Second Handbook of Research on Mathematics Teaching and Learning

Second Handbook of Research on Mathematics Teaching and Learning

A Project of the
National Council of Teachers of Mathematics

Frank K. Lester, Jr.

EDITOR

NATIONAL COUNCIL OF
TEACHERS OF MATHEMATICS

INFORMATION AGE
PUBLISHING

Information Age Publishing • Charlotte, NC • infoagepub.com

Library of Congress Cataloging-in-Publication Data

Second handbook of research on mathematics teaching and learning : a project
of the national council of teachers of mathematics / Frank K. Lester, Jr.,
editor.
     p. cm.
 Includes bibliographical references and index.
 ISBN-13: 978-1-59311-176-2 (pbk.)
 ISBN-13: 978-1-59311-177-9 (hardcover)
 1. Mathematics–Study and teaching–Research. I. Lester, Frank K. II.
Handbook of research on mathematics teaching and learning
 QA11.S365 2007
 510.71–dc22

                                            2006033783

Volume 1:
   ISBN-13: 978-1-59311-586-9 (pbk.)
   ISBN-13: 978-1-59311-587-6 (hardcover)

Volume 2:
   ISBN-13: 978-1-59311-588-3 (pbk.)
   ISBN-13: 978-1-59311-589-0 (hardcover)

Copyright © 2007 Information Age Publishing Inc.

Printed in the United States of America

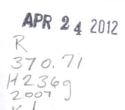

This handbook is dedicated to the memory of four eminent researchers who were authors
of chapters in the first handbook, but who have since died —

Merlyn J. Behr   1932–1995
Robert B. Davis   1923–1997
James J. Kaput   1943–2005
Alba G. Thompson   1946–1996

With their passing, the mathematics education community lost talented, dedicated researchers;
those of us who knew them lost valued friends.

.⁙ CONTENTS ⁙

Part IV: Students and Learning

Part V: Assessment

Part VI: Issues and Perspectives

⠿ PREFACE ⠿

In 1965, a bit more than 40 years ago, in response to growing interest in mathematics education research, the Board of Directors of NCTM appointed a Committee on Research in Mathematics Education (Johnson, Romberg, & Scandura, 1994). One result of the committee's deliberations was the publication in 1967 of a slim volume arguing for the need for more high-quality research (Scandura, 1967). In the preface to this volume, the editor observed that "leading citizens believe in research as never before. . . . [And] many teachers are asking for evidence to support or deny the current crop of claims demanding changes in curriculum and pedagogy. . . . Recommendations for change should be based on research" (Scandura, 1967, p. iii). In the same year, another prominent American mathematics education researcher, Robert Davis, wondered "in a society which has modernized agriculture, medicine, industrial production, communication, transportation, and even warfare as ours has done, it is compelling to ask why we have experienced such difficulty in making more satisfactory improvements in education" (Davis, 1967, p. 53).

The situation is not so different today, but the criticism has become more strident. Presently, in the middle of the first decade of the 21st century, diverse individuals and groups have been promoting a variety of old and new instructional approaches, programs, and policies for mathematics education. Researchers are being exhorted to gather and analyze data for evaluating the efficacy of various instructional approaches and curricula as never before. Individuals both within and outside of the mathematics education research community have begun to promote specific agendas, some largely for political reasons. Thus, a pressing challenge for researchers has been to reach out to its constituents by providing research-based advice about questions that concern that constituent community, and to do so in an unbiased, rational, and convincing manner. Members of the research community have discovered that the general public understands and appreciates very little about *what* researchers do or *why* they do it. They find themselves in the position of having to defend their work. Many factions—from teachers to politicians—are calling for mathematics educators to find out "what

works" in classrooms at all levels (see, for example, U.S. Department of Education, What Works Clearinghouse, www.whatworks.ed.gov, retrieved August 6, 2006). This emphasis on studies to document what works presents mathematics education researchers with a dilemma: How can they provide research results showing "what works" and at the same time advance the theoretical and model-building goals that are so important to scientific inquiry in every field?

The purpose of this Handbook is not to answer "what works" questions, although answers to many of these sorts of questions can be found in this volume. Instead, the primary purposes are to coalesce the research that has been done, to provide (perhaps) new conceptualizations of research problems, and to suggest possible research programs to move the field forward. It is only by occasionally stepping back and looking at where mathematics education research has been and suggesting where it should be going that the field is likely to ever be able to answer the questions that practitioners, policy makers, and politicians are asking.

HISTORY OF THE DEVELOPMENT OF THIS HANDBOOK

In May 2002, the Educational Materials Committee (EMC) of NCTM sent the NCTM Board of Directors a request to authorize and fund the development of a new edition of the *Handbook of Research in Mathematics Teaching and Learning*, which had been published in 1992. This request followed a review by the EMC of a proposal from Douglas Grouws and the NCTM Research Advisory Committee (RAC) regarding the revision of the Handbook. Grouws was the editor of the original Handbook and he and the RAC were convinced that, after 10 years, it was time to begin work on a new edition. In July of that year, the Board of Directors approved the EMC's request, and by September NCTM president Johnny Lott had convinced me to serve as editor. In November 2002, I had a very productive discussion with Harry Tunis, Director of Publications at NCTM and George Johnson, President of Information Age Publishing, Inc. (IAP). The result of that discussion was an agreement between NCTM and IAP to publish the new Handbook. A few days later, I met for the first and only time with the advisory board to conceptualize the new Handbook and begin to identify potential authors. The members of the advisory board included some of the most prominent members of the mathematics education community: Doug Grouws, University of Missouri; James Hiebert, University of Delaware; Carolyn Kieran, Université du Québec à Montréal; Judith Sowder, San Diego State University; and past-President of NCTM Lee Stiff, North Carolina State University. My colleague at Indiana University, Paul Kehle, was to serve as assistant editor for the project. At the end of our two-day meeting we had sketched out the overall structure of the Handbook, including identification of chapter topics to emphasize or de-emphasize, and possible authors. Although the topics remained essentially the same throughout the development of the Handbook, some of the authors changed over time for various reasons. During the ensuing four-plus years, the members of the advisory board were also available to give much-needed advice and, in general, to help out in various other ways. During the development process, authors and author teams submitted outlines of their chapters for review by at least two outside reviewers. Having received the reviews, authors proceeded to prepare an initial draft that was closely scrutinized by the same reviewers, who provided valuable constructive criticism of the drafts. Of course, I, as editor, also reviewed each draft and, when appropriate, provided additional commentary. The authors then used the reviewers' feedback to prepare a second draft. I reviewed the second drafts to determine if the authors had adequately addressed the concerns raised by the reviewers. In a few cases, additional work was called for. Once I decided that a draft was acceptable, the editorial staff at Indiana University commenced to prepare it to be sent to the publisher. On average, the process from submission of an outline to submission to the publisher took two full years.

AUDIENCE

The audience remains much the same as for the 1992 Handbook, namely, mathematics education researchers and other scholars conducting work in mathematics education.

> This group includes college and university faculty, graduate students, investigators in research and development centers, and staff members at federal, state, and local agencies that conduct and use research

within the discipline of mathematics. The intent of the authors of this volume was to provide useful perspectives as well as pertinent information for conducting investigations that are informed by previous work. The Handbook should also be a useful textbook for graduate research seminars. (Grouws, 1992, p. ix)

In addition to the audience mentioned above, the present Handbook contains chapters that should be relevant to four other groups: teacher educators, curriculum developers, state and national policy makers, and test developers and others involved with assessment. Taken as a whole, the chapters reflect the mathematics education research community's willingness to accept the challenge of helping the public understand what mathematics education research is all about and what the relevance of their research findings might be for those outside their immediate community.

SCOPE

More than 30 years ago, in an effort to define the scope of a reference volume on research in mathematics education, the Publications Committee of NCTM asked mathematics educators to answer two questions about the kind of

> reference on research in mathematics education which you would insist be on your shelf, the shelf of every doctoral student in mathematics education, and the shelf of every person responsible for research in mathematics education. What would you want such a reference to contain? What do you think such a reference need not contain? (Shumway, 1980, p. v).

The advisory board for the present Handbook considered the same two questions. A quick perusal of the previous research compendia published by NCTM (i.e., those published in 1980 [Shumway], and 1992 [Grouws]) shows that the scope of mathematics education research has expanded with each new publication. Recognizing a need "to introduce the beginner to some of the issues and problems in the research process" (Shumway, 1980, p. v), the editorial board of the 1980 volume decided to devote nearly one-third of the volume to this topic, with the balance dealing with the identification of critical, productive problems for researchers, dealing primarily with learning, teaching and curriculum.

Not only was the 1992 handbook considerably longer than Shumway's volume, its scope expanded dramatically. In addition to chapters dealing with specific aspects of mathematics learning (i.e., chapters on research on addition and subtraction, multiplication and division, rational numbers, problem solving, estimation and number sense, algebra, geometry and spatial reasoning, probability and statistics, and advanced mathematical thinking), new avenues of active inquiry appeared. For example, attention to instructional issues was quite prominent (chapters on learning and teaching with understanding, mathematics teaching practices, and grouping for instruction) and research on teachers and teacher education had become part of the mainstream (chapters on teachers' beliefs and conceptions, teachers' knowledge, teacher professionalism, and becoming a mathematics teacher). The topics of the remaining chapters of that handbook indicated that new emphasis had been placed on technology; on assessment and international comparisons; and on philosophical, social, personal, and cultural dimensions of mathematics education (one chapter each on the nature of mathematics, the culture of the mathematics classroom, ethnomathematics, affect, gender, and race, ethnicity, social class, language, and achievement in mathematics).

The current Handbook demonstrates that the research community has sustained its focus on problems of learning, teaching, teacher education, assessment, technology, and social and cultural aspects of mathematics education. But, in addition, some new areas of interest have emerged or been expanded.

The present volume has six sections. The first section, *Foundations*, contains a chapter devoted to two areas not specifically addressed in the 1992 handbook: philosophy (Cobb) and theory (Silver & Herbst). These two chapters plus a chapter on research methods (Schoenfeld) illustrate that the mathematics education research community has taken seriously the importance of thinking hard about the underpinnings of its work.

The second section, *Teachers and Teaching*, contains four chapters focusing on: (1) what knowledge mathematics teachers should have (Hill, Sleep, Lewis, & Ball), (2) how teachers are prepared and the kinds of professional development they receive (J. Sowder), (3) what teachers do in their classrooms and the culture within which teaching takes place (Franke, Kazemi & Battey), and (4) teachers' beliefs and affects (Philipp). Teachers and teaching received considerable attention in the 1992 handbook, but the chapters in section I of the new Handbook demonstrate that new perspectives and new lines of inquiry have emerged.

Section III, *Influences on Student Outcomes*, is also made up of four chapters, one of which focuses on how teaching influences what students learn (Hiebert & Grouws). Among the other three chapters in this section, one focuses on a topic that was not present in the 1992 handbook: how curricula influence student learning (Stein, Remillard, & Smith). The other two chapters comprising the balance of this section—one dealing with the notions of ethnomathematics and everyday cognition (Presmeg), the other with issues of race, class, gender, language, culture, and power (Diversity in Mathematics Education Research Center)—illustrate the ever-expanding scope of research in the field.

More than one-third of the volume (11 chapters) is contained in section IV, *Students and Learning*. Several of the chapters address topics present in the previous handbook: whole number concepts and operations (Verschaffel, Greer, & DeCorte), rational numbers and proportional reasoning (Lamon), algebra in the middle school through college levels (Kieran), problem solving (Lesh & Zawojewski), geometric and spatial thinking (Battista), and post-secondary mathematics thinking and learning (Artigue, Batanero, & Kent). Research on probability and statistics, a single chapter in 1992, has been split into two chapters—one on probability (Jones, Langrall, & Mooney), the other on statistics (Shaughnessy). There also are three chapters representing new areas of research that have burgeoned since the 1992 volume: early childhood (i.e., pre-Grade 1) mathematics learning (Clements & Sarama), early algebra and algebraic reasoning (i.e., pre-middle school; Carraher & Schliemann), and the learning and teaching of proof (Harel & L. Sowder).

The ever-increasing role of assessment in mathematics education made it appropriate to devote an entire section to this topic. The first chapter of Section V, *Assessment*, considers how classroom assessment can actually support learning goals, rather than just measure them (Wiliam). The second chapter (Wilson) deals with high-stakes, standardized tests and considers the influence these tests have had on educational practices, and the third chapter (de Lange) considers issues associated with large-scale, international assessments.

The sixth and final section contains six chapters dealing with a wide range of topics and issues. One chapter (Bishop & Forgasz) focuses on the enduring issues and perspectives that relate to equity and access, another with issues and research associated efforts to understand the impact of computer technologies on learning, teaching, and curriculum (Zbiek, Heid, Blume, & Dick), and a third describes an approach for specifying the content for any assessment of students' achievement (Webb). Two chapters consider issues associated with an area of study not previously within the purview of mathematics education research, namely educational policy. The authors of one of the policy chapters (Ferrini-Mundy & Floden) discuss studies that illustrate the formulation, status, implementation, and effects of educational policy that relate to mathematics education. The authors of the other policy chapter (Tate & Rousseau) provide a review of mathematics education research and relevant policy literature that informs school leaders who are endeavoring to build effective school learning environments for traditionally underserved students. In the final chapter of the volume, the author, Mogens Niss, reflects on the state of mathematics education research, where it seems to be going, and what sort of research would need to be done to get us to the utopian state of being able to provide final, conclusive answers to all the important questions in mathematics education.

ACKNOWLEDGEMENTS

A large number of dedicated individuals helped in the development and publication of this Handbook. As I have already noted, the members of the advisory board (Doug Grouws, James Hiebert, Carolyn Kieran, Judith Sowder, and Lee Stiff) were especially helpful in conceptualizing the volume, providing timely advice, and, on occasion, helping me deal with the inevitable frustrations involved in keeping the project on course and schedule. On behalf of the entire mathematics education community, I wish to express my sincere gratitude to them. I also extend my thanks to Paul Kehle, who, unfortunately, was not able to continue to assist me after the first year—he left Indiana University for another position. He provided valuable help during the early stages of the project. Harry Tunis, Director of Publication Services at NCTM was always helpful and supportive of the project at every stage—even when I had to tell him that we would not be able to stick to our timeline! A special thanks is due to the many reviewers who were willing to spend countless hours poring over drafts of chapters, making insightful and always helpful comments. Indeed, the Handbook exists to a great extent because of their tireless, thoughtful, largely unrecognized efforts on behalf of mathematics education.

A colleague suggested to me that when the Handbook is complete I will be better informed about contemporary mathematics education research than anyone else in the U.S. Perhaps; it is at least arguable! To the extent that it is true, I owe it to the authors and reviewers who worked so hard to make this Handbook the source authority on research in the field. I learned a tremendous amount from reading chapter drafts, considering reviewers' comments, and discussing the focus and organization of chapters. I also gained even more appreciation for my colleagues than I had before—and I already had tremendous respect for them. They are such busy folks—I really don't know how they were able to put together such thoughtful and wonderfully insightful chapters. So, to all authors and reviewers, I extend my sincerest, Thank You!

Closer to home, Indiana University doctoral students Paula Stickles and Andrea McCloskey were absolutely indispensable to the success of the project. Paula helped with the thankless job of formatting the final drafts to make them conform to APA standards. Andrea had an even more tedious task in addition to assisting with formatting: checking every reference of every chapter for correctness and completeness. I am convinced that she now knows as much or more about APA reference conventions as anyone! Both Paula and Andrea assisted me with the tedious job of checking page proofs, as did Zac Rutledge, another very able doctoral student. One other person who played a very important behind-the-scenes role in putting drafts into final form was Cheryl Burkey. Cheryl, an accomplished and very experienced copy editor, knows the *APA Publications Manual* extremely well. She copy edited almost all of the drafts in preparation for sending them to the IAP editors for final production. Thank you Paula, Andrea, and Cheryl—your professionalism, competence, and cheerful dispositions made my job much easier and rewarding. Thanks are also due to the staff of Information Age Publishing. In particular, IAP president, George Johnson, deserves special acknowledgement for his patience and understanding as the Handbook slowly became a reality. I wish also to express my thanks to Scott Suckling of MetroVoice Publishing Services for his work on page design and composition and Ann Salinger who created the indices.

Finally, I wish to thank my wife, Diana, who sometimes wondered if her often-cranky husband would ever bring the Handbook to fruition. Now we can begin to enjoy some time together when I will not have some Handbook issue in the back of my mind.

—Frank K. Lester, Jr.
Bloomington, IN

REFERENCES

Davis, R. J. (1967). The range of rhetorics, scale and other variables. *Proceedings of the National Conference on Needed Research in Mathematics Education* in the *Journal of Research and Development in Education, 1*(1), 51–74.

Grouws, D. A. (Ed.). (1992). *Handbook of research on mathematics teaching and learning.* New York: Macmillan.

Johnson, D. C., Romberg, T. A., & Scandura, J. M. (1994). The origins of *JRME*: A retrospective account. 25th Anniversary Special Issue of the *Journal for Research in Mathematics Education, 25*, 561–582.

Scandura, J. (Ed.). (1967). *Research in mathematics education.* Washington, DC: National Council of Teachers of Mathematics.

Shumway, R. J. (Ed.). (1980). *Research in mathematics education.* Reston, VA: National Council of Teachers of Mathematics.

Foundations

PUTTING PHILOSOPHY TO WORK

Coping with Multiple Theoretical Perspectives

Paul Cobb

VANDERBILT UNIVERSITY

In inviting me to write this chapter on philosophical issues in mathematics education, the editor has given me the leeway to present a personal perspective rather than to develop a comprehensive overview of currently influential philosophical positions as they relate to mathematics education. I invoke this privilege by taking as my primary focus an issue that has been the subject of considerable debate in both mathematics education and the broader educational research community, that of coping with multiple and frequently conflicting theoretical perspectives. The theoretical perspectives currently on offer include radical constructivism, sociocultural theory, symbolic interactionism, distributed cognition, information-processing psychology, situated cognition, critical theory, critical race theory, and discourse theory. To add to the mix, experimental psychology has emerged with a renewed vigor in the last few years. Proponents of various perspectives frequently advocate their viewpoint with what can only be described as ideological fervor, generating more heat than light in the process. In the face of this sometimes bewildering array of theoretical alternatives, the issue I seek to address in this chapter is that of how we might make and justify our decision to adopt one theoretical perspective rather than another. In doing so, I put philosophy to work by drawing on the analyses of a number of thinkers who have grappled with the thorny problem of making reasoned decisions about competing theoretical perspectives.

In the first section of the chapter, I follow Schön (1983) in challenging what he termed the positivist epistemology of practice wherein practical reason is construed as the application of theory. My goal in doing so is to question the repeated attempts that have been made in mathematics education to derive instructional prescriptions directly from background theoretical perspectives. I argue that it is instead more productive to compare and contrast various perspectives by using as a criterion the manner in which they orient and constrain the types of questions that are asked about the learning and teaching of mathematics, the nature of the phenomena that are investigated, and the forms of knowledge that are produced.

In the second section of the chapter, I argue that mathematics education can be productively construed as a design science, the collective mission of which involves developing, testing, and revising conjectured designs for supporting envisioned learning processes. As will become apparent, my intent in developing this viewpoint is inclusive rather than exclusive. I therefore construe designs broadly and clarify that the learning processes might, for example, be those of individual students or individual teachers, classroom communities or professional teaching communities, or indeed of schools or school districts as organizations. This perspective on our collective activity as mathematics educators gives rise to a second criterion for comparing and contrasting background theoretical positions, namely that of how they might contribute to the enterprise of formulating, testing, and revising conjectured designs for supporting mathematical learning. This second criterion is therefore concerned with the nature of the potentially useful work that different theoretical positions might do.

In section three, I draw both on the work of John Dewey and on the more recent contributions of neo-Pragmatist philosophers to question the relevance of a number of philosophical distinctions that have featured prominently in the mathematics education research literature. Foremost among these is the central problem of traditional epistemology, that of the opposition between philosophical realism and constructivist positions that deny that ontological reality is knowable. In doing so, I outline an alternative position that Putnam (1987) terms pragmatic realism and clarify its relevance to the issue of comparing and contrasting different theoretical perspectives in mathematics education research. Against this background in section four and five, I then sharpen the two criteria that I propose for comparing background theoretical positions. The first criterion concerns the nature of the phenomena that are investigated and is often framed in terms of whether a particular perspective treats activity as being primarily individual or social in character. I argue that this dichotomy is misleading in that it assumes that what is meant by *the individual* is self-evident and theory neutral. As an alternative, I propose to compare and contrast different theoretical positions in terms of how they characterize individuals, be they students, teachers, or administrators. The second criterion concerns the potentially useful work that different theoretical positions might do. In addressing this issue, I differentiate Dewey's sophisticated account of pragmatic justification and his related analysis of verification and truth from a purely instrumental approach in which methods or strategies that enable us to reach our goals are deemed to be true, and those that do not are treated as false.

In the subsequent sections of the chapter, I use these two criteria as my primary points of reference as I discuss four broad theoretical positions: experimental psychology, cognitive psychology, sociocultural theory, and distributed cognition. Experimental psychology refers to the psychological research tradition in which the primary methods employed involve experimental and quasi-experimental designs, preferably with the random assignment of subjects. The cognitive psychology tradition on which I focus involves accounting for teachers' and students' inferred interpretations and understandings in terms of internal cognitive structures and processes. So-

ciocultural theory and distributed cognition provide points of contrast with this psychological tradition in that people's activity and learning are considered to be situated with respect to the social and cultural practices in which they participate. As I clarify, sociocultural theory developed largely independently of western cognitive psychology by drawing directly on the writings of Vygotsky and Leont'ev and is concerned with people's induction into and participation in relatively broad cultural practices. In contrast, distributed cognition emerged in reaction to mainstream cognitive psychology and tends to focus on people's activity in their immediate physical, symbolic, and social environments.

I conclude from the comparison of the four perspectives that each has limitations in terms of the extent to which it can contribute to the enterprise of formulating, testing, and revising designs for supporting learning. As part of the analysis, I outline the historical origins of each perspective in order to understand adequately the types of phenomena that are investigated and the forms of knowledge produced. I note that the concerns and interests that motivated the development of these four perspectives differ from those of mathematics educators. In light of this conclusion, I propose and illustrate how we can view the various theoretical perspectives as sources of ideas to be appropriated and adapted to our purposes as mathematics educators. In the final section of the chapter, I then step back to consider the issue of competing, incommensurable theories more broadly. In doing so, I locate the approach I have taken within a philosophical tradition that seeks to transcend the longstanding dichotomy between the quest for a neutral framework for comparing theoretical perspectives on the one hand and the view that we cannot reasonably compare perspectives on the other hand.

THE POSITIVIST EPISTEMOLOGY OF PRACTICE

In his influential book *The Reflective Practitioner*, Donald Schön (1983) explicitly targets what he terms the positivist epistemology of practice wherein practical reasoning is accounted for in terms of the application of abstract theoretical principles to specific cases. As Schön (1983, 1986, 1987) notes, this epistemology is apparent in teacher education programs that (a) initially emphasize the acquisition of theoretical principles derived from fields such as psychology and sociology, (b) then focus on the process of applying these principles in subject matter methods courses before (c) allowing future teachers to engage in this application process in field experiences. This epistemology is also implicit in instructional approaches that separate students' initial acquisition of mathematical concepts, strategies, or procedures from their subsequent application of those concepts and methods. In the research literature, the positivist epistemology of practice is central to what De Corte, Greer, and Verschaffel (1996) term the first wave of the cognitive revolution, the goal of which was to formulate internal information-processing mechanisms that account for observed relations between the external stimulus environment and documented responses. Anderson, Reder, and Simon (1997) exemplified this position in an exchange with Greeno (1997) when they argued that the contributions of cognitive analyses of this type stem from the analytic power they provide for extracting principles that can generalize from one setting to another. Anderson et al. suggested that teachers should base their instruction on these principles while relying on common sense and professional experience when issues arise in the classroom that psychologists cannot yet answer. Most importantly for the purposes of this chapter, the positivist epistemology of practice is also apparent in debates about the implications of background theoretical positions for mathematics learning and teaching.

The most prominent case in which attempts have been made to derive instructional implications directly from a background theory is that of the development of the general pedagogical approach known as *constructivist teaching*. This pedagogy claims to translate the theoretical contention that learning is a constructive activity directly into instructional recommendations. As Noddings (1990) and Ball and Chazan (1994) observe, it is closely associated with the dubious assertion that "telling is bad" because it deprives students of the opportunity to construct knowledge for themselves. For his part, J. P. Smith (1996) clarifies that adherents of the pedagogy tend to frame teachers' proactive efforts to support their students' learning as interfering with students' attempts to construct meaning for themselves. As Smith demonstrates, in emphasizing what the teacher does not do compared with traditional instructional practices, the teacher's role in supporting learning is cast in relatively passive terms, thereby resulting in a sense of loss of efficacy.

It is important to note that constructivism does not have a monopoly on questionable reasoning of this type in which background theoretical contentions that are *descriptive* in nature are translated directly into instructional maxims that are *prescriptive*. For example, an advocate of symbolic interactionism might argue

that as learning involves the negotiation of meaning, students should be encouraged to continually discuss their differing interpretations. Similarly, a devotee of the distributed view of intelligence might argue that as cognition is stretched over individuals, tools and social contexts, it is important to ensure that students' mathematical activity involves the use of computers and other tools. In each of these cases, the difficulty is not with the background theory but with the relation that is assumed to hold between theory and instructional practice. As I have argued elsewhere (P. Cobb, 2002), pedagogical proposals developed in this manner involve a category error wherein the central tenets of a descriptive theoretical perspective are transformed directly into instructional prescriptions. It is noteworthy that similar instructional recommendations (e.g., the importance of small group work) are frequently derived from contrasting theoretical positions (e.g., constructivism, symbolic interactionism, and distributed intelligence). The resulting pedagogies are underspecified and are based on ideology (in the disparaging sense of the term) rather than empirical analyses of the process of students' learning and the means of supporting it in specific domains.

Later in this chapter, I will build on this critique of the positivist epistemology of practice by proposing an alternative view of the relation between theory and practice. In doing so, I will question the assumption that theory consists of decontextualized propositions, statements, or assertions that are elevated above and stand apart from the activities of practitioners. For the present, it suffices to note that work in the philosophy, sociology, and history of science, sparked by the publication of Thomas Kuhn's (1962) landmark book, *The Structure of Scientific Revolutions*, has challenged this view of theoretical reasoning in the natural sciences. In this and subsequent publications (Kuhn, 1970, 1977), Kuhn developed an analysis of both the processes by which scientists develop theory within an established research tradition and the processes by which they choose between competing research traditions. In the case of theory development within an established research tradition, he followed Michael Polanyi (1958) in questioning the assumption that scientists fully explicate the bases of their reasoning by presenting analyses of a number of historical cases. His goal in doing so was to demonstrate that the development and use of theory necessarily involves tacit suppositions and assumptions that scientists learn in the course of their induction into their chosen specialties. Kuhn went on to clarify that these implicit aspects of scientific reasoning serve as a primary means by which scientists distinguish between insiders and outsiders in the process of establishing the boundaries of scientific communities.

Kuhn (1962) extended this argument about the tacit aspects of scientific reasoning when he considered how scientists choose between competing research traditions by arguing that "there is no neutral algorithm of theory-choice, no systematic decision procedure which, properly applied, must lead each individual in the group to the same decision" (p. 200). Not surprisingly, Kuhn was accused of claiming that the process by which scientists resolve disputes is irrational. It is therefore important to stress that he was not in fact questioning the rationality of scientists. Instead, he was challenging the dominant view that scientific reasoning could be modeled as a process of applying general rules and procedures to specific cases. As Bernstein (1983) observed, "Kuhn always intended to distinguish forms of rational persuasion and argumentation that take place in scientific communities from those irrational forms of persuasion that he has been accused of endorsing" (p. 53). In responding to his critics, Kuhn (1970) subsequently clarified that

> what I am denying is neither the existence of good reasons [for choosing one theory over another] nor that these reasons are of the sort usually described. I am, however, insisting that such reasons constitute values to be used in making choices rather than rules of choice. Scientists who share them may nevertheless make different choices in the same concrete situation... [In concrete cases, scientific values such as] simplicity, scope, fruitfulness, and even accuracy can be judged quite differently (which is not to say that they can be judged arbitrarily) by different people. Again, they may differ in their conclusions without violating any accepted rule. (p. 262)

Bernstein (1983) summarized the debate incited by Kuhn's analysis of scientific reasoning by cautioning that

> one must be sensitive to and acknowledge the important differences between the nature of scientific knowledge and other forms of knowledge. But the more closely we examine the nature of this scientific knowledge that has become the paradigm of theoretical knowledge, the more we realize that the character of rationality in the sciences, especially in matters of theory-choice, is closer to those features of rationality that have been characteristic of the tradition of practical philosophy than to many modern images of what is supposed to be the character of genuine *episteme* [i.e., the application of general, decontextualized methods to specific cases]. (p. 47)

In speaking of practical philosophy, Bernstein was referring specifically to the work of Gadamer (1975). The specific phenomenon that Gadamer analyzed when he developed his philosophy of practical activity was the process by which people interpret and understand texts, particularly religious texts. In developing his position, Gadamer responded to earlier work in this area that instantiated the positivist epistemology of practice by differentiating between general methods of interpretation and understanding on the one hand, and the process of applying them to specific texts on the other hand. Gadamer rejected this distinction, arguing that every act of understanding involves interpretation, and all interpretation involves application. On this basis, he concluded that the characterization of theoretical reasoning as the application of general, decontextualized methods serves both to mystify science and to degrade practical reasoning to technical control.

The philosophical positions that Kuhn, Gadamer, and others developed to challenge the dominance of positivist epistemology of practice foreshadowed a number of more recent developments that are familiar to mathematics education researchers. For example, their argument that reasoning in all domains, including those that are typically characterized as formal and abstract, has much in common with practical reasoning anticipated the view proposed by a number of educational researchers that people's reasoning is situated with respect to their participation in specific activities or practices (e.g., Beach, 1999; J. S. Brown, Collins, & Duguid, 1989; P. Cobb & Bowers, 1999; Forman, 2003; Greeno & The Middle School Mathematics Through Applications Project Group, 1998; Saxe, 1991; Sfard, 1998). Relatedly, their characterization of people as members of intellectual communities anticipated investigations in which the community of practice has been taken as a primary unit of analysis (e.g., Lave & Wenger, 1991; Rogoff, 1990; Stein, Silver, & Smith, 1998; Wenger, 1998).

Kuhn's and Gadamer's arguments are also directly relevant to the issues that I seek to address in this chapter. For example, they call into question attempts to develop a so-called *philosophy of mathematics education* if its intent is to be normative and prescriptive by providing foundational principles for the activity of mathematics education researchers and practitioners. Instead, their arguments orient us to see research, theorizing, and indeed philosophizing as distinct forms of practice rather than activities whose products provide a viable foundation for the activities of practitioners. It is for this reason that I shy away from the approach of first surveying currently fashionable philosophical positions and then deriving implications for mathematics education research and practice from them. Instead, I take as my starting point several background theoretical positions that are already currently influential in mathematics education research: experimental psychology, cognitive psychology, sociocultural theory, and distributed cognition. At the most immediate level, my goal in discussing these theoretical positions is descriptive rather than prescriptive in that I explicate their basic tenets and consider their potential contributions to mathematics education research. This first goal gives rise to a second that is inherently philosophical in nature, that of moving beyond the realization that there is no neutral algorithm of theory choice by considering how we might sensibly compare and contrast different theoretical perspectives given our concerns and interests as mathematics educators. In my view, this latter (meta)issue is of the greatest importance given the profusion of perspectives that has become emblematic of educational research in general, and of mathematics education research in particular.

In order to address this (meta)issue, it is necessary not merely to be explicit about the criteria of comparison, but to justify them. My goal in doing so is not to formulate a general method or procedure for choosing between these and other theoretical perspectives, but to initiate a conversation in which we, as mathematics education researchers, can begin to work through the challenges posed by a proliferation of perspectives. As the neo-pragmatist philosopher Rorty (1979) clarified in his much cited book, *Philosophy and the Mirror of Nature*, philosophy so construed is therapeutic in that it does not presume to tell people how they should act in particular types of situations. Instead, its goal is to enable people themselves to cope with the complexities, tensions, and ambiguities that characterize the settings in which they act and interact. As will become apparent, such an approach seeks to avoid both the unbridled relativism evident in the contention that one cannot sensibly weigh the potential contributions of different theoretical perspectives and the unhealthy fanaticism inherent in the all-too-common claim that one particular perspective gets the world of teaching and learning right.

As an initial criterion, I propose comparing and contrasting the different theoretical perspectives in terms of the manner in which they orient and constrain the types of questions that are asked about the learning and teaching of mathematics, and thus the nature of the phenomena that are investigated and the forms of knowledge produced. Adherents to a research tradition view themselves as making progress to the extent that they are able to address questions that they judge to be important. The content of a research tradition, including the types of phenomena that are considered to be significant, therefore represents solutions to previously posed questions. However, as Jardine (1991) demonstrates by means of historical examples, the questions that are posed within one research tradition frequently seem unreasonable and, at times, unintelligible from the perspective of another tradition. In Lakatos' (1970) formulation, a research tradition comprises positive and negative heuristics that constrain (but do not predetermine) the types of questions that can be asked and those that cannot, and thus the overall direction of the research agenda. In my view, delineating the types of phenomena that can be investigated within different research traditions is a key step in the process of comparing and contrasting them. As Hacking (2000) observed, while the specific questions posed and the ways of addressing them are visible to researchers working within a given research tradition, the constraints on what is thinkable and possible are typically invisible. This initial criterion is therapeutic in Rorty's (1979) sense in that the process of comparing and contrasting perspectives provides a means both of deepening our understanding of the research traditions in which we work, and of enabling us to de-center and develop a basis for communication with colleagues whose work is grounded in different research traditions.

MATHEMATICS EDUCATION AS A DESIGN SCIENCE

The second criterion that I propose for comparing different theoretical perspectives focuses on their potential usefulness given our concerns and interests as mathematics educators. My treatment of this criterion is premised on the argument that mathematics education can be productively viewed as a design science, the collective mission of which involves developing, testing, and revising conjectured designs for supporting envisioned learning processes. As an illustration of this collective mission, the first two *Standards* documents developed by the National Council of Teachers of Mathematics (1989, 1991) can be viewed as specifying an initial design for the reform of mathematics teaching and learning in North America. The more recent *Principles and Standards for School Mathematics* (National Council of Teachers of Mathematics, 2000) propose a revised design that was formulated in response to developments in the field, many of which were related to attempts to realize the initial design. Significantly, as part of this revision process, the designers found it necessary to substantiate their proposals by developing a research companion volume (Kilpatrick, Martin, & Schifter, 2003). As a second illustration, the functions of leadership in mathematics education at the level of schools and school districts include mobilizing teachers and other stakeholders to notice, face, and take on the task of improving mathematics instruction (Spillane, 2000). Ideally, this task involves the iterative development and revision of designs for improvement as informed by ongoing documentation of teachers' instructional practices and students' learning (Confrey, Castro-Filho, & Wilhelm, 2000; Fishman, Marx, Blumenfeld, & Krajcik, 2004; Hill, 2001; McLaughlin & Mitra, 2004, April). At the classroom level, the design aspect of teaching is particularly evident in Stigler and Hiebert's (1999) description of the process by which Japanese mathematics teachers collaborate to develop and revise the design of lessons. This aspect of teaching is also readily apparent in analyses that researchers such as Ball (1993), Lampert (2001), and Simon (1995) have developed of their own instructional practices.

Although the three illustrations focus on developing, testing, and revising designs at the level of a national educational system, a school or school district, and a classroom respectively, the ultimate goal in each case is to support the improvement of students' mathematical learning. As a point of clarification, I should stress that the contention that mathematics education can be productively viewed as a design science is *not* an argument in favor of one particular research methodology such as design experiments. Rather than being narrowly methodological, the claim focuses on our collective mission and thus on the concerns and interests inherent in our work. This framing of mathematics education research therefore acknowledges that a wide spectrum of research methods ranging from experimental and quasi-experimental designs to surveys and ethnographies can contribute to this collective enterprise. Furthermore, this framing recognizes the value of theoretical analyses that result in the development of conceptual tools that can contribute to this enterprise. The fruitfulness of the framing stems from the manner in which it delineates core aspects of mathematics education research as a disciplinary activity. These core aspects include:

- Specifying and clarifying the prospective endpoints of learning
- Formulating and testing conjectures about both the nature of learning processes that aim towards those prospective endpoints and the specific means of supporting their realization

(Confrey & Lachance, 2000; Gravemeijer, 1994a; Simon, 1995)

This formulation is intended to be inclusive in that the learning processes of interest could be those of individual students or teachers, of classroom or professional teaching communities, or indeed of schools or school districts viewed as organizations.

A discussion of the relative merits of various goals or envisioned endpoints for students', teachers', and school districts' learning is beyond the scope of this chapter. It is, however, worth noting that the choice of goals and thus of what counts as improvement is not a matter of mere subjective whim or taste. Instead, our choices involve judgments that are eminently discussable in that we are expected to support them by giving reasons. As an illustration, I outline my view of the primary issues that should be considered when formulating endpoints for students' learning. I propose that, at a minimum, the process of formulating instructional goals should involve delineating central mathematical ideas in particular mathematical domains and clarifying more encompassing activities such as mathematical argumentation and modeling (Lehrer & Lesh, 2003). In addition, I contend that it is important to justify the proposed endpoints in terms of the types of future activities to which they might give students access. One important consideration is therefore the extent to which the envisioned forms of mathematical reasoning have what Bruner (1986) termed clout by enabling students to participate in significant out-of-school practices in relatively substantial ways. For example, it is apparent that public policy discourse increasingly involves the formulation and critique of data-based arguments. Students' development of the relatively sophisticated forms of statistical reasoning that are implicated in such arguments therefore has clout in that it enables students to participate in a type of discourse that is central to what Delpit (1988) termed the culture of power (cf. G. W. Cobb, 1997).

In my view, it is also important to take explicit account of the function of schools in differentiating between and sorting students when gauging the extent to which envisioned forms of mathematical reasoning have clout (cf. Secada, 1995). As an example, Moses and C. Cobb (2001) clarify that the goal of the Algebra Project is to make it possible for *all* students to have access to and to succeed in high school algebra courses that function as gatekeepers to college preparatory tracks, and thus to future educational and economic opportunities. As Moses and Cobb also indicate, cultivating students' development of mathematical interests such that they come to view classroom mathematical activities as worthy of their engagement should be an important goal in its own right. Although many students will not pursue career trajectories beyond school that involve direct engagement in mathematical activity, it is nonetheless reasonable to propose that mathematical literacy include an empathy for and sense of affiliation with mathematics together with the desire and capability to learn more about mathematics when the opportunity arises.

The importance of this latter goal becomes even more apparent when we note that an increasing appreciation for mathematics inherent in the development of mathematical interests is compatible with what D'Amato (1992) refers to as *situational significance* wherein students come to view engagement in mathematical activities as a means of gaining experiences of mastery and accomplishment, and of maintaining valued relationships with teachers and peers. D'Amato contrasts this general way in which learning mathematics in school comes to have value for students with a second that he terms *structural significance*. In this second case, students come to view achievement in mathematics as a means of attaining other ends such as entry to college and high-status careers, or acceptance and approval in household and other social networks. As he notes, not all students have access to a structural rationale for learning mathematics in school. Gutiérrez (2004) observes, for example, that many urban students do not see themselves going to college, hold activist stances, have more pressing daily concerns (e.g., housing, safety, healthcare), or do not believe that hard work and effort will be rewarded in terms of future educational and economic opportunities. As D'Amato (1992), Erickson (1992), and Mehan, Hubbard, & Villanueva (1994) all document, students' access to a structural rationale varies as a consequence of family history, race or ethnic history, class structure, and caste structure within society. Their analyses indicate that failure to give all students access to a situational rationale for learning mathematics by framing the cultivation of mathematical interests as an explicit goal of design and teaching will result in what Nicholls (1989) termed inequities in motivation.

The issues on which I have focused when discussing the formulation of goals for students' learning clearly reflect my personal view. However, the fact that they are open to critique and challenge indicates that I cannot choose goals on the basis of unarticulated whim or fancy. For example, it might be argued that in focusing on students' participation in significant out-of-school practices, I failed to consider the practices of mathematicians as my primary point of reference (cf. J. S. Brown et al., 1989). Alternatively,

it might be argued that my concern for equity in students' access to significant mathematical ideas is inadequate in that I failed to adopt social justice agenda (cf. Frankenstein, 2002; Gutstein, in press). In responding to both critiques, I would be obliged to give additional justifications to back up the types of goals I have proposed. The process of specifying and clarifying prospective endpoints for students' learning inherent in such exchanges is analogous to the type of reasoning that Kuhn (1962) argued is involved in choices between competing research traditions. In both cases, the type of rationality involved "is a judgmental activity requiring imagination, interpretation, the weighing of alternatives, and application of criteria that are essentially open" (Bernstein, 1983, p. 56). Kuhn contrasted subjectivity with the process of developing supporting reasons or warrants that satisfy communal standards of rationality. Furthermore, he argued that these standards are, as Rorty (1979) put it, "hammered out" by members of an intellectual community as they pursue their collective enterprise. Viewed in these terms, it is quite possible for mathematics educators to make rational decisions about instructional goals while acknowledging that what counts as improvement in students' mathematical learning is itself open to revision.

The view that I have proposed of mathematics education as a design science gives rise to a second criterion for comparing and contrasting background theoretical perspectives that concerns their usefulness. Stated as directly as possible, the usefulness criterion focuses on the manner in which different theoretical perspectives might contribute to the collective enterprise of developing, testing, and revising designs for supporting learning. The illustrations I gave of design at the levels of a national educational system, a school or school district, and a classroom indicate that it will be important to consider for whom particular perspectives might do useful work when comparing perspectives with respect to this criterion. For example, a perspective that does useful work for a classroom teacher responsible for supporting the mathematical learning of specific groups of students might not necessarily be valuable for a school or district administrator whose primary concern is with the learning outcomes of a relatively large student population. This second criterion reflects the view that the choice of theoretical perspective requires pragmatic justification. In a later section of this chapter, I sharpen this criterion by drawing on Dewey's account of pragmatic justification. First, however, I pause to consider traditional epistemological distinctions that have plagued discussions of philosophy in mathematics education. In doing so, I outline an alternative position that Putnam (1987) terms internal or pragmatic realism.

PRAGMATIC REALISM

Discussions of philosophical issues in mathematics education have focused primarily on epistemology in recent years. Epistemology is concerned with inquiries into both the nature of knowledge and the process by which we come to know. It therefore addresses questions that center on the origin, nature, limits, methods, and justification of human knowledge (Hofer, 2002). These discussions were initially sparked in large part by the increasing influence of radical constructivism in mathematics education research in the 1980s. The general *psychological* contention that learning is an active, constructive process is common to numerous variants of constructivism and has become widely accepted. A key distinguishing feature of radical constructivism is the manner in which it supplements this psychological contention with the *epistemological* assertion that it is impossible to check whether our ideas and concepts correspond to external reality (von Glasersfeld, 1984). It is important to emphasize that radical constructivists do not deny the existence of a pre-given external reality. Their central epistemological claim is instead that this reality is unknowable precisely because it exists independently of human thought and action. Radical constructivists therefore speak of knowledge that has proven viable as fitting with, rather than matching, external reality in that it satisfies the constraints of that reality as much as a key satisfies the constraints of a particular type of lock (von Glasersfeld & Cobb, 1984).

Viewed within a broader philosophical context, radical constructivism is a recent variant of a long line of skeptical epistemologies that can be traced back to the sophists of ancient Greece. Radical constructivism, like other skeptical positions, challenges the traditional project of epistemology, that of identifying a universal method for determining whether a particular theory or conceptual scheme matches or corresponds with external reality. The controversy that radical constructivism has evoked within the mathematics education community stems from the manner in which it, as a skeptical position, plays on the fear that unless we overcome human fallibility and achieve finality in our knowledge claims, we have achieved nothing (Bernstein, 1983; Latour, 2000). The pragmatic realist position that I will outline also seeks to end the traditional epistemological project. However, whereas radical constructivism contends that it is impossible to bridge the gulf between knowledge and reality, pragmatic realism questions a fundamental assumption common to both traditional epistemology and to skeptical positions such as radical constructivism. I can best clarify these two contrasting approaches by focusing on the notion of *reality* that underpins the arguments of both philosophical skeptics and their realist counterparts.

The reality on which realists and skeptics both focus their attention is not populated with tables and chairs, students and teachers, or differential equations and geometry proofs. It is instead an imagined, or perhaps better, an imaginary realm that has been the center of philosophical debate since the time of Plato. Putnam (1987) refers to it as Reality with a capital "R" to distinguish it from the world in which our lives take on significance and meaning. As Bernstein (1983) observes, although skeptics seek an end to the traditional epistemological project, they are also more than willing to play the traditional epistemological game. For all their differences, realists and skeptics agree on the basic image of people as knowers separated from Reality (with a capital "R"). Radical constructivists, for example, question neither the relevance of this notion of an imaginary, ahistorical Reality nor the value of debating whether it is knowable. Putnam (1987), in contrast, follows John Dewey in challenging the dichotomy between a putative external Reality on the one hand, and the concepts and ideas that people use to think about and discuss it on the other. In contrast to radical constructivism, his goal is not to offer new solutions to traditional epistemological problems, but to question the problems themselves. In his view, realists and skeptics are both enthralled by what Dewey (1910/1976) termed "the alleged discipline of epistemology." The appropriate response to both realists and skeptics is therefore not to join them in their debate, but to challenge their exchange as an academic exercise of limited relevance.

For Putnam and Dewey, and indeed for a number of influential twentieth century philosophers including Goodman (1978), Quine (1992), and Taylor (1995), the world is not a screen shutting off nature but a path into it. As Dewey (1929/1958) put it, "experience is *of* as well as *in* nature. It is not experience that is experienced, but nature" (p. 12, italics in the original). In speaking of people being in nature, Dewey displaced the traditional preoccupation with Reality with a focus on people's activities in the realities in which they actually live their lives. As Putnam clarifies, the quest for certainty is given up once this focus is adopted in favor of understanding how people are able to produce fallible truths and achieve relative security in their knowledge claims.

In Putnam's view, the traditional epistemological goal of prescribing to scientists and non-scientists alike how they ought to reason is unjustifiable self-aggrandizement. Like Dewey, his primary concern is instead with processes of inquiry as they are enacted by flesh-and-blood people. Putnam's contribution to this pragmatic line of thought is particularly relevant to my concerns in this chapter because he addresses the issue of how people cope with the multiplicity of perspectives that characterize various domains and social contexts.

As Putnam observes, contemporary science has taken

> away foundations without providing a replacement. Whether we want to be there or not, science has put us in a position of having to live without foundations. . . . That there are ways of describing what are (in some way) the "same facts" which are (in some way) "equivalent" but also (in some way) "incompatible" is a strikingly non-classical phenomenon. (p. 29)

Putnam goes on to note that scientists in many fields switch flexibly from one perspective to another and treat each set of facts as real when they do so. On this basis, he concludes that these scientists *act as* conceptual relativists who treat what counts as relevant facts and as legitimate ways of describing them as relative to the background theoretical perspective that they adopt (cf. Sfard, 1998). In making this claim, he does not focus on scientists' commentaries on the practices of their chosen disciplines. Instead, he uses the pragmatic criterion employed by Dewey, Charles Sanders Peirce, and particularly by William James, namely that the indicator of what people actually believe is not what they say about their activity (i.e., espoused belief) but the suppositions and assumptions on which they risk acting. In the case at hand, scientists risk acting on the assumption that the differing constellations of phenomena that they investigate when they adopt differing theoretical perspectives are each real (with a small "r"). In taking this stance, Putnam rejects as irrelevant the traditional epistemological project of determining which of these constellations of phenomena correspond to an imaginary Reality. He terms his position *pragmatic realism* in that he takes at face value the realities that scientists investigate. In his view, questions concerning the existence of abstract mathematical and scientific entities should be addressed not by making claims about Reality (with a capital "R") but by examining disciplinary practices. In this regard, he notes approvingly that Quine (1953)

urges us to accept the existence of abstract entities on the ground that these are indispensable in mathematics, and of microparticles and space-time points on the ground that these are indispensable in physics; and what better justification is there for accepting an ontology than its indispensability in our scientific practice? (Putnam, 1987, p. 21)

Putnam makes it clear that pragmatic realism is not limited to mathematics and science.

> [It] is, at bottom, just the insistence that realism is *not* inconsistent with conceptual relativity. One can be *both* a realist and a conceptual relativist. Realism (with a small "r") ... is a view that takes our familiar common sense [or everyday] scheme, as well as our artistic and scientific and other schemes, at face value, without helping itself to the thing 'in itself' [i.e., Reality with a capital "R"]. (p. 17, italics in the original)

He goes on to clarify that the world looks both familiar and different when we reject the dichotomy between Reality and people's ideas about it.

> It looks familiar, insofar as we no longer try to divide up mundane reality into a 'scientific image' and a 'manifest [everyday] image' (or our evolving doctrine into a 'first-class' and a 'second-class' conceptual system). Tables and chairs ... exist just as much as quarks and gravitational fields. . . . The idea that most of mundane reality is an illusion (an idea that has haunted Western philosophy since Plato . . .) is given up once and for all. But mundane reality looks different, in that we are forced to acknowledge that many of our familiar descriptions reflect our interests and choices. (Putnam, 1987, p. 37)

Putnam takes care to differentiate *conceptual relativism*, the notion that the phenomena that are treated as real differ from one perspective to another, from the view that every perspective is as good as every other. "Conceptual relativity sounds like 'relativism', but it has none of the 'there is no truth to be found . . . "true" is just a name for what a bunch of people can agree on' implications of relativism" (1987, pp. 17–18). Putnam instead characterizes truths as fallible, historically contingent, human productions that are subject to correction. His goal in adopting this view is to rehabilitate the notion of truth while simultaneously rejecting the view that Truth should be ascertained in terms of correspondence with Reality. In his view, it is imperative to preserve the notion of truth given that people make and risk acting on the basis of their truth claims in both mundane and scientific realities.

Putnam demonstrates convincingly that scientists and the proverbial person in the street both *act as* conceptual relativists. However, in the cases that he considers, the various perspectives are well established and are not subject to dispute. They therefore contrast sharply with the fluid situation in mathematics education research where theoretical perspectives continue to proliferate and are openly contested. Pragmatic realism is nonetheless relevant in that it provides an initial orientation as we begin to address the meta-issue of how we might compare and contrast different perspectives. For example, it leads us to question claims made by adherents to a particular perspective that their viewpoint gets the world of teaching and learning right—that the phenomena that they take as real correspond to Reality. Arguments of this type involve what Putnam terms the fiction that we can usefully imagine a God's Eye point of view from which we can decide which perspective matches Reality. Pragmatic realism instead orients us to acknowledge that we make choices when we adopt a particular theoretical perspective, and that these choices reflect particular interests and concerns. In addition, pragmatic realism alerts us to the danger of inferring from the conclusion that there is no neutral algorithm of theory choice that any theoretical perspective is as good as any other. In Putnam's view, we can counter this "anything goes" claim by making the criteria that we use when justifying our theoretical choices an explicit focus of scrutiny and discussion. It is to contribute to such a discussion that I have proposed two potentially revisable criteria for comparing and contrasting background theoretical perspectives in mathematical education research. The first criterion is concerned with the nature of the realities that are investigated by researchers who adopt different theoretical perspectives whereas the second focuses on the extent to which research conducted within a particular perspective can contribute to the enterprise of formulating, testing, and revising conjectured designs for supporting envisioned learning processes. In the next two sections of this chapter, I sharpen each of these criteria in turn.

THE NOTION OF THE INDIVIDUAL AS CONCEPTUALLY RELATIVE

Historically, mathematics education researchers looked to cognitive psychology as a primary source of theoretical insight. However, during the last fifteen years, a number of theoretical perspectives that treat individual cognition as socially and culturally situated have become increasingly prominent. In their historical overview of the field, De Corte et al.

(1996) speak of first wave and second wave theories to distinguish between these two types of theories. The goal of first-wave theories is to model teachers' and students' individual knowledge and beliefs by positing internal cognitive structures and processes that account for their observed activity. They portray the emergence of second-wave theories as a response to the limited emphasis on affect, context, and culture in first-wave research. Theories of this type typically treat teachers' and students' cognitions as situated with respect to their participation in particular social and cultural practices.

The distinction that De Corte et al. (1996) draw between cognitive and situated perspectives has become institutionalized within the mathematics education research community and serves as the primary way in which we compare and contrast theoretical perspectives. These comparisons focus on the extent to which particular perspectives take individual cognitive processes or collective social and cultural processes as primary. Although these comparisons throw important differences between perspectives into sharp relief, they are misleading in one important respect. As I will illustrate when I compare the four background theories of experimental psychology, cognitive psychology, sociocultural theory, and distributed cognition, the notion of the individual is conceptually relative. Adherents to these different perspectives conceptualize the individual in fundamentally different ways. Furthermore, these differences are central to the types of questions that adherents to the four perspectives ask, the nature of the phenomena that they investigate, and the forms of knowledge they produce. I will therefore concretize the first of the two criteria that I have proposed for comparing and contrasting theoretical perspectives by teasing out these differences.

I can best exemplify the difficulties that arise when the notion of the individual is taken as self evident by previewing my discussion of distributed theories of intelligence. The term distributed intelligence is perhaps most closely associated with Pea (1985; 1987; 1993). A central assumption of this theoretical perspective is that intelligence is distributed "across minds, persons, and symbolic and physical environments, both natural and artificial" (Pea, 1993, p. 47). Dörfler (1993) clarified the relevance of this theoretical perspective for design and research in mathematics education when he argued that thinking

is no longer considered to be located exclusively within the human subject. The whole system made up of the subject and the available cognitive tools and aids realizes the thinking process. . . . Mathematical thinking for instance not only *uses* those cognitive tools as a separate means but they form a constitutive and systematic part of the thinking process. The cognitive models and symbol systems, the sign systems, are not merely means for expressing a qualitatively distinct and purely mental thinking process. The latter realizes itself and consists in the usage and development of the various cognitive technologies. (p. 164)

This theoretical orientation is consistent with the basic Vygotskian insight that students' use of symbols and other tools profoundly influences both the process of their mathematical development and its products, increasingly sophisticated mathematical ways of knowing (Dörfler, 2000; Hall & Rubin, 1998; Kaput, 1994; Lehrer & Schauble, 2000; Meira, 1998; van Oers, 2000).

In developing his distributed perspective, Pea (1993) has been outspoken in delegitimizing analyses that take the individual as a unit of analysis. In his view, the functional system consisting of the individual, tools, and social contexts is the appropriate unit. Not surprisingly, Pea's admonition has been controversial. For example, Solomon (1993) responded by arguing that, in distributed accounts of intelligence, "the individual has been dismissed from theoretical consideration, possibly as an antithesis to the excessive emphasis on the individual by traditional psychology and educational approaches. But as a result the theory is truncated and conceptually unsatisfactory" (p. 111). It is worth pausing before adopting one or the other of these opposing positions to clarify what Pea might mean when he speaks of the individual. His arguments indicate that individuals, as he conceptualizes them, do not use tools and do not take account of context as they act and interact. They appear to be very much like individuals as portrayed by mainstream cognitive science who produce observed behaviors by creating and manipulating internal symbolic representations of the external environment. Pea's proposal is to equip these encoders and processors of information with cultural tools and place them in social context. This proposition is less contentious than his apparent claim that all approaches that focus on the quality of individuals' reasoning should be rejected regardless of how the individual is conceptualized. The relevant issue once we clarify what we are talking about is not whether it is legitimate to focus on individual teachers' and students' reasoning. Instead, it is how we might usefully conceptualize the individual given our concerns and interests as mathematics educators.

USEFULNESS AND TRUTH

A concern for the usefulness of different theoretical perspectives carries with it the implication that they can be viewed as conceptual tools. This metaphor fits well with the characterization of mathematics education as a design science and orients us away from making forced choices between one perspective and another. However, for this criterion to itself do useful work, we have to clarify what it means for a theoretical perspective to contribute to the collective enterprise of mathematics education. Prawat's (1995) discussion of three types of pragmatic justification identified by Pepper (1942) is relevant in this regard.

The first type of pragmatic justification is purely instrumental in that actions are judged to be true if they enable the achievement of goals. Prawat (1995) indicates the limitations of this formulation when he observes that "a rat navigating a maze has as much claim on truth, according to this approach, as the most clear-headed scientist" (p. 19). In the second type of pragmatic justification,

> it is not a successful act that is true, but the hypothesis that leads to the successful act. When one entertains a hypothesis, Pepper (1942) points out, it is in anticipation of a specific outcome. When that outcome occurs, the hypothesis is verified or judged true. (Prawat, 1995, p. 19)

As Prawat goes on to note, this is a minimalist view in that hypotheses are treated as tools for the control of nature. This portrayal sits uncomfortably with Tolumin's (1963) demonstration that the primary goal of science is to develop insight and understanding into the phenomena under investigation, and that instrumental control is a by-product.

The third type of pragmatic justification, which Pepper (1942) calls the qualitative confirmation test of truth, brings the development of insight and understanding to the fore. This type of justification builds on Dewey's analysis of the function of thought and the process of verification. Dewey viewed ideas as potentially revisable plans for action and argued that their truth is judged in terms of the extent to which they lead to a satisfactory resolution to problematic situations (Westbrook, 1991). In taking this stance, he equated experience with physical and conceptual action in a socially and culturally organized reality, and maintained that it is characterized by a future-oriented projection that involves an attempt to change the given situation (Sleeper, 1986; J. E. Smith, 1978). He therefore contended that foresight and understanding are integral to thought, the primary function of which is to project future possibilities and to prepare us to come to grips with novel, unanticipated occurrences.

These aspects of thought are central to Dewey's analysis of verification as a process in which the phenomena under investigation talk back, giving rise to surprises and inconsistencies. He clarified that some of these surprises can be accounted for relatively easily by elaborating underlying ideas, whereas others constitute conceptual impasses that typically precipitate either the reworking of ideas or their eventual rejection (Prawat, 1995). The crucial point to note is that, for Dewey, verification constitutes a context in which theoretical ideas are elaborated and modified. As Pepper (1942) put it, ideas judged to be true are those that give *insight* into what he termed the texture and quality of the phenomena that serve to verify them. In the case of mathematics education, for example, ideas are potentially useful to the extent that they give rise to conjectures about envisioned learning processes and the specific means of supporting them. However, these ideas are not simply either confirmed unchanged or rejected during the process of testing and revising designs. Instead, Dewey drew attention to the process by which ideas that are confirmed evolve as they are verified. In his formulation, *the truth of fallible, potentially revisable ideas is justified primarily in terms of the insight and understanding they give into learning processes and the means of supporting their realization.* This is the criterion that I will use when I focus on the potentially useful work that experimental psychology, cognitive psychology, sociocultural theory, and distributed cognition might do in contributing to the enterprise of formulating, testing, and revising conjectured designs for supporting envisioned learning processes.

As a point of clarification, it is important to note that Dewey's intent in analyzing verification processes was descriptive rather than prescriptive. He was not attempting to specify rules or norms for verification to which scientists ought to adhere. Instead, he sought to understand verification as it is actually enacted in the course of inquiry. He claimed that regardless of the pronouncements of philosophers who strive to identify the Method for determining Truth, verification as it is enacted by scientists is an interpretive process in which they modify and elaborate theoretical ideas. Dewey's analysis has far reaching implications in that it challenges a distinction central to the traditional project of epistemology, that between the context of discovery and the context of justification. As Bernstein (1983) explains,

> while no contemporary philosopher of science has wanted to claim that there is a determinate decision procedure or method for advancing scientific discovery, many have been firmly convinced that

there are (or ought to be) permanent procedures for testing and evaluating rival theories. This is the basis of the "orthodox" distinction between the context of discovery and the context of justification. The latter has been taken to be the proper domain of the philosopher of science. His or her task is to discover, specify and reconstruct the criteria. (p. 70)

Popper's (1972) contention that to be scientific, hypotheses and theory have to be open to falsification is the most widely cited proposal of this type. As a rough rule of thumb, Dewey would probably not have disputed Popper's claim. However, his analysis of verification indicates that he would add that the application of this rule necessarily involves wisdom and judgment. In this regard, he would concur with Bernstein's observation that

> Popper frequently writes as if we always know in advance what will count as a good argument or criticism against a conjecture. The basic idea behind the appeal to falsification as a demarcation criterion between science and nonscience is that there are clear criteria for determining under what conditions a conjecture or hypothesis is to be rejected. (p. 70)

As Dewey pointed out and as Kuhn (1962; 1977), Lakatos (1970), and Feyerabend (1975) subsequently underscored, the situation becomes less straightforward than Popper's proposal implies when we focus on the actual practices of scientists. Although the phenomena under investigation talk back, they do not serve as a jury that tell us unambiguously whether a theoretical idea should be accepted or rejected.

> We frequently do not know, in a concrete scientific situation, whether we are confronted with an obstacle to be overcome, a counterinstance that can be tolerated because of the enormous success of the theory, or with evidence that should be taken as falsifying our claim. Data or evidence do not come marked "falsification"; in part, it is we who decide what is to count as a falsification or refutation. (Bernstein, 1983, p. 71)

Furthermore, as Dewey in particular emphasized, the ideas under scrutiny are moving targets that evolve as scientists, in effect, engage in a dialogue with the phenomena under investigation.

Latour and Wollgar's (1979) analysis of activity in an organic chemistry laboratory and Pickering's (1984) of a high energy physics laboratory substantiate Dewey's analysis of verification, particularly as the work carried out in both laboratories resulted in Nobel prizes. The case studies reveal that the scientists did not simply formulate conjectures, conduct experiments to test them, and then passively let the resulting data determine which were confirmed and which were false. To be sure, the scientists had theoretical ideas and conjectures expressed in terms of those ideas. However, they also had views about how the experimental apparatus functioned, how it could be used, and so forth. As Latour and Wollgar, and Pickering both document, it was typical for the apparatus to initially not behave as the scientists expected. In these situations, the scientists attempted to adjust to the perceived anomaly in a number of ways that included revising the theory under investigation, revising their views of how the apparatus functioned, or tinkering with and rebuilding the apparatus itself (Traweek, 1988). Their goal in doing so was to achieve what Pickering (1984) terms a robust fit between these different aspects of the experimental situation. Furthermore, the scientists judged the theoretical ideas that they successfully revised and elaborated in the course of this problem-solving process to be true because they gave them insight into the phenomena they were investigating. Latour and Wollgar's and Pickering's analyses of the process by which scientists produce truths as they engage in a dialogue with nature are consistent with Dewey's contention that truths are made through the process by which they are verified (Westbrook, 1991). This pragmatic notion of truth and, relatedly, of usefulness underpins my comparison of different theoretical perspectives in terms of their potential to contribute to the enterprise of formulating, testing, and revising conjectured designs for supporting envisioned learning processes.

COMPARING THEORETICAL PERSPECTIVES

My purpose in comparing and contrasting the four theoretical perspectives is to illustrate the relevance of the two criteria I have proposed: (a) How the individual is conceptualized in the differing perspectives, and (b) the potential of the perspectives to contribute to our understanding of learning processes and the means of supporting their realization. I therefore make no pretense at providing a comprehensive overview of theoretical perspectives in mathematics education. For example, I largely ignore both symbolic interactionism (Bauersfeld, 1980, 1988; P. Cobb, Wood, Yackel, & McNeal, 1992; Voigt, 1985, 1996) and discourse and communicative perspectives (Ernest, 1994; Pimm, 1987, 1995; Rotman, 1988, 1994; Sfard, 2000a, 2000b). Experimental psychology refers to the psychological research tradition in which the primary methods employed involve experimental and quasi-experimental designs, preferably with the

random assignment of subjects. My decision to focus on this perspective has been influenced by the strong advocacy for research designs of this type by personnel in the Institute of the Educational Sciences and other U.S. government funding agencies.

My discussion of cognitive psychology is limited to theoretical orientations that involve what MacKay (1969) termed the actor's viewpoint. The goal of psychologies of this type is to account not merely for teachers' and students' observed behaviors but for their inferred interpretations and understandings in terms of internal cognitive structures and processes. This restricted focus is premised on the observation that the relationships we establish with teachers and students in the course of our work as mathematics educators frequently involve collaboration and mutual engagement. As Rommetveit (1992) and Schutz (1962) both illustrate, relationships of this type involve communicative interactions characterized by a reciprocity of perspectives typical of the actor's viewpoint. I will therefore have little to say about cognitive approaches that involve the observer's viewpoint (MacKay, 1969) and that focus on internal cognitive processes that intervene between an observed stimulus environment and observed response activity. Sociocultural theory and distributed cognition are both second wave perspectives as described by De Corte et al. (1996) and provide points of contrast with cognitive psychology by viewing individual activity as situated with respect to social and cultural practices. As I will clarify, sociocultural theory has developed largely independently of western cognitive psychology by drawing inspiration directly from the writings of Vygotsky and Leont'ev whereas distributed cognition has emerged in reaction to information-processing psychology and incorporates aspects of the Soviet work.

Experimental Psychology

In speaking of experimental psychology, I refer to the psychological research tradition whose primary contributions to mathematics education have involved the development of assessment instruments, particularly norm-referenced tests, and the findings of studies that have assessed the relative effectiveness of alternative curricular and instructional approaches. This perspective merits attention given the far reaching implication of a recent legislative initiative in the United States known as the No Child Left Behind (U.S. Congress, 2001). As Slavin (2002) observes, the Act "mentions 'scientifically based research' 110 times. It defines 'scientifically based research' as 'rigorous, systematic and objective procedures to obtain valid knowledge,' which includes research that 'is evaluated using experimental or quasi-experimental designs,' preferably with random assignment" (p. 15). Slavin goes on to note that the recently established Institute for the Educational Sciences, a major federal funding agency, is organized "to focus resources on randomized and rigorously matched experimental research on programs and policies that are central to the education of large numbers of children" (p. 15). Whitehurst (2003, April), the Institute's Director, addresses this point when he states that although "interpretations of the results of randomized trials can be enhanced with results from other methods" (p. 9), "randomized trials are the only sure method for determining the effectiveness of education programs and practices" (p. 6). In a nutshell, "randomized trials are the gold standard for determining what works" (p. 8). As Whitehurst's reference to "what works" indicates, the emphasis on a particular research method profoundly influences both the nature of the questions asked and the forms of knowledge produced. In this regard, Slavin (2004) clarifies that well-designed studies of this type are not limited to x versus y comparisons but can "also characterize the conditions under which x works better or worse than y, the identity of the students for whom x works better or worse than y, and often produce rich qualitative information to supplement the quantitative comparisons" (p. 27).

Knowledge claims of this type are premised on a particular conception of the individual. In teasing out this conception and thus the nature of the reality that is the focus of investigation, it is important to distinguish between the abstract, collective individual to whom knowledge claims refer and the individual students who participate in experiments. This collective individual is a statistical aggregate that is constructed by combining measures of psychological attributes of the participating students (e.g., measures of mathematical competence as measured by achievement test scores). As Danziger (1990) demonstrates in his seminal historical analysis of experimental psychology, the purpose of experimentation is to make predictions about how certain variations in instructional conditions affect the performance of this abstract individual as assessed by aggregating the scores of individual students. This statistically constructed individual is abstract in the sense that it need not correspond to any particular student. The construction of this collective individual enables investigators who work in this tradition to avoid the issue of individual differences and the challenges involved in accounting for the reasoning and learning of specific students (Danziger, 1990). This is accomplished by treating differences in measures of students' performance as

error variance. Students are then characterized by the extent to which their performance deviates from group norms that measure the performance of the collective individual.

This methodological approach of investigating the performance of the abstract, collective subject rests on two underlying assumptions. The first is that students are composed of discrete, isolatable attributes or qualities that vary only in degree from one student to another (Danziger, 1990). This assumption makes it legitimate to combine measures of students' performance. Students so portrayed are therefore limited to possessing larger or smaller measurable amounts of these psychological attributes. Second, the environments in which students acquire these capacities are composed of independent features that the investigator can manipulate and control directly. The implicit ontology is that of environmental settings made up of separate independent variables and students composed of collections of dependent psychological attributes. Together, these two theoretical suppositions ground investigations that seek to discern causal relationships between the manipulation of instructional conditions on the performance of the collective, abstract individual. This theoretical underpinning indicates that experimental psychology should be treated as a theoretical perspective rather than merely as a set of methodological prescriptions.

The forcefulness with which adherents of this perspective have attempted to advance their viewpoint has elicited a number of responses. One line of critique claims that the approach of casting students as objects in studies that focus on their responses to environmental manipulations is ethically dubious and potentially dehumanizing. As these critiques target the morality of investigators working in this tradition, it is worth noting with Porter (1996) that

> social quantification means studying people in classes, abstracting away their individuality. This is not unambiguously evil, though of late it has been much criticized. Much, probably most, statistical study of human populations has aimed to improve the condition of working people, children, beggars, criminals, women, or racial and ethnic minorities. (p. 77)

A second line of attack develops the argument that experimental psychology lacks theoretical depth. For example, Danziger (1990) contends that this analytic approach both separates people from the social contexts in which their actions take on significance, and eschews the study of the interpretations that they make in those contexts. Criticisms of this type miss the mark in my view because they use the theoretical commitments of alternative theoretical perspectives as criteria against which to assess experimental psychology. Judgments of theoretical depth do not transcend research traditions but are instead conceptually relative to the norms and values of particular research communities. Experimental psychologists, for example, make judgments of theoretical depth by gauging the extent to which the findings of investigations contribute to their collective understanding of the psychology of the abstract, statistically constructed individual. More generally, exchanges between experimental psychologists and their opponents typically involve people talking past each other as they point to features of the different realities that they investigate. In my view, it is more productive to assess the value of experimental psychology in terms of its potential contributions to a specific enterprise such as that of formulating, testing, and revising conjectured designs for supporting mathematical learning. There is, however, one issue that requires additional scrutiny before we consider the potential usefulness of this theoretical perspective.

Advocates of experimental psychology frequently argue with considerable vehemence that their perspective and its associated research methods are scientific, and that all other perspectives and methodological approaches are not. As the case studies conducted by Pickering (1984) and by Latour and Woolgar (1979) indicate, this claim is not based on an examination of the practices of scientists whose work is widely acknowledged to be of the highest caliber. The claim instead capitalizes on folk beliefs about what it means to be scientific and objective. In the history of the discipline, what was important

> was the widespread acceptance of a set of firm convictions about the nature of science. To be socially effective, it was not necessary that these convictions actually reflected the essence of successful scientific practice. In fact, most popular beliefs in this area were based on external and unanalyzed features of certain practices in the most prestigious parts of science. Such beliefs belong to the rhetoric of science rather than its substance. . . . Such unquestioned emblems of scientific status included features like quantification, experimentation, and the search for universal (i.e., ahistorical) truths. (Danziger, 1990, p. 120)

It is instructive to compare the features of scientific practice that advocates of experimental psychology deem to be critical with those identified by the National Research Council (2002):

Scientific Principle 1
Pose Significant Questions That Can Be Investigated Empirically

Scientific Principle 2
Link Research to Relevant Theory

Scientific Principle 3
Use Methods That Permit Direct Investigation of the Question

Scientific Principle 4
Provide a Coherent and Explicit Chain of Reasoning

Scientific Principle 5
Replicate and Generalize Across Studies

Scientific Principle 6
Disclose Research to Encourage Professional Scrutiny and Critique (pp. 3–5)

Eisenhart and Towne (2003) clarify that these principles were identified by "reviewing *actual* research programs—both basic and applied—in natural science, social science, education, medicine, and agriculture" (p. 33, italics in the original). As they note, scientifically based research is best defined not by the employment of particular research methods but by characteristics that cut across a range of methods. I would only add that the use of these or any other set of principles to assess specific research programs necessarily involves interpretations and judgments of type that all scientists including experimental psychologists make in the course of their practice.

It is ironic that the claim that experimental psychology has hegemony over what is regarded as scientific lacks empirical grounding and is contradicted by the available evidence. This claim is ideological in the pernicious sense of the term. It rests on the indefensible proposition that the methods of experimental psychology are theory neutral and constitute the only guaranteed means of gaining access to Reality with a capital "R". As I have demonstrated, experimental psychologists' use of these methods is theory laden and reflects theoretical suppositions and assumptions about the nature of the reality that they are investigating. In using these methods, they attempt to gain insights into the psychology of the statistically constructed collective individual, a character that is composed of a set of isolatable psychological characteristics and that inhabits a world made up of manipulable independent variables. This observation does not threaten the credibility of experimental psychology as a viable research tradition: The research practices that any research community establishes as normative necessarily entail theoretical commitments. The observations, however, do undermine the claim that experimental psychologists have found the Method for discerning Truth. The failure of many of their most forceful advocates to acknowledge that research in this tradition is conducted from a theoretical perspective closes down debate and, in my view, highlights the political nature of boundary disputes between science and non-science. Adherents to experimental psychology frequently state that their primary motivation in conducting investigations is to improve the students' intellectual and moral welfare. I accept these statements at face value but question the implication that insights developed in any other perspective can make at best marginal contributions to students' intellectual and moral welfare. This brings us to the issue of usefulness.

Adherents of experimental psychology repeatedly emphasize their commitment to conduct research that is pragmatically relevant. For example, Whitehurst (2003, April) stresses that "the primary focus for the Institute [of Educational Sciences] will be on work that has high consideration of use, that is practical, that is applied, that is relevant to practitioners and policy makers" (p. 4). In addressing this and similar claims, it is important to consider for whom the forms of knowledge produced by experimental psychology might be useful. Danziger's (1990) historical analysis is pertinent in this regard. He reports that

> after the turn of the [twentieth] century, psychologists' relations with teachers became increasingly overshadowed by a new professional alliance which was consummated through the medium of a new set of investigative practices. The group of educators with whom psychologists now began to establish an important and beneficial professional alliance consisted of a new generation of professional educational administrators. This group took control of a process of educational rationalization that adapted education to the changed social order of corporate industrialism. The interests of the new breed of educational administrator had little in common with those of the classroom teacher. Not only were the administrators not directly concerned with the process of classroom teaching, they were actually determined to separate their professional concerns as much as possible from those of the lowly army of frontline teachers. In this context, they emphasized scientific research as a basis for the rationalized educational system of which they were the chief architects. In the United States the needs of educational administration provided the first significant external market for the products of psychological research in the years immediately preceding World War I. (p. 103)

Danziger goes on to clarify that knowledge about the responses of the statistically constructed collective individual to different instructional conditions served the needs of administrators who managed institutions in which instruction is carried out in groups. Furthermore, the conception of learning environments as composed of manipulable independent variables was directly relevant to administrators who sought to both distance themselves from and manage classroom instructional processes. Danziger also demonstrates that the consequences of this alliance were at least as profound for experimental psychology as they were for education. For example, it was during this period that the goal of gaining insight into individual mental processes was displaced by that of putting psychological prediction at the service of administrative needs. In addition, the high value that experimental psychologists place on quantification was influenced by the usefulness of statistical constructions based on group data to administrators. On the basis of these observations, Danziger concludes that experimental psychology was transformed in large measure into what he terms an administrative science in that it investigates the reality that administrators seek to manage.

In terms of the metaphor of a theoretical perspective as a conceptual tool, Danziger's analysis indicates that experimental psychology is a tool that has been fashioned to produce forms of knowledge that fulfill administrative concerns. The resulting forms of knowledge enable administrators who are removed from the classroom and who do have little if any specialized knowledge of teaching and learning in particular content domains to make informed decisions about the curricula and instructional strategies that teachers should use. Although these forms of knowledge can be of some relevance to teachers, they are less well suited to the contingencies of classroom teaching. This is because the knowledge produced of individual students is, as Danziger (1990, p. 165) puts it, "a knowledge of strangers" who are known only through their standing in the group. It does not therefore touch on the challenges, dilemmas, and uncertainties that arise in the classroom as teachers attempt to achieve a mathematical agenda while simultaneously taking account of their students' proficiencies, interests, and needs (Ball, 1993; Davis, 1997; Fennema, Franke, & Carpenter, 1993; Lampert, 1990, 2001). I therefore question the frequent assertion made by proponents of experimental psychology that the forms of knowledge it produces are of equal relevance to teachers and administrators. It is significant that the examples that proponents cite to illustrate the need for these forms of knowledge almost invariably concern administrators and policymakers but not classroom teachers. For example, Whitehurst (2003) does not mention classroom teachers when he explains that personnel at the Institute of Educational Sciences "recently completed a survey of a purposive sample of our customers to determine what they think we ought to be doing to serve their needs. The sample included school superintendents and principals, chief state school officers, and legislative policy makers" (p. 5).

In summary, the arguments that adherents of experimental psychology advance in support of their perspective are based on the claims that it has hegemony over the production of scientific knowledge about teaching and learning, and that these forms of knowledge are pragmatically useful. I have suggested that the first of these claims is primarily ideological but that the second has merit provided we clarify for whom the resulting forms of knowledge are useful. Proponents of this perspective offer us a bold vision in which studies involving experimental and quasi-experimental designs constitute the primary basis for an ongoing process of educational improvement. It is worth noting that this vision is based on belief rather than evidence and is not supported by the historical record. There was an initial period of enthusiasm for such a vision in the early part of the last century, but this enthusiasm has since waned.

> In spite of these promising beginnings, the history of treatment-group methodology in research in education and educational psychology was not exactly a march of triumph. . . . In due course . . . a certain pessimism about the prospects of experimental research in education began to set in. The claims made on behalf of the quantitative and experimental method had undoubtedly been wildly unrealistic, and in the light of changing priorities the illusions of the early years were unable to survive. (Danziger, 1990, p. 115)

In my judgment, the claim that sustained improvements in mathematics teaching and learning can be made by relying *almost exclusively* on the findings of treatment-group studies is untenable. To be sure well-designed studies of this type can, in all probability, make an important contribution. However, advocates of experimental psychology should, in my view, moderate their rhetoric lest the history of high hopes and dashed expectations repeat itself.

Cognitive Psychology

In comparing and contrasting cognitive psychology with other theoretical perspectives, I restrict my focus to theoretical approaches that seek to account for teachers' and students' inferred interpretations and understandings in terms of internal cognitive structures and processes. Cognitive approaches of this type take on a challenge sidestepped by experimental psychology, that of accounting for specific students' and teachers' mathematical reasoning and learning. It is useful to distinguish between two general types of theories developed within this tradition. The first are theories of the *process* of mathematical learning that are intended to offer insights into students' learning in any mathematical domain, whereas the second are theories of the development of students' reasoning in specific mathematical domains. Pirie and Kieren's (1994) recursive theory of mathematical understanding serves to illustrate theories of the first type. They differentiate a sequence of levels of mathematical reasoning and model mathematical understanding as a recursive phenomenon that occurs as thinking moves between levels of sophistication. In their theoretical scheme, students who are novices to a particular mathematical domain initially make images of either their situation-specific activity or its results. The first significant development occurs when students can take such images as givens and do not have to create them anew each time. Later developments involve students noticing properties of their mathematical images and subsequently taking these properties as givens that can be formalized. One of the notable features of Pirie and Kieren's theory of the growth of mathematical understanding is its broad scope in tracing development from the creation of images to formalization and axiomatization.

Additional examples of theories of the process of mathematical learning include Dubinsky's (1991) theory of encapsulation, Sfard's (1991; Sfard & Linchevski, 1994) theory of reification, Dörfler's (1989) analysis of protocols of action, and Vergnaud's (1982) analysis of the process by which students gradually explicate their initial theorems-in-action. Each of these theorists consider the primary source of increasingly sophisticated forms of mathematical reasoning to be students' activity of interpreting and attempting to complete instructional activities, not the instructional activities themselves. Furthermore, they each characterize mathematical learning as a process in which operational or process conceptions evolve into what Sfard (1991) terms object-like structural conceptions. Proficiency in a particular mathematical domain is therefore seen to involve the conceptual manipulation of mathematical objects whose reality is taken for granted. Greeno (1991) captured this aspect of mathematical proficiency when he introduced the metaphor of acting in a mathematical environment in which tools and resources are ready at hand to characterize number sense. For her part, Sfard (2000b) describes mathematical discourse as a virtual reality discourse to highlight the parallels between this discourse and the ways in which we talk about physical reality. The intent of each of the theoretical schemes I have referenced is to provide a conceptual framework or toolkit that can be used to develop accounts of the process of specific students' mathematical learning (Thompson & Saldanha, 2000).

In contrast to these general conceptual frameworks, theories of the second type focus on the development of students' reasoning in specific mathematical domains. Examples include analyses of early number reasoning (Carpenter & Moser, 1984; Fuson, 1992; Steffe & Cobb, 1988), multiplicative reasoning (Confrey & Smith, 1995; Streefland, 1991; Thompson, 1994; Vergnaud, 1994), geometric reasoning (Clements & Battista, 1992; van Hiele, 1986), algebraic reasoning (Filloy & Rojano, 1984; Kaput, 1999; Sfard & Linchevski, 1994), and statistical reasoning (Konold & Higgins, in press; Mokros & Russell, 1995; Saldanha & Thompson, 2001). It is important to note that these domain specific frameworks do not focus on the mathematical development of any particular student but are instead, like theories of the first type, concerned with the learning of an idealized student that Thompson and Saldanha (2000) refer to as the *epistemic individual*. Researchers working in this cognitive tradition account for variations in specific students' reasoning by using the constructs that comprise their framework to develop explanatory accounts of each student's mathematical activity. This approach enables the researchers to both compare and contrast the quality of specific students' reasoning and to consider the possibilities for their mathematical development.

The common element that ties both types of theories together is their portrayal of both the epistemic individual and of specific students as active constructors of increasingly sophisticated forms of mathematical reasoning. The metaphor of learning as a process of construction can be traced to the eighteenth century Italian philosopher Giambattista Vico who was the first to advance an explicitly constructivist position when he argued that "the known is the made" (Berlin, 1976). Vico's arguments anticipated several of the major claims that the German philosopher Immanuel Kant made 80 years later. In his treatise *A Critique of Pure Reason* (Kant, 1998), Kant contended that our perceptions are always in the form of objects because our minds have a priori structures or intuitions of space and time. He also argued that in addition to being perceived in space and time, objects are experienced through four

a priori categories of understanding: quantity (e.g., plurality, totality), quality (e.g., negation), relation (e.g., causality and dependence), and modality (e.g., possibility and necessity).

Kant's philosophical analysis provided the backdrop against which the most significant contributor to constructivism in cognitive psychology, Jean Piaget, conducted his research (Fabricius, 1979). Although Piaget is typically viewed as a psychologist, he described himself as a genetic epistemologist (Piaget, 1970). In this context, the term genetic denotes genesis and encompasses both the origins and development of forms of knowledge. Piaget explained that, ideally, he would have studied the origins and subsequent evolution of foundational categories of understanding as they occurred in the history of humankind. However, as this was impossible, he attempted to gain insight into epistemological issues by studying the development of these notions in children. Significantly, the concepts that he focused on in his investigations correspond almost exactly to the fundamental categories of thought proposed by Kant 150 years earlier. His intent was to account for the development of these categories by relying on constructs such as assimilation, accommodation, and equilibration thereby showing that there is no need to posit that they are *a priori*.

The aspect of Piaget's work that has gained most attention among both psychologists and educators is his claim to have identified a sequence of invariant stages through which children's thinking progresses. The cognitive theorists I have referenced have, in contrast, focused primarily on the process aspects of Piaget's theory. Piaget drew on his early training as a biologist to characterize intellectual development as an adaptive process in the course of which children reorganize their sensory-motor and conceptual activity (Piaget, 1980). In appropriating Piaget's general constructivist orientation to the process of development, cognitive theorists have necessarily had to adapt his theoretical constructs because their concern is to gain insight into the process of students' mathematical learning rather than to address problems of genetic epistemology. The resulting characterizations that they propose of the individual differ significantly from that of experimental psychology.

As we have seen, experimental psychology focuses on how variations in instructional conditions affect the performance of the statistically constructed collective subject. In contrast, cognitive psychology focuses on how the epistemic individual successively reorganizes its activity and comes to act in a mathematical environment. Thus, whereas experimental psychologists direct their attention to an environment composed of manipulable independent variables, cognitive psychologists seek to delineate how the world of meaning and significance in which the epistemic individual acts changes in the course of development. We have also seen that experimental psychology characterizes specific students in terms of the deviation of their performance from that of the collective subject. In contrast, cognitive psychology characterizes specific students in terms of the nature or quality of their mathematical reasoning. Thus, whereas experimental psychology is premised on the assumption that students possess large or small amounts of psychological attributes, cognitive psychology is premised on the assumption that students' development involves *qualitative* changes in their mathematical reasoning.

In cases such as this and other cases where characterizations of the individual contrast sharply, it is tempting to try and determine which perspective gets people right. Following Putnam (1987), I have argued that this quest is misguided and that it is more productive to compare and contrast differing theoretical perspectives in terms of their potential usefulness. In this regard, I have suggested that the forms of knowledge produced by experimental psychology are particularly useful to administrators who are responsible for managing educational systems at some distance removed from the classroom and who have little grounding in classroom teaching and learning processes in particular content domains. Given their administrative concerns and interests, it is unlikely that they would see value in the forms of knowledge produced by cognitive psychologists working in the tradition on which I have focused. Conversely, having conducted studies that involve experimental designs (P. Cobb et al., 1991; P. Cobb, Wood, Yackel, & Perlwitz, 1992), I can attest that the resulting forms of knowledge are not well suited to the demands of instructional design at the classroom level. A fundamental difficulty is that studies involving experimental designs do not produce the detailed kinds of data that are needed to guide the often subtle refinements made when improving an instructional design at this level.

Proponents of the cognitive tradition that I have discussed frequently assume that the types of explanatory frameworks they produce constitute an adequate basis for both classroom instructional design and pedagogical decision making. I have contributed to the development of such a framework (Steffe & Cobb, 1988; Steffe, von Glasersfeld, Richards, & Cobb, 1983) and have attempted to use it to guide the development and refinement of classroom instructional designs. On the basis of this experience, I question the claim that such frameworks are, by themselves, sufficient. The primary difficulty is precisely that the forms of knowledge produced are cognitive rather than instructional and do not involve positive heuristics for design (Gravemeijer, 1994b). In my view, researchers working in this tradition have inherited perspectives and associated methodologies from cognitive psychology but have not always reflected on the relevance of the forms of knowledge produced to the collective enterprise of mathematics education. The issue of how these perspectives and methodologies might be adapted so that they can better contribute to the concerns and interests of mathematics education has rarely arisen because their sufficiency has been assumed almost as an article of faith (see Thompson, 2002, for a rare discussion of these issues).

Given that cognitive theories of mathematical learning do not, by themselves, constitute a sufficient basis for design, the question of clarifying the contributions that they can make remains. Drawing on my experience of developing and revising designs at the classroom level, I can identify three contributions of theories of development in particular mathematical domains that I do not claim are exhaustive. First, domain-specific theories typically include analyses of the forms of reasoning that we want students to develop. These analyses are non-trivial accomplishments and can serve to specify the "big mathematical ideas," thereby giving an overall orientation to the instructional design effort. Second, domain-specific frameworks can alert the designer to major shifts in students' mathematical reasoning that the design should support. Third, the designer's or teacher's use of a domain-specific framework to gain insight into specific students' mathematical reasoning can inform the design of instructional activities intended to support subsequent learning. This use of a cognitive framework is particularly evident in the successful Cognitively Guided Instruction program developed by Carpenter, Fennema, and colleagues (Carpenter & Fennema, 1992; Carpenter, Fennema, Peterson, Chiang, & Loef, 1989; Fennema et al., 1993). As I will clarify when I discuss distributed cognition, I consider this emphasis on instructional activities as the primary means of supporting students' mathematical learning to be overly restrictive. It is also worth noting that domain-specific cognitive frameworks do not orient designers to consider issues of equity in students' access to significant mathematical ideas even though the resulting designs can, on occasion, make a significant contribution in this regard (Carey, Fennema, Carpenter, & Franke, 1995; Silver, Smith, & Nelson, 1995).

In summary, general and domain-specific theories of mathematical learning are both concerned with the learning of an idealized student, the epistemic individual. Variations in specific students' reasoning are accounted for by using the constructs central to theories of this type to develop explanations of their mathematical activity. Although there are substantive differences between the various cognitive theories I have referenced, they are tied together by the characterization of mathematical learning as a constructive process in the course of which students successively reorganize their sensory-motor and conceptual activity. From the perspective of experimental psychology, this multiplicity of theories is frequently taken as evidence that the cognitive tradition on which I have focused is not scientific. It is therefore worth reiterating that experimental psychologists have achieved unanimity only by eschewing a concern for cognitive processes and structures that account for observable performance and by treating variation in performance as error variance. Whereas the forms of knowledge produced by experimental psychology are well suited to the concerns of administrators, cognitive theories can contribute to the development and improvement of classroom instructional designs.

Sociocultural Theory

At their core, the forms of knowledge produced by the cognitive theorists on whose work I have focused concern the process by which the epistemic individual successively reorganizes its activity. In contrast, the forms of knowledge produced by sociocultural theorists concern the process by which people develop particular forms of reasoning as they participate in established cultural practices. This theoretical perspective treats intellectual development and the process by which people become increasingly substantial participants in various cultural practices as aspects of a single process. Consequently, whereas cognitive theorists investigate the activity of the epistemic individual, sociocultural theorists investigate the participation of the *individual-in-cultural-practice*. In a very real sense, the two groups of researchers are attempting to understand different realities. It is necessary to clarify the origins of sociocultural theory in order to understand the reality into which it gives insight.

Contemporary sociocultural theory draws directly on the writings of Vygotsky and Leont'ev. Vygotsky (1962; 1978; 1981) made his foundational contributions to this perspective during the period of intellectual ferment and social change that followed the Russian revolution. In doing so, he was profoundly influenced by Marx's argument that it is the making and use of tools that serves to differentiate humans from other animal species. For Vygotsky, human history is the history of artifacts such as language, counting systems, and writing that are not invented anew by each generation but are instead passed on and constitute the intellectual bequest of one generation to the next. In formulating his theory of intellectual development, Vygotsky developed an analogy between the use of physical tools and the use of intellectual tools such as sign systems (Kozulin, 1990; van der Veer & Valsiner, 1991). His central claim was that just as the use of a physical tool serves to reorganize activity by making new goals possible, so the use of sign systems serves to reorganize thought. He viewed culture as a repository of sign systems and other artifacts that are appropriated by children in the course of their intellectual development (Vygotsky, 1978). It is important to stress that for Vygotsky, children's mastery of a counting system does not merely enhance or amplify an already existing cognitive capability. He instead argued that children's ability to reason numerically is created as they appropriate the counting systems of their culture. This example illustrates Vygotsky's more general contention that children's minds are formed as they appropriate sign systems and other artifacts.

In the most well known series of investigations that he conducted, Vygotsky attempted to demonstrate the crucial role of face-to-face interactions in which an adult or more knowledgeable peer supports the child's use of an intellectual tool such as a counting system (Vygotsky, 1981). He concluded from these studies that the use of sign systems initially appears in children's cognitive development on what he termed the "intermental" plane of social interaction and that, over time, the child eventually becomes able to carry out what was previously a joint activity on his or her own. On this basis, he argued that the child's mind is created via a process of internalization from the intermental plane of social interaction to the "intramental" plane of individual thought.

The central role that Vygotsky attributed to social interactions with more knowledgeable others usually features prominently in accounts of his work. However, there is some indication that shortly before his premature death in 1934, he began to view the relation between social interaction and cognitive development as a special case of a more general relation between cultural practices and cognitive development (Davydov & Radzikhovskii, 1985; Minick, 1987). This aspect of sociocultural theory was developed more fully after Vygotsky's death by a group of Soviet psychologists, the most prominent of whom was Alexei Leont'ev.

Although Leont'ev (1978; 1981) acknowledged the importance of face-to-face interactions, he saw the encompassing cultural practices in which the child participates as constituting the broader context of his or her development. For example, Leont'ev might have viewed interactions in which a parent engages a child in activities that involve counting as an instance of the child's initial, supported participation in cultural practices that involve dealing with quantities. He argued that the child's progressive participation in specific cultural practices underlies the development of his or her thinking. Intellectual development was, for him, synonymous with the process by which the child becomes a full participant in particular cultural practices. Because he considered the cognitive capabilities that a child develops to be inseparable from the cultural practices that constitute the context of their development, he viewed those capabilities to be characteristics of the child-in-culture-practice rather than the child per se.

The second contribution that Leont'ev made to sociocultural theory concerns his analysis of material objects and events. Although Vygotsky brought sign systems and other cultural tools to the fore, he gave less attention to material reality. In building on Vygotsky's ideas, Leont'ev argued that material objects as they come to be experienced by the developing child are defined by the cultural practices in which he or she participates. For example, a pen becomes a writing instrument rather than a brute material object for the child as he or she participates in literacy practices. In Leont'ev's view, the child does not come into contact with material reality directly, but is instead oriented to this reality as he or she participates in cultural practices. He therefore concluded that the meanings that material objects come to have are a product of their inclusion in specific practices. This thesis serves to underscore his argument that the individual-in-cultural-practice constitutes the appropriate analytical unit.

To clarify the distinction between the individual as characterized by cognitive and by sociocultural theorists, it is useful to contrast the role attributed to tools and social interactions in the two perspectives. Sociocultural theorists have sometimes accused proponents of the cognitive tradition on which I have focused of adopting a so-called Robinson Crusoe perspective in which they study the reasoning of socially and culturally isolated people. It is therefore important to emphasize that adherents of the cognitive tradition readily acknowledge that students' reasoning is influenced by both the tools that they use to accomplish goals and by their ongoing social interactions with others. However, tools and others' actions are considered to be external to students' reasoning, and the focus is on documenting how students interpret them. Following Vygotsky, sociocultural theorists question the assumption that social processes can be clearly partitioned off from cognitive processes and treated as external conditions for them. These theorists instead view cognition as extending out into the world and as being inherently social. They therefore attempt to break down a distinction that is basic to the cognitive perspective and indeed to experimental psychology, that between the reasoner and the world reasoned about. Furthermore, following Leont'ev, sociocultural theorists typically situate tool use and face-to-face interactions within encompassing cultural practices. From this perspective, students' actions are viewed as elements of a system of cultural practices and students are viewed as participating in cultural practices even when they are in physical isolation from others.

In considering the potential usefulness of sociocultural theory, I should acknowledge that the instructional designs that I and my colleagues have developed in recent years to support students' mathematical learning are broadly compatible with some of the basic tenets of this theoretical perspective (P. Cobb & McClain, 2002). Nonetheless, I contend that sociocultural theory is of limited utility when actually formulating designs at the classroom level. The contributions of Davydov (1988a, 1988b), notwithstanding, it is in fact difficult to identify instances of influential designs whose development has been primarily informed by sociocultural theory. This becomes understandable once we note that the notion of cultural practices employed by sociocultural theorists typically refers to ways of talking and reasoning that have emerged during extended periods of human history. This construct makes it possible to characterize mathematics as a complex human activity rather than as disembodied subject matter (van Oers, 1996). The task facing both the teacher and the instructional designer is therefore framed as that of supporting and organizing students' induction into practices that have emerged during the discipline's intellectual history. While the importance of the goals inherent in this framing is indisputable, they provide only the most global orientation for design. A key difficulty is that the disciplinary practices that are taken as the primary point of reference exist prior to and independently of the activities of teachers and their students. In contrast, the ways of reasoning and communicating that are actually established in the classroom do not exist independently of the teacher's and students' activity, but are instead constituted by them in the course of their ongoing interactions

(Bauersfeld, 1980; Beach, 1999; Boaler, 2000; P. Cobb, 2000). A central challenge of design is to develop, test, and refine conjectures about both the classroom processes in which students might participate and the nature of their mathematical learning as they do so. Sociocultural theory provides only limited guidance because the classroom processes on which design focuses are emergent phenomena rather than already-established practices into which students are inducted.

Extending our purview beyond the classroom, there are two areas where, in my judgment, sociocultural theory can make significant contributions. I introduce each by first discussing recent bodies of scholarship developed within this tradition that are relevant to mathematics educators. The first body of scholarship is exemplified by investigations that have compared mathematical reasoning in school with that in various out-of-school settings such as grocery shopping (Lave, 1988), packing crates in a dairy (Scribner, 1984), selling candies on the street (Nunes, Schliemann, & Carraher, 1993; Saxe, 1991), playing dominoes and basketball (Nasir, 2002), laying carpet (Masingila, 1994), woodworking (Millroy, 1992), and sugar cane farming (de Abreu, 1995). These studies document that people develop significantly different forms of mathematical reasoning as they participate in different cultural practices that involve the use of different tools and sign systems, and that are organized by different overall motives (e.g., learning mathematics as an end in itself in school versus doing arithmetical calculations while selling candies on the street in order to survive economically). This approach of contrasting the forms of reasoning inherent in different cultural practices bears directly on issues of equity in students' access to significant mathematical ideas.

Although an adequate treatment of cultural diversity and equity is beyond the scope of this chapter, it is worth noting that a number of investigators have documented that the out-of-school practices in which students participate can involve differing norms of participation, language, and communication, some of which might be in conflict with those that the teacher seeks to establish in the mathematics classroom (Civil & Andrade, 2002; Ladson-Billings, 1998; Zevenbergen, 2000). An emerging line of research in mathematics education draws on sociocultural theory to document such conflicts and to understand the tensions that students experience (Boaler & Greeno, 2000; Gutiérrez, 2002; Martin, 2000; Moschkovich, 2002). The value of work of this type is that it enables us to view students' activity in the classroom as situated not merely with respect to the immediate learning environment, but with the respect to students' history

of participation in the practices of particular out-of-school groups and communities. It therefore has the potential to inform the development of designs in which the diversity in the out-of-school practices in which students participate is treated as an instructional resource rather than an obstacle to be overcome (Bouillion & Gomez, 2002; Civil, 2002; Gutstein, 2002; Warren, Ballenger, Ogonowski, Rosebery, & Hudicourt-Barnes, 2001).

The second relevant body of sociocultural research centers on the notion of a community of practice. In their overview of this line of research, Lave and Wenger (1991) clarify that they consider learning to be synonymous with the changes that occur in people's activity as they move from relatively peripheral participation to increasingly substantial participation in the practices of established communities. In doing so, they also argue that the tools used by community members carry a substantial portion of a community's intellectual heritage. Franke and Kazemi (2001) and Stein et al. (1998) have used Wenger's (1998) more recent formulation of these ideas to analyze teachers' learning as they are inducted into the practices of an established professional teaching community. However, mathematics education researchers are yet to exploit the full potential of the notion of community of practice in my view. This becomes apparent when we follow Wenger (1998) in noting that this notion brings together (a) theories of social structure that give primacy to institutions, norms and rules, and (b) theories of situated experience that give primacy to the dynamics of everyday existence and the local construction of interpersonal events. These two types of theories correspond to a dichotomy in the teacher education literature between analyses that focus on the structural or organizational features of schools and analyses that focus on the role of professional development in supporting teachers' reorganization of their instructional practices (Engestrom, 1998; Franke, Carpenter, Levi, & Fennema, 2001).

As Engestrom (1998) observes, the notion of community of practice has the potential to transcend this dichotomy in the literature by providing a unit of analysis that captures social structures that are within the scope of teachers' engagement as they develop and refine their instructional practices. Elsewhere, I and my colleagues have drawn heavily on this notion to develop an analytic approach for locating mathematics teachers' instructional practices within the institutional settings of the schools in which they work (P. Cobb, McClain, Lamberg, & Dean, 2003). The goal of research of this type is to gain insight into the processes by which teachers' instructional practices are partially constituted by the institutional setting of the schools in which they work. The potential contribution of such work is indicated by the substantiated finding that teachers' instructional practices are profoundly influenced by the institutional constraints that they attempt to satisfy, the formal and informal sources of assistance on which they draw, and the materials and resources that they use in their classroom practice (Ball & Cohen, 1996; C. A. Brown, Stein, & Forman, 1996; Nelson, 1999; Price & Ball, 1997; Senger, 1999; Stein & Brown, 1997). Analyses that document these affordances and constraints can inform the development of designs for supporting teachers' learning. In particular, they orient researchers and teacher educators to consider whether their collaborations with teachers should involve concerted attempts to bring about change in the institutional settings in which the teachers have developed and revised their instructional practices.

In summary, sociocultural theory characterizes the individual as a participant in established, historically evolving cultural practices. Thus, whereas the cognitive tradition that I discussed accounts for learning in terms of the epistemic individual's reorganization of its activity, sociocultural theory does so by documenting process by which people become increasingly substantial participants in various cultural practices. I have questioned the relevance of sociocultural theory to the development of designs at the classroom level but identified two other areas in which it can potentially make contributions. The first concerns equity in students' access to significant mathematical ideas and involves analyzing the out-of-school practices in which they participate. The second involves analyses of the institutional settings in which teachers develop and refine their instructional practices that can inform the development of designs for supporting their learning. In both these areas, the central issue is that of understanding how people deal with the tensions that they experience when different practices in which they participate are in conflict. In the first case, the conflicts are between the out-of-school practice in which students participate and those established in the mathematics classroom, whereas in the second case, the conflicts are between the practices of the schools in which teachers work and those established in a professional teaching community.

Distributed Cognition

As I noted, sociocultural theory has developed largely independently of western psychology and draws inspiration directly from the writings of Vygotsky and Leont'ev. Distributed cognition, in contrast, has developed in reaction to mainstream cognitive science and incorporates aspects of the Soviet work. Mainstream cognitive science should be differentiated from the cognitive tradition on which I have focused in this chapter in that it involves the observer's rather than the actor's viewpoint and posits what Anderson (1983) terms cognitive behaviors that intervene between stimulus and observed response activity. Several of the most important contributors to distributed cognition such as John Seeley Brown (1989), Alan Collins (1992), and James Greeno (1997) in fact achieved initial prominence as mainstream cognitive scientists before substantially modifying their theoretical commitments. Whereas sociocultural theorists usually frame people's reasoning as acts of participation in relatively broad systems of cultural practices, distributed cognition theorists typically restrict their focus to the immediate physical, social, and symbolic environment. Empirical studies conducted within the distributed tradition therefore tend to involve detailed analysis of either specific people's or a small group's activity rather than analyses of people's participation in established cultural practices.

As I indicated when clarifying the claim that the notion of the individual is conceptually relative, the term distributed cognition is most closely associated with Roy Pea. Pea (1985, 1993) coined this term to emphasize that, in his view, cognition is distributed across minds, persons, and symbolic and physical environments. As he and other distributed cognition theorists make clear in their writings, this perspective directly challenges a foundational assumption of both mainstream cognitive science and of the cognitive tradition that I have discussed. This is the assumption that cognition is bounded by the skin and can be adequately accounted for solely in terms of internal processes. Distributed cognition theorists instead see cognition as extending out into the immediate environment such that the environment becomes a resource for reasoning. As a consequence, the individual is characterized in this tradition as an element of a reasoning system.

In developing this position, distributed cognition theorists have been influenced by a number of studies conducted by sociocultural researchers, particularly those that compare people's reasoning in different settings. In one of the most frequently cited investigations, Scribner (1984) analyzed the reasoning of workers in a dairy as they filled orders by packing products into crates of different sizes. Her analysis revealed that the loaders did not perform purely mental calculations but instead used the structure of the crates as a resource in their reasoning. For example, if an order called for ten units of a particular product and six units were already in a crate that held twelve units, experienced loaders rarely subtracted six from ten to find how many additional units they needed. Instead, they might realize that an order of ten units would leave two slots in the crate empty and just know immediately from looking at the partially filled crate that four additional units are needed. As part of her analysis, Scribner demonstrated that the loaders developed strategies of this type as they went about their daily business of filling orders. For distributed cognition theorists, this indicates that the system that did the thinking was the loader in interaction with a crate. From this perspective, the loaders' reasoning is therefore treated as emergent relations between them and the immediate environment in which they worked.

Part of the reason that distributed cognition theorists attribute such significance to Scribner's study and to other investigations conducted by sociocultural researchers is that they capture what Hutchins (1995) refers to as cognition in the wild. This focus on people's reasoning as they engage in both everyday and workplace activities contrasts sharply with the traditional school-like tasks that are often used to investigate cognition. In addition to questioning whether people's reasoning on school-like tasks constitutes a viable set of cases from which to develop adequate accounts of cognition, several distributed cognition theorists have also critiqued current school instruction. In doing so, they have broadened their focus beyond mainstream cognitive science's traditional emphasis on the structure of particular tasks by drawing attention to the nature of the classroom activities within which the tasks take on meaning and significance for students.

J. S. Brown et al. (1989) developed one such critique by observing that school instruction typically aims to teach students abstract concepts and general skills on the assumption that students will be able to apply them directly in a wide range of settings. In challenging this assumption, they argue that the appropriate use of a concept or skill requires engagement in activities similar to those in which the concept or skill was developed and is actually used. In their view, the well-documented finding that most students do not develop widely applicable concepts and skills in school is attributable to the radical differences between classroom activities and those of both the disciplines and of everyday, out-of-school settings. They contend that successful students learn to meet the teacher's expectations by relying on specific features of classroom activities that are alien to activities in the other settings. In developing this explanation, Brown et al. treat the concepts and skills that students actually develop in school as relations between students and the material, social, and symbolic resources of the classroom environment.

It might be concluded from the two examples given thus far, those of the dairy workers and of students relying on what might be termed superficial cues in the classroom, that distributed cognition theorists do not address more sophisticated types of reasoning. Researchers working in this tradition have in fact analyzed a number of highly technical, work-related activities. The most noteworthy of these studies is, perhaps, Hutchins' (1995) analysis of the navigation team of a navel vessel as they brought their ship into harbor. In line with other investigations of this type, Hutchins argues that the entire navigation team and the artifacts it used constitutes the appropriate unit for a cognitive analysis. From the distributed perspective, it is this system of people and artifacts that did the navigating and over which cognition was distributed. In developing his analysis, Hutchins pays particular attention to the role of the artifacts as elements of this cognitive system. He argues, for example, that the cartographer has done much of the reasoning for the navigator who uses a map. This observation is characteristic of distributed analyses and implies that to understand a cognitive process, it is essential to understand how parts of that process have, in effect, been sedimented in tools and artifacts. Distributed cognition theorists therefore contend the environments of human thinking are thoroughly artificial. In their view, the cognitive resources that people exercise in particular environments are partially constituted by the cognitive resources with which they have populated those environments. As a consequence, the claim that artifacts do not merely serve to amplify cognitive process but instead reorganize them is a core tenet of the distributed cognition perspective (Dörfler, 1993; Pea, 1993).

In contrast to sociocultural theory, distributed cognition treats classroom processes as emergent phenomena rather than already-established practices into which students are inducted. For example, distributed theorists consider that aspects of the classroom learning environment such as classroom norms, discourse, and ways of using tools are constituted collectively by the teacher and students in the course of their ongoing interactions. In my judgment, the distributed perspective therefore has greater potential than sociocultural theory to inform the formulation of designs at the classroom level. A number of design research studies have in fact been conducted from this perspective (Bowers, Cobb, & McClain, 1999; Confrey & Smith, 1995; Fishman et al., 2004; Hershkowitz & Schwarz, 1999; Lehrer, Strom, & Confrey, 2002). In investigations of this type, researchers both develop designs to "engineer" novel forms of mathematical reasoning, and analyze the process of students' learning in these designed learning environments together with the means by which that learning is supported (P. Cobb, Confrey, diSessa, Lehrer, & Schauble, 2003; Collins, Joseph, & Bielaczyc, 2004; Confrey & Lachance, 2000; Design-Based Research Collaborative, 2003; Edelson, 2002; Gravemeijer, 1994b). I noted that designs developed from the cognitive perspective on which I have focused typically emphasize the development of instructional activities. In contrast, distributed cognition theorists' characterization of students as elements of reasoning systems orients them to construe the means of supporting students' mathematical learning more broadly. The means of support that are incorporated into designs are usually not limited to instructional activities but also encompass classroom norms, the nature of discourse, and the ways in which notations and other types of tools are used. Thus, whereas design from the cognitive perspective involves developing instructional activities as informed by analyses of specific students' reasoning, design from the distributed perspective focuses on the physical, social, and symbolic classroom environment that constitutes the immediate situation of the students' mathematical learning.

Given my generally positive assessment of the usefulness of the distributed perspective in informing the development, testing, and revision of designs, it is also important to note two potential limitations. The first limitation concerns the scant attention typically given to issues of equity. For example, the focus of researchers who develop and refine designs at the classroom level usually centers on students' individual and collective development of particular forms of mathematical reasoning. Pragmatically, it is essential that students come to see classroom activities as worthy of their engagement if the designs are to be effective. However, the process of supporting students' engagement by cultivating their mathematical interests is rarely an explicit focus of inquiry (for exceptions, see diSessa, 2001; Eisenhart & Edwards, 2001, April). As a consequence, differences in students' engagement that might reflect differential access to the instructional activities used and to the types of discourse established in the classroom can easily escape notice. In my view, this limitation stems from an almost exclusive focus on the classroom as the immediate context of students' learning. This focus precludes a consideration of tensions that some students might experience between aspects of this social context and the out-of-school practices in which they participate. Adherents to this perspective can address this limitation by coordinating their viewpoint with a sociocultural perspective that enables them to see students' classroom activity as situated not merely with respect to the immediate learning environment, but also with respect to their history of participation in the practices of out-of-school groups and communities.

The second limitation concerns the manner in which the characterization of the individual as an element of a reasoning system is taken as delegitimizing cognitive analysis of specific students' reasoning. As I indicated earlier in this chapter, this view of the individual reflects the evolution of the distributed perspective from mainstream cognitive science in that it involves equipping the individual as portrayed by this latter perspective with cultural tools and placing it in social context. I consider this restriction to be a limitation of the distributed perspective because it fails to acknowledge the contributions that cognitive analyses of specific students' reasoning can make to the process of adjusting and modifying an instructional design (see P. Cobb, 1998, for a detailed discussion of this point). Furthermore, in the hands of a skillful teacher, the diversity in students' reasoning is a primary resource on which he or she can draw to support sustained classroom discussions that focus on substantive mathematical issues. Earlier in this chapter, I followed Putnam (1987) in endorsing Quine's (1953) argument that we should accept the existence of abstract mathematical entities on the grounds that this ontology is indispensable to the mathematics viewed as a practice. Similarly, I contend that we should accept the existence of specific students' mathematical reasoning because this ontology is indispensable to design and teaching in mathematics education. It is therefore necessary, in my view, to resist theoretical arguments that delegitimize this ontology. In the next section of this chapter, I indicate how it might be possible to capitalize on the potential contributions of the distributed perspective while circumventing this limitation.

In summary, the distributed perspective emerged in response to the limited attention given to context, culture, and affect by mainstream cognitive science (De Corte et al., 1996). The individual is characterized as an element of a reasoning system that also includes aspects of the immediate physical, social, and symbolic environment. In contrast to sociocultural theorists' focus on people's participation in established cultural practices, distributed theorists usually conduct detailed analyses of reasoning processes that are stretched over people and the aspects of their immediate environment that they use as cognitive resources. In my view, the distributed perspective has greater potential than sociocultural theory to contribute to the formulation of designs at the classroom level because it treats classroom processes as emergent phenomena. However, I tempered this positive appraisal by noting two limitations. These concern the limited attention given to issues of equity and the disavowal of cognitive analyses of specific students' mathematical reasoning. Taken together, these two limitations indicate the value of attempting to capitalize on the ways that multiple perspectives can contribute to the enterprise of developing, testing, and revising designs for supporting learning. I address this issue shortly.

REFLECTION

The contrasts I have drawn between the four theoretical perspectives are summarized in Table 1. The development of these contrasts involved the use of two criteria of comparison. The first criterion concerns how each perspective characterizes the individual whereas the second focus on the potential of each to contribute to our understanding of learning processes and the means of supporting their realization. Following Kuhn (1962; 1970), it is important to acknowledge that these criteria are not neutral standards but are instead values that I propose should be considered when coming to terms with the multiplicity of theoretical perspectives that characterize mathematics education research. As I indicated, these criteria reflect commitments and interests inherent in the view that mathematics education is a design science and, for this reason, are eminently debatable and are open to critique and revision. Furthermore, the process of using

them to compare theoretical perspectives necessarily involves interpretation and judgment. Consequently, mathematics educators who use these same criteria "may nevertheless make different choices in the same concrete situation" (Kuhn, 1970, p. 262). My intent in proposing criteria and illustrating how they might be used has therefore not been to shut down debate in the face of competing theoretical perspectives, but to move the debate to a meta-level at which we are obliged to give good reasons for our theoretical choices. At this meta-level, the ideological fervor with which particular perspectives are sometimes promoted is no substitute for justifications that articulate the choice criteria or values together with the interests and concerns that they reflect. The pragmatic realist stance that I have taken to the issue of multiple theoretical perspectives will have achieved what Rorty (1979) refers to as its therapeutic purpose to the extent that justifications of this sort become commonplace.

THEORIZING AS BRICOLAGE

As part of the process of discussing the four theoretical perspectives, I briefly outlined the historical origins of each. I noted, for example, that experimental psychology has been profoundly shaped by its alliance with educational administrators. In contrast, the origins of the cognitive tradition on which I focused can be traced to Piaget's interest in problems of genetic epistemology. Sociocultural theory for its part reflects Vygotsky's and Leont'ev's commitment to the notion of the "the new socialist man" and to the view of education as a primary means of bringing about this change in Soviet society (van der Veer & Valsiner, 1991).

Table 1.1   Contrasts Between Four Theoretical Perspectives

Theoretical perspective	Characterization of the individual	Usefulness	Limitations
Experimental Psychology	Statistically constructed collective individual	Administration of educational systems	Limited relevance to design at classroom level
Cognitive Psychology	Epistemic individual as reorganizer of activity	Specification of "big ideas" Design of instructional activities	Means of supporting learning limited to instructional tasks
Sociocultural Theory	Individual as participant in cultural practices	Designs that take account of students' out-of-school practices Designs that take account of institutional setting of teaching and learning	Limited relevance to design at classroom level
Distributed Cognition	Individual element of a reasoning system	Design of classroom learning environments including norms, discourse, and tools	Delegitimizes cognitive analyses of specific students' reasoning

Finally, distributed cognition emerged in reaction to perceived shortcomings of mainstream cognitive science. The concerns and interests that motivated the development of each of these perspectives differ significantly from those of mathematics educators. In terms of the metaphor of theoretical perspectives as conceptual tools that I introduced earlier in this chapter, each of these perspectives is a tool that has been fashioned while addressing problems that are not of immediate concern to most mathematics educators. It is therefore unreasonable to expect that any one of these perspectives is ready-made for the collective enterprise of developing, testing, and revising designs. Given the limitations that I have discussed of each perspective with respect to this enterprise, the question that arises is not that of how to choose between the various perspectives. Instead, it is how they can be adapted to the concerns and interests of mathematics educators.

In addressing this question, I propose to view the four perspectives as sources of ideas that we can appropriate and modify for our purposes as mathematics educators. This process of developing conceptual tools for mathematics education research parallels that of instructional design as described by Gravemeijer (1994b).

> [Design] resembles the thinking process that Lawler (1985) characterizes by the French word *bricolage,* a metaphor taken from Claude Levi–Strauss. A *bricoleur* is a handy man who invents pragmatic solutions in practical situations. . . . [T]he bricoleur has become adept at using whatever is available. The bricoleur's tools and materials are very heterogeneous: Some remain from earlier jobs, others have been collected with a certain project in mind. (p. 447)

Similarly, I suggest that rather than adhering to one particular theoretical perspective, we act as bricoleurs by adapting ideas from a range of theoretical sources.

To illustrate this approach, I take as a case an interpretive framework that several colleagues and I developed over a number of years while addressing concrete problems and issues that arose while working in classrooms (P. Cobb, Stephan, McClain, & Gravemeijer, 2001; P. Cobb & Yackel, 1996). The intent of the framework is to locate students' mathematical reasoning in the social context of the classroom in a manner that can feed back to inform instructional design and teaching. For my current purposes, it suffices to note that the framework involves the coordination of two distinct perspectives on classroom activity. One is a social perspective that is concerned with ways of acting, reasoning, and arguing that have been established as normative in a classroom community. From this perspective, an individual student's reasoning is framed as an act of participation in these normative activities. The other is a cognitive perspective that focuses squarely on the nature of individual students' reasoning or, in other words, on their specific ways of participating in communal classroom activities. Analyses developed by using the framework bring the diversity in students' mathematical reasoning to the fore while situating that diversity in the social context of their participation in communal activities.

We take the relation between the social and cognitive perspectives to be one of reflexivity. This is an extremely strong relationship that does not merely mean that the two perspectives are interdependent. Instead, it implies that neither exists without the other in that each perspective constitutes the background against which mathematical activity is interpreted from the other perspective (Mehan & Wood, 1975). For example, the collective activities of the classroom community (social perspective) emerge and are continually regenerated by the teacher and students as they interpret and respond to each other's actions (cognitive perspective). Conversely, the teacher's and students' interpretations and actions in the classroom (cognitive perspective) are not seen to exist apart from their participation in communal classroom practices (social perspective). The coordination is therefore not between individual students and the classroom community viewed as separate entities but between two alternative ways of looking at and making sense of what is going on in classrooms.

This interpretive framework is a bricolage in that the social perspective draws on sociocultural theory (Cole, 1996; Lave, 1991; Rogoff, 1990) and the cognitive perspective draws on both cognitive psychology (Piaget, 1970; Steffe & Kieren, 1994; Thompson, 1991), and distributed accounts of cognition (Hutchins, 1995; Pea, 1993; Wertsch, 1998, 2002). One of the key theoretical constructs that we use when we take a social perspective, that of a classroom mathematical practice, serves to illustrate how we have appropriated ideas to our agenda as mathematics educators. We developed this construct by adapting sociocultural theorists' notion of a cultural practice. As I indicated when discussing sociocultural theory, this idea is attractive because it makes it possible to characterize mathematics as a complex human activity rather than as disembodied subject matter. However, I also noted that the notion of a practice as framed in sociocultural theory is problematic from the point of view of design because practices are typically characterized as existing prior to and independently of the teacher's and students' activity. We therefore modified this notion by explicitly defining a classroom mathematical practice as an emergent phenomenon that is established jointly by the teacher and students in the course of their ongoing interactions. This is a non-trivial adaptation in that students are then seen to contribute to the development of the classroom norms and practices that constitute the social situation of their mathematical learning.

We also made modifications when fashioning a cognitive perspective appropriate for our purposes by adapting ideas from the cognitive perspective I have discussed and from distributed cognition. For example, we took from the cognitive perspective the notion of learning as a process of reorganizing activity. However, influenced by distributed theories of intelligence, we found it important to broaden our view of activity so that it is not restricted solely to solo sensory-motor and conceptual activity but instead reaches out into the world and includes the use of tools and symbols. The rationale for this modification is at least in part pragmatic in that our work as instructional designers involves developing notation systems, physical tools, and computer-based tools for students to use. We therefore needed an analytic approach that can take account both of the diverse ways in which students reason with tools and symbols, and of how those ways of reasoning evolve over time.

Although we saw value in distributed accounts of intelligence, we could not accept this theoretical orientation ready-made given its rejection of analytical approaches that focus explicitly on the nature of specific students' reasoning. The adaptation we made was to modify how the individual is characterized. As I have indicated, distributed theorists appear to have accepted the portrayal of the individual offered by mainstream cognitive science and propose equipping this character with cultural tools and locating it in social context. The primary adaptation we made was to characterize the individual not as needing to be placed in its immediate physical, social, and symbolic environment, but as already acting in that environment. The tools and symbols that students use are then viewed as constituent parts of their activity rather than as standing apart from or outside their activity.

Once this modification is made, what is viewed as a student-tool system from the perspective of distributed cognition becomes, in the psychological perspective we take, an individual student engaging in mathematical activity that involves *reasoning with* tools and symbols. Thus, although the focus of this psychological perspective is explicitly on the quality of individual students' reasoning, its emphasis on tools is generally consistent with the notion of mediated action as discussed by sociocultural theorists such

as Wertsch (2002). Further, as I have indicated, the remaining component of the functional system posited by distributed theorists, the classroom social context, becomes an explicit focus of attention when this psychological perspective is coordinated with the social perspective. The interpretive framework therefore characterizes students as reasoning with tools while participating in and contributing to the development of communal practices. With regard to usefulness, the framework yields analyses of students' mathematical learning that are tied to the classroom social setting in which that learning actually occurs. As a consequence, these analyses enable us to tease out aspects of this setting that served to support the development of students' reasoning. This, in turn, makes it possible to develop testable conjectures about ways in which those means of support and thus the instructional design can be improved.

In keeping with the critical stance I took to the four theoretical perspectives, I should acknowledge that the interpretive framework has at least two major limitations. These concern the isolation of classroom learning environments from the institutional settings of the schools in which they are located, and failure to address issues of cultural diversity and equity in students' access to significant mathematical ideas. As a consequence, these two areas have become a focus of our research in recent years (P. Cobb & Hodge, 2002; P. Cobb, McClain et al., 2003). It is also worth noting that the framework does not draw on the theoretical perspective of experimental psychology. This is primarily because experimental psychology views the classroom learning environment from an administrative viewpoint and characterizes it as composed of independent variables that can be manipulated from the outside. In contrast, the framework I have outlined characterizes the classroom learning environment from the inside as jointly constituted by the teacher and students. Analyses developed when using this framework will therefore be of little value to most administrators. The experimental psychology perspective might be useful when the goal is to contribute to public policy discourse about mathematics teaching and learning.

My purpose in discussing the interpretive framework has been to illustrate the process of adapting and modifying ideas appropriated from a range of theoretical sources. The pragmatic spirit of the bricolage metaphor indicates that the goal in doing so is to fashion conceptual tools that are useful for our purposes as mathematics educators. The metaphor therefore serves to differentiate relatively modest efforts of the type that I have illustrated from more ambitious projects that aim to develop theoretical cosmologies (Shotter, 1995). For example, our goal in developing the framework was to craft a tool that would enable us to make sense of what is happening in mathematics classrooms rather than to produce a grand synthesis of cognitive psychology, sociocultural theory, and distributed cognition. In my view, theorizing as a modest process of bricolage offers a better prospect of mathematics education research developing an intellectual identity distinct from the various perspectives on which it draws than does the attempt to formulate all-encompassing theoretical schemes.

INCOMMENSURABILITY

The approach I have illustrated for comparing theoretical perspectives is best viewed as a proposal about how we might cope with competing theoretical perspectives in mathematics education research. The account that I have given of the relations between different perspectives can be contrasted with analyses that portray a historical sequence of theoretical perspectives, each of which overcomes the limitations of its predecessors. Accounts of this latter type depict the history of a field as an ordered progression that typically culminates with the perspective to which the writer subscribes. As Guerra (1998) notes, these narratives are based on the implicit metaphor of theoretical developments as a *relentless march of progress.* For example, adherents of distributed cognition frequently portray their perspective as overcoming the limitations of mainstream cognitive science. Proponents of perspectives that have supposedly been superseded typically employ a different type of narrative that is based on the metaphor of *potential redemption.* For example, experimental psychologists frequently characterize the increasing prominence of competing perspectives as a period in which the issue of what counts as credible evidence has been largely ignored, and portray their perspective as offering redemption by making educational research a prestigious scientific enterprise.

The account I have given reflects an alternative metaphor, that of *co-existence and conflict.* The tension between the march of progress and potential redemption narratives indicates the relevance of this metaphor. I fleshed out this metaphor by proposing two criteria for comparing and contrasting competing perspectives. The first focused on how each perspective characterized the individual, thereby delineating the types of phenomena that proponents of the different perspectives are investigating. The second focused

on the potential contributions of each perspective to the collective enterprise of formulating, testing, and revising designs for supporting learning. In developing this second criterion, I followed Dewey (1890/1969), Pepper (1942), and Prawat (1995) in arguing that a theoretical perspective is a useful conceptual tool for mathematics educators to the extent that it gives insight and understanding into learning processes and the specific means of supporting their realization. In using these two criteria, I noted that the goals for which each of the four perspectives was originally developed differ from those of mathematics educators. On this basis, I argued that we should view the various co-existing perspectives as sources of ideas to be adapted to our purposes.

The contrasting ways in which the different perspectives characterize the individual indicate that they are incommensurable. In terms of Putnam's (1987) pragmatic realism, adherents to the differing perspectives ask different types of questions and produce different forms of knowledge as they attempt to develop insights into different realities. I followed Putnam (1987) and Kuhn (1962, 1977) in arguing that the realities that researchers investigate are conceptually relative to their particular theoretical perspectives, but rejected the claim that any of these realities is as good as any other. The approach I took is consistent with Feyerabend's (1975) claim that we cope with incommensurability both in research and in other areas of life by drawing comparisons and contrasts in the course of which we delineate similarities and differences. Feyerabend also argued that there is no single ultimate grid for comparing theoretical perspectives, and demonstrated that they can be compared in multiple ways. The primary challenge posed by incommensurability is to develop a way of comparing and understanding different perspectives. It was for this reason that I discussed the two criteria I used in some detail.

As Bernstein (1983) observed, the process of comparing incommensurable perspectives has parallels with anthropology in that the goal is to figure out what the "natives" think they are doing (Geertz, 1973). Kuhn (1977) described how he attempts to avoid merely imposing his viewpoint when he discussed his work as a philosopher and historian of science.

[The] plasticity of texts does not place all ways of reading on a par, for some of them (ultimately, one hopes, only one) possess a plausibility and coherence absent from others. Trying to transmit such lessons to students, I offer them a maxim: When reading the works of an [historically] important thinker, look first for the apparent absurdities in the text and ask yourself how a sensible person could have written them. When you find an answer, I continue, when those passages make sense, then you may find that more central passages, ones you previously thought you understood, have changed their meaning. (p. xii)

The openness inherent in this stance to incommensurability has the benefit that in coming to understand what adherents to an alternative perspective think they are doing, we develop a more sensitive and critical understanding of some of the taken-for-granted aspects of our own perspective. This understanding is critical to the justification of theoretical choices in that these justifications should, in my view, involve the specification of both the choice criteria used and the interests and concerns that they reflect.

As Geertz (1973) emphasizes, the absence of abstract, cross-cultural universals does not condemn anthropology to absolute relativism.

If we want to discover what man amounts to, we can only find it in what men are, and what men are, above all other things, is various. It is in understanding that variousness—its range, its nature, its basis, and its implications—that we shall come to construct a concept of human nature that more than a statistical shadow and less than a primitive dream has both substance and truth. (p. 52)

Similarly, the absence of a neutral framework for comparing incommensurable theoretical perspectives does not condemn us to absolute relativism in which theoretical decisions amount to nothing more than personal whim or taste. A primary purpose of philosophy for Dewey, Gadamer, Kuhn, Putnam, and Rorty is to transcend the dichotomy between an unobtainable neutral framework on the one hand and an absolute, "anything goes" brand of relativism on the other. Philosophy as they conceive it is a discourse about incommensurable perspectives and discourses. Its intent is both edifying and therapeutic in that it aims to support conversation both about and between various perspectives. My purpose in this chapter has been to advance a conversation of this type in the mathematics education research community.

REFERENCES

Anderson, J. R. (1983). *The architecture of cognition.* Cambridge, MA: Harvard University Press.

Anderson, J. R., Reder, L. M., & Simon, H. A. (1997). Situative versus cognitive perspectives: Form versus substance. *Educational Researcher, 26*(1), 18–21.

Ball, D. L. (1993). With an eye on the mathematical horizon: Dilemmas of teaching elementary school mathematics. *Elementary School Journal, 93*(4), 373–397.

Ball, D. L., & Chazan, D. (1994, April). *An examination of teacher telling in constructivist mathematics pedagogy: Not just excusable but essential.* Paper presented at the annual meeting of the American Educational Research Association, New Orleans, LA.

Ball, D. L., & Cohen, D. K. (1996). Reform by the book: What is—or might be—the role of curriculum materials in teacher learning and instructional reform? *Educational Researcher, 25*(9), 6–8,14.

Bauersfeld, H. (1980). Hidden dimensions in the so-called reality of a mathematics classroom. *Educational Studies in Mathematics, 11,* 23–41.

Bauersfeld, H. (1988). Interaction, construction, and knowledge: Alternative perspectives for mathematics education. In T. Cooney & D. Grouws (Eds.), *Effective mathematics teaching* (pp. 27–46). Reston, VA: National Council of Teachers of Mathematics and Erlbaum.

Beach, K. (1999). Consequential transitions: A sociocultural expedition beyond transfer in education. *Review of Research in Education, 24,* 103–141.

Berlin, I. (1976). *Vico and Herder: Two studies in the history if ideas.* London: Chatto and Windus.

Bernstein, R. J. (1983). *Beyond objectivism and relativism: Science, hermeneutics, and praxis.* Philadelphia: University of Pennsylvania Press.

Boaler, J. (2000). Exploring situated insights into research and learning. *Journal for Research in Mathematics Education, 31,* 113–119.

Boaler, J., & Greeno, J. G. (2000). Identity, agency, and knowing in mathematical worlds. In J. Boaler (Ed.), *Multiple perspectives on mathematics teaching and learning* (pp. 45–82). Stamford, CT: Ablex.

Bouillion, L. M., & Gomez, L. M. (2002, June). *Connecting school and community in science learning.* Paper presented at the Fifth Congress of the International Society for Cultural Research and Activity Theory, Amsterdam.

Bowers, J. S., Cobb, P., & McClain, K. (1999). The evolution of mathematical practices: A case study. *Cognition and Instruction, 17,* 25–64.

Brown, C. A., Stein, M. K., & Forman, E. A. (1996). Assisting teachers and students to reform the mathematics classroom. *Educational Studies in Mathematics, 31,* 63–93.

Brown, J. S., Collins, A., & Duguid, P. (1989). Situated cognition and the culture of learning. *Educational Researcher, 18,* 32–42.

Bruner, J. (1986). *Actual minds, possible worlds.* Cambridge, MA: Harvard University Press.

Carey, D. A., Fennema, E., Carpenter, T. P., & Franke, M. L. (1995). Equity and mathematics education. In W. G. Secada, E. Fennema, & L. B. Adajion (Eds.), *New directions for equity in mathematics education* (pp. 93–125). New York: Cambridge University Press.

Carpenter, T. P., & Fennema, E. (1992). Cognitively guided instruction: Building on the knowledge of students and teachers. *International Journal of Educational Research, 16,* 457–470.

Carpenter, T. P., Fennema, E., Peterson, P. L., Chiang, C., & Loef, M. (1989). Using knowledge of children's mathematics thinking in classroom teaching: An experimental study. *American Educational Research Journal, 26,* 499–532.

Carpenter, T. P., & Moser, J. M. (1984). The acquisition of addition and subtraction concepts in grades one through three. *Journal for Research in Mathematics Education, 15,* 179–202.

Civil, M. (2002). Culture and mathematics: A community approach. *Journal of Intercultural Studies, 23,* 133–148.

Civil, M., & Andrade, R. (2002). Transitions between home and school mathematics: Rays of hope amidst the passing clouds. In G. d. Abreu, A. J. Bishop, & N. C. Presmeg (Eds.), *Transitions between contexts of mathematical practices* (pp. 149–169). Dordrecht, The Netherlands: Kluwer.

Clements, D. H., & Battista, M. T. (1992). Geometry and spatial reasoning. In D. A. Grouws (Ed.), *Handbook of research on mathematics teaching and learning* (pp. 420–464). New York: Macmillan.

Cobb, G. W. (1997). *Mere literacy is not enough.* New York: College Entrance Examination Board.

Cobb, P. (1998). Learning from distributed theories of intelligence. *Mind, Culture, and Activity, 5,* 187–204.

Cobb, P. (2000). The importance of a situated view of learning to the design of research and instruction. In J. Boaler (Ed.), *Multiple perspectives on mathematics teaching and learning* (pp. 45–82). Stamford, CT: Ablex.

Cobb, P. (2002). Theories of knowledge and instructional design: A response to Colliver. *Teaching and Learning in Medicine, 14,* 52–55.

Cobb, P., & Bowers, J. S. (1999). Cognitive and situated perspectives in theory and practice. *Educational Researcher, 28*(2), 4–15.

Cobb, P., Confrey, J., diSessa, A. A., Lehrer, R., & Schauble, L. (2003). Design experiments in education research. *Educational Researcher, 32*(1), 9–13.

Cobb, P., & Hodge, L. L. (2002). A relational perspective on issues of cultural diversity and equity as they play out in the mathematics classroom. *Mathematical Thinking and Learning, 4,* 249–284.

Cobb, P., & McClain, K. (2002). Supporting students' learning of significant mathematical ideas. In G. Wells & G. Claxton (Eds.), *Learning for life in the 21st Century* (pp. 154–166). Oxford, UK: Blackwell.

Cobb, P., McClain, K., Lamberg, T., & Dean, C. (2003). Situating teachers' instructional practices in the institutional setting of the school and school district. *Educational Researcher, 32*(6), 13–24.

Cobb, P., Stephan, M., McClain, K., & Gravemeijer, K. (2001). Participating in classroom mathematical practices. *Journal of the Learning Sciences, 10,* 113–164.

Cobb, P., Wood, T., Yackel, E., & McNeal, G. (1992). Characteristics of classroom mathematics traditions: An interactional analysis. *American Educational Research Journal, 29,* 573–602.

Cobb, P., Wood, T., Yackel, E., Nicholls, J., Wheatley, G., Trigatti, B., et al. (1991). Assessment of a problem-centered second grade mathematics project. *Journal for Research in Mathematics Education, 22,* 3–29.

Cobb, P., Wood, T., Yackel, E., & Perlwitz, M. (1992). A longitudinal, follow-up assessment of a second-grade problem centered mathematics project. *Educational Studies in Mathematics, 23,* 483–504.

Cobb, P., & Yackel, E. (1996). Constructivist, emergent, and sociocultural perspectives in the context of developmental research. *Educational Psychologist, 31,* 175–190.

Cole, M. (1996). *Cultural psychology.* Cambridge, MA: Belknap Press of Harvard University Press.

Collins, A. (1992). Portfolios for science education: Issues in purpose, structure, and authenticity. *Science Education, 76,* 451–463.

Collins, A., Joseph, D., & Bielaczyc, K. (2004). Design research: Theoretical and methodological issues. *Journal of the Learning Sciences, 13,* 15–42.

Confrey, J., Castro-Filho, J., & Wilhelm, J. (2000). Implementation research as a means to link systemic reform and applied psychology in mathematics education. *Educational Psychologist, 35,* 179–191.

Confrey, J., & Lachance, A. (2000). Transformative teaching experiments through conjecture-driven research design. In A. E. Kelly & R. A. Lesh (Eds.), *Handbook of research design in mathematics and science education* (pp. 231–266). Mahwah, NJ: Erlbaum.

Confrey, J., & Smith, E. (1995). Splitting, covariation, and their role in the development of exponential functions. *Journal for Research in Mathematics Education, 26,* 66–86.

D'Amato, J. (1992). Resistance and compliance in minority classrooms. In E. Jacob & C. Jordan (Eds.), *Minority education: Anthropological perspectives* (pp. 181–207). Norwood, NJ: Ablex.

Danziger, K. (1990). *Constructing the subject.* New York: Cambridge University Press.

Davis, B. (1997). Listening for differences: An evolving conception of mathematics teaching. *Journal for Research in Mathematics Education, 28,* 355–376.

Davydov, V. V. (1988a). Problems of developmental teaching (Part I). *Soviet Education, 30*(8), 6–97.

Davydov, V. V. (1988b). Problems of developmental teaching (Part II). *Soviet Education, 30*(9), 3–83.

Davydov, V. V., & Radzikhovskii, L. A. (1985). Vygotsky's theory and the activity-oriented approach in psychology. In J. V. Wertsch (Ed.), *Culture, communication, and cognition: Vygotskian perspectives* (pp. 35–65). New York: Cambridge University Press.

de Abreu, G. (1995). Understanding how children experience the relationship between home and school mathematics. *Mind, Culture, and Activity, 2,* 119–142.

De Corte, E., Greer, B., & Verschaffel, L. (1996). Mathematics learning and teaching. In D. Berliner & R. Calfee (Eds.), *Handbook of educational psychology* (pp. 491–549). New York: Macmillan.

Delpit, L. D. (1988). The silenced dialogue: Power and pedagogy in educating other people's children. *Harvard Educational Review, 58,* 280–298.

Design-Based Research Collaborative. (2003). Design-based research: An emerging paradigm for educational inquiry. *Educational Researcher, 32*(1), 5–8.

Dewey, J. (1890/1969). The logic of verification. In J. A. Boyston (Ed.), *John Dewey: The early works* (Vol. 3, pp. 83–92). Carbondale: Southern Illinois University Press.

Dewey, J. (1910/1976). The short-cuts to realism examined. In J. A. Boyston (Ed.), *John Dewey: The middle works, 1899–1924* (Vol. 6, pp. 136–140). Carbondale: Southern Illinois University Press.

Dewey, J. (1929/1958). *Experience and nature.* New York: Dober.

diSessa, A. A. (2001). *Changing minds: Computers, learning, and literacy.* Cambridge, MA: MIT Press.

Dörfler, W. (1989, July). *Protocols of actions as a cognitive tool for knowledge construction.* Paper presented at the Thirteenth Conference of the International Group for the Psychology of Mathematics Education, Paris.

Dörfler, W. (1993). Computer use and views of the mind. In C. Keitel & K. Ruthven (Eds.), *Learning from computers: Mathematics education and technology* (pp. 159–186). Berlin: Springer-Verlag.

Dörfler, W. (2000). Means for meaning. In P. Cobb, E. Yackel, & K. McClain (Eds.), *Symbolizing and communicating in mathematics classrooms: Perspectives on discourse, tools, and instructional design* (pp. 99–132). Mahwah, NJ: Erlbaum.

Dubinsky, E. (1991). Reflective abstraction in advanced mathematical thinking. In D. Tall (Ed.), *Advanced mathematical thinking* (pp. 95–123). Dordrecht, The Netherlands: Kluwer.

Edelson, D. C. (2002). Design research: What we learn when we engage in design. *Journal of the Learning Sciences, 11*, 105–121.

Eisenhart, M. A., & Edwards, L. (2001, April). *Grabbing the interest of girls: African-American eighth graders and authentic science.* Paper presented at the annual meeting of the American Educational Research Association, Seattle, WA.

Eisenhart, M. A., & Towne, L. (2003). Contestation and change in national policy on "scientifically based" education research. *Educational Researcher, 32*(8), 31–38.

Engestrom, Y. (1998). Reorganizing the motivational sphere of classroom culture: An activity—theoretical analysis of planning in a teacher team. In F. Seeger, J. Voigt, & U. Waschescio (Eds.), *The culture of the mathematics classroom* (pp. 76–103). New York: Cambridge University Press.

Erickson, F. (1992). Transformation and school success: The policies and culture of educational achievement. In E. Jacob & C. Jordan (Eds.), *Minority education: Anthropological perspectives* (pp. 27–51). Norwood, NJ: Ablex.

Ernest, P. (1994). The dialogical nature of mathematics. In P. Ernest (Ed.), *Mathematics, education, and philosophy* (pp. 33–48). London: Falmer.

Fabricius, W. (1979). Piaget's theory of knowledge—Its philosophical context. *High/Scope Report, 7*, 4–13.

Fennema, E., Franke, M. L., & Carpenter, T. P. (1993). Using children's mathematical knowledge in instruction. *American Educational Research Journal, 30*, 555–583.

Feyerabend, P. (1975). *Against method.* London: Verso.

Filloy, E., & Rojano, T. (1984). Solving equations: The transition from arithmetic to algebra. *Journal for Research in Mathematics Education, 9*, 19–25.

Fishman, B., Marx, R. W., Blumenfeld, P., & Krajcik, J. S. (2004). Creating a framework for research on systemic technology innovations. *Journal of the Learning Sciences, 13*, 43–76.

Forman, E. A. (2003). A sociocultural approach to mathematics reform: Speaking, inscribing, and doing mathematics within communities of practice. In J. Kilpatrick, W. G. Martin, & D. Schifter (Eds.), *A research companion to principles and standards for school mathematics* (pp. 333–352). Reston, VA: National Council of Teachers of Mathematics.

Franke, M. L., Carpenter, T. P., Levi, L., & Fennema, E. (2001). Capturing teachers' generative change: A follow-up study of teachers' professional development in mathematics. *American Educational Research Journal, 38*, 653–689.

Franke, M. L., & Kazemi, E. (2001). Teaching as learning within a community of practice: Characterizing generative growth. In T. Wood, B. C. Nelson, & J. Warfield (Eds.), *Beyond classical pedagogy in elementary mathematics: The nature of facilitative teaching* (pp. 47–74). Mahwah, NJ: Erlbaum.

Frankenstein, M. (2002, April). *To read the world: Goals for a critical mathematical literacy.* Paper presented at the Research Presession of the annual meeting of the National Council of Teachers of Mathematics, Las Vegas, NV.

Fuson, K. C. (1992). Research on whole number addition and subtraction. In D. Grouws (Ed.), *Handbook for research on mathematics teaching and learning* (pp. 243–275). New York: Macmillan.

Gadamer, H.-G. (1975). *Truth and method.* New York: Seabury Press.

Geertz, C. (1973). *The interpretation of cultures.* New York: Basic Books.

Goodman, N. (1978). *Ways of worldmaking.* Indianapolis, IN: Hackett.

Gravemeijer, K. (1994a). *Developing realistic mathematics education.* Utrecht, The Netherlands: CD-ß Press.

Gravemeijer, K. (1994b). Educational development and developmental research. *Journal for Research in Mathematics Education, 25*, 443–471.

Greeno, J. G. (1991). Number sense as situated knowing in a conceptual domain. *Journal for Research in Mathematics Education, 22*, 170–218.

Greeno, J. G. (1997). On claims that answer the wrong questions. *Educational Researcher, 26*(1), 5–17.

Greeno, J. G., & The Middle School Mathematics Through Applications Project Group. (1998). The situativity of knowing, learning, and research. *American Psychologist, 53*, 5–26.

Guerra, J. C. (1998). *Close to home: Oral and literate practices in a transnational Mexicano community.* New York: Teachers College Press.

Gutiérrez, R. (2002). Enabling the practice of mathematics teachers in context: Toward a new research agenda. *Mathematical Thinking and Learning, 4*, 145–189.

Gutiérrez, R. (2004, August). *The complex nature of practice for urban (mathematics) teachers.* Paper presented at the Rockefeller Symposium on the Practice of School Improvement: Theory, Methodology, and Relevance, Bellagio, Italy.

Gutstein, E. (2002, April). *Roads towards equity in mathematics education: Helping students develop a sense of agency.* Paper presented at the annual meeting of the American Educational Research Association, New Orleans, LA.

Gutstein, E. (in press). "So one question leads to another": Using mathematics to develop a pedagogy of questioning. In N. S. Nasir & P. Cobb (Eds.), *Diversity, equity, and access to mathematical ideas.* New York: Teachers College Press.

Hacking, I. (2000). *The social construction of what?* Cambridge, MA: Harvard University Press.

Hall, R., & Rubin, A. (1998). There's five little notches in here: Dilemmas in teaching and learning the conventional structure of rate. In J. G. Greeno & S. V. Goldman (Eds.), *Thinking practices in mathematics and science learning* (pp. 189–236). Mahwah, NJ: Erlbaum.

Hershkowitz, R., & Schwarz, B. (1999). The emergent perspective in rich learning environments: Some roles of tools and activities in the construction of sociomathematical norms. *Educational Studies in Mathematics, 39*, 149–166.

Hill, H. C. (2001). Policy is not enough: Language and the interpretation of state standards. *American Educational Research Journal, 38*, 289–318.

Hofer, B. (2002). Personal epistemology: Conflicts and consensus in an emerging area of inquiry. *Educational Psychology Review, 13*, 353–384.

Hutchins, E. (1995). *Cognition in the wild.* Cambridge, MA: MIT Press.

Jardine, N. (1991). *The scenes of inquiry: On the reality of questions in the sciences.* Oxford, UK: Clarendon Press.

Kant, I. (1998). *A critique of pure reason.* Cambridge, UK: Cambridge University Press.

Kaput, J. J. (1994). The representational roles of technology in connecting mathematics with authentic experience. In R. Biehler, R. V. Scholz, R. Strasser, & B. Winkelmann (Eds.), *Didactics of mathematics as a scientific discipline* (pp. 379–397). Dordrecht, The Netherlands: Kluwer.

Kaput, J. J. (1999). Teaching and learning a new algebra. In E. Fennema & T. P. Carpenter (Eds.), *Mathematics classrooms that promote understanding* (pp. 133–155). Mahwah, NJ: Erlbaum.

Kilpatrick, J., Martin, W. G., & Schifter, D. (Eds.). (2003). *A research companion to principles and standards of school mathematics.* Reston, VA: National Council of Teachers of Mathematics.

Konold, C., & Higgins, T. (in press). Working with data. In S. J. Russell, D. Schifter, & V. Bastable (Eds.), *Developing mathematical ideas: Collecting, representing, and analyzing data.* Parsippany, NJ: Dale Seymour Publications.

Kozulin, A. (1990). *Vygotsky's psychology: A biography of ideas.* Cambridge, MA: Harvard University Press.

Kuhn, T. S. (1962). *The structure of scientific revolutions* (2nd ed.). Chicago: University of Chicago Press.

Kuhn, T. S. (1970). Reflections on my critics. In I. Lakatos & A. Musgrave (Eds.), *Criticism and the growth of knowledge* (pp. 231–278). Cambridge, England: Cambridge University Press.

Kuhn, T. S. (1977). *The essential tension.* Chicago: University of Chicago Press.

Ladson-Billings, G. (1998). It doesn't add up: African American students' mathematics achievement. *Journal for Research in Mathematics Education, 28,* 697–708.

Lakatos, I. (1970). Falsification and the methodology of scientific research programmes. In I. Lakatos & A. Musgrave (Eds.), *Criticism and the growth of knowledge* (pp. 91–195). Cambridge, UK: Cambridge University Press.

Lampert, M. (1990). When the problem is not the question and the solution is not the answer: Mathematical knowing and teaching. *American Educational Research Journal, 27,* 29–63.

Lampert, M. (2001). *Teaching problems and the problems of teaching.* New Haven, CT: Yale University Press.

Latour, B. (2000). *Pandora's hope: Essays on the reality of science studies.* Cambridge, MA: Harvard University Press.

Latour, B., & Wollgar, S. (1979). *Laboratory life: The social construction of scientific facts.* Beverly Hills, CA: Sage.

Lave, J. (1988). *Cognition in practice: Mind, mathematics and culture in everyday life.* New York: Cambridge University Press.

Lave, J. (1991). Situating learning in communities of practice. In L. B. Resnick, J. M. Levine, & S. D. Teasley (Eds.), *Perspectives on socially shared cognition* (pp. 63–82). Washington, DC: American Psychological Association.

Lave, J., & Wenger, E. (1991). *Situated learning: Legitimate peripheral participation.* New York: Cambridge University Press.

Lawler, R. W. (1985). *Computer experience and cognitive development: A child's learning in a computer culture.* New York: Wiley.

Lehrer, R., & Lesh, R. (2003). Mathematical learning. In W. Reynolds & G. Miller (Eds.), *Comprehensive handbook of psychology* (Vol. 7, pp. 357–391). New York: Wiley.

Lehrer, R., & Schauble, L. (2000). Inventing data structures for representational purposes: Elementary grade students' classification models. *Mathematical Thinking and Learning, 2,* 51–74.

Lehrer, R., Strom, D. A., & Confrey, J. (2002). Grounding metaphors and inscriptional resonance: Children's emerging understanding of mathematical similarity. *Cognition and Instruction, 20,* 359–398.

Leont'ev, A. N. (1978). *Activity, consciousness, and personality.* Englewood Cliffs, NJ: Prentice-Hall.

Leont'ev, A. N. (1981). The problem of activity in psychology. In J. V. Wertsch (Ed.), *The concept of activity in Soviet psychology* (pp. 37–71). Armonk, NY: Scharpe.

MacKay, D. M. (1969). *Information, mechanism, and meaning.* Cambridge, MA: MIT Press.

Martin, J. B. (2000). *Mathematics success and failure among African-American youth.* Mahwah, NJ: Erlbaum.

Masingila, J. (1994). Mathematics practice in carpet laying. *Anthropology and Education Quarterly, 25,* 430–462.

McLaughlin, M., & Mitra, D. (2004, April). *The cycle of inquiry as the engine of school reform: Lessons from the Bay Area School Reform Collaborative.* Paper presented at the annual meeting of the American Educational Research Association, San Diego, CA.

Mehan, H., Hubbard, L., & Villanueva, I. (1994). Forming academic identities: Accommodation without assimilation among involuntary minorities. *Anthropology and Education Quarterly, 25,* 91–117.

Mehan, H., & Wood, H. (1975). *The reality of ethnomethodology.* New York: John Wiley.

Meira, L. (1998). Making sense of instructional devices: The emergence of transparency in mathematical activity. *Journal for Research in Mathematics Education, 29,* 121–142.

Millroy, W. L. (1992). An ethnographic study of the mathematical ideas of a group of carpenters. *Journal for Research in Mathematics Education, Monograph No. 5.*

Minick, N. (1987). The development of Vygotsky's thought: An introduction. In R. W. Rieber & A. S. Carton (Eds.), *The collected works of Vygotsky, L.S.* (Vol. 1, pp. 17–38). New York: Plenum.

Mokros, J., & Russell, S. J. (1995). Children's concepts of average and representativeness. *Journal for Research in Mathematics Education, 26,* 20–39.

Moschkovich, J. (2002). A situated and sociocultural perspective on bilingual mathematics learners. *Mathematical Thinking and Learning, 4,* 189–212.

Moses, R. P., & Cobb, C. E. (2001). *Radical equations: Math literacy and civil rights.* Boston: Beacon Press.

Nasir, N. S. (2002). Identity, goals, and learning: Mathematics in cultural practice. *Mathematical Thinking and Learning, 4,* 213–248.

National Council of Teachers of Mathematics. (1989). *Curriculum and evaluation standards for school mathematics.* Reston, VA: Author.

National Council of Teachers of Mathematics. (1991). *Professional Standards for teaching mathematics.* Reston, VA: Author.

National Council of Teachers of Mathematics. (2000). *Principles and standards for school mathematics.* Reston, VA: Author.

National Research Council. (2002). *Scientific research in education.* Washington, DC: National Academy Press.

Nelson, B. C. (1999). *Building new knowledge by thinking: How administrators can learn what they need to know about mathematics education reform.* Cambridge, MA: Educational Development Center.

Nicholls, J. (1989). *The competitive ethos and democratic education.* Cambridge, MA: Harvard University Press.

Noddings, N. (1990). Constructivism in mathematics education. In R. B. Davis, C. A. Maher, & N. Noddings (Eds.), *Constructivist views on the learning and teaching of mathematics. Journal for Research in Mathematics Education Monograph No. 4.* (pp. 7–18). Reston, VA: National Council of Teachers of Mathematics.

Nunes, T., Schliemann, A. D., & Carraher, D. W. (1993). *Street mathematics and school mathematics.* Cambridge, MA: Cambridge University Press.

Pea, R. D. (1985). Beyond amplification: Using computers to reorganize human mental functioning. *Educational Psychologist, 20,* 167–182.

Pea, R. D. (1987). Cognitive technologies for mathematics education. In A. H. Schoenfeld (Ed.), *Cognitive science and mathematics education* (pp. 89–122). Hillsdale, NJ: Erlbaum.

Pea, R. D. (1993). Practices of distributed intelligence and designs for education. In G. Salomon (Ed.), *Distributed cognitions* (pp. 47–87). New York: Cambridge University Press.

Pepper, S. C. (1942). *World hypotheses.* Berkeley: University of California Press.

Piaget, J. (1970). *Genetic epistemology.* New York: Columbia University Press.

Piaget, J. (1980). *Adaptation and intelligence: Organic selection and phenocopy.* Chicago: University of Chicago Press.

Pickering, A. (1984). *Constructing quarks: A sociological history of particle physics.* Edinburgh, UK: Edinburgh University Press.

Pimm, D. (1987). *Speaking mathematically: Communication in mathematics classrooms.* London: Routledge and Kegan Paul.

Pimm, D. (1995). *Symbols and meanings in school mathematics.* London: Routledge.

Pirie, S., & Kieren, T. E. (1994). Growth in mathematical understanding: How can we characterize it and how can we represent it? *Educational Studies in Mathematics, 26,* 61–86.

Polanyi, M. (1958). *Personal knowledge.* Chicago: University of Chicago Press.

Popper, K. (1972). *Objective knowledge: An evolutionary approach.* Oxford, UK: Clarendon Press.

Porter, T. M. (1996). *Trust in numbers: The pursuit of objectivity in science and public life.* Princeton, NJ: Princeton University Press.

Prawat, R. S. (1995). Misreading Dewey: Reform, projects, and the language game. *Educational Researcher, 24(7),* 13–22.

Price, J. N., & Ball, D. L. (1997). 'There's always another agenda': Marshalling resources for mathematics reform. *Journal of Curriculum Studies, 29,* 637–666.

Putnam, H. (1987). *The many faces of realism.* LaSalle, IL: Open Court.

Quine, W. (1953). *From a logical point of view.* Cambridge, MA: Harvard University Press.

Quine, W. (1992). *Pursuit of truth.* Cambridge, MA: Harvard University Press.

Rogoff, B. (1990). *Apprenticeship in thinking: Cognitive development in social context.* Oxford, UK: Oxford University Press.

Rommetveit, R. (1992). Outlines of a dialogically based social-cognitive approach to human cognition and communication. In A. H. Wold (Ed.), *The dialogical alternative towards a theory of language and mind* (pp. 19–44). Oslo, Norway: Scandinavian University Press.

Rorty, R. (1979). *Philosophy and the mirror of nature.* Princeton, NJ: Princeton University Press.

Rotman, B. (1988). Towards a semiotics of mathematics. *Semiotica, 72,* 1–35.

Rotman, B. (1994). Mathematical writing, thinking, and virtual reality. In P. Ernest (Ed.), *Mathematics, education, and philosophy: An international perspective* (pp. 76–86). London: Falmer.

Saldanha, L. A., & Thompson, P. W. (2001). Students' reasoning about sampling distributions and statistical inference. In R. Speiser & C. A. Maher (Eds.), *Proceedings of the Twenty-Third Annual Meeting of the North American Chapter of the International Group for the Psychology of Mathematics Education* (Vol. 2, pp. 449–454). Columbus, OH: ERIC Clearinghouse for Science, Mathematics, and Environmental Education.

Saxe, G. B. (1991). *Culture and cognitive development: Studies in mathematical understanding.* Hillsdale, NJ: Erlbaum.

Schön, D. A. (1983). *The reflective practitioner.* New York: Basic Books.

Schön, D. A. (1986). *The design studio.* London: Royal Institute of British Architects.

Schön, D. A. (1987). *Educating the reflective practitioner.* San Francisco: Jossey-Bass.

Schutz, A. (1962). *The problem of social reality.* The Hague, The Netherlands: Martinus Nijhoff.

Scribner, S. (1984). Studying working intelligence. In B. Rogoff & J. Lave (Eds.), *Everyday cognition: Its development in social context* (pp. 9–40). Cambridge, MA: Harvard University Press.

Secada, W. G. (1995). Social and critical dimensions for equity in mathematics education. In W. G. Secada, E. Fennema, & L. B. Adajion (Eds.), *New directions for equity in mathematics education* (pp. 146–164). New York: Cambridge University Press.

Senger, E. (1999). Reflective reform in mathematics: The recursive nature of teacher change. *Educational Studies in Mathematics, 37,* 199–201.

Sfard, A. (1991). On the dual nature of mathematical conceptions: Reflections on processes and objects as different sides of the same coin. *Educational Studies in Mathematics, 22,* 1–36.

Sfard, A. (1998). On two metaphors for learning and the dangers of choosing just one. *Educational Researcher, 27(2),* 4–13.

Sfard, A. (2000a). On the reform movement and the limits of mathematical discourse. *Mathematical Thinking and Learning, 2,* 157–189.

Sfard, A. (2000b). Symbolizing mathematical reality into being. In P. Cobb, E. Yackel, & K. McClain (Eds.), *Symbolizing, communicating, and mathematizing in reform classrooms: Perspectives on discourse, tools, and instructional design* (pp. 37–98). Mahwah, NJ: Erlbaum.

Sfard, A., & Linchevski, L. (1994). The gains and pitfalls of reification – the case of algebra. *Educational Studies in Mathematics, 26,* 87–124.

Shotter, J. (1995). In dialogue: Social constructionism and radical constructivism. In L. P. Steffe & J. Gale (Eds.), *Constructivism in education* (pp. 41–56). Hillsdale, NJ: Erlbaum.

Silver, E. A., Smith, M. S., & Nelson, B. S. (1995). The QUASAR Project: Equity concerns meet mathematics education reform in middle school. In E. Fennema, W. Secada, & L.

Byrd (Eds.), *New directions for equity in mathematics education* (pp. 9–56). New York: Cambridge University Press.

Simon, M. A. (1995). Reconstructing mathematics pedagogy from a constructivist perspective. *Journal for Research in Mathematics Education, 26*, 114–145.

Slavin, R. E. (2002). Evidence-based educational policies: Transforming educational practice and research. *Educational Researcher, 31*(7), 15–21.

Slavin, R. E. (2004). Educational research can and must address "what works" questions. *Educational Researcher, 33*(1), 27–28.

Sleeper, R. W. (1986). *The necessity of pragmatism: John Dewey's conception of philosophy.* New Haven, CT: Yale University Press.

Smith, J. E. (1978). *Purpose and thought: The meaning of pragmatism.* Chicago: University of Chicago Press.

Smith, J. P. (1996). Efficacy and teaching mathematics by telling: A challenge for reform. *Journal for Research in Mathematics Education, 27*, 387–402.

Solomon, G. (1993). No distribution without individuals' cognition: A dynamic interactional view. In G. Solomon (Ed.), *Distributed cognitions* (pp. 111–138). Cambridge, MA: Cambridge University Press.

Spillane, J. P. (2000). Cognition and policy implementation: District policy-makers and the reform of mathematics education. *Cognition and Instruction, 18*, 141–179.

Steffe, L. P., & Cobb, P. (1988). *Construction of arithmetical meanings and strategies.* New York: Springer-Verlag.

Steffe, L. P., & Kieren, T. E. (1994). Radical constructivism and mathematics education. *Journal for Research in Mathematics Education, 25*, 711–733.

Steffe, L. P., von Glasersfeld, E., Richards, J., & Cobb, P. (1983). *Children's counting types: Philosophy, theory, and application.* New York: Praeger Scientific.

Stein, M. K., & Brown, C. A. (1997). Teacher learning in a social context: Integrating collaborative and institutional processes with the study of teacher change. In E. Fennema & B. Scott Nelson (Eds.), *Mathematics teachers in transition* (pp. 155–192). Mahwah, NJ: Erlbaum.

Stein, M. K., Silver, E. A., & Smith, M. S. (1998). Mathematics reform and teacher development: A community of practice perspective. In J. G. Greeno & S. V. Goldman (Eds.), *Thinking practices in mathematics and science learning* (pp. 17–52). Mahwah, NJ: Erlbaum.

Stigler, J. W., & Hiebert, J. I. (1999). *The teaching gap.* New York: Free Press.

Streefland, L. (1991). *Fractions in realistic mathematics education. A paradigm of developmental research.* Dordrecht, The Netherlands: Kluwer.

Taylor, C. (1995). *Philosophical arguments.* Cambridge, MA: Harvard University Press.

Thompson, P. W. (1991). To experience is to conceptualize: A discussion of epistemology and mathematical experience. In L. P. Steffe (Ed.), *Epistemological foundations of mathematical experience* (pp. 260–281). New York: Springer-Verlag.

Thompson, P. W. (1994). The development of the concept of speed and its relationship to concepts of rate. In J. Confrey (Ed.), *The development of multiplicative reasoning in the learning of mathematics* (pp. 181–287). Albany, NY: SUNY Press.

Thompson, P. W. (2002). Didactical objects and didactical models in radical constructivism. In K. Gravemeijer, R. Lehrer, B. van Oers, & L. Verschaffel (Eds.), *Symbolizing, modeling and tool use in mathematics education* (pp. 197–220). Dordrecht, The Netherlands: Kluwer.

Thompson, P. W., & Saldanha, L. A. (2000). Epistemological analyses of mathematical ideas: A research methodology. In M. Fernandez (Ed.), *Proceedings of the Twenty-Second Annual Meeting of the North American Chapter of the International Group for the Psychology of Mathematics Education* (Vol. 2, pp. 403–407). Columbus, OH: ERIC Clearinghouse for Science, Mathematics, and Environmental Education.

Toulmin, S. (1963). *Foresight and understanding.* New York: Harper Torchbook.

Traweek, S. (1988). *Beamtimes and lifetimes: The world of high energy physicists.* Cambridge, MA: Harvard University Press.

U.S. Congress. (2001). *No Child Left Behind Act of 2001.* Washington, DC: Author.

van der Veer, R., & Valsiner, J. (1991). *Understanding Vygotsky: A quest for synthesis.* Cambridge, MA: Blackwell.

van Hiele, P. M. (1986). *Structure and insight.* Orlando, FL: Academic Press.

van Oers, B. (1996). Learning mathematics as meaningful activity. In P. Nesher, L. P. Steffe, P. Cobb, G. A. Goldin, & B. Greer (Eds.), *Theories of mathematical learning* (pp. 91–114). Hillsdale, NJ: Erlbaum.

van Oers, B. (2000). The appropriation of mathematical symbols: A psychosemiotic approach to mathematical learning. In P. Cobb, E. Yackel, & K. McClain (Eds.), *Symbolizing and communicating in mathematics classrooms: Perspectives on discourse, tools, and instructional design* (pp. 133–176). Mahwah, NJ: Erlbaum.

Vergnaud, G. (1982). A classification of cognitive tasks and operations of thought involved in addition and subtraction. In T. P. Carpenter, J. M. Moser, & T. A. Romberg (Eds.), *Addition and subtraction: A cognitive perspective.* Hillsdale, NJ: Erlbaum.

Vergnaud, G. (1994). Multiplicative conceptual field: What and why? In G. Harel & J. Confrey (Eds.), *The development of multiplicative reasoning in the learning of mathematics* (pp. 41–59). Albany, NY: SUNY Press.

Voigt, J. (1985). Patterns and routines in classroom interaction. *Recherches en Didactique des Mathematiques, 6*, 69–118.

Voigt, J. (1996). Negotiation of mathematical meaning in classroom processes. In P. Nesher, L. P. Steffe, P. Cobb, G. A. Goldin, & B. Greer (Eds.), *Theories of mathematical learning* (pp. 21–50). Hillsdale, NJ: Erlbaum.

von Glasersfeld, E. (1984). An introduction to radical constructivism. In P. Watzlawick (Ed.), *The invented reality* (pp. 17–40). New York: Norton.

von Glasersfeld, E., & Cobb, P. (1984). Knowledge as environmental fit. *Man-Environment Systems, 13*, 216–224.

Vygotsky, L. S. (1962). *Thought and language.* Cambridge, MA: MIT Press.

Vygotsky, L. S. (1978). *Mind in society: The development of higher psychological processes.* Cambridge, MA: Harvard University Press.

Vygotsky, L. S. (1981). The genesis of higher mental functions. In J. V. Wertsch (Ed.), *The concept of activity in Soviet psychology.* Armonk, NY: M.E. Sharpe.

Warren, B., Ballenger, C., Ogonowski, M., Rosebery, A. S., & Hudicourt-Barnes, J. (2001). Rethinking diversity in learning science: The logic of everyday sense-making. *Journal of Research in Science Teaching, 38*, 529–552.

Wenger, E. (1998). *Communities of practice.* New York: Cambridge University Press.

Wertsch, J. V. (1998). *Mind as action.* New York: Oxford University Press.

Wertsch, J. V. (2002). *Voices of collective remembering.* New York: Cambridge University Press.

Westbrook, R. B. (1991). *John Dewey and American democracy.* Ithaca, NY: Cornell University Press.

Whitehurst, G. J. (2003). *Research on mathematics education.* Washington, DC: US Department of Education.

Whitehurst, G. J. (2003, April). *The Institute of Educational Sciences: New wine, new bottles.* Paper presented at the annual meeting of the American Educational Research Association, Chicago.

Zevenbergen, R. L. (2000). "Cracking the code" of mathematics classrooms: School success as a function of linguistic, social and cultural background. In J. Boaler (Ed.), *Multiple perspectives on mathematics teaching and learning.* Stamford, CT: Ablex.

AUTHOR NOTE

I am grateful to Steve Lerman, Nel Noddings, and Anna Sfard for their constructive and helpful comments on a drafts of this chapter.

Correspondence concerning this article should be addressed to Paul Cobb, Vanderbilt University, Peabody College Box 330, Nashville, TN 37203. E-mail: paul.cobb@vanderbilt.edu

2

THEORY IN MATHEMATICS EDUCATION SCHOLARSHIP

Edward A. Silver and Patricio G. Herbst

UNIVERSITY OF MICHIGAN

This chapter addresses the use of theory in mathematics education scholarship.[1] We approach our task as one of characterizing how theory is used in our field, taking a practice-oriented perspective rather than a normative one. Yet we do not advocate a passive stance toward theory development and use. Rather, we intend that the review we offer here, and the embedded framework for considering theory as central to the relationships between problems, research, and practice, will stimulate reflection, growth, and improvement in the way that theory is viewed, developed, and used in our field.

As Mason and Waywood (1996) have noted, "theory is a value-laden term with a long and convoluted history" (p. 1055). A synopsis of that history may be seen in contemporary usage of the word *theory*. The Oxford English Dictionary (OED) lists the following as nonobsolete meanings of the word:[2]

3. A conception or mental scheme of something to be done, or of the method of doing it; a systematic statement of rules or principles to be followed.

4. a. A scheme or system of ideas or statements held as an explanation or account of a group of facts or phenomena; a hypothesis that has been confirmed or established by observation or experiment, and is propounded or accepted as accounting for the known facts; a statement of what are held to be the general laws, principles, or causes of something known or observed.

b. That department of an art or technical subject which consists in the knowledge or statement of the facts on which it depends, or of its principles or methods, as distinguished from the *practice* of it.

c. A systematic statement of the general principles or laws of some branch of mathematics; a set of theorems forming a connected system: as *the theory of equations, of functions, of numbers, of probabilities*.

5. In the abstract (without article): Systematic conception or statement of the principles of something; abstract knowledge, or the formulation of it: often used as implying more or less unsupported hypothesis (cf. 6): distinguished from or opposed to *practice* (cf. 4b). *in theory* (formerly *in the theory*): according to theory, theoretically (opp. to *in practice* or *in fact*).

6. In loose or general sense: A hypothesis proposed as an explanation; hence, a mere hypothesis, speculation, conjecture; an idea or set of ideas about something; an individual view or notion. Cf. 4.

---

[1] No single chapter could cover the extensive territory that our charge encompasses, so we have made choices about how to focus the chapter in ways that we hope will complement other sources that a reader may find helpful on this topic. One excellent resource is Mason and Waywood (1996). Moreover, given the intimate connection of theory to philosophical perspectives, on the one hand, and to research methods and the nature of evidence, on the other hand, we urge readers interested in theory to also read Paul Cobb's chapter on philosophical perspectives and Alan Schoenfeld's chapter on research methods, both of which appear in this volume. Other chapters in this volume will explicate theory as it is used and developed in specific areas of research in our field.

[2] http://dictionary.oed.com, retrieved May 16, 2005.

The attentive reader of this chapter will note the use of most, perhaps all, of these senses of the word theory in this chapter.

A fundamental assumption of this chapter is that theory appears in our scholarship, especially in research in mathematics education, in many guises and at many levels. Our goal here is to describe, illustrate, and discuss a variety of these ways. But we also intend to bring some order to this diversity, both by pointing to characteristics of theory that appear to be common across the many uses and by providing a unified framework that locates theory as a central matter in the research endeavor in our field. Along the way we suggest several scopes of work where theories of different kinds can be useful. To accomplish our task we use many examples, some short, some long, which have been chosen from the literature that we know best to make key points. Although we treat many examples in this chapter, we do not profess to be comprehensive. Our omission of some theoretical perspective should not be interpreted as implied critique, nor should inclusion be seen as overt endorsement. Our examples are chosen only to help with the larger task undertaken here.

We prepared this chapter against the backdrop of two seemingly contradictory trends in mathematics education scholarship. On the one hand, there has been a growing concern for increasing the relevance and usefulness of research in our field. The press for relevance has come not only from policymakers and practitioners but also from within the community of scholars itself. On the other hand, mathematics education scholarship has become increasingly attentive to theory in recent years. Whereas the role of theory in mathematics education scholarship was usually tacit (at best) 30 years ago, it has become much more visible and important in contemporary scholarship.

What makes these two trends appear to be contradictory is an ambient skepticism in the field of mathematics education about the possibility that theory-driven research can also be relevant to educational practice. It is not uncommon to hear teachers and many other mathematics education professionals complain that mathematics education scholarship is "too theoretical." This declaration appears to rest on OED meanings 5 or 6 (above); the sense in which "true in theory" is euphemism for "not true in practice." Moreover, one can often hear researchers deride work for being "too applied," thereby identifying research that that is closely tied

to educational practice as insufficiently theoretical. Although one can debate the validity of such statements and views, there can be little debate that these views are deeply and widely held.

There can be little argument that there is a legitimate tension between some of the meanings evoked by the words *theory* and *practice* when these words are pitted against each other. In his analysis of some of the challenges faced by teachers who enter research-oriented doctoral programs in education, David Labaree has deftly captured the essence of this tension in characterizing the challenging transition that individuals make if they leave teaching careers and enter research-oriented doctoral programs: "The shift from K–12 teaching to educational research often asks students to transform their cultural orientation from normative to analytic, from personal to intellectual, from the particular to the universal, and from the experiential to the theoretical" (2003, p. 16). Nevertheless, despite the tensions that are evident when one pits the words *theory* and *practice* against each other, we contend that mathematics education scholarship can be both invested in theory and relevant to educational practice.[3] The essence of our argument is straightforward. Viewed historically, mathematics education scholarship is itself a practice—an institutional practice, done by scholars who belong to disciplines and to organizations and are thus disposed to address certain realities, to perceive certain puzzles, and to employ particular kinds of tools. As participants in a scholarly field, not only do we researchers examine educational practice, but also many in our midst come from that practice or are active reformers of that practice. Thus, we assert that theory, research, and practice in mathematics education should exist in a more harmonic relation than has been the case to date.

Although scholarship in the field of mathematics education has become more conversant with theory (more theory-friendly) over the past 35 years or so, it has done so in a variety of ways; that is, the use of theory in scholarship has become more common, yet ways of using theory are quite diverse. We offer a framework that can account for the diverse ways in which scholarship has made room for theory and how scholarly pursuits might be *improved* by being more self-conscious about matters of theory.

Given that theory has become much more evident in scholarship in our field in the past decade or so, the moment seems propitious for a serious examination of the role that theory plays and could play in the

---

[3] For another excellent treatment of the tension between theory and practice in mathematics education, see Malara and Zan (2002). Perceptive commentary on the potential connection between research and practice has also been offered by Lester and Wiliam (2002).

formulation of problems, in the design and methods employed, and in the interpretation of findings in education research. But, given increased pressures for relevance, we think that such an examination should not take an *insider* perspective only—focusing on research questions, research methods, and research findings. To be sure, theory has much to offer in regard to all three of these—helping to shape research questions, suggest research methods, and explain research findings—but we think a broader view is needed to span the perceived chasm between scholarship and practice in mathematics education. Through our consideration of the place of theory in mathematics education scholarship in this chapter, we hope to offer the field some new ways to think about the relationship between scholarship and educational practice.

Many in the policy and practice communities would argue that the growth in attention to theory in recent decades has driven the field away from central concerns of educational practice. As noted above, many view theory as a likely culprit in the perceived gap between the work done in educational research and the problems and needs of educational practice. In contrast, Sfard (2005) argued that researchers' attention to theory is not the culprit in the separation of research and practice; rather, it is researchers' careless use of theory that inhibits both the explanation and communication of findings and their implications. We hope to offer a view of theory use in mathematics education scholarship that will narrow the gulf between research and practice.

In this chapter we describe the emergence of theory (in its many forms) in mathematics education scholarship, and we offer examples of ways in which theory has contributed to, or might contribute to, our individual and collective endeavors. In our view, a careful examination of the diverse ways in which research in mathematics education has become and can continue to grow more theoretical should promote continued improvements in the formulation, examination, and justification of research problems. This examination could in turn transform debates over educational research, educational policy, and educational practice into collective deliberations on the ways in which it is or might be productive to give theory a more central role in our field.

THE EMERGENCE OF THEORY IN MATHEMATICS EDUCATION SCHOLARSHIP

In 1980 the National Council of Teachers of Mathematics (NCTM) published the first book that might legitimately be dubbed a "handbook of research in mathematics education" in the United States. Aptly titled *Research in Mathematics Education* (Shumway, 1980), this was one volume in a series of professional reference books. A close examination of that volume offers a glimpse at the state of research, and research critique, in our field at that time. Of particular relevance to this chapter are commentaries in the 1980 volume that address the status and role of theory in research in mathematics education. For example, in his chapter on "the research process," David Johnson (who had served a decade earlier as the first editor of the *Journal for Research in Mathematics Education*) decried the lack of theoretical grounding for much of the research in mathematics education at that time: "Perhaps the most significant shortcoming of much contemporary research [is] the lack of adherence to any type of theoretical or conceptual framework to direct the research efforts and to provide a basis for interpreting the results of the experimentation" (1980, p. 31).

There can be little argument that the situation has changed. Although many legitimate questions and concerns can and should be raised about the "state of the art" in using theory in our field, a perusal of articles in major research journals in our field reveals that theory is alive and well in recent work, as evidenced by the frequent appearance of the words *theory, theoretical framework,* and *theorizing.* Further support for this claim comes from an interesting analysis by Lerman and Tsatsaroni (2004) of articles published in two major research journals in our field (*Educational Studies in Mathematics* [*ESM*] and the *Journal for Research in Mathematics Education* [*JRME*]). They noted a sharp decline between 1990–1995 and 1996–2001 in the number of published articles that they judged to have no evident theoretical framing; the decline was from 10% to 5% for *ESM* and from 24% to 11% for *JRME.*

This shift in emphasis and visibility is likely due in part to the fact that the importance of theory in relation to education research in general has received considerable emphasis over the past quarter century. But other factors have likely also contributed to the growth in attention to theory in research in mathematics education.

Disciplinary Connections

An essential characteristic of the field of mathematics education is that its questions and concerns are deeply tied to matters related to the teaching and learning of mathematics. As such, the need for ties to other academic disciplines is obvious. For example, research in mathematics education is concerned with mathematics, hence its ties to such disciplines as mathematics itself and to epistemology.

In interviews conducted by Silver and Kilpatrick (1994) with leading scholars in the field of mathematics education, many researchers commented on the need for mathematics education to maintain strong ties to well-established disciplines. Some expressed a desire to tie research in mathematics education even more closely to its "mother discipline" of mathematics by considering the historical-epistemological dimension of research on student learning or by examining the learning of advanced topics in school mathematics.[4]

For those among us who consider scholarship in mathematics education as a form of development of scholarship in mathematics, the meaning given to the word *theory* could have, at least in a professional deformation sense, some roots in the discipline of mathematics. OED definition 4c (above) refers to the use of the word *theory* in the discipline of mathematics; the Wikipedia[5] expands on its usage in mathematics as follows:

> In mathematics, the word **theory** is used informally to refer to certain distinct bodies of knowledge about mathematics. This knowledge consists of axioms, definitions, theorems and computational techniques, all related in some way by tradition or practice. Examples include group theory, set theory, Lebesgue integration theory. and field theory.
>
> The term **theory** also has a precise technical usage in mathematics, particularly in mathematical logic and model theory. A theory in this sense is a set of statements in a formal language, which is closed upon application of certain procedures called rules of inference. A special case of this, an axiomatic theory, consists of axioms (or axiom schemata) and rules of inference. A theorem is a statement which can be derived from those axioms by application of these rules of inference. Theories used in applications are abstractions of observed phenomena and the resulting theorems provide solutions to real-world problems. Obvious examples include arithmetic (abstracting concepts of number), geometry (concepts of space), and probability (concepts of randomness and likelihood).

Although a direct mapping of theory from mathematics to mathematics education is not possible, definitions of *theory* associated with mathematics and other disciplines that mattered in the inception of our field may nevertheless serve as useful heuristics that help uncover what our own theories are or might be.

Other researchers in the Silver and Kilpatrick (1994) study pointed to the very fruitful connections that have been formed over many years with the field of psychology—connections that have helped illuminate many aspects of mathematics learning and performance. Yet others noted the importance of the social dimensions and organizational contexts of teaching and learning interactions and events, which underscores a need to develop ties to other disciplines such as anthropology, sociology, and linguistics.

Steiner (1985) and other scholars have pointed out that the field of mathematics education, and especially research in this field, occupies a space at the intersection of many other academic disciplines. Sierpinska et al. (1993) put it very well when they noted that our field lies at the intellectual "crossroads of many well-established domains such as mathematics, psychology, sociology, epistemology, cognitive science, semiotics, and economics" (p. 276). Although this positioning of our field sometimes leads to disputes over the "ownership" of certain problems or the "legitimacy" of certain claims and approaches, it also promotes giving attention to theoretical issues. Scholars in most social science disciplines (e.g., psychology, sociology) typically justify their research investigations on grounds of developing understanding by building or testing theories. More generally, some researchers have argued that educational research should emulate scientific research "the long-term goal of which . . . [is] to generate theories that can offer stable explanations of phenomena that generalize beyond the particular" (Shavelson & Towne, 2002, p. 3).

Diverse Perspectives

When David Johnson wrote his chapter nearly 3 decades ago, psychology was virtually alone as a source of theory for mathematics education researchers. Reflecting the prevailing view of his time, Johnson (1980) referred exclusively to "psychological theory," which he described thus,

> First, a theory serves as a means to explain observed phenomena. Second, it enables the researcher to make predictions about as yet unobserved relationships. Third, and possibly most important, a theory structures the conduct of inquiry by guiding the researcher in the process of asking questions, formulating hypotheses, and determining what key variables and relationships to investigate. (p. 31)

---

[4] For more on the historical development of mathematics education as a scholarly field, and its early ties to mathematics and psychology, see Kilpatrick (1992).

[5] http://en.wikipedia.org/wiki/Theory#Mathematics, retrieved May 16, 2006.

Johnson urged researchers in mathematics education to "*first* investigate the adaptability of various psychological theories . . . to the learning and teaching of mathematics, and [only] in the event such adaptation is not feasible, move to the creation of a new theory" (p. 32). Although psychological theories still have considerable influence in contemporary research in mathematics education, one can also see many other disciplinary roots for the theoretical perspectives taken by researchers in the field.

In their study of articles published in *ESM* and *JRME*, Lerman and Tsatsaroni (2004) noted growth during the period 1990–2001 in attention to sociological and sociocultural theories, as well as increase in the use of linguistics and semiotics. Clearly, mathematics education researchers have had opportunities in recent years to draw on a broad array of theoretical positions to support their research investigations beyond the ones available when David Johnson wrote his chapter.

The diversification of theoretical perspectives in the field is closely tied to the emergence of modes of inquiry that are more qualitative and less quantitative in nature, and to the emergence of paradigms for experimental research alternative to the experimental paradigms in psychology—such as didactical engineering (e.g., Artigue & Perrin-Glorian, 1991) and design experiments (e.g., Cobb, Confrey, diSessa, Lehrer, & Schauble, 2003). The quantitative research methods, adapted from other fields such as agriculture, that dominated education research in the 1950s and 1960s relied on experimental and quasi-experimental research designs and statistical theories as warrants for generalizable claims concerning teaching and learning. Unfortunately, though such research occasionally *demonstrated* some important phenomenon or relationship, it rarely *explained* the basis for the relationship or why a relationship existed. As researchers in mathematics education became dissatisfied with the results of such research and frustrated by the myriad questions that such research tended to leave unanswered, they began to turn away from quantitative approaches and toward more qualitative approaches. This turn toward the qualitative demanded increased attention to theory. It was through the careful use of theory that case studies and detailed analyses of small-scale interventions could generate generalizable claims.

Academic Politics

Pragmatics and politics have also undoubtedly played at least some small role in increasing attentiveness to theory. A *theoretical perspective* (as opposed to a *practical perspective*) currently dominates the process of scholarly publication in mathematics education (Hanna, 1998). Thus, it is common for research journals that publish empirical studies examining some aspect of mathematics teaching and learning to require that the papers accepted for publication provide an explicit theoretical framework. In fact, manuscripts are often rejected for journal publication if they are deemed atheoretical or if their theoretical foundations are insufficiently explicated.

It can be argued that university and academic politics may also contribute to the growth of attention to theory in our field. Most researchers in mathematics education work in university settings, holding faculty appointments in mathematics departments or in schools of education. Given that research in mathematics education often deals with the teaching or learning of precollege mathematics or content found in introductory courses in college-level mathematics, rather than with "cutting edge" mathematical ideas, university-based researchers find it useful to establish the academic quality and intellectual rigor of their work by connecting it to disciplines that have higher status within the academic community. As Lagemann (2000) and other commentators have written, this is a feature of educational research more broadly, and it has deep roots in the historical development of educational scholarship and attempts to increase its status within the academic community.

For these reasons alone, it is not surprising that mathematics educators have in recent decades formed diverse and strong disciplinary allegiances and connections to theories and methods from other disciplines and used them to undergird their own research. Some of the disciplinary connections, such as those to psychology and to mathematics itself, are longstanding; others, such as those to anthropology and sociology, are of a more recent vintage. Moreover, because research in mathematics education is a relatively young enterprise, its practitioners have wisely looked to the theories and methods from longer established disciplines as a basis for framing their work. And the proliferation of disciplinary connections has led in recent years to much greater diversity in the paradigms undergirding research reported in the leading research journals in the field. Yet another factor influencing the growth of attention to theory in mathematics education is the increasing internationalization of the field.

Internationalization

Research in mathematics education can be viewed as an international endeavor even more than a national enterprise. In fact, as has been noted by several commentators (e.g., Bishop, 1992; Fischbein, 1990; Kilpatrick, 1993), a vital international community of mathematics education researchers has developed and continues to flourish. A discussion document prepared in advance of a meeting entitled "What Is Research in Mathematics Education and What Are Its Results," which was sponsored by the International Commission on Mathematics Instruction (ICMI), pointed to the existence of diverse voices and perspectives among members of our community (Sierpinska et al., 1993). The growth of an international community of mathematics education researchers and scholars has fostered consideration of a broad array of theoretical perspectives, many of which grew naturally within the confines of the cultural and historical contexts of particular countries and then captured the attention of others in the international community as both the theories and the community itself matured.

Contemporary conversations within the international mathematics education research community often focus on matters of theory and epistemology. The contents of proceedings for conferences of the International Group for the Psychology of Mathematics Education (PME) reveal this strong tendency. Moreover, in a book intended to summarize and highlight important features of research conducted by members of the PME (Nesher & Kilpatrick, 1990), four of the seven chapters deal almost exclusively with issues of learning topics and processes in particular mathematics content domains, and the other three deal primarily with theoretical issues or a blend of epistemological and content issues. In fact, considerable progress has been made internationally in studying the learning and teaching of mathematical content and processes from a variety of interesting and important theoretical perspectives.

(RE)CONSIDERING THE PLACE OF THEORY IN MATHEMATICS EDUCATION SCHOLARSHIP

One way to begin a consideration of the role of theory in mathematics education scholarship would be to ask what the field would be like if we had no theories, or if our theories were left implicit and kept from view. Some might argue that we can easily conduct this thought experiment by simply examining the historical record of scholarship in our field.

The earlier quote from David Johnson is only one of many that could be offered as evidence of the frequent calls for greater attention to theory in mathematics education scholarship over the years.

It is not difficult to find commentary on the state of mathematics education scholarship that characterizes the field as atheoretical. One way of visualizing this state of affairs is to refer to the scholarship triangle shown in Figure 2.1.

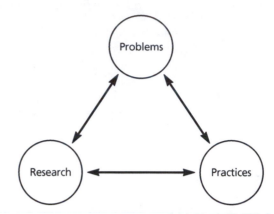

Figure 2.1 The scholarship triangle.

This triangle describes a view of scholarship that is constituted by relations between mathematics education *practices* and *research,* each of which is connected to a set of *problems.* Note that the connection between problems and practices is bidirectional, with stable practices attesting to problems and with perceptions of problems suggesting new practices. For example, certain established practices (e.g., activities in which children or adult engage with mathematics, algebra instruction in eighth-grade classrooms, children's play with wooden blocks) can pose certain kinds of problems (e.g., puzzlements about states of affairs that are different than what might be desired or surmised, undesirable differences in achievement across groups of students, nonoptimal use of resources, low yield in student achievement or motivation). The reverse can also hold. Practices may exist in response to perceived problems, such as when practitioners engage in innovative practices oriented to solving problems of certain kinds (e.g., the problem of ensuring that teachers know the mathematics they teach has inspired the practice of teaching mathematics to teachers in professional development programs, the problem of making access to higher education equitable motivates special programs for low-SES students).

Scholarly discourse about the connections between problems and practices has long been a part of mathematics education scholarship, though it has not always made explicit use of theory. In fact, an atheoretical approach is often quite appealing, offering a straightforward, commonsensical characterization of practices and problems deemed sufficient to guide research, which is the third vertex of the scholarship triangle shown in Figure 2.1.

In our view, research in which the role of theory is unspecified and apparently absent is not always lacking in merit. It often addresses problems deemed important by the public and points to critical issues. Moreover, such research can be influential and important. One need only consider how many citations of results derived from the National Assessment of Educational Progress [NAEP] (e.g., Silver & Kenney, 2000) or the trends in the Third International Mathematics and Science Study [TIMSS] (e.g., Mullis, Martin, González, & Chrostowski, 2004) are included in research articles to see the influence that research that is not theory-based can have on the field.

Such research is not restricted to large-scale surveys of national or international scope. Consider, for example, George Levine's (1972) study of the attitudes of elementary students and their parents toward mathematics and other subjects. He asked students and their parents to rate four school subjects (one of which was mathematics) in regard to their importance relative to 9 statements such as "when I was a student I did best in this subject" (p. 52, parent questionnaire) and "I do my best work in this subject" (p. 53, pupil questionnaire). He looked into the relative placement of mathematics in the ratings and found that students and their parents had a high regard for mathematics.

In terms of the scholarship triangle, Levine's (1972) research can be described as addressing the practice of studying elementary mathematics and the problem of increasing students' positive attitudes. Levine had a hunch that in addition to increasing teachers' positive attitudes toward mathematics, another driver to increase students' attitudes toward mathematics could be parental attitudes. Other than this hunch, supported to some extent by prior research, no theory was explicitly involved that justified why attitudes were important, why parents' attitudes might be important to look at, or what was the warrant for looking at mathematics in comparison with other school subjects as opposed to mathematics in comparison with other activities students might like to do.

Another interesting example to consider is the recently published study by Judson & Nishimori (2005) on the differences between American and Japanese high school students of calculus, with respect to their conceptual understanding of the derivative and their ability to use algebra in solving derivative problems. The investigators found no significant differences between the two groups of students in conceptual understanding, but they noted higher fluency on the part of the Japanese students. In terms of our triangle, we could describe Judson and Nishimori's (2005) study as addressing the practice of thinking and problem solving in calculus and the problem of describing and sizing up the gap between two school systems of economically competing nations. Prior cross-national studies like TIMSS and their lack of data for high school served as warrant to do this study, in which the authors relied on rather intuitive (though mathematically sophisticated) ideas of "conceptual understanding" and "algebraic skills" to code students' responses to problems. Additionally, no theory other than mathematics itself justified the items included in the testing, so students were compared in their understanding of the fundamental theorem of calculus or of Riemann sums without much information about what the authors considered these topics and items to be cases of, nor consideration of why particular versions of test questions were used and not others.

These two examples of interesting research are offered to underscore the observation that scholarship in our field includes some research[6] that can be legitimately justified on the basis of connections between practice and problems. Research in our field has long been done in this way—responding to problems derived from practice. That is, research has been a tool to produce specific and trustable information about problematic aspects of practice. Research has also been done to systematically describe practices originated in response to problems. In this way, research has been a tool to monitor or evaluate the impact of new practices. Research that looks into problems derived from practice has taken a variety of forms, such as comparing alternative manifestations of a given practice (e.g., when curriculum experts have done content analyses of materials being used in instruction; or when evaluators have inspected the differential achievement by comparably instructed students of different race, class, or gender). Research that looks into practices oriented to solving problems may look among other things like evaluating the practical outcomes of a problem-solving attempt (e.g., when evaluation specialists compare instructional treatments on the basis of students' achievement, when researchers have inspected the added value of novel practices such as professional

---

[6] In order to deal precisely with the role of theory in scholarship, we use the word *research* here in a rather restricted sense—as a particular research study done within certain timeframe, in response to a problem or question, and addressing a particular patch of reality. We recognize and endorse a broader use of the word *research* that is commonly used to refer to practices in which researchers also conceive problems, identify practices, and formulate theories; those activities may also be viewed as research. Our use of the word *scholarship* here is our way of capturing that larger sense of practices that might be dubbed research.

development or new materials such as manipulative materials or technology).

The arrows pointing from practices to research and from problems to research in Figure 2.1 are bidirectional because problems and practices have not just directed research but also at times drawn on research for useful or applicable results or for records of how problems have been addressed. Thus, practices may have looked into prior research in order to make policies of certain scope (e.g., as when differential achievement gains of two instructional treatments have been used to recommend policy decisions as to how to shape practice). In identifying and handling problems, the field has also looked to research to help inform decisions about the appropriate grain size for further inquiry into a problem (e.g., when reviews of the research literature have illuminated the importance to control for SES in studying differential achievement). Mathematics educators have for a long time been interested in questions like those, yet they have often proceeded with little overt attention to or investment in theory.

One could argue that neither example given above was truly atheoretical scholarship. Tacitly, each investigation relied on an organization of ideas that helped the authors make distinctions and decisions along the way. This is what Mason and Waywood (1996) referred to as *background theory*, "because every act of teaching and of research can be seen as based on a theory of or about mathematics education" (p. 1056). Though we recognize the sense in which this argument may be considered valid, we prefer to focus our attention on what Mason and Waywood dubbed *foreground theory*, in which theoretical considerations are made explicit.

Toward that end we propose the insertion of theory at the center of the scholarship triangle (formed by practice, problems, and research) as a way to capture our conception of how theory has provided or may provide added value to mathematics education scholarship (see Figure 2.2).

In Figure 2.2 theory appears as the center of gravity of the triangle, and it mediates relationships among the three vertices of the scholarship triangle shown earlier in Figure 2.1. We think that this framework can be used to describe an emerging state of affairs in mathematics education scholarship, in which theory is becoming more and more of an explicit feature in the work of mathematics education scholars, enabling and mediating connections between pairs of vertices in the triangle, often suggesting connections that may have not been visible before. In the remainder of this chapter, we will use this frame to describe and

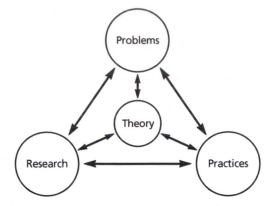

Figure 2.2 A theory-centered scholarship triangle.

exemplify the roles that theory plays in the work done in our field.

MAKING THEORY USEFUL: THEORY AS A VITAL MEDIATOR IN OUR SCHOLARSHIP

Whereas the role of theory in mathematics education scholarship was largely tacit in work published 30 years ago in our field, it has become more prominent in current scholarship. In this section we discuss how scholarship in mathematics education has related emerging uses of theory to each of the vertices of the triangle shown in Figure 2.2.

In positing that theory mediates the connection between each pair of vertices of the triangle we are proposing that one might ask: What could theory be doing there? In each of the next three subsections, we consider how theory might serve as a mediator of the dyadic interaction between vertices of the triangle in Figure 2.2 as a way to sketch possible answers to that question.

Theory As Mediator of Connections Between Research and Problems

What are some ways in which the connection between problems and research has been mediated by theory? As we note above the activity of scholarly problem posing has looked into prior research in order to inform the grain size in which further inquiry into a problem might proceed. Alternatively, the activity of addressing problems might have looked into public commentary or into theory. Here we attend to the roles that theory has played helping research get a handle on problems (see Figure 2.3).

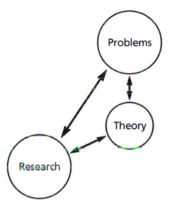

Figure 2.3 Theory mediating connections between research and problems.

Recalling the earlier quotes from David Johnson and the discussion of the close historical ties between psychology and mathematics education, we begin by noting that one way that theory has mediated the way problems appeal to research and research informs problems has often occurred through the use of intellectual formulations imported from other disciplines, such as psychology or sociology, *to interpret (or give meaning to) the results* of research in studies where these results had been produced from analysis, data, and problem formulation, which are independent of that theory. Consider for example John Clement's (1982) landmark study of college students' responses to algebra word problems in which they were asked to write equations to express relationships between two quantities, such as the following: "There are six times as many students as there are professors." Clement found that, when students were asked to write an equation that expressed the relation between S (the number of students) and P (the number of professors), many students who were otherwise competent in algebraic symbol manipulation mistakenly wrote $6S = P$. He further probed their thinking and found that the same individual who had modeled the statement using $6S = P$ might relate to the same situation in the correct way ($S = 6P$). In trying to explain what his results might mean Clement made an appeal to theory:

We call this evidence for "shifting between approaches" and hypothesize that it reflects an unobservable internal process of shifting between cognitive schemes used to deal with the problem. This provides evidence for the notion that human cognition is not always based on consistent processes—schemes that lead to contradictory results can apparently exist fairly autonomously and independently in the same individual. . . . There is an important implication of the theory that contradictory schemes for viewing problems can exist in people. The theory implies that students who are successfully taught a standard method for a mathematical skill may still possess intuitive, nonstandard methods that compete for control. (Clement, 1982, p. 28)

Thus this theory that individuals can be "of two minds" helped Clement interpret the empirical data displaying contradictory ways in which a student modeled the "students and professors" problem.

A theory, however, need not come in handy only to interpret the results of research in regard to a problem. It could also be imported from somewhere else as *a lens to look at data gathered or coded* and to proceed with the analysis. This role of theory may still presume no relationship between the nature of the lens and the nature of the problem. In work on teacher beliefs, Tom Cooney and his colleagues (see Cooney, 1985, 1994; Cooney, Shealy, & Arvold, 1998) used theories of individual intellectual development by Perry (1970) and by Belenky, Clinchy, Goldberger, and Tarule (1986) as lenses to look at interviews, observations, and artifacts pertaining to a problem that was somewhat distant from the one examined by Perry and Belenky: the individual development of beginning teachers. The notion that this development in this population was a case of the more general intellectual development in young adults described in the theories of Perry and Belenky et al. is symptomatic of this use of theory in our field: It presumes that the problems investigated are particular cases of more general problems addressed by theories generated in other social or behavioral sciences and that mathematics teaching and learning is a context of application of those theories. In one way or another, all our scholarship can display this relationship to theory, nesting our scholarship to larger fields of human inquiry, but not all our scholarship relates to theory as an imported tool to look at data collected from a homegrown problem.

Theory could also appear yet a bit earlier in the research work, as *a tool that helps give closure to the corpus of data, pointing to relevant sources of data* that complement the sources already identified by a problem, or to relevant categories that might not be represented in the data. Again, it can do so without necessarily requiring theory to play a role in the definition of the problem. Some researchers have thus used theories that describe a population of interest in order to make sure their sample to study a problem is comprehensive. At times these theories are unlikely to be identified as theories in the strict sense; they are more or less shared ways of talking about a population or a kind of data. An example of this use of theory is provided in the study by Cooney and Henderson (1972, p. 21) on the "ways mathematics teachers help students organize knowledge." The researchers set out to develop a taxonomy of the instructional moves that good teachers used in "assisting students in organizing knowledge" (p. 22). They analyzed the transcripts of 44 observed lessons in grade 7–12 classrooms, identified key instructional moves that appeared to support students' organization of knowledge, and proposed a scheme consisting of nine elements. To explain how they developed this taxonomy, the authors noted

> The taxonomy was developed through an interaction of logical considerations and analyses of the transcripts. It was felt that basically an item of knowledge is organized by incorporating it in some sort of structure. This is done by establishing one or more relations between the item and other knowledge within the structure. Hence the concept of a relation as a set of ordered pairs could be used to explicate each of the relations identified. Theoretical relations were constructed on the basis of logical considerations and the previous writings of Henderson, Smith, Ryle and others. Analysis of the transcripts suggested that modifications of the theoretical relations were necessary in order to adequately describe the classroom discourse. The present taxonomy then, was conceived in part from theoretical considerations and in part from empirical evidence gained through analyses of the transcripts. (pp. 22–23)

Although Cooney and Henderson derived their taxonomy through empirical analysis, it is interesting and important to note that the table in which they presented their data contained several cells with 0 observations at a particular grade level, and one relation (abstracting) was not observed at any grade level. Had they developed their taxonomy merely as a description of their data corpus, they would have come up with a shorter list, certainly one that did not contain abstracting. But they also used theory to guide their development of the taxonomy, particularly the theory of the logic of teaching developed by B. O. Smith and Henderson. As a result of working with both the theory and the data, and the resident tensions between the relations predicted by the theory and those observed in the data, Cooney and Henderson were able to propose and argue forcefully for a taxonomy of nine relations that could be used to characterize how teachers assist students to organize their knowledge.

A theory could also appear even earlier in mediating the connection between research and problems. A theory could provide ways of examining and transforming a problem that had initially been formulated through common sense, turning the problem into a researchable problem. The research done by Hill and Ball (2004) characterizing and measuring teachers' knowledge of mathematics for teaching provides an example. The research is grounded in the problem of gauging teacher knowledge of mathematics that makes a difference in their teaching, particularly as it regards the evaluation of professional development opportunities for teachers. The transit from that problem to the design of instruments and collection of data for the research is mediated by the development of a theory of the object of inquiry—teachers' knowledge of mathematics. The authors recounted the argument that the surprisingly little evidence of a relationship between teacher knowledge and student achievement is symptomatic of the field not having attended to the particular form of teacher knowledge that might make a difference. They built on Shulman's (1986) notion of *pedagogical content knowledge* to argue that, "at least in mathematics, how teachers hold knowledge may matter more than how much knowledge they hold" (p. 332). And they referred to Ball's notion of *knowledge of mathematics for teaching* (a particular kind of specialized knowledge of content; see Ball, Lubienski, & Mewborn, 2001) as the form of knowledge of mathematics that a teacher needs to have to be able to do the activities of teaching (e.g., explaining what an idea means, appraising students' methods for solving problems, determining whether invented student methods can be used in other problems). This theoretical framing of what can be meant by *teacher knowledge of mathematics* directed their pursuit of the research. Thus, among the assessment items that the researchers developed to measure teacher knowledge one finds, for example, an item that reads, "Imagine that you are working with your class on multiplying large numbers. Among your students' papers, you notice that some have displayed their work in the following ways" (p. 350). The item shows three different ways in which students worked out 35×25 and asks "which of these students would you judge to be using a method that could be used to multiply any two numbers?" Other items also demonstrate how the theoretical conceptualization of knowledge of mathematics for teaching directs the collection of data informing teacher knowledge by way of designing items that involve teachers in doing the mathematical work that they need to do while teaching.

A theory could also function like *a machine that produces problems as questions framed using the constructs and relations found within the theory*—problems that might or might not be readily seen as applicable or relevant to some concern in educational practice. Nevertheless, research that is spawned from theoretically generated questions can eventually help address a practical problem. An example of this use of theory can be found in Steffe's (2004) study of his own interactions with two children in the context of a constructivist teaching experiment on fractions.

Steffe's investigation was prompted by the theoretically important issue of reconciling the notion of a teacher's *hypothetical learning trajectory* (Simon, 1995) with the model of individual knowing proposed by radical constructivism (von Glasersfeld, 1995). Through a comparison between what the teacher-researcher had hypothesized to be the difference between the fractional schemes of two children and a retrospective investigation of how both children arrived at the notion of commensurate fractions, Steffe (2004) argued that a teacher's hypothetical learning trajectory regarding children's learning of fractions cannot be accounted for solely in terms of their progress in acquiring various fractional models, but rather that "learning trajectories of children must be constructed by teacher-researchers who participate first-hand in children's constructive activity" (p. 155). Steffe's inquiry into how the two children constructed the notion of commensurate fractions in interaction with the teacher-researcher emphasizes that the hypothetical learning trajectories are schemes that are themselves being adapted in the constructive activity of the teacher-researcher. Thus a theoretically generated question led to a study of teaching (outside of a classroom) in which a researcher demonstrated that hypothetical learning trajectories are less of a prediction based on existing schemes and forthcoming tasks and more of an emergent product of the interaction among individuals operating in the same milieu with different schemes.

At the risk of oversimplifying Steffe's investigation and findings to offer an example of this role of theory, we note that what appeared to be a completely theoretically generated research question seems to speak to a quite pressing pragmatic problem, that of whether any conclusive judgment as to the developmentally appropriate nature of a task can be made *a priori* for a child based solely on the child's current schemes. Thus, a question about a feature of a theory (a prediction, a necessary notion) may motivate the research, but this research may eventually translate into information on one or more publicly owned problems.

A final example of a way in which problem formulation has used theoretical considerations to draw on research that provides information is to consider the logical and conceptual structure of reviews or syntheses of past research on a problem. The background of a problem is often established on the basis of *an organized totality of the previous studies on related problems*. Not the studies themselves but the conceptual scheme that assembles the studies into a totality is a scholarly artifact that we suggest is a kind of theory in our field. To the point that the way a summary of the available research is organized shapes the extent to which some things about a problem are "known" and others are unknown and to the point that considerations such as what has or has not been sufficiently studied often drive decisions as to what else should be studied, we can assert that the organizational scheme of a review of research has functioned in mathematics education scholarship like theories function in more mature disciplines.

Some literature-organizing efforts have explicit theoretical aims, such as Anna Sfard's deft synthesis of major trends and perspectives in research on mathematics learning (Sfard, 1998) that contrasted approaches that reflected the acquisition metaphor and those that were captured by the participation metaphor. Her explicit aim was to raise theoretical issues and concerns through her analysis of prior work. Sfard's work offers an important example of a role for theory in the formulation of a problem through the organization and synthesis of prior work.

But such undertakings are not always explicitly oriented by or toward theory. Consider, for example, Begle's (1979) account of the research on mathematics teachers. He organized the empirical literature to seek insights into the relationship between teacher effectiveness and two large categories of teacher characteristics—teacher attitudes and teacher knowledge. Although Begle identified constructs, categories, and relationships, he was not explicit about the theory that connected them, perhaps because he viewed the relationships as ones to be accepted at face value: "There is no doubt that teachers play an important part in the learning of mathematics by their students. However, the specific ways in which teachers' understanding, attitudes, and characteristics affect their students are not widely understood" (p. 27).

Although Begle did not explicitly portray his literature review as an exercise in theory, and some might think it pretentious to apply the label *theory* to a summary of research, theory appeared to lurk implicitly in his review, in the form of assumptions of what research to look for and how to weigh it. For example, Begle argued the importance of the question of what role teachers played in students' learning, by considering teachers to be resources or assets for students' learning and by assuming that their differential effectiveness in promoting student achievement could be explained in relation to differences in their attitudes toward mathematics or in their knowledge of mathematics.

In the sense that a thorough, problem-oriented review of research can serve as a bridge between problems and research, insofar as a review of research posits an organization of the extant research knowledge and makes visible certain problems—and the associated categories, constructs, and relationships—a review can be argued to be a theoretical artifact, to function as a theory. Any such organization groups some things together and separates others, highlights some problems and leaves others—and the associated categories, constructs, and relationships—less evident. In the case of Begle's review, for example, the actual activity of teaching mathematics, and the differential ways in which the learning resource represented by teachers in relation to students and the mathematics to be learned might function, was at that time a minor area of research; Begle's organization of the literature did nothing to make it visible (see Ball et al., 2001; Cohen, Raudenbush, & Ball, 2003). Thus, although we have called attention to the (often implicit) theoretical nature of the organization of a literature review, unless the theoretical aspects are made explicit as in the case of Sfard's article, it may be a weak use of theory if the goal is to advance the research enterprise.

In summary, in this subsection we have described the bidirectional relationship between problems and research as one in which problems call for research to be done, and research produces information on problems. We have argued that one of the roles of theory in our scholarship is that of articulating or mediating that bidirectional relationship between research and problems. And we have identified and exemplified at least six distinct ways in which theory (or "theory") plays the role of mediator between problems and research:

- Theory as intellectual instrument imported from other disciplines to interpret results of research on a problem
- Theory as intellectual instrument imported from other disciplines as a lens to analyze data and produce results of research on a problem
- Theory as a tool that helps give closure to the corpus of data to study a problem, complementing the sources identified by the problem
- Theory as the means to transform a commonsensical problem into a researchable problem

- Theory as a generator of researchable problems that may later be translated in commonsensical terms
- (Often implicit/background) theory as systematic organization of a corpus of research on a problem

As a by-product of these different manifestations of theory we can identify a more nuanced notion of *problem*. On the one hand there are problems in the sense of undesirable or inconvenient happenstances to be fixed or managed (and research is expected to do that). On the other hand there are problems in the sense of puzzling, compelling phenomena that provoke interest and curiosity to find out more about them, to understand them (and research is expected to do that). The two notions of *problem*, the one active and the other one contemplative, are not disjoint, but the extent to which their correspondence can be worked out seems to owe to the involvement of theory in mediating the connection between problems and research. Figure 2.4 shows how the notion of a scholarship triangle accommodates this dual notion of problem.[7]

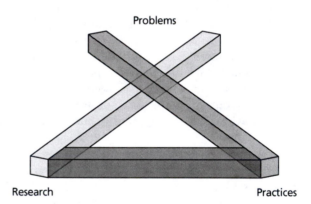

Figure 2.4 An open trilateral as a model of scholarship on two kinds of problems.

We propose those various ways in which theory mediates the connection between research and problems as a set of different entry points that have at various times been used by scholars in our field. The examples also show that these different roles of theory mediating the connection between problems and research are not neatly sorted in a historical development process. Research practices in our field display different kinds of engagements with theory as it regards posing and investigating researchable problems.

[7] We acknowledge Nicolas Balacheff's suggestion of this diagram and of the need to elaborate on the different nature of the problems.

Theory Mediating Connections Between Research and Practice

What are some ways in which the connection between research and practice has been mediated by theory? As we note above, scholarship in our field has addressed various practices: Individual mathematical thinking, teaching and learning in classrooms, and the development of mathematics teachers are three distinct practices of mainstream interest in our community (along with a number of others). Furthermore, new practices have developed in response to some research work: New curricula, pedagogical tools, or practices, even new institutions (such as summer institutes for mathematics teacher educators) have been developed that respond in some way to research. Another role of theory is that of mediating the connection between research and the practices with which our community is concerned (see Figure 2.5).

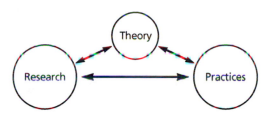

Figure 2.5 Theory mediating the connections between research and practice.

A widely held view of the role that theory plays in relating research to practice is judgmental: The more theoretical the research is, the less practical and applicable it is assumed or asserted to be. For some, perhaps many, teachers, teacher educators, and curriculum developers in mathematics education, research that is invested in theory seems to be irrelevant to practice. Although researchers might argue with this view by critiquing the critique (e.g., Sfard, 2005) or by arguing that many aspects of research have potential relevance to practice (e.g., Silver, 1990), we think it might be more productive in this chapter to look at what this view criticizes and to ask, What are the roles that theory plays and could play in connecting research to practice? What are the practices that our research connects to, for better or worse, through the use of theory? How does or could theory articulate research and practice in our scholarly community?

The word *theory* is often used to label sets of *prescriptive statements that could be used to guide the deliberate enactment of a practice.* Unlike physics, which concerns natural phenomena, teaching and learning are social phenomena. As Goffman (1974) pointed out, we humans tend to frame social events

as "guided doings"; that is, we assume that actors of the social world are in some significant sense willful agents of practices that they can significantly shape. Accordingly, the word *theory* as it relates to a social phenomenon such as educational practice is at times biased by such framing. In particular, the word *theory* is often used to indicate what, according to some vision, the agents *should* be doing—*theory* takes its everyday meaning, as an ideal or desired way of conducting social events. In that scheme of things, *practice* is used to refer to what social actors happen to or eventually are able to do, and *theory* is used to refer to what they should, or should be able to, do. We refer to these theories as prescriptive, noting that we make no distinction between forceful prescriptions and friendly suggestions, between progressive or reactionary ones—all of those "theories," no matter their force or their intention, relate to practice in similar, prescriptive or normative ways. We submit that when *theory* is understood in that prescriptive way, the relationship between research and practice is mediated by theory in at least two possible ways.

On the one hand research uses (prescriptive) theory as a norm against which to evaluate practices. An easy example of that is how mathematics itself has operated as a theory in curriculum research on the content deployed in textbooks (see for example Herbst, 1997). In such cases, mathematical knowledge as represented in texts and practices of the discipline has provided the theory—what should be the definition of a concept, what theorems should be proved, and so on. Textbooks have been looked upon as empirical material that attests to the practice of curriculum development, and the research has been one of describing, perhaps assessing the gap between theory and practice. Note also that a second example of this kind of involvement of prescriptive theories in mediating the relationship between research and practice is the use that some researchers have given to policy documents such as the NCTM *Standards* (NCTM, 1989, 2000) or derived ones in the evaluation of instructional programs.

On the other hand, practice has used research to shape prescriptive theories of the research-based "best practices" kind. Examples of this include the use that consumers of research might give to cognitive research on learning in developing a notion of best practice in teaching or, more frequently, the ways in which research often translates into "implications for practice." In both of these cases, the key theoretical operation is the assumption that relationships among constructs described or explained by research correspond to actual or desirable states of affairs in the real world that agents of a practice can willfully decide to strive towards. One could dispute the warrant of conceiving human action as amenable to voluntaristic theorizing, yet it is undeniable that the social world in which the practices of mathematics education exist (the world of schools, for example) frames those practices as guided doings. Thus agents of practices are under institutional obligations of trying harder and being better in whatever socially accepted definition of those terms, and the more specific those terms are the more they are identified as theories to be aimed to and the more amenable practices are to be certified as good by way of research. Theory as a conception of what is desirable thus helps research assist practice.

Diametrically opposed to the notion that theory's role is to prescribe or guide practice is the notion that social practices, such as mathematics teaching and learning, are complex social systems whose dynamics depends only partially on what their actors are willing to do. Rather, the conditions in which educational practices exist allow them to happen in similar ways as natural phenomena happen: Even when key agents may be unknown and some levers may not be controllable, the phenomena a practice produces are amenable for description and explanation. In that perspective, theory, rather than being the means for research to legitimize practice, is *a tool for research to understand practice.*

In particular, *theories can be languages to encode and read, that is to describe, a practice* so that researchers can examine practice according to such reading. There are many examples of descriptive theories in mathematics education research, particularly as regards the description of the practices of thinking and problem solving. The theory of conceptual fields developed by Gerard Vergnaud, for example, has been used to describe different kinds of conceptions tied to a common conceptual field. Examples worked out are those of additive structures—which include conceptions such as putting together, comparing states, or comparing transformations—and multiplicative structures—which include conceptions such as repeated addition and ratio. The theory of conceptual fields provides the means to describe the manifestations of those conceptions in the observation of children's mathematical work by requiring the observer to identify the problems, operational invariants, and signifiers involved. Much of the empirical work done by scholars like Carpenter, Fuson, Behr, and their associates concurs with Vergnaud's observations (see chapters 13 and 14, this volume, for more on this).

But theories that help research understand practice may not only aim at describing, they might also aim at explaining certain practices. It may be useful to borrow a distinction between causal and rational explanations from physical science. Likewise in mathematics education scholarship, among explanatory theories some may help research understand practice by disclosing the *causes* for practices, as when one looks for what has created a certain phenomenon. The recognition of a widespread prevalence of errors in arithmetic calculation has led researchers at various times to rely on theory to offer causal explanations for students' errors. For example, in a landmark analysis of children's subtraction errors, Brown and VanLehn (1980) proposed cognitive *repair theory* as a theoretical explanation for common errors based on children's incorrect generalization from correctly learned procedures and their attempts to perform local repairs on global procedures when problems were encountered. This gave rise to other explanatory theories that were proposed across a variety of errors (e.g., Resnick et al., 1989) to reconcile the apparent paradox that students could consciously do things that turned out to be incorrect in response to knowledge they acquired by way of doing things correctly.

Other explanatory theories may help research understand practice by providing rational arguments for why a certain phenomenon should not be surprising, or why it is plausible. An example of an explanatory theory of the rational kind can be found in the constructivist theory of learning (von Glasersfeld, 1995), which explains changes in a learner's action in terms of adaptations of schemes to fit what the learner perceives to be the situation in which she is operating. Another example may be found in Schoenfeld's (1988) article, "When Good Teaching Leads to Bad Results." In his analysis of one teacher's classroom, Schoenfeld theorized that an educational system that valued a particular kind of instruction—oriented toward having students master the skills needed to pass a test—could also encourage the relative isolation of intrinsically connected mathematical practices such as geometric construction and geometric proof. He provided a rational explanation for a practice of instruction that appeared to be dysfunctional for the development of students' mathematical dispositions. Researchers in France relied on similar observations to make the point that epistemological obstacles (i.e., items of knowledge, such as 'decimal numbers are like natural numbers but with a dot,' which may be true within a certain domain of practices but false beyond that domain) exist that could have a didactical (or instructional) origin (Brousseau, 1997). That is, these obstacles might exist because they help teacher and student get their classroom work done.

Finally, a particular case of theories that provide rational explanation is that of statistical theory, which helps research understand practice by demonstrating correlations between variables at play in a phenomenon. This last variety of theories is common in educational research at large and is sometimes (but not always) connected to theories of one or the other kind (causal or rational). Ethington and Wolfle's (1984) study of gender differences in mathematics achievement and Hembree's (1990) study of mathematics anxiety are examples of how statistical theory has been used to create explanations.

In addition to describing or explaining practice, theories that help research explain practice may also aim at predicting aspects of practice. A theory may help predicting what instances of a phenomenon may look like. For example, the repair theory of arithmetic bugs (Brown & VanLehn, 1980, 1982) not only provided a rational explanation of subtraction errors by proposing that students would engage in local repairs of a partially forgotten procedure. The theory also had the capacity to predict which repairs would more likely be done for which kind of problems (provided that students had forgotten the correct procedure), thus predicting which errors would likely be observed.

Another way in which theories may enable research to predict practice is by enabling researchers to construct probes that would perturb the stability of practices and elicit some of the norms that undergird their variability. Herbst and Brach (2006) used a theory of instructional situations proposing that mathematical work in classrooms is framed within customary systems of norms that regulate division of labor between student and teacher to propose a range of possible tasks to students. The theory allowed researchers to predict that students would react to the presumption that they be responsible for doing those tasks in their class by commenting on (and often amending) the share of labor that their teacher should be responsible for.

The distinction between theories that prescribe practice on the one hand and theories to understand practice is not as sharp as it may seem. Inasmuch as every observation is an experiment of sorts, eventually the extent to which the observed exists in a natural state is very much a construction itself. Without a scholar looking at it there would be no *phenomenon* per se, there would be people here and there doing more or less similar things (see Margolinas, 1998). That is, even the most descriptive approaches of research to practice include a prescription of what that practice should be that allows it to be visible and isolated from the rest of experience. Yet this is a different kind of prescription: Rather than prescribing what the practice should be like, it prescribes what the researcher will consider to be an instance of the practice, without necessarily expecting that agents of a practice should strive to behave as the theory says. A theory could therefore prescribe what is to be seen (how to observe something) whereas (possibly another) theory might also serve to understand what is seen.

To illustrate this use of theory we turn to Brousseau's (1997, especially pp. 137–144, 261, and chapter 5) notion of the *didactical contract*. To understand its origins, consider a scenario that might have been observed in elementary classrooms during the times of the so-called New Math. In concert with the goal of bringing school mathematics closer in content and aims to the mathematics of mathematicians, abstract mathematical ideas, particularly algebraic structures, had found their way into the prescribed curriculum for elementary school. Also, Piagetian research on the psychogenesis of knowledge had brought attention to the role of action in the formation of cognitive schemes. Into that mix were added prescriptive pedagogical theories, such as those proposed, for example, by Zoltan Dienes, that outlined how the teaching of abstract structures could be accomplished by having students actively play with different kinds of concrete materials (for example, different games), all of which had a common structure underlying the actions the child was meant to do.

From a mathematical perspective, the abstract structure to be taught could be thought of as the equivalence class of the games or structured materials that the student would play with—none of them per se being isomorphic to the structure itself (see Schoenfeld, 1987). From a psychological viewpoint, the claim could be made that the abstract structure was a viable way to describe the (second-order) model that an observer could make of the cognition of an individual student who was successfully playing any one of those games (*viability* in the sense that the model would not be falsified as long as the play was successful).[8] Pedagogical theories of the time, however, turned those games, which could be appropriate places where someone who knew the structure could see a representations of it, or the play of those games, that could be appropriate contexts for the observation of knowledge-in-action and learning, into methods of teaching. To teach a given structure to students who did not know the structure, teachers were to have

---

[8] Mathematics can certainly provide a language for an observer to describe what a student is doing—as long as the observer is clear not to equate the ideational system that the observer points to with that language with the ideational system present for the student. That is, stating that everything happens "as if" the student is thinking T is not the same as saying that the student is thinking T.

them play one game, then another one, and another one, until the students would see the structure. To characterize such instruction, Brousseau referred to the *Jourdain effect*: The student moves a set of cups of yogurt and the teacher says, "You have discovered the Klein group."[9]

For Brousseau, the observation of the Jourdain effect was much more than an amusing interlude. He saw it as a symptom of a possible breach of expectations in instruction. To characterize those expectations, Brousseau brought in the notion of the didactical contract that he had also used to explain other puzzling happenstances in student-teacher interactions (see Brousseau & Warfield, 1999).

Brousseau's accomplishment in naming the Jourdain effect a consequence of the didactical contract was not just to point at something gone wrong but also to say how it was, in a way, reasonable. After all, the student has played the game with the yogurt cups, not necessarily because it was fun or interesting but most surely because it was assigned by the teacher in school. Hence the student could expect the teacher to care to see what she has done and to give an official meaning to what she has done. Furthermore, because they are in a mathematics class, what they are doing has little chance of being officially meaningful just in its character of play; it must also be mathematical in some official way, no matter what it has meant for the individuals themselves.

The teacher knows that she has the responsibility to sanction what the action is worth, or at least she knows that the student thinks she has that responsibility. Furthermore the teacher knows that such responsibility is predicated on her supposedly different, greater knowledge of the subject, which, again, she knows she has or at least she knows that the student thinks she has. To be under the curricular prescription of having to teach abstract structures by concrete methods imposes extra constraints on the teacher that make the teacher's actions perplexing and yet also plausible, reasonable, predictable.

As this example illustrates, the point of a theory that describes and explains is not to say that something is wrong (or right for that matter) but to describe what it is and explain why it happens, that is, to shift attention away from simpleminded blame or handyman troubleshooting. One could say that the teacher's actions breach the contract at least in regard to how valid it is to label those ordinary actions after such a serious mathematical idea, but

one understands that the teacher has to provide a label for the play of the game and that she has to find indications of knowledge of the algebraic structure in students' actions. It is not unavoidable for a teacher to do those things, but if she does them one can understand that to be a predictable outcome of the didactical game the student, the teacher, and the knowledge at stake are summoned to play. In that sense our example illustrates more than just how a theory can produce a meaningful research result out of the observation of an undesirable happenstance in practice. It also illustrates how a theory can transform a commonsensical problem into a researchable problem: Being attuned to what makes the original observation meaningful enables one to look for places where other related observations could be made, and even to create experimental lessons where those phenomena could be reproduced and studied.

What theory does in this case is give the grounds to restore reason to what happened or, as Gaston Bachelard would say it, make the real appear as something that one should have thought of.[10] Theory does that without asking one to forget what happened: Theory allows one to see something as *odd* (or singular, surprising, special, funny, amusing, worthy of attention, pick your word) and as *reasonable* (plausible, expectable, predictable) at the same time. Ideologically one could continue to be disgusted by what happened and unmoved by the explanation. As we say above, if we were thinking from our positions as actors in the educational system (in addition to being researchers, both of us are parents and teacher educators) we might like classroom practices to be different, better for the sake of children, for the sake of mathematics, and for the sake of the society that receives them eventually. But theory bestows the mission to react rationally rather than ideologically; theory allows one to *address the system* that produces problems, not just attack the symptoms of those problems. Like the high fever that comes with an infection, the aberration called Jourdain effect might be less of a problem than the pedagogical theories or the curriculum on whose reaction the aberration appeared. But again, one can only ponder the merits of such assertion if one has a theory to stand on. We suggest that insofar as it affords grounds for civil discourse, insofar as it provides a way to see that people (not only children) do odd things for a reason, theory is useful. All of this is visible in the

---

[9] Guy Brousseau named the Jourdain effect after a character in an episode from the *Bourgeois Gentilhomme* by Molière. In the play, after an evidently nonpoetic verbiage by the rich and coarse Monsieur Jourdain, his private tutor tells him that what he has produced is actually prose. In naming the Jourdain effect, Brousseau reminds us that a structure imposed by an observer of a scenario may not be shared by the actor(s) in the scenario.

[10] "Le réel n'est jamais 'ce qu'on pourrait croire' mais il est toujours ce qu'on aurait dû penser" (Bachelard, 1938, p. 13).

example of the Jourdain effect, but there is more to the role of theory in mathematics education scholarship.

Another way in which theory mediates the relationship between research and practice is by providing means to find out what is generalizable or what is useful from research to practice—especially when these are not the things officially allowed by the traditional way of arguing generalizability in empirical research. Thus theories help practice read research by suggesting ways in which the results of research might orient policy-making (even if research cannot *validate* those policies). Theories may also help practice find by-products of research that practitioners may use at their own risk.

Consider, for example, the ways in which notions regarding the negotiation of task demands in classrooms have moved into the work of teachers and teacher educators. One strand of this body of scholarship begins with the work of Walter Doyle (1983, 1988) on academic tasks and flows into the work of Mary Kay Stein and her colleagues in the QUASAR project (Silver & Stein, 1996). Stein borrowed Doyle's conception of the centrality of academic tasks in creating learning opportunities for students, and she adapted and particularized it to the mathematics classroom at about the time of the introduction of the NCTM *Standards*. Although Stein did not refer to the notion of didactical contract, the theoretical and empirical contributions of this work can be seen as illuminating the origins of some contractual difficulties and entanglements.

Central to the work on academic tasks done by Stein and colleagues (e.g., Henningsen & Stein, 1997; Stein, Grover, & Henningsen, 1996; Stein, Smith, Henningsen, & Silver, 2000) has been the Mathematical Task Framework (MTF). The MTF (see Figure 2.6) underscores the important role that mathematical tasks play in influencing students' learning opportunities in the way they unfold during classroom instruction.

The MTF portrays a mathematical task as passing through phases, from the task as found in curricular materials, to the task as it is set up by the teacher in the classroom, to the task as enacted by students and their teacher interacting with one another and with the task

during the lesson. The MTF underscores the important role that mathematical tasks play in influencing students' learning opportunities. Moreover, it points to the centrality of the work that teachers do with and around mathematical tasks. Teachers' decisions and actions influence the nature and extent of student engagement with challenging tasks and ultimately affect students' opportunities to learn from their work on such tasks. The MTF and associated notions have contributed to theoretical understanding of the centrality of tasks in the academic learning that occurs in classrooms, the changes in cognitive demands that may emerge as students and teachers work on mathematics tasks in the classroom, and how changes (especially reductions) in cognitive demand can affect (usually constrain) students' opportunities to learn. The MTF was used as the basis for analyses of empirical data on the extent to which teachers succeed in maintaining the cognitive demand of challenging tasks in the classroom (Stein et al., 1996), factors that influence their success (Henningsen & Stein, 1997), and the relationship between teachers' success in maintaining the cognitive demand of tasks and the mathematics learning of their students (Stein & Lane, 1996).

The MTF has also been transformed into a tool with which to consider some of the challenges that inhere in teachers' use of complex tasks in the mathematics classroom. Recognizing that the MTF research suggests that teachers need to learn to orchestrate the work of the students while resisting the persistent urge to tell students precisely what to do, thereby removing the opportunity for thoughtful engagement, by responding to student queries and requests for information in ways that support student thinking rather than replace it, Stein and colleagues (Stein et al., 2000) constructed a set of narrative cases to illustrate observed patterns of classroom activity in which high-level task demands were maintained (or declined) during a lesson. These cases, and a second generation of narrative cases (e.g., Smith, Silver, & Stein, 2005) have been used widely by teachers in teacher professional development projects and in teacher preparation programs. This example raises to visibility the potentially important role that tools,

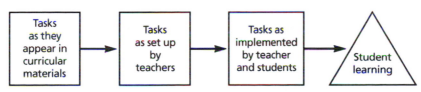

Figure 2.6 The Mathematical Tasks Framework.

such as frameworks, can play in the work of theory mediating connections between research and practice. One can think of other examples, such as the framework used in Cognitively Guided Instruction (CGI) to organize addition and subtraction problems not by operation but rather by underlying structure (Carpenter, Fennema, Franke, Levi, & Empson, 1999); this framework has had extensive use with and impact on teachers, teacher educators, and curriculum developers.

In this subsection we have argued that one of the roles of theory in scholarship is that of articulating or mediating the bidirectional relationship between research and practices. We have identified and exemplified at least four distinct ways in which theory (or "theory") plays the role of mediator between research and practice and made some distinctions within those ways:

Prescription, and elaborated that in this sense theory can be

- a system of prescriptions of what an educational practice should be like and used by research to evaluate the merit of an actual practice
- a system of "best practices" derived from research that has inspected an actual practice (usually with a criterion such as student achievement)
- a definition of what that practice consists of that allows it to be visible for the purposes of analysis and inquiry

Understanding, and elaborated that in this sense theory can be

- a language of description of an educational practice
- a conceptual system that may explain the causes of an educational practice
- a conceptual system that may establish the grounds on which the existence of an educational practice is plausible or reasonable
- a system of correlated statistical variables concurrent in an educational practice

Prediction, and elaborated that in this sense theory can be

- a system of conjectures that foretell instances of a phenomenon
- a system of assumptions that enable researchers to construct probes that perturb the stability of practices so as to elicit some of the norms that undergird their variability

Generalization, and elaborated that in this sense theory can be

- an account of which results of research might orient policy-making and why
- an organization of by-products of research that practitioners might use

Through this examination a more nuanced notion of practice also emerged. We demonstrated that researchers in mathematics education are concerned with real educational practices. But they engage in other kinds of practices, though these are often embedded in activity and thus less obvious. One is the practice of doing research itself; another is the practice of deriving implications from research (see Schoenfeld, this volume, for more on these practices). The connection between research and those practices benefit from the use of theories just as much the connection between research and educational practices does.

Theory Mediating Connections Between Practice and Problems

We initiated our consideration of the scholarship triangle (Figures 2.1 and 2.2) by arguing that scholarship inquires on problems posed by educational practices. We argued that it makes sense to deal with at least two meanings of *problem*: problem as undesirable happenstance in a practice and problem as provoking puzzle of a practice. Likewise we argued that scholarship also focuses on practices that respond to problems. We noted that new practices are often developed precisely to curb undesirable happenstances (as when a new program is established to remedy underachievement) or to capitalize on the potential of untapped resources (as when a new program is developed to teach mathematics with technology). Further, scholarship relates to other practices in addition to the educational practices that we study: The practice of doing research and the practice of shaping policies were provided as key examples. We now elaborate on how the investment of theory might help mediate the relationships between problems and practice (see Figure 2.7).

How does one know what makes sense to investigate, what is puzzling about a practice? How does one know what is undesirable or problematic about a practice or what needs improvement? How does one know what is the practice that produces a given problem? How does one know what elements need to be put together to develop a new practice? Theory seems to play a role in helping answer those questions. Theory can allow one to identify the practices that pose problems and

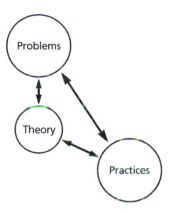

Figure 2.7 Theory mediating the connection between problems and practice.

to imagine possibilities in practice that can lead into researchable problems. In elaborating on this role of theory we attend to the more nuanced notions of problem and practice presented above.

One of the senses of *problem* is that of an inconvenience or an undesirable circumstance. All the practices that scholarship is concerned with, the educational practices studied, the novel practices studied experimentally, the practices carried out while involved in doing research, and those carried out when communicating research to policymakers and practitioners could all be problematic in that sense of the word *problem*. We suggest that theories may help each of those practices deal with those kinds of problems.

Consider first the case of the problems posed by educational practices, be those achievement gaps in K–12 education or myriad others. Even if problems were of the kind, "How can we fix this? How can we improve this?," there could be a role for theory in helping *propose a solution to a problem*. Note for example that this was the role that van Hiele's theory of levels of geometric thinking played in helping researchers come to grips with the problem of large numbers of students failing to learn the content of traditional high school courses in geometry focused on formal proof (Senk, 1983, 1989). With the assistance of the theory, scholars understood the problem as one of student readiness and sought to solve it by providing different, developmentally appropriate curricular options for students who had to take high school geometry but whose assessments did not show capacity to benefit from a course in formal geometry. The course on "informal geometry" was one of those curricular options, offered to solve a problem (see Cox, 1985; González & Herbst, 2006; Hamlin, 2006; Hoffer, 1981).

But theories also help one understand some calls for improvement as problematic. The role of theory in such cases is not to tell how to improve doing something practical but rather to help understand what the terms involved in the call for improvement mean, to analyze and characterize what it might mean to do it better, and to support attempts to explain why one or another attempted intervention succeeds or fails. For example, the increasingly well-articulated theory of mathematical knowledge for teaching (Ball & Bass, 2000; Ball, Hill, & Bass, 2005; Ball et al., 2001; Hill & Ball, 2004) offers many reasons why a simplistic call to improve mathematics teaching by having teachers take more mathematics courses is unlikely to succeed unless those courses are very carefully designed to reflect the conception of mathematics knowledge for teaching that is embodied in the theory.

Theory is pivotal in identifying practices and delimiting their scope in order to address a problem. For example, the recognition that the mathematical thinking that is of interest to the betterment of mathematics education is manifest in practices that involve communication with others (peers, teacher, or an observer), activities (such as problems), and artifacts (inscriptions, tools) has led many researchers to theorize the object of study not any more as an autonomous cognitive system but rather as a larger communicational or activity system, a discourse (see Sfard & Lavie, 2005). Thus the theoretical notion that language is not just a means for somebody to express what she thinks but actually a tool for her to think with, enlarges the practice being considered (to include, for example, the interactions between subjects and interviewer, as shown in Sfard & Lavie, 2005) and adds possible problems (e.g., understanding the register or the modality that enables or inhibits communication between interviewer and subject).

Also, theory can play an important role when one wishes to define and propose as important a problem of practice that needs improvement. Consider, for example, the subtle shift in the argument for instructional change in the NCTM *Standards* documents over a decade or so. The initial rationale for proposed changes was based on allegations of low instructional expectations, in documents such as *A Nation at Risk* (National Commission on Excellence in Education, 1983). This led initially to calls for higher standards for student achievement. But the nature of the problem was further elaborated and informed over a decade or so of discussion and argument by theoretical considerations that went beyond the policy of establishing higher standards for student achievement. Consideration of what mathematics is and what school mathematics could be about, which were nurtured by theoretical accounts of the history of mathematics and of mathematical problem solving, helped inform the posing of more complex, nuanced problems regarding notions like *opportunity to learn* and *instructional quality*. Theory and evidence played an important role in shaping the emerging argument for change, and the spotlight shifted from a focus on low expectations for students to a consideration of the ways in which schools and teachers support or inhibit student learning. As this example suggests, theory can be the ally of practitioners who wish to define problems in ways that honor their complexity and point to the inadequacy of simplistic solutions that might be suggested.

In our prior discussion of problems we noted that the word *problem* refers not only to inconveniences that need fixing but also to puzzles that appeal to the intellect. The relationship between those problems and practice is another place where there is a role for theory to play. Specifically, theory can help establish the way in which puzzling questions that matter to scholars might also matter to practice and to practitioners. For example, why is it that mathematics lessons sometimes go smoothly, exactly as planned, but often they do not? Why is it that students have little difficulty learning to add and subtract decimal fractions, but many struggle to learn to multiply and divide them?

We noted before that the practices that we are concerned with include not just the existing practices that call for scholarship to begin with, but also the new practices that result from scholarship. Some of those practices, such as the development of learning environments enhanced by various kinds of technologies, actually create opportunities to pose problems of both kinds noted above. And theory can play a role in mediating the posing of those problems as well. The emergence of dynamic geometry software has not only provided a useful tool to do constructions in classrooms but also motivated concerns such as how to maintain a role for proving in spite of the software's capacity to inspect a large number of cases. A theoretical reconsideration of the relationship between proof and perception such as that offered by Michael Otte (1998) has been instrumental not only in reaffirming the importance of proof but also in providing heuristics for the design of dynamic geometry-based tasks that create the need for proof. On the other hand, whereas all sorts of technologically enhanced learning environments may have been proposed and accepted as tools to improve practice, without necessarily posing problems that need fixing, scholars have also found ways of puzzling about them. Theory has played a role in that as well.

One such problem-as-puzzle that has been posed is that of whether and how technology changes the nature of the mathematics to be learned rather than just the way in which the mathematics is learned. In proposing the notion that technology might operate a transposition of knowledge, a computational transposition, Balacheff (1994; see also Balacheff & Kaput, 1996) was banking on the existence of a theoretical notion of didactical transposition (Chevallard, 1985) that had argued for the role of the medium (text in particular) in shaping the message for the case of disciplinary knowledge turned into academically learnable content over time.

In fact one could turn the previous observation around and notice that the development of new practices, including the development of technology-enhanced educational environments of many kinds, has allocated a role for theory in relating to the problems that call for that development. A classic example is the development of the LOGO programming language and the many interactive environments based on turtle graphics. A theory of cognition that allocated an important role to the interplay between on the one hand the analytical, differential-geometry-kind of thinking called for by the programming interface and on the other hand the synthetic, Euclidean-geometry-kind of thinking called for by the visual interface was key in the design of the language. This theory was instrumental to deploy into practice the notion that young children should exploit their intuitive knowledge of drawing to learn the properties of geometric figures.

Finally, the practices we are concerned with include not only the practices studied or created, but the very practice of doing research. Like a craftsman, the mathematics education scholar is often involved in making tools (to make tools, to make tools) to study practices. In many cases that aspect of the practice of research has been simplified by importing tools and methods from other disciplines, such as the use of experimental and quasi-experimental designs from psychology and the analyses of variance and covariance from statistics. Yet even in those cases implicit theories about the participants and about the treatments have been beneath the practice of choosing designs, tools, and methods. The emergence of case-study designs and qualitative methods for data analysis have been only the most visible evidence of a much more important development in our field. This development could be characterized as the inclusion of a more deliberate theoretical stance regarding our practice of research, by using theory to justify choices in study design, data-collection instruments, and data-analysis procedures. In particular, mathematics education researchers have developed research designs that use the word *experiment* but that do not defer to the psychological experiment for characterization. One of those is the *experimental epistemology* (also called *didactical engineering*) that Guy Brousseau and his colleagues in France used to carry out research within the theory of didactical situations, as illustrated in Brousseau's (1997, ch. 3) studies of the emergence of rational numbers and decimal numbers in fourth-grade mathematics. The qualitative-quantitative distinction is ill prepared to describe this kind of research. Brousseau's objective was to find a set of situations that could create the conditions for the emergence of specific ideas about fraction (such as commensurability). The object of study was thus not the individual cognition of a child but the system of relationships between a cognitive being and the elements of its environment relevant to the mathematical ideas at stake. In terms of design for data collection, however, the research implemented the situation under study in a real classroom.

Underlying this use of an implementation paradigm was a theory of experimentation very different from that of experimental psychology: Unlike in the experimental psychology paradigm where the implemented situation would be seen as treatment and the yield on the individual students would be seen as outcome, Brousseau's experimental research took the real classroom as the treatment or experimental probe on the situation under study and saw the yield in the robustness of the situation (in terms of its capacity to actually produce the ideas that it sought to produce). For Brousseau what was potentially variable was the *treatment,* and the possible cause of the variation in those outcomes was the variability induced by actual students impersonating the cognitive agent addressed by the situation. A theory of observation was also at play that permitted attributing the actions of each and every real child to an empirical aggregate that could illuminate the extent to which the (theoretical) student might or might not take the situation to different places.

Another example of how a theoretical stance about what research practice should be has facilitated the doing of that practice has been common in the United States: the constructivist "teaching experiment" and its generalization, the "design experiment." This form of theory use is dealt with extensively by Cobb (this volume); Cobb et al. (2003); and Kelly and Lesh (2000), so we direct the reader to these sources for more details.

In this subsection we have argued that one of the roles of theory in scholarship is that of articulating or mediating the bidirectional relationship between problems and practice. We have elaborated on this relationship by building on the multiple meanings of the words *problem* and *practice* that are relevant to our scholarship. We have proposed that theories help relate problems to practice in at least the following eight ways:

- A theory is a proposed solution to a problem of practice.
- A theory establishes the criterion according to which one state of a problem in a practice can be compared to another, so as to judge whether there has been any improvement (i.e., it establishes the meaning of *improvement*).
- A theory identifies the boundaries of the practices that pose problems and how those practices contribute to the information about the problem.
- A theory helps argue that a certain state of affairs in a practice is actually a problem (either as a circumstance that needs improvement or as a puzzling contingence that needs explanation).
- A theory helps establish the way in which puzzling questions that are problems for scholars might also be problems to practitioners or create inconveniences to the pursuit of practice.
- A theory can help envision the inconveniences and puzzles that might result from the unfolding of new practices.
- A theory can help design new practices.
- A theory can help make and justify choices (e.g., design, sampling, coding) in scholarly practices such as research, consulting, or communication.

MAKING USEFUL THEORY: THEORY MAKING AS A PART OF SCHOLARSHIP

The foregoing discussion was predicated on the premise that theory is instrumental to the relationships among practice, problems, and research. This approach seemed justified given the historical emergence of mathematics education as a field of practice and reflection on practice. Other social sciences (such as psychology and economics) also seem to have followed a similar historical developmental trajectory: Economics, for example, was first the art of doing multiple things with limited resources and only later the science of the transformations of goods and services. Mathematics education seems to be at a reasonably early stage in a similar kind of development. But it may be time to prepare for and initiate the next stage of development.

Mature social sciences grow not only in response to external demands and responses, but also from within. Their own dynamics tend to build them as solid fields of inquiry, in addition to fields of practice. And theory plays a critical role in strengthening that development from within. In fact, inward-focused development often fuels the specification of new theories and theory-based tools that offer new possibilities for application in the ways described in the previous section.

In this concluding section of the chapter we consider briefly the role of theory not so much as a mediator of relationships among practice, problems, and research, as we have in the bulk of this chapter, but rather (or also) as the collector, beneficiary, or target of that interplay in a fundamentally academic theory-making exercise. What roles could theory play when doing the job of strengthening the field as an endeavor of academic scholarship?

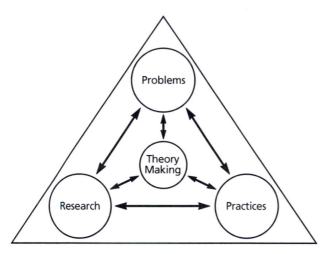

Figure 2.8 Scholarship as a theory-making endeavor.

In addition to the local theories that help mediate specific connections among the three vertices of the scholarship triangle, there are at least two kinds of theories that we scholars should aspire to build. The first could be dubbed *grand theories* of mathematics education. This type of theory responds to a need for broad schemes of thought that can help us organize the field and relate our field to other fields, much in the same way as evolutionary theory has produced a complete reorganization of biological sciences. It can also be seen as a means to aggregate scholarly production within the field. Notwithstanding the famous critique by the sociologist C. Wright Mills (1959) of such approaches in the social sciences, this has long been the goal of some pioneers in the field of mathematics education (e.g., Steiner, 1985).

According to Skinner (1985), a grand theory is any theory that tries to provide a comprehensive explanation of social life, history, or human experience. The development of a grand theory of mathematics education could be useful in providing warrants for our field's identity and intellectual autonomy within apparently broader fields such as education, psychology, or mathematics. In that sense, *a grand theory could be helpful to organize the field*, imposing something like a grand translational or relational scheme that allows a large number of people to see phenomena and constructs in places where others only see people, words, and things. A grand theory of the field of mathematics education could seek to spell out what is singular (if anything) of *mathematics education as an institutional field* or perhaps seek to spell out connections with other fields that may not be so immediately related and that establish the field as one among many contributors to an academic discipline.

This seems to have been the aim of Yves Chevallard's TAD (*Theorie Anthropologique du Didactique*, or Anthropological Theory of Didactics; see Chevallard, 1991; Chevallard, Bosch, & Gascón, 1997). In this approach, scientific production of mathematical knowledge and its communication within and across institutions (including schools, teacher preparation centers, and mathematics departments in universities), as well as the internal divisions of labor that facilitate production and communication, are turned into objects of study. Mathematics education writ large, and likely including most of what the research community does, becomes an object of study, which the TAD theory attempts to describe and explain. Its goal might be to push the limits of the theoretical approach: For example, can an anthropological framework provide an account of all the problems that the field generates and of the theoretical problems generated in addressing those?

Reciprocally, a theory that organizes the field could also take the form of *a theory of mathematics, a mathematical interpretation of the practice of mathematics education that took that practice as a repository of mathematical ideas* just as the historical documents of mathematics constitute another repository of mathematical ideas and just as (as viewed by the Piagetian studies of genetic epistemology) individual development is a repository for the mathematical ideas (such as space, time, causality, and moral judgment) from an ontogenetic perspective. The problems posed could relate to the evolution of mathematical ideas, and the theories made or tested could be epistemological ones, cultivated in educational settings much like Piagetian epistemology was an epistemological theory that used child development as the medium in which to observe ontogenesis of knowledge. Important threads that inspire this theoretical approach are found in the works of Guy Brousseau (particularly his seminal idea that for each piece of mathematical knowledge there is an *adidactical* situation, a set of relationships between a subject and an antagonist milieu, which creates the meaning of the former; see Brousseau, 1997, chapter 1).

Embodiments of mathematics are present in human actions in school and out of school. The actions in which a teacher and her students use the Pythagorean theorem are as much a cultural representation of the Pythagorean theorem as Proposition I.47 in Book I of Euclid is. It seems that the more that our culture becomes conversant with the idea of cultural embodiments of ideas and with the technologies that can capture them, the more likely we are to witness the development of an empirical epistemology of mathematics. Balacheff's (1995) proposal to identify mathematical conceptions in coherent spheres of mathematical practice provides the beginning of a theory of mathematics in practice.

A *grand theory could also be a statement of the field's concerns—a statement that identifies the field.* In this case, scholarship does its job with relative freedom by way of naming its components in some taken-as-shared vocabulary as we do when we explain what mathematics education is to scholars from outside our community. So, for example, we use expressions like "mathematical thinking," "mathematical knowledge," "mathematical communication," or "learning of mathematics" to identify objects that we take for granted to exist. Yet their theoretical nature, their constructed being, may be questioned or even denounced when we have to explain what we do to others.

One way to resolve this tension might seem to be a return to David Johnson's long-ago suggestion that we first borrow theory from other social science fields, and only if we can find nothing useful, to invent our own theories within mathematics education. But there is no guarantee that this approach will solve the problem, though it may appear to finesse it. Consider, for example, what we mean by the expression *mathematics learning.* We are mindful that students might be learning mathematical ideas in the mathematics class, but we also are aware that they might very well be learning something that is not mathematics, such as learning how to adapt to their teacher, how to get a good grade, or how to speak English. If the capacity to talk about mathematics learning were dependent on taking learning theories from elsewhere and applying them to mathematics, we might be better integrated with our colleagues in educational research yet limited in our capacity to

understand what we care about. A general theory of learning, such as one based on *communities of practice* and *legitimate peripheral participation* (Lave & Wenger, 1991), offers a compelling framework to analyze student learning as a public activity in classrooms or teacher learning as a communal activity in collegial settings such as schools (Stein, Silver, & Smith, 1998). Yet, without fundamental reconsiderations, this theory might be insufficient to distinguish, for example, the sense in which student learning is becoming a central participant in the game of school mathematics from the sense to which student learning is becoming a central participant in a game of mathematical ideas in school.

Our capacity to see the teaching and learning of mathematics where others just see teaching and learning, and yet others see only social interaction, attests to the different purposes for which individuals and organized fields construct theories. Having a grand theory of mathematics education that distinguishes constructs like *mathematics learning* from *learning,* beyond the mainstream distinction between genus and species, would be useful in making scholars in our field more resourceful and autonomous. With better theories of our own, mathematics educators will be well positioned to say things of interest to a wide variety of other fields from which we now freely borrow their theories, such as communication studies, information science, cognitive anthropology, cognitive psychology, or organizational studies.

A second type of theory grows from the need to inform a discrete variety of practices, including individual mathematical thinking, teaching and learning in classrooms, or mathematics teacher education. That is, we need what the sociologist Robert K. Merton dubbed *middle-range theories,* comparable in scope to mathematical theories (e.g., Lie theory, spectral theory), which identify subfields of study and the specific methodologies used to study those. We in mathematics education would be better able to inform outsiders about those specific practices if our scholarship contributed to the building of theories that aspired to inform, both in the sense of help understand and help improve, these practices.

We need considerable attention to middle-range theories that conceptualize patches of reality and guide empirical research, bridging the gap between grand theorizing and empiricism. Although many scholars will be attracted to the goals of grand theory outlined above, we think that more fertile ground is likely to be found if we till the soil for middle-range theories. According to Mjøset (1999), Merton studied the evolution of the social sciences and claimed that social science was still too young. When they matured, as physics already had, the body of middle range theories would converge into a system of universal laws on the model of experimental natural sciences. Until social science matured, he concluded, it should give priority to middle range theories, deliberately avoiding universal theory.[11]

Accordingly, a possible expectation for theory making in mathematics education could be not to develop a grand theory of the field but to parse the field into patches where middle-range theories could be developed as a result of the process of doing scholarship. In particular, these theories could help expose and identify things that are invisible without it, producing constructs that are operationally defined but only visible indirectly or by their consequences. Most modern science is of this kind; mathematics education scholarship has been making strides in that direction through the more extensive involvement of theory. We suspect that the building blocks for much of this kind of theory-building can be found in the other chapters in this volume.

To illustrate one example of such a middle-range theory, consider the case of mathematics instruction augmented by Brousseau's notion of *didactical contract*, which was introduced above. This idea, the didactical contract, appears to be comparable in nature to others like the *social contract* of Jean-Jacques Rousseau or *social capital* in the scholarship of Pierre Bourdieu. They are not as much precise concepts that can be defined operationally, but rather omnibus concepts that designate a perspective from which to be attuned to notice some things while looking at the real in search for phenomena and to develop finer grained systems of operational concepts that explain them.[12] We argue however that the notion of didactical contract has turned out to be a foundational contribution to the development of instruction as a theory of a middle range.

We commented above about the initial uses of didactical contract, for example pointing to the significance of the Jourdain effect, when illustrating the role that theory can play while connecting research to practices or problems. The hypothesis that a didactical contract was in place was useful to explain what was odd in the observations labeled after the Jourdain effect. And this had to do with recognizing a number of things that regulate classroom mathematical work in situations where learning a new idea was at stake. But

the notion of didactical contract itself, albeit useful to explain in general why some things happened, was not what Brousseau was interested in developing a theory of.

The key problematique Brousseau pursued asked on what conditions it was possible to communicate the ideas of the discipline of mathematics while at the same time making them function in (even emerge from) the activity of children's thinking about problems. The theory of *didactical situations*, within which the didactical contract is only a subsidiary notion, is a theory of the conditions under which the knowledge that has to be taught can emerge as a by-product of the interaction between the learner and an associated mathematically relevant environment (the milieu). This rendering stresses the affinity of Brousseau's work and that done in the United States by teacher-researchers like Deborah Ball (1993) and Magdalene Lampert (1990). One might conjecture that insofar as those independent lines of work pursued agendas for research on the enacted curriculum, both were probes on Bruner's provocative assertion that any subject can be taught to anyone at any age in "some" intellectually honest way (Bruner, 1960). Both agendas were different, however, in how they accounted for the work of the teacher and how they contributed to the development of a middle-range theory of mathematics instruction. Specifically, the contribution of those who asked the question of what it takes to teach mathematics in an intellectually honest way has been the notion that the teacher of mathematics is a player of a complex game who must select moves and strategies depending on the mathematical work that has to be accomplished, balancing commitments, some of which are unconscious and which come to the fore in the form of tensions or dilemmas that need management (Ball, 1993; Chazan, 2000; Chazan & Ball, 1999; Heaton, 1994; Lampert, 1985, 1990, 2001; Smith, 2000).

We bring this brief review up to suggest that as far as their usefulness to explain the actions of the teacher in instruction there is complementarity between the notion that a didactical contract provides entitlements and imposes responsibilities on the teacher and the notion that a teacher experiences her work as one of managing tensions and dilemmas. The role of the didactical contract in the theory of didactical situations was to identify a source of relatively ancillary regulations that needed to be managed as a precondition to engage in the experimental study

---

[11] Note that because this quote is from an online publication we cannot provide a page number.

[12] Many mathematical concepts function as omnibus concepts. For example, parallelism or area are typically not captured by precise operational definitions. Rather, they designate systems of related phenomena amenable to theories within which less important but more precise concepts can be defined (e.g., alternate interior angles, sequence of nested polygons).

of the emergence of knowledge and in order to make epistemological claims that validly referred to classroom mathematical work. In that role, the notion of contract could serve to explain why the teacher or the student might at times get out of the roles they were expected to play in the knowledge game associated with a task. Specifically what structure the didactical contract imposed on the ordinary work of a teacher and what sorts of tensions and contradictions a teacher would have to manage by way of playing the role that the didactical contract prescribes on her were not a principal part of the theoretical work in the theory of didactical situations.

The work of the teacher in the classroom has however been the object of theorization in the United States. In the 1960s Kenneth Henderson developed a relational conception of teaching into a research program that challenged then-current ways of studying teaching in educational research (which focused on personal characteristics of the teacher or on teacher behaviors). Henderson (1965) proposed a sophisticated and abstract conception of teaching as a set of relationships between and among elements of a triplet: the teacher, the student, and the knowledge at stake. Later scholars have named that relationship many things, including the "I, Thou, It, triangular relationship" (Hawkins, 1974), the *didactic system* (Chevallard, 1985), and the *instructional triangle* (Cohen et al., 2003). The important point is likely not what each chooses to call it, but rather the common, collectively acquired capacity to identify and recognize the set of relationships among those three elements and within a school institution as an object of inquiry. The notion of didactical contract has turned out to be a particularly helpful notion to turn a descriptive theory of instruction based on that relational conception of teaching into an explanatory theory (see, e.g., Herbst, 2003, 2006). The complementarity between the hypothetical obligations that the didactical contract places on any teacher and the dilemmas and tensions that individual teachers perceive is currently being exploited to study empirically the practical rationality that underlies the work of teaching (Herbst & Chazan, 2003, in press). Overall we can describe this work as one of trying to flesh out a theory of the didactical contract that can contribute a way of crafting rational explanations for instructional phenomena, hence contributing to the development of a theory of instruction of the middle-range kind.

With this theoretical insight in hand, the field is now able to grasp the idea that interaction in any mathematics classroom is not merely the projection of what an individual teacher, individual students, or the discipline of mathematics itself bring to the encounter, but rather the joint construction of a viable response to the conditions and constraints that make possible as well as affect their encounter. The emergent explanatory theory of instruction based on a relational conception of teaching is a possible example of how theory making can help sustain the identity of mathematics education as a field of inquiry. It does so by way of giving the means to stake a claim on a perspective on an object of study, by directing where some efforts should go, and by giving the means to relate to other disciplines inside and outside educational research.

The development of the notion of the didactical contract from being an omnibus concept into being a system of constructs that can afford explanation and facilitate experimentation exemplifies one way in which a relational conception of instruction has been developed. The hypothesis that a didactical contract articulates teacher and student mutual responsibilities vis-à-vis the knowledge at stake provides a way to explain how what participants actually do in classrooms is as much the joint construction of a response to the conditions in which they work together and for each other as it is an expression of who they are. What use does that have? In our opinion it allows the same kind of evolution in thinking mathematics instruction that the notion that errors have a conceptual basis had for the way researchers think about individual cognition. In the case of cognition, theory helped scholarship evolve from seeing students' mistakes as indications of a deficit to the recognition that errors are an indication of the having of a different knowledge, which is adapted to cope with a range of situations. That theoretical recasting of what errors are has served to restore rationality to the child; this has at least potentially helped promote a more respectful and empathetic treatment of learners by teachers. This has not meant that educators ignore error, any more than medicine ignores disease, but the evolving understanding of the *logic of errors* has helped support the design of better instructional treatments, in much the same way that the evolving understanding of the *logic of diseases* has helped the design of better medical treatments. Like that understanding of the nature of errors, the understanding of the nature of instructional actions provided by an explanatory theory of instruction constitutes a particular achievement that mathematics education scholarship has made and that contributes to the scholarly world writ large. Mathematical cognition in its instructional context and mathematics instruction in institutional context are two areas in which the work of our field has developed key contributions towards theories of the middle range. These contributions say things about cognition, communication, and mathematics to the educated world that no other academic field can say.

Advancing the field of mathematics education—both its scholarship and its practice—requires theoretical advances of this type. Adapting the oft-cited motto of "If you want to get ahead, get a theory" (Karmiloff-Smith & Inhelder, 1975), originally offered to capture some aspects of children's cognitive developmental trajectory, to our purposes here and applying the maxim to mathematics education professionals as a collective, we can think of no better precept both to end this chapter and to guide our future endeavors: If we want to get ahead, we need to develop theories and use them wisely!

REFERENCES

Artigue, M., & Perrin-Glorian, M. J. (1991). Didactic engineering, research and development tool: Some theoretical problems linked to this duality. *For the Learning of Mathematics, 11*(1), 13–18.

Bachelard, G. (1938). *La formation de l'esprit scientifique* [Formation of the scientific spirit]. Paris: Vrin.

Balacheff, N. (1994). La transposition informatique. Note sur un nouveau problème pour la didactique [The data-processing transposition: A new problem for the didactic one]. In M. Artigue et al. (Eds.), *Vingt ans de Didactique des Mathématiques en France* (pp. 364–370). Grenoble, France: La Pensée Sauvage.

Balacheff, N. (1995). Conception, connaissance et concept [Design, knowledge, and concept]. In D. Grenier (Ed.), *Séminaire Didactique et Technologies Cognitives en Mathématiques* (pp. 219–244). Grenoble, France: IMAG.

Balacheff, N., & Kaput, J. (1996). Computer-based learning environments in mathematics. In A. Bishop et al. (Eds.), *International handbook of mathematics education* (pp. 469–501). Dordrecht, The Netherlands: Kluwer.

Ball, D. L. (1993). With an eye on the mathematical horizon: Dilemmas of teaching elementary school mathematics. *The Elementary School Journal, 93*(4), 373–397.

Ball, D. L., & Bass, H. (2000). Interweaving content and pedagogy in teaching and learning to teach: Knowing and using mathematics. In J. Boaler (Ed.), *Multiple perspectives on mathematics teaching and learning* (pp. 83–104). Westport, CT: Ablex.

Ball, D. L., Hill, H. C., & Bass, H. (2005, Fall). Knowing mathematics for teaching: Who knows mathematics well enough to teach third grade, and how can we decide? *American Educator,* 14–17, 20–22, 43–46.

Ball, D. L., Lubienski, S. T., & Mewborn, D. S. (2001). Research on teaching mathematics: The unsolved problem of teachers' mathematical knowledge. In V. Richardson (Ed.), *Handbook of research on teaching* (pp. 433–456). Washington, DC: American Educational Research Association.

Begle, E. G. (1979). *Critical variables in mathematics education: Findings from a survey of the empirical literature.* Washington, DC: Mathematical Association of America & National Council of Teachers of Mathematics.

Belenky, M. F., Clinchy, B. M., Goldberger, N. R., & Tarule, J. M. (1986). *Women's ways of knowing: The development of self, voice, and mind.* New York: Basic Books.

Bishop, A. (1992). International perspectives on research in mathematics education. In D. A. Grouws (Ed.), *Handbook of research on mathematics teaching and learning* (pp. 710–723). New York: Macmillan.

Brousseau, G. (1997). *Theory of didactical situations in mathematics: Didactique des Mathematiques 1970–1990* (N. Balacheff, M. Cooper, R. Sutherland, & V. Warfield, Eds. & Trans.). Dordrecht, The Netherlands: Kluwer.

Brousseau, G., & Warfield, V. (1999). The case of Gael. *Journal of Mathematical Behavior, 18,* 7–52.

Brown, J. S., & VanLehn, K. (1980). Repair theory: A generative theory of bugs in procedural skills. *Cognitive Science, 4,* 379–426.

Brown, J. S., & VanLehn, K. (1982). Toward a generative theory of "bugs." In T. P. Carpenter, J. M. Moser, & T. A. Romberg (Eds.), *Addition and subtraction: A cognitive perspective* (pp. 117–135). Hillsdale, NJ: Erlbaum.

Bruner, J. (1960). *The process of education.* Cambridge, MA: Harvard University Press.

Carpenter, T. P., Fennema, E., Franke, M. L., Levi, L. & Empson, S. B. (1999). *Children's mathematics: Cognitively guided instruction.* Portsmouth, NH: Heinemann.

Chazan, D. (2000). *Beyond formulas in mathematics and teaching: Dynamics of the high school algebra classroom.* New York: Teachers College.

Chazan, D., & Ball, D. L. (1999). Beyond being told not to tell. *For the Learning of Mathematics, 19*(2), 2–10.

Chevallard, Y. (1985). *La transposition didactique: Du savoir savant au savoir enseigné* [Didactic transposition: From the knowledge of the expert to the knowledge taught]. Grenoble, France: La Pensée Sauvage.

Chevallard, Y. (1991). *La transposition didactique: Du savoir savant au savoir enseigné* (2nd ed.). [Didactical transposition: From the knowledge of the expert to the knowledge taught]. Grenoble, France: La Pensée Sauvage.

Chevallard, I., Bosch, M., & Gascón, J. (1997). Estudiar matemáticas. El eslabón perdido entre la enseñanza y el aprendizaje [Studying mathematics: The missing link between teaching and learning.] Barcelona, Spain: ICE/Horsori

Clement, J. (1982). Algebra word problem solutions: Thought processes underlying a common misconception. *Journal for Research in Mathematics Education, 13,* 16–30.

Cobb, P., Confrey, J., diSessa, A., Lehrer, R., & Schauble, L. (2003). Design experiments in educational research. *Educational Researcher, 32*(1), 9–13.

Cohen, D. K., Raudenbush, S. W., & Ball, D. L. (2003). Resources, instruction, and research. *Educational Evaluation and Policy Analysis, 25*(2), 119–142.

Cooney, T. (1985). A beginning teacher's view of problem solving. *Journal for Research in Mathematics Education, 16,* 324–336

Cooney, T. (1994). Research and teacher education: In search of common ground. *Journal for Research in Mathematics Education, 25,* 608–636.

Cooney, T., & Henderson, K. (1972). Ways mathematics teachers help students organize knowledge. *Journal for Research in Mathematics Education, 3,* 21–31.

Cooney, T. J., Shealy, B. E., & Arvold, B. (1998). Conceptualizing belief structures of preservice secondary mathematics teachers. *Journal for Research in Mathematics Education, 29*, 306–333.

Cox, P. L. (1985). Informal geometry—more is needed. *Mathematics Teacher, 88*(6), 404–405, 435.

Doyle, W. (1983). Academic work. *Review of Educational Research, 53*(2), 159–199.

Doyle, W. (1988). Work in mathematics classes: The context of students' thinking during instruction. *Educational Psychologist, 23*(2), 167–180.

Ethington, C., & Wolfle, L. (1984). Sex differences in a causal model of mathematics achievement. *Journal for Research in Mathematics Education, 15*, 361–377.

Fischbein, E. (1990). Introduction. In P. Nesher & J. Kilpatrick (Eds.), *Mathematics and cognition: A research synthesis by the International Group for the Psychology of Mathematics Education* (pp. 1–13). Cambridge, England: Cambridge University Press.

Goffman, E. (1974). *Frame analysis: An essay on the organization of experience.* Boston: Northeastern University Press.

González, G., & Herbst, P. (2006). Competing arguments for the geometry course: Why were American high school students supposed to study geometry in the twentieth century? *International Journal for the History of Mathematics Education, 1*(1), 7–33.

Hamlin, M. (2006). *Lessons in educational equity: Opportunities for learning in an informal geometry class.* Unpublished doctoral dissertation, University of Michigan.

Hanna, G. (1998). Evaluating research papers in mathematics education. In A. Sierpinska & J. Kilpatrick (Eds.), *Mathematics education as a research domain: A search for identity* (pp. 399–407). Dordrecht, The Netherlands: Kluwer.

Hawkins, D. (1974). *The informed vision.* New York: Agathon.

Heaton, R. (1994). *Teaching mathematics to the new standards: Relearning the dance.* New York: Teachers College Press.

Hembree, R. (1990). The nature, effects, and relief of mathematics anxiety. *Journal for Research in Mathematics Education, 21*, 33–46.

Herbst, P. (1997). The number-line metaphor in the discourse of a textbook series. *For the Learning of Mathematics, 17*(3), 36–45.

Herbst, P. (2003). Using novel tasks in teaching mathematics: Three tensions affecting the work of the teacher. *American Educational Research Journal, 40* (1), 197–238.

Herbst, P. (2006). Teaching geometry with problems: Negotiating instructional situations and mathematical tasks. *Journal for Research in Mathematics Education, 37*, 313–347.

Herbst, P., & Brach, C. (2006). Proving and 'doing proofs' in high school geometry classes: What is 'it' that is going on for students and how do they make sense of it? *Cognition and Instruction, 24*, 73–122.

Herbst, P., & Chazan, D. (2003). Exploring the practical rationality of mathematics teaching through conversations about videotaped episodes: The case of engaging students in proving. *For the Learning of Mathematics, 23*(1), 2–14.

Herbst, P., & Chazan, D. (in press). Producing a viable story of geometry instruction: What kind of representation calls forth teachers' practical rationality? In *Proceedings of the 28th annual meeting of the North American Chapter of the International Group for the Psychology of Mathematics Education.*

Henderson, K. (1965). A theoretical model for teaching. *The School Review, 73*(4), 384–391.

Henningsen, M., & Stein, M. K. (1997). Mathematical tasks and student cognition: Classroom-based factors that support and inhibit high-level mathematical thinking and reasoning. *Journal for Research in Mathematics Education, 28*, 524–549.

Hill, H., & Ball, D. L. (2004). Learning mathematics for teaching: Results from California's mathematics professional development institutes. *Journal for Research in Mathematics Education, 35*, 330–351.

Hoffer, A. (1981). Geometry is more than proof. *Mathematics Teacher, 84*(1), 11–18.

Johnson, D. C. (1980). The research process. In R. J. Shumway (Ed.), *Research in mathematics education* (pp. 29–46). Reston, VA: National Council of Teachers of Mathematics.

Judson, T. W., & Nishimori, T. (2005). Concepts and skills in high school calculus: An examination of a special case in Japan and the United States. *Journal for Research in Mathematics Education, 36*, 24–43.

Karmiloff-Smith, A., & Inhelder, B. (1975). If you want to get ahead, get a theory. *Cognition, 3*(3), 195–212.

Kelly, A. E., & Lesh, R. A. (Eds.). (2000). *Handbook of research design in mathematics and science education.* Mahwah, NJ: Erlbaum.

Kilpatrick, J. (1992). A history of research in mathematics education. In D. Grouws (Ed.), *Handbook of research in mathematics teaching and learning* (pp. 3–38). New York: Macmillan.

Kilpatrick, J. (1993). Beyond face value: Assessing research in mathematics education. In G. Nissen & M. Blomhøj (Eds.), *Criteria for scientific quality and relevance in the didactics of mathematics* (pp. 15–34). Roskilde, Denmark: Roskilde University, IMFUFA.

Labaree, D. F. (2003). The peculiar problems of preparing educational researchers. *Educational Researcher, 32*(4), 13–22.

Lagemann, E. C. (2000). *An elusive science: The troubling history of education research.* Chicago: University of Chicago Press.

Lampert, M. (1985). How do teachers manage to teach? Perspectives on problems in practice. *Harvard Educational Review, 55*(2), 178–194.

Lampert, M. (1990). When the problem is not the question and the answer is not the solution: Mathematical knowing and teaching. *American Educational Research Journal, 27*(1), 29–63.

Lampert, M. (2001). *Teaching problems and the problems of teaching.* New Haven, CT: Yale University.

Lave, J., & Wenger, E. (1981). *Situated learning: Legitimate peripheral participation.* Cambridge, England: Cambridge University Press.

Lerman, S., & Tsatsaroni, A. (2004, July). *Surveying the field of mathematics education research.* Paper prepared for Discussion Group 10 at the Tenth International Congress on Mathematical Education, Copenhagen, Denmark. Retrieved July 20, 2005, at http://myweb.lsbu.ac.uk/~lermans/ESRCProjectHOMEPAGE.html.

Lester, F. K., & Wiliam, D. (2002). On the purpose of mathematics education research: Making productive contributions to policy and practice. In L. D. English (Ed.), *Handbook of international research in mathematics education* (pp. 489–506). Mahwah, NJ: Erlbaum.

Levine, G. (1972). Attitudes of elementary school pupils and their parents toward mathematics and other subjects. *Journal for Research in Mathematics Education, 3*(1), 51–58.

Malara, N. A., & Zan, R. (2002). The problematic relationship between theory and practice. In L. D. English (Ed.), *Handbook of international research in mathematics education* (pp. 553–580). Mahwah, NJ: Erlbaum.

Margolinas, C. (1998). Relations between the theoretical field and the practical field in mathematics education. In A. Sierpinska, & J. Kilpatrick (Eds.), *Mathematics education as a research domain: A search for identity.* (pp. 351–356). Dordrecht, The Netherlands: Kluwer.

Mason, J., & Waywood, A. (1996). The role of theory in mathematics education and research. In A. J. Bishop, K. Clements, C. Keitel, J. Kilpatrick, & C. Laborde (Eds.), *International handbook of mathematics education* (Pt. 2, pp. 1055–1089). Dordrecht, The Netherlands: Kluwer.

Mjøset, L. (1999). Understanding of theory in the social sciences. *Arena, 99*(33). Retrieved May 16, 2006, http://www.arena.uio.no/publications/wp99_33.htm

Mullis, I.V.S., Martin, M.O., González, E.J., & Chrostowski, S.J. (2004). *TIMSS 2003 international mathematics report: Findings from IEA's Trends in International Mathematics and Science Study at the eighth and fourth Grades.* Chestnut Hill, MA: Boston College.

National Commission on Excellence in Education. (1983). *A nation at risk: The imperative for educational reform.* Washington, DC: United States Department of Education.

National Council of Teachers of Mathematics. (1989). *Curriculum and evaluation standards for school mathematics.* Reston, VA: Author.

National Council of Teachers of Mathematics. (2000). *Principles and standards for school mathematics.* Reston, VA: Author.

Nesher, P., & Kilpatrick, J. (Eds.). (1990). *Mathematics and cognition: A research synthesis by the International Group for the Psychology of Mathematics Education.* Cambridge, England: Cambridge University Press.

Otte, M. (1998, July/August). Proof and perception II. *Preuve: International Newsletter on the Teaching and Learning of Mathematical Proof.* Retrieved July 10, 2006, from www.lettredelapreuve.it/ Newsletter/980708.html.

Perry, W. G. (1970). *Forms of intellectual and ethical development in the college years: A scheme.* New York: Holt, Rinehart and Winston.

Resnick, L., Nesher, P., Leonard, F., Magone, M., Omanson, S., & Peled, I. (1989). Conceptual basis of arithmetic errors: The case of decimal fractions. *Journal for Research in Mathematics Education, 20,* 8–27.

Schoenfeld, A. H. (1987). On having and using geometrical knowledge. In J. Hiebert (Ed.), *Conceptual and procedural knowledge: The case of mathematics* (pp. 225–264). Hillsdale, NJ: Erlbaum.

Schoenfeld, A. H. (1988). When good teaching leads to bad results: The disasters of "well-taught" mathematics courses. *Educational Psychologist, 23*(2), 145–166.

Sfard, A. (1998). On two metaphors for learning and on the dangers of choosing just one. *Educational Researcher, 27*(2), 4–13.

Sfard, A. (2005). What could be more practical than good research? *Educational Studies in Mathematics, 58,* 393–413.

Sfard, A., & Lavie, I. (2005). Why children cannot see as the same what grown ups cannot see as different? Early numerical thinking revisited. *Cognition and Instruction, 23,* 237–309.

Shavelson, R. J., & Towne, L. (Eds.). (2002). *Scientific research in education.* Washington, DC: National Academy Press.

Shulman, L. S. (1986). Those who understand: Knowledge growth in teaching. *Educational Researcher, 15*(2), 4–14.

Shumway, R. J. (Ed.). (1980). *Research in mathematics education.* Reston, VA: National Council of Teachers of Mathematics.

Senk, S. (1983). *Proof-writing achievement and van Hiele levels among secondary school geometry students.* Unpublished doctoral dissertation, University of Chicago.

Senk, S. (1989). Van Hiele levels and achievement in writing geometry proofs. *Journal for Research in Mathematics Education, 20,* 309–321.

Sierpinska, A., Kilpatrick, J., Balacheff, N., Howson, A. G., Sfard, A., & Steinbring, H. (1993). What is research in mathematics education, and what are its results? *Journal for Research in Mathematics Education, 24,* 274–278.

Silver, E. A. (1990). Contributions of research to practice: Applying findings, methods, and perspectives. In T. J. Cooney (Ed.), *Mathematics teaching and learning in the 1990s* (pp. 1–11). Reston, VA: National Council of Teachers of Mathematics.

Silver, E. A., & Kenney, P. A. (Eds.). (2000). *Results from the Seventh Mathematics Assessment of the National Assessment of Educational Progress.* Reston, VA: National Council of Teacher of Mathematics.

Silver, E. A., & Kilpatrick, J. (1994). *E pluribus unum:* Challenges of diversity in the future of mathematics education research. *Journal for Research in Mathematics Education, 25,* 734–754.

Silver, E. A., & Stein, M. K. (1996). The QUASAR project: The "revolution of the possible" in mathematics instructional reform in urban middle schools. *Urban Education, 30*(4), 476–521.

Simon, M. (1995) Reconstructing mathematics pedagogy from a constructivist perspective. *Journal for Research in Mathematics Education, 26*(2), 114–145.

Skinner, Q. (1985). *The return of grand theory in the human sciences.* Cambridge, England: Cambridge University Press.

Smith, M. S. (2000). Balancing old and new: An experienced middle school teacher's learning in the context of mathematics instructional reform. *Elementary School Journal, 100*(4), 351–375.

Smith, M. S., Silver, E. A., & Stein, M. K. (2005). *Improving instruction in rational numbers and proportionality: Using cases to transform mathematics teaching and learning: Vol. 1.* New York: Teachers College Press.

Steffe, L. P. (2004). On the construction of learning trajectories of children: The case of commensurate fractions. *Mathematical Thinking and Learning, 6*(2), 129–162.

Stein, M. K., Grover, B. W., & Henningsen, M. A. (1996). Building student capacity for mathematical thinking and reasoning: An analysis of mathematical tasks used in reform classrooms. *American Educational Research Journal, 33,* 455–488.

Stein, M. K., & Lane, S. (1996). Instructional tasks and the development of student capacity to think and reason: An analysis of the relationship between teaching and learning in a reform mathematics project. *Educational Research and Evaluation, 2*(1), 50–80.

Stein, M. K., Silver, E. A., & Smith, M. S. (1998). Mathematics reform and teacher development: A community of practice perspective. In J. G. Greeno & S. Goldman (Eds.), *Thinking practices for mathematics and science education* (pp. 17–52). Hillsdale, NJ: Erlbaum.

Stein, M. K., Smith, M. S., Henningsen, M. A., & Silver, E. A. (2000). *Implementing standards-based mathematics instruction: A casebook for professional development.* New York: Teachers College Press.

Steiner, H. G. (1985). Theory of mathematics education (TME): An Introduction. *For the Learning of Mathematics,* 5(2), 11–17.

von Glasersfeld, E. (1995). *Radical constructivism: A way of knowing and learning.* Washington, DC: Falmer.

Wright Mills, C. (1959). *The sociological imagination.* New York: Oxford University Press.

AUTHOR NOTE

Several colleagues provided valuable input as we prepared this chapter, including useful critique of earlier drafts. In particular, we acknowledge the assistance of Nicolas Balacheff, James Greeno, Douglas Grouws, James Hiebert, Mary Kennedy, Frank Lester, and Anna Sfard. We also thank the members of the GRIP (Geometry, Reasoning, and Instructional Practices) research group at the University of Michigan for their valuable feedback. Nevertheless, the authors bear sole responsibility for the content of this chapter. Both authors contributed equally to the conceptualization and the writing of this chapter; the order of authorship derives from the fact that the first author was originally invited to prepare the chapter individually. Each author reserves the right to blame the other for any errors of fact or interpretation found herein! The authors also wish to thank the volume editor, Frank Lester, for his patience and support during the lengthy gestation period preceding the birth of this chapter.

3

METHOD

Alan H. Schoenfeld

UNIVERSITY OF CALIFORNIA, BERKELEY

This chapter is concerned with research methods in mathematics education and, more broadly, with research methods in education writ large. As explained below, space constraints do not allow for the detailed consideration of individual methods, or even classes of methods. Hence I have chosen to address broad metatheoretical issues in much of the chapter, which is divided into three main parts. Part 1 provides an overview of the process of conducting and reflecting on empirical research. It examines the major phases of empirical research and some of the issues researchers must confront as they conduct their studies. A main thesis underlying the discussion in Part 1 is that there is a close relationship between theory and method. I describe the process of conducting empirical research and elaborate on how researchers' theoretical assumptions, whether tacit or explicit, shape what they choose to examine, what they see and represent in the data, and the conclusions they draw from them. Part 2 presents a framework for evaluating the quality of research. In it I argue that research must be judged by at least the following three criteria: trustworthiness, generality, and importance. A range of examples is given to elaborate on the issues discussed in Parts 1 and 2. In Part 3 I try to bring together the general arguments from the first two parts of the chapter by focusing methodologically on a topic of current interest and long-term importance. As this *Handbook* is being produced, there is great pressure on educational researchers and curriculum developers in the U.S. to employ randomized controlled trials as the primary if

not sole means of evaluating educational interventions. In an attempt to move forward methodologically, I propose and discuss an educational analog of medical "clinical trials": the structured development and evaluation of instructional interventions. Part 3 offers a description of how the development and refinement of educational interventions might be conducted in meaningful ways, beginning with exploratory empirical/theoretical studies that reside squarely in "Pasteur's quadrant" (Stokes, 1997) and concluding with appropriately designed large-scale studies.

Before proceeding, I should comment about what this chapter is and is not. It is not a survey of research methods or (with the exception, in some sense, of Part 3) a "how to" guide to research. Such an approach would require a volume as large as this *Handbook* itself. Moreover, it would be largely redundant. There exist numerous handbooks of research methods in education, many weighing in at close to 1000 pages (see, e.g., Bruning & Kintz, 1987; Conrad & Serlin, 2005; Denzin & Lincoln, 2005; Green, Camilli, & Elmore, 2006; Keeves, 1997; Kelley, & Lesh, 2000; LeCompte, Millroy & Preissle, 1992; Riley, 1990; Tashakkori & Teddlie, 2002). To give just one example of the extent of the methodological domain, the *Handbook of Complementary Methods in Education Research* (Green, Camilli & Elmore, 2006) contains chapters on 35 different research methods. The methods that begin with the letters C and D alone include: case studies: individual and multiple; cross-case analysis; curriculum assessment; data modeling:

structural equation modeling; definition and analysis of data from videotape: some research procedures and their rationales; design experiments; developmental research: theory, method, design and statistical analysis; and discourse-in-use. It should be clear that even a cursory coverage of methods, much less a "how to," is beyond what can be done in this chapter.

What can and will be done is to take a bird's eye view of the terrain—to examine some overarching issues regarding the conduct of empirical research. It should be noted that from this perspective mathematics education is both special and not special. Mathematics education is special in that it is the focus of this *Handbook* and one of the best-mined fields of empirical research. All of the examples discussed in this chapter come from or serve to illuminate issues in mathematics education. At the same time, however, the issues addressed by these examples—What processes are involved in making sense of thinking, learning, and teaching? What are the attributes of high quality empirical research? How might one characterize a rigorous development and testing process for instructional interventions?—are general. The discussions in this chapter apply to all empirical research in education; indeed, to all empirical research.

PART 1
ON THE RELATIONSHIP BETWEEN THEORY AND METHOD; ON QUALITATIVE AND QUANTITATIVE METHODS; AND A FRAMEWORK FOR EXAMINING FUNDAMENTAL ISSUES RELATED TO EMPIRICAL INQUIRY

There is no empirical method without speculative concepts and systems; and there is no speculative thinking whose concepts do not reveal, on closer investigation, the empirical material from which they stem.

—Albert Einstein

All empirical research is concerned with observation and interpretation. This is the case when one is crafting "rich, thick" descriptions (Geertz, 1975) of classrooms or of aboriginal cultures; it is also the case when one is conducting randomized controlled trials of rats running mazes after being subjected to different training regimes or of students taking mathematics assessments after being taught from different curricula.

What may be less obvious, but is equally essential, is that all empirical research is concerned with and deeply grounded in (at times tacit but nevertheless strong) theoretical assumptions. Even the simplest observations or data gathering are conducted under the umbrella of either implicit or explicit theoretical assumptions, which shape the interpretation of the information that has been gathered. Failure to recognize this fact and to act appropriately on it can render research worthless or misleading.

In this opening part of this chapter I focus on issues of theory and method. First, I provide some examples to make the point that theory and method are deeply intertwined—that, as the quotation from Einstein attests, there are no data without theory and there is no theory without data. Then I proceed to put some flesh on the bare bones of this assertion. I offer a framework for conducting and examining empirical research. Readers are taken on two "tours" of this framework, one describing an example of qualitative research and one describing an example of quantitative research. A main point of the discussions is to show that divisions between the two types of research are artificial—that the same theoretical and empirical concerns apply to both.

On Framing Questions, Data Gathering, and Questions of Values

From the onset of a study, the questions that one chooses to ask and the data that one chooses to gather have a fundamental impact on the conclusions that can be drawn. Lurking behind the framing of any study is the question of what is valued by the investigators, and what is privileged in the inquiry.

For example, a recurrent issue in college level mathematics is typically posed as follows: "Is there evidence that small classes (e.g., recitation sections with thirty or fewer students) are more effective than large lecture classes?" What must be understood is that the way this question is operationalized and the choice of evidence that will be used to inform a decision are consequential.

One way to judge course effectiveness is to examine student scores on a uniform end-of-term examination. For reasons of efficiency, students in large lecture classes are often tested using skills-oriented multiple choice tests. Thus, one might decide to give such tests to students in both small and large calculus classes, and look for differences in scores.[1] It might

---

[1] How well a skills-oriented test might actually reflect what a group of students has learned, and what conclusions can be drawn from such using such tests, are serious matters. Those issues are considered in the discussion of Ridgway, Crust, Burkhardt, Wilcox, Fisher and Foster (2000) later in this chapter.

well be the case that on such a test there would be no statistically significant differences between the scores of students in large and small classes. On the basis of this evidence, the two forms of instruction could be judged equivalent. Once that judgment has been made, cost might be used as the deciding factor. The institution might opt to offer lecture classes with large enrollments.

An alternative way to evaluate course effectiveness is to look at the percentage of students in each instructional format who enroll in subsequent mathematics courses or who become mathematics majors. With that form of evaluation, small classes might produce better results. On the basis of such evidence, the institution might decide (cost factors permitting) to offer classes with small enrollments.

The point of this example is that both test scores and subsequent enrollment rates are legitimate measures of the outcomes of instruction. Each can be quantified objectively and used as the justification for policy decisions. Yet, the two measures might lead to different conclusions. A decision to use one measure or the other, or a combination of both, is a reflection of one's values—a reflection of what one considers to be important about the students' experience. In this sense, even simple quantitative data gathering and analysis are value-laden. The same is the case for qualitative analyses. Historians, for example, will decide that certain pieces of evidence in the historical record are relevant to their framing of an historical issue while others are not. These acts of selection/rejection are consequential for the subsequent representation and analysis of those data.[2]

On the Relationship Between Theory and Data

In recent years "evidence-based medicine" (see, e.g., the Cochrane Collaboration at http://www.cochrane.org/index0.htm) has been advocated by some, notably by federal administration figures such as Grover Whitehurst, director of the U. S. Department of Education's Institute for Education Sciences, as a model for how to conduct empirical research in education (see, e.g., Whitehurst, 2003). For this reason I have selected as cases in point for this discussion of the relationship between theory and data some uses of the experimental paradigm in medical research.[3] Direct connections to research

in mathematics education will be drawn after the examples have been presented.

Consider as an example the use of the "male norm" in clinical studies (Muldoon, Manuck, & Matthews, 1990; National Research Council, 1994; World Health Organization, 1998; Wysowski, Kennedy, & Gross, 1990), in which the results of male-only studies have been assumed to apply to both men and women. The March 2005 issue of the *New England Journal of Medicine* reported the results of a 10-year study of women's use of low-dose aspirin to combat heart disease. Among the findings are the following. In contrast to the situation with men, for whom taking low-dose aspirin on a daily basis has consistently been shown to lower the likelihood of heart attacks, taking a low daily dose of aspirin did not, overall, reduce the likelihood of a first heart attack or death from cardiovascular disease for women. However, there were age-specific results: Aspirin did substantially reduce the likelihood of heart attacks in women over the age of 65. Similarly, recent medical research indicates that there are differential risks of diabetes for different subpopulations of the general population.

There is a sampling point here: assuming that the results of a study (no matter how well executed) that is conducted on a subpopulation will apply to the population as a whole is not necessarily warranted. Selecting an appropriate sample is a subtle art, and unexamined assumptions may skew a sample badly. Conversely, studies that average results over an entire population may fail to reveal important information about specific sub-populations—that is, averages may mask important effects. (See, e.g., Siegler, 1987, and the discussion of Bhattachargee, 2005, below.)

This example also makes an equally important point regarding the researchers' underlying conceptual models. When "male norm" studies were paradigmatic, the assumption was that a random selection of males was a random selection of people—that gender didn't matter. That is, the experimenters did not consider gender to be a relevant variable in their experiments. This failure to conceptualize gender as a variable rendered the studies of questionable value.

In sum: Whether it is tacit or explicit, one's conceptual model of a situation, including one's view of what counts as a relevant variable in that situation, shapes data-gathering—and it shapes the nature of the conclusions that can be drawn from the data that are gathered. As will be discussed later in

---

[2] N.B. Historians, and social scientists in general, often make their cases via narrative. One must understand that narrative is a form of representation; my comments about representations apply to narrative work as well.

[3] Throughout this chapter I discuss examples of significant current interest such as controversies over randomized controlled trials as the "gold standard" for educational research. In doing so I am attempting to achieve simultaneously the dual goal of addressing enduring points of concern and clarifying current issues.

this chapter, issues such as the characteristics of the student population (e.g., what percentage of students are second-language learners?) or of the environment (e.g., is the school capable of implementing a curriculum as intended?) can be fundamental factors shaping what takes place in a learning environment. Whether and how those factors are taken into account in formulating a study and gathering data for it will shape how that study's findings can be interpreted.

A second issue, touched on in the class size example discussed above, concerns the experimenter's selection of outcomes (dependent variables) and the selection of measures to document those outcomes. To give a medical example: By the 1990s hormone replacement therapy (HRT) had become a commonly recommended treatment for some of the symptoms of menopause. When subsequent research examined an expanded set of outcomes such as the incidence of heart disease, breast cancer, and strokes, the value of HRT was called into question (see Medline Plus, 2005, for an overview). Delayed or unexpected consequences are also an issue. The devastating impact of thalidomide was not discovered until some years after the drug had been in common use.

It may seem quite a leap to compare the results of such medical studies with the results of educational interventions. However, there are direct analogues. Like medical interventions, educational interventions can have unintended and often long-term consequences. For example, a body of research in the 1970s and 1980s, which included the qualitative documentation of classroom interactions and results, documented the results of students' school mathematics experiences. These were summarized by Lampert (1990) as follows:

> Commonly, mathematics is associated with certainty; knowing it, with being able to get the right answer, quickly (Ball, 1988; Schoenfeld, 1985b; Stodolsky, 1985). These cultural assumptions are shaped by school experience, in which doing mathematics means following the rules laid down by the teacher; knowing mathematics means remembering and applying the correct rule when the teacher asks a question; and mathematical truth is determined when the answer is ratified by the teacher. Beliefs about how to do mathematics and what it means to know it in school are acquired through years of watching, listening, and practicing. (p. 32)

Let me reframe this summary in terms of contemporary discussions. As Lampert indicated, years of learning mathematics passively result in a population that tends to be mathematically passive. That population may be able, on demand, to perform some mathematical procedures—but it tends not to possess conceptual understanding, strategic competency, or productive mathematical dispositions. If the measures and descriptions of educational outcomes that are employed in empirical research fail to take into account such centrally important classes of outcomes (e.g., conceptual understanding as well as procedural competency; the ability to apply one's knowledge to novel concepts; problem-solving ability; beliefs and dispositions; drop-out rates), then researchers, teachers, and policymakers who wish to make judgments on the basis of those outcomes are potentially misinformed about the likely consequences of their decisions.

A Framework for Conducting and Examining Empirical Research

The preceding metatheoretical comments frame what is to come in this section. In what follows I set forth a framework for conceptualizing empirical work, whether that research is qualitative or quantitative, in *any* field. Figure 3.1 (modified from Schoenfeld, 2002, with permission) offers a framework within which to consider issues of method. After an introductory caveat, I briefly introduce the framework. Then I work through it in some detail.

Caveat

The discussion of Figure 3.1 proceeds in an ostensibly straightforward manner, from the "beginning" of the research process (conception and formulation of problems) to its "end" (drawing conclusions). However, the linear nature of the exposition and the representation in Figure 3.1 belie the complexity of the process, which is decidedly non-linear as it plays out in practice. Research proceeds in cycles, in which one considers and then reconsiders every aspect of the process. Even within cycles, insights (including those caused by failure or chance observation) may cause a reformulation of underlying perspective, or of what are considered salient phenomena; they may result in new representations, alternative data gathering or new ways of thinking about data that have already been gathered; and new conclusions. Specifically, Figure 3.1 is not to be taken as a linear prescription for research.

In simplest terms, empirical research is concerned with making observations of and drawing conclusions about some "real world" situation. Data are gathered and interpreted, and conclusions are drawn. That process is represented by the dotted line from Box A to Box F in Figure 3.1. The conclusions drawn are sometimes just about the situation itself ("I observed

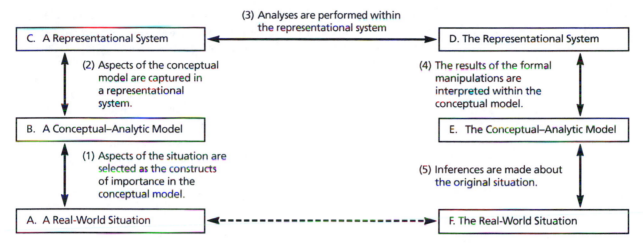

Figure 3.1 A schematic representation of the process of conducting empirical research.

the following…"), but more typically they are drawn with intimations of generality ("What happened here is likely to be the case in circumstances that resemble those described here.") and importance ("This information should shape the ways we think about X, Y, and Z."). The main purpose of Figure 3.1 is to indicate that the pathway from observations to conclusions is not as simple as it might seem.

In line with Figure 3.1, I claim that all empirical research involves the following processes:

- conceptualization, in which the situation to be analyzed is seen and understood in certain (often consequential) ways;
- the creation, use, or refinement of a conceptual-analytic framework or model, in which specific aspects of the situation are singled out for attention (and, typically, relationships among them are hypothesized);
- the creation, use, or refinement of a representational/analytic system, in which aspects of the situation singled out for attention are selected, represented and analyzed;
- the interpretation of the analysis within the conceptual-analytic framework or model; and
- attributions and interpretations from the preceding analytic process to the situation of interest (and possibly beyond).

To illustrate the main points above I consider at some length one primarily qualitative example and one primarily quantitative example.

A First Example

As a first qualitative example I discuss the decision I made, nearly 30 years ago, to explore aspects of

students' metacognitive behavior during problem solving. (Extensive detail regarding this work can be found in my 1985 book *Mathematical Problem Solving*.) The starting place for this work seemed simple. I brought students (either by themselves or in pairs) into a room near my office (my "laboratory") and asked them to solve a series of problems out loud. I was in the vicinity while they worked on the problems, and I occasionally intervened if a long time had passed without their saying something audible. I videotaped the students' solution attempts and saved their written work.

The primary sources of data for analysis were the videotapes I made of their problem-solving attempts and the written work they produced while working the problems. On the basis of those data I drew inferences about the students' decision making during problem solving and its impact on their success or failure at problem solving. I also drew inferences about the frequency and import of the students' "executive decision making" in general.

To illustrate the issues involved, I start with Box A at the lower left of Figure 3.1, and make a circuit of the figure by following the arrows up, around, and down to Box F. To begin, it should be clear that I was making a fair number of assumptions about the "real world situation" examined here—students solving problems in the laboratory. Two major assumptions were that (a) the students' problem solving behavior in the laboratory bore some relation to their problem solving behavior in other contexts; and (b) the students' overt actions bore some relation to their internal cognitive processes.

Both of these assumptions were and are controversial to some degree. Regarding (a), for example, over the years some researchers have questioned the value of laboratory studies, saying that the artificial behavior induced in the laboratory renders laboratory studies of little or no value in understanding the kinds of interactions that take place amidst (for example) the blooming complexity of the classroom. Regarding (b), for quite some time there have been controversies over the role of verbal reports as data. Retrospective reports of thought processes were roundly discredited in the early years of the 20th century, and for some years *any* reports of thought processes were deemed illegitimate. (Indeed, behaviorists banished the notion of thought processes from "scientific" explanations of human behavior.) In the 1980s Nobel prize winner Herbert A. Simon and colleague K. Anders Ericsson wrote a review (Ericsson & Simon, 1980) for *Psychological Review* and then a book (Ericsson & Simon, 1984) entitled *Verbal Reports As Data*, trying to make the case that although post hoc reports of thought processes could not be taken as veridical, "on the spot" verbalizations of what one was doing could be taken as data suggestive of the individuals' thought processes.

One could say a great deal more about assumptions (a) and (b)—teasing out what "some relation" means in each of them is a nontrivial exercise! What matters here is something simpler. Wherever one comes down with regard to assumptions (a) and (b), the fact is that they *are* assumptions, and one's stance toward them shapes how one considers the data gathered. What should be clear is that a form of naïve realism—that the videotapes and written record *directly* capture (some of) what people were thinking as they worked on the problems—is not warranted. Equally clear is that I began my work with specific assumptions about what "out loud" problem-solving protocols could reveal; I entered into the work with a set of underlying assumptions about the nature of cognition that framed the way I saw what was in the tapes. Someone who was not a cognitive scientist, or whose orientation to cognition was different, would not look for or see the same things.

When I began examining the videotapes, I knew there was *something* important about students' decision making during problem solving—something that was a factor in success or failure—but I did not know what it might be. My earlier work had focused on teaching an explicit decision making strategy, to help students use their problem-solving knowledge effectively. Now I was looking at videotapes made before the instruction, trying to identify causes of success or failure. I was looking at the tapes "from scratch" in part because the fine-grained coding schemes I had found in the literature had not seemed informative.

My research assistants and I watched a fair number of tapes, trying to figure out how to capture events of importance in a coding scheme. We started in a somewhat systematic way, looking for what we called "reasons to stop the tapes." These occurred at places in the videotapes where we saw students acting in ways that seemed to bear on the success or failure of their problem solving attempts. We made a list of such events and composed for each event a series of questions designed to trace its impact on the problem solution. This was a prototype analytic scheme. And after polishing it a bit I asked my students to try to analyze the data using it.

When we reconvened, my research assistants were unhappy. They said that the scheme we had developed was impossible to use. Far too many of our questions, which had seemed to make sense when we looked at one tape, seemed irrelevant on another. Our system had so many reasons to stop a tape, and so many unanswerable or irrelevant questions when we did, that whatever was truly important about the problem-solving episode was lost among the huge collection of questions and answers.

Confronted with this failure, I decided to begin again. I chose to look at an "interesting" tape—a tape in which it seemed that the students "should have" solved the problem but did not. My assistants and I tossed the coding scheme aside and looked at the tape afresh. As we did, I noticed one particular decision that the students in the videotape had made. They had chosen, without much deliberation, to perform a particular computation. As the solution unfolded, they spent a great deal of time on the computation, which I realized would not help them to solve the problem. As I watched them persevere in the computation, things clicked. That single decision to perform the computation, unless reversed, could result in the expenditure of so much time and energy in an unprofitable direction that the students were essentially guaranteed to fail to solve the problem.

I had the feeling I was on the trail of something important. My assistants and I looked at more tapes, this time searching for consequential "make-or-break" decisions. It turned out that these were of two kinds: paths wrongly taken and opportunities missed. These make-or-break decisions were consequential in more than half of our tapes. With this understanding, we had a new perspective on what counts as a major factor in problem solving. This new conceptual/ analytic perspective oriented us differently toward the tapes and changed our subsequent data analyses. At this point, with a conceptual model in place, we were in Box B of Figure 3.1.

[Before proceeding, I must stress that not every study involves a new conceptual model; most studies involve the use or refinement of well-established conceptual models. The point of this particular example is that any conceptual model highlights some things and obscures or ignores others; it takes some things into account and does not consider others. For example, my analyses of the videotapes of students solving problems did not, at that point, include a focus on issues of affect or belief. They did not include the detailed examination of student knowledge or knowledge organization, save for the fact that I had been careful to have the students work problems for which I had evidence that they possessed adequate knowledge to obtain a solution. (It is of little theoretical interest when a student fails to solve a problem simply because he or she lacks the knowledge that is essential to solve it.) Hence, as I was examining the problem-solving tapes, I was viewing them through a particular theoretical lens, one that focused on the impact of a particular kind of decision making. The videotapes might well have supported different kinds of analyses, but other aspects of the students' solutions were not to be seen in our analyses (and, equally important, ceased to be salient to us as we analyzed the tapes). I also note that this example demonstrates the dialectic between representational/analytic schemes and conceptual frameworks, thus illustrating the non-linear character of Figure 3.1.]

Once my research assistants and I had a first-order conceptual-analytic framework, we needed a representational scheme to capture and analyze our data. In simplest terms, we decided to parse problem-solving sessions into major chunks called "episodes," periods of consistent goal-oriented activity on the part of the problem solver. The notion of an episode was a useful device for identifying the loci of consequential decisions. The places where the direction of a solution changed were the natural boundaries between episodes, and they were often the sites of consequential decisions. It also turned out that, at a gross level, there were relatively few kinds of episodes: reading the problem, working in a structured way to analyze the problem, planning, implementing a plan, and working in a somewhat random or ill-thought-out way ("exploration").

Over time we refined this representational scheme, which was later supplemented by a more compact and informative time-line representation suggested by Don Woods (Figure 3.2, below, is an example). With a representational scheme in place, we were able to code the data. Things were straightforward. The idea was to represent the contents of a videotape (typically a 20-minute problem-solving session) by an episode diagram, which identified and labeled the episodes and the consequential decisions in a problem session.

At this point we were comfortably ensconced in Box C, working within a particular representational system. It is important to observe that the representational system reified our conceptual model. Events that did not fit into the representational scheme were not captured in the representation, and thus were not fodder for data analysis.

My research assistants and I worked together on a series of tapes, developing a communal understanding of the meanings and types of episodes, and of consequential decision making (including the failure to act on a relevant piece of information). We then coded some tapes independently. Our codings matched more than 90% of the time.

With the consistency of coding established, we were working within Box D—performing analysis with and within the representational system. Coding the sessions was still a matter of interpretation, but with practice it became a relatively straightforward task, as indicated by the high interrater reliability. The hard work had been done in the conceptualization of the scheme. Once the tapes were represented in the coding scheme, data analysis was simply a matter of counting. More than half of the students' problem-solving attempts were represented by the following schematic diagram: an episode of reading the problem followed by an episode of exploration (and failure). That is, the bar graph in Figure 3.2 represented more than half of the problem-solving sessions we coded.

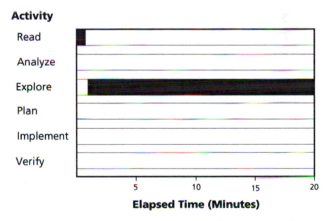

Figure 3.2 A timeline representation of a typical student solution attempt.

These data had a straightforward interpretation. More than half the time, the students—who had the knowledge required to solve the given problems—failed to solve the problems because of their poor choice of initial solution direction. As long as the students lacked effective mechanisms for reflecting on and undoing those initial decisions, they did not have the opportunity to use what they did know to solve the problems.

This interpretation placed us in Box E, and the extrapolation to Box F (and beyond) was also straightforward. We recognized, of course, that factors other than monitoring and self-regulation affected problem solving success—knowledge and problem-solving strategies among them. In our work, we had controlled for these by giving the students problems that they had the subject matter knowledge to solve; thus we saw the devastating effects of poor metacognitive decision making. We asserted that ineffective monitoring and self-regulation were significant causes of problem-solving failure, both in the laboratory and beyond it.

As noted, there were many assumptions made here, in the attribution of causality both in laboratory problem solving and in the extrapolation to more general problem solving. At this point, given the small number of students we had videotaped, our ideas about the importance of metacognitive decision making were suggestive but not yet validated. Further studies led to substantiation of those ideas. The study of accomplished problem solvers (in the laboratory) documented the ways that effective monitoring and self-regulation can be a productive force in problem solving. Subsequent research showed the robustness of the phenomena of monitoring and self-regulation, and their importance. (See, e.g., Brown, 1987; deCorte, Green, & Verschaffel, 1996; Lester, 1994.)

A second example

I now examine the issues raised by a prototypical quantitative study (Bhattachargee, 2005). Again, I focus on the kinds of choices that are made at various points in the process described in Figure 3.1 and the impact they have on the conclusions that can be drawn. Consider the task of determining whether students learn more from Curriculum X or from Curriculum Y. As many people see it, this is as straightforward as you can get: All one has to do is perform some sort of randomized controlled trial in which half the student population is randomly assigned each treatment, and then see what differences emerge on an appropriate outcome measure. Would that life were so simple.

The complexities begin in the real-world context, Box A. Even before articulating a conceptual model, there are choices at the pragmatic level. Does one randomize the instructional treatment at the student level (with two adjacent students perhaps studying different curricular materials)? At the classroom level? At the school level? Such considerations are often driven by practicalities. But then, when one goes to

the conceptual level (Box B), choices get much more complex—and consequential.

At least two kinds of conceptual issues are fundamental in shaping how one does research on curricular outcomes. The first is methodological, the second subject-matter related. A fundamental conceptual issue related to curriculum implementation is what one considers an "implemented curriculum" to be. One perspective is as follows.

Perspective 1. A curriculum is the set of instructional materials and preparation to use them that teachers are given. Whatever the teachers do with those materials in the classroom is the "implemented curriculum."

In this case, what counts as the simplest measure of the curriculum's effectiveness is the average performance of all those students who were in classrooms where that curriculum was used.

Another perspective is as follows.

Perspective 2. There is a strong degree of interaction between curriculum and context. Given different contexts or different degrees of support, there may be more or less fidelity of curriculum implementation. "Degree of fidelity of implementation" (in conformity with the intention of the designers) matters and should be taken into account in analyses of curriculum impact.

Questions of interest to people with this orientation include the following. What kinds and levels of support are necessary, in what kinds of contexts, to guarantee some level of fidelity of implementation for a particular curriculum? What is the curricular impact (measured in terms of student outcomes) when there is some degree of curriculum fidelity? When one is considering a choice between two curricula, what kinds of outcomes can one expect for each, given the resources that one has to put into implementing them?

This distinction in framing has significant consequences. Here is an example, taken from a recent issue of *Science* (Bhattachargee, 2005). In a randomized trial in a school district, three schools used Curriculum X and three used Curriculum Y. The schools were roughly matched on demographics. When one looked at overall curriculum implementation—that is, the complete sets of scores from students who had worked through Curriculum X and Curriculum Y—no statistically significant differences between outcomes were found. Is one to conclude, then, that the two curricula are equally good?

The answer depends on one's conceptual model. For those who adhere to perspective 1 as described above, the situation is straightforward. Given that the schools were randomly assigned to the treatments and the data showed no differences, it follows (within perspective 1) that the curricula are equally effective. But for those who hold perspective 2, there might be a world of difference between the two curricula.

It turns out that of the three schools that used Curriculum X, one school embraced the curriculum and implemented it in a way consistent with the designers' intentions. Students at that school outperformed those who used Curriculum Y. At a second school the implementation of Curriculum X was uneven. There, scores were not statistically different from overall scores on Curriculum Y. In a third school Curriculum X was poorly implemented, and students did poorly in comparison to Curriculum Y.

"On average" the two curricula were equally effective. The averages are uninformative, however. Another way to look at the data is as follows. When Curriculum X is implemented as intended, outcomes are superior to outcomes from Curriculum Y. Under those conditions, Curriculum X is preferable. But when Curriculum X is not implemented effectively, students would do better with Curriculum Y. Hence instructional leadership should assess the capacity of the staff at each site to implement Curriculum X—either now or with professional development—and decide on that basis whether to use it at that site. From perspective 2, then, the decision as to whether to use Curriculum X or Curriculum Y is context-dependent, depending on the school staff's current or potential capacity to implement either curriculum with some degree of fidelity. Note that this is a very different kind of conclusion than the kind of conclusion drawn by those with the "curriculum is context-independent" perspective.

Here is a relevant analogy. Suppose there are two surgical treatments for a particular condition. Treatment A returns patients to full functioning *if* they undergo a full regimen of physical therapy for a year, but the results are unsatisfactory if a patient does not. Treatment B is reasonably but not completely effective, regardless of whether the patient undergoes physical therapy. Suppose that, on average, not that many people follow through on physical therapy. On average, then, people do slightly better with Treatment B than with Treatment A.

Put yourself in the position of a patient who is facing surgery for that condition. Would you want your doctor to recommend Treatment B on the basis of the statistical average? Or would you rather have the doctor explain that Treatment A might be an option for you, but only if you commit yourself to a serious regimen of physical therapy afterward? Both statements represent legitimate interpretations of the data, within the frames of particular conceptual models. Those models make a difference. As a patient, I would much rather be offered the second choice. There is no reason to settle for the statistical average if there are reliable ways to beat that average. (One should settle for it, however, if one does not have the wherewithal to follow through with physical therapy.)

To return to the curricular example: one's conception of what is meant by "curriculum implementation" has tremendous implications for the ways that findings are reported and interpreted. One can report on data of the type discussed by Bhattachargee (2005) either by saying

(a) "There were no significant differences between Curriculum X and Curriculum Y" or

(b) "Curriculum X is superior to Curriculum Y under certain well-specified conditions; Curriculum X and Curriculum Y produce equivalent test scores under a different set of well-specified conditions; and Curriculum Y is superior to Curriculum X under yet another set of well-specified conditions."

The possibilities for *acting* on the information in (a) and (b) differ substantially.[4] I now consider conceptual models related to subject matter.

Just what does it mean to know (to have learned) some particular body of mathematics? This is not only a philosophical issue, but a practical one as well: Different conceptual models of mathematical understanding lie at the heart of the "math wars" (see Schoenfeld, 2004). One point of view, which underlies much of the "traditional" curricula and standardized assessments, is that knowledge of mathematics consists of the mastery of a body of facts, procedures, and concepts. A more current perspective, grounded in contemporary research, is that mathematical knowledge is more complex. The "cognitive revolution" (see, e.g., Gardner, 1985) produced a fundamental epistemological shift regarding the nature of mathematical understanding. Aspects of mathematical competency are now seen to include not only the knowledge base, but also the ability to implement problem-solving strategies, to be able to use what one knows effectively and efficiently, and more (deCorte, Greer, & Verschaffel, 1996; Lester,

---

[4] This idea is not new: see, e.g., Brownell, 1947.

1994; Schoenfeld, 1985a, 1985b, 1992). In elementary arithmetic, for example, the National Research Council volume *Adding It Up* (2001) described five interwoven strands of mathematical proficiency:

- *conceptual understanding*—comprehension of mathematical concepts, operations, and relations
- *procedural fluency*—skill in carrying out procedures flexibly, accurately, efficiently, and appropriately
- *strategic competence*—ability to formulate, represent, and solve mathematical problems
- *adaptive reasoning*—capacity for logical thought, reflection, explanation, and justification
- *productive disposition*—habitual inclination to see mathematics as sensible, useful and worthwhile, coupled with a belief in diligence and one's own efficacy. (p. 5)

Fine-grained analyses of proficiency tend to be aligned with the content and process delineations found in the National Council of Teachers of Mathematics' (NCTM, 2000) *Principles and Standards for School Mathematics*:

Content: Number and Operations; Algebra; Geometry; Measurement; Data Analysis and Probability;

Process: Problem Solving; Reasoning and Proof; Making Connections; Oral and Written Communication; Uses of Mathematical Representation.

These views of proficiency extend far beyond what is captured by traditional content-oriented conceptual frameworks.

In the experimental paradigm, one's view of domain competency is typically instantiated in the tests that are used as outcome measures. What view of mathematical proficiency one holds, and how that view is instantiated in the outcome measures one uses for educational interventions, can make a tremendous difference.

The issues at stake are as follows. Traditional assessments tend to focus on procedural competency, while assessments grounded in broad sets of standards such as NCTM's *Curriculum and Evaluation Standards* (1989) or *Principles and Standards* (2000) include procedural (skills) components but also assess conceptual understanding and problem solving. In a rough sense, the traditional assessments can be seen as addressing a subset of content of the more comprehensive standards-based assessments. Hence a choice of one assessment instead of another represents a value choice—an indication of which aspects of mathematical competency will be privileged when students are declared to be proficient on the basis of test scores. As the following example shows, these choices are consequential.

Ridgway, Crust, Burkhardt, Wilcox, Fisher, and Foster (2000) compared students' performance at Grades 3, 5, and 7 on two examinations. The first was a standardized high-stakes, skills-oriented test—California's STAR test, primarily the SAT-9 examination. The second was the Balanced Assessment test produced by the Mathematics Assessment Resource Service, known as MARS. The MARS tests cover a broad range of skills, concepts, and problem solving. For purposes of simplicity in what follows, scores on both tests are collapsed into two simple categories. Student who took both tests are reported below as being either "proficient" or "not proficient" as indicated by their scores on each of the examinations. More than 16,000 students took both tests. The score distribution is given in Table 3.1.

Table 3.1 Comparison of Students' Performance on Two Examinations

MARS	SAT-9	
	Not Proficient	**Proficient**
Grade 3 (N = 6136)		
Not proficient	27%	21%
Proficient	6%	46%
Grade 5 (N = 5247)		
Not proficient	28%	18%
Proficient	5%	49%
Grade 7 (N = 5037)		
Not proficient	32%	28%
Proficient	2%	38%

Unsurprisingly, there is a substantial overlap in test performance: Overall 73%, 77%, and 70% of the students at Grades 3, 5, and 7, respectively, either passed both tests or failed both tests. The interesting statistics, however, concern the students who were rated as proficient on one test but not the other.

For each grade, consider the row of Table 3.1 that reports the SAT-9 scores for those students rated "proficient" on the MARS test. At Grades 3, 5, and 7 respectively, 88%, 91%, and 95% of those students were rated proficient on the SAT-9. Thus being rated proficient on the MARS test yields a very high probability of being rated proficient on the SAT-9. That is: being declared proficient on the MARS exam virtually assures having the procedural skills required for the SAT-9.

The converse it not true. Consider the final column of Table 3.1, which indicates the MARS ratings of the students who were rated proficient on the SAT-9. Approximately 31% of the third graders, 27% of the fifth graders, and 42% of the fifth graders who were declared proficient by the SAT-9 were declared not proficient on the MARS exam. That is, possessing procedural fluency as certified by the SAT-9 is clearly *not* a guarantee that the student will possess conceptual understanding or problem-solving skills, as measured by the MARS test. Indeed, the students who were declared proficient on the SAT-9 but not the MARS test—roughly 1/3 of those declared proficient on the SAT-9—can be seen as false positives, who have inappropriately been deemed proficient on the basis of a narrow, skills-oriented examination.

Once an assessment has been given, the die has been cast in terms of data collection. One is now in Box C in Figure 3.1, where there exist standard techniques for scoring tests and representing test data. The pathway from Box C to Box D in Figure 3.1 is relatively straightforward, as are analyses within Box D. This, after all, is the province of standard statistical analysis. However, *interpretation*—the pathway to Boxes E and F—is anything but straightforward, for it depends on the conceptual models being employed.

There is suggestive, though hardly definitive, evidence (see, e.g., Senk & Thompson, 2003) that nearly all of the National Science Foundation-supported standards-based curricula have the following property. When the test scores of students who have studied from the NSF-supported curricula are compared with test scores of students who have studied from more traditional skills-oriented curricula, there tend to be no statistically significant differences between the two groups in performance on skills-oriented tests (or the skills components of broader tests). However, there tend to be large and significant differences favoring the students from the NSF-supported curricula on measures of conceptual understanding and problem solving. Thus, if appropriately broad assessments are used, comparison studies will tend to produce statistically significant differences favoring the performance of students in these standards-based curricula over the performance of students from more traditional comparison curricula. However, if skills-oriented assessments are used, no significant differences will be found. Hence at the curriculum level, the use of measures that focus on skills can result in curricular false negatives—the tests will fail to show the real differences that exist.

The fundamental point to be taken from the preceding discussion is that the specific contents of any given assessment matter a great deal. One can draw meaningful conclusions about the relative efficacy of two curricula on the basis of a particular assessment only when one knows what the assessment really assesses (that is, when a content analysis of that assessment has been done). Without a content analysis, it is impossible to interpret a finding of "no significant differences." Such a finding might occur because both curricula are equally effective. Or, it might occur because an inappropriately narrow assessment failed to pick up what are indeed significant differences in impact. For this reason, a report of a randomized controlled trial that does not contain a content analysis of the assessment employed is of no value. Indeed, the conclusions drawn from it may be false or misleading.

Ironically, this is the mistake made by the nation's most ambitious attempt to provide information about curricular effectiveness, the What Works Clearinghouse (WWC). WWC (http://www.whatworks.ed.gov/) does not conduct research itself. Rather, it was created to review and report findings from the literature. WWC searches the literature for studies that meet stringent methodological criteria. Studies that qualify for vetting by WWC must be of one of the following three types: randomized controlled trials, quasi-experiments that use equating procedures, or studies that use regression discontinuity designs. These are vetted for technical proficiency and empirical flaws. Only studies that make it through WWC's methodological filter are reported.

WWC committed the fundamental error identified above in reporting one of the few studies that did make it through its methodological filter. In a report (What Works Clearinghouse, 2004), WWC gave part of the statistical analyses in the study it examined (a quasi-experimental design with matching reported in 2001 by C. Kerstyn) full marks. Here is what WWC (2004) said about its choice of that part of the study:

> The fifth outcome is the Florida Comprehensive Assessment Test (FCAT), which was administered in February 2001. The author does not present the reliability information for this test; however, this information is available in a technical report written by the Florida Department of Education (2002). This WWC Study Report focuses only on the FCAT measures, because this assessment was taken by all students and is the only assessment with independently documented reliability and validity information.

Note that reliability and validity are psychometric properties of an assessment: They do not provide a characterization of the actual content of the examination. Neither Kerstyn nor WWC conducted content analyses of the FCAT exam. For all one knows, it could be as narrow as the SAT-9 examination discussed by Ridgway et al. (2000). The Kerstyn study reported "no significant differences"—but why? Was it because there were none, or because the narrowness of the measure used failed to reveal a significant difference that actually existed? Because of the lack of information provided by WWC, it is impossible to know. Given that WWC failed to conduct a content analysis of the FCAT, the findings reported in the WWC report are at best worthless and at worst misleading. In addition, WWC's unwillingness to conduct content analyses of the measures used in the randomized controlled trials of mathematics studies makes it impossible for WWC to achieve its core mission. WWC was created with the intention of conducting meta-analyses of the literature—to sort out through analytical means the impact of various curricula. Properly conducted, the analyses and meta-analyses are intended to reveal information such as the following: "Curriculum X tends to be strong on procedural skills and on conceptual understanding, but not especially strong on problem solving. Students tend to do well on tests of geometry, measurement, and number, but they do less well on tests of algebra and data analysis." Given that WWC has refused to conduct content analyses,[5] WWC can offer no insights of this type. Once again, what is attended to, both in conceptual models and in assessments, is highly consequential.

In sum, although one must be proficient in the application of quantitative and qualitative methods on their own (specifically, the pathway from Box C to Box D in Figure 3.1), such proficiency is no guarantee that the interpretation of the results will be meaningful or useful. A meaningful report must respect all of the pathways from Box A to Box F in Figure 3.1.

Discussion

In this section I have focused on some fundamental issues of theory and method. First, I argued that theory and method are deeply intertwined. Every empirical act of representation, analysis, and interpretation is done in the context of a (sometimes explicit, sometimes implicit) conceptual and theoretical model. The character of such models shapes the conclusions that are produced by subsequent analysis and interpretation. Second, I have presented a framework (Figure 3.1) that highlights major aspects of empirical research including conceptualization, representation, analysis, and interpretation. I remind the reader that although the figure and the linearity of prose as a medium may suggest that the process is linear, it is not: the process is cyclical, and there can be substantial give-and-take between all of the aspects of research reflected in Figure 3.1 during each cycle of research. The extensive discussion of Figure 3.1 highlighted the complexity of the process and the ways in which conceptual models can affect what one captures in data and how those data are interpreted. Third, I deliberately chose to work through one prototypically qualitative and one prototypically quantitative example to indicate that the fundamental issues of focus, data gathering, data analysis, and interpretation of findings are the same whether one is conducting qualitative or quantitative research.[6] The serious question to be considered is not, "is this research of one type or another" but "what assumptions are being made, and how strong is the warrant for the claims being made?"

Finally, I want to point to the fact that the framework outlined in Figure 3.1 can be used reflectively, both as one conducts research and as one examines research conducted by others. Each of the pathways between the boxes in Figure 3.1, and each of the boxes, represents a series of decisions made by the researcher. Thus, for example, the pathway from Box A to Box B indicated by Arrow 1 ("aspects of the situation are selected as the constructs of importance in the conceptual model") offers a reminder that any choice of focal phenomena represents a set of theoretical commitments. This provides the opportunity to reflect on the choice and implications of the conceptual model that is being (even if tacitly) employed. For example, which phenomena are not taken into account by this perspective? Which are given significant emphasis? How are those theoretical biases likely to shape the interpretation of the situation?

---

[5] I served as the Senior Content Advisor for WWC's mathematics studies (at first for middle school mathematics, then for all mathematics reports) from WWC's beginnings. I resigned in early 2005 when WWC refused to correct the flaws identified above and reneged on a commitment to publish an article in which I had discussed such issues. For details see Schoenfeld (2006a, 2006b).

[6] If space permitted I would include a third example. Suppose one wanted to conduct an ethnographic study of classrooms using different curricula, with a focus on (say) discourse structures and their impact. It is left as an exercise for the reader to work through Figure 3.1, with regard to issues such as unit of analysis, selection and form of data, outcome measures (e.g., test scores, or discussions of identity), and interpretation. All of the issues that arose in the quantitative example arise here as well.

Similarly, the pathway between Boxes B and C indicated by Arrow 2 ("aspects of the conceptual model are captured in a representational system") represents an act of *data selection and reduction* as well as representation. In historical studies, for example, whose voices are selected and heard? Or, suppose one is conducting classroom observations. Does one take field notes or make videotapes? If one tapes, what is the focus of the camera? If one takes notes, are they structured according to a predetermined system (in which case they reflect an explicit focus on particular aspects of the situation) or are they somewhat open (in which case the selection is tacit)? For example, data-gathering during the days of the process-product paradigm typically consisted of tallying certain kinds of behavior (teacher actions, student actions) and looking for correlations with educational outcomes (e.g., test scores). In contrast, many current studies of classroom discourse focus on the character of student and teacher interactions, and the results in terms of community norms, beliefs, and knowledge. Each act of data selection, reduction, and representation will have the potential to illuminate certain aspects of a situation, and to obscure others (or even render them invisible). Even if the selection is done with great fidelity to the theoretical model, an act of sampling is taking place.

The third arrow, "analyses are performed within the representational system," is deceptively simple. The key questions to ask are, What is meaningful within the representational scheme? What can be said about the quality of the inferences drawn? It should be obvious that great care must be taken in subjective analyses. But it is equally important to take comparable care in the case of ostensibly objective quantitative analyses. The results of data analyses will be no better than the quality of the data that are subjected to analysis. For example, there may be a statistically significant difference in the performance levels of two classes on an outcome measure. But is the cause a difference in the two instructional treatments, the fact that they were taught by different teachers, or (if the same teacher taught both) either the enthusiasm of the teacher for one treatment over the other or the fact that one course was taught in the morning and the other right after lunch? Many of the variables that affect performance go unmeasured in statistical analyses. I shall review the issue of *trustworthiness* of analyses in the next section.

The fourth arrow is the mirror image of the second. Just as the passage from a conceptual model to a representational system involves data selection and reduction, the return from the representational system to the conceptual model involves significant acts of interpretation. A difference in two measures might be statistically significant, for example—but is it meaningful or consequential? If so, along what lines? Or, to take a qualitative example, suppose the representational system involves coding student-to-student dialogue in classroom interactions. If the coding scheme focuses on the frequency of interactions and dialogic "take-up," one might, for example, get a picture of a highly collaborative working group. But what was the collaboration about? An analysis of the content of the interactions might or might not indicate that the group was focused productively on important mathematical issues. Thus the extrapolation from representational system to the conceptual system must be made with care.

Finally, there is the return from the conceptual model (Box E) to the "real world" situation—the original Box A, now Box F. Here too there is at least interpretation, and perhaps extrapolation. or example, what are test scores taken to mean? History has made clear the consequences of confusing test scores such as IQ tests with the traits they ostensibly represent, such as "intelligence." Likewise, whether one attributes mathematical proficiency to a good score on the SAT-9 or the Balanced Assessment tests can make a big difference. And, saying "students from Curriculum X did (or did not) outperform students from Curriculum Y on this test" is a very different thing than saying "Curriculum A is (or is not) better than Curriculum B." I address the idea of *generality* in the next section.

PART 2
ASPECTS OF RESEARCH—ISSUES OF TRUSTWORTHINESS, GENERALITY, AND IMPORTANCE

In this section I discuss three fundamental issues related to all research studies. Those issues can be posed as questions that can be asked about any study:

- Why should one believe what the author says? (the issue of trustworthiness)
- What situations or contexts does the research really apply to? (the issue of generality, or scope)
- Why should one care? (the issue of importance)

The following diagram (Figure 3.3) may be useful in thinking about the ultimate contributions made by various studies or bodies of studies:

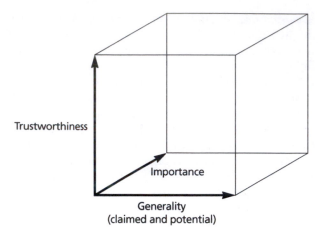

Figure 3.3 Three important dimensions along which studies can be characterized. (Reprinted, with permission, from Schoenfeld, 2002, p. 450)

As an example, a superbly written and insightful autobiographical account might score high on the trustworthiness and importance dimensions, while low on generality—although it might, by virtue of elaborating what might be seen as common experience, have intimations of generality in some ways. The same could be said of some rich descriptions of rather singular events, such as Fawcett's (1938) description of a 2-year-long geometry course. Fawcett's course served as an important and well-documented existence proof: It *is* possible to teach a course in which students develop certain kinds of understandings of the mathematical enterprise. A technically proficient comparison of two instructional treatments might rank reasonably well on the trustworthiness dimension. Such a study might or might not be important, and might or might not have potential generality, depending on the particulars of the situation. Myriad dissertation studies with conclusions of the form "students using materials that I developed scored better than students using a standard text" score low on both. However, individual and collective evaluations of some of the newer and widely used mathematics curricula begin to suggest generality and, as the findings mount, some importance (see, e.g., Senk & Thompson, 2003). Studies that are technically proficient but flawed along the vertical pathways illustrated in Figure 3.1 are *not* trustworthy. By considering the three dimensions in Figure 3.3 one can get a sense of what studies can make as their contributions. In what follows I address the three dimensions one at a time.[7]

Trustworthiness

*What did the President know,
and when did he know it?*

—United States Senator Sam Ervin,
during the Watergate hearings

The United States Senate's impeachment hearings of President Richard Nixon, known as "the Watergate hearings," were one of the defining moments of the American presidency. Richard Nixon's presidency was at stake. Time and time again, Senator Sam Ervin boiled things down to their evidentiary core. The answer to his oft-repeated question "What did the President know, and when did he know it?" would, once arrived at with a significant degree of certainty, determine Nixon's fate as president.

What mattered in Nixon's impeachment trial—what matters in all legal proceedings—is the idea of "a significant degree of certainty." Legal matters, unlike matters of mathematics, are not axiomatic. Typically, one does not resolve complex legal issues with absolute certainty; rather, the standard is whether particular claims have been substantiated "beyond a reasonable doubt." The underlying evidentiary issues are: What warrants are given for the claims being made? How believable and trustworthy are they? How robust are the conclusions being drawn from them?

Thus it is in mathematics education as well. As indicated in the previous section, once one has developed a theoretical orientation toward a situation, the core processes of empirical research are the gathering, representation, manipulation, and interpretation of data. A core question regarding the quality of the conclusions drawn from the research is, how *trustworthy* are each of those processes?

In what follows I elaborate on a number of criteria that are useful for examining the quality of empirical and theoretical research. Aspects of trustworthiness include the following, which are discussed below:

- Descriptive and explanatory power
- Prediction and falsification
- Rigor and specificity
- Replicability
- Triangulation

---

[7] The three aspects of research I have called "dimensions" are not truly independent, of course. What I am offering here is a heuristic frame and an argument that attention to all three aspects of research is essential.

Descriptive and Explanatory Power

Descriptive power denotes the capacity of theories or models to represent "what counts" in ways that seem faithful to the phenomena being described. Descriptions need not be veridical but they must focus on what is consequential for the analysis. To give a classical mathematical example, consider a typical related-rates problem that involves a ladder sliding down the side of a building. The building is assumed to be (or is explicitly stated to be) vertical and the ground horizontal. In the diagram representing the situation, the ladder, the building, and the ground are represented as *lines* that comprise parts of a right triangle. What matters for purposes of the desired analysis are their lengths, and the way the ladder is moving. That information, properly represented and analyzed, enables one to solve the given problem; that information and nothing else is represented in the diagram and the equations one derives from it. What does not matter (indeed, what would be distracting in this context) includes how many rungs are on the ladder or how much weight it might support. In a different context, of course, such things would matter quite a bit. The issue of descriptive power, then, is, does the research focus on what is essential for the analysis, in a way that is clear and compelling?[8]

Explanatory power denotes the degree to which a characterization of some phenomenon explains how and why the phenomenon functions the way it does. Before getting down to educational specifics, I note that explanatory power, like many of the properties described in this section of this chapter, is an issue in all analytic fields. In mathematics, many people will prefer a constructive proof to an existence argument that employs proof by contradiction. The latter says that something exists, but not how to build or find it; the former, in providing a "blueprint" for finding or building it, provides more detailed guidance regarding how and why it exists.

A fundamental issue here is the distinction between correlation—the fact that X and Y tend to co-occur—and what I shall call explanation at a level of mechanism, an attempt to say how and why X and Y are linked. (Note that causal or constraint-based explanations are the traditional forms of explanation at a level of mechanism, but that statistical or probabilistic models also provide such explanations.)

As one example, consider Mendelian genetics as an expression of a theory of heritability. Surely, before Mendel, the notion that traits appear to be passed from generation to generation had been observed. Indeed, the observation had some power—but that power was limited. With Mendel came a suggestion of mechanism—the idea that (at least in the case of specific traits of relatively simple plants such as peas) genes determine the characteristics of those traits. And with that suggestion came the possibility of experimentation and the refinement of the underlying theoretical notions.

Productive, data-based explanations need not be right in all detail—but they do need to be taken seriously in order for the field to progress. For example, Galileo (at great personal cost) advanced a solar-centric theory of planetary motion, which was more accountable to data and explanation than the faith-based earth-centric model it supplanted. Centuries later this work was established on a much more solid footing when Newton proposed a theory of gravitation that had the potential to take things further. His theory explained, for example, why planetary orbits are elliptical rather than circular. (To anticipate arguments made later in this section, it should also be noted that issues of prediction and falsifiability make a big difference. They allow for the evaluation of competing theories, and the refinement of explanations. And, to emphasize a point made earlier, increased explanatory power may come from a shift in conceptual-analytic frameworks.)

One example within mathematics education concerns the relationship between teachers' classroom practices and student learning. In the 1970s and 1980s a dominant approach in classroom research was the process-product paradigm, in which the data gathered focused on (a) tallies of specific classroom practices (e.g., time on task, worksheet use, asking certain kinds of questions); (b) student outcomes, typically as measured on standardized tests; and (c) the statistical relationships between (a) and (b). The results of such analyses were typically of the form "students did better when teachers did X more frequently," with the implication that it would be good for teachers to do X more often. Of course, researchers had ideas as to why some practices were more productive than others—

---

[8] As I stress in Part 1 of this chapter, a great deal of what one attends to in empirical research depends on the conceptual-analytic frameworks that orient one to the phenomena at hand. Thus, when I say that descriptive power denotes the capacity of theories or models to represent what counts, I am making a contextual statement: a characterization or representation that has substantial descriptive power has that power relative to the conceptual-analytic framework being employed. Finding a "better description" may entail finding an alternative conceptualization.

but the research methods themselves did not explore how or why they worked.[9]

In contrast, more recent studies of classroom activities focus on notions such as discourse communities, practices, and classroom norms and their impact—students' participation structures, their sense of the mathematics, and their understanding as reflected on a variety of assessments. Boaler (2002), for example, described in detail the philosophy and teaching practices at two schools, and the beliefs and understandings they are likely to engender. She interviewed students about their understandings regarding the nature of mathematics and their sense of the in-school enterprise. She examined their performance on standardized tests and on problem-solving tests of her own construction. As a result, readers understand that there are differences—and they have been provided evidence that substantiates a plausible story about how and why those differences came to be.

I should stress that I do not wish to privilege any form of explanation in this discussion. In a field that draws from the social sciences, the sciences, and the humanities, there will be various ways to try to explain how and why things happen. What is important is that the attempt be made and that claims must be held accountable to data. Thus, if one claims that teachers' decision making is based on their knowledge, goals, and beliefs (e.g., Schoenfeld, 1998), one should offer models of that decision making and enough evidence to warrant such claims; if one claims that a particular kind of action or sequence of actions on the part of a teacher supports the development of an "oppositional culture" in the classroom, one should offer evidence of the growth of opposition and link that growth in plausible ways to the teacher's actions.

Prediction and Falsification

I first describe issues of prediction and falsification in general, then with regard to educational research.

In the physical and other sciences, prediction has been the name of the game, and the potential for falsification a theoretical essential. Physics is the archetype, of course (and, alas, the source of much inappropriate positivism in education). Newton's laws, for example, say that under certain conditions, certain things will take place. Those predictions, and myriad others, serve both practice and theory. In practical terms, they allow people to build things that work consistently and reliably. Predictions are a mechanism for theoretical progress. As noted above, for example, a solar-centric explanation of planetary motion ultimately triumphed over the earth-centric view because the former explanation fit the data better than the latter. A theory of gravitation based on Newtonian mechanics and the inverse square law of gravitational attraction provides close descriptions and predictions of planetary motion, and allows for fine-grained predictions of planetary movement. The Newtonian view prevailed for centuries, but with some known anomalies—some of its predictions were not quite right. In one of the more famous incidents in the history of science, Einstein's theory of relativity predicted that under the conditions of a solar eclipse, the planet Mercury would appear to be in a different location than Newtonian theory predicted. It took some years before the eclipse took place—and Einstein's view was substantiated.

The point, as noted by Popper (1963), is that Einstein could have been proved wrong. Had Mercury not been where Einstein predicted, there would have been evidence of problems with Einstein's approach. That is, Einstein's theory was *falsifiable*. Evidence could be gathered that substantiated it, but also that cast it into doubt. According to Popper:

- Every "good" scientific theory is a prohibition: it forbids certain things to happen.
- A theory which is not refutable by any conceivable event is non-scientific.
- Every genuine test of a theory is an attempt to falsify it, or to refute it. Testability is falsifiability; …
- Confirming evidence should not count except when it is the result of a genuine test of the theory; and this means that it can be presented as a serious but unsuccessful attempt to falsify the theory;

---

[9] Indeed, subsequent studies showed some of the limitations of this kind of approach. In one study comparing highly effective teachers with other teachers, Leinhardt (1990) showed that the highly effective teachers (defined as those teachers whose students performed in the top performance range for their demographic group—whether "learning disabled," "gifted," or anything in between) uniformly established clear disciplinary routines for their classes at the beginning of the school year. In a companion study, however, Leinhardt et al. (1991) documented the limited utility of such findings. It turned out that many of the teachers had rather shaky mathematical knowledge, extending only a little bit beyond that of the curriculum; and that the standardized tests used to assess students (and teachers) were rather narrow and procedural. Thus, the other work could be reinterpreted as indicating that a high level of discipline is effective in helping teachers remain within their "comfort zone" and tends to produce students who have mastered procedures but may have little or no conceptual understanding.

- One can sum up all this by saying that the criterion of the scientific status of a theory is its falsifiability, or refutability, or testability. (Popper, 1963, p. 36)

The kind of predictions made in classical physics represent only one type of prediction, which has given rise to many misconceptions about predictions in the social sciences. Although "absolutist" arguments in domains such as physics may have face validity, theorists such as Toulmin (1958) argue that they do not apply in more complex, contextual situations in which human actions are involved. Toulmin seeks ways to mediate between absolutism on the one hand and relativism on the other; theorists such as Pickering (1995) replace notions of absolutism with concepts of scientific *practices* that are bound to the histories of the scientific communities in which they emerge. Simply put, the notions of prediction and falsification are unsettled. For that reason, I outline a range of prediction in the sciences and then education, while trying to preserve what I can of the notion of theory-testing.

As noted, the physical sciences sometimes support predictions of the type "under these conditions, the following will take place." All of the traditional laws in the sciences afford such predictions. For example, the ideal gas law $PV = nRT$ says that under certain conditions, the values of three of the variables P, V, n, and T determine the value of the fourth. Likewise, the creation of the periodic table as a theoretical characterization of atomic structure supported predictions about the existence of elements that had not yet been found. These kinds of predictions are deterministic.

The life sciences often entail predictions that are not deterministic in the sense above, but which still describe an expected state. Consider, for example, predator-prey models of animal populations. In simplest terms, predators flourish when they are few in number and there are many prey, but they diminish in number when they are densely crowded and few prey remain as food. Prey flourish when they are few in number and predators are few in number, but they diminish in number when they are overcrowded or the number of predators is large. All of these states, and population change rates, can be quantified, at which point the predator-prey model will predict changes in the sizes of both populations. The fates of individual animals are not determined in such models,

but trends are. In this sense, predictions are not absolute: A theory is not "true" or "false" in the sense that it proposes a universal, and one counterexample serves to invalidate it. Nonetheless, the theory does give rise to models, and the accuracy of the models (and the theory that generated them) can be judged by their fidelity to actual data. Similarly, Mendelian genetics predicts the percentages of offspring that will have specific traits, but not (except in the case when $p = 1$) the traits of individuals; it yields probability distributions regarding the traits of individuals. Yet, at least cumulatively, this is a strong form of prediction. (And the predictions led to refinements of the theory—for example, data that did not conform with theory led to the uncovering of linked traits.)

A weaker form of prediction has to do with *constraints*. Here, evolutionary theory is a primary example. Under ordinary circumstances, evolutionary theory cannot specify how an organism will evolve—just that it will be responsive to its environment.[10] However, evolutionary theory does impose constraints about the ways on which organisms change over time: There is, for example, greater complexity and differentiation. Thus, according to the theory, certain evolutionary sequences are plausible and others are implausible (if not impossible). Thus the theory can be challenged by empirical evidence. Every exploration of geological strata offers potential disconfirmation.

And then, of course, there is weather prediction, which is not yet a science. However, various models of climatic behavior can be assessed and refined. The main point here is not that those in the business of building climatic models "have it right"—far from it. But, because each model allows for predictions, and variations in the models support different predictions, the careful examination of the predictions made and their relation to the theory can help to improve prediction and the underlying theory.

I now turn to educational issues. Of course, only some empirical work in the social sciences or in education involves prediction; substantial bodies of empirical work in education (e.g., autobiographical reports and descriptive studies) involve no claims beyond those made about the evidence discussed. However, I note that a great deal of descriptive work contains implicit claims of generality, and thus of prediction. As soon as there is the implication that "in similar circumstances, similar things happen," one is,

---

[10] There are exceptions in simple cases of natural selection, where the relationship between certain animal traits and the environment is clear. In one classic case, for example, it was possible to predict that as pollution darkened local trees, the population of moths would darken because darker moths on trees were less visible to predators than lighter moths. (And, when pollution was reversed, the population balance changed back in the other direction.)

at least tacitly, making a prediction (see the section on *generality* below).

Within educational research, as in the examples from the sciences discussed above, there is a wide range of prediction. Randomized controlled trials offer one kind of prediction: The assumption underlying experimentation is that under conditions similar to the circumstances of experimentation, results similar to the results of experimentation will be obtained. This is not unique to statistically oriented studies, however: The same is often true of "rich, thick" anthropological descriptions. A main purpose of descriptions of productive classroom discourse structures is to explain not only how things took place (descriptive power) but why students learned what they did (explanatory power), thus enabling others to try similar things in the hope of obtaining similar results (prediction). The more that such descriptions and claims can be made rigorous, the more likely they are to have a productive impact on practice and to serve theory refinement.

Here are some illustrative examples. Brown and Burton's (1978) study of children's arithmetic "bugs" described the authors' analyses of children's errors in base-ten subtraction. Brown and Burton found that children's patterns of errors were so systematic (and rule-based) that, after giving children a relatively short diagnostic test, they could predict with some regularity the incorrect answers that those students would produce on new problems. Brown, Burton, and colleagues (Brown & Burton, 1978; Brown & VanLehn, 1982; VanLehn, Brown, & Greeno, 1984) provided well-grounded explanations of why students made the mistakes they did. But prediction added a great deal to their work. First, the data provided clear evidence of the power of cognitive models: If you can predict the incorrect answers a student will produce on a wide range of problems before the student works them, you must have a good idea of what is going on inside the student's head! Second, prediction played a role in theory refinement: If predictions do not work, then one has reason to look for alternative explanations. Third, a fact often overlooked by those who view the "buggy" work as overly simplistic and mechanistic is that this work provided clear empirical evidence of the constructivist perspective. In 1978, many people believed in the simplistic idea that one teaches something (perhaps in multiple ways) until the student "gets it," and that nothing has been "gotten" until the student has learned whatever it was to be learned. Brown and Burton showed that students had indeed "gotten" something: They had developed/learned an incorrect interpretation of what they had been taught and used it with consistency. That is, what they did

was a function of what they perceived, not simply what they had been shown. If that isn't data in favor of the constructivist perspective, I don't know what is. (See also Smith, diSessa, & Roschelle, 1993/1994.)

Another famous study, predating Brown and Burton, is George Miller's 1956 article "The Magical Number Seven, Plus or Minus Two." After numerous observations in different intellectual domains, Miller hypothesized that humans have the following kind of short-term memory limitation: We can only keep between (roughly) five and nine things in working memory at the same time. That hypothesis gives rise to simple and replicable predictions. For example, carrying out the multiplication

$$\begin{array}{r} 634 \\ \times\,857 \end{array}$$

requires keeping track of far more than nine pieces of information. According to Miller, it would be nearly impossible to look at these numbers, close your eyes, and compute their product. This has been shown time and time again—once at a talk I gave to more than a thousand mathematicians, many of whom, at least, can do their arithmetic. (It *is* possible to do such products if one knows certain arithmetic shortcuts, or if one rehearses some of the subtotals so that they become "chunked" and only use up one "slot" in short-term memory. But such instances are rare.) Note that the element of falsifiability of Miller's work is critically important. If a nontrivial fraction of the people given tasks like this succeeded at them, then Miller's claim would have to be rejected or modified.

A third class of examples consists of process models of cognitive phenomena. In physics, diSessa's models of *phenomenological primitives* and the ways they develop and shape cognition provide predictions of the ways people will interpret physical phenomena, and tests of the adequacy of the theory. In a different domain, the Teacher Model Group at Berkeley (see, e.g., Schoenfeld, 1998) constructed detailed models of a range of teachers' decision making. One theoretical claim made by the group is that a teacher's in-the-moment actions can be modeled as a function of the teacher's knowledge, goals, beliefs, and a straightforward decision procedure. Each new case explored (an inexperienced high school mathematics teacher conducting a more or less traditional lesson, an experienced high school teacher conducting an innovative lesson of his own design that largely went "according to plan," and an experienced third-grade teacher conducting a lesson in which the agenda was coconstructed with the students) tested the scope and adequacy of the theory. Again, the idea is straightforward. The more that theoretical claims can be examined and tested by data, the more there is the potential for refinement and validation.

Rigor and Specificity

At this point, little needs to be said about the need for rigor and specificity in conducting empirical research. The more careful one is, the better one's work will be. And the more carefully one describes both theoretical notions and empirical actions (including methods and data), the more likely one's readers will be able to understand and use them productively, in ways consistent with one's intentions.

Precision is essential; a lack of specificity causes problems. One historical example is the potentially useful construct of *advance organizer* introduced by Ausubel (see, e.g., Ausubel, 1960). Over the years, a large number of studies indicated that the use of advance organizers (roughly speaking, top-level introductions to the content of a body of text) improved reading comprehension. But an equally large number of studies indicated that the use of advance organizers did not make a difference. Why? On closer examination, the construct was so loosely defined that the advance organizers used in various studies varied substantially in their characteristics. Thus the results did as well. Similarly, a theory that involves terms such as *action, process, object, and schema* needs to say what those terms are, in clear ways. And, of course, appropriate detail and warrants need to be provided in discussing those terms. One example of a productively used term is the *didactical contract* as used by the French (see, e.g., Brousseau, 1997). The term has a specific meaning, which provides a useful backdrop for many such studies. In contrast, the generic use of the term *constructivist* (for example, in the meaningless term *constructivist teaching*) has not been helpful.

Note that rigor does not simply mean rigor in the use of one's data analyses (the path denoted by Arrow 3 in Figure 3.1). It means attending carefully to all of the pathways from Box A to Box F in Figure 3.1.

Replicability

Every person is different; every classroom is different. How can one possibly speak of replication in education? The idea seems strange. One might replicate experiments of some types—but how many educational researchers do experimental work? And, if every classroom is different, what does it mean to replicate someone else's research study?

One way to think about the issue of replicability is to think about generality. As noted above and as will be elaborated below, there is a tacit if not explicit aspect of generality to most studies: the expectation is that the lessons learned from a study will apply, in some way, to other situations. To that degree, some key aspects of those studies are assumed to be replicable. The issue, then, is, how does one characterize those aspects of a study—in enough detail so that readers can profit from the work in the right ways, and so that they can refine the ideas in it as well? Thus, for example, studies like those referred to in the section on prediction (Brown and Burton's studies of arithmetic bugs, Miller's article on the magic number seven plus or minus two, and the Teacher Model Group's article on teachers' decision making) all involve prediction; hence they should be potentially replicable as well.

Multiple Sources of Evidence (Triangulation)

In a well-known article published in 1962, Martin Orne introduced the concept of the *demand characteristics* of an environment via this personal anecdote:

> A number of casual acquaintances were asked whether they would do the experimenter a favor; on their acquiescence, they were asked to perform five push-ups. Their response tended to be amazement, incredulity and the question 'Why?' Another similar group of individuals were asked whether they would take part in an experiment of brief duration. When they agreed to do so, they too were asked to perform five push-ups. Their typical response was "Where?"

The idea, later abstracted as *context effects,* is that context makes a difference: People will do things in some circumstances that they might do differently (or not at all) in other circumstances. Some behavior is purely artifactual, *caused by* the particulars of the circumstances.

For this reason, the use of multiple lenses on the same phenomena is essential. In some cases, that means employing multiple methods to look at the same phenomena. Thus, observations, questionnaires, and interviews can all be used to challenge, confirm, or expand the information gathered from each other. Similarly, behavior that manifests itself in some contexts may or may not manifest itself in other contexts.[11] One should constantly be on guard with regard to such issues.

---

[11] *Context* is meant to be taken very broadly here. Orne's example is a case in point: a question asked of a friend is very different from a question asked of a voluntary participant in an experiment. Similarly, a question asked of individual experimental participants may be treated very differently by *pairs* of subjects.

Generality and Importance

A central issue with regard to any research is its scope, or generality—the question being, how widely does this idea, or theory, or finding, actually apply? A second, equally critical issue is importance. A study may or may not be warranted to apply widely. But whether it is or is not, one must ask: Does it matter? Just what is its contribution to theory and practice?

The concept of generality is slippery, because in many papers a few instances of a phenomenon are taken as an indication of the more widespread existence (or potential existence) of that phenomenon. Thus, it may be useful to distinguish the following kinds of generality:

- The *claimed generality* of a body of research is the set of circumstances in which the author of that work claims that the findings of the research apply.
- The *implied generality* of the work is the set of circumstances in which the authors of that work appear to suggest that the findings of the research apply.
- The *potential generality* of the work is the set of circumstances in which the results of the research might reasonably be expected to apply.
- The *warranted generality* of the work is the set of circumstances for which the authors have provided trustworthy evidence that the findings do apply.

There is often a significant gulf between these. And there is often slippage between evidence-based claims (warranted using the constructs within the conceptual-analytic system—the path from Box C to Box D in Figure 3.1) and explicit or implicit claims of generality (the path from Box A to Box F in Figure 3.1). Thus, for example, an experimental study may compare outcomes of a new instructional method with an unspecified control method. If the study is done by the author using a measure that he or she developed for the purpose of this study, the warranted conclusion may be "students in this context, taught by the instructor who developed the materials and tested using an assessment similar in kind to the materials, did better on that assessment than students who took a somewhat different course." The warranted generality is thus quite low. The claimed generality might be "students in the experimental group outperformed control students," with the implied generality being that other students who experience the new instructional method will similarly outperform other students. The potential generality is actually unknown—it depends on the contexts, the character of instruction in experimental and control groups, the attributes of the assessment, and more.

A key point is that the warranted generality of a study and its importance are not necessarily linked. For example, Wilbur and Orville Wright demonstrated (with substantial trustworthiness!) on December 17, 1903, that a heavier-than-air machine could take flight and sustain that flight for 59 seconds. That was the warranted generality. However, that flight was critically important as an existence proof. It demonstrated that something *could* be done and opened up fertile new territory as a result. Likewise, many studies in mathematics education can be seen as existence proofs. They may demonstrate the existence of a phenomenon worthy of investigation (e.g., early studies of metacognition or beliefs) or of instructional possibilities (e.g., early studies of Cognitively Guided Instruction). The findings of studies with existence proofs are not yet general—but there may be the potential for them to be.

Conversely, a body of research that has broad generality can turn out to be theoretically sterile and to have little practical value. One such example is a spate of studies conducted in the 1980s, which showed that a substantial proportion of those examined produced the equation $P = 6S$ instead of $S = 6P$ in the problem

> Using the letter S to represent the number of students at this university and the letter P to represent the number professors, write an equation that summarizes the following sentence: "There are six times as many students as professors at this university."

(Clement, 1982; Clement, Lochhead, & Monk, 1981; Rosnick & Clement, 1980). The phenomenon was documented repeatedly, and various attempts were made to get people to do better at such problems. However, after all was said and done, the field had no real understanding of why people made this mistake, and no effective mechanisms for preventing or fixing it. Thus—given that this body of work was general, but did not produce significant understandings or applications—we see that generality and importance are at least somewhat independent dimensions.

In what follows I briefly describe the (warranted-to-potential) generality of a number of studies and discuss the importance of those studies. Broadly speaking, the studies are clustered by their degree of generality. For the most part I have chosen studies whose trustworthiness is well established. Judgments of importance reflect the author's perspective but are also included as a reminder that importance is an essential dimension to take into account.

Studies of Limited Warranted Generality

There is a large class of studies for which there is limited warranted generality, but which are worth noting because of their current or potential importance. Such studies may be of interest because they offer existence proofs, bring important issues to the attention of the field, make theoretical contributions, or have the potential to catalyze productive new lines of inquiry.

An autobiographical report may be of interest because it is motivational, or, like other historical documents, it illuminates certain historical decisions or contexts. (For example, what was the historical context for the *Brown v. Board of Education* decision?[12]) In cases such as these there are no warranted conclusions beyond those described, but there may be lessons to be learned nonetheless. (Note that there is an implied aspect of generality to historical studies in Santayana's oft-quoted statement that "those who cannot learn from history are doomed to repeat it.")

Another singular event is the existence proof—a study that shows that something is possible and may elaborate on the means by which it became possible. For example, Harold Fawcett's classic 1938 volume *The Nature of Proof* demonstrated that it was possible to create a classroom mathematical community in which students engage meaningfully in many of the activities in which mathematicians engage (e.g., making definitions and deriving results logically from them). Early papers on Cognitively Guided Instruction (CGI) showed that it is possible to capitalize on young students' informal models and situational understandings to help them produce a wide range of meaningful solutions to word problems (see, e.g., Carpenter, Fennema, & Franke, 1996). Moll, Amanti, Neff, and Gonzalez (1992) showed how it is possible to develop instruction that builds on the kinds of cultural knowledge and traditions that students have in their out-of-school lives. Gutstein (2003) showed that it is possible to teach a mathematics course in which students do well on traditional measures of mathematical performance and become engaged as social activists as well. Boaler (in press) showed that it is possible to create a discourse community in a high school mathematics classroom with a large percentage of low-SES and ESL students in which there are high mathematical standards—and the classroom accountability structures are such that the students hold each other accountable for producing meaningful and coherent mathematical explanations.

Each of these studies had a high level of trustworthiness, in that it satisfied many of the criteria discussed in the previous section of this chapter. Each had relatively low warranted generality, in that it made the case that something had happened in specific circumstances, with a small number of students. But each showed that something could be done, opening up a previously undocumented space of possibilities. In that sense, they rate high on the potential importance scale (and may, indeed, pave the way to phenomena that become more general). Over time, some (e.g., Fawcett's study) have languished as inspirational examples not taken up on a large scale; some (e.g., Cognitively Guided Instruction) have had increasing impact.

Other studies bring important phenomena to readers' attention. They may suggest as-yet-undocumented generality, opening up arenas for investigation. Thus, for example, Cooney's (1985) case study of "Fred" showed that a teacher may espouse a set of values that sounds compatible with a particular pedagogical direction (in this case, "problem solving") but may interpret those terms in contradictory ways and thus act in ways contrary to (the normative interpretation of) those values. Cohen (1990) showed that a teacher may adopt the rhetoric of a particular pedagogical approach and believe that she is implementing that approach, while in fact assimilating (in the Piagetian sense) many of the surface aspects of that approach to her long-established pedagogical practices. Eisenhart et al. (1993) showed that a particular teacher, subject to the pressures of her environment, wound up teaching in a way differently than she intended. Each of these studies made a well-documented case that something important was happening in one specific set of circumstances. But each study also made a plausibility case that the phenomenon under discussion was more widespread.

A classic example of this genre is Heinrich Bauersfeld's 1980 article "Hidden Dimensions in the So-Called Reality of a Mathematics Classroom." Bauersfeld revisited the data from a dissertation by Shirk (1972), which had focused largely on the content and pedagogical goals of beginning teachers. Examining the same data, Bauersfeld focused on the social dimensions of the classroom. His analysis addressed four areas of classroom research that had hitherto received a negligible amount of attention: "the constitution of meaning through human interaction, the impact of institutional settings, the development of personality, and the process of reducing classroom

---

[12] In a 1954 decision that had far-reaching implications, the United States Supreme Court declared that the establishment of segregated public schools was unconstitutional.

complexity" (Bauersfeld, 1980, p. 109). Although the warranted generality of the findings was small, the face-value typicality of the classroom he explored made at least a plausibility case that the phenomena under investigation were relatively widespread. The phenomena were not (yet) claimed to be general but were seen as worthy of investigation.

In a similar way my 1988 article "When Good Teaching Leads to Bad Results: The Disasters of Well Taught Mathematics Classes" presented a trustworthy account of one classroom, in which students had developed a series of counterproductive beliefs about the nature of mathematics as a result of receiving well-intentioned but narrow instruction that focused on preparing the students with particular procedural skills to do well on a high-stakes test. By virtue of the typicality of the instruction in the focal class, and the causal explanation of how the classroom practices resulted in specific outcomes, the article had substantial potential generality.

Other papers may be important because of their theoretical or methodological contributions. For example, diSessa's 1983 chapter on *phenomenological primitives* introduced a new way to conceptualize the development of conceptual structures and reframed the theoretical debate on the nature of conceptual understanding. Brown and colleagues (e.g., Brown, 1992; Brown & Campione, 1996) introduced and elaborated on the notion of design experiments, describing nonstandard instructional practices and a novel way to think about data gathering and analysis in the context of such instruction. Yackel and Cobb's (1996) discussion of sociomathematical norms provided a theoretical tool (which will have widespread application) for examining the ways in which "taken-as-shared" classroom practices can shape individual cognition. Similarly, Cobb and Hodge's (2002) characterization of diversity as a function of differences in cultural practices rather than a measure of "spread" of some particular demographic variable, although not yet widely used, has the potential to reframe studies of diversity in productive ways. Therein lies its importance.

Studies Where Some Degree of Generality is Claimed and/or Warranted

The simplest claims of generality are statistical. Consider, for example, the following datum from Artigue (1999):

> More than 40% of students entering French universities consider that if two numbers A and B are closer than 1/N for every positive N, then they are not necessarily equal, just infinitely close. (p. 1379)

In and of itself, this statement (whose generality is warranted in statistical terms) may or may not seem important. However, this and other data (e.g., that a significant number of such students believe that the decimal number 0.9999... is infinitely close to but less than 1) help to establish realistic expectations regarding the knowledge of entering calculus students, and expectations about necessary focal points of instruction.

In a similar way, some findings from the U.S. National Assessment of Educational Progress (NAEP) pointed to serious across-the-board issues in American mathematics instruction. Perhaps the best-known single item from NAEP is the following (Carpenter, Lindquist, Matthews, & Silver, 1983):

> An army bus holds 36 soldiers. If 1128 soldiers are being bussed to their training site, how many buses are needed?

The calculation is straightforward: 36 goes into 1128 thirty-one times, with a remainder of 12. Hence 32 buses are needed. Here is the frequency of responses from the stratified nationwide sample of students who worked the problem:

23%	32
29%	31R12
18%	31
30%	other

The vast majority of students who worked the problem performed the necessary computation correctly. But then, more than a third of the students who did the computation correctly wrote down an impossible answer (the number of buses needed must be an integer, after all!), and a substantial number of others rounded down instead of up, which makes no sense in the given context.

Once again, the statistics attest to the generality of the finding. What made this finding *important* was the way it fit into an emerging body of research findings. In the early 1980s, researchers were coming to understand the importance of beliefs (Schoenfeld, 1983), both in terms of their impact on students' mathematical performance and in terms of their origins in classroom practices. Many of the arguments in support of beliefs had been local: In a specific class, students experienced these specific things; they developed the certain understandings of the mathematical enterprise as a result; and they then acted in accord with those understandings. In small-scale studies the argument had been made that students had come to experience word problems not as meaningful descriptions of mathematical situations, but as "cover stories" for arithmetic operations; they understood their task as students as uncovering the relevant mathematical numbers and operations in the cover story, performing the operations, and writing the answer down. In this view, the ostensible reality described in the cover story played no role in the problem once the numbers and operations had been extracted from it. The data on the NAEP provided evidence that the problem was a national one. Therein resides the importance of the research.

In a similar way, I first explored the role of metacognition in mathematical problem solving in the research described in part 1 of this chapter. The data in the original experiment were suggestive: more than half of the videotapes I examined were of the type shown in Figure 3.2. This finding was potentially important. The question was whether it would turn out to be general. The number of cases I had examined was small, but a priori there was nothing to suggest that they might be atypical. When the kinds of effects demonstrated in my early studies were replicated elsewhere and tied to emerging theories of the importance of metacognition in other domains (e.g., Brown, 1987), the generality and importance of the phenomenon became increasingly clear. Note that this kind of sequence is typical: Individual studies point to aspects of a potentially interesting or general phenomenon, and an expanding body of studies refine the idea over time.

A set of ideas that is in its early stages, suggesting generality but not yet fully explored or validated, has to do with the attributes of particular kinds "learner-centered" learning environments. Engle and Conant have suggested that

> productive disciplinary engagement can be fostered by designing learning environments that support (a) problematizing subject matter, (b) giving students authority to address such problems, (c) holding students accountable to others and to shared disciplinary norms, and (d) providing students with relevant resources. (2002, p. 399)

The authors provided a detailed analysis of a series of discussions in a "Fostering a Community of Learners" classroom (Brown & Campione, 1996), demonstrating how each of the four themes identified above played out in that classroom. They then briefly reexamined discourse patterns in two other types of instruction known for engaging students productively

with disciplinary content: Hypothesis-Experiment-Instruction Method Classrooms (Hatano & Inagaki, 1991; Inagaki, 1981; Inagaki, Hatano, & Morita, 1998) and the Chèche Konnen project (National Science Foundation, 1997; Rosebery, Warren, & Conant, 1992). This evidence suggests the potential generality of the findings. Readers' familiarity with other such environments—e.g., Scardamalia and Bereiter's "Knowledge Forum" classrooms (Scardamalia, 2002; Scardamalia & Bereiter, 1991; Scardamalia, Bereiter, & Lamon, 1994) or my problem-solving courses (Schoenfeld, 1985) may add to the sense of generality and importance of the findings. Time will tell.

A final kind of argument with some generality and potential importance, which is a bit further along than the Engle and Conant study, is done by aggregating over a collection of studies, each of which offers consistent data. Such an argument is made in Sharon Senk and Denisse Thompson's 2003 volume, *Standards-Based School Mathematics Curricula: What are They? What do Students Learn?* Senk and Thompson present the data from studies evaluating a dozen *Standards*-based curricula. A careful reader could find reasons to quibble with a number of the studies. For example, many of the assessments employed in the studies were locally developed.[13] Hence one could argue that some of the advantages demonstrated by the *Standards*-based curricula were due in part to assessments tailored to those curricula. In addition the conditions of implementation for the curricula (e.g., the preparation that teachers received before teaching the curriculum) may have been higher than one might expect in "ordinary" curriculum adoptions. Nonetheless, the pattern of results was compelling.

The book is divided into sections assessing curricula at the elementary, middle, and high school levels. Putnam (2003) provided the following summary of the evaluations of *Math Trailblazers, Everyday Mathematics, Investigations,* and *Number Power* (all elementary curricula):

> The four curricula … all focus in various ways on helping students develop conceptually powerful and useful knowledge of mathematics while avoiding the learning of computational procedures as rote symbolic manipulations.

> The first striking thing to note about [them] is the overall similarity in their findings.

> Students in these new curricula generally perform as well as other students on traditional measures of

---

[13] A concern about locally developed measures is that there is the potential for bias (giving an unfair advantage to the "experimental" curriculum) whenever the developers of that curriculum are closely tied to the developers of an assessment used in the evaluation of that curriculum. Whether advertently or not, items on the assessment could be designed in ways that favor the experimental curriculum.

mathematical achievement, including computational skill, and generally do better on formal and informal assessments of conceptual understanding and ability to use mathematics to solve problems. These chapters demonstrate that "reform-based" mathematics curricula can work. (p. 161)

Chapell (2003) summarized the evaluations of *Connected Mathematics, Mathematics in Context,* and *Middle Grades MATH Thematics: The STEM Project* (all middle grades curricula) as follows:

Collectively, the evaluation results provide converging evidence that *Standards*-based curricula may positively affect middle-school students' mathematical achievement, both in conceptual and procedural understanding. . . . They reveal that the curricula can indeed push students beyond the "basics" to more in-depth problem-oriented mathematical thinking without jeopardizing their thinking in either area (pp. 290–291)

Swafford (2003) examined the evaluations of the following high school curricula: *Core-Plus Mathematics Project, Math Connections,* the *Interactive Mathematics Program (IMP),* the *SIMMS Integrated Mathematics Project,* and the *UCSMP Secondary School Mathematics Program.* She concluded that:

Taken as a group, these studies offer overwhelming evidence that the reform curricula can have a positive impact on high school mathematics achievement. It is not that students in these curricul[a] learn traditional content better but that they develop other skills and understandings while not falling behind on traditional content. (p. 468)

Although many of the individual studies in Senk and Thompson (2003) may be problematic in some regard, the cumulative weight of the evidence in favor of the impact of the *Standards*-based curricula—and the absence from the literature of any comparable evaluations favoring the more traditional, skills-oriented curricula—is such that the findings, although preliminary, acquire a nontrivial degree of trustworthiness. There is a degree of generality in that all of the curricula examined, although differing substantially in style and content, all emphasized conceptual understanding, problem solving, and deeper engagement with mathematical concepts than the "traditional" curricula with which they were compared. And if students learn more, there is a prima facie case for importance.

A final example with a somewhat different kind of warrant is the claim made by Stigler and Hiebert that "teaching is a cultural activity (1999, p. 85)." As Stigler and Hiebert noted, the Third International Mathematics and Science Study (TIMSS) video study performed a random sampling of classrooms selected from the larger TIMSS study, which was carefully constructed; the final video study sample included 100 German, 50 Japanese, and 81 U.S. classrooms that the authors claimed "approximated, in their totality, the mathematics instruction to which students in the three countries were exposed" (p. 19). Classroom videos were then coded for various kinds of detail, and the codings were analyzed. The data revealed that there was much less within-country variation than across-country variation—that is, that there were relatively consistent practices in each nation that differed substantially from the practices in the other nations. It goes without saying that one might code for things other than those captured by the coding scheme developed for the TIMSS study. But the main finding, backed up by descriptions of lesson content, appears trustworthy and (by virtue of the sampling) general across the three countries in the study. Subsequent replication studies are extending the findings. But, do the results matter? In this author's opinion, the answer is a clear *yes.* Cross-national comparisons reveal what turn out to be cultural assumptions that one would not see if one kept one's eyes within one nation's borders. They reveal practices that challenge one's notion of what may be possible and thus challenge one to think about the premises underlying instruction.

Studies Where Significant Generality, if Not Universality, is Claimed or Implied

On the one hand, as Henry Pollak once said, "there are no theorems in mathematics education." On the other hand, there are some well-warranted results of significant generality. Many of these have to be with (attributions of) cognitive structures. Thus, for example, although one can certainly quarrel with major aspects of Piaget's work, phenomena such as children's development of *object permanence* and *conservation of number* are well established and well documented. Similarly, replication and follow-up studies of phenomena such as "the magic number seven plus or minus two" (Miller, 1956) have established the limitations of short-term memory as essentially universal. Cumulative bodies of literature with regard to the role and existence of schemata or their equivalents as forms of knowledge organization have established the utility of those constructs for describing people's knowledge structures, likewise for studies of beliefs and metacognition as shapers of cognition. Cumulatively, these studies provide documentation of phenomena that are robust (hence the research is trustworthy); that are essentially universal; that provide an understanding of individual cognition; and are thus clearly important.

Robust findings can, of course, be misused in applications to mathematics education. For example, the fact that people who are fluent in secondary school mathematics possess a large number of schemata for solving word problems does not imply that students should be directly taught a large set of those schemata. (See the exchange between Mayer, 1985, and Sowder, 1985, for a discussion of this issue.) Nor should a focus on cognitive structures in the preceding examples be taken in any way as suggesting that cognitive structures provide complete and coherent explanations of thinking, teaching, and learning. As noted in previous sections, many ideas from sociocultural theory have great promise in helping to unravel the complex interactive processes in which all humans engage. The arena is important, and many individual findings are trustworthy, but the field is still in its genesis. One expects to see many examples with a sociocultural emphasis—studies of discourse patterns, identity formation, etc.—in this category of studies when the third edition of this *Handbook* is published.

Finally, I note another category of potential universals: large-scale comparative studies of various types. As examples, cross-national studies such as TIMSS (see, e.g., Mullis et al., 1998; Mullis et al., 2000) and PISA (see, e.g., Lemke et al., 2004), provide trustworthy comparative information at nationwide levels about student performance. However, it should be remembered that any assessment is only as good as the items it uses (cf. Part 1 of this chapter). Moreover, any assessment reflects the mathematical values of its developers. Specifically, PISA and TIMSS emphasize different aspects of mathematical competence (PISA being more "applied"). Nations that do well on one assessment do not necessarily do well on the other. Other comparative studies (e.g., Lee, 2002) reveal trends in similarities and differences in the mathematical performance of various subgroups of larger populations. The same caveats apply.

Discussion

A major challenge in the conduct of educational research is the tension between the desire to make progress[14] and the dangers of positivism and reductivism. The point of Part 1 of this chapter was that

reductivism, at least, comes with the territory: One's explicit or implicit theoretical biases frame what one looks at, how one characterizes it, how one analyzes it, and how one interprets what one has analyzed. It is easy to slip into positivism as well.

It is in that spirit that I refer to the ideas in Part 2. The question addressed here is, what makes for good empirical work? I have argued that research must be examined along three somewhat independent dimensions. The first is trustworthiness—the degree of believability of the claims made in a piece of research. The core issue is: if claims are made, do the warrants for them ring true? As discussed above, there are various criteria for the trustworthiness of empirical research: a study's descriptive power; its explanatory power; whether the claims made are in some sense falsifiable; whether the study makes predictions and, if so, how well those predictions fare; how rigorous and detailed the work is; whether the work has been described in ways that allow for attempts at replication and, if so, whether the findings are duplicated or extended; and whether the study offers multiple lenses on the same phenomena and multiple lines of evidence that converge on the same conclusions. Of course, not every criterion is relevant for every study; the nature of a study will determine how each of the relevant criteria is applied. But trustworthiness is an essential quality of good research, and attending to the criteria discussed in Part 2 will help one to make informed judgments about it.

Trustworthiness is not enough, however. Simply put, a study may be trustworthy (and thus publishable in the sense that it meets the appropriate methodological criteria) but trivial, along one or both of the two other dimensions: generality (scope) and importance. Assessing the generality of a result is often a delicate matter. Typically authors imply the generality of a phenomenon by tacitly or explicitly suggesting the typicality of the circumstances discussed in the study. Implying generality is one thing, however, and providing solid evidence for it is another. For that reason I have introduced the notions of *claimed, implied, potential,* and *warranted generality* as ways to think about the scope or generality of a study.

Importance is, of course, a value judgment. But it is an essential one, to be made reflectively.

---

[14] If one defines *progress* in educational research as either clarifying/adding to the field's knowledge or producing information or materials that help people do something more effectively, then just about all research is aimed at progress in some way. I intend this broad a definition.

PART 3
FROM IDEAS TO IMPLEMENTATION: A RECONSIDERATION OF THE CONCEPT OF "CLINICAL TRIALS" IN EDUCATIONAL RESEARCH AND DEVELOPMENT

My purpose in Part 3 of this chapter is to describe a sequence of research and development activities employing both qualitative and quantitative methods that is intended to serve as a mechanism for the improved development and effectiveness of educational interventions. This effort is motivated both by a need, in general, for mechanisms to improve the curriculum design and evaluation process (Burkhardt, 2006; Burkhardt & Schoenfeld, 2003) and the wish to bring together the statistical and mathematics education communities in profitable ways (Scheaffer, 2006). I shall try to draw upon the best of three traditions that are often seen as in conflict with one another: educational research and design as craft, fundamental research in mathematics education, and calls for the scientific validation of curricular effectiveness through experimental means.

This part of this chapter differs substantially from the two that preceded it. Earlier in this chapter, my goal was generality. Part 1 offered a scheme for characterizing and reflecting on all empirical research, and Part 2 offered a framework that can be used to assess the quality of all empirical research studies. In a sense, the issues addressed in Parts 1 and 2 are timeless—questions of how to conduct research of increasingly high quality will always be with us. Here my focus is more narrow. In a chapter that addresses issues similar to those addressed here, Clements (2002) poses this question: "Why does curriculum development in the United States not improve?" (p. 599) One of my goals is to help solve this practical problem. The problem is timely for practical and political reasons, as will be discussed below. It is far from ephemeral, however. The general issue of experimentalism in education, and the scientism that bedevils it, have been problematic since long before I entered the field.

I begin with some caveats, in the hope of avoiding misinterpretations of what follows. I am *not* proposing that curriculum development is the solution to instructional problems in mathematics education; nor am I suggesting that it is the only (or even the primary) way in which mathematics instruction can be improved. A curriculum is a tool, and a tool is only effective in the hands of those who can wield it effectively. The fact that I do not focus here on issues of professional development, or on strategies for improving instruction via professional development, should not be taken as an indication that I underestimate the value of professional development as a vehicle for improving instruction. In theoretical terms, I do not propose that curriculum development is the solution to the problem of "travel"—the ability of materials developed in one context to be used profitably in another (National Academy of Education, 1999). The more narrow issue that I address is, how can we refine the curriculum development process, so that the curricular tools that are developed are more likely to be effective (when used in the right ways)? The approach I take is to try to develop a serious educational analog to the idea of clinical trials in the development of medicines and medical treatments. The general approach outlined here should be applicable to all educational interventions.

I note by way of preface that I have been engaged in educational research for long enough to see a number of pendulum swings in what appears to be the eternal (and erroneous) dialectic between emphases on qualitative and quantitative methods. In the 1960s and 1970s, an emphasis on experimental and statistical methods was stifling the field. The problem was often a misapplication of good statistical methods. In many cases a researcher employed a "Treatment A *versus* treatment B" experimental design to compare curricular materials developed by the researcher with some form of control treatment. The assessment measures employed were typically standardized tests or measures developed by the author, each of which have their own problems (recall the discussion in Part 1 of this chapter). Sometimes the experimenter taught both the experimental and control classes, in which case the most significant (and unmeasured!) variable may have been teacher enthusiasm for the experimental materials, or the fact that one course was taught before lunch and one after. Sometimes different teachers taught the experimental and control classes, in which case teacher variation (again unmeasured) may have been the most important variable shaping outcomes. A number of scholars, including myself (see also Kilpatrick, 1978), have argued that there was as much scientism as science in educators' attempts to adopt scientific methods in educational research—for example, I called for a moratorium on the use of factor analyses on "mathematical abilities" until researchers could explain what the factors determined by statistical methods actually meant.

Over the 1980s and 1990s a flowering of the cognitive and sociocultural perspectives led to a wide range of methods and approaches. Attention to these, combined with a reaction against the sterility of the earlier decades of "scientific" work, led to a general abandonment of quantitative research in the field. This too is problematic, unless there is a firm commitment on the part of researchers to provide the warrants that allow research to be deemed truly trustworthy (cf. Part 2 of this chapter). In a column leading up to the American Educational Research Association annual program in 1999, program chair Geoffrey Saxe and I (1998) reproduced the following quote from a letter I had received:

> At [Annual Meetings] we had a hard time finding rigorous research that reported actual conclusions. Perhaps we should rename the Association the American Educational Discussion Association. . . . This is a serious problem. We serve a profession that has little regard for knowledge as the basis for decision making. By encouraging anything that passes for inquiry to be a valid way of discovering answers to complex questions, we support a culture of intuition and artistry rather than building reliable research bases and robust theories. Incidentally, theory was even harder to find than good research. (p. 33)

At the federal level, at least, the pendulum has swung back in the direction of experimentalism in recent years. Here is an excerpt from the U.S. Department of Education's (2002) Strategic Plan for 2002–2007:

> Strategic Goal 4: Transform education into an evidence-based field

> Unlike medicine, agriculture and industrial production, the field of education operates largely on the basis of ideology and professional consensus. As such, it is subject to fads and is incapable of the cumulative progress that follows from the application of the scientific method and from the systematic collection and use of objective information in policy making. We will change education to make it an evidence-based field. We will accomplish this goal by dramatically improving the quality and relevance of research funded or conducted by the Department, by providing policy makers, educators, parents, and other concerned citizens with ready access to syntheses of research and objective information that allow more informed and effective decisions, and by encouraging the use of this knowledge. (U.S. Department of Education, 2002, p. 51)

Grover Whitehurst, Director of the Department of Education's Institute of Education Sciences, has followed through on that agenda. IES has funded the What Works Clearinghouse, some of whose limitations are discussed in Part 1 and in Schoenfeld (2006a, 2006b). Early in his tenure, Whitehurst made his agenda clear in a presentation (2002) to the American Educational Research Association. Referring to Donald Stokes's argument in *Pasteur's Quadrant* (Stokes, 1997; see below), Whitehurst argued strongly in favor of his Institute's mission to support fundamentally applied work, in "Edison's quadrant":

> Yes, the world needs basic research in disciplines related to education, such as economics, psychology, and management. But education won't be transformed by applications of research until someone engineers systems and approaches and packages that work in the settings in which they will be deployed. For my example of massed versus distributed practice, we need curricula that administrators will select and that teachers will follow that distributes and sequences content appropriately. Likewise, for other existing knowledge or new breakthroughs, we need effective delivery systems. The model that Edison provides of an invention factory that moves from inspiration through lab research to trials of effectiveness to promotion and finally to distribution and product support is particularly applicable to education.

> In summary, the Institute's statutory mission, as well as the conceptual model I've just outlined, points the Institute toward applied research, Edison's quadrant. (Whitehurst, 2002; unpaginated.)

While paying homage to other forms of research, Whitehurst goes on to state his stance in clear terms: "Randomized trials are the only sure method for determining the effectiveness of education programs and practices."[15] (Whitehurst, 2002)

But *sure* in what sense? As noted in the section of Part 1 called "a second example," how one interprets curricular effectiveness depends on how (if at all) one defines and attends to context. And where and when does one move from investigative studies to documentation of effectiveness? Must they be separate? In this section I address these issues.

Because much of the move toward evidence-based research in education is based on comparable work in the health sciences (see, e.g., the work of the Cochrane Collaboration, at http://www.cochrane.org/), I shall make my proposal by way of analogy to the model of clinical trials favored in medical and pharmaceutical research. It should be understood, however, that in using this analogy I am not accepting the reductive

---

[15] Whitehurst goes on to say that randomized controlled trials (RCTs) are not appropriate for all research questions, and that other methods are appropriate to supplement the findings of RCTs. However, it is clear that "questions of what works are paramount for practitioners; hence randomized trials are of high priority at the Institute."

metaphor of curricular "treatments" in education as being anything like "simple" drug treatments for medical conditions. In what follows I wish to honor the complexity of the human and social processes that characterize the educational enterprise. What I wish to show is that the pharmaceutical enterprise is more complex than many of those who would analogize to education would have it. There is much to learn from the systematic study of proposed medical interventions, *if* the proper analogies are made, fully respecting the complexity of the educational process. If they are not, however—if steps in the process are skipped, or if proper attention is not paid to the complexities of educational research discussed in Part 1 of this chapter—then there is the potential for reductivism or scientism to do more harm than good.

I begin, with some prefatory comments about *Pasteur's quadrant*. Stokes (1997) observed that, traditionally, basic and applied research have been seen as being at opposite ends of the research spectrum. This is captured by the one-dimensional picture in Figure 3.4.

Basic Research ⟷ Applied Research

Figure 3.4 Basic and applied research viewed as polar opposites.

Moreover, basic research has normally been thought of as preceding applied research, which precedes large-scale applications in practice. Stokes argued, however, that the basic/applied dichotomy was too simple. Although some research is almost exclusively basic (the work of physicist Niels Bohr is a classic example) and some is almost exclusively applied (Thomas Edison's being a case in point), some critically important research has been framed in ways that allowed it to simultaneously make fundamental contributions to theory *and* to the solution of important practical problems. Pasteur's research, which advanced the field of microbiology while helping to cure problems of food spoilage and disease, is the generic example of such work. Stokes proposed replacing the scheme in Figure 3.4 with a two-dimensional version, shown in Figure 3.5.

Whereas Whitehurst argued that an emphasis on Edison's quadrant—in essence, focusing on "What Works" without necessarily asking why—is essential for making progress on curricular issues in mathematics education and other fields, I shall argue that a more balanced approach, with much more significant attention to work in Pasteur's quadrant, will have a much larger payoff. To do this I shall make a direct analogy with clinical studies in evidence-

Research is inspired by:

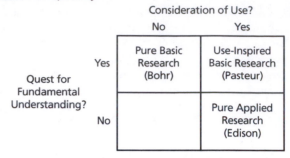

Figure 3.5 A 2-D representation of "basic" and "applied" considerations for research (Stokes, 1997, p. 73).

based medicine. The arguments made here have been influenced by a long-term collaboration with Hugh Burkhardt. An approach parallel to the one discussed here has been called the "engineering approach" to education (see Burkhardt, 2006; Burkhardt & Schoenfeld, 2003).

The U.S. Food and Drug Administration (FDA, 2005) described four "steps from test tube to new drug application review": preclinical research and three phases of clinical studies (see http://www.fda.gov/cder/handbook/). In preclinical drug (or indeed, any medical treatment) research, a wide range of experimentation is done with animals, in order to determine that proposed interventions are likely to be useful and cause no harm to humans. The FDA then defines Phase 1, 2, and 3 studies as follows:

Phase 1 Clinical Studies. Phase 1 includes the initial introduction of an investigational new drug into humans. These studies are closely monitored and may be conducted in patients, but are usually conducted in healthy volunteer subjects. These studies are designed to determine the metabolic and pharmacologic actions of the drug in humans, the side effects associated with increasing doses, and, if possible, to gain early evidence on effectiveness.... The total number of subjects included in Phase 1 studies varies with the drug, but is generally in the range of twenty to eighty. (http://www.fda.gov/cder/handbook/phase1.htm, August 14, 2004)

Phase 2 Clinical Studies. Phase 2 includes the early controlled clinical studies conducted to obtain some preliminary data on the effectiveness of the drug for a particular indication or indications in patients with the disease or condition. This phase of testing also helps determine the common short-term side effects and risks associated with the drug. Phase 2 studies are typically well-controlled, closely monitored, and conducted in a relatively small number of patients, usually involving several hundred people. (http://www.fda.gov/cder/handbook/phase2.htm, August 14, 2004)

Phase 3 Clinical Studies. Phase 3 studies are expanded controlled and uncontrolled trials. They are performed after preliminary evidence suggesting effectiveness of the drug has been obtained in Phase 2, and are intended to gather the additional information about effectiveness and safety that is needed to evaluate the overall benefit-risk relationship of the drug. Phase 3 studies also provide an adequate basis for extrapolating the results to the general population and transmitting that information in the physician labeling. Phase 3 studies usually include several hundred to several thousand people. (http://www.fda.gov/cder/handbook/phase3.htm, August 14, 2004)

I have abstracted the general approach to phases 1, 2, and 3 on the left-hand side of Figure 3.6. My argument is that there are analogues of such studies in education, and that it would be useful to follow through with them. The educational analogues are given on the right-hand side of Figure 3.6.

In what follows I flesh out the details of Figure 3.6.

Preliminary and Phase 1 Studies in Educational R&D

Research in each of Bohr's, Edison's, and Pasteur's quadrants makes significant contributions as preliminary and Phase 1 studies in educational R&D.

Although the emphasis here is on curricular design, it is worth recalling that many curriculum ideas have their origins in Bohr's quadrant (research with a focus on fundamental understanding)—and in addition that such research encompassed a wide range of methods, from experimental studies in the laboratory to classroom ethnographies. Basic research in the 1970s and 1980s resulted in a fundamental reconceptualization of the nature of mathematical thinking, along multiple dimensions. Theoretical considerations in the field at large resulted in an epistemological shift away from an acquisitionist theory of knowing to a constructivist view, laying the foundation for a different conception of student engagement with mathematics. Laboratory studies revealed the mechanisms by which problem-solving strategies could be elaborated and learned, and demonstrated that poor metacognition could hamper problem solving. Classroom observations and laboratory studies demonstrated the negative impact of counterproductive beliefs on mathematical performance, and classroom studies suggested the origins of such beliefs. All of this work informed NCTM's (1989) *Curriculum and Evaluation Standards for School Mathematics,* which served as a catalyst for the National Science Foundation to issue a series of requests for proposals (RFPs) for curriculum development in the early 1990s. Hence basic research

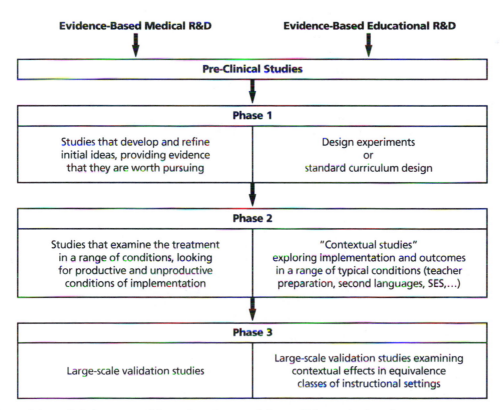

Figure 3.6 Potential parallels between evidence-based research in medicine and education.

played a fundamental role in shaping curriculum development over the final quarter of the 20th century, and in shaping the criteria by which curricula should be evaluated.

Once the goals and theoretical orientation had been established, fundamentally applied work played a major role in curriculum development. The teams that produced curricula in response to the National Science Foundation's curriculum RFPs differed in character, with some being university-based and some housed in organizations known for the production of instructional materials. However, all shared the property that they drew on whatever sources they could—knowledge from research where available, but equally if not more important given the state of the art, prior experience in curriculum development and, more generally, what has been called the wisdom of practice (see, e.g., Shulman, 2004). Given the timetable for curriculum development and refinement—typically, the grant for the development of an *n*-year curriculum had a term of *n* years!—the process of curriculum development and refinement necessarily resided squarely in Edison's quadrant.

Over the same time period, work in Pasteur's quadrant—work that simultaneously addressed theoretical issues and pressing issues of practice in fundamental ways—was beginning to become a productive reality. Early exemplars, before the concept of design experiment had been named, were called *apprenticeship environments* (Collins, Brown, & Newman, 1989). These included my (1985a, 1985b) courses in mathematical problem solving, Palincsar and Brown's (1984) work on reciprocal teaching, and Scardamalia and Bereiter's (1985) work on facilitating the writing process. Related concepts within the mathematics education community have included teaching experiments (Cobb, 2000, 2001; Steffe & Thompson, 2000), classroom-based research (Ball,

2000; Ball & Lampert, 1999; Lampert, 2001), and curriculum research as design (Battista & Clements, 2000; Roschelle & Jackiw, 2000).

The term *design experiment* was introduced into the literature by Ann Brown (1992) and Allan Collins (1992). The authors intended for both of the terms in the name to be taken with great seriousness. Design experiments were exercises in instructional *design*: they involved the creation of (novel) instructional environments (contexts and materials) that played out as real instruction for real students. They were also true *experiments*—not necessarily in the statistical sense, though data were often gathered and analyzed statistically, but in the scientific sense, in that theory-based hypotheses were made, measurement tools were established, a deliberate procedure was rigorously employed, outcomes were measured and analyzed, and theory was revisited in light of the data.

Brown (1992) encapsulated this complexity as shown in Figure 3.7. It is worth noting that the double arrow between "contributions to learning theory" and "engineering a learning environment" represents the dualism inherent in working in Pasteur's quadrant.

The meaning of design experiments has not been settled in the literature. Related but clearly distinct formulations may be found in Collins (1999), the Design-Based Research Collaborative (2003), and Cobb, Confrey, diSessa, Lehrer, and Schauble (2003). Collins provided a contrast between traditional psychological methods and the context and methods of typical design experiments. In the latter case one gives up a significant degree of control and straightforward experimental prediction for a much more complex set of human interactions, and the need to develop rich characterizations of those interactions. The Design-Based Research Collaborative noted the interaction of theoretical and empirical goals, and the cyclic nature of design, enactment, analysis, and redesign;

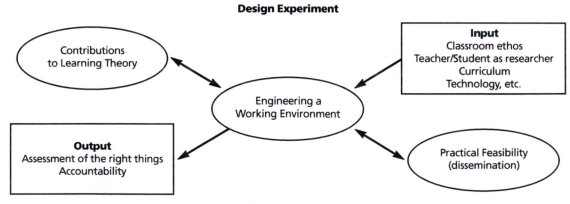

Design Experiment

Figure 3.7 The complex features of design experiments. (From Brown, 1992, p. 142. With permission)

they called for methods to "document and connect processes of enactment to outcomes of interest." Cobb et al. identified a wide range of studies that involve design and fit squarely into Pasteur's quadrant: small-scale versions of a learning ecology that enable detailed studies of how and why they function the way they do (e.g., Steffe & Thompson, 2000); collaborations between a teacher and a research team to construct, conduct, and assess instruction (e.g., Cobb, 2000, 2001; Confrey & Lachance, 2000; Gravemeijer, 1994); preservice teacher development experiments (Simon, 2000); collaborations between researchers and in-service teachers to support the development of a professional community (Lehrer & Schauble, 2000; Silver, 1996; Stein, Silver, & Smith, 1998); and systemic interventions such as in New York City's Community School District #2 (High Performance Learning Communities, 2003). They then delineated the following common features of those efforts:

- Purpose: "to develop a class of theories about both the process of learning and the means that are designed to support that learning."
- "Highly interventionist methodology: Design studies are typically test beds for innovation."
- Theory testing: "Design experiments create the conditions for developing theories yet must place those theories in harm's way."
- "Iterative design."
- "Pragmatic roots: theories developed during the process of experiment are humble not merely in the sense that they are concerned with domain-specific learning experiences, but also because they are accountable to the activity of design." (Cobb et al., 2003, pp. 9–10)

Despite the differences in emphasis among those who theorize about design experiments, there is certainly an agreement that one desirable outcome of design experiments is a "road tested" design for an instructional environment or curriculum that shows promise in terms of student learning and does not have any major downside risks.

However one arrives at them—inspired by basic research (origins in Bohr's quadrant), drafted by skilled designers on the basis of their intuitions or produced by a commercial process (designed or refined in Edison's quadrant), or produced by a process that attends to both theory and materials development (Pasteur's quadrant)—the road-tested design described in the previous paragraph has all of the desired properties that a Phase 1 product should have: promise, feasibility, and no obvious risks. So: What next?

One possibility, consistent with much of the current political zeitgeist, is to move directly to randomized controlled trials. After all, if you have a product that you think "works," why not "prove it," using what has been called the gold standard of experimental methods? In what follows I shall argue that such an approach, although well intended, represents a mistaken parallel to evidence-based medicine. The next step in clinical trials is to move to Phase 2, mid-scale exploratory studies. There is an appropriate educational analogue, which has not been explored by the field at large.

Phase 2 Studies in Educational R&D

I begin with a story that illustrates the main point underlying the approach suggested in this section. I had been following the use of a standards-based curriculum in local classrooms. Implementation was uneven at times, but what I saw in local middle schools, across the boards, confirmed the impression I had from the literature. In comparison with students who experienced the "traditional" curriculum (represented by two state-adopted textbooks in California), students who experienced the standards-based curriculum performed respectably with regard to skills, and they were better off when it came to concept acquisition and problem solving. At that point, I was leaning toward a blanket recommendation that the newer standards-based curricula be adopted. It appeared that the floor for such curricula was higher than the floor for traditional curricula.

Then I visited an off-track algebra class in a local high school. This was a second semester algebra course being taught in the first semester. All of the students in the class were there because they had had some difficulty with algebra—they were either repeating the course or had entered it from a tracked course that had covered one semester of algebra over the course of a year. By and large, the students were disaffected and saw themselves as poor mathematics learners.

In this school as in many urban schools, privilege played a role in course assignments. The "best" teaching assignments, such as calculus, went to those with the most seniority and clout; the most problematic assignments, of which this was one, went to those who had little choice and sometimes little motivation to teach them. The person assigned to teach this course was a computer science teacher, and the students were assigned to his room—which had computers bolted to each desk.

The curriculum calls for a great deal of group work. Roughly speaking, students are expected to work 10 problems a day in class, largely in groups; this work prepares them for the 10 problems they are assigned each day for homework. The in-class work is essential, because the homework from this curriculum looks somewhat different from the work in the traditional curriculum. Parents or other caregivers are less likely than usual to be able to help with homework.

Simply put, the class was dysfunctional. The teacher assigned students to groups but provided neither monitoring nor assistance; some students listened to headphones, some drummed basketballs or otherwise indicated their disaffectedness. With computers bolted to their desks, even those students who wanted to work found themselves in an environment that was crowded, noisy, and inhospitable to such work. (The teacher reacted to an early complaint about the level of chaos by saying that the disruptive students would soon drop out, and that the classroom would then be quieter.) Generally speaking, students were either unwilling to work in class or found it nearly impossible to do so. This caused serious problems for those who tried. Unable to work in class, they needed to do 20 problems (the in-class assignment *and* the homework assignment) each night, often without support, in order to keep up.

The fact is that (short of replacing the teacher with a more attentive and competent one, which was not likely to happen) these students would have been better off taking a course that used the traditional curriculum, for the following reasons. Being given straightforward presentations, even those aimed at procedural learning, would have been better than receiving no instruction at all. Seatwork on standard problems might have been doable for a larger percentage of the students. And many of them could have received homework help from parents or caregivers. In sum, in this somewhat pathological context, using a *Standards*-based curriculum actually lowered the floor. In this kind of context the students would not have been well served, but they would have been better served, by the traditional curriculum.

This kind of story will come as no surprise to those who are familiar with urban classrooms. However, it illustrates an essential point that seems to have been missed by many. A curriculum is not a "thing" that is "given" to students, with consistent effects. A curriculum plays out differently in different contexts. The same instructional materials are unlikely to result in the same patterns of learning in inner-city schools that have broken-down facilities and a low proportion of credentialed staff that they produce in suburban schools that have modern facilities and a well-credentialed staff. Some curricula will work well with, or be easily adaptable to, a student body that has a high proportion of English language learners;

others, because of their high threshold of formal or academic language, may prove difficult for such students. Some may require a significant amount of teacher preparation before implementation can be successful; some may not. If one is in the position of choosing a curriculum, one wants to be aware of these *contextual effects* and choose a curriculum that has the potential to provide the best results given the real or potential affordances of the present context.

This is precisely the kind of information provided by Phase 2 studies in medical research. The pharmaceutical industry has long understood, for example, that drugs may have differential impacts (some positive, some negative) depending on: the health of the patient and what other conditions he or she may have; on what other medicines the patient may be taking; and on whether the medicine is taken on a full or empty stomach, or with particular foods. Thus, for example, Medline Plus, a web server made available by the U.S. National Library of Medicine and the National Institutes of Health at http://www.nlm.nih.gov/medlineplus/druginformation.html, has standard categories of information related to various medical treatments. A typical page describing a drug or treatment regimen includes categories such as the following:

- *Before Using This Medicine*
  This category includes information about allergies, special situations such as pregnancy or breast feeding, appropriateness for particular populations, and interactions with other conditions and/or medicines.
- *Proper Use of This Medicine*
  This category includes information about necessary conditions for proper use. This includes a drug or physical therapy regimen, but also information such as "to be taken on a full stomach" or even as specific as "do not drink grapefruit juice with this medicine."
- *Precautions While Using This Medicine*
  This category includes descriptions of downside risks to taking the medicine, possible complications, and things that should be avoided (typically, taking alcohol or other medicines that might interact with the one being considered).

In pharmacology, the point of Phase 2 studies is to uncover this kind of information, *before* proceeding to large-scale testing via randomized controlled trials. In broad terms, the issues are, What needs to be in place before the proposed approach stands a decent chance of being effective? Which outcomes should be tracked, in what ways? What makes effectiveness unlikely? What kinds of collateral benefits or costs might one encounter, with what frequency, for what populations? Only after such questions have been answered—when one has a decent sense of where and when something works—should one set about the large-scale documentation *that* it works in those circumstances.

In what follows I explore the educational analogue. Lest I be accused of reductiveness at the start, I should point out that public health may be a better analogy than pharmacology for instruction-in-context. The rough parallel is that curricula play a role in the educational system (where there is room for anything from almost total neglect to faithful implementation of what is proposed) that is analogous to the role that medicines play in public health implementation (where there is room for anything from almost total neglect to faithful implementation of what is proposed). What I am proposing here is somewhat speculative and (like all work in Pasteur's quadrant) will require a good deal of adaptation and midcourse correction to be fully effective.

Suppose, then, that one has developed some instruction,[16] possibly as part of a larger learning environment. Assume that the instruction extends over a substantial period of time—perhaps a semester, a year, or some number of years. The path toward the development of the instruction may have been any of those described in the section "Preliminary and Phase 1 Studies"—it may have been inspired by research, may have emerged from a team of skilled designers, or may be the product of one or more design experiments. The instruction has presumably been used with a relatively small number of students and found promising. What next?

It seems clear that the instruction should be implemented and examined in a range of contexts that are representative of the wide range of circumstances in which it might be adopted. (These might include suburban schools, inner city schools, schools with low and high proportions of English language learners, schools with low and high proportions of credentialed mathematics teachers, and so on.) This implementation and analysis should be systematized, perhaps as follows.

Although no two schools are the same, it might be possible to identify a collection of "Phase 2 schools" that, in some way, represent equivalence classes of schools. That is, suppose one selects a collection of schools such that each school has a collection of demographic and perhaps administrative attributes that typify a significant number of schools. It is not unreasonable to expect that the implementation in a Phase 2 school that represents an equivalence class would be at least somewhat similar to implementation in other schools in that equivalence class, with somewhat similar results. Hence implementation over a range of schools selected from different equivalence classes would suggest what might happen if the instruction were implemented on a wider scale. It would also indicate which kinds of contexts are likely to experience beneficial results, and which are not.

In each of the Phase 2 schools, one would investigate questions such as the following:

- Under what conditions can the instruction be made to work (and what does one mean by work)?
- What measures, both of process (e.g., classroom dynamics, and discourse patterns) and product (mathematical or other outcomes) can be used or developed to help understand and calibrate the impact of such instruction?
- How does the instruction get operationalized in different contexts, and with different kinds of support systems? What are the patterns of uptake, of interaction?
- How does implementation in different contexts reflect on the theoretical underpinnings of the design?
- How might the curriculum be refined in light of what is revealed by the attempts to implement it?

In the early stages of Phase 2 work in these varied contexts (Phase 2a), the work would resemble a family of coordinated design experiments. Understanding and describing instructional impact in different contexts would still be a contribution to fundamental research at that point; a wide range of methods would still be appropriate to explore the implementation, and to find out what support structures, in different environments, improved outcomes. This kind of work could provide "user's guides" to the curricula. (If a district has a particular profile, it might expect certain patterns of outcomes. Thus it might want to consider

---

[16] In what follows I use the term *instruction* because it is somewhat generic. I intend a very broad interpretation in what follows. If, for example, the intended instruction requires certain classroom configurations (e.g., the kind of "jigsawing" used by Brown and Campione, 1996, in their FCL work) or certain technology (as in Scardamalia and Bereiter's, 2002, Knowledge Forum work), those are considered to be part of the instruction.

modifying certain administrative features of the environment and providing professional development along certain lines to its staff.)

One would, of course, expect some aspects of this kind of work to be more efficient than in a first design experiment: Some of the key phenomena would have been noted in prior R&D, and some of the student assessment tools would have been developed. Indeed, in the interests of efficiency and with an eye toward later studies, one would want to have or develop similar observational and assessment tools for as many of these contexts as possible.

Once the basic research in the different contexts has been done, efforts turn toward systematization (Phase 2b). Are there features of implementation fidelity that are essential to produce certain kinds of student outcomes? Looking across the various contexts in which the instruction has been implemented, are there features of those contexts that seem consistently to produce certain kinds of effects? Are there certain contextual variables that, having been identified in the more detailed studies, can be captured more efficiently in summary fashion once one knows to look for them? (For example, early studies of accountability—to what and to whom are students accountable, and how does that play out in terms of learning and identity?—can be expected to be highly descriptive and detailed. After some time, however, one can imagine that the character of accountability structures could be codified in straightforward ways.)

A major goal of the work in Phase 2b is to develop the instrumentation that supports the meaningful conduct of Phase 3 studies. This includes the ability to capture central aspects of the context as described in the previous paragraph. It includes broader contextual variables found to have an impact on implementation or outcomes. (These might, for example, include larger aspects of the context, such as whether the state or district has high-stakes accountability tests of particular types. Given that many teachers and schools depart from planned curricula for substantial periods of the school year in order to focus on test preparation, the presence of such tests might have a significant impact and should be included as a variable.) It includes characterizations of implementation fidelity. And it includes a range of outcome measures. Content-related measures would, in the best of circumstances, be independent of the curriculum and standards-based. (Recall that a major

problem with early curricular evaluations such as those found in Senk and Thompson, 2003, is that many of the student assessments were locally developed, and thus not comparable to other such measures.) Other measures might well include characterizations of the learning community, of student beliefs, and of other aspects of the students' mathematical identities. Some comparative studies would be conducted during either Phase 2a or 2b, in order to assess the relative impact of the instruction and gauge the plausibility of the claim that the instruction is indeed worth adopting. To be explicit in summary, much of Phase 2 research as imagined here would reside squarely in Pasteur's quadrant—contributing to basic understandings, but also laying the foundations for the kind of applied work done in Phase 3.

Phase 3 Studies in Educational R&D

The main focus of Phase 2 was to understand how and why instruction "works," to revise and improve that instruction, to identify relevant variables, and to develop relevant instrumentation. With that established, Phase 3 can be both straightforward and informative. One can imagine two kinds of Phase 3 studies: (1) expanded studies replicating the effects discovered in Phase 2, and (2) comparative in-depth evaluations of any number of promising curricula. The two kinds of studies can be done independently, or studies of type 1 (which are in essence comparative studies of two treatments, the instruction and a control) can be done as subsets of studies of type 2.[17]

A key aspect of such studies is the random assignment of students to instructional treatments. That said, the rest of the work should be reasonably straightforward, in conceptual terms.[18] That is, it should be, *if* the proper work has been done in Phase 2, and the right perspective is taken toward the analysis of the data. As noted above, major goals for research in Phase 2 are to identify contextual variables and be able to code for them, and to use appropriately independent and meaningful assessments of student performance. With such tools at one's disposal, the relevant issue becomes one of interpretation.

Recall the second example discussed in Part 1, which illustrated two different ways to think about the results of randomized controlled trials. In this author's opinion, aggregating across all contexts—"on average, students who studied from Curriculum A

---

[17] For a broad discussion of comparative studies, as part of the "engineering approach," see Burkhardt (2006). For an extended treatment of some of the relevant statistical issues, see Scheaffer (2006).

[18] Of course, the devil is in the details. To say statistical analyses are conceptually straightforward is one thing; to carry them out correctly can be something else altogether. Even issues of the appropriate unit of analysis are non-trivial.

outperformed students who studied from Curriculum B"—is of little value. Such an approach provides evaluative information in summary, but it provides little useful information for anyone faced with a choice of instructional materials. (What if, for example, Curriculum A outperformed Curriculum B in general, but Curriculum B outperformed Curriculum A by a significant margin in schools that are similar to the ones for which instructional materials must be chosen? In that case Curriculum B is the right choice, even though Curriculum A is better on average. Or, what if the research reveals that Curriculum B is superior under present circumstances, but that a program of professional development will equip your staff to employ Curriculum A to better effect?) In sum, the most beneficial role of randomized controlled trials in curricular evaluation would be to investigate the hypotheses raised in Phase 2 studies—to determine which contextual variables have which effects on implementation fidelity of a wide range of curricula, and what the effects of those curricula are in those conditions. That kind of information would be truly useful for decision makers.

CODA

I began doing research in mathematics education in the mid-1970s. At the time, the field's primary research methods (in the United States, at least) were statistical, but their use was often unsophisticated, and the field as a whole suffered somewhat from a reductive form of what has been called "science envy." The field of cognitive science had not yet coalesced (the first annual meeting of the Cognitive Science Society was held in 1978), and many of the fields that would ultimately contribute both methods and perspectives to our current understanding of mathematical thinking and learning (among them anthropology, artificial intelligence, linguistics, philosophy, psychology, and sociology) were at best peripheral to the enterprise. When I began my research on problem solving, the cutting edge of research consisted of attempts to make sense of factors that shaped the success or failure of individuals working problems in isolation in the laboratory. Major constructs that have now come to be cornerstones of our understanding (the very notion of cognitive modeling; the roles of metacognition and beliefs; the concepts of identity and of communities of practice) and the methods that help to elaborate them were only beginning to emerge, if they were on the horizon at all.

The progress that has been made since that time is nothing short of phenomenal. In little more than a quarter century there have been major epistemological shifts, accompanied by a flourishing of tools, techniques, and the theoretical perspectives that underlie them. The cognitive sciences and sociocultural research within mathematics education have both matured and become increasingly robust; what at first seemed almost like thesis and antithesis have, over the past decade or so, begun a synthesis that seems increasingly promising in terms of its possibility to help explain issues of (mathematical) thinking, teaching, and learning. The same is the case for the artificial distinction between quantitative and qualitative methods, which becomes less and less important as one begins to ask these central research questions: "What assumptions are being made? What claims are being made? What warrants are being offered for those claims?" This is remarkable progress, and one hopes and expects to see more.

REFERENCES

Artigue, M. (1999). The teaching and learning of mathematics at the university level: Crucial questions for contemporary research in education. *Notices of the American Mathematical Society, 46*, 1377–1385.

Ausubel, D.P. (1960). The use of advance organizers in the learning and retention of meaningful verbal material. *Journal of Educational Psychology*, 51, 267–272.

Ball, D. (1988). *Knowledge and reasoning in mathematical pedagogy: Examining what prospective teachers bring to teacher education*. Unpublished doctoral dissertation, Michigan State University.

Ball, D. (2000). Working on the inside: Using one's own practice as a site for studying teaching and learning. In A. E. Kelley & R. Lesh (Eds.), *Handbook of research design in mathematics and science education* (pp. 365–402). Mahwah, NJ: Erlbaum.

Ball, D., & Lampert, M. (1999). Multiples of evidence, time, and perspective: Revising the study of teaching and learning. In E. Lagemann & L. Shulman (Eds.), *Issues in education research: Problems and possibilities* (pp. 371–398). New York: Jossey-Bass.

Battista, M., & Clements, D. (2000). Mathematics curriculum development as a scientific endeavor. In A. E. Kelley & R. Lesh (Eds.), *Handbook of research design in mathematics and science education* (pp. 737–760). Mahwah, NJ: Erlbaum.

Bauersfeld, H. (1980). Hidden dimensions in the so-called reality of a mathematics classroom. *Educational Studies in Mathematics, 11*(1), 109–136.

Bhattachargee, Y. (2005). Can randomized trials answer the question of what works? *Science, 25*, 1861–1862.

Boaler, J. (2002) *Experiencing school mathematics* (Rev. ed.). Mahwah, NJ: Erlbaum.

Boaler, J. (in press). *Promoting relational equity in mathematics classrooms—Important teaching practices and their impact on student learning*. Text of a "regular lecture" given at the 10th International Congress of Mathematics Education (ICME

X), 2004, Copenhagen. To appear in the *Proceedings of the 10th International Congress of Mathematics Education.*

Brousseau, G. (1997). *Theory of didactical situations in mathematics: Didactique des mathématiques 1970–1990.* (N. Balacheff, M. Cooper, R. Sutherland, & V. Warfield, Eds. &Trans.). Dordrecht, The Netherlands: Kluwer.

Brown, A. (1987). Metacognition, executive control, self-regulation, and other more mysterious mechanisms. In F. Reiner & R. Kluwe (Eds.), *Metacognition, motivation, and understanding* (pp. 65–116). Hillsdale, NJ: Erlbaum.

Brown, A. (1992) Design experiments: Theoretical and methodological challenges in creating complex interventions in classroom settings. *Journal of the Learning Sciences, 2*(2), 141–178.

Brown, A., & Campione, J. (1996) Psychological theory and the design of innovative learning environments: On procedures, principles, and systems. In L. Schauble & R. Glaser (Eds.), *Innovations in learning: New environments for education* (pp. 289–325). Mahwah, NJ: Erlbaum.

Brown, J. S., & Burton, R. R. (1978). Diagnostic models for procedural bugs in basic mathematical skills. *Cognitive Science, 2*(2), 155–192.

Brown, J.S., & VanLehn, K. (1982). Toward a generative theory of bugs. In T. Carpenter, J. Moser, & T. Romberg (Eds.), *Addition and subtraction: A cognitive perspective* (pp. 117–135). Hillsdale, NJ: Erlbaum.

Brownell, W. (1947). An experiment on "borrowing" in third-grade arithmetic. *Journal of Educational Research, 41*(3), 161–171.

Bruning, J.L., & Kintz, B.L. (1987). *Computational handbook of statistics.* (3rd edition). Glenview, IL: Scott, Foresman.

Burkhardt, H. (2006). From design research to large-scale impact: Engineering research in education. In J.Akker, K. Gravemeijer, S. McKenney, & N. Nieveen (Eds.), *Design and development research: Emerging perspectives.* Manuscript in preparation.

Burkhardt, H., & Schoenfeld, A. H. (2003). Improving educational research: Toward a more useful, more influential, and better funded enterprise. *Educational Researcher, 32*(9), 3–14.

Carpenter, T. P., Fennema, E., & Franke, M. L. (1996). Cognitively guided instruction: A knowledge base for reform in primary mathematics instruction. *Elementary School Journal, 97*(1), 1–20.

Carpenter, T. P., Lindquist, M. M., Matthews, W., & Silver, E. A. (1983). Results of the third NAEP mathematics assessment: Secondary school. *Mathematics Teacher, 76*(9), 652–659.

Chapell, M. (2003). Keeping mathematics front and center: Reaction to middle-grades curriculum projects research. In S. Senk & D. Thompson (Eds.), *Standards-based school mathematics curricula: What are they? What do students learn?* (pp. 285–298). Mahwah, NJ: Erlbaum.

Clement, J. (1982). Algebra word problem solutions: Thought processes underlying a common misconception. *Journal for Research in Mathematics Education, 13*(1), 16–30.

Clement, J., Lochhead, J., & Monk, G. S. (1981). Translation difficulties in learning mathematics. *American Mathematical Monthly, 88*(3), 286–290.

Cobb, P. (2000). Conducting teaching experiments in collaboration with teachers. In A. E. Kelly & R. A. Lesh (Eds.), *Handbook of research design in mathematics and science education* (pp. 307–333). Mahwah, NJ: Erlbaum.

Cobb, P. (2001). Supporting the improvement of learning and teaching in social and institutional context. In S. M. Carver & D. Klahr (Eds.), *Cognition and instruction: Twenty-five years of progress* (pp. 455–478). Mahwah, NJ: Erlbaum.

Cobb, P., Confrey, J. diSessa, A., Lehrer, R., & Schauble, L. (2003). Design experiments in educational research. *Educational Researcher, 32*(1), 9–13.

Cobb, P., & Hodge, L. L. (2002). A relational perspective on issues of cultural diversity and equity as they play out in the mathematics classroom. *Mathematical Thinking and Learning, 4*(2&3), 249–284.

Cohen, D. (1990). A revolution in one classroom: The case of Mrs. Oublier. *Education Evaluation and Policy Analysis, 12*(3), 311–329.

Collins, A. (1992). Toward a design science of education. In E. Scanlon & T. O'Shea (Eds.), *New directions in educational technology* (pp. 15–22). Berlin, Germany: Springer.

Collins, A. (1999). The changing infrastructure of educational research. In E. Lagemann & L. Shulman (Eds.), *Issues in education research: Problems and possibilities* (pp. 289–298). New York: Jossey-Bass.

Collins, A., Brown, J.S., & Newman, S. (1989). Cognitive apprenticeship: Teaching the craft of reading, writing, and mathematics. In L. B. Resnick (Ed.), *Knowing, learning, and instruction: Essays in honor of Robert Glaser* (pp. 453–494). Hillsdale, NJ: Erlbaum.

Confrey, J., & Lachance, A. (2000). Transformative reading experiments through conjecture-driven research design. In A. E. Kelly & A. Lesh (Eds.), *Handbook of research design in mathematics and science education* (pp. 231–266). Mahwah, NJ: Erlbaum.

Conrad, C., & Serlin, R. (Eds.). (2005). *The Sage handbook for research in education: Engaging ideas and enriching inquiry.* Thousand Oaks, CA: Sage.

Cooney, T. (1985). A beginning teacher's view of problem solving. *Journal for Research in Mathematics Education, 16,* 324–336.

deCorte, E., Greer, B., & Verschaffel, L. (1996). Mathematics teaching and learning. In D. Berliner & R. Calfee (Eds.), *Handbook of Educational Psychology,* 491–549. New York: MacMillan.

Denzin, N.K., & Lincoln, Y.S. (Eds.). (2005). *The Sage handbook of qualitative research.* Thousand Oaks, CA: Sage.

Design-Based Research Collective. (2003). Design-based research: An emerging paradigm for educational inquiry. *Educational Researcher, 32*(1), 5–8.

diSessa, A. (1983). Phenomenology and the evolution of intuition. In D. Gentner & A. Stevens (Eds.), *Mental models* (pp. 15–34). Hillsdale, NJ: Erlbaum.

Eisenhart, M., Borko, H., Underhill, R., Brown, C., Jones, D. & Agard, P. (1993). Conceptual knowledge falls through the cracks: Complexities of learning to teach mathematics for understanding. *Journal for Research in Mathematics Education, 24,* 8–40.

Engle, R., & Conant, F. (2002) Guiding principles for fostering productive disciplinary engagement: Explaining emerging argument in a community of learners classroom. *Cognition and Instruction, 20*(4), 399–483.

Ericsson, K.A., & Simon, H.A. (1980). Verbal reports as data. *Psychological Review, 87*(3), 215–251.

Ericsson, K. A., & Simon, H. A. (1984). *Protocol analysis: Verbal reports as data.* Cambridge, MA: MIT Press.

Fawcett, H. (1938). *The nature of proof.* (Thirteenth Yearbook of the National Council of Teachers of Mathematics.) New York: Teachers College.

Gardner, H. E. (1985). *The mind's new science : A history of the cognitive revolution.* New York: Basic Books.

Geertz, C. (1975). On the nature of anthropological understanding. *American Scientist, 63*, 47–53.

Gravemeijer, K. (1994). Educational development and developmental research. *Journal for Research in Mathematics Education, 25*, 443–471.

Green, J. L., Camilli, G., & Elmore, P. B. (Eds.). (2006). *Handbook of complementary methods in education research.* Mahwah, NJ: Erlbaum.

Gutstein, E. (2003). Teaching and learning mathematics for social justice in an urban, Latino school. *Journal for Research in Mathematics Education, 34*, 37–73.

Hatano, G., & Inagaki, K. (1991). Sharing cognition through collective comprehension activity. In L. B. Resnick, J. M. Levine, & S. D. Teasley (Eds.), *Perspectives on socially shared cognition* (pp. 331–348). Washington, DC: American Psychological Association.

High performance learning communities project (HPLC). (n.d.). Retrieved August 9, 2003, from http://www.lrdc.pitt.edu/hplc/.

Inagaki, K. (1981). Facilitation of knowledge integration through classroom discussion. *Quarterly Newsletter of the Laboratory of Comparative Human Cognition, 3*(2), 26–28.

Inagaki, K., Hatano, G., & Morita, E. (1998). Construction of mathematical knowledge through whole-class discussion. *Learning and Instruction, 8*, 503–526.

Kerstyn, C. (2001). *Evaluation of the I CAN Learn® mathematics classroom: First year of implementation (2000–2001 school year).* (Available from the Division of Instruction, Hillsborough County Public Schools, Tampa, FL)

Keeves, J. (Ed.). (1997). *Educational research, methodology and measurement: An international handbook.* Amsterdam: Elsevier.

Kelley, A. E., & Lesh, R. A. (2000). *Handbook of research design in mathematics and science education.* Mahwah, NJ: Erlbaum.

Kilpatrick, J. (1978). Variables and methodologies in research on problem solving. In L. Hatfield (Ed.), *Mathematical problem solving* (pp. 7–20). Columbus, OH: ERIC.

Lampert, M. (1990). When the problem is not the question and the solution is not the answer. *American Educational Research Journal, 27*(1), 29–63.

Lampert, M. (2001). *Teaching problems and the problems of teaching.* New Haven, CT: Yale University Press.

LeCompte, M., Millroy, W., & Preissle, J. (Eds.). (1992). *Handbook of qualitative research in education.* New York: Academic Press.

Lee, J. (2002). Racial and ethnic achievement gap trends: Reversing the progress toward equity? *Educational Researcher, 31*(1), 3–12.

Lehrer, R., & Schauble, L. (2000). Modeling in mathematics and science. In R. Glaser (Ed.), *Advances in instructional psychology: Educational design and cognitive science* (pp. 101–159). Mahwah, NJ: Erlbaum.

Leinhardt, G. (1990). A contrast of novice and expert competence in math lessons. In J. Lowyck, & C. Clark (Eds.), *Teacher thinking and professional action* (pp. 75–97). Leuven, Belgium: Leuven University Press.

Leinhardt, G., Putnam, R. T., Stein, M. K., & Baxter, J. (1991). Where subject knowledge matters. In J. Brophy (Ed.), *Advances in research on teaching* (Vol. 2, pp. 87–113). Greenwich CT: JAI Press.

Lemke, M., Sen, A., Pahlke, E., Partelow, L., Miller, D., Williams, T., et al. (2004). *International outcomes of learning in mathematics literacy and problem solving: PISA 2003 results from the U.S. Perspective.* (NCES 2005-003). Washington, DC: U.S. Department of Education, National Center for Education Statistics.

Lester, F. (1994). Musings about mathematical problem-solving research: 1970–1994. *Journal for Research in Mathematics Education, 25*(6), 660–675.

Lucas, J., Branca, N., Goldberg, D., Kantowsky, M., Kellogg, H., & Smith, J. (1980). A process-sequence coding system for behavioral analysis of mathematical problem solving. In G. Goldin & E. McClintock (Eds.), *Task variables in mathematical problem solving* (pp. 345–378). Columbus, OH: ERIC.

Mayer, R. (1985). Implications of cognitive psychology for instruction in mathematical problem solving. In E.A. Silver (Ed.), *Learning and teaching mathematical problem solving: Multiple research perspectives* (pp. 123–138). Hillsdale, NJ: Erlbaum.

Medline Plus. (2005). Hormone replacement therapy. Retrieved May 15, 2005, from http://www.nlm.nih.gov/medlineplus/hormonereplacementtherapy.html

Miller, G. A. (1956). The magical number seven, plus or minus two: Some limits on our capacity for processing information. *The Psychological Review, 63*, 81–97.

Moll, L. C., Amanti, C., Neff, D., & Gonzalez, N. (1992). Funds of knowledge for teaching: Using a qualitative approach to connect homes and classrooms. *Theory Into Practice, 31*(2), 132–141.

Muldoon, M.F., Manuck, S.B., & Matthews, K.A. 1990. Lowering cholesterol concentrations and mortality: A quantitative review of primary prevention trials. *British Medical Journal, 301*, 309–314.

Mullis, I. Martin, M., Beaton, A., Gonzalez E., Kelly, D., & Smith, T. (1998). *Mathematics and science achievement in the final year of secondary school.* Boston: The International Association for the Evaluation of Educational Achievement, at Boston College.

Mullis, I., Martin, M., Gonzalez, E., Gregory, K., Garden, R., O'Connor, K., et al. (2000). *TIMSS 1999: Findings from IEA's repeat of the third international mathematics and science study at the eighth grade. International mathematics report.* Boston: The International Association for the Evaluation of Educational Achievement, at Boston College.

National Academy of Education. (1999). *Recommendations regarding research priorities: An advisory report to the National Educational Research Policy and Priorities Board.* Washington, DC: National Academy of Education.

National Council of Teachers of Mathematics. (1989) *Curriculum and evaluation standards for school mathematics.* Reston, VA: Author.

National Council of Teachers of Mathematics. (2000). *Principles and standards for school mathematics.* Reston, VA: Author.

National Research Council. (1994). *Women and health research: Ethical and legal issues of including women in clinical studies, Vol. 1.* (A.C. Mastroianni, R. Faden, & D. Federman, Eds.). Washington, DC: Institute of Medicine, National Academy Press.

National Research Council. (2001). *Adding it up: Helping children learn mathematics.* Washington DC: National Academy Press.

National Science Foundation. (1997). *Foundations: The challenge and promise of K–8 science education reform, Vol. 1* (NSF 97-76). Washington, DC: Author.

Orne, M.T. (1962). On the social psychology of the psychological experiment: With particular reference to demand characteristics and their implications. *American Psychologist, 17*(11), 776–783.

Palincsar, A. S., & Brown, A. L. (1984). Reciprocal teaching of comprehension-fostering and comprehension-monitoring activities. *Cognition and Instruction, 2,* 117–175.

Pickering, A. (1995). *The mangle of practice: Time, agency, and science.* Chicago: University of Chicago Press.

Popper, K. (1963). *Conjectures and refutations: The growth of scientific knowledge.* London: Routledge, Keagan & Paul.

Putnam, R. (2003). Commentary on four elementary mathematics curricula. In S. Senk & D. Thompson (Eds.), *Standards-based school mathematics curricula: What are they? What do students learn?* (pp. 161–178). Mahwah, NJ: Erlbaum.

Ridgway, J., Crust, R., Burkhardt, H., Wilcox, S., Fisher, L., & Foster, D. (2000). *MARS report on the 2000 tests.* San Jose, CA: Mathematics Assessment Collaborative.

Riley, J. (1990). *Getting the most from your data: A handbook of practical ideas on how to analyse qualitative data.* Bristol, England: Technical and Educational Services Ltd.

Roschelle, J., & Jackiw, N. (2000). Technology design as educational research: Interweaving imagination, inquiry, and impact. In A. E. Kelley & R. Lesh (Eds.), *Handbook of research design in mathematics and science education* (pp. 777–798). Mahwah, NJ: Erlbaum.

Rosebery, A., Warren, B., & Conant, F. (1992). Appropriating scientific discourse: Findings from language minority classrooms. *Journal of the Learning Sciences, 2,* 61–94.

Rosnick, P., & Clement, C. (1980). Learning without understanding: the effect of tutoring strategies on algebra misconceptions. *Journal of Mathematical Behavior, 3*(1), 3–27.

Saxe, G., & Schoenfeld, A. (1998). Annual meeting, 1999. *Educational Researcher, 27*(5), 33.

Scardamalia, M. (2002). Collective cognitive responsibility for the advancement of knowledge. In B. Smith (Ed.), *Liberal education in a knowledge society* (pp. 67–98). Chicago: Open Court.

Scardamalia, M., & Bereiter, C. (1985). Fostering the development of self-regulation in children's knowledge processing. In S. F. Chipman, J. W. Segal, & R. Glaser (Eds.), *Thinking and learning skills: Research and open questions* (pp. 563–577). Hillsdale, NJ: Erlbaum.

Scardamalia, M., & Bereiter, C. (1991). Higher levels of agency for children in knowledge building: A challenge for the design of new knowledge media. *Journal of the Learning Sciences, 1,* 37–68.

Scardamalia, M., Bereiter, C., & Lamon, M. (1994). The CSILE project: Trying to bring the classroom into World 3. In K. McGilley, (Ed.), *Classroom lessons: Integrating cognitive theory and classroom practice* (pp. 201–228). Cambridge, MA: MIT Press.

Scheaffer, Richard L. (Ed.) (2006). Guidelines for reporting and evaluating mathematics education research. *Proceedings of an NSF-Supported workshop series on the uses of statistics in mathematics education research.* Manuscript in preparation.

Schoenfeld, A. H. (1983). Beyond the purely cognitive: Belief systems, social cognitions, and metacognitions as driving focuses in intellectual performance. *Cognitive Science, 7,* 329–363.

Schoenfeld, A. (1985a). *Mathematical problem solving.* New York: Academic Press.

Schoenfeld, A. (1985b). Metacognitive and epistemological issues in mathematical understanding. In E. A. Silver (Ed.), *Teaching and learning mathematical problem solving: Multiple research perspectives* (pp. 361–380). Hillsdale, NJ: Erlbaum.

Schoenfeld, A. H. (1988) When good teaching leads to bad results: The disasters of well taught mathematics classes. *Educational Psychologist, 23*(2), 145–166.

Schoenfeld, A. H. (1992). Learning to think mathematically: Problem solving, metacognition, and sense-making in mathematics. In D. Grouws (Ed.), *Handbook for research on mathematics teaching and learning* (pp. 334–370). New York: MacMillan.

Schoenfeld, A. H. (1998). Toward a theory of teaching-in-context. *Issues in Education, 4*(1), 1–94.

Schoenfeld, A. H. (2002). Research methods in (mathematics) education. In L. English (Ed.), *Handbook of International Research in Mathematics Education* (pp. 435–488). Mahwah, NJ: Erlbaum.

Schoenfeld, A. H. (2004). The math wars. *Educational Policy, 18*(1), 253–286.

Schoenfeld, A. H. (2006a). What doesn't work: The challenge and failure of the What Works Clearinghouse to conduct meaningful reviews of studies of mathematics curricula. *Educational Researcher, 35*(2), 13–21.

Schoenfeld, A. H. (2006b). Reply to comments from the What Works Clearinghouse on "What Doesn't Work." *Educational Researcher, 35*(2), 23.

Senk, S., & Thompson, D. (Eds.). (2003). *Standards-based school mathematics curricula: What are they? What do students learn?* Mahwah, NJ: Erlbaum.

Shirk, G. B. (1972). *An examination of the conceptual frameworks of beginning mathematical teachers.* Unpublished dissertation. Urbana-Champaign: University of Illinois.

Shulman, L. S. (2004). *The wisdom of practice: Essays on teaching, learning, and learning to teach.* San Francisco: Jossey-Bass.

Siegler, R. S. (1987). The perils of averaging data over strategies: An example from children's addition. *Journal of Experimental Psychology: General, 116,* 250–264.

Silver. E. (1996). Moving beyond learning alone and in silence: observations from the QUASAR project concerning communication in mathematics classrooms. In L. Schauble & R. Glaser (Eds.), *Innovations in learning: New environments for education* (pp. 127–159). Mahwah, NJ: Erlbaum.

Simon, M. A. (2000). Research on the development of mathematics teachers: The teacher development experiment. In A. E. Kelly & R. A. Lesh (Eds.), *Handbook of research design in mathematics and science education* (pp. 335–359). Mahwah, NJ: Erlbaum.

Smith, J., diSessa, A. & Roschelle, J. (1993/1994). Misconceptions reconceived: A constructivist analysis of knowledge in transition. *The Journal of the Learning Sciences, 3,* 115–163.

Sowder, L. (1985). Cognitive psychology and mathematical problem solving: A discussion of Mayer's paper. In E.A. Silver (Ed.), *Learning and teaching mathematical problem solving: Multiple research perspectives* (pp. 139–146). Hillsdale, NJ: Erlbaum.

Steffe, L., & Thompson, P. (2000). Teaching experiment methodology: Underlying principles and essential elements. In A. E. Kelley & R. Lesh (Eds.), *Handbook of research design in mathematics and science education* (pp. 267–306). Mahwah, NJ: Erlbaum.

Stein, M. K., Silver, E. A., & Smith, M. S. (1998). Mathematics reform and teacher development: A community of practice perspective. In J. G. Greeno & S. V. Goldman (Eds.), *Thinking practices in mathematics and science learning* (pp. 17–52). Mahwah, NJ: Erlbaum.

Stigler, J., & Hiebert, J. (1999). *The teaching gap.* New York: Free Press.

Stodolsky, S. S. (1985). Telling math: Origins of math aversion and anxiety. *Educational Psychologist 20,* 125–133.

Stokes, D. E. (1997). *Pasteur's quadrant: Basic science and technical innovation.* Washington, DC: Brookings.

Swafford, J. (2003). Reaction to high school curriculum projects' research. In S. Senk & D. Thompson (Eds.), *Standards-based school mathematics curricula: What are they? What do students learn?* (pp. 457–468). Mahwah, NJ: Erlbaum.

Tashakkori, A., & Teddlie, C. (Eds). (2002). *Handbook of mixed methods in social and behavioral research.* Thousand Oaks, CA: Sage.

Toulmin, S. (1958). *The uses of argument.* Cambridge, MA: Cambridge University Press.

U.S. Department of Education (2002). *Strategic plan.* Washington, DC: U.S. Department of Education.

U.S. Food and Drug Administration. (2004) *Definitions of Phase 1, Phase 2, and Phase 3 clinical research.* Retrieved August 14, 2005, from http://www.fda.gov/cder/handbook/phase1.htm, http://www.fda.gov/cder/handbook/phase2.htm, and http://www.fda.gov/cder/handbook/phase3.htm.

VanLehn, K., Brown, J.S., & Greeno, J.G. (1984). Competitive argumentation in computational theories of cognition, In W. Kintsch, J. R. Miller, & P. G. Polson (Eds.), *Methods and tactics in cognitive science* (pp. 235–262). Hillsdale, NJ: Erlbaum.

What Works Clearinghouse. (2004). Detailed study report: Kerstyn, C. (2001). Evaluation of the I CAN LEARN Mathematics Classroom: First year of implementation (2000–2001 school year). Unpublished manuscript. Retrieved February 27, 2005, from http://www.whatworks.ed.gov/.

Whitehurst, G. (2002, April). *The Institute of Education Sciences: New Wine, New Bottles.* Presentation at the annual meeting of the American Educational Research Association, New Orleans, LA. Retrieved from http://www.ed.gov/rschstat/research/pubs/ies.html.

Whitehurst, G. (2003, June 9). *Evidence-based education.* [Powerpoint presentation]. Retrieved November 25, 2005, from http://www.ed.gov.

World Health Organization. (1998). Gender and health: Technical paper 98.16 (Reference WHO/FRH/WHD/98.16) New York: World Health Organization.

Wysowski, D.K., Kennedy, D.L., & Gross, T.P. (1990). Prescribed use of cholesterol-lowering drugs in the United States, 1978 through 1988. *Journal of the American Medical Association 263*(16), 2185–2188

Yackel, E., & Cobb, P. (1996). Sociomathematical norms, argumentation, and autonomy in mathematics. *Journal for Research in Mathematics Education, 27*(4), 458–477.

AUTHOR NOTE

I am most grateful to Jim Greeno, Randi Engle, Tom Carpenter, Natasha Speer, Ellen Lagemann, Andreas Stylianides, Cathy Kessel, Dor Abrahamson, and proximal and distal members of the Functions Group including Markku Hannula, Mara Landers, Mari Levin, Katherine Lewis and Daniel Wolfroot for their penetrating and thought-provoking comments on earlier drafts of this chapter. Frank Lester has been an ideal editor, demonstrating remarkable patience and attention to detail. The final version of this chapter is much improved as a result of their help.

Teachers and Teaching

4

ASSESSING TEACHERS' MATHEMATICAL KNOWLEDGE

What Knowledge Matters and What Evidence Counts?

Heather C. Hill, Laurie Sleep, Jennifer M. Lewis,
and Deborah Loewenberg Ball

UNIVERSITY OF MICHIGAN

For more than two hundred years, teachers—and what they know—have been objects of scrutiny. Teachers have been tested, studied, analyzed, lauded, and criticized. In short, both they and their performance have been assessed to an extent rare in other professions. Both the purposes and methods for assessing teachers have varied. With some assessments, the goal has been to evaluate individuals' qualifications for the work of teaching, while with others, the focus has been to analyze the knowledge teachers use to do that work. Most assessments have been evaluative, either explicitly or implicitly, seeking to appraise the adequacy of individual teachers' knowledge or the quality of their performance. Some assessments have contributed to building evidence about the knowledge needed for teaching, thus helping to establish criteria for professional qualifications and the methods for certifying them. What is assessed differs across approaches, and how it is assessed varies: Some are indistinguishable from a test that could be given to students, while others pose tasks special to the work of teaching. Teachers have been interviewed and observed; they have been queried, given tasks, and asked to construct portfolios representing their work. The variation in purposes, test content, and assessment methods has increased over the past 30 years, as teacher testing has become more commonplace and the number of teacher tests has multiplied.

Developing a more coherent approach to the assessment of teachers' knowledge, and in particular their knowledge of mathematics, is now both possible and necessary. Three important contemporary pressures call out for a system of assessment that is not only rigorous, but professionally relevant and broadly credible. The first is a political environment that demands that students be taught by "highly qualified" teachers. Too many students, especially those in under-resourced schools, currently are assigned to teachers who are not certified to teach mathematics (Darling-Hammond & Sykes, 2003; Ingersoll, 1999). Still, what counts as "qualified" and how this could be measured and certified remains a significant problem. A second pressure is the growing need to establish evidence on the effects of teacher education on teachers' capacity, and of teachers' knowledge and skill on their students' learning. Many question the effectiveness of professional training and argue that it is unnecessary; by this logic, all that is needed to be qualified to teach is a mathematics major and experience. Evaluating this claim requires empirical evidence about teachers' knowledge and skill. This, in turn, necessitates a method for appraising such capacities. Third is the need to distinguish what makes teaching professional—that is, a domain of professional knowledge and skill not possessed by just any educated adult. Is the knowledge mastered by someone who majors in mathematics sufficient content knowledge for teaching? If not, then what exactly distinguishes mathematical knowledge for teaching and how could this distinction be established? Given the confluence of these three forces, the field has an opportunity to advance both the tools to assess teachers' mathematical knowledge and our collective understanding of the notion of mathematical knowledge *for teaching*. A set of agreed-upon reliable and valid methods of assessing teachers' mathematical knowledge would afford the capacity to gather and analyze the sorts of evidence needed to make progress on these issues.

Toward that end, this chapter reviews and seeks to appraise the myriad ways in which U.S. teachers' knowledge of mathematics has been assessed. It asks: What is measured on tests of teachers' mathematical knowledge? What *should* be measured? *How* should it be measured? And how has the evolution of both assessment methods and scholarly thinking about teachers' mathematical knowledge influenced these tests? Answering these questions has taken us into territories that have often remained disconnected. On one hand, the history of teacher testing consists largely of assessments designed to *certify* teachers, that is, to attest to the adequacy of teachers' knowledge to teach mathematics. On the other hand, the history of research on teaching and teacher education reveals a stream of systematic efforts to *investigate* what teachers know, and to associate that knowledge with their professional training and their instructional effectiveness. Although these two lines of work—certification and investigation—have both focused on the assessment of teachers' mathematical knowledge, they have proceeded almost independently of one another. They differ in the knowledge they seek to assess, the methods they use to do so, and the conclusions they aim to make. In this chapter, we assemble these lines of work under a common light in order to consider how teachers' knowledge might be responsibly assessed. Important to note, however, is that the assessment

of teachers is hotly contested terrain. Our goal is to contribute resources that might help move the debate from one of argument and opinion to one of professional responsibility and evidence.

The chapter is organized in four parts. First we review the history of U.S. teacher certification testing, beginning in the nineteenth century, and trace its evolution and re-emergence in the 1980s. In the second section, we examine early *scholarly* research involving the measurement of teachers' mathematical knowledge. Both early certification exams and scholarly studies were constructed in the absence of any elaborated theory about the elements of mathematical knowledge for teaching, a fact that led to important limitations on the interpretations of this early work. We then show, through extended example, that in contrast to the views of teaching mathematics implicit in both early teacher certification tests and early scholarly research, teaching requires knowing more than simply how to solve the problems a student might solve. In the third section, we describe measurement methods in studies that explore the mathematical knowledge as it is *used* in teaching. This section provides a review of the tools and instruments developed to study teachers' mathematical knowledge. In the fourth section, we return once again to teacher certification exams, considering the extent to which modern exams have taken up the ideas and methods that appear in scholars' study of mathematical knowledge for teaching.

Throughout, we observe that the mathematical assessments used in each period are linked both to contemporary methods of assessment and ideas about the mathematical knowledge teachers need for success with students. We do not explore ideas about teachers' mathematical knowledge in their own right, but instead describe how these ideas have been instantiated in and influenced the development of assessments.[1] We also limit our analysis to U.S. teacher assessments, as reviewing international teacher assessments is beyond the scope of a single chapter.

TESTING TEACHERS: HISTORICAL APPROACHES TO ASSESSING TEACHERS' MATHEMATICAL KNOWLEDGE

The history of assessing teachers' mathematical knowledge begins with the history of teacher examinations for certification. For the majority of U.S. history, in fact, the main route to teaching was through

---

[1] For a fuller treatment of mathematical knowledge for teaching, see Ball, Lubienski, and Mewborn (2001).

taking an exam, usually offered at the local or county level, that certified those with passing scores as eligible for classroom work (Angus, 2001; Haney, Madaus, & Kreitzer, 1987). It was only at the turn of the twentieth century that the completion of professional education programs eclipsed certification exams as a pathway into teaching, though by the late twentieth century, certification exams had returned full force. In this section, we examine the earliest exams, describe their decline and subsequent rebirth in the 1980s. In doing so, we focus closely on the content of these assessments in order to examine the mathematical knowledge thought to be required for teaching.

Early teaching exams were notable for what they included—not only the three "R's", but also questions designed to assess the moral fiber, ability to manage a classroom, and, possibly, religious affiliations of those tested (Angus, 2001). Over the course of the nineteenth century, exams lengthened to include a wider array of subject matter and, by the 1850s, some pedagogical questions as well. This period was also marked by a change in exam format, from an oral exam often administered by a local district board to written exams administered by county or state officials.

The mathematics sections of these written assessments reflected closely the curriculum of the day. The state of Michigan, for instance, maintained a vigorous teacher examination system through the early twentieth century, issuing tests in physics, literature, geography, arithmetic, history, civics, botany, physiology, and the "theory and art" of teaching, among other topics. An 1895 exam[2] contained the following arithmetic problems:

1. A pole 63 feet long was broken in two unequal pieces, and 3/5 of the longer piece equaled 3/4 of the shorter. What was the length of each piece? Give a good solution.

2. Two men hire a pasture for $20. The one puts in 9 horses, and the other puts in 48 sheep. If 18 sheep eat as much as three horses, what must each man pay?

3. A boat whose rate of sailing in still water is 14 miles an hour, was accelerated 3 1/2 miles per hour in going down stream, and retarded the same distance per hour in coming up. It was five hours longer in coming up a certain distance than in going down. What was the distance?

4. (a) How do you read any decimal? (b) How do you express decimally any common fraction?

(Michigan Department of Public Instruction, 1896, pp. 297–298)

The state exam also contained the following algebra problems:

1. There is a number consisting of three digits, the first of which is to the second as the second is to the third; the number itself is to the sum of its digits as 124 to 7; and if 594 be added to the number, the digits will be reversed. What is the number?

2. Find the value of x:

$$\frac{\sqrt{3x+1}+3}{\sqrt{3x+1}-3} = \frac{\sqrt{7x+8}+4\frac{1}{5}}{\sqrt{7x+8}-4\frac{1}{5}}$$

(Michigan Department of Public Instruction, 1896, p. 297)

Candidates answering these and other problems were granted a state certificate for teaching any grade. Michigan also maintained, through this period, a separate county examination system in which graded certificates were available to candidates in rural areas. This system reflected the tradition of local control over such exams, but more important, also accommodated the realities of the rural teacher labor market. Candidates in rural areas could choose to sit for an exam leading to a general certificate, which authorized its holder to teach at any grade level in the county. These general certification exams were substantially similar in length and content to the state exam excerpted above. Records show that the vast majority of rural teacher candidates, however, chose instead to sit for exams leading to primary (K–4) certificates. These exams did not have a formal algebra section, and occasionally had less computationally intensive arithmetic problems. County education commissioners and legislators argued that requiring more difficult tests for the rural population would have led to fewer certified teachers, a chronic problem in underpopulated counties. In fact, even the primary-grade exams boasted a 50% failure rate in some counties (Michigan Department of Public Instruction, 1897, p. 152). One observer noted that, "The chief cause for failure is a want of knowledge of the subject; and this is due to poor teaching" (Michigan Department of Public Instruction, 1897, p. 153).

Beyond illustrating how mathematics was imagined to be used in daily life at the turn of the twentieth century, these exams and their attendant materials reveal the views of teacher knowledge held during this period. To start, many believed teachers should know more difficult mathematics than they were expected to teach. On the typical primary certification exam, for instance, candidates were asked to solve classic middle school arithmetic problems—rates, proportions, and percents:

---

[2] Throughout the chapter, the item numbering is our own, and does not necessarily reflect the item numbering on the actual tests.

1. (a) Explain as to a class the difference between the simple and local value of figures.

   (b) Write in figures fourteen million, one thousand and five hundredths; three trillion, two hundred one and one thousand seventy billionths; also write in words 7504306.040521/4.

2. I own a horse and a farm; one-fourth the value of the farm is four times the value of the horse. Both taken together are worth $1,700. Find the value of each. Write out a complete analysis.

3. A man sold a lot for $84 and by so doing gained 1/5 of what it cost. What % would he have gained if he had sold for $100? Analyze.

4. A field of 5 acres in form of a square is to be surrounded by a fence 4 1/2 feet high, to be built of boards 6 inches wide, placed horizontally. The lower board is to be four inches above the ground, and there is to be a space of 5 inches between the boards. What will be the cost of the boards required at $18 per M?

5. A merchant gets 500 barrels of flour insured for 75% of its cost, at 2 1/2%, paying $80.85 premium. For how much per barrel must he sell the flour to make 20% upon cost price?

6. A man devotes 40% of his income for household purposes, 35% for the education of his children, 16% of the remainder to charitable purposes, and saves the remainder, which is $705.60. What is his income?

7. One-half of a stack of hay will keep a cow for 20 weeks, and 3/4 of the stack will keep a horse 120 days. How many weeks will the whole stack of hay keep both the cow and the horse?

(Michigan Department of Public Instruction, 1896, p. 315)

Although it is possible that these topics were taught in the primary grades during this era, this is probably not why this content was included. In their annual reports, state superintendents of the time asserted that the "idea that a teacher should know very much more than he is expected to teach is so generally agreed upon that no discussion here is necessary" (Michigan Department of Public Instruction, 1898, p. 10). Mathematics certification exams throughout this period were also marked by the use of numerically complex examples. Whereas the examiners could have chosen mathematics problems that yielded easily to mental computation or shortcuts, few did. This implies that successful teacher candidates of this time would not only understand the general principles for solving such problems, but also have a solid foundation in facts, procedures, and patience.

According to state documents, the difficulty of these exams was in part a political move, the state bureaucracy's attempt to ratchet up the quality of rural teachers by making the test more difficult and removing the possibility of political favoritism. Defending an increase in the difficulty of test questions, the Superintendent of Public Instruction Henry R. Pattengill wrote in 1896: "The rural schools have received the benefit of a better prepared, more mature, and more broadly educated class of teachers. . . . Sympathy, politics, sect, 'pull', should play no part in the choice of teacher. It is not understood that the teacher need be examined year after year in the same studies, but every teacher should forever be a student" (p. 7). Another state report of the time presaged current political slogans: "Better qualified teachers in the country schools should be our motto" (Michigan Department of Public Instruction, 1897, p. 10).

The exams and their materials reveal other key assumptions about teacher knowledge. In Michigan, nearly all the arithmetic problems were what would today be called "real-world," ones that asked teachers to engage in the practical calculations facing businesses, farms, and banks. Problems of a purely mathematical nature—for instance, those that explored the conceptual underpinnings of arithmetic—were not included. The problems included were based on the mathematics curriculum that teachers were expected to transmit to their all-male students (Michalowicz & Howard, 2003), and suggested that test writers viewed teachers' mathematical knowledge as the knowledge they were responsible for instilling in future business owners, farmers, and bankers.

And, despite the applied focus of the exams, some Michigan state superintendents advocated what might today be called conceptual knowledge. In 1896, for instance, graders were cautioned that "in arithmetic, a knowledge of principles and general accuracy in method shall be considered not less than three times as important as obtaining a correct answer" (Michigan Department of Public Instruction, 1896, p. 313). At an 1896 conference on teacher testing, one county commissioner stated, "Our examinations should be of such a character as to demand less of the memory and more of the reason; less rote work, more sequence, more analysis, and more simplicity" (Michigan Department of Public Instruction, 1897, p. 409). Without evidence from the actual grading of exams, it is difficult to determine the extent to which this sentiment was implemented. However, the existence of these admonitions foreshadows the debates in the twentieth century over procedural fluency and conceptual understanding.

Finally, these tests also illuminate nascent political struggles about what should be known in order to teach. These struggles continue into the present; in fact, the words Pattengill wrote in 1894 could well have been written today:

The question of teachers' examinations has always been a perplexing one. No matter how excellent the questions for such examinations may be, nor how thoroughly and honestly conducted, it will still be true that examinations alone are an inadequate test of a candidate's ability to teach school. The qualifications requisite for good school teaching cannot be ascertained by any set of questions whether oral or written; and yet, where the teachers must be employed for a large extent of territory and for many schools, no other plan has been suggested better than the plan of examinations combined by the supervision exercised by the country school commissioner.

The first essential element in a teacher is good scholarship. No amount of tact, or method, or skill in the use of devices will make up for the deficiency in scholarship. In thus emphasizing the value of this factor, we by no means overlook the value of tact and good method. If, however, we are to do without one of these, we think we could more safely risk the teacher with scholarship than the one with method and poor scholarship. (Michigan Department of Public Instruction, 1894, pp. 1–2)

In Pattengill's view, and in the exams of the early 1890s, content knowledge was separate from knowledge of the "theory and art" of teaching. Which is more important to student learning remains both an unanswered question and a topic of hot debate.

In contrast to the business-focused mathematics on the exams of the 1890s, the period after Pattengill's tenure saw an increase in teaching-specific mathematics questions on these exams. For instance, while half the 1900 state exam contained problems similar to those found in previous years, the arithmetic section also carried the following:

1. In what arithmetical processes is drill the essential element? In what operation is memory a factor?

2. When may the algebraic equation be used to advantage in arithmetic?
   Is it advisable to teach its transformations in eighth grade arithmetic?

3. How do you present the table of Surveyors' Long Measure so that link and chain stand for ideas in the mind of a child? Write the table.

4. Outline a first lesson on the subject of ratio.

(Michigan Department of Public Instruction, 1901, p. 70)

Without information on the grading of specific answers, it is difficult to tell exactly what types of knowledge test developers desired to assess; however, it seems likely that these state assessments were shaped by Michigan's growing class of professional teacher educators, some of whom perhaps called attention to the mathematical work teachers needed to do once inside classrooms. In fact, in 1900 the State Board of Education passed control of the state-level certification exams to the president of the normal school system (Michigan Department of Public Instruction, 1901, p. 69).

These results refer to the exams in one state; without similar and detailed analyses of teacher exams in other state archives, it would be difficult to tell whether Michigan's tests were typical. However, we think it likely that they were, and that the major issues regarding test content were common to other states, as well. It is especially striking that these major issues remain unresolved today: how difficult the exams should be—and by extension, how many prospective teachers should be excluded on the basis of lack of knowledge; whether it is content knowledge or knowledge of methods that make a good teacher; whether conceptual or procedural knowledge should be assessed. The passage of 125 years has done little to bring closure to these important questions.

These exams are also revealing in how they *measured* teachers' knowledge. Nineteenth-century teacher exams were designed to draw inferences about *individuals* and their capacity to teach children. Yet it would be nearly a half century before modern psychometrics, with its concerns for reliability and validity, began to influence the development of teacher certification exams. Meanwhile these assessments were used to make high-stakes decisions about individuals' careers without actual evidence of their effectiveness in this regard.

In the interim, however, the use of teacher exams shrank rapidly. As reported in Angus (2001), the late nineteenth century saw a growing class of professional teacher educators begin to criticize these exams, primarily on the grounds that they were too easy, thus allowing incompetent teachers to serve. Histories of teacher certification also note that local control of these exams encouraged favoritism and left open the possibility that examiners would know less than the teachers they tested. In response, teacher educators successfully convinced state bureaucracies and legislatures to approve what was then an alternative form of certification: completion of professional training, usually at a normal school or teachers' college. By 1937, 28 states had abolished teacher testing (Wilson & Youngs, 2005), requiring instead the completion of a professional preparation program. These programs, according to Wilson and Youngs, came to include courses in educational history, psychology, educational foundations, teaching methods, and assessment. Although teacher educators would soon face criticism from other academic disciplines, the first half of the twentieth century saw an increasing claim to professional knowledge in education, and the assent of state legislatures to allow teacher preparation to occur in professional schools.

The early history of teacher testing, in fact, might help explain the aversion of many teacher educators to it. Angus (2001) describes this aversion as a byproduct of professionalization: teacher educators sought to legitimize their product and accumulate resources by claiming a codified body of knowledge and basing entry to their profession on the completion of coursework designed to cover that body of knowledge. Teacher tests like Michigan's, by virtue of being "too easy" and a "back door" into teaching, threatened teacher preparation programs. These tests were also largely controlled not by educational professionals but by either bureaucrats or elected officials. These twin problems—the perceived ease of teacher tests and the lack of control over testing by educational professionals—would afflict the profession in even greater ways later in the century, after teacher certification exams regained political favor.

Enabled by the development of standardized tests and test firms, and in response to teacher surpluses and urban school officials' concerns about how to hire the "best" candidates, certification exams reemerged in the second half of the 1900s. The National Teacher Examination (NTE), established in 1940, eventually became the most widely administered—and most studied—teacher examination in the U.S. The NTE saw three major growth periods. The first two were fueled less by concerns about underqualified teachers than by Southern white attempts to block efforts to reduce racial discrimination. As Baker (2001) describes, school districts in the South typically paid comparably educated African American teachers less than whites through the 1940s. Challenged by NAACP litigation, Southern districts eliminated this practice from official policy. To continue a de facto policy of discriminatory pay, many turned to using the early NTE, on which African Americans tended to score lower than whites. Ben Wood, the major author of the NTE, aggressively promoted the test for this purpose to Southern school officials, touring the South to spread the word. As Baker recounts, "When school leaders asked about how white and black teachers might score on the NTE, [Wood] informed them that on previous administrations of the test the average score of blacks was 'at the lower fifth percentile' of whites" (p. 326). During the second period of NTE expansion in the 1960s, Arthur L. Benson, director of teacher testing at the Educational Testing Service (ETS), which assumed control of the NTE during the 1940s, repeated Wood's strategy. This time, the issue was the desegregation of

schools; Benson sold the NTE to Southern states by again pointing out that African American and white teachers performed differently on the exam. By 1968, nine Southern states required the exam of at least some teacher candidates. Criticism of this practice led ETS in 1971 to issue guidelines on the proper use of the NTE and to attempt to eliminate test bias, but African Americans continued to score lower—and thus be more likely to fail the certification tests—through the 1980s.

The third wave of state adoptions of the NTE occurred in the 1980s in response to *A Nation at Risk* (National Commission on Excellence in Education, 1983). This wave of adoptions was, by all reports, driven by public suspicion about teacher quality. The most popular among these tests were basic skills assessments (Haney et al., 1987). As Wilson and Youngs (2005) report, 37 states had adopted basic skills testing as a requirement for certification by 2002. Over the last two decades of the twentieth century, the percentage of teachers taking one of these assessments prior to entry into either preservice education programs or teaching increased dramatically. As in earlier periods, professional educators and professional education associations (e.g., the National Educational Association) remained opposed to the tests, and perhaps with good reason, as pressure mounted from non-education quarters, mainly economists, political scientists, and policy-makers who questioned the value of professional education in teaching (Hess, 2002; Walsh, 2001). The solution many non-educators propose is to return to a system in which passing an exam certifies a teacher to work in classrooms, regardless of the candidate's level of formal professional training or experience. Such proposals have met with strenuous objection from some involved in teacher education (see, for example, Berliner, 2005; Darling-Hammond & Youngs, 2002).

In this debate, much hinges on the quality of the test itself—what it assesses, and whether what it assesses is reasonably related to successful classroom teaching performance. The NTE tested mathematical knowledge among elementary and secondary teachers. All teachers took mathematics "items," as mathematics problems are referred to in modern assessment terms, as part of the "Core Battery" test, which assessed communication skills, general knowledge (including mathematics), and professional knowledge. Elementary teachers also took mathematics items as part of the "Elementary School Specialty Area Test," an examination that measured subject area knowledge and pedagogical knowledge across the elementary school curriculum.

Secondary teachers intending to be certified to teach mathematics took a stand-alone mathematics test.

Several items from an elementary-level core battery test illustrate the nature of this assessment:

1. An elevator operator is not allowed to carry more than 10 passengers in the elevator at one time. If there are 35 people waiting on the ground floor, what is the minimum number of trips up that the elevator must make in order to transport these 35 people?

   a) 3   b) $3\frac{1}{2}$ c) 4   d) 5   e) $5\frac{1}{2}$

2. Statement: "The product of any two numbers is always greater than or equal to either of those numbers."

   Which of the following examples proves the statement above FALSE?

   a) $1 \times 1 = 1$ b) $3 \times 4 = 12$ c) $5 \times 1 = 5$ d) $\frac{5}{2} \times 4 = 10$ e) $\frac{1}{2} \times 4 = 2$

3. Which of these is NOT a correct way to find 75% of 40?

   a) 75.0×40 b) $(75 \times 40) \div 100$ c) $\frac{75}{100} \times 40$ d) $\frac{3}{4} \times 40$ e) 0.75×40

4. A pattern requires $1\frac{7}{8}$ yards of a certain fabric. If five remnants of the fabric are on sale and their lengths are as follows, which of these lengths will provide enough fabric and result in the least waste?

   a) $1\frac{1}{2}$ yd   b) $1\frac{3}{4}$ yd   c) $1\frac{15}{16}$ yd   d) 2 yd   e) $2\frac{1}{16}$ yd

5. Written as a percent, 2 =

   a) 0.02%   b) 0.2%   c) 2%   d) 20%   e) 200%

(ETS, 1984, pp. 68–73)

Two observations about this assessment stand out. In contrast to the early teacher tests we uncovered, the NTE Core Battery contains a mixture of basics (e.g., item #5 above), disciplinary knowledge (e.g., item #2 above, where candidates must recognize a counterexample), and word problems (e.g., #4 above) that reflect, like the early tests, the ways mathematics was imagined to be used in contemporary society. This new mix of items reflected, perhaps, the changing nature of mathematical knowledge in society.

Second, the items are arguably easier than those on both nineteenth century tests and contemporary tests. Unlike the historical tests, for instance, few items extend beyond the formal elementary school curriculum. And few problems are, like those on the nineteenth century assessments, numerically complex; in many cases, teacher candidates likely did not have to put pencil to paper to correctly answer the item. Further, quite a number of the items are extremely basic. Item #2, for instance, might assess mathematical reasoning and proof from one per-

spective, but from another, it simply asks candidates to recognize a case that violates the statement made in the problem situation. One reason for the easier assessments might be the methods adopted for testing teacher candidates, described below. Another might be that these were conceived in many quarters as basic skills tests. Nevertheless, the items contained on this assessment constitute a very different theory about the substance of teacher knowledge than earlier teacher certification exams, and from the certification exams to come.

Interestingly, some items from the NTE Elementary School Specialty Area test in the 1980s seemed to tap into what today we might call "mathematical knowledge for teaching." In fact, taken as a whole, these elementary exams presage what Shulman would soon name as "pedagogical content knowledge" (Shulman, 1986): the items convey a broad sense that teachers need a specialized knowledge, something beyond what other educated adults know. The following is an example of one such item (ETS, 1987, p. 19):

$$
\begin{array}{ccccc}
521 & 348 & 863 & 508 & 108 \\
-386 & -187 & -37 & -43 & -26 \\
\hline
245 & 261 & 836 & 545 & 182 \\
\end{array}
$$

A pupil's work on five subtraction problems is shown above. Which of the following is the most appropriate diagnosis and suggestion about the pupil's work in arithmetic?

a) The pupil has not had enough practice in doing this type of problem. The teacher should assign the pupil at least ten such problems each day for a few weeks.

b) The pupil shows no recognizable error pattern in the problems shown; the poor performance is probably due to lack of interest. The teacher should seek some means of motivating the pupil to give more attention to class work.

c) The pupil does not understand place value as it relates to subtraction problems. The teacher should have the pupil use concrete materials to explore this concept.

d) The pupil has no understanding of the subtraction concept. The teacher should have the pupil review basic subtraction facts with the help of concrete materials.

e) The pupil probably understands the subtraction concept but has not memorized the basic subtraction facts. The teacher should use drill activities and games to help the pupil memorize those facts.

Answering correctly by choosing option (c) requires knowledge of subtraction and the standard procedure for subtracting multi-digit numbers. It also emulates the reasoning that teachers do *in practice*—sizing up student work and deciding what to do next.

Another item focuses on a common mistake that students make when learning about the multiplication of rational numbers:

Ralph has worked with decimal representations of numbers in class. In response to a decimal exercise posed by his teacher, Ralph wrote .2 × .4 = .8 on his paper. Ralph's work is best evaluated as

a) correct, and this fact can be verified by multiplication of fractions

b) correct, but this fact cannot be verified by multiplication of fractions

c) incorrect, and multiplication of fractions can be used to show why it is incorrect

d) incorrect, and decimal addition can be used to derive the correct answer

e) incorrect, but neither multiplication of fractions nor decimal addition can be used to demonstrate why it is incorrect

It takes only basic mathematical knowledge to know that answer choices (a) and (b) are incorrect, because .2 × .4 is .08. It would be possible to distinguish between answer choices (c), (d), and (e) using purely mathematical knowledge as well, but most people other than teachers are unlikely to have done such work or have occasion to do so. One would need to know that the model of multiplication as repeated addition (2 × 4 is the same as 4 + 4) does not apply here: .2 × .4 is not the same as .4 + .4. That in fact would yield .8, which is incorrect. This knowledge is not the sole province of teachers, but teachers need to know the meanings of operations for teaching in a way that the common educated person does not. The answer (c) is correct, because one could show that

$$\frac{2}{10} \times \frac{4}{10} = \frac{8}{100}$$

by multiplying across numerators and denominators. Again, this is not knowledge unavailable to adults other than teachers, but teachers would have the most facility working with such ideas about meanings of operations and how they can be modeled.

The next item is similarly constructed to gauge mathematical knowledge that is used in teaching multiplication (ETS, 1987, p. 23):

In developing concepts of multiplication, the following types of exercises should be included:

(1) 2 × 23   (3) 4 × 13

(2) 3 × 20   (4) 5 × 35

Which of the following is the most appropriate sequencing of these exercises?

a) 1, 2, 4, 3
b) 1, 3, 4, 2
c) 2, 1, 3, 4
d) 2, 3, 4, 1
e) 3, 4, 1, 2

This item draws upon pure mathematical knowledge (multiplication of a one-digit by a two-digit number), but having such knowledge is not sufficient to identify the correct answer. This item also demands knowledge of how this topic is learned by children—the logical sequencing, the typical errors children make and the obstacles likely to be encountered by a child who is new to learning this skill. Thus, it encompasses basic mathematical competence but goes far beyond into mathematical territory that teachers traverse in their work with children.

However, test items that tap mathematical knowledge for teaching can be difficult to construct. As later generations of item writers discovered, particularly as scholars moved into measuring knowledge in this domain, items tend to suffer from three types of problems. One is the construction of an unambiguously correct answer. Teaching is heavily contextualized—teachers work with particular materials, and with particular children. It is also highly variable: textbooks have different approaches to teaching specific topics, and different developments of those topics over time. And students have unique learning and instructional histories. This contextualization and variability work against writing items in this domain. For instance, in the item above, arguments can be made for a sequence that starts with 3 × 20, as it yields easily to mental mathematics and modeling via a number line. Also, a teacher who plans to develop related number facts (3 × 2, 3 × 10, 3 × 20) to logically reason about the problem might choose one of the options that begins with 3 × 20 (b or c). However, a teacher who plans to begin with a partial product method might choose 23 × 2 over 20 × 3:

$$
\begin{array}{r}
20 \\
\times 3 \\
\hline
0 \\
60 \\
\hline
60
\end{array}
\qquad
\begin{array}{r}
23 \\
\times 2 \\
\hline
6 \\
40 \\
\hline
46
\end{array}
$$

As one can see, 20 × 3 yields a 0 in the first partial product, which might be confusing to children just learning this method. Items without clear correct answers were not uncommon occurrences in the tests we examined.

The following item illustrates a second problem associated with these tests—items that appear to be contextualized in teaching but that, in fact, do not require special professional knowledge (ETS, 1987, p. 28):

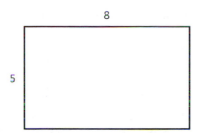

To find the perimeter of the rectangle above, a child can add 5 and 8 and multiply by 2 [2(5 + 8)] or multiply each number by 2 and then add [(2 × 5) + (2 × 8)]. This example illustrates an application of which of the following properties?

a) Distributive property of multiplication over addition
b) Commutative property of addition
c) Commutative property of multiplication
d) Associative property of addition
e) Associative property of multiplication

Although this is important knowledge to have, and a kind of mathematical problem teachers are likely to come across in their work with children, this item could be solved simply by knowing the mathematics (properties of addition and multiplication). The school learning context makes no further knowledge demands for determining the correct answer to the item; it is, for testing purposes, window dressing. Note this contrast with the previous two items, in which specialized knowledge of children and learning in the domain of mathematics was required to determine the correct answer.

Third, some attempts to measure the mathematical knowledge used in teaching produced items that that depended more on ideology than professional skill. For instance (ETS, 1987, p. 15):

In teaching a third-grade class regrouping with two-digit numbers for the first time, which of the following would provide the best means for a teacher to demonstrate the concept?

a) Manipulatives

b) Illustrations drawn on the chalkboard

c) A chart

d) Examples from an approrpiate mathematics textbook

e) Examples from a local newspaper

The test materials identify (a) as the correct answer. Yet we might argue that several of these possibilities are plausible, and that it is not clear on what evidence any one answer is best. For example, a diagram (b) that showed the regrouping of tens and ones and linked it to the symbolic form might help pupils to link the representation to the written procedure more clearly than with manipulatives. It is unlikely that there is any one best approach in this example, and

as we pointed out above, good teaching would likely depend on the particulars of the situation. Here ideology about teaching—that concrete materials are always best—seems to override judgment.

These items demonstrate how an earlier version of teacher tests had an underlying sensibility about specialized knowledge for teaching, one that predates the formalized naming of "pedagogical content knowledge." Second, the items show the difficulty of constructing measures that tap into this knowledge domain with precision. We return to both issues below.

In both NTE tests, mathematical knowledge at the elementary level is subsumed under "general knowledge"—it is one of several areas (literature, fine arts, social science) in which the successful teacher candidate was expected to be conversant. Scores were not returned for subjects separately. In this view of teacher knowledge, general proficiency across a wide range of topics is needed for successful teaching.

In addition, the methods for measuring teacher knowledge underwent a significant shift in the years since our nineteenth-century examples. With the development of more sophisticated methods for scaling items and scoring responses, test developers could better assure candidates and education officials of at least the reliability of the measures. These new methods—and related technology, such as computer-scannable response sheets—changed the face of the tests. Multiple choice items came into vogue. More items were put on the assessments to increase the accuracy of individual-level scores. And, perhaps to increase the precision of measurement among less knowledgeable teachers, easier items were added.

One area of trouble for the NTE, however, was its predictive validity—that is, whether teachers' performance on the assessment correlated with the quality of their classroom instruction or their students' learning. As Wilson and Youngs (2005) describe, six studies addressed the relationship between NTE score and student learning; all but one failed to find a relationship. Official documents make clear that ETS never claimed predictive validity, however, instead proffering their test construction process as proof of validity. The tests were written at the request of educators, to specifications drafted and reviewed by educators, and based on an analysis identifying what teacher candidates should know in a particular area, the tests must represent the knowledge needed to teach well.

The claims to validity made by ETS and other large teacher testing firms lead back to the issue of control over these professional tests. One hallmark of a profession, in fact, is control of barriers to entry by members of the profession itself (Abbott, 1988).

Yet through the 1980s, and even today, there has been only a weak claim to professional control over teacher assessments (Haney et al., 1987). Even after control of teacher testing shifted from local and state bureaucracies to large testing firms, the representation of professionals—in this case, mathematics educators, teachers, and mathematicians—on key decision-making panels remained more nominal than real. Certainly testing firms have included teachers, school officials, disciplinary experts, and others in the test construction process. On the whole, however, the leaders in the field have not tended to play a role or to be consulted in this work. Instead, control over the tests remains to a large extent in the hands of psychometricians, other testing experts, and officials from the state departments of education that mandate the assessments.

Throughout U.S. history, then, teacher tests have been largely controlled by those outside the education profession: local officials, state legislators and bureaucrats, and testing firms. By many accounts (Berliner, 2005; Haney et al., 1987; Wilson & Youngs, 2005), these tests have tended to capture basic skills rather than more complex and job-specific competencies associated with helping students learn.

At the same time, rather than contributing expertise to the process, many educators condemn any form of standardized testing. In a 2002 issue of *English Education*, nearly all articles railed against the very idea of standardized testing for teachers:

> Virtually all of the criticisms leveled against testing in schools also apply to the quick and dirty attempt to demand accountability in testing teachers. Timed tests given to children are really evaluating speed rather than thoughtfulness, and the same is true when they're given to adults. Multiple choice tests and contrived open response items are not meaningful ways of assessing how much students understand, and neither are they particularly effective in telling us how well educators can educate. (Kohn, quoted in Appleman & Thompson, 2002, p. 96)

Others take the view that educators need to assume control over licensing exams, moving toward new forms of assessment and focusing on professionally relevant knowledge and skill. We review some of these new assessments in the last section of this chapter. First, however, we review *scholarly* approaches to *studying* teachers' mathematical knowledge—studies that measure teacher knowledge not for high-stakes decisions, but instead for the purpose of uncovering what teachers do know, understanding the relationship between teacher knowledge and student learning, and,

eventually, for understanding the nature of knowledge and knowing in teaching. Insights from this body of work have opened new possibilities for the assessment of teachers' mathematical knowledge.

EARLY SCHOLARLY RESEARCH INVOLVING THE MEASUREMENT OF TEACHERS' MATHEMATICAL KNOWLEDGE

Studies measuring teachers' knowledge date to the 1960s, originating in what is now known as the "educational production function" literature. The main goal of this research program was to predict student achievement on standardized tests as a function of the resources held by students, teachers, schools, and others. Key resources were seen to include students' family background and socioeconomic status, district financial commitments to teacher salaries, teacher–pupil ratios, other material resources, and teacher and classroom characteristics (Hanushek, 1981; Greenwald, Hedges, & Laine, 1996). Published mainly in economic journals, these studies have been influential in shaping American public opinion and policy on how best to foster school improvement. In particular, educational economists' studies of teachers and their knowledge and skills have been especially influential.

One approach taken in this literature involves using data on teachers' preparation, coursework, and experience to predict student achievement. Key measures included overall teacher education level, certification status, number of post-secondary subject matter courses taken, number of teaching methods courses taken, and years of experience in classrooms. By using such measures, researchers assumed a connection between formal schooling and employment experiences and the more proximate aspects of teachers' knowledge and performance that produce student outcomes. This makes sense, given teacher educators' assertion that formal preparation provides the knowledge needed for teaching. However, reviews of educational production function studies have disputed the extent to which variables like teacher preparation and experience in fact contribute to student achievement (Begle, 1972, 1979; Greenwald et al., 1996; Hanushek, 1981, 1996), with conflicting interpretations resting on the samples of studies and methods used for conducting meta-analyses. Wayne and Youngs (2003) argue that when studies use subject-matter specific markers of teacher preparation—for instance, certification in mathematics rather than general certification—results are a bit more positive. This increasing subject-matter specificity is one characteristic of studies conducted in the educational production function tradition.

A second approach measures teacher knowledge by looking at teachers' performance on certification exams or other tests of subject-matter competence. The first study to do so was the Coleman report, *Equality of Educational Opportunity*, completed in 1966. Coleman and colleagues measured teacher knowledge via a multiple-choice questionnaire, then used teachers' scores to predict student achievement in both reading and mathematics. They found that teacher scores did indeed predict student achievement in both subjects, and that this relationship grows stronger in the higher grades. Notably, however, none of the items on their measure focused specifically on mathematics; instead, the questionnaire asks teachers to complete a "short test of verbal facility" with items such as:

Dick apparently had little _____ in his own ideas, for he desperately feared being laughed at.

   a) interest

   b) depth

   c) confidence

   d) difficulty

   e) continuity

Thus the first study to identify a positive relationship between teacher knowledge and student mathematics achievement did not actually measure teachers' mathematical knowledge at all. Instead, it measured teacher knowledge via a vocabulary test—close, some would argue, to a test of general intelligence. Similar studies have taken advantage of available data from the NTE to create composite measures of teacher knowledge to compare to student achievement scores (Strauss & Sawyer, 1986; Summers & Wolfe, 1977). And more recently, Ferguson (1991) found that the mix of reading skills and professional knowledge captured on the Texas Examination of Current Administrators and Teachers, an exam used in the 1980s, was positively related to student achievement.

By the 1990s, some studies began to focus on how teachers' *mathematical* knowledge related to student gains in mathematics achievement (Harbison & Hanushek, 1992; Mullens, Murnane, & Willett, 1996; Rowan, Chiang, & Miller, 1997). This move toward subject-matter-specific measures coincided with increasing indications that teachers' effectiveness was related to their knowledge of subject matter rather than their general or pedagogical knowledge. Rather than use proxy measures such as degrees or mathematics coursework, however,

the strategy was to administer (or obtain teachers' scores on) problems they might assign students. Harbison and Hanushek (1992), for instance, administered a fourth grade student assessment to both teachers and students, using scores from the first group to predict performance among the second. Mullens et al. (1996) used teachers' scores recorded on the Belize National Selection Exam, a primary-school leaving exam administered to all students seeking access to secondary school. In both cases, the authors provided little information about the actual content of the assessment; Harbison and Hanushek (1992) listed the mathematical topics tested (number recognition, measurement, multiplication and division, rational numbers, unit measures, four operations, story problems) but did not include any actual items. Because these articles described assessments given to students, however, it seems probable that both studies measured teachers' competency with basic computation and mathematical procedures, rather than knowledge of teaching those topics.

Rowan et al. (1997) employed a somewhat different approach, one that began an increased effort to capture the mathematical knowledge used in classrooms, a trend we will take up in more detail in the next section of the paper. In this study, the authors used a one-item assessment to explore how teacher knowledge related to high-school student gains in the National Education Longitudinal Study. The item, developed by the Teacher Education and Learning to Teach study (Kennedy, Ball, & McDiarmid, 1993), reads:

Your students have been learning how to write math statements expressing proportions. Last night you assigned the following:

A one-pound bag contains 50 percent more tan M&Ms than green ones. Write a mathematical statement that represents the relationship between the tan(t) and green(g) M&Ms, using t and g to stand for the number of tan and green M&Ms.

Here are some responses you get from students:

Kelly: $1.5t = g$
Lee: $.50t = g$
Pat: $.5g = t$
Sandy: $g = 1/2g = t$

Which of the students has represented the relationship best? (Mark ONE)

  ❑ All of them
  ❑ Kelly
  ❑ Lee
  ❑ Pat
  ❑ Sandy
  ❑ None of them. It should be _____
  ❑ Don't know.

When stripped of the teaching context, this might easily be a problem that would appear on an end-of-chapter test or a student exam. The problem is different in that it is situated in a common task of teaching. However, like items on the NTE test, the teaching situation can be considered window-dressing.

Studies that directly measure teachers' verbal or mathematical knowledge typically show a positive relationship to student mathematics achievement (e.g., Boardman, Davis, & Sanday, 1977; Ferguson, 1991; Hanushek, 1972; Harbison & Hanushek, 1992; Mullens et al., 1996; Rowan et al., 1997; Strauss & Sawyer, 1986; Tatto, Nielsen, Cummings, Kularatna, & Dharmadasa, 1993; for an exception, see Summers & Wolfe, 1977; for reviews, see Greenwald et al., 1996; Hanushek, 1996; Wayne & Youngs, 2003). By using such measures, these studies assume a relationship between teacher content knowledge as measured by such assessments and the kinds of teaching performances that produce improved student achievement. Yet there is little evidence in this literature for what is actually measured by these assessments, and how what is measured relates to classroom performance. These tests were not developed, for the most part, by specifying domains of teacher knowledge and developing an assessment from this map. Instead, the claim to validity is that these assessments covered material in the K–12 curriculum, and which should be known by the students themselves at the end of schooling. This is true—but may miss important elements of the knowledge that makes teachers successful. And using these theoretically impoverished tools also limits the conclusions that could be drawn from these studies. Is it that teachers who perform better on these student-level tests simply make fewer mistakes in classroom teaching? Or, do teachers who perform better offer a qualitatively different kind of mathematics instruction? Measuring quality *teachers* through performance on tests of verbal or mathematics ability may overlook key elements in what produces quality *teaching*.

The conjecture that teachers may need to know subject matter differently than their students or non-teachers has become the subject of intensive research over the last two decades. In his presidential address to the American Educational Research Association, Shulman (1986) argued for the centrality of subject matter in teaching, drawing attention to the particular ways that teachers must know and use content knowledge in their work. He introduced the term "pedagogical content knowledge," as a special kind of teacher knowledge that intertwines content and pedagogy. According to Shulman (1986, 1987) and colleagues (Wilson, Shulman, & Richert, 1987), pedagogical content knowledge includes understanding which topics students find interesting or difficult, the common

misconceptions that students have, and what forms of representation are useful for teaching particular topics. More recently, scholars who focus on studies of mathematics teachers and teaching have hypothesized that in addition to topic-specific knowledge of students and teaching, teachers might in fact use subject-matter knowledge in ways unique to teaching. Our research group, for instance, has developed the notion that teachers have mathematical knowledge that is "specialized" for the work of teaching. This specialized knowledge of the content can be seen in the demands of providing explanations for mathematical procedures or ideas, representing mathematical phenomena, and working flexibly with non-standard solution methods (Ball & Bass, 2003; Ball, Thames, & Phelps, 2005; Hill, Schilling, & Ball, 2004). In her study of Chinese teachers, Ma (1999) points to the depth and coherence of these teachers' knowledge both over the range of mathematics topics and over time, a kind of knowledge that she labeled "profound understanding of fundamental mathematics." And Ferrini-Mundy, Floden, McCrory, Burrill, and Sandow (2005) argue that teaching algebra requires teaching-specific mathematical practices, among them "bridging" between what students know and disciplinary knowledge, and "trimming" disciplinary knowledge to levels that are understandable by students, but that are also still mathematically accurate.

To illustrate what it means to know mathematics *for teaching*, not just to know mathematics, we turn to an example from a topic in the upper elementary curriculum: division of fractions. By considering the ways teachers might need to know the mathematics behind this topic, we offer conceptual proof that there is more to be known than has been captured to date in most of the educational production function studies described above. This conceptual proof also serves as an outline of the mathematics-specific teaching knowledge used by teachers. Following the example, we then turn in the next section to measurement methods used in the literature that uncovered this knowledge.

We begin with the following problem:

$$\frac{5}{6} \div \frac{1}{3}$$

A common way to calculate the answer is using the infamous "invert-and-multiply" algorithm:

$$\frac{5}{6} \div \frac{1}{3} = \frac{5}{6} \times \frac{3}{1} = \frac{15}{6} = \frac{5}{2} = 2\frac{1}{2}$$

To teach division of fractions to students, teachers must of course be able to carry out the procedure correctly themselves. Thus one aspect of knowing mathematics for teaching is being able to do the mathematics that one is teaching one's students. Yet being able to divide fractions oneself is far from sufficient. After all, teaching mathematics is not simply "knowing" in front of students. Teaching requires making the content accessible, interpreting students' questions and productions, and being able to explain or represent ideas and procedures in multiple ways. For example, a student might ask a teacher why it works to invert and multiply, or why the second fraction is "flipped over" rather than the first. Or, a student might be perplexed because the answer, 2½, is larger than both 5/6 and 1/3—isn't dividing supposed to make the answer smaller? To respond to these questions, teachers must understand and be able to explain why the algorithm works, and be able to explain why and in what cases the quotient is larger than both the dividend and the divisor.

Or, instead of questioning why the standard procedure works, a student might claim that she doesn't even need to learn how to invert and multiply because she has found an easier way to divide fractions—namely, to just divide the numerators and divide the denominators:

$$\frac{5}{6} \div \frac{1}{3} = \frac{5 \div 1}{6 \div 3} = \frac{5}{2} = 2\frac{1}{2}$$

This situation is not uncommon; students often develop their own methods of computation—some valid and some not—and teachers must be able to size up these alternative approaches. In this case, the teacher would have to be able to judge whether the student's method for dividing fractions is mathematically valid or whether the correct answer was simply a coincidence. And, if the method is valid for this particular problem, would it work in general to divide any two fractions, or does it only work in special cases? Generating and testing examples and counterexamples in teaching involves mathematical reasoning. In fact, even the disposition to consider a method's generalizability and efficiency is in itself an example of the type of mathematical knowledge used in teaching.

What else arises in teaching division of fractions? Students, of course, do not always solve problems correctly. For example, suppose a teacher sees the following calculation on a student's paper:

$$\frac{5}{6} \div \frac{1}{3} = \frac{6}{15} = \frac{3}{5}$$

Spotting this as an incorrect answer is insufficient. A teacher must be able to figure out what steps the student might have taken to produce this error, as well as the reasons it might have been made, and then, in light of this mathematical analysis, determine an appropriate response.

After analyzing the above error, for example, a teacher might decide to pose a word problem that corresponds to the calculation to encourage the student to think about whether 3/5 is a reasonable answer. Or, the teacher might decide to use a number line to represent the calculation:

Choosing representations such as story problems or diagrams involves mathematical reasoning and skill. For example, to use the number line representation above, a teacher needs to be able to map each element of the computation to the representation: Where is the 5/6, the 1/3, and the 2 ½? Why does this representation model division? In this case, the representation is using a measurement interpretation of division (i.e., How many 1/3s are in 5/6?). Five-sixths is marked on the number line and each loop represents "measuring off" 1/3. Together, the loops represent the number of 1/3s that "go into" 5/6: Two full loops and a half a loop, or two and a half 1/3s. In addition to making these correspondences, explaining this representation requires an understanding of central mathematical ideas such as the meaning of division and attention to the unit. The teacher must also recognize that, although this representation models division with fractions and shows why the quotient is greater than the dividend and the divisor, it does not readily explain the invert-and-multiply algorithm.

For this purpose, a teacher might rely on the inverse relationship between multiplication and division, noting that dividing by a number is equivalent to multiplying by its reciprocal. Alternatively, a teacher could appeal to students' knowledge that multiplying by 1 does not change the value of a number. In this case, the complex fraction

$$\frac{\frac{5}{6}}{\frac{1}{3}}$$

is multiplied by a convenient form of 1,

$$\frac{\frac{3}{1}}{\frac{3}{1}},$$

to show that $\frac{5}{6} \div \frac{1}{3} = \frac{5}{6} \times \frac{3}{1}$:

$$\frac{5}{6} \div \frac{1}{3} = \frac{\frac{5}{6}}{\frac{1}{3}} = \frac{\frac{5}{6}}{\frac{1}{3}} \times 1 = \frac{\frac{5}{6}}{\frac{1}{3}} \times \frac{\frac{3}{1}}{\frac{3}{1}} = \frac{\frac{5}{6} \times \frac{3}{1}}{\frac{1}{3} \times \frac{3}{1}} = \frac{\frac{5}{6} \times \frac{3}{1}}{1} = \frac{5}{6} \times \frac{3}{1}$$

Another means for explaining the invert-and-multiply algorithm again turns to the meaning of division, but this time using a partitive interpretation. In the case of

$$\frac{5}{6} \div \frac{1}{3},$$

the partitive interpretation asks: What number is 5/6 one-third *of*? Like the measurement interpretation, the partitive interpretation can also be represented on a number line. If 5/6 is one-third of the length of the whole, then the length of whole can be found by putting together three lengths of 5/6, in other words, multiplying 5/6 by 3, which is equivalent to 5/6 × 3/1, which is precisely the invert-and-multiply step:

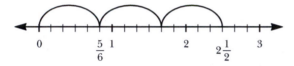

Understanding these subtleties requires a robust understanding of the different meanings of operations and the ways they can be represented. And, as this example shows, teachers not only need to be able to map between a representation and the problem or concept being modeled, they must also determine which representation or explanation would be most appropriate for a particular instructional goal. Thus, and significantly for the assessment of teachers' mathematical knowledge, an item on a test for teachers that asks for the correct answer to

$$\frac{5}{6} \div \frac{1}{3}$$

might reveal knowledge of basic mathematics, but not *all* the possible types of knowledge needed for teaching this topic.

Clearly, global and verbal assessments do not capture the sort of mathematical knowledge and reasoning required to navigate the situations of teaching sketched in the example above. At the time when most of the education production function studies were conducted, however, few alternatives existed; scholars needed to make inferences about teachers' mathematical knowledge based on proxies and indirect indicators. Aligning teachers' and students'

scores on tests was a good bet. Yet these studies can only make inferences about the value of the specific knowledge domains that appear on the assessments: teachers' knowledge of the content they are directly responsible for teaching. The same can be said of the teacher certification exams reviewed to this point: although teaching-specific mathematics questions crept in and out of the tests of the nineteenth and twentieth centuries, they never captured much real estate. Why is this important? Without broader conceptualizations of teacher knowledge and without assessments that reflect this broader conceptualization, we lack precision about the kinds of teacher knowledge that matter. Is it, as some have concluded from analyzing the educational production function literature, simple competency in the topics teachers teach students? Or do teachers need to know "advanced" knowledge, a view common in the 1800s and again today (Education Trust, 1999)? Or, do teachers need specialized knowledge of the topics they teach, illustrated by the teaching tasks in our example? If we had assessments that reliably and validly measured these different conceptions of mathematical knowledge for teaching, and could construct models of student achievement using teachers' scores on such assessments, we could specify more precisely the nature of the knowledge that makes a difference for instructional quality and student learning. Without this precise specification, we cannot design preservice and in-service learning experiences that equip teachers with this knowledge.

Through the 1980s and 1990s, researchers began to approach these questions from another perspective: studying teachers' knowledge by studying teaching and teachers. Through this work, researchers began to hypothesize about the discipline-specific knowledge of content, students, and instruction required for the work of teaching. The studies they conducted focused on teachers' knowledge in practice situations; nearly all involved measuring or making inferences about teacher knowledge in some way. We turn next to a discussion of the strategies used by researchers engaged in this line of research.

METHODS FOR MEASURING *PROFESSIONALLY SITUATED* MATHEMATICAL KNOWLEDGE

Sparked in part by Shulman and colleagues' conceptualization of pedagogical content knowledge, and by a perceived need to assess the content knowledge needed for teaching as a basis for certification and licensure decisions, researchers in the 1980s and 1990s sought to identify what teachers know (or should know) in teaching classroom mathematics. From the start, these studies suggested that teaching requires more than the ability to do the mathematics in the school curriculum. Teaching mathematics, in this view, is not the same as standing at the board and doing mathematics in front of students; it entails additional mathematical knowledge, competencies, and skills—what we call "mathematical knowledge for teaching" (Ball & Bass, 2003). This knowledge, as shown in the division of fractions example above, is multifaceted, including not only teachers' ability to solve the problems their students are expected to solve, but also to understand the content in the particular ways needed for teaching it, to understand what students are likely to make of the content, and to craft instruction that takes into account both students and the mathematics.

In this section of the chapter, we discuss the methods used to study—and later measure—mathematical knowledge for teaching. To do so, we shift from making inferences about conceptions of teachers' mathematical knowledge based on analyses of assessments, as we did with historical teacher exams and the NTE, to exploring the tools used to develop a theory of mathematical knowledge for teaching. We do not provide results from these studies, or chronicle the growth of this concept, because others have done so (Ball et al., 2001). Instead, we focus on the methods used, in part to provide readers with a sense for the array of methods and instruments that can measure mathematical knowledge for teaching. We also discuss the inferences that can be made from each type of research and consider each method's advantages and disadvantages.

We have organized this section of the chapter both by method and chronology. We first discuss observations of teaching practice, beginning with efforts in the 1980s to uncover aspects of mathematical knowledge for teaching and later, post-2000, as they began to be used as a method for quantifying the mathematical characteristics of classroom instruction. We next discuss written tasks and interviews, linked methods used in the 1980s and beyond to assess the extent to which teachers held mathematical knowledge in forms useful to instruction. As concern over the level of teachers' mathematical knowledge grew, so did programs designed to improve such knowledge—and measures to evaluate the results of these programs. We then describe some of the results of efforts to develop such evaluation tools—multiple-choice and other methods designed to assess teacher learning.

The assessment methods used in (and lessons learned from) the studies described below are related only distantly to the methods used in early teacher certification exams and to the assessments given to teachers in the educational production function studies. While they are related more closely to the newer teacher certification exams described in the next section, development of the methods and content for both types of assessment have proceeded, to some degree, on separate tracks. One result is that lessons from one field (e.g., large-scale certification testing) do not easily penetrate another field (e.g., close studies of teacher knowledge). We observe here and in the next section the extent to which the separation of these paths has inhibited the development of stronger measures.

Uncovering Mathematical Knowledge for Teaching: Observations of Teaching Practice

Perhaps the earliest and most widely adopted technique for measuring teachers' mathematical knowledge is one that, on its face, does not appear to be a measurement strategy at all: direct or videotaped observations of teachers' mathematical instruction. However, because scholars followed these observations by careful analysis and explication of the quality and characteristics of the mathematics delivered to students, and because scholars often made inferences about teachers' knowledge from such evidence, we argue that this strategy verged on measurement. We describe some of these early efforts below.

Early observational research focused on teachers' mathematical knowledge had some common elements. Researchers typically collected tens, if not hundreds, of hours of observations or videotapes; published research, though, typically focused on a tiny fraction of the data, often even just a few minutes. Analysis that appeared in print was primarily qualitative, with researchers using methods and coding systems tailored specifically to the mathematical topics and questions at hand. Scholars typically combined observational records with other sources of data to gain insight into mathematical knowledge for teaching.

One classic example of a study in this tradition is Leinhardt and Smith's (1985) study of expertise in mathematics instruction. To explore the relationship between teacher behavior and subject matter knowledge, the authors studied eight teachers intensively, collecting three months of observational field notes from their mathematics lessons, ten hours of videotaped lessons, interviews about the videotaped lessons and other topics, and responses to a card sort task. They used the classroom observations of instruction to construct a measure, ranking teachers' knowledge as high, medium, or low based on "in-class discussions over 3 years and by considering their presentations and explanations as well as their errors" (p. 251). This strategy—sorting teachers by their actual in-class mathematical performance—is rare in the literature, and, unfortunately, not well explicated in their published work. The authors then examined teachers' knowledge in light of performance on interview tasks and examined three teachers' teaching of fractions in much more depth by intensive description of single lessons on simplifying fractions. This method—thick description (Geertz, 1973)—would prove a mainstay in probing the mathematical knowledge needed for teaching.

Borko et al. (1992) provide another example of the observational method for measuring teachers' knowledge of mathematics. Their article focuses on a few minutes of an hour-long review lesson, moments in which a student teacher was asked by a student to explain the division of fractions algorithm. Audiotapes of the lesson were augmented by fieldnotes taken by live observers, and allowed the construction of a dependent variable of sorts: an assessment of this teacher's capacity to provide a conceptually-based justification for the standard algorithm. The authors showed that this teacher's capability in this area was poor—she used a concrete model that actually represented multiplication rather than division. They then used data from interviews, her performance on open-ended mathematics problems, and records from this teacher's teacher education program to explain her in-class performance.

These early observational studies began to record elements of mathematical knowledge for teaching: presentations, representations, explanations, linking between key mathematical elements in instruction, flexibility, and other uses of content knowledge that could be seen in observations of teachers' work. As a result, many of these studies were largely descriptive; rather than inferring any *individual's* overall level of knowledge, classroom practice, or learning, these studies served as explorations of the *territory* of mathematical knowledge for teaching. Borko et al., for instance, did not make global generalizations about their teacher's level of mathematical knowledge for teaching or mathematics instruction per se, or attempt to use their observations to inform any high-stakes (e.g., licensure) decision. This is entirely appropriate, given the limitations of the method.

More recently, several standardized protocols for observing the mathematical quality of classroom instruction (or videotaped records) have emerged in response

to the need to study teacher performance, learning, and change. These protocols are designed to make more generalizable inferences about particular teachers' classroom mathematical work, and thus differ from the early observational studies in the sense that they are intended to evaluate, with some degree of confidence, how well a teacher or group of teachers can teach mathematics. Leaving aside instruments that focus solely on the pedagogical aspects of teaching mathematics—e.g., the degree to which students work in groups, work on extended investigations, or answer questions—a number of instruments combine descriptions of the nature of classroom work with estimates of teachers' skill and knowledge in teaching: the *Reformed Teaching Observation Protocol* (RTOP) (Sawada & Pilburn, 2000), *Inside the Classroom Observation and Analytic Protocol* (Horizon Research, 2000), and the *Learning Mathematics for Teaching: Quality of Mathematics in Instruction (LMT-QMI)* instrument (LMT, 2006a).[3] These instruments vary in the extent to which they infer the *quality of the mathematics in instruction,* as opposed to the *quality of mathematics instruction.* All three instruments, for instance, ask for ratings of the extent to which content is presented accurately. All three ask whether the content presented to students is mathematically worthwhile. And all three ask about some elements of what some consider "rich" instruction—the use of representations, explanations, and abstractions, for instance. As such, all three contain elements that some might use to infer the *quality of the mathematics in instruction,* or the accuracy of content, richness of representation and explanation, and connectedness of classroom tasks to mathematical principles. We argue that while not a measure of teacher knowledge per se, the quality of the mathematics in instruction is a product of, and thus closely related to, teachers' mathematical knowledge for teaching.

RTOP and Horizon's instruments, however, both embed the ratings of teacher knowledge in larger scales intended to measure the extent to which classroom instruction aligns with the National Council for Teachers of Mathematics standards. For instance, RTOP has the following prompts:

- In this lesson, student exploration preceded formal presentation.
- This lesson encouraged students to seek and value alternative modes of investigation or of problem-solving.
- Connections with other content disciplines and/or real world phenomena were explored and valued.

---

[3] We also surveyed the TIMSS-R Video Math Coding Manual (2003). The coding effort here appears more focused on the quality of the problems presented and enacted in classrooms, rather than making inferences about teacher knowledge.

- Students were actively engaged in thought-provoking activity that often involved the critical assessment of procedures.

The issue here is not whether these are desirable lesson characteristics, but instead what is being measured. These two instruments are designed, according to their materials, to measure the quality of mathematics instruction, where quality is defined as both the richness and correctness of mathematical content *and* the way material is conveyed to students. In these instruments, no direct measure of teacher knowledge is available. Instead, teacher knowledge is estimated as a component of how mathematical material is presented to students. The RTOP and Horizon instruments are also designed for rating both science and mathematics lessons, which limits the specificity with which they can ask about particular mathematical practices.

The LMT-QMI instrument was intended to serve as a means to validate multiple-choice measures of teachers' knowledge for teaching mathematics. As such, it is catholic with respect to teaching style, but places heavy emphasis on ways in which teachers' mathematical knowledge might appear in instruction. It includes many of the elements common to the three instruments, but at a finer level of specification. The accuracy of mathematical content, for instance, is assessed over a broad domain of prompts, including use of mathematical language, computational errors, explanations, and notation. The rubric also asks about the presence and accuracy of mathematically rich elements of instruction, including representations, links between multiple representations, explanations, justifications, and the explicit development of mathematical practices. As such, the LMT-QMI measures the quality of the mathematics in instruction.

Observations of teaching have high validity, if the goal is measuring mathematical knowledge for teaching, because classrooms are where mathematical knowledge for teaching is expressed—in teachers' classroom moves, explanations, representations, and computations. However, all three instruments suffer from two main problems. The first involves language and interpretation. Put simply, prompts such as "the mathematics/science content was significant and worthwhile" are just a collection of words; their interpretation will be shaped by the coders' knowledge and views of mathematics itself, and will vary significantly across individuals. In our own work using LMT-QMI, for instance, we have found that the rating of an explanation as accurate or inaccurate—or even as an explanation at all—is shaped by observers' knowledge of the specific mathematical topic, knowledge of what constitutes explanation within mathematics, knowledge of mathematics education, and interpretation of the instrument. The three instruments attempt to eliminate this problem in various ways: Horizon researchers, for instance, spent two days in training and were given an annotated coding manual prior to conducting actual observations. LMT coders spent nearly two years coming to agreement on the coding scheme itself, and on how particular video clips should be rated. And RTOP offers online training for those interested in using their instrument.

The second problem relates to generalizability and the logistics of measurement in the context of research projects: one must observe teachers for multiple lessons before reaching any conclusions about their level of mathematical knowledge for teaching (Rowan, Harrison, & Hayes, 2004; Stein, Baxter, & Leinhardt, 1990). On any particular day, a teacher might be working on a topic with which she is unfamiliar, making a generalization from that day's performance to her overall level of knowledge invalid; there is also natural variation in teachers' knowledge across content areas (number, operations, geometry) that may bias results from a small sample of lessons. Clearly, making many observations over a large sample of teachers participating in preservice or in-service coursework would be a significant burden on any study; the hundreds of teachers required to model statistically student achievement effectively prohibits this strategy. Thus while classroom and videotape observations can be generative for answering fine-grained questions about particular teachers working with students around particular topics, this technique is not amenable to studies that formally test the effects of preservice or teacher development programs, or that link different conceptualizations of teacher knowledge to student achievement.

Exploring Teacher Knowledge: Mathematical Interviews and Tasks

A second method used widely for investigating mathematical knowledge for teaching is mathematical tasks and interviews. These mathematical tasks are different from mathematics assessments, which we consider below, by virtue of the fact that they are not designed to yield generalizable inferences about individual participants' knowledge, but to help scholars understand the nature and extent of teachers' knowledge. In other words, the focus of inference is a group of teachers' performance on the task itself, often to extend our understanding of the task or of teacher knowledge very generally, rather than the performance of individual teachers (e.g., drawing inferences about mathematical knowledge as one component of suitability for teaching). Tasks can be paper-and-pencil problems or included as components of interviews exploring teachers' mathematical knowledge (Borko et al., 1992; Ma, 1999; Simon, 1993; Tirosh & Graeber, 1990) and have typically been designed to study the nature of teachers' mathematical knowledge in specific content areas such as multiplication, rational numbers, division, geometry, and functions (Ball et al., 2001). Some of the tasks that have been used to study teachers' knowledge, such as those derived from elementary textbooks or from studies of children's thinking, could also be used to assess students' mathematical knowledge. Other tasks have been explicitly designed to investigate mathematical knowledge beyond what students would be expected to be able to do, such as explaining what misconception led to a student's alternative method or generating a representation that could be used to teach a particular concept. These tasks are often based on situations that arise in teaching such as analyzing a student's error or answering a student's question. Providing a detailed analysis of all of the mathematical tasks and interview probes that have been used to study teachers' mathematical knowledge for teaching is beyond the scope of this chapter; however, we describe below some of the tasks that have been used in one content area, division, as a representative sample of the types of tasks that have been used.

Some tasks used on written assessments and interviews were, literally, the same tasks given to students. For instance, Graeber, Tirosh, and Glover (1989) investigated the extent to which prospective elementary teachers held misconceptions about multiplication and division similar to those commonly held by children—namely, that multiplication always yields larger numbers and division always yields smaller numbers. They used a written test constructed by slightly modifying 26 problems that had been administered by Fischbein, Deri, Nello, and Marino (1985) in their study of adolescents' misconceptions about operations used to solve multiplication and division word problems. The test presented various word problems and asked respondents not to perform the calculation, but "to write an expression in the form of 'a number, an operation, and a number' that would lead to the solution of the problem" (Graeber et al., 1989, p. 96). It was administered to 129 prospective elementary teachers; 33 follow-up interviews were conducted. Interviewees were given a problem similar to one that had been answered incorrectly on their written form. If they still answered incorrectly, they were asked to explain their answer and show how they would check their solution. Upon realizing (or being told) that the original

expression was incorrect, interviewees were asked to explain what they thought might have caused them to write an incorrect expression. This example shows the ways tasks and interviews might be combined. It also demonstrates one major strength of this method: questions can be tailored to the answers given by specific respondents, often on the spot, to probe the reasons for misconceptions or the support for understanding. It also illustrates a central finding of the studies that used this method: some prospective teachers had serious gaps in their understanding of the mathematics they would be expected to teach.

In another study, Tirosh and Graeber (1989) investigated whether prospective teachers' misconceptions about multiplication and division were explicitly held or just implicitly influenced their calculations. A written instrument was administered to 136 prospective elementary teachers, and then 71 were interviewed. The test asked respondents to label each of the following statements as "true" or "false" and to provide a justification for their response:

A. In a multiplication problem, the product is greater than either factor.

B. The product of .45 x 90 is less than 90.

C. In a division problem, the quotient must be less than the dividend.

D. In a division problem, the divisor must be a whole number.

E. The quotient for the problem 60/.65 is greater than 60.

F. The quotient for the problem 70 ÷ ½ is less than 70.

There were also questions that asked respondents to write expressions for word problems, as well as perform calculations that served as counterexamples to the statements above. This example demonstrates a class of problems slightly removed from actual classroom teaching itself, but that is not knowledge used exclusively in teaching. Few teachers would give the above examples A–F directly to students; yet a firm foundation in operations with rational numbers, and in mathematical proof and reasoning, is necessary for teaching children successfully. Building on this work, Tirosh and Graeber (1990) then explored whether cognitive conflict could be introduced to prompt prospective teachers to confront their misconceptions, in particular, the misconception that in division the quotient must be less than the dividend. Twenty-one respondents who correctly computed 3.75 ÷ .75 but agreed that "the quotient must be less than the dividend," were selected for interviews. The interview protocol was designed to point out the conflict between these two responses and to consider the source of this inconsistency. This study reinforced the predominant findings of other research—i.e., that some teacher candidates do not adequately know the mathematics they will teach—but also used tasks and interviews as a means of studying teacher learning.

With the work of Ball (1990), the field turned more directly toward examining knowledge of mathematics that is specialized to teaching. Ball studied elementary and secondary prospective teachers' knowledge of division in three contexts: division with fractions, division by zero, and division with algebraic equations. The following example tasks required respondents to generate and explain representations, and interview probes were designed to gather information about their notions of what counts as a mathematical explanation:

1. People have different approaches to solving problems involving division with fractions. How would *you* solve this one:

$$1\tfrac{3}{4} \div \tfrac{1}{2}$$

2. Sometimes teachers try to come up with real-world situations or story problems to show the meaning or application of some particular piece of content. This can be pretty challenging to do. What would you say would be a good situation or story for $1\tfrac{3}{4} \div \tfrac{1}{2}$—something real for which $1\tfrac{3}{4} \div \tfrac{1}{2}$ is the appropriate mathematical formulation?

3. Suppose that a student asks you what 7 divided by 0 is. How would you respond? Why is that what you'd want to say?

4. Suppose that one of your students asks you for help with the following:

$$\text{If } \frac{x}{0.2} = 5, \text{then } x =$$

How would you respond? Why is that what you'd do?

These four items illustrate different aspects of what Ball and her colleagues (Ball & Bass, 2003; Ball, Thames, et al., 2005) call specialized mathematical knowledge. Using fractions to calculate the answer to real-life problems is not uncommon among mathematically literate adults; however, constructing real-life problems to illustrate fractional computation is likely limited to teachers. Responding to students' questions about division by zero requires more than the knowledge that "this cannot be done"—it requires mathematical knowledge of the reason this is undefined. And the fourth item requires insight into how to make the procedures of algebraic manipulations comprehensible to learners, a different kind of understanding from simply being able to do the procedure oneself.

These items, written by Ball and others working on the Teacher Education and Learning to Teach project (TELT) (Kennedy et al., 1993), have been widely used in the field. Ma (1999), for instance, compared U.S. teachers in the TELT sample with Chinese teachers,

analyzing responses to the items. She found that many Chinese teachers had a depth of understanding, for example how mathematical topics are related/connected, not present in U.S. teachers.

Another example of an instrument that taps specialized knowledge for teaching is described by Simon (1993), who investigated prospective teachers' knowledge of division using written open-response items and individual interviews. The questions were designed to assess the connectedness of prospective teachers' knowledge and their understanding of units:

1. Write three different story problems that would be solved by dividing 51 by 4 and for which the answers would be, respectively:

   a) $12\tfrac{3}{4}$ b) 13   c) 12

2. Write a story problem for which $\tfrac{3}{4}$ divided by $\tfrac{1}{4}$ would represent the operation used to solve the problem.

3. In the long division carried out as in the example below [the actual item shows the problem 715 divided by 12 being calculated with the standard long division algorithm], the sequence *divide, multiply, subtract, bring down* is repeated. Explain what information the *multiply* step and the *subtract* step provide and how they contribute to arriving at the answer.

Like Ball's items, these ask respondents to perform tasks that are unique to teaching: writing story problems that take into account the different interpretations of remainders, and explaining how portions of the standard long division algorithm work. The entire written instrument was administered to 33 prospective elementary teachers, and another eight were interviewed about a subset of the problems. Interviewees were asked to "think aloud as they solved the problem" and their comments were probed by the interviewer. Simon's results echoed previous studies, finding that, in general, prospective teachers' knowledge was fragmented and procedural.

The division tasks described above are examples of the types of written questions and interview prompts that have been used to better understand teachers' knowledge of mathematics. In developing the tasks, researchers had to grapple with questions about what mathematical knowledge is needed for teaching and how it might be assessed. It makes sense that the source of many such tasks were problems students encounter in K–12 schools, for teachers certainly must be able to solve the problems they give to their students. But the tasks also try to tap into mathematical knowledge needed for teaching beyond what students need to know, for example, knowing not only how to perform an algorithm, but also why it works and how it can be represented to students. Thus, designing these tasks required researchers to reconceptualize what it means to know elementary mathematics in ways that are central to teaching and enabled scholars to develop a more refined conception of the mathematical knowledge needed for teaching.

The use of these tasks helped explicate what knowledge for teaching might look like—in essence, making the case for specialized, professional knowledge. These studies have also brought attention to problems with the quality of mathematical instruction delivered in the U.S. and can be used to inform the content and methods of teacher education courses. And, from a measurement standpoint, there are certain benefits to using these types of measures. Once tasks are designed, they are fairly low cost and easy to implement. There is also high face validity, at least as compared to standard mathematics tests and/or multiple-choice exams.

However, their use is also constrained in ways similar to direct observations. Analyzing or grading teachers' responses is, at scale, a lengthy and expensive process. Further, these tasks and interviews cannot be used to make inferences about particular individuals' (or even groups of individuals, we argue) level of knowledge, for they have seldom been studied to provide information regarding their validity, reliability, and generalizability. Contrary to some beliefs in mathematics education research, using an open-ended task or interview format does not free researchers from an obligation to examine the measurement qualities of the assessments used. Instruments with only a few mathematics problems, or with a very narrowly defined band of mathematics content, encounter the same generalizability problem that using only a handful of observations engenders, that of making a broad claim about knowledge from only a few samples of teachers' work. Open-ended tasks and interview probes are not always reliable, either; teachers' performance on one open-ended task designed by Ball—asking teachers to provide a story that represents 1¾ divided by ½—has been used as evidence that U.S. teachers lag in mathematical knowledge for teaching. A pilot of the multiple-choice version of this measure, however, suggests that part of the item fails to discriminate reliably between highly knowledgeable and less knowledgeable teachers.

Moreover, although these tasks are often situated in teaching contexts, they do not necessarily indicate how a teacher uses mathematical knowledge in practice. A teacher may perform well on the tasks in a clinical or instructional setting but not be able to access this knowledge during a lesson (Borko et al., 1992). Observations of teaching have revealed differences in teachers' mathematical knowledge that were not evident in interview tasks, such as their skill in selecting and using representations and their emphasis of key concepts during lessons (Leinhardt & Smith, 1985).

Assessing Teacher and Student Learning: A New Generation of Teacher Assessments

As researchers uncovered elements of mathematics knowledge unique to teaching, scholars and evaluators faced new measurement needs. One concerned the development of tools to test the relationship between mathematical knowledge for teaching and student achievement; the other involved the development of efficient tools to track change in teachers' mathematical knowledge over time. In both cases, researchers required tools that could be used to make valid and reliable inferences about individuals' or groups' mathematical knowledge for teaching. And researchers needed these tools at scale. Although they are not used for teacher certification, they are often used to deliver information about dozens if not hundreds of teachers, often at multiple time points.

These issues spurred the development of what many consider to be an unusual choice given the history of antipathy between professional educators and teacher assessments: pencil-and-paper tests. In some cases, such as the Study of Instructional Improvement/Learning Mathematics for Teaching (SII/LMT) measures (LMT, 2006b) and the SimCalc rate and proportionality teaching survey (Shechtman et al., 2006), these tests come in multiple-choice format. In other cases, such as the Knowledge for Algebra Teaching (KAT) measures (Knowing Mathematics for Teaching Algebra Project, 2006) and the Diagnostic Teacher Assessments in Mathematics and Science (DTAMS) measures (Bush, Ronau, Brown, & Myers, 2006) the instruments contain both multiple-choice and short open responses—often simply the answer to a computation problem, but sometimes short explanations for problems or procedures. All of the above tests, however, emphasize the ability to measure teachers' knowledge at scale, and with known reliability and validity. Teachers' answers to the items are marked as correct or incorrect, and a score is calculated, typically in standard deviation units. Thus these assessments are easily scored, with the ease of scoring a major factor in the selection of this format.

One reason for this selection was the scope of the research projects and intended uses of these measures. Having new insight into mathematical knowledge for teaching, scholars now turned to modeling its growth and contributions to the educational production function. For instance, the SII/LMT measures grew out of a study designed to examine and compare the performance of elementary schools, teachers, and students participating in one of three whole-school reforms. Teacher learning was one potential target of these reforms, and also a mediator of student outcomes (Hill, Rowan, & Ball, 2005). Later, the instrument's authors received continued funding from the National Science Foundation to pilot additional items and extend the measures upwards to middle school. The SimCalc project began as a curriculum development project, one aimed at introducing the mathematics of change and variation in the early grades, typically with some technological component. The SimCalc instrument grew out of interest in teachers' learning as a result of engaging with their professional development and curriculum, and the desire to model student achievement as a function of teacher learning (Shechtman et al., 2006). KAT began as a project to explore the nature of knowledge used in teaching algebra and progressed to instrument development. Instrument developers hope to use these measures in models of student achievement. DTAMS is solely a measures development project, but has designed measures to be used not only to diagnose the status of middle school teachers' mathematical knowledge, but to track teacher learning and student achievement as a function of teacher knowledge.

This new generation of pencil-and-paper tests shares some common attributes. All, for instance, contain items intended to represent the mathematics problems encountered in teaching, rather than only "common" content knowledge, or knowledge of mathematics that is common across professions and available in the public domain. As such, items focus on teachers' grasp of representations of content, unusual student work or mathematical methods, student errors, mathematical explanations, teaching moves, and other raw materials of classroom mathematics. This sets these instruments apart from historical teacher tests, which focused only on asking teachers to solve the mathematics problems they would be teaching to students, or in some cases, mathematics problems somewhat more advanced than the curriculum they would be teaching. The new instruments also focus on relatively narrow mathematical domains, rather than returning an overall score across all mathematical content. And, critically in our view, all these tests have been developed by experts in mathematics, in the teaching of mathematics, and in psychometrics. This collaboration has led to explicit interest in the multidimensionality of the measures. Rather than attempting to write measures that represent "pure" mathematical or pedagogical knowledge, as many test development firms do for technical reasons, these

measures embrace the idea that teachers' knowledge has many facets that must be included in any instrument. Project psychometricians, for their part, have been willing to work with this specification and, in turn, lent generous expertise in gauging the reliability and validity of these instruments. Reliability and validity analyses, which we shall discuss in more detail below, is relatively rare anywhere in the educational assessment enterprise (Messick, 1989).

One major difference among these pencil-and-paper assessments is in instrument goals and, by extension, the interpretation of scores. Two projects, KAT and DTAMS, criterion referenced their exams—DTAMS to nine policy documents describing both middle school student *and* teacher knowledge and KAT to a rubric that outlines the categories of teacher knowledge and the content involved in teaching algebra. By choosing to criterion reference and assuming defensible cutoffs for different performance levels, scores on this measure can be concretely interpreted vis-à-vis this standard—e.g., teacher Y knew X% of the mathematics needed according to standards. A third project, SII/LMT, initially designed its measures to discriminate accurately among teachers, in essence ordering them as correctly as possible relative to one another and to the underlying trait being assessed, mathematical knowledge for teaching. Although this design allows users to measure change over time as teachers learn, raw frequencies have no meaning relative to any criterion, such as "the mathematical knowledge teachers need to know." Finally, the fourth project, SimCalc, designed its measures to gauge curriculum-specific mastery of content—whether teachers grasp key principles in rate and proportionality, principles needed to teach using the SimCalc curriculum.

It is in the arena of underlying theory, however, that these instruments differ most. Despite claiming to cover roughly the same terrain, these projects have strikingly different approaches to specifying domains for measurement—in essence, different approaches to organizing what is "in" mathematical knowledge for teaching. SII/LMT has four knowledge domains; SRI has two; DTAMS has four; and KAT's conceptual framework crosses three aspects of knowledge for teaching and four mathematical domains. None of these conceptual maps match one another exactly, and none match Shulman's original formulation of pedagogical content knowledge. From one view, these differences might reflect ways the content itself (algebra, number) influences the knowledge needed for teaching. But at least to those interested in building theoretical coherence around mathematical knowledge for teaching, the variety of approaches is distressing.

Some examples from the actual assessments help elaborate these theoretical differences. Three-quarters of DTAMS' items measure important types of mathematical knowledge that teachers are charged with developing in students:

1. *Memorized facts and skills:* Which of the following numbers is the least common denominator of 5/9 and 7/12? (a: 108; b: 36; c: 72; d: 3)

2. *Conceptual understanding:* Name two numbers between 1.35 and 1.36. (a: $1.3\overline{5}$ and $1.\overline{35}$; b: 0.351 and 0.352; c: 1.345 and 1.354; d: There are no numbers between them.)

3. *High-order thinking:* Mr. Short is three paper clips in height. Mr. Tall is five buttons in height. Two buttons laid end-to-end are the same height as three paper clips laid end-to-end. Find the height of Mr. Tall in paper clips. Justify your solution.

These are relatively common problems in the "public domain," so to speak, of mathematical knowledge. Other than the fact that teachers are more likely than the general population to be solving such problems, none represent knowledge specific to teaching. One quarter of the items measure mathematically situated pedagogical knowledge:

1. A student said, "Whenever you add two non-zero integers the sum is always between the two integers." Explain why this is incorrect. Explain how you would use a diagram or model to help the student understand she is incorrect.

2. A student claims that all squares are congruent to each other because they all have four right angles. Why is this claim incorrect? Explain how you would help the student understand the error in her thinking.

These pedagogical items, which like the higher-order thinking items are in short-answer format and hand-scored, mimic the mathematical judgments teachers must make during classroom work—understanding common mathematical errors and designing instruction to correct them.

The SII/LMT items, by contrast, are in multiple-choice format and cover three domains of teacher knowledge, with one domain sub-divided into two constituent parts:

Content Knowledge:

Common content knowledge, or the mathematical knowledge teachers are responsible for developing in students:

Ms. Dominguez was working with a new textbook and she noticed that it gave more attention to the number 0 than her old book. She came across a page that asked students to determine if a few statements about 0 were true or false. Intrigued, she showed them to her sister who is also a teacher, and asked her what she thought.

Which statement(s) should the sisters select as being true? (Mark YES, NO, or I'M NOT SURE for each item below.)

	Yes	No	I'm not sure
0 is an even number	1	2	3
0 is not really a number. It is a placeholder in writing big numbers.	1	2	3
The number 8 can be written as 008.	1	2	3

Items in this category are similar to DTAMS' memorized facts and skills and conceptual understanding. They tap knowledge that is in the public domain, including that used in other professions (e.g., writing 8 as 008 as in computer science).

Specialized content knowledge, or mathematical knowledge that is used in teaching, but not directly taught to students:

Imagine that you are working with your class on multiplying large numbers. Among your students' papers, you notice that some have displayed their work in the following ways:

Student A	Student B	Student C
35	35	35
×25	×25	×25
125	175	25
+75	+700	150
875	875	100
		+600
		875

Which of these students would you judge to be using a method that could be used to multiply any two whole numbers?

	Method would work for all whole numbers	Method would *not* work for all whole numbers	I'm not sure
Method A	1	2	3
Method B	1	2	3
Method C	1	2	3

Here, the teacher must inspect the three student methods to determine what, in fact, is occurring. Student A, for instance, has multiplied 5×25 and then 30×25; student B is using a version of the standard U.S. algorithm; and student C has used a partial product method. If the teacher understands these explanations for the students' methods, she must then make a determination about whether each method generalizes to the multiplication of other whole numbers, perhaps by referencing the commutative or distributive properties of multiplication. This work is entirely mathematical, but it is not mathematical work done by many non-teaching adults.

Knowledge of Content and Students

Knowledge of content and students, or the amalgamated knowledge that teachers possess about how students learn content:

Takeem's teacher asks him to make a drawing to compare ³/₄ and ⁵/₆. He draws the following:

and claims that ³/₄ and ⁵/₆ are the same amount.

What is the most likely explanation for Takeem's answer? (Mark ONE answer.)

a) Takeem is noticing that each figure leaves one square unshaded.

b) Takeem has not yet learned the procedure for finding common denominators.

c) Takeem is adding 2 to both the numerator and denominator of ³/₄, and he sees that that equals ⁵/₆.

d) All of the above are equally likely.

Here, a teacher must look beyond the standard method for comparing fractions (finding common denominators) to see that Takeem has focused on the amount that is missing from each whole. This knowledge is likely held by teachers who have worked with children learning to compare fractions; it also corresponds to Shulman's "common student misconceptions" portion of pedagogical content knowledge.

Knowledge of Content and Teaching

Knowledge of content and teaching, or mathematical knowledge of the design of instruction, includes how to choose examples and representations, and how to guide student discussions toward accurate mathematical ideas.

While planning an introductory lesson on primes and composites, Mr. Rubenstein is considering what numbers to use as initial examples. He is concerned because he knows that choosing poor examples can mislead students about these important ideas. Of the choices below, which set of numbers would be best for introducing primes as composites? (Mark one answer.)

	Primes	Composites
a)	3, 5, 11	6, 30, 44
b)	2, 5, 17	8, 14, 32
c)	3, 7, 11	4, 16, 25
d)	2, 7, 13	9, 24, 40
e)	All of these would work equally well to introduce prime and composite numbers.	

In this item, the teacher must first know that students are likely to think that all prime numbers are odd; including 2 is important given this misconception. Students are also likely to think that odd numbers cannot be composite—making choice (d), which tests both misconceptions, the best of the four options. This SII/LMT category is the furthest from pure mathematical knowledge, as it involves reasoning about both students and teaching. It corresponds with another component of Shulman's pedagogical content knowledge (choosing the "best representation") and to DTAMS' pedagogical knowledge category.

The schematic for item development used by the Knowledge of Algebra for Teaching is more detailed, consisting of 24 separate cells that are the product of two algebraic content areas, three types of algebra knowledge for teaching, and four domains of mathematical knowledge. The first side of this three-dimensional matrix names two central algebraic topics: expressions, equations, and inequalities; and functions and their properties. The second side names algebra knowledge for teaching, which in this view consists of knowledge of school algebra, or the mathematics in the intended curriculum that *students* should learn; advanced knowledge, or college-level mathematics that gives teachers perspectives on the trajectory of mathematics beyond middle or high school; and teaching knowledge, or mathematical knowledge beyond that directly taught to students, but which is useful in instruction. This last category includes the mathematics involved with student misconceptions, knowledge of materials and texts, and ways to develop mathematics within and across lessons with particular goals in mind. Finally, the third side of the matrix encompasses four domains of mathematical knowledge, including "core" content knowledge, or declarative or substantive knowledge; representations of mathematical content such as number lines, tables, graphs, and area models as well as the materials that can be used to represent content during instruction (e.g., algebra tiles); applications and contexts, such as situations that can be modeled by linear functions; and reasoning and proof, or knowledge of how truth is established in the discipline.

Four items illustrate aspects of this domain specification. The first item comes from the category of school algebra, or the topics that teachers would directly teach students. It illustrates an "application" problem from KAT's four domains of mathematical knowledge, modeling the ways in which exponential functions might be used in real-life situations:

Which of the following situations can be modeled using an exponential function?

i.   The height *h* of a ball *t* seconds after it is thrown into the air.

ii. The population *P* of a community after *t* years with an increase of *n* people annually.

iii. The value *V* of a car after *t* years if it depreciates *d*% per year.

A. i only

B. ii only

C. iii only

D. i and ii only

E. ii and iii only

This problem is similar, perhaps, to those found in high school or college textbooks. This item would fall in the SII/LMT common content knowledge category, and in DTAMS' conceptual knowledge category.

The next KAT item is from the category of advanced knowledge or knowledge beyond the actual middle or high school curriculum, and focuses on functions and their properties:

For which of the following sets S is the following statement true?
For all a and b in S, if ab = 0, then either a = 0 or b = 0.

i.   the set of real numbers

ii. the set of complex numbers

iii. the set of integers mod 6

iv. the set of integers mod 5

v.   the set of 2 × 2 matrices with real number entries

A. i only

B. i and ii only

C. i, ii, and iv only

D. i, ii, iii, and iv only

E. i, ii, iii, iv, and v

Answering this item properly requires an important type of mathematical reasoning. Teachers must be sensitive to the fact that familiar properties from algebra are not necessarily true in general, but, rather, depend on the set being referenced. In the item above, the statement *if ab = 0, then either a = 0 or b = 0* is true when working with all the choices save for the set of real 2 × 2 matrices and integers mod 6.[4] So although integers mod 6 is not in the K–12 curriculum, teachers must not lead high school students to believe that the original statement is universally true. This category is unique to the KAT measures and represents one

very traditional line of thinking about the mathematical knowledge teachers need—that a teacher should know "very much more than he is expected to teach" (Michigan Department of Public Instruction, 1898, p. 10; see also Education Trust, 1999, p. 4).

The third item is open-ended, tapping "teaching knowledge" of student errors, and is also classified as "core content knowledge" by the KAT project:

A student solved the equation 3(n – 7) = 4 – n and obtained the solution n = 2.75.

What might the student have done wrong?

To answer correctly, teachers must know how to correctly solve this equation for themselves, and they must also know a common student error, namely that students will forget to distribute the 3 over the 7 as well as the n. This corresponds to the SII/LMT category named knowledge of students and content.

The last item is also open-ended and draws from the "teaching knowledge" category, focusing on representations of expressions:

Hot tubs and swimming pools are sometimes surrounded by borders of tiles. The drawing at the right shows a square hot tub with sides of length s feet. This tub is surrounded by a border of 1 foot by 1 foot square tiles.

How many 1-foot square tiles will be needed for the border of this pool?

a. Paul wrote the following expression:
$$2s + 2(s + 2)$$
Explain how Paul might have come up with his expression.

b. Bill found the following expression:
$$(s + 2)^2 - s^2$$
Explain how Bill might have found his expression.

c. How would you convince the students in your class that the two expressions above are equivalent?

Here, teachers must "see" how the same representation might yield very different expressions, mapping closely between the representation and the construction of each expression. Teachers must also make a judgment about how best to prove these two expressions are equivalent, a question that grapples with the complex intersection between content, students, and teaching. This item would fall, in the SII/LMT scheme, into the "specialized" and "knowledge of content and teaching" categories.

---

[4] For example, when working with the set of integers mod 6, 4 times 3 equals 12, which is 0 mod 6, but neither 4 nor 3 is equal to 0 mod 6.

Finally, the SimCalc materials have no similar domain map, instead using a domain map for *student* learning of their "mathematics of change" curriculum materials. Nevertheless, their items do reveal a theory of teacher knowledge. The item below, for instance, asks teachers to analyze an unusual solution method—and in so doing, to draw on underlying knowledge of proportional relationships as represented in such a table:

Students are working on creating tables of proportional relationships. During the lesson, Mr. Lewis has the following table up on the board.

x	y
3	12
2	8
4	16
8	32

He asks the students to come up with another (x,y) pair that could fill in the blank cells. One student says that you can create a new (x,y) pair by taking the averages of the last two rows in the table. That is, the new x is

$$\frac{4+8}{2}=6,$$

and the new y is

$$\frac{16+32}{2}=24.$$

Mr. Lewis had hoped that the students would pick a value for x, then multiply by 4 to get a value of y. He is intrigued by this student's solution to the problem, however, and wonders whether it will always work.

This method will produce a valid x,y pair: (Mark ALL that apply.)

a) Only when x and y are always whole numbers.

b) Only when the last two rows have even numbers.

c) Only when the last row is exactly double the next-to-last row.

d) Always, as long as the table represents a proportional relationship.

e) Always, as long as the table represents a linear function.

In addition, because this novel curriculum asks students to explain, interpret, and represent linear functions, proportional relationships, and rates, teachers are often asked to do the same:

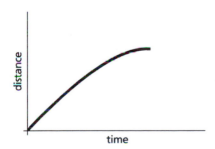

Ms. Chopra asks her students to explain why the graph above is NOT proportional. Her students' explanations are given below.

Which of the following student statements explains why the graph is NOT proportional? (Mark ALL that apply.)

a) You can't have a curve in a graph.

b) The relationship between the distance and time is not always constant.

c) The graph is not labeled correctly.

d) The ratio of rise over run is different at different points on the curve.

e) At every point on the curve, the d/t is not always the same ratio.

f) None of the above.

This item blurs the distinction between common and specialized knowledge set out by the SII/LMT instrument developers; here, both students and teachers are asked to know mathematics in ways that probe the underlying meanings and to provide mathematical explanations to counter a common misconception.

However, many more of the SimCalc items ask teachers to solve the same types of problems they would be asking students to solve:

Mr. Aneesh presents the following graph to his class. The graph represents the entire race run by Adrianne, Bettina, Corinna, and Destiny.

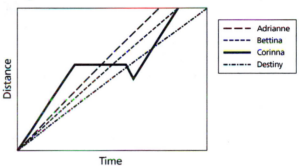

Mr. Aneesh asks the students to describe what is happening during the race.

Which of the following student statements are FALSE? (Mark ALL that apply.)

a) Corinna and Bettina finished at the same time.

b) All runners finished the race at the same time.

c) Adrianne had the highest average speed.

d) Adrianne led the race from start to finish.

These four projects have progressed the furthest in instrument development, in the sense that all have taken care to specify domain maps, completed pilot work with large groups of teachers, conducted psychometric analyses, and begun work on validation of the measures. However, many other pencil-and-paper tests exist. Most faculty responsible for teaching mathematics content and methods courses to preservice teachers, for instance, write end-of-unit or end-of-course assessments. Research projects interested in studying the development of teacher knowledge design their own instruments, tailored to their own program or purpose. Schweingruber and Nease (2000) describe a content knowledge assessment used to gauge learning in a teacher professional development program at Rice University. Phillip et al. (in press) describe and make available a test that combines an assessment of preservice elementary teachers' content knowledge with their knowledge of students. In it, preservice teachers are presented with student work, asked whether that work is correct and, if not, to explain the student error. Some of the correct student work presents unusual mathematical solutions or representations, such as non-standard algorithms or unusual representations of number and content. Many of the student errors are drawn from the literature on student learning and error analysis.

As this review shows, the field has responded strongly to the need for pencil-and-paper assessments that can be used to study teacher and student learning. Many build on the types of items and mathematics described in the initial work on teachers' knowledge identified in the beginning of this section, and thus come closer to tests of professional knowledge than have past assessments, particularly those used in the educational production function literature. Although development is not complete on any of these assessments, and such assessments are far from perfect vehicles for this work, significant progress has been made. One reason for this is the interest on the part of funders, particularly the federal government, in developing such measures; three of the projects described in detail above were funded by the National Science Foundation, and one by the U.S. Department of Education. Measures development is expensive, and a commitment on the part of major funding agencies is critical. Another reason might be a climate of increased accountability, not only for districts, schools and students, but also for the professional development providers who serve these educators.

Like any form of assessment, the instruments reviewed here have significant drawbacks. To start, few would agree that instruments that use multiple choice or short-answer responses can represent completely the mathematical knowledge teachers use in their work. First, the construct of mathematical knowledge for teaching is still emergent. Without better theoretical mapping of this domain, no instrument can hope to fully capture the knowledge and reasoning skills teachers possess. Second, teaching any subject matter involves complex professional judgments, and such

judgments are not always amenable to testing in a format with single correct answers. While there may be some regularities in student errors, for instance, the errors of any particular student are shaped by his or her prior learning experiences and the task at hand; either may lead to idiosyncrasies. And while base ten materials may generally work well to represent multiplication of decimals, whether they will work well in a particular situation depends in part on the teacher's skill in using them, students' exposure to the use of base ten materials to model whole number operations, and so forth. Teachers' reasoning about these situations is not well reflected in right/wrong formats. Problems that draw only on mathematics avoid some of these complications, but also fail to gauge other aspects of knowledge and reasoning that are key to classroom teaching.

A third problem with multiple-choice/short-answer assessments involves teachers' reactions to such instruments. Several of the projects described here note teacher resistance to assessments and, in particular, to multiple-choice forms of assessments. Teachers, whose work is about helping others learn, are well aware of the problems inherent in assessing complex learning outcomes. They know that many important achievements are difficult to measure, and that multiple-choice formats are often inadequate in capturing learning. Using multiple-choice formats to assess teachers' own knowledge suggests a limited view of the complexity of professional knowledge and thus conflicts with professionals' reasonable experience with and distrust of test questions framed in this form.[5]

Fourth, validity work on these tests has generally lagged behind assessment development itself. Test developers are most likely to have conducted standard psychometric analyses, such as construct identification via factor analyses and reliability analyses. Reliabilities for these assessments meet or exceed industry standards for the specific test goals listed, such as diagnosing teacher knowledge or tracking groups of teachers over time. However, for the most part, validation work has been limited. The DTAMS and the SII/LMT projects both conducted content validity checks, ensuring that the test reflected key elements of the mathematics curriculum for students, and elements of teacher knowledge suggested by the literature described above. KAT is currently conducting a discriminant validity study, giving their assessment to college mathematics majors, for instance, to see whether this group scores high on advanced mathematics knowledge but low, as expected, on teaching mathematics

---

[5] That most multiple-choice tests are disappointing in their capacity does not rule out the possibility of designing new types of items that take advantage of the format rather than distorting the nature of the knowledge being assessed.

knowledge. Construct identification, however, is only one form of validity. Another important form is predictive—whether teachers' performance on the assessment is related to the mathematical quality of instruction, and to student learning. Currently, only the SII/LMT test developers have performed this work, and even there, more work remains to be done.

Finally, using multiple-choice/short answer assessments as part of an evaluation or study requires at least basic statistical expertise—a t-test or ANOVA to examine teacher gains, for instance. In some cases, such as the SII/LMT measures, scores are returned not in raw frequencies but in standard deviation units. Users also have the option of creating their own forms from the item pool. This contrasts with the analytic demands of many of the other methods described here. Eventually, many of these pencil-and-paper assessments may be put online, a condition under which these assessments will likely be scored automatically by computer.

Other Methods

New technologies may also allow improvements on pencil-and-paper assessments, at least for small-to-moderate scale research projects. Kersting (2004) and associates at LessonLab have developed an online measure that gauges teacher knowledge by examining responses to ten short video clips of teaching. Teachers' responses are credited according to the amount they discuss features of the mathematics, student thinking and understanding, teaching strategies, and according to the amount that the discussion of the clips embeds these observations in a cause/effect framework. Like interviews or open-ended tasks, this method may be cumbersome to score. However, the technical properties of the instrument appear promising, and the assessment format itself holds three appealing features. First, it complements current professional development methods, particularly the use of videotape for teacher analysis and reflection. Second, an analysis of mathematics teaching may be less threatening to teachers than a pencil-and-paper assessment. Finally, the assessment can be embedded in the flow of teachers' work in a video-based professional development setting.

Other embedded assessment techniques also exist. For instance, Sowder, Phillip, Armstrong, and Shappelle (1998) effectively used discourse analysis—or studying what teachers say and how they talk about mathematics—to trace growth in teachers' knowledge of division of fractions. One benefit of this method is that it requires no statistical expertise. Discourse analysis, however, does require close and specialized train-ing. It is a promising approach because teachers' use of mathematical language is both an indicator and target of growth in knowledge and skill; teachers need, in classrooms, to be able to use mathematical terms accurately and precisely. But more, they also need to be able to "translate" between children's home language, informal mathematical language, and disciplinary language (Ball, Masters-Goffney, & Bass, 2005; Pimm, 1987). Our own observations suggest that this linguistic capacity may be held variably in today's teaching population.

The drawbacks to discourse analysis include a lack of generalizability past the particular professional development (or other) setting that participants are in and difficulties separating individuals' skill with language from group skill with language. Further, such measurements do not scale easily, and they require finely calibrated frameworks for interpreting and analyzing discourse patterns. That said, this remains an intriguing possibility, one that should be investigated more formally.

Finally, some key studies have used self-reports of improvements in knowledge and skill as proxy measures for actual growth (e.g., Garet, Porter, Desimone, Birman, & Yoon, 2001; Horizon Research, 2002; NCES, 1999). However, the degree to which these are related to actual teacher knowledge growth is largely unknown.

In sum, scholars have developed a variety of methods for studying mathematical knowledge for teaching, from those intended to help understand the construct itself to those designed to provide evidence of teachers' learning in this domain. Our review of these methods has taught us several things. First, developers in this field report that there is in fact considerable interplay between assessment and theoretical development. Writing test items, for instance, helps illuminate aspects of mathematical knowledge specific to teaching; conducting a discourse analysis can help illuminate the understandings teachers have of students and mathematics. Second, every method has both advantages and drawbacks; there is no one perfect method, although there may be better and worse methods for particular research questions. In fact, more diversity in assessments and assessment formats should be encouraged, to allow better matching between the subjects of inquiry and instruments used.

Third, these assessments should be widely available, and widely used. This requires broader support than exists now in the mathematics education community. For instance, there is currently no clearinghouse through which scholars "shopping" for a measure can collect and compare information about each measure's content, design, and technical features.

Also, there is no systematic method for disseminating measures or for training individuals in their use. Instead, scholars and evaluators sit at the mercy of the research projects and agencies developing these measures, whose capacity to disseminate varies according to funding and the availability of personnel. But broader use of these measures in the study of teacher knowledge and learning would surely benefit the field, as results from different research projects would be more directly comparable and, we hope, more rigorously evaluated.

Finally, each of the efforts described here have made an attempt to bring the measurement of teachers' mathematical knowledge closer to the actual practice of teaching itself. In many cases, the developers of these assessments explicitly state their desire to measure the mathematical knowledge used in teaching, although the frameworks for what those measures might be vary. One would expect that this improves the validity of the assessments, although as we noted above, most projects do little predictive validity work on their measures. A similar effort to move the assessment of mathematical knowledge closer to actual teaching has been occurring in the field of teacher certification testing, a topic to which we now turn.

CONTEMPORARY APPROACHES TO TESTING TEACHERS FOR CERTIFICATION

In the last decades of the twentieth and into the twenty-first centuries, three elements converged to spur the development of new forms of teacher certification tests. First, policy-makers and the public made increasing demands for accountability on the part of teachers and teacher educators; as we have seen, the amount of required teacher testing grew markedly over the period 1980–2000. Second, mathematics educators began to understand the nature of mathematical knowledge for teaching, investing heavily in the idea that beyond knowing content, teachers must also have profession-specific mathematical knowledge about content, students, and teaching. Finally, this era saw emergent interest on the part of professional educators in testing teachers; although opposition to teacher testing still runs high in some quarters, at least two national professional organizations have been established to promote new forms of teacher assessment. In this way, the historic antipathy between teaching and teacher educator organizations and testing agencies has begun to crumble. Whether the professional organizations can gain control over widespread teacher testing, and whether the tests produced are satisfactory, remains to be seen.

In this section, we describe in detail the mathematical elements of the three most prominent of these efforts. Two of the assessments are for beginning teachers: the Praxis Series (the successor to the NTE), and the Interstate New Teacher Assessment and Support Consortium's (INTASC) portfolio assessment system. Although INTASC is not now formally in use for certification in any state, Connecticut's second-stage licensure system uses a portfolio assessment based on INTASC principles, and INTASC is looked to as a leader in the field. The third assessment is an endorsement for experienced practitioners: the National Board for Professional Teaching Standards (NBPTS) certification. We examine each of these assessments along four dimensions: its history and development; a description of the test and how it is used; an analysis of sample mathematics items; and, based on our analysis of available items, some conjectures about the test's implicit views of teaching, of mathematics, and of the mathematics needed for teaching. We also briefly review other tests used in the U.S., including the California Subject Examinations for Teachers (CSET), and the tests developed by the American Board for Certification of Teacher Excellence (ABCTE). Each assessment plays a unique role in the current debates over how to certify teachers.

Before beginning, it is important to note what these modern certification tests are designed to do. Unlike the mathematical assessments common in the scholarly literature, the tests discussed below are designed to allow the user to draw inferences about a candidate's suitability for teaching. As this implies, there are often great stakes attached to teachers' performance on these exams: pay raises, at a minimum, and more often the ability to compete for jobs in one's chosen profession. The high-stakes nature of many of these assessments helps explain why there has been so much debate over them and so much effort devoted to ensuring their accuracy. We begin with Praxis, the most commonly used teacher certification assessment in the U.S.

The Praxis Series

The Praxis Series began to replace the National Teacher Examination (NTE) in 1993. Currently, 43 states use at least one portion of the series for beginning teacher certification (ETS, 2006a). In the wake of increasing calls for teacher testing, and building on new advances in psychology and research on teaching, Praxis was to take a more robust view of teaching and the knowledge needed to teach. Praxis takes a three-stage approach to teacher certification that includes tests of prerequisite content knowledge, subject-specific content knowledge and pedagogical knowledge, and interactive teaching skills. Each construct in Praxis would be measured separately, and at different stages in a new teacher's career.

In many states, teachers begin by taking Praxis I, a multiple-choice test of general academic knowledge, before admission to teacher education programs. This test is similar to the SAT and in some states the tests are used interchangeably. If the study guide questions are representative, the mathematics questions in Praxis I are similar to the SII/LMT "common content knowledge" category, meaning mathematics that is common across professions and, in many cases, in the public domain. This makes sense, given that this assessment is administered prior to the start of formal teacher education programs.

Praxis II is a range of multiple-choice and constructed-answer assessments that test general academic and pedagogical knowledge as well as subject-specific knowledge for teaching. Praxis II is generally taken at the completion of a teacher education program. Elementary teachers might see mathematics questions on one of three separate tests: the content knowledge test for elementary teaching, a widely used assessment, containing roughly 30 multiple-choice items on mathematics; the curriculum, instruction, and assessment test, where roughly 20% of 110 multiple choice items ask about mathematics; and the content area exercises, a test that includes as one of its four prompts a problem that asks teachers to design mathematics instruction or respond to a student error. At the secondary level, ETS maintains several mathematics-specific Praxis II assessments, including content knowledge; proofs, models, and problems; general mathematics; and pedagogy. The mathematics pedagogy assessment is similar to the elementary content area exercise test, in that it asks teachers to respond to prompts about how best to design or implement mathematics instruction.

Praxis III is an inventory of teaching skills that is conducted by observing the beginning teacher in her classroom. Only two states are currently using Praxis III, and none require it for initial licensure. Praxis III is used instead in some districts and states for professional development.

Each state that uses Praxis requires its own combination of tests, and each state determines its own passing rates for licensure. So, for example, Idaho requires that its teacher candidates take a single subject-area Praxis II test for licensure; Hawaii requires the Praxis I test and two Praxis II tests, the Principles of Learning and Teaching as well as a subject-matter area test. Because of the prevalence of Praxis II, and its claim to measure subject-matter-specific disciplinary knowledge, we focus on it for much of the discussion below.

Praxis II, like other modern teacher certification tests, stakes its claim to validity on the idea that it captures the knowledge needed to successfully teach. Two aspects of assessment design buttress this claim. First, the Praxis II tests were developed through a process designed to include key stakeholders in the education and certification of teachers. In mathematics, this included subject-matter specialists (mathematicians, mathematics educators), teachers, mathematics coordinators in school districts, psychometricians, and psychologists. Second, the items on the assessment were developed following what the test developers call a "job analysis," or a review of the mathematical knowledge teachers need in their jobs. This "job analysis" is not an ethnographic study of a mathematics classroom with special focus paid to the teacher and the knowledge employed in teaching. Rather, by "job analysis," the developers mean that subject-matter specialists examined earlier tests such as the NTE, textbooks, the NCTM Standards documents, curricular documents, then drew up lists of topics that teachers would need to know for teaching mathematics. This list was then shared with teachers and other school personnel who made suggestions and changes and, when approved, translated into assessment items.

Because ETS does not release active Praxis items, it is difficult to characterize the results from this process with any degree of certainty. If Praxis study materials are any guide, however, the problems prospective teachers face on the content area tests are largely dependent on the Praxis II assessment their state requires. For both elementary and secondary content knowledge tests, among the most commonly used, the items resemble those that have comprised teacher assessments for generations. The following items are taken from study guides and test preparation materials (ETS, 2005b, p. 11; ETS, 2006c, p. 4):

1. $15(4 + 3) = 15 \times 4 + 15 \times 3$

The equation above demonstrates which of the following?
   a) The distributive property of multiplication over addition
   b) The commutative property of multiplication
   c) The associative property of multiplication
   d) Additive inverse and additive identity

2. Riding on a school bus are 20 students in 9th grade, 10 in 10th grade, 9 in 11th grade, and 7 in 12th grade. Approximately what percent of the students on the bus are in 9th grade?
   a) 23%   b) 43%   c) 46%   d 76%

3. Which of the following is equal to 8^4?
   a) 4,032   b) 4,064   c) 4,096   d) 4,128

4. For which of the following values of k does the equation $x^4 - 4x^2 + x + k = 0$ have four distinct real roots?
   I. −2

   II. 1

   III. 3

   a) II only   b) III only   c) II and III only   d) I, II, and III

Items 1–3 are from elementary materials; item 4 is from secondary test preparation materials. These items ask teachers to solve problems that contain mathematical content common to the public and other professions, including mathematics as a discipline. And they are not far from the types of items found in most teacher certification exams for the past two centuries, with changes in mathematical content and item design to allow the tests to stay current with the content taught in elementary and secondary schools.

By contrast, the small number of mathematics items on the elementary curriculum, instruction, and assessment test appear aimed at assessing prospective teachers' ability to work with the mathematics in instruction (ETS, 2005a, pp. 5–6):

1.

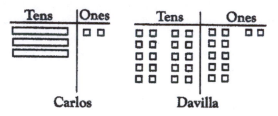

The examples above are representative of a student's work. If the error pattern indicated in these examples continues, the student's answer to the problem 9/11 − 1/7 will most likely be

a) $\frac{10}{4}$ b) $\frac{8}{7}$ c) $\frac{8}{4}$ d) $\frac{9}{18}$

2.

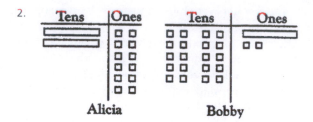

The illustrations above show how four students, Alicia, Bobby, Carlos, and Davilla, used base ten blocks to represent the number 32. Which of the students used the blocks to represent the number 32 in a way that reveals an understanding of the underlying concepts of a numeration system based on powers of ten?

a) Alicia   b) Bobby   c) Carlos   d) Davilla

The first item asks teachers to recognize or reason through students' errors; the second asks about how materials represent concepts. Taken together, these items seem to draw from the kinds of thinking described in Shulman's "pedagogical content knowledge" and the finer subdivisions of this construct by modern academic test developers (e.g., KAT's "teaching knowledge" or SII/LMT's "knowledge of content and students" and "knowledge of content and teaching"). These items are also the modern version of those on the NTE's elementary school specialist area test.

Note that the second item again illustrates the difficulty of writing defensible multiple-choice items in this arena. According to test materials, Carlos' representation is best. However, Alicia might also have a solid understanding of base ten numeration, as evidenced by the fact that she showed 2 tens and 12 ones—a representation of 32 commonly used for subtraction with regrouping. In fact, many teachers use base ten materials to explicitly teach the equivalence of Carlos' and Alicia's representations of 32, and we hazard that some educators might go as far as to argue that Alicia has a better understanding than Carlos for particular purposes (e.g., subtraction with regrouping). Although the presence of this item on Praxis released materials might reflect a simple mistake, it may also indicate a significant flaw in the construction of this assessment, in that we cannot imagine many mathematics educators would approve of such an item. If mathematics educators do not have a voice in the construction of this assessment, it again calls into question the degree of professional control over the assessment. It also harkens to the precursor to the Praxis II, the National Teacher Examination, which carried other such ambiguous items.

Moving further toward practice, two Praxis II tests (elementary content area exercises and secondary mathematics pedagogy assessment) ask teachers to answer essay questions involving mathematical scenarios like the following:

> You are teaching a unit on solving quadratic equations. You have already taught the students how to solve quadratics by taking square roots and by factoring. In your next lesson, you plan to teach the students how to solve quadratic equations by completing the square.
>
> Design a homework assignment for your students to complete after the lesson on solving quadratic equations by completing the square. The homework assignment should consist of 5 problems that review previously taught skills and concepts while also providing practice in the newly introduced material.
>
> Briefly explain your rationale for including the skills and concepts that the problems illustrate.
>
> (ETS, 2003, p. 106)

Here, and in every other sample secondary-level prompt available to us, teachers must engage in lesson design. Other prompts include "Describe an investigation you would have your students make to discover the difference between 'regular' and 'equilateral' [polygons]. Your investigation may involve the use of any type of manipulative or any software package" (ETS, 2003, p. 114). And "Describe a strategy, using pictures or manipulatives, that you could use to help foster the students' conceptual understanding of equivalent fractions" (p. 110). On their face, these tasks do appear to tap mathematical knowledge as it is used in teaching; responses are scored based on not only demonstrated understanding of the mathematics, but also on the basis that the instruction described therein is "very likely to achieve the desired goals" (ETS, 2005, p. 2). However, it does so in a particular way, assuming that good teaching equates with constructing lessons from scratch. Left out in this type of assessment is the kind of mathematical knowledge used when adapting curriculum materials, or using these materials well.

Finally, like some of the professional organizations we describe below, the Praxis has moved into the realm of measuring teaching in the classroom. The Praxis III assessment, now used in two states, evaluates teachers during their first and second years in the classroom via observations and interviews by trained assessors. It assesses knowledge in 19 criteria in four interrelated domains: organizing content knowledge for student learning; creating an environment for student learning; teaching for student learning; and teacher professionalism. Although not specific to mathematics instruction, the prominence of content knowledge is noteworthy. The description of this domain begins with the following: "Knowledge of the content to be taught underlies all aspects of good instruction. Domain A focuses on how teachers use their understanding of students and subject matter to decide on learning goals, to design or select appropriate activities and instructional materials, to sequence instruction in ways that will help students to meet short- and long-term curricular goals, and to design or select informative evaluation strategies" (Dwyer, 1998, p. 182). Nonetheless, the specifications for content knowledge observed in Praxis III are generic, so how and whether teachers' mathematical knowledge for teaching is observed and appraised remains an open question. Whether mathematical knowledge is closely evaluated also depends in large part upon whether the observer is knowledgeable in this area.

Interstate New Teacher Assessment and Support Consortium (INTASC) Portfolio Assessments

Another effort to reformulate teacher certification has taken place under the auspices of the Interstate New Teacher Assessment and Support Consortium (INTASC), a consortium of state education agencies and national educational organizations sponsored by the Council of Chief State School Officers. INTASC was created in 1987 to provide a forum for states to learn collaboratively about and develop new accountability requirements for teacher preparation programs, new assessments for teacher licensing and evaluation, and new professional development programs. Following the publication of its *Model Standards in Mathematics for Beginning Teacher Licensing & Development: A Resource for State Dialogue* (1995), INTASC began to develop standards-based licensing tests for teachers. On the view that no single test could adequately assess a prospective teacher's ability to teach, INTASC recommended three types of licensing tests: 1) a test of content knowledge; 2) a test of teaching knowledge (pedagogy); and 3) an assessment of actual teaching. The first two types of tests—content knowledge and teaching knowledge—are intended for the conclusion of teacher preparation programs as a requirement for a provisional licensure.

To qualify for a permanent license, teachers would complete a third, performance-based test in the initial years of teaching. INTASC used a portfolio format for its assessment, "a collection of documents that tell the story of a candidate's teaching as it develops over a period of time" (INTASC, 2006). Portfolios had been growing in use as K–12 assessments through the 1980s, and their extension to teacher assessments made sense: they can provide evidence of the development of a teacher's practice and include instructional materials, student work samples, video records of teaching, and written commentaries. Like their model standards, INTASC's portfolio assessments are intended to be a resource that can be adopted or adapted, for example, by schools of education as a requirement for program completion, or by states as a requirement for certification or completion during the induction years. We focus on INTASC's portfolio assessment both because of its novel format and its view of teacher knowledge.

The INTASC mathematics teaching portfolio documents approximately 8 to 12 hours of a beginning teacher's practice, organized around a central mathematical topic. Portfolio entries include a description of the teaching context; a sequence of lesson plans highlighting central features of the lessons such as the mathematics content, tasks, and accommodations for three focal students; two featured lessons each documented with 20–30 minutes of video, assessments, and student work samples; a cumulative student assessment and scoring criteria, accompanied by student work marked with feedback; and an analysis of teaching and professional growth. Along with these records of practice, each entry also requires a written commentary with a rationale for and evaluation of the teacher's instructional choices. The following excerpt describes part of the commentary required for the series of lessons:

The Mathematics: Identify your goals and expectations for student learning across this series of lessons. Begin your commentary with a statement of these goals. Continue with a description of the broad mathematical concept(s) or idea(s) that unifies the lessons. Why is this concept or idea important to the study of mathematics? How do the lessons you have designed build a cohesive set of plans to address this concept or idea? Explain the mathematical connections across the lessons. Provide one or two examples of how the series of lessons builds on mathematics the students have already learned. Give one or two examples of how the series of lessons serves as a foundation for the mathematics students will learn later.

Provide a general description of your goals and expectations for students to reason mathematically, solve problems, communicate mathematically, and see connections within mathematics, connections to other disciplines, and connections to the real world across these lessons. Provide specific examples of how your lessons will promote reasoning, problem solving, communication, and connections through the mathematical concept or idea that is the focus.

(INTASC, 1996, p. 25)

Completed portfolios are evaluated based on five interpretive categories—tasks, discourse, learning environment, analysis of learning, and analysis of teaching. Guiding questions for each category illustrate aspects of teachers' performance that could be considered in the collection and analysis of evidence.

The INTASC portfolio assessment is thus very different from prior teacher certification exams in that teacher knowledge of mathematics is assessed through the tasks, discourse, and analysis of student learning presented in the portfolio. Three of the evaluation framework's guiding questions are identified as specifically capturing facets of mathematical knowledge used in teaching: the appropriateness of the tasks for the instructional goals; the teacher's use of notation, language, and representation; and the accuracy of the teacher's interpretation and evaluation of information about students. Other areas of the portfolio might also demonstrate a teacher's use of mathematical knowledge in teaching, for example, in making decisions about when to probe students or in the ability to maintain the cognitive demand of a task during its implementation. Thus, the INTASC assessment has the potential for capturing the type of professionally situated knowledge identified by researchers—mathematical knowledge as it is deployed in the work of teaching. However, recognizing a teacher's skillful task selection in mathematics or imprecise use of mathematical language requires portfolio readers with significant mathematical knowledge. Thus, although the INTASC portfolio has the possibility of capturing mathematical knowledge as it is used, or misused, in teaching, its ability to do so would depend on the reader.

It is also worth noting that INTASC materials embrace the particular vision of mathematics instruction that is articulated in the INTASC standards and NCTM's *Professional Standards for the Teaching of Mathematics* (1991). Although the assessment handbook says that "there is no singular right way to teach mathematics" (INTASC, 1996, p. 36), teachers are required to demonstrate evidence of particular instructional formats and types of mathematics tasks. For example, one of the featured videotaped lessons must show students engaged in "mathematical problem solving and reasoning" where problem solving is defined as "students exploring new ideas and trying to pull prior learning together to address non-routine problems, not simply applying an algorithm to a new context," and reasoning is defined as "students making and testing conjectures, constructing arguments, judging the validity of arguments, formulating counterexamples, etc." (INTASC, 1996, p. 27). In addition to problem solving and reasoning, the featured lessons must, between them, capture other aspects of instruction such as both whole group and non-whole group formats, student-to-student discourse, and the introduction or development of a mathematical concept. Whereas historical teacher tests kept knowledge of pedagogical techniques and knowledge of content separate, they are here intertwined; teachers must exhibit specific instructional skills. While this is part of a larger story about changes in "pedagogical" knowledge over the last two centuries, it warrants noting here because of its implications for the orientations and skills novice teachers must possess.

National Board for Professional Teaching Standards (NBPTS) Certification

In contrast to both the Praxis and INTASC assessments, the National Board for Professional Teaching Standards (NBPTS) certification is a voluntary endorsement for teachers who already possess the minimum state requirements for teacher certification and who have at least three years of experience. Certification is currently available in 24 areas (NBPTS, 2006b)—including Early Childhood/Generalist, Middle Childhood/Generalist, and Early Adolescence/Mathematics—and is valid for ten years, after which teachers can apply for renewal. There are specific standards for each certificate that describe "what an accomplished teacher should know and be able to do" (NBPTS, 2006a). In fact, the National Board Standards informed the development of INTASC standards, and, like INTASC, the NBPTS is endorsed both by governance (e.g., National Governors' Association; National School Boards Association) and teacher education associations (e.g., NCATE); NBPTS is also backed by the National Education Association.

NBPTS certification has two components: assessment center exercises and portfolio entries. These two types of performance-based assessments are designed to measure not only teachers' content knowledge, but also the skills and judgment they must deploy routinely in practice. Although the specifics and standards vary, the process is the same across different types of certificates.

Assessment center exercises focus primarily on content knowledge and are designed to measure what accomplished teachers should *know*, rather than questions that can be studied for. Teachers complete six open-ended response exercises at designated testing centers, located across the nation. Candidates are given up to 30 minutes to respond to each exercise. Organized around "challenging teaching issues," exercises require teachers to use their content knowledge in pedagogical situations, for example, to identify student misconceptions and plan an appropriate instructional strategy, or to describe a learning experience based on a foundational concept or in response to a student's inquiry. NBPTS contracted with ETS, the same organization that publishes Praxis, to develop its scoring systems.

All six assessment center exercises for the Early Adolescence/Mathematics certificate are focused on mathematics topics: algebra and functions, connections, data analysis, geometry, number and operation sense, and technology and manipulatives. For both the Early and Middle Childhood/Generalist certificates, however, only one of the six exercises explicitly involves mathematics. The Early Childhood/Generalist scoring guide describes such an exercise:

> Candidates will demonstrate their ability to identify mathematical misconceptions or difficulty in a student's work, to state the fundamental prerequisites needed by this student in order to learn this particular mathematical concept, and to plan an instructional strategy based on real-world applications. They will also be asked to choose the materials and to provide a rationale for their choice of these materials that will be used to teach these prerequisites. (NBPTS, 2005b, p. 9)

The scoring guide provides the example in Figure 4.1.

In this example, candidates must identify and explain a common student misunderstanding—failing to recognize non-equilateral triangles as triangles—then design instruction to remedy this problem. This exemplar can be characterized as "pedagogical content knowledge," for it asks teachers to solve mathematics-specific problems involving student misunderstandings and teaching strategies. However, because there is only one mathematics-specific task on the generalist certification assessments, it is unlikely that *mathematical* pedagogical content knowledge is the construct that is contained in the overall score teachers receive. Instead, teachers' scores on all of the assessment center exercises and portfolio entries, across content areas, are weighted and combined to make a final assessment about their level of expertise.

The NBPTS portfolio allows for in-depth examination of practice. It requires four entries, three of which document specific aspects of a teacher's practice

Stimulus

A kindergarten student is having difficulty with a mathematical concept:

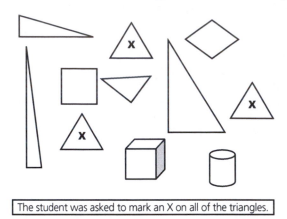

The student was asked to mark an X on all of the triangles.

Prompts

1. Identify the mathematical misconception/difficulty of this student work sample.
2. What fundamental concepts are prerequisites for this student at this grade level in order to learn this skill?
3. Based on real-world applications, state your goal for an instructional strategy or learning experience to help this student. Plan a learning experience or instructional strategy based on this goal that would further the student's understanding of this mathematical concept.
4. What materials would you use to teach this mathematical concept to this child? Provide a rationale for your choice of materials.

Figure 4.1 Example in the Early Childhood/Generalist scoring guide. (Reprinted with permission from page 84 of the National Board for Professional Teaching Standards, Early Childhood/Generalist Scoring Guide, 2005. All rights reserved.)

using records such as videotapes, student work samples, and related instructional tasks and classroom assessments.[6] In addition to creating these records, teachers write reflective commentaries that describe, analyze, and evaluate their practice. Similar to the assessment center exercises, the Early and Middle Childhood/Generalist certificates require only one of the four portfolio entries to focus on mathematics. And in fact, this entry requires teachers to document their integration of mathematics and science, as described in the 308-page portfolio instructions document for the Middle Childhood/Generalist certificate:

In this entry you will demonstrate how you help students better understand a "big idea" in science using relevant science and mathematics knowledge. You will engage students in the discovery, exploration, and implementation of these science and mathematics concepts, procedures, and processes by integrating these two disciplinary areas. This entry is designed for

you to provide evidence of your ability to plan, describe, illustrate, assess, and reflect on your teaching practice.

For this entry, you must submit the following:

- Written Commentary (14 pages maximum) that contextualizes, analyzes, and evaluates the teaching and integration of the math and science instruction.

- One Video Recording (15 minutes maximum) that demonstrates how you engage two small groups of students, with at least three students per group, in a science lesson that integrates mathematics.

- Three Instructional Materials (no more than a total of 6 pages) related to the lesson featured on the video recording that will help assessors understand what occurred during the lesson. (NBPTS, 2006d, p. 169)

Each of the portfolio entries is evaluated independently according to a detailed rubric addressing selected NBPTS standards. For example, the portfolio entry described above is scored for standards such as knowledge of students, knowledge of content and curriculum, meaningful applications of knowledge, and reflection. For the two mathematics-specific certificates (Early Adolescence/Mathematics and Adolescence and Young Adulthood/Mathematics), the four portfolio entries each focus on a different aspect of mathematics teaching: developing and assessing mathematical thinking and reasoning, whole class mathematical discourse, small group mathematical collaborations, and contributions to student learning.

Our inspection of the NBPTS documents suggests several stances toward the mathematical knowledge needed for teaching, and toward the measurement of this construct. First, the NBPTS assessments do offer the opportunity to capture how teachers use their knowledge of mathematics in practice. For example, the mathematics-related assessment center exercise for the Middle Childhood/Generalist certificate describes a high scoring response as follows:

The response offers clear, consistent, and convincing evidence of the ability to demonstrate pedagogical and content knowledge of math by accurately identifying the math misconception/error, providing an instructional strategy to assist student understanding of concepts/skills needed to accurately solve the mathematical problem, and providing a rationale that supports the instructional strategy. (NBPTS, 2005c, p. 53)

Like the INTASC standards, the NBPTS assessment presumes that teachers need intertwined knowledge

---

[6] The fourth focuses on accomplishments outside the classroom.

of content, students, and instructional design—and fluency with that knowledge, as evidenced by the short span of time allotted for teachers to answer the assessment center exercises. In this way, both the assessment center exercises and the portfolio entries reflect two aspects of teachers' mathematical work in classrooms: using integrated knowledge, and using it in the moment, rather than in a context more removed from actual classroom work. Thus, the NBPTS assessments better measure professionally situated mathematical knowledge than did past teacher certification exams, but, like the INTASC portfolios, are dependent on the samples of practice the teacher submits, as well as the evaluator's ability to recognize a teacher's use of mathematical knowledge in practice.

Second, there are a number of measurement-related concerns, some more pressing than others. While both INTASC and NBPTS have provided evidence for the validity of their measures, evidence on their reliability and other technical properties is more difficult to come by. Such evidence should be broadly available. Also, the small number of mathematics/science questions within K–8 exercises and prompts raises concerns about the degree to which the sampled tasks generalize to teachers' overall level of expertise in teaching. This is complicated by the fact that teachers can choose lessons for portfolio entries, and those lessons may not represent their everyday practice (Porter, Youngs & Odden, 2001). And this is even further complicated in the NBPTS generalist certificates by the combination of mathematics and science in the same portfolio task. Because NBPTS is interested in measuring accomplished general practice, they chose not to design separate measures of proficiency in teaching these two content areas. But accomplished practice might be discipline-specific. This leaves under-assessed the specific mathematical qualities of a teacher's instructional practice.

Third, both the NBPTS and INTASC portfolios seem as much designed to capture teachers' *reflection* on practice as the practice itself. For instance, the written commentary guidelines for the NBPTS include prompts such as: *What evidence of inquiry, intellectual engagement, discussion, and content is demonstrated in your video recording? How did you further students' knowledge and skills and engage them intellectually? How does the discussion and/or activity featured on the video recording reveal students' reasoning and understanding? Describe a specific example from this lesson as seen on the video recording that shows how you ensure fairness, equity, and access for students in your class* (NBPTS, 2006d, p. 173–174). This is a view of teaching that is heavily tailored toward ideas about teachers as "reflective practitioners" (Schön,

1995). It is worth considering, however, whether accomplished reflection on teaching is required for accomplished teaching; if it is not, then NBPTS might be missing a class of proficient educators. Moreover, the relationship of articulate reflection to effective instruction is not clearly established; yet the NBPTS makes this a central demand of its assessment strategy. Are teachers who can describe, explain, and reflect on their work better teachers than those who are less able to articulate their designs, purposes, or analyses? Possibly they are, for it may be that the ability to articulate one's practice is an indicator of deliberateness, and that the ability to write cogent reflections an indicator of analytic capacity, both of which may predict student achievement. But the nature of the evidence for this remains unexplained.

Finally, like INTASC, with its very specific view of mathematics instruction, and Praxis, with its instructional design tasks, the view of instruction presented in the NBPTS precludes some types of teachers from being recognized as accomplished—for instance, those who follow scripted curriculum materials, or those who do not teach mathematics using the inquiry/discussion/engagement methods suggested above. In the scoring guide, for instance, teachers whose portfolio is rated poorly are urged to consider whether it "provide(s) evidence that you have created a classroom environment that promotes active learning by all your students" and "provide(s) evidence that you were able to engage students in an effective classroom discussion or inquiry appropriate to the goals of your teaching" (NBPTS, 2005a, p. 15). Thus, there are examples of competent teaching that might not be recognized by NBPTS. Should a teacher who controls most classroom talk, but who also gives clear explanations of mathematical procedures or careful demonstrations of proofs be recognized as accomplished? A read of NBPTS documents suggests not: the organization bases its scoring system on what we have called the *quality of mathematics instruction*, or teachers' mathematical knowledge *and* pedagogical choices, considered together. However, other stances are also possible.

Other Certification Tests

Several other certification tests affect a large number of prospective teachers.

American Board for Certification of Teacher Excellence (ABCTE)

The certification program designed by the ABCTE is meant to attract career-changing adults into teaching, allowing these individuals to circumvent teacher education programs and other state requirements for entry into the profession. As part of ABCTE certification, teachers must possess a bachelor's degree, pass "rigorous examinations of subject area and professional teaching knowledge" (ABCTE, 2006a), and submit materials for a federal background check. This process brings to mind certification routes from the eighteenth and nineteenth centuries, in which teacher candidates possessed some minimum level of schooling, passed an exam, and were assessed for moral fiber. As we shall see below, many of the debates evoked by the ABCTE program echo those from the late nineteenth century.

In mathematics, as in most other ABCTE subjects, the exam is entirely multiple-choice and administered via computer. Test content is based on ABCTE standards, which were developed in turn by reviewing state and professional standards. Standards for elementary teachers suggest that test developers expect teachers to know not only the content in the K–6 curriculum, but also content their students will encounter in middle or high school. For instance, in the area of algebra, teachers certified via the elementary exam must know the symbolic language of mathematics, "including applying proportional reasoning, defining the inverse function and performing arithmetic operations on functions, solving systems of linear and quadratic equations and inequalities, and extending and recognizing linear patterns" (ABCTE, 2006b). Many of these topics are typically taught in middle or high school. The middle/high school standards extend into content often taught in post-secondary education, such as least squares regression, matrix algebra, and calculus topics such as limits, derivatives, integrals, and convergence of sums. Secondary test items reflect this emphasis:

1. Twenty gallons of solution A, which is 35% acid, is mixed with 30 gallons of solution B. The resulting mixture contains 20% acid.

   What percent of solution B is acid?
   a) 5%   b) 10%   c) 12.5%   d) 15%

2. The following list of ordered pairs represents a functional relationship:

   $$(1,1)\quad(2,4)\quad(3,9)\quad(4,16)$$

   What principle of functional relationships supports this statement?
   a) The x values are all different.
   b) The relationship between each pair of numbers is reciprocal.
   c) The pairs are listed in consecutive order.
   d) The relationship between each pair of numbers is the same.

3. What is $\lim\limits_{x \to 2}\dfrac{x^3 - 8}{x - 2}$?

   a) 0   b) 4   c) 8   d) 12

The first item is perhaps a modern-day equivalent of the types of problems that appeared on the 1890s Michigan teacher exams; the second taps more conceptual knowledge, investigating teachers' understanding of the definition of functions; the third item taps a topic typically not taught until college. Overall, this is a mathematically demanding assessment, with items that are computationally intensive, designed to push on candidates' misconceptions, and to push into territory beyond that typically taught in high school.

To date, the ABCTE certification process has not been widely used. Although the program is funded by two U.S. Department of Education grants totaling $40 million, it certified only 135 individuals between its founding in 2001 and April of 2006 (Schimmel, personal communication, April 19, 2006). One reason for the slow start is that the ABCTE certification must be recognized by states before teachers can gain employment; in mid-2006, only five states recognized the ABCTE certification. Another reason might be the proliferation of state and district-based alternative certification programs, which reduce incentives for teachers to seek the ABCTE stamp of approval, and also for states to recognize the organization's certification process.

The tepid response by states and teachers has not dulled the uproar over the ABCTE certification process. For the most part, critiques center around the assumption that teachers can be certified via a multiple-choice exam, absent professional training and field experiences. As Berliner (2005) writes of the ABCTE and similar tests, "it is a near impossibility to adequately assess quality teaching through pencil-and-paper tests of professional knowledge. . . . These tests fail in part because of the complexity of classroom environments and the near impossibility of capturing that reality in pencil-and-paper formats" (p. 211). Educators also assert that in other professions, such as law, medicine and architecture, the certification process includes not only an exam but also extended professional learning and apprenticeship.

Thus the ABCTE has reopened the debate over who is qualified to teach, and how best to assess individuals' skills and capacity for this work. Teacher educators, many of whom oppose any form of alternative certification, have singled out this test-based certification method for particular opprobrium, much as educators at the turn of the twentieth century railed against the teacher tests administered by local school boards and officials. Many issues are the same. Teacher educators claim there is a professional body of knowledge learned primarily through preservice preparation and apprenticeship, and that tests provide a "back door" for under-qualified individuals to enter teaching. ABCTE supporters argue that teachers' subject matter knowledge is of the foremost importance, and that there needs to be a method to identify knowledgeable individuals to help alleviate teacher shortages, shortages caused in part by the bureaucratic restrictions most states have placed on the entry to teaching. ABCTE supporters also take the view that teachers need to know more advanced content than they are expected to teach students, much as the Michigan superintendent (and test promoter) Henry R. Pattengill did over 100 years earlier. Teacher educators, for their part, tend to focus more on promoting deep knowledge of the specific curriculum teachers will teach students.

California Subject Examinations for Teachers

The California Subject Examinations for Teachers (CSET) are reviewed here briefly as an example of both a state-specific test, and as an example of a customized exam developed by National Evaluation Systems (NES, 2006) specifically for one state's use. NES is the second-largest supplier of teacher certification assessments, providing tests to not only California but also Texas and a number of smaller states. In California, NES customized tests to align with California curriculum frameworks and student standards.

Two sample items from the elementary-level (multiple subjects) exam provide some insight into the composition of the test:

1. If the number 360 is written as a product of its prime factors in the form $a^3 b^2 c$, what is the numerical value of $a + b + c$?

   a) 10   b) 16   c) 17   d) 22

2. The problem below shows steps in finding the product of two two-digit numbers using this standard multiplication algorithm. The missing digits in the problem are represented by the symbol □.

$$\begin{array}{r} □9 \\ \times\, 36 \\ \hline 29□ \\ +\, □4□□ \\ \hline □□□□ \end{array}$$

   What is the hundreds digit in the product of the two numbers?

   a) 1   b) 4   c) 6   d) 7

These and other items assess upper-elementary content (e.g., slope; proportionality). They also appear designed to draw on both candidates' basic skills and on their ability to reason through novel mathematics problems. On this test, candidates' mathematics and science ability is reported as a separate subtest score, allowing test users to gauge individual competency in both areas combined. Praxis II, in contrast, only reports overall scores to employers.

Candidates for secondary mathematics positions must take two mathematics examinations, one focusing on number theory and algebra and the other focusing on geometry and probability and statistics. To teach advanced placement classes or classes in analysis and calculus, a candidate must take a third mathematics examination covering the history of mathematics and calculus. The examinations were designed by NES to follow the mathematics content standards for California Public Schools, and these are expressed as lists of topics to know for teaching. This parallels the development of Praxis, where items were developed to reflect mathematical knowledge represented by lists of topics. The example below shows the kind of background CSET provides to explain the undergirding of its mathematics items:

1.4   Linear Algebra

a. Understand and apply the geometric interpretation and basic operations of vectors in two and three dimensions, including their scalar multiples and scalar (dot) and cross products.

b. Prove the basic properties of vectors (e.g., perpendicular vectors have zero dot product)

c. Understand and apply the basic properties and operations of matrices and determinants (e.g., to determine the solvability of linear systems of equations)

(Mathematics Content Standards for California Public Schools, Algebra I: 9.0; Algebra II: 2.0; Mathematical Analysis: 1.0; Linear Algebra: 1.0–12.0)

One can see the direct connection between this policy statement and the item below:

If vectors $\vec{a} = (a_1, a_2)$ and $\vec{b} = (b_1, b_2)$ are perpendicular,

$$A = \begin{pmatrix} \cos\theta & -\sin\theta \\ \sin\theta & \cos\theta \end{pmatrix},$$

and if we identify vectors with column matrices in the usual manner, then show that the vectors $A\vec{a}$ and $A\vec{b}$ are perpendicular for all values of θ.

Like the elementary exam, the secondary exam appears to emphasize reasoning and proof, at least in its constructed response items. The exam also appears quite challenging; the following items draw from post-secondary content:

1. Which of the following statements refutes the claim that $GL_R(3)$, the set of 3×3 invertible matrices over the real numbers, is a field?

   a) There exist elements A and B of $GL_R(3)$ such that $AB \neq BA$.

   b) There exist elements A and B of $GL_R(3)$ such that $\det(AB) = \det(A)\det(B)$.

   c) If A is an element of $GL_R(3)$, then there exists a matrix A^{-1} such that $A^{-1}A = I$.

   d) If A is an element of $GL_R(3)$, then there exists a matrix A such that $\det(A) \neq 1$.

2. Show that the subset of complex numbers of the form $a + bi$ with a and b rational numbers satisfies the axioms of a field under the operations of addition and multiplication of complex numbers.

In all, however, many of the items on the CSET, ABCTE, and secondary Praxis II look quite similar. This overlap is one characteristic of current secondary teacher certification tests: aside from different content emphases and more or less difficult items, there is less variation than one would think, given the proliferation of current tests.

The Future of Certifying Teachers of Mathematics

At no time in this nation's history have there been so many routes to becoming a certified teacher; teachers can do everything from taking multiple-choice content exams to constructing elaborate records of their teaching practice. Where does this leave those interested in moving teacher certification toward an assessment of more job-relevant, discipline-specific professional skills? We begin by discussing the benefits and drawbacks of current certification exams, then in our conclusion consider some of the issues that might lead to the next generation of certification tests.

Clearly, most new forms of teacher licensure testing have significant benefits as compared to past forms and, in some cases, even as compared to scholarly methods for assessing mathematical knowledge for teaching. To start, unlike their historical counterparts, most of the current tests attend carefully to issues of reliability and, increasingly, validity. For instance, a number of studies comparing NBPTS teacher candidates who have received certification and those who have not have revealed measurable differences between NBPTS-certified and regularly certified teachers and their impact on student learning and growth (Bond, Smith, Baker, & Hattie, 2000; Cavalluzzo, 2004; Goldhaber, Perry & Anthony, 2004; Vandevoort, Amrein-Beardsley, & Berliner, 2004; for a counterexample, see Sanders, Ashton, & Wright, 2005). This is a key question centering on predictive validity: Does the assessment provide information about the candidate's actual effectiveness in the classroom? If not, the assessment is of little value.

Second, most current certification tests explicitly intend to assess professionally specific knowledge of mathematics, such as pedagogical content knowledge. And, in the case of performance-based assessments— such as observations, portfolios, the Praxis content exercises, or the NBPTS assessment exercises—not only does the content of the assessments more accurately reflect ways that teachers use mathematics in practice,

the assessments themselves emerge from or are closely tied to the actual work teachers do. Given that sociologists typically define professions by the presence of a codified body of knowledge with professionally controlled barriers to entry for new practitioners, this is a significant advance. In fact, efforts such as INTASC and NBPTS actually contribute to this development, for their standards attempt to articulate a coherent and comprehensive vision of the knowledge, skills, and dispositions needed for teaching.

Another benefit of performance-based assessments is they may actually stimulate teacher reflection and learning, thus providing an opportunity for professional development. For instance, individuals who seek ABCTE certification take a self-assessment and, if necessary, work with a learning advisor to refresh and/or learn content areas prior to taking the actual exam. Teachers have reported that the NBPTS certification process inspires growth and self-evaluation (Rotberg, Futrell, & Lieberman, 1998) and encourages teachers to become more reflective about their practice (Haynes, 1995; Letofsky, 1999; Sumner, 1997; Tracz et al., 1995).

However, new forms of teacher certification also have drawbacks. To start, the closer to practice the assessment, the less likely it is to be used. The secondary content knowledge Praxis II, for instance, was taken by six times more examinees than the secondary mathematics pedagogy exam (ETS, 2006b, p. 11). Only 2 of the 43 states that use Praxis I and/or II for certification require Praxis III. And only small numbers of teachers seek NBPTS certification; one recent survey found that only 2% of middle grades teachers and 3% of high school teachers had applied for certification in the past three years (Horizon Research, 2002).

A second issue centers on equity: Research routinely suggests certification testing has adverse impacts on teachers from minority groups, who for instance have historically scored lower on both the NTE and Praxis.[7] Minority candidates may successfully complete teacher education coursework and internships but not meet their states' requirements for a passing score on Praxis II (Albers, 2002). This raises questions about the construction of the Praxis tests themselves. Some suggest the content and the format privilege the knowledge base of certain groups: for example, that Praxis tests in English favor white literature over minority literatures, and therefore disadvantage minority teacher education students. It has also been suggested that the curricula and style of learning predominant

---

[7] See Haney et al. (1987) for a discussion of the NTE. For Praxis, Wakefield (2003) examined statistics released from the Georgia Professional Standards Commission. These statistics revealed that in 2001, European Americans had a 93% pass rate on Praxis II, while African Americans had a 71% pass rate (p. 383).

in historically black universities and colleges is not reflected in standardized tests in general and in Praxis in particular (Albers, 2002).

There also have been equity concerns about NBPTS certification. Some believe that the process creates an unnecessary hierarchy within the field of teaching. Even though a growing array of grants, scholarships, and no-cost loans are now available to cover the cost of participation in the certification process, some argue that the $2,500 fee may still limit who can participate in the process (King, 1994; Marshall, 1996). Others critique the lack of inclusion of "culturally sensitive pedagogies" in the framework (Hamsa, 1998; Irvine & Fraser, 1998). Furthermore, while the number of minority applicants is proportionate to their representation in the teaching work force (17%), their achievement of certification is lower (only 11% of NBPTS-certified teachers are minority). NBPTS and other researchers are critically examining the role of cultural bias in these results as well as support mechanisms that might address this situation (Kraft, 2001; SRI, 2004; Serafini, 2002).

A third concern involves reliability. How accurately can the assessments determine how much mathematical knowledge for teaching an individual has? Standardized assessments, such as Praxis, return indices of reliability that can be used to assess the accuracy of the test. For more novel formats, such as portfolios and constructed response tasks, the determination of the degree of accuracy is more complex—dependent on the selection of specific tasks, scoring rubrics, and also on the ability of the person scoring the assessment. Reliability between scorers must be reached, but there must also be some assurance, in our view, that if one took another sample of a teacher's mathematical practice, the same overall judgment regarding certification would be reached. With any stakes attached to the assessment, this is a critical issue.

Yet another issue is the lack of evidence on the validity of many of these assessments, or the degree to which these certification tests predict what teachers can do in classrooms to help students learn. Aside from the NBPTS and some of the scholarly measures, few assessments have been validated by anything more than a review of the literature and signatures from the practicing teachers and mathematics educators who serve on standards development and assessment review panels. The need for more evidence on predictive validity in particular—that is, whether teachers who pass specific certification tests produce better-educated students than those who do not—is dire.

Fourth, it is interesting to note how the mathematical topics to be tested have changed over time. Today there is an increasing emphasis on algebra in some exams: fully 40% of the CSET items for mathematics certification in California are algebra problems, and the Praxis II examination for secondary mathematics teachers has 16–18% algebra items, up from 12–14% on the NTE. Interest in geometry has grown as well: in the days of the NTE, geometry items represented 16–18% of the entire pool of mathematics items, and the Praxis II exams, which replaced the NTE in most states, now have 22–26% geometry out of all math items. What counts as mathematically important knowledge for teaching seems to vary over time and by test.

So where does this leave us? Clearly, there is currently no ideal certification assessment, at least in mathematics. However, we do stand in a better position than just twenty years ago. Basic research into mathematical knowledge for teaching has supplied a vision for what might be measured on these assessments, one that contrasts with the "teachers should know a little more than their students" view taken during most of U.S. history. And advances in assessment theory—both psychometrics and test theory in general—have encouraged the adoption of better formats and improved test characteristics. Yet much remains to be done. Despite the alternative vision of what should be measured, we do not know whether this is the mathematics that actually matters to student learning. Pattengill's old question—whether teachers need to know content, or teaching methods, or both—survives today, although in slightly different form. And the items on this list are complicated by the fluid nature of professional knowledge, and by debates over mathematics teaching more broadly. Some of what counted as profession-specific mathematical knowledge on the 1900 certification exams does not count as knowledge today; what counts as knowledge today might not count as knowledge tomorrow.

CONCLUSION

Teacher examinations have a complicated history and pose a number of serious contemporary dilemmas. Even 150 years after the first written teacher tests emerged in the U.S., there is little agreement on what, whom, and how to measure, and for what purpose. Pressing questions—such as the balance of knowledge of content *and* knowledge of pedagogy, the nature of content knowledge useful for teaching, and the "content" of pedagogical knowledge — have not been answered. Meanwhile, the number of teacher exams continues to increase. And pressures for more accountability, both among the programs that prepare teachers and among teachers themselves, continue to mount.

Given these pressures, it is not likely that teacher testing will subside. The challenge, perhaps, is to create the best tests possible. Measuring teacher knowledge, even using standardized modes of assessment, *can* be done in ways that honor and define the work of teaching, ratify teachers' expertise, and help to ensure that every child has a qualified teacher. Doing so requires carefully constructed instruments that take seriously the work of teaching and that can be used at scale. In this concluding section, we turn to consider possible ways to contribute to the improvement of teacher assessment, both standardized and other forms. Our remarks take the form of proposed directions and foci, based on our review of practice in this domain.

Measure Mathematical Knowledge for Teaching

It is no surprise that we argue for measuring the mathematical knowledge used in teaching. Valid teacher assessments should not be remote from what teachers are asked to do in classrooms, with real students, materials, and content. The extended example of division of fractions in the middle of this chapter illustrates some of our own bets about what kind of knowledge matters; others have different bets, and there will no doubt be some evolution toward the "correct" mix of problem types, tasks, and items that should be contained on teacher assessments. With effort, there can also be an evolution toward appropriate formats for the assessment, through in-person observations, pencil-and-paper assessments, portfolios, or other formats. These directions can bridge more effectively the scholarly work on teacher knowledge and the policy-related work on assessing teacher quality.

However, we are worried about two trends. One is this field's tendency to conflate teachers' knowledge of mathematics for teaching with other types of knowledge or skill. For instance, many of the assessments we studied conflated knowledge of instructional approaches (e.g., using manipulatives; involving students in meaning-making; teaching in line with the NCTM standards) with knowledge of mathematics. This is reminiscent of our distinction between the *quality of mathematics instruction*, which we defined as inclusive of instructional approaches, and the *quality of the mathematics in instruction*, which we defined as focused specifically on the actual mathematics deployed in the course of a lesson. Other assessments of mathematical knowledge for teaching required teachers to design lessons, a possibly separate skill. And still other assessments appear to be scored based on a teacher's ability to reflect about her instruction. Such assessments are not necessarily invalid; however, these other elements of teacher knowledge should be recognized as a crucial component of what is being measured, and test scores should be interpreted appropriately.

Mathematical knowledge for teaching should also not be represented as only one element in a composite score. Mathematics teaching is important in its own right, and students deserve teachers who have qualified for their jobs based on knowing mathematics in the ways it should be known for teaching—not demonstrating preparedness in many areas, including mathematics. This is especially true, in our analysis, for elementary-school certification exams, where mathematics is likely to be combined with other areas (e.g., science, in the elementary CSET and NBPTS) or with all subjects (e.g., elementary content area assessment of the Praxis II).

Measure with Care

No test is perfect; as we have shown, every format, whether multiple choice, observational, or performance based, has advantages and disadvantages. Recognizing this, the task facing educators and policy-makers is to carefully and responsibly link the research/policy question being investigated to the most appropriate form of assessment.

Two other concerns are worthy of mention. Assessments can send unintended messages about their purpose. The format of the assessment may serve, inadvertently, as a model of assessment that teachers believe should then be used in assessing their students. Multiple-choice exams are vulnerable to many criticisms, yet their use for the testing of teachers may suggest to teachers that this is a preferred assessment method. Even when the content of the multiple choice questions is focused on robust expressions of what teachers know to teach, the format may speak louder and have greater implications than intended by the developers. Similarly, mathematics educators have worked hard to counter the misconception that mathematical competence is demonstrated by quick solutions to routine mathematical problems. Certain testing formats, multiple-choice in particular, may solidify this misconception among teachers.

This field also needs to recognize how the technical requirements for standardized testing affect test content. In his scathing critique of standardized testing in history, Wineburg (2004) notes that the vagaries of modern psychometrics conspire to hide what students know about history rather than measure what they *do* know. Wineburg describes how analytic techniques such as biserial correlation disadvantage those whose knowledge base is different than the mainstream, and that only items that produce a good "discrimination index" (that is, that differentiate between high achievers and low achievers) are included in multiple-choice exams, regardless of the intrinsic value of their content. He writes, "If ETS statisticians determined during pilot testing that most students could identify George Washington, 'The Star-Spangled Banner,' Rosa Parks, the dropping of the bomb on Hiroshima, slavery as a main cause of the Civil War, the purpose of Auschwitz, Babe Ruth, Harriet Tubman, the civil rights movement, the 'I Have a Dream' speech, all those items would be eliminated from the test, for such questions fail to discriminate among students" (2004, p. 1409). Measures of mathematical knowledge for teaching are vulnerable to the same dilemma.

Use Multiple Approaches

One way to effect well-considered improvements in this enterprise is to avoid repeating the mistakes of the past. One of the problems with assessing professional knowledge has been reliance on traditional test formats. In this chapter, we have examined both standard assessments formats and more novel forms of assessment, including discourse analysis, video analysis, and portfolio assessment. The most widely used teacher test today is the Praxis series, which offers multiple-choice and open-ended items in the paper-and-pencil tests that measure inert knowledge for teaching, as well as real-time observation of teaching that is designed to assess knowledge in action. Only two states are currently using the observation-based Praxis III for certification, but we think this combination of testing formats is promising. New uses of multimedia for documenting teaching offer other avenues for comprehensive appraisals of teaching work, as discussed earlier in the chapter.

Meet Professional Standards of Rigor in Assessment

Another important step is to confront the challenge of rigor, and what "rigor" might appropriately mean. New assessments should be held to more rigorous standards than old assessments, which were amazingly often validated only with reference to the content they covered, despite their use in making inferences about practice and its quality. New assessments should be rooted in theory, and scores should be validated using a number of methods. In our own work, paper-and-pencil measures of the mathematical knowledge for teaching are validated by close analyses of the mathematical work of teachers that we videotaped. Is the mathematical knowledge that teachers demonstrate on a paper-and-pencil test expressed in their teaching? Can we show that students of highly qualified teachers, as defined by this paper-and-pencil test, learn more mathematics? If all mathematics assessments for teachers answered such questions, tests could bring real value to our efforts to improve educational quality. Rigorous standards for assessment already exist in the form of American Educational Research Association/American Psychological Association guidelines (1999) and the field can readily use them.

Learn from Other Measurement Methods

In the introduction to this chapter, we noted that seldom do lessons from one line of work measuring teachers' mathematical knowledge "travel" to other lines of work. Qualitative researchers have much to learn from large-scale test developers; large-scale test developers need lessons learned in qualitative research to succeed. More crossover needs to occur between these separate enterprises.

Attend to Issues of Equity

Attention to equity is crucial. As we saw, exams have a long and disturbing history of excluding people of color from the teaching profession. That these assessments regularly discriminated among candidates on the basis of characteristics unrelated to the knowledge and skill needed for teaching was a serious problem, and a threat to the validity of the exams themselves. Doing well on the tests must be related to success in teaching, not demographic characteristics. When patterns of performance are predicted by race, examinations must be scrutinized to uncover what it is about the test questions that may be producing this result. Tests that are invalid across groups in such ways must be challenged publicly; moreover, their use also serves to inhibit the development of a diverse teaching force.

Investigate the Relationship among Mathematical Knowledge for Teaching, Other Domains of Teaching Knowledge, and Student Learning

Much work remains to be done to specify what we are calling "mathematical knowledge for teaching." Important to understand is how this domain intersects with other domains of knowledge for teaching, and how these together affect student learning. We know, for example, that mathematical knowledge for teaching intersects knowledge for equitable teaching. Being explicit about mathematical practices makes mathematics learning accessible for wider populations. Mathematical knowledge for teaching may also act upon (or be acted upon by) teachers' ability to motivate students to learn, to organize classrooms for productive instruction, and to produce learning gains.

Increase Professional Role and Control

The history of U.S. teacher testing in mathematics has been one in which the tests have largely been developed by those outside the profession: local officials, bureaucrats, and later, professional test developers. In the coming years, professionals in education and in mathematics must develop the earned authority to control the identification of professional knowledge and the means of certifying it reliably. The judgments involved require specialized insight and perspective, rooted in evidence that is tied to instructional effectiveness—that is, instructional practices that reliably produce student achievement. Improvement in assessing teachers depends on close involvement of experts in mathematics instruction.

Where does the tour on which we have been lead? We have seen that teacher assessment was originally narrow, closely bound to the content of the curriculum. Teacher qualification was judged by teachers' ability to solve difficult word problems of the sort they might teach their students. Over the last 100 years, approaches to assessment have multiplied. What is assessed has broadened dramatically. How it is assessed has similarly expanded. Even within the same testing firm, different assessments are developed. The field has moved from a narrow and underspecified conception of the knowledge necessary for teaching to a broadband approach which lacks the precision and common agreement about requirements that one would expect of a profession. The agenda for assessing teacher knowledge is clear: The need to prepare and certify knowledgeable and skilled teachers who can be reliably effective with a wide range of students has perhaps never been greater. With the expansion of interest in and approaches to assessment, the field needs to move toward clearer precision, specification and agreement of measures, and broader skilled use of reliable and valid tools for appraising teachers' knowledge, skill, and performance. Research on teaching and teaching effectiveness is ready for this step; the current state of policy and practice demands it.

REFERENCES

Abbott, A. D. (1988). *The system of the professions: An essay on the division of expert labor.* Chicago: University of Chicago Press.

Albers, P. (2002). Praxis II and African American teacher candidates (or, is everything black bad?). *English Education 34*(2), 105–125.

American Board for Certification of Teacher Excellence. (2006a). *About passport to teaching.* Retrieved June 21, 2006, from http://www.abcte.org/passport

American Board for Certification of Teacher Excellence. (2006b). *Multiple subject exam.* Retrieved June 21, 2006, from http://www.abcte.org/standards/mse

American Education Research Association/American Psychological Association. (1999). *Standards for educational and psychological testing.* Washington, DC: AERA.

Angus, D. L. (2001). *Professionalism and the public good: A brief history of teacher certification.* Washington, DC: Thomas Fordham Foundation.

Appleman, D., & Thompson, M. J. (2002). "Fighting the toxic status quo": Alfie Kohn on standardized tests and teacher education. *English Education 34*(2), 95–103.

Baker, R. S. (2001). The paradoxes of desegregation: Race, class, and education, 1935–1975. *American Journal of Education, 109*(3), 320–343.

Ball, D. L. (1990). The mathematical understandings that prospective teachers bring to teacher education. *Elementary School Journal, 90,* 449–466.

Ball, D. L., & Bass, H. (2003). Making mathematics reasonable in school. In G. Martin (Ed.), *Research compendium for the Principles and Standards for School Mathematics* (pp. 27–44). Reston, VA: National Council of Teachers of Mathematics.

Ball, D. L., Lubienski, S., & Mewborn, D. S. (2001). Research on teaching mathematics: The unsolved problem of teachers' mathematical knowledge. In V. Richardson (Ed.), *Handbook of research on teaching* (4th ed., pp. 433–456). Washington, DC: American Educational Research Association.

Ball, D. L., Masters-Goffney, I., & Bass, H. (2005). The role of mathematics instruction in building a socially just and diverse democracy. *The Mathematics Educator, 15*(1), 2–6.

Ball, D. L., Thames, M. H., & Phelps, G. (2005, April). *Articulating domains of mathematical knowledge for teaching.* Paper presented at the annual meeting of the American Educational Research Association, Montréal, Quebec.

Begle, E. G. (1972). *Teacher knowledge and student achievement in algebra* (SMSG Rep. No. 9). Palo Alto, CA: Stanford University.

Begle, E. G. (1979). *Critical variables in mathematics education: Findings from a survey of the empirical literature.* Washington, DC: Mathematical Association of America and National Council of Teachers of Mathematics.

Berliner, D. (2005). The near impossibility of testing for teacher quality. *Journal of Teacher Education, 56*(3), 205–213.

Boardman, A. E., Davis, O. A., & Sanday, P. R. (1977). A simultaneous equations model of the educational process. *Journal of Public Economics, 7,* 23–49.

Borko, H., Eisenhart, M., Brown, C. A., Underhill, R. G., Jones, D., & Agard, P. C. (1992). Learning to teach hard mathematics: Do novice teachers and their instructors give up too easily? *Journal for Research in Mathematics Education, 23,* 194–222.

Bond, L., Smith, T., Baker, W., & Hattie, J. (2000). *The certification system of the National Board for Professional Teaching Standards:*

A construct and consequential validity study. Greensboro: University of North Carolina, Greensboro, Center for Educational Research and Evaluation.

Bush, W. S., Ronau, R., Brown, T. E., & Myers, M. H. (2006, April). *Reliability and validity of diagnostic mathematics assessments for middle school teachers.* Paper presented at the American Educational Association Annual Meeting, San Francisco.

Cavalluzzo, L. (2004, November). *Is National Board certification an effective signal of teacher quality?* Alexandria, VA: The CNA Corporation.

Coleman, J. (1966). *Equality of educational opportunity.* Washington, DC: U.S. Department of Health, Education, and Welfare.

Darling-Hammond, L., & Sykes, G. (2003, September 17). Wanted: A national teacher supply policy for education: The right way to meet the "highly qualified teacher" challenge. *Education Policy Analysis Archives, 11*(33). Retrieved June 21, 2006, from http://epaa.asu.edu/epaa/v11n33/.

Darling-Hammond, L., & Youngs, P. (2002). Defining "highly qualified teacher": What does "scientifically-based research" actually tell us? *Educational Researcher, 31*(9), 13–25.

Dwyer, C. A. (1998). Psychometrics of Praxis III: Classroom performance assessments. *Journal of Personnel Evaluation in Education, 12*(2), 163–187.

Education Trust. (1999). Not good enough: A content analysis of teacher licensing exams. *Thinking K–16, 3*(1).

Educational Testing Service. (1984). *A guide to the NTE core battery tests: Communication skills, general knowledge, professional knowledge.* Princeton, NJ: Author.

Educational Testing Service (1987). *NTE: A guide to the Education in the Elementary School Specialty Area Test.* Princeton, NJ: Author.

Educational Testing Service. (2003). *Praxis study guide for the mathematics tests.* Princeton, NJ: Author.

Educational Testing Service. (2005a). *Elementary education: Curriculum, instruction, and assessment (0011).* Princeton, NJ: Author.

Educational Testing Service. (2005b). *Mathematics: Content knowledge (0061).* Princeton, NJ: Author.

Educational Testing Service. (2005c). *Mathematics: Pedagogy test at a glance.* Princeton, NJ: Author.

Educational Testing Service (2006a). *State requirements.* Princeton, NJ: Author. Retrieved June 27, 2006, from http://www.ets.org/portal/site/ets/menuitem.22f30af61d34e9c39a77b13bc3921509/vgnextoid=d378197a484f4010VgnVCM10000022f95190RCRD.

Educational Testing Service (2006b). *Understanding your Praxis scores, 2005–2006.* Princeton, NJ: Author. Retrieved June 15, 2006, from http://www.ets.org/Media/Tests/PRAXIS/pdf/09706PRAXIS.pdf

Educational Testing Service. (2006c). *Elementary education: Content knowledge (0014).* Princeton, NJ: Author.

Ferguson, R. F. (1991). Paying for public education: New evidence on how and why money matters. *Harvard Journal on Legislation, 28,* 458–498.

Ferrini-Mundy, J., Floden, R., McCrory, R., Burrill, G., & Sandow, D. (2005). *A conceptual framework for knowledge for teaching school algebra.* East Lansing, MI: Authors.

Fischbein, E., Deri, M., Nello, M. S., & Marino, M. S. (1985). The role of implicit models in solving verbal problems in multiplication and division. *Journal for Research in Mathematics Education, 16*(1), 3–17.

Garet, M. S., Porter, A. C., Desimone, L., Birman, B. F., & Yoon, K .S. (2001). What makes professional development effective? Lessons from a national sample of teachers. *American Educational Research Journal, 38*(4), 915–945.

Geertz, C. (1973). *The interpretation of cultures: Selected essays.* New York: Basic Books.

Goldhaber, D., Perry, D., & Anthony, E. (2004). *National Board certification: Who applies and what factors are associated with success?* (Working Paper). Washington, DC: The Urban Institute, Education Policy Center.

Graeber, A. O., Tirosh, D., & Glover, R. (1989). Preservice teachers' misconceptions in solving verbal problems in multiplication and division. *Journal for Research in Mathematics Education, 20*(1), 95–102.

Greenwald, R., Hedges, L.V., & Laine, R. D. (1996). The effect of school resources on student achievement. *Review of Educational Research, 6,* 361–396.

Hamsa, I. S. (1998). The role of the National Board for Professional Teaching Standards. *Education, 118*(3), 452–458.

Haney, W., Madaus, G., & Kreitzer, A. (1987). Charms talismanic: Testing teachers for the improvement of American education. *Review of Research in Education, 14,* 169–238.

Hanushek, E. A. (1972). *Education and race: An analysis of the educational production process.* Lexington, MA: D. C. Heath.

Hanushek, E. A. (1981). Throwing money at schools. *Journal of Policy Analysis and Management, 1,* 19–41.

Hanushek, E. A. (1996). A more complete picture of school resource policies. *Review of Educational Research, 66,* 397–409.

Harbison, R. W., & Hanushek, E. A. (1992). *Educational performance for the poor: Lessons from rural northeast Brazil.* Oxford, England: Oxford University Press.

Haynes, D. (1995). One teacher's experience with National Board assessment. *Educational Leadership, 52,* 58–60.

Hess, F. (2002). Break the link. *Education Next, 2*(1), 22–28.

Hill, H.C., Rowan, B., & Ball, D.L. (2005). Effects of teachers' mathematical knowledge for teaching on student achievement. *American Educational Research Journal, 42,* 371–406.

Hill, H. C., Schilling, S. G., & Ball, D. L. (2004). Developing measures of teachers' mathematics knowledge for teaching. *Elementary School Journal, 105,* 11–30.

Horizon Research. (2000). *Inside the classroom observation and analytic protocol.* Chapel Hill, NC: Horizon Research.

Horizon Research. (2002). *The 2000 national survey of science and mathematics education: Compendium of tables.* Chapel Hill, NC: Horizon Research.

Ingersoll, R. (1999). The problem of underqualified teachers in American secondary schools. *Educational Researcher, 28*(2), 26–37.

Interstate New Teacher Assessment and Support Consortium. (1995). *Model standards in mathematics for beginning teacher licensing and development: A resource for state dialog.* Washington, DC: Author.

Interstate New Teacher Assessment and Support Consortium. (1996). *INTASC mathematics teacher performance assessment handbook.* Washington, DC: Author.

Interstate New Teacher Assessment and Support Consortium. (2006). *INTASC portfolio development.* Retrieved June 21, 2006, from http://www.ccsso.org/projects/Interstate_New_Teacher_Assessment_and_Support_Consortium/Projects/Portfolio_Development/

Irvine, J. J., & Fraser, J. W. (1998). Warm demanders: Culturally responsive pedagogy of African American teachers. *Education Week, 17*(35), 42.

Kennedy, M. M., Ball, D. L., & McDiarmid, G. W. (1993). *A study package for examining and tracking changes in teachers' knowledge* (Technical Series 93–1). East Lansing, MI: The National Center for Research on Teacher Education.

Kersting, N. (2004, April). *Assessing what teachers learn from professional development programs centered around classroom videos and the analysis of teaching: The importance of reliable and valid measures to understand program effectiveness.* Paper presented at the annual meeting of the American Educational Research Association, San Diego, CA.

King, M. B. (1994). Locking ourselves in: National standards for the teaching profession. *Teaching and Teacher Education, 10*(1), 95–108.

Knowing Mathematics for Teaching Algebra Project. (2006). *Survey of knowledge for teaching algebra.* East Lansing: Michigan State University. Retrieved June 21, 2006, from http://www.msu.edu/~kat/

Kraft, N. P. (2001). *Standards in teacher education: A critical analysis of NCATE, INTASC, and NBPTS (A conceptual paper/review of the research).* (ERIC Document No. ED462378), 1–29.

Learning Mathematics for Teaching Project. (2006a). *A coding rubric for measuring the quality of mathematics in instruction.* Ann Arbor, MI: Author.

Learning Mathematics for Teaching Project. (2006b). Measures of teachers' knowledge for teaching mathematics. Ann Arbor, MI: Author. Retreived June 15, 2006, from www.sitemaker.umich.edu/lmt.

Leinhardt, G., & Smith, D. A. (1985). Expertise in mathematics instruction: Subject matter knowledge. *Journal of Educational Psychology, 77*, 247–271.

Letofsky, J. (1999). National Board certification. *Center-space, 13*(2), 1–5.

Ma, L. (1999). *Knowing and teaching elementary mathematics: Teachers' understanding of fundamental mathematics in China and the United States.* Mahwah, NJ: Erlbaum.

Marshall, M. (1996). Familiar stories: Public discourse, National Board standards and professionalizing teaching. *English Education, 28*, 39–67.

Messick, S. (1989). Validity. In R. L. Linn (Ed.), *Educational measurement,* (3rd ed., pp. 13–103). New York: American Council on Education and Macmillan.

Michalowicz, K. D., & Howard, A. C. (2003). Pedagogy in text: An analysis of mathematics texts from the nineteenth century. In G. M. A. Stanic & J. Kilpatrick (Eds.), *A history of school mathematics* (Vol. 1, pp. 77–109). Reston, VA: National Council of Teachers of Mathematics.

Michigan Department of Public Instruction. (1894). *Fifty-seventh annual report of the superintendent of public instruction, Michigan.* Lansing: State Printers. Bentley Historical Library, University of Michigan.

Michigan Department of Public Instruction. (1896). *Fifty-ninth annual report of the superintendent of public instruction, Michigan.* Lansing: State Printers. Bentley Historical Library, University of Michigan.

Michigan Department of Public Instruction. (1897). *Sixtieth annual report of the superintendent of public instruction, Michigan.* Lansing: State Printers. Bentley Historical Library, University of Michigan.

Michigan Department of Public Instruction. (1898). *Sixty-first annual report of the superintendent of public instruction, Michigan.* Lansing: State Printers. Bentley Historical Library, University of Michigan.

Michigan Department of Public Instruction. (1901). *Sixty-fourth annual report of the superintendent of public instruction, Michigan.* Lansing: State Printers. Bentley Historical Library, University of Michigan.

Mullens, J. E., Murnane, R. J., & Willett, J. B. (1996). The contribution of training and subject matter knowledge to teaching effectiveness: A multilevel analysis of longitudinal evidence from Belize. *Comparative Education Review, 40*, 139–157.

National Board for Professional Teaching Standards. (2005a). *Handbook on National Board Certification.* Retrieved June 29, 2006, from http://www.nbpts.org/UserFiles/File/scoringhandbook.pdf

National Board for Professional Teaching Standards. (2005b). *NBPTS early childhood generalist scoring guide.* Retrieved July 20, 2006, from http://www.nbpts.org/for_candidates/certificate_areas?ID=17&x=49&y=6

National Board for Professional Teaching Standards. (2005c). *NBPTS middle childhood generalist scoring guide.* Retrieved July 20, 2006, from http://www.nbpts.org/for_candidates/certificate_areas?ID=27&x=48&y=6

National Board for Professional Teaching Standards. (2006a). Retrieved July 20, 2006, from http://www.nbpts.org/.

National Board for Professional Teaching Standards. (2006b). *Certification areas.* Retrieved on June 29, 2006 http://www.nbpts.org/for_candidates/certificate_areas

National Board for Professional Teaching Standards. (2006c). *NBPTS early childhood generalist portfolio instructions.* Retrieved July 20, 2006, from http://www.nbpts.org/for_candidates/certificate_areas?ID=17&x=49&y=6

National Board for Professional Teaching Standards. (2006d). *NBPTS middle childhood generalist portfolio instructions.* Retrieved July 20, 2006, from http://www.nbpts.org/for_candidates/certificate_areas?ID=27&x=48&y=6

National Center for Education Statistics. (1999). *Teacher quality: A report on the preparation and qualifications of public school teachers.* (NCES 1999-080). Washington, DC: U.S. Department of Education/OERI.

National Commission on Excellence in Education. (1983). *A nation at risk: The imperative for educational reform: a report to the nation and the Secretary of Education.* Washington, DC: U.S. Department of Education.

National Council of Teachers of Mathematics. (1991). *Professional standards for teaching mathematics.* Reston, VA: Author.

National Evaluations Systems, Inc. (2006). *California subject examinations for teachers.* Amherst, MA: Author.

Phillip, R.A., Ambrose, R., Lamb, L.C., Sowder, J.R., Schappelle, J.C., Sowder, L., et al. (in press). Effects of early field experiences on the mathematical content knowledge and beliefs of prospective elementary school teachers: An experimental study. *Journal for Research in Mathematics Education.*

Pimm, D. (1987). *Speaking mathematically: Communication in mathematics classrooms.* London: Routledge.

Porter, A.C., Youngs, P., & Odden, A. (2001). Advances in teacher assessments and their uses. In V. Richardson (Ed.), *Handbook of research on teaching,* (4th ed., pp. 259–297). Washington, DC: American Educational Research Association.

Rotberg, I. C., Futrell, M. H., & Lieberman, J. M. (1998). National Board certification: Increasing participation and assessing impacts. *Phi Delta Kappan, 79*(6), 462–466.

Rowan, B., Chiang, F., & Miller, R. J. (1997). Using research on employees' performance to study the effects of teachers on students' achievement. *Sociology of Education, 70*, 256–284.

Rowan, B., Harrison, D., & Hayes, A. (2004). Using instructional logs to study elementary school mathematics: A close look at curriculum and teaching in the early grades. *Elementary School Journal, 105*, 103–127.

Sanders, W. L., Ashton, J. J., & Wright, S. P. (2005, March 7). *Comparison of the effects of NBPTS certified teachers with other teachers on the rate of student academic progress.* Retrieved June 21, 2006, from http://www.nbpts.org/pdf/sas_final_report.pdf

Sawada, D., & Pilburn, M. (2000). *Reformed teaching observation protocol.* (Tech. Rep. No. IN00-1). Arizona Collaborative for Excellence in the Preparation of Teachers: Arizona State University.

Schön, D. A. (1995). *The reflective practitioner: How professionals think in action.* Aldershot, England: Arena.

Schweingruber, H. A., & Nease, A. A. (2000, April). *Teachers' reasons for participating in professional development programs: Do they impact program outcomes?* Paper presented at the annual meeting of the American Educational Research Association, New Orleans, LA.

Serafini, F. (2002). Possibilities and challenges: The National Board for Professional Teaching Standards. *Journal of Teacher Education, 53*(4), 316–327.

Shechtman, N., Roschelle, J., Knudsen, J., Vahey, P., Rafanan, K., Haertel, G., et al. (2006). SRI *Teaching survey: Rate and proportionality.* Unpublished manuscript, SRI International, Menlo Park, CA.

Shulman, L. S. (1986). Those who understand: Knowledge growth in teaching. *Educational Researcher, 15*, 4–14.

Shulman, L. S. (1987). Knowledge and teaching: Foundations of the new reform. *Harvard Educational Review, 57*, 1–22.

Simon, M. (1993). Prospective elementary teachers' knowledge of division. *Journal for Research in Mathematics Education, 24*(3), 233–254.

Sowder, J. T., Phillip, R. A., Armstrong, B. E., & Shappelle, B. P. (1998). *Middle-grade teachers' mathematical knowledge and its relationship to instruction.* Albany, NY: SUNY Press.

SRI International. (2004, May). *Exploring differences in minority and majority teachers' decisions about and preparation for NBPTS certification* (SRI Project—P12209). Arlington, VA: Author.

Stein, M. K., Baxter, J. A., & Leinhardt, G. (1990). Subject-matter knowledge and elementary instruction: A case from functions and graphing. *American Educational Research Journal, 27*(4), 639–663.

Strauss, R. P., & Sawyer, E. A. (1986). Some new evidence on teacher and student competencies. *Economics of Education Review, 5*, 41–48.

Summers, A. A., & Wolfe, B. L. (1977). Do schools make a difference? *American Economic Review, 67*, 639–652.

Sumner, A. (1997). The toughest test. *Techniques, 71*, 28–31, 65.

Tatto, M. T., Nielsen, H. D., Cummings, W., Kularatna, N. G., & Dharmadasa, K. H. (1993). Comparing the effectiveness and costs of different approaches for educating primary school teachers in Sri Lanka. *Teaching and Teacher Education, 9*, 41–64.

TIMSS-R video math coding manual. (2003). Retrieved June 26, 2006, from http://www.lessonlab.com/TIMMS/download/TIMSS%201999%20Video%20Coding%20Manual.pdf

Tirosh, D., & Graeber, A. O. (1989). Preservice elementary teachers' explicit beliefs about multiplication and division. *Educational Studies in Mathematics, 20*, 79–96.

Tirosh, D., & Graeber, A. (1990). Evoking cognitive conflict to explore preservice teachers' thinking about division. *Journal for Research in Mathematics Education, 21*(2), 98–108.

Tracz, S. M., Sienty, S., Todorov, K., Snyder, J., Takashima, B., Pensabene, R., et al. (1995, April). *Improvement in teaching skills: Perspectives from National Board for Professional Teaching Standards field test network candidates.* Paper presented at the annual meeting of the American Educational Research Association, San Francisco.

Vandevoort, L. G., Amrein-Beardsley, A., & Berliner, D. C. (2004). National Board certified teachers and their students' achievement. *Education Policy Analysis Archives, 12*(46). Retrieved March 10, 2006, from http://epaa.asu.edu/epaa/v12n46/

Wakefield, D. (2003). Screening teacher candidates: Problems with high-stakes testing. *The Educational Forum, 67*(4), 380–388.

Walsh, K. (2001). *Teacher certification reconsidered: Stumbling for quality.* Baltimore: Abell Foundation.

Wayne, A. J., & Youngs, P. (2003). Teacher characteristics and student achievement gains: A review. *Review of Educational Research, 73*, 89–122.

Wilson, S., & Youngs, P. (2005). Research on accountability processes in teacher education. In M. Cochran-Smith & K. M. Zeichner (Eds.), *Studying teacher education: The report of the AERA panel on research and teacher education* (pp. 591–644). Mahwah, NJ: Erlbaum.

Wilson, S. M., Shulman, L. S., & Richert, A. (1987). 150 different ways of knowing: Representations of knowledge in teaching. In J. Calderhead (Ed.), *Exploring teachers' thinking* (pp. 104–124). Sussex, England: Holt, Rinehart & Winston.

Wineburg, S. (2004). Crazy for history. *The Journal of American History, 90*(4), 1401–1414.

AUTHOR NOTE

Work on this chapter was supported by grants from the National Science Foundation REC-0207649, EHR-0233456, and EHR-0335411. We would like to acknowledge the assistance of Richard Askey, Timothy Boerst, Catherine Brach, and Seán Delaney. We also thank Pamela Moss, Judith Sowder, Suzanne Wilson and Peter Youngs for their critical and helpful reviews of an earlier draft of this chapter. Errors and omissions remain the property of the authors.

5

THE MATHEMATICAL EDUCATION AND DEVELOPMENT OF TEACHERS

Judith T. Sowder

SAN DIEGO STATE UNIVERSITY

The nation can adopt rigorous standards, set forth a visionary scenario, compile the best research about how students learn, change textbooks and assessment, promote teaching strategies that have been successful with a wide range of students, and change all the other elements involved in systemic reform—But without professional development, school reform and improved achievement for all students will not happen. Unless the classroom teacher understands and is committed to standards-based reform and knows how to make it happen, the dream will not be realized.

—Principles for Professional Development, American Federation of Teachers, 2002, p. 2

Teachers matter. A variety of recently published documents support the notion that the key to increasing students' mathematical knowledge and to closing the achievement gap is to put knowledgeable teachers in every classroom. For example, in a survey designed by Lou Harris and undertaken in 1998 by the Recruiting New Teachers organization, a carefully chosen sample of a cross-section of U.S. adults was questioned about how to improve America's schools (Haselkorn & Harris, 1998). Of the 2525 people interviewed, an overwhelming majority agreed that "improving the quality of teaching is the most important way to improve public education" (p. 1). Other documents essentially repeating this message include the 1998 report *Every Child Mathematically Proficient: An Action Plan* from the Learning First Alliance, the 2000 report *Before It's Too Late: A Report to the Nation* from the National Commission on Mathematics and Science Teaching for the 21st Century as requested by the U.S. Department of Education, the 1996 and 2003 reports *What Matters Most: Teaching for America's Future* and *No Dream Denied* from the National Commission on Teaching and America's Future, the 2000 report *The Mathematical Education of Teachers* (MET) from the Conference Board of the Mathematical Sciences, the 2001 act *No Child Left Behind* from the U.S. Department of Education (2002), and the 2003 report *Mathematical Proficiency for All Students* from the RAND Mathematics Study Panel.

The purpose of this chapter is to address what it means to prepare teachers of mathematics and to provide them, while they progress in their careers, with the professional learning opportunities they need to lead their students to succeed in learning mathematics.

In this chapter I attempt to answer, as much as is possible on the basis of theory and research evidence, the following questions:

1. Why has professional development become a priority in realizing current goals for mathematics education?

2. What are the goals of professional development?

3. What principles can be used to guide the design of professional development?

4. How do teachers learn what they need to know for teaching mathematics?

5. How do teachers learn from their professional communities about teaching mathematics?

6. What can teachers learn by investigating their own teaching of mathematics?

7. What can be learned from research on teacher change?

8. What can be learned from research about the preparation of teachers to teach mathematics?

9. What do researchers know about issues that affect professional development, such as support and accountability, preparing teacher leaders, evaluating professional development, and policy issues?

10. Where do we stand now, and what comes next?

In the ten major sections of this chapter, I address these ten questions.

Before beginning to address these questions, I raise six caveats that will clarify some decisions I made in organizing and writing this chapter. First, because this volume also contains chapters on teacher knowledge, teaching, and teacher beliefs, I address these topics only to the extent needed to provide a basis for much of the ensuing discussion. Second, the more general literature on professional development and teacher education provides a context from which to interpret the literature within mathematics education. Thus, the two bodies of literature are intermingled in this chapter. Third, many of the claims made by authors and quoted here are based on reviews of literature, and, thus, for understandable reasons, the original literature on which they base their claims is not usually included here. Fourth, much of the research discussed here could easily fit under more than one heading; that is, the categories I use are not disjoint. My placement decisions are apt to be different from those of some readers. Fifth, some authors make distinctions among teacher education, professional development, and teacher change. For example, Lerman (2001) and Ponte (2001) associated teacher education with preservice teacher preparation, that is, courses and experiences designed for students preparing to be teachers, whereas professional development and teacher change are linked to in-service programs and teachers' changing roles and could occur in taking courses, completing projects, reading, reflecting, sharing experiences, and so on. That distinction is not always made here because much of what is true for professional development is also true for teacher preparation. For example, an inquiry approach can be used in both; strategies such as focusing on children's thinking or studying cases can be used in both; changes of beliefs about mathematics and about learning mathematics can be expected from both, even though these approaches are discussed in sections on professional development. Likewise, topics such as mentoring are discussed within the section on teacher preparation. Indeed, the National Science Foundation now uses the term *professional development* to refer both to teacher preparation and to the development of practicing teachers. That said, some issues, such as the effects of methods courses and field experiences, relate to teacher preparation but not to professional development of practicing teachers. For this reason, I devote a section to only the preparation of teachers. And, sixth, although the term *reform* has been both overused and often abused, it so thoroughly permeates the literature that I realized that I could not completely avoid its use here. When possible, however, I use the term *principled knowledge* in the manner articulated by Spillane (2000) as a way of thinking about the goals of school mathematics reform:

> Reformers want principled mathematical knowledge, as distinct from procedural knowledge, to receive more attention in school work. Whereas *procedural knowledge* centers on computational procedures and involves memorizing and following predetermined steps to compute answers, *principled knowledge* focuses on the mathematical ideas and concepts that undergird mathematical procedures. Procedural knowledge has dominated the K through 12 curriculum. Reformers also propose that students develop a more sophisticated appreciation for doing mathematics including framing and solving mathematical problems, articulating conjectures, and reasoning with others about mathematical ideas: Students need to appreciate mathematical activity as more than computation. (p. 144)

In this chapter I focus on what researchers can tell us about teaching mathematics for principled knowledge and the implications this information has for the design of programs of teacher education and development. Changing the focus from procedural to principled knowledge is a new frontier for many teachers, administrators, and policy makers. The knowledge demanded is not an add-on; for many, this change can mean a complete reconceptualization of such fundamental questions as: What does it mean to teach mathematics? What is mathematics? What mathematics is important for teachers to know and why? What does teaching mathematics entail? How does one acquire knowledge? How can teachers productively assess that knowledge and use assessment data to plan instruction? The answers to all these questions change from those that underlie teaching for procedural knowledge.

ADDRESSING QUESTION 1: WHY HAS PROFESSIONAL DEVELOPMENT BECOME A PRIORITY IN REALIZING CURRENT GOALS FOR MATHEMATICS EDUCATION?

The standards movement of the 1990s coupled with a call for school reform, both set against a backdrop of performance-based accountability demands that include high-stakes testing, conspire to change the ways in which schools operate. The standards movement was heralded by the publication of the *Curriculum and Evaluation Standards for School Mathematics* by the National Council of Teachers of Mathematics (NCTM) in 1989; this document set the stage for standards in other K–12 disciplines. NCTM, too, continued to publish standards for teaching mathematics (1991) and assessing mathematics learning (1995) and in 2000, published the *Principles and Standards for School Mathematics*. States have also developed standards for teaching mathematics, some based on the NCTM Standards and some not.

The accountability movement is more recent and results from recognition that all students have the ability to learn—and to learn more than they are currently learning—even while the student population becomes more diverse. To some, *accountability* means that all should know their basic skills, but others recognize that "the future of education clearly presses beyond the basics to more demanding forms of academic learning" (Sykes, 1999, p. 153). The accountability movement is perhaps best represented in the United States by the 2001 *No Child Left Behind* Act, which, by law, calls for vast changes that are intended to result in all children's becoming proficient in key areas.

Education reformers realize that instructional improvement and increased student achievement depend on the professional development of teachers and administrators (e.g., Ball & Cohen, 1999; Elmore & Burney, 1999; Nelson & Hammerman, 1996; Sykes, 1999; C. L. Thompson & Zeuli, 1999). Hawley and Valli (1999) provided evidence from a number of sources that "one of the most persistent findings from research on school improvement is, in fact, the symbiotic relation between professional development and school improvement efforts" (p. 129). Sykes (1999) noted that the view that professional development can act as a "significant lever for education improvement" (p. 151) is fairly recent and has occurred as the result of three major societal changes: systemic reform efforts, new research knowledge about instruction, and the escalation of the commitment to quality education for all. Systemic reform has been a major policy influence.

Recognition of the need to change the way in which mathematics is taught and learned is international in scope—"Across the world, . . . mathematics teacher education programmes are preparing teachers to work with and promote reform in the practice of school mathematics" (Adler, 2000, p. 205). International researchers view teachers as the key figures on which the success of current reforms depends (Ponte, 1994). In the United States, the escalation of the commitment to providing quality education to all students, at a time when the student population grows more diverse, has led to higher demands on teachers, who in turn need the support provided through appropriate professional development. In mathematics education, "mathematics for all" has been a foundation for all volumes of NCTM *Standards* (1989, 1991, 1995, 2000).

Even with increased recognition that teacher professional development must be a priority, the professional development offered in mathematics often does not meet teachers' needs. According to Ball and Cohen (1999), most professional development funding is spent on

> sessions and workshops that are often intellectually superficial, disconnected from deep issues of curriculum and learning, fragmented, and noncumulative (Cohen and Hill, 1997; Little, 1994). Rarely do these in-services seem based on a curricular view of teachers' learning. Teachers are thought to need updating rather than opportunities for serious and sustained learning of curriculum, students, and teaching. (pp. 3–4)

In many school districts a "one size fits all" mentality limits the type of professional assistance teachers receive (Lord, 1994). Some frame the issue even more dramatically—Miles (1995) called current professional development a "joke"—"pedagogically naive, a demeaning exercise that often leaves its participants more cynical and no more knowledgeable, skilled, or committed than before" (p. vii). Others (e.g., Elmore, 2002a; Hargreaves, 1995; Middleton, Sawada, Judson, Bloom, & Turley, 2002) have been milder in stating their criticism of current professional development efforts but, nonetheless, have found them to be ineffective in terms of current reform and accountability demands.

The confirmation of the belief that much conventional professional development is ineffective and wasteful was found by Hawley and Valli (1999) to be one of four converging developments that appeared to provide a consensus among stakeholders that major changes in how professional development is conceptualized and conducted are needed. A second development they noted was the body of research that showed how change is related to professional development. Schools cannot change unless teaching is improved. And schools will not be improved for children unless schools also become places for teachers to learn. A third development was a "growing agreement that students should be expected to achieve much higher standards of performance, which include a capacity for complex and collaborative problem solving" (p. 128). The view of teaching as transmission of knowledge is at odds with the type of performance now expected of children. The fourth development they noted was that the research on how people learn leads to strategies for teaching and assessment that differ from those most commonly used. Many of teachers' core beliefs need to be challenged before change can occur.

Case studies have become a useful way to understand and evaluate the effectiveness of professional development or intervention into the teacher preparation process. Large-scale research projects in which professional development has been a major factor often use case studies as a means to study the effects of the professional development on individual teachers. The results can be dismaying, as exemplified in a case study by D. K. Cohen (1990). His portrayal of Mrs. Oublier has attained the status of a study that can be used as a generic description of a class of teachers who have misinterpreted the principles underlying the professional development they received. Mrs. Oublier, an elementary school teacher in California, willingly participated in professional development that included familiarization with the 1985 *Mathematics Framework for California Public Schools* (California Department of Education), a framework much different from California's current framework. She eagerly adopted new curriculum materials and activities introduced to her during professional development in which she participated. She believed that she had revolutionized her practice. Cohen noted that she had simply adapted these new materials and activities to her traditional teaching style. She continued to teach mathematics as though there were only right and wrong answers, and she discouraged exploration that would reveal students' understanding. Cohen observed many tensions that lay beneath the surface of the work in Mrs. Oublier's classroom. Because her own understanding of mathematics was still superficial, she could not have been successful with the give and take of open discourse. Cohen reflected on his observations of Mrs. Oublier, stating that the kind of change professional developers believe may be taking place is illusory.

> The teachers and students who try to carry out such change are historical beings. They cannot simply shed their old ideas and practices like a shabby coat, and slip on something new. . . . As they reach out to embrace or invent a new instruction, they reach out with their old professional selves, including all the ideas and practices comprised therein. The past is their path to the future. Some sorts of mixed practice, and many confusions, therefore seem inevitable. (p. 339)

Mrs. Oublier had little opportunity for sustained guidance and support. She had much to unlearn, but no one to help her do this unlearning. The lessons here for the need for sustained professional development and mentoring are significant.

These developments and stories show that the quality of preparation and professional development of teachers must improve well beyond current levels if teachers are to become "capable of far more sophisticated forms of practice" (Sykes, 1999, p. 153). Exploration of how this improvement can come about permeates the literature of teacher education.

ADDRESSING QUESTION 2: WHAT ARE THE GOALS OF PROFESSIONAL DEVELOPMENT?

To teach mathematics for principled knowledge, teachers need an extensive knowledge base, technical teaching skills, and epistemological stances that should all be goals of teacher preparation and professional development programs (Elmore, 2002a; Sykes, 1999). To meet these goals, professional development must provide opportunities for professional growth on the part of teachers and motivate them to develop the knowledge, skills, and dispositions they need to teach mathematics well. Related outcomes of professional development include changing teachers' understanding of how students learn mathematics and of the nature of mathematics, of mathematical knowledge, and of teaching mathematics well. Ideally, this new understanding will result in mathematics teachers equipped to choose appropriate curricula, plan instruction, and organize their classrooms to promote and support learning for *all* the students they teach. Professional growth is thus marked by change in teachers' knowledge, beliefs, and instructional strategies. Professional development should be designed to address the needs of teachers so that they can teach for principled knowledge in a manner envisioned by researchers and educators. Several educators have addressed the issue of what these needs are. Ball and Cohen (1999) suggested that teachers need to become serious learners of practice rather than learners of strategies and activities. For teachers to become such learners would entail their coming to understand well the mathematics they teach and what it means to reason mathematically. They need to learn to attend to their students in insightful ways, a skill that "requires expertise beyond what one gathers from one's own experience" (p. 9). They need to "develop and expand their ideas about learning, [which would require that] longstanding beliefs and assumptions about learning would need to be examined" (p. 9). They would need to better understand pedagogy and how to establish a classroom culture that supports learning goals. Ball and Cohen noted that even when these needs are addressed, the complexity of interactions and situations in classrooms also affect the ways in which teachers teach.

Borasi and Fonzi (2002) also identified several needs of teachers that could serve as a focus for professional development. In addition to the needs identified by Ball and Cohen, they included the needs to develop a vision and commitment to school mathematics reform, to become familiar with exemplary curriculum materials, to understand equity issues and their implications for the classroom, to cope with the emotional aspects of engaging in reform, and to develop an attitude of inquiry towards one's practice.

In this section I address these teacher needs by first grouping them into six, sometimes overlapping, goals and then considering literature that informs and supports each goal. These goals are to develop (a) a shared vision for mathematics teaching and learning, (b) a sound understanding of mathematics for the level taught, (c) an understanding of how students learn mathematics, (d) deep pedagogical content knowledge, (e) an understanding of the role of equity in school mathematics, and (f) a sense of self as a mathematics teacher.

Goal 1. Developing a Shared Vision

When professional development providers plan, they themselves are guided by a vision of what they want teachers to know and be able to do. These visions will differ according to the providers' views of the role of school mathematics, how children learn, and how instruction should be designed to best help children learn. Those who believe that the goal of school mathematics is to lead students to develop principled knowledge of the mathematics they need to function in today's society will formulate their goals and structure their programs in ways that reflect this vision.

Several documents published over the past few decades have contributed to the formation of a vision of what teaching and learning mathematics encompasses. The NCTM Standards documents of 1989, 1991, 1995, and 2000 have provided images of classroom teaching and learning that are far different from those familiar to most teachers. For example, the *Principles and Standards for School Mathematics* (NCTM, 2000) provided a vision of the mathematics all students ought to know and be able to do. The Teaching Principle is a recognition of the role of teachers in attaining this vision: "Effective mathematics teaching requires understanding what students know and need to learn and then challenging and supporting them to learn it well" (p. 16). Accordingly,

> teachers need to know and use "mathematics for teaching" that combines mathematical knowledge and pedagogical knowledge. They must be information providers, planners, consultants, and explorers of uncharted mathematical territory. They must adjust their practices and extend their knowledge to reflect changing curricula and technologies and to incorporate new knowledge about how students learn mathematics. They must also be able to describe and explain why they are aiming for particular goals. (p. 370)

Other sources such as *Adding It Up* (National Research Council [NRC], 2001a), with its intertwined strands of abilities and dispositions representing mathematical proficiency; *Making Sense: Teaching and Learning Mathematics With Understanding* (Hiebert et al., 1997), with its descriptions of what learning with understanding entails; *Reconstructing Mathematics Education* (Schifter & Fosnot, 1993) and *What's Happening in Math Class* (Schifter, 1995b), with their stories of teachers who transformed their instruction; *Relearning to Teach Arithmetic,* with its videotapes showing children's approaches to multiplication and division;

the *Integrating Mathematics and Pedagogy* (Philipp, Cabral, & Schappelle, 2004, 2005) with its rich sets of videos of children's thinking, organized by mathematics topic (within whole number and rational number areas), grade level, and other categories; *Learning and Teaching Linear Functions* (Seago, Mumme, & Branca, 2004), with its video cases of real classroom lessons; *Fostering Algebraic Thinking* (Driscoll, 1999), with its grounded assistance for teaching algebra beginning in Grade 6; *Mathematics Teaching Cases* (Barnett, Goldenstein, & Jackson, 1994) and *Windows on Teaching Math* (Merseth, 2003), with their cases of classroom teaching prepared for discussion by teachers—these and many other sources reflect a common vision of teaching and learning mathematics and provide professional developers with assistance in their planning and materials for presentations and workshops. Some of this literature was based on research; some, on shared values and beliefs. Cooney (1994) reminded us that teachers' beliefs about teaching "are shaped by social situations and therefore can only be reshaped by social situations" (p. 109). Professional development can provide the social situations teachers need to develop and clarify their values and beliefs while they learn new content and skills.

Ferrini-Mundy (1997) quoted from Fullan in her discussion of vision as a process rather than a product: "Shared vision, which is essential for success, must evolve through the dynamic interaction of organizational members and leaders" (Fullan, 1993, p. 28). In her project, designed to initiate reform of mathematics at many schools, Ferrini-Mundy learned that vision is "evolutionary and open-ended" (p. 124) and that in her project, teachers and administrators "became more and more articulate and definite about the goals and nature of the mathematics changes as they accumulated experience and evidence from their practice" (p. 124).

Including a shared vision as a goal of professional development does not necessarily require the visible focus one maintains for the remaining goals discussed here. A teacher's vision of learning and teaching mathematics unfolds while the other goals are being addressed.

Goal 2. Developing Mathematical Content Knowledge

Mathematics continues to be an essential component in the work of professionals in a variety of disciplines, some of which have existed only for the past few decades and many of which, though not new, have changed the ways in which they use mathematics. Technology has forever changed the way math-

ematics is used not only by professionals but also by workers in many fields. Accordingly, expectations for school mathematics are also changing—students will need to know more and different mathematics to be successful. And, in turn, expectations for teachers of school mathematics are different than they were in the 20th century. In the RAND report (2003), these expectations are stated in terms of urgency—"as both a matter of national interest and a moral imperative, the overall level of mathematical proficiency must be raised, and the differences in proficiency among societal groups must be eliminated" (p. xi).

Yet many policy makers underestimate the knowledge required for teaching mathematics. More than a decade ago, Ball and Wilson (1990) noted and countered two common but mistaken assumptions about obtaining knowledge needed for teaching: that traditional mathematics content needed by teachers can be best obtained through traditional undergraduate mathematics courses and that content knowledge is the only professional knowledge teachers need. Many policy makers continue to exhibit these assumptions, even though many educators believe that what mathematics knowledge is needed to teach well remains uncertain (Ball, Lubienski, & Mewborn, 2001). However, welcome attempts have been made (and continue to be made) to determine the content knowledge needed to teach mathematics, and these needs have been expressed in terms of a deep understanding that requires at least some specially designed coursework or programs. The quality of the mathematical knowledge also needs to be addressed. At the 2003 Department of Education Secretary's Summit on Mathematics, Ball, in her presentation, said that the qualities of mathematics knowledge needed for effective instruction included being "respectful of the integrity of the discipline, able to be extended and opened up for learners—*unpacked*, justified and reasoned, connected within and across domains, building on earlier ideas and anticipating more advanced topics, and organized psychologically as well as logically."

The need for teachers to know mathematics differently than mathematicians do has been recognized by educators for a very long while. More than a century ago Dewey addressed this issue:

Every study or subject thus has two aspects: one for the scientist as a scientist; the other for the teacher as a teacher. These two aspects are in no sense opposed or conflicting. But neither are they immediately identical. For the scientist, the subject matter represents simply a given body of truth to be employed in locating new problems, instituting new researches, and carrying them through to a verified outcome. . . . The problem of the teacher is a different one. . . . What concerns him as teacher is the ways in which that subject may become part of experience, what there is in the child's present that is usable with reference to it; how such elements are to be used; how his own knowledge of the subject-matter may assist in interpreting the child's needs and doings, and determine the medium in which the child should be placed in order that his growth may be properly directed. He is concerned, not with the subject-matter as such, but with the subject-matter as a related factor in a total and growing experience. (Dewey 1902, pp. 29–30)

The mathematics community is only now beginning to understand the need for teachers to know appropriate mathematics, as described by Dewey. Recent documents from mathematical organizations have addressed the mathematical needs of teachers. *The Mathematical Education of Teachers*, published by the Conference Board of the Mathematical Sciences (CBMS, 2000), provided detailed information about the appropriate mathematics teachers need to know to teach at elementary, middle, and secondary school levels. The work of the writing committee was influenced by the then recent publication by Ma (1999). Her study of Chinese and American elementary school teachers' knowledge, in which she described the "Profound Understanding of Elementary Mathematics (PUFM)" needed to teach elementary grades, led many mathematicians to understand that the mathematics of elementary grades is not trivial and that teachers need more and different preparation than has been common. Additionally, the writers recognized that the ways present and future teachers need to know mathematics differs from the ways other college mathematics students need to know mathematics.

A second document that includes the U.S. mathematics community's content recommendations for the preparation of teachers is the *Undergraduate Programs and Courses in the Mathematical Sciences* (Mathematical Association of America [MAA], 2003), prepared by the Committee on the Undergraduate Programs in Mathematics (CUPM), a standing committee of the MAA. The teacher education recommendations in this document are informed by the MET recommendations and are embedded in a set of recommendations for the mathematical education of all undergraduates.

Substantial evidence indicates that too few classroom teachers have the mathematical knowledge described above (Borko & Putnam, 1995; RAND, 2003). Preservice teachers at all levels have been found to know rules and procedures but lack knowledge of concepts and reasoning skills (Wilson, Floden, & Ferrini-Mundy, 2001). The current efforts to focus school mathematics instruction on students' developing principled knowledge of mathematics must go hand-in-hand with teachers' developing principled knowledge of mathematics also. Expanding principled knowledge calls for a shift from content as a focus to "content in relation to student learning and context" (Elmore, 2001, p. 11).

Professional development provides an opportunity for teachers to learn more mathematics, even when the focus is on student thinking or curriculum or classroom events. The projects described later in this chapter resulted in increased teacher knowledge of mathematics even though knowledge development was not their primary focus.

Goal 3. Developing an Understanding of How Students Think About and Learn Mathematics

Schifter (2001) has reminded the mathematics education community that additional mathematical skills needed for teaching may evolve not from a focus on mathematical content but from

> attending to the mathematics in what one's students are saying and doing, assessing the mathematical validity of their ideas, listening for the sense in children's mathematical thinking even when something is amiss, and identifying the conceptual issues on which they are working. (p. 131)

Attending to students' reasoning has become a focus of many professional development programs. The Cognitively Guided Instruction (CGI) work is but one example of using student thinking in professional development. The literature on CGI is extensive; several studies (e.g., Carpenter & Fennema, 1992; Carpenter, Fennema, Peterson, & Carey, 1988; Carpenter, Fennema, Peterson, Chiang, & Loef, 1989; Fennema, Carpenter, Franke, & Carey, 1992; Fennema et al., 1996; Franke, Carpenter, Levi, & Fennema, 2001; Franke, Carpenter, Fennema, Ansell, & Behrend, 1998; Franke, Fennema, & Carpenter, 1997; Jacobson & Lehrer, 2000) over a number of years pointed to teachers' changes of instructional practice that led to substantial gains in student achievement. This research is based on the hypothesis that if teachers listen to children, understand their reasoning, and teach in a manner that reflects this knowledge, they can and will provide children with a mathematics education better than if they did not have this knowledge. To this end, the researchers developed a framework of children's thinking about early number problems—problem types and solution strategies—and used this framework as the core of a professional development program. This research has provided evidence of the efficacy of focusing professional development on student thinking and learning.

Developing an understanding of how students think about mathematics is a goal of professional development—it is also a strategy used in professional development. When teachers examine work of their own students, they gain insight into their students' thinking and understanding. When teachers examine the work of other students, in company with other teachers, they discuss the kinds of strategies used and hypothesize about the students' knowledge, the kind of instruction received, and the instructional changes needed to improve understanding and achievement.

Student thinking can also be thought of as an interpretive lens that helps teachers think about their students, the mathematics they are learning, the tasks that are appropriate for the learning of that mathematics, and the questions that need to be asked to lead them to better understanding. As with any interpretive lens, development takes place over time, with many experiences with individual children and with whole classes of students. Professional development that has as a goal teachers' coming to understand how children think about and learn mathematics can awaken teachers to the power of understanding children's thinking and could lead them to develop a listening capacity that deepens over time.

A later section of this chapter is devoted to the exploration of projects that have focused on using student thinking as an approach to teacher learning. These studies confirm the powerful effect that the study of student reasoning can have on the ways in which teachers think about mathematics learning and gear their own instruction to account for this new knowledge.

Goal 4. Developing Pedagogical Content Knowledge

The grade-band content recommendations described in *The Mathematical Education of Teachers* (CBMS, 2000) are buttressed with descriptions of mathematical content as it relates to pedagogy. They focus on providing teachers with robust knowledge of mathematics that is conceptual in nature and appropriate for teaching. Teachers

> need to see the topics they teach as embedded in rich networks of interrelated concepts, know where, within those networks, to situate the tasks they set their students and the ideas these tasks elicit; . . . they must be able to appraise and select appropriate activities, and choose representations that will bring into focus the mathematics on the agenda. (p. 55)

The knowledge described in this document aligns well with the notion of *pedagogical content knowledge,* that is, the knowledge of a subject one needs to teach it, which Shulman (1986) described as including

> the most useful forms of representations of those ideas, the most powerful analogies, illustrations, examples, explanations, and demonstrations—in a word, ways of representing and formulating the subject that make it comprehensible to others. [It] also includes understanding of what makes the learning of specific topics easy or difficult; the conceptions and preconceptions that students of different ages and backgrounds bring with them to the learning of those most frequently taught topics and lessons. (p. 9)

This description is reminiscent of the quote from Dewey earlier in this section—the knowledge of the scientist (or mathematician, in this case) and the knowledge of the teacher are quite different because their goals are so very different.

Grossman's (1990) description of the four central components of pedagogical content knowledge is helpful to those developing teacher education programs and professional development opportunities for mathematics teachers. Her four categories (here changed to refer to mathematics) are (a) an overarching knowledge and belief about the purposes for teaching (mathematics); (b) knowledge of students' understandings, conceptions, and potential misunderstandings (in mathematics); (c) knowledge of (mathematics) curriculum and curricular materials; and (d) knowledge of the instructional strategies and representations for teaching particular topics (in mathematics).

The categories are not independent of one another. According to Borko and Putnam (1995), having an overarching conception of mathematics teaching (the first category above)

> serves as a "conceptual map" for instructional decision making, as the basis for judgments about classroom objectives, instructional strategies and student assignments, textbooks and curricular materials, and the evaluation of student learning. . . . Teachers' overarching conceptions are a particularly salient component of the professional knowledge base. (p. 47)

Effective teachers can often predict what mathematics students will understand, how they will understand it, and some of the potential for misunderstandings (Grossman's second category above). They know, for example, that children's work with whole numbers often leads them to believe that multiplication "makes bigger," which is untrue when their world of numbers is extended to include fractions and decimal numbers.

These teachers can plan more effectively because they can anticipate students' difficulties. They know what prior knowledge must be present to understand something new. They know how to scaffold knowledge to assist students in developing understanding. They know how to listen to students. Much of this knowledge comes from practice, but teachers who themselves have poor understanding of mathematics are unlikely to develop this type of knowledge, particularly when the mathematics in the curriculum becomes more sophisticated, such as it does when moving from additive to multiplicative situations or from Cartesian coordinates to polar coordinates.

Effective mathematics teachers think about mathematics curricula (Grossman's third category above) in terms of *big ideas,* such as proportional reasoning or the mathematics of change, around which to structure instruction. According to Kennedy (1997), attending to the central ideas rather than to minutiae is one way to define *conceptual understanding.* Effective teachers recognize the particular strengths and weaknesses of the textbooks and materials they are using. For example, a sixth-grade teacher studied by Philipp, Flores, Sowder, and Schappelle (1994) expressed astonishment that the mathematics text his district had adopted "allocated only three pages to perimeter and area, which indicated to him that the authors thought these difficult concepts could be learned by looking at a few textbook examples of regular figures and then learning formulas" (p. 165). Many effective teachers have (and often own) an array of materials they use when teaching mathematics. Another teacher interviewed by Philipp et al. (1994) said, "I have been known for a long time as having an incredible resource library. It's too bad I didn't buy stock in Dale Seymour and Creative Publications"[1] (p. 173). This teacher and others interviewed and observed by these researchers were able to use manipulatives and other materials effectively because they knew exactly what mathematics children would learn from any particular manipulatives when appropriate scaffolding was provided.

Knowledge of the strategies and representations for teaching particular topics (Grossman's fourth category above) can be characterized as "an extensive repertoire of powerful representations and the ability to adapt these representations in multiple ways in order to meet specific goals for specific sets of learners" (Borko & Putnam, 1995, pp. 48–49). Grouws and Schultz's (1996) description of *pedagogical content knowledge* in mathematics as including, but not limited to, "useful representations, unifying ideas, clarifying examples and counterexamples, helpful analogies, important relationships, and connections among ideas" (p. 444) embodied this knowledge component. The work of Sowder et al. (1998), on the analysis of the mathematical content knowledge necessary to teach multiplicative structures in the middle grades also fits with this fourth component. Many teachers fail to develop this pedagogical knowledge of mathematics because they lack the fundamental mathematical knowledge that undergirds such pedagogical content knowledge. According to Borko and Putnam (1995), teachers come to understand the various messages about changing instruction through their existing knowledge and practices, and teachers' knowledge and beliefs serve as "powerful filters through which learning takes place" (p. 60).

The teacher knowledge characterized by Grossman as combining many types of knowledge is, of course, the sort of knowledge teachers need, and the knowledge components must be integrated in practice, even though researchers at times, rightly or wrongly, separate them theoretically so that they might be more easily studied.

Simon (1997), too, characterized the knowledge teachers need as the combination of aspects of knowledge. The eight areas (overlapping and often interdependent) he identified illustrate this intertwining of facets of knowledge needed by teachers.

1. Knowledge of and about mathematics . . . [that] defines what is worth learning and the activities that are appropriate to the enterprise.

2. A personally meaningful model of mathematics learning . . . [that includes a] teacher's concepts of how mathematical knowledge is developed.

3. Knowledge of students' development of relevant concepts [that allows teachers] to make sense of the students' mathematics.

4. Relationship to students' mathematics . . . [that] encompasses commitment to understanding his or her students' mathematics, ability to elicit and probe their mathematics, and ability to analyze students' mathematical activity and form useful models of their knowledge.

5. A personally meaningful model of mathematics teaching . . . is itself recursively embedded . . . [and] is constantly in a state of construction and renovation.

---

[1] These companies sold, among other things, manipulative materials for teaching mathematics.

6. Ability to define appropriate learning goals for students, . . . [and] longer range goals as well as goals for spontaneous interventions, [implying] the ability to identify key mathematical ideas of the mathematics being considered, as well as to make use of knowledge of students' concept development and knowledge of that teacher's own students' mathematics.

7. Ability to anticipate how students' learning might ensue, . . . learning to anticipate how student learning might progress

8. Ability to construct lessons consistent with one's model of teaching . . . [by] constituting new forms of practice both by adapting new techniques and by rethinking the role of familiar ones. (pp. 81–82)

Although progress has been made, "the nature and development of [teachers'] knowledge is only beginning to be understood by the present generations of researchers in teaching and teacher education" (Munby, Russell, & Martin, 2001, p. 877). Limited knowledge of key mathematical ideas can lead to "missed opportunities for fostering meaningful connections between key concepts and representations" (Borko & Putnam, 1995, p. 44). Fundamental changes are needed in what we consider to be the nature of teacher knowledge. We now have at least some knowledge about what teachers need to know to be effective teachers, and this knowledge should guide us in developing programs for teacher education and professional development.

Goal 5. Developing an Understanding of the Role of Equity in School Mathematics

The NCTM Standards documents (1989, 1991, 1995, 2000) advocated exemplary mathematics instruction for *all* students. In fact, in the 2000 Standards the Equity Principle was quite deliberately listed as the first of the six principles to describe a high-quality mathematics education: "Excellence in mathematics education requires equity—high expectations and strong support for all students" (NCTM, 2000, p. 11). This principle was the core element of the vision of the Standards and required that "all students, regardless of their personal characteristics, backgrounds, or physical challenges, must have the opportunities to study—and support to learn—mathematics" (p. 12). This sentiment can be found in other documents; for example, the Holmes Group (1990), an influential consortium of deans of the nation's colleges of education, claimed that "schools fail students if they deny them a passport to the mainstream; they also fail if students graduate with a sense that their own culture is worthless" (p. 22).

But what exactly is *equity*? And how is it incorporated into the mathematical education of teachers? Weissglass (1997) had defined it thus:

Equity is the ongoing process (not a product) of increasing our own and society's capacity and commitment to

- completely respect individuals as complex thinking and feeling humans with different social, cultural, gender, and class backgrounds and values, and

- provide the necessary resources to assist people in learning. This includes overcoming the effects of any mistreatment on the ability to learn, whether it be at the hands of individuals or institutions. (pp. 120–121)

Equity must be a goal of professional development because teachers teach in school settings marked by diversity, schools in which people of different racial, ethnic, language, and socioeconomic backgrounds coexist. "Professional development must attend to teachers' unique circumstances, particularly in those contexts where social justice and educational equity are most needed" (Quiroz & Secada , 2003, p. 104). The difficulty with dealing with everyone equitably is that we are many times unaware that we are not doing so. Or, as Delpit (1995) has so aptly stated: "We all interpret behaviors, information, and situations through our own cultural lenses; these lenses operate involuntarily, below the level of conscious awareness, making it seem that our own view is simply 'the way it is'" (p. 151).

Reviews of research on teacher knowledge (Tate, 1994) have not focused on teachers' understanding of mathematics as it relates to "society's democratic and economic processes" (p. 56). Rather, "research on teachers' subject matter knowledge has been built on the assumption that mathematics is a neutral, objective, abstract, culture-free discipline" (p 56). Yet according to Secada (1989), an equitable curriculum for teachers would include

real contexts that reflect the lived realities of people who are members of equity groups, and unless those contexts are as rich in the sorts of mathematics which can be drawn from them as from others, we are likely to stereotype mathematics as knowledge that belongs to a few privileged groups. (p. 49)

In Sleeter's (1997) review of literature of mathematics, multicultural education, and teaching, she noted many connections between multicultural education and mathematics. However, she found that professional development in multicultural education does not usually help teachers make this connection. But professional development on these issues

> can help them grapple with some concepts and issues that are fundamental to multicultural mathematics, such as culturally relevant pedagogy, pedagogy for language differences, a deep-level understanding of culture, and the operation of institutional discrimination both inside and outside schools. (p. 693)

Professional developers and teachers have to avoid what Cahnmann and Remillard (2002) called the pitfall of relying on "token symbols of so-called culture that do not of themselves enhance students' learning opportunities. Embracing culturally contextualized practices is not simply a matter of incorporating ethnic symbols and artifacts into the mathematics classroom" (p. 199).

Tate (1994) reviewed other literature to make the point that teachers need to be prepared to effectively teach diverse populations. He described two multicultural approaches to teacher education. The ultimate goal for the first, the Teaching the Exceptional and Culturally Different (TECD) approach, is the assimilation of all students into the American mainstream so that they can compete in the job market. Teaching is viewed as a static process of passing on information; mathematics is viewed as a set of facts and procedures to be learned. In the second, the Education that is MultiCultural and Social Reconstructionist (EMCSR) approach, the preparation of students to critically analyze social conditions is advocated. Tate, borrowing from Dewey (1920) and Ernest (1991), described this approach as based on three interconnected sets of beliefs: "All knowledge is tentative and fallible; . . . education should prepare students for democracy; . . . and education should be built on the child's interests and experiences" (p. 61). Those using this approach address diversity by "promoting critical thinking by situating students in realistic problem contexts found in our multicultural society and having them analyze and solve problems" (p. 61). Teachers using this approach would encourage development of a community of learners and must "facilitate genuine discussion between the students and the teacher" (p. 61).

One particular area in which attention to equity sometimes fails is with the type of learning expected of students in diverse classrooms. Teachers of students from low-income families or second-language families often expect too little of their students, teach mathematics procedurally, and focus on basics. Yet a large body of research indicates that when teachers of students in diverse classrooms focus on meaning-making and on mathematics content, students do learn more and better mathematics than when they are taught only procedures and skills (e.g., Boaler, 2002; Knapp et al., 1995; Ladson-Billings, 1994; Silver, Smith, & Nelson, 1995; Zeichner, 1992). This is the good news.

The bad news is that few teachers in these schools can teach in this manner. "Teaching in urban schools demands a different set of skills and abilities and requires people who themselves are committed to protecting learners and learning" (Ladson-Billings, 1999a, p. 233). Many teachers lack these skills. Further, many teachers in urban schools are the youngest and least skilled, ready to move on when they have achieved enough seniority. Sykes (1999) turned to teacher preparation for an answer: "One clear implication is that teachers in the future must be capable of far more sophisticated forms of practice than in any prior era. The stakes around their learning are now much higher" (p. 153).

Thus we have a problem—teachers of the future are unlikely to be capable of these forms of practice without major changes in our teacher preparation and professional development programs. Zeichner (1992) pointed out that demographic composition of the teaching corps is unlikely to change significantly; thus "the problem of educating teachers for diversity will continue to be one of educating white, monolingual, mostly female teacher education students" (p. 1) to teach an increasingly diverse population with backgrounds and life experiences very different from their own. According to Ladson-Billings (1999b), applicants to teacher preparation programs who are most apt to have the background and skills that would make them good urban teachers are too often screened out of programs. In short, the education community is only beginning to acknowledge and deal with the challenges presented by the NCTM Equity Principle.

Goal 6. Developing a Sense of Self As a Teacher of Mathematics

This particular goal, like the goal of developing a shared vision, can be a byproduct of sound, ongoing professional development. Dewey (1916/1944) has told us that the self is "something that is in continuous formation through choice of actions" (p. 351). A knowledge of self develops when teachers regularly engage "in reflection, in, on, and about their values, purposes, emotions, and relationships" (Day & Sachs, 2004, p. 9). Dewey.1916/1944, p. 351)This goal was addressed in the NCTM *Professional Standards for Teaching Mathematics* (1991) as part of its standard on developing as a teacher of mathematics:

> Being a teacher of mathematics means developing a sense of self [as] a teacher. Such an identity grows over time. It is built from many different experiences with teaching and learning. Further, it is reinforced by feedback from students that indicates they are learning mathematics, from colleagues who demonstrate professional respect and acceptance, and from a variety of external sources that demonstrate recognition of teaching as a valued profession. Confident teachers of mathematics exhibit flexibility and comfort with mathematical knowledge and commitment to their own professional development within the larger community of mathematics educators. (p. 161)

A sense of self as a confident teacher of mathematics is particularly difficult for elementary teachers to develop because of the mathematical anxiety many of them experience. Much of this anxiety is due to their lack of understanding of the mathematics they teach—in fact, many have had negative experiences with mathematics from the time they were in elementary school, in part because their own teachers were probably "math anxious" and did not understand well the mathematics they were teaching. Professional development that allows teachers to experience learning as an intellectual endeavor in which mathematics is explored together with others can lead them to have confidence in their abilities to develop understanding of other mathematical content (Nelson & Hammerman, 1996; Sowder, Philipp, Armstrong, & Schappelle, 1998). That is, developing the ability to explore mathematics intellectually can be generative. Strong, ongoing professional development can lead teachers to this point, and, further, teachers should be helped to understand that this type of learning can also be generative for their students. "It literally creates understanding in the mind of the thinker" (C. L. Thompson & Zeuli, 1999, p. 346).

Shedding anxiety about the teaching of mathematics can lead to a sense of empowerment. Research reviewed by Hargreaves (1995) showed that when teachers become empowered, they

> work collaboratively rather than individually, take more risks (Little, 1987), commit to continuous rather than episodic improvement (Rosenholtz, 1989), tend to be more caring with students and colleagues alike (Nias, Southworth, & Yeomans, 1989; Taafaki, 1992), have stronger senses of teaching efficacy (Ashton & Webb, 1986), are more assertive in relation to external pressures and demands (A. Hargreaves, 1994), experience greater opportunities to learn and improve

from one another (Woods, 1990), and have access to more feedback (Lortie, 1975) and opportunities for reflection Grimmett & Crehan, 1991. (p. 19)

In spite of this empowerment, teachers are not entirely ready to cast off old practices and beliefs. Like teachers in other fields, teachers of mathematics need to be convinced that new teaching practices work before they are willing to adopt them (Guskey, 1995). Contrary beliefs are sometimes held simultaneously. Only when the inconsistencies of certain beliefs, such as a view of teaching exclusively as presentation versus teaching as mutual exploring, are brought to the forefront can teachers address them. And, as is to be expected, new anxieties can result in such exposures.

Other evolving aspects of a teacher's sense of self as a mathematics teacher are the deep emotions and passions that are "often at the very heart of teacher commitment and desire" (Hargreaves, 1995, p. 24). Teachers who share the vision of mathematics as a place in which students can become intellectually alive are likely to be passionate about teaching mathematics. They are also more likely to want to share this passion with their colleagues, to feel a commitment toward empowering others to teach principled mathematics. They come to think that their roles as mathematics teachers have a moral component. In a chapter on conceptualizing teacher development Cooney (2001) said: "I suggest that there remains a philosophical perspective that suggests that reform-oriented classroom is more consistent with the kind of society most of us would embrace. I do so under the assumption that the teaching of mathematics, or any subject for that matter, is ultimately a moral undertaking" (p. 11).

Coping with change is certain to have emotional aspects. Professional development providers need to recognize this aspect and help teachers address it. One tends to think of change as an outcome of professional development, but new teachers can also benefit from attending to the emotional aspect of teaching during their teacher preparation. For them, "learning how to handle the emotional responses . . . was as important as learning how to conduct tasks, meet new experiences, make judgments, build relationships, or assimilate new knowledge (Tickle, 1991, p. 320).

A sense of self as a mathematics teacher not only will differ among teachers but also will change for individual teachers while they progress in their profession. Day and Sachs (2004) described five career phases through which teachers pass:

1. Launching a career: initial commitment (easy or painful beginnings).

2. Stabilization: find commitment (consolidation, emancipation, integration into peer group).

3. New challenges, new concerns (experimentation, responsibility, consternation).

4. Reaching a professional plateau (sense of mortality, stop striving for promotion, enjoy or stagnate).

5. The final phase (increased concern with pupil learning and increasing pursuit of outside interests, disenchantment, contraction of professional activity and interest). (p. 11)

Finally, a sense of self can be thought about in terms of *identity formation,* which Wenger (1998) reminded readers is not only or always an individual process but is a process that can involve the communities in which teachers practice.

Identity formation is a lifelong process whose phases and rhythms change as the world changes. From this perspective, we need to think about education not merely in terms of an initial period of socialization into a culture, but more fundamentally in terms of rhythms by which communities and individuals continually renew themselves. Education thus becomes a mutual developmental process between communities and individuals, one that goes beyond mere socialization. It is an investment of a community in its own future, not as a reproduction of the past through cultural transmission, but as a formation of new identities that can take its history of learning forward. (pp. 263–264)

In the discussion of various types of professional development in the remainder of this chapter, the effects of the career stages of individuals are important to consider.

Summary and Discussion of Goals for Professional Development

The six goals stated and discussed here are not independent of one another. For example, if mathematics learning is to be useful in the classroom, it cannot be separated from the learning of pedagogical knowledge. Knowing mathematics and having the associated pedagogical knowledge to teach mathematics will lead teachers to develop a sense of self as a mathematics teacher. Learning about children's thinking about mathematics will affect teachers' visions of mathematics teaching (D. K. Cohen & Hill, 2001).

The goals listed here will not be achieved if teachers are simply told what to do or how to teach differently. Rather,

Teachers themselves must make the desired changes. To do so, they must acquire richer knowledge of subject matter, pedagogy, and subject-specific pedagogy; and they must come to hold new beliefs in these domains. Successful professional development efforts are those that help teachers to acquire or develop new ways of thinking about learning, learners, and subject matter, thus constructing a professional knowledge base that will enable them to teach students in more powerful and meaningful ways. (Borko & Putnam, 1995, p. 60)

ADDRESSING QUESTION 3: WHAT PRINCIPLES CAN BE USED TO GUIDE THE DESIGN OF EFFECTIVE PROFESSIONAL DEVELOPMENT?

The quality and effectiveness of professional development programs are, as noted in the introduction of this chapter, of interest not only to educators and policy makers but also to the public at large. Various educators have prepared lists of principles of effective professional development, some based on reviews of research and some based on years of success as a professional developer. Success is usually measured by teacher changes in knowledge of mathematics, beliefs about mathematics, and instructional practice and by increased student learning. These four aspects of success are somewhat easier to measure, but no more important, than other goals of professional development described earlier: a vision for mathematics teaching and learning, an understanding of the role of equity in the mathematics classroom, and a sense of self as a mathematical teacher.

Basing their synthesis of research on professional development that successfully contributed to student learning, Hawley and Valli (1999) identified eight characteristics that both embody this research and take into account recent "calls for action" (p. 137). A similar cut on the literature on successful professional development, that is, professional development that improves student learning, has been provided by Elmore (2002a). Elmore called his list a consensus view based on professional literature and academic research. Both syntheses apply more generally to professional development and both are based on literature about professional development that leads to student success. Clarke (1994) has compiled 10 important principles of mathematics professional development on the basis of both mathematics education research literature and his own years of providing professional development.

Table 5.1 A Comparison of Three Lists of Elements of Successful Professional Development

Hawley and Valli (1999, p. 138)	Elmore (2002a, p. 7)	Clarke (1994, p. 38)
Driven, fundamentally, by analyses of the differences between (a) goals and standards for student learning and (b) student performance.	Focuses on a well-articulated mission or purpose anchored in student learning of core disciplines and skills; derives from analysis of student learning of specific content in a specific setting.	Addresses issues of concern and interest, largely (but not exclusively) identified by the teachers themselves.
Involves learners (such as teachers) in the identification of their learning needs and, when possible, the development of the learning opportunity and the process to be used.	Focuses on specific issues of curriculum and pedagogy derived from research and exemplary practice; connected with specific issues of instruction and student learning of academic disciplines and skills in the context of actual classrooms.	Solicits teachers' conscious commitment to participate actively; involves a degree of choice for participants.
Is primarily school based and integral to school operations.	Involves active participation of school leaders and staff.	Involves groups of teachers from a number of schools; enlists support from administrators, students, parents, and broader community.
Provides learning opportunities that relate to individual needs but for the most part are organized around collaborative problem solving.	Develops, reinforces, and sustains group work using collaborative practice within schools and networks across schools.	Enables participating teachers to gain a substantial degree of ownership by their involvement in decision making and by being regarded as true partners in the change process.
Is continuous and ongoing, involving follow-up and support for further learning, including support from sources external to the school.	Sustains focus over time—projects an expectation of continuous improvement.	Recognizes that change is a gradual, difficult, and often painful process and affords opportunities for ongoing support.
		Recognizes that changes in teachers' beliefs about teaching and learning derive largely from classroom practice and, thus, such changes will follow the opportunity to validate, through observing positive student learning, information supplied by professional development programs.
Incorporates evaluation of multiple sources of information on outcomes for students and processes involved in implementing the lessons learned through professional development.	Uses assessment and evaluation for active monitoring of student learning and providing feedback on teacher learning and practice.	Encourages participants to set further goals for their professional growth.
Provides opportunities to develop a theoretical understanding of the knowledge and skills to be learned.	Embodies a clearly articulated theory or model of adult learning.	Allows time and opportunities to report successes and failures and to share the wisdom of practice.
Is integrated with a comprehensive change process that deals with the full range of impediments to and facilitators of student learning.		Recognizes and address the many impediments to teachers' growth at the individual, school, and district levels.
	Models effective practice delivered in schools and classrooms and is consistent with the message.	Models desired classroom approaches to project a clearer vision of the proposed changes.

As can be seen in Table 5.1, Hawley and Valli's (1999), Elmore's (2002a), and Clarke's (1994) lists of what makes professional development successful have considerable overlap. The first two included recognition of the importance of a theoretical base for professional development. Other researchers have agreed that however professional development is designed, it will be ineffective unless it is grounded in sound theories of learning, particularly adult learning (Knapp, 2003; Knight, 2002; Mewborn, 2003).

Clarke's (1994) list lacks the emphasis on theory found in the other two lists, but his list addressed issues in more practical ways. All three lists provide guidance for those planning professional development and could be used as a means for considering the success of a professional development program. (The fit of different elements of each list into a table is somewhat forced but allows for some comparisons.)

Primary commonalities include the role of determining the purpose of a professional development program, the role of teachers in deciding on foci (or, at least, a focus on issues relating to teachers' needs), the need to have support from other constituencies (e.g., administrators, peers, parents) to undertake changes in instruction, the important role of collaborative problem solving, the need for continuity over time, the necessity of modeling the type of instruction expected, and the need for assessment that provides teachers with feedback they need to grow. Hawley & Valli's (1999) list seems to have been developed with a focus on the needs of teachers (see, in particular, the second component); the second (Elmore, 2002a) was more focused on student learning. Both are the result of literature reviews, in contrast to Clarke's (1994) list, which was based on experience with long-term professional development of mathematics teachers in Australia and is more pragmatic in nature. Even with these differences, the three lists have a great deal in common.

Others who study professional development have made some of the points found in the first three lists and some new ones. For example, Borasi and Fonzi (2002), in their monograph on professional development prepared at the request of the National Science Foundation (NSF), reviewed successful professional development programs in mathematics education and, on the basis of their review, recommended that programs be sustained and intensive, be informed by how people learn best, focus on the critical activities of teaching and learning rather than on abstractions and generalities, offer a rich set of diverse experiences, and foster collaboration. Also, in a report of work in NSF Teacher Enhancement projects, Friel and Bright's (1997) recommendations for professional development in mathematics included many of the elements above; new in their list were a focus on the need to understand children's thinking, the link between curriculum and teachers' decision making, the need to develop teacher leadership, the consideration of ways to deal with the tensions of change experienced by teachers, and the important role of becoming mathematics learners themselves and developing an understanding of the mathematics they are expected to teach. Knapp's (2003) review of literature on professional development that "is likely to deepen teachers' knowledge and skill *as well as* prompting application of this knowledge and skill in the classroom" led to recommendations including opportunities that

(a) "challenge teachers intellectually and are built around powerful images of teaching and learning;" (b) engage teachers as active learners with "concrete images of what high-quality practice looks like;" (pp. 120–121), (c) provide reinforcement through long-term exposure to ideas and interactions with other teachers; and (d) address specific problems of practice. This set of recommendations reinforces the need for teachers to be challenged by powerful images of good practice.

Ball and Cohen's (1999) examination of the needs of teachers were discussed previously in this chapter.

Missing in all but Friel and Bright's (1997) and Ball and Cohen's (1999) recommendations is a focus on developing a strong content-knowledge base through professional development. Perhaps this need was assumed; also, content knowledge is viewed as less important in some areas (e.g., reading) than in other areas (e.g., mathematics). Wilson and Berne (1999), in their review of high-quality professional development programs in mathematics, found that providers in these programs "acknowledged that professional teaching knowledge might include, at the very least, knowledge of subject matter, of individual students, of cultural differences across groups of students, of learning, and of pedagogy" (p. 177). All the projects they reviewed involved communities of collegial learners who were redefining their teaching practice and teacher learning, leading to their reconceptualizing both teaching and professional development. Wilson and Berne also noted that teachers often undertake professional development with preconceived ideas of what they will find helpful and relevant and that these ideas undergo, at times, radical revision, a point to consider in the light of the recommendations of others (above) that teachers should identify their learning needs and participate in planning for their professional development.

Professional developers might do well to also consider, as did Hargreaves (1995) in a literature review of professional development programs, reasons some professional development is *not* effective: Teachers are likely to reject knowledge and skill requirements when (a) the requirements are imposed or encountered in the context of multiple, contradictory, and overwhelming innovations; (b) teachers (except for those selected for design teams) are excluded from the development; (c) professional development is packaged in off-site courses or one-shot workshops that are alien to the purposes and contexts of teachers' work; or (d) teachers experience them alone and are afraid of being criticized by colleagues or of being seen as elevating themselves on pedestals above them. Hargreaves concluded that

The reason that knowledge about how to improve teaching is often not well utilized by teachers is not just that it is bad knowledge (though sometimes it is), or even badly communicated and disseminated knowledge. Rather, it does not acknowledge or address the personal identities and moral purposes of teachers, nor the cultures and contexts in which they work. (p. 14)

Brown and Borko (1992) noted that because learning to teach requires the acquisition of both a knowledge base and observable teaching behavior, cognitive psychologists are drawn to research on teachers' knowledge and their development of teaching skills. Indeed, much of the literature on professional development is produced by cognitive psychologists. Borko and Putnam (1995, 1996; also Brown & Borko, 1992) reviewed the literature on professional development programs, much of it in mathematics education, and concluded that there was substantial evidence showing that teachers in these programs did experience significant changes in their instructional practices. These changes depended on having "opportunities for teachers to construct knowledge of subject matter and pedagogy in an environment that supports and encourages risk taking and reflection" (Borko & Putnam, 1995, p. 59). Thus, in their view teachers can be considered to be both the objects and agents of change.

The above reviews on what makes professional development effective or ineffective were based on examinations of many studies. A final research study discussed here is one in which effective professional development was investigated through a large-scale, empirical data-collection and analysis. Garet, Porter, Desimone, Birman, and Yoon (2001) examined a national probability sample of more than a thousand mathematics and science teachers who had participated in professional development sponsored by the Eisenhower Professional Development Program. The dimensions of the study included the forms of professional development activity; the duration; the collective participation of groups of teachers from the same department, school, or district; and the core features, including the content focus, the opportunities for active learning, and the degree to which the activities were coherent and aligned with teachers' goals, state standards, and assessments. The researchers found that sustained and intensive professional development was more likely to be effective, as reported by teachers, than was shorter professional development. Their results also indicated that professional development that was focused on academic subject matter (content), was based on providing teachers opportunities for "hands-on" work (active

learning), and was integrated into the daily life of the school (coherence) was more likely to produce enhanced knowledge and skills than programs lacking these features. Thus they confirmed the importance of professional development that is focused on mathematics content, and their results support literature that said that coherence and collective participation were related to improvements in teacher knowledge and practice.

In one way or another, these reviews of the literature (except the last one) on successful professional development used student learning as a criterion for success. But strengthening the connection between teacher learning and student learning is difficult because reform efforts do not always match the system in which the reform is to take place. Moreover, teacher professional development (TPD) that

pursues more ambitious learning goals, requiring, for example, conceptual understanding of subject matter or application of knowledge to novel problems and new situations of use, is unlikely to exert much influence even if the TPD itself is well designed and supported. State and district curriculum frameworks, instructional materials, tests, and not least, longstanding traditions will combine to subvert teachers' efforts to change instructional practice. . . . [These portrayals] yield a paradox. If TPD is to be successful, it must fit with the regularities in place, but if it fits, it is unlikely to exert much beneficial influence on teacher or student learning. (Sykes, 1999, p. 160)

Sykes (1999) did provide suggestions for ways to strengthen the teacher-student learning connection. First, this connection should be used as a criterion for the selection and design of teacher professional development. Planners must have a theory about how the learning opportunities can affect the eventual learning by students. Second, professional development should be embedded in the content of the student curriculum. Teachers need opportunities to develop a deeper understanding of the curriculum they teach; they need to develop ways of presenting the content to their students; and they need to understand how students learn the curriculum. Third, teachers need to better understand student learning and the multiple ways to examine that learning. Fourth, teachers need opportunities to fully understand new implementations, whether they be changes in curriculum, in testing, or in instructional practice. "Relatively few schoolwide innovations have been validated on the basis of their impact on student learning" (p. 168). And fifth, an evaluation process that has student learning as an outcome is needed to develop better accountability.

A SEGUE TO DISCUSSING THE ACQUISITION OF PROFESSIONAL KNOWLEDGE

Professional development is an umbrella term for many types of activities and settings. Loucks-Horsley and her colleagues (Loucks-Horsley, Love, Stiles, Mundry, & Hewson, 2003) listed several: (a) aligning and implementing curriculum including curriculum-replacement units and selecting instructional materials; (b) participating in collaborative structures such as professional networks, study groups, and partnerships with mathematicians in business, industry, and universities; (c) examining teaching and learning through collaboration and action research, case discussions, examination of student work and thinking, scoring of assessments, and lesson study; (d) immersion in inquiry in mathematics and into the world of mathematicians; (e) practice teaching, including coaching, demonstration lessons, and mentoring; and (f) participating in workshops, institutes, courses, and seminars that can include developing professional developers and using technology for professional development. My organization of professional development activities and settings in this chapter is different from the organization of Loucks-Horsley et al., but, overall, I discuss most of the types they listed.

In the following three sections (and subsections), I focus on various types of professional development in which teachers' knowledge needed for teaching well is considered important, under the premise that without well-prepared teachers, students will not achieve to their capacities. Because teacher knowledge can be acquired in many ways and because the literature on professional development is vast and diverse, I have organized this literature in a way that helps me discuss it and that readers can, I hope, follow. I found the types of teacher knowledge suggested by Cochran-Smith and Lytle (1999), *knowledge-for-practice, knowledge-in-practice,* and *knowledge-of-practice,* to be a useful organizational tool. These authors described *knowledge-for-practice* as acquired by learning from formal professional development programs and university coursework. The emphasis here is that this is shared knowledge, already known by others such as those who provide teacher education and professional development experiences. *Knowledge-in-practice* is the practical knowledge of teaching known by competent teachers "as it is embedded in practice and in teachers' reflections on practice" (p. 250). Here teaching is treated as a *craft.* Teachers acquire knowledge-in-practice when they deliberate on their practice, "consciously reflecting on the flow of classroom action and invention of knowledge in action in order to take note of new

situations, intentionally and introspectively examining those situations, and consciously enhancing and articulating what is tacit or implicit" (p. 268). This kind of learning is the "stuff" of communities of practice, of lesson study. Finally, when teachers use their own classroom and school sites to investigate learning, knowledge, and theory, they generate *knowledge-of-practice;* they are learning from practice. "Teachers learn when they generate local knowledge *of* practice by working within the contexts of inquiry communities to theorize and construct their work and to connect it to larger social, cultural, and political issues" (p. 250). This kind of knowledge is acquired when teachers practice inquiry of teaching, whether or not they undertake research in a more formal sense.

Certainly this method of categorizing the ways in which teachers acquire knowledge is not a perfect one. The primary disadvantage of using this organizational tool is that, in reality, the examples of each are not clear cut, and in many cases the examples I have selected to characterize a particular type of professional development could also be used to characterize other types of professional development. But the purpose of this categorization is to provide examples of the different types of professional development (many of which are also relevant to teacher preparation) rather than to fully describe each project that serves as an example.

ADDRESSING QUESTION 4: HOW DO TEACHERS LEARN WHAT THEY NEED TO KNOW TO TEACH MATHEMATICS? THAT IS, HOW DO THEY ACQUIRE KNOWLEDGE-FOR-PRACTICE?

The literature on formal professional development, that is, on the contexts and strategies that provide the knowledge needed for the practice of teaching mathematics, are considered here within four categories: approaches that focus on student thinking; approaches that focus on curriculum, whether student textbooks or curriculum designed expressly for professional development; approaches that focus on classroom activities and artifacts; and, finally, approaches that make use of formal coursework focused on the knowledge needed for teaching well. I next discuss each of these approaches and provide examples of professional development programs and projects within each of these categories. The projects cited are not, by any means, exhaustive of those that have been successful in providing teachers with the professional knowledge they need to teach mathematics well. Those that are included have been the subject of research reports, although not even all of those can be discussed here. My purpose is rather to provide examples of an array of ways that each approach can be used.

A Focus on Student Thinking As an Approach to Teacher Learning

A review of research by staff members at the Eisenhower National Clearinghouse (2000) on examining student work and thinking led to suggestions for using this strategy productively. Although the strategy appears easy to apply, professional developers need to ensure that the tasks examined are appropriate and that student work varies in terms of the types of responses and the quality of explanations. Teachers' motivation for studying student work can vary from curiosity about disappointing student performance on standardized tests to the desire to develop a rubric for examining student work on particular problems (e.g., on a performance-based evaluation of learning), to the desire to design instruction that is based on what students know. Teachers can examine student work on their own, but having someone with better pedagogical knowledge of the mathematics being examined can frequently lead teachers to a deeper understanding of children's thinking and reasoning processes.

Little (2004) has described three complementary ways in which written student work can be used in professional development. First, it can be used as a resource for strengthening teacher knowledge, providing evidence of "what works" in the classroom, and developing the teachers' confidence in their own instructional practices. Second, student work can catalyze a spirit of inquiry and provide a basis for developing a professional community. Third, student work can provide evidence of student learning to be used for external accountability, thus assuring that instruction is aligned with standards.

It is precisely this constellation of purposes that makes "looking at student work" of particular interest in the landscape of teachers' continuing professional development (CPD). Embedded in the arguments for looking closely and collectively at student work, and embodied in the conception and design of specific approaches, these purposes reflect certain prevailing interests and emerging tensions in teachers' work and teacher development more generally. In this regard, "looking at student work" is a case emblematic of contemporary professional development in the USA. (p. 95)

Warfield (2001) found that elementary school teachers who have a research-based knowledge of children's thinking and deep understanding of the mathematics they teach "can extend what [they] learn about the mathematical thinking of their children and the ways in which they use what they learn to inform their practice" (p. 137). She described several ways this extension can take place. These teachers could

(a) pose questions that go beyond asking children to describe their solution strategies to asking them to think more deeply about the mathematics underlying those strategies; (b) understand students' mathematical thinking that differs from what might be expected based on the research-based information on children's thinking; (c) critically examine that thinking to determine whether it is mathematically valid; and (d) use what they learn about their students' thinking to create tasks that enable students to extend their understanding. (p. 137)

Some examples of projects that use student thinking as a means of developing teacher knowledge are next discussed.

Example: Cognitively Guided Instruction (CGI)

This primary-grades program was briefly described in the Goals section as an example of meeting the goal of developing understanding of how students think and learn about mathematics. An extensive body of literature on the effectiveness of using student thinking to guide instructional decision making was cited. This body of research is based on the hypothesis that if teachers listen to children, understand their reasoning, and teach in a manner that reflects this knowledge, they can and will provide children with a better mathematics education than if they did not have this knowledge. The research began in the late 1980s and continues to generate new research studies. In the first publications of their work (e.g., Carpenter et al., 1988; Carpenter et al., 1989), the CGI researchers reported on their study of 40 first-grade teachers—20 randomly assigned to a treatment group and 20 to a control group. The treatment consisted of a month-long workshop during which the participants studied children's thinking about addition and subtraction problems. Earlier work by Carpenter and Moser (1983) provided a detailed analysis of this domain of addition and subtraction problems and was used to familiarize the treatment-group teachers with the processes children use to solve problems. The treatment consisted of a summer workshop followed by a year of teaching after the workshop. During the second year, the teachers were observed teaching, and treatment teachers were found to listen to their students while the students were solving problems and to encourage children to solve problems significantly more than control teachers. The treatment teachers overcame their reluctance to place a greater emphasis on problem solving and found that their students achieved better than students of the control teachers.

In other reports (e.g., Carpenter & Fennema, 1992; Fennema, Carpenter, Franke, & Carey, 1992), the researchers provided detail on the teachers' use of knowledge of student thinking and the changes in teachers' beliefs and knowledge about children's abilities to solve problems. They undertook case studies of individual teachers to illustrate the CGI child-focused instruction and noted how two teachers who had access to the principled knowledge offered through CGI differed in how they used this knowledge in teaching. They studied the extent to which teachers in the workshop internalized the knowledge of children's thinking and then adapted instruction to meet student's needs, which varied from teacher to teacher.

In a longitudinal study (Fennema et al., 1996) of using children's mathematical thinking in primary classrooms with 21 CGI teachers, a pattern of increasingly greater gains was found for students who had been in CGI classrooms for more than one year. Overall, the gains in students' concepts and ability to solve problems were substantial. The descriptions of teacher gains and of these researchers' further study (Franke et al., 1997) of the self-sustaining, generative change that resulted from teachers' work with CGI are more appropriately discussed in later sections.

Examples: Teaching to the Big Ideas (TBI) and Developing Mathematical Ideas (DMI)

In the Teaching to the Big Ideas project, Schifter and colleagues (Schifter 1998; Schifter, Russell, & Bastable, 1999) examined the role of student thinking in elementary school together with exploration of mathematical content as avenues for teacher development, reflecting their belief that an analysis of student thinking provided a powerful site for continued mathematics learning. They acknowledged the difficulties inherent in moving from a presentation type of instructional practice to teaching as engaging students in mathematical discourse through understanding the student thinking that prompts students' questions and responses during classroom discussions of mathematics. "In a practice in which student thinking takes center stage, classroom practices become much harder to manage and much less predictable" (Schifter, 1998, p. 56). The elementary school teachers in the TBI project participated as mathematics learners in sessions with project staff. The lessons addressed the content of elementary school mathematics and were

directed toward mathematical issues that would challenge adult learners. The teachers in the classes struggled with the ideas. "Though most adults, including most elementary teachers, learned years before how to operate with fractions, few developed an articulated sense of the conceptual distinctions between whole numbers and fractions" (p. 78). They also dealt with these ideas in their own classrooms, which became resources for learning about children's thinking.

An important part of the work of the TBI seminar involved learning how to use children's perplexities to make such deep conceptual issues visible. The TBI teachers began their investigations into children's mathematical thinking by analyzing other teachers' students, studying videotapes of clinical interviews and classroom discourse as well as written materials illustrating student work. . . . However, the teaching practice TBI was working to help these teachers construct requires a facility that goes beyond an understanding of basic principles of children's mathematical thinking in general. Rather, it involves developing a new ear, one that is attuned to the mathematical ideas of one's own students. (pp. 78–79)

In addition to helping the participants work though mathematical topics by engaging in the mathematics and by examining student thinking, the providers also intended for teachers to develop a "disposition to inquiry. . . . [They] must come not only to expect, but to seek, situations in their own teaching in which they can view the mathematics in new ways, especially through the perspective that their students bring to the work" (Schifter, 1998, p. 84).

The TBI was a professional development and research project that led the authors to develop curriculum materials for teachers of Grades K–6—the Developing Mathematics Ideas (DMI) materials. The seven volumes of DMI, each of which focuses on a different mathematical topic, are being used at many sites for professional development seminars. The materials are heavily based on children's thinking. S. Cohen (2004) described this use of the materials:

> During seminar meetings teachers analyze the mathematical thinking of the children depicted in the cases (both video and print), listening to the students for the logic of their thinking, the parts of their reasoning and their understanding that are strong, and the mathematical issues that the children are still working on. (p. 32)

Three aspects of DMI that, together, "help describe DMI's particular place within the new genre of professional development" (S. Cohen, 2004, p. 34) are "(1) solidity and complexity of the mathematics under study, (2) the concurrent examination of teachers' and students' mathematics; and (3) the parallel between the seminar's pedagogy and elementary classroom pedagogy as envisioned by both national Standards and the DMI designers" (pp. 33–34).

S. Cohen (2004) reported on a research project she undertook to examine teachers' development of understanding both of mathematics and of children's thinking about mathematics over the course of a year of participation in DMI seminars. On the basis of teacher reports, she described the benefits to children:

> Children began to experience themselves, the subject matter, and the classroom community differently. The teachers see increased student engagement in the classroom, and it shows up in four ways. (1) The students enjoy mathematics, (2) they feel like they can do it, (3) they go further, more deeply into the subject matter than the teachers had thought was possible, and (4) a new kind of community is built. (p. 245)

Cohen found that these changes, described by the teachers, paralleled the changes she and the other researchers had found among the teachers. She noted the supportiveness, agenda, and rigor of the seminar community. "Professional development [groups] such as these are vibrant, rigorous, respectful intellectual communities that support teacher learning in all of its complexity" (p. 293).

Example: Integrating Mathematics Assessment (IMA)

This IMA project (Gearhart & Saxe, 2004; Saxe, Gearhart, & Nasir, 2001) was also designed to help elementary school teachers understand children's mathematical thinking and learn to use the knowledge gained to guide children to a deeper understanding of the mathematics. The researchers worked with a group of teachers who were using the *Seeing Fractions* "replacement unit" that had been designed for California teachers for use in place of textbook pages on fractions. The teachers were participants in the researchers' Integrating Mathematics Assessment program. This program was built on the assumptions that (a) when children solve problems, they make use of what they already know to create opportunities for developing new understandings, and (b) for an instructional approach to be effective, it had to help children build new knowledge based on prior knowledge. Thus, a curriculum alone was not necessarily sufficient for constructing understandings; the role of the teacher was to support students' development. Ongoing assessment was a critical factor. The professional development with these teachers consisted of cycles of activity; during each cycle, the first focus was on the teachers' understanding of the mathematics

(with the teacher as learner), then on understanding children's mathematics (with the teacher as researcher), and then on implementation of the lesson (with the teacher as educator). Assessment was critical to the third phase, and several ways to assess children's understanding were practiced.

The researchers compared the students of the IMA teachers to both students of a second group of teachers who had participated in professional development not focused on children's understanding of mathematics and students of a third group of teachers who had not participated in professional development. The IMA students had better conceptual understanding than students from other groups and had as much or more procedural proficiency. The researchers concluded that the replacement curriculum was not sufficient for students' improved learning—teachers needed to build activities and assessments to move these students toward understanding. In this project, the curriculum-replacement unit was used in the professional development to help teachers understand the mathematics and to learn what to expect from students using that curriculum.

Summary

Providing opportunities for teachers to examine student work has been quite productive, as indicated by these projects. The major focus of each was teachers' coming to understand student thinking about mathematics at their grade levels so that their instructional practice might improve and student learning might be enhanced. Learning the mathematics of the grades taught and how students learned this mathematics was essential to the success of all projects. The teachers had opportunities to "practice" on their own students, and the students' reactions were powerful reinforcers of the value of teaching for understanding.

A Focus on Curriculum As an Approach to Teacher Learning

Many researchers and professional developers believe that school curriculum materials should be designed with both teachers and students in mind. For example, Schifter, Russell, and Bastable (1999) argued that "innovative materials can aid teachers in rethinking their mathematics programs; indeed, we are convinced that curriculum must be thought of as a vehicle supporting ongoing teacher development, especially with regard to teachers' mathematical understanding" (p. 31).

The authors of *Adding It Up* (NRC, 2001a) also suggested that instructional material, including textbooks, could assist teachers in understanding mathematical concepts and student errors and in learning about effective pedagogy. Ball and Cohen (1996), too, spoke of the usefulness of curriculum materials, when integrated into professional development, in generating learning. The belief that student curriculum materials can be useful in a professional development environment is not new. Indeed, in the preface of the 1977 edition of *The Process of Education* (1960/1977), Bruner wrote, "A curriculum is more for teachers than it is for pupils. If it cannot change, move, perturb, inform teachers, it will have no effect on those whom they teach. It must first and foremost be a curriculum for teachers" (p. xv). According to Brown, Smith, and Stein (1996), research has shown that well-designed curricula can provide strong professional development for teachers of mathematics.

Many efforts over the past decade or so have been aimed at providing well-designed curricula for school mathematics. Soon after the publication of the *Curriculum and Evaluation Standards for School Mathematics* by NCTM in 1989, the National Science Foundation funded several school mathematics curriculum projects, for example, *Investigations in Number, Data, and Space* (Russell, Tierney, Mokros, & Economopoulos, 2004) at the elementary level; *Connected Mathematics* (Fey, Fitzgerald, Friel, Lappan, & Phillips, 2006) at the middle school level; and *Core-Plus Mathematics Project* (Coxford et al., 2003) at the secondary level. Each has spawned an industry of professional development workshops and conferences focused on helping teachers prepare to use the materials in their classrooms. The National Science Foundation has funded centers to study of the effects on the elementary curricula (ARC Center in Lexington, MA, http://www.comap.com/elementary/projects/arctest/), the middle school curricula (Show-Me Center at University of Missouri, http://showmecenter.missouri.edu), and the secondary curricula (COMPASS at Ithaca College http://www.ithaca.edu/compass/). Each center's website Home Page cites studies on the effectiveness of the various curricula, related primarily to student achievement but also to teacher learning.

In this section both children's textbooks and curricula designed for professional development are explored.

Example: Studying the Manner in Which Teachers Used Adopted Textbooks for Professional Development

Remillard and her colleagues (e.g., 1999, 2000; Remillard & Bryans, 2004; Remillard & Geist, 2002) have studied the effectiveness of using curriculum materials for professional development. In a 1999 study, Remillard described two levels of curriculum development. The first is what curriculum writers do when they produce curriculum materials for use in the classroom. The second is what teachers do as they "alter, adapt, or translate textbook offerings to make them appropriate for their students" (p. 318). In this research report, Remillard reviewed literature on recent teacher development projects, concluded that teachers need to have well-designed curriculum guidance, and suggested that researchers need to be more in tune with the role textbooks play in teachers' larger curricular agendas. In a 2000 study, she addressed the question of whether curriculum materials can support teachers' learning. By observing the manner in which two 4th-grade teachers "constructed curriculum" in their classrooms, she analyzed the nature and contexts of their learning over an academic year when they used a newly adopted reform-oriented textbook. Using her data, she was able to consider the manner in which teachers *read* the textbook, with *reading* referring "to a constructive and dynamic process of meaning making through engaging written text, such as would be found in a textbook" (p. 335). She also found and interpreted the manner in which teachers read both students and the tasks with which the students engage. She found that substantial learning occurred during these three types of reading. The teachers reexamined their beliefs and understandings during this process and, in so doing, influenced the curriculum. Remillard (2000) concluded that new textbooks alone will not affect teaching unless teachers are provided support to learn to construct curriculum from text in good ways. "To promote productive use of curriculum materials, professional development opportunities need to foster teachers' reading and decision making [and] deepen and broaden their mathematical knowledge" (Remillard, 2000, p. 347). Even when professional development was provided, however, Remillard and Bryans (2004) found that the perspectives and orientations teachers bring to their teaching of a particular curriculum can greatly affect the *enacted* curriculum in their classrooms. But when curriculum is well designed, it can be useful in professional development for many teachers and can contribute to larger instructional agendas.

Curriculum developers have also recognized the need for appropriate professional development and have suggested that good curriculum materials,

> when developed through careful, extended work with diverse students and teachers, when based on sound mathematics and on what we know about how people learn mathematics, are a tool that allows the teacher to do her best work with students. (Russell, 1997, p. 248)

Russell advocated using innovative curriculum materials with both preservice and in-service teachers and suggested that the curriculum can provide a forum for thinking about both mathematics content and students' mathematical thinking. When teachers can explore the depth of the mathematics and the changes in instruction called for in some of the new curriculum materials, such opportunities can lead to a commitment to the curriculum and to changes in teaching practices needed to be successful with the curriculum (Loucks-Horsley et al., 2003). Schoen, Cebulla, Finn, and Fi (2003) also found that professional development of secondary school teachers specifically aimed at preparing to teach the Core-Plus curriculum was important for teachers and led to higher student achievement.

A curriculum might exist not in traditional textbook format but rather as a computer microworld in which students can learn a narrow set of related concepts. Bowers and Doerr (2001) presented a microworld focused on concepts related to mathematical change to secondary preservice and in-service mathematics teachers in classes on two campuses. As students, the participants found value in conceptual explanations and agreed that "teaching with technology involves rethinking the format of activities and what counts as an acceptable explanation and solution" (p. 126) and they recognized "tradeoffs in having exploration before or after symbolization" (p. 126). As teachers, they learned "the value of building on students' incorrect explanations" (p. 130) and the "influence of hidden supports and constraints of technology on students' mathematical activities" (p. 131). The mathematical insights were all related to their increased conceptual understanding of mathematical change.

Another way in which curriculum has been extended is the deliberate development of instructional units to replace part of an adopted curriculum that falls short of expectations. In the IMA project described above, the curriculum-replacement unit was used in the professional development to help teachers understand the mathematics in the unit and to anticipate what to expect from students using that curriculum.

In professional development, the curriculum can sometimes be a set of materials or experiences designed to provide teachers with specific experiences and ways of learning. The two projects described next fit within this category.

Example: Introducing Mathematics Teachers to Inquiry

In this project, Borasi and Fonzi (1999) developed a set of materials that could be used to offer professional development or to provide experiences valu-

able to prospective teachers, primarily at the middle school level. The goals included helping teachers in "becoming mathematical problem solvers/inquirers, understanding the nature of mathematics and the 'big ideas' in mathematics, and developing mathematical confidence" (p. 45). Their package of materials, to be used by professional providers, incorporated an inquiry approach both to learning and to teaching. Their framework included a sequence of activities:

> beginning to develop a community of learners, experiencing inquiry as learners, reflecting on the inquiry experience, developing images of inquiry-based mathematics classes, articulating the principles of an inquiry approach, scaffolding a first "experiences as a teacher" of an inquiry approach, moving towards teaching mathematics through inquiry throughout the curriculum, and looking back and ahead. (p. 63)

(Each of these components was explained in more detail than provided here and was accompanied with a list of suggested activities.) The supporting print and electronic media materials provided professional developers with a detailed plan for implementing their program.

Borasi and her colleagues (Borasi, Fonzi, Smith, & Rose, 1999) described using the three mathematical-inquiry units they had designed with two groups of teachers of heterogeneous backgrounds. Working on the instructional units, which provided scaffolding, the teachers experienced inquiry learning of mathematics that they could take back to their classrooms. The mathematics topics (e.g., tessellations) could be used at various grade levels. The teachers' backgrounds varied—some were teaching special education classes. Participants were positive about the experience, and many found that the help they had received through their own inquiry-based learning led them to want to teach other mathematics using this approach.

Borasi and Fonzi (2002) described their units and the ways they can be used. Their analysis of results of their program led them to claim that it was "successful in initiating a long-term process of rethinking one's pedagogical beliefs and practices, as well as in promoting some immediate instructional change" (pp. 196–197). They found that follow-up was a key to success.

Example: Relating Mathematical Knowledge and Teaching

In this 3-year research study Sowder and colleagues (Sowder, Philipp et al., 1998; Sowder & Schappelle, 1995) investigated the relation of middle grade teachers' understanding of mathematics at their grade levels to their teaching practices. The researchers and selected teachers met weekly for a year, then once or twice a month during the second year. The 3-hour meetings focused on rational number, quantity, and proportional reasoning, with prolonged discussions about the concepts involved—an informal curriculum designed around teachers' content needs. Teachers were observed several times over the 2-year period. The results of the study were presented in terms of the effects of the seminars on teacher knowledge, case studies of the changes in teachers' instructional practices over the 2 years, and evidence of change in student learning during the second year. The teachers' conversations at the beginning of the first year were profoundly different from the conversations at the end of the first year and during the second year. Although the researchers chose initial topics and mathematical problems, the teachers controlled the agenda during the class period while they mulled over and persisted in trying to fully understand the mathematics. The positive effects of the seminars on the teachers were measured by pretests and posttests on content and by written comments from teachers.

The first year was difficult for these teachers. They struggled to understand while they came to realize that the mathematics they taught was not trivial; in fact, it was quite complex. But the teachers' changing understanding of content and their comfort level with content grew noticeably stronger, sometimes dramatically so, and influenced mathematics instruction in their classrooms. Student learning in their classes during the second year was measurably better than during the first year. The teachers changed their expectations of their students' capabilities when they learned that students could understand mathematics in ways they themselves had not understood before. Their views about the role of curriculum materials changed as they came to recognize the difficulty of teaching good mathematics without access to good instructional materials. The manner of classroom discourse also changed. Teachers began to probe for understanding and were delighted to find that students wanted to share their solutions with others. When asked why they now focused on questioning, even though questioning had never been a topic of discussion during the seminars, the teachers said that they had come to understand the importance of questioning when reflecting on the seminars and the manner in which the researchers' questioning them had helped them in developing their own mathematical understanding. The case studies of these teachers illustrated the need for support over an extended period of time for teachers to make meaningful change.

Example: The NCTM Academy

For approximately 4 years, The National Council of Teachers of Mathematics offered a series of professional development workshops through their Academy for Professional Development. The primary focus was the content of the *NCTM Principles and Standards of School Mathematics* (2000), as portrayed in the *Navigations* series published by NCTM. Each year several long-weekend academies were offered in different parts of the country, for elementary, middle, and secondary school teachers. Instructors were carefully selected and teachers reported that they found the workshops useful for implementing Standards-based instruction in their classrooms. The NCTM Academy was quite different from the other projects in that each academy offered short, concentrated programs on one topic.

Summary

All the research projects described in this subsection focused on developing, through the use of curricular materials, teachers' understanding of the content the teachers taught. But the approaches were quite varied. Borasi and Fonzi (1999, 2002) introduced mathematics to teachers through a set of carefully selected problems that were conducive to an inquiry approach to learning mathematics and were materials that the teachers would use in their own classrooms. Remillard and her colleagues (e.g., 1999, 2000; Remillard & Bryans, 2004; Remillard & Geist, 2002) studied the use of actual elementary school curricula in professional development and concluded that teachers need to develop curricula from textbooks, and to do so, they had to understand the mathematics in the textbooks. Sowder et al. (1998) led teachers to explore mathematical content in ways that introduced them to the challenges of the mathematics they taught, using a somewhat Socratic approach. The mathematics of elementary school *is* challenging (Ma, 1999) and needs to be a major focus of professional development. The fundamental message of the first three projects is that teachers need assistance in coming to understand well the mathematics they teach but that when they become confident of their own abilities to learn, the need for outside assistance diminishes. The purpose of the NCTM Academies was somewhat different, but they were attractive to teachers because they provided opportunities to explore Standards-based instruction by focusing on limited content (for example, geometry in Grades 3–5) and were of short duration.

A Focus on Case Studies As an Approach to Teacher Learning

Studying cases as a way of developing knowledge within a profession has long been used in law and medicine. Shulman, in his 1985 American Educational Research Association Presidential Address (published in 1986), made the case that education could also profit from the study of cases. He described cases as providing "knowledge of specific, well-documented, and richly described events" (p. 11). In 1992 he advocated that the study of cases be included in postgraduate programs much as they are in areas such as law and business. Cases are now frequently used in education to assist teachers in examining their practice and their students' reasoning and understanding.

When cases are used in formal professional development, a facilitator usually guides the case discussion and can influence the focus, the progress, and the outcomes of the discussion. A facilitator can play the role of a devil's advocate to encourage teachers to consider their claims in a careful fashion and to model the manner in which one can respectfully challenge the assumptions and ideas of others in the group (Barnett, 1998). Levin (1999) has noted other research that supported the important role of a facilitator in challenging misinterpretations and assuring that the discussions are fruitful, particularly with preservice teachers. She noted, "Recurring evidence that case discussions held in small groups followed by processing in the larger group may mean more active involvement and more attention by the participants than cases read but never discussed, or discussed without a facilitator" (p. 112).

In this section I consider how teachers can come to hear about, read about, and sometimes see how other teachers teach and the effects these experiences have on their professional development. Three types of cases are discussed: written cases, videocases, and multimedia cases.

Examples: Studying Written Cases of Classroom Events

Richert (1991) noted three components of cases that are important for teacher learning: They are descriptive in nature (even though they may contain analytical elements); they describe teaching practice; and they are "situated in a way that is significant for thinking of them as 'texts' for teacher learning (Brown, Collins, and Diguid, 1989), they are about particular students in a particular setting, at a particular time, for a particular purpose" (p. 116) and represent knowledge "as it is embedded in the practice of teachers' work" (p. 121). Furthermore:

> [Cases] present the dilemmas of teaching, the tradeoffs, the uncertainties. They capture teacher actions as they exist in the uncertain context where those actions occur. . . . [Cases] provide the potential for connecting the act of teaching with the cognitions and feelings that both motivate and explain that act. They offer a vehicle for making the tacit explicit. (p. 117)

Others have noted these and other qualities of case studies. Lee and Yarger (1996) found the merit in case studies to be that they provided rich detail and were explicit about contextual factors.

Cases provide opportunities for teachers to make judgments about what is worthwhile, to develop critical analysis of teaching and learning that is student centered, to analyze situations and weigh the effectiveness of various alternatives, to exchange perspectives with peers, to reflect on their own practices, and, in so doing, to extend their pedagogical content knowledge and become empowered in ways that lead to changes in beliefs about teaching (Barnett, 1991, 1998; Barnett & Friedman, 1997; Merseth, 1996, 2003). Teachers can examine what other teachers think and do, together with their rationales, then compare these findings with how they themselves think and what they do, together with their own rationales (Richert, 1991). In addition to the opportunities to explore teachers' and students' thinking and coming to understand mathematics in new ways, cases offer a safe environment for teacher inquiry because they provide teachers opportunities to explore and discuss instruction that is not their own or that of peers (Merseth, 2003).

Merseth (1996), in her review of research on cases and case methods, found that studying cases showed that skillful teachers do not operate "from a set of principles or theories, but rather build, through experience in contextualized situations, multiple strategies for practice" (p. 724). Studying cases, by helping teachers become more reflective about their own practices, allows them to develop this skill. Merseth (2003) has used case methods with secondary school mathematics teachers. Barnett, whose work has focused on case studies in mathematics with elementary school teachers, reviewed research that showed that case studies, as a method of professional development, can bring about changes in practices, beliefs, and awareness of student learning (Barnett & Friedman, 1997). Barnett's work is based on the theoretical framework of Spiro and colleagues (Spiro, 1993) in which a case-based curriculum

> that is effective in helping people acquire advanced knowledge in complex, context-dependent, and ill-structured domains . . . presents opportunities to reexamine a variety of ideas from different vantage points and in new contexts, thus offering the possibility of establishing multiple connections among cases or experiences that on the surface may seem dissimilar. (Barnett & Friedman, 1997, p. 391)

In later work, Barnett (1998) used aspects of cases and case discussions to illustrate four areas of development. The case studies seemed to be used successfully to promote a deeper understanding of mathematics. First, transcripts and conversations with the teachers illustrated how the discussions help teachers "extend and resolve their own understandings of the mathematics" (p. 87). Second, teachers learned to "consider student thinking as a feedback loop that is used to continually reevaluate, revise, and learn from teaching experiences" (p. 88); Barnett believed that this process also influenced the ways in which teachers selected tasks for their students. Third, the case studies helped teachers to understand that concepts they thought were difficult were actually quite easy, and, just the opposite, mathematics they thought was easy was often more complex than they had realized. Fourth, the opportunities to critically examine alternative views helped them reconsider their own understandings and beliefs. Barnett concluded by noting two hypotheses: that case methods such as these

> demonstrate the extraordinary value of collective inquiry and critical reflection [and that] an individual teacher's decision to pursue an idea that has been carefully examined through the case discussion process is more likely to endure the tests of classroom practice. (p. 92)

Four pivotal experiences "help move teachers beyond their own perceptions and beliefs" while participating in case studies (Barnett & Friedman, 1997, pp. 389–390): First, the cases are written by other classroom teachers and provide opportunities for peers to collaborate in formulating approaches to the case. Second, the discussions model a way of teaching; the facilitator is a guide, not an expert. Third, the issues are usually unresolved and leave teachers with a desire to continue the inquiry. Fourth, teachers have opportunities to see classroom experiences through the eyes of students. Barnett and Friedman argued that "reflection about specific student thinking in the context of classroom situations is perhaps the most potent experience a teacher can have" (p. 390).

Stein, Smith, Henningsen, and Silver (2000) referred to cases in work such as that of Barnett and Merseth as *dilemma-driven* because each case focuses on a pedagogical problem, thus helping teachers realize that teaching is inherently dilemma-driven and that they face trade-offs when selecting a course of action in the classroom. In contrast, Stein et al. called their own cases *paradigm* cases because they "embody certain principles or ideas related to the teaching and learning of mathematics" (p. 33); they are designed to help teachers understand tasks and the associated cognitive demands, and they require that teachers critically reflect on their own practice. They claimed

that teachers often do not know how to think about their own practices. By learning to critically examine and reflect on their practices, using the Mathematical Tasks Framework the authors designed, teachers can "bring meaning to the myriad actions and interactions that constitute classroom activity" (p. 34). This framework focused on student learning as the outcome of mathematical tasks (a) as they appear in curricular materials, (b) as they are set up by teachers, and (c) as they are implemented by students. Stein et al. used *cases*, that is, episodes of classroom practice, as a way to situate the abstract ideas of their Tasks Framework. They explained, "Once teachers begin to view our cases as *cases of* various patterns of task enactment, they can begin to reflect on their own practices through the lens of the cognitive demands of tasks and the Mathematical Tasks Framework" (p. 34). The Mathematical Tasks Framework and the use of tasks and cases themselves were developed as part of the Quantitative Understanding: Amplifying Student Achievement and Reasoning (QUASAR) project.

Other researchers have also used cases successfully as a form of professional development that supports teacher change. For example, Walen and Williams (2000) designed a research study that was based on their belief that "reflection on the practice of teaching is vital to professional growth [and] as a data source in the study of teaching" (p. 5). Their review of research on case methods led them to conclude that cases used in group settings can lead to a community of learners because of the power of shared inquiry. They collected data on two cases used during a 2-day teacher meeting with 115 teachers. Teachers began the first case by passively accepting their situation and ended by challenging the system. "As individuals became aware of similarities of their ideas, teachers expressed confidence in their ability to find ways to change the system. Teachers genuinely enjoyed the opportunity that this case provided for discussion" (p. 15). The authors situated their work within Merseth's (1996) conceptual framework—but it differs from other studies because of the manner in which Walen and Williams viewed the role of reflection:

> We see reflection not as one more among many psychological processes, subject to development and training, but as a fundamental ontological fact of human existence: Reflection is a movement among modes of being, a movement from the unreflective ready-to-hand mode to the more abstract unready-to-hand or present-to-hand modes. For us, reflection is intimately tied to individual concerns and is always a movement from within a context of concern. Thus, we see reflection as bringing out for each teacher what is of individual concern, allowing those concerns to be

examined and made sense of within the context of their own lived experience." (pp. 22–23)

Case-based studies have been reported in other countries. For example, constructing cases as part of school-based professional development in Taiwan helped teachers to become more aware of and help students who had difficulties learning mathematics (Lin, 2002).

Examples: Video Case Studies

The study of videos of teachers teaching is another form of professional development using cases. Video cases offer an opportunity for teachers to share a common experience of teaching practice and can act as a scaffold for developing theory from practice and applying theory to practice.

> The potential that video has for making practice accessible in manageable sized chunks, and the opportunity video offers to pause, rewind, and replay practice is worthy of attention. It has the possibility of portraying the complexity of practice and at the same time enables a focus into the mass of action. (LeFevre, 2002, p. 13)

But developing video cases is expensive, and they are more difficult to use than written cases because of the need for appropriate and dependable technology, both drawbacks that are disappearing, however. Thus, more of these cases are expected in the future. The video-cases studies I discuss here are only a sample of such studies.

Classroom video of American, German, and Japanese teachers taped as part of the Third International Mathematics and Science Study (TIMSS) have been widely used in professional development because of their power to draw teachers into deep discussions of instructional practice (U.S. Department of Education, 1997). The videos clearly illustrate very different teaching styles unique to each country (Stigler & Hiebert, 1997); they provoke thoughtful analysis of the differences and of how these instructional differences might be related to achievement differences.

Seago and Mumme (2002; Mumme & Seago, 2002), in their Videocases for Mathematics Professional Development project (VCMPD) have developed videocases as tools for K–12 mathematics teacher development. They deliberately avoided creating the image that the teachers in the videos were exemplary teachers because they were not convinced that exemplary images would help teachers understand and reason about the practice unless they could relate to the class and teacher being observed. They believed that "video provides an opportunity to study an instance of teaching that is emotionally distant enough to create a safe place to scrutinize practice carefully" (Seago & Mumme, 2002, p. 5). They hypothesized four ways in which teachers could learn from videocases:

1. "Videocases afford teacher opportunities to learn mathematics embedded in teaching" (p. 11); that is, teachers can learn *useable* mathematics within discussions of the videocases.

2. "The cross-analysis of multiple and varied video segments can provide teachers the opportunity to generalize mathematical and pedagogical ideas" (p. 13).

3. "Videocases afford the opportunity for teachers to develop a more complex view of teaching" (p. 15). Seeing real teaching that is "messy" and complex provides teachers opportunities for identifying and extracting what is relevant and important.

4. "Videocases can provide teachers opportunities to develop new norms of professional discourse and a more precise language of practice" (p. 17). Teachers need ways to critically analyze teaching practice and to make claims that they can support; they need to learn accepted ways of describing what they are seeing and thinking (e.g., terms such as *classroom culture* and *teacher directed* become part of their professional language).

The VCMPD videos of cases through which to study classrooms are not unique. The CGI and DMI projects (both discussed earlier) have video cases to be used in professional development; both of these sets of videos can also be useful in teacher preparation courses. The difficulty is not so much in locating good video to use in professional development; it is rather that the video should provide needed lessons for teachers and that the provider using the videos should have the background to help teachers profit from the viewing of video.

Some video case studies described here focused on the learning of preservice teachers whereas others focused on practicing teachers. The manner in which these two groups engaged in cases, whether video or not, differed. Richardson and Kille (1999) pointed out that preservice teachers "do not have the cognitive schema or practical knowledge that are acquired from teaching experience to be able to ask appropriate questions and reconstruct their own sense of theory and practice" (pp. 121–122). This difference carries over to many types of professional experiences educators provide for both groups. Sowder, Philipp et al. (1998) concluded from their work that teachers contextualized, in terms of their own students, almost every mathematical task the group undertook, whereas preservice teachers were unable to think in terms of what children can and cannot do. Videocases do, however, provide a context through which preservice teachers can explore classroom situations.

Examples: Multimedia Case Studies

These case studies usually include some video but complement the video with other types of media. In an intensive year-long study, Lampert and Ball (1998) used videotapes of themselves teaching third graders (Ball) and fifth graders (Lampert) to create records of teacher practices that could communicate the complexity of teaching. The focus was on instruction over an entire school year rather than on isolated lessons or even units of mathematics. In addition to the videos of most lessons and some audiotapes of the remaining lessons, they collected field notes written by observers of every lesson, their own written reflections after each lesson taught, and records of student work. A curriculum was developed around these materials and then used in their teacher preparation program. The researchers' aim was to

> put beginning teachers in a position where they have the capacity to knit the resources of theory with those of practice as they explore "big ideas" in knowing teaching. Just as place value and functions are big ideas in knowing mathematics, discourse, curriculum, the teacher's role, and classroom culture are all big ideas in knowing teaching. (p. 79)

Prospective teachers carried out investigations on a topic or question. An investigation might have included examples of student work, excerpts from the teacher's journal, and segments of video. Lampert and Ball analyzed 68 investigations; most analyses focused closely on the teaching. For example, one project focused on "What does Deborah do when a student gives the wrong answer?" (p. 96). The analyses led the investigators "to think that the multimedia environment offered significant potential as a site for a new pedagogy of teacher education" (p. 109) but that challenges remained to explore in future projects: The investigators did not know what the teacher education students actually took from their work in these classes, how to manage the interplay between big ideas and the contexts and particulars of practice, and what types of policies or programs would be most appropriate for supporting teacher education programs.

The difficulties in finding a proper field placement for preservice teachers and the challenges of trying to use student thinking in instructional practice led Masingila and Doerr (2002) to design multimedia case studies that could be used to explore "the complexities of teaching and the complexities of teaching *about* teaching" (p. 238). They used a case study of a middle-school teacher teaching a four-day sequence of lessons on the use of data; the case included a video overview of the school setting, the teacher's lesson plans, video of the class lessons, students' written work, the teacher's reflections and anticipations on each lesson in a video journal, and the transcripts of all the included video. They found that this rich video source served as a site in which the preservice teachers could investigate, analyze, and reflect on the practice of the teacher and the effects of her instruction on student learning.

Van Es and Sherin (2002), building on multimedia case studies, developed and successfully used a software tool, Video Analysis Support Tool (VAST), to support secondary intern teachers in developing abilities to "notice and interpret aspects of classroom practice that are important to reform pedagogy" (p. 8). However, they did not investigate how the participants' teaching might have been affected by the experience.

The possibility of interactive media increases the usefulness of technology in teacher development. Sullivan and Mousley (2001) suggested that interactive media can be a tool used to develop both "student teachers' and teachers' understandings of the complexities of mathematics classrooms as well as about the many roles that they can play. It allows moments that have the potential for active classroom decision-making to be 'frozen,' discussed, and explored" (p. 162).

Summary

Case-study methods include the study of written cases, the study of video cases, and the study of multimedia. The study of classroom events and activities is formal in the sense that the events and activities studied are selected by a facilitator who then guides discussion by teachers who have read about the events or watched videos of the events. At times, teachers' writings about events in their own classrooms are shared with the group, and usually the discussion is still coordinated by a facilitator. Using case studies such as these, teachers can talk about instruction in a way that is nonthreatening and rewarding. Cases also preserve the complexity of teaching while providing opportunities to examine it. The mathematics inherent in these events provides opportunities to increase content knowledge. During discussions of pedagogy teachers examine the practices of others and reflect on the cases and take away lessons, both mathematical and pedagogical, that they can apply to their own instruction.

Formal Course Work

Many teachers seek out professional development opportunities in continued university coursework, such as master's degree programs, certificate programs, or focused courses on mathematical topics they feel unprepared to teach. Through this type of professional development, teachers earn university credit toward credential programs, fulfill an obligation for continuing education, obtain higher salaries, and participate in structured learning experiences that are otherwise unavailable to them. At the elementary level, degree programs are sometimes offered for cohort groups, thus allowing teachers to form communities of support. But frequently teachers, particularly secondary teachers, are not in cohorts and have classes with, for example, master's-degree students who are from many school sites or who are not even educators but rather students needing mathematics for some other purpose. Teachers in such situations are less likely than those in cohort groups to form communities of support.

Courses offered in programs for teachers should be based on sound principles of professional development (Loucks-Horsley et al., 2003), but often they are not. For example, graduate-level courses in mathematics are often taught in a lecture format only. Some institutions offer teachers separate courses that focus on developing pedagogical knowledge together with content knowledge. The disadvantage of separating teachers from other mathematics students, particularly at the secondary level, is that some mathematicians believe that the mathematics taught to prospective or practicing teachers is less rigorous than the mathematics in regularly offered courses and are more likely to consider these students to be second-class citizens in the department.

Acquiring Knowledge-for-Practice: Discussion

Acquiring *knowledge-for-practice,* that is, knowledge that belongs to someone else who shares it with teachers, is a common form of professional development and is what most people think of when the term *professional development* is used. Professional development provided by school districts or funding agencies is usually offered in one or more of the forms discussed here.

As noted in the introduction to this section, the categorization scheme (using student thinking as an approach to teacher learning, using curriculum materials as an approach to teacher learning, using classroom activities and artifacts as an approach to teacher learning, and formal courses) used to organize the many projects and studies that fit under *acquiring knowledge-for-practice* can lead to a false implication that any particular project was narrowly focused in terms of its delivery of professional development. For example, the study of written cases of classroom events is placed in the subsection on using classroom activities and artifacts, but the study of written cases is bound to focus on the student thinking that is exemplified in the case. Thus this work could have been described as using student thinking as an approach to teacher learning.

Some Common Features

There were commonalities across projects; two in particular deserve attention. First, all the professional development projects described here provided teachers with opportunities to themselves learn mathematics. For example, in the DMI project (Schifter 1998; Schifter, Russell, & Bastable, 1999) the learning of mathematics by the teachers was overt and taken seriously. And in the IMA project (Gearhart & Saxe, 2004; Saxe, Gearhart, & Nasir, 2001) the first focus was on the teachers' understanding of the mathematics, with the teacher as learner. In fact, one might question the success of any mathematics professional development in which the teachers do not themselves learn mathematics, whether or not their mathematics learning is a stated goal.

A second commonality was that all the projects described in this section fostered an *inquiry approach* to learning and teaching; that is, the providers emphasized teachers' "engagement in problem-solving and theory-building about important situations and concepts" (Lloyd, 2002, p. 1). For example, Borasi and Fonzi (1999), in their *Introducing Math Teachers to Inquiry* project, consciously set out to provide teachers with opportunities not only to learn mathematics but to do so by experiencing learning themselves. The teachers were placed in situations in which they worked through the instructional units they would be using, an experience that required them to adopt an inquiry stance toward learning, and eventually, to teaching. As with the opportunities to learn mathematics, one might question whether professional development that does not foster inquiry as a necessary component of mathematics learning could be considered successful.

Mention of such commonalities serves as a reminder that projects selected as examples of a particular approach could have been selected differently, because most projects used more than one approach.

Thus the examples should not be construed as being limited to the particular focus under which they are placed. The placement here is simply one means of looking at how different approaches were realized in different projects and studies.

ADDRESSING QUESTION 5: HOW DO TEACHERS LEARN FROM THEIR PROFESSIONAL COMMUNITIES ABOUT TEACHING MATHEMATICS? THAT IS, HOW DO THEY ACQUIRE KNOWLEDGE-IN-PRACTICE?

Recall that the second category of knowledge described by Cochran-Smith and Lytle (1999) was *knowledge-in-practice*, that is, the practical knowledge of teaching "embedded in practice and in teachers' reflections on practice" (p. 250). Lesson study (e.g., Lewis, 2000), which could be considered as a special type of case study, is discussed here as an example of learning-in-practice because in lesson study teachers deliberate on the practices they observe with others. Communities of practice and professional development schools also provide settings for learning from practice. Learning often takes place in settings in which teachers join with other teachers—making learning a communal process. These settings can be thought of as sites of practice that "seem promising as contexts through which teachers can learn mathematics, in particular the mathematics it takes to teach well" (Ferrini-Mundy, 2001, p. 125).

Professional Communities

Many professional communities into which teachers might be drawn to participate do not necessarily use the strategies described in the previous section, but teachers profit from them professionally. Many of these communities are informal—a teacher might meet occasionally with other teachers who are teaching the same grade level within their school to develop an instructional plan that meets state standards. Sometimes pinpointing exactly what teachers are learning is difficult—at times teachers exchange stories that might seem like gossip but could be regarded as " learning about the moral principles that guide one another's work" (Hargreaves, 1995, p. 22).

The professional communities in which teachers seek membership are often outside their school districts. We have much yet to learn about the effects of participation in professional associations such as state and national organizations (Little, 1993), although we know that effects are real. For many teachers, attend-

ing a national NCTM conference, for example, is a heady experience. These and other teacher members subscribe to journals and website offerings, purchase books and instructional materials, and attend conferences and professional academies of NCTM. State organizations of mathematics teachers often have strong memberships and offer conferences that are well received by teachers. Thus, teachers do participate in and take advantage of opportunities offered by professional societies, but this type of professional activity is nebulous and thus its effect on teachers is more difficult to determine.

A professional community shares a common base of specialized knowledge, and that specialized knowledge in a community of teachers is of the types discussed in earlier sections of this chapter—content and pedagogical knowledge and understanding how students think. The formation of teacher communities is inherently more difficult than community formation in other professions because participation is voluntary: Teachers most often experience teaching as individuals behind closed doors and do not feel the need for outside assistance. Also, the work of these communities often takes place primarily outside the workday and thus are seen by some as taking away from time with family or other personal pursuits.

Yet many teachers feel the need to participate in a professional community; they seek to find or establish such a community because, instead of being extraneous to their lives within a school, "constructing and participating in such professional communities (Little & McLaughlin, 1993) in schools is itself a vibrant form of teacher and school development [in which the] members are constantly searching for ways to improve practice" (Hargreaves, 1995, p. 19).

In a review of literature on teacher communities, Grossman, Wineburg, and Woolworth (2001) found that "a key rationale for teacher community is that it provides an ongoing venue for teacher learning" (p. 947). Although teacher learning is an important outcome, it is not the only incentive for teachers to join together in communities of practice. Wenger (1998) argued that "issues of education should be addressed first and foremost in terms of identities and modes of belonging, and only secondarily in terms of skills and information" (p. 263). Members of a true community share a sense of purpose, usually with an understanding that they are responsible to one another for achieving their goals; they coordinate their efforts to assure student learning, learn together to improve their practice, and share responsibilities for decision making about matters pertaining to their group (Secada & Adajian, 1997). Lord (1994) has listed seven other characteristics that help define professional communities of teachers; some overlap with those of Secada and Adajian:

1. Teacher ownership or, at the very least, increased partnership for teachers in decisions regarding professional development activities.

2. A collective commitment to acquiring and using new knowledge in the subject areas, especially knowledge that could be characterized as 'cutting edge.'

3. A reliable connection to resource-rich institutions, organizations, or associations independent of the school or school district, e.g., university education, liberal arts, and science departments, as well as libraries, museums, theaters, businesses and industries, and civic agencies.

4. Intensive, and in some cases long-term, professional relationships among participants.

5. A perspective on the profession of teaching that extends beyond the four walls of the school and beyond the duration of individual teachers' careers.

6. A greater commitment to 'lateral accountability' within the teaching profession, i.e., the critical review of teachers' practices by other teachers.

7. High levels of teacher involvement in the reform of system wide structures. (p. 199)

Lord (1994) advocated working with communities beyond the school because they provide access to intellectual resources that often do not exist in a small community at the school level, even though external communities commonly "operate at the margins of local school district life" (p. 201). Some teachers seek out communities in which they can find these intellectual resources and establish professional identities. Philipp et al. (1994) found that a common element in the lives of four extraordinary teachers whom they studied was long-term, active participation in professional communities outside their schools, which provided them with identities as professional mathematics teachers.

From another viewpoint, Grossman et al. (2001) suggested that summer workshops and professional days away from the school are often viewed as activities that take place during free time rather than as part of an ongoing professional life. They argued that professional communities located in the school offer "the possibility of individual transformation as well as the transformation of the social settings in which individuals work" (p. 948), noting that the expectation of a teacher's teaching for 9 months then learning for 2 weeks in the summer is like "having a marathoner train all week long and eat only on weekends" (p. 993). But the ways in which professional communities within a school can affect the instructional process should not be underestimated—they can have a profound effect on classroom practices and student learning.

A community of practice is considered by some to be more than the professional communities just described. Communities of practice "create, expand, and exchange knowledge about their practice" (Gallucci, 2003, p. 15). They vary in terms of their strength and their openness to engage with new ideas about their practices. These variations can "affect teachers' interpretations of standards-based reform policies, their engagement with learning opportunities, and, subsequently, the kinds of changes they make in classroom practice" (p. 14). They are more likely to remain intact if the membership is stable and if the members think that they are profiting professionally by their participation.

Mathematics education researchers have, in recent years, begun to study professional communities of teachers and communities of practice in mathematics. Some examples follow.

Example: Communities of Practice among Mathematics Specialists

Nickerson and Moriarty (2005) worked with teachers in eight low-performing, urban elementary schools in which the district had placed mathematics specialists who had 60–90 minutes each day of professional development time together. The teachers were all taking mathematics and pedagogy classes to earn a mathematics-specialist certificate. In some schools, teachers took advantage of their time together to develop a community of practice within their school. They developed a shared sense of purpose, and their discussions focused on how to provide students the best opportunities possible to learn mathematics. But at other schools, the teachers used the time for grading, individual lesson planning, and discussion of their students' shortcomings. Where teachers formed a community of practice, they were more apt to have good relationships with other teachers in the school and with the principal; they respected and accessed the particular knowledge of one another; their mathematical content knowledge was strong enough that they felt comfortable in their classrooms; and they were familiar with the culture and language of their students.

The next two examples provide instances in which researchers have taken theoretical stances that helped them study communities of teachers in new ways.

Example: QUASAR

The QUASAR project researchers set out to explore the possibility of changing student achievement in low-performing middle schools at several locations around the United States. The project was focused on providing assistance to the staff of the middle schools and then evaluating the effects, using the site as the unit of analysis of change. Instead of focusing on one teacher at a time, the researchers (Stein & Brown, 1997) studied the "process that takes advantage of the synergy, support, and motivation supplied when a 'critical mass' of teachers undertakes reform for all students in a given school" (p. 156). By changing the focus of attention from the individual to the group, they defined *learning* at each site by a transformation in the ways teachers participated in the collaborative activities and in the community that was being formed.

In a QUASAR study of teachers in one school, Stein and Brown (1997) used the *framework of change* to define, as well as measure, *teacher learning* as movement from the peripheral to a fuller form of participation. The researchers visited the school a number of times over several years and so had to consider changes in teaching staff along with other data collected. They decided to examine teachers' participation by measuring *learning* as breadth of participation in all the activities in which members participated. That is, a long-time staff member was expected to participate in more activities than a newcomer, and each teacher was expected to increase his or her range of participation over the course of the project—thus both breadth and depth of activity were measured. These measures proved to be useful indices of participation and, thus, of learning.

This learning encompassed the gradual appropriation of a complex set of knowledge and skills related to valued classroom instructional practices as well as professional work associated with building, sustaining, and explaining a reform program of mathematics instruction. Moreover, teachers were developing the attitudes, values, and orientations that underlie the central tenets of the reform movement. The learning that occurred did so without the teachers being explicitly taught in any sort of structured manner, but rather through their ongoing exposure to old-timers and the work practices of the community." (p. 170)

The teachers were motivated to work collaboratively on projects they valued. Each teacher's contribution depended on the level of expertise of that teacher, and all were able to move forward and to learn.

In a second study, in a different school, the researchers (Stein & Brown, 1997) found that the framework of the study had to be altered because reform goals and joint productive activity had not yet taken hold. They turned to work by Tharp and Gallimore (1988), for whom *learning* was viewed as assisted performance, which "within the social organization of schools helps to explain the organizational and interpersonal factors necessary for teacher learning to occur and the obstacles that must be overcome in most schools in order for reform communities to become established" (Stein & Brown, 1997, p. 171). *Learning*, though again viewed as a process of transformation of participation, was defined in terms of the *zone of proximal development* (ZPD, a term used by Vygotsky, 1978), and the learning of teachers was studied while these teachers moved from assisted performance to unassisted performance.

The framework Tharp and Gallimore (1988) used had four stages. In the first stage, performance was assisted by others; in the second, performance was assisted by oneself; in the third, performance was independent and automated; in the fourth, a stage of review and revision, the first three stages were revisited. The researchers examined lines of influence, which they called *chains of assessment*, and the manner in which they became *chains of assistance*. They were led to study the larger social system in which the teachers worked. Principals arranged for expert and experienced people to help the teachers and worked alongside the teachers during these sessions.

In the beginning of this study, resource partners were, together with administrators, sources of assistance. (The school had committed to using resource partners as part of the agreement with the QUASAR project.) The district, however, held to its schedule of testing, and consequently the district administrators, the principal, and teachers at this middle school had conflicting goals. Also, the common planning time funded by QUASAR was not used productively. Thus, little progress was made during the first two years. In the third and fourth years, however, the district provided more leeway to the school mathematics teachers, and they progressed toward the QUASAR goals of reformed teaching of mathematics. The common planning time was used as an "effective activity setting for teachers learning through assisted performance" (Stein & Brown, 1997, p. 184). The researchers began to see movement through the ZPD while teachers worked toward unassisted performance.

Stein and Brown (1997) noted that their two studies were sociocultural in nature and "channeled attention away from the cognitive attributes and instructional practices of individual teachers toward the collaborative interactions that occur as teachers (and others) attempt to develop and improve their school's mathematics instructional programs" (p. 185). Yet this channeling of attention played out differently in the two schools:

> In the communities-of-practice framework, this transformation of participation is characterized as movement from *peripheral to fuller* forms of participation in the overall work activities of the community. In the framework advocated by Tharp and Gallimore, the transformation is characterized as movement from *assisted to unassisted* performance in specific activity settings. (p. 185)

In another study, Stein, Smith, and Silver (1999) provided detailed examples of how professional development evolved over time at two QUASAR sites and the manner in which, at the first site, the teachers took from the university partners the lead in designing professional development. The evolution of the work at this site led teachers to form a community of teachers with a shared vision of the mathematical competence they wanted their students to attain. These teachers decided that they no longer needed the expertise provided by the outside developers. Apparently because these two outside providers never taught in or even observed their classes, the teachers concluded that the outsiders did not really understand the work of the teachers. A lesson to be learned from this experience is that when mathematicians become involved in providing professional development for teachers, it is imperative that they visit classrooms and that they do so as learners. Only then will they understand the constraints under which teachers work, the teachers' students, and the type of help teachers need.

Example: Situating Work in Instructional Practices

To study teachers' instructional practices, Cobb, McClain, Lamberg, and Dean (2003) also situated their work in the instructional setting. The authors developed "an interpretive perspective that seeks to situate teachers' instructional practices with respect to the affordances and constraints of the schools and districts in which they work" (p. 13). They noted that most research on teachers has focused, on the one hand, on the role of professional development that leads to teachers' reorganizations of their teaching practices and, on the other hand and independently of the first, around the structural and organizational features of schools. They viewed the school in which they worked as a "lived organization" consisting of interconnected communities of practice, and they developed what they called an *analytic* approach to research that they have used to study teaching as a distributed activity and to situate teachers' instructional practices "within the institutional settings of the schools and school districts in which they work" (p. 13). The notion of interconnected communities of practice led them to distinguish among three types of interconnections: (a) *boundary encounters* occurred when the teachers' community joined in activities with another community (e.g., the district mathematics leaders); (b) interaction with *brokers* who could "bridge activities of different communities by facilitating the translation, coordination, and alignment of perspectives and meanings (Wenger 1998)" (p. 19); and (c) *boundary objects,* interconnections among communities of practice based on *reification* (a term also attributed to Wenger) that "play a significant role in enabling members of different communities to coordinate their activities even when they are used differently and have different meanings" (p. 19). They used the example of a district pacing guide, used by the teachers, as a boundary object.

Three general aspects of the Cobb et al. (2003) analytic approach, taken together, provide a general method of understanding specific situations and thus make a useful way to study teachers. First, using their approach provides a way to account for different interpretations held by different communities to "acknowledge and account for the frustrations and antagonism that they experience" (p. 21). Second, people's actions, as situated within a school "as a lived organization, constitute resources that might enable members of particular communities to move beyond viewing members of other communities merely as impediments to their agendas" (p. 21). Third, this type of analysis can "support the formulation of strategies for institutional change that involve the creation of new tools as prospective boundary objects as well as the orchestration of boundary encounters and the development of brokers" (p. 21). Together, these three aspects constituted a "*general* method for understanding the specific settings in which particular groups of teachers work" (p. 21) and allowed the research group to more effectively collaborate with teachers.

In the next part of this section I turn to another type of professional community connecting schools and universities.

Professional Development Schools (PDSs)

Some school sites link with universities to form professional communities. PDSs are an outgrowth of the work of the Holmes Group, a consortium of research universities dedicated to improving teacher education. The Holmes Group prepared three sets of goals: In 1986, they published goals for tomorrow's teachers; in 1990, goals for tomorrow's schools; and in 1995, goals for tomorrow's schools of education.

One goal stated in the first report was "to connect our own institutions to schools" (Holmes Group, 1986, p. 4). The second report introduced Professional Development Schools to provide "continuing learning by teachers, teacher educators, and administrators" (Holmes Group, 1990, p. 7) with a major commitment of "overcoming the educational and social barriers raised by an unequal society" (p. 7). A PDS would "devise for itself a different kind of organizational structure, supported over time by enduring alliances of all the institutions with a stake in better professional preparation for school faculty" (p. 7). In the third report the Holmes Group called for connecting "professional schools of education with professionals directly responsible for elementary and secondary education at local, state, regional, and national levels to coalesce around higher standards" (Holmes Group, 1995, p. 20) . . . so that a PDS would function "as a place where prospective and practicing educators from the school and the university immerse themselves in a sea of inquiry in pursuit of ever more effective learning" (p. 20). By 1999, some 250 professional development schools were located across the United States (Cochran-Smith & Lytle, 1999).

Lanier (1994) referred to PDSs as places of "ongoing invention and discovery; places where school and university faculty together carry on the applied study and demonstration of the good practice and policy the profession needs to improve learning for young students and prospective educators" (p. ix). These schools provide opportunities for school faculty and university faculty to join together in research and in rethinking instructional practice (Darling-Hammond, 1994a). PDSs, from the beginning, have had equity as a central focus (Valli, Cooper, & Frankes, 1997). They are viewed as places that produce professionals. Because "professionalism starts from the proposition that thoughtful and ethical use of knowledge must inform practice," PDSs, by focusing on "knowledge and expertise, ethical commitment to clients, and responsibility for setting standards" (Darling-Hammond, 1994a, p. 4), are professionalizing the field of teaching.

The critical connection between teacher education and professional development, beginning with teacher induction, has been recognized by school districts, universities, and states, and PDSs are recognized as a formal way of acknowledging and addressing this need (NRC, 2001b). PDSs have been established at a large number of sites, and the movement has gained support from the American Federation of Teachers, the National Education Association, and the American Association of Colleges for Teacher Education (Book, 1996; Darling-Hammond, 1994b). In fact, Maryland state law requires that teacher education include a one-year internship in a PDS. The NRC (2001b) study of PDSs was a literature review in which more than 600 PDS sites were identified. By far the majority of PDSs studied reported having teachers and university faculty working together to improve teacher education. Teachers who participated in a PDS showed greater understanding of diversity and the needs of children than teachers who had not had this experience.

Mathematics learning and teaching are rarely singled out in descriptions of PDSs. Exceptions are the PDS described by Whitford (1994), in which the increased cooperation led teachers to reorganize their instruction around the NCTM *Standards,* and the PDS described by Valli, Cooper, and Frankes (1997), in which a university professor guided a group to teachers in implementing the NCTM *Standards.* PDSs, because of their comprehensive natures, often include mathematics, a core subject taught in K–12 schools. The NRC Committee on Science and Mathematics Teacher Preparation (NRC, 2001b) considered PDSs to be an exemplary approach to teacher education, places where "teachers, students, and university faculty create new knowledge and experiment with, evaluate, and revise practices" (p. 76).

Establishing a PDS is not an easy task. Neither a university nor a school can begin this type of collaboration unless it builds on prior work together. "Building coherence requires building shared understandings, which takes dialogue, time, and compromise" (Arends & Winitzky, 1996, p. 551). Financing a PDS is also problematic. Schools rarely allocate much funding for professional development, and universities are often unwilling to expand budgets for teacher education to include the type of collaboration necessary for a PDS partnership to exist. A third problem is the existing reward structure for university faculty. Work in a PDS is often viewed as service, which counts little toward promotion, tenure, or merit pay. "For PDSs to survive, universities will have to take seriously this kind of research and count it in the reward structure for faculty" (Book, 1996, p. 206). Only when issues of how collaboration is undertaken, when all those involved benefit, when trust and respect exist, when the focus is clear to all involved, and when the agenda is manageable can PDSs hope to succeed as a means of furthering the professionalization of teaching.

Lesson Study

Hiebert, Gallimore, and Stigler (2002) claimed that U.S. "teachers rarely draw from a shared knowledge base to improve their practice" (p. 3) and that specific research information has had little effect. Literature they reviewed indicated that the type of *craft* knowledge needed for teaching does not follow from the work of educational researchers. This finding led them to consider requirements for building a professional knowledge base for teaching from practitioner knowledge and to reflect on whether this approach would be more successful than trying to build a professional knowledge base from research. "Observation and replication across multiple trials can produce dependable knowledge as well [as experimental methods can]" (p. 12). They considered lesson study to be a large-scale system that could facilitate this type of transformation.

Lesson study, a recent phenomenon in this country, is a basic practice in Japanese schools at all levels and in all subject areas. Recognizing what U.S. educators could learn from Japanese lesson study led to a U.S.–Japan Teacher Development Workshop following the 2000 ICME-9 (International Congress on Mathematics Education) conference in Japan. The conference proceedings were published by National Academy Press (Bass, Usiskin, & Burrill, 2002). Chapters by Shimizu and by Yoshida provided background on Japanese lesson study, and these chapters are the basis of the information in the next paragraph.

Japanese educators at this conference explained how lesson studies, which are of different levels and different types, operate in Japan (Shimizu 2002; Yoshida, 2002). Intraschool lessons are held a few times a year within a particular school. Lesson studies also occur at the city level, the school-district level, and even the national level. The three basic parts of a lesson study consist of choosing a particular topic and designing a lesson around that topic, teaching the lesson to students with teachers observing the particular lesson, and discussing the lesson with the observing teachers. The discussion, usually conducted by the lesson teacher, begins with comments. The process is cyclic, and each lesson study serves as preparation for the next. Lesson studies are the primary means of education for novice teachers and of professional development for experienced teachers. An obvious outcome is that participating teachers learn much about the skill of teaching. A less obvious, but important, outcome is that the participating teachers learn a great deal of content knowledge through lesson-study discussions.

Studied lessons are often called *research lessons.* They are set apart from everyday lessons in that "they are observed by other teachers; . . . they are carefully planned, usually in collaboration with one or more colleagues; . . . they are focused; . . . they are recorded; . . . (and) they are discussed" (Lewis & Tsuchida, 1998, pp. 12–14). National research lessons are disseminated all over the country. At the national level, thousands of teachers might participate in a lesson study.

The first two steps of lesson study—preparing and implementing a lesson—are not new to U.S. teachers (although these activities rarely involve other teachers), but the third step— analyzing and evaluating the lesson—is uncommon. Ball (2002) listed three skills needed to undertake lesson study. First, a teacher must have the ability to pay attention to and teach every student in the class, knowing what each child understands or does not understand. Second, the teacher must know mathematics well enough to be able to organize the content, create appropriate activities, and make adjustments as needed during class. Third, a teacher must be able to work with others on developing knowledge for teaching. Hiebert and Stigler (2000) have also described lesson study:

> The knowledge being shared through lesson study is not just collections of lesson plans that teachers can pull off the shelves and use. The goal of lesson study is not just to produce lessons that can be copied but to produce knowledge about teaching upon which colleagues can build (see Ball & Cohen, 1996). Such a knowledge base grows as a teacher reflects on and improves what others have done, working to understand the basis for the improvements. (p. 12)

The impact of research lessons can be substantial. Lewis (2000) listed nine ways in which these lessons contribute to Japanese instruction: They (a) provide professional development; (b) help teachers understand student thinking (or "see children"), (c) spread knowledge of new content and new approaches, (d) help individual teachers connect their practices to school goals and broader goals, (e) allow competing views to be heard, (f) create a demand for improved instruction, (g) shape national educational policy, and, last but not least, (h) honor the central role of teachers in the development of instruction. Lewis (2002) has worked with teachers in American schools to implement lesson study. To be successful, she advocated that American educators must answer the following questions: "What are the essential features of lesson study that must be honored when lesson study is conducted in the US? How do educators improve instruction through lesson study? What supports will be needed for lesson study in the US, given its system and culture?" (pp. 6–7). Lewis then provided tentative answers to these questions and noted that the educational systems in the United States and Japan differ greatly and that lesson study must be thoughtfully adapted rather than "borrowed."

Acquiring Knowledge-In-Practice: Summary and Discussion

The process of acquiring knowledge-in-practice can take many forms. Communities of practice established within a school can have a powerful influence on the mathematics instruction and student learning in that school. However, establishing such a community is not at all easy, and maintaining the community is difficult, particularly because teachers move in and out of the school. Several positive characteristics of communities of practice were listed, and examples of communities of practice were described. Lesson study, originating in Japan, is being studied in this country as a form of practice in which teachers can reflect on and learn from their own practices.

Professional development schools continue to be a source for learning, but, for many reasons, they are difficult to establish. Where they exist, they provide professional development for both teachers and university colleagues. PDSs serve as venues for forming collaborations, but other groups have collaborated successfully without actually establishing a PDS. Clark et al. (1996) provided an example of professional development consisting of collaboration in the form of dialogue in which teachers and researchers engage. *Dialogue* was considered to be

a means of achieving parity in collaboration while facilitating mutual reflection, growth, and change. Instead of teachers and researchers trying to take on the *work* of one another as their own—an approach which may be particularly *disempowering* for teachers—reciprocity is achieved dialogically, in part by talking through the hidden assumptions which teachers and researchers bring about the "other." (p. 228)

These collaborations centered on evaluation of portfolios in English and science classrooms but might also be a useful technique in mathematics, particular as a way of introducing mathematicians into the work of teachers.

ADDRESSING QUESTION 6: WHAT CAN TEACHERS LEARN BY INVESTIGATING THEIR OWN TEACHING OF MATHEMATICS? THAT IS, HOW DO THEY ACQUIRE KNOWLEDGE-OF-PRACTICE?

Cochran-Smith and Lytle (1999) called their third type of teacher knowledge *knowledge-of-practice*. This type of knowledge can be generated when teachers investigate learning and teaching in their own classrooms and school sites. Their work is located within

the "contexts of inquiry communities to theorize and construct their work and to connect it to larger social, cultural, and political issues" (p. 250). Stated more formally, *knowledge-of-practice* is

constructed in the context of use, intimately connected to the knower, and, although relevant to immediate situations, also inevitably a process of theorizing. From this perspective, knowledge is not bound by the instrumental imperative that it be used in or applied to an immediate situation; it may also shape the conceptual and interpretive frameworks teachers develop to make judgments, theorize practice, and connect their efforts to larger intellectual, social, and political issues as well as to the work of other teachers, researchers, and communities. (pp. 272–273)

Practitioner Research

In the 1940s Kurt Lewin introduced research that teachers undertake to better understand their own teaching and its effects. It has evolved into "an ongoing process of systematic study in which teachers examine their own teaching and students' learning through descriptive reporting, purposeful conversation, collegial sharing, and reflection for the purpose of improving classroom practice" (Eisenhower National Clearinghouse, 2000, p. 18). According to Holly (1991), "Action research, at root, constitutes participative learning for teachers. Who better to analyze teaching and learning than the teachers themselves through the agency of action research?" (p. 135). Hiebert et al. (2002) advocated the development of a knowledge base for teaching, beginning with practitioner knowledge:

The teacher-as-researcher movement has oriented teachers to studying their own practice, thereby making it more public and testing its effectiveness (Berthoff, 1997; Burnaford, Fisher, and Hobson, 1996; Cochran-Smith and Lytle, 1993; 1999). During the same time that the movement has been increasing educators' awareness of the richness of teachers' personal knowledge, it also has focused attention on the kind of teacher learning that is required to teach more effectively. (p. 11)

Much of the early work in teacher research in mathematics, commonly referred to as *action research,* took place in Australia and South Africa and has been described by Crawford and Adler (1996), who viewed research activities on the part of teachers as a prerequisite for changes in the quality of mathematics education. They claimed that teachers who act in "generative research-like ways . . . may learn about the teaching/learning process, and about mathematics, in ways that empower them to better meet the needs of their students" (p. 1187). Borasi and Fonzi (2002) considered action research to offer "an ideal way for teachers to learn more about teaching and learning mathematics and to apply the results immediately to their own practice" (p. 100). Middleton et al. (2002) described research that moved from the study of systems to a new model of teacher as professional: "Under such a model, teachers are seen as knowledge producers, curriculum designers, and policy analysts with a unique configuration of knowledge, skills, and practices that have merit in the larger order of knowledge production, curriculum design, and policy analysis" (p. 411).

Action research has multiple forms—teachers might work alone to pursue a research interest, they might work together in inquiry teams, or they might work with university researchers. They might produce their own data or undertake a study of existing data on a question of interest. The defining factors are that teachers themselves formulate or contribute to the formulation of the research questions; they collect data to answer the questions; they use an action-research cycle of planning, acting, observing, and reflecting; they work collaboratively when possible; they have access to outside sources of knowledge and stimulation; and they document and share their research (Loucks-Horsley et al., 2003). According to Cochran-Smith and Lytle (1999), teacher research, particularly when collaboratively undertaken, has the potential to

> profoundly alter the cultures of teaching. . . . In work of this kind, the image of knowledge as collectively constructed is particularly striking; knowledge emerges from the conjoined understandings of teachers and others committed to long-term highly systematic observation and documentation of learners and their sense making. To generate knowledge that accounts for multiple layers of context and multiple meaning perspectives, teachers draw on a wide range of experiences and their whole intellectual histories in and out of schools. (p. 275)

Teachers who undertake research are likely to use the results, and they gain power over decision making that guides their practice. "The change process—with all its emotional undertow—is internalized and personalized to the point where the teacher action researcher becomes 'hooked' on his or her change agenda" (Holly, 1991, p. 153). This process was exemplified in a project in Arizona in which a group of teachers worked with researchers studying children's reasoning using the "model of teacher-as-researcher as

a mechanism for supporting their unique needs and questions of interest but also for disseminating their understandings across a larger audience" (Middleton et al., 2002, p. 411).

Some concerns have been raised about the validity of action research. Jaworski (1998), working with a group of teacher researchers focusing on the ways research interfaced with reflective practice, had the following to say about this matter:

> In making public the outcomes of their research, teachers contribute to research rigour. It is hard to validate the research in terms of substantive outcomes. For example, what Julie [a teacher engaged in action research] learned about mathematical talk must be seen in the context of Julie's classroom and personal teaching perspective. However, in offering the outcomes of her research to other teachers, she shares her own insights and invites responses. Validity rests in the possibility of others seeing and making sense of Julie's work and using it to illuminate their own work and thinking. This is a rather different way of regarding research rigour, but nevertheless important to the use of research as a vehicle for developing teaching. (p. 27)

Obstacles need to be overcome in the process of planning and undertaking teacher research (ENC, 2000). Undertaking research always demands a significant amount of time and is not always recognized as a form of professional development. When possible, teachers arrange to have this work count toward district-required professional development. Also, teachers have different goals and priorities, and not all want to become involved in research. If teachers are, for example, teaching from a new curriculum or participating in other professional development activities, they often cannot, at the same time, undertake research that requires additional time and effort. For others, however, action research provides an avenue to develop professionally. The ENC (2000) report provided an example of a teacher from New York who wanted to study how interactions with computers affected learning in her mathematics class. Her research led to a collaborative effort with another teacher and a university researcher and to a sizable grant from the National Science Foundation. She remarked, "Action research lends itself to classroom use through professionally inquisitive teachers. When this happens, the classroom becomes a learning environment for the teacher as well as the student. It involves doing what comes naturally to teachers, questioning classroom practices" (p. 20).

Knowledge-Building Schools

Whereas *action research* often refers to one or a few teachers in a school undertaking research in their classrooms, *knowledge-building schools* are schools in which research is embedded into the school structure. Groundwater-Smith and Dadds (2004) described a school-improvement project in which teachers from seven schools in Australia "sat together and discussed the possible formation of a coalition of knowledge-building schools . . . constructed and examined in ways which illuminate understanding rather than [serving] as a means of proving a particular case" (p. 247). The participants wanted to embed inquiry practices into their teaching and develop a workplace-learning culture, believing that "professional learning is not an exclusively individualistic enterprise but that learning and growth can take place at the organizational or corporate level" (p. 248).

The Inquiry of Teaching, Or Practical Inquiry

The *inquiry of teaching*, that is, "the ongoing every-day inquiry into their students' learning that teachers are expected to conduct" (Hammer & Schifter, 2001, p. 471) is different from research on teaching and learning as practiced by researchers, whether or not the researchers are teachers. Inquiry of teaching could be considered a form of research, and, in fact, the boundary between inquiry of teaching and teacher research is indistinct. Hammer and Schifter have examined this border by distinguishing between the inquiry of teaching and the inquiry of research: "Both . . . inquire into students' knowledge and reasoning, but they do so under different conditions and with different opportunities and constraints" (pp. 475–476). Some teachers are uninterested in research but do want to know more about their teaching and their students and undertake daily analysis of their students' thinking. Their study is distinguished from research in terms of the audience, the scope and purpose of the inquiry, and the manner in which teachers identify themselves, whether as researchers or as teachers who study their students' thinking. Inquiry of teaching is often prompted by the use of innovative textbooks and other instructional tools that lead to teachers' experimenting with instructional practices. "Curriculum reform requires teacher inquiry into students' knowledge, reasoning, and participation" (p. 442).

Cognitively Guided Instruction researchers have used a different label, *practical inquiry*, for what seems closely related to inquiry of teaching. These researchers have recently turned their attention to the under-standing of teachers' self-sustaining, generative change in the context of professional development (Franke et al., 1998). The authors claimed that for change to occur, teachers must be engaged in *practical inquiry*, that is, inquiry that helps the teachers understand their students, the contexts in which they work, and the practices of the community in which the teachers work. It can be viewed as a teacher's "questioning and reflecting about his/her practice with a specific focus. The focus of a teacher's practical inquiry determines what a teacher sees as critical, and what constitutes an opportunity for reflection" (p. 68). In undergoing this type of change, teachers are not learning new procedures but, instead, are making basic epistemological changes in their views of learning and teaching.

In the first of two levels of reflecting on practice, one's aim is to see "what works." Knowing some practices that can be used successfully can encourage the teacher to try to find other such activities or practices. This type of reflecting on practice is self-sustaining because it supports continuation of using these practices, but at this first level, teachers do not know *why* a practice is effective. When teachers also engage in detailed analysis of their practices, but now in relation to their own thinking, they come to understand the "principled ideas that drive their practice and their continued practical inquiry" (Franke et al., 1998, p. 68). This second level is both self-sustaining and *generative* because when teachers understand the relation between their practices and student learning, they have a basis for generating new insights about effective instruction. Thinking in terms of these two levels provides fodder for those individuals who provide professional development for teachers.

> A concern for self-sustaining, generative change shifts the focus of professional development from the factors that initiate change to principles that make it possible for teachers to continue to grow and learn. . . . In order to understand the effects of a teacher development program, it is necessary to understand how the participants in the program construe and implement the principles. Teachers, as well as students, do not simply assimilate knowledge. Teachers in teacher development programs may construct interpretations of a program's first principles that are quite different from the principles intended by the designers of the program. (p. 68)

Acquiring Knowledge-of-Practice: Summary and Discussion

"The knowledge-of-practice conception turns on the assumption that the knowledge teachers need to teach well emanates from systematic inquiries about teaching, learners and learning, subject matter and curriculum, and schools and schooling" (Cochran-Smith & Lytle, 1999, p. 274). Whether this inquiry is viewed by the teacher as research or as practical inquiry depends on whether the inquiry has some sort of research question and employs research methods.

One way to distinguish knowledge-of-practice from knowledge-in-practice is to consider the knowledge gained in lesson study, described in the previous section. From the standpoint of the teachers participating in the lesson study as observers and discussants, they are acquiring knowledge-in-practice by engaging in a community discussion of the lesson. But the person who teaches the lesson and participates in discussions about the class is more likely to be acquiring knowledge-of-practice. (Recall that the second category of knowledge described by Cochran-Smith and Lytle [1999] was *knowledge-in-practice*, that is, the practical knowledge of teaching "embedded in practice and in teachers' reflections on practice" [p. 250].)

ADDRESSING QUESTION 7: WHAT CAN BE LEARNED FROM RESEARCH ON TEACHER CHANGE?

Recent research and thinking about teaching has led to a "radically revised picture of what kind of instruction should take place in the classroom" (Goldsmith & Schifter, 1997, p. 20). A climate of inquiry and sense making, a change in the role of the teacher as the sole authority, and a focus on reasoning and problem solving call for significant changes in a teachers' beliefs about mathematics, about children's learning, and about teaching. On the part of the teacher these changes call for a fundamental reconceptualization of his or her role in the educational process— not an easy change to make. According to Nelson and Hammerman (1996), researchers must come to understand

what it is in innovative professional development projects that makes this process of change possible for teachers, and why it works. . . . We must specify the nature of the change that is at issue and advance a theory of the process by which teachers make these changes. (p. 4)

Instructional change can occur in several ways (Richardson & Placier, 2001). Change that is mandated rather than initiated by teachers may be negative and, more often than not, leads to teachers' becoming recalcitrant and resistant to implementation of changes they feel forced to make. Such change, usually minimal and superficial, is the type referred to by Loucks-Horsley and Stiegelbauer (1991) in the following:

Too often policy makers, administrators, even teachers assume that change results from a policy mandate, an administrative requirement, adoption of new curricula, or a revised procedure. They assume that a teacher will put aside tried-and-true practice and immediately use a new one with great facility. In the most absurd situations, there is the companion assumption that the new program can soon be evaluated for its impact on students. In reality, change takes time and is accomplished only in stages. (p. 17)

Because teacher change is a process rather than an event, it must be considered in terms of continuous growth over time. Darling-Hammond (1990) said that the reason change is slow and difficult is that in addition to perseverance, "it requires investments in those things that allow teachers, as change-agents, to grapple with transformations of ideas and behavior: time for learning about, looking at, discussing, struggling with, trying out, constructing and reconstructing new ways of thinking and teaching" (p. 240).

One type of experience necessary for teachers to undertake instructional change is the opportunity to see how innovative teaching practices can enhance their students' learning, particularly when they see that their students become more confident and more engaged in learning, with resulting higher achievement. "Then, and perhaps only then, a significant change in teachers' attitudes and perceptions is likely to occur" (Guskey, 1989, p. 446).

Ways of Describing Teacher Change

Researchers have tried, in several ways, to describe the change process in terms of differences in instructional practice. For example, the manner in which teachers question their students can provide insight into their orientations toward learning and instruction (A. G. Thompson, Philipp, P. W. Thompson, & Boyd, 1994), and the study of the manner in which teachers question students can become a way to measure change. These authors claimed that teachers who have a *calculational orientation* are focused on procedures for obtaining answers. Such teachers speak almost exclusively in the language of numbers and operations. Teachers who have a *conceptual orientation* focus students' attention on the "rich conception of situations, ideas, and relationships among ideas. [They] strive for conceptual coherence, both in their pedagogical actions and in students' conceptions" (p. 86). Teachers' implementation of new curricula is particularly influenced by their orientations toward mathematics teaching.

Other researchers have described stages that occur in teacher growth. On the basis of studies of teachers in professional development programs designed to help teachers change their teaching practices, Schifter (1995a) proposed four stages of conceptions of mathematics teaching:

(1) an ad hoc accumulation of facts, definitions, and computational routines;

(2) student-centered activity, but with little or no systematic inquiry into issues of mathematical structure and validity;

(3) student-centered activity directed toward systemic inquiry into issues of mathematical structure and validity;

(4) systematic mathematical inquiry organized around investigation of "big" mathematical ideas. (p. 18)

Each of these broadly defined stages "entails an understanding of what counts as 'doing mathematics,' of the extent to which mathematical results are interconnected, and where mathematical authority resides and how it is established" (Schifter, 1995a, p. 18). The model was developed to assess change on the part of program participants, but Schifter and her colleagues found it more useful as a pedagogical heuristic that could clarify goals, interpret practice, and guide the design of professional development programs. These stages are similar to ones she and Fosnot used in their SummerMath program (Schifter & Fosnot, 1993).

The Cognitively Guided Instruction researchers also developed a way of documenting growth on the part of teachers (Fennema et al., 1996). They categorized teachers as belonging to one of four levels, similar but not identical to those proposed by Schifter (1995a). In Level 1, instruction focused on the learning of procedures and was guided by an adopted textbook. In Level 2, instruction was similar to that of teachers in Level 1 but was enriched by problems similar to those studied during the teacher workshops. Teachers were exploring new ways of teaching, but within a model of fixed routines. Classrooms of teachers in Level 3 were characterized as those in which children were engaged in solving and reporting solutions to problems that were not found in standard textbooks. The teachers talked about specific children's thinking. By Level 4, teachers based instructional decisions (to a greater extent than those at lower levels) on the kinds of problems children were capable of solving, the strategies children use, and the kinds of communication children are capable of using. At the end of their longitudinal study, 90% of the teachers were at Level 3 or 4; Level 4 was the researchers' category for

the most cognitively guided instruction they observed. They concluded, "Starting with an explicit, robust, research-based model of children's thinking, as we did, enabled almost all teachers to gain knowledge, change their beliefs about teaching and learning, and improve their mathematics teaching and their students' mathematics learning" (Fennema et al., 1996, p. 433). The levels and patterns of change were further discussed in a book chapter (Franke et al., 1997) in which the authors noted that teachers at lower levels were able to change instructional practice without substantially changing beliefs, but that for progress from Level 3 to Level 4, changes in teachers' beliefs were essential.

Schifter's (1995a) and Franke et al.'s (1997) descriptions of stages of development were based on their work with teachers. Goldsmith and Schifter (1997) used theories of cognitive development in addition to personal observations to provide descriptions of the understandings of individuals over time, while teachers' mental organizations were being reconstructed to allow for more complex thought and action. They believed that studying these stages could be fruitful avenues for research on teachers and teaching. Their first stage of development is characterized by instruction that is organized around the transfer of knowledge; thus teachers' instruction at this stage was focused on providing opportunities for children to automatize and generalize the demonstrated procedures, as happens in the majority of mathematics classrooms. In subsequent stages, teachers become "less intent on helping students acquire facts and procedures, and more involved in building on what (and how) their students understand" (p. 23). The final stage is one in which teachers' instruction is aligned with that envisioned in the NCTM Standards. Although this last stage has different interpretations, one constant is that teachers focus on understanding what their children are capable of achieving and take this information into account when planning instruction. The authors spoke of stages as providing important information about the process of change—noting that the development of pedagogical content knowledge can be seen as orderly but that it usually is not. Thus, research on models of mathematics teaching might profitably focus on the orderliness of change. "Research directed toward mapping the issues teachers confront as they enact their new beliefs and understandings in the classroom will help to create a fuller picture of how teachers move through the terrain of creating a reformed mathematics practice" (p. 38).

Goldsmith and Schifter (1997) encouraged attention to both psychological and sociocultural mechanisms in research on teacher development and suggested that each view of learning will lead to a different way of viewing teacher change. They emphasized the importance of language in the process of "becoming acculturated to new forms of mathematics teaching and new ways of communicating those thoughts and ideas" (p. 45). Teachers often need new vocabulary to talk about the changes in their instruction and in student learning and new images of good instruction:

> Teachers' efforts to construct new forms of mathematics teaching can be affected considerably by the images they have of good teaching, the nature of collegial relationships, the criteria and procedures for job evaluations, and the kinds of professional development that prevail in their schools and districts. (p. 45)

Goldsmith and Schifter also suggested that interactions with students affect teachers' development and that research centered around these interactions could prove useful. Finally, they said that still to come is "a well-developed account of what sustains people as they undertake the often difficult and frustrating task of restructuring their understanding" (p. 46). Their earlier work indicated that "courage is an important ingredient in the motivation equation" (p. 46).

The difficulties of moving through these stages were acknowledged by Goldsmith and Schifter (1997). They spoke of teachers' having "one foot in each of two conflicting paradigms" (p. 28) when they move from one set of beliefs to another, conflicts that they resolve over time. Even while attitudes and perceptions are changing, teachers experience a certain amount of anxiety and reluctance to change instruction until they are certain that the changes will be successful in their classrooms (Borasi & Fonzi, 2002; Guskey, 1995). At times teachers' changing perceptions can bring about a belief that their past teaching practices actually harmed their students, a belief that can bring with it a great deal of pain on the part of a conscientious teacher (Sowder, Philipp et al., 1998). Schifter and Fosnot (1993) found that for their SummerMath teachers the process of transformation was "highly affect-laden" (p. 195). Teachers who have transformed their mathematics teaching and want to help others sometimes encounter professional jealousy, which, in turn, can cause tension and pain (Koch, 1997). Professional developers must be aware of and able to help teachers deal with the affective issues related to change.

Psychological and Sociological Frameworks for Studying Teacher Change

Studies of mathematics teachers' development have generally followed two broad frameworks within which to develop and explore theories about teacher change. In the first, based on a psychological perspective, individual teachers are themselves the objects of study—how they learn, how they teach, and how they transform their practices, usually as a result of some professional development intervention. The Goldsmith and Schifter (1997) stages discussed above are based on a psychological perspective. Case studies of individual teachers—and there are many such studies—fit within this genre of research. The bulk of the studies reviewed here fit within this framework.

The second type of framework is based on a sociocultural perspective in which teachers' learning is conceptualized as a process of *transformation of participation* in a community of practice (Stein & Brown, 1997). Whereas in the psychological framework the study of teacher change is focused on changes in the individual teachers, their knowledge, beliefs and practices, one using a sociological perspective instead focuses on the learning processes of teachers by looking "beyond formally structured events to other times and places in which individuals learn; . . . the focus on analysis shifts away from pedagogical activity toward an analysis of the community's activities and learning resources" (Stein & Brown, 1997, p. 162).

Stein and Brown (1997) further delineated this sociocultural perspective into two frameworks. One is based on Lave and Wenger's (1991) "theory of learning though legitimate peripheral participation in communities of practice" (Stein & Brown, 1997, p. 155) and the other on Tharp and Gallimore's (1988) "model of learning as movement from assisted performance to unassisted performance through a Zone of Proximal Development" (Stein & Brown, 1997, p. 155).

Although these are the two most commonly used frameworks, other frameworks, too, provide ways to study teacher change. For example, Drake, Spillane, and Hufferd-Ackles (2001) used the construct of *identities* as a framework for studying teacher change. They claimed that a teacher's sense of self, or sense of identity, reveals the way that a teacher knows himself and his life and that teachers' stories "can serve as a lens through which teachers understand themselves personally and professionally and through which they view the content and context of their work, including any attempts at instructional innovation" (p. 2). Using this framework, Drake et al. identified three types of elementary mathematics teachers: *Turning-point teachers*, who had had mostly negative experiences with mathematics learning but had had a recent positive experience and who focused on using manipulatives and on the affective components of reform; *failing teachers*, who had only negative experiences with learning and teaching mathematics, were traditional in their teaching, and focused only on superficial aspects of reform; and *roller-coaster teachers,* who had had both positive and negative mathematics learning experiences and now focused on both the content and affective components of reform. Drake et al. noted, "It is not surprising to find that individual teachers exposed to identical reform programmes will respond differently depending in part on the dispositions and beliefs which are embedded in their identities as teachers and as learners" (p. 3). The authors claimed that a better understanding of teachers' positions in terms of their mathematics identities could be used to "inform the design of future policies, curricula, and teacher development activities intended to foster instructional innovation" (p. 17).

Another framework that has been used to study professional development of elementary school mathematics teachers is the *Diffusions of Innovations* framework of Rogers (1995). Diffusion research can be used to understand how innovations are shared over time among the members of a social system. Jacobs and Raynes (2001, 2002) used this framework to study professional development being offered by teacher leaders in a small school district. They found four diffusion principles useful in guiding their work. First, the characteristics of an innovation can help determine how quickly it will be adopted. "Innovations that do not require radical changes but have highly visible benefits over current practices can have an inherent advantage" (Jacobs & Raynes, 2002, p. 4). Second, an adoption of an innovation is determined by information gathering and decision making, both of which require time and opportunities to explore the innovation. Third, those who communicate about the innovations are more effective if they have "personal or social characteristics that make them similar to but also respected by those considering adoption" (p. 5). Fourth, "the social system in which an innovation would be used has structures and norms that encourage its members to act in particular ways . . . [and that can] increase or decrease the rate of diffusion" (p. 5).

In yet another way to understand how teachers develop their practices, Simon (2001) intertwined two lines of study, the empirical work of conducting research on mathematics teacher development and the theoretical work of elaborating theory of mathematics teaching and learning, as the basis for his work on the Mathematics Teacher Development project. Teachers' development "can be thought of generally as evolution from more traditional practices toward practices more consistent with current mathematics education reforms" (pp. 160–161). Simon and his research group developed two constructs, a perception-based perspective and a conception-based perspec-

tive, which they used to describe the mathematical experiences of the students with whom they worked. The *perception-based perspective* can be thought of as "best achieved by students examining for themselves mathematical situations in which the relevant mathematics can be seen; that is, they learn best through first-hand experience with the mathematics" (p. 163). Simon viewed the teacher's role as a guide to "afford students' direct apprehension of the mathematics" (p. 163). The *conception based-perspective* refers to "an array of perspectives that share the following assumptions:

1. Mathematics is a product of human activity. Humans have no access to mathematics that is independent of their ways of knowing.

2. What individuals see, understand, and learn are afforded and constrained by their current conceptions.

3. Mathematics learning is a process of transformation of learners' knowing and ways of acting. (p. 164)

Someone with a perception-based perspective assumes "that everyone is looking at the same mathematics; what is seen is assumed to be unvarying from person to person" (p. 164). Someone with a conception-based perspective makes "no assumption that what is seen matches with what others see" (p. 164) in any particular situation. Simon argued that theoretical frameworks of mathematics teaching and learning can promote discourse and advance understanding of our field.

A Role for Researchers in the Study of Teaching and Teacher Change

Steps for professional development design and research were suggested by Borko (2004) in her AERA presidential address, in which she called upon researchers "to play a leadership role in providing high-quality professional development for all teachers" (p. 13) by addressing two major questions: "What do we know about professional development programs and their impact on teacher learning? (and) What are important directions and strategies for extending our knowledge?" (p. 3). She noted that in her discussion of successful research programs on teaching and teacher change her choices were constrained by the limited availability of such programs. She suggested design experiments to bring researchers and professional developers together to adapt existing successful professional development to new subject areas by considering a successful program and engaging "in multiple design/research cycles to refine the program and study its impact on the development of professional community and the learning of individual teachers" (p. 12). Just as educational innovations are changed by users, so is professional development program changed by the manner in which it is implemented. She thus suggested research "investigating the balances and tradeoffs between fidelity and adaptation, and consider which elements of a program must be preserved to ensure the integrity of its underlying goals and principles" (p 13). Finally, she noted the need to develop instruments to study "change over time in teachers' subject matter knowledge for teaching and instructional practices, and analytic tools that can separate out the influences of various program, school, and individual factors on teacher and student learning" (p. 13).

Several successful professional development programs have been described in this chapter. Borko's suggestions lay out terrain for conducting research on professional development and teacher change in mathematics.

This discussion of teacher change is based on a vision of mathematics teaching as described in my first of six goals and found in the above descriptions of final stages of change (Fennema et al., 1996; Goldsmith & Schifter, 1997; Schifter, 1995a). An enhanced vision of mathematics teaching and learning develops hand-in-hand with a developing knowledge base for teaching.

A Few Words About the Role of Reflection in Teacher Change

Throughout the research discussed in this chapter, the topic of reflection and its role in furthering teachers' professional development and education is a constant in the discussions. The importance of teachers' reflecting on their practice is a recurring theme in research on teaching, and the degree and kind of reflection is often used to describe teacher change. Schön (1983, 1987) is most often credited for this focus on reflection, although recognition of its importance can be found in Dewey's work (e.g., 1910/1977). In Mewborn's (1999) research on reflective thinking among preservice teachers, she noted that the term *reflection* is used in different ways but that three features are common in all uses. First, reflection is always considered to be qualitatively different from recollection or rationalization. Because the object of reflection when applied to teaching is problematic in nature, the teaching itself is recognized as problematic. Second, "action is an integral part of the reflective process. . . . Reflection and action together are seen as a bridge across the chasm between educational theory and practice" (p.

317). Third, reflection requires introspection but also requires outside probing; thus, it is "both an individual and [a] shared experience" (p. 317).

The ability to reflect on practice is considered a necessity for effective instruction.

> Teachers need to be able to analyze and reflect on their practice, to assess the effects of their teaching, and to refine and improve their instruction. They must continuously evaluate what students are thinking and understanding and reshape their plans to take account of what they have discovered. (Darling-Hammond, 1998, p. 8)

The notion of teacher inquiry is at the heart of reflective practice and plays a central role in "many of the prevailing conceptions of teacher learning including critical reflection, reflection in and on action, personal and pedagogical theorizing, narrative inquiry, action research, and teacher research" (Barnett, 1998, p. 81). In their work with four "extraordinary" teachers, Philipp et al. (1994) were struck by the quality of the reflectiveness of these teachers and their tendencies to, as A. G. Thompson (1992) said, "think about their actions, vis-à-vis their beliefs, their students, the subject matter, and the specific context of instruction" (p. 139).

But important issues need to be considered when we ask or expect teachers to become reflective about their practices. First, Hargreaves (1995) noted that although

> reflection is central to teacher development, the mirror of reflection does not capture all there is to see in a teacher. It tends to miss what lies deep inside teachers, what motivates them most about their work. However conscientiously it is done, the reflective glance can never quite get to the emotional heart of teaching. (p. 21)

Second, time is needed for developing the ability and habit of reflecting on practice. Reflection rarely occurs when time is not a resource available to teachers (Pugach & Johnson, 1990). Third, "like many enduring educational concepts, [reflection's] rhetorical appeal has outstripped the evidence. . . . Reason only has limited access to that which drives our actions, which rather limits the promise of reflection for improving practice" (Knight, 2002, p. 232). Within the classroom setting, teachers rarely have opportunities to reflect—they must constantly be observing students and making on-the-spot inferences and decisions. Only unusual teachers will carve out the time needed to reflect on their own instruction and students' learning unless support is given in terms of time and strong collegial relationships.

ADDRESSING QUESTION 8: WHAT CAN BE LEARNED FROM RESEARCH ABOUT THE PREPARATION OF TEACHERS TO TEACH MATHEMATICS?

Many of the goals, types, and strategies we have described for professional development are also applicable to teacher preparation. For example, teacher educators use videos that illustrate student thinking; school curriculum materials; case studies; and, most certainly, formal course work in their teacher preparation programs. Still, some aspects of teacher preparation need to be considered apart from professional development of practicing teachers. Here I discuss the development of a new frame of reference for teaching, ways of improving teacher preparation, and the particular role of mathematicians in preparing teachers.

Developing a New Frame of Reference for Teaching

A prospective teacher undergoes a lengthy *apprenticeship of observation* (Ball & Cohen, 1999; Ebby, 2000, attributed to Lortie, 1975; Kennedy, 1999) during his or her K–12 education. In mathematics, this apprenticeship can be described as experiencing mathematics classes as an immutable sequence beginning with checking homework, asking questions about the homework, watching the teacher demonstrate how to do new problems, then receiving an assignment to work problems similar to those demonstrated (Romberg & Carpenter, 1986). University course experiences are likely to reinforce this framework. Thus to improve the way in which teachers are prepared to teach mathematics, the mathematics education community must attend to changing this familiar model of a mathematics class. "The kind of teaching that reformers envision requires teachers to shift their thinking so that they have different ideas about what they should be trying to accomplish, interpret classroom situations differently, and generate different ideas about how they might respond to these situations" (Kennedy, 1999, p. 56). According to Kennedy (1999), unless this shift is initiated during teacher preparation, teachers develop habits of teaching based on their pre-preparation experiences and solidify these practices, making change all the more difficult. The demands of the job are intense when teachers begin teaching, so to find that new teachers often fall back on practices they know best is not surprising. Dewey (1904) expressed this concern a century ago:

> The student adjusts his actual methods of teaching, not to the principles which he is acquiring, but to what he sees succeed and fail in an empirical way from moment to moment; to what he sees other teachers doing who are more experienced and successful in keeping order than he is; and to the injunctions and directions given him by others. In this way the controlling habits of the teaching finally get fixed with comparatively little reference to principles in the psychology, logic, and history of education. (p. 24)

The influence of the apprenticeship-of-observation framework is more prevalent at the secondary level than at the elementary or middle school level. Secondary school teachers have experienced many more mathematics classes, most, if not all, taught in a traditional fashion. They are likely to have taken far fewer courses dealing with education or pedagogy. "Teachers need to know and use 'mathematics for teaching' that combines mathematical knowledge and pedagogical knowledge" (NCTM, 2000, p. 370). Providing secondary school teachers with a new frame of reference for how mathematics should be taught is particularly challenging.

Improving Mathematics Teacher Preparation

What makes teacher preparation effective? In this section I explore what many researchers and educators have described as components of a strong teacher preparation program and changes needed to make programs more effective.

A concern for better preparation of teachers of mathematics and science led the National Research Council, in 1998, to appoint a committee to study teacher preparation. The committee was comprised of mathematicians and scientists, mathematics and science educators including K–12 teachers, and a business representative. The committee's report, *Educating Teachers of Science, Mathematics, and Technology* (NRC, 2001b), was written to identify "critical issues in existing practices and policies for K–12 teacher preparation in mathematics and science" (p. xiii). Several recommendations were made about the characteristics that teacher education programs in mathematics and science should exhibit. The recommendations are broad and call for interactions among all groups concerned with K–12 teacher preparation. They recommend that programs should have the following features:

- Be collaborative endeavors developed and conducted by mathematicians,[2] education faculty, and practicing K–12 teachers with assistance from members of professional organizations and mathematics-rich businesses and industries . . . ;
- Help prospective teacher to know well, understand deeply, and use effectively and creatively the fundamental content and concepts of the disciplines that they will teach . . . ;
- Unify, coordinate, and connect content courses in mathematics with methods courses and field experiences . . . ;
- Teach content through the perspectives and methods of inquiry and problem solving . . .;
- Present content in ways that allow students to appreciate the applications of mathematics . . . ;
- Provide learning experiences in which mathematics is related to and integrated with students' interests, community concerns, and societal issues . . . ;
- Integrate education theory with actual teaching practice, and knowledge from mathematics teaching experience with research on how people learn mathematics;
- Provide opportunities for prospective teachers to learn about and practice teaching in a variety of school contexts and with diverse groups of children . . . ;
- Encourage reflective inquiry into teaching through individual and collaborative study, discussion, assessment, analysis, [and] classroom-based research and practice; and
- Welcome students into the professional community of educators and promote a professional vision of teaching. (pp. 69–71)

Howey (1996) also addressed the elements needed to make teacher preparation programs effective. Though many of his suggestions were too general to be appropriate for this chapter, he did suggest eight understandings, abilities, and dispositions that should "permeate a preservice program in the form of interrelated pedagogical activities designed to promote them" (p. 163) and that could profitably be the focus of a discussion by those who prepare mathematics teachers. He offered a different slant from that of the National Research Council on ways to make teacher preparation effective.

Understand and celebrate individual and cultural diversity; understand the subject matter to be taught and be able to represent it in multiple ways pedagogically; reflect on the moral and ethical consequences of policy and classroom practice; analyze and justify teaching practice; engage in teaching as a shared responsibility; monitor student understanding and foster conceptual learning; engage learners in active, self-monitoring learning tasks; and relate experiences in school to critical issues in society. (Howey, 1996, p. 163)

Carrying out these recommendations requires new ways of thinking about teacher preparation. Wilson and Ball (1996) suggested three ways that teacher education programs can provide models of teaching different from those the majority of entering teachers have experienced. One way is to sort out what needs to be preserved and what needs to be changed. Being able to manage a large group of students will always be something teachers are required to do. But new components of teacher education might need to be added—how to elicit student talk, how to ask good questions, how to undertake classroom discourse. A second way to provide models of teaching is to create opportunities for prospective teachers to view the sort of instruction that is needed to teach for understanding. Observing good teachers can be productive, but probably only if sufficient guidance is provided. For example, watching a teacher work with groups in a classroom should not leave the impression that the teacher is not teaching. Models might exist as videos that can be discussed in a teacher education class. Third, new teachers will need to be more prepared to "reason in the face of a host of dilemmas of practice" (p. 133). Reasoning in and about practice requires new ways of watching and hearing children. It means learning to reason about children's development, about instructional tasks, and about the interaction between the two.

Lampert and Ball (1999) also suggested changes in teacher preparation that might result in teachers better prepared for teaching mathematics. They suggested orienting teacher education around investigations of practices of teaching and learning instead of focusing teacher education programs on providing knowledge and skills for teaching. Making this change would require a focus on "inquiring into concrete phenomena of practice" (p. 43).

Investigations draw on but are not restricted to current theories and formal knowledge about teaching and learning in general or a particular kind of teach-

---

[2] Throughout the recommendations, science and technology were included in the original document but were omitted here for the purposes of this chapter.

ing and learning. Instructors help teacher education students access and use ideas of the field, and they also encourage students to make novel conjectures and advance possible novel interpretations. Engaging with situated questions of practice can support a context for the intertwining of knowledge use and knowledge construction. (Lampert & Ball, 1999, p. 44)

Few teacher preparation programs currently focus on investigations, although individual professors might make investigations a focus in individual courses. Programs that are associated with a Professional Development School might more readily accept this challenge.

The Role Children's Thinking Can Play in Teacher Preparation

Preparing teachers to teach in new ways requires that they be provided opportunities to reason in and about practice and to learn to listen to, hear, and watch students. The opportunity to observe and study children thinking about mathematics has proved to be a powerful component of professional development, as noted earlier, but the study of children's thinking can also be a productive component of teacher preparation. For example, the Cognitively Guided Instruction model most often used in professional development of practicing teachers has also been used successfully with preservice teachers (Philipp, Armstrong, & Bezuk, 1993; Vacc & Bright, 1999). The Developing Mathematical Ideas (DMI) program, in which teachers examine how children develop understanding of the big ideas of elementary school mathematics, has been successfully used with preservice teachers (Schifter, Russell, & Bastable, 1999). Studying mathematics cases such as those of Barnett et al. (1994) for elementary and middle grades and Merseth (2003) for secondary school can also provide insight into children's ways of understanding mathematics, together with new ways to think about teaching.

The Integrating Mathematics and Pedagogy (IMAP) project (Philipp et al., in press) also focused on children's thinking during teacher preparation. Prospective teachers in their first mathematics content course who volunteered to become involved in the project were randomly assigned (as much as was possible) to one of five treatments: two that provided opportunities to study children's thinking through interviewing children or through watching videos of interviews of children, two that visited classrooms either of exemplary teachers or of teachers unknown to the researchers, and a control group. Belief changes were measured using a project-developed, web-based beliefs-assessment instrument on which students whose beliefs reflected better understanding of children's

thinking and of the nature of school mathematics had higher scores. The treatment groups who studied children's mathematical thinking had significantly higher gain scores on the beliefs-assessment instrument than either the group who observed traditional teachers or the control group, but not than the group who had observed weekly in classes taught by carefully selected teachers.

Others have also provided opportunities for prospective teachers to work with children and observe their thinking. Nicol (1998) used "prospective teachers' weekly interactions with students as springboards for investigations of mathematics, teaching, and learning" (p. 45). Using a framework of questioning, listening, and responding, she found that the prospective teachers began to pose questions and listened and responded in ways they had not at the beginning of the project.

At universities with large numbers of prospective teachers, opportunities for preservice teachers to work with K–12 students one-on-one are difficult to arrange. Videos of children doing and explaining mathematics provide a different, but also effective, way to study children's mathematical reasoning if instructors help teachers to understand the children's thinking.

Methods Courses, Field Experiences, Induction, and Mentoring

Most prospective teachers of mathematics take a course (or more) on methods of teaching mathematics. This course is the most likely place that Lampert and Ball's (1999) suggested changes begin. Mewborn (1999, 2000) studied four preservice teachers during a field-based mathematics methods class so that she could examine what these future teachers found problematic and how they resolved their problems. She found that they viewed the knowledge they gained in professional education classes as rules or prescriptions to apply to classrooms, and they became disillusioned when these rules were ineffective. Thus she encouraged the prospective teachers to analyze their experiences in a way that would lead to reflective judgments rather than evaluative judgments. She concluded that "teacher education that is conducted in a setting that promotes investigation and inquiry into the problems of mathematics teaching seems to hold promise for assisting preservice teachers in becoming inquiring, reflective mathematics teachers" (Mewborn, 1999, p. 339). Ebby (2000), too, studied ways of thinking about relating coursework and fieldwork. Her study of the very different ways prospective teachers make connections between their coursework in teacher preparation and their fieldwork led her to conclude that teacher educators need to rethink how methods courses are organized and delivered. Rather than focusing only on developing new knowledge and beliefs, methods courses should help future teachers develop habits of mind so that they can learn from the classroom—"methods courses need to be explicitly oriented towards learning from fieldwork" (Ebby, 2000, p. 94). Howey's (1996) previously described recommendations might be useful to consider in light of Ebby's research.

In the Mewborn (1999) and Ebby (2000) studies, methods courses had field-experience components. However, such is not always the case. When separated, the field experience can be either a positive or negative experience. A review of research on teacher preparation commissioned by the U.S. Department of Education (Wilson, Floden, & Ferrini-Mundy, 2001) addressed issues of school experiences as part of teacher preparation programs. Both experienced and newly certified teachers considered clinical experiences, such as student teaching, as the most powerful element in their preparation to teach. Research showed significant shifts in attitudes during clinical experiences, but unclear is whether these changes enhanced the quality of the experiences. Too often school experiences, particularly student teaching, are disconnected from university-based components of teacher preparation. Problems with appropriate placement are common, and this problem is magnified by the fact that classroom teachers have a powerful influence on prospective teachers. Across several studies "one theme that emerges is that field experiences led to more significant learning when activities are focused and well structured" (p. 19). Teachers who had fifth-year induction internships were more satisfied with being a teacher than teachers who lacked this induction experience.

According to Darling-Hammond, Berry, Haselkorn, and Fideler (1999), research indicates that 30% of new teachers leave within 5 years, that beginning teachers of mathematics leave the field at an even higher rate, and that secondary school teachers leave at a higher rate than elementary or middle school teachers. Higher salaries in other professions are one inducement to leave teaching. Inadequate preparation is another. "The disproportionate placement of underprepared teachers in disadvantaged schools is especially problematic given the mounting research evidence that fully qualified teachers . . . are more effective than those who do not possess this knowledge and skill; especially in contexts where skillful teaching is most needed" (p. 189). The importance of strong teacher preparation followed by induction that includes mentoring and further professional development can reduce attrition and, at the same time,

strengthen teachers' abilities to be effective. Darling-Hammond and her colleagues suggested that all our reform efforts will be at risk if our teaching force is not prepared to raise student achievement. The need for new programs that better prepare teachers is a call to

> envision the professional teacher as one who learns from teaching rather than as one who has finished learning to teach, and the job of teacher education as developing the capacity to inquire systematically and sensitively into the nature of learning and the effects of teaching. [This approach to knowledge production is] one that empowers teachers with greater understanding of complex situations rather than seeking to control them with simplistic formulas or cookie-cutter routines. (Darling-Hammond, 1998, p. 9)

Mentoring new teachers is considered by many to be a crucial element in reducing attrition. The role of a mentor was well described by authors of *Mentoring in Mathematics Teaching* (Jaworski & Watson, 1994):

> The mentoring process offers an intensive opportunity to develop the philosophy and practice for all concerned. Mathematics teaching is a particularly interesting field in which to mentor because schools can exhibit vastly different styles of teaching and make contrasting demands on teachers. (p. 10)

A great deal of knowledge-in-practice is acquired during the first years of teaching. A master-teacher mentor who has content knowledge, knowledge of teaching, and knowledge of how to reflect on practice (Cochran-Smith & Lytle, 1999), together with time for the beginning teacher to learn from the mentor teacher, can effectively influence the knowledge-in-practice gained by the new teacher.

As educators study the manner in which mathematics teachers are prepared, consideration must also be given to alternative ways of obtaining certification to teach. Wilson, Floden, and Ferrini-Mundy (2001), in their study of teacher preparation programs, also considered the success of high-quality, alternative certification programs. They found that a more diverse pool of teachers has entered these programs but that the programs have a mixed record in terms of the quality of teachers recruited and trained.

What Teachers Carry from the University into the Classroom

The move from the university to the classroom was the focus of studies by researchers in the *Learning to Teach Mathematics* project (Borko et al., 1992, Eisenhart et al., 1993). These studies carry important messages for those trying to reform teacher preparation. The researchers examined the progress of a small number of prospective middle school teachers through their final year of teacher preparation and their first year of teaching. The researchers' goal was to "describe and understand the novice teachers' knowledge, beliefs, thinking, and actions related to the teaching of mathematics" (Eisenhart et al., 1993, p. 11). The 1992 report focused on a classroom lesson in which the novice teacher, Ms. Daniels, was unsuccessful in providing meaningful justification for the division-of-fractions algorithm, although she had the most extensive mathematics background in the group and believed that good mathematics teaching included making mathematics relevant and meaningful for students. In interviews at the beginning of her student teaching, she was unable to give meaning to division of fractions. She developed some knowledge about division of fractions during student teaching but was unable to use this knowledge to help students understand division of fractions. The methods course did not challenge her beliefs, and, in fact, she was disappointed in this class because it lacked the examples and explanations she had expected to learn.

The authors of the first study (Borko et al., 1992) noted two major implications of the study. "First, prospective teachers must be given the opportunity in their university course work to strengthen their subject matter knowledge (p. 219)," not just by increasing courses but through improvement in the curriculum and instruction offered in university mathematics departments. Second, university course work must provide prospective teachers opportunities to develop the "concepts and language to draw connections between representations and applications on the one hand and algorithms and procedures on the other" (p. 220). Prospective teachers' fundamental beliefs about teaching and learning must be challenged; they need more opportunities to strengthen their pedagogical content knowledge.

The second study in this project (Eisenhart et al., 1993) focused on the complex process of teaching for understanding. The data collected for the study were again the interviews of eight prospective teachers when they completed their professional coursework and began student teaching. Ms. Daniels was again the focus of the study. The researchers noted that "teaching for conceptual knowledge was a major theme of the mathematics education course work, an expressed commitment of the placement schools, and a part of the student teachers' own beliefs about good teaching" (p. 10). Yet although the teachers and teacher educators they observed "struggled to teach for conceptual knowledge, they often appeared to emphasize procedural knowledge instead" (p. 10).

Eisenhart et al. (1993) examined the tensions created over learning to teach for conceptual understanding in their university courses and the procedural knowledge required for their daily responsibilities in their placement classrooms. "Tensions felt by the student teachers seemed to blind them to the conceptual knowledge, and ways of teaching for conceptual knowledge, that the instructor was trying to help them learn" (p. 29). Additional tension came from the fact that administrators appeared to want teachers to teach for both conceptual understanding and procedural knowledge yet held teachers accountable only for the procedural knowledge. As a result, teachers in the schools taught primarily for procedural knowledge. Ms. Daniels received no support for the ideas she had learned in her methods course.

The researchers (Eisenhart et al., 1993) concluded by offering two recommendations. First, student teachers must be placed with teachers who will reinforce the importance of teaching for understanding. Second, the university must find a way to provide time and opportunities for the prospective teachers and the instructors "to explore and develop approaches to teaching that are considered desirable" (p. 38). This study illustrates the need to find ways for prospective teachers to develop a framework that will be strong enough to combat the apprenticeship-of-observation framework that not only is brought with them into teacher education but also is sometimes reinforced by what they observe even while in a teacher education program.

When the same questions are examined in more than one country, the results can be surprisingly similar. For example, Oldham, van der Valk, Broekman, and Berenson (1998) analyzed lesson plans of prospective teachers from four countries. All the prospective teachers had been told to prepare a lesson on area, and all were interviewed about the lesson. The researchers found that all lesson plans had some procedural approaches, and some also used conceptual approaches. Many plans were completely oriented around the teacher as the leader, and a smaller number had some student activities. The researchers believed that this model provided them with robust frameworks and insights across cultures concerning the role of the teacher.

A Model for Learning to Teach

Some researchers are proposing new ways to think about teacher preparation. The *experiment model* for learning to teach, developed by Hiebert, Morris, and Glass (2003), is a way of thinking about the objectives of coursework during teacher preparation. The authors first acknowledged and described the tools that teachers need to become effective mathematics teachers—personal mathematical proficiency and the knowledge, competencies, and dispositions to learn to teach for mathematical proficiency. They then proposed that "as a way of making some aspects of teachers' routine, natural activity more systematic and intensive" (p. 207), teachers need to learn to treat lessons as experiments. When designing a lesson, a teacher should be guided by the learning goals for the lesson and by hypotheses about the instructional activities planned to achieve the learning goals. "Lessons-as-experiments require, on the one hand, constructing local theories regarding the relationships between teaching and learning and, on the other hand, the tying of the theories to *this* learning goal in *this* context" (p. 209). Data are collected and analyzed in terms of the learning goals and are used to make informed decisions about future lessons. Following this model leads prospective teachers to acquire "the identity and dispositions of a professional teacher" (p. 211) with the obligation to draw from and contribute to a shared knowledge base. The authors suggested that "the principles and processes that are proposed for the generation of knowledge needed to improve teaching are the same as those that are proposed for the generation of the knowledge needed to improve teacher preparation" (p. 213).

Mathematicians' Role in Teacher Preparation

The mathematics community has, for a number of years, actively promoted more involvement by mathematicians in the preparation of teachers. At the time the first NCTM Standards document was released in 1989, the National Research Council released *Everybody Counts: A Report to the Nation on the Future of Mathematics Education* (1989). *Everybody Counts* was a consensus document of the Mathematical Sciences Education Board, the Conference Board of the Mathematical Sciences, and the Committee on the Mathematical Sciences in the Year 2000. Mathematicians were well represented on these three boards. The document authors spoke forcefully about the role of mathematicians in the education of teachers:

> Undergraduate mathematics is the linchpin for revitalization of mathematics education. Not only do the sciences depend on strong undergraduate mathematics, but also all students who prepare to teach mathematics acquire attitudes about mathematics, styles of teaching, and knowledge of content from their undergraduate experience. No reform of mathematic education is possible unless it begins with revitalization of undergraduate mathematics in both curriculum and teaching style. (NRC, 1989, p. 39)

In 1991 the Committee on the Mathematical Education of Teachers, representing the Mathematical Association of America (MAA), published *A Call for Change: Recommendations for the Mathematical Preparation of Teachers of Mathematics* (Leitzel, 1991). The authors believed that the substantive changes in mathematics required corresponding changes in the preparation of teachers. They emphasized that "although skills in mathematics must be acquired, processes, concepts, and understanding should take precedence" (p. 39) and called for a rethinking of the collegiate curriculum in mathematics that would allow future teachers to grapple with their own learning of mathematics. The most recent version of the report from the MAA's (2003) Committee on the Undergraduate Program in Mathematics (CUPM) and the Conference Board of the Mathematical Sciences' report (2000) *The Mathematical Education of Teachers* (discussed in the earlier section on what mathematics teachers need to know) both addressed the mathematical and pedagogical preparation of teachers and the role of the mathematics community in this work.

Another example of mathematicians' recognition of their role in preparing teachers is the MAA initiative Preparing Mathematicians to Educate Teachers (PMET). Funding was received from the National Science Foundation for three summers of workshops for college and university mathematics and mathematics education faculty to prepare them to offer better courses for prospective teachers at the elementary, middle, and secondary levels.

The NCTM's 2000 version of standards, *Principles and Standards for School Mathematics*, was developed with a great deal of advice from professional associations of mathematicians in the form of recommendations from various Association Review Groups that included groups of mathematicians. The considerable influence of faculty in 2-year and 4-year colleges and universities was apparent and was welcomed. Hyman Bass, in his role as the president of the American Mathematical Society, said in an interview (2001) that "mathematicians and scientists have a special responsibility that extends their traditional roles in research and education at the university level to concerns for K–12."

ADDRESSING QUESTION 9: WHAT DO RESEARCHERS KNOW ABOUT ISSUES THAT AFFECT PROFESSIONAL DEVELOPMENT?

The people who provide professional development play, of course, a key role in whether the six previously discussed goals of professional development are met. Who are these providers? How do they become providers? What professional development and other experiences do *they* need and receive? What role is played by school site and district administrators? What resources do they provide? If they have a vision of what should be happening in their classrooms that is different from that of teachers, how is this difference resolved? How does it affect teacher evaluation? What is the effect of district, state, and national policies on professional development? How is the success of professional development, at any level, determined? These issues are explored in this section.

Developing Professional Developers

Those responsible for providing professional development fall into several categories. In one category are university educators, particularly those in schools and colleges of education, who are frequently engaged with professional development and who may or may not receive university recognition for this work. Most mathematics departments do not reward work with teachers; thus, mathematicians who work in professional development and teacher preparation are usually established and find themselves interested in school mathematics, often because of their children's experiences in mathematics or because they are approached to teach mathematics courses for teachers. To be effective, these providers must spend time in classrooms at the level of the teachers whom they teach. In another category are those whose careers are built around professional development, such as professional providers associated with the Education Development Center or Marilyn Burns Associates. In a third category of providers are those who become teacher leaders within a school or district—they may act as resource teachers, mentors, coaches, consultants, curriculum developers, and professional development providers. In some cases they fill this role by choice—a teacher who wants more challenges might enter a master's program that provides them with knowledge and skills needed to lead. Some teachers may be recognized for their potential by administrators, other teachers, or outside collaborators and may be asked to take on additional responsibilities with coaching, mentoring, providing workshops, or working as resource teachers. Teachers as professional development providers need their own professional development (Fonzi, 2003; Sugrue, 2004). Spencer, Moon, Miller, and Elko (2000) have developed a casebook to support the development of teachers who wish to become teacher leaders.

Loucks-Horsley et al. (2003) identified three key elements for professional developers, each of which must be obtained through their own professional development opportunities. These elements (from pp. 230–231) are summarized here:

1. *Effective professional developers must have diverse knowledge and skills,* including a deep understanding of the appropriate level of school mathematics; knowledge of how both children and adults learn; knowledge of school organization, structures, and culture; self-awareness and the ability to be self-critical; willingness and ability to learn from mistakes and successes; and an understanding of the process of implementing and evaluating changes.

2. *Professional developers use a repertoire of professional development strategies,* and the strategies needed depend partly on the role played. If a new curriculum is being implemented, the role of the professional developer might include workshops focusing on the curriculum. If a professional developer is charged with providing assistance to new teachers, mentoring and coaching skills will be required. If the content knowledge of teachers within a school is weak, the professional developer might need to focus on developing understanding of the mathematics being taught.

3. *Professional developers have their own learner community* with whom to collaborate and share. They need access to many resources and information. They need their own professional development through courses, workshops, and attendance at state and national meetings. They should not be expected to work in isolation.

The knowledge and skills needed can be accessed in a variety of ways. Colleges and universities offer courses on leadership and management. Workshops offered by, for example, the Association for Supervision and Curriculum Development (ASCD), the American Federation of Teachers (AFT), and the National Council of Teachers of Mathematics (NCTM) can provide teachers leaders with the kinds of information and skills they need. Sometimes state funding, such as for the California Mathematics Project, provides opportunities for teachers both to develop as leaders and to participate in a professional community. In large school districts, special provisions and opportunities are offered to potential leaders. In small school districts, a teacher might become a leader by being mentored by a recognized teacher leader.

Potential leaders are often recognized through their interest in mathematics, their voluntary participation in professional development opportunities of many kinds, their interest in their students' reasoning and thinking, and their abilities to work with others. "Leadership is not a personality trait but an attribute of self-development in social relationships (Grossman et al., 2001, p. 996).

Schifter and Lester (2002) studied the role of facilitators of a professional development seminar (DMI), asking whether facilitators should play an active role and, if so, what they should do and what they must understand to be able to intervene appropriately. They built on the work of Remillard and Geist (2002), who had observed DMI seminars and were drawn to study the instances that "required facilitators to make judgments about how to guide the discourse. [These moments, they argued,] arose from conflicts among the goals and commitments of the facilitators, the expectations of the participants, and the agenda of the curriculum" (Schifter & Lester, 2002, p. 6). These moments were ultimately referred to as *openings in the curriculum.*

Schifter and Lester (2002) argued for an active role by facilitators of professional development (in this case, of DMI) and said that if "instances of discontinuity between participants' ideas or beliefs and the goals of the curriculum are to provide fruitful opportunities for learning, then the facilitator must take determined action to exploit them" (p. 15). The authors described the knowledge facilitators must have to be able to exploit such moments and the type of support DMI facilitators are provided.

> Responding to openings for teacher learning is not just a matter of having the right cognitive dispositions. Just as important, effective facilitation requires courage—courage to challenge the thinking of other adults; courage to redirect a discussion that is moving in an unproductive direction; and courage to face the agitation, and sometimes even the tears, that result when firmly held ideas begin to crack. This form of facilitation also demands a stance of respect for and commitment to the teachers being supported and the ideas to be explored. Perhaps this disposition is best reflected in one facilitator's injunction to herself and her colleagues: "We can do better—go deeper—than where we are now." (p. 21)

A story of teachers' becoming professional development leaders can be found in the work of Jacobs and Raynes (2001, 2002). These researchers worked with a group of primary-grades teachers in schools in low-socioeconomic communities. The professional development offered was voluntarily undertaken by teachers who were not satisfied with their own mathematics instruction. Even though some did not even like mathematics when they began, they found that the focus on listening to children and using student thinking as a basis for instruction was compelling. During the third year of the project, teachers took on leadership roles and began to offer professional development to other teachers in their schools.

> The one underlying principle that all of the teacher leaders embraced was that participation in the learning opportunities they designed needed to be voluntary. They recognized that real change is not always smooth. It requires risk taking and a willingness to critically examine one's own practices—often practices that have been taken for granted for many years. They were convinced that this type of professional development could only succeed if a teacher wanted to change. Therefore, they targeted their support toward teachers who volunteered to participate and work to improve their mathematics instruction. On the other hand, the teacher leaders also made sure that their learning opportunities did not preclude other teachers from initially participating peripherally and then becoming more deeply involved at a later date. (Jacobs & Raynes, 2001, pp. 4–5)

Teachers become leaders through a number of routes (Fonzi, 2003). Many begin simply, by honing their own practice or presenting at a local or state meeting for teachers of mathematics. When a teacher first presents at a meeting or conference, she or he has probably been prompted to do so by someone such as a university professor or project director. Some teachers are asked to coach new teachers or act as supervisors of prospective teachers while they are concurrently themselves teaching. Others become engaged in action research and want to share their results. When teachers become engaged in projects and work with other teacher leaders, they gain self-confidence while they are gaining more knowledge and are being asked to provide leadership in many ways. Some will become full-time leaders, working with a school district as a mathematics coordinator or as a resource teacher for a number of schools.

University educators also need to develop new ways of providing professional development. Many of those ways have been described in various sections of this chapter.

> Not only will professional developers need to build their repertoires beyond workshops and courses, they will also need to learn how to manage these repertoires in relation to their goals and the context within which they are working. Not only will professional developers need to develop teachers as individuals, they

will also need to learn how to develop whole communities of practice. And, finally, not only will professional developers be held accountable for adding new techniques or skills to teachers' repertoires, but also for teachers' successful enactments of valid practices that raise student achievement. Never before have the stakes been higher or the learning demands greater. (Stein et al., 1999, pp. 267–268)

Supporting Professional Development

Professional development is the key to improving education and effecting changes in student achievement (e.g., Adler, 2000; Guskey, 1995), but only when it is offered in a form that is relevant to teachers and addresses their specific needs and concerns (Guskey, 1995). In addition, effective professional development must "provide regular opportunities for participants to share perspectives and seek solutions to common problems in an atmosphere of collegiality and professional respect" (Guskey, 1995, p. 121). The authors of *Adding It Up* (NRC, 2001a) also advocated that when teachers come together to work on practice, they should be provided with opportunities that will result in serious and substantive professional development and that these opportunities can result only when those providing professional development have strong mathematical backgrounds and understand students' thinking, students' learning, and teachers' thinking about both mathematics and their own instructional practices.

Resources for and of Professional Development

Professional development, particularly in mathematics and science, can lead to changes that result in teaching for understanding; professional development can also be viewed as a resource for change. "Schools can support teacher change by responding to teacher learning" (Gamoran, 2003, p. 66). Gamoran and his colleagues studied six schools in which teachers and researchers collaborated to investigate factors that brought about teaching for understanding. They found evidence that resources that support teaching for understanding include "time and other material conditions, expertise from outsiders and insiders, and communities of teachers within schools that facilitated the professional development collaboratives" (p. 68). They also found that teaching for understanding in mathematics and science can itself generate resources that can help sustain such teaching. The social and human resources generated included the development of professional communities and the increased knowledge of teachers. Material resources generated in their study included curriculum materials and addi-

tional resources that could be used to facilitate teachers' work in the classroom.

For professional development to be "high quality, sustained, and systematically designed and deployed to help all students develop mathematical proficiency" (NRC, 2001a, pp. 409–410), teachers need the time and resources to engage in sustained efforts to improve their practices.

The Roles of Administrators

Administrators have a major role in bringing about change that can foster teaching for understanding.

> Leaders who wish to support teaching for understanding must not only allocate resources, but also foster new ways of generating resources; and they must deal with not only material resources (time, materials, and compensation),but also human resources (teachers' knowledge, skills, and dispositions) and social resources (relations of trust and collaboration among educators). Although educators in positions of authority (e.g., principals and curriculum coordinators) have special roles to play in responding to these tasks, they cannot provide all the leadership necessary. Instead, leadership must be distributed throughout the organization, including teachers who take on leadership roles that support their learning and that of their colleagues. Developing capacity to support teaching for understanding involves recognizing new leadership demands and engaging in a distributed response. (Gamoran, Anderson, & Ashmann, 2003, p. 107)

Principals in the Gamoran et al. (2003) study did not play a major role in articulating a vision for the mathematics and science teaching in their schools beyond fostering such general visions as "all students can learn." Administrators who played a role in articulating this vision were more often district specialists and outside researchers. Teachers themselves usually began professional development with visions of what they intended to gain, but the tempering of these visions by what they learned and by their work with other teachers led to clarification of how to enact what had been learned. "Thus the professional development groups developed their own visions of what teaching for understanding meant in their particular settings" (p. 115).

Nelson (1998), in her Lens on Learning project, found that many administrators view mathematical knowledge as hierarchical, with facts and operational facility coming first for students and with the possibility of students' being taught and learning concepts by melding them with practice of math facts. Thus, parents' demands for instruction on skills and facts are seen as reasonable. When faced with calls for reforming mathematics instruction, administrators do not understand that teachers need deeper understanding of mathematics; rather, they think that the purpose of professional development is to teach teachers new techniques and that their role is to support teachers in trying these new techniques. Other administrators, who view mathematics as a way of knowing and thinking, can explain to parents that mathematics focused on understanding might seem less efficient but is more enduring. These administrators understand that beliefs must change before practice can change, that fundamental instructional changes are called for, and that this process takes time. They recognize that professional development is a process rather than an event.

In Nelson's (1998) study, administrators "developed views of what a professional development project that aimed at reconceptualization might look like" (p. 204). They came to understand that the reforms being introduced in schools would require a deep examination of beliefs and ideas about learning, teaching, and the nature of mathematics and mathematical knowledge. Just as teachers and children need opportunities to investigate new ideas and restructure their conceptual understanding, so do administrators. Some of Nelson's work is reflected in the statements made in the NCTM (2000) Standards' comments on the role of administrators:

> Administrators at every level have responsibilities for sharing the instructional mission in their jurisdictions, providing for the professional development of teachers, designing and implementing policies, and allocating resources. . . . When administrators themselves become part of the mathematics learning community, they develop deeper understandings of the goals of mathematics instruction. They can understand better what they are seeing in mathematics classrooms (Nelson, 1999) and can make more informed judgments about the curricular, technological, and pedagogical resources teachers need. (p. 376)

Knapp (2003), too, has noted that administrators need professional development to fully understand what improving learning, teaching, and instructional development entails.

The Challenges Faced

Challenges to change go hand-in-hand with support needed for change to take place. Gallucci (2003) advocated that schools be organized so that vision setting and professional learning are encouraged and teachers are involved in communities of practice. He recognized that the most serious barriers to the formation and longevity of such communities include time and other institutional structures. Overcoming these barriers invariably translates into money. The major challenge is then, to no one's surprise, cost (Garet et al., 2001).

The nature of challenges that faced school professionals in the Gamoran et al. (2003) study were described in the chapter by Anderson (2003):

- Providing resources for classroom teaching. For mathematics these resources include tools and teaching materials, ability to understand and respond to students' reasoning, understanding of mathematics and how children learn mathematics, understanding of differences among students, and ability to develop in classroom learning communities social norms that promote learning;
- Aligning purposes, perceptions, and commitments. Teachers, parents, and students often value traditional schools because they are safe, predictable, and do not make excessive demands. "Yet is seems unlikely that teaching for understanding can become a widespread practice without changes in professional cultures" (p. 18). Complete transformation of schools is unlikely; the alternatives are conflict over resources, standards, and practice; *compromise,* which is accommodation instead of transformation, or *coexistence,* which is described as accommodation together with transformation.
- Sustaining teaching for understanding. To sustain such teaching requires both collective and individual efforts that depend on working interdependently rather than dependently, with leadership from professional communities.

Additionally, even teachers and administrators who support teaching for understanding often do not share researchers' notions of what teaching for understanding entails.

Finally, a seemingly trivial but an actually widespread and serious challenge was noted by Guskey (1995)—that many teachers think any reform is once again a fad that will pass, and they share this view with younger colleagues who express concerns over implementing new programs.

Evaluating Professional Development

Although teachers' satisfaction with professional development is desirable, more information is needed to judge how much teachers have learned, whether teachers are likely to change their instructional practices, and whether student achievement is likely to be affected (Guskey, 1995; Mumme & Seago, 2002; Wilson & Berne, 1999). One effective way to evaluate professional development is to consider what elements research has shown to be successful—for example, Table 5.1 from earlier in this chapter provides three such lists (one by Hawley and Valli, 1999; a second by Elmore, 2002a; and a third by Clarke, 1994) that contain similar elements expressed in different ways.

The most comprehensive guidelines for evaluating professional development (Guskey, 2000) are quite recent. Understanding Guskey's models requires understanding his view that professional development must be systemic to be effective:

> Experience has taught us that you cannot be successful in any endeavor designed to improve student learning by focusing only on professional development (Sparks, 1999). Too many other things in the system affect student learning, such as the curriculum, assessment, school organization, materials, support, and leadership. The dynamic interaction of these elements requires a systemic approach (p. 69)

Guskey presented two evaluation models, one focused on improvement of student learning and the other on the five critical levels in professional development evaluations.

The need to clarify the "precise nature of the relationship" (Guskey, 2000, p. 72) between professional development and student achievement has become more pressing, particularly because government funding agencies consider this relationship imperative in their consideration of future funding. Guskey's first model includes three factors that influence the quality of professional development: content characteristics, process variables, and context considerations. The first two of these factors have been discussed in previous sections of this chapter. The context characteristics "refer to the 'who,' 'when,' 'where,' and 'why' of professional development" (p. 74). All three factors are important to the success of professional development, primarily because "most content and processes must be adapted, at least in part, to the unique characteristics of the setting (Firestone & Corbett, 1987; Fullan, 1985; Huberman & Miles, 1984)" (Guskey, 2000, p. 74) but also because if any one of these three factors is neglected, the effectiveness of the entire experience will be reduced.

Guskey's (2000) second model consists of five levels of information that address a broad range of *what* questions related to professional development but do not provide much information that contributes to learning *why*. Still, because each higher level builds on the previous levels, this model can be particular-

ly useful in evaluating professional development in mathematics. At Level 1, participants' reactions are considered. Collecting information at this level is relatively easy and usually consists of asking questions about comfort level, such as the temperature of the room. Guskey referred to these reactions as *happiness quotients*. At Level 2, the focus changes to measuring the knowledge, skills, and activities in which participants engaged and, at times, the attitudes changed by the experiences. Level 3 focuses on the organization—what support is provided and what is changed. For example, professional development may not result in desired change because of organizational characteristics. Gathering such information is not easy but is necessary to inform future change initiatives. Level 4 focuses on participants' use of new knowledge and skills. The central question at this level is whether what participants learned made any difference in their professional practices. "The key to gathering relevant information at this level rests in the clear specification of indicators that reveal both the degree and quality of implementation. In other words, how can you tell if what participants learned is being used and being used well?" (p. 84). At Level 5, the level of most sophisticated teaching, student learning outcomes are measured. Both intended and unintended learning are important to consider, and, thus, multiple measures are necessary. Measures from student records should be augmented by affective indicators, and information on factors such as "student self-concepts, study habits, school attendance, homework completions rates, and classroom behaviors" (p. 86) could be included.

Guskey (2000) warned that "evaluation at any of these five levels can be done well or poorly, convincingly or laughably" (p. 86) and that effectiveness at one level tells no story about effectiveness at the next level.

Evaluators of professional development in mathematics can also profit from Borasi and Fonzi's (2002) identification of nine learning needs of teachers. The authors, in an NSF-requested review of professional development that supports school mathematics, analyzed promising professional development experiences by discussing each experience in terms of each of these needs, some of which were identified earlier but are listed in full here because they can be used as an inventory for evaluating professional development at both a formative and a summative level.

1. Developing a vision and commitment to school mathematics reform,
2. Strengthening one's knowledge of mathematics,
3. Understanding the pedagogical theories that underlie school mathematics reform,

4. Understanding students' mathematical thinking,

5. Learning to use effective teaching and assessment strategies,

6. Becoming familiar with exemplary instructional materials and resources,

7. Understanding equity issues and their implications for the classroom,

8. Coping with the emotional aspects of engaging in reform, and

9. Developing an attitude of inquiry toward one's practice.

No single model emerged from their analysis of successful professional development programs—many programs met these nine needs, but in different ways.

Policy Issues

Policy matters such as how schools are structured, how curriculum is adopted and implemented, how teachers are recruited, how teachers are evaluated, and how testing influences what and how mathematics is taught are relevant, in some manner, to the professional development of teachers and to teacher preparation. But *policy matters* can also refer to the fact that policy *matters*—schools and universities cannot provide professional development and teacher education independent of local, state, and national policies. Many of us in mathematics education have for too long ignored the policy implications of what we do.

> Teachers teach from what they know. If policymakers want to change teaching, they must pay attention to teacher knowledge. And if they are to attend to teacher knowledge, they must look beyond curriculum policies to those policies that control teacher education and certification, as well as ongoing professional development, supervision, and evaluation. (Darling-Hammond, 1990, p. 240)

Policy makers have the power to create an infrastructure to support professional development in ways that involve both school-level staff as well as external groups (Knapp, 2003). Small districts in particular often lack personnel to undertake meaningful, long-term professional development. In California, during a time of budget surplus, the state funded a Professional Development Institute (PDI) with sites at many areas across the state where teachers could participate in summer and year-long professional development, primarily offered by universities and primarily focusing on increasing content knowledge and pedagogical content knowledge. Teachers received attractive sti-

pends and in many cases also earned university credit. The PDI was intended to continue indefinitely but was abandoned in 2004 when the budget surplus vanished. Before that time many hundreds of teachers took advantage of the professional development offered by PDI. This initiative serves as an example of how policy makers at the state level can provide a large-scale infrastructure for professional development.

District- and School-Level Policies

Teachers work in policy environments, whether or not they think about their work in this way. Policy makers at the district level play an important role in professional development by "creating a variety of venues and occasions for learning, helping to establish a focus and incentives for professional learning, and enabling teachers to connect their learning experiences over time" (Knapp, 2003, p. 147). At the site level, policy decisions can affect the allocation of resources, the scheduling of classes, the extra-curricular duties of teachers, and other such matters. Hargreaves (1995) suggested that some site-level professional development focus on political agendas and concerns: "Teachers who have been introduced to the micro-politics of schooling in relation to their own institutions can achieve breakthroughs in insight, action, and effectiveness that help them secure support and resources for their students (Goodson & Fliesser, 1992)" (p. 18). The Urban Mathematics Collaboratives funded by the Ford Foundation in the early 1990s serve as an example of teachers' becoming empowered after coming to understand how they could affect decisions regarding their work.

At the district level, administrators make decisions regarding curriculum, school structures, assessment, and opportunities for professional development. Teachers are expected to make changes in their instructional practices to align with state and national standards, but if they have few opportunities and few incentives to learn how to change practice, they are less likely to make the changes dictated by policy (Spillane, 2000). Policies that call for changing instruction require far more than mandates.

> Coming to know involves the reconstruction of existing knowledge rather than the passive absorption of knowledge (Anderson & Smith, 1987; Confrey, 1990). Moreover, sense making is influenced by the social, physical, and cultural contexts of the sense maker. (J. Brown, Collins, & Duguid, 1989; Resnick, 1991). Applying these ideas to policy implementation, we notice that implementers construct what it is policy asks of them, and it is these understandings they respond to in implementing policy. Thus, the policy stimulus is not all that matters: Implementers' beliefs, knowl-

edge, and experience, as well as their situation, also influence the ideas they come to understand from policy. (Spillane, 2000, p. 146)

If policies are to lead to the making of needed changes, those policies must be aligned with one another. Teachers and administrators might ask questions such as

Does a new curriculum framework stress "implementation of texts," thereby describing a passive teacher and student role, or are teachers assumed to participate in the construction of practices that start with the students' experiences and needs and aim at student outcome standards? Does an assessment system evaluate student understanding, or does it test for rote recall of disconnected facts? Do teacher evaluation systems look for teaching behaviors aimed at keeping students quiet or for practices that engage students actively in their learning? Do administrator licensing standards require that principals know how students learn and how teachers teach for understanding, or do they stress noninstructional matters? Do school accountability requirements enforce current, highly fragmented bureaucratic structures and uses of time, or do they allow for more integrated and student-centered forms of staffing and fund allocations? (Darling-Hammond & McLaughlin, 1996, p. 213)

If high quality professional development is taken seriously as the most important mechanism for improving teachers, then schools must make a major financial commitment to providing it. But schools can rarely afford to provide high-quality professional development to all the teachers in a district, and choices need to be made. In a report of a study by Garet et al. (2001), the researchers were blunt about their findings:

A major challenge to providing . . . high-quality professional development is cost. Schools and districts understandably feel a responsibility to reach large numbers of teachers. But a focus on breadth in terms of number of teachers served comes at the expense of depth in terms of quality of experience. Our results indicate a clear direction for schools and districts: in order to provide useful and effective professional development that has a meaningful effect on teacher learning and fosters improvement in classroom practice, funds should be focused on providing high-quality professional development experiences. This would require schools and districts either to focus resources on fewer teachers, or to invest sufficient resources so that more teachers can benefit from high-quality professional development. (p. 937)

One way to think about this challenge is to consider seriously, at least in Grades 5 and up, that teachers specialize, that they receive high-quality professional development in one or two areas but are not expected to be experts in all the subject areas in the elementary school curriculum. Hiring of mathematics specialists has been frequently advocated in elementary school for teachers of Grades 5 and above (CBMS, 2000; Learning First Alliance, 1998; NCTM, 2000; NRC, 1989).

Evaluating Teachers

The manner in which teachers are evaluated is usually a site-based matter undertaken with some guidelines from the district. All too often, the on-site evaluation of teachers works against the very reforms that a district is attempting to make (Spillane, 2000). Spillane distinguished *form-focused understanding*, which refers to "pedagogical forms of learning activities, including students' work, instructional materials, and grouping arrangements" (p. 154), from *functional understandings*, which focus on "what counts as mathematical knowledge, doing mathematics, and learning and knowing mathematics" (p. 154). When teachers receive professional development that leads them to attempt to teach for functional understanding but are then evaluated by administrators who focus on form, the mismatch can lead to discouragement and probably to their abandoning teaching for functional understanding. Of course, the situation could be reversed—a teacher who is form-focused may be evaluated on her students' success in achieving functional understanding, confusing the teacher who does not know how to teach for functional understanding. Darling-Hammond and McLaughlin (1996) extended these ideas to consider administrative leadership:

To support teaching for understanding and the professional development necessary to it, new forms of teacher evaluation will need to emphasize the *appropriateness* of teaching decisions to the goals and contexts of instruction and the needs of students, rather than focusing on teachers' adherence in prescribed routines. Evaluation must be conceived not as a discrete annual event conducted by supervisors making brief visits bearing checklists but as a constant feature of organizational and classroom life for practitioners. . . . A critical role for administrative leadership involves creating and sustaining settings in which teachers feel safe to disclose aspects of their teaching, admit mistakes, try and possibly fail, and continue to seek out advice and counsel. (p. 215)

State- and National-Level Policies

A great deal of variance is found in the ways that state policies affect teachers. If the policies regarding professional development, teacher preparation, teaching expectations, curriculum, and testing are aligned, achievement scores are likely to increase. For example, Wilson, Darling-Hammond, and Berry (2001) undertook a case study of a 15-year policy in Connecticut to recruit, prepare, and support teachers; this policy, combined with strong and consistent leadership and investment in research and development, led to success in creating effective teachers and increasing achievement. The intention of the Connecticut State Department of Education was to move the state's educational system toward a "conception of teaching grounded in strong disciplinary and pedagogical content knowledge as well as in a broad research base about teaching and learning" (p. 25). The Department staff realized that continued school improvement depended on a coherent system of continued educator improvement, which required aligning teacher preparation, teacher evaluation, and professional development with the state's student and teaching standards. The primary focus of all professional development is to provide teachers with the skills, knowledge, and abilities to improve student learning. One factor in the success in Connecticut has been the focus on improving teacher preparation and supporting beginning teachers. Licensing requirements were changed to strengthen clinical experiences for beginning teachers. All teacher preparation programs were required to demonstrate that their graduates understood the state's student standards and teaching standards. A strong support system for new teachers was put into place. "By systematically educating mentors who support and evaluate novice teachers, the support system has increased capacity in beginners and veterans alike" (p. 4). The system has high standards for teachers, with strong competition for openings. Only top-quality teachers are hired.

Another large-scale study of state education reform in mathematics was undertaken by D. K. Cohen and Hill (2001) in California. In 1985 a new state framework for mathematics turned from a focus on basics to a focus on teaching for understanding. The reforms advocated in this framework were continued by the state department of education when they contracted outside agencies to develop replacement units that could be used for particular curriculum topics, such as fractions. A state adoption of textbooks forced publishers to conform (as much as was possible) to the 1985 framework. Large-scale professional development was undertaken. Cohen and Hill studied this reform beginning in the early 1990s by focusing on professional development as a way to examine the relations between policy and practice. They found that only a small portion of teachers were affected by the

reform, partly because of the fragmentation of educational governance and partly because of missed opportunities on the part of the reformers. Their analysis did show that when assessment was consistent with professional development, "teachers' opportunities to learn pay off in students' math performance . . . [showing] the significant role that the education of professionals can play in efforts to improve public education" (p. 148). They pointed out that the opportunities to study policy as they had done are rare because states and districts are often unwilling to mandate curriculum and thus are unable to focus professional development on the curriculum. Unfortunately, the policy makers and reformers in California

> made no attempt to learn systematically about how the reforms played out in schools and classrooms. As a result, they had no evidence to inform either their work or public debate, and when skeptics charged that the reforms were diluting instruction to basic math skills, state officials and professionals had no basis on which to investigate or respond. (p. 187)

One could disagree with this analysis. California state officials and professionals did attempt to establish ways of recording learning by developing first the California Assessment Program (CAP), which suffered a line-item veto by the governor, then the California Learning Assessment System (CLAS), with items in reading and mathematics that led to a public outcry before also being abandoned. Both of these state examinations were based on the state framework, which had many similarities with the NCTM Standards, and the tests were held in high regard by most mathematics educators who knew about them. New policy makers in California believed that mathematics reform had gone too far, and much has changed on the educational scene. However, the need for professional development continued to be recognized, and in 2000 the state legislature funded the Mathematics Professional Development Institutes previously described. Hill and Ball (2004) evaluated these Institutes by developing pretests and posttests to measure teacher learning; these tests were administered at most sites. Their report provided evidence that statewide, large-scale, funded professional development is beneficial in terms of teachers' learning of both content knowledge and pedagogical content knowledge.

Today, throughout the country, a major factor influencing what teachers teach is the student-testing program within the state, further complicated by the federal rulings regarding the testing for Academic Yearly Progress (AYP) as mandated by the No Child Left Behind legislation and the threat of sanctions for schools that do not meet the required increases in scores. These new demands have been met with criticism—severe in some cases. The federal requirements of upward change each year ignore much of what is known about testing. For example, Linn, Baker, and Betebenner (2002) noted that "interpretations of the law also should recognize the volatility of school-level results from year to year and provide states with latitude to identify ways of reducing that volatility" (p. 16). After examining the psychometric work pertaining to national testing, Tucker (2004) had this to say.

A central tenet of the No Child Left Behind Act is that educational improvement at a school can be measured by comparing student scores on standards-based tests from one year to the next. An important question about such a strategy that has gotten surprisingly little attention, is this: How accurate are such year-to-year comparisons? The answer is that they are much less accurate than people assume—in some cases, wildly inaccurate.... Most standards-based tests are based on a technical psychometric methodology called Item Response Theory, or IRT. Item response theory makes many critical assumptions, both of a practical nature—in getting all the technical details of test development right—and of a theoretical nature—in its one-dimensional model for assessing student's knowledge. Few states have the resources to implement IRT-based tests with the attention to detail they require. Such tests should be reliable for assessing standard procedural skills. such as solving a quadratic equation. Unfortunately, the more thoughtful and thus unpredictable a test, the more likely equating methods will misperform.

In addition to the psychometric problems of large-scale testing, equity is an issue, primarily because low-performing schools are most likely to have second-language students and students of color. Elmore (2002b) argued that testing is not the way to solve problems of equity.

The threat of such measures is supposed to motivate students and schools to ever higher levels of achievement. In fact, this is a naïve view of what it takes to improve student learning. Fundamentally, *internal* accountability must precede *external* accountability.... Giving test results to an incoherent, atomized, badly run school doesn't automatically make it a better school. A school's ability to make improvements has to do with the beliefs and practices that people in the organization share, not with the kind of information they receive about their performance. *Low-performing schools aren't coherent enough to respond to external demands for accountability.* (p. 4)

Beyond the problems of psychometrics and equity, the phenomenon of tests driving what is taught

is not necessarily bad. Instead, testing can be one way to drive desired changes that are not otherwise taken seriously.

D. K. Cohen and Hill (2001) advocated taking more seriously the need for better research about education policy and practice: "Such work could contribute to efforts to improve instruction and to understanding improvement. Research of this sort could not determine decisions, but it can inform understanding, debate, and, in come cases, decisions" (p. 189).

ADDRESSING QUESTION 10: WHERE DO WE STAND NOW, AND WHAT COMES NEXT? PERSONAL THOUGHTS

The recognition that mathematics-teacher change is a process, rather than an event, is being successfully used in some places to justify the need for resources for multiyear, coherent programs of professional development. Policy makers, including district and state personnel, are recognizing that workshops lacking continuity and direction have little effect on teachers. Yet large numbers of teachers remain untouched. The efforts described in this chapter provide evidence that good professional development *can* make a difference. However, "if we are to move past these and other 'pockets of wonderfulness,' we must advance to another level of sophistication as mathematics educators and change facilitators" (C. Seeley, personal e-mail, September 6, 2004). Scale-up of successful professional development programs remains the greatest challenge we educators face.

I once thought of teacher preparation as a solution to this problem. Teachers who are well prepared at universities, I believed, could lessen the need for professional development, which seemed almost too enormous a task to undertake. With so many teachers close to retirement, the education system would have new, better prepared teachers to replace them. I came to realize, of course, that this solution was untenable. Preparing teachers to teach mathematics in four short years, often beginning with students who have little real understanding of mathematics, is not possible. Rather, our goal should be to prepare them for future learning, in part because at the university we can focus only on learning-for-practice (and not enough of that), and we know they have much more to learn *in* practice while teaching. Even the arena of preparation that belongs to the university is an undertaking made doubly difficult by transient graduate-student instructors, by mathematicians who fail to model good instruction, by methods courses taught as though all subjects are taught in the same way, and by what sometimes seems to be random placement for student teaching. Even one of these problems can have a significant influence on a teacher's preparation to teach. Few future teachers have an idea of what understanding mathematics for teaching entails, and, unfortunately, few have sufficient opportunities during their university coursework to change this state of knowledge of mathematics. The career stages described by Day and Sachs (2004) earlier in this chapter need to have a zero stage before the first stage—it is "Preparing to become learners of teaching practice."

Unfortunately, professional developers, whether coordinators, university professors, or private providers, tend to reinvent the wheel time and time again. Yet according to Elmore (2001),

> We know enough about teacher learning, its connection to practice, and the connection of practice to student performance to design some fairly powerful interventions that would reduce variability in practice and increase the consistency with which teachers influence student learning in powerful [ways]. The main constraint is not knowledge; [it is] the *social construction of teaching*. That is, most of our theories about student learning can't be deployed because the people who teach don't believe them—in fact, actively oppose them–and the organizations in which teaching occurs—schools and school systems—are designed in ways that make it virtually impossible to use them. (p. 10)

Why do teachers not believe theories about student learning? As researchers, we have not taken seriously our job to relate research to practice, to work with teachers on classroom research that builds on and tests the theories about student learning.

Research on teachers' knowledge, beliefs, preparation, and professional development is a fairly new phenomenon. Nelson (1997) noted that the mental life of teachers has been a topic of study only since the mid 1970s. For many years, research on teaching used a process-product paradigm that focused on teacher behaviors rather than on teachers' thinking. The 1991 Professional Standards from NCTM was not a research document, but it provided researchers with a clear agenda for undertaking research that could confirm or disconfirm the ideas in that document, and it moved researchers beyond process-product research into new paradigms for professional development, many of which have been discussed in this chapter.

One paradigm in particular illustrates ways that professional development has evolved to encompass features quite different from the traditional workshop type of professional development (Stein et al., 1999). These features (summarized from pp. 239–241) include

- *Teacher assistance embedded in or directly related to the work of teaching.* Teachers need assistance that focuses on their day-to-day efforts to teach in these new and demanding ways. Such assistance is best provided by support directly focused on an individual teacher's practice . . . or by guided group discussions surrounding selected authentic artifacts from practice, such as student work, instructional tasks, or cases of instructional practice written by and for teachers (Ball and Cohen, 1999).

- *Teacher assistance grounded in the content of teaching and learning.* A characteristic of new calls for reform focus on meeting high standards associated with the major school subjects (Little, 1993). . . . Teachers frequently need to encounter the discipline as learners themselves before grappling with how to teach it (Ball, 1991). Professional development should give them the opportunity to do so.

- *Development of teacher communities of professional practice.* The new paradigm for professional development encourages collegiality among teachers to counter the isolation typical of teaching. . . . A new set of goals has been placed under the purview of professional development, including community building; the development of teachers' capacities to explain, challenge, and critique the work of peers (Lord, 1994); and the development of teacher leaders (Lieberman, Saxl, & Miles, 1988).

- *Collaboration with experts outside the teaching community.* Teachers cannot be expected to be knowledgeable about all aspects of school reform, subject-matter standards, or professional practice. . . . Outside experts—often university-based educators—bring fresh perspectives, ideas about what has proven successful elsewhere, and an analytic stance toward the school-improvement process (Little, 1993).

- *Consideration of organizational context.* Teachers perform their work within multiple contexts, including schools, districts, communities, and states; the values and established procedures each have an impact on classroom practice. Professional developers must carefully analyze the constraints and alternatives offered by each of these contexts, ranging from the unwritten cultural norms to explicit regulations and policies.

This emerging paradigm moves from providing workshops and focusing on techniques to building teachers' capacities, including strong content knowledge and a variety of ways to build practice on children's thinking and clear learning goals. The new paradigm requires "a new conceptual grounding, one that also incorporates theories of how teachers learn within organizations and through interactions with others (Stein & Brown, 1997)" (Stein et al., 1999, p. 265).

Too little has been said here about connecting teachers with research. Some action research was discussed, but here I mean something more. How do teachers come to know about relevant research? What is the responsibility of researchers in sharing their research? What is the responsibility of the researcher in undertaking research that will be relevant to teachers? What is the responsibility of teachers in seeking out research that can help them be better teachers? What is the responsibility of teachers in sharing with researchers their issues and concerns and questions that could be addressed by researchers? What is the responsibility of researchers *and* teachers in mutually identifying and undertaking meaningful research? In a 2003 *Journal for Research in Mathematics Education* editorial, Silver used the metaphor of border crossing to describe the work of researchers and teachers—each traveling across a border to meet the other, with the values of each group as the currency to be exchanged.

> In the research community, the valued currency is theory. . . . Work that contributes to the development or refinement of theory is highly valued. In contrast, across the border in the land of educational practice, the valued currency is practical application. Work has value in this community to the extent that it can be directly applied to the improvement of some important domain of practice. (p. 183)

The metaphor of border crossing can be thought of as a cyclic activity, wherein additional knowledge is created when additional conversations are held between researchers and practitioners. Ruthven (2002) described these cycles in phases.

> In one phase of this cycle, scholarly knowledge is (re)contextualized and activated within teaching, stimulating (re)construction of craft knowledge. In the complementary phase, craft knowledge is elicited and codified through researching, stimulating (re)construction of the scholarly knowledge. In both phases, conversation involves the filtering and reformulating of knowledge: only certain derivatives of scholarly knowledge will prove capable of being productively incorporated within craft knowledge;

equally, only some derivatives of craft knowledge will prove able to be fruitfully appropriated as scholarly knowledge. (p. 595)

Situations that facilitate these conversations are needed. At times, a researcher seeks access to classrooms to undertake research. At other times, teachers approach a researcher for assistance in answering some question. These are not ideal situations—in the first, the teachers have probably had little or no input into the research; in the second, the particular question asked is not a research question. Both sides must be open and seek opportunities for conversations. At times, outside agencies bring teachers and researchers together and provide opportunities for sharing.

On a final note, in the introduction to this chapter I focused on the national need for more and better teacher education. Education, however, is a global enterprise, being shaped by government policy universally (Day & Sachs, 2004). Although different countries have different educational systems, the recognition that teachers must be well prepared is universal. At the time I began to write this final section of this chapter, I received notice of the Fifteenth International Congress on Mathematics Instruction (ICMI, 2004) Study, The Professional Education and Development of Teachers of Mathematics. The study focus was selected because "the continued development of teachers is key to students' opportunities to learn mathematics" (p. 1). The call for participants noted that the past decade "has seen substantial increase in scholarship on mathematics education and development" (p. 3), but much remains to be learned about the "knowledge, skills, personal qualities and sensibilities that teaching mathematics entails, and about how such professional resources are acquired" (p. 3). This invitation to participate provided lists of topics and questions that could serve as a research agenda throughout the world. A list of topics dealing with teacher preparation included the structure of teacher preparation, recruitment and retention, curriculum of teacher preparation, the early years of teaching, the most pressing problems of preparing teachers, and the history and change in teacher preparation. In the strand on professional development questions rather than topics are proposed:

> What sorts of learning seem to emerge from the study of practice? In what ways are practices of teaching and learning mathematics made available for study? What kinds of collaboration are practiced in different countries? What kinds of leadership help support teachers in learning from the practice of mathematics teaching? What are crucial practices of learning from

practice? How does language play a role in learning from practice? (p. 7).

The answers are often universal, and we in the United States are too often insular and not open to answers found in other countries. Thus, more than one border needs to be crossed while we, researchers, administrators, policy makers, and teachers, seek to better educate teachers for the purpose of improving student learning.

REFERENCES

Adler, J. (2000). Conceptualising resources as a theme for teacher education. *Journal of Mathematics Teacher Education, 3,* 205–224.

American Federation of Teachers. (2002). *Principles for professional development.* Washington, DC: Author.

Anderson, C. W. (2003). How can schools support teaching for understanding in mathematics and science? In A. Gamoran, C. W. Anderson, P. A. Quiroz, W. G. Secada, T. Williams, & S. Ashmann (Eds.), *Transforming teaching in math and science: How schools and districts can support change* (pp. 3–21). New York: Teachers College Press.

Arends, R., & Winitzky, N. (1996). Program structures and learning to teach. In F. B. Murray (Ed.), *The teacher educator's handbook* (pp. 526–556). San Francisco: Jossey-Bass.

Ball, D. L. (2002). Setting the stage. In H. Bass, Z. P. Usiskin, & G. Burrill. (Eds.), *Studying classroom teaching as a medium for professional development. Proceedings of a U. S.-Japan workshop* (pp. 49–52). Washington, DC: National Academy Press.

Ball, D. L. (2003). What mathematics knowledge is needed for teaching mathematics? *Secretary's Mathematics Summit,* Feb 6, 2003, Washington, DC. Retrieved March 15, 2005 from http://www.ed.gov/rschstat/research/progs/mathscience/summit.html

Ball, D. L., & Cohen, D. K., (1996). Reform by the book: What is—or might be—the role of curriculum materials in teacher learning and instructional reform? *Educational Researcher, 25*(9), 6–8.

Ball, D. L., & Cohen, D. K., (1999). Developing practice, developing practitioners: Toward a practice-based theory of professional education. In L. Darling-Hammond & G. Sykes (Eds.), *Teaching as the learning profession: Handbook of policy and practice* (pp. 3–32). San Francisco: Jossey-Bass.

Ball, D. L., Lubienski, S. T., & Mewborn, D. S. (2001). Research on teaching mathematics: The unsolved problem of teachers' mathematical knowledge. In V. Richardson (Ed.), *Handbook of research on teaching* (4th ed, pp. 433–456). Washington, DC: American Educational Research Association.

Ball, D. L, & Wilson, S. M. (1990). *Knowing the subject and learning to teach it: Examining assumptions about becoming a mathematics teacher* (Research Report 90–7). East Lansing: National Center for Research on Teacher Learning, Michigan State University.

Barnett, C. (1991). Building a case-based curriculum to enhance the pedagogical content knowledge of mathematics teachers. *Journal of Teacher Education, 42,* 263–272.

Barnett, C. (1998). Mathematics teaching cases as a catalyst for informed strategic inquiry. *Teaching and Teacher Education, 14,* 81–93.

Barnett, C., & Friedman, S. (1997). Mathematics case discussions: Nothing is sacred. In E. Fennema & B. S. Nelson (Eds.), *Mathematics teachers in transition* (pp. 381–399). Mahwah, NJ: Erlbaum.

Barnett, C., Goldenstein, D., & Jackson, B. (Eds.). (1994). *Mathematics teaching cases: Fractions, decimals, ratios, and percents: Hard to teach and hard to learn?* Portsmouth, NH: Heinemann.

Bass, H. (2001). Presidential views: Interview with Hyman Bass. *Notices of the AMS, 48,* 312–315.

Bass, H., Usiskin, Z. P., & Burrill, G. (Eds.). (2002). *Studying classroom teaching as a medium for professional development.* Proceedings of a U.S.–Japan workshop. Washington, DC: National Academy Press.

Boaler, J. (2002). Learning from teaching: Exploring the relationship between reform curriculum and equity. *Journal for Research in Mathematics Education, 33,* 239–258.

Book, C. L. (1996). Professional development schools. In J. Sikula, T. J. Buttery, & E. Guyton (Eds.), *Handbook of research on teacher education* (2nd ed., pp. 194–210). New York: Macmillan.

Borasi, R., & Fonzi, J. (1999). *Introducing math teachers to inquiry: Framework and supporting materials to design professional development.* Report prepared for the National Science Foundation for Grants TPE-9153812 & DUE-9254475. Rochester, NY: University of Rochester.

Borasi, R., & Fonzi, J. (2002). *Professional development that supports school mathematics reform.* Foundations series of monographs for professionals in science, mathematics and technology education. Arlington, VA: National Science Foundation.

Borasi, R., Fonzi, J., Smith, C. F., & Rose, B. J. (1999). Beginning the process of rethinking mathematics instruction: A professional development program. *Journal of Mathematics Teacher Education, 2,* 49–78.

Borko, H. (2004). Professional development and teacher learning: Mapping the terrain. *Educational Researcher, 33* (8), 3–15.

Borko, H., Eisenhart, M., Brown, C. A., Underhill, R. G., Jones, D., & Agard, P. C. (1992). Learning to teach hard mathematics: Do novice teachers and their instructors give up too easily? *Journal for Research in Mathematics Education, 23,* 194–222.

Borko, H., & Putnam, R. (1995). Expanding a teachers' knowledge base: A cognitive psychological perspective on professional development. In T. Guskey & M. Huberman (Eds.), *Professional development in education: New paradigms and practices* (pp. 35–65). New York: Teachers College Press.

Borko, H., & Putnam, R. T. (1996). Learning to teach. In D. C. Berliner & R. C. Calfee (Eds.), *Handbook of educational psychology* (pp. 673–708). New York: Macmillan.

Bowers, J., & Doerr, H. J. (2001). An analysis of prospective teachers' dual roles in understanding the mathematics of change: Eliciting growth with technology. *Journal of Mathematics Teacher Education, 4,* 115–137.

Brown, C. A., & Borko, H. (1992). Becoming a mathematics teacher. In D. A. Grouws (Ed.), *Becoming a mathematics teacher. Handbook of research on mathematics teaching and learning* (pp. 209–239). New York: Macmillan.

Brown, C. A., Smith, M. S., & Stein, M. K. (1996, April). *Linking teacher professional development to enhanced classroom instruction.* Paper presented at the annual meeting of the American Educational Research Association, New York.

Bruner, J. (1960/1977). *The process of education.* Cambridge, MA: Harvard University Press.

Cahnmann, M. S., & Remillard, J. T. (2002). What counts and how: Mathematics teaching in culturally, linguistically, and socioeconomically diverse urban settings. *Urban Review, 34,* 179–204.

California Department of Education. (1985). *Mathematics framework for California public schools, kindergarten through Grade 12.* Sacramento: Author.

Carpenter, T. P., & Fennema, E. (1992). Cognitively guided instruction: Building on the knowledge of students and teachers. *International Journal of Educational Research, 17,* 457–470.

Carpenter, T. P., Fennema, E., Peterson, P. L., & Carey, D. A. (1988). Teachers' pedagogical content knowledge of students' problem solving in elementary arithmetic. *Journal for Research in Mathematics Education, 19,* 385–401.

Carpenter, T. P., Fennema, E., Peterson, P. L., Chiang, C. P., & Loef, M. (1989). Using knowledge of children's mathematics thinking in classroom teaching: An experimental study. *American Educational Research Journal, 26,* 499–531.

Carpenter, T. P., & Moser, J. M. (1983). The acquisition of addition and subtraction concepts. In R. Lesh & M. Landau (Eds.), *The acquisition of mathematics concepts and processes* (pp. 7–44). New York: Academic Press.

Clark, C., Moss, P., Goering, S., Herter, R., Lamar, B., Leonard, D., et al. (1996). Collaboration as dialogue: Teachers and researchers engaged in conversation and professional development. *American Educational Research Journal, 33,* 193–231.

Clarke, D. (1994). Ten key principles from research on the professional development of mathematics teachers. In D. B. Aichele & A. F. Coxford (Eds.), *Professional development for teachers of mathematics* (pp. 37–48). Reston, VA: National Council of Teachers of Mathematics.

Cobb, P., McClain, K., Lamberg, T., & Dean, C. (2003). Situating teachers' instructional practices in the institutional setting of the school and district. *Educational Researcher, 32*(6), 13–24.

Cochran-Smith, M., & Lytle, S. (1999). Relationships of knowledge and practice: Teacher learning in communities. In A. Iran-Nejad & P. D. Pearson (Eds.), *Review of research in education* (Vol. 24, pp. 249–305). Washington, DC: American Educational Research Association.

Cohen, D. K. (1990). A revolution in one classroom: The case of Mrs. Oublier. *Educational Evaluation and Policy Analysis, 12,* 327–345.

Cohen, D. K., & Hill, H. C. (2001). *Learning policy: When state education reform works.* New Haven, CT: Yale University Press.

Cohen, S. (2004). *Teachers' professional development and the elementary mathematics classroom: Bringing understandings to light.* Mahwah, NJ: Erlbaum.

Conference Board of the Mathematical Sciences (CBMS). (2000). *The mathematical education of teachers.* Washington, DC: Mathematical Association of America.

Cooney, T. J. (1994). On the application of science to teaching and teacher education. In R. Biehler, R. W. Scholz, R. Sträßer, & B. Winkelmann (Eds.), *Didactics of mathematics as a scientific discipline* (pp. 103–116). Dordrecht, The Netherlands: Kluwer.

Cooney, T. J. (2001). Considering the paradoxes, perils, and purposes of conceptualizing teacher development. In F. L. Lin & T. Cooney (Eds.), *Making sense of mathematics teacher education* (pp. 9–31). Dordrecht, The Netherlands: Kluwer.

Coxford, A.F., Fey, J.T., Hirsch, C.R., Schoen, H.L., Burrill, G., Hart, E.W., et al. (2003). *Contemporary mathematics in context: A unified approach.* Columbus, OH: Glencoe/McGraw-Hill.

Crawford, K., & Adler, J. (1996). Teachers as researchers in mathematics education. In A. Bishop, K. Clements, C. Keitel, J. Kilpatrick, & C. Laborde (Eds.), *International handbook of mathematics education* (Vol. 2, pp. 1187–1205). Dordrecht, The Netherlands: Kluwer.

Darling-Hammond, L. (1990). Instructional policy into practice: "The power of the bottom over the top." *Educational Evaluation and Policy Analysis, 12,* 233–241.

Darling-Hammond, L. (1994a). Developing professional development schools: Early lessons, challenge, and promise. In L. Darling-Hammond (Ed.), *Professional development schools: Schools for developing a profession* (pp. 1–27). New York: Teachers College Press.

Darling-Hammond, L. (Ed.). (1994b). *Professional development schools: Schools for developing a profession.* New York: Teachers College Press.

Darling-Hammond, L. (1998). Teacher learning that supports student learning. *Educational Leadership, 55,* 6–11.

Darling-Hammond, L., Berry, B. T., Haselkorn, D., & Fideler, E. (1999). Teacher recruitment, selection, and induction. In L. Darling-Hammond & G. Sykes (Eds.), *Teaching as the learning profession: Handbook of policy and practice* (pp. 183–232). San Francisco: Jossey-Bass.

Darling-Hammond, L., & McLaughlin, M. W. (1996). Policies that support professional development in an era of reform. In M. W. McLaughlin & I. Oberman (Eds.), *Teacher learning: New policies, new practices* (pp. 202–218). New York: Teachers College Press.

Day, C., & Sachs, J. (2004). Professionalism, performativity, and empowerment: Discourses in the politics, policies, and purposes of continuing professional development. In C. Day & J. Sachs (Eds.), *International handbook on the continuing professional development of teachers* (pp. 3–32). Berkshire, England: Open University Press.

Delpit, L. (1995). *Other people's children: Cultural conflict in the classroom.* New York: New Press.

Dewey, J. (1902). *The child and the curriculum.* Chicago: University of Chicago Press.

Dewey, J. (1904). The relation of theory to practice in education. In C. Murray (Ed.), *The third NSSE (The National Society for the Study of Education) yearbook: Part I* (pp. 15–25). Chicago: University of Chicago Press.

Dewey, J. (1910/1997). *How we think.* Boston: Heath.

Dewey, J. (1916/1944). *Democracy and education.* New York: The Free Press.

Dewey, J. (1920). *Reconstruction in philosophy.* New York: Holt.

Drake, C., Spillane, J. P., & Hufferd-Ackles, K. (2001). Storied identities: Teacher learning and subject-matter context. *Journal of Curriculum Studies, 33*(1), 1–23.

Driscoll, M. (1999). *Fostering algebraic thinking: A guide for teachers Grades 6–10.* Portsmouth, NH: Heinemann.

Ebby, C. B. (2000). Learning to teach mathematics differently: The interaction between coursework and fieldwork for preservice teachers. *Journal of Mathematics Teacher Education, 3,* 69–97.

Eisenhart, M., Borko, H., Underhill, R., Brown, C., Jones, D., & Agard, P. (1993). Conceptual knowledge falls through the cracks: Complexities of learning to teach mathematics for understanding. *Journal for Research in Mathematics Education, 24,* 8–40.

Eisenhower National Clearinghouse (ENC) for Mathematics and Science Education. (2000). *Ideas that work: Mathematics professional development.* Columbus, OH: Author.

Elmore, R. F. (2001, April). *Content-focused professional development: An issue of policy and practice in large-scale school reform.* Invited presentation to the Special Interest Group on Research in Mathematics Education at the annual meeting of the American Educational Research Association, Seattle, WA.

Elmore, R. F. (2002a). *Bridging the gap between standards and achievement: The imperative for professional development in education.* Washington, DC: Albert Shanker Institute.

Elmore, R. F. (2002b). Testing trap. *Harvard Magazine, 105,* [September–October], retrieved March 15, 2004, from http://www.harvardmag.com/on-line/0902140.html

Elmore, R. F., & Burney, D. (1999). Investing in teacher learning. In L. Darling-Hammond & G. Sykes (Eds.), *Teaching as the learning profession: Handbook of policy and practice* (pp. 263–291). San Francisco: Jossey-Bass.

Fennema, E., Carpenter, T. P., Franke, M. L., & Carey, D. A. (1992). Learnng to use children's mathematical thinking: A case study. In R. Davis & C. Maher (Eds.), *Schools, mathematics, and the world of reality* (pp. 93–117). Needham Height, MA: Allyn & Bacon.

Fennema, E., Carpenter, T. P., Franke, M. L., Levi, M., Jacobs, V., & Empson, S. (1996). A longitudinal study of learning to use children's thinking in mathematics instruction. *Journal for Research in Mathematics Education, 27,* 403–434.

Ferrini-Mundy, J. (1997). Reform efforts in mathematics education: Reckoning with the realities. In S. N. Friel & G. W. Bright (Eds.), *Reflecting on our work: NSF teacher enhancement in K–6 mathematics* (pp. 113–132). Lanham, MD: University Press of America.

Ferrini-Mundy, J. (2001). Where are we? Moderator's summary. In the National Research Council's *Knowing and learning mathematics for teaching* (pp. 125–126). Washington, DC: National Academy Press.

Fey, J. T. , Fitzgerald, W. M., Friel, S. N., Lappan, G, & Phillips, E. D. (2006). *Connected Mathematics 2.* Englewood Cliffs, NJ: Prentice Hall.

Fonzi, J. (2003). *Understanding the initial stages of initiating lead teacher positions in mathematics.* Unpublished doctoral dissertation, University of Rochester, NY.

Franke, M. L., Carpenter, T., Fennema, E., Ansell, E., & Behrend, J. (1998). Understanding teachers' self-sustaining change in the context of professional development. *Teaching and Teacher Education, 14*(1), 67–80.

Franke, M. L., Carpenter, T. P., Levi, L., & Fennema, E. (2001). Capturing teachers' generative change: A follow-up study of professional development in mathematics. *American Educational Research Journal, 38,* 653–689.

Franke, M. L, Fennema, E., & Carpenter, T. P. (1997). Teachers creating change: Examining evolving beliefs and classroom practice. In E. Fennema & B. S. Nelson (Eds.), *Mathematics teachers in transition* (pp. 255–282). Mahwah, NJ: Erlbaum.

Friel, S. N., & Bright, G. W. (1997). Common components and guiding principles for teacher enhancement programs. In S. N. Friel & G. W. Bright (Eds.), *Reflecting on our work: NSF teacher enhancement in K–6 mathematics* (pp. 7–21). New York: University Press.

Fullan, M. G. (1993). *Change forces: Probing the depths of educational reform.* New York: Falmer Press.

Gallucci, C. (2003). *Theorizing about responses to reform: The role of communities of practice in teacher learning.* Seattle: Center for the Study of Teaching and Policy, University of Washington.

Gamoran, A. (2003). Access to resources. In A. Gamoran, C. W. Anderson, P. A. Quiroz, W. G. Secada, T. Williams, & S. Ashmann (Eds.), *Transforming teaching in math and science: How schools and districts can support change* (pp. 65–86). New York: Teachers College Press.

Gamoran, A., Anderson, C. W., & Ashmann, S. (2003). Leadership for change. In A. Gamoran, C. W. Anderson, P. A. Quiroz, W. G. Secada, T. Williams, & S. Ashmann (Eds.), *Transforming teaching in math and science: How schools and districts can support change* (pp. 105–126). New York: Teachers College Press.

Garet, M. S., Porter, A. C., Desimone, L., Birman, B. F., Yoon, K. S. (2001). What makes professional development effective? Results from a national sample of teachers. *American Educational Research Journal, 38,* 915–945.

Gearhart, M., & Saxe, G. B. (2004). When teachers know what students know: Integrating assessment in elementary mathematics teaching. *Theory Into Practice, 43,* 304–313.

Goldsmith, L., & Schifter, D. (1997). Understanding teachers in transition: Characteristics of a model for developing teachers. In E. Fennema & B. S. Nelson (Eds.), *Mathematics teachers in transition* (pp. 19–54). Mahwah, NJ: Erlbaum.

Grossman, P. L. (1990). *The making of a teacher: Teacher knowledge and teacher education.* New York: Teachers College Press.

Grossman, P., Wineburg, S., & Woolworth, S. (2001). Toward a theory of teacher community. *Teachers College Record, 103,* 942–1012.

Groundwater-Smith, S., & Dadds, M. (2004). Critical practitioner inquiry: Towards responsible professional communities of practice. In C. Day & J. Sachs (Eds.), *International handbook on the continuing professional development of teachers* (pp. 238–263). Berkshire, England: Open University Press.

Grouws, D. A., & Schultz, K. A. (1996). Mathematics teacher education. In J. Sikula, T. J. Buttery, & E. Guyton (Eds.), *Handbook of research on teacher education* (2nd ed., pp. 442–458). New York: Macmillan.

Guskey, T. R. (1989). Attitude and perceptual change in teachers. *International Journal of Educational Research, 13,* 439–453.

Guskey, T. R. (1995). Professional development in education. In T. R. Guskey & M. Huberman (Eds.), *Professional development in education: New paradigms and practices,* (pp. 114–131). New York: Teachers College Press.

Guskey, T. R. (2000). *Evaluating professional development.* Thousand Oaks, CA: Corwin Press.

Hammer, D., & Schifter, D. (2001). Practices of inquiry in teaching and research. *Cognition and Instruction, 19,* 441–478.

Hargreaves, A. (1995). Development and desire: A postmodern perspective. In T. R. Guskey & M. Huberman (Eds.), *Professional development in education: New paradigms and practices* (pp. 9–34). New York: Teachers College Press.

Haselkorn, D., & Harris, L. (1998). *The essential profession: American education at the crossroads.* Belmont, MA: Recruiting New Teachers.

Hawley, W. D., & Valli, L. (1999). The essentials of effective professional development: A new consensus. In L. Darling-Hammond & G. Sykes (Eds.), *Teaching as the learning profession: Handbook of policy and practice* (pp. 127–150). San Francisco: Jossey-Bass.

Hiebert, J., Carpenter, T. P., Fennema, E., Fuson, K. C., Wearne, D., Murray, et al. (1997). *Making sense: Teaching and learning mathematics with understanding*. Portsmouth, NH: Heinemann.

Hiebert, J., Gallimore, R., & Stigler, J. W. (2002). A knowledge base for the teaching profession: What would it look like and how can we get one? *Educational Researcher, 31*(5), 3–15.

Hiebert, J., Morris, A. K., & Glass, G. (2003). Learning to learn to teach: An "experiment" model for teaching and teacher preparation in mathematics. *Journal of Mathematics Teacher Education, 6*, 201–222.

Hiebert, J., & Stigler, J. W. (2000). A proposal for improving classroom teaching: Lessons from the TIMSS video study. *Elementary School Journal, 101*(1), 3–20.

Hill, H. C., & Ball, D. L. (2004). Learning mathematics for teaching: Results from California's Mathematics Professional Development Institutes. *Journal for Research in Mathematics Education, 35*, 330–351.

Holly, P. (1991). Action research: The missing link in the creation of schools as the centers of inquiry. In A. Lieberman & L. Miller (Eds.), *Staff development for education in the '90s: New demands, new realities, new perspectives* (pp. 133–157). New York: Teachers College Press.

Holmes Group. (1986). *Tomorrow's teachers*. Retrieved March 7, 2005, from http://www.holmespartnership.org/Tomorrows _Teachers.pdf

Holmes Group. (1990). *Tomorrow's schools: Principles for the design of professional development schools*. Retrieved March 7, 2005, from http://www.holmespartnership.org/Tomorrows_Schools .pdf

Holmes Group. (1995). *Tomorrow's schools of education: A report of the Holmes Group*. Retrieved March 7, 2005, from http://www.holmespartnership.org/Tomorrows_Schools %20of%20Education.pdf

Howey, K. (1996). Designing coherent and effective teacher preparation programs. In J. Sikula, T. J. Buttery, & E. Guyton (Eds.), *Handbook of research on teacher education* (2nd ed., pp. 143–170). New York: Macmillan.

International Commission on Mathematics Instruction (ICMI). (2004). *The fifteenth ICMI study: The professional education and development of teachers of mathematics*. Retrieved March 15, 2005 from http://www.ams.org/mathcal/info/2005_ may15-21_s.html

Jacobs, V. R., & Raynes, D. (2001, April). *Teacher-directed school-based professional development in elementary mathematics*. Paper presented at the annual meeting of the American Educational Research Association, Seattle, WA.

Jacobs, V. R., & Raynes, D. (2002, April). *The importance of teacher—teacher talk in the adoption of professional development ideas in elementary mathematics*. Paper presented at the annual meeting of the American Educational Research Association, New Orleans, LA.

Jacobson, C., & Lehrer, R. (2000). Teacher appropriation and student learning of geometry through design. *Journal for Research in Mathematics Education, 31*, 71–88.

Jaworski, B. (1998). Mathematics teacher research: Process, practice, and the development of teaching. *Journal of Mathematics Teacher Education, 1*, 3–31.

Jaworski, B., & Watson, A. (1994). *Mentoring in mathematics teaching*. London: Falmer Press.

Kennedy, M. (1997). *Defining optimal knowledge for teaching science and mathematics (Research Monograph 10)*. Madison: National Institute for Science Education, University of Wisconsin.

Kennedy, M. (1999). The role of preservice teacher education. In L. Darling-Hammond & G. Sykes (Eds.), *Teaching as the learning profession: Handbook of policy and practice* (pp. 54–85). San Francisco: Jossey-Bass.

Knapp, M. S. (2003). Professional development as a policy pathway. *Review of Research in Education, 27*, 109–157.

Knapp, M. S., Adelman, N. E., Marder, C., McCollum, H., Needels, M. C., Padilla, C., et al. (1995). *Teaching for meaning in high-poverty classrooms*. New York: Teachers College Press.

Knight, P. (2002). A systemic approach to professional development: Learning as practice. *Teaching and Teacher Education, 18*, 229–241.

Koch, L. C. (1997). The growing pains of change: A case study of a third-grade teacher. In J. Ferrini-Mundy & T. Schram (Eds.), *The recognizing and recording reform in mathematics education project: Insights, issues, and implication. Journal for Research in Mathematics Education, Monograph Series, No. 8* (pp. 87–109). Reston, VA: National Council of Teachers of Mathematics.

Ladson-Billings, G. (1994). *The dreamkeepers: Successful teachers of African American children*. San Francisco, Jossey-Bass.

Ladson-Billings, G. (1999a). Preparing teachers for diverse student populations: A critical race-theory perspective. In A. Iran-Nejad & P. D. Pearson (Eds.), *Review of research in education* (Vol. 24, pp. 211–247). Washington, DC: American Educational Research Association.

Ladson-Billings, G. (1999b). Preparing teachers for diversity: Historical perspectives, current trends, and future directions. In L. Darling-Hammond & G. Sykes (Eds.), *Teaching as the learning profession: Handbook of policy and practice* (pp. 86–123). San Francisco: Jossey-Bass.

Lampert, M., & Ball, D. L. (1998). *Teaching, multimedia, and mathematics: Investigations of real practice*. New York: Teachers College Press.

Lampert, M., & Ball, D. L. (1999). Aligning teacher education with contemporary K–12 reform visions. In L. Darling-Hammond & G. Sykes (Eds.), *Teaching as the learning profession: Handbook of policy and practice* (pp. 33–53). San Francisco: Jossey-Bass.

Lanier, J. T. (1994). Foreword. In L. Darling-Hammond (Ed.), *Professional development schools: Schools for developing a profession* (pp. ix–xii). New York: Teachers College Press.

Lave, J., & Wenger, E. (1991). *Situated learning: Legitimate peripheral participation*. Cambridge, England: Cambridge University Press.

Learning First Alliance. (1998). *Every child mathematically proficient: An action plan of the Learning First Alliance*. Washington, DC: Author.

Lee, O., & Yarger, S. J. (1996). Modes of inquiry in research on teacher education. In J. Sikula, T. J. Buttery, & E. Guyton (Eds.), *Handbook of research on teacher education* (2nd ed., pp. 14–37). New York: Macmillan.

LeFevre, D. M. (2002, April). *Video records of practice as a tool for learning practice: What is to be learned?* Paper presented at the annual meeting of the American Educational Research Association, New Orleans, LA.

Leitzel, J. R. C. (Ed.). (1991). *A call for change: Recommendations for the mathematical preparation of teachers of mathematics.* Washington, DC: Mathematical Association of America.

Lerman, S. (2001). A review of research perspectives on mathematics teacher education. In F. L. Lin & T. J. Cooney (Eds.), *Making sense of mathematics teacher education* (pp. 33–52). Dordrecht, The Netherlands: Kluwer.

Levin, B. B. (1999). The role of the facilitator in case discussions. In M. A. Lundeberg, B. B. Levin, & H. L. Harrington (Eds.), *Who learns what from cases and how? The research base for teaching and learning with cases* (pp. 101–115). Mahwah, NJ: Erlbaum.

Lewis, C. C. (2000, April). *Lesson study: The core of Japanese professional development.* Invited presentation to the Special Interest Group on Research in Mathematics Education at the annual meeting of the American Educational Research Association, New Orleans, LA.

Lewis, C. C. (2002). Does lesson study have a future in the United States? *Nagoya Journal of Education and Human Development, 2002,* 1–23.

Lewis, C. C., & Tsuchida, I. (1998). A lesson is like a swiftly flowing river: How research lessons improve Japanese education. *American Educator, 22*(Winter), 12–17 & 50–52.

Lin, P. (2002). On enhancing teachers' knowledge by constructing cases in classrooms. *Journal of Mathematics Teacher Education, 5,* 317–349.

Linn, R. L., Baker, E. L., & Betebenner, D. W. (2002). Accountability systems: Implications of requirements of the No Child Left Behind Act of 2001. *Educational Researcher, 31*(6), 3–16.

Little, J. W. (1993). Teachers' professional development in a climate of educational reform. *Educational Evaluation and Policy Analysis, 15*(2), 129–151.

Little, J. W. (2004). 'Looking at student work' in the United States: A case of competing impulses in professional development. In C. Day & J. Sachs (Eds.), *International handbook on the continuing professional development of teachers* (pp. 94–118). Berkshire, England: Open University Press.

Lloyd, G. (2002). Mathematics teachers' beliefs and experiences with innovative curriculum materials. In G. Leder, E. Pehkonen, & G. Toerner (Eds.), *Beliefs: A hidden variable in mathematics education* (pp. 1–11). Dordrecht, The Netherlands: Kluwer.

Lord, B. (1994). Teachers' professional development: Critical colleagueship and the role of professional communities. In N. Cobb (Ed.), *The future of education: Perspectives on national standards in America* (pp. 175–204). New York: College Board.

Loucks-Horsley, S., Love, N., Stiles, K. E., Mundry, S., & Hewson, P. W. (2003). *Designing professional development for teachers of science and mathematics.* Thousand Oaks, CA: Corwin Press.

Loucks-Horsley, S., & Stiegelbauer, S. (1991). Using knowledge of change to guide staff development. In A. Lieberman & L. Miller (Eds.), *Staff development for education in the '90s: New demands, new realities, new perspectives* (pp. 15–36). New York: Teachers College Press.

Ma, L. (1999). *Knowing and teaching elementary mathematics: Teachers' understanding of fundamental mathematics in China and the United States.* Mahwah, NJ: Erlbaum.

Masingila, J. O., & Doerr, H. M. (2002). Understanding pre-service teachers' emerging practices through their analyses of a multimedia case study of practice. *Journal of Mathematics Teacher Education, 5,* 235–263.

Mathematical Association of America (MAA). (2003). *Undergraduate programs and courses in the mathematical sciences.* Washington, DC: Author.

Merseth, K. (1996). Cases and case methods in education. In J. Sikula (Ed.), *Handbook of research on teacher education* (2nd ed., pp. 722–744). New York: Macmillan.

Merseth, K. (2003). *Windows on teaching math: Cases of middle and secondary classrooms.* New York: Teachers College Press.

Mewborn, D. S. (1999). Reflective thinking among preservice elementary mathematics teachers. *Journal for Research in Mathematics Education, 30,* 316–341.

Mewborn, D. S. (2000). Learning to teach elementary mathematics: Ecological elements of a field experience. *Journal of Mathematics Teacher Education, 3,* 27–46.

Mewborn, D. S. (2003). Teaching, teachers' knowledge, and their professional development. In J. Kilpatrick, W. G. Martin, & D. Schifter (Eds.), *A research companion to Principles and Standards for School Mathematics* (pp. 45–52). Reston, VA: National Council of Teachers of Mathematics.

Middleton, J., Sawada, D., Judson, E., Bloom, I., & Turley, J. (2002). Relationships build reform: Treating classroom research as emergent systems. In L. English (Ed.), *Handbook of international research in mathematics education* (pp. 409–431). Mahwah, NJ: Erlbaum.

Miles, M. G. (1995). Foreword. In T. R Guskey & M. Huberman (Eds.), *Professional development in education* (pp. vii–ix). New York: Teachers College Press.

Mumme, J., & Seago, N. (2002, April). *Issues and challenges in facilitating videocases for mathematics professional development.* Paper presented at the annual meeting of the American Educational Research Association, New Orleans, LA.

Munby, H., Russell, T., & Martin, A. K. (2001). Teachers' knowledge and how it develops. In V. Richardson (Ed.), *Handbook of research on teaching* (4th ed., pp. 877–904). Washington, DC: American Educational Research Association.

National Commission on Teaching and America's Future. (1996). *What matters most: Teaching for America's future.* New York: Author.

National Commission on Teaching and America's Future. (2003). *No dream denied: Pledge to America's children.* New York: Author.

National Council of Teachers of Mathematics. (1989). *Curriculum and evaluation standards for school mathematics.* Reston, VA: Author.

National Council of Teachers of Mathematics. (1991). *Professional standards for teaching mathematics.* Reston, VA: Author.

National Council of Teachers of Mathematics. (1995). *Assessment standards for school mathematics.* Reston, VA: Author.

National Council of Teachers of Mathematics. (2000). *Principles and standards for school mathematics.* Reston, VA: Author.

National Research Council (NRC). (1989). *Everybody counts: A report to the nation on the future of mathematics.* Washington, DC: National Academy Press.

National Research Council (NRC). (2001a). *Adding it up: Helping children learn mathematics.* J. Kilpatrick, J. Swafford, & B. Findell (Eds.), Mathematics Learning Study Committee, Center for Education, Division of Behavioral and Social Sciences and Education. Washington, DC: National Academy Press.

National Research Council, Committee on Science and Mathematics Teacher Preparation. (2001b). *Educating teachers of science, mathematics, and technology: New practices for the new millennium.* Washington, DC: National Academy Press.

Nelson, B. S. (1997). Learning about teacher change in the context of mathematics reform: Where have we come from? In E. Fennema & B. S. Nelson (Eds.), *Mathematics teachers in transition* (pp. 3–15). Mahwah, NJ: Erlbaum.

Nelson, B. S. (1998). Lenses on learning: Administrators' views on reform and the professional development of teachers. *Journal of Mathematics Teacher Education, 1,* 191–215.

Nelson, B. S., & Hammerman, J. K. (1996). Reconceptualizing teaching: Moving toward the creation of intellectual communities of students, teachers, and teacher educators. In M. W. McLaughlin & I. Oberman (Eds.), *Teacher learning: New policies, new practices* (pp. 3–21). New York: Teachers College Press.

Nickerson, S. D., & Moriarty, G. (2005). Professional communities in the context of teachers' professional lives: A case of mathematics specialists. *Journal of Mathematics Teacher Education, 8,* 113–140.

Nicol, C. (1998). Learning to teach mathematics: Questioning, listening, and responding. *Educational Studies in Mathematics, 37,* 45–66.

Oldham, E. E., van der Valk, A. E., Broekman, H. G. B., & Berenson, S. B. (1998). Preservice teachers' ideas about teaching mathematics: Contributions to frameworks for teacher education courses. In S. Berenson, K. Dawkins, M. Blanton, W. Coulombe, J. Kolb, K. Norwood, et al. (Eds.), *Proceedings of the twentieth annual meeting of the North American Chapter of the International Group for the Psychology of Mathematics Education* (Vol. 2, pp. 710–716). Columbus, OH: ERIC.

Philipp, R. A., Ambrose, R., Clement, L. L., Sowder, J. T., Schappelle, B. P., Sowder, L., et al. (in press). *The effects of early field experiences on the mathematical content knowledge and beliefs of prospective elementary school teachers: An experimental study.* Manuscript submitted for publication.

Philipp, R. A., Armstrong, B. E., & Bezuk, N. S. (1993). A preservice teacher learning to teach mathematics in a cognitively guided manner. In J. Becker & B. Pence (Eds.), *Proceedings of the fifteenth annual meeting of the PME-NA* (Vol. 2, pp. 159–165). Pacific Grove, CA: San Jose State University.

Philipp, R., Cabral, C., & Schappelle, B. (2004). *Integrating mathematics and pedagogy to illustrate children's reasoning.* Upper Saddle River, NJ: Prentice Hall.

Philipp, R., Cabral, C., & Schappelle, B. (2005). *Searchable IMAP video collection: Children's mathematical thinking clips.* San Diego, CA: Center for Research in Mathematics and Science Education, San Diego State University.

Philipp, R. A., Flores, A., Sowder, J. T., & Schappelle, B. P. (1994). Conceptions and practices of extraordinary mathematics teachers. *Journal of Mathematical Behavior, 13,* 155–180.

Ponte, J. P. (1994). Mathematics teachers' professional development. In J. P. da Ponte & J. F. Matos (Eds.), *Proceedings of the eighteenth international conference for the Psychology of Mathematics Education* (Vol. 1, pp. 195–210). Lisbon, Portugal: University of Lisbon.

Ponte, J. P. (2001). Investigating mathematics and learning to teach mathematics. In F. L. Lin & T. Cooney (Eds.), *Making sense of mathematics teacher education* (pp. 53–72). Dordrecht, The Netherlands: Kluwer.

Pugach, M. C., & Johnson, L. J. (1990). Developing reflective practice through structured dialogue. In R. T. Clift, W.

R. Houston, & M. C. Pugach (Eds.), *Encouraging reflective practice in education: An analysis of issues and programs* (pp. 186–207). New York: Teachers College Press.

Quiroz, P. A. & Secada, W. G. (2003). Responding to diversity. In A. Gamoran, C. W. Anderson, P. A. Quiroz, W. G. Secada, T. Williams, & S. Ashmann (Eds.), *Transforming teaching in math and science: How schools and districts can support change* (pp. 87–104). New York: Teachers College Press.

RAND Mathematics Study Panel. (2003). *Mathematical proficiency for all students.* Santa Monica, CA: RAND.

Remillard, J. T. (1999). Curriculum materials in mathematics education reform: A framework for examining teachers' curriculum development. *Curriculum Inquiry 29*(3). 315–342.

Remillard, J. T. (2000). Can curriculum materials support teachers' learning? Two fourth-grade teachers' use of a new mathematics text. *The Elementary School Journal, 100,* 331–350.

Remillard, J. T., & Bryans, M. B. (2004). Teachers' orientations toward mathematics curriculum materials: Implications for teacher learning. *Journal for Research in Mathematics Education, 35,* 352–388.

Remillard, J. T., & Geist, P. K. (2002). Supporting teachers' professional learning by navigating openings in the curriculum. *Journal of Mathematics Teacher Education, 5,* 7–34.

Richardson, V., & Kille, R. S. (1999). Learning from videocases. In M. A. Lundeberg, B. B. Levin, & H. L. Harrington (Eds.), *Who learns what from cases and how? The research base for teaching and learning with cases* (pp. 121–136). Mahwah, NJ: Erlbaum.

Richardson, V., & Placier, P. (2001). Teacher change. In V. Richardson (Ed.), *Handbook of research on teaching* (4th ed., pp. 905–947). Washington, DC: American Educational Research Association.

Richert, A. E. (1991). Using teacher cases or reflection and enhanced understanding. In A. Lieberman & L. Miller (Eds.), *Staff development for education in the '90s: New demands, new realities, new perspectives* (pp. 113–132). New York: Teachers College Press.

Rogers, E. M. (1995). *Diffusions of innovations* (4th ed.). New York: The Free Press.

Romberg, T. A., & Carpenter, T. P. (1986). Research on teaching and learning mathematics: Two disciplines of scientific inquiry. In M. C. Wittrock (Ed.), *Handbook of research on teaching* (3rd ed., pp. 850–873.) New York: Macmillan.

Russell, S. J. (1997). The role of curriculum in teacher development. In S. N. Friel & G. W. Bright (Eds.), *Reflecting on our work: NSF teacher enhancement in K–6 mathematics* (pp. 247–254). New York: University Press.

Russell, S. J., Tierney, C., Mokros, J., & Economopoulos, K. (2004). *Investigations in number, data, and space.* Glenview, IL: Pearson Scott Foresman.

Ruthven, K. (2002). Linking researching with teaching: Towards synergy of scholarly and craft knowledge. In L. D. English (Ed.), *Handbook of international research in mathematics education* (pp. 581–598). Mahwah, NJ: Erlbaum.

Saxe, G. B., Gearhart, M., & Nasir, N. S. (2001). Enhancing students' understanding of mathematics: A study of three contrasting approaches to professional support. *Journal of Mathematics Teacher Education, 4,* 55–79.

Schifter, D. (1995a). Teachers' changing conceptions of the nature of mathematics: Enactment in the classroom. In B. S. Nelson (Ed.), *Inquiry and the development of teaching: Issues in the transformation of mathematics teaching* (pp. 17–25). [Center for the Development of Teaching Paper Series.] Newton, MA: Center for the Development of Teaching, Educational Development Center.

Schifter, D. (1995b). *What's happening in math class?* New York: Teachers College Press.

Schifter, D. (1998). Learning mathematics for teaching: From a teachers' seminar to the classroom. *Journal of Mathematics Teacher Education, 1,* 55–87.

Schifter, D. (2001). Learning to see the invisible: What skills and knowledge are needed to engage with students' mathematical ideas? In T. Wood, B. S., Nelson, & J. Warfield (Eds.), *Beyond classical pedagogy: Teaching elementary school mathematics* (pp. 109–134). Mahwah, NJ: Erlbaum.

Schifter, D., & Fosnot, C. T. (1993). *Reconstructing mathematics education: Stories of teachers meeting the challenge of reform.* New York: Teachers College Press.

Schifter, D., & Lester, J. B. (2002, April). *Active facilitation: What do specialists need to know and how might they learn it?* Paper presented at the annual meeting of the American Educational Research Association, New Orleans, LA.

Schifter, D., Russell, S. J., & Bastable, V. (1999). Teaching to the big ideas. In M. S. Solomon (Ed.), *The diagnostic teacher: Constructing new approaches to professional development* (pp. 22–47). New York: Teachers College Press.

Schoen, H. L., Cebulla, K. J., Finn, K. F., & Fi, C. (2003). Teacher variables that relate to student achievement when using a Standards-based curriculum. *Journal for Research in Mathematics Education, 34,* 228–259.

Schön, D. (1983). *The reflective practitioner: How professionals think in action.* New York: Basic Books.

Schön, D. (1987). *Educating the reflective practitioner: Toward a new design of teaching and learning in the professions.* San Francisco: Jossey-Bass.

Seago, N., & Mumme, J. (2002, April). *The promises and challenges of using video as a tool for teacher learning.* Paper presented at the annual meeting of the American Educational Research Association, New Orleans, LA.

Seago, N., Mumme, J., & Branca, N. (2004). *Learning and teaching linear function: Video cases for mathematics professional development.* Portsmouth, NH: Heinemann.

Secada, W. G. (1989). Agenda setting, enlightened self-interest, and equity in mathematics education. *Peabody Journal of Education, 66*(Winter), 22–56.

Secada, W. G., & Adajian, L. B. (1997). Mathematics teachers' change in the context of their professional communities. In E. Fennema & B. S. Nelson (Eds.), *Mathematics teachers in transition* (pp. 193–219). Mahwah, NJ: Erlbaum.

Shimizu, Y. (2002). Lesson study: What, why, and how? In H. Bass, Z. P. Usiskin, & G. Burrill. (Eds.), *Studying classroom teaching as a medium for professional development. Proceedings of a U. S.-Japan workshop* (pp. 53–57). Washington, DC: National Academy Press.

Shulman, L. S. (1986). Those who understand: Knowledge growth in teaching. *Educational Researcher, 15*(2), 4–14.

Silver, E. A. (2003). Border crossing: Relating research and practice in mathematics education. *Journal for Research in Mathematics Education, 34,* 182–184.

Silver, E. A., Smith, M. S., & Nelson, B. S. (1995). The QUASAR project: Equity concerns meet mathematics reform in the middle school. In W. G. Secada, E. Fennema, & L. B. Adajian (Eds.), *New directions in equity in mathematics education* (pp. 9–56). New York: Cambridge University Press.

Simon, M. (1997). Developing new models of mathematics teaching: An imperative for research on mathematics teacher development. In E. Fennema & B. S. Nelson (Eds.), *Mathematics teachers in transition* (pp. 55–86). Mahwah, NJ: Erlbaum.

Simon, M. A. (2001). Two intertwined bodies of work: Conducting research on mathematics teacher development and elaborating theory of mathematics teaching/learning. In T. Wood, B. S., Nelson, & J. Warfield (Eds.), *Beyond classical pedagogy: Teaching elementary school mathematics* (pp. 157–169). Mahwah, NJ: Erlbaum.

Sleeter, C. E. (1997). Mathematics, multicultural education, and professional development. *Journal for Research in Mathematics Education, 28,* 680–696.

Sowder, J., Armstrong, B., Lamon, S., Simon, M., Sowder, L., & Thompson, A. (1998). Educating teachers to teach multiplicative structures in the middle grades. *Journal of Mathematics Teacher Education, 1,* 127–155.

Sowder, J. T., Philipp, R. A., Armstrong, B. E., & Schappelle, B. P. (1998). *Middle-grade teachers' mathematical knowledge and its relationship to instruction: A research monograph.* Albany, NY: SUNY Press.

Sowder, J. T., & Schappelle, B. P. (Eds.). (1995). *Providing a foundation for teaching mathematics in the middle grades.* Albany, NY: SUNY Press.

Spencer, D. S., Moon, J., Miller, B., & Elko, S. (2000). *Teacher leadership in mathematics and science: Casebook and facilitator's guide.* Portsmouth, NH: Heinemann.

Spillane, J. P. (2000). Cognition and policy implementation: District policymakers and the reform of mathematics education. *Cognition and Instruction, 18,* 141–179.

Spiro, R. J. (1993, April). *Theoretical issues in situation-sensitive construction of understanding.* Paper presented at the annual meeting of the American Educational Research Association, Atlanta, GA.

Stein, M. K., & Brown, C. (1997). Teacher learning in a social context: Integrating collaborative and institutional processes with the study of teacher change. In E. Fennema & B. S. Nelson (Eds.), *Mathematics teachers in transition* (pp. 155–191). Mahwah, NJ: Erlbaum.

Stein, M. K., Smith, M. S., Henningsen, M. A., & Silver, E. A. (2000). *Implementing standards-based mathematics instruction: A casebook for professional development.* New York: Teachers College Press.

Stein, M. K., Smith, M. S., & Silver, E. A. (1999). The development of professional developers: Learning to assist teachers in new settings in new ways. *Harvard Educational Review, 69,* 237–269.

Stigler, J. W., & Hiebert, J. (1997, September). Understanding and improving classroom mathematics instruction: An overview of the TIMSS video study. *Phi Delta Kappan, 79,* 14–21.

Sugrue, C. (2004). Rhetorics and realities of CPD across Europe: From cacophony towards coherence? In C. Day & J. Sachs (Eds.), *International handbook on the continuing professional development of teachers* (pp. 67–93). Berkshire, England: Open University Press.

Sullivan, P., & Mousley, J. (2001). Thinking teaching: Seeing mathematics teachers as active decision makers. In F. L. Lin & T. Cooney (Eds.), *Making sense of mathematics teacher education* (pp. 147–163). Dordrecht, The Netherlands: Kluwer.

Sykes, G. (1999). Teacher and student learning: Strengthening the connection. In L. Darling-Hammond & G. Sykes

(Eds.), *Teaching as the learning profession: Handbook of policy and practice* (pp. 151–179). San Francisco: Jossey-Bass.

Tate, W. F. (1994). Diversity, reform, and professional knowledge: The need for multicultural clarity. In D. B. Aichele & A. F. Coxford (Eds.), *Professional development for teachers of mathematics* (pp. 55–66). Reston, VA: National Council of Teachers of Mathematics.

Tharp, R. G., & Gallimore, R. (1988). *Rousing minds to life: Teaching, learning, and schooling in social context.* Cambridge, England: Cambridge University Press.

Thompson, A. G. (1992). Teachers' beliefs and conceptions: A synthesis of the research. In D. A. Grouws (Ed.), *Handbook of research on mathematics teaching and learning* (pp. 127–146). New York: Macmillan.

Thompson, A. G., Philipp, R. A., Thompson, P. W., & Boyd, B. A. (1994). Calculational and conceptual orientations in teaching mathematics. In D. B. Aichele & A. F. Coxford (Eds.), *Professional development for teachers of mathematics* (pp. 79–92). Reston, VA: National Council of Teachers of Mathematics.

Thompson, C. L. & Zeuli, S. J. (1999). The frame and the tapestry: Standards-based reform and professional development. In L. Darling-Hammond & G. Sykes (Eds.), *Teaching as the learning profession: Handbook of policy and practice* (pp. 341–375). San Francisco: Jossey-Bass.

Tickle, L. (1991). New teachers and the emotions of learning teaching. *Cambridge Journal of Education, 21,* 319–329.

Tucker, A. (2004, August 11). In testing, how reliable are year-to-year comparisons? *Education Week.* Retrieved April 10, 2005, from http://www.edweek.org/ew/articles/2004/08/11/44tucker.h23.html?querystring=standards-based%20tests.

U.S. Department of Education. (1997). *Attaining excellence: A TIMSS resource kit.* Washington, DC: Office of Educational Research and Improvement.

U.S. Department of Education. (2000). *Before it's too late: A report to the nation from the National Commission on Mathematics and Science Teaching for the 21st Century.* Washington, DC: Author.

U.S. Department of Education. (2002). *No child left behind.* Retrieved March 15, 2005, from http://www.ed.gov/nclb/

Vacc, N. N., & Bright, G. W. (1999). Elementary preservice teachers' changing beliefs and instructional use of children's mathematical thinking. *Journal for Research in Mathematics Education, 30,* 89–110.

Valli, L., Cooper, D., & Frankes, L. (1997). Professional development schools and equity: A critical analysis of rhetoric and research. In M. W. Apple (Ed.), *Review of research in education* (Vol. 22, pp. 251–304). Washington, DC: American Educational Research Association.

van Es, E. A., & Sherin, M. G. (2002, April). *Learning to notice: Scaffolding new teachers' interpretations of classroom interactions.* Paper presented at the annual meeting of the American Educational Research Association, New Orleans, LA.

Vygotsky, L. S. (1978). *Mind in society: The development of higher psychological processes.* M. Cole, V. John-Steiner, S. Scribner, & E. Souberman (Eds.). Cambridge, MA: Harvard University Press.

Walen, S. B., & Williams, S. R. (2000). Validating classroom issues: Case method in support of teacher change. *Journal of Mathematics Teacher Education, 3,* 3–26.

Warfield, J. (2001). Where mathematics content matters: Learning about and building on children's mathematical thinking. In T. Wood, B. S. Nelson, & J. Warfield (Eds.), *Beyond classical pedagogy: Teaching elementary school mathematics* (pp. 135–155). Mahwah, NJ: Erlbaum.

Weissglass, J. (1997). *Ripples of hope: Building relationships for educational practice.* Santa Barbara, CA: Center for Educational Change in Mathematics and Science.

Wenger, E. (1998). *Communities of practice: Learning, meaning, and identity.* New York: Cambridge University Press.

Whitford, B. L. (1994). Permission, persistence, and resistance: Linking high school restructuring with teacher education reform. In L. Darling-Hammond (Ed.), *Professional development schools: Schools for developing a profession* (pp. 74–97). New York: Teachers College Press.

Wilson, S. M., & Ball, D. L. (1996). Helping teachers meet the standards: New challenges for teacher educators. *Elementary School Journal, 97,* 121–138.

Wilson, S. M., & Berne, J. (1999). Teacher learning and the acquisition of professional knowledge: An examination of research on contemporary professional development. In A. Iran-Nejad & P. D. Pearson (Eds.), *Review of research in education* (Vol. 24, pp. 173–209). Washington, DC: American Educational Research Association.

Wilson, S. M., Darling-Hammond, L., & Berry, B. (2001). *A case of successful teaching policy: Connecticut's long-term efforts to improve teaching and learning.* Seattle: Center for the Study of Teaching and Policy, University of Washington.

Wilson, S. M., Floden, R. E., & Ferrini-Mundy, J. (2001). *Teacher preparation research: Current knowledge, gaps, and recommendations.* Seattle: Center for the Study of Teaching and Policy, University of Washington.

Yoshida, M. (2002). Framing lesson study for U. S. participants. In H. Bass, Z. P. Usiskin, & G. Burrill (Eds.), *Studying classroom teaching as a medium for professional development. Proceedings of a U. S.-Japan workshop* (pp. 58–64). Washington, DC: National Academy Press.

Zeichner, K. M. (1992). *Educating teachers for cultural diversity.* (SR 2/93). East Lansing: National Center for Research on Teacher Learning, Michigan State University.

AUTHOR NOTE

I am deeply indebted to Raffaella Borasi of the University of Rochester, Vicki Jacobs of San Diego State University, and Deborah Schifter of the Education Development Center for their many helpful comments and suggestions through two draft versions of this chapter. I also wish to acknowledge the editorial assistance of Bonnie Schappelle.

6

MATHEMATICS TEACHING AND CLASSROOM PRACTICE

Megan Loef Franke

UCLA

Elham Kazemi

UNIVERSITY OF WASHINGTON

Daniel Battey

ARIZONA STATE UNIVERSITY

Ms. Michaels began her third-grade mathematics lesson by posing a problem for all her students to solve. The problem drew on an idea students raised the previous day regarding the equal sign. Ms. Michaels continued to pose a series of problems intending to build and extend students' ideas. She sequenced the problems carefully; each one built mathematically on the previous one. Ms. Michaels reminded the students about what they had discussed the previous day and tried to get them re-engaged with the same mathematical ideas. As the lesson went on, she asked students general questions about how they would solve the problems and why. Students shared their ideas, usually with the whole class, while Ms. Michaels focused them on the important mathematical ideas.

In another classroom, at the very same school, Ms. Jimenez's third-grade class was working on a similar set of mathematical ideas. Ms. Jimenez also created a series of problems to pose. The sequence of problems she created did not develop or build on a coherent mathematical idea. The students were asked to solve the first problem and discuss it with their partner to see if they agreed and could explain their solutions. As Ms. Jimenez circulated the room and listened to students explain to each other, she asked questions that focused on prompting students to further detail their thinking. She asked students to share publicly, and again she and her students asked questions about detail and about why a student had taken a particular approach.

Student participation varied greatly in these two classrooms. As Ms. Michaels worked through the sequence of problems, an increasing number of students stopped participating in both solving and sharing, while in Ms. Jimenez's class, a majority of students continued to participate throughout the sequence, both in solving the problems and in the sharing and discussion that followed.

Advances in theory and research would characterize aspects of the classroom practice depicted in both cases as productive. The teachers listened to students' mathematical thinking and tried to use it to encourage conversation that revolved around the mathematical ideas in the sequenced problems. Yet, these brief vignettes only depict the skeleton of what occurred in the classrooms. They convey little about what shaped student participation, how students' interactions arose from the social space of the classroom, and how the many discursive storylines fed into the way students engaged with one another and the teacher. Clearly, teaching is not just about starting with mathematically rich problems, even ones connected to what students are thinking. And it is also not just about listening to students and asking them to describe their thinking. But these are central parts of the work, and although researchers in the field are beginning to agree on what constitutes central features of classroom practice, we have much to do to elaborate the details. In particular, we need to consider the relationship between particular classroom practice and opportunities for students to engage. How is student participation negotiated in the classroom? In what ways are students able to participate across various classroom practices? This work is critical because students' ways of being and interacting in classrooms impact not only their mathematical thinking but also their own sense of their ability to do and persist with mathematics, the way they are viewed as competent in mathematics, and their ability to perform successfully in school.

In reviewing the research on classroom practice in mathematics, we have a dual focus. We attend to what teachers like Ms. Michaels and Ms. Jimenez do to structure opportunities for students to understand and learn mathematics. In the vignettes show, the focus is on the teacher and how we understand the teacher's role in classroom mathematical work. Our intent is to also consider classroom practice as experienced by students, attending to what students bring and take from their participation in the social context of the classroom.[1] In maintaining this dual focus, we will summarize what we do know about teaching and classroom practice while attending to what remains underdeveloped or underspecified.

OVERVIEW OF CHAPTER

This chapter begins with making explicit the views of teaching that guided our review. We discuss in detail the theoretical perspectives about teaching and learning that frame our views of classroom practice. We then use what we know from the research literature about mathematics teaching and classroom practice to situate our review of the three features of classroom practice we see as currently most central in helping us come to understand the teaching and learning mathematics: (a) creating mathematical classroom discourse, (b) developing classroom norms that support opportunities for mathematical learning, and (c) building relationships that support mathematical learning.

CONCEPTUALIZING CLASSROOM PRACTICE

Teaching Mathematics

What constitutes good teaching[2] is consistently controversial and will remain controversial. Our goal in this chapter is not to resolve the controversy but rather focus on what research and theory can currently tell us about classroom instruction with the intent of (a) making explicit current findings (drawing widely on both research and theory); (b) highlighting where future research can contribute to our ongoing understanding of classroom practice; (c) attending to how research takes into account or raises questions about supporting student engagement, participation, and learning in classroom practice; and (d) characterizing how current findings and conceptions of classroom practice relate to issues of equity.

Teaching mathematics in classrooms, no matter what the level, engages students, teachers, administrators, and schools in contexts that vary from day to day in ways that make it difficult to create a formula, a set of guidelines, or even a set of practices that all teachers should engage in. We can lay out a theory that can help provide a base for thinking about mathematics classrooms and a set of principles that we can link to outcomes under particular conditions (so we can provide some research-based evidence about what would happen if someone else were to engage under similar circumstances). Our intention is that as researchers and teachers read this chapter, they will finish with the feeling they know something about effective mathematics classroom practices and

---

[1] The students' perspective is not evident in either vignette and often not evident in research on classroom practice.

[2] We consider the act of teaching as involving a number of different practices, with different people, which go beyond classrooms. Teachers teach when they work with parents, talk with the principal, and engage in professional development. This chapter is about one aspect of teacher work, that which occurs as teachers work with students in the context of the classroom. In this chapter, we focus on teaching and classroom practice, recognizing it is not all of teachers' work, nor separate from the other aspects of the work, but it is a place where opportunities for student and teacher learning are created and developed.

the kinds of questions to ask to help shape their teaching or studies of teaching.

As we consider classroom practices here, we are interested in practices that have been shown to positively impact both students and teachers. And although we recognize that we must focus on classroom practices that impact students' mathematical understanding, we must also be clear that mathematical understanding involves students' relation to the mathematics—how they see themselves as doers of mathematics. We also recognize that productive classroom practices create ongoing opportunities for teacher and student learning, where students and teachers see themselves as comfortable, confident, and knowledgeable in their abilities to engage in mathematics.

We also examine the work on teaching and mathematics classroom practice through the lens of equity. We present research that considers the opportunity for each student to become successful, to become a part of ongoing engagement, to be a full participant, where assumptions about particular reasons for a student's participation and success or failure are challenged. We recognize that focusing on individual children and creating opportunities in relation to individuals is not enough to ensure equity. We consider how teaching and classroom practice can create opportunities that support and inspire changes in longstanding classroom, school, and district practices that lead to inequities; challenge assumptions on which instruction is consistently built; and support the development of students' identities, particularly disenfranchised students, as competent doers of mathematics.

Claims about Teaching

We want to recognize up front that making claims about teaching and classroom practice in general is difficult. Classrooms involve people, the lives of people over time, people who work in social, cultural, and political contexts that shape how they do their work and how that work gets interpreted. So as we put forward claims about what is known about teaching and classroom practice, we must remember that any aspect of classroom practice may evolve differently depending on the classroom, the teacher, the student, and the broader social, cultural, and political context. However, we do not want these concerns to so override the interpretations of research that we become unable to make progress in our understanding. We ask that the reader consistently consider that the statements we make must be taken as interpretable within particular local contexts (and as always there exists variability; see Goldenberg & Gallimore, 1991).

A Conception of Teaching[3]

Teaching is relational. Teachers, students, and subject matter can only be understood in relation to one another. The teacher works to orchestrate the content, representations of the content, and the people in the classroom in relation to one another. Students' ways of being, their forms of participation, and their learning emerge out of these mutually constitutive relationships. Teaching is also multidimensional. Within research on classroom practice, several images of this multidimensionality emerge, images that overlap with one another but that are often emphasized in one line of work. Teaching is creating an environment for learning mathematics, orchestrating participation so that students relate to representations of subject matter and to one another in particular ways (Ball & Bass, 2000a, 2000b; Civil, 2002; Cobb, 2000; Lampert, 2001; Moschkovich, 2002). Teaching is eliciting and interpreting what students do and know. It is detailing students' learning trajectories (Fennema et al., 1996; Kazemi & Franke, 2004; McClain, 2000; Simon, 1997; Simon, Tzur, Heinz, & Kinzel, 2004). Teaching is principled decision-making that emerges from complex interactions between teachers' knowledge, beliefs, and goals (Schoenfeld, 1998, 1999a, 2000, in press). Teaching is learning in and from practice in ways that continually supports students (Franke, Carpenter, Levi, & Fennema, 2000; Kazemi & Franke, 2004; Lave, 1996).

Specifying a View of Teaching

Lampert has written extensively and in detail about the teaching of mathematics. She makes explicit her view of teaching and the theory and research underlying that view. We draw on her work here to provide the foundation for understanding classroom practice in mathematics. Our hope is that the portrait she provides will allow those who share her view as well as those who may differ to see where the similarities and differences exist and how a particular view of teaching can be used to make sense of the literature around classroom practice.

Lampert views teaching practice not as a collection of actions that the individual teacher completes but as the evolution of relationships. "Teaching practice is what teachers do, but it is more than how teachers behave with students or the actions of individual

---

[3] Throughout the chapter we draw on a range of theoretical perspectives. We draw on cognitive and sociocultural theories of learning as well as critical theory and critical race theory. Our intention is to draw on work from these theoretical perspectives that taken together can help shape our understanding of teaching and classroom practice in mathematics.

teachers; action is behavior with meaning, and practice is action informed by a particular organizational context" (2004, p. 2). For Lampert, then, teaching is about building relationships—between students and the teacher and among students themselves around mathematics—and engaging together in constructing mathematical meaning. Teaching involves orchestrating the content, the representation, and the people in relation to one another. It is about making decisions in the moment that serve the individuals and the collective. It is about understanding the students, who they are as individuals and as a group, and continuing to learn more about that as one engages in the process of learning. It is about working together to negotiate meaning.

Supporting learning comes from knowing the students, the situation, and the content and then making decisions that support interaction that productively engages students in moving their ideas forward. Teaching is about knowing about the various tools and the meanings students make of them and how they relate to the content and the development of students' thinking as well as knowing how this plays out in the context in which one teaches. Teaching is deliberate work, but it is deliberate work that takes into account the interaction among people and ideas and context. Teaching requires adaptation and learning for teachers and students.

Within this view the complexity and layers of teaching emerge. Rather than teaching being a summation of individual acts, it becomes multifaceted in that relationships among people and content are constantly negotiated. As this negotiation takes place, teachers must learn and adjust to the changing nature of individual and collective learning.

Conceptions of Student and Teacher Learning

Classrooms need to be places where teachers and students are engaged in rigorous mathematics in ways that both parties learn. Opportunities must exist in classrooms for teachers to fine-tune their skills and knowledge as they learn what works, for whom, and when. Opportunities must also exist for teachers to engage in sense-making about their own practice in relation to the mathematics and the students. We have chosen in this chapter to review the work that enables us to discuss mathematics classroom practice that affords learning for teachers and students—classroom practice that makes possible ways for teachers and students to participate together without marginalization.

Emerging Consensus on Student Learning

The field appears to be moving toward some consensus about the focus of student learning (as evidenced by *Adding it Up*, see other chapters in this volume). This emerging consensus includes the idea that all students need opportunities to develop both concepts and skills, to develop flexibility in their abilities to engage with mathematical ideas, and to engage in what some may call higher order or critical thinking. Developing these ideas is about rigor in both depth and breadth and is how we frame understanding in this chapter.

The authors of *Adding it Up* (National Research Council, 2002) defined *understanding* as consisting of five interwoven strands of mathematical proficiency: conceptual understanding, procedural fluency, strategic competence, adaptive reasoning, and productive disposition. We find it quite important that these five strands include one that focuses on the disposition of the learner, highlighting that understanding is more than a set of knowledge and skills; understanding includes one's perception of oneself in relation to the mathematics. In the words of Dutro and colleagues,

> Students learn to locate themselves within the dichotomies of schooling. They narrate themselves and are narrated by others into storylines of success and failure, competence and incompetence, participant and non-participant, included and excluded, etc. In addition, students bring other storylines from home, the media, and from their own particular...ways of negotiating the world. (Dutro, Kazemi, & Balf, 2006, pp. 25–26)

These storylines, as they emerge and become reinforced through classroom practice, shape students' ways of being in the classroom and their stances towards learning and doing mathematics.

In *Adding it Up*, the focus on the interwoven strands highlights that these ideas about oneself and the acquisition of knowledge and skills are not developed in a linear fashion. One does not, the authors claimed, develop mathematical skill first then the others follow or develop conceptual understanding and then the others follow but rather all of the aspects of understanding must be addressed together over time.

These perspectives on understanding mathematics are shaped and further defined as one relates them to views of learning. Learning here is seen as social and shared, where teachers and students bring histories and identities to the interactions, where participation is the focus. Knowledge and skills help shape who we are and how we see ourselves (Greeno & MMAP, 1998; Lave & Wenger, 1991; Rogoff, 1990; Wenger, 1998; Wertsch, 1991). They help shape the ways we interact with others in new and existing settings. Language and tools mediate these interactions. Learning is not about receiving information; it is about engaging in sense-making as we participate together.

Connecting Views of Learning Mathematics with the Challenge for Teaching

Within the field of mathematics education, researchers seem to agree in principle that classrooms that support mathematical proficiency would be places where students are encouraged to be curious about mathematical ideas, where they can develop their mathematical intuition and analytic capabilities, where they learn to talk about and with mathematical expertise. Difficulties arise as we have left these ideas underspecified. In particular, helping teachers, who have not experienced such learning environments themselves, to adapt their classroom practice in significant ways to meet these goals for all students requires helping teachers understand the details of what it would mean to engage in this kind of mathematical work and how to support students in coming to participate in this way.[4] We attempt in this chapter to provide some of that specificity and give direction to future inquiry in the field.

We know from studies like the Third International Mathematics and Science Study (TIMMS) that, in the United States, most mathematics instruction is not consistent with current reform ideas (Hiebert et al., 2003; Hiebert & Stigler, 2000; Stigler & Hiebert, 1997, 1999). Most U.S. mathematics classrooms maintain an initiation-response-evaluation (IRE) interaction pattern, where the evaluation move on the part of the teacher focuses on students' answers rather than the strategies they use to arrive at them. The teacher assumes responsibility for solving the problem while student participation often involves providing the next step in a procedure. TIMMS has further reported that U.S. students had little opportunity to discuss connections among mathematical ideas and reason about mathematical concepts. A number of studies support the TIMMS report, and agreement exists that currently many mathematics classrooms do not provide sufficient opportunities for students to develop mathematical understanding.

Researchers also know that, even in classrooms where teachers are attempting to teach for understanding as specified by the NCTM *Standards* (and other reform documents), teachers often maintain the IRE pattern and maintain many existing mathematics practices inconsistent with the reform (Hiebert & Stigler, 2000). This was even the case when the teachers were using conceptually rich curricula (RAND Mathematics Study Panel, 2002; Spillane & Zeuli, 1999). Changing existing mathematical classroom practice has proven to be challenging, particularly inasmuch as the change requires not simply adopting a set of practices or curricula but must include a change in how teachers engage around the practice with their students. For instance, simply using manipulatives, putting students in cooperative groups, or asking higher order questions does not lead to classrooms that support the development of mathematical understanding (Stigler & Hiebert, 1997, 1999; Webb, Nember, & Ing, 2006). How teachers and students engage with higher order questions, engage students in cooperative groups, or use manipulatives matters.

A growing body of research is beginning to provide insight into mathematical classroom practice that supports mathematical proficiency. Many mathematics reform projects found that practicing procedures and asking students to solve many problems does not support the development of understanding (Carpenter, Ansell, Franke, Fennema, & Weisbeck, 1993; Carpenter, Franke, Jacobs, & Fennema, 1998; Fennema et al., 1996; Hiebert & Wearne, 1992). Decreasing time on conventional procedural skill does not decrease students' proficiency on routine problems (Carpenter, Fennema, Peterson, Chiang, & Loef, 1989; Hiebert & Wearne, 1992). Spending more time on a problem supports a greater degree of reflective and analytic thought (Hiebert & Wearne, 1992) but the nature of discourse around a problem is critical (Cobb, Wood, & Yackel, 1993; Cobb et al., 1991; Hiebert & Wearne, 1992, 1993, Stein, Grover, & Henningsen, 1996; Yackel, Cobb, Wood, Wheatley, & Merckel, 1990). Teachers who know about their students' mathematical thinking can support the development of mathematical proficiency. Knowing about students' mathematical thinking supports opportunities for question asking linked to students' ideas, eliciting multiple strategies, drawing connections across strategies, and so on. When teachers ask more questions and ask for more than recall, providing students' opportunities to express their ideas and actions verbally, students develop understanding. When students are required to describe their strategies in detail and why they work, they develop understanding. These initial studies, as they supported the development of aspects of reform-based classroom practice, provided knowledge about aspects of teachers' work that researchers can leverage to better understand classroom practice.

---

[4] We draw attention throughout the chapter to "detailing" practice. Details matter. Describing practices in our research and our work in classrooms allows for unpacking, supports conversations about meaning, and helps us be explicit about agreement and disagreement.

Summary of Stance Towards Classroom Practice Organizing this Chapter

Maintaining our dual focus on what teachers do and how that is experienced by students; keeping equity central; drawing on a current, coherent conception of teaching and learning; and attending to the limits and strengths of the reform studies leads us to a particular approach to our review of classroom practice. Research suggests the value of listening to students' mathematical ideas and building on them, that attending to the details of student thinking and the mathematics (particularly the mathematical task) supports student learning, and creating opportunities for a range of ideas to be a part of the classroom discourse matters for students and teachers alike. Lampert's conception of teaching parallels these ideas and suggests that knowing the development of students' mathematical thinking, understanding the mathematical tasks, and orchestrating conversation around the particulars of student thinking and mathematics are all a part of teaching for understanding. In addition, Lampert's conception would require that teachers know who their students are within the social environment of the classroom (and potentially outside) and who the students are as an evolving intellectual community. Our equity focus requires we attend to the multiple aspects of who students are and how they participate. These understandings lead us to elaborate three features of classroom practice (a) shaping classroom mathematical discourse, (b) developing classroom norms that support engagement around the mathematical ideas, and (c) developing relationships with students and the class in a way that supports opportunities for participation in the classroom's mathematical work.[5] We now turn to an extended discussion of each of these features of classroom practice.

SUPPORTING DISCOURSE FOR DOING AND LEARNING MATHEMATICS

Shaping mathematical discourse is a significant aspect of a teacher's work. How teachers and students talk with one another in the social context of the classroom is critical to what students learn about mathematics and about themselves as doers of mathematics. What it means to do and learn mathematics is enacted through the discursive practices that form in the classroom. As we argued at the beginning of this chapter,

teaching is relational. Through classroom discourse, one can see how students, the teacher, and subject matter interact and what the consequences are for students. Developing mathematical understanding requires that students have the opportunity to present problem solutions, make conjectures, talk about a variety of mathematical representations, explain their solution processes, prove why solutions work, and make explicit generalizations. A number of researchers have been engaged in research to better understand the discourse practices that support the development of students' mathematical understanding. Although in its relative infancy, this work has already made a number of contributions to what researchers know about mathematics classroom practice.

In this section of the chapter, we contrast the predominant discourse structure of initiation–response–evaluation, or IRE, with structures that enable discussion and deliberation of mathematical ideas. We review how researchers have conceptualized and documented these alternative discursive structures and discuss both the role of the teacher in managing and enabling such structures and how these structures position students in relation to one another, the teacher, and the subject matter. Issues of equity, diversity, and social justice are again both implicitly and explicitly central in understanding the impact of classroom discourse practices. Mathematics classrooms are necessarily cultural and social spaces that can either perpetuate social inequities by privileging certain forms of discourse and ways of reasoning or reorganize them by positioning multiple and hybrid forms of learning and knowing as "having clout" (Cobb & Hodge, 2002; Gutierrez, Baquedano-Lopez, & Tajeda, 1999; Gutstein, in press; Lee, 2002). Although researchers know that the role the teacher plays in supporting discourse practices is critical to their success and consequences, our review claims that we know little about what teachers need to do to best support classroom discourse in a way that opens participation and supports the development of students' knowledge and identities. For instance, we do not know whether argumentation, where students are aligned in opposition to one another, is productive for all students' development of mathematical understanding.

The IRE Pattern and Its Alternatives

Communication in the traditional mathematics classroom can be characterized by teacher talk: teach-

---

[5] All of this mathematical work within classrooms requires teachers understand the key mathematical ideas of the work they do with students and a particular stance towards that engagement. Chapters in this volume delineate that in detail. Here we focus on the classroom teaching work, not what is supporting it.

ers explaining procedures, giving directions, explaining mistakes in ways that require very little student-to-student or even student-to-teacher talk (Hiebert et al., 2003; Porter, 1989; Silver & Smith, 1996; Stigler & Hiebert, 1997). The most dominant classroom discourse pattern has been the IRE, where the lesson follows a teacher-dominated pattern of teacher-*initiated* question, student *response*, and teacher *evaluation* (Cazden, 2001; Doyle, 1985; Mehan, 1985). The IRE pattern is a well-documented part of mathematics classrooms in the United States and elsewhere, especially for students from economically disadvantaged backgrounds (Silver, Smith, & Nelson, 1995). In mathematics classrooms students are typically asked to listen and remember what the teacher said. Usually, little emphasis has been placed on students' explaining their thinking, working publicly through an incorrect idea, making a conjecture, or coming to consensus about a mathematical idea. Even in classrooms where teachers are attempting to teach for understanding they often maintain this pattern. Spillane and Zeuli (1999) found in their study of reform-minded teachers that they predominately engaged in procedure-bound discourse, rarely asked students to do more than provide the correct answer, focused on procedural rather than conceptual knowledge, and engaged students in memorizing procedures to calculate answers (see also RAND Mathematics Study Panel, 2002).

Over the last 15 years, a growing body of research has documented and theorized alternatives to the IRE discourse structure. New conceptions of classroom discourse focus on how conversations foster mathematical argument so students can come to understand the different forms of mathematical explanation, create a public knowledge base that can be used by the class as a resource, align students with one another and the content, develop students' mathematical identities, and generally foster higher level thinking (Ball & Bass, 2000b; Boaler & Greeno, 2000; Cobb et al., 1993; Lampert, 2001; Yackel et al., 1990). O'Connor and Michaels (1993) described classroom conversations as providing a site for aligning students with one another and the content of the academic work while also socializing them into ways of engaging in mathematics. In addition teachers, through classroom discourse, can draw on a range of student resources, particularly for English language learners (ELL), and build on them in a way that attends to the strengths of students, not their deficits (Moschkovich, 1999, 2000). Creating these opportunities takes more than simply posing the problem and asking the children how to solve it within a non-threatening environment.

Although the kind of discourse we have just described seems like such a necessary and productive part of classroom practice, what teachers must do to support such opportunities is complex and not well characterized in the literature. The way teachers support mathematical discourse matters. Even in what may seem like the simplest form of classroom discourse—large-group discussion to foster students' participation in thinking through a problem—the teacher must attend to many issues. The teacher must attend to who is participating, how they are participating, the mathematical ideas being pursued, the students' linguistic and mathematical backgrounds, the students' current understandings, and the attitudes and identities the students bring to the conversation (Lampert, 2001). The teacher must give each student the opportunity to participate in working through the problem while simultaneously encouraging each student to attend to the solution paths of others, in ways that she can orchestrate opportunities for students to build one another's thinking. While attending to this, the teacher must also actively take a role in making certain that the class gets to the implicit and explicit goals. She needs to make judgments about what to avoid, navigate through solution paths that do not always work, respond to incorrect statements, and watch out for those not participating. She must also find a way to make explicit the underlying mathematical similarities and differences in the solutions in a way that makes sense to her students. All these actions and decisions must of course fit within the given period of time of a lesson, a unit, and a school year.

Teachers are expected to pose problems but not provide answers (Lampert, 1990), stop or slow down the discussion to provide access to more students (Rittenhouse, 1998), model the academic discourse for the students (Ball, 1993; Lampert, 1990; Rittenhouse, 1998), comment and elaborate on student ideas (Rittenhouse, 1998), and question student reasoning so as to foster certain habits of mind (Lampert, 1990; Lampert, Rittenhouse, & Crumbaugh, 1996; Rittenhouse, 1998). Thus, as Ball (1993) pointed out, the teacher is responsible for the students' learning of mathematical content and, at the same time, for fostering a discourse environment that both supports students and helps to create, among them, new identities that include a favorable disposition towards mathematics. It is no wonder IRE remains prevalent.

Conception of Discourse in Mathematics Education

Researchers and theorists alike are calling for a focus on mathematical communication. How communication is defined and why we need it in classrooms varies. However, all perspectives include social engagement, through talk and shared representation, with a focus on developing mathematical ideas. Some researchers highlight communication as creating publicly shared knowledge that can serve as a future resource; others see the opportunities for multiple ways to participate with the content, while still others see communication as a way to further develop the practices of mathematics: explanation, argument, and so on. Some researchers focus on students' sharing their solutions and ideas surrounding those solutions, while others focus on developing mathematical arguments. Some see mathematical conversation as another opportunity to develop further mathematical knowledge, while others look to the social and individual benefits.

So although researchers engaged in understanding mathematical conversations and their benefits may think differently about what to look for in a conversation and what the benefits may be, what seems critical here is that to engage in the type of mathematics required by such documents as *Adding it Up* (NRC, 2002) and the NCTM *Standards* (1989, 2000) in a way that makes use of what is known about learning in a social context, mathematical conversations are necessary. Yet, just talking is not enough. Cazden (2001) and Nemirovsky, Barros, Noble, Schnepp, and Solomon (2005) argued that teachers need a repertoire of ways to engage students in mathematical conversation with the understanding of what a particular conversational move may afford or constrain and how to support it in the classroom with a range of students. Questions then become about what types of conversational moves support student participation and learning in mathematical conversations. How can teachers support opportunities for engagement in mathematical conversations in ways that are productive for students?

Engaging in Classroom Conversations

Much of our initial learning about engaging students in mathematical conversations comes out of the research on cooperative groups. Increasing the level of discourse in cooperative groups often produces greater student learning; explaining to other students is positively related to achievement outcomes, even when controlling for prior achievement (Brown & Palincsar, 1989; King, 1992; Peterson, Janicki, & Swing, 1981; Saxe, Gearhart, Note, & Paduano, 1993; Webb, 1991; Yackel et al., 1990). However, just getting students to talk was not enough; what they talked about mattered. Often students in talking with a peer shared a procedure with little discussion about con-

cepts or strategies (Webb, 1991; Webb et al., 2004). Webb and others have found that most often in these cases the students are mirroring what teachers do as they engage in conversations with students. However, although talking could often be productive for the student sharing, it is typically less productive for the student hearing it. The learning of the students hearing an explanation is dependent on the teacher and the support the teacher provides that allows students to learn together.

Consistent throughout the research on classroom talk, whether from the cooperative learning work or from work on classroom discourse, is the fact that details matter (Webb et al., 2004). Sfard and Kieran (2001a) found that students were neither precise nor explicit in their talk (see also Nathan & Knuth, 2003). Often the language students used to describe their ideas was not taken-as-shared by those listening, referents students made were unclear, and they generally did not provide enough detail. This is much like the cooperative group literature that differentiates non-elaborated from elaborated help or specific questions about a problem solution from questions about following a procedure. Elaborated help and specific questions, both more detailed, led to more student learning. Kieran (2002) stated

> it is the way we make our thoughts available that is critical, it is not just about making them 'available,' it is how…utterances that were neither complete nor ever expanded upon seemed much less conducive to the emergence of mathematical thought for both participants. (p. 219)

So one of the most powerful pedagogical moves a teacher can make is one that supports making detail explicit in mathematical talk, in both explanations given and questions asked.

The most prominent finding about creating opportunities for mathematical conversations in classrooms is that what the teacher does to structure these opportunities matters. Teachers need to scaffold, monitor, and facilitate discourse around the mathematical ideas in ways that support student learning (Kieran & Dreyfus, 1998). Wood (1998) contrasted discourse patterns in which teachers' questioning *funneled* student responses to discourse patterns in which teachers' questioning *focused* student mathematical thinking. In the former, the teacher's questions funnel the conversation so that the teacher actually does the bulk of the intellectual work. In the latter, the teacher's questions focus student attention on important mathematical ideas but place the responsibility of the intellectual work on students. Sherin (2002) described how the teacher used a "filtering approach" to focus students' attention on particular mathematical ideas. After soliciting multiple solutions from students during which they listened and evaluated one another's ideas, the teacher intentionally filters the ideas, choosing which ones to pursue with the whole class in order to advance particular instructional goals. Although researchers have long recognized the *potential* of teacher practices to foster meaningful conversation and student learning in classrooms (e.g., Cazden, 1986; Forman, 1989, 1993), researchers have only begun to study ways of changing classroom discourse from traditional recitation patterns in which the teacher dominates classroom exchanges to more balanced and student-centered communication in which students take a more active role in classroom discourse (Carpenter, Fennema, Franke, Levi, & Empson, 1999; Cobb & Bauersfeld, 1995; Forman & Ansell, 2002; Hufferd-Ackles, Fuson, & Sherin, 2004; Leinhardt & Steele, 2005; Martino & Maher, 1999; Nathan & Knuth, 2003; Sherin, 2002; Wood, Williams, & McNeal, 2006; Yackel, 2002).

In working to create opportunities for rich conversation a number of researchers have found ways teachers can provide support. From Palinscar and Brown's (1984) *reciprocal teaching* to Hogan and Pressley's (1997) *instructional scaffolding*, researchers have developed structured ways for teachers to engage with students that have proven to produce more conceptually based conversation as well as productive student outcomes.

Researchers have in mathematical classrooms also learned something about the general ways teachers can support rich mathematical conversations. Teachers have found certain strategies to be effective in supporting student engagement and discourse. Yackel, Cobb, and Wood (Yackel, Cobb, & Wood, 1991; Yackel et al., 1990) found two strategies that teachers used to develop norms around students' explaining their methods. First, teachers used specific situations that arose spontaneously during group work as a springboard for whole-class discussion, and second, the teacher invented specific situations that they could use to discuss participation with the class. These teachers also intervened in small-group work directly to help renegotiate student participation. It took significant attention on the teachers' part to support the development of different forms of participation.

Silver and Smith (1996; Silver, 1996) found through their QUASAR work that teachers need to (a) give students sufficient time to generate and explore their own ideas and discuss their solutions, (b) value understanding solutions by pressing for justification and explanation, (c) refrain from stepping in and providing a solution strategy for the student, and (d) create op-

portunities for competent performance in sharing to be modeled by a capable student (see also Stein et al., 1996). Forman, Larreamendy-Joerns, Stein, and Brown (1998) drew on a study of a teacher engaged in creating opportunities for argumentation. They found that the teacher orchestrated discourse by recruiting attention and participation from students, aligning students with one another on the basis of their mathematical ideas, highlighting positions through repetition, and pointing out implicit but important aspects of an explanation as she expanded on what was shared. These findings are similar to those of Strom, Kemeny, Lehrer, and Forman (2001). In addition, they highlight the teacher's stance and excitement toward the collective work. She made her investment in the work explicit for the students. In all of these cases teachers' moment-to-moment participation acts to structure the conversations in ways that allow for all students to participate in meaningful mathematical work.

Beyond these general ideas about mathematical discourse a few promising core ideas and practices are emerging. The four ideas we touch on briefly highlight the detail at which we can come to understand mathematical discourse and the direction we can go in developing additional knowledge to support teachers and students as they engage in this work. We look at the research around revoicing as it represents an area where there is accumulating evidence about the role teachers play in the particulars of how they respond to students within a mathematical conversation. Second, we look at the work discussed earlier on the ways tasks play a role in the development of mathematical discourse. Third, we look particularly at mathematical discourse for English language learners. Fourth, we propose that a discourse practice drawn from science classrooms, interrogating meaning, as defined by Rosebery and Warren, can further efforts to make equity central in thinking about mathematical discourse and participation. We end with emerging work about students' own experiences of mathematical discourse and participation.

Revoicing

Revoicing, most simply, is reuttering of someone's talk. This reuttering can involve repetition, expansion, rephrasing, or reporting what a student says (O'Connor & Michaels, 1993, 1996, drew on the work of Goffman in their elaboration of revoicing). Revoicing can accomplish a range of goals and can either support or limit productive discourse. Revoicing serves to clarify or amplify an idea and allows the teacher to substitute mathematical vocabulary for everyday words or redirect the conversation. Beyond these goals, revoicing can communicate a way of thinking about doing mathematics, a respect for student ideas, and an encouragement of students' developing mathematical voice. Here revoicing can frame the content discussion and the social roles. In addition, most recently researchers have investigated the use of revoicing as a pedagogical move to position students in relation to one another and to the mathematical argument (Forman et al., 1998). The teacher can use revoicing as a way to align students to a particular argument or way of thinking about the mathematics. Revoicing then is a way of orchestrating the conversation.

Studies of teachers in the mathematics classroom that examine both the types of revoicing that occur and the impact of the revoicing on students have found that often revoicing plays a supportive role in terms of both mathematical support for ideas and the development of students' identities around the learning of mathematics (Forman et al., 1998; O'Connor & Michaels, 1993, 1996; Strom et al., 2001). These case studies of teacher and classrooms provide support for the ways revoicing structures student participation as students work together and connects students to the mathematics. Work like that of Forman and her colleagues or Strom and her colleagues is also beginning to clarify the particulars around how revoicing can support student learning and how that revoicing brings together the mathematics, forms of notation and representation, and the students.

However, this type of language socialization into the discourse practices of academic argumentation is not without its difficulties. First, it creates teaching dilemmas. For example, conflicts can arise between the teacher's content objectives and a pedagogy that honors the intellectual process, particularly when it means leaving incorrect mathematical ideas on the table (Ball, 1993; Lampert, 1990). Second, because academic argumentation can easily deteriorate into oppositional or confrontational talk, there is a risk of interpersonal conflicts that spill over the intellectual content or even the boundaries of the classroom (Lampert et al., 1996; O' Connor, 1998). Thus, it is particularly important for teachers to attend to creating a set of social norms suitable for classroom debate alongside the sociomathematical norm about what is accepted as an adequate mathematical answer or justification (Horn, in press; Yackel & Cobb, 1996). Finally, these discourse practices may be aligned with, or in conflict with, practices from the students' home or the practices from other communities that they participate in outside of school (Gallego, Cole, & The Laboratory of Comparative Human Cognition, 2001; O'Connor, 1998; Walkerdine, 1990).

Although the work on revoicing has stimulated detailed look at supporting classroom discourse in mathematics and there is evidence that revoicing occurs and can productively support student engagement and learning, researchers do not yet know when particular aspects of revoicing are productive and when they might lead to disengagement or learning. There is enough evidence to argue that the teacher's role in orchestrating through revoicing and other means is crucial, and that continued research in this area can support both the development of classroom practice and the learning of teachers.

The Role of Tasks

Silver and his colleagues in their focus on mathematical tasks describe how worthwhile mathematical problems can "give students something to talk about" (Silver, 1996; Stein & Lane, 1996). These worthwhile tasks from their perspective are ones that engage students in thinking and reasoning about important mathematical ideas. They drew on the NCTM *Standards* and further described worthwhile tasks that can be solved in multiple ways, involve multiple representations, and require students to justify, conjecture, and interpret (see also Engle & Conant, 2002, for discussions of productive disciplinary engagement).

Stein and Lane (1996), drawing on the work of the QUASAR project, reported that the highest learning gains on a mathematics performance assessment were related to the extent to which tasks were set up and implanted to engage students in high-level cognitive thinking and reasoning. Selecting and setting up a high-level task, though, does not guarantee students' high-level engagement. However, starting with a good task is necessary for providing opportunities to engage students in high-level thinking. So starting with a cognitively demanding task allows the teacher to engage students in sharing their thinking, comparing different approaches, making conjectures, and generalizing (Silver & Smith, 1996).

Oftentimes as teachers begin to think about engaging their students in mathematical discourse they draw upon the tasks they have been using, and these tasks often lend themselves to using a single procedure and do not lend themselves to the types of conversations that stimulate rich discourse. Supporting productive discourse would be easier for teachers if they worked from mathematical tasks that allowed for multiple strategies, connected core mathematical ideas, and were of interest to the students.

Classroom Discourse for English Language Learners

There is significant evidence that students bring to the classroom ways of participating in conversations;

language socialization patterns develop in the home environment. The teacher, through classroom discourse practices, plays a role in helping to socialize students with potentially varying notions about participating in mathematical conversations. Revoicing is a way to begin to understand this, but the challenge becomes even greater when the students do not speak English as their native language (O'Connor, 1998). However, to date, much of this research has been conducted in monolingual classrooms, led by exemplary teachers.

Moschkovich (2002a, 2000b) advocated a perspective for ELL that focuses on *participation in mathematical discourse*. The focus on participation expands and complicates the typical explanations of how ELL students learn mathematics and what the difficulties are. According to Moschkovich (2002b), the focus on participation in mathematical discourse starts with the assumption that knowing and doing mathematics is not merely an issue of vocabulary or semantics. Instead, it leads educators to identify the resources that ELL students can bring to mathematical discussions and the ways to build on these strengths to construct meaningful mathematical understandings (Gutierrez, 2002b). This implies that ELL students should be allowed to use their primary language (particularly in small groups) to talk about mathematics. With the help of the teacher, who can explicitly link these ideas to how they can be expressed in the English mathematical register, students will be able to bridge their mathematical knowledge and their mathematical language proficiency in English (Moschkovich, 2002b).

Nemirovsky et al. (2005) in their studies of high school mathematics classrooms built on the work of Moschkovich (1999) and found that the relationship between everyday language and the mathematical register is one of continuity. This continuity implies that the mathematical meanings are normally expressed in the combined use of everyday language and the specialized register and that, far from being "dangerous," the open discussion of multiple word usages, including those that tend to be mistaken from a mathematical point of view, enriches mathematics learning.

Revoicing can be one such strategy that supports the participation of students with limited academic English proficiency in mathematical discussions (Moschkovich, 1999). Rather than adopting the typical perspective that sees students' limited English proficiency as a barrier to understanding, teachers can use revoicing to frame students' first language as a resource for the construction of meaning (Moschkovich, 2000). Through revoicing, teachers acknowledge and demonstrate respect for students' contributions, while also potentially elaborating, clarifying, or formalizing students' thinking. These types of friendly uptakes of student voices may lead to a greater degree of student participation in the discourse of mathematics. Therefore, revoicing contributes to a new model for classrooms, in which students learn academic English and mathematics at the same time.

Enyedy and his colleagues (Enyedy, Rubel, Castellon, Mukhopadhyay, & Esmond, 2006) studied revoicing in a multilingual high school algebra classroom. They found that revoicing was not being used to initiate arguments between students (as described by Forman et al., 1998). However, revoicing was a frequently used tool. The teacher in this classroom used revoicing to support elaboration, clarification, and positioning with respect to mathematical tasks, goals, and roles. Even though the students did in this class engage in making mathematical claims and justifying them, the teacher did not use revoicing as a means to create alignments and oppositions in an argument. So, little debate occurred. Enyedy and his colleagues (2006) argued then that it may be more productive to conceptualize revoicing as a "flexible rhetorical form that can take on many functions" beyond facilitating student debate.

Interrogating Meaning

An additional discourse practice with promise for supporting mathematical learning and providing opportunities for students from different backgrounds to participate is raised in the work of Rosebery, Warren, and their colleagues (Rosebery, Warren, Ballenger, & Ogonowski, 2005; Rosebery, Warren, & Conant, 1992). In attempting to create opportunities for students to engage in discussion around science inquiry, Rosebery and Warren have developed an approach they term *interrogating meaning* (Rosebery et al., 2005). Here students interrogate situated meanings, familiar and unfamiliar in light of one another, and in the process make explicit the assumptions of use, purpose, and context. Engaging in interrogating situated meaning engages students in the ways language and symbolic practices are used more generally in a discipline and allows the teacher to be explicit that this is not separate from engagement with the knowledge of that discipline. Interrogating meaning engages students in an analytic stance toward the ways of seeing. In Rosebery and Warren's work students were encouraged by the teacher to interrogate (ask questions of, challenge, wonder about) "deceptively familiar experiences, such as motion down a ramp and ideas such as 'faster and faster', 'speed' or 'darkness' in relation to situations of use and perspective" (p. 66). As students and teachers engaged with the often-conflicting ways students made sense of the ideas, the students made explicit to themselves and to the teacher their conceptualizations and stances toward those conceptualizations in ways that supported ongoing learning as students made connections between what they knew and what they were learning.

What is appealing about Rosebery and Warren's conception of interrogating meaning is the opportunities this move provided for participation. Interrogating meaning creates a norm around the expected nature of questioning and challenging one's ideas. The focus is not then on who was wrong, or whose explanation was incomplete, but on unpacking every response in ways that support learning.

Although we know little to date about interrogating meaning in a mathematics classroom, the conceptualization of the pedagogical move and the research in science education by Rosebery and Warren leads us to see promise in promoting both the focus on mathematics we hope to support and the opening of participation across students.

Considering Students' Experiences of Classroom Discourse

As our review of the classroom discourse literature indicates, there has been growing attention to the role teachers play in managing classroom discourse. A set of studies is also emerging that considers the experiences students have in classroom discussions. Some of this work includes students' own accounts of these experiences, whereas others are interpretations of students' experiences. Attention to both ways of considering students' experience of classroom discursive practices is clearly a fertile area for research. Importantly, this body of work has often focused on marginalized students. We summarize a number of articles here that reflect this promise.

Empson (2003) presented an in-depth analysis of two of the lowest performing students in a first-grade unit on fractions. Her examination of classroom interactions provided new insights into the learning gains made by these students. In particular, the teacher leveraged certain interactional mechanisms to support the students to learn mathematics and enhance their dispositions towards doing mathematics: (a) Tasks were used to elicit students' understanding, not just their confusions, and (b) teachers positioned the students, through interaction, as mathematically competent. The teacher often supported the students to work through problematic reasoning and then positioned them as authorities on these strategies to the group. In instances when the two students' problematic reasoning was resolved during group discussions, the teacher most often used revoicing strategies to repair

their claims or positioned the students as evaluators of other students' comments or alternative strategies. Rarely did the group conversations cause the focal students to opt out. As Empson's work has continued, she is working with colleagues to use these observations of students to create a set of conjectures about how teachers can take proactive roles to use discourse strategies and participation structures to position students in ways that allow them to *display* understanding and competence (Empson, Turner, Dominguez, & Maldonado, 2006).

The teacher's ability to draw out, connect with, and frame student achievement and competence is clearly linked to her ability to know students, cultivate norms for generative interactions, and build relationship with students, ideas that we will discuss in depth in the major sections of this chapter. By analyzing student participation Moschkovich (2002) found that students brought different ways of talking about mathematical objects and mathematical situations to the discussions. None of these ways of talking should be seen as privileged over the other; instead each way of talking can contribute in its own way to the mathematical discussion and bring resources to the conversation (p. 12). Instead of concluding that the students were wrong or lacked vocabulary, Moschkovich underscored that they were bringing a different point of view or using different language than the teacher expected. Moschkovich concluded that taking a discourse approach to mathematical learning means considering the different ways of talking and different points of view students bring to the discussion.

In much of her work, Boaler (Boaler, 2003; Boaler & Greeno, 2000) intentionally elicits the feelings and interpretations students have of their own classroom experiences. The results have been striking and have repeatedly shown that the practices of learning mathematics enacted in classrooms defined the mathematical knowledge that students learned and their orientations and dispositions toward mathematics.

Lubienski (2002) found in her case study of a seventh-grade class that the high SES and low SES students experienced mathematical discussions differently. Whereas the higher SES students claimed to feel confident in contributing to and in sorting out the ideas presented, the lower SES students found discussions frustrating and wanted the teacher to show them how to do the problems. Lubienski's case study stands in contrast to much work, including the other two studies reviewed above, that have documented the benefits of engaging students in collective mathematical discussions. This case and others (see, e.g., Civil, Baxter, Woodward, & Olson, 2001) begin to open up the complications that can arise in classroom practice that centers on mathematical discussions. Much research remains in understanding how classroom practices are negotiated so that they do not have differentially negative impacts on any particular group of students.

Summary

The nature of mathematical discourse is a central feature of classroom practice. If we take seriously that teachers need opportunities to learn from their practice, developing mathematical conversations allows teachers to continually learn from their students. Mathematical conversations that center on students' ideas can provide teachers a window into students' thinking in ways that students' individual work cannot do alone. Mathematical conversations provide opportunities for teachers to hear regularly from their students and to learn about the range of ideas students have about a particular mathematical idea, the details supporting students' ideas, the values students attach to those ideas, and the language students use to express those ideas. The knowledge teachers gain from engaging with their students in conversations is essential for teaching for understanding. This is a place where teachers can learn and hone their skills.

Courtney Cazden (2001) in her second edition of *Classroom Discourse* pointed out that teachers are being asked more often to add nontraditional discussions to their repertoires to better support the development of students' higher level thinking. She also pointed out that the "challenges of deciding, planning, and acting together across differences of race, ethnicity and religion are growing" (p. 5). So more than ever educators need to pay attention to who speaks and how opportunities are provided for varied participation. Cazden suggested, as do other educators, that when engaging in different forms of discourse the rules and practices for students be made explicit. Cazden claimed that educational purpose and equitable opportunity to learn remain the most important design principles for classroom discourse. This remains true within mathematics education. The classroom discourse literature consistently finds that (a) many forms of discourse support student participation and learning, (b) the teacher plays a pivotal role in supporting productive discourse and participation, (c) details within the discourse matter, and (d) issues of equity remain.

So much of the work we reviewed addresses the concerns raised by Cazden around both enriching repertoires of different opportunities for talk as well as considering issues of equity. Revoicing, interrogating meaning, and attending to detail all provide opportunities to continue to challenge the community of teachers to examine how we support mathematical discourse, how the way we structure talk supports the development of students' mathematical identities in productive ways, and how our practices as teachers make room for only limited opportunities for participation.

The limited citations here clearly indicate that the research in these areas has only just begun. What is reassuring is that the work being done is providing productive ways of detailing classroom practice in ways that attend to teacher learning. In addition this work provides a way of thinking about how to continue to engage in research around classroom discourse. It also, taken together with the studies of cooperative group work, demonstrates that we have evidence about how to support student discourse and that these forms of discourse are productive for student learning.

ESTABLISHING NORMS FOR DOING AND LEARNING MATHEMATICS

As the research on managing classroom discourse shows, much of teachers' classroom work involves negotiating—with students—the context for learning, or in other words, supporting the development of classroom norms. Research has shown mathematics classrooms to be particular kinds of social contexts where the structures of activity within them afford and constrain what is learned, how it is learned, and which students learn it (Boaler, 1997; Yackel & Cobb, 1996). As Lampert et al., (1996) argued, for children to learn to formulate mathematical claims in school, the terms and structure of discourse need to be reconfigured. And beyond that, a sense of what counts as mathematical work in school needs to be reconfigured. Lampert (2001) described it as "establishing and maintaining norms of action and interaction within which the teacher can teach and students can study." (p. 51) The teacher plays a critical role in nurturing the mathematical climate of the classroom. Wood, Cobb, and Yackel (1991) describe the teacher's role as guiding mathematical dialogue among students to describe their solutions. The teacher uses these contributions to ask more questions and support the learning of the individual and the group.

In this chapter, we will describe the relationship between managing classroom discourse and the demands it places on the teacher to establish, maintain, and continually negotiate with students the norms that govern classroom interactions and mathematical work. Quoting Goffman (1974), Wood defined norms as the "interlocking networks of obligations and expectations that exist for both the teacher and students [that] influence the regularities by which students and teacher interact and create opportunities for communication to occur between the participants" (Wood, 1998, p. 170). These "hidden regularities guide the actions of participants in the classroom . . . [and] become the taken-for-granted ways of interacting that constitute the culture of the classroom" (Wood, 1998, p. 170). Key to the developing understanding of the centrality of norms is Yackel and Cobb's work on distinguishing social norms from sociomathematical norms. They have drawn researchers' attention not only to what is generally valued in classroom interaction but also to how the teacher and students engage with one another mathematically (Yackel & Cobb, 1996). Several other bodies of work have provided further contours to this work. Lampert, we will show, theorized extensively about her moves to create a particular set of norms related to doing mathematics in the classroom. Work by Civil, Boaler, and others has also underscored the challenges teachers face in supporting the cultivation of particular norms, especially in light of the complex ways students' social and schooling histories position them in relation to one another as well as to mathematics.

While teachers create opportunities that structure participation, they do not do this alone. Schools and communities have long cultural histories that shape what happens in classrooms. Families and students bring their own histories and identities that shape participation. Yet, together, the teacher, students, and families can shape who learns and how. Who gets to talk, when, about what, who participates and how, the kinds of solutions and questions considered acceptable, and so on are determined by those participating and the experiences they bring to the classroom. Shaping participation is not always accomplished explicitly but can be driven by implicit goals, beliefs, and identities of the teacher, school, and community. Boaler and Greeno (2000) reported that the practices that constitute mathematics learning in one classroom may be vastly different from one another and foster different understandings of mathematics. Shaping learning opportunities in the kinds of classrooms we are advocating requires explicit attention to who gets to participate and how. The ways norms are shaped influences which students learn, what they learn, and how they learn it.

Relating Classroom Norms to Doing Mathematics

Students come to mathematics classrooms with definite notions about what it means to "do" mathematics.

In an early study of first-grade classrooms, Franke and Carey (1996) found students to be quite articulate about what it meant to do mathematics and do it successfully in their classrooms. The first graders clearly said they knew their answer was wrong if the teacher asked them a question about it. They explained either that they had to be fast at math to be good at it or that everyone who had a strategy was good at math. And student responses within classrooms were quite consistent.

Students, like these first graders, develop definite notions of what constitutes participation in a mathematics classroom, and often they come to mathematics classrooms expecting participation that includes working in textbooks, solving pages of problems, listening to teachers explaining, finishing, and finishing quickly. Students typically do not see explaining their thinking as a part of doing mathematics, and they certainly do not typically consider justification of their mathematical ideas as part of the process. This makes creating classroom cultures that support the development of understanding through ongoing mathematical discourse challenging, particularly for those students who are already in middle and high school with a great deal of traditional school experiences (Silver, Kilpatrick, & Schlesinger, 1990; Silver & Smith, 1996).

Depending on one's theoretical perspective, the discussion of creating opportunities for participation can focus on rule-setting, creating a context for learning, or developing a set of classroom norms. In each case, the concern is with shaping the learning environment in a way that supports student learning and with developing certain dispositions toward doing mathematics. However, given the goals of creating a particular type of mathematics classroom where sense-making and discourse are central and where different forms of participation are acceptable and expected (so each student is included), detailing the development of norms is critical. Cobb et al. (1993) emphasized that it is essential to adopt a perspective that focuses on the culture of the classroom community even if one's primary concern is to understand individual students' construction of mathematical knowledge, as individual mathematical learning is influenced by both the mathematical practices and the norms negotiated by the classroom community.

Distinguishing Social From Sociomathematical Norms

In considering norms for classrooms that support mathematical understanding, one must consider the regularities in classroom social interactions that constitute the grammar of classroom life. These general classroom norms apply to any subject matter area and are not unique to mathematics but sustain classroom cultures characterized by explanation, justification, and argumentation(Cobb, Yackel, & Wood, 1989; Yackel et al., 1991). The classroom norms, interactively negotiated by teachers and students, set guidelines for acceptable participation and constrain individual actions. From the perspective of general social norms, these regularities include how participants engage together—how participants use tools, work together, listen, respect one another's ideas, and determine language to support participation.

Adding it Up describes particular norms for creating mathematics classroom communities. The norms proposed, based on theory and research, include (a) value placed on ideas and methods, (b) student autonomy in choosing and sharing their problem solving methods, (c) appreciation of the value of mistakes as sites for learning for everyone, and (d) renegotiation of authority for whether something is correct or sensible so that authority lies in the logic and structure of the subject and not with the status of the teacher or popularity of the person making the argument. These norms shape opportunities for students to share their thinking, engage in discussion around ideas, value and learn different ways of thinking about problems, and build mathematical arguments.

Lampert's (2001) description of evolving norms, although similar to those put forth in *Adding it Up*, are more specific. She focused on what it means to investigate meaning in public. Lampert (2001) argued that names accord importance and, by labeling them, make particular ideas, skills, and habits usable. She articulated three explicit activities related to mathematical problem solving that she names and works to establish with students: (a) finding and articulating the "conditions" or assumptions in problem situations that must be taken into account in making a judgment about whether a solution strategy is appropriate, (b) producing "conjectures" about elements of the problem situation including the solution, and (c) revising conjectures based on mathematical evidence and the identification of conditions (Lampert, 2001, p. 66). She describes what it means to display alternative legitimate responses and how to disagree. She works on how to use tools, specifically the use of journals. She works at developing norms for small-group as well as large-group interaction and for written communication (Lampert, 2001). Central to all of these norms is mathematics and mathematical work.

Yackel and Cobb (1996) argued that within mathematics classrooms norms are developed that are particular to the mathematical activity and guide the mathematical work, norms that go beyond the general norms for managing discourse. These norms would be more closely aligned with Lampert's ideas of what it means to engage in mathematical disagreement or to revise a mathematical explanation. Yackel and Cobb termed these *sociomathematical norms*. Knowing that one is expected to explain one's thinking is a social norm; knowing what counts as an acceptable mathematical explanation is a sociomathematical norm. Sociomathematical norms include such norms as those around what counts as mathematically different, mathematically sophisticated, mathematically efficient, or mathematically elegant explanations.

Considering the Consequences of Norms for Student Learning and Dispositions

Attention to sociomathematical norms is critical because it allows one to explicitly address the mathematical aspects of teachers' and students' work in the mathematics classroom. In particular, it allows one to make explicit the implicit ways in which the mathematical work is being driven by particular ways of considering sophistication, elegance, and difference in mathematical ideas. Sociomathematical norms can also be a way to make explicit the teacher's role in relation to both the students and the mathematics. Moreover, attention to norms enables one to think about the values that are being enacted as students engage in the work of mathematics with one another.

Yackel and her colleagues have pointed out that implicit in the inquiry approach to mathematics are social norms that can foster students' development of social autonomy (Cobb et al., 1991; Cobb et al., 1989; Kamii, 1985; Nicholls, Cobb, Wood, Yackel, & Patashnick, 1990). The sociomathematical norms that are negotiated within inquiry-based mathematics classrooms can foster the development of intellectual autonomy: "students who are intellectually autonomous in mathematics are aware of, and draw on, their own intellectual capabilities when making mathematical decisions and judgments as they participate in these (classroom) practices" (Yackel & Cobb, 1996, p. 473).

Lampert (2001) also discussed the broader goals she has for developing norms for articulating conditions, making conjectures, and revising ideas. She described these activities as central to the work of doing mathematics in the classroom, but she also discussed at length her efforts to see that these practices also help students to become "the kinds of people who study in the classroom and who expect others to do so as well" (p. 265) and "who are academic resources for themselves and for one another" (p. 266). She commits herself to helping students learn to practice Polya's *intellectual virtues*, those of intellectual courage, intellectual honesty, and wise restraint. For Lampert, developing norms for what constitutes mathematical reasoning (developing and revising conjectures and conditions) should also lead students to learn certain intellectual virtues.

Studying the Emergence and Negotiation of Norms

Classroom norms develop through ongoing negotiation among the participants. Mathematics is continually negotiated and institutionalized by a community of knowers. Mathematical activity in the classroom occurs within mathematical practices that have been institutionalized by the classroom community and are taken-as-shared by its members. According to Cobb et al., (1993), "the background for mathematical activity and dialogue is constituted by the results of prior negotiations of mathematical meanings as well as by current social norms" (p. 112). As Lampert (2001) asserted, developing classroom mathematical norms happens at the beginning of the year but continues throughout the year as norms are continually renegotiated.

Creating norms does not occur solely as a result of the teacher's statements or actions. As teachers and students work together on mathematics, ways of participating in those practices, and the language supporting them, develop. Regularities in these interactions shape the norms and support learning. Thus, the development of norms can be constantly renegotiated. The cause for renegotiation, according to Much and Shweder (1978), often occurs when either the teacher's or child's expectations are not fulfilled or when there is a perceived breach of a social norm.

Cobb, Yackel, and Wood studied the norms that developed in the classrooms in their elementary school project. They documented four types of norms, norms around (a) working together, (b) students' thinking through ideas for themselves, (c) who decides on acceptable explanations, and (d) what types of mathematical reasons are acceptable. They found that acceptable explanations were mathematics-based and not status-based. Acceptability of explanations was not decided by the teachers' response; students were expected to show mathematically why a solution was correct. In fact, project students were significantly more likely to believe that one succeeds in mathematics by attempting to make sense of things and significantly less likely to believe that success stems from conforming to the teacher's or other children's solution methods, from attempting to be superior to classmates, or from being lucky, neat, or quiet (Nicholls et al., 1990; Yackel et al., 1991). In addition, rather

than providing a procedural explanation of the steps to a solution without reference to the quantities or the mathematical ideas, students expected to reference the mathematics.

McClain and Cobb (2001) studied the process through which sociomathematical norms of acceptable mathematical solutions, mathematical difference, and mathematically efficient solutions were established in a first-grade classroom. They focused particularly on the teacher's proactive role. They found that one of the teacher's primary challenges was to overcome the dichotomy between providing all the direction and structure for students when discussing solutions and following the students' own interests and interpretations. In addition, they found a relationship between the emergence of particular sociomathematical norms. As the class worked on what counted as a mathematically different solution, the teacher leveraged those ideas to support the development of what counted as an efficient or sophisticated solution. Further, the norms for what counted as an efficient solution gave direction to students' learning about arithmetic and supported the development of flexible reasoning, in particular towards algebraic understandings of properties of operations.

Significant in McClain and Cobb's (2001) study of the emergence of sociomathematical norms was the role teacher's notation played in supporting the development of norms of mathematical efficiency and sophistication (see also McClain, 2000). This finding adds to the understanding that talk is not the only contributor to the development of sociomathematical norms. For example, how the teacher chose to represent and notate students' strategies provided lasting records of ideas students used as they made judgments about efficient solutions. In middle school classrooms, Hershkowitz and Schwarz (1999) also noted that the use of nonverbal interactions between students and tools, in this case multi-representational software and multiphase activities, contributed to the development of sociomathematical norms. In their study, how students engaged with the tools themselves, as well as with the activity structure, supported sociomathematical norms of what counts as evidence and what counts as a good hypothesis.

Kazemi and Stipek (2001) found four sociomathematical norms distinguished classrooms in which there was a press for conceptual learning and understanding: (a) explanations were supported by mathematical reasons, (b) mistakes created opportunities to engage further with mathematical ideas, (c) students drew mathematical connections between strategies, and (d) consensus was reached through mathematical argumentation. They concluded that when teachers created a high press for conceptual thinking, resulting in the development of these four types of norms, mathematics drove not only the activity but the students' explanations as well, thus increasing student understanding.

Silver et al., (1995) found that creating an atmosphere of trust and mutual respect in QUASAR middle school classrooms was critical for the development of mathematical discourse communities. Specifically, they reported that in supporting students to share alternative explanations and question each other's ideas, it was important to avoid criticizing and blaming and to focus on respecting each other's ideas (Silver et al., 1995). They pointed to the importance of norms developed around attention to language. Here they talked about supporting and attending to the language of the students, developing a shared vocabulary, respecting a range of ways of using language to share ideas, and supporting the sharing of information in a form everyone can understand (Silver, 1996). They found it challenging to get students to participate together in ways that challenged one another's thinking and justified their ideas. They suggested that developing the social norms might first need to occur prior to developing those norms focused particularly on the mathematics (Silver & Smith, 1996).

Emerging Complications in the Negotiation of Norms

A number of studies have also drawn attention to the impact that traditional mathematics instruction has on students and the challenges created for teachers who attempt to create new classroom norms based on mathematical reasoning. These studies point to the need to attend to the dispositions and values that students learn as classroom norms are negotiated.

Boaler (2001) found that if students encountered textbook situations that departed from their expectations, they became confused not only because of the limited extent of their mathematics knowledge but also because of the regularities of the mathematics classrooms to which they had become accustomed (Greeno & MMAP, 1998). When students engaged in assessments that asked questions in a different way than in their classrooms, they tried to draw upon their classroom practices and found that these practices were limited in helping them respond to the task. They could not (or did not) use the mathematics they learned. Students' practices were highly specific to the mathematics classrooms, and when they were in different situations, even exams, the students became confused because they tried to follow the cues they had learned in class and found that those practices did not help. Boaler (2001) concluded that students learn a set of practices "and these come to define their knowledge" (p. 9). In class, students who adapted and applied the mathematical methods and who discussed ideas and situations with different people could use a variety of practices and more readily found ones that could be used in the "real world."

In interviewing students Boaler and Greeno (2000) found that many of the students who talked about disliking mathematics felt so not because of the cognitive demands of the mathematics or because they were unsuccessful (often they were quite successful) but rather because they wanted to pursue subjects they felt offered opportunities for expression, interpretation, and agency (they saw mathematics as about receiving knowledge). These findings underscore the sometimes invisible impact of classroom practices that promote traditional approaches to mathematics as a discipline. The findings depart markedly from virtues promoted by Lampert, Cobb, Yackel, and others of developing students' intellectual autonomy and virtues, discussed above, of the importance of intellectual autonomy, intellectual courage, intellectual honesty, and wise restraint.

Civil and Planas (2004) documented the challenges to creating sociomathematical norms for mathematical deliberation in classrooms, particularly for low-income students of color. Students in the classes studied were not accustomed to analyzing one another's work, especially within the mathematics class (Civil, 2002). As teachers worked to establish a set of sociomathematical norms (involving students engaging together around mathematical thinking and exchanging ideas), the researchers found that forms of participation were influenced by the organizational structures at school and the emerging memberships they created. Civil and Planas reported that when they tried to open up the patterns of participation in the classroom, power and status structures were deeply engrained. Students who were athletes or from a gifted class had the status and power in the class that made it difficult for others to participate. The athletes and "gifted" students were listened to and their ideas given credence over others regardless of the mathematical argument put forth. Other students, without this status, did not see it their place, or worth their time, to participate. These existing forms of participation, often created by structures outside the classroom, made creating particular types of norms challenging at best. Civil's work (1995, 1998, 2002) leads to a consideration of how to support teachers as they attempt to recognize and address the complexity involved in negotiating such norms.

Developing Norms and Teacher Learning

We have argued throughout the chapter that one aspect of designing classroom practice is to consider teachers' learning through practice. Cobb and his colleagues have pointed to how teachers in their project learned through their interaction with their students and how this learning was shaped also by the norms that were negotiated (Cobb et al., 1989; Wood et al., 1991). This is particularly true for the development of sociomathematical norms. If teachers are expected to facilitate discussion, teachers need the opportunity to make sense of a wide array of student solutions. This need returns us to the importance of teachers' developing deep knowledge of students' thinking and subject matter (Yackel, 2002). Teachers must learn how to identify increasingly sophisticated solutions and know how to support students as they detail them and justify them through conversations. The kind of sociomathematical norms that teachers attempt to develop in the classroom will reflect how they are able to make the mathematics central as they attend to the sharing, make sense of a justification, or decide about the sophistication of a solution. Thus, teachers' learning opportunities are shaped by the sociomathematical norms that are developed in the classroom. If only social norms are developed, such as sharing solutions so that ideas are on the table, and the mathematical differences among strategies are not consistently part of the conversation, teachers do not have to listen for them, focus on them, or learn from them.

Summary

In this section, we have argued that as teachers plan and carry out instruction, they must attend to the norms that govern interactions in general and interactions around mathematics in particular. The social and sociomathematical norms that emerge in classrooms have consequences for what students learn about particular mathematical ideas as well as what it means to do mathematics. As many researchers have also argued, the norms that define the classroom culture also impact the kinds of intellectual virtues that students learn. Negotiating norms for engaging in mathematical deliberation together is not without its challenges and complications. In this section, we again drew on research of students' experiences in classrooms to raise questions about how the negotiation of norms positions students with one another and with the subject matter. Negotiating sociomathematical norms demands, then, that the teacher know not only how to manage discourse but also how to draw on subject-matter knowledge and knowledge of students' mathematics, how to use tools strategically, and how to build relationships with students. These activities also imply that teachers need to learn from their practice as they engage in this work.

BUILDING RELATIONSHIPS FOR DOING AND LEARNING MATHEMATICS

Developing norms that open opportunities for participation in ways that support the development of mathematical discourse can be challenging (Civil & Planas, 2004; Silver, 1996; Silver & Smith, 1996) and requires building relationships with students. Building relationships with students allows teachers to challenge assumptions about who students are and what they bring to the mathematics classroom in a way that supports creating opportunities for participation and mathematical learning. Building relationships with students involves understanding their identities and the experiences that have shaped them as well as allowing the students to get to know one another and the teacher. Attending to understanding mathematics requires that teachers attend to the details of the students' work, the various identities students develop in relation to the mathematics and the classroom community, and the cultural histories that shape students' identities and participation.

Within mathematics education, significant work has focused on getting to know students' mathematical thinking. Knowing about students' mathematical thinking involves knowing the details of how one's students are making sense of particular mathematical ideas and at the same time knowing in general how students develop in relation to a particular mathematical idea. Knowing students' mathematical thinking provides the substance for the mathematical conversations. The benefits of knowing students' mathematical thinking and the trajectories that describe particular mathematical development is well documented (National Research Council, 2002).[6]

Simon (1997; Simon et al., 2004) emphasized the detailed knowledge of student trajectories teachers need to support their classroom practice. The work of the teacher is embedded in the decisions teachers make around students' trajectories. Teachers' knowledge of the cognitive learning trajectories of each student in

---

[6] We cannot stress enough how critical this aspect of teachers' work is. We do not detail research on knowing students' mathematical thinking here as it is detailed both in other chapters in this volume and in a number of content-based reviews.

the class provides the basis for choosing tasks that support students' reflective abstraction. These trajectories provide a way to think more clearly about the mechanisms underlying students' acquisition of particular knowledge. Mechanisms are critical here. Understanding the mechanisms enables teachers to choose activities or tasks that provide opportunity for learning.

Cognitively Guided Instruction: A Detailed Example

Cognitively Guided Instruction has much to say about classroom practice that supports the development of students' mathematical understanding. Although the focus of the CGI work was not detailing instruction, much can be drawn from the body of work that can inform classroom practice. The work focused on supporting teachers and learning with teachers as they engaged with research-based information about students' trajectories within particular whole number domains. The data from the experimental study of first-grade classrooms, the qualitative studies of six teachers over a year, the longitudinal study of first-through third-grade teachers, and the follow-up study to the longitudinal studies are quite consistent (Carpenter, Fennema & Franke, 1997; Carpenter et al., 1999; Carpenter et al., 1989; Carpenter et al., 1998; Fennema et al., 1996; Fennema, Carpenter, Franke, & Carey, 1993; Franke et al., 2000). Together these studies provide links between teacher learning and student outcomes, and between teacher learning and changes in classroom practice. Teachers who listened to their students' explanations and asked for multiple strategies to problems had higher achieving students on written measures of skill and problem solving. Spending twice as much time learning number facts did not mean students could recall their number facts better; actually, students in classrooms where teachers spent more time on problem solving and half as much time learning their number facts were more successful using recall to solve problems. Teachers who could report accurately and in detail about their students' mathematical thinking had higher achieving students. These teachers were also ones who continued to learn through their practice.

In the CGI longitudinal study of first-through third-grade teachers and their students and in the follow-up study, 4 years after the professional development work ended, CGI researchers found that although many teachers figured out how to ask their students about their thinking and share strategies it was more difficult for teachers to engage in discussions that required students to compare strategies (Fennema et al., 1996; Franke et al., 2000). They also found that teachers who were able to detail the development of student's mathematical thinking, who asked students how they solved problems and made use of what they heard from students in their instruction, were those who saw themselves as continuing to learn from their students. These teachers, engaged in what CGI researchers termed *generative growth*, believed that they had a way to make sense of what they learned from interacting with their students, a way of making sense mathematically of what they heard from students. Here teachers had both knowledge to support their practice and a stance toward that knowledge, a stance that they could add to and create ongoing understanding for themselves and their students.

CGI provides evidence that teachers' classroom practice that (a) includes eliciting and making public student thinking, (b) involves eliciting multiple strategies, (c) focuses on solving word problems, and (d) uses what is heard from students to make instructional decisions led to the development of student understanding. Further, teachers whose classroom practice drew on their (a) detailed knowledge of the development of students' mathematical thinking within a domain, (b) organization of student thinking in relation to the mathematical content, and (c) notions that they could continue to learn from their practice contributed to the development of student learning. Clearly, this could be interpreted simply and pronouncements could be made about "just get teachers to ask their students how they solved a problem," but it is important to note that factors about the teachers themselves and their classroom practice evolve together, and it is that coming together that drives how teachers engage in their classroom practices.

Beyond Knowing Children's Thinking

Research then provides substantial evidence that building relationships with students around the specifics of the mathematical work is critical to developing understanding. However, we argue that relationships based on getting to know students' mathematical thinking alone are limited. Teaching to open opportunities for participation requires getting to know students' identities, histories, and cultural and school experiences, all in relation to the mathematical work.

> Educators need to know what happens in the world of the children with whom they work. They need to know the universe of their dreams, the language with which they skillfully defend themselves from the aggressiveness of their world, what they know independently of the school and how they know it. (Freire, 1998, p. 72)

When the classroom environment and pedagogical practice are brought into line with the lived experiences of students, student performance is positively influenced (see reviews by Dilworth & Brown, 2001 and Gallego, Cole & Laboratory of Comparative Human Cognition, 2001). In addition, without attention to students' cultural and racial backgrounds, mathematics classrooms will continue to play a role in producing and maintaining the inequities that exist in society (e.g., Gutstein, 2003; Moses & Cobb, 2001; Oakes, Muir, & Joseph, 2003).

We purposely focus this section on building relationships as we intend to push beyond seeing teachers' work as eliciting students' thinking in mathematics and consider what it means for teachers and students to get to know one another in ways that lead to different opportunities for participation in mathematics for teachers and students, challenge the status quo, and allow for the possibility of the distribution of student and teacher voices.

Drawing on Ideas of Identity and Culture[7]

Embedded in current definitions of mathematical understanding are notions of students' perceptions of themselves in relation to the mathematics. Notions of identity are included in understanding not only because educators want students to like mathematics for the purposes of engagement and motivation over time but also because developing a particular stance towards the mathematics is part of understanding the mathematics (National Research Council, 2002). The relationship students develop with mathematics—how they see themselves in relation to the mathematics—influences how they participate, how they make sense of the mathematics, and the different ways they persist. This is all a part of understanding mathematics. This places one's mathematical identity not as something peripheral to doing mathematics but as central.

Gee (2001) described identity as "being recognized as a certain 'kind of person' in a given context." Wenger (1998) described identity as the "pivot between the social and the individual" (p. 145). Both would argue that identities are constantly being formed and re-formed in the dialectic between social structures and individuals' lived experiences. Identities reflect histories of engagement in sociocultural communities, institutions, and practices. As Stuart

Hall (1996) explained, "[This view] accepts that identities are never unified and, in late modern times, increasingly fragmented and fractured; never singular but multiply constructed across different, often intersecting and antagonistic, discourses, practices, and positions" (p. 4).

These conceptions of identity illustrate the importance of thinking about classrooms as places where identities are constructed and reconstructed through discourse. In many mathematics classrooms the identities constructed through discourse practices perpetuate beliefs about who is mathematically literate, and often this construction differentiates students on racial lines (Boaler, 2002; Martin, 2000).[8] A number of researchers have begun to explore how understanding identities in the context of a classroom's discursive practices can help us understand students in a new way and see how to challenge existing inequities. For example, students are often located and locate themselves within dichotomous identities of achievement: successful versus struggling. Students' own sense of struggle may be hidden behind what they find appropriate to share and divulge about themselves to their peers or their teacher (see, e.g., Dutro et al., in press).

Researchers have recently begun to examine how individuals are positioned and position themselves within the mathematics classroom context (Gresalfi, 2004; Martin, 2000; Nasir, & Saxe, 2003). Just as we researchers have learned a great deal about the different ways that males and females may participate in and with mathematics, we are now just beginning to understand how students from different races and cultural backgrounds may also have different participation histories that influence how they engage in mathematics. Often minority students believe that they must choose between a strong academic and a positive ethnic identity (Nasir & Saxe, 2003). How the tensions continue to play out matters for how students engage in school and in mathematics. Martin (2000, 2003, 2006, in press) studies the co-construction of African American and mathematics identities, highlighting the challenges that many African Americans face in negotiating positive identities as mathematics learners. Martin has shown that mathematically successful African American students not only maintain their sense of self as African Americans and as students but also construct a number of other positive identities;

---

[7] We draw heavily here on work outside of mathematics education. This work has been reviewed in a number of places (see Dilworth & Brown, 2001) and not only serves as a basis for conceptualizing what can happen in mathematics classrooms but also provides details we see useful in mathematics classrooms.

[8] We recognize that race is not the only way students are differentiated (e.g., language, gender, sexual orientation). However we focus here specifically on issues of race as we see the need for mathematics classrooms to attend to race more explicitly. Research on race and mathematics teaching is limited. See equity chapter in this volume for a more detailed examination of these issues.

they are not "acting white." He frames mathematics teaching and learning as *racialized forms of experience*—where meanings for race and racism emerge in the day-to-day struggles for mathematics literacy among the African American adolescents and adults he has studied. In all of their stories of mathematics teaching and learning, race was salient. Martin (2006, in press) connects the identities developed by the African Americans in his study to the sociohistorical and structural forces that shaped the kinds of mathematics practices in which students participated. Martin (in press) argued for the need to

> acknowledge that mathematics learning and participation, like many other areas in life can be viewed as *racialized forms of experience*, that is, as experiences where race and the *meanings for race* emerge as highly salient in structuring (a) the way that mathematical experiences and opportunities to learn unfold and are interpreted and (b) who is perceived to be literate, and who is not.

The complexity of issues of identity for students of color in the United States highlights the importance of understanding how minority students structure and manage emerging tensions as they construct and negotiate ethnic and academic identities in the course of their everyday activities.

Drawing on Cultural Practices

Identities are shaped, constructed, and negotiated through cultural practices (Holland, Lachoitte, Skinner, & Cain, 1998; Lave & Wenger, 1991; Martin, 2000; Moll & Gonzalez, 2004; Nasir, 2002; Wenger, 1998). Understanding cultural practices becomes essential to understanding students' evolving identities, both the cultural practices students engage in outside of school and those of the classroom.

Leacock's (1971) definition of culture

> refers to the totality of a group's learned norms for behavior and manifestations of this behavior. This includes the technological and economic mechanisms through which a group adapts to its environment, its related social and political institutions and the values, goals, definitions, prescriptions, and assumptions which define and rationalize individual motivation and participation (p. 35).

So individuals' ways of participating are shaped by the norms, values, and goals of the group. As teachers and students together create a new set of norms they must understand the existing norms for doing and talking to one another. Understanding the cultures—the

groups students have participated in—can help the teacher build relationships based on a depth of understanding and not on assumptions or stereotypes.

A challenge to developing relationships with students that draw on their histories and cultural experiences involves the deficit views of families that frequently exist. These deficit views often permeate mathematics classrooms (Oakes et al., 2003; Spencer, 2006). A number of researchers argue for the need to change deficit views of families and communities and look to draw on the strengths that exist (Brice-Heath, 1982; Civil, 2002; Cole, Griffin, & LCHC, 1987; Moll & Gonzalez, 2004). Researchers drawing on the *cultural difference* perspective rather than the *cultural deficit* perspective argue that teachers' work requires that they find ways to reduce the cultural mismatch between the home and classroom cultures. The goal of reducing the mismatch is to use what the children already know, along with the associated cultural practices, as resources for understanding in the classroom (Moll & Greenberg, 1990, cited in Gallego et al., 2001). Many researchers have investigated and developed strategies to reduce mismatch, and Gallego and her colleagues in analyzing this work stated that, "these examples make it clear that a strategy of local accommodations of school culture to home culture can be educationally productive" (p. 979).

Moll and his colleagues documented how households contain accumulated *funds of knowledge*, the skills, abilities, and practices that support the functioning of the household. Moll and Gonzalez (1997) pressed teachers to come to know the cultures of their students. They argued that teachers need to become qualitative researchers in some ways as they come to understand students in their home environments, detailing the strengths and skills students use in the home that can be used to understand participation (including language use) in school. This work explicitly rejects the notion that the problem of underachievement either is located within the students or is due to students' culture, and it shifts responsibility for underachievement to understanding what students do bring to school and how schools can draw on the wealth of knowledge and experience that the student brings (Moll, 1992).

Civil has taken these ideas into the mathematics classroom through the BRIDGE Project. The explicit goal of the project was the development of mathematics teaching that builds on students' backgrounds and experiences (Civil, 2002; González, Andrade, Civil, & Moll, 2001). Although this work points to the potential of bringing parents and families into schools so teachers see students' strengths, it also makes explicit the challenges of actually building on the everyday mathematics knowledge and experiences of students in schools. Civil found that often the mathematics in everyday activity was hidden: People did not see themselves as doing mathematics and often rejected the idea that what they were doing was mathematics (Gonzalez et al., 2001). And although this work is challenging it does begin to point to the fact that students' competencies identified outside of the classroom are often not used as a resource in school.

Hand (2003) analyzed classroom dynamics in two reform-driven algebra classrooms in an urban high school. She found what she has termed a *participation gap* between the two classrooms. In one classroom any activity that was not directly related to the lesson was treated as a potential distraction to students' learning and thus prohibited. In contrast, in the other classroom, the teacher searched for links between mathematics and the knowledge and practices that the children brought with them to the classroom. In the class where the teacher validated students' contributions, greater participation was evident. Erickson and Mohatt (1982) found that when teachers used language patterns that approximated the students' home cultural patterns or balanced these patterns with the Anglo instructional patterns typically used in school, they were more successful in improving student academic performance. Here the goal was to include student culture in the classroom as authorized knowledge. In work with African American students researchers have found that teachers have successfully built on forms drawn from African American churches. The teachers built relationships in their classrooms that were marked by social equality, egalitarianism, and mutuality stemming from a group rather than an individual ethos. The focus was on collective and personal responsibility (Henry, 1992; Hollins, 1982; Foster, 1993, all as cited in Dilworth & Brown, 2001).

The work of Lipka and the Alaskan Natives, a group of indigenous people of Alaska whose language and culture has been marginalized within the educational system by its emphasis on English-only language and middle-class White American cultural practices (Lipka, 1994, 2002, 2005; Lipka, Wildfeuer, Wahlberg, George, & Ezran, 2001), draws upon the knowledge of the Yup'ik elders in making curricular decisions. Yup'ik elders, teachers, bilingual aides, students, school-board members, and university faculty created a working environment where both insider and outsider knowledge was valued. In this way the teaching of mathematics became a community project that drew on and respected both local knowledge and more traditional mathematics. The researchers have documented that the achievement of students whose teachers used the curriculum after receiving professional development surpassed that of comparable groups of students (Lipka, 1994, 2002; Lipka et al., 2005; Lipka, Wildfeuer, Wahlberg, George, & Ezran, 2001).

Moschkovich (2002), through her work on discourse and language, has pushed for making the mathematical practices of different groups accessible to students in classrooms. She has stated that it is not the using of everyday practices that is itself important but rather making connections between familiar practices and academic mathematical practices. She suggested that there needs to be a balance between the everyday and academic mathematics so students can be encouraged to engage in the study of mathematics while also providing them the discursive practices necessary for pursuing mathematics. The goal of the work that draws on students' cultural experiences is to understand and use the resources students bring to the classroom, thereby allowing students to accept and affirm their cultural identity while at the same time developing a critical perspective that can challenge the existing accepted practices of schooling. This work points to the value of teachers' understanding students' identities and cultural practices in building classroom relationships. Bringing these identities within the classroom walls legitimizes different forms of participation and allows for more diverse mathematical knowledge.

Teaching As Building Relationships Around Identity and Culture

Although we can make a theoretical argument and provide a number of existence proofs that attending to both identity and cultural practices in building relationships with students makes a difference in students' participation, the field continues to try to identify what exactly it means to attend to identity and cultural practices in mathematics classrooms. We argue here that educators can learn from culturally relevant pedagogy (and related literatures) and the critical stance it takes in providing opportunities for students of color and use it as a way to think about the types of relationships that can be built with students, families, and communities to shape the teaching and learning of mathematics.

Culturally relevant teaching, from Ladson-Billings's (1995) perspective, must develop students academically, support and nurture cultural competence, and develop a sociohistorical or critical consciousness. Culturally relevant teachers organize their classrooms and interactions with their students in ways that are guided by the following principles: (a) Treat students as competent and they will likely demonstrate competence, (b) provide instructional scaffolding so students can move from what they know to what they need to know, (c) focus the classroom on instruction, (d) extend students' thinking and abilities, (e) gain in-depth knowledge of both students and the subject matter, and (f) link student understanding to meaningful cultural referents. How these principles are instantiated matters greatly. Although some may argue that these principles signify good mathematics teaching, what differs here is the focus on the relationships with students, and how students are respected for who they are and what they bring to the classroom (Ladson-Billings, 1995). Ladson-Billings has argued that culturally relevant teaching fosters student achievement and provides ways for students to maintain cultural integrity, develop academic success, and recognize, understand, and critique current social inequities (Ladson-Billings, 1995).

Many scholars have pursued the notion of culturally relevant teaching (see reviews by Gallego et al., 2001; Mercado, 2001). And a growing body of research shows how African American, Latino, and Native American students draw on their cultural and community knowledge to help them succeed in mathematics (Gutstein, 1997, 2003; Martin, 2000; Nasir, 2002).

Gutstein, Lipman, Hernandez, and de los Reyes (1997), in a multiyear study of mathematics teaching in a Mexican-American community, focused on three components they saw as critical to culturally relevant teaching: thinking critically in general and about mathematics, building on both students' informal math knowledge and their cultural and experiential knowledge, and gearing teachers' orientations to students' cultures. In their observations of mathematics lessons, the researchers watched teachers encourage students to explain and justify their answers, push students for multiple interpretations, and encourage mathematical communication (p. 720). They observed teachers' using students' out-of-school knowledge to help students understand mathematical concepts. And teachers came to understand and acknowledge students' cultures by building relationships with students and families.

Gutstein (2003) continued this work through a 2-year study in an urban, Latino seventh/eighth grade mathematics classroom where he taught using these principles of culturally relevant pedagogy along with reform curricula. Gutstein worked to help students develop sociopolitical consciousness, a sense of agency, and positive social and cultural identities while also supporting them to "read" the world using mathematics, developing students' mathematical power, and changing their disposition towards mathematics. He found that, although it was challenging to merge his various goals, attending to both sets of goals supported student participation and learning.

Like Ladson-Billings and Gutstein, Frankenstein (1990, 1995, 1997) has looked for ways to connect to students' lived experiences so that students develop mathematical understanding while learning to challenge social inequities. Frankenstein operated from the position that effective curricular choices build upon students' existing funds of knowledge. She found that students were often reluctant to acknowledge—despite the gap in their formal mathematical knowledge base—that they are already quite capable logical thinkers in their everyday lives as consumers and workers. In addition, Frankenstein found that several of her students had a very particular vision of what counted as legitimate school knowledge, and as such she faced some resistance to addressing social justice issues through mathematics. For Frankenstein, mathematics can be a tool for grappling with social issues. She argued that the incorporation of ethnomathematics into mathematics curricula can be both culturally and intellectually empowering for students and can lead to increased participation in mathematics.

Most of this work studies exceptional teachers trying to challenge social inequities. However, Gau Bartell engaged more typical teachers in learning to teach mathematics for social justice. Gau Bartell (2005) engaged eight secondary mathematics teachers through the creation, implementation, observation, and revision of a mathematics lesson that incorporated social justice goals. She found that the teachers struggled in balancing the mathematics and social justice. In one group, mathematics "trumped the social justice," as teachers never addressed social justice issues, and for the other group, the mathematics was often not connected with social justice. In this group, teachers were so focused on guiding students to a particular idea that they supported conclusions that the mathematics did not. Although this work speaks to the difficulty for teachers to change their practice to incorporate social justice issues, researchers are just beginning to develop these ideas. It represents a fruitful area for future inquiry.

Summary

Within this section on building relationships the theory and literature point to the importance of understanding students beyond how they think about mathematics. Although understanding how students think about mathematics is critical and has been shown to make a difference for teachers' and student learning, understanding students in terms of their race, cultural histories, and previous experiences enables teachers to build relationships that challenge assumptions and open opportunities.[9]

Just as knowing how individual students' thinking is supported by knowing trajectories of students within particular content domains, knowing students' identities in relation to school and mathematics is supported by knowing the communities in which they have participated. If teaching involves orchestrating and negotiating participation, as we have argued here, teachers need to understand students' participation histories. Teachers need to recognize the forces inside and outside that shape the multiple identities for students as they engage in mathematics.[10] Teachers need to know how to draw students' identities into the mathematical work, support them to evolve in how they participate, honor different forms of participation, and structure opportunities that allow for different participation forms. We have described some work that begins to document what this can look like in a mathematics classroom. Yet researchers still know little; we particularly know little in terms of the type of critical stance some would argue for in creating classroom cultures and supporting participation. Teachers clearly, though, hold many assumptions about how certain groups of students do or should participate that can narrow forms of participation (often due to their own experiences), and opportunities are needed within the context of teaching to learn to challenge those assumptions in ongoing ways so that each student is challenged to participate in and understand the mathematics.

"Bringing the lived experiences" of students into the mathematics classroom is both challenging and controversial. It is challenging in that researchers who have begun to try to accomplish this have found resistance from a variety of sources, including the students themselves. It is controversial because including lived experiences can be accomplished in a way that focuses on deficit and not strength. The challenging and controversial nature of the work only points to the need for building relationships in classrooms that allow for a multidimensional understanding of each other and the mathematical work.

CONCLUSION

We began this chapter with two brief scenarios drawn from the mathematical teaching practices of Ms. Michaels and Ms. Jimenez. Both Ms. Michaels and Ms. Jimenez engaged in some aspects of practice supported by research findings that emerged from large-scale reform projects. They posed a sequenced set of problems, elicited student thinking, asked follow-up questions, facilitated classroom discussion around mathematical ideas, and used what they knew about students' mathematical thinking. They each engaged in these practices to different degrees and in somewhat different ways, and those differences had an impact on student participation. What is unknown from the short description of a single lesson in these teachers' classrooms are the details about how the teachers orchestrated mathematical conversation, the norms established in each of the classrooms around what it means to do mathematics, and the kinds of relationships teachers had developed with their students. Knowing more about each of these would enable us to have a better sense of how and why the practices in these two classrooms played out in the way they did and how they shaped student participation. We argued throughout this chapter that there have been important advances in research on classroom practice since the publication of the last *Handbook*. We can articulate and detail some critical features of classroom practices. We know what to begin to look for in Ms. Michaels's and Ms. Jimenez's classrooms. We recognize the need to know more about what surrounds those classroom practices, and this is illustrated by what we do not know about Ms. Michaels's and Ms. Jimenez's classrooms, no matter how much detail we provide about the interaction from one day. We believe that the field is in a productive place to move understanding of classroom practice forward by attending more closely to the relationship between the already articulated critical features of classroom practices and the surrounding discourse, norms, and relationships.

Three Features of Classroom Practice

We chose to focus the chapter on three features of teachers' classroom practice: discourse, norms, and building relationships. We chose to focus on these three aspects of mathematics classroom practice because consensus is building that students need opportunities in classrooms to share their mathematical thinking, discuss alternative approaches to solve problems, use mathematical tools flexibly, and so on. Providing students these types of opportunities requires

---

[9] One needs to be careful here in avoiding essentializing by drawing on narrow preconceived notions of cultures. Notions and understandings about cultural histories need to be continually challenged to support the development of a complex set of understandings about culture and the student in relation to his or her culture.

[10] We see this as part of what Ball, Goffney, and Bass (2005) called for in having teachers bring an *equity awareness* and willingness to act on that awareness to the mathematics classroom.

that teachers and students develop new ways to engage together around mathematical ideas—specifically, it requires new forms of mathematical discourse. Creating opportunities for discourse demands attention to the negotiation of both social and sociomathematical norms. The negotiation of these norms to support discourse requires building relationships between teachers and students that include attention to the identities students bring, their participation histories, and the norms and cultural practices of their communities. All of this is needed so that assumptions are continually challenged and renegotiated around what it means to participate and be literate in mathematics. And within this, issues of race, class, and gender need to be made salient.

The research on classroom discourse and specifically mathematics discourse has been accumulating over a number of years, and educators are in the position of beginning to understand the kinds of foci that can facilitate understanding mathematics classroom practice. The work on revoicing, interrogating meaning, and supporting language provides details around what teachers can attend to in managing discourse. And although researchers are accumulating evidence about the details of when and how to use these ideas to manage discourse, further research is needed, particularly to relate how teachers manage discourse to the norms that govern discourse and the teacher-student relationships that are built and enacted.

We also have evidence that classroom norms that attend to the details of working through mathematical ideas are critical for creating classroom cultures where students can develop mathematical understanding. The work on sociomathematical norms has pushed the field to consider the kinds of norms needed to support mathematical work and mathematical reasoning. We also have evidence that creating the norms that expect students to share their mathematical thinking, challenge and test ideas publicly, and work to make sense of one another's ideas is not easy; it involves issues of power and positionality as well as issues of identity that develop both inside and outside the classroom and are shaped by the societal, community, and school structures. This all makes building relationships with students critical to the success of developing classroom norms that support the development of mathematical understanding.

Evidence clearly exists to support the need for developing relationships with students, relationships built upon an understanding of students' past experiences, their cultural histories, their race, and their participation histories. Teachers who attend to the cultural backgrounds of their students, who build relationships with students as well as their families

and communities, make a difference in the success of students in their classrooms. Assumptions need to be challenged about who can do mathematics, what counts as mathematical argument, and what it means to participate in mathematical work. These assumptions are constructed around long-standing personal and societal beliefs about "groups" that shape interactions with individuals. Developing relationships with students that push teachers to get to know the multiple identities students bring, how those identities have been shaped, and the communities that they learn from provides teachers with a different way to see students and opens opportunities for what it means to participate in the mathematics classroom. We recognize that building this type of relationship with students will not resolve all of the equity issues in mathematics classrooms. We recognize the societal, structural basis for much inequity. However, we (as do others) see that building different kinds of relationships and opening different opportunities for participation and practice can lead to using mathematics to help transform what happens for students of color, English language learners, students living in poverty, and other marginalized groups.

Advancing Research on Classroom Practice

In the conclusion to this chapter, we would like to propose the direction needed in order to advance research on classroom practice. To do this, we draw on Lampert's (2005) and Graziani's (2005) recent work studying Italian language instruction in Rome. Following their lead, we argue that the three features of classroom practice we have discussed, although important in helping us think about the work of teachers and the experiences of students, still need to be assembled together and studied as they emerge, unfold, and lead to student learning and empowerment. We propose that the field needs to identify routines of practice and study how the three features of classroom practice discussed in this chapter work together in such routines. By *routines of practice* we mean that we need to agree on core activities (within each mathematical domain and at appropriate grade levels) that could and should occur regularly in the teaching of mathematics. We propose that such routines could become the central hub for teachers' practice and should become the focus of the study of classroom practice. Research would clarify how these routines of practice support teacher and student learning in different contexts with different teachers and students.

To connect to Simon's writing about learning trajectories (1997; Simon et al., 2004), these core routines of practice should be set within an explicit framework for understanding the development of students' mathematical competence in any particular domain. In addition to our knowledge of students' thinking, we also want to underscore the role of mathematics. As seen through the other chapters in this volume, researchers are accumulating significant knowledge about particular domains of mathematics and how the content itself makes a difference in how ideas evolve in classrooms and in work with teachers (see Franke, Carpenter, & Battey, in press). Mathematics itself needs to be one of the fundamental considerations in both identifying the core practices and analyzing the surround—in how to consider how these core practices play out in classrooms. We cannot ignore that orchestrating mathematical discourse around algebraic thinking may need to look different than orchestrating discourse around addition of whole numbers. So although we argue for focus on routines of practice, we want to be careful and explicit in detailing how these routines of practice are embedded within particular mathematical domains. It is with this detail that we feel we will gain the information we need to support teachers and advance research.

In recent writings, we see advances in specifying routines. We would like to consider a few of these in order to begin to compare the level of specificity in which researchers are beginning to tackle this issue and to lay some of the groundwork for what work lies ahead. What Lampert (2005) and Graziani (2005) described in their analyses is that the Italian language school has a sophisticated framework for understanding linguistic competency. Within this framework, key instructional activities or routines of practice have been identified and in fact extensively specified. As teachers learn these routines of practice, they engage simultaneously in understanding teachers' instructional moves and students' contributions and understandings. Set within the framework for understanding linguistic competence, the entire set of instructional routines allows teachers to develop their knowledge and pedagogy in a principled way. The spare set of routines also enables teachers to develop their practices around common objects of study. We can imagine that researchers in mathematics education, within various mathematical domains, could also marry the research on student thinking with classroom practice by beginning to identify core practices that teachers could follow. The routines of practice serve as a set of focal points (all conceptually connected) in which to embed our developing understanding of the interrelations of the features of classroom practice discussed in this chapter.

Stein and colleagues, in their current project, have built one possible model for framing the work of teachers when planning and enacting lessons for the purpose of "moderating the degree of improvisation required by the teacher during a discussion," (Stein et al., 2006). They propose five practices:

1. anticipating likely student responses to rich mathematical tasks

2. monitoring students' responses to the tasks during the explore phase

3. selecting particular students' responses to present their mathematical responses during the discuss-and-summarize phase

4. purposefully sequencing the students' responses that will be displayed

5. helping the class make mathematical connections between different students' responses. (pp. 7–8)

As Stein and colleagues discuss in their paper, teachers must make many decisions within each of these practices, and clearly the work that is summarized in this chapter about the work of teaching is implicated in each of these five practices. However, drawing teachers' attention to the goals and routines for bringing mathematical reasoning into the public space of the classroom and helping students engage with it would advance knowledge of classroom practices that support students' intellectual and dispositional growth. We see the promise of connecting the fourth and fifth practices proposed by Stein and colleagues to a recent close examination of interactional patterns in different types of reform-oriented classes studied by Wood et al., (2006). In *strategy-reporting* classrooms, children present different strategies. "Children presenting may be asked to provide more information about *how* they solved the problem by the teacher but rarely by other student listeners" (p. 224). In *inquiry/argument* classrooms, children also share strategies, but the goal of sharing is for other listeners (both teachers and students) to ask questions for further clarification and understanding. These discussions, "often include a challenge or disagreement from student listeners or teachers, which initiates an exchange that in turn prompts the thinking of justification in support of children's ideas" (p. 224). Their extensive analysis of the patterns of interaction in these classrooms provides nuanced images of discourse patterns that enable participants to (a) pull mathematical ideas together, (b) identify and explain mathematical errors, and (c) develop robust mathematical arguments and reasoning.

What the field needs is to define these routines of practice so that the research and teacher education communities can engage with them, unpack them, and study them.[11] This first step in identifying routines would involve both conceptual and empirical work. How should we define a *routine*? What important mathematical ideas do students work with in particular mathematical domains that would help us converge on particular routines? To what degree should we specify the routine? How do we make visible teachers' work in managing discourse, establishing norms, and building relationships with students as we specify and study the enactment of routines? The routines of practice should be examined in relation to the critical features of classroom practice raised throughout this chapter, connecting the details of practice to the surround in which they are embedded. Understanding the relationships in the details of practice, discourse, norms, and personal relationships allows researchers to look both within and outside of classrooms. This would potentially serve to connect what happens in classrooms to the social and political structures that influence what occurs. As with Civil's work (Civil & Planas, 2004), we would begin to see how the structures of schools and the social and political nature of communities shape what happens for teachers and students in mathematics classrooms.

We recognize that the research directions we propose have already been initiated by many of those cited throughout this chapter who are looking closely at classroom practice and how teachers and students together shape what occurs. We also recognize that this type of research is challenging for a number of reasons. First, the work is challenging because one needs to either find or create classrooms in which the identified routines of practice exist. Second, the data needed to understand the routines of practice in relation to the surround is complicated—it requires detailed descriptions of classroom interactions over time, interviews with teachers and students, and an understanding of the school culture and students' communities. It also requires attention to race, class, gender, and language, and other ways students and teachers may be marginalized within the school and community. Despite these complications, we see tremendous potential for advancing our understanding of classroom practices if researchers take on such challenging work and create a richer, synthetic view of how discourse, sociomathematical norms, and student-teacher relationships contribute to the enactment of classroom practices that support student learning and empowerment.

REFERENCES

Ball, D. L. (1993). With an eye on mathematical horizon: Dilemmas of teaching elementary mathematics. *The Elementary School Journal, 93* (4), 373–397.

Ball, D. L., & Bass, H. (2000a). Interweaving content and pedagogy in teaching and learning to teach: Knowing and using mathematics. In J. Boaler (Ed.), *Multiple perspectives on mathematics teaching and learning* (pp. 83–104). Westport, CT: Ablex.

Ball, D. L., & Bass, H. (2000b). Making believe: The collective construction of public mathematical knowledge in the elementary classroom. In D. Phillips (Ed.), *Yearbook of the national society for the study of education, Constructivism in education.* (pp. 193–224). Chicago: University of Chicago Press.

Ball, D. L., Goffney, I., & Bass, H. (2005). The role of mathematics instruction in building a socially just and diverse democracy. *Mathematics Educator, 15*(1), 2–6.

Boaler, J. (1997). *Experiencing school mathematics: Teaching styles, sex and setting.* Buckingham, U.K.: Open University Press.

Boaler, J. (2001, October). Opening the dimensions of mathematical capability: The development of knowledge, practice, and identity in mathematics classrooms. *Proceedings of the 20th Annual Meeting of the North American Chapter of the Psychology of Mathematics Education Conference* (Vol. 1, pp. 3–21). Snowbird, Utah.

Boaler, J. (2002). *Experiencing school mathematics.* NJ: Erlbaum.

Boaler, J. (2003). Learning from teaching: Exploring the relationship between reform curriculum and equity. *Journal for Research in Mathematics Education, 33*(4), 239–258.

Boaler, J. (in press). *Promoting relational equity: The mixed ability mathematics approach that taught students high levels of responsibility, respect, and thought.* Unpublished manuscript, Stanford University, CA.

Boaler, J. & Greeno, J. G. (2000). Identity, agency and knowing in mathematical worlds. In J. Boaler (Ed.), *Multiple perspectives on mathematics teaching and learning* (pp. 171–200). Westport, CT: Ablex.

Brice-Heath, S. (1982). Questioning at home and at school: A comparative study. In G. Spindler (Ed.), *Doing the ethnography of schooling: Educational anthropology in action* (pp. 102–131). Prospect Heights, IL: Waveland.

Brown, A.L., & Palinsar, A.S. (1989). Guided, cooperative learning, and individual knowledge acquisition. In L.B. Resnick (Ed.), *Knowing, learning and instruction: Essays in honor of Robert Glaser* (pp. 393–451). Hillsdale, NJ: Erlbaum.

Carpenter, T. P., Ansell, E., Franke, M. L., Fennema, E., & Weisbeck, L. (1993) Models of problem solving: A study of kindergarten children's problem-solving processes. *Journal for Research in Mathematics Education, 24,* 427–440.

Carpenter, T., Fennema, E., & Franke, M. (1997). Cognitively Guided Instruction: A knowledge base for reform in primary mathematics instruction. *Elementary School Journal, 97,* 3–20.

Carpenter, T. P. Fennema, E., Franke, M. L., Levi, L. W., & Empson, S. B. (1999). *Children's mathematics: Cognitively Guided Instruction.* Portsmouth, NH: Heinemann

Carpenter, T. P., Fennema, E., Peterson, P. L., Chiang, C. P., & Loef, M. (1989). Using knowledge of children's mathematics

---

[11] There may be different routines of practice across elementary, middle, and high school mathematics.

thinking in classroom teaching: An experimental study. *American Educational Research Journal, 26,* 499–531.

Carpenter, T.P, Franke, M.L., Jacobs, V.R., & Fennema, E. (1998). A longitudinal study of invention and understanding in children's multidigit addition and subtraction. *Journal for Research in Mathematics Education, 29* (1), 3–20.

Cazden, C. (1986). Classroom discourse. In M. Wittrock (Ed.), *Handbook of Research on Teaching* (3rd, ed., pp. 432–463). New York: Macmillan.

Cazden, C. (2001). *Classroom discourse: The language of teaching and learning (2nd ed.).* Portsmouth, NH: Heinemann.

Civil, M. (1995, July). Listening to students' ideas: Teachers interviewing in mathematics. *Proceedings of the Nineteenth Annual International Conference for the Psychology of Mathematics Education,* Recife, Brazil.

Civil, M. (1998). Mathematical communication through small group discussions. In M. Bartolini-Bussi, A. Sierpinska, & H. Steinbrig (Eds.), *Language and communication in the mathematics classroom* (pp. 207–222). Reston, VA: National council of Teachers of Mathematics.

Civil, M. (2002). Culture and mathematics: A community approach. *Journal of Intercultural Studies, 23,* 133–148.

Civil, M., & Planas, N. (2004). Participation in the mathematics classroom: Does every student have a voice? *For the Learning of Mathematics, 24* (1), 7–12.

Cobb, P. (2000). The importance of a situated view of learning to the design of research and instruction. In J. Boaler (Ed.), *Multiple perspectives on mathematics teaching and learning* (pp. 45–82). Westport, CT: Ablex.

Cobb, P., & Bauersfeld, H. (Eds.). (1995). *Emergence of mathematical meaning: Interaction in classroom cultures.* Hillsdale, NJ: Erlbaum.

Cobb, P., & Hodge, L.L. (2002). A relational perspective on issues of cultural diversity and equity as they play out in the mathematics classroom. *Mathematical Thinking and Learning, 4*(2&3), 249–284.

Cobb, P., Wood, T., & Yackel, E. (1993). Discourse, mathematical thinking, and classroom practice. In E. A. Forman, N. Minick, & C. A. Stone (Eds.), *Contexts for learning: Sociocultural dynamics in children's development* (pp. 91–119). New York: Oxford University Press.

Cobb, P., Wood, T., Yackel, E., Nicholls, J., Wheatley, G., Trigatti, B., et al. (1991). Assessment of a problem-centered second-grade mathematics project. *Journal for Research in Mathematics Education, 22,* 3–9.

Cobb, P., Yackel, E., & Wood, T. (1989). Young children's emotional acts while doing mathematical problem solving. In D. B McLeod & V. M. Adams (Eds.), *Affect and mathematical problem solving: A new perspective* (pp. 117–148). New York: Springer-Verlag.

Cole, M., Griffin, P., & Laboratory of Comparative Human Cognition. (1987). *Contextual factors in education.* Madison: Wisconsin Center for Educational Research.

Dilworth, M.E., & Brown, C.E. (2001). Consider the difference: Teaching and learning in culturally rich schools. In V. Richardson (Ed.), *Handbook of Research on Teaching* (4th ed., pp. 643–667). Washington, DC: American Educational Research Association.

Doyle, W. (1985). Classroom organization and management. In M.C. Wittrock (Ed.), *Handbook of research on teaching* (3rd ed., pp. 392–431). New York: Macmillan.

Dutro, E., Kazemi, E., & Balf, R. (2006). Making sense of "The Boy Who Died": Tales of a struggling successful writer. *Reading and Writing Quarterly, 22,* 1–36.

Empson, S. (2003). Low performing students and teaching fractions for understanding: An interactional analysis. *Journal for Research in Mathematics Education, 34,* 305–343.

Empson, S., Turner, E., Dominguez, H., & Maldonado, L. (2006, April). *"Because it's kind of:" Lower achieving elementary students' participation in problem-based mathematics.* Paper presented at the annual meeting of the National Council of Teachers of Mathematics, St. Louis, MO.

Engle, R. A., & Conant, F. R. (2002). Guiding principles for fostering productive disciplinary engagement: Explaining an emergent argument. *Cognition and Instruction, 20,* 399–483.

Enyedy, N., Rubel, L., Castellon, V., Mukhopadhyay, S., & Esmond, I. (2006). Revoicing in a multilingual classroom: Learning implications of discourse. *Mathematical Thinking and Learning.* Manuscript submitted for publication.

Erickson, F., & Mohatt, G. (1982). Cultural organization of participation structures in two classrooms of Indian students. In G. Spindler (Ed.), *Doing the ethnography of schooling: Educational anthropology in action* (pp. 132–175). Prospect Heights, IL: Waveland Press.

Fennema, E., Carpenter, T., Franke, M., Levi, L, Jacobs, V., & Empson, S. (1996). A longitudinal study of learning to use children's thinking in mathematics instruction. *Journal for Research in Mathematics Education, 27*(4), 403–434.

Fennema, E., Franke, M. L., Carpenter, T. P., & Carey, D. A. (1993). Mathematical pedagogical content knowledge in use. *American Educational Research Journal, 30,* 555–583.

Fennema, E., Carpenter, T. P., Franke, M., & Carey, D. (1993). Learning to use children's mathematical thinking: A case study. In R. B. Davis & C. A. Maher (Eds.), *Schools, mathematics, and the world of reality* (pp. 93–117). Boston: Allyn & Bacon.

Forman, E. (1989). The role of peer interaction in the social construction of mathematical knowledge. In N.M Webb (Ed.), *Peer interaction, problem-solving, and cognition: Multidisciplinary perspectives* (pp. 55–70). Oxford, U.K: Pergamon.

Forman, E. (1993). *Contexts for learning.* New York: Oxford University Press.

Forman, E. A., & Ansell, E. (2002). Orchestrating the multiple voices and inscriptions of a mathematics classroom. *The Journal of the Learning Sciences, 11,* 251–274.

Forman, E. A., Larreamendy-Joerns, J., Stein, M. K. & Brown, C.A. (1998). You're going to want to find out which and prove it: Collective argumentation in a mathematics classroom. *Learning and Instruction, 8*(6), 527–548.

Franke, M., Carpenter, T., & Battey, D. (in press). Content matters: The case of algebraic reasoning in teacher professional development. In J. Kaput, & M. Blanton (Eds.), *Exploring early algebra.* Hillsdale, NJ: Erlbaum.

Franke, M.L., & Carey, D.A. (1996). Young children's perceptions of mathematics in problem solving environments. *Journal for Research in Mathematics Education, 28*(1), 8–25.

Franke, M. L., Carpenter, T.P., Levi, L., & Fennema, E. (2000). Capturing teachers' generative growth: A follow-up study of professional development in mathematics. *American Educational Research Journal, 38,* 653–689.

Frankenstein, M. (1990). Incorporating race, gender, and class issues into a critical mathematical literacy curriculum. *Journal of Negro Education, 59*(3), 336–347.

Frankenstein, M. (1995). Equity in mathematics education: Class in a world outside of class. In W.G. Secada, E. Fennema, & L.B. Adajian (Eds.), *New directions for equity in mathematics education* (pp. 165–190). New York: Cambridge University Press.

Frankenstein, M. (1997). In addition to the mathematics: Including equity issues in the curriculum. In J. Trentacosta & M. Kenny (Eds.), *Multicultural and gender equity in the mathematics classroom.* Reston, VA: National Council of Teachers of Mathematics.

Freire, P. (1998). *Teachers as cultural workers: Letters to those who dare to teach.* Boulder, CO: Westview Press.

Gallego, M.A, Cole, M., & The Laboratory of Comparative Human Cognition. (2001). Classroom cultures and cultures in the classroom. In V. Richardson (Ed.), *Handbook of research on teaching* (pp. 951–997). Washington, DC: American Educational Research Association.

Gau Bartell, T. (2005). *Learning to teach mathematics for social justice.* Unpublished doctoral dissertation. University of Wisconsin, Madison.

Gee, J. P. (2001). Identity as an analytic lens for research in education. In W. Secada (Ed.), *Review of research in education, 25.* Washington, DC: American Educational Research Association.

Goldenberg, C., & Gallimore, R. (1991). Local knowledge, research knowledge, and educational change: A case study of early Spanish reading improvement. *Educational Researcher, 20*(8), 2–14.

González, N., Andrade, R., Civil, M., & Moll, L.C. (2001). Bridging funds of distributed knowledge: Creating zones of practices in mathematics. *Journal of Education for Students Placed at Risk, 6,* 115–132.

Graziani, F. (2005, April). *The pedagogical practice of the authoritative professional in professional education.* Paper presented at the annual meeting of the American Educational Research Association, Montreal.

Gresalfi, M. (2004). *Taking up opportunities to learn: Examining the construction of participatory mathematical identities in middle school students.* Unpublished doctoral dissertation, Stanford University, CA.

Greeno, J. G., & MMAP. (1998). The situativity of knowing, learning and research. *American Psychologist, 53*(1), 5–26.

Gutierrez, K.D., Baquedano-Lopez, P., & Tajeda, C. (1999). Rethinking diversity: Hybridity and hybrid language practices in the third space. *Mind, Culture, and Activity, 6*(4), 286–303.

Gutierrez, R. (2002). Beyond Essentialism: The complexity of language in teaching mathematics to Latina/o students. *American Educational Research Journal, 39*(4), 1047–1088

Gutstein, E. (1997). Culturally relevant mathematics teaching in a Mexican American context. *Journal for Research in Mathematics Education, 28*(6), 709–738.

Gutstein, E. (2003). Teaching and learning mathematics for social justice in an urban, Latino school. *Journal for Research in Mathematics Education, 34*(1), 37–73.

Gutstein, E. (in press). "So one question leads to another": Using mathematics to develop a pedagogy of questioning. In Cobb (Ed.), *Diversity, equity, and access to mathematical ideas.* New York: Teachers College Press.

Gutstein, E., Lipman, P., Hernandez, P., & de los Reyes, R. (1997). Culturally relevant mathematics teaching in a Mexican American context. *Journal for Research in Mathematics Education, 28,* 709–737.

Hall, S. (1996). Introduction: Who needs identity? In S. Hall & P. Du Gay (Eds.), *Questions of cultural identity,* (pp. 1–17). London: Sage

Hand, V. (2003). *Reframing participation: Meaningful mathematical activity in diverse classrooms.* Unpublished doctoral dissertation, Stanford University, CA.

Hershkowitz, R., & Schwarz, B. (1999). The emergent perspective in rich learning environments: Some roles of tools and activities in the construction of sociomathematical norms. *Educational Studies in Mathematics, 39,* 149–166.

Hiebert, J. (1992). Reflection and communication: Cognitive considerations in school reform. *International Journal of Educational Research, 17*(5), 439–456.

Hiebert, J., Gallimore, R., Garnier, H. Givving, K.B., Hollingsworth, H., Jacobs, J., et al. (2003). *Teaching mathematics in seven countries: Results from the TIMSS 1999 video study,* NCES (2003-013), U.S. Department of Education. Washington, DC: National Center for Education Statistics.

Hiebert, J., & Stigler, J. (2000). A proposal for improving classroom teaching: Lessons from the TIMSS Video Study. *The Elementary School Journal, 101,* 3–20.

Hiebert, J., & Wearne, D. (1992). Links between teaching and learning place value with understanding in first grade. *Journal for Research in Mathematics Education, 23*(2), 98–122.

Hiebert, J., & Wearne, D. (1993). Instructional tasks, classroom discourse, and students' learning in second-grade arithmetic. *American Educational Research Journal, 30,* 393–425.

Hogan, K., & Pressley, M. (1997). Scaffolded scientific competencies within classroom communities of inquiry. In K. Hogan & M. Pressley (Eds.), *Scaffolded student learning* (pp. 74–107). Albany, NY: SUNY Press.

Horn, I.S. (in press). Accountable argumentation as a participation structure to support learning through disagreement. In A.H. Shoenfeld (Ed.), *A study of teaching: Multiple lenses, multiple views. JRME Monograph series.* Reston, VA: National Council of Teachers of Mathematics.

Hufferd-Ackles, K., Fuson, K.C., & Sherin, M.G. (2004). Describing levels and components of a math-talk learning community. *Journal for Research in Mathematics Education, 35*(2), 81–116.

Kamii, C. (1985). *Young children reinvent arithmetic: Implications of Piaget's theory.* New York: Teachers College Press.

Kazemi, E., & Franke, M. L. (2004). Teacher learning in mathematics: Using student work to promote collective inquiry. *Journal of Mathematics Teacher Education. 7,* 203–235.

Kazemi, E., & Stipek, D. (2001). Promoting conceptual understanding in four upper-elementary mathematics classrooms. *Elementary School Journal, 102,* 59–80.

Kieran, C. (2002). The mathematical discourse of 13 year old partnered problem solving and its relation to the mathematics that emerges. In C. Kieran, E. Forman, & A. Sfard. (Eds.), *Learning discourse.* Dordrecht, The Netherlands: Kluwer.

Kieran, C., & Dreyfus, T. (1998). Collaborative versus individual problem solving: Entering another's universe of thought. In A. Olivier, & K. Newstead (Eds.), *Proceedings of the 22nd International Conference for the Psychology of Mathematics Education,* (Vol. 3, pp. 112–119). Stellenbach, South Africa

King, A. (1992). Facilitating elaborative learning through guided student-generated questioning. *Educational Psychologist, 27,* 111–126.

Ladson-Billings, G. (1994). *The dreamkeepers: Successful teachers of African American students.* San Francisco: Jossey-Bass.

Ladson-Billings, G. (1995). Toward a theory of culturally relevant pedagogy. *American Educational Research Journal, 32*(3), 465–491.

Lampert, M. (1990). When the problem is not the question and the solution is not the answer: Mathematical knowing and teaching. *American Educational Research Journal,* 27, 29–63.

Lampert, M. (2001). *Teaching problems and the problems of teaching.* New Haven, CT: Yale University Press.

Lampert, M. (2004, August). Response to Teaching Practice/ Teacher Learning Practice Group. In J. Spillane, P. Cobb, & A. Sfard (organizers). Investigating the Practice of School Improvement: Theory, Methodology, and Relevance, Bellagio, Italy.

Lampert, M. (2005, April). *Training in instructional routines: Learning to "listen" and constructing responses.* Paper presented at the annual meeting of the American Educational Research Association, Montreal.

Lampert, M., Rittenhouse, P., & Crumbaugh, C. (1996). Agreeing to disagree: Developing sociable mathematical discourse. In D. R. Olson & N. Torrance (Eds.), *Handbook of education and human development: New methods of learning, teaching, and schooling* (pp. 731–764). Oxford, U.K.: Blackwell Publishing

Lave, J. (1988). *Cognition in practice.* Cambridge, U.K.: Cambridge University Press.

Lave, J. (1996). Teaching, as learning, in practice. *Mind, Culture, and Activity, 3,* 149–164.

Lave, J., & Wenger, E. (1991). *Situated learning: Legitimate peripheral participation.* New York: Cambridge University Press.

Leacock, E.B. (1971). Introduction. In E.B. Leacock (Ed.), *The culture of poverty: A critique* (pp. 9–37). New York: Simon & Schuster.

Lehrer, R., Jacobson, C., Kemeny, V., & Strom, D. (1999). Building on children's intuitions to develop mathematical understanding of space. In E. Fennema & T.A. Romberg (Eds.), *Mathematics classrooms that promote understanding* (pp. 63–87). Mahwah, NJ: Erlbaum.

Leinhardt, G., & Steele, M. D. (2005). Seeing the complexity of standing on the side: Instructional dialogues. *Cognition and Instruction, 23,* 87–163.

Lipka, J. (1994). Culturally negotiated schooling: Toward a Yup'ik mathematics. *Journal of American Indian Studies, 33*(3), 1–12.

Lipka, J. (2002). *Schooling for self-determination: Research on the effects of including native language and culture in the schools.* ERIC Digest.

Lipka, J. (2005). Math in a cultural context: Two case studies of a successful culturally based math project. *Anthropology & Education Quarterly, 36*(4), 367–385.

Lipka, J., Wildfeuer, S., Wahlberg, N., George, M., & Ezran, D. (2001). Elastic geometry and storyknifing: A Yup'ik Eskimo example. *Teaching Children Mathematics, 7*(6), 337–343.

Lubienski, S. T. (2002). Research, reform, and equity in U.S. mathematics education. *Mathematical Thinking and Learning, 4*(2&3), 103–125.

Martin, D. B. (2000). *Mathematics success and failure among African-American youth: The roles of sociohistorical context, community forces, school influence, and individual agency.* Mahwah, NJ: Erlbaum.

Martin, D. (2003). Hidden assumptions and unaddressed questions in the *Mathematics for All* rhetoric. *The Mathematics Educator, 13*(2), 7–21.

Martin, D. B. (2006). Mathematics learning and participation as racialized forms of experience: African American parents speak on the struggle for mathematics literacy, *Mathematical Thinking and Learning, 8*(3), 197–229.

Martin, D. (in press). Mathematics learning and participation in African American context: The co-construction of identity in two intersecting realms of experience. In N. Nasir & P. Cobb (Eds.), *Diversity, equity, and access to mathematical ideas.* New York: Teachers College Press.

Martino, A.M. & Maher, C.A. (1999). Teacher questioning to promote justification and generalization in mathematics: What research practice has taught us. *Journal of Mathematical Behavior, 18*(1), 53–78.

McClain, K. (2000). The teacher's role in supporting the emergence of ways of symbolizing. *Journal of Mathematical Behavior, 19,* 189–207.

McClain, K., & Cobb, P. (2001). An analysis of the development of sociomathematical norms in one first-grade classroom. *Journal for Research in Mathematics Education, 32,* 236–266.

Mehan, H. (1985). The structure of classroom discourse. In T.A. Van Dijk (Ed.), *Handbook of discourse analysis,* (Vol. 3, pp. 119–131). London: Academic Press.

Mercado, C. (2001). The learner: "Race," "Ethnicity," and linguistic difference. In V. Richardson (Ed.), *Handbook of research on teaching,* (pp. 668–694). Washington, DC: American Educational Research Association.

Moll, L. (1992). Bilingual classroom studies and community analysis: Some recent trends. *Educational Researcher, 21*(2), 20–24.

Moll, L., & Gonzalez, N. (2004). Engaging life: A funds-of-knowledge approach to multicultural education. In J. A. Banks & C. A. M. Banks (Eds.), *Handbook of research on multicultural education* (2nd ed., pp. 699–715). San Francisco: Jossey-Bass.

Moll, L., & Greenberg, J. (1990). Creating zones of possibilities: Combining social contexts for instruction. In L. C. Moll (Ed.), *Vygotsky and education* (pp. 319–348). Cambridge, England: Cambridge University Press.

Moschkovich, J.N. (1999). Supporting the participation of English language learners in mathematical discussions. *For the Learning of Mathematics, 19*(1), 11–19.

Moschkovich, J.N. (2000) Learning mathematics in two languages: Moving from obstacles to resources. In W. Secada (Ed.), *Changing the faces of mathematics: Vol. 1. Perspectives on multiculturalism and gender equity* (pp. 85–93). Reston, VA: National Council of Teachers of Mathematics.

Moschkovich, J. N. (2002a). An introduction to examining everyday and academic mathematical practices. In M. Brenner & J. Moschkovich (Eds.), *Everyday and academic mathematics: Implications for the classroom. Journal for Research in Mathematics Education* [Monograph No. 11], 1–11.

Moschkovich, J. N. (2002b). A situated and sociocultural perspective on bilingual mathematics learners [Special issue]. *Mathematical Thinking and Learning, 4*(2/3), 189–212.

Moses, R., Kamii, M., Swap, S.M., & Howard, J. (1989). The Algebra Project: Organizing in the spirit of Ella. *Harvard Educational Review, 59*(4), 423–443.

Moses, R. P., & Cobb, C.E. (2001). *Radical equations: Civil rights from Mississippi to the Algebra Project.* Boston: Beacon Press.

Much, N.C., & Shweder, R.A. (1978). Speaking of rules: The analysis of culture in the breach. *New Directions for Child Development, 2,* 19–39.

Nasir, N. S. (2002). Identity, goals, and learning: Mathematics in cultural practice. *Mathematical Thinking and Learning, 4*(2&3), 213–248.

Nasir, N. S., & Saxe, G. (2003). Ethnic and academic identities: A cultural practice perspective on emerging tensions and their management in the lives of minority students. *Educational Researcher, 32*(5), 14–18.

Nathan, M., & Knuth, E.J. (2003). A study of whole classroom mathematical discourse and teacher change. *Cognition & Instruction, 21*, 175–207.

National Council of Teachers of Mathematics. (1989). *Curriculum and evaluation standards for school mathematics.* Reston, VA: Author.

National Council of Teachers of Mathematics. (2000). *Principles and standards for school mathematics.* Reston, VA: Author.

National Research Council. (2002). *Adding it up: Helping children learn mathematics.* J. Kilpatrick, J. Swafford, & B. Findell (Eds.). Mathematics Learning Study Committee, Center for Education, Division of Behavioral and Social Sciences and Education. Washington, DC: National Academy Press.

Nicholls, J., Cobb, P., Wood, T., Yackel, E., & Patashnick, M. (1990). Dimensions of success in mathematics: Individual and classroom differences. *Journal for Research in Mathematics Education, 21*, 109–122.

Nemirovsky, R., Barros, A., Noble, T., Schnepp, M., & Solomon, J. (2005). Learning mathematics in high school: Symbolic places and family resemblances. In T. Romberg, T. Carpenter, & F. Dremock (Eds.) *Understanding mathematics and science matters* (pp. 185–208). Mahwah, NJ: Erlbaum.

Oakes, J., Joseph, R., & Muir, K. (2003). Access and achievement in mathematics and science: Inequalities that endure and change. In J. A. Banks & C.A. Banks, (Eds.), *Handbook of research on multicultural education* (2nd ed., pp. 69–90). San Francisco: Jossey-Bass.

O'Connor, M. C., & Michaels, S. (1993). Aligning academic task and participation status through revoicing: Analysis of a classroom discourse strategy. *Anthropology and Education Quarterly, 24*, 318–335.

O'Connor, M.C., & Michaels, S. (1996). Shifting participant frameworks: Orchestrating thinking practices in group discussion. In D. Hicks (Ed.), *Discourse, learning, and schooling* (pp. 63–103). New York: Cambridge University Press.

O'Connor, M.C. (1998). Language socialization in the classroom. In M. Lampert & M.L. Blunk (Eds.), *Talking mathematics in school: Studies of teaching and learning* (pp. 15–55). Cambridge, U.K.: Cambridge University Press.

Palincsar, A. S., & Brown, A. L. (1984). Reciprocal teaching of comprehension-fostering and comprehension-monitoring activities. *Cognition and Instruction, 1*, 117–175.

Peterson, P., Janicki, T.C. & Swing, S.R. (1981). Ability x treatment interaction effects on children's learning in large-group and small-group approaches. *American Educational Research Journal, 18*, 453–473.

Porter, A. C. (1989). A curriculum out of balance: The case of elementary school mathematics. *Educational Researcher, 18*(5), 9–15.

RAND Mathematics Study Panel, (2002). *Mathematics proficiency for all students: Toward a strategic research and development program in mathematics education.* Santa Monica, CA: RAND Foundation.

Rittenhouse, P.S. (1989). The teacher's role in mathematical conversation: Stepping in and stepping out. In M. Lampert & M.L. Blunk (Eds.), *Talking mathematics in school: Studies of teaching and learning* (pp. 163–189). Cambridge, U.K.: Cambridge University Press.

Rogoff, B. (1990). *Apprenticeship in thinking: cognitive development in social context.* New York: Oxford University Press.

Rosebery, A., Warren, B., Ballenger, C., & Ogonowski, M. (2005). The generative potential of students' everyday knowledge in learning science. In T. Romberg, T. Carpenter, & F. Dremock (Eds.), *Understanding mathematics and science matters.* (pp. 55–80). Mahwah, NJ: Erlbaum.

Rosebery, A., Warren, B., & Conant, F. (1992). Appropriating scientific discourse: Findings from language minority classrooms. *The Journal of the Learning Sciences, 2*, 61–94.

Saxe, G. B., Gearhart, M., Note, M., & Paduano, P. (1993). Peer interaction and the development of mathematical understanding. In H. Daniels (Ed.), *Charting the agenda: Educational activity after Vygotsky* (pp. 107–144). London: Routledge.

Schoenfeld, A.H. (1998). Toward a theory of teaching in context. *Issues in Education, 1*, 1–94.

Schoenfeld, A.H. (1999). Looking toward the 21st century: Challenges of educational theory and practice. *Educational Researcher, 28* (7), 4–14.

Schoenfeld, A.H. (2000). Models of the teaching process. *Journal of Mathematical Behavior, 18*(3), 243–261.

Schoenfeld, A.H. (2002). A highly interactive discourse structure. In J. Brophy (Ed.), *Social constructivist teaching: Its affordances and constraints* (Vol. 9, pp. 131–170). Amsterdam: JAI Press.

Schoenfeld, A.H. (in press). On modeling teachers' in-the-moment decision-making. In A. H. Schoenfeld (Ed.), *A study of teaching: Multiple lenses, multiple views. JRME Monograph series.* Reston, VA: National Council of Teachers of Mathematics.

Sfard, A., & Kieran, C. (2001a). Preparing teachers for handling students' mathematical communication: Gathering knowledge and building tools. In F.L. Lin & T. Cooney (Eds.), *Making sense of mathematics teacher education,* Dordrecht, The Netherlands: Kluwer.

Sfard, A., & Kieran, C. (2001b). Cognition as communication: Rethinking learning-by-talking through multi-faceted analysis of students' mathematical interactions. *Mind, Culture, and Activity, 8*(1), 42–76.

Sherin, M.G. (2002). A balancing act: Developing a discourse community in a mathematics classroom. *Journal of Mathematics Teacher Education, 5*, 205–233.

Silver, E. A. (1996). Moving beyond learning alone and in silence: Observations from the QUASAR Project concerning some challenges and possibilities of communication in mathematics classrooms. In L. Schauble & R. Glaser (Eds.), *Innovations in learning: New environments for education* (pp. 127–159). Hillsdale, NJ: Erlbaum.

Silver, E. A., & Smith, M. S. (1996). Building discourse communities in mathematics classrooms: A worthwhile but challenging journey. In P. Elliott (Ed.), *Communication in mathematics, K–12 and beyond* (pp. 20–28) [1996 yearbook of the National Council of Teachers of Mathematics]. Reston, VA: National Council of Teachers of Mathematics.

Silver, E. A., Kilpatrick, J., & Schlesinger, B. (1990). *Thinking through mathematics.* New York: College Entrance Examination Board.

Silver, E. A., Smith, M. S., & Nelson, B. S. (1995). The QUASAR project: Equity concerns meet mathematics education reform in the middle school. In E. Fennema, W. Secada, &

L. B. Adajian (Eds.), *New directions in equity in mathematics education* (pp. 9–56). New York: Cambridge University Press.

Simon, M. (1997). Developing new models of mathematics teaching: An imperative for research on mathematics teacher development. In E. Fennema & B. Nelson. (Eds.) *Mathematics teachers in transition* (pp. 55–86). Hillsdale, NJ: Erlbaum.

Simon, M., Tzur, R., Heinz, K., & Kinzel, M. (2004). Explicating a mechanism for conceptual learning: Elaborating the construct of reflective abstraction. *Journal for Research in Mathematics Education, 35*(5), 305–329.

Spencer, J. (2006). *Balancing the equation: African American students' opportunity to learn mathematics with understanding in two central city middle schools.* Unpublished doctoral dissertation, University of California, Los Angeles.

Spillane, J. P., & Zeuli, J. S. (1999). Reform and teaching: Exploring patterns of practice in the context of national and state mathematics reforms. *Educational Evaluation and Policy Analysis, 21*, 1–27.

Stein, M. K., Engle, R. A., Hughes, E. K., & Smith, M. S. (2006). *Orchestrating productive mathematical discussions: Helping teachers learn to better incorporate student thinking.* Manuscript submitted for publication.

Stein, M. K., Silver, E. A., & Smith, M. S. (1998). Mathematics reform and teacher development: A community of practice perspectives. In J. G. Greeno & S. V. Goldman (Eds.), *Thinking practices in mathematical and science learning* (pp. 17–52). Mahwah, NJ: Erlbaum.

Stein, M. K., Grover, B. W., & Henningsen, M. (1996). Enhanced instruction as a means of building student capacity for mathematical thinking and reasoning. *American Educational Research Journal, 33*, 455–488.

Stein, M. K., & Lane, S. (1996). Instructional tasks and the development of student capacity to think and reason: An analysis of the relationship between teaching and learning in a reform mathematics project. *Educational Research and Evaluation, 2*, 50–80.

Stigler, J. & Hiebert, J. (1997). Understanding and improving classroom mathematics instruction. *Phi Delta Kappan, 79*, 14–21.

Stigler, J & Hiebert, J. (1999). *The teacher gap, best ideas from the world's teachers for improving education in the classroom.* New York: Free Press.

Strom, D., Kemeny, V., Lehrer, R., & Forman, E. (2001). Visualizing the emergent structure of children's mathematical argument. *Cognitive Science, 25*, 733–773.

Walkerdine, V. (1990). Difference, cognition, and mathematics education. *For the Learning of Mathematics, 10*(3), 51–56.

Webb, N. (1991). Task-related verbal interaction and mathematics learning in small groups. *Journal for Research in Mathematics Education, 22*, 366–389.

Webb, N., Ing, M., Kersting, N., & Nemer, K.M. (1994, April). *The effects of teacher discourse on student behavior in peer-directed groups.* Presented at the annual meeting of the American Educational Research Association, San Diego, CA.

Webb, N. M., Nemer, K. M., & Ing, M. (2006). Small-group reflections: Parallels between teacher discourse and student behavior in peer-directed groups. *Journal of the Learning Sciences, 15*(1), 63–119.

Wenger, E. (1998). *Communities of practice: Learning, meaning and identity.* Cambridge, U.K.: Cambridge University Press.

Wertsch, J. V. (1991). *Voices of the mind: A sociocultural approach to mediated action.* Cambridge, MA: Harvard University Press.

Wood, T., Williams, G., & McNeal, B. (2006). Children's mathematical thinking in different classroom cultures. *Journal for Research in Mathematics Education, 37*, 222–255.

Wood, T., & Cobb, P. (1991). Change in teaching mathematics: A case study. *American Education Research Journal, 28* (3), 587–616.

Wood, T., Cobb, P., & Yackel, E. (1991). Change in teaching mathematics: A case study. *American Educational Research Journal, 28*, 587–616.

Wood, T. (1998). Alternative patterns of communication in mathematics classes: Funneling or focusing? In H. Steinbring, M. G. B. Bussi & A. Sierpinska (Eds.), *Language and communication in the mathematics classroom* (pp. 167–178). Reston, VA: National Council of Teachers of Mathematics.

Yackel, E. (2002). What can we learn from analyzing the teacher's role in collective argumentation. *Journal of Mathematical Behavior, 21*, 423–440.

Yackel, E., & Cobb, P. (1996). Sociomathematical norms, argumentation, and autonomy in mathematics. *Journal for Research in Mathematics Education, 27*, 458–477.

Yackel, E., Cobb, P., & Wood, T. (1991). Small-group interactions as a source of learning opportunities in second-grade mathematics. *Journal for Research in Mathematics Education, 22*, 390–408.

Yackel, E., Cobb, P., & Wood, T., Wheatley, G., & Merckel, G. (1990). The importance of social interaction in children's construction of mathematical knowledge. In T.J. Cooney, & C.R. Hirsch (Eds.), *Teaching and learning mathematics in the 1990s,* (pp. 12–21). Reston, VA: National Council of Teachers of Mathematics.

AUTHOR NOTE

We would like to thank Thomas Carpenter and Magdalene Lampert for their thoughtful and ongoing feedback on the chapter. We are also grateful for the invaluable assistance of Angela Chan and Julie Kern Schwerdtfeger in gathering literature. The material in this chapter is based in part on work supported by the National Science Foundation under Grant No. ESI-0119732 to the Diversity in Mathematics Education Center for Learning and Teaching. Any opinions, findings, and conclusions or recommendations expressed in this material are those of the author(s) and do not necessarily reflect the position, policy, or endorsement of the National Science Foundation.

7

MATHEMATICS TEACHERS' BELIEFS AND AFFECT

Randolph A. Philipp

SAN DIEGO STATE UNIVERSITY

Be careful how you interpret the world: It is like that.
—Erich Heller (Also attributed to Helen Keller)

A colleague once complained to me that after his daughter had successfully completed her last required mathematics course from a highly regarded university, she reported to him that as a result of that course she hated mathematics. My colleague was particularly upset because his daughter had enjoyed mathematics up to that point, and he was worried that the negative effects of her last mathematics class might linger long into her future. For many students studying mathematics in school, the beliefs or feelings that they carry away *about* the subject are at least as important as the knowledge they learn *of* the subject. A 2005 Associated Press poll (AP–AOL News, 2005) showed that nearly 40% of the adults surveyed said that they had hated mathematics in school, and although those polled also acknowledged hating other subjects, twice as many people said that they hated mathematics as said that about any other subject. While students are learning mathematics, they are also learning lessons about what mathematics is, what value it has, how it is learned, who should learn it, and what engagement in mathematical reasoning entails.

To understand students' experiences with school mathematics, one must understand a central factor in their experience: mathematics teachers. Two de-

cades have passed since Lee Shulman (1986) rejected George Bernard Shaw's infamous statement "He who can, does. He who cannot, teaches" as overly simplistic because it was not reflective of the highly specialized knowledge required of teachers. He introduced the now famous term *pedagogical content knowledge* to refer to the complex knowledge that lies at the intersection of content and pedagogy and that teachers must possess to make the curriculum accessible to their students. Shulman ended his landmark article by suggesting that a more apropos saying would be "Those who can, do. Those who understand, teach" (p. 14). Researchers studying teachers' knowledge, beliefs, and affect related to mathematics teaching and learning are still trying to tease out the relationships among these constructs and to determine how teachers' knowledge, beliefs, and affect relate to their instruction.

The focus of this chapter is a consideration of what researchers have to say about teachers' beliefs and affect. A saying by Heller, "Be careful how you interpret the world: It is like that," indicates that the way one makes sense of his or her world not only defines that person for the world but also defines the world for that person. Beliefs might be thought of as lenses through which one looks when interpreting the world, and affect might be thought of as a disposition or tendency one takes toward some aspect of his or her world; as such, the beliefs and affect one holds surely affect the way one interacts with his or her world. Although few researchers have examined the relationship between mathematics teachers' affect and their instruction, the existing research shows that the feelings teachers experienced as learners carry forward to their adult lives, and these feelings are important factors in the ways teachers interpret their mathematical worlds.

ORGANIZATION OF CHAPTER

I begin this chapter with a brief review of two chapters from the first *Handbook of Research on Mathematics Teaching and Learning* (Grouws, 1992): one on teachers' beliefs, written by Alba G. Thompson, and a general chapter on affect, written by Doug McLeod. After summarizing the state of research on mathematics teachers' beliefs and affect at the time of publication of the first handbook chapter, I review the research conducted on teachers' beliefs since that time in two sections. In the first section, I consider what beliefs are, how they are measured, what stances are taken on the role of inconsistent beliefs, and how they are changed. In the second section, I review emerging areas of research related to mathematics teachers' beliefs by looking to research on teachers' beliefs about students' mathematical thinking, teachers' beliefs about curriculum, teachers' beliefs about technology, and teachers' beliefs about gender. I then review the research on teachers' affect. I begin with a brief review of the relationship between affect and achievement and then summarize studies that address teachers' affect. Because little research specifically addresses the topic of mathematics teachers' affect and several researchers have proposed frameworks on students' affect that might be helpful for considering teachers' affect, I follow the research on teachers' affect with a review of some of the student-focused frameworks. After providing a review of the research on teachers' beliefs and affect, I consider researchers who have taken a broader look at the construct of beliefs by considering such constructs as *teachers' orientations, teachers' perceptions,* and *teacher identity,* and I highlight how studying identity provides one promising path by which to integrate teachers' beliefs and affect. I end with some final comments.

WORKING DEFINITIONS/DESCRIPTIONS OF TERMS

Many of the terms in this chapter are not used in the literature in a uniform way. However, I recognize that readers can become confused about the relationships among terms, and so I have attempted to distill meanings that capture distinctions that emerge in usage by researchers, and I list them in Figure 7.1. I call these *definitions/descriptions* because they are based upon a combination of the literature usage and the dictionary definitions. These definitions/descriptions are not intended to stand alone as definitions, but instead I provide them to support the reader in drawing distinctions among the commonly used meanings of terms in this chapter. Each of these terms is discussed in the chapter, but I provide them at the beginning (see Figure 7.1) so that the reader can refer to them as needed.

THE STATE OF RESEARCH ON MATHEMATICS TEACHERS' BELIEFS AND AFFECT AT THE PUBLICATION OF THE FIRST HANDBOOK OF RESEARCH ON MATHEMATICS TEACHING AND LEARNING

Brief Summary of A. G. Thompson's 1992 Handbook Chapter on Teachers' Beliefs and Conceptions

Twice in her review of the literature on teachers' beliefs and conceptions, A. G. Thompson (1992) noted the importance for researchers studying mathematics teachers' beliefs to make explicit to themselves and to others the perspectives they hold about teaching, learning, and the nature of mathematics, because these perspectives greatly affect researchers' approaches to and interpretations of their work. In keeping with her own advice, A. G. Thompson stated her stance that researchers must consider the discipline of mathematics and the relationship between what a teacher thinks about mathematics and how the teacher teaches. She embraced a view of mathematics "as a kind of mental activity, a social construction involving conjectures, proofs, and refutations, whose results are subject to revolutionary change and whose validity, therefore, must be judged in relation to a social and cultural setting" (p. 127). The conception of mathematics teaching associated with this view of mathematics and reflected in several documents A. G. Thompson cited from the 1980s was "one in which

Affect—a disposition or tendency or an emotion or feeling attached to an idea or object. Affect is comprised of *emotions*, *attitudes*, and *beliefs*.

> **Emotions**—feelings or states of consciousness, distinguished from cognition. Emotions change more rapidly and are felt more intensely than attitudes and beliefs. Emotions may be positive (e.g., the feeling of "aha") or negative (e.g., the feeling of panic). Emotions are less cognitive than attitudes.
>
> **Attitudes**—manners of acting, feeling, or thinking that show one's disposition or opinion. Attitudes change more slowly than emotions, but they change more quickly than beliefs. Attitudes, like emotions, may involve positive or negative feelings, and they are felt with less intensity than emotions. Attitudes are more cognitive than emotion but less cognitive than beliefs.

Beliefs—Psychologically held understandings, premises, or propositions about the world that are thought to be true. Beliefs are more cognitive, are felt less intensely, and are harder to change than attitudes. Beliefs might be thought of as lenses that affect one's view of some aspect of the world or as dispositions toward action. Beliefs, unlike knowledge, may be held with varying degrees of conviction and are not consensual. Beliefs are more cognitive than emotions and attitudes. (I do not indent this definition under affect because, although beliefs are considered a component of *affect* by those studying affect, they are not seen in this way by most who study teachers' beliefs.)

Beliefs System—a metaphor for describing the manner in which one's beliefs are organized in a cluster, generally around a particular idea or object. Beliefs systems are associated with three aspects: (a) Beliefs within a beliefs system may be primary or derivative; (b) beliefs within a beliefs system may be central or peripheral; (c) beliefs are never held in isolation and might be thought of as existing in clusters.

Conception—a general notion or mental structure encompassing beliefs, meanings, concepts, propositions, rules, mental images, and preferences.

Identity—the embodiment of an individual's knowledge, beliefs, values, commitments, intentions, and affect as they relate to one's participation within a particular community of practice; the ways one has learned to think, act, and interact.

Knowledge—beliefs held with certainty or justified true belief. What is knowledge for one person may be belief for another, depending upon whether one holds the conception as beyond question.

Value—the worth of something. A belief one holds deeply, even to the point of cherishing, and acts upon. Whereas beliefs are associated with a true/false dichotomy, values are associated with a desirable/undesirable dichotomy. Values are less context-specific than beliefs.

Figure 7.1 Working definitions/descriptions of terms (listed alphabetically, except that *emotion, attitude,* and *belief* are listed from least to most cognitive component of *affect*. I accept sole responsibility for these definitions while crediting many).

students engage in purposeful activities that grow out of problem situations, requiring reasoning and creative thinking, gathering and applying information, discovering, inventing, and communicating ideas, and testing those ideas through critical reflection and argumentation" (p. 128). The role of teachers' mathematical conceptions was a recurring theme underlying her 1992 review chapter, no surprise to readers familiar with her research (e.g., A. G. Thompson, 1984; A. G. Thompson, Philipp, Thompson, & Boyd, 1994; P. W. Thompson & Thompson, 1994).

The term *belief* is so popular in the education literature today that many who write about beliefs do so without defining the term. A. G. Thompson (1992) found, "For the most part, researchers have assumed that readers know what beliefs are" (p. 129). Further, many educators contend that distinguishing between knowledge and belief is unimportant for research, but investigating how, if at all, teachers' beliefs and knowledge affected their experience is important. Although Thompson used both *beliefs* and *conceptions* in the title of her chapter, in most of her chapter, she seemed to

think of beliefs as a subset of conceptions, and her definition of *conceptions* included *beliefs*. And yet, at times she seemed to use the terms interchangeably. She referred to teachers' conceptions "as a more general mental structure, encompassing beliefs, meanings, concepts, propositions, rules, mental images, preferences, and the like" (p. 130). When using the term *conceptions*, Thompson, recognizing the important relationship between knowledge and beliefs, seemed less interested in drawing distinctions between these terms, and she stated, "To look at research on mathematics teachers' beliefs and conceptions in isolation from research on mathematics teachers' knowledge will necessarily result in an incomplete picture" (p. 131). However, early in her chapter she recognized two distinctions between beliefs and knowledge, *conviction* and *consensuality*; these distinctions are generally referred to in the field.

First, beliefs can be held with varying degrees of conviction, whereas knowledge is generally not thought of in this way. For example, whereas one might say that he or she believed something strongly, one is less likely to speak of knowing a fact strongly. Second, beliefs are not consensual, whereas knowledge is. One is generally aware that others may believe differently and that their stances cannot be disproved, whereas with respect to knowledge, one finds "general agreement about procedures for evaluating and judging its validity" (A. G. Thompson, 1992, p. 130).

A. G. Thompson defined a *belief system* as "a metaphor for examining and describing how an individual's beliefs are organized" (1992, p. 130). Drawing upon the work of Green (1971) and Rokeach (1960, cited in A. G. Thompson, 1992), she highlighted three aspects of beliefs systems identified by Green, all related to the notion that because a belief is never held in total isolation from other beliefs, considering how beliefs are held in relation to one another is useful. First, some beliefs serve as the foundation for other beliefs in a quasi-logical structure, meaning that some beliefs might be thought of as *primary beliefs* whereas others serve as *derivative beliefs*. Thompson's example of a primary belief was a teacher's belief that clearly presenting mathematics to students is important, and an associated derivative belief that might follow is that teachers should be able to readily answer any questions asked by students. A second dimension of beliefs systems is that beliefs can be either *central*, which means strongly held, or *peripheral*, which means less strongly held and more susceptible to change. Green contended that primary beliefs might not necessarily be more central than the associated derivative beliefs. In the example above, the derivative belief that a teacher must be prepared to answer questions may be more central (strongly held) to the teacher than the primary belief that a lesson must be clear. A teacher might even, perhaps as a result of a professional development opportunity, change her primary belief about presenting clear, sequential lessons without affecting the derivative, more central, belief about being able to answer any questions posed. Green's third dimension of beliefs systems is that beliefs are held in *clusters* that are more or less in isolation from other clusters. One outcome of holding beliefs in this manner is that people can avoid confrontations between belief structures. Another outcome is that these beliefs systems may appear contradictory or inconsistent from the point of view of an observer.

The major portion of her chapter was a review of the literature about topics of study that, at that time, A. G. Thompson (1992) considered relatively new: teachers' conceptions of mathematics, their conceptions of mathematics teaching and learning, and the relationship between these conceptions and instructional practice. She ended the section by discussing changing teachers' conceptions. I briefly summarize these topics below.

She first defined a *teacher's conception of the nature of mathematics* as "that teacher's conscious or subconscious beliefs, concepts, meanings, rules, mental images, and preferences concerning the discipline of mathematics" (A. G. Thompson, 1992, p. 132). Her examples of ways that scholars have highlighted teachers' conceptions included Ernest's (1988, cited in A. G. Thompson, 1992) three conceptions of mathematics as (a) a dynamic, problem-driven discipline, (b) a static, unified body of knowledge, or (c) a bag of tools. She described Lerman's (1983, cited in A. G. Thompson, 1992) *absolutist* view of mathematics as universal, absolute, certain, value-free mathematical knowledge and the *fallibilist* view of mathematics as developing through conjectures, proofs, and refutations. She described Perry's (1970, cited in A. G. Thompson, 1992) scheme of intellectual and moral development and Copes's (1979, cited in A. G. Thompson, 1992) attempt to adapt this general scheme to mathematical knowledge. She also described Skemp's (1978) ideas of instrumental and relational understanding. Her look at teachers' conceptions of mathematics teaching and learning focused upon models of mathematics teaching, and she drew upon the work of Kuhs and Ball (1986, cited in A. G. Thompson, 1992) to highlight four views: learner-focused, content-focused with an emphasis on conceptual understanding, content-focused with an emphasis on performance, and classroom-focused. In considering the relationship between teachers' conceptions and their instructional practices, Thompson noted that although some researchers found a strong relationship whereas others found more variability and inconsistency, in general, a higher degree of consistency was reported by researchers studying the relationship between teachers' conceptions of mathematics and their instructional practices than by researchers studying teachers' conceptions about teaching and learning and their instructional practices. She cautioned that inconsistencies between professed beliefs and instructional practice raise a methodological concern related to how beliefs or conceptions are measured, and she suggested that researchers must go beyond teachers' professed beliefs and at least "examine teachers' verbal data along with observational data of their instructional practice or mathematical behavior" (A. G. Thompson, 1992, p. 135).

In the last major section of her chapter, on changing teachers' conceptions, A. G. Thompson (1992) reported studies of preservice teachers who showed little change, and she noted that teachers often assimilate new ideas to fit their existing schemata instead of accommodating their existing schemata to internalize new ideas. She suggested that studies providing in-depth, detailed analyses would be required to better explicate teachers' difficulties in accommodating new ideas into their existing schemata. She described a promising paradigm and indicated that teachers' beliefs and practices underwent large changes when teachers learned about children's mathematical thinking (Carpenter, Fennema, Peterson, Chiang, & Loef, 1989).

A. G. Thompson (1992) ended her chapter with a few suggestions. First, she was concerned that too many researchers thought of beliefs systems as static entities to be uncovered whereas she viewed beliefs systems as dynamic mental structures that were susceptible to change in light of experience. She challenged researchers to look less at how beliefs affected practice and more at the dialectic between beliefs and practice. She reiterated the position put forth by some researchers that distinguishing between teachers' knowledge and teachers' beliefs is not useful and that, instead, researchers should investigate teachers' conceptions encompassing both beliefs and any relevant knowledge—including meanings, concepts, propositions, rules, or mental images—that bears on the experience. She also noted that virtually no researchers had investigated the extent to which teachers' and students' conceptions interact during instruction.

Brief Summary of McLeod's 1992 Handbook Chapter on Affect

McLeod defined the *affective domain* as referring "to a wide range of beliefs, feelings, and moods that are generally regarded as going beyond the domain of cognition" (McLeod, 1992, p. 576), and he identified several specific terms that make up the affective domain. *Emotions, attitudes,* and *beliefs* were the three terms to which he devoted most attention. Emotions change more rapidly than attitudes and beliefs; they are also less cognitive and are felt more intensely than attitudes and beliefs. An example of a negative emotion is the feeling of panic experienced by some students when doing, or even thinking about doing, mathematics, whereas an example of a positive emotion is the satisfactory feeling of "Aha!" experienced by students who have an insight during mathematical problem solving. McLeod noted that emotional reactions had not been major factors in research on affect in mathematics education, and he conjectured that the reason for this lack of attention was that emotions, being prone to changing quickly, lack the stability necessary for researchers to think that they can reliably gather questionnaire data about them.

Attitudes refer to "affective responses that involve positive or negative feelings of moderate intensity" (McLeod, 1992, p. 581). Attitudes are more cognitive in nature and more stable than emotions; they are felt less intensely. Examples of attitudes students might experience toward, for example, geometry are liking, disliking, being curious about, or being bored by the subject. One connection between emotions and attitudes is that repeated emotional reaction to an experience related to mathematics can result in automatizing that emotion into an attitude toward that experience. McLeod provided the example of a student who has repeated negative experiences with geometry proofs. The initial emotional response the student experiences will, with time, become more automatic and less physiologically arousing and will eventually lead to the student's forming a more stable response, an attitude, that can be measured by use of a questionnaire. For example, a student who is struggling in a geometry class may feel discomfort, or even illness, when faced with devising geometry proofs, and with time these feelings may result in the more general attitude that the student dislikes geometry, and, eventually, mathematics.

Beliefs are more cognitive in nature than attitudes (and, hence, also than emotions), are generally stable, and are experienced with a lower level of intensity than emotions or attitudes. McLeod concluded that beliefs tend to develop gradually and that cultural factors play a key role in their development. Four categories of beliefs addressed in McLeod's chapter are beliefs about mathematics, beliefs about self, beliefs about mathematics teaching, and beliefs about the social context. These discussions were almost exclusively about students' beliefs, not teachers' beliefs.

Although his major focus on affect was on emotions, attitudes, and beliefs, McLeod addressed several other affect-related concepts, such as confidence, self-concept, mathematics anxiety, and learned helplessness. He also addressed concepts more closely related to cognition, such as autonomy, intuition, metacognition, and social context.

A major theme identified by McLeod in his 1992 chapter was that "all research in mathematics education can be strengthened if researchers will integrate affective issues into studies of cognition and instruction" (p. 575), and he drew heavily from Mandler's theory that affective factors arise out of emotional responses to interrupted plans. According to Mandler, one approaches a task with a schema for how the task will be completed, and if the anticipated sequence of actions cannot be completed, the result is a physiological response. For Mandler, these affective factors were connected to one's knowledge and beliefs, because interpretations of interruptions vary from individual to individual. McLeod offered the example of a sixth-grade student solving a story problem. If the student believed that all mathematics problems should be solvable in a couple minutes but the student was unable to solve the problem in that period of time, the student might experience an arousal that he or she would interpret as negative. If these experiences were repeated frequently, the student might develop a negative attitude toward story problems, and, in many cases, such negative attitudes toward one aspect of mathematics generalize to negative attitudes toward mathematics in general or, perhaps even worse, toward the student's view of himself or herself as a mathematical learner. If, however, students believe that story problems can challenge even good problem solvers and require a longer period of time to solve, then arousal at an inability to quickly solve the problem might not be interpreted as negative. The student's interpretation of the experience, not the experience itself, determines the outcome.

In the upcoming section on affect, I relate the research since the publication of McLeod's chapter to two conclusions he drew. First, little research in mathematics education integrated cognitive and affective factors, and McLeod called for more integration, which, unfortunately, is not generally reflected in the recent research. At the end of the chapter, however, I describe some promising directions for such research.

Second, although McLeod addressed affect generally, he noted that teacher affect in mathematics education had been studied little, a trend that seems to have continued. However, the emotional responses felt by students learning mathematics and the associated attitudes that develop are believed to linger into adulthood and may have important implications for teachers. I look to some research on comparison of children's and adults' affective responses, for example, research showing that mathematics test anxiety can equally affect children and prospective teachers, and I consider implications.

Summary of the State of Research on Beliefs and Affect at the Publication of the First Handbook

A. G. Thompson (1992) found that most research on teachers' beliefs and conceptions was interpretive in nature, employed qualitative methods of analysis, and was comprised of in-depth case studies with small numbers of subjects. She suggested that more such in-depth studies were needed to explain why teachers did not accommodate new conceptions into their existing schemata. Methodological approaches used at the time for these in-depth studies included "interviews, classroom observations, stimulated-recall interviews, linguistic analysis of teacher talk, paragraph-completion tests, responses to simulation materials such as vignettes describing hypothetical students in classroom situations, and concept generation and mapping exercises such as the Kelly Repertory Grid Technique" (p. 131). Those who conducted experimental or pseudoexperimental studies of beliefs used Likert-type instruments, and they tended to measure beliefs isolated from knowledge.

McLeod, (1992) in his chapter on affect, addressed beliefs as a cognitive component of affect and called for affect to be included in the study of cognition and, therefore, of beliefs but found that little research had been conducted on teachers' affect. A. G. Thompson (1992), in her chapter on teacher beliefs, addressed the importance of considering beliefs together with knowledge and referred to this construct as *teachers' conceptions*. Thompson did not address affect. This lack of attention to affect was common throughout the first handbook. With one exception, all index references to *affect* from the first National Council of Teachers of Mathematics (NCTM) *Handbook of Research on Mathematics Teaching and Learning* (Grouws, 1992) were to the McLeod chapter on affect in mathematics education. The one exception was Schoenfeld's (1992) mention, in his chapter on learning to think mathematically, that space constraints precluded his addressing the topic of affect and that readers should refer to the work by McLeod for "authoritative starting points" (p. 358). Absent from both chapters was the topic of technology, either as a tool to support the study of teachers' beliefs and affect or as an object of study about teacher's beliefs or affect with respect to technology.

Setting the Context for Mathematics Teacher Education Research Since 1992

Much has changed since 1992, when the first handbook was published. I note five occurrences that, at least indirectly, have affected research in mathematics education in the United States and, to some extent, abroad. These five occurrences are the general acceptance and infusion of ideas from the NCTM Standards' documents (NCTM, 1989, 1991, 1995) into the educational arena; the increased number of outlets for publishing research in mathematics education; the increased politicization of United States education in general and educational research in particular; technological advances affecting the manner in which we obtain and report research and capture, edit, and post video for use when assessing teachers' beliefs or affect; and the emergence of sociocultural and participatory theories of learning.

Occurrence 1: The Acceptance and Infusion of Ideas From the NCTM Standards' Documents Into the Educational Arena

Between 1989 and 1995 the National Council of Teachers of Mathematics published three important documents: *Curriculum and Evaluation Standards for School Mathematics* (1989), *Professional Standards for Teaching Mathematics* (1991), and *Assessment Standards for School Mathematics* (1995). (Although NCTM subsequently published the *Principles and Standards for School Mathematics* [2000], this document appeared too late to influence much of the research reviewed for this chapter.) The *Curriculum and Evaluation Standards* document, which emphasized mathematics as problem solving, communication, reasoning, and making connections, particularly affected curriculum development in the United States and, as a result, also affected research that was conducted (Hiebert, 1999). Although this document had been published before the first *Handbook* (Grouws, 1992), it had had no substantial influence on either practice or research by that time. However, in the next decade, that situation was to change, with publication of hundreds of research studies about mathematics education reform (Ross, McDougall, & Hogaboam-Gray, 2002). One might conclude from the abundance of studies on reform that schools were engaged in important and fundamental change. However, a peek into randomly selected American classrooms has led to the conclusion that the reform movement in the United States has not led to widespread change in mathematics instruction (Hiebert et al., 2005; Stigler & Hiebert, 1999). Whereas reform documents have had an effect on mathematics education research, they have had a much smaller effect on what takes place in American schools. When Lerman (2002) wrote that "a reform along the lines of that in the USA could not take place in the UK because of the current dominance of the official pedagogical field" (p. 237), he could, to some extent, also have been writing about the reform's effect in U.S. schools (Becker & Jacob, 2000; Gregg, 1995; Smith, 1996). One reason that schools change so little is that teaching is cultural (Stevenson & Stigler, 1992), and major commitments are required to enact meaningful school-based change. I am reminded of the response of a mathematics curriculum specialist in the mid 1990s to my question about whether the reform movement had affected the teaching of mathematics in his district: He had ensured that every school received a copy of the *Standards* (NCTM, 1989). But

even if major reform initiatives designed to implement innovative curriculum were funded, meaningful and lasting changes in schools would not occur without sustained professional development designed to change teachers' beliefs. In their review of research studies published about mathematics education reform, Ross, McDougall, and Hogaboam-Gray (2002) concluded that the main obstacle to implementation was teachers' beliefs about mathematics teaching. During the past 10 years, many other researchers have studied teachers' beliefs as they relate to reform (e.g., Lloyd & Wilson, 1998; Steele, 2001; Sztajn, 2003), and some have developed instruments to measure teachers' beliefs as they relate to reform (e.g., Ross, McDougall, Hogaboam-Gray, & LeSage, 2003). Although others studying teachers' beliefs may not explicitly mention reform, their focus reflects beliefs that are in the spirit of the reform.

Occurrence 2: The Increased Number of Publishing Outlets

A second notable occurrence to affect research in mathematics education over the past decade has been the increased number of outlets for publishing research in mathematics education in general and research on mathematics teachers in particular. Journals that have begun publishing since publication of the 1992 *Handbook* (Grouws) include *Hiroshima Journal of Mathematics Education; International Journal for Mathematics Teaching and Learning; The International Journal for Technology in Mathematics Education; Canadian Journal of Science, Mathematics, & Technology Education; International Journal of Science & Math Education; Journal of Mathematics Teacher Education; Mathematics Education Review; Nordic Studies in Mathematics Education;* and *Zentralblatt Für Didaktik Der Mathematik (International Reviews on Mathematical Education)*. To indicate the effect such journals have had on research on teachers' beliefs and affect, I consider more closely the *Journal of Mathematics Teacher Education*, first published in 1998 by Kluwer Academic Publishers. This journal is devoted to publication of research, critical analyses, and critiques related to mathematics teachers' learning and development at all stages of their professional development. The journal provided a new outlet for research on teachers' affect and beliefs, and many studies on these topics have been published in this journal. In his final editorial as the first editor for this journal, Cooney (2001), after reviewing the first four volumes of the journal, wondered whether the articles reflected a field that is generating subsequent research questions but failing to provide the necessary foundation for substantial progress. In acknowledging that mathematics teacher education is inherently a field of practice, Cooney was concerned that we teacher educators not allow the practical orientation of our work to preclude the development of powerful and useful new constructs:

> The problems we study may be inherently practical, but surely their solutions (in the sense of greater wisdom) lie in our ability to see teaching and teacher education outside the confines of the acts themselves. This makes life difficult. It is relatively easy for us to see teaching from a more abstract level if we are not part of the act. In teacher education, however, teacher educators are indeed part of the act. So how is it that we can educate ourselves to engage in Dewey's notion of reflective thinking or to follow von Glasersfeld's notion of reflection which requires us to step out of ourselves and see our actions from a vantage point beyond our actions themselves? It is only from this meta-vantage point that we can begin developing schemes and constructs that capture the webs in which we are all entangled. Simply put, we cannot engage in an analysis of our own activity using only the language of that activity itself. (pp. 256–257)

Cooney emphasized the need to engage in theoretical case-study research that goes beyond reporting anecdotes to develop constructs that might be applied to help make sense of teaching and learning environments.

Occurrence 3: The Increased Politicization of United States Education and Educational Research

Questions about the quality of research in mathematics education have not been isolated to concerns about mathematics-teacher education. Even the broader field of educational research has come into question. Some contend that educational research tends to lack influence or usefulness, partially because it does not lead to practical advances that can be readily applied by practitioners (Burkhardt & Schoenfeld, 2003). By comparing education to other applied fields, Burkhardt and Schoenfeld suggested that the research-based development of tools and processes used by practitioners found in, say, the medical field are lacking in education. Mathematics educators who raise questions about research in our field often take as given that qualitative research is the appropriate means by which to conduct research, and their questions are about the methodological, analytical, or interpretive issues related to qualitative research (Simon, 2004). However, with the political agendas of granting agencies often driving the research that is conducted (Silver & Kilpatrick, 1994), a call for "scientifically based research" has affected funding agencies (Eisenhart & Towne, 2003). Sadly, during the 1990s, many states in the United States experienced a "Math War" (Becker & Jacob, 2000) that pitted mathematics educators calling for innovation and reform against those who believed that the best way to proceed was through a "back to basics" approach. Although published research has generally not yet reflected this recent politicization of educational research, it is a third notable occurrence since 1992 to influence subsequent research on teachers' beliefs and affect.

Occurrence 4: Technological Advances That Support Obtaining and Reporting Research, and Capturing, Editing, and Posting Video for Use When Assessing Teachers' Beliefs and Affect

Many researchers conducting literature reviews today begin the task with their computers. Many articles are available through online journals or online libraries. Furthermore, authors routinely post presentations and papers on their websites or send pdf versions of papers upon request. These changes speed the process of conducting research but do not fundamentally change it. However, other technological advances have changed the process of assessing teachers' beliefs and affect. For example, one group of researchers developed a web-based survey with embedded video requiring open-ended responses to be used when assessing elementary school teachers' beliefs (Integrating Mathematics and Pedagogy, 2003), together with a user's manual also available on-line (Integrating Mathematics and Pedagogy, 2004). Those completing the survey respond online, and their data are downloaded to a spreadsheet. Use of this survey enabled the researchers to conduct a large-scale experimental study (Philipp et al., 2005) that would have been impossible 20 years ago.

Occurrence 5: The Emergence of Sociocultural and Participatory Theories of Learning

A fifth development to affect the research on teachers' beliefs and affect has been the increasing adoption by researchers of emerging theories of learning. By 1992, when the first handbook was published, theoreticians (e.g., Brown, Collins, & Duguid, 1989; Lave, 1988) had already raised questions about psychological theories for explaining the complexities associated with learning, and during the 1990s, sociocultural (Cobb, 1995) and participatory (Wenger, 1998) frameworks were increasingly adopted by mathematics education researchers. The researchers investigating teachers' beliefs and affect since 1992 have increasingly adopted these theoretical frameworks.

RESEARCH ON TEACHERS' BELIEFS CONDUCTED SINCE 1992

I address, in the first part of the section on mathematics teachers' beliefs, four important areas: defining *teachers' beliefs*, measuring them, considering inconsistencies in beliefs, and changing beliefs. In the second part of the beliefs section, I review emerging areas of research related to mathematics teachers' beliefs and how their beliefs are changed by looking to beliefs about students' mathematical thinking, beliefs about curriculum, beliefs about technology, and beliefs about gender. Although all these areas were studied before publication of the first handbook chapter, they have received substantially more attention from researchers since that time.

What Are Beliefs?

The construct *belief* is of great interest to those attempting to understand mathematics teaching and learning. An ERIC-database search for 1990s' articles based on the terms *teacher* and *belief* located 3,105 documents, and although the numbers of documents for the 1980s (1,382) and the 1970s (745) were smaller, clearly the subject of teacher beliefs has been popular.[1] When the term *mathematics* was added to the terms *teacher* and *belief*, the 23, 123, and 407 items located for the 1970s, 1980s, and 1990s, respectively, though far fewer than the number of generic teacher-beliefs documents, still reflect an area of growing importance in our field. In all these studies, the term *belief* was used. However, as popular as the term *belief* is in education research, it is often undefined by those who study it (Pajares, 1992; A. G. Thompson, 1992). Since publication of A. G. Thompson's 1992 *Handbook* chapter, the amount of research on mathematics teachers' beliefs has increased, and the research has included efforts to define *belief*. However, in general, no clear agreement about the definition has been reached. Two constructs often closely related to beliefs are values and knowledge.

Belief Versus Values

If you must tell me your opinions, tell me what you believe in. I have plenty of doubts of my own.

—Johann Wolfgang von Goethe

In the 1950s, Edward R. Murrow hosted a radio show titled "This I Believe," for which Americans wrote about the "rules they live by, the things they have found to be the basic values in their lives" (Murrow, 1951/2005). National Public Radio recently revived this show with modern-day Americans from all walks of life reading short essays about the core beliefs guiding their daily lives. Topics included belief in the power of love to transform and to heal; belief in one's personal responsibility to positively affect society; belief in empathy; and belief that the presence of a person who often feels perplexed, torn by issues, and unsure of his or her beliefs can be an asset, especially "in periods of crisis, when passions are high and certainty runs rabid" (Gup, 2005). These statements of belief are examples of creeds people believe *in*. Murrow (1951/2005) referred to these as "basic values," and I take that stance: A belief *that* is about beliefs, but a belief *in* is about values.

In a review of the literature on values in mathematics teaching, Bishop, Seah, and Chin (2003) addressed the differences and the similarities found in the ways researchers think about beliefs and values. One difference they identified is that beliefs tend to be associated with a true/false dichotomy whereas values are often associated with a desirable/undesirable dichotomy. They contended that beliefs are, therefore, more context-dependent than values, because whereas a true/false judgment must be made in reference to some object, desirable/undesirable dichotomies are associated with more general, less context-dependent, attributes. They provided the example of the belief that mathematics is fun, a true/false judgment made about a particular subject. This judgment must be made in context, because holding this belief does not imply that one finds all subjects, or even all mathematics, fun. One who views fun as a personally desirable quality in a more universal way is likely to seek fun throughout his or her life. For a person who values fun, the belief that mathematics is fun whereas literature is not fun would be more pertinent than for a person who values something else, say usefulness, more than he values fun. To summarize this distinction between beliefs and values, beliefs are true/false statements about constructs whereas the choice of the particular constructs one finds desirable or undesirable represents one's more context-independent values. These more context-independent values are often viewed as more internalized than beliefs and, hence, harder to change.

---

[1] This search was conducted in 2001, before the United States Department of Education closed the ERIC database in an effort to restructure the search engine to include only documents that met a particular standard of "evidence-based" research.

In another approach to distinguishing between beliefs and values, some researchers view values as a subset of beliefs. Rokeach (1973, cited in Bishop et al., 2003) viewed values as enduring beliefs, and Clarkson and Bishop (1999, cited in Bishop et al., 2003) viewed values as beliefs in action. Raths, Harmin, and Simon (1987, cited in Clement, 1999) presented several attributes of beliefs that, if present, would constitute a value. They viewed values as beliefs that one chooses freely from among alternatives after reflection and that one cherishes, affirms, and acts upon. Because life situations may invoke incompatible values, many researchers relax the constraint that one must act upon one's values and instead suggest that so long as one is committed to a particular belief, it might be said that that belief is a value for the person (Clement, 1999). As such, values influence, rather than determine, the choice of possible actions available (Bishop et al., 2003).

The similarities in researchers' views on values and beliefs may be greater than the differences, and the two terms are often used interchangeably by mathematics educators (Bishop et al., 2003). Just as people hold incompatible values, so too do they hold beliefs that may conflict. Values exist within more complex values systems, and individual values alone seldom determine one's decisions, actions, and responses (Bishop et al., 2003). Because beliefs also exist within systems (A. G. Thompson, 1992), in assessing a teacher's values or beliefs, one must recognize the role played by a person's incompatible or conflicting values and beliefs. Consider, for example, a teacher who values students' development of mathematical proficiency. She will want to support their developing a combination of conceptual understanding, procedural fluency, and problem-representing and -solving approaches, all built upon the capacities to reason logically about relationships and justify their reasoning and upon the disposition to view mathematics as useful, worthwhile, and attainable (National Research Council, 2001). If this teacher also values raising her students' test scores as measured by a standardized test focused primarily upon procedural fluency, then she is likely to find herself in a quandary. If this teacher also teaches in a low-performing school, subject to strict local or federal guidelines (like those currently applicable in U.S. schools), then the requirement to immediately improve students' test scores is likely to cause the teacher to feel pressured into pursuing a course of action different from the one she believes will lead her students to develop mathematical proficiency. This teacher will have to prioritize among competing values. These priorities will vary across situations and contexts (Bishop et al., 2003); for example, for some time before the standardized test, the teacher may feel more constrained to alter her usual curricular and pedagogical approaches to focus upon developing procedural fluency. Simply observing this teacher's performance is unlikely to provide one an accurate assessment of her values or beliefs. Furthermore, this teacher may be unaware of how these conflicting forces are affecting her curricular and pedagogical decisions; thus, teasing out the relationships among her values and beliefs and her thoughts and actions would be difficult, even in conversation.

Belief Versus Knowledge

Not . . . what opinions are held, but . . . how they are held: instead of being held dogmatically, [liberal] opinions are held tentatively, and with a consciousness that new evidence may at any moment lead to their abandonment.

—Bertrand Russell, Unpopular Essays, 1950

Across disciplines of anthropology, social psychology, and philosophy, one finds general agreement in viewing *beliefs* as "psychologically held understandings, premises, or propositions about the world that are felt to be true" (Richardson, 1996, p. 103). The notion that a belief is thought to be true raises one of the more common distinctions drawn between belief and knowledge (Pajares, 1992; A. G. Thompson, 1992), with researchers often viewing *knowledge* as "belief with certainty" (Clement, 1999). This definition is similar to one that is more than 2,300 years old, Plato's definition of *knowledge* as "justified true belief" (McDowell, 1987, cited in Furinghetti & Pehkonen, 2002, p. 42), but with one important difference. By using the term *true*, Plato implied the existence of an external reality of which one could know and be certain, and von Glasersfeld's observation that for many people today "there reigns the conviction that knowledge is knowledge only if it reflects the worlds as it is" (von Glasersfeld, 1984, p. 20) captures the sense that the view of knowledge put forth by Plato continues to prevail.

The notion that knowledge must be associated with truth has influenced many definitions of *knowledge* and, therefore, the distinction between knowledge and belief. For example, Scheffler (1965, cited in M. S. Wilson & Cooney, 2002) viewed both *belief* and *knowledge* as cognitive constructs but viewed knowledge as a stronger condition than belief. For Scheffler, one knows Q if three conditions hold: One believes Q; one has the right to believe Q; and Q is, in fact, true. Under this definition, a belief that is true is not knowledge for an individual unless one has a warrant for believing it, that is, unless one's believing is based upon evidence. Furthermore, under this definition, a belief is not knowledge unless it is true. This truth requirement alarms many, because the means by which truth is determined is debatable, and the history of science indicates that what was taken as true at one point in time was generally modified or subsumed into more encompassing theories at other times. In von Glasersfeld's (1993) view of constructivism, the notion that knowledge is a representation of reality is rejected. Instead, the most one can say about ability to predict physical phenomena is that knowledge has proved to be viable under particular circumstances. Cobb, Yackel, and Wood (1992) noted important implications for teachers' holding the position that to know is to accurately represent what is outside the mind. Teachers who hold this view are likely to believe that they can use materials as a means for presenting readily apprehensible mathematical relationships as if there is a direct mapping between the materials and the mathematics, and all that is necessary for students to construct this mapping is for a teacher to clearly present the mathematics by using relevant materials (Cobb et al.). For example, a teacher might believe that the place-value relationships that students so often struggle to comprehend may be obvious if the students use base-ten blocks.

Some are troubled by the view of radical constructivists, who reject the notion that one has access to truth. For example, Goldin (2002b), who thought that the radical constructivist epistemology was deeply flawed and was "entering the realm of passé" (p. 205), feared that defining *knowledge* as justified true belief is a grave mistake made by "cultural relativists" (Goldin, 2002a, p. 64), because doing so "leaves no convenient word to distinguish beliefs that are *in fact* true, correct, good approximations, valid, insightful, rational, or veridical, from those that are *in fact* false, incorrect, poor approximations, invalid, mistaken, irrational, or illusionary" (Goldin, 2002a, p. 65).

This debate about the relationship between knowledge and belief is unlikely to cease, and I suspect that researchers will continue to disagree about the affordances and constraints of various relationships between the two. More important for researchers is to take clear stances on how they are viewing beliefs, stances that are reflected in their operationalization of the construct. For example, M. S. Wilson and Cooney (2002) dropped the truth condition and modified Scheffler's second condition so that they suggested that *X* knows *Q* if *X* believes *Q* and *X* has reasonable evidence to support *Q*. M. S. Wilson and Cooney acknowledged that although in using this definition, they dodged the claim to know reality, the major question of knowing what constitutes evidence is unresolved. Using this definition, researchers can,

however, acknowledge that what is evidence for one may not be evidence for another.

Educational researchers attempting to understand teachers' knowledge and beliefs can benefit by recognizing how teachers look at these constructs and by drawing distinctions between knowledge and belief according to the world view of the subjects holding the knowledge or belief. The important question for researchers studying mathematics teachers' beliefs and knowledge is not whether some conception is true in an ontological way but how a teacher views the conception. As a researcher, I have found the following stance useful when I attempt to understand how a person holds a particular conception: A conception is a *belief* for an individual if he or she could respect a position that is in disagreement with the conception as reasonable and intelligent, and it is *knowledge* for that individual if he or she could not respect a disagreeing position with the conception as reasonable or intelligent. By this definition, agreement upon what constitutes "a reasonable, intelligent position" is unnecessary; instead, the principles are left for the subject to apply and for the researcher to reveal. Under this definition, one person's belief may be another person's knowledge. For researchers to know *how* one holds a notion may be as important as knowing *what* one holds as the notion. Two people who hold contradictory *beliefs* about something may have more in common with each other than with a person who holds one of those conceptions as *knowledge*. For example, consider the following three people and the notions they hold about students' learning of concepts and procedures: Person A holds as *belief* the notion that students are better served by being taught concepts before procedures; Person B holds as *belief* the notion that students are better served by being taught procedures before concepts; Person C holds as *knowledge* the notion that students are better served by being taught concepts before procedures. On the one hand, Persons A and B disagree about the beliefs they hold, but because they hold the notions as beliefs, they may find important points of agreement in a discussion, because they could imagine reasonable, disagreeing positions. On the other hand, Persons A and C, who hold the same notion but one questions it and the other does not, may experience unexpected difficulty when holding a meaningful discussion about their positions. The real roadblocks to meaningful dialogue are created when at least one of party of a disagreement holds a notion as knowledge and, hence, does not respect the position of those who disagree as reasonable or intelligent. Unfortunately, in mathematics education, this lack of ability to find common ground for discussion has fueled the fires in what have come to be known as The Math Wars (Becker & Jacob, 2000).

Summary

General distinctions have been drawn among values, beliefs, and knowledge, with values and knowledge often defined in terms of beliefs. Values are generally viewed as a type of belief to which one is deeply committed. Values are also viewed as preferences that are not associated with truth values, whereas beliefs are held by an individual as true/false dichotomies with the understanding that others may disagree. If one takes the ontological view that truth exists and people have access to it, then knowledge might be viewed as true belief. If one takes a view that truth, though it may exist, is not accessible to humans and instead the best one can hope for is viability, then knowledge is belief with certainty. Although for researchers to understand *what* beliefs teachers hold is important, perhaps equally important is understanding *how* teachers hold these beliefs.

A Comment About Terminology

A. G. Thompson, in her handbook chapter (1992), used the terms *conception* and *belief,* and although at times she seemed to use the terms interchangeably, more often she used *conception* to refer to a general construct that included beliefs as a component. Other researchers have used the term *conception* in their work, but with little agreement on its meaning. For example, Tirosh (2000) used *conception* as interchangeable with knowledge whereas Knuth (2002) used the term to represent knowledge and belief "in tandem" (p. 85). Andrews and Hatch (2000) noted that for some the term *conception* is essentially cognitive whereas for others it is affective. Although attempts to unify the definitions of constructs might seem critical for moving the field forward, I am not convinced of this need, because the constructs that researchers apply when studying the integration of teachers' beliefs, knowledge, and practices are operationalized as much by the research methodologies used as by a particular definition. Throughout this chapter, I use the terms as the researchers used them in their studies and attempt to tease out similarities and differences in their approaches.

Measuring Beliefs

However one chooses to define (or not define) *beliefs,* when studying the beliefs of others, researchers must draw inferences from what people do or say (Pajares, 1992). Mathematics education researchers have typically approached the study of teachers' beliefs in one of two ways: by using case-study methodology or by using beliefs-assessment instruments. Perhaps the more common approach has been to use case-study methodology to provide detailed descriptions of the beliefs of a small number of teachers by relying upon rich data sets that include some combination of classroom observations, interviews, surveys, stimulated-recall interviews, concept mapping, responses to vignettes or videotapes, and linguistic analyses. These data are often collected over a period of time and triangulated. These rich data sets are important for theory building, inasmuch as they enable researchers to consider interrelationships in the complex world of teachers. Such studies also enable researchers to meet the challenge put forth by A. G. Thompson (1992) to investigate the dialectic relationship between teachers' conceptions, including their beliefs and knowledge, and teachers' practices. Researchers in most studies I examined took this case-study approach to studying beliefs.

Likert-Scale Surveys

The case-study approach to studying teachers is powerful for building theory, but testing theory often requires tools for measuring the beliefs of larger groups of teachers. Typically, beliefs of large numbers of mathematics teachers are measured using Likert scales, and I present four examples of Likert scales used for measuring mathematics teachers' beliefs.

As part of their long-term study of teachers' use of knowledge about children's mathematical thinking, referred to as Cognitively Guided Instruction, Carpenter, Fennema, and colleagues developed a beliefs survey, the Mathematics Belief Scale, comprised of 48 Likert-scale items with four subscales to assess Role of the Learner, Relationship Between Skills and Understanding, Sequencing of Topics, and Role of the Teacher (Fennema, Carpenter, & Loef, 1990; Peterson, Fennema, Carpenter, & Loef, 1989). The Role of the Learner subscale measured the belief that children are able to construct their own knowledge instead of being receivers of knowledge. The Relationship Between Skills and Understanding subscale measured the belief that skills should be taught in relation to understanding and problem solving rather than in isolation. The Sequencing of Topics subscale measured the belief that children's natural development, rather than the logical structure of formal mathematics, should guide the sequencing of topics. The Role of Teacher subscale measured the belief that instruction should facilitate children's construction of knowledge rather than consist of teachers' presenting materials. An example of an item from The Role of the Learner subscale is "It is important for a child to be a good listener in order to learn how to do mathematics." This beliefs survey has been used by researchers working with prospective teachers (Vacc & Bright, 1999) and with practicing teachers (Fennema et al, 1996).

Zollman and Mason (1992) developed a beliefs instrument based upon the National Council of Teachers of Mathematics (1989) *Curriculum and Evaluation Standards for School Mathematics*. Their instrument was designed to assess teachers' beliefs about the *Standards*, and they selected 16 items to reflect 16 of the 54 standards included in the NCTM document. The items were stated as nearly direct quotes or the inverse of direct quotes from the *Standards* document, and each item met the criteria that the answer should not be "intuitively obvious" (Zollman & Mason, 1992, p. 359) and the item could be incorporated into a single sentence easily stated in a positive or negative manner. During their pilot work developing the survey, the authors found that respondents appeared to emphasize aspects of the items that were distracting, and so to ensure that respondents focused attention on the intent of the item, select words were capitalized. An example of a positively stated item is "Students should share their problem-solving thinking and approaches WITH OTHER STUDENTS"; a negatively stated item is "Children should be encouraged to justify their solutions, thinking, and conjectures in a SINGLE way" (Zollman & Mason, 1992).

Researchers seeking to use Likert-scale beliefs surveys to study mathematics teachers' beliefs use existing surveys, as Vacc and Bright (1999) used the survey developed by Fennema et al. (1990), develop a new survey, as Zollman and Mason (1992) did, or create a hybrid, drawing upon an existing survey. Hart (2002) administered a three-part beliefs' survey to prospective elementary school teachers before and after they had completed an integrated content/methods course; the first two parts were adapted from existing instruments, and the third part was Hart's creation. The first part was a form of Zollman and Mason's (1992) *Standards'* Belief Instrument; the second was adapted from Schoenfeld's (1989, cited in Hart, 2002) Problem-Solving Project. A positively stated item from this second part, which Hart used to assess change in teachers' beliefs about teaching and learning mathematics within and outside the school setting is "Good mathematics teachers show students lots of different ways to look at the same question"; an item that was negatively stated is "Good math teachers show you the exact way to answer the math question you will be tested on." Hart's third part was comprised of two items on teacher efficacy, with the first stating, "I am very good at learning mathemat-

ics," and the second stating, "I think I will be very good at teaching mathematics."

Enochs, Smith, and Huinker (2000) developed a Mathematics Teaching Efficacy Belief Instrument designed to be used with prospective teachers. Their beliefs survey was comprised of 21 Likert-scale items, 13 on the Personal Mathematics Teaching Efficacy subscale and 8 on the Mathematics Teaching Outcome Expectancy subscale. An item on the former subscale was "Even if I try very hard, I will not teach mathematics as well as I will most subjects," and one from the latter was "The mathematics achievement of some students cannot generally be blamed on their teachers."

One concern about self-report surveys is whether teachers' reports are accurate. Although Ross et al. (2003) provided evidence that the self-report data for elementary school teachers' degrees of implementation of standards-based mathematics teaching in Ontario, Canada, were reliable and valid, whether teachers' self-reports of their practices are different from their self-reports of their beliefs remains an open question. Is a Likert-scale survey a valid measure on which to base inferences about teachers' beliefs? In the next section, I present an example from a group of researchers whose concern about the validity of Likert-scale beliefs surveys led them to develop an alternative survey that can be administered to large numbers of subjects.

An Alternative to Likert-Scale Beliefs Surveys for Large-Scale Data Collection

A group of researchers engaged in a large-scale research and development project referred to as IMAP (Integrating Mathematics and Pedagogy, 2003; see, also, Ambrose, Clement, Philipp, & Chauvot, 2004) developed a web-based survey requiring open-ended responses to overcome three problems they had identified with use of Likert-scale surveys. They noted, first, that inferring how a respondent interprets the words in Likert-scale items is difficult. For example, in the item "It is important for a child to be a good listener in order to learn how to do mathematics" (Fennema et al., 1990), how do researchers know how the respondent interprets the idea of *good listener*? Might a respondent think differently about this statement depending upon whether she was thinking of a child listening to a teacher demonstrating a procedure, a teacher presenting a problem situation, or a child sharing a solution strategy? Individual respondents are not able to explain their responses for Likert items, so researchers can only infer what considerations may have guided the respondents. A second difficulty the IMAP researchers identified with Likert items is that responses provide no information for determining the importance of the issue to respondents (Ambrose et al., 2004). McGuire (1969) stated, "When asked, people are usually willing to give an opinion even on matters about which they have never previously thought" (p. 151). For example, although teachers who build their instruction around having children solve problems and share solutions with other students are likely to agree strongly with the statement "Students should share their problem-solving thinking and approaches WITH OTHER STUDENTS," teachers who have not thought much about basing instruction upon their students' sharing their thinking with other students may, when reading this statement, think that it makes sense and may agree or even strongly agree. The third concern the IMAP researchers raised is that Likert scales tend to provide little, and often no, context. For example, Collier (1972) developed a Likert-scale survey containing the item "In mathematics, perhaps more than in other fields, one can find set routines and procedures." Respondents might interpret such a comment in multiple ways, and their responses give no indication of their interpretations. Some respondents may differentiate between middle-school algebra and primary-grade arithmetic and believe that one has set routines and procedures and the other does not. Because no context is associated with the item, respondents may feel compelled to consider mathematics in general when, in fact, they view different levels of mathematics in different ways.

The IMAP team set out to create a beliefs survey by identifying four characteristics of beliefs identified in the literature as accounting for the critical role that beliefs play in teaching and learning and, thus, were important for the way the researchers attempted to measure the beliefs. First, beliefs influence perception (Pajares, 1992). That is, beliefs serve to filter some complexity of a situation to make it comprehensible, shaping individuals' interpretations of events (Grant, Hiebert, & Wearne, 1998). Beliefs might be thought of as serving as a type of constructed model (von Glasersfeld, 1993) or theory (Nisbett & Ross, 1980) *through* which one looks, so that beliefs affect what one sees or notices (Mason, 2002). The IMAP team addressed this issue in their web-based survey by providing respondents with complex situations to interpret (Ambrose et al., 2004). Second, beliefs are viewed as drawing one toward a position or, at least, predisposing one in a particular direction. Allport (cited in McGuire, 1969) considered a belief as a readiness to respond, and Rokeach (1968) thought of a belief as a disposition to action. The notion that beliefs are dispositions toward action has played a key role in some important research on beliefs (e.g., Cooney, Shealy, & Arvold,

1998), because one's beliefs can be inferred by attending to the manner in which one is disposed to act in a particular situation. Because the IMAP team intended their beliefs survey to be used by practicing and prospective teachers, they took into account the fact that many prospective teachers are not yet positioned to act and thus many means available to researchers for measuring practicing teachers' dispositions toward action are unavailable for use with prospective teachers. They addressed this issue in their beliefs survey by providing respondents with scenarios in which they were asked to make teaching decisions. The prospective and practicing teachers' dispositions to act in these situations provided evidence from which to infer their beliefs. A third characteristic the IMAP team identified as pertinent for their work was that beliefs are not all-or-nothing entities; they are, instead, held with differing intensities (Pajares, 1992, citing Rokeach, 1968). To address this characteristic in their survey, they provided tasks with multiple interpretation points; the order in which questions were posed was carefully constructed to elicit views that were important to the respondents before "giving away" the "preferred" answers. For example, in one task, after being shown a video clip, respondents were asked, "What stands out to you in this video?" before they were asked, "What are weaknesses of the teaching in the video?" Those who identified weaknesses *before* being prompted for weaknesses were scored as holding the designated belief more strongly than those who did not mention weaknesses until asked to do so. Furthermore, because of the electronic administration of the survey, respondents could not return to change their responses to previous items. To allow for the differing intensities with which individuals hold beliefs, the team devised the scoring rubrics to differentiate among *strong evidence, evidence, weak evidence,* and *no evidence* for a respondent's holding a belief (Integrating Mathematics and Pedagogy, 2004). Furthermore, they were careful not to claim that an individual lacked a particular belief but instead stated that they found no evidence for the belief in the responses the individual provided. Fourth, beliefs tend to be context specific, arising in situations with specific features (Cooney et al., 1998). The IMAP team addressed this issue by situating survey segments in contexts, and they inferred a respondent's belief on the basis of his or her interpretation of the context (Ambrose et al., 2004).

This beliefs survey was used in an experimental study of prospective elementary school teachers enrolled in a mathematics course and randomly assigned to (a) concurrently learn about children's mathematical thinking by watching children on video or working directly with children, (b) concurrently visit elementary school classrooms of conveniently chosen or specially selected teachers, or (c) a control group that had no relevant additional instructional activities outside the mathematical course (Philipp et al., 2005). Those who studied children's mathematical thinking while learning mathematics developed more sophisticated beliefs about mathematics, teaching, and learning and improved their mathematical content knowledge more than those who did not study children's thinking. Furthermore, beliefs of those who observed in conveniently chosen classrooms underwent less change than the beliefs of those in the other groups, including those in the control group. The researchers concluded that by developing a context-based open-ended beliefs survey, they were able to measure prospective elementary school teachers' beliefs related to mathematics and to mathematics learning and teaching years before these prospective teachers were in the position to act as teachers. Beliefs of PSTs could be changed over the course of a semester, although the authors did not speculate whether these changes would remain with time (Philipp et al., 2005).

Summary

The predominant approach used by researchers to measure beliefs is to employ qualitative measures, including teacher interviews, classroom observations, responses to simulated materials, concept generation and mapping, paragraph completion, or Kelly Repertory Grid techniques. Although these qualitative approaches provide rich data sets, they are expensive to employ across large numbers of participants; researchers who need to collect large data sets often rely upon less expensive approaches, such as Likert-scale surveys. Some have identified problems with Likert-scales surveys, and one example was provided of an alternative beliefs survey that was developed and successfully used to survey large numbers of prospective elementary school teachers without relying upon Likert-scale responses.

Inconsistent Beliefs and the Role of Context

Because of the complexity of teachers' beliefs systems, researchers may find that teachers hold beliefs that appear to be inconsistent with their teaching practices. One approach taken by researchers to explain inconsistencies is to examine whether particular beliefs within a beliefs system are more central or primary, and hence play a greater role in influencing practice, than other beliefs. Another approach is to study whether a teacher's perspective on his or her practice might help explain the apparent contradiction. In this section I share examples of both ap-

proaches and conclude by considering the stances researchers take regarding inconsistent beliefs.

In a study of six novice elementary school teachers, Raymond (1997) investigated the inconsistency between one teacher's mathematics beliefs and teaching practice. For each teacher, data, collected during the 10-month period between March of their first year of full-time teaching and December of their second year, included an introductory phone interview, six 1-hour audiotaped interviews, five monthly classroom observations, an analysis of several samples of lesson planning, a concept-mapping activity in which teachers presented their views of the relationships between mathematics beliefs and practice, and a take-home questionnaire on mathematics beliefs and factors that influence teaching practice. Raymond analyzed these data within four categories—teachers' beliefs about the nature of mathematics, teachers' beliefs about learning mathematics, teachers' beliefs about teaching mathematics, and teachers' mathematics-teaching practices—on a 5-level scale ranging from *traditional* to *nontraditional* (traditional, primarily traditional, even mix of traditional and nontraditional, primarily nontraditional, and nontraditional). The focus of her case study was a fourth-grade teacher, Joanna, selected, first, because her expressed beliefs were similar to those of the other five teachers and, second, because her case was dramatic in that although her beliefs about the nature of mathematics were fairly traditional, her beliefs about learning and teaching mathematics were placed at the *nontraditional* end of the scale.

Joanna expressed the view that students should discover mathematics without being shown and that students learn mathematics better when solving problems. Joanna also stated the beliefs that teachers could be effective without following a textbook and that they should use various activities from various sources; she was a strong advocate of using manipulatives to achieve hands-on learning. Raymond noted a major inconsistency between Joanna's nontraditional beliefs about mathematics teaching and learning and her traditional practices. Joanna's class was run in an orderly way with her students' working exclusively at their desks in a quiet atmosphere, reflecting strict discipline. Joanna established herself as the authority, presenting information as teacher-directed instruction; although some teacher-student dialogue took place, on no occasion was student-to-student dialogue observed. Joanna rigidly followed her mathematics textbook, and the students worked quietly on problems from the textbook without access to manipulatives.

Joanna's beliefs about mathematics were categorized as traditional—at the opposite end of the continuum from her beliefs about mathematics teaching and learning but on the same end as her practice. Joanna identified her own negative experiences as a mathematical learner as the primary influence on her beliefs about the nature of mathematics but her own teaching practice as the main influence on her beliefs about teaching and learning. Despite her negative view of mathematics as a student, and perhaps because of her negative experiences, Joanna wanted her own students to experience mathematics in a more positive light than she had experienced, a desire that may account for her nontraditional beliefs about learning and teaching mathematics. But if Joanna held these nontraditional beliefs she espoused, why was her practice traditional?

Raymond (1997) concluded that this inconsistency arose from the effects on Joanna's teaching of time constraints, scarcity of resources, concerns over standardized tests, and students' behavior. As a novice teacher, Joanna was particularly concerned with classroom management and discipline. Raymond reported also that, in at least one case, Joanna seemed to view her mathematics-teaching practice in terms of what she wanted to do, or thought she should do, rather than what she accomplished. She claimed that her belief in teaching mathematics through manipulatives led to her using them, but the only time Joanna actually used manipulatives was during a class she described as chaotic.

Raymond (1997) concluded, further, that her study provided evidence that the teacher's beliefs and practice were not wholly consistent and that her practice was more closely related to her beliefs about mathematics content than to her beliefs about mathematics learning and teaching. She also suggested that during teacher education, beliefs of prospective elementary school teachers might be more explicitly addressed so that teachers become aware of the beliefs with which they enter and attend to how these beliefs begin to change during this important period of growth.

Hoyles (1992) reflected upon how her changing view of beliefs affected her thinking about inconsistent beliefs. During her work in the early 1980s on a research project designed to identify characteristics of good secondary-school teaching practice, she and her colleagues approached teachers' beliefs as a construct that researchers might access from outside the classroom. Assuming that beliefs could be assessed in a decontextualized manner and that a linear relationship existed between teachers' beliefs and their classroom practices, the researchers first assessed teachers' beliefs and then observed teachers in the classroom. By the early 1990s, after observing the changes teachers experienced while interacting with a classroom innovation involving computers, Hoyles drew upon work in situated cognition (Brown et al., 1989) and subscribed to the notion that situations coproduce beliefs through activities. She came to view beliefs as situated and dialectically constructed from the relationships among activity, context, and culture and, thus, teachers' beliefs about mathematics or mathematics teaching and learning as being affected by factors such as grade level, students' level, textbook, or computer use.

For Hoyles (1992), viewing all beliefs as situated reconciled the apparent inconsistencies between teachers' beliefs and actions reported by researchers. Once the embodied nature of beliefs is recognized, the notion of inconsistent beliefs is an irrelevant consideration that can be replaced by considering the circumstances and constraints within settings. When Hoyles reconsidered data from the 1980s study with this new lens, she noted "the remarkable harmony between what the teacher wished to 'deliver' and 'what the students wanted'—an example of beliefs constructed in practice" (p. 40). Hoyles continued, "Such harmony exemplifies *a* good practice—a different blend of variables in the setting might produce alternative pictures of success. The notion of situated beliefs allows us to cope with this diversity" (p. 40).

Skott (2001) also considered the influence of context on teachers' beliefs, and although his conclusion about context differed from Hoyles's, he shared her conclusion about inconsistencies. He investigated the relationship between teachers' beliefs about mathematics and mathematics teaching and learning and the teachers' classroom practices. Skott introduced the term *school-mathematics images* to "describe teachers' idiosyncratic priorities in relation to mathematics, mathematics as a school subject and the teaching and learning of mathematics in schools" (p. 6). On the basis of responses to a questionnaire comprised of open and closed questions, administered to Danish student teachers two months before their graduation, Skott selected 11 student teachers, representing a variety of school-mathematics images, to interview after graduation. Four teachers who presented school-mathematics images inspired by current reform efforts in school mathematics were videotaped teaching their mathematics classes for 2 to 3 weeks during their first 18 months of teaching; they participated in informal discussions after each lesson and in a comprehensive interview after the videotaped lessons were completed. Skott selected one of the four teachers, Christopher, as the subject of a research study.

Christopher's responses to the questionnaire and the first interview indicated that his school mathematics images were in contrast to those he remembered experiencing as a high school student in the Danish gymnasium, in which teachers lectured at the board. He favored process aspects of mathematics; he viewed his role as initiating and supporting investigative activities so that his students could assume responsibility for their own learning. Christopher was committed to school-mathematics reform but also to his students' developing broad educational, nonmathematical skills: learning to plan, reflecting on their own learning, working cooperatively, relating critically to information, and independently finding solution strategies and models.

After graduation, as a music and mathematics teacher in a typical Danish *folkeskole* (a municipal school for Grades 1–10), Christopher enjoyed freedom in his choice of teaching methods and curricula. He was observed teaching his Grade 6 mathematics class, in which he had introduced a textbook different from the traditional book used by the previous teacher. Because as a novice teacher Christopher felt overwhelmed with lesson preparation and grading, he relied upon the textbook for the aims, the content, and the tasks for instruction, but he strongly influenced the flow of the classroom by the methodology he chose. He lectured little and addressed the whole class primarily to initiate small-group or individual work. He sought student explanations and elaborations, and he encouraged students to explore in the noisy, but productive, classroom.

Skott (2001) described two visits to Christopher's classroom. The first visit took place during the second day of a unit on area; the students were cutting congruent triangles to make parallelograms, cutting parallelograms to make rectangles, and measuring areas of desks, blackboards, and other items. When two students called for help with a problem, Christopher listened to the weaker student's approach and asked the student questions, although he recognized that the approach would be nonproductive, before he asked whether they could approach the problem another way. He then pursued the stronger student's approach with the two students. Skott viewed this episode as being consistent with Christopher's school-mathematics image, because he attempted to initiate and support the students' learning by letting them devise suggestions and by refraining from deflecting suggestions that might be wrong or insufficient.

The second visit took place during the next lesson. When two other boys asked for help to find the areas of rectangular lawns drawn with different scales, Christopher intervened and led the two students through a series of computationally oriented questions in such a way as to deplete the task of any mathematical challenge and replace it with carrying out simple, prescribed computations. Skott

(2001) viewed this episode as being inconsistent with Christopher's school-mathematics image and Christopher's actions as replacing his general reformist school-mathematics priorities with traditional values in mathematics education.

When confronted with the roles he had played in the two episodes, Christopher, who was often critical of his own teaching, was satisfied with both. Skott (2001) suggested that although Christopher's second interaction appeared to reflect inconsistencies both with his action during the first class and with his stated beliefs, the relationship between Christopher's school-mathematics images and his actions could explain his seemingly inconsistent behavior. Christopher's evaluation of each episode encompassed aspects of his approach to teaching beyond students' mathematical learning, and Skott identified three such aspects that became apparent in the two episodes and in the interview. First, Christopher consistently mentioned the need to think of the children as learners generally, not solely as mathematical learners, for example, as needing to develop self-confidence in mathematics and in general. Second, Christopher encouraged and expected his students to interpret and solve tasks on their own and to discuss strategies and results among themselves. In keeping with this goal, he provided minimal support before they started working, and as a result, the students were in constant need of assistance from the teacher during group work. Third, Christopher's constructivist view of learning did not lead to a particular method for teaching, and he spoke about the need to play multiple roles when teaching. Understanding these three aspects helped Skott understand Christopher's actions. In the first episode, he delved into the first student's suggestion because he believed that the student's self-confidence was an important consideration. In the second episode, he ensured that the students arrived at the correct answer because of his concern for the general management of the class, realizing that one of the two boys was extremely demanding and would require excessive attention unless the two boys believed that they had solved the problem. Skott concluded that teachers have "multiple and sometimes conflicting educational priorities, and that in this case the priorities related to Christopher's SMIs [school-mathematics images] were dominated by others concerned with managing the classroom with broader educational issues" (p. 18).

Skott (2001) concluded that Christopher's school-mathematics image shaped and filtered the objects of his reflection and framed his interpretation of what he saw but was insufficient for explaining all his actions, sometimes being overshadowed by the more general educational priorities of building students' confidence and managing the class. Skott also noted how the teacher's classroom actions tended to replicate themselves. Christopher used the same calculational approach (A. G. Thompson et al., 1994), seemingly incompatible with his school mathematics image, that he had used to move the class along by appeasing these two students with a number of other groups of students, even when the motives for which it was originally used, building student confidence and managing the class, no longer were issues. The way the teacher coped with the original problem became a "prototypical action type" (Skott, 2001, p. 22) that was applied even when the original reasons for applying the action type were no longer present.

Skott's (2001) conclusion that the relationship between the teacher's school-mathematics images and his classroom practices varied according to the classroom situation seems to support Hoyle's (1992) notion that beliefs are situated. Although both Skott and Hoyles considered context critical to assessment of teachers' beliefs, they differed in their views on the situativity of beliefs. Skott contended that Christopher's beliefs did not change with the situation but that Christopher continued to believe in the importance of mathematical processes and the teacher's role in initiating and monitoring student activity to facilitate both the processes and the products. The change was in his specific goals during the two episodes—from mathematical learning of the two students in the first to class management in the second. Skott (2001) elaborated upon how this view differed from that of Hoyles (1992):

> It is not the situation that produces a new set of school mathematical priorities. Situations that are outwardly different—for example, a research interview as opposed to a classroom setting—do not necessarily create or radically change a teacher's beliefs. Rather, the situations, that is, very specific situations each concerned with a specific interaction between student(s) and teacher, create or co-produce competing objects and motives of the teacher's activity. It is these competing objects and motives that form the basis of what apparently, but only apparently, is a new set of priorities produced under new and slightly different circumstances. In other words, the contextual embeddedness of Christopher's activity did not necessarily lead to a similar contextualisation of his school mathematical priorities. (Skott, 2001, pp. 24–25)

Skott acknowledged one general conclusion *about* beliefs research: No general conclusions should be expected *in* beliefs research. Furthermore, because teachers and the contexts in which they work differ, the relationship between classroom practices and teachers' beliefs may require more detailed, long-term investigations of teachers' approaches in specific classrooms over several studies to begin to generate preliminary understandings of theoretical constructs.

Sztajn (2003) also considered the relationship between teachers' beliefs and practices as mediated by the contexts in which teachers were working, but instead of analyzing one teacher in two contexts as Skott had done, she analyzed two teachers who held similar beliefs but taught in different contexts. She observed each of two elementary school teachers a week at a time for four weeks spaced throughout the semester and conducted five semistructured interviews with each teacher. When visiting the teachers, she observed throughout the day, including during lunch, breaks, and staff meetings. Other classroom data included the teachers' lesson plans, grading books, students' report cards, mathematics worksheets and completed assignments, and materials from activities in other subjects. To triangulate her data related to the class context, Sztajn interviewed the principal, other teachers, and parents from each school.

Because she set out to study beliefs that go beyond mathematics and mathematics teaching and learning experiences, Sztajn (2003) initially coded the data using seven descriptive categories: the teacher, the students, the classroom, the school, the parents, general educational goals, and mathematics. The two public-school teachers she investigated taught classes in small, midwestern towns, one a third-grade and the other a fourth-grade class; each class was comprised only of Caucasian children. The teachers held similar beliefs about mathematics, believing that problem solving and basic skills were important but that basic skills had to be mastered before children could engage in problem solving. They believed that some children bring basic-skills knowledge from home but that the others lack both basic and social skills. In spite of the similarity of their beliefs about mathematics, their instruction differed; to explain the differences, Sztajn turned to their beliefs that go beyond mathematics.

Sztajn (2003) found that the concept of students' needs, which included teachers' beliefs about children, society, and education, emerged as an important factor in the teaching of both teachers and accounted for the differences in their instruction, because the two schools differed with respect to the manner in which the children were viewed. The students of the third-grade teacher, Theresa, were at a lower socioeconomic level than the students of the fourth-grade teacher, Julie. At Theresa's school, 40% of the students received free or reduced-cost lunches, compared to only 10% of the students at Julie's school. The educational levels of the parents, as estimated by the principals and teachers, differed; most parents of the students at Theresa's school held low-income, manual-labor jobs whereas in Julie's school, many middle-income parents were doctors, lawyers, and university professors. Although Theresa believed in the importance of students' engaging in problem solving and developing higher order thinking skills, because she saw her students as coming from unstable, chaotic homes—poor environments for children—basic facts, drill, and practice were at the core of what she thought her students needed to prepare for their roles in the workplace. Most important for Theresa, whose teaching goal was to form responsible citizens, was that "learning and following rules in a responsible and organized way is what her students need in order to find their place in society" (p. 64). The fourth-grade teacher, Julie, held beliefs about mathematics and about students' needs that were similar to Theresa's, but because she thought that she had to deal with few problems and worried little about teaching rote basics, she structured her classroom around problem solving and projects. In Julie's class, even for routine mental-computation tasks, students shared various solution strategies. Although at the time of the study Julie and Theresa approached teaching differently, the approach Julie described having used previously, at a school in which children were more like Theresa's students, was very similar to Theresa's approach. Both teachers believed that teaching social skills was a main role of school. For example, Julie said that responsibility was important to teach, but because she thought that her students learned responsibility at home, they needed less practice on social skills and could explore more sophisticated mathematical content.

Sztajn (2003) contended that both teachers

are the heroes of the stories presented—never the villains. They are not, in a Machiavellian way, trying to hold students back, lessen their chances of going to college, or ensure that they will remain where they are in the socioeconomic spectrum. Quite the contrary, both of them use their best judgments when defining what their students need, and once they form this concept, they direct their work and all their efforts toward meeting these needs. (p. 71)

Sztajn concluded that because these students came from different socioeconomic backgrounds, they were learning different mathematics. It was the teachers' beliefs about children, society, and education, not their beliefs about mathematics, that accounted for the differences in instruction.

Summary

Researchers who have investigated apparent inconsistencies between teachers' beliefs and practices have attempted to explain them. Raymond found two factors that accounted for such inconsistencies. First, a teacher's practices were more in line with her beliefs about mathematics than her beliefs about mathematics teaching and learning. Second, as important as beliefs about mathematics were, observed inconsistencies between a teacher's espoused beliefs and her practice were explained by general educational issues, such as time constraints, resources, standardized tests, and students' behavior.

Hoyles (1992) found that when she viewed beliefs as decontextualized, she observed inconsistencies, but when her view of beliefs became more contextualized and situated, she could explain apparent inconsistencies by considering the circumstances and constraints within settings.

Skott (2001) found that a teacher's beliefs about mathematics and about mathematics teaching and learning were important for understanding the teacher's practice but that these beliefs were often overshadowed by the more general educational priorities of building students' confidence and managing the class. But unlike Hoyles, who concluded that beliefs were situated, Skott concluded that not the teacher's beliefs but the teacher's goal for the particular activity changed, and after understanding the goal, the researcher could explain apparent contradictions between the teacher's beliefs and actions.

Sztajn (2003) studied two teachers who held similar beliefs about mathematics but taught in very different contexts and differed in their teaching. She found that attending to the teachers' beliefs about mathematics was insufficient for explaining the teachers' practices; only after considering the teachers' beliefs about children, society, and education was she able to account for the differences in the teachers' practices.

A Stance on Inconsistent Beliefs

What are we mathematics education researchers to make of inconsistencies? Is resolving inconsistencies important for the work we do? Within mathematics, consistency is so central that, historically, when mathematicians have faced a new way to consider some aspect of mathematics that seemed to them unimaginable or even untrue, they would, often after much resistance, accept this new way so long as no inconsistencies arose. As a classic example, 19th-century mathematicians realized that rejecting Euclid's fifth postulate led to consistent, albeit previously unimaginable, possibilities, resulting in mathematicians' accepting the findings and inventing new non-Euclidean geometries that led to productive ways for considering the physical world. Other examples are the expansion of the number systems to include irrational numbers, negative numbers, and complex numbers and Cantor's theory of transfinites, which engendered new ways of thinking about infinity. Although inconsistencies are unacceptable in logical domains, people, in general, have come to accept that inconsistencies abound in human affairs. So what are the implications for how we choose to consider inconsistencies when studying teachers' beliefs?

Most researchers cited in this section found that the inconsistencies that arose when documenting teachers' beliefs and actions ceased to exist after they better understood the teachers' thinking about some aspects of their contexts. I propose that as a research stance in studying teachers and their beliefs, we researchers assume that contradictions do not exist. Taking this stance when we observe apparent contradictions, we would assume that the inconsistencies exist only in our minds, not within the teachers, and would strive to understand the teachers' perspectives to resolve the inconsistencies. Inconsistencies should still present problems, but for researchers instead of teachers. I do not propose that this position is ontological; that is, I do not suggest that inconsistencies are nonexistent or can all be explained away. I suggest, instead, that researchers who assume that inconsistencies do not exist will attempt to better understand teachers' beliefs systems and that the circumstances surrounding the teachers' practices will often, as for the researchers cited in this section, lead to resolution of the inconsistencies.

Changing Beliefs

Two general approaches are considered for the relationship between the change in beliefs and the change in behavior. One notion, supported by research, is that to change behavior, one must first change one's beliefs, because beliefs act as filters that affect what one sees (Pajares, 1992) and people have difficulty seeing what they do not already believe. However, if beliefs must change before behavior, then how do beliefs change? Can one change his or her beliefs merely by reflecting upon them? Or might reflection and action form a dialectic that can contribute to teachers' learning and growth? If Socrates was right that the unexamined life is not worth living, perhaps it is equally true that the unlived life is not worth examining.

Guskey (1986) presented an alternative model predicated on the notion that "significant change in teachers' beliefs and attitudes is likely to take place only *after* changes in student learning outcomes are evidenced" (p. 7). He offered a linear model, pro-

posing that teachers who implement an instructional change that leads to students' success will then alter their related beliefs. In 1993, von Glasersfeld, in stating, "If one succeeds in getting teachers to make a serious effort to apply some of the constructivist methodology, even if they don't believe in it, they become enthralled after five or six weeks" (p. 29), also observed that changes in teachers' beliefs follow changes in their instruction.

In his research on teacher change, Guskey (1986) examined classroom teachers' success in implementing what he described as a relatively minor alteration in their practices. Grant, Hiebert, and Wearne (1998) wondered how Guskey's model might hold for substantive changes to teachers' practices, such as those called for in the effective-schools literature. They noted that implementers of past programs had often incorrectly assumed that simply *telling* teachers how they should teach would lead to the desired changes in practice and wondered whether, instead of telling, *showing* teachers models of teaching would be effective? What would teachers notice when observing reform-oriented instruction applied with their own students, and would what the teachers noticed be related to the beliefs they held? To investigate these questions, nine teachers observed reform-oriented lessons for 6–12 weeks. The teachers' beliefs about what mathematics should be learned and how to teach mathematics were placed on a continuum (from *skills/teacher-responsibility* end to the *process/student-responsibility* end). The researchers then compared the positions of the nine observing teachers' beliefs along the continuum with their evaluations of the alternative instruction, and they discovered a clear relationship between the teachers' beliefs and what they observed during the instruction. Beliefs of four of the nine teachers were on the skills/teacher-responsibility end of the beliefs continuum, and these four teachers tended to focus on one particular aspect of the reform-oriented lesson or to notice features that the researchers believed were tangential, not the intended point of the instruction. For example, whereas the intended purpose for using manipulatives was to provide opportunities for teachers to promote multiple strategies that might link representations for the students, the teachers on the skills/teacher-responsibility end of the continuum thought that their purpose was to present the *only* way to solve problems.

Three of the nine observing teachers, whose beliefs were placed in the middle of the beliefs continuum, recognized some features of the instruction but did not connect these features with the larger intended goals. For example, one teacher noted the importance of explaining strategies, doing story problems, and engaging in critical thinking, but she juxtaposed these against her view of the teacher's role as the provider and demonstrator of strategies. This teacher seemed to be driven by an image of a teacher as one who removes obstacles from her students' paths. When working with story problems, she tried to support her students' success by helping the students focus on key words. One teacher who was classified in *the middle-ground* category later taught the place-value unit she had observed. She explained that except for making a little change, she "pretty well stuck to their lesson plans." The change she made was that instead of encouraging students to solve the problems as they chose, she explained the rule of solving by adding from the one's column.

Beliefs of two of the nine observing teachers were placed at the process/student-responsibility end of the beliefs continuum; these teachers recognized the features of instruction important to the researchers and understood the underlying goals of the instruction. For example, even after hearing other teachers say that the approach was based on use of manipulatives, one of these two teachers noted that the approach was less about manipulatives and more about having children describe their thinking processes and solve problems in different ways.

The authors concluded that their study provided evidence for the position that the beliefs teachers hold filter what they see and, consequently, what they internalize. Teachers whose beliefs are at odds with a particular instructional approach may not change their thinking or their practices solely by observing the approach. The researchers suggested that observations combined with other activities designed to support the teachers' reflecting upon the experience, such as collegial discussions of the instruction, are more likely to lead to change.

In a study of 14 third-grade teachers undergoing professional development, Borko, Mayfield, Marion, Flexer, and Cumbo (1997) found a similar result. When teachers' beliefs were incompatible with the goals of the staff developers, the teachers generally either ignored the new ideas or inappropriately assimilated the ideas into their existing practices. Borko et al. concluded that beliefs served as filters through which ideas were perceived, and teachers needed to be challenged to reflect upon their beliefs.

Although Benbow (1995) acknowledged that beliefs act as filters through which teachers interpret their school experiences and influence their classroom practices, he rejected the linear model—that beliefs change must precede change in practice—in investigating the relationship between prospective elementary school teachers' mathematical beliefs and

their classroom teaching in the context of an early field experience. On the basis of data from three measures of beliefs, two Likert-type surveys and one open-ended questionnaire, he concluded that the preexisting beliefs held by the preservice teachers played key roles in their planning and implementation decisions for their mathematics lessons. At the same time, the preservice teachers' interpretations of the instructional outcomes affected their beliefs and, therefore, their subsequent practices. The early field experience had little effect on the preservice teachers' beliefs about mathematics, but it did affect their beliefs about pedagogical issues, such as their becoming more open to alternative points of view. The most dramatic belief change measured by the researcher was in the preservice teachers' *personal mathematics-teaching efficacy,* that is, their confidence in their own abilities to help students learn mathematics. Benbow concluded that the preservice teachers' core beliefs underwent little change and that, overall, the early field experience was primarily a confirming experience for the participants.

Reflection and Changing Beliefs

Cooney et al. (1998) also found that reflection played an important role in the growth of prospective secondary-school teachers over their last year in an undergraduate teacher preparation program. They applied a rich theoretical perspective, drawing heavily from the work of Green (1971, cited in Cooney et al., 1998) to attend not only to what beliefs the prospective teachers held but also to the ways in which the beliefs were held. They considered the three aspects of Green's belief structures: Beliefs can be held as primary or derivative beliefs; they then can be held with differing degrees of strength; and they can be held more centrally or more peripherally in relation to other beliefs. Another important factor for Green was the role that evidence plays in the ways one holds beliefs, because nonevidentially held beliefs are impervious to change through reason or evidence whereas beliefs held evidentially can be changed through teacher reflection. A related construct is *teacher authority,* because for one who holds beliefs nonevidentially, the role of an outside authority figure is important for the construction of beliefs. Cooney et al. applied two frameworks for considering authority, one based upon Perry's (1970, cited in Cooney et al., 1998) scheme, which was developed by investigating how college males think and come to know, and the other based upon women's ways of knowing (Belenky, Clinchy, Goldberger, & Tarule, 1986, cited in Cooney et al., 1998). Cooney et al. "were interested in capturing the meanings each of the teachers ascribed to his or her teacher-education experiences" (p. 315), and doing so required them to assess each preservice teacher's beliefs structures by collecting multiple data sources and searching for disconfirming evidence across the data. Beginning with a written survey administered to all the students in the class, the researchers referred to the students' responses to the survey along with classroom-observation field notes and written assignments over the first few class sessions to select four students who initially held a range of positions toward mathematics teaching and learning. The survey questions, some rather unconventional and designed to assess the preservice teachers' views of mathematics, included this item:

> Consider the following similes: A mathematics teacher is like a news broadcaster; entertainer; doctor; orchestra conductor; gardener; coach. Choose the simile that you believe best describes a mathematics teacher and explain your choice. Choose a simile that you believe does not describe a mathematics teacher and explain your choice. (p. 314)

They administered a similar written survey later in the year and conducted individual interviews at five points over the course of the year: The initial interview was conducted in November during the integrated content and pedagogy course, and the fifth interview was in August near the end of the post-student-teaching seminar. Other data sources included observation field notes that were discussed weekly by the research team. Their results were published as four separate case studies in which relationships were drawn between the types of beliefs changes each student experienced and the manners in which the students held their beliefs. On the one hand, Greg's beliefs changed over the course of two quarters; he had begun the program "adamantly opposed to using technology in the classroom" (p. 317) but later stated that in looking for a job, he would look for a school with a Mac lab. Henry, on the other hand, was initially opposed to using technology in education, and his belief did not change over the same period of time. Cooney et al. (1998) attributed Greg's shift to his own involvement in technologically rich learning environments, but these were the same learning environments in which Henry was involved. The authors suggested that not the beliefs the two students began with but the ways they held their beliefs had made the difference.

Greg was open to others' perspectives and enjoyed the exchange of diverse opinions, whereas Henry tended to either accept or reject an opinion, holding tightly to his own, and felt frustrated when his beliefs were challenged. He held a more *authoritarian view* in which assimilation, but not accommodation, may be possible. Greg held a *reflective stance* toward the world whereby he was able to hold onto his core ideas while reformulating his core beliefs when faced with perturbations. The two students who changed the most were the two who were most reflective and who were open to teaching as a problematic activity.

The authors presented four characterizations for how the preservice teachers held their beliefs, characterizations that were further elaborated by Cooney (1999): isolationist, naive idealist, naive connectionist, and reflective connectionist. Henry was classified as an *isolationist*, one who

> tends to have beliefs structured in such a way that beliefs remain separated or clustered away from others. Accommodation is not a theme that characterizes an isolationist. For whatever reason, the isolationist tends to reject the beliefs of others at least as they pertain to his/her own situation. (Cooney, 1999, p. 172)

The *naive idealist* "tends to be a received knower in that, unlike the isolationist, he/she absorbs what others believe to be the case but often without analysis of what he/she believes" (Cooney, 1999, p. 172). The last two characterizations are considered to be connectionist positions that emphasize reflection and attention to the beliefs of others as compared to one's own. "The naive connectionist fails to resolve conflict or differences in beliefs whereas the reflective connectionist resolves conflict through reflective thinking" (Cooney, 1999, p. 172). Cooney et al. (1998) considered Greg a *reflective connectionist*, and they suggested that this position set the stage for the preservice teachers to emerge into reflective practitioners (Schön, 1983). The researchers concluded that a goal of teacher education was to produce reflective connectionists, but they could not conceive of some of the students in their study moving in that direction. They suggested that understanding the belief structures of students might help teacher educators make decisions about effective activities:

> We submit that understanding the belief structures as evidenced by these four young teachers can provide a basis for conceptualizing the nature and activities of teacher education programs. The inculcation of doubt and the posing of perplexing situations would seem to be central to the promotion of movement from being a naive idealist or even isolationist to being a connectionist. Inciting doubt and making the previously unproblematic problematic can have significant impact on a person's world and lead to varied and perhaps unsettling responses. It is not enough to make mathematics and teaching problematic for teachers. We need to understand the effect of this inculcation of doubt and also understand the kind of

support teachers need to make sense of it. (Cooney et al., 1998, pp. 330–331)

The authors ended by suggesting that the research community needs to develop a better understanding of the linkages between what teacher educators provide to preservice teachers and the effect these activities have on the preservice teachers' beliefs systems.

Mewborn (1999) investigated reflection and its effect on four preservice elementary school teachers in a field-based mathematics methodology course. She noted that reflection and action must go together, and she used the terms *verbalism* and *activism* to refer to reflection without action and action without reflection, respectively. Mewborn created a community of learners among four preservice elementary school teachers, herself, and a fourth-grade teacher in whose class they observed and worked. Over the course of the 11-week experience, the preservice teachers observed and discussed lessons in the fourth-grade class, observed and discussed videotaped examples of children's thinking, interviewed children and discussed the experience, and taught small groups of children and discussed the experience. Mewborn studied what the four prospective teachers found problematic, and she classified the concerns into four somewhat-ordered categories of issues related to classroom management and the classroom context; issues related to teaching mathematics; issues related to children's mathematical thinking; and, to a lesser extent, mathematics content. She found that the preservice teachers were able to engage reflectively, but they needed support in learning to observe mathematics teaching and learning environments and, in particular, in developing an internal locus of authority for pedagogical ideas. Initially, the students held an external locus of authority whereby they would turn to the classroom teacher or to Mewborn for answers. Later, they moved to a second stage in which the locus of authority was both internal and external, and during this stage, they moved beyond asking questions and began to state problems and generate hypotheses. During the third stage, which was reached while the students immersed themselves with children's mathematical thinking, the locus of authority was internal, and the preservice teachers became fully reflective when they not only presented hypotheses but also searched for evidence by which to test their hypotheses. Mewborn identified five elements of the design of the field experience that she considered critical to successfully helping the students become reflective about teaching and learning mathematics: (a) The field experience was approached from an inquiry perspective; (b) the students, teacher, and teacher educator participated

as a community of learners; (c) the community was nonevaluative; (d) the preservice teachers were given time to reflect; and (e) the field experience was subject specific.

Many have noted the need to support teachers in their becoming reflective practitioners, but we educators must be careful lest we think that this support, alone, will be sufficient to create major change in our schools. The organization and structure of schools present a formidable challenge to the kind of change called for in the reform, even if we create reflective preservice teachers. Gregg (1995), in a case study of a beginning high school teacher's acculturation into the school mathematics tradition, concluded that encouraging teachers to view their current practices as problematic may not be sufficient to promote reform, because the school culture has provided explanations designed to institutionalize the problems as matters to be coped with, not resolved. Gregg noted, for example, that the school mathematics tradition has separated the acts of teaching and learning so that a teacher can be considered to have met the obligation to teach even though the students may not have met their obligation to learn. This view is perpetuated by the commonly accepted conception of ability as capacity, because the teacher cannot be accountable for failing to successfully teach students who are unable to learn. Gregg provided the example of designating a test as "too hard" or certain questions as "unfair" because many students scored poorly on them, so that teachers can blame the test, even if they constructed it, instead of blaming their instruction; blaming their instruction might lead to questioning other of their fundamental beliefs and practices related to the teaching and learning of mathematics. Gregg doubted that "a teacher in the school mathematics tradition would question or reflect on the taken-as-shared beliefs and practices of this tradition as a result of students' poor test performance" (p. 463). He concluded that to be successful, a reform movement "must challenge the classroom, school, and societal obligations that characterize teachers' roles in the school mathematics tradition" (p. 463). In other words, Gregg raised the prospect that looking to change teachers' beliefs and knowledge within the current school mathematics tradition will, for the vast majority of teachers, fail.

Two Obstacles to Changing Beliefs: Teachers' Caring and Teachers' Belief That Teaching is Telling

Cooney (1999) suggested that to help teachers change their beliefs, teacher educators must find ways to support teachers to become more reflective while they unravel their notions about teaching and rebuild them in a rational way. But he identified two major roadblocks to attaining this effect: teachers' overwhelming propensities to be caring teachers, and teachers' orientations toward teaching-as-telling. Cooney recommended supporting preservice teachers while they transform their notions of caring and telling into notions that encourage attention to context and reflection by helping them move beyond their concerns for children's personal comfort levels to consider students' intellectual needs and by integrating content and pedagogy in mathematics teacher education to support the transition of teachers toward reflection. These two obstacles are related, because the challenges teachers face when they change their role in the class affect their perceived sense of their effectiveness in meeting the needs of their students.

Nearly all mathematics educators seem to agree that effective teaching is more than telling, a view reflected in the U.S. national reform documents (NCTM, 1989, 2000). However, even expert instructors grapple with the difficulties of developing and maintaining a view of teaching as more than telling. In a study during which a researcher and a teacher collaborated on the curriculum that they each taught to separate ninth-grade mathematics classes, Romagnano (1994), the researcher, described difficulties he experienced when he encouraged his students to solve problems independently without first being told what to do, whereas the teacher with whom he was working avoided such difficulties by being more directive with the class. Neither Romagnano's nor the other teacher's approach was without challenges, because whereas the teacher's approach removed much of the conceptual mathematics for the students, Romagnano's approach often led his students to frustration so that they avoided participating in classroom activities. He concluded that the differences between his and the teacher's views of mathematics, of how it is learned, and of the role of the teacher in the process led to his facing the Ask Them or Tell Them dilemma, whereas the teacher faced no such dilemma.

Chazan and Ball (1999) described challenges they experienced when trying not to "tell" while teaching a high school algebra lesson and a third-grade class, respectively.

Chazan described how in the algebra lesson he had taught to a lower track high school class the students disagreed on whether, when determining the average raise given to employees of a company, they should include a person who received no raise. When the discussion became heated and several students became excited, Chazan found himself wondering how to strike a balance between his desire for his students to develop greater confidence in their abilities to discuss their way through a mathematical problem and

his desire for the students to reach some common ground from which they could stand back and reflect upon the mathematics. Chazan wondered whether, or how, he should intervene. During a lesson Ball taught to third graders, the students seemed to agree on an incorrect way of labeling the number line, and when her attempts to induce disequilibrium were unsuccessful, she was left wondering how best to support the students' learning. Whereas Chazan was concerned that his students might have engaged in an unproductive disagreement, Ball was concerned that her students were engaged in an unproductive agreement. Chazan and Ball were aware that telling can take different forms and that the type of telling must be considered with the context. For example, a teacher's telling through demonstration of a mathematical procedure for the students to practice is different from a teacher's telling by attaching conventional mathematical terminology to a distinction the students are already making. In spite of this knowledge and their vast experience teaching without simply telling, Chazan and Ball, two expert teachers, faced difficult challenges in their lessons. They warned that the commonly accepted reform exhortation not to tell is a statement about what *not* to do, but it contributes nothing toward supporting teachers while they consider what they might do. They described three factors they balanced in considering how to proceed during the class discussion: the mathematics under consideration, the nature and direction of the class discussion, and the social and emotional climate of the class.

If expert teachers face difficulties with teaching without telling, one could expect that many teachers struggle even more. Smith (1996) suggested that a *teacher's efficacy*, originally defined as "the extent to which the teacher believes he or she has the capacity to affect student performance" (Tschannen-Moran, Hoy, & Hoy, 1998, p. 202), is being challenged by a reform movement that calls for teachers to approach teaching from an entirely different perspective. Smith pointed to the existence of a core set of beliefs that characterize how most mathematics teachers in the United States view mathematics, mathematics learning, and mathematics teaching. For most teachers, school mathematics is a fixed set of facts and procedures for determining answers, and the authority for school mathematics resides in the textbooks, with the teacher serving as the intermediary authority between the textbooks and the students. For them, teaching mathematics requires telling, or providing clear, step-by-step demonstrations of these procedures, and students learn by listening to the teachers' demonstrations and practicing these procedures. These beliefs support teachers' senses of efficacy, because the conception of mathematics as a fixed set of facts and procedures restricts the content teachers must know; thus they can think that they have mastered the necessary content. The notion of teaching-as-telling provides a detailed but attainable model that teachers can hope to master. Telling students how to perform procedures also supports teachers' senses of efficacy, because the conventional nature of procedures is such that students cannot be expected to know them until the teacher shows them, and so the students' successes in mastering the procedures can be attributed to the teacher. Furthermore, the belief that students learn by listening, watching, and practicing clearly specifies to the students and the teacher what the students should be doing in class.

> Teachers of mathematics, like all teachers, need to believe that their teaching actions have significant causal impact on their students' learning. Telling, irrespective of its pedagogical strengths and weaknesses, provides a clear model for teachers of mathematics to develop a sense of efficacy. (Smith, 1996, p. 393)

Smith went on to suggest that the reform movement has substantially changed what teachers must know and do to succeed in helping their students learn. How can teachers who have learned in a telling environment achieve a sense of efficacy consistent with the reform? Two things must happen. First, teachers must learn about the reforms and attempt to change their practices accordingly; a growing body of research on teachers' knowledge, beliefs, and practices in the context of reform has addressed this need. Second, according to Smith, teachers' success in making the changes to their practices must bring about a reconceptualization of their senses of efficacy, and this transformation has received no explicit focus. Now, nearly a decade since Smith's appeal, I think that the movement to consider teachers' identity, to be discussed later in the chapter, might address this call for research into teacher efficacy.

Summary

Researchers, teacher educators, and professional developers are generally not interested in just measuring teachers' beliefs; they want to change beliefs. The research indicates that the beliefs teachers hold affect their views of the instruction they observe, complicating attempts to answer the question of whether beliefs' change precedes or follows change in instruction. Determining which changes first is less important than supporting teachers to change their beliefs and practices in tandem, and reflection is the critical factor for supporting teachers' changing beliefs and practices. Through

reflection, teachers learn new ways to make sense of what they observe, enabling them to see differently those things that they had been seeing while developing the ability to see things previously unnoticed. While teachers are learning to see differently, they challenge their existing beliefs, leading to associated beliefs change. Change in teachers' beliefs may not lead to change in their practices, or vice versa, but I conjecture that the most lasting change will result from professional development experiences that provide teachers with opportunities to coordinate incremental change in beliefs with corresponding change in practice. Teacher educators and professional developers must better understand not only *what* beliefs teachers hold but also *how* they hold them, because the ways that teachers hold their beliefs affect the extent to which existing beliefs can be challenged. Two impediments to changing teachers' beliefs are concern for the well-being of children that often inhibits teachers' willingness to challenge students and difficulty in overcoming the classroom challenges that derive from moving beyond their role as the teacher as one whose job it is to tell students how to be successful.

TEACHERS' BELIEFS, PART II: FOUR AREAS OF RESEARCH

Three major areas of research on teachers' beliefs are their beliefs about students' mathematical thinking, about curriculum, and about technology. These three areas are important because of their potential role in changing teachers' beliefs. One other area I briefly review in this section is teachers' beliefs about gender.

Teachers' Beliefs Related to Students' Mathematical Thinking

For more than 25 years, mathematics educators have gained substantial knowledge about children's mathematical thinking within specific content domains, and this research has served as an important base for supporting professional development of prospective and practicing teachers (Grouws, 1992; National Research Council, 2001). Prospective elementary school teachers care fundamentally about children, but not necessarily about mathematics (Darling-Hammond & Sclan, 1996), so helping prospective teachers learn about children's mathematical thinking has been an important aspect of teacher development. Likewise, professional development based on children's thinking can help teachers create rich instructional environments that promote mathematical inquiry and understanding, leading to documented improvement in student performance (S. M. Wilson & Berne, 1999). I now turn to recent research on teachers' beliefs about children and the relationship of teachers' beliefs about mathematics, teaching, and learning to teachers' understanding of children's mathematical thinking.

Elementary School Teachers' Beliefs Related to Students' Mathematical Thinking

Fennema et al. (1996) conducted a longitudinal study of 21 primary-grade teachers over a 4-year period during which the teachers participated in a Cognitively Guided Instruction (CGI) (Carpenter, Fennema, Franke, Levi, & Empson, 1999) professional development program focused on helping teachers understand the development of children's mathematical thinking by interacting with a specific research-based model. They studied changes in the beliefs and in the instruction of the teachers and growth in the children's learning. Data sources for measuring teachers' beliefs were comprised of audiotape transcriptions of classroom observations, interviews, Mathematics Belief Scale (Fennema et al., 1990) scores, and field notes from many informal interactions with the teachers. As noted previously, the beliefs survey was comprised of 48 Likert-scale items assessing four subscales: Role of the Learner, Relationship Between Skills and Understanding, Sequencing of Topics, and Role of the Teacher. The Role of the Learner subscale measures the belief that children are able to construct their own knowledge instead of being merely receivers of knowledge. The Relationship Between Skills and Understanding subscale measures the belief that skills should be taught in relation to understanding and problem solving rather than in isolation. The Sequencing of Topics subscale measures the belief that children's natural development, rather than the logical structure of formal mathematics, should guide the sequencing of topics. The Role of the Teacher subscale measures the belief that instruction should facilitate children's construction of knowledge rather than consist of teachers' presenting materials.

The researchers engaged in an extensive 2-year process to define levels of teachers' instruction and levels of teachers' beliefs. These two multilevel scales (that later were collapsed into one) were scored on a scale of 1, 2, 3, 4-A, and 4-B. The belief levels were based on the extent to which teachers held the belief that children can solve problems without instruction and the belief that what the teachers know about their students' thinking should inform the teacher's curricular decisions. Level 1, the lowest level, reflected the belief that children cannot solve problems without instruction. Level 2 teachers were struggling with the belief that children can solve problems without instruction.

Level 3 reflected the belief that children can solve problems without instruction, but the teacher believed only in a limited way that his or her students' thinking should be used to make instructional decisions. The highest level was divided into Levels 4-A and 4-B; Level 4-A reflected the belief that children can solve problems without instruction in specific domains and that teachers should use knowledge of their students to guide interactions with them; Level 4-B reflected the belief that children can solve problems without instruction across mathematics content domains and that teachers should use knowledge of their individual children's thinking to inform their decision making, regarding both interactions with the students and curriculum design.

The researchers used the same Levels (1, 2, 3, 4A, and 4B) to characterize levels of cognitively guided instruction. Level 1, the lowest level of cognitively guided instruction, was characterized by the teacher's providing few or no opportunities for students to engage in problem solving and share their thinking. The highest level, Level 4-B, was characterized by the teacher's providing opportunities for children to engage in various problem-solving activities and eliciting children's thinking, attending to the thinking children shared, and adapting instruction according to what was shared. That is, at Level 4B, instruction was driven by the teacher's knowledge of individual children.

Results (Fennema et al., 1996) indicated that over the course of the longitudinal study, 18 of the 21 teachers increased their levels in terms of sophistication of their beliefs, and 18 of the 21 teachers increased their levels in terms of cognitively guided instruction. Overall, 17 teachers increased on both levels. At the beginning of the study, 2 teachers were categorized as holding beliefs at Level 4 and 2 teachers were categorized for instruction at Level 4 (I do not know whether these were the same teachers), and by the end of the study, 11 of the 21 teachers were categorized as holding beliefs at Level 4, and 7 of the 21 teachers were characterized for instruction at Level 4. The authors noted that in spite of the obvious relation between levels of instruction and levels of belief, they were unable to compare relationships, because a teacher's beliefs and instruction were not always categorized at the same level, and they found no overall pattern as to whether a teacher's level was higher for beliefs or for instruction. In short, the researchers found no consistent relationship between change in beliefs and change in instruction. Regarding the 17 teachers whose final ratings were higher than their initial ratings on both beliefs and instruction, 6 teachers' beliefs changed first, 5 teachers' instruction changed first, and the change occurred simultaneously for 6 teachers. The authors stated that the exploration of the relationships between change in beliefs and change in instruction was complex and could be understood only in terms of specific teachers. In stating their conclusions, the authors noted that although many factors contributed to the teachers' change, the two that seemed most critical were that the teachers learned the specific research-based model about children's thinking and used that model in their classrooms.

One other important finding in this study (Fennema et al., 1996) was that "gains in students' concepts and problem-solving performance appeared to be directly related to changes in teaches' instruction" (p. 430). The authors attributed the changes in the students' achievement to the facts that teachers provided the students more opportunities to engage in problem solving, encouraged their students to share their thinking, and adapted instruction to the problem-solving abilities of the students. Although the authors expected that more problem-solving emphases by the teachers would lead to an increase in problem-solving performance by their students, they found striking that the shift in emphasis from skills to concepts and problem solving did not result in a decline in students' computational-skill performance.

In a follow-up to the original CGI study (Peterson, Fennema, Carpenter, & Loef, 1989), Knapp and Peterson (1995) conducted phone interviews with 20 of the original 40 CGI teachers 3–4 years after their initial introduction to CGI; they found that all but one of the teachers reported using CGI in some way in their current classroom practices. They identified three patterns of CGI use: (a) Ten teachers reported "steadily, if often gradually, developing their use of CGI" (p. 47) and currently using it as the main basis for their instruction; (b) 6 teachers reported having used CGI more extensively when they first learned about it but currently using it only supplementally or occasionally; and (c) 4 teachers reported having never used CGI more than supplementally or occasionally. Knapp and Peterson found that "the relationship between changes in teachers' beliefs and their classroom practices was often interactive, and in many cases, beliefs seem to have followed practice" (p. 61). They reported that the teachers' beliefs in CGI principles began to develop during the workshop but deepened over time when their students generated solutions to complex mathematical problems.

In a longitudinal study of 496 German elementary school students in 27 classrooms, Staub and Stern (2002) also found a relationship between teachers' beliefs as measured by the CGI Belief Scale (Fennema et al., 1990) and students' achievement. Staub and Stern found that students of third-grade teachers who scored higher on the CGI Belief Scale displayed higher achievement in solving mathematical word problems than did students of third-grade teachers who scored lower on the CGI Belief Scale. Furthermore, as in the study by Fennema et al. (1996), they found that the students of teachers who scored higher on the CGI Belief Scale did no worse on mathematics-facts achievement than the students of teachers who scored lower on the survey. In fact, the authors concluded, "Teachers with a cognitive constructivist view of learning tend to be more successful in fostering math fact achievement" (Staub & Stern, 2002, p. 354).

Preservice Elementary School Teachers' Beliefs Related to Students' Mathematical Thinking

Vacc and Bright (1999) investigated the effects that introducing Cognitively Guided Instruction had on preservice elementary school teachers' changing beliefs about teaching and learning and on their instructional use. Thirty-four preservice teachers enrolled in a 2-year undergraduate cohort for preservice teachers participated in the 2-year study starting with the beginning of the students' junior year when they began the professional development program and continuing through the end of student teaching at the end of their senior year. On four occasions all students completed the CGI Belief Scale (Fennema et al., 1990); Vacc and Bright (1999) also conducted eight on-site, formal observations of each prospective teacher. In-depth case studies of two students were conducted, and data for these case studies comprised reflective journal entries written during the mathematics methods course and student teaching, videotapes of four mathematics lessons conducted during the student teaching, and three open-ended interviews. Vacc and Bright applied the Levels of Instruction framework and Levels of Belief framework used by Fennema et al. (1996) for determining the levels of the two student teachers.

Results of the study (Vacc & Bright, 1999) indicated that although the preservice teachers' beliefs-scale scores changed little during the first year of the 2-year program of professional course work, which included a 10-hour/week internship in a professional development school, the scores increased significantly during the third semester during their enrollment in the mathematics methodology course, about one third of which was devoted to instruction related to CGI. The students' belief-scale scores continued to increase significantly during the fourth semester during their student-teaching experiences. On the basis of their analysis of the case-study data, Vacc and Bright categorized one of the two preservice teachers at the end of student teaching at Level 3 for belief and Level 3 for instruction; that is, they found that she believed that children can solve problems without instruction and, in a limited way, that children's thinking should be used to make instructional decisions; also, she was putting these beliefs into practice. The other teacher was categorized at between Levels 2 and 3 for belief and at Level 2 for instruction; therefore, she was providing limited opportunities for her children to engage in problem solving or share their thinking, and she was, in a limited way, eliciting and attending to children's thinking. The authors concluded that counter to previous research indicating that preservice teachers' beliefs are resistant to change, this study provided evidence that preservice teachers' beliefs did change. The researchers were cautious about drawing conclusions regarding either the depth of the changes in beliefs or the contributions of various factors to the change. They concluded, "The data indicate the possibility that intensity of experience and a focus on children's thinking in the mathematics methods course may be keys for helping preservice teachers change their views" (p. 108).

Ambrose (2004) studied the beliefs of prospective elementary school teachers engaged, in Southern California, in an early field experience linked to a mathematics course. The purpose of the early field experience, described by Philipp, Thanheiser, and Clement (2002), was to motivate students to recognize the importance of learning mathematics by exploring the complexities involved when interviewing and tutoring elementary school children. In her theoretical framework for beliefs, Ambrose (2004) considered two primary sources for beliefs: emotion-packed experiences and cultural transmission. An example of an emotion-packed experience, described later in this chapter in the section on affect, is the prospective elementary school teacher Jo's negative childhood experiences with timed multiplication tests (Walen & Williams, 2002). The second source of beliefs mentioned by Ambrose (2004), cultural transmissions, might be thought of as the hidden curriculum in which people are engaged every day; beliefs created as a result of these experiences are held at a subconscious level. Examples of hidden curricula at the secondary school level were provided by Schoenfeld (1988): He found that a "well-taught" high school geometry course perpetuated the views that (a) all problems can be solved in just a few minutes; (b) the answer is what counts; (c) students are passive consumers of other people's mathematics; and (d) the world of deductive geometry is separate from the world of constructive geometry. Ambrose (2004) noted that beliefs have a filtering effect on people's thinking, so that, through time,

beliefs become more resilient. Ambrose provided the example shared by a colleague, who, in an effort to help her prospective elementary school teachers understand that children come to school with a great deal of knowledge on which teachers can build, had them interview kindergarten children during the second week of school. These preservice teachers were amazed at some of the problem-solving skills the students demonstrated, but instead of taking these skills as evidence that students bring much informal knowledge to school, some mentioned how impressed they were with how much the teacher had taught the children during the first week of school! The students' belief that one's mathematical knowledge is a result of school learning colored their view of the experience in such a way that they misinterpreted it.

Ambrose (2004) presented four mechanisms by which students' beliefs systems might change; the first two of these were mentioned as mechanisms for creating new beliefs:

> (a) they can have emotion-packed, vivid experiences that leave an impression; (b) they can become immersed in a community such that they become enculturated into new beliefs through cultural transmissions; (c) they can reflect on their beliefs so that hidden beliefs become overt; (d) they can have experiences or reflections that help them to connect beliefs to one another and, thus, to develop more elaborated attitudes. (p. 95)

Ambrose (like Cooney, 1999, mentioned earlier) noted that the two beliefs central to prospective teachers related to teaching as explaining and teaching as caring about students; the study was designed to provide opportunities for the prospective teachers to challenge their belief about teaching-as-telling by building upon their caring of children.

The 15 prospective elementary school teachers who participated in the study were simultaneously enrolled in a mathematics course and in a Children's Mathematical Thinking Experience. Data sources were surveys, interviews, prospective teachers' written work, field notes, and a computerized beliefs survey completed at the beginning and end of the semester. Ambrose (2004) conducted an intense analysis of one student, Donna, and compared the emergent themes from Donna's case with experiences of the other 14 students. Ambrose learned that for Donna the most powerful experiences were her three sessions working with a fifth-grade student on fractions, assessing the child during the first session and trying to expand the child's fraction understanding during the second and third sessions. The second session was a high point for most of the prospective teachers, because they felt confident that they had successfully taught the children something about fractions, but during the third session, most were surprised to discover that the children with whom they had previously worked experienced difficulty with the same fractions topic the prospective teachers thought they had taught during the previous lesson. The language the preservice teachers used captured the intensity of the experience; they described feelings of being *stoked* or *excited* by the events of the second session, whereas words they used to describe the third session included *shocked, disappointed,* and *aggravating.* Ambrose concluded that for many of the prospective teachers, their initial beliefs were undifferentiated and were thus held in an unreflective manner. The intense experiences in working with the students resulted in their beliefs' becoming more salient to them. Whereas many teacher educators hold belief reversal as the goal for work with prospective teachers, Ambrose found that instead of letting go of their old beliefs, the prospective teachers in her study held onto their old beliefs while forming new ones, and hence the prospective teachers engaged in actions with their children that reflected a combination of beliefs about mathematics teaching and learning. Ambrose suggested that the students' experiencing a "failed teaching experiment" may have been an important component in their learning experience, and she concluded by stating, "Providing prospective teachers with intense experiences that involve them intimately with children poses a promising avenue for belief change" (p. 117).

D'Ambrosio and Campos (1992) studied five preservice teachers enrolled in their senior year of the mathematics program at the Catholic University of São Paulo, Brazil, and, like Ambrose (2004), they found that providing preservice teachers with opportunities to learn about children's mathematical thinking led to their reflecting upon conflicting situations that arose. These situations created a state of disequilibrium for the preservice teachers, leading them to question normally accepted instructional practices and generally to develop characteristics of reflective practitioners (Schön, 1983).

Secondary School Teachers' Beliefs Related to Students' Mathematical Thinking

Most researchers investigating the relationship between teachers' beliefs and students' thinking have focused at the elementary school level. Nathan and Koedinger (2000b), however, studied 67 secondary school mathematics teachers from the United States and 35 mathematics education researchers to determine their predictions of problem-solving difficulty for a set of arithmetic and algebra problems. Nathan

and Koedinger constructed six types of problems using three presentation formats (story problems, word-equation problems, and symbolic-equation problems), and for each format, they wrote two problems, one with the start unknown and the other with the result unknown. Their previous research had indicated that students found result-unknown story problems and result-unknown word equations to be the easiest; result-unknown equations, start-unknown story problems, and start-unknown word equations of easy-to-medium difficulty; and start-unknown equations to be the most difficult of the problems. On the one hand, Nathan and Koedinger suggested that this research supported the view that the development of students' verbal reasoning precedes the development of their symbolic reasoning, and they developed a model they named a *Verbal-Precedence Model* to reflect the student data. On the other hand, the teachers incorrectly predicted that the students' symbolic-reasoning skills would develop first, with word-problem-solving ability developing later, and the researchers referred to the model they developed to reflect the teachers' view as the *Symbolic-Precedence Model.* On the basis of their data, they concluded, "High school mathematics teachers hold beliefs that cause them to systematically misjudge students' symbolic- and verbal-reasoning abilities" (Nathan & Koedinger, 2000a, p. 212). The researchers found that in two commonly used mathematics-textbook series the development seemed to match the teachers' *Symbolic-Precedence Model,* and they suggested that through repeated exposure to textbooks, "teachers internalize the symbolic precedence view as a basis for their predictions of problem difficulty for students" (Nathan & Koedinger, 2000b, p. 181).

In a follow-up study, Nathan and Koedinger (2000a) asked 107 Grade K–12 teachers to rank order the six mathematics problems used in the previous study (Nathan & Koedinger, 2000b); each teacher also completed a 47-item Likert-scale assessment designed to assess each of six constructs related to teachers' views of mathematics, mathematics instruction, and student learning:

1. Algebraic procedures are "best" for effectively solving mathematical problems.

2. Invented solution methods are effective for solving problems.

3. Arithmetic problems are easier than algebra problems and should be presented first, and mathematics problems presented in words are most difficult and need to appear later in the curriculum (symbolic-precedence view).

4. Students may enter the classroom with valid ways of reasoning, and teachers should encourage invented strategies.

5. Correct answers are more important than reasoning processes.

6. Alternative solution strategies (such as arithmetic, guess-and-check, and other nonsymbolic methods) demonstrate gaps in students' knowledge.

Results indicated that across grade levels, the teachers reflected recent mathematical reform views, and they tended to disagree with views that challenged reform-based views. The elementary school teachers were most likely and high school teachers were least likely to agree with reform-oriented views. Teachers across all grade levels correctly predicted that result-unknown problems would be easier than start-unknown problems. High school teachers incorrectly ranked symbolic equations as easier than verbal problems, and the rank orders of the elementary school teachers were similar to those of the high school teachers, although elementary school teachers were more likely to rank problems on the basis of algebraic and arithmetic structure than on presentation format. Middle school teachers stood out as a group because their rank orderings were most consistent with the research on students' performance. That is, "middle school teachers were most accurate in predicting students' problem-solving performance in contrast to the view held by many high school teachers that symbolically presented problems were easier to solve than verbally presented problems (Nathan & Koedinger, 2000a, p. 226). In their discussion, the authors suggested that middle school teachers, who work with students who have not typically received formal algebra training, may have occasions to observe students' use of more informal and invented methods, and thereby the teachers may have more opportunities to observe the transition from arithmetic to algebraic reasoning. They suggested, further, that elementary school teachers, who held the strongest reform-based views of learning and teaching, are relatively unfamiliar with students' algebraic reasoning and may have therefore expected that students would operate with a traditional vision of algebra curriculum and instruction. High school teachers were least aware of the efficacy of students' invented algebra solution strategies, and the authors suggested that because high school teachers have greater expertise in their content areas, they are personally more distant from their novice students' difficulties. The authors introduced the term *expert blind spot,* a feature they attributed to teachers with high levels of content knowledge but low levels of

awareness of alternative interpretations for symbolic equations.

Nathan and Petrosino (2003) studied 48 prospective secondary school teachers enrolled in a "nationally acclaimed teacher education program" (p. 911) in the United States to test their *expert-blind-spot hypothesis,* stipulated as follows:

> Educators with advanced subject-matter knowledge of a scholarly discipline tend to use the powerful organizing principles, formalisms, and methods of analysis that serve as the foundation of that discipline as guiding principles for their students' conceptual development and instruction, rather than being guided by knowledge of the learning needs and developmental profiles of novices. (p. 906)

Participants were categorized in one of three ways, on the basis of their subject-matter knowledge: (a) 13 were placed in a "basic math" category because they had not completed a precalculus course; (b) 16 from the general population of teacher education students who were pursuing licensure to teach elementary school but who had completed at least one calculus course were placed into the HiMath knowledge group; (c) 19 who were enrolled in a specialized program for mathematics and science majors and had completed many higher level mathematics courses were placed in the MathSci category. Using problems similar to those used in previous studies (Nathan & Koedinger, 2000a, 2000b), Nathan and Petrosino (2003) asked all participants to rank order six problems as they would predict beginning-level algebra students would experience ease or difficulty in solving them. The participants also responded to the 47-item beliefs survey used in the previous study (Nathan & Koedinger, 2000a). Results of the predictions of problem difficulty (Nathan & Petrosino, 2003) indicated that the preservice teachers with more advanced mathematics knowledge (the MathSci and HiMath groups) were far more likely to follow the symbolic-precedence view of algebra development than the preservice teachers with less mathematics knowledge (the Basic Math group). Results of the beliefs survey supported this finding, with a greater percentage of those preservice teachers in the two higher mathematics knowledge groups agreeing that using algebraic formalisms is best for solving complex problems, whereas students in the Basic Math group were more likely to believe that instruction should build on students' intuitions and invented methods. Nathan and Petrosino (2003) conjectured,

> Educators with greater subject-matter knowledge tend to view student development through a domain-centric lens and consequently tend to make judgments

about student problem-solving performance and mathematical development that differ from actual performance patterns in predictable ways. (p. 918)

Summary

Researchers measuring teachers' beliefs about children's mathematical thinking have correlated these beliefs to teachers' instruction and to students' learning. The Cognitively Guided Instruction group (Carpenter et al., 1999) developed an instrument to measure teachers' beliefs about children's abilities to solve problems without instruction and about teachers' beliefs about the role that knowledge of children's thinking should play when one makes instructional decisions, and they used this instrument in a study in which they found that teachers' beliefs were related to instruction reflecting a focus on children's mathematical thinking, and, furthermore, teachers' instructional changes were related to gains in students' understanding and problem-solving performance (Fennema et al., 1996). Knapp and Peterson (1995) found that most of the original CGI teachers who had made instructional changes to focus upon children's mathematical thinking continued to apply this focus several years later. Staub and Stern (2002) used the same beliefs survey and found a direct link between teachers' beliefs about children's mathematical thinking and students' achievement. Preservice elementary school teachers provided with opportunities to learn about children's mathematical thinking can change their beliefs, and the intensity of the experience associated with focusing upon children's thinking was identified as a cause for these changes (Ambrose, 2004; Vacc & Bright, 1999). Although teachers' beliefs about children's mathematical thinking have been studied more extensively at the elementary school level than at the secondary school level, Nathan and his colleagues have investigated secondary school mathematics teachers' beliefs about students' mathematical thinking. Nathan and Koedinger (2000b) found that the development of students' verbal reasoning skills preceded the development of their symbolic reasoning but that high school mathematics teachers believed just the opposite, and they coined the term *expert blind spot* (Nathan & Koedinger, 2000a) to refer to the relationship between teachers' higher levels of content knowledge and lower levels of awareness of students' understanding. Nathan and Petrosino (2003) found that the expert-blind-spot characterization also applied to prospective secondary school teachers. Although researchers studying teachers' beliefs about students' mathematical reasoning have noted the complex relationship between beliefs and practices, several studies provide evidence that teachers' beliefs often change as a result of observing children's mathematical reasoning (Ambrose, 2004; D'Ambrosio & Campos, 1992; Knapp & Peterson, 1995; Vacc & Bright, 1999).

Teachers' Beliefs Related to or Changed by Use of Mathematics Curricula

Any mathematics class has two curricula: the intended curriculum and the enacted curriculum. The intended curriculum, that which is stated or explicit, is comprised of the materials of a class, including textbooks, curriculum guides, and course descriptions. Although many think of these materials as defining the curriculum, one cannot understand the experience students have in a class without attending to how materials are enacted by a teacher, with students, in a particular context (Ball & Cohen, 1996). Just as the knowledge and beliefs teachers hold about mathematics, teaching, and learning affect the ways in which they enact curricula, so too should we expect that teachers might learn from the curricula they use. That is, teachers not only adapt and change curricula but also are changed by the curricula they use. The relationships between teachers' beliefs and the innovative curricula they use are addressed in this section.

Lloyd (2002) highlighted two aspects of reform-oriented curricula designed to support teacher learning. First, by focusing upon the exploration of solutions to real-world problems, reform-curricula designers emphasize a more inquiry-oriented approach to mathematics than is found in traditional curricula, and the materials are formatted to support these mathematical and pedagogical differences. Second, reform-oriented curricula include more extensive information for teachers than traditional curricula. For example, they provide historical information, details about different representations, and examples of students' reasoning. Lloyd (1999, 2002) found that using the reform-oriented Core-Plus curriculum over several years supported a high-school teacher's learning or relearning of mathematical subject matter. Furthermore, as a result of becoming more familiar with the curriculum and interacting with students, this teacher's beliefs about the graphical representations of functions became richer and more nuanced.

Collopy (2003) reported on a study designed to investigate two elementary school teachers' changes in beliefs related to teaching and learning mathematics and their changes in instructional practices resulting from their first year's use of *Investigations in Number, Data, and Space*, a reform-oriented curriculum. The two teachers were experienced elementary school teachers working in similar contexts, schools comprised of a high percentage of at-risk students described as "below average in mathematics." Both teachers sat on their district mathematics committees, both had recently used a district-mandated traditional Addison-Wesley mathematics textbook, and both had volunteered to pilot the *Investigations* curriculum.

Collopy (2003) collected data about the teachers' beliefs and practices in three stages. During the first stage, she collected data on the teachers' backgrounds and baseline data on their beliefs and knowledge in an extended formal interview and on their initial use of the curriculum through 7 or 8 classroom observations. During the second stage, she focused on investigating the teachers' decisions about curriculum use and changes in beliefs and practices through six days of observations, in two-day sets, a month apart. During the third stage of data collection, Collopy collected data on teachers' reflections; use of the curriculum; and beliefs, knowledge, and instructional practices by observing each class for two consecutive weeks during the final weeks of the school year and conducting a final interview of each teacher. Collopy conducted two data analyses: a thematic analysis and a segment analysis. In the thematic analysis of the interviews and field notes, she developed categories of codes to explore the stability, changes, and relationships among the teachers' beliefs about mathematics, students, pedagogy, curriculum, and themselves as learners and teachers. She presented her interpretations during each final interview; both teachers concurred with her interpretations. In the segment analysis, Collopy tracked the format and focus of each teacher's practice across the year by coding each observed lesson into segments and coding each segment for length in minutes, instructional format, the teacher's role and focus, student behavior, and teacher's expectations for students' thinking.

From her detailed case studies, Collopy (2003) concluded that the two teachers differed dramatically in what they had learned from the materials and ways they engaged with the materials as a support for learning. Ms. Clark held a tightly integrated set of beliefs about mathematics, student learning, the teacher's role, the purposes of mathematics instruction, and her own mathematical efficacy, and her instructional practice was consistent with these beliefs. She believed that computational speed and accuracy distinguished successful students from unsuccessful students, that her students needed to learn the rules of mathematics, and that mathematics topics should be presented systematically from easier to harder. Ms. Clark agreed with the emphasis in the *Investigations* materials on students' understanding mathematical concepts but not with the definition of *understanding* as "familiarity with

the magnitude of numbers, mathematical relations, and the meaning of mathematical operations and situations" (p. 296); Ms. Clark defined *understanding* as the memorization and correct execution of standard algorithms. Eventually she became frustrated with the *Investigations* materials, in which speed and rote memorization were downplayed, and she adapted lessons to compensate, changing their essence by omitting alternative problem-solving strategies and mathematics discussions she considered distractions; her frustration level rose until January, when she shelved the *Investigations* curriculum and returned to her more traditional curriculum. Collopy's segment analysis indicated that in Ms. Clark's use of the *Investigations* curriculum, 61.8% of the observed time was devoted to procedures and correctness, 35.4% to organizational and management issues, and 2.8% (consisting of 24 minutes during one lesson) to conceptual understanding.

The other teacher, Ms. Ross, approached the *Investigations* curriculum differently than Ms. Clark, significantly changing her approach to instruction in shifting from her early focus on procedures and correctness (94% of the early observations to 21% for November through May) to conceptual understanding and mathematical reasoning (from no early focus to more than 75% in later observations). Unlike Ms. Clark, whose actions were driven by beliefs about mathematics as a set of facts and procedures that one learns through exposure to direct and clear instruction, Ms. Ross held beliefs about the importance of developing students' confidence for learning mathematics by building upon their prior knowledge. Ms. Ross saw no clear structure for mathematics, so she relied upon her curriculum materials and objectives for the state's standardized tests to direct the content of and emphasis in her instruction. Ms. Ross followed the curriculum carefully, reading and using the student notes, and she learned mathematics while challenging her beliefs. For example, in the process of asking students to demonstrate, write, and discuss their ideas, she came, for the first time, to see that problems can be solved in many ways.

Although the two teachers in this study had similar backgrounds, taught in similar situations, and willingly piloted the same reform-oriented curriculum, their enactment of the curriculum provided different opportunities for them to learn, differences Collopy (2003) explained in terms of the beliefs most integral to the teachers' identities. Ms. Clark's identity as a teacher and learner of mathematics was centered in her belief that mathematics is a set of rules and procedures that students learn by being shown, incompatible with the beliefs underlying the curriculum, so in her curriculum adaptation, she lost most opportunities to learn from the materials. The beliefs most integral to Ms. Ross's identity as a teacher were not about mathematics but, instead, were about developing students' confidence for learning mathematics and were compatible with those espoused in the curriculum. Thus, she was able to use the materials as they were intended and was, thereby, poised to learn much from their use. Collopy's findings show that while considering how to build opportunities for teachers to learn from their curriculum, writers must also consider the beliefs held by teachers using the materials.

Remillard and Bryans (2004) extended the work of Collopy (2003) by studying the way the reform-oriented *Investigations* curriculum supported the learning of urban elementary school teachers in one school over several years. In this study, seven of the eight teachers who volunteered to participate and attend monthly study-group meetings and related research activities attended for two years. During the first year, each teacher was observed 2–4 times and interviewed at least twice; during the second year, six of the seven teachers were observed 7–8 times and interviewed three times. Two of the teachers' classrooms were designated as focus classrooms, which were observed daily over 2-week periods four times during the second year; each focus teacher was interviewed approximately five times. Data were comprised of audiotapes of the lessons and written field notes, which were later used to complete a predesigned observation instrument. The researchers conducted segment analyses of these data, documenting the time devoted to each task, the teacher's aim and focus of the task, and ways the students engaged with the task.

Three broad categories of teachers' use of the *Investigations* curriculum were identified by Remillard and Bryans (2004): *intermittent and narrow, adopting and adapting,* and *thorough piloting.* Teachers whose use was *intermittent and narrow* used the materials minimally and relied primarily on their own teaching routines and other resources to guide their curricula over the year. *Adopting and adapting* described use of the materials as a guide for determining what topics to teach and how to sequence them, but with teachers' adapting the materials to fit their familiar strategies and approaches to teaching. *Thorough piloting* referred to use of all parts of the curriculum guides. Although teachers whose use was in this category continued to draw upon activities and strategies from other sources, when they used the *Investigations* curriculum, these teachers followed the lessons as suggested in the guide and studied and grappled with most of the information provided for the teacher.

To address each teacher's relationship with the *Investigations* curriculum during the first 2 years of use, Remillard and Bryans (2004) introduced the construct *orientation toward curriculum,*

> a set of perspectives and dispositions about mathematics, teaching, learning, and curriculum that together influence how a teacher engages and interacts with a particular set of curriculum materials and consequently the curriculum enacted in the classroom and the subsequent opportunities for student and teacher learning. (p. 364)

They presented a dynamic model designed to capture the interactive relationships among teachers' perspectives and beliefs, the use of curriculum materials, the enacted curriculum, and consequent opportunities for teachers to learn.

To illustrate orientation toward curriculum, Remillard and Bryans (2004) described three veteran teachers, Jackson, Reston, and Kitcher, who first used the *Investigations* curriculum during the study. Jackson held beliefs about mathematics teaching and learning incompatible with those underlying the curriculum, and his curriculum use was categorized as intermittent and narrow. Jackson not only seemed intent upon using his own materials to emphasize procedural skills but also, in his use of the curriculum, failed to use the suggestions to facilitate students' work and to have students explain their reasoning or discuss their strategies. The other two veteran teachers, Reston and Kitcher, shared views about mathematics teaching and learning that seemed compatible with the curriculum, and so one might have expected their use of the materials to differ from Jackson's. Kitcher's use was as intended (thorough piloting), but Reston's use, in spite of her views of mathematics, teaching, and learning, was similar to Jackson's (intermittent and narrow). Remillard and Bryans found that Reston's "general mistrust of published materials trumped any potential compatibility between her beliefs about mathematics and those represented in the book" (p. 366). That is, her orientation toward curriculum, not her beliefs about mathematics, teaching, and learning, explained her skeptical approach to the curriculum and led her to use the curriculum only as a collection of useful activities.

Use by four of the eight teachers in the study was categorized as thorough piloting of the curriculum, and although these four teachers shared beliefs about mathematics, teaching, and learning that were consistent with those reflected in the *Investigations* curriculum, others in the study did as well, leading the authors to conclude that their shared ideas about the role of curriculum materials, not their beliefs, were the most significant factor in shaping their orientations toward the curriculum. The teachers viewed curriculum resources as guides or partners in their teaching, and they were all committed to learning about the curriculum and giving it a fair trial. They viewed curriculum as more than a source of activities for students: They saw it as a resource for themselves.

The authors (Remillard & Bryans, 2004) noted that, except for Kitcher (with 25 years' experience), those thoroughly piloting the curriculum were the teachers with the least teaching experience— Graves, Larson, and Schwarz (4, 3, and 3 years, respectively)—who, unlike the other five teachers, lacked well-established routines in their practices and appreciated the structure and learning opportunities available in the curriculum. Graves and Larson used the curriculum to develop deeper insights into students' thinking, and they used those insights to reconstruct their roles as teachers to create more opportunities to attend to their students' thinking. Larson and Schwarz placed substantial emphasis on exploring the mathematical ideas embedded in the units to extend their mathematical knowledge. The authors concluded,

> Standards-based curricula are not a panacea, but they can play an important role in fostering reform-based practices. However, without additional support for teachers, the impact of these curriculum materials is likely to be unpredictable and varied. Teachers using curricula like Investigations would benefit from opportunities to explore the content of the materials and have conversations with others about how they use them. (p. 386)

Clearly, curricula may be an important site of learning for practicing teachers. What of prospective teachers? Spielman and Lloyd (2004) attempted to use reform-oriented curricula to change prospective elementary school teachers' beliefs. They arranged for two classes of mathematics for elementary school to be taught by the same instructor using different curriculum models. The *textbook section* was taught using a popular mathematics textbook and pedagogical strategies the researchers believed reflected the philosophy of the textbook authors. The class met weekly for 3 hours, and a typical lesson began with a homework review during which groups of students (who were prospective teachers) were assigned to discuss problems then present solutions to the class, followed by the instructor's comments on the homework review and lecture on the new material. Following the lecture, the instructor showed a short video or students worked, individually or in groups, on a problem or activity at their tables, often with manipulatives available. Student volunteers shared outcomes of their work with the class until a consensus

was formed. During the final hour of class, after the instructor made final comments or summary statements, students worked alone or in groups on homework, with the teacher available to answer questions.

The *curriculum-materials section* used units from two reform-oriented middle school curricula, *Mathematics in Context* and *Connected Mathematics Project*, selected to correspond to the mathematical topics in the textbook section (Spielman & Lloyd, 2004). Whenever possible, instructional decisions were made to support the intent of the authors of the curricula. For example, the curricula materials rarely offered rules or solutions to sample problems, and the instructor refrained from providing rules or solutions. A typical lesson began, as in the textbook section, with the students' reviewing homework by working in groups and sharing their solutions. However, instead of providing solutions or explanations when students had questions, the instructor referred the students to other students for support, often redirecting or rephrasing questions. During the lively homework discussions, students responded to one another's questions without being called upon, and although the students expressed frustration with the instructor's refusal to offer rules, procedures, or solutions, "the students appeared to become accustomed to this type of classroom" (p. 35). After the homework review, the students worked together at their tables on problems or activities, and they would occasionally watch video. Although the videos were the same for both classes, the problems and activities differed both in type and presentation. In the curriculum materials section, no instructor presentation or lecture preceded work on problems, and the instructor's role was to clarify and pose questions. During the last hour of the class students worked on their homework, but instead of answering students' questions, the instructor posed other questions designed to facilitate students' finding their own solutions.

Students completed a Likert-scale teaching-beliefs survey and a mathematics-content instrument as pretests and posttests. No between-group differences were found on the content test, but differences were found on the beliefs survey. The students in the textbook section saw the instructor as more of a classroom authority than did the students in the curriculum-materials section. When asked whether the textbook, other students, or the instructor had been most beneficial to learning, only about 10% of the students in either group selected the textbook, but more than 68% of the students in the textbook section selected the instructor whereas more than 82% of the students in the curriculum-materials section selected other students. A paired-sample *t*-test comparing students' pretest and posttest responses on the percentage of time students thought should be devoted to teacher lecture and explanation or to group work and discussions showed significant differences for the students in the curriculum-materials group, but no significant differences were found for students in the textbook section. Over the course of the semester, the amount of time students in the curriculum-materials group thought should be devoted to teacher lecture and explanation decreased (from 30.6% to 22.2%) whereas the amount of time they thought should be devoted to group work and discussion increased (from 54.1% to 63.7%). Spielman and Lloyd (2004) concluded that prospective elementary school teachers' beliefs about textbooks, teaching, and learning changed as a result of using reform-oriented middle school textbooks to learn mathematics.

Summary

Several mathematics curricula have been designed to promote a reform-oriented approach (NCTM, 2000; NRC, 2001) to mathematics teaching and learning. The curriculum designers of these materials recognized the role that curriculum might play in supporting teachers' learning, and they included instructor-support materials designed to provide guidance for classroom instruction. These curricula are said to speak *to* teachers, rather than *through* them (Remillard, 2000). As important as curriculum materials are for supporting teachers in the classroom, assuming that teacher-proof curriculum can, or should, be used would oversimplify the matter; the research indicates that if teachers' beliefs about mathematics, teaching, and learning are not consistent with the beliefs that serve as the foundation of the reform-oriented curriculum, the teachers do not use the materials as they were intended (Collopy, 2003; Remillard & Bryans, 2004). Holding particular reform-oriented beliefs is a necessary, but not sufficient, condition for teachers to use a curriculum as intended, because Remillard and Bryans (2004) found that another factor, teachers' orientation toward curriculum, played a significant role in curriculum use even for teachers who held reform-oriented beliefs about mathematics, teaching, and learning. According to Collopy (2003), a teacher who viewed mathematics as a set of procedures and rules was less open to implementing the curriculum as intended than was the teacher who lacked such a clear structure for mathematics. The less experienced teachers who had yet to develop classroom routines for teaching mathematics were more likely than the experienced teachers to read and learn from the supportive materials in a study by Remillard and Bryans, indicating that providing these curricular materials to teachers early in their careers may be a promising means

of supporting their growth as teachers. Spielman and Lloyd (2004), studying prospective teachers enrolled in a course of mathematics for elementary school, found that prospective teachers need not yet be in their own classrooms to accrue the educational benefits of reform-oriented curricula; beliefs of students in a course based on a reform-oriented curricula changed more than the beliefs of students who used a conceptually oriented college textbook.

Teachers' Beliefs About Technology

Prospective teachers believe that computers are important in education, and they want to use them in their preservice programs, but except at a few institutions, students do not learn to use technology effectively (Willis & Mehlinger, 1996). This failure to prepare teachers is reflected in the fact that elementary school teachers generally do not like to teach with computers, and when they do use them, they most commonly do so for drill and practice activities (Cummings, 1998). Using a Likert-scale instrument to survey 33 elementary school teachers about their uses of technology, Cummings found that although most teachers thought that they possessed expertise about using computers, they cited time spent preparing for computer instruction as the greatest barrier to computer use.

Schmidt (1998) investigated beliefs of teachers in Grades 4–6 about the use of calculators and concluded that they reflected their more traditional views about mathematics. In a follow-up study, Schmidt (1999) provided the teachers with an in-service during which they spent one week using calculators to explore fifth- and sixth-grade concepts and applications. After the in-service, during which teachers kept reflective journals; evaluated teaching materials; read and discussed journal articles about a variety of topics, including research about calculators; and participated in and led discussions, the teachers continued their professional development by engaging in action-research projects and sharing the results with colleagues, attending the state mathematics conference, and cohosting a conference for teachers and principals about using calculators. Before and after the in-service, 32 teachers completed a two-part questionnaire: In the first part, they gave personal and professional characteristics and answered three open-ended questions related to what would facilitate the integration of calculators into mathematics, what most hindered the use of calculators in mathematics, and how students in their classes currently used calculators. In Part 2, comprised of 29 Likert-items, respondents provided one of six responses (strongly agree, agree, slightly agree, slightly disagree, disagree, and strongly disagree) to items such as "If calculators are used in school mathematics programs, students no longer need to know paper-pencil computing techniques" (Schmidt, 1999, p. 24), a statement with which the teachers strongly *disagreed* on both administrations of the questionnaire. On another item, "Calculators are tools that allow students to focus more attention on mathematics concept development and understanding" (p. 25), the teachers strongly *agreed* on both the pretest and posttest. After finding little change in the teachers' beliefs about calculator use as a result of the in-service, Schmidt suggested that the "teachers' perspective and philosophy of teaching and learning mathematics constrained their beliefs about calculators" (1999, p. 32).

Wiegel and Bell (1996) investigated the integration of computers into six mathematics content classes for prospective elementary school teachers—three sections each of the first and second courses. Data for the first course consisted of precourse surveys and a topic-specific survey, part of a required but ungraded assignment, for which students wrote about their attitudes and beliefs about mathematics and the integration of computer activities into the mathematics content course. Data for the second course consisted of one essay, weekly reflections from all students, and anecdotal comments collected during the computer labs. Used in the first mathematics course were two interactive microworlds, one for geometry and the other for fractions, created as part of a research grant directed by Les Steffe and John Olive. The students in the second mathematics course used a spreadsheet, a probability microworld, LOGO, and Geometer's Sketchpad. The authors concluded that the students and instructors all rated the overall experience positively, and that the classroom atmosphere, which was tense at the beginning of the semester, seemed to relax when students used computers, working in small groups and communicating with one another. The students reported appreciating the opportunity to use the computer, an activity they came to see as a welcome change in the class routine. Although the authors of this mid-1990s' study found that many of the students needed to learn to operate computers and manipulate programs and that often the students' attention to the basic mechanical operation of the computer detracted from their learning the mathematics, college students today would have more expertise with computers. Also, because their computer work was to be ungraded, the students wanted to stop using the computers toward the end of the semester to prepare for the paper-and-pencil items they expected on the final examination.

Tharp, Fitzsimmons, and Ayers (1997) conducted a study designed to determine whether providing teachers with instruction on the use of graphing calculators would have an effect on the teachers' beliefs about computers or on the teachers' instruction. Five 3-hour, monthly in-services designed to help teachers integrate graphing calculators into their instruction were provided to 261 Grades 6–12 teachers (168 mathematics teachers, 72 science teachers, and 21 teachers of other subjects). Topics covered during these in-services included solving linear equations; studying mathematical modeling, exponential growth, projectile motion, and parametric equations; and connecting multiple (graphic, tabular, and numeric) representations. Teachers watched videotaped lessons offered by experienced teachers using calculators in their classes. All the teachers were required to provide instruction using graphing calculators in their own classrooms. Data included pre and post questionnaires to assess how teachers' attitudes toward graphing calculators and mathematics shifted because of the instruction and the use of technology. In journals, teachers described their initial use of the graphing calculators during instruction. Likert-scale items, such as "Calculators should 'only' be used to check work," and "When doing mathematics, it is only important to know how to do a process and not why it works," were used to measure the teachers' views of calculator use and mathematics learning, teachers' access to calculators, teachers' efficacy, student efficacy, and administrative support. Results indicated that teachers' beliefs about the use of calculators changed significantly to favor using calculators as an integral part of instruction. A significant positive correlation was found between teachers' views of mathematics and their views on the use of calculators in the classroom, with those teachers holding more rule-based views of mathematics tending to hold the view that calculators do not enhance instruction and those with less rule-based views of mathematics more likely to view calculators as integral for mathematics and science instruction. After coding and analyzing all the teachers' journal entries, the authors found that although all the teachers attempted to integrate the use of the graphing calculator into their instruction, the rule-based teachers tended to quickly return to limiting the amount and type of calculator use in their classrooms. Whereas the rule-based teachers tended to focus upon the students' emotional reactions, the non-rule-based teachers were more likely to concern themselves with the students' conceptual understanding and thinking. The authors concluded that technology can be used to support changes in teachers' instruction, followed by changes in teachers' beliefs.

Fleener (1995) used a Likert scale to measure the attitudes toward the use of technology held by 233 elementary, intermediate, and high school classroom teachers and by 78 preservice teachers enrolled in a mathematics methodology course for K–3 teachers. Items on the 29-item scale included "Calculators make mathematics fun" and "Students should not be allowed to use calculators until they have mastered the concept." Fleener's previous research had shown that a critical issue (which she labeled the *mastery issue*) that divided teachers related to whether they believed that students should be allowed to use calculators prior to having mastered the concept. In the 1995 study, Fleener investigated whether teachers with different views on the mastery issue also differed with respect to their responses to other items on the survey. Results showed that 55% of the preservice teachers were in the *mastery group* (i.e., believed that students should attain conceptual mastery before being allowed to use calculators), and Fleener found no differences between the mastery views of preservice and practicing teachers. She did, however, find differences in the ways both preservice and in-service teachers in the mastery group tended to respond to other items compared with teachers not in the mastery group. Finally, she concluded that the mastery issue may be key for affecting teachers' beliefs about using calculators.

Walen, Williams, and Garner (2003) investigated the relationship between students' use of calculators in college mathematics courses for prospective elementary school teachers and their views of appropriate use of calculators in elementary school classrooms. Although technology has been identified as an essential component of teaching and learning mathematics (NCTM, 2000), Walen et al. (2003) reported that many college mathematics faculty limit the use of calculators for fear that their students will grow to depend upon them or because calculators enable students to "get answers that they don't 'know'" (p. 447). The researchers contended that this position, held by many faculty, reflects a view of mathematics as primarily comprised of a set of rule-governed calculations and procedures. They studied four college mathematics classes in which calculators were incorporated into a collaborative setting dedicated to students' learning with understanding (NCTM, 2000). They hoped that because calculators were readily available for students, they would use them "naturally and unreflectively to accomplish mathematical tasks" (Walen et al., p. 449). To determine whether college students who learned mathematics in such an environment developed be-

liefs about how calculators should be used in elementary school classrooms different from the beliefs of others, Walen et al. (2003) collected informal participant-observation notes, classroom artifacts, and questionnaire responses from students in four sections of a mathematics content course designed for prospective elementary school teachers. They also conducted four case studies of students enrolled in the course. Their results indicated that when asked to solve computational tasks, the students used more mental approaches or paper-and-pencil approaches than calculator solutions. In other words, their students did not overuse the calculator by using it on simple problems. When analyzing the students' responses to three open-ended questions, the researchers found that the three most common responses to the question "When do you use a calculator?" were when accuracy or confidence in the answer was needed, when the problem was large or complex, or when trying to save time, with each answer given by about one third of the respondents (although a person may have given more than one answer). Overall the researchers concluded that their students reported positive experiences using the calculators in their mathematics classes.

The researchers (Walen et al., 2003) had hoped that the classroom experiences provided to their students would positively influence their attitudes toward calculator use in elementary classrooms and their decisions to model the experiences by using calculators with their subsequent students. However, they were disappointed to see that their "success towards this goal was limited" (p. 457). The second open-ended question was "When do you think students should use a calculator?" The most common answer, given by about half the students (32 of 66), was that students should use the calculator only after they knew the operations; 8 students responded that children should never use calculators because they will handicap students' learning. The most common response (by 42 of 66) to the third open-ended question—"Is there a time when a calculator shouldn't be used?"—was that calculators should not be used while students are learning the basics. The researchers concluded that although some students' responses regarding their subsequent teaching with calculators paralleled their descriptions of their positive experiences using calculators, many students' responses did not. Instead, the prospective teachers in this study held "seemingly contrasting views that it is acceptable for them to use a calculator to do an arithmetic problem, but it is not acceptable for their students" (p. 459). The researchers suggested that their students found themselves in two different worlds, one in which, as doers of mathematics, they should use calculators whenever the need arises and the other in which, as teachers helping children learn mathematics, they juxtapose the use of calculators against a set of beliefs about mathematics teaching: "that basic arithmetic skills must be learned; that the task of their future students was not to *do* mathematics, but to *learn* mathematics"; and the calculator was neither an integral nor welcome part of this context. The students' view of mathematics as something that must first be learned and then done caused them to view calculators as inappropriate for children before they learn the mathematics.

Summary of Teachers' Beliefs About Technology

Teachers need support to effectively use computers or calculators in their classes, but even when teachers are comfortable using computers or calculators for their own learning, they may not believe that using technology with their students is appropriate. Teachers' beliefs about appropriate use of technology for children are constrained by their beliefs about mathematics (Tharp et al., 1997; Walen et al., 2003) and by their beliefs about teaching and learning mathematics (Fleener, 1995; Schmidt, 1999). One issue that appeared in several studies was the belief that calculators should be used only after students have learned the mathematics (Walen et al., 2003) or have mastered the concepts (Fleener, 1995).

Mathematics Teachers' Beliefs Related to Gender

Results of the Third International Mathematics and Science Study (TIMSS) indicated that, although performance of boys and girls worldwide is comparable at the 4th- and 8th-grade levels, by the last year of senior high school, in 18 of 21 countries participating in the testing, boys perform significantly better on mathematics assessments than girls, and even in the remaining 3 countries (including the United States), the differences favored males (Mullis, Martin, Fierros, Goldberg, & Stemler, 2000). In the United States, males and females take similar mathematics classes at the K–12 level, but males participate in mathematics more than females after high school (Levi, 2000). Although boys and girls at the early elementary school level have differed little in achievement-test scores, when researchers extended the focus of their investigations to the types of strategies students used, gender differences between boys and girls appeared as early as first grade, with boys consistently using more abstract strategies than girls used to solve problems (Fennema, Carpenter, Jacobs, Franke, & Levi, 1998). Such early differences may explain why major differences appear at the senior high school level, because students who

learn to make sense of mathematics early on by virtue of applying mental computation and estimation strategies are demonstrating the kind of understanding of numbers and operations that will help them be more successful when studying more advanced mathematics (Sowder, 1998).

Some proposed reasons boys consistently outperform girls in mathematics at the senior high school level and some differences favoring boys that appear even much earlier relate to the views that students have about mathematics. For example, both female and male students in the United States, Australia, and Japan have stereotyped mathematics as a male domain, and the males held this stereotype more than the females (Keller, 2001). Might these students' beliefs be affected by teachers' beliefs? Although much remains to be learned about how teachers' verbal and nonverbal behaviors affect students, the finding that teacher expectancy affects pupils' performance, sometimes known as the *educational self-fulfilling prophecy,* has been clearly established (Rosenthal, 2002). Less clear, however, is what the research on teachers' beliefs about boys, girls, and mathematics has shown.

Li (1999), in a review of literature on teachers' beliefs related to gender, found that although results were inconclusive, the literature indicated that teachers had different beliefs about male and female students and that they tended to stereotype mathematics as a male domain. Teachers tended to overrate male students' mathematics capabilities and had more positive attitudes about male students. Li found no differences between male and female teachers' beliefs.

Helwig, Anderson, and Tindal (2001) found little evidence of a relationship between third- and fifth-grade teachers' ratings of their students' mathematics skills and gender. They also agreed with an earlier conclusion by Brophy (1983, cited in Helwig et al., 2001) that teachers' perceptions of their students were accurate.

In urban and rural Kansas, Leedy, LaLonde, and Runk (2003) used a survey to assess mathematics attitudes of 4th-, 6th-, and 8th-grade girls and boys participating in a regional mathematics contest; their parents; and their teachers. The students were selected on the basis of their interest in the mathematics competition and their mathematical abilities. Using 12 Likert-type items drawn from the Fennema-Sherman Scales (Fennema & Sherman, 1976), the authors assessed the students on six factors; they modified the items to assess parents' and teachers' views of mathematics as a male domain and usefulness of mathematics. The teachers were comprised of two groups, those mathematics teachers who served as coaches for the mathematics teams and those who did not. Results indicated that the boys, fathers, parents of sons, and noncoaching mathematics teachers more strongly supported the belief that mathematics is a male domain than the girls, mothers, parents of daughters, and mathematics coaches. The most strongly endorsed individual statement related to gender bias was "Men are naturally better at mathematics than women," whereas the most strongly endorsed nonbiased statement was "It is just as appropriate for women to study mathematics as for men" (p. 289). The authors questioned whether agreement with the first statement is an indication of gender bias or of "real, inherent differences between girls and boys, men and women, in how they learn and think about mathematics" (p. 289).

The belief that mathematics is a male domain was also found among teachers in Switzerland. Keller (2001) drew from TIMSS data and used a Likert-type scale she designed to assess teachers' and 6th-, 7th-, and 8th-grade students' aptitude concepts, such as mathematics as a male domain. Results indicated that students and teachers stereotyped mathematics as a male domain, and evidence showed that teachers' views affected the students' stereotyping. Keller concluded that the assumption, put forth by Fennema, "that teachers transmit their views of mathematics during instruction and that students adopt such views is more plausible than the assumption that teachers adopt the views of students" (p. 171).

Tiedemann (2000) studied the beliefs of Grades 3 and 4 German elementary school students and their teachers. The teachers were each asked to name three girls and three boys from their classes, one of each gender performing in mathematics at the low, medium, and high levels, and then the teachers completed a questionnaire for each identified child. The 21-item questionnaire addressed six categories, including estimation of the child's competence, attribution of improvement, and presumed self-concept of the child. The results indicated that even for equally achieving, average boys and girls, teachers rated mathematics as more difficult for the girls than the boys. Teachers attributed unexpected failures more to low ability and less to lack of effort for girls than for boys. Tiedemann also reported similar results for third- and fourth-grade Russian teachers, who thought of mathematics as more a male domain than a female domain. Tiedemann concluded, "If teachers' beliefs are important influences on how they interact with and teach students, these teachers' beliefs could be seen as an influence on the development of gender differences in mathematics" (p. 205).

Summary of Teachers' Beliefs Related to Gender

Researchers raise more questions than they answer related to teachers' beliefs about boys, girls, and mathematics and the effects of these beliefs on children's performance. Apparently, teachers continue to hold the stereotype that mathematics is a male domain, and evidence indicates that this teacher belief affects students' beliefs. Most research in this area has been conducted using Likert scales, although case-study research has also proven to be useful.

RESEARCH ON TEACHERS' AFFECT SINCE 1992

One can better understand the state of the research on teacher affect in 1992 by considering the intersection of McLeod's handbook chapter on affect and A. G. Thompson's handbook chapter on teachers' beliefs and conceptions. Neither author wrote about teachers' affect. In writing about affect, McLeod focused on students' affect when solving mathematics problems, and in her chapter, Thompson focused on teachers' beliefs and conceptions but not on their affect. McLeod called for researchers to begin to incorporate affect into research on cognition, but most of those responding have not focused specifically on teachers.

I begin by sharing some of the little research that has addressed teachers' affect as it relates to mathematics teaching and learning and then turn to work that addresses either affect in general or affect-related frameworks developed to describe children's experiences, because I contend that the frameworks may be applicable for work with prospective and practicing teachers learning or engaged in mathematics. But first, I address a more general question: Does affect matter in the learning of mathematics? I report two reviews of research that clearly indicate that, at least for students, affect is related to achievement.

Relationship Between Affect and Achievement

The relationship between affective factors and achievement appears to be important for researchers in mathematics education. Ma and Kishor (1997) conducted a meta-analysis of 113 studies to investigate this relationship between students' attitudes toward mathematics and their achievement in mathematics. Their results indicated an overall mean effect size of 0.12, which they concluded was "statistically significant but not strong for educational practice" (p. 39). They found that gender differences were too weak to have practical implications and that whereas the relationship between attitude toward mathematics and achievement in mathematics was not strong at the elementary school level, it "may be strong enough for practical considerations at the secondary school level" (p. 40). They concluded that the junior high school level may be the most important period during which students shape their attitudes toward mathematics as these attitudes relate to students' mathematics achievement and that by high school, many students' attitudes are fixed or stable, which may result in their having less effect on or being less affected by achievement in mathematics. They also found a relationship between attitudes toward mathematics and the means by which researchers measured achievement in mathematics. The researchers found that attitudes toward mathematics could predict achievement measured using complex conceptual and procedural mathematics tasks, such as applications and problem solving, better than achievement measured using less mathematically complex tasks. They concluded that researchers should work toward substantially refining their measures for affect. They also called for studies in which researchers differentiate among students' ability levels, take into account students' grade levels and ethnicities, and examine the relationships between varying school characteristics (e.g., school size, school mean socioeconomic status, school policies) and mathematics achievement.

Ma (1999) conducted a meta-analysis of 26 studies narrowly focused on the relationship between anxiety toward mathematics and achievement in mathematics and reported a significant correlation of –.27. Although several definitions for *mathematics anxiety* were used by the authors of the studies analyzed, Ma drew upon McLeod's (1992) definitions to conclude that *affect* was a general construct and *mathematics anxiety* was a distinct construct generally associated with the more intense feelings students exhibit in mathematics classrooms, including tension, helplessness, dislike, worry, and fear. Ma also summarized theoretical models put forth for the negative relationship between mathematics anxiety and mathematics achievement. According to one model, when arousal states are either below or above an optimal state, performance decreases. Under this inverted-U–curve model depicting a curvilinear relationship between anxiety and performance, some arousal is beneficial to performance, but when the arousal level increases too much, performance drops. Although this model is popular with arousal theorists, Ma found that most researchers tended to consider the relationship between arousal and achievement to be linear, with increased arousal associated with decreased achievement. Ma cited researchers who have

identified this linear relationship among children at the elementary and secondary level, among adults in general, and particularly among college students. A primary reason posited for the negative relationship between mathematics anxiety and mathematics achievement stems from the theory of test anxiety. Researchers who adhere to this notion regard mathematics anxiety "as a kind of subject-specific test anxiety" (Ma, 1999, p. 522). In his meta-analysis, Ma quantified the potential improvement in mathematics achievement when anxiety toward mathematics is reduced, reporting that when an average student who is highly anxious about mathematics is able to decrease that level of anxiety, improvement from the 50th to the 71st percentile in mathematics achievement may occur. Ma thought that this meta-analysis resulted in more definitive results than the study of the relationship between attitude and achievement (Ma & Kishor, 1997) because the instruments used to measure mathematics anxiety are more effective than those used to measure the more general construct of attitude, perhaps because "anxiety toward mathematics is easier to measure than attitude toward mathematics in that it is more operationally definable for researchers and more verbally expressible for students" (Ma, 1999, pp. 533–534). Although Ma identified several measures of mathematics anxiety, almost half the studies (12 of 26) used the same one—the Mathematics Anxiety Rating Scale (MARS) (Richardson & Suinn, 1972); he found that the research results were similar regardless of the anxiety-measurement instrument. As in the 1997 study, Ma found that the type of instrument the researchers used to measure mathematics achievement made a difference. Researchers who used standardized achievement tests tended to report weaker relationships than those who used researcher-made achievement tests or mathematics teachers' grades. No gender differences were found in this study.

Mathematics Teachers' Affect

The meta-analyses described previously indicated that affect in general and mathematics anxiety in particular impede students' mathematical learning. Clearly, this is an important finding about students, but if the anxiety dissipates by the time these students enroll in college, we need not concern ourselves with mathematics anxiety of prospective or practicing teachers. Unfortunately, this dissipation does not occur, as the following studies indicated.

Harper and Daane (1998) investigated the causes of mathematics anxiety among prospective elementary school teachers enrolled in a U.S. midsized southeastern university before and after they completed their mathematics methodology courses. Fifty-three prospective teachers enrolled in three methodology courses completed three assessments. Mathematics anxiety was assessed using the 98-item Mathematics Anxiety Rating Scale (MARS) (Richardson & Suinn, 1972) test as a pretest and a posttest. The MARS test is a self-rating Likert-scale comprised of items the respondent rates in terms of the degree of anxiety aroused. Items include "Thinking about beginning a math assignment," "Figuring out your monthly budget," and Doing a word problem in algebra" (Bessant, 1995, pp. 334–335). Factors influencing mathematics anxiety (FIMA) were assessed with a 26-item checklist designed by the authors and used only as a pretest. Respondents completing the FIMA assessment were asked to either agree or disagree with each item on the assessment, and if they agreed, they then indicated whether they believed that the factor had caused them to experience any mathematics anxiety. Examples of items on this survey are "There was an emphasis on drill and practice," "I lacked an interest in math," and "I had embarrassing or negative experiences in past math classes." To further assess the influence of the methods course, the authors developed a 7-item methods-course reflection prospective teachers completed at the end of the course; it was designed to assess effects on students of (a) working with a partner, (b) working in cooperative groups, (c) working with small groups or in centers, (d) using manipulatives, (e) working on problem-solving activities, (f) writing about mathematics in journals, and (g) doing field-work in local schools. Finally, 11 of the 53 students were interviewed. Results indicated that mathematics anxiety persists in prospective elementary school teachers and that, often, the anxiety originated in elementary school. Causes for these students' mathematics anxiety included an emphasis on right answers and the right method, fear of making mistakes, insufficient time, and word problems and problem solving. The mathematics methods course led to decreased levels of mathematics anxiety for 44 of the 53 prospective teachers, but 9 of them experienced increased levels of mathematics anxiety during the course.

Mapolelo (1998) conducted a case study of one Botswanian preservice teacher's beliefs and attitudes toward mathematics. In Botswana, mathematics teaching was, at that time, procedurally oriented, and many students had difficulty with it in elementary school. These difficulties continued into teacher training to the extent that more than one third of the 490 prospective teachers at the teacher-training colleges failed the mathematics final examination. Mapolelo selected the case-study student because she expressed encountering difficulty in mathematics and had previous teaching experience. Dudu was 22 years old and an average student with "less than impressive overall class performance, particularly in mathematics" (p. 339) and two years of primary-level teaching experience in a school in a remote area. Dudu, like most of the preservice teachers, reported that she had had unpleasant experiences with mathematics at the junior and senior secondary levels. Results indicated that Dudu had been frustrated by an educational system in which memorizing facts and procedures without understanding the memorized procedures was emphasized. She came to view mathematics as something that was right or wrong, and she saw little place for reasoning when engaged in mathematics. Dudu was obsessed with failing examinations and attributed passing any examination to luck. Mapolelo concluded that to develop the better mathematics teachers needed in Botswana, mathematics educators needed to attend to the beliefs and affect of preservice teachers.

Philippou and Christou (1998) reported that Greek prospective teachers "brought very negative attitudes to Teacher Education" (p. 196). For example, 24% of the prospective teachers agreed with the statement "I detest mathematics and avoid using it at all times," and 62% agreed with the statement "I do not think mathematics is fun, but I have no real dislike of it" (p. 197). Efforts to change these attitudes were successful over the course of the program, during which prospective teachers completed two mathematics content courses based on the history of mathematics (with particular attention paid to the contributions of the Greek culture to mathematics) and a methods course. Attitudes that changed were the prospective teachers' satisfaction from mathematics and their views of its usefulness. The prospective teachers' deeply rooted anxieties about mathematics did not seem to change, and the researchers concluded that students' emotions are resistant to change.

Bessant (1995) studied attitudes of 173 Canadian university students enrolled in the introductory statistics courses in the mathematics, psychology, or sociology department. The author administered a 35-item mathematics-attitude scale the first week of class, followed a week later by a slightly reduced 80-item version of the Mathematics Anxiety Rating Scale (MARS) (Richardson & Suinn, 1972). He also administered a questionnaire designed to identify students' learning approaches. Using factor analysis, the author identified six factors that accounted for 54.9% of the total variance: general evaluation anxiety, everyday numerical anxiety, passive-observation anxiety, performance anxiety, mathematics-test anxiety, and problem-solving anxiety. The most important factors were general evaluation anxiety, referring to students' apprehension over the completion of mathematics-related tasks, and mathematics-test anxiety from thinking about a test one day, one hour, or five minutes before it began and waiting for the return of the test. The other factors were "statistically less significant, but they are of theoretical value" (p. 335).

In their study designed to investigate students' emotional responses to timed tests, Walen and Williams (2002) provided a deeper look into issues that might explain Bessant's findings related to the role of general test anxiety. In the only study that I found to establish a relationship between affect experienced by children and adults, Walen and Williams conducted two case studies of college students and one of a third-grade child They built their work on Mandler's (1989) framework of emotion as resulting from cognitive analysis and physiological response. In particular, visceral arousals can be produced by some discrepancy or the interruption of some ongoing action, and these emotional responses can substantially affect cognitive functioning. One experiencing such stress tends to focus on aspects of a situation deemed important by that individual, but Walen and Williams (2002) pointed out that although this focus may appear to be helpful, often "what an individual considers important in a given context may not lead to a resolution of the perceived stress" (p. 364). Timed mathematics tests are, for some people, such a context.

Walen and Williams juxtaposed this notion of stress induced by interruptions or blockage of activity, an occurrence that varies from individual to individual, with the more culturally based view that people have of time and temporality. Whereas in the Western culture time is seen as "objective, continuous, universal, linear, and infinitely reducible" (2002, p. 364), many in *polychronic* cultures view time as more subjective and variable in speed. For such people, "time is not something life is measured against, but rather something that conforms to life" (p. 364). Walen and Williams cited the American Indian culture as including a view of time that often clashes with the Western view. For example, whereas in the American Indian culture, careful, slow, well-considered responses are rewarded, this approach does not serve well for one required to provide quick answers on timed tests.

Walen and Williams (2002) reported case studies of Pat and Jo, two women enrolled during different semesters in the same mathematics course for elementary school teachers, and a case study of Em, a third grader. Pat, of Asian descent, and Jo, of American Indian descent, were nontraditional students returning to school after spending time as mothers and home-

makers. Pat reported past experiences of failure that caused her to feel anxious and insecure about doing mathematics, but she also reported that she enjoyed mathematics. This seemingly contradictory comment left the instructor (the first author) unsure of Pat's meaning. During problem-solving, small-group activities, the instructor came to see Pat as confident and articulate and not at all anxious or insecure. The instructor was convinced that Pat was learning the mathematics well and so was baffled when Pat failed the first 1-hour exam. As a matter of practice, the instructor informally interviewed each student after each test; during her interview, Pat explained that she had felt that she was progressing satisfactorily on the test until she realized that she had only 20 minutes left to complete the exam and panicked. She explained, "My heart pounded. I saw spots. For me, the test was over. I turned in the exam and left." The instructor offered Pat the opportunity to take as much time as she needed on the subsequent test; Pat responded with great relief to the offer. On the next exam, Pat required an extra 30 minutes to complete the exam; the instructor noted, however, that she saw Pat begin to panic when she saw the first student turn in the exam. On the remaining exams, Pat needed no additional time, and she told the instructor, "When I have all the time I need, I don't even need the whole hour" (p. 369).

Jo's story is almost identical to Pat's. She appeared to understand in class, but during the first exam, the instructor noted that Jo "paled, then flushed, then broke out with beads of sweat" (Walen & Williams, 2002, p. 369). Although Jo assured the instructor that she was okay, the instructor did not believe her. Jo failed the exam, and she reflected later on her experiences taking tests in elementary school: "It was those tests . . . those tests that we took on multiplication. I'm so slow. I just couldn't do it. I understand, but I just can't do it when there's *time*" (p. 369). The instructor arranged for Jo to have extra time on subsequent tests, but Jo, like Pat, did not need it. When turning in the next exam, "she smiled and said, 'I did it. I did it in the hour. I don't need more time. Because I could have more, I didn't need it. I didn't even feel a bit scared'" (p. 370).

Em's story provides a glimpse into the possible development of Pat's and Jo's feelings of anxiety and fear about mathematics tests. Em was described as a bright, articulate third-grade girl who was "particularly successful in mathematics" (Walen & Williams, 2002, p. 370). She was mathematically curious, and she seemed conceptually oriented. However, she had difficulty memorizing her multiplication facts and responding to the 100 facts in 5 minutes, and this frustration left her "dreading every single Friday" (p. 370).

She seemed aware of other children in the class who also struggled with these multiplication timed tests, and she distinguished between one student who was not very good in mathematics and other students who, like her, were. Her teacher had told Em that she was aware that Em knew her facts but that these timed tests had to be administered because the other classes were taking them. The culminating experience for Em was a Friday on which only the students who had mastered all their tables could attend a banana-split party. Em described feeling humiliated by the prospect of being excluded and stayed home to have her own banana-split party. She said that had she gone to school that day but not attended the party, the other students would think "that I wasn't very good at mathematics" (p. 374).

In their analysis of these three case studies, Walen and Williams (2002) noted that these students experienced neither math anxiety nor even test anxiety; instead, their enemy was time. All three students were comfortable doing mathematics, and they

> had trouble with timed mathematics tests precisely because they were willing to view mathematics as we wanted them to—as personal, meaningful, and contextual. It was not the mathematics, but rather the imposition of a foreign temporal structure over their mathematical knowing, that led to their discomfort and eventual paralysis in mathematical testing situations. (p. 375)

This research is particularly important for thinking about teachers' affect, not only because it provides two rich case studies of prospective teachers' affect but also because the researchers draw connections between the anxiety experienced by a child doing mathematics and that experienced by adults. Pat and Jo were both studying to be teachers. I am left wondering how many adult students like Pat and Jo are teaching in the classroom now? Worse yet, how many of them feel constrained to perpetuate practices that they recognize as harmful to students because that is what they believe is expected of them?

Summary of Studies of Elementary School Teachers' Mathematics Anxiety

Results of these studies indicate that mathematics anxiety among prospective elementary school teachers is of worldwide concern. Causes of mathematics anxiety have been identified as fear of failure, general test anxiety, and emphasis on right answers and right methods instead of on developing ways of reasoning about mathematics. Efforts to help prospective elementary school teachers reduce their mathematics anxiety are mixed, depending upon the approaches

taken. Simply providing more time during test taking was sufficient for some prospective elementary school teachers to overcome serious anxiety experienced during exams. Coursework promoting mathematical sense making in less rigid environments supported reduced mathematics anxiety among prospective elementary school teachers, although in one case the changes that occurred supported the students' sense of the usefulness of mathematics but not their deeply rooted anxieties toward mathematics.

Perhaps the most compelling finding is that the causes of negative affect toward mathematics or mathematics learning tend to go to prospective teachers' experiences as learners of mathematics. The process of preparing mathematics teachers starts long before prospective teachers take college courses; it begins when students enter kindergarten and first begin to learn mathematics in school. Therefore, I turn now to several studies that addressed affect, but not teacher affect. I consider these studies important because understanding the affect of K–12 students may help us understand the affect of our prospective and practicing teachers. Further, these studies all extend Mandler's widely used theory of affect.

Frameworks for Considering Affect: Extending Mandler's Framework

Hannula (2002) extended the work of Mandler (1989) by incorporating the influence of less intensive emotional states not addressed in Mandler's theory. Whereas Mandler focused upon interruptions that occur only while students are in the process of engaging in mathematics, Hannula argued that he considered students' reactions when they are thinking about mathematics, both generally and specifically. He presented a framework in which the mathematics-related emotions students experience were classified into four evaluative processes: (a) associations when generally thinking about the concept *mathematics*, (b) expectations when thinking about engaging in mathematics, (c) situational emotions when actually doing mathematics, and (d) values when considering the role of mathematics in relationship to other goals. For Hannula, each is a process that produces an expression of an evaluation of mathematics. The first evaluation is a nearly automatic association students make when, for example, they are asked how they feel about mathematics; it depends solely upon their prior experiences. Because it is not related to a particular context, it is an evaluation commonly assessed using questionnaires. The second evaluation is the most cognitive of the four; it goes deeper than the first because whereas for the first, students react generally to their image of *mathematics,* for the second, students imagine a specific mathematical situation and the consequences involving some emotions that would follow. The third evaluation occurs when a student, while engaged in doing mathematics, engages in a continuous, often unconscious, evaluation of the situation with respect to her or his personal goals. This evaluation is represented as an emotion. The fourth evaluation is based on, often unconscious, cognitive analysis in which a student evaluates his or her life and the value given to different goals in it.

In a study of Finnish students in Grades 7–9, Hannula (2002) applied his framework to the case of Rita. At the beginning of the year, Rita held negative associations about mathematics in general and about word problems in particular. In spite of these negative associations and negative expectations, Rita exerted much effort when studying mathematics, a fact Hannula interpreted as a reflection of Rita's values related to mathematics. Hannula described experiences Rita had while learning mathematics in a group of three students; two goals she held during the first task were the cognitive goal of understanding the solution of the task and the social goal of effectively interacting with her two peers. Rita experienced frustration with the mathematics and with her feelings of being left out of the group discussions. Later when Rita began to think that she understood the mathematics, her attitude toward mathematics improved. Hannula found that Rita's evaluation of mathematics changed for the better "regardless which of the four evaluations (emotion, expectation, association, values) we consider" (p. 41), and he concluded that the proposed framework provided a useful way to describe attitudes and their change. For Hannula, this case study indicated that attitudes can change over a relatively short period of time and, finally, that Rita's negative attitude toward mathematics provided a "successful defence strategy of a positive self-concept" (p. 42).

Gómez-Chacón (2000) presented a model for the study of the interaction between cognition and affect in mathematics. She subdivided affect related to mathematics into two constructs, *local affect,* which referred to the states of change of feelings or emotional reactions one experiences while engaged in mathematical activity, and *global affect,* which is the concept of self and the beliefs about mathematics and learning one holds. The global affect results from "routes followed in the local affect" (p. 151); that is, global affect might be thought of as the beliefs and identity that one develops as a result of the feelings and emotional reactions one experiences while engaged in mathematics within the individual's sociocultural context. Fear of mathematics is an example of global affect. Gómez-

Chacón developed instruments for measuring the cognition-affect interaction, and she applied those instruments in a study of 23 cabinetmaking students in their late teens enrolled in a job-training workshop in Spain.

One instrument Gómez-Chacón (2000) developed was a *problem mood map*, in which codes are used by students to reflect their emotional reactions while engaged in mathematics. At the end of each mathematical activity, the students were asked to select from a list of emotions they might have experienced while working on the activity. The 14 emotions and descriptors offered, selected as a consensus of the most relevant emotional reactions the students in this study exhibited in the classrooms, were curiosity, cheerful, despair, calm, hurry, boredom, bewilderment/confused/puzzlement, brain teaser, liking, indifference, amusement, confidence, blocked, and *De abuty*, which is a Spanish term meaning "just great." Gómez-Chacón contended that students' coding of their moods while engaged in mathematics could foster awareness of their emotional reactions, and, as a result, they might feel more in control of their emotions; also, teachers could gather information about their students' affective reactions at different stages of a lesson.

In applying her framework to a student named Adrián, Gómez-Chacón (2000) concluded that using her instrument to assess both local and global affect was a valuable way to describe Adrián's affect. For example, although Adrián exhibited the global affective trait of fear toward mathematics, he exhibited a variety of local affective traits, including anxiety, satisfaction, and surprise, while solving problems. She also found that Adrián's negative emotions were associated with a lack of understanding of the mathematics. Gómez-Chacón concluded that although the model for affect put forth by Mandler (1989), whereby students experience affect because of interruptions in problem-solving situations, was valid, the model needed to be enhanced to account for both local and global affect.

Hannula (2002) and Gómez-Chacón (2000), like Mandler (1989), considered affective factors experienced by students while engaged in mathematical problem solving. However, they both believed that Mandler's framework needed to be expanded to consider additional factors. Hannula extended Mandler's framework by considering associations, expectations, and values one holds toward mathematics, and Gómez-Chacón extended the framework by considering the global affective characteristics. Both also found in their case studies that a student's affect was inversely correlated with the student's mathematical understanding.

In her dissertation, Malmivuori (2001) presented a theoretical analysis centered on affect but integrated with cognition and beliefs around particular social environments. She addressed, in addition to affect, beliefs, metacognition, self-regulation and self-perceptions, motivation, and the influence of context on all these factors; she built her framework upon constructivist and sociocognitive theories, and I contend that she attempts to synthesize constructs in such a way as to address teacher identity. Although considering all these elements simultaneously is difficult, rather than isolate these constructs, she consistently drew connections between and among them in an extraordinarily ambitious review, analysis, and reconceptualization based upon 600 references! She wrote that the main point of her study was

> to constitute the theoretical and dynamic linkages between the often applied constructs and educational research results, as well as of the mathematics education research results with affect, that would also apply to and clarify self-regulated learning processes or the dynamic interplay of affect and cognition more generally. (p. 299)

Envisioning a study that utilizes her rich theoretical framework is difficult; still, her work is important for those rethinking affect, and Malmivuori's work is an important contribution to the future of research on affect.

Mathematical Intimacy, Integrity, and Meta-Affect

Goldin (2002a) expressed concern that research in mathematics education has focused primarily on cognition and far less on affect; he attributed this situation, in part, to the popular myth that mathematics is a purely intellectual endeavor in which emotion plays little role. Yet, he noted that, contrary to this myth, negative emotion is widespread in mathematics, and he recommended that further attention "be devoted to the psychology of developing effective affect in students" (Goldin, 1998, p. 154). According to Goldin, "When individuals are doing mathematics, the affective system is not merely auxiliary to cognition—it is central" (2002a, p. 60). He, like McLeod, subcategorized affect as emotions, attitudes, and beliefs, but he included a fourth category comprised of values, ethics, and morals. *Emotions* are rapidly changing states of feeling usually embedded in context. *Attitudes* are moderately stable predispositions toward ways of feeling in classes of situations that involve a balance of affect and cognition. *Beliefs* are internal representations to which the holder attributes truth,

validity, or applicability, and these representations are usually stable, highly cognitive, and highly structured. Also like McLeod, Goldin (2002a) viewed emotions as rapidly changing and highly affective; he viewed attitudes as more stable and cognitive than emotions, and he viewed beliefs as the most stable and cognitive of the three. The *values, ethics, and morals* comprising his fourth category are deeply held preferences that might be characterized as personal truths, are stable, may be highly structured, and are highly affective and highly cognitive.

Goldin, working with DeBellis, identified several other affective constructs related to mathematics; two of these constructs are *mathematical intimacy* and *mathematical integrity*. *Mathematical intimacy* is used to describe a psychological relationship between an individual and mathematics that connects with the individual's sense of and value of self. DeBellis and Goldin (1999) provided the example of a child, Jerome, experiencing difficulty solving a problem in which he moved 10 jelly beans from Jar 1 containing 100 green jelly beans into Jar 2 containing 100 orange jelly beans, and then, after mixing up the jelly beans, he moved 10 jelly beans from Jar 2 (now containing 100 orange and 10 green jelly beans) back to Jar 1. The child was puzzled to discover that there were the same number of the orange jelly beans in Jar 1 as green jelly beans in Jar 2, and this relationship stayed constant regardless of how many jelly beans he moved from Jar 1 if he moved the same number back to Jar 1. The mathematical intimacy of the child was reflected in the child's concern about, and involvement in, the problem.

> Jerome's interactions, from which we infer intimate engagement, include his close proximity to the jelly beans when performing the experiments, his raised voice, his deep breaths, the gesture of brushing his hand through his hair, his shrugging of shoulders, his smiling, and the silent pauses. He sits back in his chair as if to push himself away from the experiment, to distance himself when the outcome contradicts his expectation. (p. 253)

Mathematical intimacy might be thought of as the extent to which a person working on a problem is engaged with the problem.

Mathematical integrity "describes an individual's affective psychological posture in relation to when mathematics is 'right,' when a problem is solved satisfactorily, when the learner's understanding is sufficient, or when mathematical achievement is deserving of respect or commendation" (DeBellis & Goldin, 1999, p. 253). Mathematical integrity is associated with a learner's desire to want to understand, with one's discomfort toward inconsistencies, and with a focus on justifying one's reasoning. Often students confronted with tests or other evaluations are encouraged, by being awarded partial credit for giving responses even though they know that they do not understand, to bluff. This behavior, though it may lead to a higher test score, compromises the student's mathematical integrity because instead of focusing on whether the submitted work makes sense, the student focuses on the grade that results from bluffing.

DeBellis and Goldin (1999) contended that a problem solver who possesses both mathematical intimacy and mathematical integrity will experience more mathematical power and perseverance than one who lacks these attributes. They noted that a relationship exists between mathematical intimacy and mathematical integrity. Absence of integrity is a major obstacle to intimacy, because the experience of pretending or bluffing one's way through a problem not only blocks the individual's understanding of the problem but also prevents one from experiencing intimacy in relation to the mathematics. Conversely, if one lacks mathematical intimacy, the need for mathematical integrity is reduced.

Goldin (2002a) also proposed the construct of *meta-affect* to refer to "affect about affect, affect about and within cognition that may again be about affect, the monitoring of affect, and affect itself as monitoring" (p. 62). Goldin suggested that most affective experiences exist in a context that contributes to the interpretation of the affective experience and provided the example of a person riding a roller coaster because he or she seeks the experience of fear, which, in that context, is pleasurable. "The cognition that the person is 'really safe' on the roller coaster permits the fear to occur in a meta-affective context of excitement and joy" (p. 62). However, the experience can change entirely if, for example, the rider hears an unfamiliar noise and thinks that the roller coaster may be malfunctioning. The change in the rider's meta-affect creates a different experience for the rider, whether or not anything has actually changed. Goldin wrote, "It might seem that the 'cognitive' belief, that the ride is in fact safe, is the main essential to the joyful meta-affect. In this sense, the belief stabilizes the meta-affect" (p. 62). He explained that other beliefs and values play stabilizing roles for meta-affect. Applying this notion to mathematics might help explain how a particular experience, such as difficulty solving a mathematics problem, might be interpreted in different ways depending upon the beliefs and values held by the problem solver. Whereas one student might interpret the difficulty as an indication that he or she is a failure, another might view the difficulty

with anticipation for a feeling of satisfaction at the expected success. The ways students view the context surrounding the task, together with the beliefs or values held by the student, are integrally related to the interpretation of the experience. "Powerful affective representation that fosters mathematical success inheres not so much in the surface-level affect, as it does in the meta-affect" (p. 63).

Summary of Teachers' Affect

Because so little literature focuses on the intersection between teachers and affect, I have included studies of student affect. Goldin and DeBellis introduced the constructs of *mathematics intimacy, mathematical integrity* (DeBellis & Goldin, 1999), and *meta-affect* (Goldin, 2002a), and Hannula (2002) and Gómez-Chacón (2000) extended Mandler's theory by including *less intensive states* and *global states*, respectively. Malmivuori (2001) presented a framework centered on affect but integrated with cognition, metacognition, beliefs, self-regulation and self-perceptions, motivation, and the influence of context on all these factors.

Research indicates that attitudes are less amenable to change among students in high school (Ma & Kishor, 1997) and college (Hembree, 1990) than among those in elementary and middle grades. One might, therefore, reasonably assume that prospective and practicing teachers' mathematics affect or anxiety may have been formed when the teachers were precollege students, and the research supports this assumption. Although the view that the affect of students provides insight into the affect of teachers is a conjecture, evidence shows that similarities exist between adults' and children's feelings about mathematics.

I know of no research linking teachers' affect about mathematics and mathematics learning to their classroom instructional decisions, but the prospect that the two may be linked is sufficient reason to think more carefully about what can be done to support prospective and practicing teachers in developing more positive affect toward mathematics and mathematics teaching and learning. In an age of accountability, can one reasonably assume that teachers who enjoy mathematics will teach more of it, or teach it with a deeper sense of intimacy and integrity (DeBellis & Goldin, 1999)? Will those who have more positive affect toward mathematics be more positive in their approaches to the teaching of mathematics? If teachers are able to share enthusiasm for mathematics, will their students develop more positive impressions of mathematics?

Consideration of the relationship between fear of failure and mathematical learning is important in the preparation of mathematics teachers. In-service work with practicing teachers generally does not include evaluative components, and some in-service providers attribute their success to the nonevaluative aspect of the experience. But prospective elementary school teachers generally must complete mathematics courses at the university level, and these courses are difficult for some students. The tension between the gatekeeper role of such courses and the affect promoted as a result of the students' concern seems to be worth additional consideration. For example, although we educators may believe that an important component of understanding mathematics is knowing when we do not understand something, students may feel compelled to hide their lack of understanding lest they receive an unsatisfactory grade and fail to meet the requirements for becoming teachers; students may learn that the best way to pass their courses is to hide their lack of understanding from the teacher, and possibly from themselves, thereby compromising the development of mathematical integrity (DeBellis & Goldin, 1999). Or, perhaps worse, if we help prospective teachers learn the mathematics that we consider important for them to learn but they feel that the experience was unpleasant and prefer to avoid mathematics, as practicing teachers, they may teach mathematics in a way that is consistent with their negative affect. Can a position be found between the evaluative, gatekeeper model generally applied with preservice teachers and the nonevaluative, collaborative model generally applied with practicing teachers, a position that would maintain the academic integrity required by universities but would also incorporate the long-range benefits valued by those engaged with in-service teachers' professional development? Even though providing professional development for practicing teachers and preparing preservice teachers raise different sets of issues, turning to professional development collaboratives that take as their primary goal the teaching of mathematics to practicing teachers may be useful as a context for considering innovative approaches to teaching mathematics to prospective teachers in ways that also foster prospective teachers' development of positive affect (see Sowder, this volume).

When we researchers in the field of mathematics education strive to make the field a scientific discipline, we draw heavily from other disciplines, such as psychology and sociology, gaining more theoretical perspectives in our field (*Educational Studies in Mathematics* [ESM], 2002). One result of a young and healthy discipline is that "it has become the norm rather than the exception for researchers to propose their own conceptual framework rather than adopting or refining an existing one in an explicit and disciplined way" (*ESM*, 2002 p. 253). The editors of *Educational Studies in Mathematics* (2002) and the *Journal of Mathematics Teacher Education* (Cooney, 2001) raised the possibility that the existence of these journals may have contributed to the proliferation of local theories that may not sufficiently push the field forward. This tendency to develop new conceptual frameworks seems strong in the area of research on affect in mathematics education; each study seems to introduce a new theory about affect. Mandler's theory provided a much-needed starting point around which others could build, and most researchers studying affect in mathematics education at least reference Mandler. We should keep in mind Mandler's (1989) caution that although a search for consensus on definitions of *affect* may be futile, we should take care in how we use the term. Defining our terms is certainly not a problem limited to the study of affect, as we saw in research on teachers' beliefs.

BEYOND BELIEF AND AFFECT?

Some researchers contend that teachers' beliefs should not be isolated but, instead, should be considered a component of more encompassing constructs. In this section, I look to such constructs, teachers' *orientations* and teachers' *perspectives,* and to a new area of research that holds possibilities for integrating research on teachers' cognitions, beliefs, values, and affect: communities of practice and teachers' identities.

Teachers' Orientations

In her handbook chapter, A. Thompson (1992) argued for the position that researchers ought not attempt to separate the study of teachers' beliefs from teachers' knowledge, and she used the term *conception* to refer to both. Later, she and colleagues (A. G. Thompson et al., 1994) used the term *orientation* to refer to teachers' views of mathematics and mathematics teaching. They introduced the terms *conceptual orientation* and *calculational orientation* to refer to the images teachers hold toward pedagogical tasks and the goals they served. Although teachers with conceptual and calculational orientations may agree upon the long-term goal of helping students develop problem-solving skills, their approaches differ in important ways. Actions of a teacher with a conceptual orientation are driven by an image of a system of ideas and ways of thinking she intends her students to develop; an image of how these ideas and ways of thinking can be developed; ideas about features of materials, activities, and expositions and the students' engagement with them that can orient students' attention in productive ways; and an expectation and insistence that students be intellectually engaged in tasks and activities. Although a teacher with a calculational orientation may share the general view that solving problems is important, the actions of such a teacher are driven by a fundamental image of mathematics as the application of calculations and procedures for deriving numerical results. Associated with a calculational orientation is a tendency to speak exclusively in the language of number and numerical operations, a predisposition to cast solving a problem as producing a numerical solution, an emphasis on identifying and performing procedures, and a tendency to disregard context and to calculate upon any occasion to do so. Whereas a conceptually oriented teacher can identify the important concepts embedded in a problem, a calculationally oriented teacher tends to treat problem solving as flat; that is, nothing about the problem stands out, except that the answer is most important.

Conceptual and calculational orientations lead to differences in classroom discussions. Consider the problem "Susan drives 240 miles in 5 hours. What is the average speed of her trip?" The question "How did you solve this?" usually elicits from students the calculations they performed: "I divided 240 by 5, and the answer is 48." A teacher with a calculational orientation may consider this an appropriate answer, but a teacher with a conceptual orientation might question further to direct students toward the underlying conceptual ideas, the quantities in this problem, and the relationships among those quantities: "When you divided 240 by 5 and got 48, what is 48 a number of? That is, to what does 48 refer in this situation?" Whereas a calculationally oriented teacher might ask, "What did you do?" and mean "What calculation did you perform?", a conceptually oriented teacher might look for an explanation of the students' reasoning, not for the students' calculations, and ask, "What were you thinking?" or "What are you trying to find when you do this calculation (in the situation as you currently understand it)?"

Teachers who wish to become more conceptually oriented toward mathematics and mathematics teaching must reflect deeply about their goals for, and image of, mathematics and mathematics teaching. To shift their fundamental image from mathematics as doing things to mathematics as reasoning in particular ways is difficult. Furthermore, teachers who embark on such a journey frequently encounter their old patterns and goals, for example, in finishing a conceptual lesson by presenting the standard procedure instead of having students use notation to represent the conceptual methods they produced for solving one or several problems. Too little textbook support for conceptual teaching is provided, and far too little support is available in terms of state frameworks for mathematics instruction.

A. G. Thompson and P. W. Thompson (1994, 1996) applied the constructs of conceptual and calculational orientations in analyzing 2 one-on-one teaching experiments in which different instructors worked with the same child. The first instructor, a mathematics teacher named Bill who possessed a strong and elaborate understanding of the concept of rate but who experienced difficulty in teaching the concept of rate to one of the stronger students in his sixth-grade class, acted like other good quantitative reasoners who are unable to use arithmetic to reason quantitatively.

> [Bill had] come to use arithmetic in two ways simultaneously—as a representational system and as a formulaic system to express an evaluation. What we did not foresee was the shortcoming of this development in regard to teaching. Bill's quantitative conceptualizations appeared to be encapsulated in the language of numbers, operations, and procedures. He thus had no other means outside the language of mathematical symbolism and operations to express his conceptualizations. The language of arithmetic served him well as a personal representational system, or as a system for communicating with other competent quantitative reasoners. Yet, . . . that language served him poorly when trying to communicate with children who knew the tokens of his language but had not constructed the meanings and images that Bill had constructed to go along with them. (P. W. Thompson & Thompson, 1994, p. 300)

In the follow-up article, A. G. Thompson and Thompson (1996) described a subsequent lesson in which the same child worked with an instructor who had what the authors referred to as *knowledge for conceptual teaching*, comprised of "clear images of understanding a mathematical idea conceptually, how those images might be expressed in discourse, and what benefits might accrue to students by addressing the conceptual sources of their difficulties" (p. 3). In less than 30 minutes, the instructor had supported the child in her initial development of the concept of *speed* as the "mutual accrual of distance and time." The teacher who had struggled with the child on the previous day observed the lesson but failed to notice that a key aspect of the successful lesson was helping the child attend to the proportionality of corresponding quantities, distance and time, by attending to the segments representing each of these two quantities—even though he reasoned proportionally to solve these problems himself. Teacher knowledge was not enough to support the child's reasoning, and only the second instructor's ability to coordinate his understanding of the content and the child's thinking and to use language that oriented the child toward the important underlying concepts enabled the child to develop the concept of rate.

The construct *orientation* as used by A. G. Thompson et al. (1994) incorporated teachers' knowledge, beliefs, and values about mathematics and mathematics teaching. In using the term, they described teachers' images, views, intentions, goals, and tendencies, and they operationalized the construct by attending to teachers' language and actions. Although they addressed more than beliefs, they did not explicitly address affective issues. In the next section I describe an attempt to expand the investigation of teachers' beliefs to include affective factors.

Teachers' Perspectives

When studying teachers, researchers apply their conceptual lenses to interpret teachers' knowledge, beliefs, and practices. Simon and Tzur (1999) highlighted two approaches taken by researchers studying teachers. Some researchers apply their conceptual frameworks, which encompass the current knowledge from the field, to tell their own stories about teachers, providing, for example, deficit studies focused primarily on what teachers lack; the teachers' perspectives, aspects of the teachers' practices that are critical to the teacher, are generally missing from the story. In the other approach, the researcher tells the story from the teacher's perspective, but the researcher's conceptual framework may be missing, and consequently the story may not add to the theoretical development of the field. Simon and Tzur (1999), in their methodology for studying mathematics teacher development, focus upon the teacher's approach by using their conceptual framework to provide accounts of the teacher's perspective. A key principle guiding their work is that "every teacher's approach is rational and coherent from his or her perspective" (p. 261) and that the job of the researcher is to understand the teacher's underlying perspectives that might coherently account for the approach without separating aspects of that teacher's perspectives into disparate components, such as beliefs, knowledge, and methods of questioning. They have contended that the term *teacher's practice* includes not only everything the teacher does that contributes to his or her teaching, including planning, assessing, or interacting with students, but also the teacher's values, skills, intuitions, and feelings about those aspects of the practice. These

epistemological commitments are associated with particular methodological commitments, including the view that one cannot assess the relationships among teachers' beliefs, knowledge, values, intuitions, feelings, and practices without gathering rich data. They refer to the unit of analysis for their qualitative work as a *set*, comprised of at least two consecutive related mathematics lessons and interviews before, between, and after the lessons.

Applying this methodological framework to study a group of teachers engaged in mathematics education reforms, they (Simon, Tzur, Heinz, Kinzel, & Smith, 2000; Tzur, Simon, Heinz, & Kinzel, 2001) considered these teachers in transition in terms of two perspectives commonly referred to in the literature: conception-based perspectives and traditional school-mathematics perspectives. The authors referred to *conception-based perspectives* as emergent and constructivist perspectives based on three assumptions: (a) Mathematics is created through human activity, and humans have no access to a mathematics that is independent of their ways of knowing; (b) individuals' currently held conceptions constrain and afford what they see, understand, and learn; and (c) mathematical learning is a process of transforming one's knowing and ways of acting. They contrasted this perspective with the *traditional school-mathematics perspective,* in which mathematics is viewed as existing independently from human experience and students are believed to passively receive mathematical knowledge by listening to and watching others. Teachers holding the perception-based perspective behave as if the mathematical relationships were properties of the objects being considered instead of as a function of the knowledge of the perceiver. Simon et al. (2000) provided a metaphor to capture the difference between the perception-based and conception-based perspectives:

> We think of *perception* as looking through a lens. The lens represents what the perceivers bring to the situation, which structures their perception. Perceivers who hold a perception-based perspective do not consider that there are lenses affecting what they see. Rather, they assume that what they see is what is "out there," that what is out there enters *as is* through their senses. . . . In contrast, perceivers who hold a conception-based perspective consider that they can see only that that results from looking through particular lenses and that they have no way to compare their perceptions (through their lenses) with lens-free perceptions. As a consequence, they make no claims as to what is out there, and they attempt to understand what is perceived as a contribution to understanding the perceiver. The analogy serves to emphasize that those who we infer hold a perception-based perspective are not rejecting a

conception-based perspective. They have never considered such a perspective. (p. 594)

The researchers found that neither perspective adequately characterized the views of these teachers in transition because the teachers did not view children as passively receiving mathematical knowledge, but, although they viewed mathematics as an interconnected and understandable body of knowledge, they also viewed mathematics as existing independently from human activity, and therefore "accessible *as is* to all learners" (Simon et al., 2000, p. 593).

Communities of Practice and Teacher Identity

When learning theorists evolved during the twentieth century beyond the stimulus-response of behaviorism to consider the contents of people's minds, their research focus changed from teachers' actions to understanding of teachers' thought processes, including their planning, their interactive thoughts and decisions, and their theories and beliefs (Clark & Peterson, 1986). The preponderance of research on mathematics teachers' beliefs is focused upon understanding teachers' beliefs, investigating the relationship between teachers' beliefs and practices, and changing teachers' beliefs. However, even before the publication of the first NCTM *Handbook* (Grouws, 1992), researchers on learning questioned a focus solely upon the contents of a person's mind without attending to the contexts in which learning occurred (Brown et al., 1989; Lave, 1988). By the early 1990s, during further evolution, some mathematics educators had adopted sociocultural perspectives, and a related theoretical framework that was to affect research on teachers' professional development, *communities of practice* (see Sowder, this volume), emerged.

Lave and Wenger (1991) suggested that by shifting their focus from the individual learner to groups of learners, they could account for the learning that results from changing the way one participates with others who share common goals. Wenger (1998) further elaborated the relationship between a social community's practices and individuals' identity construction resulting from participating in these communities. Many researchers studying teacher professional development have adopted the *communities-of-practice* framework and the construct of *identity*; I provide three examples of research in which mathematics educators applied this theoretical lens to their work. Two examples are drawn from major professional development projects based around reform-oriented approaches to teaching and learning mathematics. In

the CGI project (Carpenter et al., 1999) elementary school teachers learned about children's mathematical thinking as a means for changing their beliefs and their practices; the QUASAR project (Silver & Stein, 1996) focused upon inner-city middle school reform through professional development of teachers. The third example is work of Forman and Ansell (2001) with a single teacher.

In work related to the CGI project, Kazemi and Franke (2004) applied a transformation-of-participation framework to describe the learning of a group of teachers engaged in professional development focused on children's mathematical thinking. Building upon Wenger's (1998) framework, they studied ways in which the participation of the group of teachers changed over time, because "shifts in participation do not merely mark changes in activity or behavior. Shifts in participation involved a transformation of roles and the crafting of new identities" (Kazemi & Franke, 2004, p. 205). Their contribution in writing about this work was "to provide an analytic frame for understanding teacher learning as shifts in participation" (p. 206). They met with 10 cross-grade teachers at an inner-city elementary school, after school once a month throughout the year, to discuss students' mathematical thinking. Teachers adapted common problems for their students and brought student-work examples to discuss at each meeting. The study covered the first year of the teachers' participation; data analyzed were seven workgroup-meeting transcripts from audio recordings, written teacher reflections, copies of student work shared by the teachers, and end-of-the-year teacher interviews. The authors analyzed their notes and the transcripts to understand how the teachers talked about student work, noting the mathematical and pedagogical issues that were raised, and identified other descriptive themes that consistently emerged across the year, such as teachers' sharing successes from their practices or teachers' generating questions about their practices.

Kazemi and Franke (2004) noted two major shifts in teachers' workgroup participation. First, early in the year, teachers new at eliciting children's thinking tended to underestimate the children's abilities and reported students' unsuccessful attempts to solve problems, but by the third session, they had begun to interact with their students about their strategies and expressed amazement at their students' innovative strategies. The teachers helped one another by suggesting ways they might select students to share strategies so as to observe something identified as interesting.

Second, the teachers began to notice mathematical issues related to place value in the children's strategies; these issues led them to reconsider their mathematical goals for their students. Some teachers came to redefine place value less in terms of students' memorizing the names of units (ones, tens, hundreds) and more related to their using groups of 10 to solve problems (e.g., to find the difference between 48 and 111, one teacher shared that two of her students reasoned that 48 + 60 is 108, and 3 more is 111, so the answer is 63.)

Kazemi and Franke (2004), although noting the importance of traditional components of studies of teacher change, explained that the framework they had applied enabled them to focus upon shifts in the participation of the study-group teachers:

> We do not argue that examining individual teachers' developing knowledge and beliefs is unimportant. In fact, these are key resources for a developing community. However, we believe that by attending to shifts in participation, we can understand the following aspects of teacher learning: (a) how teachers working together supported the development of each other's thinking and the practices they used in their classrooms; (b) how and when teachers asked each other for help and contributed to discussions in the workgroup because of their own experimentation in the classroom; and (c) how teachers looked at the strategies students in other classrooms used and then used those as markers for what to expect from their students. (p. 231)

Stein and Brown (1997) drew upon sociocultural theories to explain teacher learning in the QUASAR project, suggesting that the "location of the phenomenon of learning" (p. 159) changed from being "located in the cognitive structures and mental representations of individual teachers" (p. 159) to being "situated in the fields of social interactions between and among individuals" (p. 159). Stein and Brown explained that when the unit of analysis shifted from the individual to the group, teacher change was defined with respect to the changing roles of the teachers as transformation of participation. For them, the communities-of-practice framework, found useful by many, lacked specificity as an analytic tool. To apply the framework in their study, Stein and Brown (1997) examined the range of the community's work practices and determined that the breadth of teachers' participation increased to include working on portfolio systems, planning curriculum, involvement with parents, making presentations to others, and working on articulation issues across other programs. Further they observed how teachers' depth of participation, as measured by movement from peripheral to more central roles, changed. For example, participants moved from peripheral involvement in "old-timer" colleagues' presentations to giving mini-presentations in the local community or copresenting to external audiences to the point of giving presentations on their own. Stein and Brown concluded that the community-of-practice framework provided a lens for viewing collaboration as a source for teacher learning and development.

In the same paper, Stein and Brown (1997) reported on a QUASAR site where a reform community with common goals and joint productive activity had not become established. The community-of-practice perspective was inappropriate for explaining teacher learning in this situation, so the authors relied instead upon Tharp and Gallimore's (1988, cited in Stein & Brown, 1997) *assisted-performance perspective.*

Stein and Brown (1997) concluded that "studies of teacher change in mathematics tend to overlook noncognitive variables such as affect or motivation for change" (p. 186). They found that the sociocultural frameworks they applied enabled them to "situate the learning of new knowledge and skills in the context of the larger meaning-making capacities and affective characteristics of individuals" (pp. 186–187) and, thus, to "integrate the study of cognitive change with the examination of goals, motivations, and identity" (p. 187).

Forman and Ansell (2001), like the authors of the previous two studies, drew from the communities-of-practice framework to view learning as a social activity, but their focus was on one teacher, Mrs. Porter, not a group of teachers. To support their theoretical framework, they adopted a methodological approach in which their unit of analysis bridged the individual and the social. Initially, in a top-down approach, they sampled major events in a third-grade mathematics classroom over several months before making a more fine-grained analysis of classroom-discourse patterns. They were surprised to find that the nature of the talk when the class was discussing standard strategies differed from their talk about student-invented strategies. During discussions of student-invented strategies, the teacher "emphasized sense making, risk-taking, persistence, being logical and finding increasingly efficient strategies" (p. 131), but in discussions of standard algorithms, she spoke in a different voice, emphasizing the confused and complicated nature of standard algorithms. Forman and Ansell (2001) traced these contrasting voices to personal experiences of Mrs. Porter. Notice the central role of teacher affect in their comments:

> The voice of her earliest experiences (both as a student and a teacher) spoke of alienation, low self-confidence, passive acceptance of authority, lack of initia-

tive, and a sense of incompetence and confusion, especially in mathematics. This first voice seemed to be connected, in part, to an instructional approach that relied upon students memorizing algorithms without understanding why they work ... [and] that ignored or discounted students' own attempts to make sense of their strategies for solving problems. The voice of her more recent experiences spoke of excitement, the enjoyment of learning from others through discussion, risk-taking, a passionate commitment to fostering students' sense-making, and an intense involvement in helping her students express their ideas and reflect on them. This voice seemed associated with an instructional approach that emphasized careful attention to students' different ways of solving problems and their informal knowledge base. Both of these voices also seemed to be linked to the broader social and institutional contexts of her classroom. (p. 134)

What is an *identity*? Collopy (2003) wrote, "A teacher's identity is the constellation of interconnected beliefs and knowledge about subject matter, teaching, and learning as well as personal self-efficacy and orientation toward work and change" (p. 289). Van Zoest and Bohl (2005) also viewed *teachers' identities* as embodying their knowledge, beliefs, commitments, and intentions, but they included, also, all the ways teachers have learned to think, act, and interact. For Van Zoest and Bohl, a teacher may hold many identities, but a given identity is not context specific—individuals carry it with them when they move from context to context. Wenger's (1998) view of identity cannot be captured in a brief definition; he devoted one third of his book to discussing it, so I state what identity is not for Wenger instead of what it is:

> [Identity] is not equivalent to a self-image; it is not, in its essence, discursive or reflective. We often think about our identities as self-images because we talk about ourselves and each other—and even think about ourselves and each other—in words. These words are important, no doubt, but they are not the full, lived experience of engagement in practice. (p. 151)

Sfard and Prusak (2005) took issue with Wenger's notion that "there is something beyond one's actions that stays the same when the actions occur, and also that there is a thing beyond discourse that remains unchanged, whoever is talking about it" (p. 16). They sought a definition of *identity* that is "operational, immune to undesirable connotations, and in tune with the claim about identities as man-made and collectively shaped rather than given" (p. 16). They thought of identity as narratives that are told by an author, about an identified person, to a recipient; the author, identified person, and recipient may be different people or

the same person. They distinguished between *actual identities,* consisting of stories about the actual state of affairs (e.g., "I am an army officer" or "I am a good driver"), and *designated identities,* consisting of narratives presenting an anticipated state of affairs (e.g., "I want to be a doctor" or "I have to be a better person"). Sfard and Prusak argued that identities are crucial to learning because they may act as self-fulfilling prophecies, and they suggested that this perspective on identity might serve as the missing link between learning and sociocultural context.

Summary

Researchers have considered constructs related to beliefs, with *orientation* (A. G. Thompson et al., 1994), *perspective* (Simon & Tzur, 1999), and *identity* (Wenger, 1998) serving as three such terms. An important issue for the future of research on teachers' beliefs is how researchers negotiate between constructivist and sociocultural perspectives. For example, Rogoff (1997) argued that the assumptions underlying the theoretical perspectives that explain how individuals acquire knowledge are fundamentally different from the assumptions underlying the transformation-of-participation perspective associated with focusing upon groups of people, and she argued that because no individual exists in isolation or out of cultural context, all learning, even reading a book, is social and should be seen from this perspective. Cobb (1995) presented the position that psychological and sociological perspectives should be viewed together:

> It is tempting to respond to the conflicting assumptions of the sociocultural and constructivist perspectives by claiming that one side or the other has got things right. However, I will instead argue that the two perspectives evolved to address different problems and issues and they are complementary in several respects. (p. 379)

P. Thompson (in P. Thompson & Cobb, 1998) agreed that psychological and social perspectives depend upon each other, but he chose to view the psychological perspective as more fundamental because "it aligns more explicitly with what I take as our fundamental goal of making a positive, lasting difference in students' lives after they leave our classrooms" (p. 19).

Lerman (1998, cited in Van Zoest & Bohl, 2005) suggested that in choosing a unit of analysis for educational research, one must be able to zoom in and out, changing focus to account for the full spectrum of locations of cognitive development. P. Thompson and Cobb (1998) applied the same metaphor, arguing, "To achieve a unification of psychological and social perspectives would mean that we become able to 'zoom' out or in with respect to a set of problems without losing sight of where we started" (p. 4).

Researchers who take a psychological approach by focusing on an individual teacher's beliefs will continue to inform the field as will those who take a sociocultural approach, by, for example, focusing on communities of practice. We in the field must balance these approaches to gather data useful for answering the types of questions driving our work. Furthermore, when we become comfortable with both approaches, the questions we consider asking will change.

SOME THEMES AND FINAL THOUGHTS

Changing Beliefs, Infusing Affect

One noteworthy difference between research on teachers' beliefs and affect is that whereas research on teachers' beliefs has been extensive and subsumed into almost all areas of research on mathematics teaching and learning, the study of teachers' affect has not. Mathematics educators generally agree on what beliefs are; we now face a greater challenge than defining *beliefs:* how to change teachers' beliefs. If beliefs are lenses through which we humans view the world, then the beliefs we hold filter what we see; yet what we see also affects our beliefs—creating a quandary: How do mathematics educators change teachers' beliefs by providing practice-based evidence if teachers cannot see what they do not already believe? The essential ingredient for solving this conundrum is reflection upon practice. When practicing teachers have opportunities to reflect upon innovative reform-oriented curricula they are using, upon their own students' mathematical thinking, or upon other aspects of their practices, their beliefs and practices change. Furthermore, although prospective teachers seldom are embedded in practice-based environments, when they are provided opportunities to learn about students' mathematical thinking and reflect upon the experiences, their beliefs change.

Researchers have found that for some teachers, beliefs change before practice, whereas for others, changes in practice precede changes in belief. I suspect that ultimately research will show that the most meaningful changes take place when teachers' beliefs and practices change together, but additional research in this area will lead to a better understanding of the relationship between teachers' changing beliefs and practices.

Judging by the attention to teachers' affect in the mathematics education literature, one might conclude that teachers' affect is not nearly as important as teachers' beliefs. Yet teachers' affect is critically important! If prospective or practicing teachers are to develop deeper content knowledge and richer beliefs about mathematics, teaching, and learning, then positive affect must be considered. I have asked several groups of prospective secondary mathematics teachers, the majority of whom had previously completed their undergraduate mathematics major, to define *mathematical proficiency.* Generally they talk about learning procedures, understanding concepts, and problem solving; a smaller number mention proofs or reasoning. Few students mention affective issues. The definition of *mathematical proficiency* presented in the consensus document *Adding It Up* (NRC, 2001) includes five interrelated strands that, together, comprise proficiency. The first four are *conceptual understanding, procedural fluency, strategic competence* (the ability to formulate, represent, and solve mathematical problems), and *adaptive reasoning* (the capacity to think logically and to informally and formally justify one's reasoning). The fifth strand, *productive disposition,* is "the tendency to see sense in mathematics, to perceive it as both useful and worthwhile, to believe that steady effort in learning mathematics pays off, and to see oneself as an effective learner and doer of mathematics" (NRC, 2001, p. 131). The inclusion of confidence in the ability to learn mathematics as a component of mathematical proficiency is important as a statement that proficiency in mathematics has affective aspects. Teachers make important decisions about the manner in which they teach mathematics, and elementary school teachers often decide how much time to devote each day to mathematics. If the mathematics courses we educators offer to prospective teachers address the first four strands of mathematical proficiency without addressing mathematical disposition, I suspect that we will continue to produce many teachers who lack the positive dispositions associated with creating positive mathematics-learning experiences for students. Although secondary school teachers generally have more positive affect than elementary school teachers toward mathematics, we must help secondary school teachers to consider not only what mathematics they are teaching but also the experiences they create for their students.

We can support prospective and practicing teachers by helping them attend to their own affect toward mathematics. Do they suffer from mathematics anxiety? Are they affected by timed tests? When they experience what Mandler referred to as *interrupted plans,* how do they interpret the experiences? Do they have experiences of mathematical intimacy? Do they strive for *mathematical integrity,* that is, the ability to know whether they understand—or have years of being falsely assured that they understand mathematics undermined their abilities to view their own mathematical understanding objectively? When they publicly reflect upon their own experiences, they may also reflect upon the experiences of their colleagues, and in so doing, they may begin to understand the wide range of experience represented by their colleagues and, thus, by the students they one day will teach.

Constructs, Measurement, and to the Future

As researchers, we do not study beliefs or affect in general; we study them in context. Researchers studying teachers' beliefs or affect are generally careful to make explicit what beliefs or affect they are studying, and this practice has facilitated convergence upon common definitions in the field. Many researchers view *knowledge* as belief with certainty and *values* as deeply held, even cherished, beliefs. Researchers studying affect tend to include beliefs as a component of affect, but the majority of research on teachers' beliefs has not included affect.

Ma (1999) found more definitive results in a meta-analysis about mathematics anxiety than he and a colleague found in a meta-analysis about attitudes toward mathematics (Ma & Kishor, 1997), and he attributed the difference to the fact that researchers have more effectively operationalized instruments for measuring mathematics anxiety than for measuring the general construct of attitude (Ma, 1999). Clearly, a field of research is extended by researchers' applying careful definitions of their constructs. But so too is a field extended when researchers consider emergent constructs that encapsulate new ways to consider old constructs. The epistemological commitments entailed by considering teachers' conceptions, perspectives, orientations, identities, or other general constructs that subsume beliefs or affect are associated with particular methodological commitments. For example, Simon and Tzur (1999) argued that studying teachers' perspectives involves the investigation of the relationships among teachers' beliefs, knowledge, values, intuitions, feelings, and practices, and to gather such rich data required a unit of analysis referred to as a *set*, comprised of observation of at least two consecutive related mathematics lessons and interviews conducted before, between, and after the lessons. In-depth qualitative research, in which data sources are triangulated using a variety of tools, is important for theory building, but for theory testing, researchers need methodologies that can be used with large numbers of subjects. Self-report data are widely used for measuring affect and beliefs of prospective and practicing teachers because such measures are easy to administer and score. However, because people may, at times, be unaware of their beliefs (Furinghetti & Pehkonen, 2002) and may offer opinions on matters about which they have given little thought (McGuire, 1969), Likert scales are of limited use. One promising alternative for assessing beliefs and affect is to employ internet technologies, such as video streaming and data downloading, in developing surveys that require open-ended responses to provide data richer than those gathered in Likert-scale instruments (Integrating Mathematics and Pedagogy, 2003).

We mathematics education researchers must continue to pursue fundamental theoretical questions that move our field forward, often through qualitative research. However, education is a political endeavor, and the past 10 years have found the mathematics education community embroiled in a political struggle for relevance. Until we produce more quantitative studies designed to test theories, too many people in positions of power will continue to ignore our research.

I accepted the task of reviewing the literature on teachers' beliefs and affect for this volume, and although teachers' beliefs and affect are closely related, I was unable to integrate the review of these two areas. Perhaps in the future, teachers' beliefs and affect will converge in such a way as to make an integration more natural. To review all the research on teachers' beliefs and affect since the publication of the last NCTM *Handbook* (Grouws, 1992) is beyond the scope of a single chapter, and with the increases in the size of the field and the number of publication outlets, I expect that for a third handbook, the amount of research to summarize will be far greater than the current body. What would such a volume contain—separate chapters on teachers' knowledge, teachers' beliefs and affect, teacher education and professional development, and teachers and teaching? Might all these chapters be integrated into one chapter under a common theme, say, teacher identity?

Teachers know; teachers believe; teachers feel; teachers participate; teachers belong. The choice researchers make about which constructs to apply in their work is one of the most important research decisions made. With new constructs come new questions and new methodologies. One way to think about how new constructs emerge and fit into existing constructs is to turn to Bernstein (1999, cited in Lerman, 2002), who distinguished vertical knowledge structures from horizontal knowledge structures. Science provides an example of a *vertical knowledge structure* in which each new theory can subsume its predecessors, whereas mathematics education is an example of a *horizontal knowledge structure* because new theories sit alongside their predecessors. New theories come with their own language and epistemological commitments, and, as such, they cannot be used to refute other theories, but instead they introduce different questions and new methodologies for answering these questions. The research on teachers' beliefs and affect has grown from psychological theories of learning that focused upon the contents of the minds of individuals, and the field has now adopted sociocultural theories through which researchers look at the world anew. I expect that the newer theories will neither replace nor subsume the old theories but, instead, will sit alongside them. Researchers will continue to study emerging sociocultural constructs, such as teacher identity, that include teachers' knowledge, beliefs, affect, and more. But researchers will also continue to study these components in isolation, and I think that this work will continue to be important.

REFERENCES

Ambrose, R. (2004). Initiating change in prospective elementary school teachers' orientations to mathematics teaching by building on beliefs. *Journal of Mathematics Teacher Education, 7*, 91–119.

Ambrose, R., Clement, L., Philipp, R., & Chauvot, J. (2004). Assessing prospective elementary school teachers' beliefs about mathematics and mathematics learning: Rationale and development of a constructed-response-format beliefs survey. *School Science and Mathematics, 104*, 56–69.

Andrews, P., & Hatch, G. (2000). A comparison of Hungarian and English teachers' conceptions of mathematics and its teaching. *Educational Studies in Mathematics, 43*, 31–64.

AP–AOL News. (2005). *Math: The most unpopular school subject.* Retrieved May 30, 2006, from http://mathforum.org/kb/thread.jspa?threadID=1204282&tstart=210

Ball, D. L., & Cohen, D. K. (1996). Reform by the book: What is—or might be—the role of curriculum materials in teacher learning and instructional reform? *Educational Researcher, 25*(9), 6–8,14.

Becker, J. P., & Jacob, B. (2000). The politics of California school mathematics: The anti-reform of 1997–99. *Phi Delta Kappan, 81*, 529–537.

Benbow, R. M. (1995, October). *Mathematical beliefs in an "early teaching experience."* Paper presented at the annual conference of the North American Chapter of the International Group for the Psychology of Mathematics Education, Columbus, OH.

Bessant, K. C. (1995). Factors associated with types of mathematics anxiety in college students. *Journal for Research in Mathematics Education, 26*, 327–345.

Bishop, A., Seah, W. T., & Chin, C. (2003). Values in mathematics teaching—The hidden persuaders? In A. J. Bishop, M. A. Clements, C. Keitel, J. Kilpatrick, & F. K. S. Leung (Eds.), *Springer international handbooks of education: Vol. 10. Second international handbook of mathematics education* (pp. 717–765). Dordrecht, The Netherlands: Kluwer.

Borko, H., Mayfield, V., Marion, S., Flexer, R., & Cumbo, K. (1997). Teachers' developing ideas and practices about mathematics performance assessment: Successes, stumbling blocks, and implications for professional development. *Teaching and Teacher Education, 13*, 259–278.

Brown, J. S., Collins, A., & Duguid, P. (1989). Situated cognition and the culture of learning. *Educational Researcher, 18*(1), 32–41.

Burkhardt, H., & Schoenfeld, A. H. (2003). Improving educational research: Toward a more useful, more influential, and better-funded enterprise. *Educational Researcher, 32*(9), 3–14.

Carpenter, T. P., Fennema, E., Franke, M. L., Levi, L., & Empson, S. (1999). *Children's mathematics: Cognitively Guided Instruction.* Portsmouth, NH: Heinemann.

Carpenter, T. P., Fennema, E., Peterson, P. L., Chiang, C.-P., & Loef, M. (1989). Using knowledge of children's mathematics thinking in classroom teaching: An experimental study. *American Educational Research Journal, 26*, 499–531.

Chazan, D., & Ball, D. (1999). Beyond being told not to tell. *For the Learning of Mathematics, 19*(2), 2–10.

Clark, C. M., & Peterson, P. L. (1986). Teachers' thought processes. In M. C. Wittrock (Ed.), *Handbook of research on teaching* (3rd ed., pp. 255–296). New York: Macmillan.

Clement, L. L. (1999). The constitution of teachers' orientations toward teaching mathematics. (Doctoral dissertation, University of California, San Diego, and San Diego State University, 1999). *Dissertation Abstracts International, 60*(06), 1949A.

Cobb, P. (1995). Cultural tools and mathematical learning: A case study. *Journal for Research in Mathematics Education, 26*, 362–385.

Cobb, P., Yackel, E., & Wood, T. (1992). A constructivist alternative to the representational view of mind in mathematics education. *Journal for Research in Mathematics Education, 23*, 2–23.

Collier, C. P. (1972). Prospective elementary teachers' intensity and ambivalence of beliefs about mathematics and mathematics instruction. *Journal for Research in Mathematics Education, 3*, 155–163.

Collopy, R. (2003). Curriculum materials as a professional development tool: How a mathematics textbook affected two teachers' learning. *Elementary School Journal, 103*, 287–311.

Cooney, T. J. (1999). Conceptualizing teachers' ways of knowing. *Educational Studies in Mathematics, 38*, 163–187.

Cooney, T. J. (2001). Editorial: Theories, opportunities, and farewell. *Journal of Mathematics Teacher Education, 4*, 255–258.

Cooney, T. J., Shealy, B. E., & Arvold, B. (1998). Conceptualizing belief structures of preservice secondary mathematics teachers. *Journal for Research in Mathematics Education, 29*, 306–333.

Cummings, C. A. (1998). *Teacher attitudes and effective computer integration.* Unpublished master's research paper, University of Virginia, Charlottesville.

D'Ambrosio, B., & Campos, T. M. M. (1992). Pre-service teachers' representations of children's understanding of mathematical concepts: Conflicts and conflict resolution. *Educational Studies in Mathematics, 23*, 213–230.

Darling-Hammond, L., & Sclan, E. M. (1996). Who teaches and why: Dilemmas of building a profession for twenty-first century schools. In J.P. Sikula, T. J. Buttery, & E. Guyton (Eds.), *Handbook of research on teacher education: A project of the Association of Teacher Educators* (2nd ed., pp. 67–101). New York: Macmillan.

DeBellis, V. A., & Goldin, G. A. (1999). Aspects of affect: Mathematical intimacy, mathematical integrity. In O. Zaslavsky (Ed.), *Proceedings of the 23rd conference of the International Group for the Psychology of Mathematics Education* (Vol. 2, pp. 249–256). Haifa, Israel.

Educational Studies in Mathematics. (2002). Reflections on educational studies in mathematics: The rise of research in mathematics education [Editorial]. *Educational Studies in Mathematics, 50*(3), 251–257.

Eisenhart, M., & Towne, L. (2003). Contestation and change in national policy on "scientifically based" education research. *Educational Researcher, 32*(7), 31–38.

Enochs, L. G., Smith, P. L., & Huinker, D. (2000). Establishing factorial validity of the Mathematics Teaching Efficacy Beliefs Instrument. *School Science and Mathematics, 100*, 194–202.

Fennema, E., Carpenter, T. P., Franke, M. L., Levi, L., Jacobs, V. R., & Empson, S. B. (1996). A longitudinal study of learning to use children's thinking in mathematics instruction. *Journal for Research in Mathematics Education, 27*, 403–434.

Fennema, E., Carpenter, T. P., Jacobs, V. R., Franke, M. L., & Levi, L. W. (1998). A longitudinal study of gender differences in young children's mathematical thinking. *Educational Researcher, 27*(5), 4–13.

Fennema, E., Carpenter, T. P., & Loef, M. (1990). *Mathematics Beliefs Scales*. Madison: University of Wisconsin-Madison.

Fennema, E., & Sherman, J. A. (1976). Fennema-Sherman Mathematics Attitude Scales. *JSAS: Catalog of Selected Documents in Psychology, 6*(1). (Ms. No. 1225).

Fleener, M. J. (1995). The relationship between experience and philosophical orientation: A comparison of preservice and practicing teachers' beliefs about calculators. *Journal of Computers in Mathematics and Science Teaching, 14*, 359–376.

Forman, E., & Ansell, E. (2001). The multiple voices of a mathematics classroom community. *Educational Studies in Mathematics, 46*, 115–142.

Furinghetti, F., & Pehkonen, E. (2002). Rethinking characterizations of beliefs. In G. C. Leder, E. Pehkonen, & G. Törner (Eds.), *Beliefs: A hidden variable in mathematics education* (pp. 39–57). Dordrecht, The Netherlands: Kluwer.

Goldin, G. A. (1998). Representational systems, learning, and problem solving in mathematics. *Journal of Mathematical Behavior, 17*, 137–165.

Goldin, G. A. (2002a). Affect, meta-affect, and mathematical belief structures. In G. C. Leder, E. Pehkonen, & G. Törner (Eds.), *Beliefs: A hidden variable in mathematics education* (pp. 59–72). Dordrecht, The Netherlands: Kluwer.

Goldin, G. A. (2002b). Representation in mathematical learning and problem solving. In L. D. English (Ed.), *Handbook of international research in mathematics education* (pp. 197–218). Mahwah, NJ: Erlbaum.

Gómez-Chacón, I. M. (2000). Affective influences in the knowledge of mathematics. *Educational Studies in Mathematics, 43*, 149–168.

Grant, T. J., Hiebert, J., & Wearne, D. (1998). Observing and teaching reform-minded lessons: What do teachers see? *Journal of Mathematics Teacher Education, 1*, 217–236.

Green, T. F. (1971). *The activities of teaching*. New York: McGraw-Hill.

Gregg, J. (1995). The tensions and contradictions of the school mathematics tradition. *Journal for Research in Mathematics Education, 26*, 442–466.

Grouws, D. A. (1992). *Handbook of research on mathematics teaching and learning*. New York: Macmillan.

Gup, T. (2005, September 12). In praise of the 'Wobblies' [Radio broadcast]. In B. Gordemer (Producer/Director), *Morning Edition*. Washington, DC: National Public Radio. Retrieved September 12, 2005, from http://www.npr.org/templates/story/story.php?storyId=4837776

Guskey, T. R. (1986). Staff development and the process of teacher change. *Educational Researcher, 15*(4), 5–12.

Hannula, M. S. (2002). Attitude towards mathematics: Emotions, expectations and values. *Educational Studies in Mathematics, 49*, 25–46.

Harper, N. W., & Daane, C. J. (1998). Causes and reductions of math anxiety in preservice elementary teachers. *Action in Teacher Education, 19*(4), 29–38.

Hart, L. C. (2002). Preservice teachers' beliefs and practice after participating in an integrated content/methods course. *School Science and Mathematics, 102*, 4–14.

Helwig, R., Anderson, L., & Tindal, G. (2001). Influence of elementary student gender on teachers' perceptions of mathematics achievement. *School Science and Mathematics, 95*, 93–102.

Hembree, R. (1990). The nature, effects, and relief of mathematics anxiety. *Journal for Research in Mathematics Education, 21*, 33–46.

Hiebert, J. (1999). Relationships between research and the NCTM Standards. *Journal for Research in Mathematics Education, 30*, 3–19.

Hiebert, J., Stigler, J. W., Jacobs, J. K., Givvin, K. B., Garnier, H., Smith, M., et al. (2005). Mathematics teaching in the United States today (and tomorrow): Results from the TIMSS 1999 video study. *Educational Evaluation and Policy Analysis, 27*, 111–132.

Hoyles, C. (1992). Mathematics teaching and mathematics teachers: A meta-case study. *For the Learning of Mathematics, 12*(3), 32–44.

Integrating Mathematics and Pedagogy. (2003). *IMAP Web-Based Beliefs Survey*. San Diego, CA: Center for Research in Mathematics and Science Education, San Diego State University. Retrieved September 15, 2005, from http://www.sci.sdsu.edu/CRMSE/IMAP/survey/

Integrating Mathematics and Pedagogy. (2004). *IMAP Web-Based Beliefs-Survey manual*. San Diego, CA: Center for Research in Mathematics and Science Education, San Diego State University. Retrieved September 15, 2005, from http://www.sci.sdsu.edu/CRMSE/IMAP/pubs.html

Kazemi, E., & Franke, M. L. (2004). Teacher learning in mathematics: Using student work to promote collective inquiry. *Journal of Mathematics Teacher Education, 7*, 203–235.

Keller, C. (2001). Effect of teachers' stereotyping on students' stereotyping of mathematics as a male domain. *The Journal of Social Psychology, 14*, 165–173.

Knapp, N. F., & Peterson, P. L. (1995). Teachers' interpretations of "CGI" after four years: Meanings and practices. *Journal for Research in Mathematics Education, 26*, 40–65.

Knuth, E. J. (2002). Teachers' conceptions of proof in the context of secondary school mathematics. *Journal of Mathematics Teacher Education, 5*, 61–88.

Lave, J. (1988). *Cognition in practice*. Cambridge, U.K.: Cambridge University Press.

Lave, J., & Wenger, E. (1991). *Situated learning: Legitimate peripheral participation*. New York: Cambridge University Press.

Leedy, G. M., LaLonde, D., & Runk, K. (2003). Gender equity in mathematics: Beliefs of students, parents, and teachers. *School Science and Mathematics, 103*, 285–292.

Lerman, S. (2002). Situating research on mathematics teachers' beliefs and on change. In G. C. Leder, E. Pehkonen, & G. Törner (Eds.), *Beliefs: A hidden variable in mathematics education* (pp. 233–243). Dordrecht, The Netherlands: Kluwer.

Levi, L. (2000). Gender equity in mathematics education. *Teaching Children Mathematics, 7*, 101–105.

Li, Q. (1999). Teachers' beliefs and gender differences in mathematics: A review. *Educational Research, 41*, 63–76.

Lloyd, G. M. (1999). Two teachers' conceptions of a reform-oriented curriculum: Implications for mathematics teacher development. *Journal of Mathematics Teacher Education, 2*, 227–252.

Lloyd, G. M. (2002). Mathematics teachers' beliefs and experiences with innovative curriculum materials: The role of curriculum in teacher development. In G. C. Leder, E. Pehkonen, & G. Törner (Eds.), *Beliefs: A hidden variable in mathematics education* (pp. 149–159). Dordrecht, The Netherlands: Kluwer.

Lloyd, G. M., & Wilson, M. S. (1998). Supporting innovation: The impact of a teacher's conceptions of functions on his implementation of a reform curriculum. *Journal for Research in Mathematics Education, 29*, 248–274.

Ma, X. (1999). A meta-analysis of the relationship between anxiety toward mathematics and achievement in mathematics. *Journal for Research in Mathematics Education, 30*, 520–540.

Ma, X., & Kishor, N. (1997). Assessing the relationship between attitude toward mathematics and achievement in mathematics: A meta-analysis. *Journal for Research in Mathematics Education, 28*, 26–47.

Malmivuori, M.-L. (2001). *The dynamics of affect, cognition, and social environment in the regulation of personal learning processes: The case of mathematics.* Unpublished doctoral dissertation. University of Helsinki, Finland.

Mandler, G. (1989). Affect and learning: Reflections and prospects. In D. B. McLeod & V. M. Adams (Eds.), *Affect and mathematical problem solving* (pp. 237–244). New York: Springer-Verlag.

Mapolelo, D. C. (1998). Pre-service teachers' beliefs about and attitudes toward mathematics: The case of Dudu. *International Journal of Educational Development, 18*, 337–346.

Mason, J. (2002). *Researching your own practice: The discipline of noticing.* London: RoutledgeFalmer.

McGuire, W. J. (1969). The nature of attitudes and attitude change. In G. Lindzey & E. Aronson (Eds.), *The handbook of social psychology* (pp. 136–314). Reading, MA: Addison-Wesley.

McLeod, D. B. (1992). Research on affect in mathematics education: A reconceptualization. In D. A. Grouws (Ed.), *Handbook of research on mathematics teaching and learning* (pp. 575–596). New York: Macmillan.

Mewborn, D. S. (1999). Reflective thinking among preservice elementary mathematics teachers. *Journal for Research in Mathematics Education, 30*, 316–341.

Mullis, I. V. S., Martin, M. O., Fierros, E. G., Goldberg, A. L., & Stemler, S. E. (2000). *Gender differences in achievement: IEA's Third International Mathematics and Science Study (TIMSS).* Chestnut Hill, MA: TIMSS International Study Center, Lynch School of Education, Boston College.

Murrow, E. R. (1951/2005, April 4). The 1951 introduction to *'This I believe.'* [Radio broadcast]. In D. Gedimen & J. Allison (Producers), *This I believe.* Washington, DC: National Public Radio. Originally broadcast in 1951. Retrieved September 12, 2005, from http://www.npr.org/templates/story/story.php?storyId=4566554

Nathan, M. J., & Koedinger, K. R. (2000a). An investigation of teachers' beliefs of students' algebra development. *Cognition and Instruction, 18*, 209–237.

Nathan, M. J., & Koedinger, K. R. (2000b). Teachers' and researchers' beliefs about the development of algebraic reasoning. *Journal for Research in Mathematics Education, 31*, 168–190.

Nathan, M. J., & Petrosino, A. (2003). Expert blind spot among preservice teachers. *American Educational Research Journal, 40*, 905–928.

National Council of Teachers of Mathematics. (1989). *Curriculum and evaluation standards for school mathematics.* Reston, VA: Author.

National Council of Teachers of Mathematics. (1991). *Professional standards for teaching mathematics.* Reston, VA: Author.

National Council of Teachers of Mathematics. (1995). *Assessment standards for school mathematics.* Reston, VA: Author.

National Council of Teachers of Mathematics. (2000). *Principles and standards for school mathematics.* Reston, VA: Author.

National Research Council. (2001). *Adding it up: Helping children learn mathematics.* Washington, DC: National Academy Press.

Nisbett, R., & Ross, L. (1980). *Human inference: Strategies and shortcomings of social judgment.* Englewood Cliffs, NJ: Prentice-Hall.

Pajares, M. F. (1992). Teachers' beliefs and educational research: Cleaning up a messy construct. *Review of Educational Research, 62*, 307–332.

Peterson, P. L., Fennema, E., Carpenter, T. P., & Loef, M. (1989). Teachers' pedagogical content beliefs in mathematics. *Cognition and Instruction, 6*, 1–40.

Philipp, R. A., Ambrose, R., Lamb, L. C., Sowder, J. T., Schappelle, B. P., Sowder, L., et al. (in press). Effects of early field experiences on the mathematical content knowledge and beliefs of prospective elementary school teachers: An experimental study. *Journal for Research in Mathematics Education.*

Philipp, R. A., Thanheiser, E., & Clement, L. L. (2002). The role of a children's mathematical thinking experience in the preparation of prospective elementary school teachers. *International Journal of Educational Reform, 37*, 195–210.

Philippou, G. N., & Christou, C. (1998). The effects of a preparatory mathematics program in changing prospective teachers' attitudes toward mathematics. *Educational Studies in Mathematics, 35*, 189–206.

Raymond, A. M. (1997). Inconsistency between a beginning elementary school teacher's mathematics beliefs and teaching practice. *Journal for Research in Mathematics Education, 28*, 550–576.

Remillard, J. T. (2000). Can curriculum materials support teachers' learning? Two fourth-grade teachers' use of a new mathematics text. *Elementary School Journal, 100*(4), 331–350.

Remillard, J. T., & Bryans, M. B. (2004). Teachers' orientations toward mathematics curriculum materials: Implications for teacher education. *Journal for Research in Mathematics Education, 35*, 352–388.

Richardson, F. C., & Suinn, R. M. (1972). The Mathematics Anxiety Rating Scale: Psychometric data. *Journal of Counseling Psychology, 19,* 551–554.

Richardson, V. (1996). The role of attitudes and beliefs in learning to teach. In J. P. Sikula, T. J. Buttery, & E. Guyton (Eds.), *Handbook of research on teacher education: A project of the Association of Teacher Educators* (2nd ed., pp. 102–119). New York: Macmillan.

Rogoff, B. (1997). Evaluating development in the process of participation: Theory, methods, and practice building on each other. In E. Amsel & K. A. Renninger (Eds.), *Change and development: Issues of theory, method, and application* (pp. 265–285). Mahwah, NJ: Erlbaum.

Rokeach, M. (1968). *Beliefs, attitudes, and values: A theory of organization and change.* San Francisco: Jossey-Bass.

Romagnano, L. (1994). *Wrestling with change: The dilemmas of teaching real mathematics.* Portsmouth, NH: Heinemann.

Rosenthal, R. (2002). Covert communications in classrooms, clinics, courtrooms, and cubicles. *American Psychologist, 57,* 838–849.

Ross, J. A., McDougall, D., & Hogaboam-Gray, A. (2002). Research on reform in mathematics education, 1993–2000. *Alberta Journal of Educational Research, 48,* 122–138.

Ross, J. A., McDougall, D., Hogaboam-Gray, A., & LeSage, A. (2003). A survey measuring elementary teachers' implementation of standards-based mathematics teaching. *Journal for Research in Mathematics Education, 34,* 344–363.

Schmidt, M. E. (1998). Research on middle grade teachers' beliefs about calculators. *Action in Teacher Education, 20*(2), 11–23.

Schmidt, M. E. (1999). Middle grade teachers' beliefs about calculator use: Pre-project and two years later. *FOCUS—On Learning Problems in Mathematics, 21*(Winter), 18–34.

Schoenfeld, A. H. (1988). When good teaching leads to bad results: The disasters of "well taught" mathematics courses. *Educational Psychologist, 23,* 145–166.

Schoenfeld, A. H. (1992). Learning to think mathematically: Problem solving, metacognition, and sense making in mathematics. In D. A. Grouws (Ed.), *Handbook of research on mathematics teaching and learning* (pp. 334–370). New York: Macmillan.

Schön, D. A. (1983). *The reflective practitioner: How professionals think in action.* New York: Basic Books.

Sfard, A., & Prusak, A. (2005). Telling identities: In search of an analytic tool for investigating learning as a culturally shaped activity. *Educational Researcher, 34*(4), 14–22.

Shulman, L. S. (1986). Those who understand: Knowledge growth in teaching. *Educational Researcher, 15*(2), 4–14.

Silver, E. A., & Kilpatrick, J. (1994). E pluribus unum: Challenges of diversity in the future of mathematics education research. *Journal for Research in Mathematics Education, 25,* 734–754.

Silver, E. A., & Stein, M. K. (1996). The QUASAR project: The "revolution of the possible" in mathematics instructional reform in urban middle schools. *Urban Education, 30,* 476–521.

Simon, M. A. (2004). Raising issues of quality in mathematics education research. *Journal for Research in Mathematics Education, 35,* 157–163.

Simon, M. A., & Tzur, R. (1999). Explicating the teacher's perspective from the researchers' perspectives: Generating accounts of mathematics teachers' practice. *Journal for Research in Mathematics Education, 30,* 252–264.

Simon, M. A., Tzur, R., Heinz, K., Kinzel, M., & Smith, M. S. (2000). Characterizing a perspective underlying the practice of mathematics teachers in transition. *Journal for Research in Mathematics Education, 31,* 579–601.

Skemp, R. (1978). Relational understanding and instrumental understanding. *Arithmetic Teacher, 26*(3), 9–15.

Skott, J. (2001). The emerging practices of a novice teacher: The roles of his school mathematics images. *Journal of Mathematics Teacher Education, 4,* 3–28.

Smith, J. P. (1996). Efficacy and teaching mathematics by telling: A challenge for reform. *Journal for Research in Mathematics Education, 27,* 387–402.

Sowder, J. T. (1998). Perspectives from mathematics education. *Educational Researcher, 27*(5), 12–13.

Spielman, L. J., & Lloyd, G. M. (2004). The impact of enacted mathematics curriculum models on prospective elementary teachers' course perceptions and beliefs. *School Science and Mathematics, 104*(1), 32–42.

Staub, F. C., & Stern, E. (2002). The nature of teachers' pedagogical content beliefs matters for students' achievement gains: Quasi-experimental evidence from elementary mathematics. *Journal of Educational Psychology, 94,* 344–355.

Steele, D. F. (2001). The interfacing of preservice and in-service experiences of reform-based teaching: A longitudinal study. *Journal of Mathematics Teacher Education, 4,* 139–172.

Stein, M. K., & Brown, C. A. (1997). Teacher learning in a social context: Integrating collaborative and institutional processes with the study of teacher change. In E. Fennema & B. S. Nelson (Eds.), *Mathematics teachers in transition* (pp. 155–192). Mahwah, NJ: Erlbaum.

Stevenson, H., & Stigler, J. W. (1992). *The learning gap.* New York: Summit Books.

Stigler, J. W., & Hiebert, J. (1999). *The teaching gap.* New York: The Free Press.

Sztajn, P. (2003). Adapting reform ideas in different mathematics classrooms: Beliefs beyond mathematics. *Journal of Mathematics Teacher Education, 6,* 53–75.

Tharp, M. L., Fitzsimmons, J. A., & Ayers, R. L. B. (1997). Negotiating a technological shift: Teacher perception of the implementation of graphing calculators. *The Journal of Computers in Mathematics and Science, 16,* 551–575.

Thompson, A. G. (1984). The relationship of teachers' conceptions of mathematics teaching to instructional practice. *Educational Studies in Mathematics, 15,* 105–127.

Thompson, A. G. (1992). Teachers' beliefs and conceptions: A synthesis of the research. In D. A. Grouws (Ed.), *Handbook of research on mathematics teaching and learning* (pp. 127–146). New York: Macmillan.

Thompson, A. G., Philipp, R. A., Thompson, P. W., & Boyd, B. A. (1994). Calculational and conceptual orientations in teaching mathematics. In D. B. Aichele & A. F. Coxford (Eds.), *Professional development for teachers of mathematics* (pp. 79–92). Reston, VA: National Council of Teachers of Mathematics.

Thompson, A. G., & Thompson, P. W. (1996). Talking about rates conceptually, Part II: Mathematical knowledge for teaching. *Journal for Research in Mathematics Education, 27,* 2–24.

Thompson, P., & Cobb, P. (1998). On relationships between psychological and sociocultural perspectives. In S. Berenson, K. Dawkins, M. Blanton, W. Coulombe, J. Kolb, K. Norwood, et al. (Eds.), *Psychology of Mathematics Education*

(Vol. 1, pp. 3–26). North Carolina State University: ERIC Clearinghouse for Science.

Thompson, P. W., & Thompson, A. G. (1994). Talking about rates conceptually, Part I: A teacher's struggle. *Journal for Research in Mathematics Education, 25*, 279–303.

Tiedemann, J. (2000). Gender-related beliefs of teachers in elementary school mathematics. *Educational Studies in Mathematics, 41*, 191–207.

Tirosh, D. (2000). Enhancing prospective teachers' knowledge of children's conceptions: The case of division of fractions. *Journal for Research in Mathematics Education, 31*, 5–25.

Tschannen-Moran, M., Hoy, A. W., & Hoy, W. K. (1998). Teacher efficacy: Its meaning and measure. *Review of Educational Research, 68*, 202–248.

Tzur, R., Simon, M. A., Heinz, K., & Kinzel, M. (2001). An account of a teacher's perspective on learning and teaching mathematics: Implications for teacher development. *Journal of Mathematics Teacher Education, 4*, 227–254.

Vacc, N. N., & Bright, G. W. (1999). Elementary preservice teachers' changing beliefs and instructional use of children's mathematical thinking. *Journal for Research in Mathematics Education, 30*, 89–110.

Van Zoest, L. R., & Bohl, J. V. (2005). Mathematics teacher identity: A framework for understanding secondary school mathematics teachers' learning through practice. *Teacher Development: An International Journal of Teachers' Professional Development, 9*, 315–345.

von Glasersfeld, E. (1984). An introduction to radical constructivism. In P. Watzlawick (Ed.), *The invented reality* (pp. 17–40). New York: W. W. Norton.

von Glasersfeld, E. (1993). Questions and answers about radical constructivism. In K. Tobin (Ed.), *Constructivism: The practice of constructivism in science education* (pp. 23–38). Washington, DC: American Association for the Advancement of Science.

Walen, S. B., & Williams, S. R. (2002). A matter of time: Emotional responses to timed mathematics tests. *Educational Studies in Mathematics, 49*, 361–378.

Walen, S. B., Williams, S. R., & Garner, B. E. (2003). Pre-service teachers learning mathematics using calculators: A failure to connect current and future practice. *Teaching and Teacher Education, 19*, 445–462.

Wenger, E. (1998). *Communities of practice: Learning, meaning, and identity.* Cambridge, U.K.: Cambridge University Press.

Wiegel, H. G., & Bell, K. (1996). *Pre-service elementary teachers' affective responses to computer activities in mathematics content courses* (Unpublished report). Athens: The University of Georgia.

Willis, J. W., & Mehlinger, H. D. (1996). Information technology and teacher education. In J.P. Sikula, T. J. Buttery, & E. Guyton (Eds.), *Handbook of research on teacher education: A project of the Association of Teacher Educators* (2nd ed., pp. 978–1029). New York: Macmillan.

Wilson, M. S., & Cooney, T. (2002). Mathematics teacher change and development. In G. C. Leder, E. Pehkonen, & G. Törner (Eds.), *Beliefs: A hidden variable in mathematics education* (pp. 127–147). Dordrecht, The Netherlands: Kluwer.

Wilson, S. M., & Berne, J. (1999). Teacher learning and the acquisition of professional knowledge: An examination of research on contemporary professional development. In A. Iran-Nejad & P. D. Pearson (Eds.), *Review of Research in Education, 24*, 173–209.

Zollman, A., & Mason, E. (1992). The Standards' Beliefs instrument (SBI): Teachers' beliefs about the NCTM standards. *School Science and Mathematics, 92*, 359–364.

AUTHOR NOTE

I would like to thank Vicki Jacobs, Lisa Clement Lamb, Gilah Leder, Doug McLeod, and Bonnie Schappelle for their insightful comments and constant encouragement, and I would like to additionally recognize Bonnie Schappelle for her editorial expertise.

Influences on Student Outcomes

.ıI 8 ꭦ

HOW CURRICULUM INFLUENCES STUDENT LEARNING

Mary Kay Stein

UNIVERSITY OF PITTSBURGH

Janine Remillard

UNIVERSITY OF PENNSYLVANIA

Margaret S. Smith

UNIVERSITY OF PITTSBURGH

The purpose of this chapter is to review research on how mathematics curriculum[1] influences student learning. Interestingly, there was not a similarly named chapter in the 1992 *Handbook of Research on Mathematics Teaching and Learning* (Grouws, 1992), suggesting that the mathematics curricula of the '70s and '80s and their relationship to student learning were not viewed as a significant object of scholarly investigation at the time. Most likely, this stance reflected the era more broadly, a time during which arithmetic—primarily computation—comprised the lion's share of work in the elementary years, and algebra—principally procedures for manipulating symbolic expressions—formed the majority of high school work in mathematics. In this milieu, textbooks were viewed primarily as a resource for problem sets:

"Very few had references to mathematical concepts or principles and virtually none had problems from everyday life or from other fields" (Senk & Thompson, 2003 p. 9). The absence of a chapter on curriculum in the 1992 *Handbook* may also reflect the period in time—a time during which reformers focused on helping teachers create lessons based on activities found *outside* of commercial textbooks.[2] This tendency may have reflected a general disillusionment among mathematics educators with the textbooks available at the time most of which were developed by publishing companies and did not draw on research on how students learn mathematics (Ball & Feiman-Nemser, 1988).

Current interest in how curriculum materials influence student learning has been fomented by sev-

---

[1] As described later, we use the term curriculum broadly to include mathematics curriculum materials and textbooks, curriculum goals as intended by teachers, and the curriculum that is enacted in the classroom.

[2] Examples include Cognitively Guided Instruction (Carpenter, Fennema, Peterson, Chiang, & Loef, 1989), replacement units, and manipulative-based programs.

eral policy shifts in mathematics education and educational research and practice more generally. First, the field has seen a sharp rise in curriculum development activity since the writing of the 1992 Handbook chapter—much of it in response to publication of the *Curriculum and Evaluation Standards for School Mathematics* (National Council of Teachers of Mathematics, 1989). As states adopted standards that reflected the NCTM's vision, the publishing industry moved quickly to make adaptations to their textbooks. At the same time, the National Science Foundation (NSF) funded extensive curriculum development projects that produced an entirely new set of curriculum materials that entered the market in the mid- to late-nineties (referred to as "standards-based curricula"). This activity dwarfed even the explosion of curriculum development that occurred during the new math era of the late '50s and '60s (Usiskin, 1997).

These materials have generated particular interest of researchers because they embody an approach to mathematics teaching and learning that is qualitatively different from textbooks or instructional resources previously available. The standards-based curricula had a common set of design specifications that included alignment with the *Curriculum and Evaluations Standards for School Mathematics* (NCTM, 1989), heavy use of non-numeric representations (e.g., diagrams, manipulatives), an expanded content base (e.g., in elementary curricula, topics that go beyond traditional arithmetic to include statistics and graphing, geometry, and pre-cursors to algebraic reasoning), and extensive use of calculators. Moreover, in order to meet new goals for student learning (i.e., mathematical thinking, reasoning, problem solving, connecting, communicating, seeking evidence, and constructing arguments to make predictions and support conclusions), these new curricula de-emphasized paper-and-pencil skills and focused on students' active construction of and communication about solutions to challenging problems. A subset of these curricula also incorporated technologies (e.g., intelligent tutors) that were not previously available.

The kind of instructional experience supported by the standards-based curricula represented a substantial departure from conventional practice in the U.S. As summarized by Fey (1979) from three studies commissioned by the National Science Foundation (Stake & Easley, 1978; Suydam & Osborne, 1977; Weiss, 1978) and corroborated by the recent TIMSS study (NCES, 2003; Stigler & Hiebert, 2004), the majority of mathematics lessons in the U.S. involved extensive, teacher-directed explanation of new material followed by student seatwork on paper-and-pencil assignments with little or no discussion or exploration of concepts. As

such, reformers cast these new curricula in the role of change agent with the hope that they would help teachers to transform their instruction from a focus on basic skills to conceptually based problems that require thinking, reasoning, problem solving and communication (Senk & Thompson, 2003). To others, however, these new curricula represented a swing of the pendulum too far toward an emphasis on the processes of doing mathematics to the detriment of time spent developing computational efficiency (Wu, 1997).

The standards-based era of curriculum development has been associated with increased vigor in the mathematics education community and an upsurge in research activity. Because the new curriculum materials and the standards documents on which they were based, offered a radically different vision of what it means to learn mathematics and consequently how it should be taught in the classroom, they prompted a substantial amount of research on the role of teachers in the standards-based classroom, as well as research on the relationship between teachers' beliefs and knowledge and their instructional practice. All of these studies are critical to the central argument of this chapter and our response to its driving question—How do curriculum materials influence student learning? As we detail below, the influence of curriculum materials is mediated by teachers and students interacting with those materials in classroom contexts.

A second policy shift, occurring in the late 1990s and early 2000s, catalyzed another research emphasis that raised different kinds of questions about the influence of curriculum materials on student learning: "Do these new materials work?" Because the passage of the *No Child Left Behind Act* restricted the use of federal monies to those programs backed by scientific evidence of student learning (NCLB, 2002), many curriculum developers became eager to "prove" the effectiveness of their materials. In addition, harsh criticism of education research and educational practices as not scientifically based led to calls for research on the effectiveness of educational programs in general, including innovative curricula such as the NSF-sponsored materials (Mosteller & Boruch, 2002; NRC, 2002; 2004). Finally, in response to perceived needs of practitioners to distinguish effective from non-effective programs, the US Department of Education established the *What Works Clearinghouse*. This Internet-based website features the results of research, that qualifies as scientifically based, on the outcomes of educational programs and curriculum materials, as measured by gains in student achievement. This climate of accountability led to a substantial number of studies that aimed to discern what students who were exposed to different kinds of curriculum materials learned. In this chapter, we review both these effectiveness studies and the broader field of research on how teachers and students use curricula.

CONCEPTUAL ISSUES, DEFINITIONS, AND BOUNDARIES

The term *curriculum* has different meanings in different contexts. Therefore, we begin with a brief discussion of the term and its varied applications in order to clarify how we are using it in this chapter. In so doing, we present the framework that structures our discussion. We also discuss how the construct of *curriculum materials* or *textbooks* has evolved and shifted in educational discourse and, particularly in mathematics education. Finally, we specify the boundaries of our review.

Multiple Meanings of Curriculum

Very broadly, curriculum refers to the substance or content of teaching and learning—the "what" of teaching and learning (as distinguished from the "how" of teaching). Those who study curriculum frequently examine planned and unplanned components of what is taught or experienced in classrooms (Jackson, 1992). However, among educational decision makers, the term curriculum is often used to refer to expectations for instruction laid out in policy documents or frameworks. Currently, and in direct response to the recent flurry of curriculum material development, many mathematics education researchers and practitioners use the term *curriculum* to refer to the material resources designed to be used by teachers in the classroom, such as "standards-based curricula."

Research on teaching and curriculum, however, has revealed that a substantial difference exists between the curriculum as represented in instructional materials and the curriculum as enacted in the classroom by teachers and students. Curriculum theorists use a number of terms to distinguish between the curriculum outlined in a guide or set of materials and that enacted in the classroom. *Formal* (Doyle, 1992) or planned curriculum (Gehrke, Knapp, & Sirotnik, 1992), for example, refer to the goals and activities outlined by school policies or designed in textbooks or by teachers. The objectives set out in curriculum frameworks or state standards as well as those specified in scope and sequence charts in textbooks also represent formal curricula, sometimes referred to as

the *institutional* or *intended* curriculum. The *enacted* curriculum refers to what actually takes place in the classroom (Gehrke et al., 1992). In order to identify the impact that the enacted curriculum has on students, researchers use terms such as the *experienced* (Gehrke et al., 1992) or *attained* (Valverde, Bianchi, Wolfe, Schmidt, & Houang, 2002) curriculum. It is worth noting that distinctions are infrequently made between the curriculum as outlined by policy makers or curriculum designers and the curriculum interpreted or intended by the teacher.

In our effort to examine the influence that curriculum materials have on student learning, we have found all these meanings of curriculum to be significant and interrelated, yet uniquely important. In short, the influence of curriculum materials on student learning is not straightforward and cannot be understood without examining the curriculum as designed by teachers and as enacted in the classroom. Drawing on our earlier work (Remillard,1999; Stein, Grover, & Henningsen, 1996), we have conceptualized these various meanings of curriculum as unfolding in a series of temporal phases from the printed page (*the written curriculum*), to the teachers' plans for instruction (*the intended curriculum*),[3] to the actual implementation of curricular-based tasks in the classroom (the *enacted* curriculum). (See Figure 8.1.)

Within and between certain of these phases, interpretative and interactive processes transform the curriculum. *Between* the written and intended phases, teachers bring their prior understandings, beliefs, and goals to bear on the written curriculum and, in the process, transform it into a form that they believe will be workable in the classroom. *Within* the enactment phase, the teacher and the students, in interaction with each other, bring the curriculum to life and, in the process, create something different than what could exist on the pages of the book or in the teacher's mind or lesson plan. While all of the phases have a bearing on student learning (the final triangle), the classroom activities that occur during the enactment phase most directly influence how students experience mathematics and what they learn (Carpenter & Fenemma, 1988; Wittrock, 1986). We have added return arrows from the enacted curriculum and student learning to the transformation processes between the written and intended curriculum to indicate that the enacted curriculum and teachers influence teachers' future interactions with written curriculum.

The oval in Figure 8.1 identifies possible factors that mediate the interpretive and interactive processes

---

[3] Our use of the term, "intended" differs from TIMSS' use of the term. TIMSS used "intended" to refer to what we call the "written" curriculum.

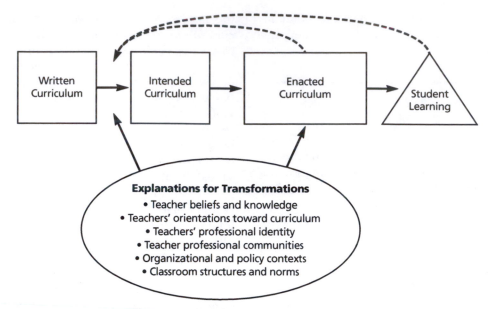

Figure 8.1 Temporal phases of curriculum use.

that occur within and between phases. A variety of studies, some specifically focused on curriculum use, others on the implementation of reform in general, have identified one or more of these factors as influencing how curriculum use unfolds. A number of researchers, for example, have explored the manner in which individual teacher characteristics and capacities, such as subject matter knowledge (Brophy, 1991, 2001; Stein, Baxter, & Leinhardt, 1990), beliefs (Cohen, 1990; Jennings, 1996; Putnam, 1992; Remillard, 1992; Spillane, 1999; Spillane & Jennings, 1997), and professional identity (Remillard & Bryans, 2004) influence how teachers understand and enact reform. Increasingly, studies—inside and outside of mathematics reform—have begun to suggest explanations for transformations that are external to the individual teacher. For example, research suggests that strong professional communities can help teachers to learn complex new ways of teaching (Cobb, McClain, de Silva Lamberg & Dean, 2003; Little & McLaughlin, 1993; Louis, Marks, & Kruse, 1996; Stein, Silver, & Smith, 1998) whereas teachers working in isolation are apt to continue to practice in the manner they always have. In addition, implementation of new programs and practices at the school level is influenced by organizational and policy contexts such as, the level of support from the principal (Berends, Kirby, Naftel, & McKelvey, 2001; Bodilly, 1998; Datnow, Hubbard, & Mehan, 2002; Fullan, 1991; Kirby, Behrends, & Naftel, 2001). Finally, factors central to the social organization of the school and classroom, including students, classroom structures, and norms (Doyle, 1983) can influence the curriculum transformation process.

The framework depicted in Figure 8.1 provides an organizing structure for the studies reviewed in this chapter. In *Section One,* we review research that has addressed the impact of curriculum materials on student learning. Many scholars and policymakers have (implicitly) conceptualized these studies as testing a (causal) relationship between the written curriculum (the first box in Figure 8.1) and student learning (the final triangle in Figure 8.1). The driving equation is, "Does curriculum X work?" with an emphasis on the written materials (not their enactment) as the intervention and a focus on student outcomes (usually achievement scores) as the indicator of what it means to work. After reviewing the findings of this research, we highlight the pros and cons of viewing the relationship between curriculum and student learning in this way; we then turn to the mediating relationships depicted in the figure to identify *how* curriculum influences student learning.

The studies reviewed in the next two sections begin to unpack how curriculum influences student learning by focusing on the transformations written curricula undergo as teachers engage with and use them in the classroom. In *Section Two* we review research on how teachers interpret curriculum materials which can be viewed as a discussion of the relationship between the written curriculum and the intended curriculum (the first and second boxes in Figure 8.1). In *Section Three* we review studies that have focused on what happens when curricular tasks are unleashed inside the classrooms; as such, this section focuses on studies falling within the box labeled enacted curriculum. The research reviewed in *Section Four* deals with studies that aim to identify reasons for the variations uncovered in Sections Two and Three (i.e., why do teachers interpret and enact curriculum in different ways?) and thus can be seen as a discussion of the explanations that appear in the oval. Finally, in *Section Five* we review research on the relationship between the enacted curriculum and student learning (the final box and the final triangle of Figure 8.1). As such, Section Five also examines the question, "Does curriculum X work?" but this time the intervention is broadly conceived to include how teachers and students engage with the curriculum resource as well as the curriculum itself.

When considered together, the research conducted on the phases of curriculum use paints a complex picture of how curriculum influences student learning. More specifically, it points to the fallacy of assuming that the materials themselves are the primary agent in shaping opportunities for student learning and instead uncovers the important role played by the interpretive and interactive influences of teachers and students. Because they often represent the starting point for instruction, the materials are important and their influence merits substantial investigation; thus, they remain a focal point of this chapter. However, tracing *how* they are important necessarily takes one into the realm of teachers, students, and classrooms. Consequently, the reader will find overlaps between the research examined in this chapter and the chapter by Hiebert and Grouws on classroom teaching.

Curriculum Materials: An Evolving Construct

Because curriculum materials are a focal point of this chapter, we begin with a brief examination of what mathematics textbooks or curriculum materials are and how they and the way they are perceived have changed throughout the history of U.S. public education. Over the years texts and curriculum materials have varied in focus, style, philosophy and degree of comprehensiveness and have taken on different roles in the minds of teachers and administrators. While we use the terms *curriculum materials* and *textbook* (along with instructional resources and guides) somewhat interchangeably to refer to printed or electronic, often published, materials designed for use by teachers and students before, during, and after mathematics instruction, many teachers and mathematics educators draw sharp delineations between the two.

During the latter part of the 20th century, mathematics *textbooks* common to classrooms in the U.S.

were viewed as a source of explanations and exercises for students to complete. The textbook played a central role in most mathematics classrooms and instruction was aimed at teaching students what they needed to know in order to complete the exercises and answer the questions in their books (Jackson, 1968).

Unlike the term mathematics textbook, the term mathematics *curriculum materials* often refers to instructional guides that place substantial emphasis on both pedagogy (the how of teaching) and mathematics (the what of teaching). To many, the term curriculum materials was used to connote something akin to an "anti-textbook" because these resources offered programs of instruction that rejected the notion that learning mathematics involved completing decontextualized exercises in a book. In contrast to textbooks, which were developed and marketed by commercial publishing companies, curriculum materials tended to be designed by mathematics experts and mathematics education researchers and, prior to the late 1990s, were sold independently to a fairly small market. For most standards-based curriculum materials, students' work during instruction involves investigative projects instead of exercises found on the pages of a "student textbook." Student textbooks are replaced by thin, often consumable, student workbooks that are designed to support students' investigative work. The centerpiece of most lessons is the thinking that is required to grapple with the investigative task; student work books are designed to support that thinking by providing a basis for recording, summarizing or reflecting on one's actions and thinking.

Curriculum material development played a central role in the "modern mathematics" period of the late 1950s and early '60s. During this time, numerous scientists and mathematicians were solicited to design instructional materials that would prepare the next generation of citizens to be scientifically competitive. The NSF played a major role in funding the development of the majority of these programs, often referred to as the "New Math."[4] The curriculum products of these content experts sought to place equal emphasis on "computational skill" and "understanding the basic concepts of mathematics and of their interrelationships, i.e., the structures of mathematics" (Begle, 1970, p. 1). Like texts of the past, these materials were designed with students (specifically talented students), not teachers, in mind. The consideration these developers gave to the teacher's role is less clear, but it is understood that they grossly "overestimated

---

[4] Specifically, the term New Math refers to the curriculum materials developed in the 1960s by the School Mathematics Study Group (SMSG). It is often used to refer to all reform-oriented mathematics materials produced during the period of modern mathematics reform.

the materials' independent educative power" and underestimated the influence of the teacher in their use (Cohen & Barnes, 1993, p. 215). As a result, the curriculum materials developed during the modern math period are frequently labeled as attempts to be "teacher proof."

Like the New Math materials, curriculum materials developed during the period of standards-based reform initiated in the early 1990's are also seen as a potential means for shifting instructional practices. Unlike the New Math, however, the role of the teacher is generally explicitly acknowledged in these materials; most include detailed pedagogical guidance for teachers and some include resources to support teacher learning. For example, *Investigations in Data, Number and Space*, an NSF-supported elementary curriculum, includes "Notes to Teachers" that explain the mathematical significance of the ideas that are featured in particular lessons and how students might think about those ideas and "Dialogue Boxes" that illustrate how a student discussion surrounding those ideas might unfold in the classroom. As such, these materials can be viewed as being written for teachers and their learning, rather than being written for direct consumption by students. Remillard (2000) refers to this approach to curriculum design as curricula speaking *to*, as opposed to speaking *through*, the teacher.

The expectation by NSF that curriculum developers distribute their materials through established commercial publishers has had far-reaching influences. Most standards-based materials have experienced unprecedented visibility and success in the commercial markets. At the same time, the final products reflect compromises reached between developers and publishers. And, ironically, curriculum materials developed independently are marketed and sold along side, and sometimes as extensions of, instructional materials and textbooks developed "in house." Thus, another result of the way standards-based curriculum materials have been marketed may be a blurring of distinct lines that—at one time—clearly differentiated them from conventional texts.

The role of curriculum materials has also been conceptualized differently—even within the same period of time. Some individuals view curriculum materials as a direct blueprint for instruction—a plan to be unerringly implemented. According to this view, material that appears on the written page of the curriculum is meant for faithful execution by teachers and students. Others view curriculum materials as a *resource* to be drawn upon by teachers as they construct lessons and units. This view suggests that complete fidelity of implementation is impossible because teachers will always bring their own frames of understanding and their knowledge of the local context to bear on how they use curricular materials. Inevitably, this view argues, teachers and students construct their own unique version of the curriculum—a version that necessarily springs from their goals, needs, and understandings (Remillard, 2005).

Literature Selection and Boundaries of this Review

In preparing this chapter, we reviewed research on the effects of mathematics curriculum materials or textbooks, both standards-based and conventional, on students and teachers available in peer-reviewed journals in addition to other sources such as research reports and edited volumes. A substantial portion of the research reviewed in this chapter was spurred by the wave of curriculum development supported by NSF and initiated after the publication of the NCTM *Standards* (1989). As noted earlier, the introduction of these curriculum materials to the public catalyzed substantial research activity—activity that was not matched by studies of conventional textbooks. The marked differences in the number of studies conducted on NSF-supported in contrast to conventional textbooks can be attributed to at least three reasons: a) NSF-supported curriculum developers were required by the funder to conduct evaluations and were sometimes provided with the funds to do so; b) the majority of NSF-supported curriculum developers were university faculty or researchers whose graduate students and colleagues conducted studies on the use of the new materials; and c) the research conducted by commercial publishing companies tended to be market driven, focusing on purchasing decisions rather than on measures of student outcomes or teacher use (NRC, 2004, pp. 25–28).

In an effort to orient the reader to the range of curriculum materials available in the U.S. during the time this chapter was written, we have identified the most well-known curriculum materials in Table 8.1. The first column identifies the full name of the text; the second column identifies the grade band for which the materials were developed. In the third column we identify the funder (if any and if known).

In the final column we indicate the category or group to which we have assigned the curriculum for

Table 8.1 Mathematics Curricula Commonly Used in U.S. Public Schools

Curriculum	Level	Funder (if any and if known)	Standards-based or Conventional
*Everyday Mathematics; SRA/McGraw Hill	Elementary	NSF	Standards-based
*Investigations in Number, Data, and Space; Scott Foresman	Elementary	NSF	Standards-based
*Math Advantage K–6; Harcourt Brace	Elementary		Conventional
Math in My World; McGraw Hill	Elementary		Conventional
*Math K–5; Scott Foresman/Addison Wesley	Elementary		Conventional
Math Land	Elementary		Standards-based
*Math Trailblazers; Kendall/Hunt	Elementary	NSF	Standards-based
Number Power (supplemental)	Elementary	NSF	Standards-based
Opening Eyes to Mathematics; Visual Mathematics	Elementary		Standards-based
Saxon Math	Elementary		Conventional
Silver Burdett Ginn Math	Elementary		Conventional
SRA Math; McGraw Hill	Elementary		Conventional
*Connected Mathematics Project (CMP); Prentice Hall	Middle School	NSF	Standards-based
Heath Mathematics Connections; DC Heath & Co.	Middle School		Conventional
Math Advantage; Harcourt Brace	Middle School		Conventional
Math Alive; Visual Mathematics	Middle School		Standards-based
*Mathematics: Applications & Connections; Glencoe/McGraw Hill	Middle School		Conventional
*Mathematics in Context (MiC); Holt, Rinehart, & Winston	Middle School	NSF	Standards-based
Mathematics Plus; Harcourt Brace & Co.	Middle School		Conventional
*MathScape: Seeing and Thinking Mathematically; Glencoe/McGraw Hill	Middle School	NSF	Standards-based
*MathThematics (STEM); McDougal Littell	Middle School	NSF	Standards-based
Middle Grades Math; Prentice Hall	Middle School		Conventional
Middle School Math; Scott Foresman/Adison Wesley	Middle School		Conventional
*Middle School Mathematics Through Applications Project (MMAP; unpublished)	Middle School	NSF	Standards-based
Passport Series; McDougal-Littell	Middle School		Conventional
Pre-Algebra: An Integrated Transition to Algebra and Geometry; Glencoe/McGraw Hill	Middle School		
*Saxon Math	Middle School		Conventional
Cognitive Tutor	Middle/High School		Standards-based
College Preparatory Mathematics (CPM)	High School		Standards-based
*Contemporary Mathematics in Context (Core-Plus)	High School	NSF	Standards-based
*Interactive Mathematics Program (IMP); Key Curriculum Press	High School	NSF	Standards-based
*Larson Series, Grades 9–12; Houghton Mifflin/McDougal Littell	High School		Conventional
*MATH Connections: A Secondary Mathematics Core Curriculum Grades 9–12; IT'S ABOUT TIME, Inc.	High School	NSF?	Standards-based
*Mathematics: Modeling Our World (MMOW/ARISE); W.H. Freeman & Co.	9–12	NSF	Standards-based
*Systemic Initiative for Montana Mathematics and Science (SIMMS) Integrated Mathematics; Kendall/Hunt	High School	NSF	Standards-based
*University of Chicago School Mathematics Project Integrated Mathematics, Grades 7–12; Prentice Hall	Middle/High School	NSF, Amoco Foundation, Carnegie Corp. of New York, General Electric Foundation	Standards-based

*= Curriculum included in NRC panel report, On Evaluating Curricular Effectiveness.

the purposes of this chapter. Curriculum materials are identified as either "standards-based" or "conventional." *Standards-based curriculum materials* include the NSF-funded materials, curriculum materials that were inspired by documents that *preceded* the 1989 *Curriculum and Evaluation Standards* (e.g., AAAS's *Benchmarks for Scientific Literacy*; and NRC's *Everybody Counts*, and the draft version of the *Curriculum and Evaluation Standards for School Mathematics* that circulated in the mid-eighties), and, finally, curriculum materials that were driven by the ideas in the NCTM *Standards*, but not funded by the National Science Foundation (e.g., the *Cognitive Tutor*). *Conventional curriculum materials* include commercially developed textbooks with earlier editions that had been on the market prior to the release of the *Curriculum and Evaluation Standards*, editions that were not influenced by the earlier reform documents mentioned above. It should be noted that many of these publishers attempted to align the content of subsequent editions with the NCTM Standards after they appeared in 1989; however, this alignment was primarily accomplished through backward-mapping as opposed to using the *Standards* as a design template from the start, thereby giving these textbooks a substantially different "look and feel" than those materials that we label standards-based.

When we cite studies that explicitly involve one or more of the materials that appear in Table 8.1, we identify the curriculum by name. When overall statements are made regarding the characteristics or efficacy of various kinds of materials, we refer to the two general categories of standards-based or conventional.

Finally, the focus of our examination of the influence of curriculum on students is on classrooms—teachers and students. Thus, our discussion of curriculum materials centers on the resources and materials that teachers typically interact with as they plan and carry out their lessons. We have not included policy documents or the curriculum frameworks developed by districts or states in our review. In taking this approach, we do not, however, ignore the role that policy and organizational designs play in curriculum use. We recognize that state and local policy shape the manner in which teachers think about and use curriculum materials. We account for these contextual features primarily in our discussion of the explanations of how the curriculum is transformed between and within phases of enactment (see oval in Figure 8.1).

SECTION ONE: RESEARCH ON CURRICULUM MATERIALS AND STUDENT LEARNING

The *Curriculum and Evaluation Standards for School Mathematics* (NCTM, 1989) were created based on a broad consensus of professionals (teachers, mathematics educators, mathematicians, users of mathematics, etc.) and were framed in broad, easy-to-agree-with terms (EEPA, 1990). The curricular resources that were developed to embody them, however, offered much more detailed specification regarding what should be taught and how it should be taught, thereby providing concrete targets for critics of the Standards. As these new standards-based curriculum materials began to penetrate the market in the mid- to late-nineties, critics charged that they watered down important mathematical concepts, relied too heavily on real-life problems at the expense of pure mathematical problem solving and discovery, provided too little attention to paper-and-pencil calculations, and were not suited for college-level preparation (Wu, 1997).

Despite the fact that conventional textbooks used at the time had little or no evidence of their effectiveness, the release of these new curriculum materials was soon followed by demands for "proof" that they were helping, and not harming, students. Kilpatrick (2003) noted that these demands should not have been surprising:

> Anyone proposing a new school mathematics curriculum faces the task of justifying its adoption. Teachers, parents, and students themselves may be dissatisfied with the current situation, but that does not mean they necessarily welcome change of the sort the new curriculum might demand. Curriculum developers are always in the position of "selling" their product by convincing their potential clients that the change it entails is both manageable and for the better. Research showing improved performance, or at least performance that is no worse than at present, is often used to bolster arguments being made on other grounds. (p. 477)

The fact that the standards-based curriculum materials challenged the status quo by embodying a radically different set of goals for student learning, however, meant that they had a particularly steep hurdle to overcome (Romberg, 1992). Moreover, the demands for "proof" were fueled by intense controversy, embodied by disagreements between two opposing camps (composed primarily, though not entirely, of mathematics educators in one group and university-based mathematicians in the other) that became so strident that they were dubbed the "math wars" (Schoen, Fey, Hirsch, & Coxford, 1999). As this controversy played out in newspaper editorials, Internet discussions, and articles in popular periodicals, its polarized nature often made it difficult for the general public to distinguish ideology from empirical fact.

In this section, two kinds of studies are reviewed: content-based studies of curriculum materials and evaluations of what and how students learn from curriculum materials. These two kinds of studies share the underlying assumption that curriculum materials—in and of themselves—matter. Although some authors acknowledge the fact that curriculum materials are not self-enacting, these studies do not highlight the enactment process but rather focus on the materials themselves or on the relationship between the adoption of written materials and student outcomes. That is, most studies do not attempt to examine how teachers interpret curriculum materials or how teachers and students interact with the materials in the classroom. Returning to Figure 8.1, the reader might imagine a solid arrow drawn from "Written Curriculum" to "Student Learning" as a way to characterize the studies cited in this section.

Research on Content of Curriculum Materials

The majority of mathematics teachers rely on curriculum materials as their primary tool for teaching mathematics (Grouws, Smith, & Sztajn, 2004). If a topic is not included in their curriculum materials, there is a good chance that teachers will not cover it.[5] And, as noted by Hiebert and Grouws (this volume), one of the best substantiated findings in the literature on classroom teaching and student learning is that students do not learn content to which they are not exposed. Thus, the identification of *what* mathematical topics are covered in a given set of curriculum materials is of fundamental importance.

Others argue, however, that "the mere presence of content in a textbook does not ensure that students will learn that content. For real learning to take place, textbooks must focus sound instructional strategies on the ideas and skills that students are intended to learn" (Project 2061, n.d., part 2). The myriad decisions that curriculum developers make regarding issues such as how material should be presented reflect the developers' theory of how students learn mathematics. These

dimensions, which we discuss under the label of *how* content is presented (as opposed to *what* content is presented), are important because they set into motion different pedagogical approaches and different opportunities for student learning.

Finally, some curriculum materials aim to promote teacher learning in addition to student learning. If one believes that students' opportunities to learn are the product of ongoing interactions among the text, the student and the teacher, then what students learn will be influenced by their teachers' understanding of and presentation of the material in the text. Given the plethora of findings about the impoverishing effect that limited content knowledge has on teachers' interpretation and use curriculum materials, especially novel and highly demanding materials such as standards-based curriculum materials (see studies reviewed in Section Four), we argue that a third dimension worthy of investigation is the extent to which the curricular materials are educative (Davis & Krajcik, 2005) for teachers. Each of these approaches to examining students' opportunities to learn—What Content is covered? How is Content Presented? And with what support for teacher learning?—is reviewed below.

What Content is Covered?

Content analysts normally compare selected curriculum materials against a set of external criteria to determine depth and/or breadth of coverage. Analysts in the United States typically use standards, frameworks, or other countries' curricula as their external criteria (NRC, 2004). Over the past decade, the external criteria used by researchers to analyze curriculum materials have varied widely, reflecting the various values held by the individuals who have conducted them.

Not surprisingly, these variations have produced different results. In their review of 36 content analyses, the NRC (2004) found that the ratings of many curricular programs vacillated from strong to weak depending on who had reviewed them and according to what criteria. Differences in values have not and cannot be decided by empirical analyses. However, once one's values are clear and learning goals compatible with those values have been specified, questions regarding what curriculum is most effective (for achieving those goals) can be answered empirically (Hiebert, 1999). As concluded by the NRC panel (2004), curriculum decision makers for schools and districts should select

---

[5] In recent years, the chances of students not being exposed to a topic if it is not in the curriculum have been somewhat mitigated by state standards. Most districts check their curricula for alignment to state standards. If a topic is found to be in the standards but not in their textbooks, they frequently find supplementary materials to address the gap.

reviews whose underlying dimensions are compatible with their values and learning goals.

It is in this spirit that we report the approaches and findings from three of the most prominent content analyses performed over the past decade: Project 2061 of the American Association for the Advancement of Science (AAAS), the ratings of curricula by the US Department of Education, and the work of Mathematically Correct.

Project 2061. One of the first, systematically documented analyses of the mathematical content of curricula was conducted in the late nineties by the American Association for the Advancement of Science [AAAS]) of middle school (and subsequently of algebra) textbooks. According to their website, the emergence of the NCTM Standards in 1989 along with AAAS's benchmarks for scientific literacy (which were closely aligned with the NCTM Standards) provided—for the first time—a solid, widely acknowledged conceptual basis for evaluating textbooks based on what students should learn. In their evaluation of 13 middle school curricula, they used a "relatively small but carefully chosen set of benchmarks to identify the strengths and weaknesses of the curricula." These benchmarks, they claimed, "deal with concepts and skills that nearly everyone would agree are important for middle school students to achieve." Another important aspect of their methodology was the decision to conduct in-depth examinations of a relatively small—but important—number of topics as opposed to examining more topics in a more superficial manner.[6]

The benchmarks used for their review of middle-school curricula included the following:

Number Concepts: The expression a/b can mean different things: a parts of size 1/b each, a divided by b, or a compared to b.

Number Skills: Use, interpret, and compare numbers in several equivalent forms such as integers, fractions, decimals, and percents.

Geometry Concepts: Some shapes have special properties: Triangular shapes tend to make structures rigid, and round shapes give the least possible boundary for a given amount of interior area. Shapes can match exactly or have the same shape in different sizes.

Geometry Skills: Calculate the circumferences and areas of rectangles, triangles, and circles, and the volumes of rectangular solids.

Algebra Graph Concepts: Graphs can show a variety of possible relationships between two variables. As one variable increases uniformly, the other may do one of the following: increase or decrease steadily, increase or decrease faster and faster, get closer and closer to some limiting value, reach some intermediate maximum or minimum, alternately increase and decrease indefinitely, increase or decrease in steps, or do something different from any of these.

Algebra Equation Concepts: Symbolic equations can be used to summarize how the quantity of something changes over time or in response to other changes. (Project 2061, n.d., part 1)

Results are presented in a matrix form for each of the 13 curricula that were reviewed. These matrices provide a content rating for each of the above criteria at one of three levels: most content, partial content, and minimal content. Page numbers and sections that provide evidence for meeting each of the content levels are provided.

The results of this review have been summarized as follows:

> Most middle grades textbooks do a credible job addressing number benchmarks. However, only the best ones develop meanings for fractions, for example, by having students measure, build models, use number lines, and compare fractions to acquire a full understanding. Almost all textbooks present the formulas in geometry. But even some of the best ones fail to relate geometry skills to real-life ideas, such as the triangular structures used in bridges or the relationship between the distance around a city park and the area enclosed. . . . Few textbooks do a good job teaching how graphs show relationships, and instead focus on simple, linear graphs. The best series involve students with data collection in situations like a bicycle tour, giving them first-hand experience connecting concepts, such as time and distance, with tables and graphs. With this solid foundation, variables and equations are used naturally and with understanding (Kulm, 1999).

Only four of the thirteen middle-grades curricula—all standards-based—were rated satisfactory; that is, high enough to be confident that students would learn the content of the selected benchmarks. These curricula were, ranked in order, *Connected Mathematics, Mathematics in Context, MathScape,* and *Middle*

---

[6] This speaks to the criticism of American curricula and programs as being a "mile wide and an inch deep" (Schmidt, McKnight, & Raizen, 1997) by focusing on the identification and treatment of "big ideas," the understanding of which should place students in good stead for future mastery of more advanced mathematical content.

Grades Math Thematics. None of the conventional curricula were rated as satisfactory. These textbooks were judged to be lacking in their coverage of important mathematics, and to provide little development in sophistication from grades 6 to 8. (See Appendix A for ratings of various curricula by the AAAS,'s Project 2061, the Department of Education, and Mathematically Correct.)

U.S. Department of Education. In 1999, the U.S. Department of Education conducted a review of mathematics curricula to identify promising and exemplary curricula. According to the NRC panel (2004), this evaluation was guided by eight criteria structured in the form of questions (page 68):

1. Are the program's learning goals challenging, clear, and appropriate; is its content aligned with its learning goals?

2. Is it accurate and appropriate for the intended audience?

3. Is the instructional design engaging and motivating for the intended student population?

4. Is the system of assessment appropriate and designed to guide teachers' instructional decision making"

5. Can it be successfully implemented, adopted, or adapted in multiple educational settings?

6. Do its learning goals reflect the vision promoted in national standards in mathematics?

7. Does it address important individual and societal needs?

8. Does the program make a measurable difference in student learning?

Although *The Curriculum and Evaluation Standards* of the NCTM are not specifically mentioned, they were used (and assumed to be used) as the national standards referred to in question #6. Not surprisingly, once again, the standards-based curricula fared better under these criteria than did more conventional curricula, with the *Connected Mathematics Project* (CMP), the *Middle School Mathematics through Applications Project* (MMAP), the *Cognitive Tutor, College Preparatory Mathematics* (CPM), *Contemporary Mathematics in Context* (Core-Plus), and the *Interactive Mathematics Program* (IMP) being named exemplary and *Everyday Mathematics, MathLand, Number Power,* and the *University of Chicago School Mathematics Program* (UCSMP) *Integrated Mathematics 7–12* being labeled promising.

Shortly after these evaluations became public, the criteria used by the Department of Education came under attack by opponents of NCTM *Standards* and Standards-based reform. For example, Richard Askey, a professor of mathematics at the University of Wisconsin and one of the authors of an open letter (signed by 200 professional mathematicians) sent to prominent national newspapers and to *Education Week* stated, "Some of the recommended programs are very careless with the mathematical core." Furthermore, he questioned the role of the NCTM *Standards,* saying, "To put in as one of the criteria that curriculum materials should be aligned with the NCTM Standards is inappropriate. We want good mathematics programs, and these do not necessarily have to be based on the NCTM Standards." Other prominent mathematicians defined their concerns with specific features of the "exemplary" curriculum materials, such as their light treatment of standard algorithms and heavy use of calculators (e.g., Klein, 2000).

Mathematically Correct. The organization of concerned mathematics professionals and parents known as Mathematically Correct have provided the third set of noteworthy content reviews in the past decade. Their reviews of second, fifth and seventh-grade mathematics textbooks (both standards-based and conventional textbooks) are guided by the concept of preparing all students for pre-algebra by eighth grade and by various sets of standards. In the seventh grade review (presented here for comparison to the criteria used by AAAS in their review of middle school textbooks), this organization evaluated "a set of key topics, representing an array of algebra readiness concepts, knowledge, skills, and problem-solving applications. Key benchmarks within these areas were drawn from various standards documents and used as the content basis for the review of each mathematical topic" (Mathematically Correct, n.d.a, p.1) The content areas were: Properties, order of operations; Exponents, squares, roots; Fractions; Decimals; Percents; Proportions; Expressions and equations-simplifying and solving; Expressions and equations-writing; graphing; Shapes, objects, angles, similarity, congruence; Area, volume, perimeter, distance.

Under each of these topic areas, detailed specifications were provided regarding what students should learn. Compared to the AAAS criteria for middle school curricula, the Mathematically Correct criteria focus more heavily on mastering procedural manipulations and rules. For example, under the first topic area, " Properties, order of operations" they state:

At this level students should master the rules of order of operations and the properties of the real number system (commutative, associative, distributive, identity and inverse) and be able to apply them, with justification, with all four operations in calculations involving fractions, decimals, percents, positive and negative numbers, and in the simplification of simple roots and powers. This should include the simplification of numerical expressions, the evaluation of expressions with substitution of numerical values and the solution of simple one and two step algebraic equations (Mathematically Correct, n.d.a, p.2).

The concern of Mathematically Correct is mastery of what they term, the mathematical core, the essential elements of knowledge and skills that students need to succeed in more advanced mathematics. Their criteria for mastering this core include procedural fluency, conceptual understanding and the ability to apply newly learned information to novel situations—criteria that most subscribers to the NCTM Standards would endorse as well. As we shall see in the next section, one of the main differences between NCTM Standards subscribers and Mathematically Correct members can be found in their respective notions of how—and in what order—students should learn the mathematical core.

Under the Mathematically Correct evaluations, conventional textbooks fared better than did standards-based curricula. (Appendix A identifies the curricula that have been rated by Mathematically Correct and the overall "grade" they received.) The highest "grade" received by a standards-based curriculum was a "C" for *Everyday Mathematics*; all others received a "D" or "F." Meanwhile, all but two of the conventional curricula received an "A" or "B" (Mathematically Correct, n.d.b)

In sum, the three content reviews cited in this section make clear the need for transparency regarding the criteria used for content review of curriculum materials. Using very different sets of criteria, AAAS and the US Department of Education reviews produced recommendations that were nearly the opposite of the recommendations produced by reviews conducted by Mathematically Correct. This suggests that consumers and decision makers must first define what *they* value in terms of student learning and then select a review that encompasses their values as closely as possible (Hiebert, 1999).

Another way of framing the issue is in terms of tradeoffs. For example, one feature that distinguishes the AAAS review from the Mathematically Correct review is the extent to which each focused on depth versus breadth. Whereas AAAS examined a few topics in great depth (reflecting their belief that students

need to develop multi-faceted, in-depth understanding of a few, salient mathematical topics or ideas), Mathematically Correct's reviews reflected the comprehensive range of skills that they believe students need to succeed in advanced mathematics courses. Each approach can be seen as having both virtues and drawbacks. Being aware of the criteria used in content reviews allows decision makers to surface and articulate this and other tradeoffs (some of which will be illuminated in the next section) and make informed decisions regarding what best fits with their K–12 instructional program, their goals for student learning, and their institutional resources and constraints.

How is Content Presented?

The above analyses primarily focused on *what* is included (and excluded) in various curriculum materials. We turn now to a discussion of *how* that content is presented, that is, analyses that more sharply focus on pedagogical intent. Similar to the analyses of content coverage discussed above, the criteria that researchers have used to make judgments about the pedagogical intent of various curricula are necessarily related to personally held views regarding the nature of mathematics and how students learn it. If one believes that mathematics is best learned through student construction based on active exploration, one set of criteria will be selected to guide one's analyses. On the other hand, if one believes that mathematics is best learned through direct instruction and skills practice, then review criteria will be of a different sort.

A host of curricular features have the potential to influence how teachers use and thus how students experience curriculum materials. These include pre-tests to assess prior student knowledge, embedded assessments, suggestions for tailoring the material for students at different levels, the inclusion of group work activities, and the encouragement of student discussion of ideas. In this section, we begin with a discussion of several overarching design features that can have potentially far-reaching effects on how students learn mathematics. We then briefly review how two organizations, AAAS and Mathematically Correct rated curriculum materials with respect to how they presented content.

Overarching features of curricula. Most standards-based and conventional curricula intend for students to learn concepts, skills, applications, problem-solving and efficient procedures. They differ, however, with regards to the order and manner in which these elements are presented, the balance that is struck among various elements, and organizational style. Each of these is discussed below.

1. *Order and manner of presentation.* Conventional curricula tend to rely on direct explication of the to-be-learned material as well as careful sequencing and accumulation of lower-level skills before presenting students with the opportunity to engage in higher-order thinking, reasoning and problem solving with those skills. Students typically are expected to master definitions and standard algorithmic procedures *before* they are exposed to opportunities to apply their knowledge. In contrast, standards-based materials rarely explicate the to-be-learned concepts *for* the students; instead they rely on student engagement with well-designed tasks to surface the concepts. Active exploration of the concept-to-be-learned is viewed as allowing students to induce important facets of the concept based on their own reasoning and thinking. After the concept has been surfaced and its features explored by students, the curriculum and teacher step in to apply definitions, standard labels, and (sometimes) standard procedural techniques related to the concept.

To illustrate the above contrast, consider two ways in which curricula might approach the concept of the mathematical average. Conventional curricula would begin with the provision of an explicit definition of the concept of average (i.e., a measure of central tendency) and would then demonstrate the conventional algorithm for finding the average of a set of numbers (taking the sum of the numbers in the set and dividing the sum by the number of addends). After practicing this procedure on several sets of numbers, students would be asked to apply what they have learned by finding the average in the context of word problems. In contrast, standards-based curricula would not begin with an explicit definition of the concept of average but rather would immerse students in a task that requires that they experience what an average means, for example, as is done with the task shown in Figure 8.2 (found in *Visual Mathematics/ Math Alive*).

As students grapple with this task, they would be approaching the concept of average through a frame of "leveling" and thus be encouraged to recognize what an average does, i.e., it evens out the extremes of a given population, a recognition that is important when deciding which measure of central tendency might be the most valid representation of a population given a particular question. Students could also be drawn to recognize that the average can be a number that is not represented in the population, as is the case in this problem. Only after exploring these more qualitative, conceptual features of the mathematical average would students be introduced to the conventional procedure for finding the average.

The theory of knowledge and of how students learn is radically different in these two approaches.

The pairs of numbers in a–d below represent the heights of stacks of cubes to be leveled off. On grid paper, sketch the front views of columns of cubes with these heights before and after they are leveled off. Write a statement under the sketches that explains how your method of leveling off is related to finding the average of the two numbers.

a) 9 and 5   b) 16 and 7   c) 7 and 12   d) 13 and 15

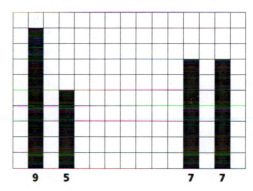

By taking 2 blocks off the first stack and giving them to the second stack, I've made the two stacks the same. So the total # of cubes is now distributed into 2 columns of equal height. And that is what average means.

[Taken from *Visual Mathematics*, (Bennett & Foreman, 1989)]

Figure 8.2 Task used in *Visual Mathematics* to develop students' understanding of the concept of average.

In the more conventional approach, knowledge is viewed as a Gagne-like (1977) hierarchy of skills and learning is viewed as proceeding from lower-level to higher-level skills. Without the prerequisite lower-level skills, this approach argues, students will not be able to master higher-order skills. In the standards-based curricula, on the other hand, learning is viewed as the development of understanding through the construction of increasingly detailed relationships between concepts. In this perspective on learning, students are viewed as making meaning of new information by relating it to their prior knowledge; learning is said to have occurred when students develop an understanding of the relationships between and among concepts and between concepts and procedures (Hiebert et al., 1997).

2. *Balance.* Curriculum materials also differ with respect to the balance that they strike among a number of elements including a) between concepts and procedures, b) between calculator-aided vs unaided computation, and c) among various kinds of representations including numerical data, manipulatives, tables, graphs, and equations. Again, the particular balance that is struck reflects the developers' philosophy regarding how students learn mathematics. For example, in summarizing reviews of algebra curricula, Clopton, McKeown, McKeown, and Clopton from Mathematically Correct (1998) expressed concern re-

garding the impact that calculators have on the development of computational fluency and tended to discourage their use. Many researchers favorable to the NCTM Standards, on the other hand, applaud the introduction and use of calculators, because of their ubiquity in present-day society and their belief that off-loading computational burdens from students "frees up" cognitive space for higher order thinking and reasoning. With regard to the balance between concepts and procedures, not surprisingly, conventional materials tend toward a balance that weighs heavily toward practice on procedures while standards-based curriculum materials are weighted more toward the development of concepts and problem solving.

3. *Organizational style.* Finally, conventional versus standards-based curriculum materials differ in terms of their overall organization. Most conventional textbooks are organized into units and chapters that follow topics that have been identified as appropriate for each of the particular grade levels. Many are also organized as a "spiral," meaning that specific topics are introduced and re-introduced throughout a set of grades with increasing levels of sophistication each time that students encounter them. Students are not expected to master the topic the first time it is introduced, but rather are expected to deepen their understanding and attain eventual mastery over time as the topic is revisited. Stein and Kim (2006) refer to spiral and other tightly sequenced curricula as *integral* curricula. In integral curricula, knowledge and skills are tightly woven into the fabric of the curriculum (i.e., they cannot be easily separated out) and they must be taught in a specified sequence over the years.

Integral curriculum materials can be contrasted with *modular* approaches in which the subcomponents can be separated and even recombined into new and varied configurations with little loss to the overall effectiveness in attaining curricular goals (Stein & Kim, 2006). Many standards-based curriculum materials are modularized with each module designed to represent a big idea or integrating conceptual theme. For example, *Bits and Pieces*, a sixth grade module from *Connected Mathematics*, subtitled "Understanding Rational Numbers" focuses on moving between and among fractions, decimals, and percents. Typically, students complete 6–12 modules per school year.

By and large, research has been silent on the trade-offs associated with integral versus modular forms of curricular organization. Stein and colleagues (Stein & Kim, 2006) have proposed that integral curricula place more demands on the social cohesion of a school's faculty (which they conceptualize as social capital) because they require that all teachers buy into one curriculum and dutifully carry out their responsibilities with respect to teaching the requisite skills at each grade level. Modular approaches, on the other hand, can afford to be taught by some teachers, but not all. Such approaches can also be used in variable sequences (within a given grade level or sometimes across two consecutive grades). Moreover, because the to-be-learned knowledge and skills are more easily identifiable and viewed as separable from the curricular fabric, modular curriculum materials are more amenable to progressive implementation within a reform effort (i.e., implementing only a partial set of modules for the first year, a few more the second year, and so forth).

In sum, the above discussion of design features illustrates the range of decisions that developers make (either implicitly or explicitly) as they design materials for student consumption. Moreover, the features discussed above—order and manner of presentation, balance, and organizational style—are not features that typically appear on textbook review checklists. Rather they are subtle attributes whose presence becomes felt over time as teachers use the materials and observe their impact on how students and their colleagues engage with the materials. Surfacing these distinctions and their potential implications for how students learn mathematics would be a significant move forward for decision makers in most schools and districts.

Reviews of how content is presented. The reviews conducted by AAAS and Mathematically Correct illustrate the selection of two very different sets of dimensions for judging the pedagogical quality of mathematics curricula. AAAS claims to have based their review of middle school curricula on criteria "derived from research on learning and teaching and on the craft knowledge of experienced teachers" (Project 2061, n.d., part 1, p. 3). Reviewers were trained to rate the extent to which textbooks explicitly included information with respect to 24 instructional criteria ranging from the specification of prerequisite knowledge to the guiding of interpretation and reasoning. The 24 criteria were organized into seven categories (identifying a sense of purpose, building on student ideas about mathematics, engaging students in mathematics, developing mathematical ideas, promoting student thinking about mathematics, assessing student progress in mathematics, and enhancing the mathematics learning environment), each of which was justified by findings from research on learning and teaching.[7] The criteria

---

[7] Sources that are cited include *Science for All Americans* (AAAS, 1990), *Benchmarks for Science Literacy;* (AAAS, 1993) and the *Handbook of Research on Mathematics Teaching and Learning* (Grouws, 1992).

and organizational categories are also cited as being compatible with NCTM's 1989 Standards.

One example of an organization category will illustrate the view of teaching and learning embodied by the AAAS approach. Category IV, labeled "Developing Mathematical Ideas" is justified as such:

> Mathematics literacy requires that students see the link between concepts and skills, see mathematics itself as logical and useful, and become skillful at using mathematics. Six criteria are used to determine whether the material justifies the importance of benchmark ideas, introduces terms and procedures only as needed, represents ideas accurately, connects benchmark ideas, demonstrates/models procedures and provides practice.

The above rationale cites the need for practice and accuracy, but also stresses the importance of students viewing mathematics as meaningful and useful; moreover, much of the emphasis is on the making of connections and the introduction of terms and procedures *as needed*. As such, the view of teaching and learning appears to be closely aligned with a constructivist, meaning-making approach to mathematics.

AAAS's analysis of instructional strategies revealed stark contrasts among the textbooks with respect to the adequacy of instruction for the specific benchmarks. Four curricula are highly rated for their engagement of students, development of mathematics concepts, and support of teachers, the same four that were rated high for content: *Connected Mathematics, Mathematics in Context, MathScape,* and *Middle Grades Math Thematics.* The remaining curricula (all conventional) were judged to be unsatisfactory with respect to instructional support. They were judged to be particularly unsatisfactory in offering a purpose for learning mathematics, taking account of student ideas, and promoting student thinking.

Mathematically Correct based their reviews of pedagogical worthiness on examination of two main dimensions: "quality of presentation" and "quality of student work." For "quality of presentation," the following were examined: "clarity of objectives within each lesson and within a topic as a whole, clarity of presentation, a sufficient number and appropriate examples, reasonable guided practice or scaffolding of the early stages of implementation of new knowledge, and a general sense that students would have a high probability of mastering the material at the level at which it was taught" (Mathematically Corect, n.d.a, p. 5). For "quality of student work," they examined the following: "Is there enough chance to practice what is taught, is the work well designed, and is the work at an appropriately high level? Importantly, a

range depth and scope of student work should be provided, appropriately building from the simpler, less abstract cases to the more difficult ones" (Mathematically Correct, n.d.a, p. 5).

Using these criteria for their review of seventh grade curricula, the Mathematically Correct reviewers gave the two standards-based curricula (*Connected Mathematics* and *Math Thematics*) the lowest scores. According to their website (Mathematically Correct, n.d.c, p. 5), "these books fall notably short in content, have ill-defined goals for each lesson, as least in the eyes of the reviewers, and have insufficient or poorly designed student work." The curricula that we have called conventional fared much better, notably those published by Saxon. Unlike the other conventional curricula which the reviewers found to be compromised by a "fusion" of teacher-directed lessons and activities based on "discovery learning and the constructivist theory of knowledge," (Mathematically Correct, n.d.c, p. 4) the Saxon curricula were found to have clarity of focus, regular practice on important topics, and to be concise enough to discourage teacher variation with regards to decisions regarding coverage.

The reader can begin to get a sense of how these two sets of evaluations privilege different approaches to teaching and learning mathematics. Not surprisingly—as with the content analyses—the same curriculum fares very differently depending on which set of criteria are applied. Again, decisions can be viewed as tradeoffs set against the backdrop of consumers' beliefs and values, combined with an analysis of their institutional resources and constraints. Potential consumers need to pose to themselves and answer questions such as the following: What kinds of knowledge and skills will prepare our students for their most likely futures? Do our students need to become procedurally sharp by building on their successes through repetition and diligent practice? Or is our ultimate goal students who recognize and begin to deal with complexity—students who, nonetheless, may have to deal with frustration along the way because they will not always arrive at a correct solution in a neat and timely manner? Are our teachers up to the task of teaching in more complex ways and/or following an integral curriculum to assure a coordinated instructional program across grade levels? The posing and answering of these questions depends on awareness of one's values, an accurate picture of one's institutional context, and the ability to assess curricula according to criteria that are transparent to consumers.

The Support of Teacher Learning

As noted earlier, the standards-based curricula tend to embed much of the mathematics learning implicitly within the tasks themselves. Students are viewed as learning mathematics by "doing mathematics" as opposed to having mathematical concepts and procedures explicitly and directly taught to them by the curricular materials (and the teacher). This student-centered approach to the learning of mathematics places considerable demand on teachers who must organize the physical and normative environment to be conducive to student inquiry, scaffold student actions and thinking as they grapple with cognitively challenging tasks, deal with student ideas that are off-track, and, finally, bring closure to a lesson that is rooted in many different student-designed methods of approaching a given task and a variety of levels and kinds of understanding of the to-be-learned concept(s).

Recently, researchers have begun to identify ways in which curricular materials might be designed so as to assist teachers to learn subject matter themselves—a prerequisite for this kind of teaching—as well as to learn how to better assist student learning of subject matter (Ball & Cohen, 1996; Davis & Krajcik, 2005). These features comprise elements that are added on top of the "base curriculum," that part of the curricular materials that is designed for direct student consumption. Educative curricula incorporate elements that are meant specifically for *teacher* learning and are distinguished from typical teacher guides that steer teachers' actions but not their thinking.

This is a relatively new area of research but is becoming increasingly important as student-centered materials gain a foothold in the marketplace and as teachers struggle to learn a new way of teaching mathematics. The most developed framework for judging how well curricula promote teacher learning identifies five high-level guidelines for educative curricula (Davis & Krajcik, 2005, pp. 5–6):

1. Educative curricula could help teachers learn how to anticipate and interpret what learners may think about or do in response to instructional activities.

2. Educative curricula could support teachers' learning of subject matter.

3. Educative curricula could help teachers consider ways to relate units during the year.

4. Educative curricula could make visible curriculum developers' pedagogical judgments.

5. Educative curricula could promote teachers' pedagogical design capacities or their ability to use personal resources and the supports embedded in the materials to adapt curriculum to achieve productive instructional ends (as opposed to performing "lethal mutations") (Brown & Campione, 1996, p. 291).

Building on this work, Stein and Kim (in press) analyzed two standards-based curricula (*Everyday Mathematics* and *Investigations*) in order to assess the extent to which they (a) made visible the curriculum developers' rationales for specific instructional tasks and/or for particular learning pathways advanced in the base curricula; and (b) helped teachers to anticipate how students might approach these tasks. Referring to the visibility of developers' judgments as "transparency," Stein and Kim reasoned that teachers who had a clear idea of where a particular task was headed mathematically would be better equipped to guide the classroom enactment of the task, especially when it did not go as planned. The ability to anticipate student thinking in response to the task, they reasoned, would allow teachers to plan ahead for how they might deal with the diversity of student responses that open-ended tasks often generate. They found differences between the two curricula with *Investigations* offering more rationales regarding the mathematical significance of tasks and more support for teachers to anticipate student thinking. Preliminary evidence regarding the impact of these differences on teacher implementation of these two curricula suggests that the open-ended tasks of *Investigations* can be enacted with considerable fidelity when teachers make use of the teacher support material whereas the relatively more constrained tasks of *Everyday Mathematics*[8] were nonetheless difficult to implement with fidelity when teachers were unsure of the mathematical purpose of the tasks or how students might respond to them (Stein, Kim, & Seeley, 2006).

Brown (in press) distinguishes two different principles that can be used to organize teacher materials: resource-centric and procedure-centric. A resource-centric approach to the design of teacher materials emphasizes the key building blocks of lessons and tries to make visible the pedagogical affordances of the building blocks, but leaves up to the teacher how, when, and whether to implement the various building blocks. A procedure-centric approach, on the other hand, focuses on the actions involved in carrying out the lessons. Stein and Kim (in press) view

---

[8] The majority of *Investigations* tasks were coded as "doing mathematics" (open-ended, unstructured), whereas the majority of *Everyday Mathematics* tasks were coded as "procedures with connections to underlying concepts and meaning." See Stein et al. (1996) for an explanation of these levels of cognitive demand.

the teacher materials that accompany *Everyday Mathematics* as procedure-centric and the teacher materials that accompany *Investigations* as resource-centric. They stress that how and where to implement each depends on the nature of the base curriculum and the institutional context.

In this sub-section (Research on Content of Curriculum Materials), we have reviewed research that has focused directly on curriculum materials, framing our review in terms of the opportunities for student learning afforded by different curricula. Our basic finding is that students' opportunity to learn is in the eye of the beholder. An examination of Appendix A illustrates what may now seem to be predictable disagreements between Mathemtically Correct and AAAS/Department of Education reviewers (see, for example, the disparate ratings for *Connected Mathematics*). If one adopts the view of teaching and learning espoused by the NCTM standards (that served as a design blueprint for most standards-based curricula) and consults reviews that base their criteria on the NCTM Standards and like-minded sources, one finds that the standards-based curriculum materials are highly rated. On the other hand, if one believes that mathematics is better taught and learned through more explicit, direct instruction and practice of what is taught and consults reviews that build their criteria from sources such as Mathematically Correct, one finds that the standards-based curriculum materials come up considerably short.

A second finding, which is obvious but nevertheless worth noting, is that standards-based and conventional curriculum materials really do appear to provide very different opportunities for student learning. Although the various reviewers used different criteria, they used them consistently suggesting that standards-based curriculum materials provide more opportunities for students to learn how to think and reason and fewer opportunities to learn and practice basic skills. Conventional textbooks, on the other hand, appear to provide more systematic opportunities to learn definitions and practice demonstrated skills but fewer opportunities to learn when and how to apply those skills.

Finally, our review suggests that standards-based curriculum materials differ with respect to the kind of support materials that they provide *for teachers*. We also suggest that this issue is more complex than discovering that one approach (e.g., the resource-centric approach) is more effective than another. Rather, the kind of teacher-support materials that are most helpful will most likely depend on the nature of the base curricula, as well as the experience level and instructional capacities of the teachers who will be implementing the materials. The issue of teacher learning from curriculum materials opens up an entire new set of design considerations that developers must attend to and that researchers must learn to evaluate.

Our examination of curriculum materials with respect to students' *opportunities to* learn represents only the first step—albeit an important first step—to investigating curricula's actual impact on student learning. In the next sub-section, we turn to studies that collected data on what students learned when they were exposed to these curriculum materials in the classroom.

Examination of Student Learning from Mathematics Curriculum Materials

All of the NSF-supported curriculum projects were required to conduct both formative and summative evaluations of their materials (NSF, 1989). Virtually all developers conducted pilot tests of their materials as various units were under construction or in different stages of completion. A key element of these formative assessments was to uncover what students were able to do and understand as a result of their exposure to the curricula's tasks. Developers also conducted summative evaluations, many of which are reported in an edited volume by Senk and Thompson (2003) entitled, *Standards-Based Mathematics Curricula: What Are They? What Do Students Learn?* Some developers also received additional funding once their materials had been completed to conduct additional studies of their impact.

On the whole, the early evaluations of standards-based curricula as reported in the Senk and Thompson volume were promising. After reviewing syntheses of findings for the four elementary curricula, Putnam (2003) stated:

> The first striking thing to note . . . is the overall similarity of their findings. Students in these new curricula generally perform as well as other students on traditional measures of mathematics achievement, including computational skill, and they generally do better on formal and informal assessments of conceptual understanding and ability to use mathematics to solve problems (p. 161).

Similarly, Chappell (2003) notes that, collectively, the evaluation results provided for three middle school curricula provide "converging evidence that standards-based curricula may positively affect middle school students' mathematics achievement, in both conceptual and procedural understanding." (p. 290). And, finally, Swafford (2003) reviewed syntheses of findings across five high school curricula and found that students in the reform programs performed as well as students in traditional programs on tests designed to measure traditional content and that they outperformed them on tasks designed to measure the content and processes emphasized in their materials (pp. 458, 460). Given the very different goals and approaches of the new standards-based curricula, these initial reports suggesting that students taught with these materials demonstrated at least comparable levels of computational learning as students taught with conventional curricula and superior understanding of concepts and problem solving were greeted with enthusiasm by their proponents.

That said, none of the approaches or methodologies used in the studies reported in the Senk and Thompson volume was without flaws. One of the biggest concerns with the studies reported in this volume was the fact that curriculum designers were the primary researchers on most of the studies which inevitably raises conflict-of-interest issues. On the positive side, however, developers who also conduct research on their curriculum materials are in the position to design assessment tasks that tap the knowledge that students should acquire from exposure to their curriculum due to their deep understanding of the material. For example, Mokros (2003) reports on a developer-designed assessment comprised partially of individual interviews that assessed the quality of explanations and use of representations and tools, prominent learning goals associated with the *Investigations* curriculum. These kinds of assessments should yield better information on the extent to which students have learned from the specific curriculum than will standardized tests.[9]

Overall, however, the curriculum designers did not appear to take full advantage of their "insider knowledge" and design such assessments. In one of the concluding chapters of the Senk and Thompson volume, Kilpatrick (2003) noted that many of the most distinguishing features of the curricula that were studied and reported on in this volume were not assessed in terms of the student learning outcomes that would be expected. Rather, the most common instruments used were standardized tests, items from national and international assessments, and the PSAT or SAT. According to Kilpatrick, it appeared as those these measures were being used "more to reassure parents and the public that the new curricula were not harming students than to assess common learning goals" (p. 479).

Other concerns that have been raised about drawing conclusions based on these early studies as reported in the Senk and Thompson volume include the often limited variability in conditions of the studies. For example, the study of the *Trailblazers* curriculum included primarily teachers who had received professional development from the curriculum developers themselves. Also, in many of the early studies, the materials were not always in their final forms, and, in some cases, the curriculum materials had not been used long enough to establish stable implementations. Finally, issues can be raised about the comparability of the groups to which some of the curricula were compared. This leaves one with limited ability to make strong conclusions based on any one of the above studies, although the number of studies and the diversity of approaches represented in the Senk and Thompson volume might—for some readers—increase confidence in the general patterns that emerged.

The above studies were undertaken during the early phases of the dissemination of standards-based curricula. As such, the studies reported in the Senk and Thompson volume might be best understood as "proof of concept" studies, indicating that the standards-based curricula "are working in classrooms in ways their designers intended for them to work" (Kilpatrick, 2003, p. 472). It should also be noted, however, that these studies did not satisfy skeptics who demanded that independent researchers develop more rigorous methodologies with which to examine student learning, a discussion to which we now turn.

Comparative Studies Conducted by External Researchers

Although it may seem simple to evaluate whether or not a new curriculum is better, worse or no different from existing approaches, designing a study to do so is fraught with pitfalls and complications, leading an NRC panel convened to examine evaluations of mathematics curricula to conclude that "comparative evaluation study is an evolving methodology" (2004, p. 96). To start, credible outcomes measures must be identified or created. This can be particularly problematic with standards-based curricula that focus on conceptual understanding, thinking, reasoning, and problem solving—outcomes for which the field has limited measures. Two accessible measures that assess conceptual understanding, reasoning and problem solving, the New Standards Reference Exam and Balanced Assessment, have not been widely used (Schoenfeld, 2006). Many researchers use their own measures whose psychometric properties have not been established (Kilpatrick, 2003).

---

[9] On the other hand, a limitation of most of the items that were developed to measure student learning by curricular developers is that their psychometric properties were not reported, that is, no reliabilities were reported and often curricular validity was the only type established.

Another challenge involves coming to grips with the issue of the differing goals of standards-based vs. conventional curricula. As noted earlier, both would agree to the need for developing some conceptual understanding and some procedural competence; however, the emphasis placed on each varies dramatically between the two kinds of curricula, as does the amount of emphasis placed on reasoning, problem solving, and communication. In addition, there are differences in the topics covered, with, for example, elementary standards-based curricula covering topics not typically found in conventional textbooks including geometry, statistics, and algebraic reasoning. In a curricular evaluation, the question thus becomes: Does one assess only the goals and topics that the two kinds of curricula have in common or does one assess all students on all goals and topics, regardless of which curriculum they were exposed to?

Third, despite being at the heart of this methodology, the creation of comparison groups can be surpisingly difficult. For example, comparison groups are most often selected based on comparability of student samples. Despite the fact that the comparability of teachers is often not considered, experience suggests that teachers in experimental vs. control groups often differ in their levels of enthusiasm, recent professional development experiences, and in their ability to attract a certain kind of student. The NRC report (2004) raises another common concern: studies that compare classrooms using standards-based curriculum with classrooms whose instruction represents a potpourri of instructional styles and approaches. Even if the two sets of classrooms (i.e., the experimental and control groups) are comparable on student background characteristics, the cleaner, more deliberate instructional approach represented by the standards-based classrooms almost always "wins."

A final issue is determining how the curriculum materials were enacted in the classroom. One cannot say that a curriculum is or is not associated with a learning outcome unless one can be reasonably certain that it was implemented as intended by the curriculum developers. Moreover, there is ample evidence that teachers vary in how they implement the same curricular task (Stein et al., 1996; Tarr, Chávez, Reys, & Reys, in press). Assessing implementation quality using observational methods is extremely resource intensive, especially in large-scale studies, and credible self-report of implementation on survey measures is notoriously difficult to achieve (Ball & Rowan, 2004). Nevertheless, claims about the effectiveness of one curriculum over another are commonly made citing student achievement data, but providing little or no data on degree of implementation.

Given the limitations associated with comparative studies, the question naturally arises: What claims can be made about the impact on student learning of various curricula? This question sits at the heart of the charge given to the NRC panel and their conclusion was:

> . . . the corpus of evaluation studies as a whole across the 19 programs studied does not permit one to determine the effectiveness of individual programs with a high degree of certainty, due to the restricted number of studies for any particular curriculum, limitations in the array of methods used, and the uneven quality of the studies (2004, p. 3).

The report goes on to caution that the inconclusive finding of the panel should not be interpreted to mean that these curricula are ineffective either, but rather that problems with the data and/or study designs prevented the panel from making confident judgments about their effectiveness. (The curricula marked with an asterisk in Table 8.1 were included in the NRC review).

Incontrovertible evidence of curricular effectiveness can be provided only by studies that randomly assign students to standards-based vs. conventional curricula, the so-called "gold standard" of the randomized controlled trail (Mosteller & Boruch, 2002; National Research Council, 2002). Although the NRC panel identified no randomized controlled designs for the 19 programs that they studied, the What Works Clearinghouse (http://www.whatworks.ed.gov/PDF/Topic/math_topic_report.html) has produced a Topic Report on middle school curriculum-based interventions that identifies four randomized controlled studies. Two of these studies yielded significant effects: a study by Morgan and Ritter (2002) that examined the effectiveness of the *Cognitive Tutor Algebra 1* and a study by Kirby (2004) on the effectiveness of *I CAN Learn.* Both curricula are standards-based and computer-mediated. The study by Kirby (2004) consisted of one teacher who taught both *I CAN Learn* and a conventional curriculum and used the Georgia Criterion-Referenced Competency Test, a state test, as the outcome measure. The study by Morgan and Ritter (2002) involved nearly 400 students and 18 teachers across five schools and used the Educational Testing Service's (ETS) Algebra I End-of-Course Assessment as the primary outcome measure. In both studies, students who used the standards-based curriculum outperformed the comparison group that was taught with a conventional curriculum (McDougal Littell's *Heath Algebra* in the case of the *Cognitive Tutor* and an unspecified "traditional" curriculum in the case of *I CAN Learn*). The other two randomized controlled studies cited by the

WWC failed to find significant differences in student learning favoring either the standards-based (the *Expert Mathematician/ UCSMP*) or conventional curricula (*Transition Math* or *Saxon*).

Turning to studies that do not meet the gold standard of the randomized controlled trail, one finds less-certain evidence for any one specific curriculum, but, nevertheless interesting patterns across findings from several large-scale studies that compared achievement in classrooms using a variety of standards-based curricula to achievement in classrooms using conventional curricula. Despite the differences in grades and curriculum studied, and the methodologies employed, many of these studies have produced fairly consistent findings. The first is that students taught using standards-based curriculum, compared with those taught using conventional curriculum, generally exhibited greater conceptual understanding and performed at higher levels with respect to problem solving (e.g., Boaler, 1997; Huntley, Rasmussen, Villarubi, Sangtong, & Fey, 2000; Thompson & Senk, 2001). Second, these gains did not appear to come at the expense of those aspects of mathematics measured on more traditional standardized tests. Compared with students taught using conventional curricula, students who were taught using standards-based curricula generally performed at approximately the same level on standardized tests that assess mathematical skills and procedures (e.g., Riordan & Noyce, 2001; Thompson & Senk, 2001). The differences that occurred were usually not significant, and some show the "standards-based curriculum" students doing slightly better, while others show the "conventional curriculum" students doing slightly better. For example, students in the Core Plus Mathematics Project (CPMP) (a standards-based curriculum) outperformed others on tests of algebraic concepts set in real world contexts, but the students taught using more traditional texts outperformed those in CPMP on tests of algebraic skills set in questions without contexts that did not allow calculators (Huntley et al., 2000). Unsurprisingly, students tend to do well on tests that match the approaches through which they have learned.

It is striking to note the similarity between the patterns of findings reported in the Senk and Thompson volume and the findings of the larger-scale comparative studies conducted by external reviewers. In both cases, students taught using standards-based curricula tended to hold their own on tests of computational skills and to outperform students taught with conventional curricula on tests of thinking, reasoning, and conceptual understanding. This pattern of findings—not the findings of any one study—has prompted some to point to the overall efficacy of standards based curricula (e.g., Schoenfeld, 2002).

But, efficacy for what? It is important to note that students tended to perform best on tests that aligned with the approaches by which they had learned, repeating the well-worn finding that students learn what they are taught. Combined with the findings from the analyses of curriculum materials cited earlier, the research examined here suggests that students taught using conventional curricula will learn to compute better, but not to think and reason better than students taught using standards-based curricula. Students taught using standards-based curricula, on the other hand, will learn to think and reason better and (maybe) to compute as well as their counterparts.

Finally, readers must look beyond sound-bite findings in order to ascertain the level of learning actually achieved by students—both in conventional and standards-based interventions. For example, what does it mean to perform "as well as" one's counterparts? In some studies cited above that meant scoring in the lower percentiles of mathematics achievement (e.g., see Morgan & Ritter, 2002; Kirby, 2004). And, studies finding "no difference" do not always mean that students performed dismally. For example, the What Works Clearinghouse reported "no effect" for a standards-based program (the Expert Mathematician) because there were no significant student achievement differences between students exposed to it vs. a more conventional curriculum. Nevertheless, students made learning gains in both groups (Viadero, 2004 as reported by Schoenfeld, 2006).

The above discussion harkens back, once again, to values. Should schools and teachers primarily be held accountable for developing students' capacities for basic proficiencies or for the development of students' abilities to think, reason and problem solve? If both, how should the two be combined?

Methodologically, we acknowledge the superior power of randomized controlled studies to determine the effectiveness of one curriculum over another and we join the National Research Council in calling for more experimental studies. At the same time, we feel that it is important to acknowledge the questions that randomized controlled studies cannot answer and to make a plea for other kinds of studies as well. Returning to the framework depicted in Figure 8.1, the majority of the comparative studies cited above did not collect data on how the various curricula were enacted in the classroom (the third box in our framework—see Figure 8.1); none examined the way in which teachers engaged with the materials to create the intended curriculum (the second box in our framework). When implementation measures are included, the findings can become less clear. For example, Boaler and Staples (in press) studied algebra classes across three high schools, approximately half of which used conventional methods, and half of which used standards-based methods. In tests of algebra at the end of their first year of high school, students in the two groups performed at the same level, both in terms of skills and conceptual understanding. However, this was not an indication that the curricular approach did not matter, only that there was a need to look beyond curriculum to the ways in which teachers implemented them. Indeed the most significant factor in comparisons of algebra achievement in this study was the teacher, with large variations between classes *within* the same curriculum. The importance of individual teachers and their particular teaching decisions was also reported by Huntley et al. (2000) who found that different classes in the same curriculum varied to a large degree in classroom implementation and student achievement. And, finally, using multiple, non-observational measures of curricular implementation,[10] Balfanz, Mac Iver, and Byrnes (2006) found differences in levels of curriculum implementation that were associated with differences in student achievement with classrooms that exhibited higher levels of implementation averaging higher achievement gains.

As noted earlier, many scholars and policymakers have (implicitly) conceptualized the question behind student-outcome studies as the testing of a (causal) relationship between curricular materials (the written curriculum—see first box of Figure 8.1) and student learning (the final triangle in Figure 8.1). Our framework, together with findings associated with most of the above studies, suggests, however, that such a conceptualization—although useful for some reasons—is incomplete. Although the above studies that focused on student outcomes can reveal *whether* a particular curriculum achieved superior outcomes, they cannot shed light on *how*. However, knowledge of how an effect was achieved is crucial, both for enhancing the field's understanding of teaching and learning mathematics as well as for others who wish to implement the curriculum and for designers hoping to improve the curriculum. Thus, a review of *only* these studies is insufficient. Without accompanying knowledge of how the various curricula were implemented, students' opportunities to learn remain shrouded. In the best case, this limits our un-

derstanding of how desirable effects were achieved; in the worst case, it raises questions about the extent to which the achieved effects on student learning can be attributed to a particular curriculum. Thus, we call for studies that are designed to include both large-scale tests of curricular effectiveness and smaller, but embedded, observational studies of instructional practice.

There are too few studies of this kind, two notable exceptions not withstanding. Tarr, Reys, et al. (in press) compared classrooms that used either standards-based or conventional curricula, collecting both achievement data and observational data on the extent to which the classrooms exhibited a "standards-based learning environment (SBLE)." These observations documented the extent to which teachers created learning environments consistent with the tenets of the NCTM Standards and were not tied to any particular curriculum. They found that students in the classrooms using standards-based curricula that also had high levels of SBLE implementation outperformed students who were in classrooms using standards-based curricula but having lower levels of SBLE implementation. In the aforementioned longitudinal study of three California high schools, Boaler and Staples (in press) utilized a range of qualitative and quantitative methods including student assessments, interviews, and questionnaires, and over 600 hours of classroom observations. As such, they were able to report not only that students in Railside high school (where instruction was built around standards-based curricula) achieved at higher levels,[11] but also the ways that the teachers brought this about through their responses to student thinking, their questioning, and the allocation of time. More studies like this—that attend to both student learning outcomes and the classroom-based mechanisms that appear to relate to those outcomes—are needed.

Despite few studies that examine student learning and instructional practice together, there are many research studies that have investigated the mechanisms and processes by which teachers engage curricula, both in their planning and during instruction. These studies provide an important complement to comparative studies that primarily focus on student outcomes. This brings us to the second area of research that will be reviewed: what happens when teachers engage with and interpret curriculum materials.

---

[10] The intervention consisted of the University of Chicago School Mathematics Program middle school curriculum supported by professional development, coaching and a surrounding whole-school reform program.

[11] The Railside students achieved at higher levels on all measures (researcher-designed and district examinations) except the standardized test used by the state of California.

SECTION TWO:
HOW TEACHERS ENGAGE WITH AND
INTERPRET CURRICULAR MATERIALS

This section examines teachers' use of curriculum materials as a factor in student learning from curriculum. As our framework suggests, teachers have a substantial role in interpreting and using written curriculum materials and enacting them in the classroom. In this section, we focus on the first step in this process that is labeled in our framework as the relationship between the written curriculum and the intended curriculum.

As discussed earlier, the written or formal curriculum refers to the curriculum represented in curriculum materials or other teaching resources. The intended curriculum refers to the teacher's intentions, that is, the curriculum the teacher plans to enact. As our model suggests, the intended curriculum is often a derivative of the written curriculum; however, there are distinct differences between the two. On a conceptual level, written and intended curricula are different kinds of representations of curriculum. The former is a set of written guidelines and resources aimed at leading and supporting teachers to enact what curriculum developers imagine taking place in the classroom. The latter is comprised of a set of ideas and plans held in the teacher's head. In this sense, the intended curriculum is more closely aligned with an image of classroom practice than the written curriculum.

A second distinction between the two is that the intended curriculum represents the written curriculum as transformed by the teacher. In other words, the process of reading and using curriculum materials necessarily involves interpretation and meaning making on the part of the teacher. In this way teachers do not merely read and follow written curriculum. Just as reading itself is a process of interpretation (Mailloux, 1982), teachers draw on their own experiences and knowledge to make meaning of what they read in curriculum guides. In this sense, they transform the written curriculum from generic, written description and guidelines to both an image of classroom practice and something that has specific meaning to them.

Scholarly interest in this transformation process and the factors that influence it is motivated, in part, by questions about the extent to which curriculum materials can influence teachers and, subsequently, impact students. Research that examines teachers' use of cur-

ricula assumes that teachers play a mediating role in how materials influence student learning and seeks to provide clarity about a) the processes by which teachers use, interpret, and transform curricula, b) the factors that influence it, and c) how conditions and contexts can impact these processes.[12] In this section, we explore how researchers have framed and studied the relationship between the written and intended curriculum and the stances toward conceptualizing the teacher-curriculum relationship represented in the literature. The factors and conditions that influence how teachers engage with curricular materials (b and c above) are explored in Section Four.

Framing of the Relationship between Written and Intended Curriculum

The framework guiding this chapter assumes that written and intended curriculum differ in significant, although sometimes, subtle ways. This view, however, has not always been embraced by curricular scholars. Historical studies of school curricula, for instance, relied heavily on textbooks of the period to reconstruct the contents of classroom practice (Love & Pimm, 1996; Walker, 1976). By highlighting ways that teachers adapted or resisted unfamiliar curriculum programs developed during the period of curriculum reform in the late 1950s and 1960s, researchers such as Sarason (1982), Stake and Easley (1978), and Berman and McLaughlin (1973, 1978) challenged conventional assumptions that teachers merely followed their textbooks and that texts alone represented classroom instruction. These studies frame teachers as active users of curriculum, rather than conduits of it and opened the door to questions about the various ways that teachers make use of curriculum materials in shaping the curriculum they intend to teach. The following discussion examines the variety of ways researchers have described and categorized teachers' approaches to curriculum use.

Content Coverage

As part of a year-long study of how content is determined in fourth grade classrooms, Freeman, Belli, Porter, Floden, Schmidt, and Schwille (1983) undertook case studies of seven teachers using mathematics textbooks published in the 1970s. Their analysis produced what the authors identified as four styles of textbook use that focus on the mathematical content

---

[12] There are only a handful of studies that examine the processes by which teachers interpret curriculum exclusively; typically, such research is incorporated into studies of how the curriculum is enacted in the classroom. As a result, some of the studies cited in this section are also cited in Section Three.

covered. They used the term *textbook bound* to refer to an approach to textbook use in which the teacher followed the guide lesson by lesson, beginning at the first chapter and, more than likely, not completing the last few chapters before the close of the academic year. The *selective omission* style, was similar to textbook bound, however, the teacher skipped some chapters entirely. Teachers who were identified as *basics focused*, placed priority on teaching the four operations and used their textbook selectively to assist them in doing so. Freeman and colleagues also identified a subcategory of these teachers who included the topics of measurement as a basic skill. The final style of use, identified as *management by objectives*, involved following a set of objectives imposed by district and school policy and using the textbooks as tools to attend to the goals set out by these policies.

It is likely evident to most readers that the categories proposed by Freeman et al. (1983) reflect the conventional mathematics curricula that dominated the educational scene at the time. However, what stands out about the approach of these researchers is their focus on content coverage over pedagogical strategies or other dimensions of textbook use.

In their examination of 39 middle school teachers' use of both standards-based and conventional textbooks, Tarr, Reys, et al. (2006) also focused on mathematical content coverage. These researchers were interested in determining how closely teachers followed the content offered in their mathematics curriculum materials, particularly in terms of the distribution of mathematical topics. By comparing the proportion of each program devoted to number and operations, algebra, geometry and measurement, and data analysis and probability to the proportion of lessons taught devoted to each topic, they found that all teachers made both omissions and additions to the material found in the text in favor of number and operations. Lessons on data analysis and probability were omitted most frequently.

Components of the Curriculum

Other ways of categorizing curriculum material use place primary emphasis on differences in *how* teachers use the same lesson. Remillard and Bryans (2004), for example, studied 8 elementary teachers during the first two years of using *Investigations*, a standards-based curriculum program. In their analysis, which drew on interviews and observations, they characterized teachers with respect to the extent to and thoroughness with which they used the structure of the curriculum, its mathematical content, and the pedagogical suggestions and information offered.

The two teachers who used the curriculum *narrowly and intermittently* relied primarily on their own teaching routines and other resources to guide their curriculum choices over the year. When they did use the standards-based curriculum program adopted in their school, they tended to use it by selecting tasks that seemed familiar and using the repertoires they had developed over years of teaching when enacting them in the classroom. The two teachers identified as *adopting and adapting* the curriculum used the materials as a guide for the general structure and content of their mathematics curriculum. They regularly adopted mathematical tasks from the curriculum guides, but drew on their own strategies and approaches to enact them in the classroom, adapting them to fit their familiar approaches to teaching. The authors identified four of the eight teachers as *thorough piloting* the curriculum. These teachers used the materials as their primary guide in teaching mathematics, which included structuring the curriculum, selecting tasks, and facilitating students' work on those tasks. They tended to read and use all parts of the curriculum guides and sought to follow the lessons as suggested.

An important distinction made in Remillard and Bryans' (2004) categorization system is between teachers who read and follow pedagogical guidance offered by the curriculum and those who relied on their own repertoires when using tasks provided by the curriculum. In other words, the same tasks can be used differently by different teachers. Moreover, because the most recent generation of curriculum materials places substantial emphasis on pedagogical practices, considering how and whether teachers use pedagogical information in the materials is critical to understanding the relationship between the written and intended curriculum.

In his study of middle school science teachers using a standards-based science curriculum, Brown (2002) also focused on *how* teachers used the curriculum guide, emphasizing the shifting nature of this relationship. He identified three different ways that teachers used the materials—offloading, adapting, or improvising. *Offloading* refers to when teachers offload curriculum design responsibilities onto the curriculum and follow it fully. *Adapting* refers to when teachers use suggestions offered in the materials but adapt them to suit their particular needs or preferences. *Improvising* refers to when teachers move away from the curriculum suggestions substantially and design their own.

These types of curriculum use are similar to those offered by Remillard and Bryans (2004) in that they focus on the extent to which the teacher used the guidance offered by the authors. Brown, however, suggests that teachers tend to engage in different kinds of use at different times. The same teacher might offload in one instance, allowing the curriculum to carry the weight of curriculum design. In another instance, she might improvise, drawing on the curriculum minimally. Although his study was of middle school science teachers, Brown's finding that teachers' approach to curriculum use can vary from lesson to lesson depending on a number of factors contributes to what is understood about the relationships between the written and intended curriculum.

Program Philosophy

Lambdin and Preston (1995) studied middle school teachers using *Connected Mathematics* (CMP), a standards-based curricula. Using interviews and observations, these researchers developed three classifications of teachers to represent the patterns they observed across teachers studied. These classifications highlight the teacher's stance in relation to the underlying philosophy of the program. *Frustrated methodologists,* for example, were teachers whose established classroom routines were in conflict with the goals of the curriculum and were reluctant to change. These teachers' use of the curriculum was mechanical in nature and "characterized by disjointed often superficial use of curriculum materials" (p. 135). The frustrated methodologists described by Lambdin and Preston were often torn between two philosophies of teaching, theirs and that underlying the curriculum. The *teacher on the grow,* describes a group of teachers "open to change and anxious to learn" (p. 136) in order to use the new curriculum more successfully. These teachers often experienced initial reluctance or doubt about the intent of the new curriculum, but, overtime, developed appreciation for its intent and approach. Lambdin and Preston described the final group of teachers as *standards bearers.* These teachers had established classroom routines and pedagogical approaches closely aligned with the philosophy of the program.

Lambdin and Preston's (1995) categories for teachers in relation to curriculum materials focus on the degree of match between the teacher's philosophy of teaching mathematics and the pedagogical stance represented in the curriculum. These researchers found that confidence in one's mathematical knowledge was critical in determining how teachers were classified. Of equal importance, however, was their understanding of the program's epistemological stance or philosophy. In fact, they found more evidence of teachers increasing mathematical knowledge and gaining confidence in their knowledge through use of the curriculum than they found of teachers coming to understand and appreciate new pedagogical approaches.

Looking across these differing schemes for categorizing how teachers use curriculum materials, we see significant interest in and reason for identifying and examining the relationship between the curriculum as written and as intended by the teacher. Indeed, there is overwhelming evidence that teachers use and enact curriculum in substantially different ways. The varieties of kinds of use examined hint at the complexity of the relationship between written curriculum materials and teachers' uses of them. Moreover, these varied categorization schemes represent some of the possible ways that curriculum use can differ (e.g., in content coverage, in extent and nature of reliance on various curriculum components, in epistemological and philosophical stance). An examination of the kinds of use identified in these studies as a set can offer insight into the dimensions along which curriculum use can vary. As a field, we ought to take as much interest in these dimensions as we do in the variations within each study. Considering these dimensions as a set can shed light on the complexity of curriculum use. While most studies tend to focus on a single dimension, such as philosophical stance, attending to a set of them can offer a fuller understanding of what matters in teachers' curriculum use. Further, attending to these dimensions raises questions for further research. Are there other dimensions along which variation in use occurs that matter? What are the differential consequences of these kinds of variation for classroom activity? Do some variations matter more than others?

A second reason to examine the dimensions of variation in material use identified by researchers is to uncover points of convergence across studies. The different foci taken by researchers along with the introduction of new categorical schemes and terms to label kinds of use points to a need for clarity in how curriculum use is examined and identified. Given the reality that the predominant methods for studying curriculum use involve intensive observation over extended time frames, it is critical that the field establishes structures that enable aggregation of findings across studies.

A final point to consider when examining studies of curriculum material use by teachers is the extent to which the dimensions of use identified parallel the dimensions used to frame analyses of the curriculum materials themselves. The studies above that examined content coverage by teachers (e.g., Freeman et al., 1983, and Tarr, Chavez et al., in pres) can, to some extent, be viewed as mirroring analyses of curriculum materials that focus primarily on content and topic coverage (see pages 327–330 in this chapter). On the other hand, the research cited above by Remillard and

Bryans (2004), Brown (2002), and Lambdin and Preston (1995) that focused on teachers' use of pedagogical and philosophical dimensions of curriculum materials align with studies of curriculum materials that focus on how the content is presented, both to students and teachers (see pages 330–335 in this chapter). We believe that the knowledge base associated with how curriculum impacts student learning would be advanced if the field established common structures for examining both the materials themselves and the dimensions along which researchers examine teachers' use of those materials.

We now turn to different conceptualizations of curriculum use in the literature and how these stances shed light on the manner in which teachers interact with curriculum materials in shaping the intended curriculum.

Conceptualizations of Curriculum Use

In an analysis of the literature on teachers' use of mathematics curriculum materials Remillard (2005) identified several different ways curriculum use was studied in the literature; three of these approaches represent differing stances on what it means to use curriculum. As a result, they reveal different dimensions of the relationship between the written and intended or enacted curriculum. These three stances on curriculum use include: use as following or subverting, use as interpretation, and use as participating with. The following descriptions are accompanied by illustrative examples from the literature.

Curriculum Use as Following or Subverting

Studies that frame curriculum use as following or subverting view the written curriculum as embodying discernible and complete images of practice and examine the degree to which teachers follow these guidelines with fidelity. These studies implicitly contend that the degree of match between the written curriculum materials and enacted curriculum is measurable. As a result, researchers undertaking such studies concern themselves with how curriculum writers and others might achieve greater clarity and closer guidance for the teacher. These studies that seek to capture "treatment integrity," as called for by the NRC (2004) report, are critical to assessing the impact of curriculum materials on student learning.

This perspective is evident in studies of teachers using conventional curricula as well as in studies of teachers using standards-based curricula. In these studies, researchers found many more cases of subverting the curriculum than of close following. As a result, they tended to make recommendations that might

result in greater fidelity, such as increased specificity within curriculum materials and enhanced professional development. It is worth noting that in almost all cases, researchers found that teachers' deviations from the written curriculum reflected tendencies toward conventional pedagogical approaches and representations of mathematics.

Studies by Freeman et al. (1983), Freeman and Porter (1989), Stodolsky (1989), and Sosniak and Stodolsky (1993) all provide examples of analyses of teachers' use of conventional curricula. Their findings revealed that teachers used texts selectively, focusing on basic skills instruction and core mathematical topics, and demonstrated, in Freeman and Porter's words, that the conviction that textbooks determined the mathematics curriculum reflected "a narrow view of teacher decision making" (p. 404).

Studies of teachers using innovative curriculum materials since the 1980s (e.g., Donovan, 1983; Manouchehri & Goodman, 1998; Stephens, 1982; Tarr, Chavez, et al., in press) examined teachers using curriculum materials that reflect unfamiliar views of mathematics and approaches to mathematics teaching. These researchers all found that teachers tended to use mathematical topics and tasks from the curriculum guides, but made alterations as well, often subverting the designers' intentions, transforming the tasks to fit conventional approaches to teaching and classroom management structures.

In a study of 39 teachers from six different states, 17 using standards-based and 22 using commercially produced texts, Tarr, Reys et al. (in press) examined textbook fidelity and its relationship to student outcomes. They measured fidelity by considering: (a) the extent to which the teacher covered topics and lessons in the textbook, (b) the amount of time spent on topics in the textbook, (c) in the case of standards-based curriculum, the extent to which the classroom activities reflected a "standards-based learning environment." They found that, on average, teachers used 60 to 70% of the lessons in the curriculum guides, regardless of the type of curriculum. In fact, the greatest influence on whether lessons were omitted was their topic. Teachers of both kinds of curriculum materials were more likely to omit lessons on data and probability than on number and operations. While the use of standards-based curriculum materials was not a predictor of the presence of a standards-based learning environment (SBLE), measured on a ten point scale, only one classrooms using commercially generated curriculum materials had a SBLE rating over 5. Half of the classrooms using standards-based curriculum materials were rated between 6 and 10 on the SBLE scale.

As a group, studies that examine the closeness of match between a written curriculum and actual practice shed light on the aspects of materials teachers tended to follow as well as the variety of ways they adapt and subvert suggestions in these materials. While teachers are likely to follow the mathematical topics offered in curriculum materials, they do so differentially, favoring topics such as number and operations over probability and statistics. Moreover, they are much less likely to follow the pedagogical recommendations or embrace the philosophical stance of the materials.

Curriculum Use as Interpretation

Research grounded in a view of curriculum use as interpretation assumes that fidelity between written plans in a teacher's guide and classroom action is impossible (Remillard, 2005). This view holds that teachers bring their own beliefs and experiences to their encounters with curriculum to create their own meanings, and that by using curriculum materials teachers interpret the intentions of the authors. In her book on teachers' encounters with curriculum materials, Ben-Peretz (1990) argues that teachers draw on personal knowledge and experience to "assign meaning to the curriculum materials they use daily in their classrooms" (p. 71). Research from this point of view investigates the nature of teachers' interpretations, the factors that influence them, and the resulting classroom practices.

Researchers who view curriculum use as a process of interpretation have applied this view to a range of initiatives intended to influence teaching, including educational policy. In their studies of the relationships among state-level policy, district and classroom practices, researchers with the Educational Policy and Practice Study (EPPS)[13] maintained that policy and its multiple instantiations, such as textbooks and tests, are open to interpretation. Beginning in 1988, EPPS used case-study methods to consider how elementary teachers in California learned about, understood, and acted on state-level mathematics policy in the late 1980s. The state's initial efforts to alter mathematics instruction placed heavy emphasis on approved textbooks to communicate its message of reform. Hence, how teachers interpreted and used their new textbooks was central to this work (e.g., Ball, 1990; Cohen, 1990; Heaton, 1992; Putnam, 1992; Remillard, 1992; Weimers-Jennings, 1990; S. M. Wilson, 1990).

A particularly striking finding from this research with respect to the interpretation of policy and curriculum materials was the conviction with which all of the participating teachers believed that their teaching reflected the ideas of the new policy's reform intentions as a result of their faithful use of a particular textbook or curriculum program. However, their interpretations of the goals of the reform initiative and their uses of their texts varied tremendously. Variations included subtle and not so subtle adaptations of the plans suggested in the text as well as diverging interpretations of what it means to engage students in problem solving or discuss their solution strategies.

A number of researchers studying teachers using standards-based curriculum materials have focused their analyses on teachers' interpretations of the unfamiliar curricula as well as the NCTM *Standards* themselves. Collopy (2003) studied two teachers using the *Investigations* curriculum for the elementary grades and found significant differences in their uses. The most extreme case of this kind of variation was in how the teachers used the illustrative dialogues provided in the curriculum. These scripted conversations, presented in the teacher's guide like the dialogue in a play, illustrated an example of a discussion a class might have about a focal concept or phenomenon. One teacher read them in order to anticipate the ideas that might come up during a class discussion while the other teacher used them as scripts and had students read the different student parts aloud. Collopy attributed these dissimilar interpretations to the teachers' contrasting views of curriculum and the degree to which they had firmly established pedagogical repertoires and curriculum structures.

Like those that frame curriculum use as following or subverting, these studies also highlight variations in how teachers use curriculum resources. They also go a step further in illuminating the process by which these variations occur. By framing curriculum use as an interpretive process, researchers make visible the ways that teachers' meaning making processes shape the intended curriculum. The factors that influence teachers meaning making are discussed in Section Four of the chapter.

Curriculum Use as Participating With

A third perspective taken by researchers studying teachers and curriculum materials focuses on the teacher-text relationship or the activity of using the text. This perspective treats curriculum material use

---

[13] This group began its work in 1988 on a study of the relationship between state-level mathematics policy and classroom practice in California. Later, in 1992, as it expanded its focus to include three states and policy in two subjects (mathematics and reading), the group took the title EPPS.

as collaboration with the materials (Remillard, 2005). Central to this perspective is the assumption that teachers and curriculum materials are engaged in a dynamic interrelationship that involves participation on the parts of both the teacher and the text. While there is significant overlap between this perspective and the stance that focuses on use as interpretation, the core difference is the focus of the analyses. Indeed, researchers in either category might view curriculum use as a process of interpretation. However, the research in this group seeks to study and explain the nature of the participatory relationship. In other words, the distinguishing characteristic of this perspective is its focus on the activity of using or participating with the curriculum resource and on the dynamic relationship between the teacher and curriculum. Studies taking this perspective consider the ways that particular features of the curriculum resource, as well as the teacher's interpretations of them, work together to shape the intended curriculum.

The view that curriculum material use involves a dynamic interchange between teacher and curriculum is reflected in Lloyd's (1999) study of two high school teachers using *CORE Plus*, a standards-based curriculum. The study investigated "how and why two teachers encountered particular successes and difficulties as each implemented a set of novel curriculum materials for the first time" (p. 229). The author examined the teachers' conception of the curriculum and of key ideas central to it, including exploration and cooperation, and the teachers' resulting mathematics teaching. She argued that "curriculum implementation consists of a dynamic relation between teachers and particular curricular features" (p. 244), but also suggested that this relationship can be strained by tensions between the structure outlined in the curriculum guide and the teachers' need to construct curriculum that is responsive to students.

Remillard's (1999, 2000) study of two teachers using a textbook that represented a commercial publisher's response to the NCTM *Standards* (1989) also examined teachers' participation with curriculum. The study illustrates how two fourth grade teachers interacted with the same text in different ways to construct contrasting opportunities for student learning. The analysis highlights the ways the teachers read the textbook and explores the factors that contributed to different approaches to reading. Not only did the two teachers read entirely separate features of the text (exercises for students versus supplementary activities), they also read for different purposes (potential activities and assignments versus big ideas to guide planning). These contrasting uses of the textbook resulted in different plans from the same lesson in the text.

Studies like these that frame and examine curriculum use as an interactive process not only capture how teachers interpret curriculum materials, but focus attention on the role the materials play in this process. They illustrate the complexity of transformation from the written to the intended curriculum.

While highlighting different aspects of the teacher-curriculum relationship, the above variations in the way curriculum use is conceptualized in the literature also point to fundamental differences in theoretical and conceptual assumptions underlying the research. These differences, when not made explicit, further complicate efforts to examine findings across studies and illustrate the need for greater clarity in how these constructs are conceptualized.

The relationship between the written and intended curriculum (the first and second boxes of Figure 8.1) has been neglected in most studies of how curriculum influences student learning. Yet the set of studies reviewed here point to important variations in curriculum use that are introduced even before the lesson begins. These studies are important because they challenge the assumption that curriculum materials tend to be implemented uniformly across teachers, schools, and districts. If research suggests that the curriculum is altered the minute that a teacher begins to read it, then the assumption that curriculum materials are single-handedly—or even primarily—responsible for student performance becomes questionable and the need to broaden the logic to include a discussion of how teachers interpret the curriculum becomes apparent. Thus, we see this set of studies as laying the first stone in the foundation of a chain of reasoning, evidence and argument that must underlie future studies of how curriculum influences student learning.

Throughout this chapter, we call for this chain to be broader, more inclusive, better-articulated, and more internally coherent than what currently exists. As we have already noted, the studies reviewed in Sections One and Two reveal the need for greater consistency with respect to the dimensions that are examined when analyzing the curriculum materials (phase one of the framework) and curriculum interpretations (phase two of the framework). Another point raised for us by these studies concerns the consistency with which the term *fidelity of implementation* is used (fidelity is an important concept in the "following vs. subverting" approach to studying curriculum use). More specifically, might "fidelity" mean something different when one is discussing a standards-based versus conventional curriculum materials? If conventional materials—as described earlier—present content and exercises in a straightforward manner, might not the attainment of fidelity be, not only easier to achieve, but also easier to measure? One would examine the extent to which teachers provided the correct definitions, examples, and procedures and whether they monitored student practice accurately. Standards-based curricula, as noted earlier, tend to present information indirectly, relying on tasks—and students' work on those tasks—to expose the concepts to be learned, thereby placing considerable demand and discretion on teachers to help students correctly surface and explicate concepts and desirable ways of thinking and reasoning. Here, one would need to examine the extent to which teachers' ways of assisting students was compatible with this indirect style of teaching, the extent to which the teacher listens to and "pulls from" students important ideas, and, if and how the students correctly surface the intended material. In the first case, one appears to be measuring fidelity to a given set of teacher procedures, in the second, the fidelity is to an underlying philosophy and theory of learning.

SECTION THREE: THE ENACTMENT OF CURRICULA IN CLASSROOMS

This section examines the ways in which curricular tasks are set up and enacted during instruction. Here our focus moves beyond teachers' intentions to the examination of what actually occurs as curriculum materials are enacted in the classroom—the third phase of curricular use shown in Figure 8.1. We begin this section by discussing the various ways in which curriculum enactment has been studied and providing a rationale for the approach taken herein—a focus on mathematics tasks. We continue by exploring the source and nature of mathematical tasks that are used as the basis of instruction in our nation's classrooms. We then turn our attention to research that illuminates the ways in which mathematical tasks are set up and enacted by classroom teachers and their students and the processes that influence the ways in which tasks actually unfold in the classroom.

Ways in Which Curriculum Enactment Has Been Studied

Studies that can shed light on the enactment of curriculum in the classroom vary with respect to the aspects of instructional practice to which they have attended. Over the past decade, researchers have conducted studies focused on classroom discourse (e.g., O'Conner, 2001), questioning (e.g., Boaler & Brodie, 2004), intellectual authority (e.g., Wilson & Lloyd, 2000) and on normative practices (e.g., Boaler & Sta-

ples, in press) in classrooms in which standards-based and conventional curricula were being used. While not focusing directly on curricular enactment, these studies provide insight into the challenges and possibilities of orchestrating teaching and learning interactions around standards-based curricular tasks.

In this section, we organize our discussion of these and other studies through the conceptual lens of mathematical instructional tasks. We have made this decision for several reasons. First, classroom instruction is generally organized and orchestrated around mathematical instructional tasks. That is, students' day-to-day work in mathematics classrooms consists of working on a tasks, activities, or problems. For example, an analysis of 100 eighth grade lessons revealed that the delivery of content "was accomplished primarily by working through problems" (NCES, 2003, p. 144). Second, the tasks with which students engage determines what they learn about mathematics and how they learn it. According to Doyle (1983, p.161), "tasks influence learners by directing their attention to particular aspects of content and by specifying ways of processing information." Finally, tasks are a way of framing how researchers have looked at the enactment of curriculum (e.g., Doyle 1983 1988). And, because a discussion of tasks typically implicates many of the dimensions discussed above (i.e., discourse, questioning, authority structures, normative practices), we are able to discuss many of these attributes of implementation under the overall umbrella of task.

The Source and Nature of Mathematical Tasks

We define mathematical tasks as classroom activities the purpose of which is to focus students' attention on a particular mathematical idea. Tasks include expectations regarding what students are expected to produce, how they are expected to produce it, and the resources available for so doing (Doyle, 1988; Stein et al., 1996). Curricular materials or textbooks are the main source of mathematical tasks used by teachers for classroom instruction. An analysis of responses to teacher questionnaires, administered under the auspices of the National Assessment of Educational Progress in 2000, revealed that more than two-thirds of students at grades 4 and 8 had teachers who reported that students worked on problems from the textbook on a daily basis (Grouws et al., 2004). An additional 20 percent of students at both grades 4 and 8 had teachers who reported using tasks from the textbook once or twice a week. Although these percentages represented a significant decrease in textbook use from 1992 (Grouws et al., 2004), these data continue to

Task A Martha's Carpeting Task	Task B The Fencing Task
Martha was recarpeting her bedroom, which was 15 feet long and 10 feet wide. How many square feet of carpeting will she need to purchase?	Ms. Brown's class will raise rabbits for their spring science fair. They have 24 feet of fencing with which to build a rectangular pen to keep the rabbits. a. If Ms. Brown's students want their rabbits to have as much room as possible, how long would each of the sides of the pen have to be? b. How long would each of the sides be if they had only 16 feet of fencing? c. How would you go about determining the pen with the most room for any amount of fencing? Organize your work so that someone else who reads it will understand it.

Figure 8.3 Two different mathematical tasks that involve the calculation of area (Stein et al., 2000, pp. 1–2)

make salient the centrality of textbooks in classroom instruction. Data from the most recent TIMSS video study provide additional evidence of the centrality of textbooks. Specifically, 98% of the lessons from the U.S. analyzed in this study used a textbook or a worksheet (NCES, 2003).

As we have discussed earlier in this chapter, curricular materials vary considerably with respect to the nature of the mathematical tasks that are found within them. For example, in comparison to conventional curricula, standards-based curricula have more contextual problems, more tasks that can be solved by using a range of strategies, and fewer exercises that require only application of procedures (Senk & Thompson, 2003). Making distinctions between tasks is important because all tasks do not provide the same opportunities for students' thinking and learning. Tasks that ask students to perform a memorized procedure in a routine manner lead to one type of opportunity for student thinking; tasks that demand engagement with concepts and that stimulate students to make connections lead to a different set of opportunities for student thinking.

Consider, for example, the two tasks shown in Figure 8.3. Task A (Martha's Carpeting Task) requires the simple application of the formula for finding the area of a rectangle (i.e., $A = l \times w$, where l and w refer to the length and width of a rectangle respectively). There is no ambiguity about what needs to be done in order to solve this problem or how one should go about doing it. Therefore it can be viewed as a task with low-level cognitive demands. By contrast in order to solve Task B (The Fencing Task), students must do more than apply a formula in order to be successful. They must generate and test different pen configurations in order to identify the pen that yields the maximum area using two different lengths of fencing (first 24 and then 16 feet). Students must then generalize beyond the two cases in order to determine the dimensions of the pen that will yield the largest area for *any* amount of fence. As a result, Task B can be viewed as a task with high-level cognitive demands. Students must construct their own approach for solving the task as none is obvious or prescribed. Although the area formula may be used to determine the area of various pen configurations, the task also requires students to engage in additional mathematical thinking and reasoning that includes looking for patterns, making and testing conjectures, and finding a generalization that can be explained and justified.

Hence, the tasks with which students become engaged in the classroom form the basis of their opportunities to learn *what* mathematics is and *how* one does it (Doyle, 1983, 1988). What students can learn from problems such as Martha's Carpeting Task is significantly different from what they learn from the Fencing Task. Over time, the cumulative effect of classroom-based tasks is students' implicit development of ideas about the nature of mathematics—about whether mathematics is something that they personally can make sense of, and how long and how hard they should have to work to do so.

Since the introduction of the *Curriculum and Evaluation Standards for School Mathematics* (NCTM, 1989) there has been an attempt to describe the types of tasks that promote and develop students' understanding of mathematical concepts and ideas. Although these descriptions differ with respect to the actual language used to describe the nature of the activities in which students should engage, they share a common focus on developing students' capacity for non-algorithmic thinking. For example, *worthwhile mathematical tasks* (NCTM, 1991) are described as "ones that do not separate mathematical thinking from mathematical concepts or skills, that capture students' curiosity, and that invite them to speculate and to pursue their hunches (p. 25)." Hiebert and his colleagues (Hiebert et al., 1997) have argued that in order for students to build mathematical understandings, the tasks with which they engage "must allow the students to treat the situations as problematic, as something they need to think about rather than as a prescription they need to follow" (p. 18). Such experiences, in their view, leave behind important insights into the structure of mathematics and strategies for solving classes of problems. More recently, Horn (2005, p. 22) has used the term "groupworthy" to describe tasks that "illustrate important mathematical concepts, allow for multiple representations, include tasks that draw effectively on the collective resources of a groups, and have several possible solution paths." While similar in some ways to the task descriptions provided by NCTM and Hiebert and his colleagues, Horn's notion of groupworthy tasks also makes salient the need to provide students who are working collaboratively—an organizational structure often found in classrooms that use standards-based curricula—with tasks that allow for multiple entry points so that group members who bring different strengths to the table can contribute to solving the task at hand.

Stein and her colleagues (Henningsen & Stein, 1997; Smith & Stein, 1998; Stein et al., 2000; Stein et al., 1996; Stein & Lane, 1996; Stein & Smith, 1998) have provided a taxonomy of mathematical tasks based on the kind and level of thinking required to solve them. From this perspective, mathematical tasks are viewed as placing *high-level* cognitive demands on students when they appear to engage students in the processes of active inquiry and validation (what Stein and colleagues refer to as "doing mathematics") or encourage them to use procedures (broadly defined to include standard and nonstandard procedures, formulas, and algorithms) in ways that are meaningfully connected to concepts or understanding. Tasks that encourage students to use procedures, formulas, or algorithms in ways that are not actively linked to meaning; or that consist primarily of memorization or the reproduction of previously memorized facts are viewed as placing *lower-level* cognitive demands on students. Examples of tasks that would be classified at each of the four levels of cognitive demand in the Stein et al. taxonomy are shown in Figure 8.4. In the TIMSS Video Study (NCES, 2003; Stigler & Hiebert, 2004) a similar distinction was made between low-level tasks that can be solved with basic computation and procedures (*using procedures*) and high-level tasks that focus on concepts and connections among mathematical ideas (*making connections*).

In addition to providing a taxonomy for classifying mathematical tasks, Stein and her colleagues also provide a framework for tracking the cognitive demands of mathematical tasks as they unfold during instruction and for exploring the connection between instruction and student learning. The Mathematical Tasks Framework (MTF) distinguishes three phases through which tasks pass as they unfold during a lesson (Stein et al, 1996): First, as they appear in curricular or instructional materials (i.e., on the printed pages of textbooks, ancillary materials, or as created by teachers); next, as they are set up or announced by the teacher; and finally, as they are actually enacted by students and the teacher in the classroom—in other words, the way in which students actually go about working on the task. All of these, but especially the enactment phase, are viewed as important influences on what students actually learn.

This research suggests that the cognitive demands of a task can change as the task unfolds. For example, the task that appears in curricular materials is not always identical to the task that is set up by the teacher in the classroom (reflecting the interpretive processes discussed in Section Two) and, this task, in turn, is not always identical to the task that the students actually do. Between the set-up and implementation phases, tasks can transform for a variety of reasons, many of which have to do with the press of the classroom environment. It is interesting to note, however, that tasks that are set-up at a low-level are nearly always implemented as intended, a point to which we will return (Stein et al., 1996).

In the remainder of this section, we use the Mathematical Tasks Framework as an organizing frame to discuss large and small-scale studies that provide insights into the ways in which mathematics tasks are enacted in the classroom.

Setting Up and Implementing Mathematical Tasks

Research has shown that maintaining high-level cognitive demands during implementation is much more difficult than selecting and setting up high-level tasks. In a study involving 144 lessons selected to be representative of instruction at four middle schools that participated in the QUASAR Project (Stein et al., 1996; Henningsen & Stein, 1997), three-quarters of the tasks were set-up in ways that demanded that students either engage in the active processes of "doing mathematics" or use procedures with connections to concepts, meaning, or understanding (Stein et al., 1996). The study also showed, however, that the higher the demands that a task placed on students at the set-up phase, the less likely it was for the task to be carried out at that level during the implementa-

Lower-Level Demands	Higher-Level Demands
Memorization What is the rule for multiplying fractions? *Expected Student Response:* *You multiply the numerator times the numerator and the denominator times the denominator* OR *You multiply the two top numbers and then the two bottom numbers.*	**Procedures with Connections** Find 1/6 of 1/2. Use pattern blocks. Draw your answer and explain your solution. *Expected Student Response:* 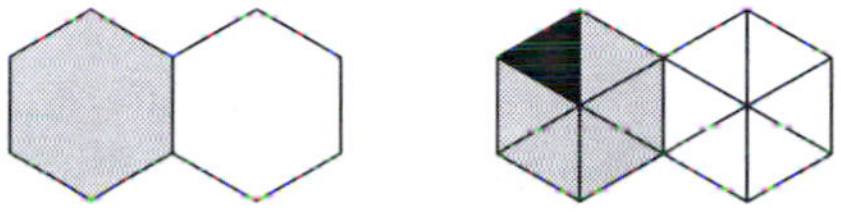 *First you take half of the whole which would be one hexagon. Then you take one-sixth of the half. So I divided the hexagon into six pieces which would be six triangles. I only needed one-sixth so that would be one triangle. Then I needed to figure out what part of the two hexagons one triangle was and it was 1 out of 12. So 1/6 of 1/2 is 1/12.*
Procedures without Connections Multiply: $\frac{2}{3} \times \frac{3}{4}$ $\frac{5}{6} \times \frac{7}{8}$ $\frac{4}{9} \times \frac{3}{5}$ *Expected Student Response:* $\frac{2}{3} \times \frac{3}{4} = \frac{2 \times 3}{3 \times 4} = \frac{6}{12}$ $\frac{5}{6} \times \frac{7}{8} = \frac{5 \times 7}{6 \times 8} = \frac{35}{48}$ $\frac{4}{9} \times \frac{3}{5} = \frac{4 \times 3}{9 \times 5} = \frac{12}{45}$	**Doing Mathematics** Create a real-world situation for the following problem: $\frac{2}{3} \times \frac{3}{4}$ Solve the problem you have created without using the rule and explain your solution. *One Possible Student Response:* *For lunch, Mom gave me three-fourths of the pizza that we ordered. I could only finish two-thirds of what she gave me. How much of the whole pizza did I eat?* *I drew a rectangle to show the whole pizza. Then I cut it into fourths and shaded three of them to show the part Mom gave me. Since I only ate two thirds of what she gave me, that would be only two of the shaded sections.* 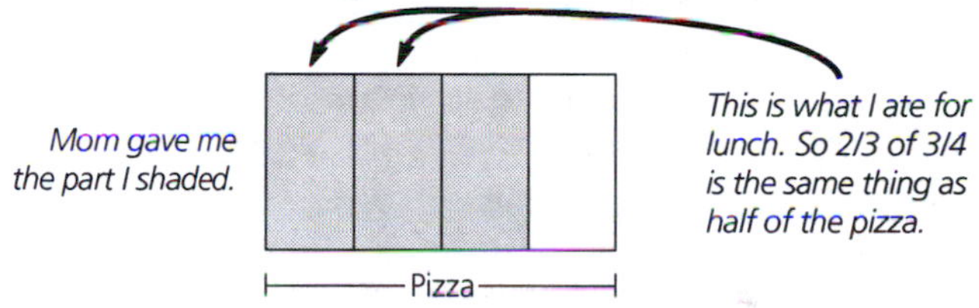 Mom gave me the part I shaded. This is what I ate for lunch. So 2/3 of 3/4 is the same thing as half of the pizza. ⊢——— Pizza ———⊣

Figure 8.4. Examples of tasks at each of the four levels of cognitive demand.

tion phase. Indeed the kind of tasks that reformers have suggested as most essential for building students' capacities to think and reason mathematically (high-level tasks) appear to be the very tasks that students and teachers have the most difficulty carrying out in a consistent manner. Overall, tasks that were set up to require a high level of cognitive engagement were found to decline into less demanding student engagement more than half of the time (in 62 out of 107 tasks or 58%).

The difficulty teachers in the United States have enacting tasks at a high level is echoed in the results of the recent TIMMS video study. In this study, a random sample of 100 8th grade mathematics classes from each of seven countries was videotaped during the 1999 school year. Although 17 percent of the problems used by teachers in the U.S. were coded as high level (*making connections*), none of these problems was implemented as intended (NCES, 2003; Stigler

& Hiebert, 2004). Instead, most of the making-connections problems were transformed into procedural exercises. Hence, the authors concluded, U.S. 8th grade students spend most of their time in mathematics classrooms practicing procedures regardless of the nature of the tasks they are given.

The *Inside the Classroom* study, conducted by Horizon Research, provides additional insight into the nature of mathematics instruction in the United States (Weiss & Pasley, 2004). In a study of 364 lessons, drawn from representative K–12 mathematics and science classrooms, only 15 percent of the lessons "were structured and implemented in a manner that engaged students with important mathematics or science concepts" and were viewed as "very likely to enhance student understanding of these concepts and to develop their capacity to do mathematics or science successfully" (p. 25). The majority of lessons (59%) were rated as low quality and were deemed unlikely to enhance students' understanding of mathematics or science concepts. Interestingly, the authors also report that the quality of lessons was not associated with a particular approach to teaching—both reform oriented and traditional lessons were found in each of the three categories (i.e., low quality, medium quality, high quality).

Case studies of teachers attempting to use challenging tasks and curricula in their classrooms provide insights into what actually occurs as these tasks are enacted by teachers and students during instruction. In their study of the early efforts of teachers to implement mathematics instructional reform in California, EPPS researchers observed an odd combination of old ways and new practices coexisting (Cohen et al., 1990). This was most evident in the case of Mrs. O who on the surface appeared to embrace reform, but a closer look revealed that not much had changed (Cohen, 1990). While Mrs. O used innovative instructional materials and activities that were designed to help students make sense of mathematics, she used them in a way "that conveyed a sense of mathematics as a fixed body of knowledge of right answers rather than as a field of inquiry" (Cohen, 1991, p. 19). Students were not asked to explain correct or incorrect solutions or the strategies used to derive them and there was no communication among students.

The mélange of old methods and new materials that characterized Mrs. O's practice highlights the challenges faced by teachers as they have endeavored to use new tasks and curricula in their classrooms and to take on new roles in the classroom. Other studies provide insights into the nature of the challenges faced by teachers who have attempted to use new tasks in new ways. For example, teachers struggle with relinquishing authority in the classroom (Wilson & Goldenberg, 1998; Wilson & Lloyd, 2000; Wood, Cobb, & Yackel 1991), ensuring that students feel successful as they work on more challenging mathematical tasks (Smith, 2000), knowing when to ask questions and when to provide information (Romagnano, 1994), and providing an appropriate amount of support and structure (Lloyd, 1999). These studies and others make it clear that taking on these new roles and responsibilities is not easy for teachers. As EPPS researchers (Cohen et al., 1990, p.163) have concluded, "changing one's teaching is not like changing one's socks"—but rather it requires a deep-seated change in belief about what it means to teach and learn mathematics." In Section Four, we discuss the role that teachers' beliefs and other mediating factors play in influencing the transformation of the written curriculum to the intended curriculum and the intended curriculum to the enacted curriculum.

Investigating Processes Involved in Task Implementation

Here we turn our attention to classroom processes that influence what teachers and students actually *do* during classroom instruction. What students do during instruction depends, to some extent, on the normative practices that have been agreed upon either explicitly or implicitly by the students and teacher. According to Carpenter and Lehrer (1999, p. 26), "although the selection of appropriate tasks and tools can facilitate the development of understanding, the normative practices of a class determine whether they will be used for that purpose." Hence we confine our discussion here to what happens during task implementation—the actions and interactions of teachers and students as they work on mathematical tasks. In Section Four, we discuss the factors (shown in the oval in Figure 8.1) that influence these interpretive processes and in so doing provide plausible explanations for the transformations that often occur during the task implementation phase.

In the previously cited study of QUASAR Project middle schools (Henningsen & Stein, 1997; Stein et al., 1996), a handful of patterns emerged that capture characteristic ways in which high-level tasks unfold in the classrooms of teachers who are attempting to reform their instruction so as to be more compatible with the *Standards* (NCTM, 1989, 2000). These patterns and the processes associated with them are described below.

Some tasks that were set up to place high levels of cognitive demand on student thinking were indeed implemented in such a way that students thought and reasoned in complex and meaningful ways. When this happened, teachers generally drew on a number of processes in order to support students' engagement in a way that maintained the demands of the task. As shown in the second column of Table 8.2, these processes include scaffolding of student thinking (i.e., assisting student thinking in a manner that preserves task complexity), modeling of high-level thinking and reasoning by the teacher or more capable peers, pressing for explanation and meaning, and selecting tasks that built on students' prior knowledge.

Other tasks that were set up to place high levels of cognitive demand on students' thinking, however, exhibited declines in terms of how students actually went about working on them. When tasks declined, a different set of processes tended to be operating in the classroom environment (see first column in Table 8.2). These processes involved a variety of teacher, student, and task-related conditions, actions, and norms. Although tasks can decline in several different ways,

Table 8.2 Processes Associated with Maintenance and Decline of High-Level Cognitive Demands

Processes Associated with the Decline of High-Level Cognitive Demands	Processes Associated with the Maintenance of High-Level Cognitive Demands
Routinizing problematic aspects of the task	Scaffolding of student thinking and reasoning
Shifting the emphasis from meaning, concepts, or understanding to the correctness or completeness of the answer	Providing a means by which students can monitor their own progress
Providing insufficient time to wrestle with the demanding aspects of the task or so much time that students drift into off-task behavior	Modeling of high-level performance by teacher or capable students
Engaging in high-level cognitive activities is prevented due to classroom management problems	Pressing for justifications, explanations, and/or meaning through questioning, comments, and/or feedback
Selecting a task that is inappropriate for a given group of students	Selecting tasks that build on students' prior knowledge
Failing to hold students accountable for high-level products or processes	Drawing frequent conceptual connections
	Providing sufficient time to explore

one common pattern is into tasks in which students use procedures without connections to meaning. (See Henningsen & Stein, 1997 for a examples of patterns of maintenance and decline of high level tasks.)

Instead of engaging deeply and meaningfully with the mathematics, students ended up utilizing a more procedural, often mechanical and shallow approach to the task. In this type of decline, one of the most prevalent factors operating in the environment was teachers' "taking over" and doing the challenging aspects of the tasks for the students. High-level tasks tend to be less structured, more difficult, and longer than the kinds of tasks to which students are typically exposed. Students often perceive these types of tasks as ambiguous and/or risky because it is not apparent what they should do, how they should do it, and how their work will be evaluated (Doyle 1988; Romagnano, 1994). In order to deal with the discomfort that surrounds this uncertainty, students often urge teachers to make these types of tasks more explicit by breaking them down into smaller steps, specifying exact procedures to be followed, or actually doing parts of the task for them. Should the teacher succumb to such requests, the challenging, sense-making aspects of the task are reduced or eliminated, thereby robbing students of the opportunity to develop thinking and reasoning skills and meaningful mathematical understandings.

A review of the processes in Table 8.2 suggests the central role of teachers' questions to the outcome of a lesson. Asking questions that scaffold or support students' continued engagement with a task and that press students to explain and justify their thinking are key to sustaining the cognitive demands of mathematical tasks. Alternatively, failure to ask questions that go beyond the correct answer or that communicate accountability for thinking and reasoning are associated with the decline of high level tasks. The nature of teachers' questioning and its impact on students' opportunities to learn are highlighted in both large and small-scale studies. For example, in the study of 350 representative mathematics and science lessons, Weiss and Pasley (2004) indicated that effective questioning was key in helping students develop an understanding of mathematical concepts and in making connections, yet rare in the observed classes. According to the authors, such scaffolding allows teachers to monitor students' understanding and encourages students to thinking more deeply and critically. The study suggests, however, that teachers more frequently asked questions that focus on getting the correct answer rather than on helping students make sense of mathematics concepts.

In a comparative study of mathematics teachers in three high schools using standards-based vs. conventional curricula, Boaler and Brodie (2004) found that teachers using standards-based curricula asked a broader range of questions than those who used a conventional curricula. In particular, nearly all of the questions (95%) asked by the teachers using conventional curricula fell into the category of procedural questions. While teachers using standards-based curricula also asked many procedural questions (60–75% of the questions asked were categorized as such) they also asked a large percentage of questions that were classified as conceptual and probing. Boaler and Brodie (2004, p. 780) argue that teachers' questions play an important role "in shaping the nature of classroom environments and the mathematical terrain that is traversed . . . and the cognitive opportunities offered to students."

Several case studies also point to the critical relationship between questioning and maintaining the integrity of the task (e.g., Smith, 2000; Wood et al., 1991). For example, in a study of a second grade teacher who was implementing standards-based mathematics materials, Wood et al. (1991) note that initially the teacher was unsure how to deal with the various procedures, solutions, and incorrect answers that students came up with during their explorations. The teacher's initial attempt to deal with this challenge resulted in beginning with a student's incorrect response and leading the student through a series of questions that would ultimately lead to the answer the teacher had in mind. According to Wood et al. (1991), "as the teacher reflected on the use of this approach later, she recognized that attempting to guide students in this manner was just another way for the teacher to ensure students get the right answer, whether they understand how to or not" (p. 604).

Related to the issue of what *type* of question to ask is the challenge of determining *when* to let students struggle without teacher input, *when* to ask leading questions, and *when* to tell students something directly (NCTM, 1991). These decisions are at the heart of what Romagnano (1994) refers to as "The Ask Them or Tell Them Dilemma." This dilemma grew out of the pull and tug of when to withhold and when to provide information to students. Romagnano found that it was the frustration and lack of clarity resulting from both the nondirective questions and the use of more complex problems that caused students to disengage. While consistent with the findings regarding the implementation of novel tasks (Doyle, 1988), this presented a dilemma to Romagnano, the teacher with whom he worked, and to many other teachers who are engaged in such work. Romagnano indicated that, "being directive was a problem because I knew that when I was that way, I reduced the chances that any significant mathematics would be learned" (1994, p. 123).

These studies raise important issues about the teacher's role in "discussion-intensive teaching" (Chazan & Ball, 1999) that is often intertwined with the use of the kind of high-level tasks found in standards-based curricula. Although teachers are told *not* to tell, they are given no explicit direction regarding what they could or should do instead. This leaves teachers struggling to determine *how* to support students without taking away the challenge *before* they completely disengage from activity in frustration. Chazan and Ball (1999, p. 9) argue that for teachers to direct a discussion productively, they must have "a repertoire of ways to add, stir, slow, redirect the class's work."

In their study of teachers at Railside, Boaler and Staples (in press) indicate that a feature of the curriculum—one that could not be seen in a review of the curricular materials themselves—was the follow-up questions that teachers asked students. According to the authors, these questions "shaped the course of implementation" (p. 19). The questions, written prior to teaching the lesson, provided teachers with specific ways of focusing students on the key mathematical ideas in the lesson leaving less "in the moment" decision-making regarding the next instructional move.

In this section, we have reviewed the changes that often occur once curricular tasks are unleashed in real classroom settings. By adding new dimensions to our understanding of how curricula are transformed over time, they provide further building blocks in our evolving chain of logic regarding how students learn from curriculum. The studies reviewed in this section suggest, once again, that the introduction of human interaction with curriculum materials brings variation to the information and styles of learning to which students will be exposed. In the previous section, the variation was introduced by the interpretive processes of the teacher "reading" the curriculum materials. In this section, variation is introduced by the flux of the classroom life—a space comprised of students as well as teachers. Instead of just one person interpreting the curriculum (the teacher as he or she plans), there is now a roomful of students interpreting the curriculum tasks and materials and bouncing off of each others' interpretations. As the teacher responds to these interpretations, the task is often transformed into a learning opportunity that is quite different from what was intended by the curriculum developers.

Interestingly, although not the specific focus on any study of which we are aware, the conceptual ideas and empirical findings presented herein suggest that one might expect less variation in the classroom-based enactment of conventional as opposed to standards-based curricula. Most of the studies cited above were conducted in classrooms in which teachers were using standards-based curricula or similarly inspired reform materials. When teachers set up instructional tasks in the way outlined by the materials, the stage was set for the study of how teachers and students enact high-level task (since these kinds of tasks comprise the majority of activities in standards-based materials). However, as noted earlier, low level tasks tended to implemented as intended (Stein et al., 1996). This is because there is much less ambiguity regarding what to do and how to do it. Thus, one would expect less variation in how teachers enact tasks that are found in conventional curricula (since these kinds of tasks—procedural and definitional activities—comprise the majority of conventional curricula).

This observation reinforces the point made at the end of the previous section regarding the concept of fidelity. There, we noted that the implementation of curricular tasks with fidelity would be expected to be more attainable and measurable with conventional as opposed to standards based curricula. Here we have extended the idea to encompass the notion that enactment of standards-based curricula can logically be expected to lead to greater variability than the enactment of conventional curricula. This is important because—as noted throughout this chapter—it is not the materials themselves but rather how students' experience the materials in the classroom that determines what students will learn. If standards-based classrooms have more variable implementation, then predicting what students will learn from them becomes more difficult.

SECTION FOUR: EXPLAINING TRANSFORMATIONS WITHIN AND BETWEEN DIFFERENT PHASES OF CURRICULUM USE

The preceding sections detail some of the transformations that occur between the written and intended curriculum and within the enactment phase (see Figure 8.1). Research on mathematics teaching and teachers' use of curriculum materials highlights the ways teachers read and interpret written curriculum materials differently (Section Two) and planned tasks and instructional designs are reconstructed by students and teachers as they unfold in real-time lessons (Section Three). In addition to identifying curricular transformations, much of the research seeks to explain them by identifying factors that influence transformations. The teacher figures most prominently in these explanations. However, much of this literature recognizes that the reasons underlying the ways teachers and students transform curriculum are multi-faceted and complex. In addition to individual teacher characteristics, scholars have identified students, the teaching context, and the curriculum itself to be influential in these processes. In the following sections, we discuss the kinds of factors that have been identified in the research. Rather than providing an extensive review of this research, our aim is to delineate those factors that appear influential across the body of studies. The specific studies cited are offered as illustrations.

The Teacher Matters

Studies that examine the way that teacher-related factors influence how teachers use materials in planning and enacting curriculum are most prominent in the literature. The factors most studied include teachers' beliefs about mathematics, teaching, and learning; their knowledge of mathematics; and their experiences as teachers and as students. More recently researchers have begun to conceptualize additional influencing factors, including teacher identity and their stance or orientation toward curriculum materials.

Beliefs and Knowledge

The research that focuses on the role that teachers' beliefs and knowledge play in teachers' curriculum use is part of a larger body of research on the influences of beliefs and knowledge on teaching practices more generally. In fact, the 1992 *Handbook* devoted an entire chapter to the role of teachers' beliefs in mathematics teaching (Thompson, 1992) and a second chapter to the role of teacher knowledge (Fennema & Franke, 1992). The research reviewed in these two chapters demonstrates the critical role teachers' beliefs about mathematics, teaching, and learning and their knowledge of mathematics and students' learning play in influencing their pedagogical and curricular decisions. We limit our discussion here to the portion of this research that is specific to curriculum use. A substantial portion of this research focuses on a) teachers' beliefs about the nature of mathematics and the nature of knowledge, b) teachers' goals for students and their beliefs about purposes of school, and c) teachers' mathematical knowledge. In most cases, these beliefs are tightly intertwined with one another and cannot be examined in isolation (Putnam, Heaton, Prawat, & Remillard, 1992). In the following sections we discuss examples of studies that illustrate how these beliefs influence curriculum use.

Examinations of teachers' beliefs about mathematics, how it is learned, and the nature of mathematical knowledge have figured prominently in studies of teachers' use of standards-based curriculum because these materials offer views of mathematics that conflict with those typically held in mainstream culture. Indeed, much of this research illustrates the ways that teachers' beliefs about mathematics and how it is learned influence how they interpret and use curriculum materials. This influence is particularly evident when conflicts exist between teachers' beliefs and the ideas embraced by the curriculum designers. A number of researchers have pointed to the deeply rooted nature of teachers' beliefs about mathematics (Wilson & Goldenberg, 1998) as an explanation for both substantial departures from and resistance to suggestions offered by standards-based curricula (Chavez, 2003; Lloyd & Wilson, 1998; Romberg, 1997). Other studies have attributed different teachers' uses of the same curriculum to substantial differences in the teachers' beliefs about mathematics and how students learn it (e.g., Collopy, 2003; Lloyd, 1999; Remillard, 1999).

While discussions of teachers' beliefs about mathematics permeate research on teachers' use of curriculum resources, there is general agreement that they should not be viewed in isolation. Putnam et al., (1992) used case studies of elementary teachers' responses to new curricula and new state policy, to illustrate how teachers' beliefs about mathematics were tightly intertwined with their beliefs about knowledge, teaching, and learning more generally. Other researchers have identified additional aspects of curriculum materials that conflict with beliefs commonly held amongst teachers. Stephens (1982) and Donovan (1983), for instance, illustrated how conflicts between teachers' beliefs about classroom control and management and purposes of school and those ideas represented in an innovative curriculum led teachers to ignore or change key features in the curriculum.

Limits in teachers' mathematical knowledge are frequently used to explain the ways that teachers use curriculum materials, particularly materials designed to foster understandings of concepts and relationships underlying standard mathematical procedures. Heaton (1992) and Cohen (1990) provide two examples of elementary teachers failing to appreciate the mathematical complexity underlying activities suggested in curriculum. Both teachers they report on used interactive mathematics tasks drawn from standards-based curriculum programs without developing the underlying mathematical ideas. Similarly, in their two-year study of two seventh grade mathematics teachers using a standards-based, middle-school curriculum, Manouchehri and Goodman (1998) found the teachers' mathematical knowledge to be the greatest influence on how they used the curriculum.

Orientation

Factors other than teachers' beliefs about mathematics teaching and learning have been found to influence their use of materials. Some studies indicate that teachers' perceptions of curriculum materials or textbooks as a particular genre of teacher resource influence how they engage and use them. In a study of two teacher education programs, Ball and Fei-

man-Nemser (1988) found that the majority of pre-service teachers tended to associate textbooks with traditional and undesirable teaching practices. As a result, they were reluctant to draw on these resources when planning mathematics lessons. Remillard and Bryans (2004) used the term "orientation" to refer to a teacher's stance toward curriculum materials in teaching. These researchers found this orientation to be influential in how and whether teachers used the curriculum regardless of the match or mismatch of the ideas in the curriculum to the teacher's view about mathematics teaching.

Professional Identity

Researchers have also found that teacher's professional identity, or how they see themselves and their roles as teachers, influences curriculum use. Spillane (2000) refers to teachers' identity as "who they are, their sense of self, and their habits of mind . . . an individual's way of understanding and being in the world" (p. 308). The concept of identity builds on and extends the large body of work that considers how teachers' knowledge and beliefs influence their curricular decision making and practice. Identity, as a construct, integrates what one knows, feels and is inclined to do in a particular context into an inseparable whole, making activity, rather than knowledge, central. While knowledge is often viewed as static, identity is active. Further, the notion of identity highlights the contextual and social nature of knowing and being. Identity is a social construct that develops in relation to others and particular contexts.[14] "Identities develop in and through social practices" (Boaler & Greeno, 2000, p. 173). That is, teachers' professional identities are produced in relation to their work as teachers in classrooms with students. And, as Spillane (2000) and Drake and Sherin (2006) suggest, identities are subject-matter specific. The role the particular subject matter plays in a teacher's professional identity is, perhaps, best illustrated by Spillane's case of an elementary teacher who routinely facilitated inquiry-oriented discussions amongst her fifth graders in the area of reading, writing, and grammar. Her mathematics instruction, on the other hand, tended to focus on directing students to memorize and follow procedural rules.

A significant component of teachers' professional identity is how they construct their roles in relation to students. Many teachers, for instance, associate good teaching with telling and showing students what to do,

---

[14] For organizational purposes, we have placed professional identity under the heading of the teacher. However, as this discussion suggests, identity is a social and context-specific phenomenon and, thus, embraces influential factors discussed in the following sections, including students, context, and curriculum materials.

whereas, standards-based curriculum materials outline a much less didactic role for the teacher. J. Smith's (1996) analysis of the fundamental conflicts between factors that tend to positively influence teachers' sense of their own efficacy, as outlined by self-efficacy research, and the view of teaching promoted by the NCTM Standards helps to explain teachers' curriculum transformations. Frykholm (2004) found that using curriculum materials that conflicted with their ideas about good teaching resulted in considerable discomfort for teachers. He argued that using standards-based curriculum materials required teachers to develop tolerance for this discomfort. Lloyd (1999) observed a different kind of conflict between a teacher's identity and standards-based curriculum materials. One of the teachers in her study of teachers using a high school curriculum viewed herself as an innovative, project-based mathematics teacher and, consequently, found the curriculum overly restricting and difficult to follow closely.

Students Matter

A number of studies have revealed that students' responses to standards-based curricula appear to be influential in how teachers use new materials. Teachers tend to be concerned about the high level of independent thought, problem solving, and self monitoring demanded of students by the tasks found in standards-based curricula, expect that students cannot manage these demands, and consequently restructure or adapt the lessons to make them less complex and more readily accessible to students. As described previously, Stein et al. (1996) observed teachers reducing the complexity of tasks during instruction in response to students' struggles with high demand tasks. M. Smith (2000) associated the tendency to reduce task demand to teachers' concerns about student success. As the previous discussion on self-efficacy indicated, teachers' reluctance to allow students to struggle is associated with how they define and assess their role as teacher.

Some researchers have suggested that teachers' responses to challenging curriculum materials can be associated with perceived rather than actual concerns about students' difficulties. Wilson and Lloyd (2000) also observed that teachers' curricular decisions were often influenced by the "invisible hand of students" (Borasi, 1990). However, they found no evidence that the students in the classrooms they observed were having the kind of difficulties identified by the teachers. They suggest that "the invisible hand was teacher *perception* of student resistance, not student resistance per se" (p. 28).

A number of researchers have identified the challenges teachers face in sharing intellectual authority with students as another influential factor in how teachers use curriculum materials. As indicated earlier, standards-based curriculum materials include tasks and teaching recommendations that offer students a degree of autonomy and decision making that is not typical in traditional teaching practices in the United States. Many teachers, however, are not comfortable sharing authority with students and implement curricular suggestions in ways that are more structured and directive than recommended (Wilson & Goldenberg, 1998; Wilson & Lloyd, 2000) and in ways that provide a level of support and structure they deem necessary (Lloyd, 1999; Stein et al., 1996).

The Context Matters

Many studies have also identified the teaching context as an important factor in teachers' use of curriculum materials. Most of these studies highlight the ways features of the context constrain teachers from using the curriculum as intended by the developers or as fully as they would like. Some studies demonstrate the ways that elements of the context enable curriculum use. Some contextual features that figure prominently in this body of research include time, local cultures, and the extent and nature of teacher support.

Time

Limits on time for planning and to devote to mathematics instruction have both been identified as factors that influence how teachers use curriculum materials. Many of the standards-based curricula assume that a significant amount of time will be devoted to mathematics lesson—90 minutes or more. Moreover, the complexity of the mathematics and the instructional recommendations require that teachers spend more time preparing than they are accustomed to. In a study of middle school teachers using *CMP*, Keiser and Lambdin (1996) identify a number of issues related to time that influence how teachers use their materials. They identify conflicts between the time available for instruction and planning and the amount of time required. They also examined how teachers used time both in planning and during instruction. They argue that using standards-based curricula as intended requires flexibility in class scheduling and timing, a luxury infrequently afforded teachers.

Local Cultures

A number of studies have pointed to the ways in which the departmental, district, school, or community culture influences teachers' use of curriculum. This is particularly the case for standards-based curricula where a mismatch with the local norms is likely. As mentioned above, these conflicts emerge around students' and teachers' roles during mathematics class, sharing authority with students, and norms for behavior (Floden et al., 1981; Lloyd, 1999; Manoucheri & Goodman, 1998; Wilson & Lloyd, 2000; Wood et al., 1991; Cobb, & Yackel 1991; Floden, et al., 1981).

Some researchers have focused attention on the norms of the school as a workplace for teacher learning. As noted earlier, the introduction of standards-based materials challenges what teachers know and how they think about mathematics, as well as their thinking regarding how students learn mathematics and what their role as teachers should be in the classroom. Research on teacher change suggests that teachers are more apt to change their approach to instruction when supported by a group of colleagues who are also struggling to learn innovative approaches to instruction (Coburn, 2001; Spillane, 1999). Some researchers who have explored the role of teacher-teacher interaction in supporting teacher reform in mathematics have utilized sociocultural concepts such as "communities of practice" to explain the ways in which groups of teachers influence each others' learning and adaptation to new forms of teaching (Cobb et al., 2003; Franke & Kazemi, 2001; Stein et al., 1998) More recently, Stein, Coburn and colleagues have utilized social capital as a theoretical construct to gain purchase on how teacher-teacher interaction influences teachers' experiences with the implementation of innovative curricula (Stein & Coburn, 2005).

Finally, since the introduction of NCLB, various policy-related factors have impacted what gets taught and how. The consequences associated with performing poorly on state-mandated tests, for example, have led to the introduction of pacing guides and other methods that McNeil along with others have suggested "narrow" the curriculum (McNeill, 2002). In addition, many districts have either begun with the reform of literacy instruction or have invested the greatest proportion of their resources in literacy. As such, the improvement of mathematics instruction often takes a back seat in overall school and district operations (Stein, Toure, Acquerelli, & Fergeson, 2003).

Teacher Support

A number of studies of teachers using standards-based curriculum materials report on the role of professional support in enabling and influencing teachers' work with the curriculum (Davenport, 2000; Remillard, 2000; Van Zoest & Bohl, 2002). Most of these studies align with what is known about the features of effective professional development in general. For example, effective professional development is content-based, promotes active learning, and is perceived by

teachers to be part of an overall coherent program of teacher learning (Garet, Porter, Desimone, Birman, & Yoon, 2001). An additional feature that is often cited as critical in mathematics teacher professional development is the opportunity for teachers to engage in mathematics thinking and reasoning as learners, often exploring the same or similar tasks to what appears in the curriculum materials for students (Smith, 2001).

The Curriculum Matters

One factor that has the potential to influence transformations between written, intended, and enacted curriculum that has received minimal attention in the literature is the curriculum itself. How do characteristics or features of the particular curriculum influence how a teacher uses it? In order to consider this question, researchers must begin with analyses of curriculum materials themselves. This is an underdeveloped area of research. However, some of the ways curriculum materials have been characterized for the purpose of analysis include a) broad types, such as conventional versus standards-based, b) particular features or characteristics of the materials, and c) how they are designed to communicate with and educate teachers.

Conventional versus Standards-based Curricula

There is some evidence that teachers use standards-based curriculum materials differently than they use traditional curriculum, although there are few studies that compare use of two different curricula by the same teacher. Herbel-Eisenmann, Lubienski, and Id Deen (in press) undertook a comparative case study of a middle school math teacher using *Connected Mathematics,* a standards-based curriculum, with one class of students and a conventional textbook with another class. They found that the nature of the instruction was notably different across these two classes with respect to classroom organization and the types of tasks students were asked to do. However, these differences seemed closely related to the teacher's beliefs that students and their parents should have a say in the kind of instruction they received. In this setting, students opted into a traditional or reform track. Respecting their choices, this teacher adjusted her teaching practices to fit the curriculum she was using.

Many teachers are less flexible in their practices and find it difficult to use a curriculum that conflicts with their beliefs about mathematics and how it should be taught. In his analysis of seven studies of teachers using standards-based curriculum, Romberg (1997) noted that the mismatch between the teachers' underlying beliefs about mathematics and learning and those implicit in the curriculum was a primary factor in how teachers used the curriculum. In his study of 53 teachers using either standards-based or conventional curricula, Chavez (2003) found that teachers using standards-based curricula felt less free to make adaptations and were less likely to see themselves as able to determine course goals and objectives.

Curriculum Features

Recently, scholars have questioned the value of applying such general classifications (standards-based or conventional) to analysis of influential factors in curriculum use, arguing that they mask substantial differences across curricula within categories (NRC, 2004)[15] and fail to provide reliable insights into how particular curricular features influence teachers' use (Remillard, 2005). Brown (2002) argues that understanding how a teacher uses curriculum resources and the resulting classroom practices requires an integrated analysis of the particular curriculum resources, the resources the teacher brings, and how they interact. Little research has been undertaken that examines these interactions, however the evidence provided here suggests this to be a fruitful area for future research.

The importance of taking into account the interactions between teachers and curriculum features is illustrated by Lloyd's (1999) findings from a study of two high school teachers using Core-Plus. One teacher found the curriculum too unstructured and struggled to navigate a curriculum where students' ideas were central. As a result, he tended to provide more support for students and more structure in the lesson than the curriculum suggested. The other teacher found the curriculum too structured for her taste.

Another feature frequently considered to be influential in how teachers use curriculum materials is sequencing—the way the material is ordered in the book. A commonly held assumption is that lessons near the end were more likely to be eliminated than those placed earlier in the book. To consider this hypothesis, Tarr et al., (2006) compared the lessons taught and omitted by 9 teachers using the same conventional textbook, *Glencoe,* and found that placement in the text was less influential in omission decisions than the particular topic of the lesson. Teachers omitted lessons throughout the book. Moreover, they omitted lessons on data analysis and probability at a significantly greater rate than lessons on number and operations or algebra.

Beyond content and sequencing, curriculum designers make a number of critical decisions about how the materials they produce are structured and communicated to the teacher. Many of the standards-based materials are designed to look substantially different from the conventional texts which were a mainstay in most classrooms in the U.S. We know little about how teachers engage these varied offerings.

Educative Curriculum

One characteristic common to the design of a number of the standards-based curriculum materials is their intent to communicate directly with teachers. Because these materials are relatively new, we know little about how teachers respond to or use these features. There are a few examples in the literature of teachers using these features to enhance their understanding of the purpose of particular lessons and to anticipate students' responses (Davenport, 2000; Rodriguez, 2000). There are also examples of teachers ignoring these features (Remillard & Bryans, 2004; Sherin & Drake, n.d.) or using them in different ways than were intended by the authors (Collopy, 2003). Drawing on research on teacher learning, Davis and Krajcik (2005) offer a set of design principles that might guide the development of educative curriculum materials. However, these principles are yet to be tested in practice.

The work cited in this section suggests possible explanations for why teachers interpret and enact curriculum in particular ways. Briefly, support can be found for factors that reside within the individual teacher, within students (or teachers' perceptions of students), within the context, and within the curriculum itself. These studies and their findings are important because they suggest additional pieces of the conceptual landscape that might be incorporated into both large and small-scale studies of curricular effectiveness.

These studies are also important from a practical perspective. The explanations that are offered for how teachers interpret and enact curriculum have primarily been generated from studies in which teachers have been attempting to adjust their instruction to be aligned with standards-based approaches to teaching and learning. Thus, if variations in curricular effectiveness are related to these factors, they suggest potential leverage points for interventions aimed at assisting teachers to enact standards-based curricula in ways that are more aligned with developers' intentions.

Finally, we believe that the study of curriculum materials and the manner in which they interact with the resources that teachers bring to the table (Brown,

---

[15] See Stein and Kim (2006) for an exception; this study investigates similarities and differences between two standards-based curricula.

2002) deserves more attention. In this perspective, curriculum materials are viewed as tools that can enable or constrain teachers' thinking and actions (Wertsch, 1998). Research conducted from this perspective in science education has uncovered properties of curriculum materials that appear to influence instruction in important ways (Brown, 2002; Schneider & Krajcik, 2002). Stein and her colleagues (in press) have adopted this perspective in their research on elementary curriculum materials and are exploring how particular features of the materials interact with certain teacher characteristics (e.g., knowledge of mathematics for teaching), and with characteristics of the school environment (e.g., amount of teacher turnover; level and kind of social capital to which teachers have access). In doing so, their work attempts to change the question from, "Does Curriculum X work?" to the question: "Which curriculum works best under which conditions?"

SECTION FIVE:
HOW THE ENACTED CURRICULUM
INFLUENCES STUDENT LEARNING

In this section we examine the impact of curriculum on student learning by looking at the relationship between the enacted curriculum (the third box in Figure 8.1) and what students appear to learn from their instructional experiences (the final triangle in Figure 8.1). We have limited our discussion to studies that involve observation as a measure of instructional practices rather than those that rely solely on self-report data. Using surveys alone to provide insights into classroom practices is problematic for several reasons. As Ball and Rowan point out (2004, p.4), "key descriptions of practice used in survey instruments are seldom understood uniformly by respondents." In addition, it is difficult for teachers to accurately remember classroom events and interactions (NCES, 2003), and this can lead to their inadvertently misrepresenting their practice (Ball & Rowan, 2004). This issue is made salient by the discrepancy between teachers' responses to questionnaire items and actual classroom observations in the TIMMS video study (NCES, 2003). For example, 86% of the U.S. eighth grade teachers surveyed reported that the videotaped lesson was consistent with current ideas about teaching and learning mathematics yet the observations suggest that the goal of the observed lessons was to learn and use procedures (NCES, 2003).

Although there have been many studies in the past 15 years (discussed earlier in this chapter) that have linked the use of particular curricula to student achievement or have analyzed the ways in which teachers and their students have enacted mathematical tasks or curricula, few studies have connected the curriculum (or tasks) *as enacted* with student learning or achievement. Three studies cited below provide evidence that the cognitive demands of the mathematics tasks in which students engage are related to student learning.

Evidence gathered across scores of middle school classrooms in four QUASAR middle schools has shown that students who performed the best on project-based measures of reasoning and problem solving were in classrooms in which tasks were more likely to be set up <u>and</u> implemented at high levels of cognitive demand (Stein & Lane, 1996; Stein, Lane, & Silver, 1996). In these classrooms a teacher's success in enacting tasks in ways that maintained the rigor of the tasks was related to the processes identified in the right-hand column of Table 8.2. For students in these classrooms, having the opportunity to work on challenging tasks in a supportive classroom environment translated into substantial learning gains on an instrument specially designed to measure high level thinking and reasoning processes. It is worth noting that the school that achieved the highest student learning gains (Site A), and the greatest percentage of tasks that were set-up and enacted at a high level, was using the standards-based *Visual Mathematics* curriculum. Results from QUASAR also show that students who had the lowest performance on project assessments were in classrooms where they had limited exposure to tasks that required thinking and reasoning (Stein & Lane, 1996).

The results of the 1999 TIMSS video study provide additional evidence of the relationship between the cognitive demands of mathematical tasks and student achievement. In this study, a random sample of 100 8th grade mathematics classes from each of six countries (Australia, the Czech Republic, Hong Kong, Japan, the Netherlands, Switzerland) and the United States, were videotaped during the 1999 school year. The six countries were selected because each performed significantly higher than the U.S. on the TIMSS 1995 mathematics achievement test for eighth grade (Stigler & Hiebert, 2004). The 1999 study revealed that the higher-achieving countries implemented a greater percentage of *making connections* tasks in ways that maintained the demands of the task. With the exception of Japan, higher-achieving countries did not use a greater percentage of high-level tasks than in the U.S. All other countries were, however, more successful in not reducing these tasks into procedural exercises. Hence, the key distinguishing feature between instruction in the U.S. and instruction in high achieving countries is that students in U.S. classrooms "rarely spend time engaged in the serious study of mathematical concepts" (Stigler & Hiebert, 2004, p. 16).

In a recent study Boaler and Staples (in press) report the results of a five-year longitudinal study of 700 students in three high schools. Students at one high school, Railside, used a standards-based curriculum designed by teachers around key concepts (e.g., What is a linear function?) and featuring groupworthy tasks drawn from curricula such as *College Preparatory Mathematics* (CPM) and the *Interactive Mathematics Program* (IMP) and a textbook of activities that use algebra manipulatives. The students at the other two high schools used conventional curricula. The researchers report that students at Railside achieved at higher levels than those at other schools. In particular, by the end of the second year Railside students significantly outperformed all other students in a test of algebra and geometry.

Boaler and Staples (in press) indicate that one factor contributing to the success of students at Railside was the high cognitive demand of the curriculum and the teachers' ability to maintain the level of demand during enactment through questioning. According to the authors, "the support that teachers gave to students did not serve to reduce the cognitive demand of the work, even when students were showing signs of frustration. . . . At Railside, teachers were highly effective in interacting with students in ways that supported their continued thinking and engagement in the core mathematics of the problems" (p. 27). In addition, the authors point to other sources of the Railside success story that include the strong commitment of teachers to the advancement of equity, the development of a curriculum that enhances success for *all* students by including problems that value a range of abilities, and the accountability placed on students for their own learning and that of their peers. Hence, Boaler and Staples conclude that although the curriculum played a part of the success of Railside students, "at the heart of this system is the work of the teachers, and the numerous different equitable practices in which they engaged" (p. 31).

The results of a study of middle school curriculum being undertaken by the Center for the Study of Mathematics Curriculum[16] provides additional evidence of the importance of the learning environment in student achievement. In this study, Tarr et al. (in press) investigated the impact of three factors on student achievement: curriculum type (conventional vs. standards-based), fidelity of curriculum implementation, and the nature of the learning environment.[17] The study included more than 4200 students in grades 6–8 from 11 middle schools across 6 states. The curricula included both standards-based curricula (i.e., *Connected Mathematics, Mathematics in Context,* and *Math Thematics*) and conventional curricula from commercial publishers (e.g., Addison Wesley, Glencoe, Harcourt Brace).

The study's findings suggest that student achievement in mathematics cannot be predicted solely by the type of curriculum used or by the fidelity of implementation of a curriculum. However, student achievement in mathematics could be predicted by the nature of the classroom environment. Specifically, a standards-based learning environment was associated with higher performance on an assessment of thinking, reasoning and problem solving regardless of the curriculum being used. A standards-based learning environment, however, was more frequently found in classrooms that used standards-based curricula. Particularly interesting is the finding that achievement was highest among students who experienced a standards-based curriculum in a standards-based learning environment over two consecutive years.

The findings of this study suggest that the learning environment is a critical factor in student learning and that standards-based curricula are most effective when the normative practices are in place that promote understanding, that is, learning is viewed as problem solving, alternative strategies and perspectives are discussed publicly, and explanations are given to support conjectures and approaches.

In addition to the above studies that have linked observed instructional practice and student achievement across many classrooms, there have been several smaller scale studies, focused on students' performance in arithmetic in the primary grades, that provide additional evidence of the connection between in-

---

16 The Center for the Study of Mathematics Curriculum (CSMC) is one of the Centers for Learning and Teaching Initiative funded by the National Science Foundation. The goal of this initiative was to encourage cross-institutional collaboration on the preparation of future K–12 STEM leaders and on research on critical issues in the field. CSMC is a collaboration of the University of Missouri, Michigan State University, University of Chicago, Horizon Research, Columbia Public Schools, Grand Ledge Public Schools, and Kalamazoo Public Schools.

17 The authors use the term "Standards-Based Learning Environment" (SBLE) to describe classrooms where students make conjectures and explain responses and teachers use students' thinking as the basis for instruction and encourage multiple perspectives and strategies.

struction and learning (Carpenter et al, 1989; Hiebert & Wearne, 1993; Wood et al., 1991). These studies show that students in classrooms that used standards-based curricula that included on-going opportunities to engage with high-level tasks outperform students who are exposed to more conventional curricula and instructional approaches. For example, in a study of instruction and learning in second grade classrooms, Hiebert and Wearne (1993) report that students in classrooms that used an alternative to the more conventional textbook program (i.e., a standards-based curricula) showed higher levels of performance than students in classrooms that used a conventional curricula. The alternative program used contextualized problems, encouraged the use of different representational forms and different solution strategies, and allowed for a public discussion and sharing of students solution strategies (Hiebert & Wearne, 1993). Thus the authors conclude that, "instructional tasks and classroom discourse mediate the relationships between teaching and learning" (p. 420).

One question raised by these studies is 'how do mathematical tasks and discussions around those tasks influence learning?" One explanation is that students learn mathematics by solving problems and by listening to what the teacher and their peers tell them. More likely, however, is that teaching influences students' cognitive processes of thinking which in turn influences their learning (Carpenter & Fennema, 1988; Hiebert & Wearne, 1993; Wittrock, 1986). According to Hiebert and Wearne (1993, p. 421) "certain kinds of instructional tasks and discourse encourage more productive ways of thinking."

The studies reviewed in Section One and this section can be seen as complementing one another. By focusing on tasks, discourse, and students' opportunities to learn, the studies reviewed in this section help to uncover the mechanisms by which curricula might improve student performance. In this way these studies begin to explain why two different classrooms using the same curriculum might result in different levels of student performance. The larger scale studies in Section One, on the other hand, are useful for suggesting what teachers on average might be expected to accomplish using the newer curricula and thus we learn something about the feasibility of implementing them at scale.

SUMMARY AND CONCLUSIONS

We begin by summarizing and in some cases elaborating on the major points that were surfaced in Sections One–Five.

Curricula Differ in Significant Ways

In Section One, we concluded that the analyses of curriculum materials completed to date suggest that the mathematics curriculum materials available today differ significantly from each another, with the most consistent differences found between those that we have labeled standards-based vs. conventional. Standards-based curricula embody an approach to learning that focuses on the students' active construction of important ideas and concepts while conventional based curricula, by and large, present content directly and expect the teacher to explicitly teach students the skills, concepts and procedures that are the goal of the lesson. Our review also pointed to the need for evaluators to clearly articulate the criteria by which they judge the quality of curricula. As shown by the ratings in Appendix A, the same curriculum fares very differently depending on the criteria by which it is judged.

These Differences Impact Student Learning

The pattern of findings associated with the comparative, mostly quantitative, research discussed in Section One suggests that the above differences matter, at least with respect to how much computational and conceptual knowledge students gain relative to each other when one group is taught using standards-based and the other using conventional curricula. Students taught using standards-based curricula tend to "keep up" with their conventionally taught counterparts with respect to computational knowledge but to surpass them in conceptual knowledge and their ability to solve non-routine problems. However, all students tend to do best on tests that align with the way they were taught, giving conventionally taught students a slight edge on traditional standardized tests and standards-taught students a considerable edge on measures of thinking, reasoning, and problem solving.

Kilpatrick (2003) has suggested, and we agree, that findings such as these appear to be oriented toward reassuring the public that the new standards-based curricula are not harming students more than assessing what students truly appear to be learning from them. Regardless, the findings that curricula differ in significant ways and that these differences impact student learning point to the role that healthy discussions of values (what kind of mathematics should students be learning?) must play in decisions to adopt one curriculum over another.

From a research—or knowledge-building—perspective, these findings feel somewhat unsatisfying. Knowing how much students learn may be less important than knowing what and how they learn. For example, an emerging insight from some of the studies cited in Section One that included qualitative data gathering (e.g., Huntley et al., 2000) is that students who learn and practice symbolic skill manipulation under the auspices of a standards-based curriculum may learn such skills in a qualitatively different way than do students who are taught skills in a direct, sequenced, and structured manner. We need investigations to understand more about the forms of knowledge and understanding that students develop in standards-based versus conventional curricular approaches, and about how the particular approaches support this development. Students most likely do not simply develop more or less knowledge but rather acquire knowledge, beliefs and understandings that differ in important ways including how they become available for use at later points.

No Curriculum is Self-Enacting

The studies reviewed in Sections Two and Three reveal the variation to the written curriculum that is introduced once human interaction enters the picture. Section Two focused on the interpretive processes by which teachers "read" curriculum; Section Three focused on the transformative forces that are unleashed once curricular materials are actually used in the classroom. The research and conceptual ideas introduced in both sections suggest that—while not self-enacting—curricula that focus on the development of specified skills and procedures offer less room for interpretation and transformation than do curricula that focus on reasoning, problem solving, and strategy development and depend on the actions of both students and teachers to bring their designs to fruition.

Counterarguments that have or might be made to this statement include the observation by Mathematically Correct that some of the so-called conventional curricula have become so overlaid with suggestions for problem solving activities and group work that teachers now have to make decisions regarding what to cover and how to prioritize their time between teacher-directed instruction and these "constructivist"-based activities. On the other hand, some might argue that technology-enhanced standards-based curricula such as the *Cognitive Tutor* offer a narrower berth for teacher interpretation because of the large amount of instructional time that occurs directly between the student and the computer.

Standards-Based Curricula are Challenging to Enact Well

Along with their unstructured learning space and room for variable enactment, standards-based curricula offer more enactment challenges to the average teacher. The studies cited in Section Three demonstrate the ways in which tasks from standards-based curricula can be distorted such that they no longer embody the vision or goals that the developer intended. Using these curricula in ways that unlock the potential that their developers envisioned requires considerable knowledge and time on the part of the teacher, as well as a philosophical orientation toward teaching and learning that aligns with that of the developers. Other factors that influence the level of success that can be expected from using standards-based curricula are identified and discussed in Section Four.

The high degree of variation associated with the enactment of standards-based curricula point to the need to interpret studies of student learning from such curricula with caution. Given the high degree of "slippage" associated with the enactment of high-level tasks, it is unclear whether limited student learning should be attributed to the materials or the manner in which they were implemented in the classroom. This is less often the case with interpreting studies of conventional curricula.

The Success of Standards-Based Curricula is Influenced by Multiple Factors

Sections Four and Five identify factors within and outside the classroom that impact the ease with which a successful implementation of a standards-based curriculum can be achieved. The studies in these sections are important to teachers because they uncover the variety of contextual factors that often accompany successful implementation of standards-based curricula but that are sometimes not overtly articulated as supporting factors. The studies that looked closely at particular teaching interactions in the classroom provide a level of detail that is needed by teachers who wish to implement tasks from standards-based curricula. From a research perspective, more fine-grained studies of classroom implementation provide a clearer, more complete picture of how and under what conditions standards-based curriculum lead to improvements in students' learning.

* * *

We began this chapter with the introduction of a framework that illustrated various phases of curriculum use. Uncovering how curriculum influences student learning, we argued, would not be possible without reviewing what is known about how the curriculum is mediated before and in the process of making its way to students. By reviewing and summarizing the implications of studies that lay at various markers along this temporal pathway, we have identified and described—not only whether—but also *how* curricula influence student learning. This review of the research on how curriculum influences student learning has illuminated the vast conceptual territory that lies between the curriculum as a designed object and student learning.

Our claims rely on studies that we have patchworked together to form a longer chain of reasoning regarding the phases through which curricula traverse. In the patchwork process, we have had to gloss over inconsistencies in conceptualizations, definitions of terms, and points of focus that existed across the various studies. In the chapter we call for the field to establish more agreement regarding what to focus on and how so that studies can be accumulated and build toward a solid understanding of how curricula are transformed and eventually influence student learning. Studies that are conducted to examine the entire chain—from materials analyses to teacher interpretation to curricular enactment to student learning—would also add considerably to our knowledge base.

Looking toward the future—including studies that might help curriculum developers improve their materials—perhaps the most pressing question is how to combine the best of both conventional and standards-based curricula into a more unified and balanced approach. How can teachers effectively teach procedural skills in the context of the open-ended, problem-based forms of instruction that characterize standards-based curricula? Standards-based lessons that encourage a variety of ideas, solution strategies, and forms of representation have the advantage of engaging students in thinking and learning to justify and reason. However, they also have the disadvantage of not converging neatly into a common core of knowledge that all students share and know—especially not in any predetermined time frame, be it a lesson, a grading period, a year, or in time for a high-stakes test. Making sure that all students eventually reach clarity and closure on important mathematics ideas and procedures is not trivial in this form of instruction (National Research Council, 2001).

The limitations of conventional curricula that focus too narrowly on procedural skill have been well documented; yet there are proficiencies that students who are exposed to those curricula gain. Learning how to incorporate elements of that learning into the curricula of tomorrow will be an important task. Given the ambition of the standards-based curricula—both in terms of the depth of student learning for which they aim and the numbers of teachers they hope to influence—we need to get much smarter about how we design studies to help the continuing evolution of those curricula.

Although we feel as though this chapter brings some conceptual order to the territory that lies between curriculum materials and student learning, the territory of the factors represented in the oval in Figure 8.1 is much less well conceptualized. As discussed in Section Four of this chapter, these factors have been supported by a host of studies, however, most studies looked at one, or at most two, factors in isolation from the others. Missing is research that systematically relates the factors to one another in a conceptually coherent manner. For example, we know that teachers' knowledge and beliefs influence how they interpret curricular materials and how they enact them in the classroom. However, we know little about how this knowledge and beliefs are influenced by context, including the informal communities in which teachers talk about mathematics and, most importantly, negotiate the meaning of new curricula. Moreover, we know little about how different features of curricula (how transparent their goals are; how much information they provided about possible student responses to their tasks) are related to various levels and kinds of teacher knowledge.

The research reviewed herein points to the challenges that most teachers face in learning to implement standards-based curricula well. There is beginning to be a body of research that examines *how teachers learn* through implementing these new curricula. We have not reviewed this research as it did not directly bear on the question of how curricula influences student learning. Nevertheless, we feel that it is an important component of the conceptual landscape that sits between the design of materials and student learning. If teachers are the mechanism through which curriculum becomes transformed and goes on to influence students, well-designed studies on what and how teachers learn from using these materials would form an important part of the overall picture.

On a related note, because these new standards-based curricula have been placed into the role of change agent, another kind of conceptual work is suggested as well: How teacher learning from curricular materials might be expected to happen at scale within schools and districts. Cobb et al. (2003) have pointed to a bifurcation of the research literature in this regard with one body of literature on professional development and another body on organizational/structural facilitators and constraints to instructional improvement. With few studies that bridge the two, as a field we are left with decontextualized knowledge about effective professional development and knowledge about organizational supports for change that fails to make contact with the particular needs of teacher learning in mathematics. Thus, we need more studies that systematically integrate schools and districts as settings for curriculum implementation and what we know about the kind of teacher learning required to enact these curricula well.

REFERENCES

American Association for the Advancement of Science. (1990). *Science for all Americans: A project 2061 report.* New York: Oxford University Press.

American Association for the Advancement of Science. (1993). *Benchmarks for science literacy.* New York: Oxford University Press.

Balfanz, R., Mac Iver, D.J., & Byrnes, V. (2006). The implementation and impact of evidence-based mathematics reforms in high-poverty middle schools: A multi-site, multi-year study. *Journal for Research in Mathematics Education, 37*(1), 33–64.

Ball, D. L. (1990). Reflections and deflections of the framework: The case of Carol Turner. *Educational Evaluation and Policy Analysis, 12*(3), 247–260.

Ball, D.L., & Cohen, D. K. (1996). Reform by the book: What is—or might be—the role of curriculum materials in teacher learning and instructional reform. *Educational Researcher, 25*(9), 6–8.

Ball, D. L., & Feiman-Nemser, S. (1988). Using textbooks and teachers' guides: A dilemma for beginning teachers and teacher educators. *Curriculum Inquiry, 18*(4), 401–423.

Ball, D. L., & Rowan, B. (2004). Introduction: Measuring instruction. *Elementary School Journal, 105*(1), 3–10.

Begle, E.G. (Ed.). (1970). *Mathematics education: The sixty-ninth yearbook of the National Society for the Study of Education.* Chicago: The National Society for the Study of Education.

Ben-Peretz, M. (1990). *The teacher-curriculum encounter: Freeing teachers from the tyranny of texts.* Albany, NY: SUNY Press.

Berends, M., Kirby, S. J., Naftel, S. N., & McKelvey, C. (2001). *Implementation and Performance in New American Schools: Three Years into Scale-Up.* Santa Monica, CA: RAND.

Berman, P., & McLaughlin, M. W. (1973). *Implementing innovations: Revisions for an agenda for a study of change agent programs in education. Study of change agent programs: A working note* (Descriptive No. WN-8450-1-HEW). Santa Monica, CA: RAND.

Berman, P., & McLaughlin, M. W. (1978). *Federal* (No. R-1589/8-HEW). Santa Monica, CA: RAND.

Boaler, J. (1997). *Experiencing school mathematics: Teaching styles, sex, and setting.* Buckingham, U.K.: Open University Press.

Boaler, J., & Brodie, K. (2004). *The importance of depth and breadth in the analysis of teaching: A framework for analyzing teacher questions.* Proceedings of the 26th meeting of the North America Chapter of the International Group for the Psychology of Mathematics Education. Toronto, Ontario, Canada.

Boaler, J., & Greeno, J. G. (2000). Identity, agency, and knowing in mathematics worlds. In J. Boaler (Ed.), *Multiple perspectives on mathematics teaching and learning* (pp. 171–200). Westport, CT: Ablex.

Boaler, J., & Staples, M. (in press). Creating mathematical futures through an equitable teaching approach: The case of Railside School. *Teachers College Record.*

Bodilly, S. J. (1998). *Lessons from new American schools' scale-up phase: Prospects for bringing designs to multiple schools.* Santa Monica, CA: RAND.

Borasi, R. (1990). The invisible hand operating in mathematics instruction: Students' conceptions and expectations. In R. Cooney (Ed.), *Teaching and learning mathematics in the 1990s* (pp. 174–182). Reston, VA: National Council of Teachers of Mathematics.

Brophy, J. (Ed.). (1991). *Advances in research on teaching: Teachers' knowledge of subject matter as it relates to their teaching practice.* Greenwich, CT: JAI Press.

Brophy, J. (Ed.). (2001). *Subject-specific instructional methods and activities.* Amsterdam: JAI Press.

Brown, M. W. (2002). *Teaching by design: Understanding the interactions between teacher practice and the design of curricular innovation.* Northwestern University, Evanston, Illinois.

Brown, M. W. (in press). Toward a theory of curriculum design and use: Understanding the teacher-tool relationship. In J. T. Remillard, G. M. Lloyd, & B. Herbel-Eisenmann (Eds.), *Teachers' use of mathematics curriculum materials: Research perspectives on the relationship between teachers and curriculum.*

Brown, A., & Campione, J. (1996). Psychological theory and the design of innovative learning environments: On procedures, principles, and systems. In L. Schauble & R. Glaser (Eds.), *Innovations in learning: New environments for education* (pp. 289–325). Mahwah, NJ: Erlbaum.

Carpenter, T. P., & Fennema, E. (1988). Research and cognitively guided instruction. In E. Fennema, T.P. Carpenter, & S. Lamon (Eds.), *Integrating research on teaching and learning mathematics* (pp. 2–19). Madison: University of Wisconsin, National Center for Research in Mathematical Sciences Education.

Carpenter, T. P., Fennema, E., Peterson, P. L., Chiang, C.-P., & Loef, M. (1989). Using knowledge of children's mathematics thinking in classroom teaching: An experimental study. *American Educational Research Journal, 26*(4), 499–531.

Carpenter, T.P., & Lehrer, R. (1999). Teaching and learning mathematics with understanding. In E. Fennema & T. Romberg (Eds.). *Mathematics classrooms that promote understanding* (pp. 19–32). Mahwah, NJ: Erlbaum.

Chappell, M. (2003). Keeping mathematics front and center: Reaction to middle-grades curriculum projects research. In S.L. Senk & D.R. Thompson (Eds.), *Standards-based school mathematics curricula: What are they? What do students learn?* (pp. 285–296). Mahwah, NJ: Erlbaum.

Chazan, D., & Ball, D. (1999). Beyond being told not to tell. *For the Learning of Mathematics, 12,* 2–10.

Chavez, O. L. (2003). *From the textbook to the enacted curriculum: Textbook use in the middle school mathematics classroom.* Unpublished manuscript, University of Missouri, Columbia.

Clopton, P., McKeown, E., McKeown, M., & Clopton, J. (1998). *Mathematically correct algebra 1 reviews.* Retrieved November 15, 2005, from http://mathematicallycorrect.com/algebra.htm

Cobb, P., McClain, K., de Silva Lamberg, T., & Dean, C. (2003). Situating teachers' instructional practices in the institutional setting of the school and district. *Educational Researcher, 32*(6), 13–24.

Coburn, C.E. (2001). Collective sense making about reading: How teachers mediate reading policy in their professional communities. *Educational Evaluation and Policy Analysis, 23*(2), 145–170.

Cohen, D. K. (1990). A revolution in one classroom: The case of Mrs. Oublier. *Educational Evaluation and Policy Analysis, 12*(3), 327–345.

Cohen, D. K. (1991, Fall). Revolution in one classroom: Or then, again, was it? *American Educator,* 16–48.

Cohen, D.K., & Barnes, C. (1993). Pedagogy and policy. In D. Cohen, M. McLaughlin, & J. Talbert (Eds.), *Teaching for understanding: Challenges for policy and practice* (pp. 207–239). San Francisco: Jossey Bass.

Cohen, D., Peterson, P., Wilson, S., Heaton, R., Remillard, J., & Weimers, N. (1990). *Effects of state-level reform of elementary school mathematics curriculum on classroom practice* (Elementary Subjects Center Series No. 25). Michigan State University, Center for the Learning and Teaching of Elementary Subjects and National Center for Research on Teacher Education.

Collopy, R. (2003). Curriculum materials as a professional development tool: How a mathematics textbook affected two teachers' learning. *Elementary School Journal, 103,* 287–311.

Datnow, A., Hubbard, L., & Mehan, H. (2002). *Extending educational reform: From one school to many.* London: Routledge Falmer.

Davenport, L. R. (2000). *Elementary mathematics curricula as a tool for mathematics education reform: Challenges of implementation and implications for professional development.* Newton, MA: Unpublished manuscript.

Davis, E. A., & Krajcik, J. S. (2005). Designing educative curriculum materials to promote teacher learning. *Educational Researcher, 34*(3), 3–14.

Donovan, B. F. (1983). *Power and curriculum in implementation: A case study of an innovative mathematics program.* Unpublished doctoral dissertation, University of Wisconsin, Madison.

Doyle, W. (1983). Academic work. *Review of Education Research, 2* (53), 159–199.

Doyle, W. (1988). Work in mathematics classes: The context of students' thinking during instruction. *Educational Psychologist, 23* (2), 167–180.

Doyle, W. (1992). Curriculum and pedagogy. In P. W. Jackson (Ed.), *Handbook of research on curriculum* (pp. 486–516). New York: Macmillon.

Drake, C., & Sherin, M. G. (2006). Practicing change: Curriculum adaptation and teacher narrative in the context of mathematics education reform. *Curriculum Inquiry, 35*(2), 153–187.

EEPA. (1990). *Educational Evaluation and Policy Analysis* [Special Issue], *12*(1), 233–353.

Fennema, E., & Franke, M. L. (1992). Teachers' knowledge and its impact. In D. A. Grouws (Ed.), *Handbook of research on mathematics teaching and learning* (pp. 147–164). New York: MacMillan.

Fey, J. (1979). Mathematics teaching today: Perspectives from three national surveys. *The Arithmetic Teacher, 27,* 10–14.

Floden, R. E., Porter, A.C., Schmidt, W.H., Freeman, D.J., & Schwille, J.R. (1981). Responses to curriculum pressures: A policy-capturing study of teacher decisions about content. *Journal of Educational Psychology, 73*(2), 129–141.

Franke, M., & Kazemi, E. (2001). Teaching as learning within a community of practice: Characterizing generative growth. In T. Wood, B. Nelson, & J. Warfield (Eds.), *Beyond classical pedagogy in teaching elementary mathematics: The nature of facilitative teaching* (pp. 47–74). Mahwah, NJ: Erlbaum.

Freeman, D.J., Belli, G.M., Porter, A.C., Floden, R.E., Schmidt, W.H., & Schwille, J.R. (1983). The influence of different styles of textbook use on instructional validity of standardized texts. *Journal of Educational Measurement, 20*(3), 259–270.

Freeman, D. J., & Porter, A. C. (1989). Do textbooks dictate the content of mathematics instruction in elementary schools? *American Educational Research Journal, 26*(3), 403–421.

Frykholm, J.A. (2004). Teachers' tolerance for discomfort: Implications for curricular reform in mathematics. *Journal of Curriculum and Supervision, 19*(2), 125–149

Fullan, M. (1991). *The new meaning of educational change* (2nd ed.). New York: Teachers College Press.

Gagne, R. (1977). *The conditions of learning* (3rd ed.). New York: Holt, Rinehart, & Winston.

Garet, M.S., Porter, A.C., Desimone, L., Birman, B.F., & Yoon, K.S. (2001). What makes professional development effective? Results from a national sample of teachers. *American Educational Research Journal, 38*(4), 915–945.

Gehrke, N.J., Knapp, M.S., & Sirotnik, K.A. (1992). In search of the school curriculum. *Review of Research in Education, 18,* 51–110.

Grouws, D. A. (Ed.). (1992). *Handbook of research on mathematics teaching and learning.* New York: MacMillan.

Grouws, D. A., Smith, M. S., & Sztajn, P. (2004). The preparation and teaching practices of U.S. mathematics teachers: Grades 4 and 8. In P. Kloosterman & F. Lester (Eds.), *The 1990 through 2000 mathematics assessments of the National Assessment of Educational Progress: Results and interpretations* (pp. 221–269). Reston, VA: National Council of Teachers of Mathematics.

Heaton, R. M. (1992). Who is minding the mathematics content? A case study of a fifth-grade teacher. *Elementary School Journal, 93,* 153–162.

Henningsen, M., & Stein, M. K. (1997). Mathematical tasks and student cognition: Classroom-based factors that support and inhibit high-level mathematical thinking and reasoning. *Journal for Research in Mathematics Education, 29,* 524–549.

Herbel-Eisenmann, B. A., Lubienski, S. T., & Id Deen, L. (in press). *Reconsidering the study of mathematics instructional practices: The importance of curricular context in understanding local and global teacher change.*

Hiebert, J. (1999). Relationships between research and the NCTM Standards. *Journal for Research in Mathematics Education, 30,* 3–19.

Hiebert, J., & Wearne, D. (1993). Instructional tasks, classroom discourse, and students' learning in second-grade arithmetic. *American Educational Research Journal, 2*(30), 393–425.

Hiebert, J., Thomas P., Carpenter, T., Fennema, E., Fuson, K., Wearne, D., et al. (1997). *Making sense: Teaching and learning mathematics with understanding.* Portsmouth, NH: Heinemann.

Horn, I. S. (2005). Learning on the job: A situated account of teacher learning in two high school mathematics departments. *Cognition & Instruction, 23*(2), 207–236.

Huntley, M.A., Rasmussen, C.L., Villarubi, R.S., Sangtong, J., & Fey, J.T. (2000). Effects of standards-based mathematics education: A study of the Core-Plus Mathematics Project algebra and functions strand. *Journal for Research in Mathematics Education, 31*(3), 328–361.

Jackson, P.W. (1968). *Life in classrooms.* New York: Holt, Rinehart and Winston.

Jackson, P.W. (1992). Conceptions of curriculum and curriculum specialists. In P.W. Jackson (Ed.), *Handbook of research on curriculum* (pp. 3–40). New York: MacMillan.

Jennings, N. E. (1996). *Interpreting policy in real classrooms: Case studies of state reform and teacher practice.* New York: Teachers College Press

Keiser, J. M., & Lambdin, D. V. (1996). The clock is ticking: Time constraint issues in mathematics teaching reform. *Journal of Educational Research, 90*(1), 23–30.

Kilpatrick, J. (2003). What works? In S.L. Senk & D.R. Thompson (Eds.), *Standards-based school mathematics curricula: What are they? What do students learn?* (pp. 471–488). Mahwah, NJ: Erlbaum.

Kirby, P.C. (2004). *Comparison of I CAN LEARN and traditionally-taught 8th grade student performance on the Georgia Criterion-Referenced Competency Test.* Unpublished manuscript.

Kirby, S. N., Berends, M., & Naftel, S. (2001). *Implementation in a longitudinal sample of new American schools: Four years into scale-up.* Santa Monica, CA: RAND.

Klein, D. (2000, April). Math problems: Why the U.S. Department of Education's recommended math programs don't add up. *American School Board Journal,* 52–57.

Kulm, G. (1999). Evaluating mathematics textbooks. *Basic Education, 43*(9). Retrieved November 2, 2006, from http://www.project2061.org/publications/articles/articles/cbe.htm.

Lambdin, D., & Preston, R. (1965). Caricatures in innovation: Teacher adaptation to an investigation-oriented middle school mathematics curriculum. *Journal of Teacher Education, 46*(2), 130–140.

Little, J. W., & McLaughlin, M. W. (Eds.). (1993). *Teachers' work: Individuals, colleagues, and contexts.* New York: Teachers College Press.

Lloyd, G. M. (1999). Two teachers" conceptions of a reform-oriented curriculum: Implications for mathematics teacher development. *Journal of Mathematics Teacher Education, 2*(3), 227–252.

Lloyd, G. M., & Wilson, M. (1998). Supporting innovation: The impact of a teacher's conceptions of functions on his implementation of a reform curriculum. *Journal for Research in Mathematics Education, 29*(3), 248–274.

Louis, K. S., Marks, H. M., & Kruse, S. (1996). Teachers' professional community in restructuring school. *American Educational Research Journal, 33*(4), 757–798.

Love, E., & Pimm, D. (1996). "This is so": A text on texts. In A. J. Bishop, K. Clements, C. Keitel, J. Kilpatrick, & C. Laborde (Eds.), *International handbook of mathematics* (Pt. 1, pp. 371–409). Boston: Kluwer.

Mailloux, S. (1982). *Interpretive conventions: The reader in the study of American fiction.* Ithaca, NY: Cornell University Press.

Manouchehri, A., & Goodman, T. (1998). Mathematics curriculum reform and teachers: Understanding the connections. *Journal of Educational Research, 92*(1), 27–41.

Mathematically Correct. (n.d.a). *Mathematics program reviews: Methods for seventh grade program reviews.* Retrieved October 18, 2005, from http://www.mathematicallycorrect.com/books7x.htm

Mathematically Correct. (n.d.b) *Mathematics program reviews for grades 2, 5, and 7: Summary of overall ratings by publisher.* Retrieved January 13, 2006, from http://www.mathematicallycorrect.com/booksy.htm

Mathematically Correct. (n.d.c) *Mathematics program reviews: Comparative summary for seventh grade.* Retrieved January 14, 2006, from http://www.mathematicallycorrect.com/books7y.htm

McNeill, L. (2002). *Contradictions of school reform: Educational costs of educational testing.* New York: Routledge.

Mokros, J. (2003). Learning to reason numerically: The impact of Investigations. In S.L. Senk & D.R. Thompson (Eds.), *Standards-based school mathematics curricula: What are they? What do students learn?* Mahwah, NJ: Erlbaum.

Morgan, P., & Ritter, S. (2002). *An experimental study of the effects of Cognitive Tutor Algebra I on student knowledge and attitude.* Pittsburgh, PA: Carnegie Learning. Retrieved November 2, 2006, from http://www.carnegielearning.com/research/research_reports/morgan_ritter_2002.pdf.

Mosteller, F., & Boruch, R. (2002). *Evidence matters: Randomized trials in education research.* Washington, DC: Brookings Institution Press.

National Center for Education Statistics (2003). *Teaching mathematics in seven countries: Result form the TIMSS video study.* Washington, DC: U.S. Department of Education.

National Council of Teachers of Mathematics. (1989). *Curriculum and evaluation standards for school mathematics.* Reston, VA: Author.

National Council of Teachers of Mathematics. (1991). *Professional standards for teaching mathematics.* Reston, VA: Author.

National Council of Teachers of Mathematics (2000). *Principles and standards for school mathematics.* Reston, VA: Author.

National Research Council. (1989). *Everybody counts.* Washington, DC: National Academy Press.

National Research Council. (2001). *Adding it up: Helping children learn mathematics.* Washington, DC: National Academy Press.

National Research Council. (2002). *Scientific research in education.* Washington, DC: National Academy Press.

National Research Council (2004). *On evaluating curricular effectiveness: Judging the quality of K–12 mathematics evaluations.* Washington, DC: National Academy Press.

National Research Council. (2004). *On evaluating curricular effectiveness: Judging the quality of K–12 mathematics evaluations.* Washington, DC: The National Academies Press.

National Science Foundation. (1989). *Materials for mathematics instruction: Program solicitation.* Arlington, VA: Author, Division of Materials Development, Research, and Informal Science Education.

No Child Left Behind Act of 2001 (2002). Pub.L. No. 107–110, 115, Stat. 1425.

O'Connor, M.C. (2001). Can any fraction be turned into a decimal? A case study of a mathematical group discussion. *Educational Studies in Mathematics , 46,* 143–185.

Project 2061. (n.d.). *Middle grades mathematics textbooks: A benchmarks-based evaluation.* Retrieved October 18, 2005, from http://www.project2061.org/publications/textbook/mgmth/report.htm

Putnam, R. T. (1992). Teaching the "hows" of mathematics for everyday life: A case of a fifth-grade teacher. *Elementary School Journal, 93*(2), 163–177.

Putnam, R. T. (2003). Commentary on four elementary mathematic curricula. In S.L. Senk & D.R. Thompson (Eds.), *Standards-based school mathematics curricula: What are they? What do students learn?* (pp. 161–178). Mahwah, NJ: Erlbaum.

Putnam, R.T., Heaton, R. M., Prawat, R., Remillard, J.T. (1992). Teaching mathematics for understanding: Discussing case studies of four fifth-grade teachers. *Elementary School Journal, 93*(2), 213–228.

Remillard, J. T. (1992). Teaching mathematics for understanding: A fifth-grade teacher's interpretation of policy. *Elementary School Journal, 93*(2), 179–193.

Remillard, J. T. (1999). Curriculum materials in mathematics education reform: A framework for examining teachers' curriculum development. *Curriculum Inquiry, 100*(4), 315–341.

Remillard, J.T. (2000). Can curriculum materials support teachers' learning? Two fourth-grade teachers' use of a new mathematics text. *Elementary School Journal, 100*(4), 331–350.

Remillard, J.T. (2005). Examining key concepts in research on teachers' use of mathematics curricula. *Review of Educational Research, 75*(2), 211–246.

Remillard, J.T., & Bryans, M. (2004). Teachers' orientations toward mathematics curriculum materials: Implications for teacher learning. *Journal of Research in Mathematics Education, 35*(5), 352–388.

Riordan, J.E., & Noyce, P.E. (2001). The impact of two standards-based mathematics curricula on student achievement in Massachusetts. *Journal for Research in Mathematics Education, 32*, 368–398.

Rodriguez, B. (2000, April). *An investigation into how a teacher uses a reform-oriented mathematics curriculum* Paper presented at the annual meeting of the American Educational Research Association, New Orleans, LA (ERIC Document No. ED 440868)

Romagnano, L.R. (1994). *Wrestling with change: The dilemmas of teaching real mathematics.* Portsmouth, NH: Heinemann.

Romberg, T.A. (1997). Mathematics in context: Impact on teachers. In E. Fennema & B. S. Nelson (Eds.), *Mathematics teachers in transition* (pp. 357–380). Mahwah, NJ: Erlbaum.

Romberg, T.A. (1992). Problematic features of the school mathematics curriculum. In P.W. Jackson (Ed.), *Handbook of research on curriculum* (pp. 749–788). New York: Macmillan.

Sarason, S. (1982). *The culture of the school and the problem of change* (2nd ed.). Boston: Allyn and Bacon.

Schmidt, W.H., McKnight, C.C., & Raizen, S.A. (1997). *Splintered vision: An investigation of U.S. mathematics and science education.* Norwel, MA: Kluwer.

Schneider, R., & Krajcik, J. (2002). Supporting science teacher learning: The role of educative curriculum materials. *Journal of Science Teacher Education, 13*(3), 221–245.

Schoen, H.L., Fey, J.T., Hirsch, C.R., & Coxford, A.F. (1999, February). Issues and options in the math wars. *Phi Delta Kappan,* 444–453.

Schoenfeld, A. (2002). Making mathematics work for all children: Issues of standards, testing, and equity. *Educational Researcher, 31*(1), 13–25.

Schoenfeld, A. (2006). What doesn't work: The challenge and failure of the What Works Clearinghouse to conduct meaningful reviews of studies of mathematics curricula. *Educational Researcher, 35*(2), 13–21.

Senk, S. L., & Thompson, D.R. (Eds). (2003). *Standards-based school mathematics curricula: What are they? What do students learn?* Mahwah, NJ: Erlbaum.

Sherin, M., & Drake C. (2004). *Identifying patterns in teachers' use of a reform-based elementary mathematics curriculum.* Manuscript submitted for publication.

Smith, J. P., III. (1996). Efficacy and teaching mathematics by telling: A challenge for reform. *Journal for Research in Mathematics Education, 27*(4), 387–402.

Smith, M. S. (2000). Balancing old and new: An experienced middle school teacher's learning in the context of mathematics instructional reform. *Elementary School Journal, 100*(4), 351–375.

Smith, M.S. (2001). *Practice-based professional development for teachers of mathematics.* Reston, VA: National Council of Teachers of Mathematics.

Smith, M. S., & Stein, M. K. (1998). Selecting and creating mathematical tasks: From research to practice. *Mathematics Teaching in the Middle School, 3*(5), 344–350.

Sosniak, L. A., & Stodolsky, S. S. (1993). Teachers and textbooks: Materials use in four fourth-grade classrooms. *Elementary School Journal, 93*(3), 249–275.

Spillane, J. C. (1999). External reform initiatives and teachers efforts to reconstruct their practice: The mediating role of teachers' zones of enactment. *Journal of Curriculum Studies, 31*(2), 1–33.

Spillane, J.C. (2000). A fifth-grade teacher's reconstruction of mathematics and literacy teaching: Exploring interactions among identity, learning, and subject matter. *Elementary School Journal, 100*(4), 307–330.

Spillane, J. P., & Jennings, N. E. (1997). Aligned instructional policy and ambitious pedagogy: Exploring instructional reform from the classroom perspective. *Teachers College Record, 98*(3), 439–481

Stake, R. E., & Easley, J. (1978). *Case studies in science education.* Urbana: University of Illinois.

Stein, M.K., Baxter, J.A., & Leinhardt, G. (1990). Subject matter knowledge for elementary instruction: A case from functions and graphing. *American Educational Research Journal, 27*(4), 639–663.

Stein, M.K., & Coburn, C.E. (2005, April). *Districts' use of instructional guidance systems.* Paper presented at the annual meeting of the American Educational Research Association, Montreal.

Stein, M. K., Grover, B. W., & Henningsen, M. (1996). Building student capacity for mathematical thinking and reasoning: An analysis of mathematical tasks used in reform classrooms. *American Educational Research Journal, 33*(2), 455–488.

Stein, M.K., & Kim, G. (in press). The role of mathematics curriculum in large-scale reform: An analysis of demands and opportunities for teacher learning. In J. T. Remillard, G. M. Lloyd, & B. Herbel-Eisenmann (Eds.), *Teachers' use of mathematics curriculum materials: Research perspectives on the relationship between teachers and curriculum.*

Stein, M.K., Kim, G., & Seeley, M. (2006 April). *The enactment of reform mathematics curricula in urban settings: A comparative analysis.* Paper presented at the annual meeting of the American Educational Research Association, San Francisco.

Stein, M. K., & Lane, S. (1996). Instructional tasks and the development of student capacity to think and reason: An analysis of the relationship between teaching and learning in a reform mathematics project. *Educational Research and Evaluation, 2,* 50–80.

Stein, M. K., Lane, S., & Silver, E.A. (1996). *Classrooms in which students successfully acquire mathematical proficiency: What are the critical features of teachers' instructional practice.* Paper presented at the annual meeting of the American Educational Research Association, New York.

Stein, M. K., Silver, E.A., & Smith, M.S. (1998). Mathematics reform and teacher development: A community of practice perspective. In J. Greeno & S. Goldman (Eds.), *Thinking practices in mathematics and science learning* (pp. 17–52). Hillsdale, NJ: Erlbaum.

Stein, M. K., & Smith, M. S. (1998). Mathematical tasks as a framework for reflection: From research to practice. *Mathematics Teaching in the Middle School, 3*(4), 268–275.

Stein, M. K., Smith, M. S., Henningsen, M. A., & Silver, E. A. (2000). *Implementing standards-based mathematics instruction: A casebook for professional development.* New York: Teachers College Press.

Stein, M.K., Toure, J., Acquerelli, K., & Fergeson, S. (2003, April). *Infusing mathematics reform into a literacy saturated environment: A challenge of content-driven systemic reform.* Paper presented at the annual meeting of the American Educational Research Association, Chicago.

Stephens, W. M. (1982). *Mathematical knowledge and school work: A case study of the teaching of developing mathematical processes.* Unpublished doctoral dissertation, University of Wisconsin, Madison.

Stigler, J.W., & Hiebert, J. (2004) Improving mathematics teaching. *Educational Leadership, 61*(5), 12–16.

Stodolsky, W. M. (1989). Mathematical knowledge and school work: A case study of the teaching of developing mathematical processes. In P. Jackson & S. Haroutunia-Gordon (Eds.), *From Socrates to software: The teacher as text and the text as teacher* (pp. 159–184). Chicago: University of Chicago Press.

Suydam, M., & Osborne, A. (1977). *The status of pre-college science, mathematics, and social studies education: 1955–1975.* Columbus: Ohio State University Center for Science and Mathematics Education.

Swafford, J. (2003). Reaction to high school curriculum projects. In S.L. Senk & D.R. Thompson (Eds.), *Standards-based school mathematics curricula: What are they? What do students learn?* (pp. 457–468). Mahwah, NJ: Erlbaum.

Tarr, J. E., Chávez, O., Reys, R. E., Reys, B. J. (2006). From the written to enacted curricula: The intermediary role of middle school mathematics teachers in shaping students' opportunity to learn. *School Science and Mathematics, 106*(4), 191–201.

Tarr, J.E., Reys, R.E., Reys, B.J., Chavez, O., Shih, J., & Osterlind, S.J. (in press). The impact of middle grades mathematics curricula on student achievement and the classroom learning environment. *Journal for Research in Mathematics Education.*

Thompson, A. (1992). Teachers' beliefs and conceptions: A synthesis of the research. In D. A. Grouws (Ed.), *Handbook of research on mathematics teaching and learning* (pp. 127–146). New York: Macmillan.

Thompson, D.R., & Senk, S.L. (2001). The effects of curriculum on achievement in second-year algebra: The example of the University of Chicago School Mathematics Project. *Journal for Research in Mathematics Education, 32*(1), 58–84.

Usiskin, Z. (1997). Applications in the secondary school curriculum: A generation of change. *The American Journal of Education, 106,* 62–84.

Valverde, G. A., Bianchi, L. J., Wolfe, R. G., Schmidt, W. H., & Houang, R. T. (2002). *Using TIMSS to investigate the translation of policy into practice through the world of textbooks.* Dordrecht, The Netherlands: Kluwer.

Van Zoest, L.R., & Bohl, J.V. (2002). The role of reform curriculum materials in an internship: The case of Alice and Gregory. *Journal of Mathematics Teacher Education, 5*(3), 265–288.

Viadero, D. (2004, August). Researchers question Clearinghouse choices. *Education Week,* 30–32.

Walker, B. F. (1976). *Curriculum evolution as portrayed through old textbooks.* Terre Haute: Indiana State University, School of Education.

Weimers-Jennings, N. J. (1990). Transformation and accommodation: A case study of Joe Scott. *Educational Evaluation and Policy, 12,* 281–292.

Weiss, I.(1978). *Report of the 1977 National Survey of Science, Mathematics, and Social Studies Education.* Research Triangle Park, NC: Research Triangle Institute.

Weiss, I. R., & Pasley, J.D. (2004). What is high-quality instruction? *Educational Leadership, 61*(5), 24–28

Wertsch, J. V. (1998). *Mind as Action.* New York: Oxford University Press.

Wilson, M., & Goldenberg, M. P. (1998). Some conceptions are difficult to change: One middle school mathematics teacher's struggle. *Journal of Mathematics Teacher Education, 1*(3), 269–293.

Wilson, M. R., & Lloyd, G. M. (2000). The challenge to share mathematical authority with students: High school teachers' experiences reforming classroom roles and activities through curriculum implementation. *Journal of Curriculum and Supervision, 15,* 146–169.

Wilson, S. M. (1990). A conflict of interests: The case of Mark Black. *Educational Evaluation and Policy Analysis, 12,* 293–310.

Wittrock, M.C. (1986). Students' thought process. In M.C. Wittrock (Ed.), *Handbook of research on teaching* (pp. 297–314). New York: Macmillan.

Wood, T., Cobb, P., & Yackel, E. (1991). Change in teaching mathematics: A case study. *American Educational Research Journal, 28*(3), 587–616.

Wu, H. (1997). The mathematics education reform: Why you should be concerned and what you can do. *American Mathematical Monthly, 104*(10), 946–954.

AUTHOR NOTE

We wish to thank a number of individuals for their helpful comments on earlier drafts of the chapter. Jere Confrey, James Hiebert, and Iris Weiss provided insightful comments at a critical point during the development of the chapter and Frank Lester shared generously of his time and advice through the entire process of preparing the chapter. We would also like to thank Valerie Klein for her assistance in gathering and formatting references. An earlier version of this paper was presented at the research pre-session of the annual meeting of the National Council of Teachers of Mathematics, Anaheim, CA, April 7, 2005. The work herein was supported in part by grants from the National Science Foundation (IERI Grant REC-0228343 and Grant REC-9875739). The content or opinions expressed herein do not necessarily reflect the views of the National Science Foundation or any other agency of the U.S. government.

Appendix A

Ratings of Various Curricula by the U.S. Department of Education, AAAS, and Mathematically Correct

DOE: Curricula identified for recognition as Promising or Exemplary
AAAS: Middle school curricula identified as Satisfactory or Unsatisfactory
MC: Overall ratings of curricula

Curriculum	Level	Standards-Based (SB) or Conventional (C)	DOE	AAAS	MC
Everyday Mathematics; SRA/McGraw Hill	EI	SB	Promising		2nd grade: C 5th grade: C–
Investigations in Number, Data, and Space; Scott Foresman	EI	SB			2nd grade: F 5th grade: F
Math Advantage K–6; Harcourt Brace	EI	C			2nd grade: B 5th grade: B–
Math in My World; McGraw Hill	EI	C			2nd grade: B+ 5th grade: B–
Math K–5; Scott Foresman/Addison Wesley	EI	C			2nd grade: B+ 5th grade: B–
MathLand	EI		Promising		
Number Power (supplemental)	EI	SB	Promising		
Saxon Math	EI	C			2nd grade: B 5th grade: B+
Silver Burdett Ginn Math	EI	C			2nd grade: B 5th grade: B
SRA Math; McGraw Hill	EI	C			2nd grade: A 5th grade: A–
Connected Mathematics Project (CMP); Prentice Hall	MS	SB	Exemplary	Satisfactory	F
Heath Mathematics Connections; DC Heath & Co.	MS	C		Unsatisfactory	
Math Advantage; Harcourt Brace	MS	C		Unsatisfactory	B
Mathematics: Applications & Connections; Glencoe/McGraw Hill	MS	C		Unsatisfactory	B
Mathematics in Context (MiC); Holt, Rinehart, & Winston	MS	SB		Satisfactory	
Mathematics Plus; Harcourt Brace & Co.	MS	C		Unsatisfactory	
MathScape: Seeing and Thinking Mathematically; Glencoe/McGraw Hill	MS	SB		Satisfactory	
MathThematics (STEM); McDougal Littell	MS	SB		Satisfactory	D+
Middle Grades Math; Prentice Hall	MS	C		Unsatisfactory	B
Middle School Math; Scott Foresman/Adison Wesley	MS	C		Unsatisfactory	B
Middle School Mathematics Through Applications Project (MMAP; unpublished)	MS	SB	Exemplary		

Curriculum	Level	Standards-Based (SB) or Conventional (C)	DOE	AAAS	MC
Passport Series; McDougal-Littell	MS	C		Unsatisfactory	Passport to Algebra & Geometry: A Passport to Math: C
Saxon Math	MS	C		Unsatisfactory	Alg ½: A Math 87: C
Transition Mathematics (Scott Foresman)	MS			Unsatisfactory	
Cognitive Tutor	MS/HS	SB	Exemplary		
College Preparatory Mathematics (CPM)		SB	Exemplary		
Contemporary Mathematics in Context (Core-Plus)	HS	SB	Exemplary		
Interactive Mathematics Program (IMP); Key Curriculum Press	HS	SB	Exemplary		
University of Chicago School Mathematics Project Integrated Mathematics, Grades 7–12; Prentice Hall	HS	SB	Promising		

9

THE EFFECTS OF CLASSROOM MATHEMATICS TEACHING ON STUDENTS' LEARNING

James Hiebert

UNIVERSITY OF DELAWARE

Douglas A. Grouws

UNIVERSITY OF MISSOURI

TEACHING MATTERS . . .
BUT *HOW* IS NOT EASY TO DOCUMENT

We begin with the following claim: The nature of classroom mathematics teaching significantly affects the nature and level of students' learning. Such a claim seems obvious. Everyone can remember teachers who were especially effective. Surely, teaching makes a difference. Systematically collected data support most people's personal experience by showing that different teachers produce different levels of student achievement gains (Nye, Konstantopoulos, & Hedges, 2004; Sanders & Rivers, 1996; Sanders, Saxton, & Horn, 1997). In fact, the cumulative effect over several years of effective teachers is substantial. Having good teachers really does make a difference.

But what makes mathematics teachers effective? This question does not have an obvious or easy answer. The answer is not found by searching the reports cited above on teacher effectiveness. Many reasons could explain why some of the teachers in these studies were more effective than others. Classrooms are filled with complex dynamics, and many factors could be responsible for increased student learning. This is not one of those pedantic educational issues that researchers get tangled up in by splitting hairs and that, in the end, simply demonstrate what everyone already knows. This is a very central and difficult question to answer. It is, in a real sense, the core question of education. What is it, exactly, about teaching that influences students' learning? Although laypersons in the street often have quick answers based on recollections of their favorite teachers, documenting particular features of teaching that are consistently effective for students' learning has proven to be one of the great research challenges in education.

Our aim in this chapter is to tackle directly the issue of teaching effectiveness—why it has been so hard to document, what is known about it, and how the mathematics education community can learn more. We first examine why it has been so difficult to establish robust links between teaching and learning, then we present a few claims that organize and structure the literature on teaching effects in what we hope are helpful ways, and finally we outline a set of goals and strategies to guide future work in this central and urgent research domain. Although we believe the claims we make about links between mathematics teaching and learning are well supported, we do not present an exhaustive review of the research literature to substantiate them. Rather, we develop arguments to support our claims, select a few studies that illustrate them, and refer the reader to other, more extensive reviews of relevant research.

Before considering some of the difficulties facing researchers who try to document the effects of teaching, we need to define what we mean by *teaching*. Simple definitions, presented in some classic treatments of teaching, are a good starting point. Thorndike (1906) defined *teaching* as the methods used to help students achieve the learning goals valued by society. Gage (1978) defined *teaching* as "any activity on the part of one person intended to facilitate learning by another" (p. 14). These definitions are incomplete because, although they contain much of what we mean by *teaching*, they treat teaching as a one-way relationship—teachers acting on students. In reality, teaching is influenced by students and has a bidirectional quality. Cohen, Raudenbush, and Ball (2003) captured this quality by saying that "instruction consists of interactions among teachers and students around content" (p. 122). If we use Cohen et al.'s definition but add back the goal-directedness explicit in Thorndike's definition and restrict the environments to school classrooms, we have a working definition that sets the boundaries for this chapter: Teaching consists of classroom interactions among teachers and students around content directed toward facilitating students' achievement of learning goals.

Why has it been so difficult to identify features of teaching, as defined here, that show consistent and strong effects on students' learning? Examining this question sets the context for both the claims we make about links between mathematics teaching and learning and the strategies we propose to advance this central research domain in mathematics education. Because the issues surrounding the difficulty of linking teaching with learning are not unique to mathematics, and because they have been addressed in the broader literature, we examine the question both within this broader context and within contexts specific to mathematics.

A Claim That Appears Obvious is Strikingly Difficult to Specify

Despite more than 100 years of work on specifying effective teaching, the field still is searching for reliable, demonstrable recommendations of teaching methods. This is not to say that recommendations have not been proposed. The education literature is filled with suggestions, recommendations, and sure-to-succeed descriptions of teaching. Pestalozzi, one of the great educators of the past, was said to have claimed discovery of "the ultimate method of effective instruction" in the early 1800s (Reusser, 2001, p. 1). A perusal of more recent analyses indicates that Pestalozzi's optimism is alive but not widespread.

Reviews of the teaching-learning literature indicate considerable differences of opinion, not just about which teaching methods are effective, but even about how much is known about whether any methods are particularly effective. Some researchers portray a field that is making good progress in documenting links between teaching and learning, but other researchers see less progress. The third edition of the *Handbook of Research on Teaching* includes two chapters that review, with considerable confidence, the connections between particular teaching behaviors and students' learning (Brophy & Good, 1986; Rosenshine & Stevens, 1986). More recently, Brophy (1999) reinforced this optimism by identifying 12 principles of effective teaching based on accumulated research.

In contrast to these optimistic claims, Romberg and Carpenter (1986) argued that research on teaching and research on learning had constituted two separate fields in mathematics education with few attempts to connect learning and teaching. Consequently, there was little to say about specific links between the two. Romberg and Carpenter's less optimistic view is found in the same third edition of the *Handbook* that contains the positive treatments by Brophy and Good (1986) and Rosenshine and Stevens (1986).

Observations about the absence of evidence-based connections between teaching and learning can be traced back to at least the middle of the last century. A report released by the American Educational Research Association Committee on the Criteria of Teacher Effectiveness concluded:

> The simple fact of the matter is that, after forty years of research on teacher effectiveness during which a vast number of studies have been carried out, one can point to few outcomes that a superintendent of schools can safely employ in hiring a teacher or granting him [*sic*] tenure, that an agency can employ in certifying teachers, or that a teacher education faculty can employ in planning or improving teacher education programs. (Barr et al., 1952, cited in Duffy, 1981, p. 113)

Two decades later, several reviews suggested that efforts to link particular teaching features with students' learning still had produced few results (Berliner, 1976; Heath & Nielson, 1974).

Even with continuing attention to the effects of classroom teaching on learning since these early reviews, researchers still are not uniformly persuaded by the evidence (Shuell, 1996). Countering Brophy's (1999) optimistic view, the two chapters in the most recent fourth edition of the *Handbook of Research on Teaching* that deal explicitly with the effects of teaching on learning are decidedly cautious about claims that link teaching with learning (Floden, 2001; Oser & Baeriswyl, 2001).

The uncertainty in the research community about what is known regarding the effects of teaching on learning can be explained by some of the challenges facing researchers engaged in this work and the ways in which the challenges have (and have not) been met. Before examining these challenges, we must point out that this uncertainty in the field has its costs. Among them is the perception by key U.S. policymakers that research does not inform classroom practice:

> Education has not relied very much on evidence, whether in regard to how to train teachers, what sort of curriculum to use or what sort of teaching methods to use. The decisions have been based on professional wisdom or the spirit of the moment rather than on research (Grover Whitehurst, U.S. Assistant Secretary of Education, as quoted by J. Traub in the *New York Times*, Nov. 10, 2002).

Pressures are increasing to provide evidence-based descriptions of effective mathematics teaching. Consequently, the focal issue of this chapter is not only fundamental, from a research perspective, but urgent, from a policy perspective.

The first step in making further progress in establishing connections between teaching and learning is understanding why this is so hard to do. In the following sections, we outline a number of difficulties that face researchers who try to connect teaching with learning. These are not excuses; they are severe obstacles that must be resolved. Indeed, after studying the following challenges, we expect that readers might change their question from "Why is it so hard to connect teaching with learning?" to "Is it even possible to establish reliable connections between particular features or methods of teaching with students' learning?" Anticipating this discouragement, we continue the chapter by developing several claims about features of teaching that affect learning, claims that we believe have accumulated considerable empirical support in spite of the challenges.

Useful Theories of Teaching Are in Short Supply

Making progress in any scientific field is difficult without explicit theories. In fact, some researchers would argue that fields are scientific to the extent that empirical research is guided by theory (National Research Council, 2002). Theories are useful because they direct researchers' attention to particular relationships, provide meaning for the phenomena being studied, rate the relative importance of the research questions being asked, and place findings from individual studies within a larger context. Theories suggest where to look when formulating the next research questions and provide an organizational scheme, or a story line, within which to accumulate and fit together individual sets of results. Karmiloff-Smith and Inhelder (1974) titled a provocative article, "If You Want to Get Ahead, Get a Theory." They were describing children's efforts to develop their understanding of physical phenomena, but the analogy holds equally well for researchers developing an understanding of connections between teaching and learning.

Robust, useful theories of classroom teaching do not yet exist. Theories that consider connections between classroom teaching and students' learning are even less developed (Floden, 2001; Oser & Baeriswyl, 2001). Within mathematics, theories of learning have been more clearly articulated than theories of teaching. Although theories of learning provide some guidance for research on teaching, they do not translate directly into theories of teaching and cannot be used, by themselves, to stimulate or coordinate research agendas on the effects of teaching on learning. Components or features of teaching that are likely to play important roles in useful working theories have been identified, and beginnings of theories of mathematics teaching have been proposed (e.g., Brousseau, 1997; Freudenthal, 1973; Gravemeijer, 1994; Leinhardt & Greeno, 1986; National Research Council, 2001; Schoenfeld, 1998; Simon, Tzur, Heinz, & Kinzel, 2004). These efforts signal important progress. But theories that specify the ways in which the key components of teaching fit together to form an interactive, dynamic system for achieving particular learning goals have not been sufficiently developed to guide research efforts that can build over time.

Without guiding theories, researchers are feeling their way through murky waters. Studies that have focused on teaching-learning relationships often are chosen because they align with the personal interests of researchers. When researchers pursue different research questions and use different measures of teaching and of learning, their findings often remain isolated from one another. The data do not easily accumulate to reveal patterns and general principles. To surmount these obstacles and move ahead, researchers need to construct useful theories. Those who attempt to develop theories face at least three major conceptual challenges.

Challenge 1: Different Teaching Methods Might Be Effective for Different Learning Goals

There is no reason to believe, based on empirical findings or theoretical arguments, that a single method of teaching is the most effective for achieving all types of learning goals. Perhaps some methods of teaching are more effective for, say, memorizing number facts whereas other methods of teaching are more effective for deepening conceptual understanding and still other methods are more effective for acquiring smooth execution of complex procedures. In addition, some methods of teaching might be especially effective for showing gains in the short-term whereas others might be better for retaining and increasing gains in the long-term. Because a range of goals might be included in a single lesson, and almost certainly in a multi-lesson unit, the best or most effective teaching method might be a mix of methods, with timely and nimble shifting among them.

One consequence of recognizing that different teaching methods might be effective for different learning goals is the fact that empirical studies that compare one teaching method to another using blunt learning measures are difficult to interpret. For what kinds of learning are the teaching methods claimed to be effective?

Challenge 2: Teaching is a System of Interacting Features

A teaching method consists of multiple features that interact with one another in almost countless ways (Cohen et al., 2003; Design-Based Research Collective, 2003; Stigler & Hiebert, 1999). Theories of teaching must account for this fact by treating teaching as a system of interacting features rather than a collection of independent and interchangeable features. This means that the effects of features of teaching on students' learning cannot be measured independently of the system in which they operate. The effects of a particular feature on students' learning likely depend on its interactions with other features in the system.

Two simple examples illustrate the system nature of teaching. One of the striking results from the 1995 Video Study of the Third International Mathematics and Science Study (TIMSS) was the difference in teaching eighth-grade mathematics found between Japan and the United States. In general, lessons in Japan challenged students to solve difficult mathematics problems whereas lessons in the United States emphasized the practice of lower level skills (Stigler & Hiebert, 1999). Casual observation of the lessons suggests that Japanese teachers asked more higher level questions than U.S. teachers, but a closer look showed that the major difference was in the *kinds* of higher order questions that were asked and *when* the questions were asked (Kawanaka & Stigler, 1999). Japanese teachers asked most of their higher order questions about student-presented solution methods for solving problems whereas U.S. teachers asked most of their higher order questions about methods they, themselves, were demonstrating. Japanese teaching was distinguished not just by the existence of high-level questions but in how the questions fit within and reinforced the system of teaching being employed.

A second example of the system nature of teaching comes from research on a feature of teaching often associated with current recommendations—asking students to work in small groups. Clearly whether small groups function productively to help students achieve learning goals depends on many surrounding factors, including the knowledge and skill students acquire for working collaboratively and the kinds of tasks they are assigned (Good, Mulryan, & McCaslin, 1992; Webb, 1991; Webb, Troper, & Fall, 1995). While the first author of this chapter was reviewing videotaped lessons as part of the TIMSS, he observed one eighth-grade teacher asking students to break into their small groups and work together to answer the question "What is the name for a 12-sided object?" Students quizzed each other quickly on whether anyone knew the name and then visited for the remaining time about non-mathematical topics. This task did not lend itself to collaborative investigation. The effect of small-group work is shaped by the system of which it is a part.

A consequence of the system nature of teaching is that the effects on students' learning of individual features of teaching, like higher-order questions or the use of small groups, are difficult to isolate. Systems of teaching, not single features of teaching, affect students' learning. This is not to say that all features within a system are equally responsible for their effects on student learning. Particular features, or combinations of features, might account for more of the system's effects than other features. But the effect of any particular feature is influenced by related and interacting features. Theories of teaching must account for this fundamental complexity.

Challenge 3: The Influence of Mediating Variables

The effects of teaching on learning are complicated, not just by the features of teaching interacting with each other, but by factors that mediate the effects of the system. In a classic analysis, Wittrock (1986) argued that teaching does not affect learning directly. The effects of teaching are mediated by students' thinking—their attentiveness during instruction, their interpretations of the teacher's presentations and of the tasks they are assigned, their entry knowledge and skills, and so on. Wittrock was not the first to comment on the important role that students play in determining the effects of teaching on learning (Berliner, 1976; Doyle, 1978; Winne & Marx, 1980). Berliner (1976), for example, cited "a student's active time-on-task" (p. 10) as a complicating factor in connecting teaching and learning. More recently, Weinert, Schrader, and Helmke (1989) proposed an even stronger claim by saying that "the activity and cognitive orientation of the learners are as important as the activities and the behaviors of the teachers" (p. 899) when specifying the relationships between teaching and learning.

The set of variables related to students' thinking is not the only factor that mediates the effects of teaching on learning. A large number of contextual conditions undoubtedly influence the nature and level of students' learning (Berliner, 1976; Dunkin & Biddle, 1974). These include conditions both inside and outside of the classroom—number of students in class, length of daily lessons, home support for completing homework, and so on.

Some of the variables that mediate the effects of teaching on learning exert a stronger influence on the relationship than simply redirecting or moderating the impact of teaching on learning. These variables can reverse the causal flow so that they change the nature of teaching. For example, students can push back and alter the teacher's intended approach in both subtle and not-so-subtle ways (Cohen et al., 2003; Cooney, 1985). This is why we adjusted Thorndike's and Gage's definitions of *teaching*, as described earlier, to include two-way interactions rather than just one-way interactions from the teacher to the students.

Developing theories of teaching and, in particular, theories that connect teaching with learning is clearly a challenging task. The lack of robust widely-accepted theories is not surprising. Although the development of such theories would enable a more quickly accumulating research base, researchers would not be wise to wait for such theories to pursue empirical studies. Theoretical and empirical work can boost each other. Unfortunately, as we argue in the next section, conducting empirical studies on teaching-learning connections is fraught with a host of additional challenges. As the reader will notice, however, not all of these challenges are unique to empirical research; some are

related to challenges we have already identified that confront theory development.

Documenting the Effects of Teaching on Learning Is Methodologically Difficult

Collecting trustworthy data on the effects of teaching on learning is hampered by a number of methodological difficulties. We place three kinds of difficulties at the top of the list for posing critical threats to research efforts and accounting for the challenge of linking teaching and learning. In Gage's (1978) terms, the first considers what variables to study and the second asks how to measure them. The third kind of difficulty—individual differences among students—has reappeared in various forms throughout the history of research on teaching and learning and remains a significant issue.

Accounting for Relevant Factors

Many, many factors, inside and outside of school, influence what and how well students learn. In addition to quality of teaching, influencing factors include things like the level of home support (health care, education values and resources, and so forth), peer pressures and social relationships, students' motivation, curricular materials, and assessments. To determine the effects of teaching in the midst of all the competing forces, researchers need to account for these forces in some way. Ideally, researchers would conduct experiments in which all of these forces are equalized across comparison groups, or at least are randomly distributed across the groups. Given the nature of school constraints, randomly assigning students to treatments for sufficiently long periods of time is often impossible. The consequence is that the results of studies can have many alternative interpretations. Numerous factors, other than teaching methods or in combinations with them, can explain the learning outcomes.

If researchers are able to conduct studies with random assignment of students to teaching methods, a problem remains of implementing the same method in a consistent way across classrooms. As we have noted, teaching interacts with students (as well as other contextual variables, such as the teacher's experience and expertise) and is likely to change somewhat from classroom to classroom. It can be difficult to know whether the critical features of teaching are being implemented in the same way across all sites. Limiting the research design to a single classroom or a single teacher does not solve the problem because it leaves to later replications the question of whether the findings will apply to other classrooms and teachers.

A final thorny issue arises when testing the relative benefits of particular features of teaching. As noted earlier, teaching is a system of interacting elements. Isolating particular features of the system is difficult and can even distort the effects of teaching being compared. As an example, consider the question of whether concrete manipulative materials aid students' learning.[1] First, one would need to decide how students' learning would be measured. This, in itself, is a challenging issue because different kinds of learning might be possible with manipulative materials than without such materials. Suppose this issue was resolved and treatments were designed with and without materials. Should the treatments contain the same teaching features except for the use of materials? Or should two different treatments be designed—one to provide the most effective possible approach with materials and another without materials? The first choice might lead to rather contrived treatments for one or both conditions because, under normal conditions, it is unlikely that teaching methods would be identical except for the presence or absence of materials. The second choice would lead to results that would be difficult to interpret—between-group differences might be due to the use of materials or, instead, to other differences between the methods that arose as a consequence of changing other features in the teaching system.

Accounting for all the relevant factors when designing studies or interpreting results on relationships between teaching and learning is a serious methodological challenge. As we will argue later, this does not mean that planned, comparative research is impossible, but it does suggest that the simple isolation of variables within teaching systems is unrealistic.

Creating Appropriate Measures

Developing instruments to measure students' learning and instruments to describe classroom teaching that are reliable *and* valid is a continuing methodological challenge (Ball & Rowan, 2004; Berliner, 1976), especially when using the instruments with large samples (Hamilton et al., 2003; Rowan, Correnti, & Miller, 2002). Describing teaching is, in many ways, more challenging than measuring students' learning due, in part, to its bewildering complexity and, in part, to the relatively less attention it has received.

Attempts to describe classroom teaching must consider what to measure and how to measure it. The question of what to measure is tied closely to our earlier discussions of developing theories that point

researchers to important features of teaching and discussions of accounting for the relevant factors in the teaching system. We focus here on the issue of *how* to describe teaching, once the "what" is determined.

Two methods for describing teaching have been used most heavily: classroom observations (Dunkin & Biddle, 1974) and survey questionnaires (Rowan et al., 2002). Both methods suffer from limitations that must be accounted for when designing studies of teaching. Classroom observations can be influenced by biases of the observer. Establishing and maintaining inter-observer reliability on subtle but potentially important features of teaching is difficult. Observation protocols can be worked out in advance to reduce observer bias, but this requires deciding in advance what to observe. Deciding what to observe, in turn, presumes theories (at least hypotheses) of relationships between teaching and learning. Teacher questionnaires have even more serious problems, especially when the questionnaires are administered on a single occasion and ask teachers to make judgments about teaching approaches and emphases (Hiebert & Stigler, 2000; Malara & Zan, 2002; Mayer, 1999; Wubbels, Brekelmans, & Hooymayers, 1992). Teachers have a tendency to over-report features of teaching that they believe are expected. In other words, self-report surveys might indicate how teachers think they should be teaching, but the validity of these reports is suspect for describing teaching practices.

Alternative instruments to describe teaching are being developed. Video provides numerous advantages over observations and questionnaires (Stigler, Gallimore, & Hiebert, 2000) both in terms of reliability and validity. In addition, it allows teaching to be analyzed by different researchers using different approaches over long periods of time, and it affords serendipitous discoveries. When researchers can afford the expense of collecting and analyzing videotape samples of teaching, it offers many methodological benefits. Less expensive alternatives that can be used routinely on a large scale include redesigned and carefully targeted questionnaires (Ross, McDougall, Hogaboam-Gray, & LeSage, 2003; Rowan et al., 2002), combinations of measures such as questionnaires and case studies (Stecher & Borko, 2002), teacher logs that include frequent diary-like notes by teachers (Porter, Floden, Freeman, Schmidt, & Schwille, 1988; Rowan, Harrison, & Hayes, 2004), and classroom artifacts that include information regarding teachers' use of instructional materials and teaching methods, classroom activities, and students' work (Borko, Stecher,

---

[1] See Hiebert (2003) for a similar analysis with respect to the effects of using calculators on students' computational skill.

Alonzo, Moncure, & McClam, 2003). Many of these instruments are currently in the development and testing phase, so it is too early to evaluate their ultimate usefulness.

One methodological issue that cuts across questions of what to measure and how to measure it is the unit of analysis. What constitutes a meaningful unit of teaching? Much of the research that focused on connections between specific teaching features and student outcomes, often referred to as process-product research (Dunkin & Biddle, 1974), used relatively narrow or local units, such as the type of questions a teacher asked or the amount of time spent on each instructional task. Because we believe that teaching is best viewed as a system of interacting features, we propose using broader units that preserve the potentially important interactions. The typical daily lesson is one such unit (Stigler & Hiebert, 1999). The daily lesson has the advantage of being large enough to include key interactions among teaching features and small enough to be thoroughly analyzable. A disadvantage of using the daily lesson as the unit is that students' learning does not occur in lesson chunks. Sequences of lessons defined by full treatment of particular concepts or topics constitute other possible units of teaching, but these can become too large to digest and analyze in detail. Clearly, the tradeoffs must be weighed carefully.

Teaching Often Gets Interpreted as Teachers

To conclude the discussion of why the obvious sounding claim of "teaching affects learning" is so difficult to specify, we focus on a conceptual issue. Questions of how teaching affects learning often get muddled with questions of how teachers affect learning. Some might view this distinction as less substantive and more pedantic than the theoretical and methodological challenges posed in the earlier sections, but we believe the confusion between teachers and teaching is important because it can interfere with theoretical and methodological efforts to establish clearer connections between teaching and learning. If the differences between teachers and teaching are not clarified, research findings are misinterpreted and misunderstood.

To reiterate, we focus on teach*ing* in this chapter—classroom interactions among teachers and students around content directed toward facilitating students' achievement of learning goals. This is different from teach*ers*—their beliefs, subject matter knowledge, years of experience, personality traits, and so on. Characteristics of teachers surely can influence their teaching, but these characteristics *do*

not determine their teaching. Teachers with different characteristics can teach in essentially the same way and vice versa. No doubt this a major reason for the absence of clear links between teachers' personal and background characteristics and student achievement (Dunkin & Biddle, 1974).

We chose to focus on teaching in this chapter, with a targeted and constrained definition, not only to make our task manageable but also because the teaching we define is largely under the control of the teacher. Some characteristics of teachers, such as personality traits, are quite unalterable, but the methods used to teach can be changed. This is not to say that changing teaching is easy, but it is something that teachers' control. Teaching is something that can be altered under supportive conditions and that can benefit from research-based information.

The distinction between teachers and teaching has been recognized by researchers for some time, but the distinction often disappears. In their classic 1974 book, Dunkin and Biddle recognized the distinction and even gave teacher characteristics a special name—*presage* variables. But the distinction has not always been made, perhaps because of a mistaken assumption that teacher characteristics determine, or nearly determine, the nature of teaching. The confusion is especially apparent, and nearly pervasive, in the public press and in many policy documents. Most news reports and policy statements, for example, that critique education and call for better teaching point to characteristics of teachers, not methods of instruction. The annual report on the condition of U.S. education sponsored by the Pew Charitable Trust has a section on "Teaching Quality" but reports no information on teaching methods, only characteristics of teachers such as certification programs and years of experience (Olson, 1997). In the first two sentences of the 2004 annual report of the U.S. Secretary of Education, *teacher* and *teaching* are equated and then used interchangeably with improved teaching apparently assumed to result from changing teacher characteristics (Paige, 2004). The policy document *Before It's Too Late* produced by the highly visible commission chaired by U.S. Senator John Glenn includes four key points in the foreword about improving mathematics and science teaching in the 21st century (National Commission on Mathematics and Science Teaching for the 21st Century, 2000). The third of the four points actually focuses on teaching, and suggests that improving teaching is a key to improving students' learning. But, consistent with Secretary Paige's report, the point is developed by recommending upgrading the quality of teachers as measured by professional and personal characteristics, not by recommending upgrading the quality of teaching methods.

One explanation for blurring the distinction between teachers and teaching is the absence of a shared, accepted knowledge base linking teaching with learning. If Secretary Paige or the Commission chaired by Senator Glenn wanted to recommend improved teaching methods, what would they say? Perhaps discussions of teaching quickly shift to teachers because not enough is known about teaching methods that support particular kinds of learning. At the same time, we believe that ignoring or blurring the conceptual distinction between teachers and teaching hinders progress in building a knowledge base. Confusion between teachers and teaching undermines research efforts to link teaching with learning.

Despite the Difficulties, Research-Based Claims Are Emerging

Given the challenges we have outlined, it now might seem surprising that empirically supported claims linking teaching with learning exist at all. However, despite the difficulties, researchers have constructed a variety of local hypotheses to guide their efforts and have used a variety of methods to investigate connections between teaching and learning. Out of this array of theoretical and empirical work, a few general patterns that link features of teaching with particular kinds of learning can be detected. This is a bold claim because it says there is sufficient empirical data, not just plausible arguments, to link some features of teaching with some types of learning.

Because the goal of this chapter is to describe empirically based connections between teaching and learning, we reviewed the literature with two primary criteria in mind. First, we wanted to select learning outcomes that were educationally significant across grade levels and specific mathematical topics. We were not concerned with detailing the kinds of teaching that promote learning of particular competencies, like solving quadratic equations or subtracting whole numbers with regrouping. Rather, we were interested in more general outcomes, like skill efficiency or problem-solving ability or conceptual understanding. If we could find links between teaching and more general outcomes, the findings would have wider applicability.

The second criterion we used to guide the review was to select only studies that described classroom teaching with some detail, that included well-described measures of students' learning, and that were designed to examine connections between the nature of the teaching and what and how well students learned. Studies that addressed well at least some of the concep-

tual and methodological challenges reviewed earlier were of special interest. Note that requiring studies to describe the nature of the classroom teaching eliminates a number of important areas of research on students' learning. For example, a well-established principle of learning is that the extent to which new material is learned well is influenced by what is already known (National Research Council, 1999). But few studies in mathematics education have examined these effects in the classroom by varying systematically the extent to which teaching connected with students' current knowledge or ways of thinking. Although the principle still has important implications for teaching mathematics, its empirical support does not come primarily from classroom studies of teaching and learning mathematics. As another example of excluded literatures, a rapidly increasing set of studies exists on the effects of different curricula on students' learning (National Research Council, 2004). At this point, however, most of these studies do not include detailed descriptions of classroom teaching so they are not included here. The interested reader will find the literature on the effects of curricula on students' learning reviewed by Stein, Remillard, and Smith (this volume).

In the next sections of the chapter, we present several patterns that connect mathematics teaching and learning. We begin by identifying one of the most firmly established but most general connections between teaching and learning. Commonly referred to as "opportunity to learn," this claim says that students learn best what they have the most opportunity to learn. We then increase the lens magnification slightly and identify two educationally significant learning outcomes in mathematics that have attracted considerable attention from researchers studying teaching effectiveness: skill efficiency and conceptual understanding. We identify the kinds of teaching that are empirically related to these learning outcomes.

The patterns we describe were induced by looking across many studies with an eye toward identifying convergent findings. We believe they are the patterns that enjoy the strongest empirical support at this point in our research history. Other patterns are emerging but do not yet have as much support at this point. For example, considerable work has been conducted on classroom discourse and its effects on learning (see, for example, Forman, 2003; Lampert & Cobb, 2003), but, in our judgment, the empirical links between particular kinds of discourse and particular kinds of student learning are not yet as developed as the patterns we describe. Parenthetically, we include in our review some studies that report data on classroom discourse, but we interpret the results to support a kind of teaching that includes more than just a particular kind of discourse.

Our reading of the literature indicates that the patterns we identify are sufficiently documented that they warrant action. That is, we believe they can be accepted with high enough levels of confidence that they can be used to inform policy and influence classroom practice. Saying this yet another way, they provide substantive empirically-based connections between teaching and learning that Secretary Paige or the Glenn Commission on Teaching could have used to formulate recommendations regarding improved teaching. In addition, the patterns point to fertile areas for future research. They suggest ingredients for modest theories of relationships between teaching and learning, and they map out a territory for promising empirical studies.

OPPORTUNITY TO LEARN: STILL THE KEY ENABLING CONDITION

"Opportunity to learn is widely considered the single most important predictor of student achievement" (National Research Council, 2001, p. 334). Defined by the National Research Council (2001) as "circumstances that allow students to engage in and spend time on academic tasks . . ." (p. 333), the concept of opportunity to learn has been proposed for some time as an explanation for differences in achievement among students (Floden, 2002). It has been used to account for differences in mathematics learning across a variety of contexts, including between-country differences on international comparisons. Fletcher (1971), for example, in a reanalysis of mathematics achievement data from the First International Study of Mathematics Achievement (Husen, 1967) found opportunity to learn a key explanatory construct. He summarized, "the inescapable conclusion, therefore, seems to be that reported 'achievement' is virtually synonymous with 'coverage' across countries (p. 145)." Carroll's (1963) influential model for school learning included time as well as topic coverage as an important variable and promoted the concept's popularity. Researchers from inside and outside of mathematics education continue to include opportunity to learn on the short list of well-documented findings that link teaching and learning (Brophy, 1999; Grouws, 2004; National Research Council, 2001).

Opportunity to learn is not entirely a function of teaching or entirely under the control of the teacher. The curriculum the teacher is required to use, for example, surely influences students' opportunity to learn (Stein et al., this volume). But teaching, as we have defined it, plays a major role in shaping students' learning opportunities. The emphasis teachers place on different learning goals and different topics, the expectations for learning that they set, the time they allocate for particular topics, the kinds of tasks they pose, the kinds of questions they ask and responses they accept, the nature of the discussions they lead—all are part of teaching and all influence the opportunities students have to learn.

Opportunity to learn as a concept that links teaching and learning is best viewed as something more nuanced and complex than simply exposure to subject matter. Put simply, students who are exposed to a topic obviously have a better chance of learning it than students who are not. But, opportunity to learn can mean something more interesting and useful. Consider first graders exposed to a lesson on calculus. Do they have an opportunity to learn calculus? Only in an overly simplistic sense. In reality, first graders would not have a good opportunity to learn calculus because they are not sufficiently prepared to learn the material and cannot engage in the tasks and discussions that would support such learning. They might learn something by sitting through the lesson, like how to sit quietly and listen to things they do not understand. Why? Because that is what the conditions afford. It might not be the intended learning, but it is, in a real sense, what they have an opportunity to learn. So, "opportunity to learn" is not the same as "being taught." Opportunity to learn includes considerations of students' entry knowledge, the nature and purpose of the tasks and activities, the likelihood of engagement, and so on.

Vygotsky's (1978) well-known concept of the *zone of proximal development* captures the general principles behind the example of first graders being taught calculus. The zone of proximal development refers to the space within which learning can be expected, under supportive conditions, given the person's current level of functioning. It describes what is possible, given appropriate opportunities.

Opportunity to learn can be a powerful concept that, if traced carefully through to its implications, provides a useful guide to both explain the effects of particular kinds of teaching on particular kinds of learning and improve the alignment of teaching methods with learning goals. If conceptual understanding, for example, is a valued learning goal, then students will need opportunities to develop such understanding. Similar statements can be made for other learning goals, such as smooth execution of procedures, problem-solving skill, and so on. These kinds of learning do not happen automatically; students need continuing opportunities to achieve them.

Although the concept is useful, the fact that opportunity to learn still is the best link between teaching and learning signals that the research community has not advanced very far in teasing out more specific connections. In the next section, we take a small step in identifying more specific connections between features of mathematics teaching and students' learning. The two patterns we describe can be thought of as instances of opportunities to learn: opportunities to develop skill efficiency and opportunities to develop conceptual understanding.

TWO PATTERNS: TEACHING FOR SKILL EFFICIENCY AND TEACHING FOR CONCEPTUAL UNDERSTANDING

Two of the most valued learning goals in school mathematics are, and have been for some time, skill efficiency and conceptual understanding (Resnick & Ford, 1981). By *skill efficiency*, we mean the accurate, smooth, and rapid execution of mathematical procedures (see, for example, *skill learning* in Gagne, 1985). We do not include the flexible use of skills or their adaptation to fit new situations. By *conceptual understanding*, we mean mental connections among mathematical facts, procedures, and ideas (Brownell, 1935; Davis, 1984; Hiebert & Carpenter, 1992). This definition suggests that conceptual understanding grows as mental connections become richer and more widespread. As we mentioned earlier, considerable research has been conducted to identify the kinds of teaching that facilitate students' achievement of these two learning goals.

An initial and rather obvious finding is that some kinds of teaching support skill efficiency and other kinds of teaching support conceptual understanding. That different kinds of teaching facilitate different kinds of learning is a direct consequence of opportunity to learn. Different kinds of teaching provide different opportunities to learn that, in turn, yield different kinds of learning. This notion has been expressed for some time. Here is what Edward Thorndike had to say in 1912, "It is possible, by focusing attention upon immediate facility, to choose methods of teaching that are excellent for that, but defective for the more important service of arousing in a pupil the desire and power to educate himself [*sic*]" (p. 197). Gagne (1985) refined the notion by detailing different instructional techniques appropriate for different learning outcomes. And Brophy (1988) reminded educators that the work he reviewed linking teaching behaviors to students' learning mostly had used measures of routine skill learning. Other teaching features might have been identified if other learning outcomes had been used.

The story gets more complicated when one notices that neither theory nor empirical data indicate a simple correspondence between one method of teaching and skill efficiency and between another method of teaching and conceptual understanding. The best way to express current knowledge in the field is to describe some features of teaching that facilitate skill efficiency and some features of teaching that facilitate conceptual understanding, and to indicate where these features overlap or intersect.

At this point we should recognize an important historic distinction that is related to but different from the distinction in learning goals we are making here. In a classic book on learning and teaching, Ausubel (1963) proposed a 2×2 matrix with rote versus meaningful learning on one axis and discovery versus expository teaching on the other axis. Ausubel contended that these dimensions were independent. Expository teaching, said Ausubel, does not necessarily produce rote learning, and discovery teaching does not necessarily produce meaningful learning. We agree with Ausubel and raise the distinction here for two reasons. First, we want to alert readers that we are not necessarily linking skill efficiency with expository teaching. Based on our definition, skills might be executed efficiently under rote or meaningful learning conditions. Conceptual understanding, on the other hand, would require meaningful learning.

A second reason for raising Ausubel's (1963) distinction is that we do not expect the features of teaching that facilitate skill efficiency and conceptual understanding to fall neatly into categories such as "expository" or "discovery." In fact, the features of teaching we describe do not fit easily into any of the categories frequently used to describe teaching: direct instruction versus inquiry-based teaching, student-centered versus teacher-centered teaching, traditional versus reform-based teaching, and so on. Although these categories and labels have been useful for some purposes because they capture constellations of features and treat teaching as systems of interacting components, they also can be misleading because they group together features in ill-defined ways and connote different kinds of teaching to different people. For our purposes, avoiding past connotations is important. In fact, we will argue that most of these categories, distinctions, and labels are now more confusing than helpful, and further advances in research as well as clarity of policy recommendations will benefit from abandoning these labels.

But abandoning the usual labels for methods or systems of instruction in favor of identifying features of teaching that link with learning appears to violate a principle we espoused earlier—teaching is a system of interacting elements, and the effects of an individual feature are determined by the system of which it is a part. Can these two points of view be reconciled? Can researchers single out features of teaching that facilitate learning and still believe that the specific effects of individual features depend on the systems of teaching in which they operate? We think so, for two reasons. First, by looking for patterns across multiple studies, we are searching for features that replicate their effects through different implementations of similar systems and even through different systems. We suspect that the exact effects of target features vary somewhat from system to system, but we expect some of these features to emerge as sufficiently robust across systems to appear as detectable patterns. Second, we assume that when we identify specific features of teaching we actually are identifying clusters of features. The features we single out in the following sections probably operate within a supportive set of less apparent features. The regularity that can be detected in their effects on students' learning might be the result of clusters of features that function well within a range of systems of teaching. We single out specific features of teaching because the empirical data do not yet allow a more precise interpretation. But the hypothesis of systems of teaching predicts that these features are simply the most salient aspects of surrounding clusters. Perhaps the development of sharper research tools will afford more precise descriptions in the future.

Becoming Efficient in Executing Skills

No empirical studies that we are aware of set out to examine which features of teaching support skill efficiency and which support conceptual understanding. Studies have contrasted methods of teaching that emphasize skills versus understanding (e.g., Brownell & Moser, 1949; Cobb et al., 1991; Heid, 1988; Hiebert & Wearne, 1993), but no single study or set of studies aimed to optimize these two learning outcomes and sort out the features of teaching associated with each. In addition, different researchers used different measures of teaching and learning, so patterns across studies must be abstracted by interpreting designs and results and inducing consistencies, not simply by adding up the findings.

The large set of studies conducted within the process-product paradigm provides the best evidence

from which to induce patterns for links between teaching and skill efficiency. The literature includes studies in mathematics and in other subjects, and, for the most part, the features of teaching that correlate with skill efficiency are similar across subjects. The most problematic aspect of interpreting this work is characterizing the nature of the dependent measures. Many studies used standardized achievement tests to assess students' learning. Although these tests often are composites of a range of items, they include a heavy skill component, require relatively quick responses, and are restricted to closed-ended, multiple choice formats. Consequently, we believe we can appropriately infer that skill efficiency, as we defined it earlier, is captured by these outcome measures.

Recall that process-product studies attempt to identify relationships between what teachers do in the classroom (the process) and what students learn as a result of this instruction (the product). To provide trustworthy data, these studies must include large sample sizes, use reliable classroom observation instruments to measure classroom processes, and assess student learning in reliable ways. We review several studies that meet these criteria and provide insights into one set of teaching strategies that is related to mathematical skill learning as measured by standardized achievement tests. The interested reader can examine a comprehensive summary of this line of research and its implications across many subject matter fields in the work of Brophy and Good (1986).

In a naturalistic study of mathematics teaching, Good and Grouws (1977) examined the teaching performance of more than 100 third- and fourth-grade teachers over a 2-year period. An initial finding was that the impact of many teachers on students' mathematics learning, as measured on a standardized achievement test, was not very stable over time. This finding suggests that their teaching methods might have changed from year to year, perhaps due to the changing composition of their classes. The authors were able, however, to identify a group of teachers who consistently facilitated better than statistically predicted achievement results over 3 consecutive school years[2] and teachers who were less successful than would have been expected during the same time period. Results from the study indicated that teaching effectiveness was associated with the following behavioral clusters: whole-class instruction[3]; general clarity of instruction; a nonevaluative, task-focused en-

---

[2] The teachers were identified by using achievement data over a two year period and their success was confirmed for the third year during the course of an observational study.

[3] Whole class instruction was the organizational mode in the high classrooms (and low classrooms) with teachers employing instruction via groups in the middle of the distribution.

vironment; higher achievement expectations (more homework, fast pace); student initiated behavior; and classrooms relatively free of behavioral problems. The faster pace finding was also identified in a study of fifth-grade classrooms conducted by McDonald (1976) and singled out by Leinhardt (1986) as a characteristic of teaching associated with standardized achievement gains. Other common findings in the literature are illustrated in the next process-product study we examine.

Using the process-product paradigm, Evertson, Anderson, Anderson, and Brophy (1980) studied junior high school teachers of mathematics ($N = 29$) and English ($N = 39$). They used both low-inference and high-inference observational coding schemes and measured student learning using the California Achievement Test (CAT). Effective mathematics teachers asked more questions than did less effective teachers with most of the questions lower-order product questions. The authors summarized that "the most effective [mathematics] teachers were active, well organized, and strongly academically oriented" (p. 58). Further, as in the previous described study, the effective teachers tended to emphasize whole-class instruction with some time devoted to seatwork. In general, they actively managed their class efficiently. Although the terminology used in this study was different from that in the Good and Grouws (1977) study, some obvious parallels exist in their findings.

A series of studies conducted by researchers at the Far West Regional Laboratory for Educational Research and Development during the 1970s focused on linking teacher behavior to student achievement in mathematics and reading. This research became known as the Beginning Teacher Evaluation Study (BTES), although in practice the studies involved teachers with at least 3 years of experience and concentrated on research rather than evaluation. Interestingly, none of the teacher behaviors measured was a significant predictor of student learning in both mathematics and reading. This finding lends credence to the position that some important teaching actions are content specific.

The results from the BTES about mathematics teaching and learning tend to agree with the previously cited studies but add some interesting findings when data are examined at the individual student level. The studies found a positive relationship between student achievement on a standardized achievement test and high success rates when engaged in mathematical work during classroom lessons (correct answers except for occasional errors). Conversely, there was a negative relationship when students were engaged in activities with low success rates. The researchers

considered this an important enough finding to refine the concept of opportunity to learn to mean "the amount of time a student spends engaged in an academic task [related to the intended learning] that s/he can perform with high success" (Fisher et al., 1980, p. 8). Interestingly, the attention given to high success rates in these studies provided the rationale for Good and Grouws emphasizing the importance of students being well prepared for seatwork before working independently. An overview of the studies in the BTES can be found in Powell (1980) with a detailed account provided in the five volume final report by McDonald and Elias (1976).

The features of teaching that emerge from the studies just reviewed can be summarized as follows: teaching that facilitates skill efficiency is rapidly paced, includes teacher modeling with many teacher-directed product-type questions, and displays a smooth transition from demonstration to substantial amounts of error free practice. Noteworthy in this set of features is the central role played by the teacher in organizing, pacing, and presenting information to meet well-defined learning goals.

The features of teaching identified above facilitate students' skill efficiency as measured by standardized tests. What effects these features have on students' conceptual development is unclear. Because conceptual understanding is of great interest in mathematics education and infuses the recently conceived goal of mathematical proficiency (National Research Council, 2001), we devote considerable space in the next section to examining links between teaching and students' conceptual understanding.

Developing Conceptual Understanding

We defined conceptual understanding earlier as the mental connections among mathematical facts, procedures, and ideas (Brownell, 1935; Davis, 1984; Hiebert & Carpenter, 1992). This simple definition works well for our purposes but is not without controversy. Some researchers argue that this definition places too much emphasis on the internal or mental representations of individuals and that social/cultural factors should be considered (Lave, 1988; Wenger, 1998). From this sociocultural view, understanding can be seen as an activity of participating in communities of people who are becoming competent in the practices of the trade. In our case, this would mean participating in a community of people who are becoming adept at doing and making sense of mathematics as well as coming to value such activity. We believe that both the mental connections and the cultural participation views of understanding can be useful (Cobb,

1994; Sfard, 1998) and that the claims we make in the following sections are reasonable under either definition. However, because the research examples we review were conducted within the then-prevailing mental connections view, this definition should be kept in mind when reading the next sections.

Teaching That Promotes Conceptual Development: Two Key Features

Educators always have been interested in promoting conceptual understanding, at least in their rhetoric, but there has been a recent explosion of interest in the topic. On the day we drafted this paragraph, amazon.com found 90,448 book entries for the terms: teaching for understanding. Among the best-selling titles were *Teaching for Musical Understanding, Strategies That Work: Teaching Comprehension to Enhance Understanding*, and *Shakespeare Without Fear: Teaching for Understanding*. Two well-known books that contain material relevant for this chapter are *Teaching for Understanding* edited by Cohen, McLaughlin, and Talbert (1993) and *Mathematics Classrooms That Promote Understanding* edited by Fennema and Romberg (1999). Clearly, the topic does not lack for analyses and recommendations.

The research base that links features of teaching with students' understanding is much thinner than the current attention might suggest. Looking across the landscape of research connecting teaching and learning, we agree with Brophy's (1997) observation that the research conducted provides less insight into how students construct understanding during instruction than in how they develop skill efficiency. This is due, in part, to the challenge of creating measures that assess conceptual understanding and the fact that much past work has used standardized tests and other similar measures that are more skill oriented. Nevertheless, we can identify two features that emerge from the literature as keys to promoting conceptual understanding. We view our presentations of both features as plausible conclusions drawn from past research *and* as descriptions of domains in which future research could be especially rich and useful.

Feature 1: Teachers and Students Attend Explicitly to Concepts

A clear pattern across a range of empirical studies is that students can acquire conceptual understandings of mathematics *if* teaching attends explicitly to concepts—to connections among mathematical facts, procedures, and ideas (Gamoran, 2001; Hiebert, 2003; National Research Council, 2001). By *attending to concepts* we mean treating mathematical connections in

an explicit and public way. Brophy (1999) described such teaching as infused with coherent, structured, and connected discussions of the key ideas of mathematics. This could include discussing the mathematical meaning underlying procedures, asking questions about how different solution strategies are similar to and different from each other, considering the ways in which mathematical problems build on each other or are special (or general) cases of each other, attending to the relationships among mathematical ideas, and reminding students about the main point of the lesson and how this point fits within the current sequence of lessons and ideas.

To elaborate our claim and to illustrate the findings that support it, we will review several studies that typify those upon which the claim is based. More extensive reviews of this literature can be found in *Adding It Up* (Chapters 6, 7, and 8, National Research Council, 2001), in *A Research Companion to Principles and Standards for School Mathematics* (Kilpatrick, Martin, & Schifter, 2003), in *Mathematics Classrooms That Promote Understanding* (Fennema & Romberg, 1999), and in a chapter by Hiebert and Carpenter (1992) in a previous handbook for teaching and learning mathematics.

In many ways, the claim that students acquire conceptual understanding of mathematics if teaching attends explicitly to mathematical concepts is a restatement of the general observation that students learn what they have the best opportunity to learn. At a surface level, the claim seems rather obvious and, perhaps, not even very interesting. But it becomes more interesting when one discovers that the claim is supported across a wide range of research designs and holds true across different instructional treatments or systems. The studies we review, and the larger literature from which these studies are drawn, display a remarkable variation. Variability across studies can be described in the following ways:

- The effects appear in studies of varying designs—from tightly controlled studies of shorter duration to more loosely designed studies of longer duration;
- Both teacher-centered and student-centered teaching systems or methods (using the old labels) that have attended to conceptual development have shown higher levels of students' conceptual understanding than similar methods that have not attended to conceptual development as directly;
- The ways in which concepts are developed in classrooms can vary—from teachers actively directing classroom activity to teachers taking

less active roles, from methods of teaching that highlight special tasks and/or materials to those that highlight special forms of classroom discourse to those that highlight student invention of solution strategies.

We believe the evidence does not justify a single or "best" method of instruction to facilitate conceptual understanding. But we believe the data do support a feature of instruction that might be part of many methods: explicit attention to conceptual development of the mathematics.

To support and clarify the claim, we chose the following first set of studies to review because they focused on the same mathematical topic—multidigit addition and subtraction in the primary grades—but implemented quite different treatments using different research designs. Fuson and Briars (1990) investigated the effects of a particular conceptual instruction approach on students' acquisition of the standard addition and subtraction algorithms. Specially designed instruction on addition was given to 169 first and second graders, and 75 of the second graders also received special instruction on subtraction. Instruction included careful demonstrations of the algorithm using base-10 blocks, with special attention to the pattern of the regrouping as the numbers increased to three and four digits. Connections were explicitly drawn between the physical and written representations. Some teachers used whole-class formats with teacher demonstrations and class discussions, and other teachers asked students to work in small groups and complete some problems on their own. In all classrooms, the focus was on understanding the standard algorithm, not inventing new strategies. Before instruction, and then after 3–6 weeks of instruction on place value and addition, plus an additional 2–4 weeks for those receiving instruction on subtraction, students were assessed using a traditional written skills test, a specially designed written test on underlying concepts, and individual interviews. As might be expected for students this young, pretest performance was quite low. For example, only 9 of the 169 students correctly regrouped when completing a multidigit addition task. In contrast, performance was uniformly high on the posttest. For example, 160 of the 169 students correctly regrouped on a similar posttest item. Even more striking were the ceiling levels of performance on conceptually based items, including those on the interviews asking students to explain why and how regrouping works in the multidigit algorithm. Given the high levels of performance across the full sample, the particular ways in which the teachers implemented the instruction (e.g., whole-class demonstrations versus small group discussions) apparently did not substantially affect the outcomes. The key seemed to be the explicit attention to the conceptual underpinnings of the algorithm through connecting written and physical representations.

The absence of a control group in the Fuson and Briars (1990) study leaves many questions unanswered. These are answered, in part, as the authors point out, by the fact that the level of performance by the first graders and the (lower achieving) second graders in their study was higher than that of many third and fourth graders on similar items reported on the National Assessment of Educational Progress. This comparatively high performance supports claims about the effectiveness of the instruction. In addition, the findings were replicated by the authors (reported in the same article), with only slightly lower levels of performance, with 2,723 second graders in the Pittsburgh City Schools. Clearly, the effects of this instruction were not an artifact of a special sample. In sum, the findings can be viewed as an existence proof that attending to conceptual underpinnings—to connections among mathematical ideas and representations—can facilitate students' conceptual understanding. The results also show strong skill learning, a finding that we will explore in a later section of the chapter.

A second study, conducted 40 years earlier by Brownell and Moser (1949), employed control groups and fills in another part of the picture on the effects of instruction that attends explicitly to conceptual development. Brownell and Moser studied the effects of teaching multidigit subtraction algorithms "meaningfully" versus "mechanically." About 1,400 third graders in 41 schools were involved in the study. Two different algorithms, equal additions (adding equal amounts to the minuend and subtrahend to keep the difference the same) and decomposition (the current standard U.S. algorithm of regrouping), were crossed with two teaching approaches (meaningful and mechanical) to create four conditions. Teachers in the 41 schools who volunteered to participate indicated which of the approaches they preferred. Classrooms were assigned to conditions by taking into account teachers' preferences and by trying to equalize students' entry abilities across conditions. Each lesson for each condition during the 15-day experimental period was prescribed, and all lessons were quite teacher-directed. The meaningful or conceptual approach included the use of physical objects (bundles of sticks) to illustrate grouping by tens, writing numerals in expanded notation and connecting written and physical representations, and delaying the step-by-step algorithm until after examining the meaning of the number manipulations. The mechanical approach began with a demonstration of the step-by-step algorithm and did not include a rationale for the procedure. The time saved by this approach was used for practicing the procedures. Students' learning was assessed immediately after the instruction and 6 weeks later using written computation tests and individual interviews. Items measured speed, accuracy, explanations of procedures, and transfer to new kinds of subtraction problems. The results are complicated, but, in general, the following conclusions can be drawn. On the immediate measures of speed and accuracy, the mechanical approach produced higher scores. On measures of retention and transfer, the meaningful approach produced higher scores. The meaningful approach worked better for the decomposition algorithm than the equal additions algorithm and for students who had not already memorized a subtraction algorithm before they began instruction. In general, if good explanations, retention, and transfer are valued learning goals (Brownell, 1948), then the meaningful approach showed the most promise, regardless of which algorithm was taught.

Critics might argue that the results reported by Brownell and Moser (1949) are not conclusive because the mechanical approach was highly constrained and the large sample of teachers, without classroom observations to suggest otherwise, could have implemented the approaches in quite different ways. The criticisms are appropriate and should be considered when interpreting the results. Still, we can reasonably conclude that explicit attention to conceptual underpinnings resulted in higher levels of conceptual understanding. This reinforces the conclusion drawn from Fuson and Briars (1990) using a much different research design and separated in time by more than 40 years.

A third study on multidigit addition and subtraction fills in another part of the picture. Hiebert and Wearne (1993) examined the effects of a mixed teacher- and student-centered form of conceptually-based instruction on conceptual understanding (and procedural skill). The 147 second graders in the six classrooms of one school were assigned to one of two treatment conditions for the 12 weeks normally devoted to multidigit addition and subtraction by the teachers. Conceptually-based instruction focused on building a foundation of place value understanding by connecting written and physical (base-10 blocks) representations of numbers and then asked students to invent their own procedures for solving addition and subtraction problems. Teachers demonstrated the standard algorithms if they were not presented by students. Traditional instruction followed the textbook as ordinarily used by the teachers. The focus in

these classrooms was on demonstrating and practicing the standard algorithms. Four classrooms (one high achieving and three average-low achieving) received the conceptually based instruction taught by specially trained teachers and two classrooms (one high achieving and one average-low achieving) received the traditional instruction taught by the classroom teachers. Weekly observations and audiotapes recorded the tasks used during the lessons and the talk by teachers and students in all the classrooms. Instruction, as implemented, showed differences between the number and kind of tasks presented and the nature of the classroom discourse. In general, students receiving the conceptually based instruction spent a greater percentage of time working on the rationale for procedures and examining the legitimacy of invented procedures. Students receiving the traditional instruction spent a greater percentage of time practicing taught procedures. On the basis of beginning-of-the-year and end-of-the-year tests, students in the conceptually based classrooms gained more on skill items and, especially, on conceptual items such as explaining the rationale for computation procedures and transferring them to new kinds of problems. By the end of the year, the performance profile of the lower achieving classroom that received the conceptually based instruction looked much like the higher achieving traditionally instructed classroom.

The use of specially trained teachers for the conceptually based instruction and the relatively small sample size in Hiebert and Wearne (1993) lead to less-than-conclusive results but for different reasons than in Fuson and Briars (1990) or in Brownell and Moser (1949). Each study contains some design deficiencies, but the deficiencies lie in different places. That is why we avoid drawing conclusions about the links between the specific kinds of instruction on multidigit addition and subtraction described in each study and the acquisition of specific skills and understandings. But we do see a pattern across these studies, and others like them, in the general feature of attending to conceptual development. We conclude that when teaching attends explicitly and directly to the important conceptual issues students are more likely to develop important conceptual understandings. This conclusion is reinforced further by considering a few additional and very different kinds of studies.

The evidentiary base supporting conceptual teaching is not restricted to research in the computational skills domain. In a summary report of three separate experimental studies of mathematics learning defined broadly in classrooms at Grades 4, 6, and 8, respectively, Good, Grouws, and Ebmeier (1983) found substantial support for an instructional program that was teacher directed, tightly controlled, and emphasized attention to conceptual development. To define (conceptual) development, the authors drew on the early work of Brownell (1947) on meaningful learning and the subsequent recapitulation of this work, termed *relational understanding*, by Skemp (1976). They summarized their use of the term *development* as "that part of a lesson devoted to increasing comprehension of skills, concepts, and other facets of the mathematics curriculum" (p. 203).

Common to the set of studies reported by Good et al. (1983), and quite different from the preceding studies reviewed in this section, was the use of a treatment that asked teachers to follow a prescribed, highly structured daily teaching plan that employed whole class instruction and included a brief introductory segment (attend to homework, short review, mental computation practice), a development component (about one half the remaining class period), seatwork, lesson closure, and assignment of homework. In the studies, intact classrooms were randomly assigned to treatment and control conditions. Control group teachers were asked to teach using their usual teaching methods.

A number of unique features of these studies are worth noting when interpreting the findings. First, total test scores and subtest scores of the mathematics portions of standardized achievement tests were the main dependent measures used to determine student learning, although as will be mentioned later, supplementary content measures were used in some studies. Second, analysis of variance on residualized gain scores were used to examine the pre-post test differences on each measure. Third, class means were used as the unit of analysis in the statistical tests employed. Finally, interpretation of results took account of treatment fidelity using data from various combinations of classroom observations, teacher diaries, and self-report questionnaires.

In the fourth-grade study of 40 classroom teachers, the treatment teachers exhibited substantially more of each of the requested teaching features than did the control teachers. Furthermore, the scores of the students of the treatment teachers showed significant improvement from pretest to posttest moving from the 27th percentile to the 57th percentile on national norms, significantly higher than the student scores from the control group ($p = .003$) (Good & Grouws, 1979). On a specially designed content test developed by an outside consultant to measure specific content in the curriculum including skills, concepts, and verbal problem solving, the treatment group again outperformed the control group, but the statistical difference was more modest ($p = .10$) with only small differences between the treatment group and the control group on the verbal problem-solving items.

In the sixth-grade study involving 36 teachers and in the eighth-grade junior-high study involving 19 teachers (41 classes), the treatment was modified to give regular, systematic attention to verbal problem solving. In both studies, the treatment students outperformed the control students on the standardized achievement test and on a research-constructed measure of verbal problem solving. The differences on both measures, however, were not as large as in the fourth-grade study. The authors hypothesized that the smaller effects in the sixth-grade study might have been due to smaller differences between teaching methods in the treatment and control classrooms than in the original study. They attributed the lack of larger teaching differences to the possibility of sixth-grade teachers visiting with and learning from the fourth-grade treatment teachers who participated in the study the previous school year. The somewhat diminished fidelity of Grade 8 teachers to the teaching expectations requested in the treatment might be due, according to the authors, to a shorter training period than in previous studies and to a possible treatment request overload—the teachers were asked to implement all the teaching requests in the previous study plus the requests associated with increased attention to verbal problem solving. In spite of less fidelity to treatment than in the previous studies, the researchers did find a statistically significant difference between treatment and control classrooms on the problem-solving measure at the sixth-grade level when 7 poorly implementing teachers were removed from the analysis.

The researchers are quick to point out that they "are *not* suggesting that the instructional program used in the study is the best approach to take for facilitating the mathematics achievement of students" (Good et al.,1983, p. 91), but rather that the ideas within the program including more attention to conceptual development, systematic attention to review, and utilization of appropriate homework assignments might be of considerable value to teachers who use a whole-class organizational pattern for teaching mathematics. The important message from these studies for the current discussion is the recurring link between explicit attention to the development of mathematical connections among ideas, facts, and procedures and the increased conceptual learning by students. The studies not only support this contention but also seem to show that such instruction facilitates the development of important mathematical skills beyond those associated with computational skills, a point to which we will return in a later section.

In contrast to the rather tightly controlled and precisely targeted designs of the studies reviewed above, Fawcett (1938) and Boaler (1998) reported multi-year studies on the cumulative effects of instruction at the secondary level. Fawcett described a 2-year experimental course in high-school geometry for which instruction was redesigned to begin with the reasoning of students and build toward conventional deductive forms of reasoning through, in part, analyzing the logic of real-life arguments. Conceptual aspects of this instruction are apparent in the explicit attention to connections among mathematical objects—definitions, arguments, proofs, claims, and so on. Performances of the 25 students in this course were compared to those of a similar number of students in a more traditional (2-year) course in geometry. On conventional measures of geometric content, on retention of content, and on transfer measures involving the analysis of reasoning in nonmathematical situations, students in the experimental group outperformed students in the control group. Although many design features undermine the comparative claims (e.g., the absence of information about the instruction provided in the more traditional course), a fact acknowledged explicitly by Fawcett (1938), it does seem clear that students in the experimental group engaged with important mathematical ideas and developed important understandings.

Boaler's (1998) study, 50 years later, addressed some of the design deficiencies of Fawcett's (1938) study and reached similar conclusions. Boaler followed a cohort of students at two different secondary schools for 3 years. The schools were chosen because they claimed to be implementing different forms of instruction. After observing regularly lessons in both schools, Boaler described the instruction in one school as an efficient and well-ordered textbook-based approach to procedural competence. Teachers demonstrated procedures and students practiced them. Instruction in the other school was directed toward more conceptual learning goals and revolved around open-ended projects. Students often worked cooperatively, sharing and explaining solution strategies. Entry measures showed that the approximately 200 students in the first school and the approximately 100 students in the second school performed similarly when they began. But, after 3 years, students in the second school, emphasizing conceptual problems and discussions, performed better than their peers on both specially designed conceptual measures and on standardized tests.

In summary, the findings from these studies typify those found in much of the literature that assesses the effects of instruction that explicitly attends to conceptual development of mathematics. Students receiving such instruction develop conceptual understandings, and do so to a greater extent than students receiving instruction with less conceptual content. As noted earlier, this claim might not be surprising because it is a straightforward instantiation of the principle of opportunity to learn. It becomes more interesting, however, when one notes that conceptual development of the mathematics can take many pedagogical forms. Concepts can be developed through teacher-centered and highly structured formats as in some of Fuson and Briar's (1990) classrooms and in Good et al.'s (1983) classrooms or through student-centered and less structured formats as in Fawcett's (1938) classroom. The claim becomes even more interesting and important when making two additional observations that will be developed later in the chapter. First, as we have noted, the findings from several of the studies we reviewed suggest that instruction emphasizing conceptual development facilitated skill learning as well as conceptual understanding. Second, although instructional attention to conceptual development might seem like an obvious feature of "good" teaching and might be expected in most classrooms, it is a feature consistently absent in descriptions of U.S. mathematics classrooms. Before exploring these two issues, we present the second feature of instruction that is associated consistently with increased conceptual understanding.

Feature 2: Students Struggle with Important Mathematics

Our interpretation of the literature on teaching for conceptual understanding points to a second feature of teaching that consistently facilitates students' conceptual understanding: the engagement of students in struggling or wrestling with important mathematical ideas. Unlike the first feature, this second feature might not be obvious to readers, so we first clarify what we mean by *struggle*, then elaborate the theoretical connection between struggling with and understanding mathematics, and finally review a small sample of empirical studies from this perspective.

We use the word *struggle* to mean that students expend effort to make sense of mathematics, to figure something out that is not immediately apparent. We do *not* use *struggle* to mean needless frustration or extreme levels of challenge created by nonsensical or overly difficult problems. We do not mean the feelings of despair that some students can experience when little of the material makes sense. The struggle we have in mind comes from solving problems that are within reach and grappling with key mathematical ideas that are comprehendible but not yet well formed (Hiebert et al., 1996). By struggling with important mathematics we mean the opposite of simply being presented information to be memorized or being asked only to practice what has been demonstrated.

Dewey (1910, 1926, 1929) devoted a good deal of attention to elaborating the processes and consequences of this kind of struggle. For Dewey, the process of struggle was essential for constructing deep understandings. The process begins, said Dewey (1910) with "some perplexity, confusion, or doubt (p. 12)." It continues as students try to fit things together to make sense of them, to work out methods for resolving the dilemma. Traditional school instruction, noted Dewey (1926), is plagued by a press for quick answers. This short-circuits opportunities to think about the conceptual aspects of the problem. Deep knowledge of the subject, said Dewey (1929), "is the fruit of the undertakings that transform a problematic situation into a resolved one" (p. 242–243).

Other cognitive theorists have developed similar arguments that connect the process of struggling to make sense of the subject with deeper understandings of the subject. Festinger's (1957) theory of cognitive dissonance developed the notion of cognitive perplexity as a central impetus for cognitive growth. In a more targeted way, Hatano (1988) identified cognitive incongruity as the critical trigger for facilitating the development of reasoning skills that display conceptual understanding. For Hatano, cognitive incongruity places the student in the situation described by Dewey as filled with perplexity, confusion, or doubt. Applying a similar theoretical construct to school classrooms, Brownell and Sims (1946) contrasted the repetitiveness of most school tasks with the processes they argued are essential for developing understanding—students must feel a need to resolve a problematic situation and then be allowed "to muddle through" (p. 40). What kinds of tasks or problems pose appropriate problematic situations? Vygotsky's (1978) zone of proximal development provides a guide for answering this question because it can be interpreted as the space within which a student's struggle is likely to be productive. That is, novel problems near the boundaries of a student's zone of proximal development are problems that are within reach but that present enough challenge, so there is something new to figure out.

Mathematicians and others who are concerned about how students interact with the discipline of mathematics have alluded to struggling with key mathematical ideas as a natural part of doing mathematics. In his classic book on problem-solving heuristics, Polya (1957) echoed themes of Dewey (1910) by opening the preface to the book's first printing with the following:

A great discovery solves a great problem but there is a grain of discovery in the solution of any problem. Your problem may be modest; but if it challenges your curiosity and brings into play your inventive faculties, and if you solve it by your own means, you may experience the tension and enjoy the triumph of discovery (p. v).

Polya continued by noting that experimentation and struggle are an integral part of developing the discipline of mathematics and suggested that students should experience this side of mathematics as well as the systematic, deductive side. Brown (1993) explicitly brought together Dewey's and Polya's views around the notion of struggling with mathematics, a phenomenon Brown called "confusion." He suggested that struggling with real problems is quite different than the tidy curriculum students usually receive, and he argued that appropriate struggle (or confusion) reveals to students a key aspect of mathematics. "We seem incapable of appreciating what it means to be confused, what virtues may follow from it and what kind of pedagogy may make use of it" (Brown, 1993, p. 108).

The logic that connects struggle with understanding, the learning goal of interest in this section, can be expressed using either of the definitions of *understanding* presented earlier. If understanding is defined as the mental connections among mathematical facts, ideas, and procedures, then struggling is viewed as a process that reconfigures these things. Relationships among facts, ideas, and procedures are re-formed when new information cannot easily be assimilated or when the old relationships are found to be inadequate to make sense of a new problem (Piaget, 1960; Skemp, 1971). Struggle results in restructuring one's mental connections in more powerful ways. If understanding is defined as participating in a community of people who are becoming adept at doing and making sense of mathematics, then struggling is vital because it can be an essential aspect of personal meaning-making within the community (Handa, 2003). It is central to what mathematicians do (Brown, 1993; Polya, 1957). In summary, struggling with important mathematics is implicated in both definitions of understanding by identifying the common processes hypothesized to develop understanding.

Although little, if any, research has tried to isolate and test the effects of struggle on the development of students' conceptual understandings, many findings suggest that some form of struggle is a key ingredient in students' conceptual learning. In a provocative review of cognitive training studies (outside of mathematics education), Robert Bjork (1994) pointed out a curious feature in many of the findings: When trainees experienced difficulties and could not smoothly master the targeted skills, they developed deeper or more useful competencies in the long run. The difficulties were introduced either intentionally or accidentally into the training procedure and took different forms—changing the context of problems, withholding feedback, introducing interfering distractions, and so on. The key condition was that trainees faced some obstacles along the way. In these cases, the trainees generalized their learning more appropriately and transferred it more effectively to novel tasks. Apparently the trainees had to think harder about what they were learning. Bjork (1994) concluded:

> Whatever the exact mixture of manipulations that might turn out to be optimal [for learning], one general characteristic of that mixture seems clear: It would introduce many more difficulties and challenges for the learner than are present in typical training routines (p. 189).

This conclusion is not surprising, said Bjork, because it simply affirms the observation, now widely accepted in cognitive psychology, that learners construct their own interpretations and memories of activities rather than simply absorbing them from the environment. When students struggle (within reason), they must work more actively and effortfully to make sense of the situation, which, in turn, leads them to construct interpretations more connected to what they already know and/or to reexamine and restructure what they already know. This yields content and skills learned more deeply.

No series of classroom studies parallels those of the laboratory studies in the training literature reviewed by Bjork (1994), but, when viewed through the lens of students' struggle, the findings from a number of classroom studies are suggestive. Again, we review only a few studies here to illustrate the nature of these findings. First, in a study outside of mathematics, Capon and Kuhn (2004) found that students in a graduate MBA course who learned a new business concept through attempting to solve problems rather than listening during a lecture and discussion could apply what they learned to more effectively explain a related concept. The authors accounted for the results by arguing that working to solve problems promoted more sense making and integrating of information, in other words more conceptual understanding.

We now review several studies in mathematics education that support the claim that some features of teaching can facilitate students' opportunities to struggle, in the sense that we have defined it, and when this happens students develop greater conceptual under-

standing. Although Carpenter, Fennema, Peterson, Chiang, and Loef (1989) did not interpret their data from this perspective, they reported findings that are consistent with our claim. The authors compared the teaching and learning in 40 first-grade classrooms. Half of the teachers from these classrooms were randomly selected to receive a 4-week summer course (called Cognitively Guided Instruction, or CGI) that focused on young children's knowledge and skills of simple addition and subtraction. Classroom observers recorded the kinds of teaching that occurred for addition and subtraction the following year, and the researchers administered a variety of student achievement measures. Teachers who had received the 4-week course spent a higher percentage of instruction time on word problems (often thought to be more challenging for students) whereas the control teachers spent a higher percentage of instructional time on number facts problems and on review. In addition, the CGI teachers posed more arithmetic problems to the whole class, listened more often to the solution procedures presented by students, and gave feedback less often on whether students' answers were correct. All these features of teaching are consistent with the hypothesis that students in the classrooms of CGI teachers had more opportunities to wrestle with the mathematics involved in adding and subtracting whole numbers. On many of the achievement measures, there were no significant differences between students in the two groups. However, students in the classes taught by CGI teachers performed better on the "complex addition and subtraction problems," problems requiring more than routine knowledge of arithmetic. Interestingly, students in these same classrooms also demonstrated significantly higher recall of number facts during individual interviews than students in the control classrooms. Although only suggestive, the results reported by Carpenter and colleagues (1989) are consistent with our primary claim in this section.

A study that addresses more directly the claim that when teaching promotes constructive struggle with mathematics students' understanding increases was conducted as part of the QUASAR Project (Silver & Stein, 1996; Stein & Lane, 1996). In particular, the results of the study suggest that mathematical tasks that make higher cognitive demands on students facilitate greater conceptual development (Stein, Grover, & Henningsen, 1996; Stein & Lane, 1996). The mathematical tasks presented to middle-school students were examined for the cognitive effort, or "thinking processes," the tasks were likely to elicit. The most challenging tasks were presumed to require reasoning mathematically (problem solving, pattern finding, conjecturing) and connecting procedures with concepts. The least challenging tasks were presumed to require executing procedures without considering connections to other ideas, memorizing facts and rules, and exploring unimportant mathematics or exploring in a haphazard, unpurposeful way. The level of challenge of a sample of 144 tasks presented in 3 classrooms in each of 4 middle schools over 3 years was correlated with the achievement gains by sixth, seventh, and eighth graders tested at the beginning and end of each school year. Achievement was measured using a specially designed pool of 36 open-ended mathematics problems that allowed tracing changes in students' performance over several years. Results are quite complicated but clearly indicate that students attending schools in which teachers presented *and* faithfully implemented more challenging problems were more likely to develop increased conceptual understanding of the mathematics.

The study by Hiebert and Wearne (1993), reviewed earlier, supplements the findings of the QUASAR studies because teachers in the classrooms that facilitated greater understanding asked a higher percentage of questions that required explanations and analyses, and students in these classrooms responded with longer utterances. These forms of discourse require more intellectual work from students and can reveal students who are struggling to make sense of mathematics and to express their emerging understandings.

That students can make sense of classroom discourse that displays genuine cognitive and mathematical struggle was documented nicely by Inagaki, Hatano, and Morita (1998). Fourth and fifth graders in 11 classrooms were observed during discussions of alternative methods for adding fractions with unlike denominators. Although the discussions were filled with confusions and incorrect conjectures along with correct analyses, most students made sense of the discussions and improved their understanding as demonstrated through both their verbal statements and written work.

Case studies of classrooms in which teachers engage students in struggling with important mathematics add support to the claim that struggle is a key enabler for conceptual understanding. Detailed descriptions of classrooms in which authors reflect on their own experiences as teachers are especially compelling (Ball, 1993; Fawcett, 1938; Heaton, 2000; Lampert, 2001; Schoenfeld, 1985). Classrooms in which this kind of struggle is noticeably absent and, in turn, students do not improve their understanding (Schoenfeld, 1988) complement the positive cases.

A number of questions regarding classrooms in which students are engaged in constructive struggle with mathematics remain unanswered. In contrast to the first feature of explicit attention to concepts, whether struggle can be incorporated into a wide range of pedagogical forms is unclear. Struggle is usually associated with student-centered or student inquiry approaches. But we can imagine teacher-centered approaches that provide targeted and highly structured activities during which students are asked to solve challenging problems and work through challenging ideas. In fact, it seems plausible that students' struggle should be sufficiently bounded and directed so that it centers on the important mathematical ideas. This requires some level of teacher guidance. There are few principles, however, from theory or data, that speak to the degree of structure needed to facilitate (or undermine) the productive effects of students' struggle. Parenthetically, the possibility that appropriate struggle can be built into teacher-centered approaches is one reason why we think the old labels of student-centered versus teacher-centered instruction can be so misleading.

A striking contrast should be apparent between teaching that encourages students to struggle and teaching for skill efficiency, described earlier, that ensures high success rates. Although these are not completely contradictory features, they would likely be found in quite different systems of teaching. On the one hand, this is consistent with what one might expect—different features and clusters of features are effective for promoting skill efficiency than are effective for conceptual understanding. But what explanation can account for the fact that skill learning was relatively high in classrooms where features of teaching associated with conceptual development were being implemented? We turn next to consider this question.

Teaching Features that Promote Conceptual Understanding also Promote Skill Fluency

We noted earlier that no simple correspondence exists between features of teaching and learning outcomes. One place in which the complex nature of teaching and learning becomes apparent is in the effects of conceptually oriented teaching on skill learning. Many of the reports on the conceptual development of students also indicate that their skills increased at a level equal to or greater than students in the control groups (Boaler, 1998; Fawcett, 1938; Fuson & Briars, 1990; Good et al., 1983; Hiebert & Wearne, 1993; Stein & Lane, 1996). Apparently, it is not the case that only one set of teaching features facilitates skill learning and another set facilitates conceptual learning. In this case, two quite different kinds of features both seem to promote skill learning.

One way to explain this observation is that the *nature* of skill learning might be somewhat different under the two instructional conditions. Under one condition, instruction is quickly paced, teachers ask short-answer targeted questions, and students complete relatively large numbers of problems during the lesson with high success rates. Under the other condition, instruction is more slowly paced, teachers ask questions that require longer responses, and students complete relatively few problems per lesson. Using Wittrock's (1986) argument, the cognitive mechanisms that students are likely to use under these different conditions would be different, leading to different skill competencies. In fact, Bjork's (1994) review seemed to show exactly that: Trainees who mastered skills under conditions that stimulated more conceptual work acquired different competencies—they were better able to adapt their skills to solve new kinds of tasks. Following the authors of *Adding It Up* (National Research Council, 2001), we use the word *fluent* (rather than *efficient*) to describe this kind of competence.

The foregoing analysis produces the following conjecture. The two features we identified for developing conceptual understanding are promising candidates for facilitating a more ambitious learning goal—mathematical proficiency (National Research Council, 2001). *Mathematical proficiency* is the proficiency achieved through integrating five kinds of competencies, including conceptual understanding and procedural fluency.[4] Because mathematical proficiency has not been defined long enough to serve as the outcome measure for empirical studies, no unambiguous data are available to test the relationships between teaching and students' mathematical proficiency. But the findings of improved conceptual understanding *and* procedural fluency related to the two features of teaching we described above make plausible the conjecture that these two features can facilitate the development of mathematical proficiency.

Summary

We believe that two features of classroom mathematics teaching facilitate students' conceptual development (and perhaps mathematical proficiency)—explicit attention to connections among ideas, facts, and procedures, and engagement of students in struggling

with important mathematics. We believe both features are justified by not only the theoretical arguments but also the empirical data. We have not identified all of the teaching actions that might call students' attention to relevant connections or engage them in productive struggle, but we have described a sample of strategies (e.g., posing problems that require making connections and then working out these problems in ways that make the connections visible for students). The centrality of the teacher was noted when reviewing the teaching features that facilitate the acquisition of skill efficiency. On the basis of the current literature, systems of teaching that facilitate the development of understanding seem to operate with the teacher playing both more and less active roles.

Our reading of the literature suggests that other features of teaching that sometimes are associated with conceptual development (e.g., use of concrete materials, asking higher order questions) are too specific and tied too closely to particular classroom conditions to support claims that apply across classrooms. That is, not enough evidence exists to conclude that these more specific features, which work well under certain conditions or within particular systems of teaching, will work well under other conditions or in other systems. The two features we identified are more general and seem to operate effectively across a range of contexts and teaching systems.

We have not commented on whether both teaching features—attending to connections and struggling with important mathematics—are necessary for promoting conceptual understanding, whether either one is sufficient, whether there are interactions between the two, and so on. We also have not speculated about how these features might function as part of clusters of features. Perhaps, for example, each of these two features is part of a cluster that travels, as a cluster, across different systems of teaching. Increased conceptual understanding might be due to a cluster of teaching features rather than just the single highlighted feature. We believe these issues, together with the specific teaching actions that are required and the mechanisms that are responsible for the effects of these teaching features, define promising research agendas for the future. We also believe that they warrant investigation using outcome measures of the re-

---

[4] In brief, mathematical proficiency is the simultaneous and integrated acquisition of five kinds of mathematical competencies, or "strands:

- *conceptual understanding*—comprehension of mathematical concepts, operations, and relations
- *procedural fluency*—skill in carrying out procedures flexibly, accurately, efficiently, and appropriately
- *strategic competence*—ability to formulate, represent, and solve mathematical problems
- *adaptive reasoning*—capacity for logical thought, reflection, explanation, and justification
- *productive disposition*—habitual inclination to see mathematics as sensible, useful, and worthwhile, coupled with a belief in diligence and one's own efficacy" (National Research Council, 2001, p. 116).

cently defined and more ambitious learning goal of mathematical proficiency (National Research Council, 2001). After commenting about the state of current classroom teaching with respect to these features, we pick up on this claim by identifying a number of principles to guide research agendas that would advance our knowledge of key relationships between teaching and learning.

Absence of Features That Support Conceptual Development from U.S. Mathematics Teaching

Before leaving the section on "Developing Conceptual Understanding," we feel compelled to point out that U.S. mathematics classrooms show a striking absence of the very two features we identified as facilitating conceptual understanding. Ever since U.S. classroom practices have been studied systematically, these two features rarely have been found. Hoetker and Ahlbrand (1969) surveyed early descriptions of classroom teaching, across all subject fields, and reported that, since records have been kept, U.S. school teaching often has focused on lower level skills using a tightly controlled and curtailed question-and-answer routine called *recitation*. A series of studies in the 1970s confirmed this general observation for mathematics classrooms. The authors described classrooms focused on developing routine skills with a basic pattern of teaching that was quite consistent across classrooms (Fey, 1979; National Advisory Committee on Mathematics Education, 1975; Stake & Easley, 1978). Welch (1978) described a typical mathematics lesson this way:

> First, answers were given for the previous day's assignment. A brief explanation, sometimes none at all, was given of the new material, and problems were assigned for the next day. The remainder of the class was devoted to students working independently on the homework while the teacher moved about the room answering questions. The most noticeable thing about math classes was the repetition of this routine (pp. 5–6).

Recent surveys have shown that this lesson pattern continues today and that instruction rarely includes explicit attention to conceptual development or engages students in struggling with key mathematical ideas (Hiebert et al., 2003; Rowan et al., 2004; Stigler,

Gonzales, Kawanaka, Knoll, & Serrano, 1999; Weiss, Pasley, Smith, Banilower, & Heck, 2003).

The mathematics Video Study of the 1999 TIMSS provides recent and direct information on the nature of U.S. mathematics teaching (Hiebert et al., 2003). To compare eighth-grade mathematics teaching in the United States with six higher achieving countries in Europe and Asia (Australia, Czech Republic, Hong Kong, Japan, Netherlands, and Switzerland), the 1999 Video Study collected a nationally representative sample of about 100 videotaped lessons from each country. An international team of researchers developed more than 75 codes to analyze the structure and organization of lessons, the mathematical content of the lessons, and the way in which content was worked on during the lessons. Only codes that could be applied with at least 85% interrater agreement were used.

One code from the 1999 Video Study that is especially instructive is the nature of the mathematics problems presented and the way in which the problems were worked on with the students.[5] On the basis of the written materials or the oral presentation by the teacher, all problem statements could be classified into one of three categories defined by the mathematical processes that were apparently intended: using procedures, stating concepts, and making connections. "Making connections" is of most interest here because making connections (among ideas, facts, and procedures) lies at the heart of our definition of conceptual development. A making connections problem could be "Graph the equations $y = 2x + 3$, $2y = x - 2$, and $y = -4x$, and examine the role played by the numbers in determining the position and the slope of the associated lines." Problem statements classified as stating concepts asked students to recall or illustrate definitions or properties. These problems seldom involved conceptual development because they simply asked students to recall or repeat concepts or propositions. A stating concepts problem could be "Show the point (3, 2) on the coordinate plane." Using procedures problems asked students to use apparently known procedures to solve problems, for example, "Solve $2x + 3 = 5$." Results showed that 17% of the problems in an average U.S. lesson indicated an intent to make connections; this was within the range of 15% to 24% evident in all other countries except Japan (Japan was an outlier with 54%).

The picture changes, however, when considering the way in which teachers worked with students on problems. All problems that had some public discussion or visible work were coded a second time into one

---

[5] This code was developed by Margaret Smith (2000) and built upon earlier work by Mary Kay Stein and colleagues (Stein, Grover, & Henningsen, 1996).

of the same three categories. A fourth category, giving only the answer, was needed for those cases when no mathematical work was done with students except reading the answer. A question of special interest is, "What happened to the making connections problems when they were implemented in the classroom?" In all countries except for Australia and the United States, 37% to 52% of these problems were worked on so that the connections implied by the problem statement were made explicit with the students—through examining the problem, comparing solution methods, justifying why the solution methods worked, and so on. In Australia, 8% of the making connection problems were worked on in this way. In the United States, so few making connection problems were worked on in this way that the percentage rounded to 0% (Hiebert et al., 2003). This nationally representative sample of eighth-grade mathematics lessons shows that students in the United States are not receiving instruction that supports conceptual development.

Before speculating about the reasons for the absence of these teaching features from U.S. classrooms, we point out the degree to which researchers in the Video Study had to probe the actual work of teachers in these classrooms to detect the differences among countries. If coding had stopped with the kinds of mathematics problems presented to students, U.S. classrooms would have looked similar to several other countries. Describing teaching—how students and teachers interact about content—required moving beyond the curriculum to describing the details of how teachers use the curriculum. Documenting connections between teaching and learning requires describing teaching with considerable detail and precision. Notice that this fact calls into question findings that connect opportunity to learn with student achievement when opportunity to learn is defined solely in terms of content coverage (Fletcher, 1971; Gamoran, Porter, Smithson, & White, 1997; Porter et al., 1988). To the extent that teachers transform problems, as they did in the TIMSS 1999 Video Study, the opportunity to learn indicated by an analysis of the curriculum might not be the opportunity to learn experienced by the students. This becomes especially critical if one hopes to understand the mechanism lying behind statistical relationships and develop policy recommendations based on the findings. Simply put, curriculum changes are unlikely to lead directly to changes in students' opportunity to learn.

A number of factors might explain the long-standing absence from U.S. mathematics classrooms of teaching features that support conceptual understanding. These include the difficulty of changing practices that are embedded in a culture, the lack of subject matter and pedagogical knowledge that teachers need in order to teach in different ways, the challenge of enacting complex skills even when they are known by teachers, the limiting nature of common professional development activities, the pressure of district- and state-mandated tests, and the absence of a useful and accessible knowledge base for teachers to improve their practices. We will not develop the arguments that justify each of these explanations, but we will use the last of the explanations, the absence of a knowledge base, as the springboard for the final section of the chapter. The question we tackle in the final section concerns the kind of research agendas that could stimulate further progress in linking teaching and learning. What kind of information will be most helpful and how can it be produced?

FUTURE DIRECTIONS FOR RESEARCH THAT CONNECTS TEACHING AND LEARNING

In this section, we identify several principles that we believe will elevate the quality of research that aims to link mathematics teaching and learning. Some of these principles can be applied immediately, and some will require development themselves. We believe that attending to each of the principles, even if they are not fully developed, will increase the chances that research findings will contribute to a growing, coherent, and useful base of knowledge for improving classroom teaching.

Be Explicit About Learning Goals

The distinction we elaborated earlier, between teaching features that support skill efficiency and those that support conceptual understanding, foreshadows the first principle of conducting research on teaching/learning relationships: Researchers must be clear and explicit about the kinds of learning they will study. Talking about teaching, in general terms, as "effective" or "not effective" is no longer helpful. Teaching is effective (or not) for helping students achieve particular kinds of learning goals. Teaching is effective *for* something and the something must be described and measured as precisely as possible.

We are not saying that a simple correspondence exists between specific features of teaching and specific learning outcomes. This empirical question has yet to be answered. Of course, the data we reviewed earlier indicate already that the correspondence will not be straightforward. The features of teaching that support conceptual understanding might also support skill fluency. Regardless of how these particular relationships are resolved, one lesson from past work is that further progress in understanding connections between teaching and learning depends on being clear about the learning outcomes to be measured and on developing assessments that measure these outcomes directly.

We recommend that the learning goal of mathematical proficiency (National Research Council, 2001) become part of some research agendas in the future. This ambitious goal, which is defined as the integration of five strands of knowledge, has not yet been operationalized through instrument development. Considerable effort will be needed to construct tasks that measure the integration of these strands. Measuring each strand individually misses the point of the construct. It is the integration of the strands that characterizes this promising goal. How to handle these conceptual and methodological challenges presents an immediate task worth pursuing.

Build and Use Theories

Once learning goals are chosen and instruments developed to measure them, theories must guide the research process. Ideally, theories will guide key aspects of the *entire* process: (1) the generation of testable hypotheses about the connections between teaching and learning, (2) the selection and modification of methodological designs that are appropriate for the questions, and (3) the analyses and interpretation of the data. Earlier, we identified the lack of theories of teaching and learning relationships as a major cause for the absence of better claims. Now, we remind the reader that the existence of theories, even local theories, has great benefits for guiding the design of individual studies and for connecting the findings from individual studies to build a useful knowledge base for teaching.

We underscore one benefit of working from theories—they allow researchers to *understand* what they are studying. Theories of teaching provide a framework within which to understand the interactions among features of teaching and the reasons that features of teaching might be more or less effective, within different systems, for facilitating particular kinds of learning. Unless researchers, and teachers, understand why teaching influences learning in a particular way, the knowledge is of little use (Rowan et al., 2002). Understanding the reasons for the effects of teaching allows teachers to adapt the features of interest to their own setting without fatal changes (Brown & Campione, 1996). This makes knowledge useful. When connections between teaching and learning are not under-

stood, isolated findings of "what works" can become rules or prescriptions for practice that are applied too rigidly or changed inappropriately. Users need to know how to modify or adjust the prescriptions to fit the expected differences from classroom to classroom. This benefit of working from theories becomes critical for our discussion of policy in the concluding section of the chapter.

The process through which theories of mathematics teaching and learning relationships are best developed is not yet clear. Ruthven (2002) described two different processes for building theories of teaching, attributed to French and Italian researchers, respectively. A first approach is characterized by top-down activities in which researchers specify the hypotheses to be tested and teachers implement the tests as prescribed. The second approach is characterized by bottom-up activities in which the teachers experiment with various instructional strategies and, together with researchers, detect more general principles that gradually emerge. We believe it is too early in the history of mathematics education research to tell whether one or the other, or a third, approach will be most productive for generating useful theories. All approaches should be considered.

A call to build theories is, fundamentally, a call for researchers to be clear about the hypotheses they use when conducting empirical studies. Hypotheses are researchers' expressions of how well they understand the phenomena under investigation. Explicit statements of hypotheses provide a measure for how the field is advancing in its collective understanding of the phenomena. Thus, hypotheses and the theories they generate not only aid researchers in conducting empirical studies, they provide an essential way for researchers to communicate with one another regarding their own understandings of the teaching-learning relationships. As specific, local hypotheses are made explicit, are refined, and are accumulated, theories can emerge that tie together individual studies and enable the construction of a useful knowledge base.

Set Realistic Expectations for What a Knowledge Base Can Do

A third principle that can inform future research agendas is recognition of the appropriate role of research in setting policy and recommending practice. To define an appropriate role for research on relationships between teaching and learning, consider first Gage's (1978) observations. Gage argued that to develop a science of teaching defined by high predictability and control would be unrealistic. The reasons are found in the challenges presented earlier in documenting straightforward connections between teaching and learning. But, said Gage, a scientific basis for the art of teaching is realistic. It is realistic to expect an increasingly sound theoretical and empirical basis for making informed (although still uncertain) decisions. Gage quoted an 1891 observation by the philosopher Josiah Royce:

> It is vain that the inadequacy of science is made a sufficient excuse for knowing nothing of it. The more inadequate science is when alone, the more need of using it as a beginning when we set about our task. (p. 20)

A realistic expectation, in our view, is that a knowledge base can inform policy and practice and can be used to make decisions with some level of confidence. The level of confidence rises as the knowledge base becomes more detailed, richer, and more coherent. Because research is further advanced in some areas than others, decisions about teaching practices that support some learning goals will be made with more confidence than decisions about achieving other learning goals.

The uncertainty of research claims is quite similar to that found in many other professional fields that involve the complexities of human behavior. Consider the field of health and nutrition. Scientific research on the effects of exercise, diet, and drugs informs recommendations regarding healthy lifestyles. Fruits and vegetables are good for you in a variety of ways. A small daily dose of aspirin reduces the risk of stroke. A high-fiber diet reduces the risk of colon cancer. But, like in education, most of these recommendations are made with some level of confidence, not with certainty. And as the specificity of the recommendation increases, the confidence level declines. Fruits and vegetables are good, but which ones are best, and exactly how many servings should everyone have? Should everyone take a small dose of aspirin each day? What about young people, and what about those who have mild stomach reactions to aspirin? Like education, the problem is that so many interactions influence the effects of any particular treatment—physiology of the individual, amount of exercise, weight, age, and so on. Specifying optimal diet, drug treatments, and lifestyle with certainty is impossible. But a knowledge base that informs decisions with increasing levels of confidence can be built.

One final parallel between decisions about physical health and about education is important. Readers surely have noticed that, as new information about nutrition and drug treatments becomes available, recommendations change, sometimes dramatically.

Should you eat butter or margarine? For years, margarine was the healthy alternative to butter. With new knowledge about trans-fatty acids, butter might be the better choice. Recommendations regarding more serious issues, like hormone treatments and breast cancer, also have witnessed considerable changes, even reversals. This does not mean that the medical profession is being silly or lazy or is conducting thoughtless and flawed research. Rather, the problems are extremely complicated, and sorting out all of the interactions and conditions under which treatments affect outcomes is enormously difficult. It also is noteworthy that people generally believe they benefit from the latest and currently best information that the health and medical professions can provide, even if it will change in the future. The same should be true in education with respect to relationships between teaching and learning.

One implication of this analysis is that educators, and the public, should not expect critical experiments in education. They should not expect single studies, regardless of how large or how well conceived, to answer the question of what kind of teaching is best for achieving particular learning goals. Well-designed studies that yield clear results can increase the level of confidence about which features of teaching to recommend, but the recommendations always are subject to change when better information becomes available or when conditions change.

We conclude this section by reminding the reader that not all decisions in education are, or should be, based on research. Learning goals, for example, are statements about what society most values and should be selected, in part, through public debates about values (Hiebert, 2003). How important is conceptual understanding? Should it be a priority or are other competencies more important? Research can inform society about what is possible but not about what is most valued. Once learning goals have been selected and made explicit, based at least partly on value judgments, then researchers can examine the features of teaching that facilitate the achievement of such goals.

Account for the Costs and Benefits of Different Research Approaches

The fourth principle we propose to elevate the quality of research on classroom teaching and learning relationships is to carefully consider the advantages and disadvantages of different research designs and strategies when conducting empirical studies and interpreting data. We elaborate this principle by considering the issues related to three particular contexts: research designs that compare the effects of instructional methods; research designs that correlate features of teaching with student learning; and the trade-offs between small-scale qualitative studies and large-scale quantitative studies.

Comparing the Effects of Different Instructional Methods

Research designs that compare students' learning under two or more instructional methods often are quasi-experiments (Campbell & Stanley, 1963). Some of the methodological features are consistent with usual conventions of scientific experiments (controlling some variables while studying differences among others) but some are not (lack of random assignment of students to instructional method). One advantage of these comparative studies is that they often have some ecological validity. That is, they are conducted in real classrooms with all of the usual complexities. Instructional methods are implemented as systems, with all of the usual interactions. A second advantage is that these studies have the opportunity to control some of the most obvious alternative explanations for the results. For example, pretests can show similar entry knowledge and skills by students across treatment groups, content can be similar across groups except in ways dictated by the different teaching methods, and classroom observations can ensure faithful implementation of methods.

Some of the disadvantages of the quasi-experiments that compare teaching methods are inherent in the design and are difficult for researchers to control. The ecological validity that assures a realistic implementation of a teaching method also guarantees that the method will not be exactly replicable in other settings. As noted earlier, many classroom conditions, including the students, influence the way in which methods are implemented and, consequently, influence the learning outcomes. The challenge for researchers is to determine whether robust features of teaching within the system yield particular kinds of outcomes regardless of the naturally occurring changes in the system from classroom to classroom. This requires multiple replications, a strategy not frequently employed by educational researchers. But replication is one of the most powerful tools available to sort out which features of teaching make a difference across different implementations of a system of teaching, and even across different systems.

Another disadvantage of these comparative studies of teaching is easier for researchers to address. Reports of comparative studies often provide insufficient descriptions of the teaching features or methods studied. Without full descriptions of the teaching features

being compared, the results cannot be interpreted in ways that connect particular kinds of teaching with particular kinds of learning. And without these kinds of interpretations, it is impossible to *understand* the relationships that are documented. Superior learning reported during a thinly described "student-centered" approach compared with a thinly described "direct teaching" approach is of no value. There is no way to develop more specific hypotheses about what, exactly, in the student-centered approach facilitated students' learning. And it is these ill-understood connections that can lead to "lethal mutations" when implementing the teaching features in other conditions (Brown & Campione, 1996).

Thinly described teaching methods often are special deficiencies in reports of studies that include a control condition. The experimental method usually is described in some detail, but the control method can be described as cryptically as "traditional instruction was provided" or "the textbook was followed closely." Thinly described control methods undermine the educational significance of a claim that the experimental method yielded better results. This is because the power of such a claim depends as much on the quality of the control method as the quality of the experimental method. When benefits are claimed for one teaching method over another, regardless of the methods being compared, the claim is only as educationally significant as the quality of the weaker method. If the weaker method is extremely weak, or not sufficiently described to judge its quality, then results showing that the alternative method yielded better results are of little importance. To evaluate comparative claims between teaching methods, all the methods must be fully described.

We conclude this section with a suggestion about research questions. History shows that hypotheses in social science research are almost always confirmed (Greenwald, Pratkanis, Lieppe, & Baumgardner, 1986). This is especially true for predictions regarding the superiority of an experimental condition over a control condition. Several well-known research biases, such as the Hawthorne effect, probably explain the frequent confirmation of hypotheses. But, as Greenwald and colleagues (1986) argued, this tendency can and should be countered by changing the nature of the research question. Rather than asking "Is Method A better than Method B?" researchers should ask "Under what conditions is Method A better than Method B?" Translating the suggestion into the situation of interest here, we recommend asking more questions of the kind "Under what conditions will teaching features of Set A facilitate achievement of particular mathematics learning goals more effectively than teaching features of Set B?" Indeed, this is exactly the characteristic of Brownell and Moser's (1949) study that received applause from Cronbach (1986). Cronbach echoed Greenwald and colleagues' (1986) recommendation by noting the benefits of documenting the conditions under which particular instruction approaches are effective rather than claiming one approach is superior to another.

Correlating Features of Teaching with Students' Learning

A common research approach to connecting teaching with learning has been to measure a range of specific features of teaching and gains in students' achievement and then, through correlation techniques, to identify the features of teaching that correspond best with students' learning. Known as *process-product research* (Brophy & Good, 1986; Dunkin & Biddle, 1974), this approach often focuses on counting the frequency of teacher behaviors and correlating these with student gains on standardized achievement tests. The approach was primarily responsible for the patterns described earlier between features of teaching and students' skill efficiency.

Advantages of the correlation approach center around the ability to sort out features of teaching that might be key facilitators of learning and those that are part of the implementation but might not be critical. For example, teaching might vary along dimensions of asking students to present alternative solution methods and time spent working in groups developing shared solution methods. One of these features (or an interaction between the two) might be more important than the other for students' conceptual development. Well-designed correlation research could provide useful information. Schoen, Cebulla, Finn, and Fi (2003) presented an example of how correlations can be used to sort out key instructional variables in classrooms using reform curricula. Although the traditional process-product approach selected a specific kind of teacher behavior and used standardized tests to measure learning, the correlational approach is not limited to these independent and dependent variables. The approach is especially useful to create initial maps of the terrain when few hypotheses exist to suggest which features of teaching might facilitate which kinds of learning.

A potential danger of the correlation approach is the flip-side of its greatest advantage—the assumption that teaching consists of a collection of interchangeable features and that the effects of individual features can be measured independently of their interactions with other features in the system. Although many re-searchers have been explicit that "any teaching act is meaningful only in relationship to other instructional behaviors that occur in the classroom" (Good et al., 1983, p. 20), their warnings have often been overlooked in summaries and syntheses of research and when research findings are incorporated into professional development activities for teachers. Without considering the systems of which features are a part, recommendations from process-product relationships can be as simple and misguided as "Teachers should engage in process [X] more often." Thus, as we noted earlier, when discussing the challenges faced in developing adequate theories of teaching and learning relationships, teaching must be conceived as a system of interacting features and overly simplistic notions of teaching must be avoided.

In spite of the danger, we believe that correlation approaches have a role to play in linking teaching with learning. They can help to identify features of teaching that might have major effects on learning and deserve further study, and they can provide descriptive data to inform the development of hypotheses regarding teaching-learning relationships. Equally important, especially in the complex classroom environment where school learning takes place, correlational studies can assist with decisions on which variables will likely play major roles and which will play minor roles in particular learning situations and in this way help develop testable hypotheses and assist with the design of manageable experiments.

Balancing the Benefits of Small-Scale Qualitative Studies and Large-Scale Quantitative Studies

In the current research climate, there are increasing pressures to quantify regularities through large-scale experimental studies that use random assignment and seek to sort out cause and effect between classroom practice and student learning (Coalition for Evidence-Based Policy, 2003). Partly in response to these pressures, the National Research Council sponsored a report that proposed guidelines for what counts as "scientific" research in education (National Research Council, 2002). A number of researchers have argued that a balanced approach between qualitative and quantitative approaches has the best chance of moving the field forward (see, for example, Maxwell, 2004). We believe that small-scale qualitative approaches and large-scale quantitative approaches and mixes between the two all have useful roles to play in advancing our understanding of the relationships between teaching and learning. Gage's (1978) simple observation is a good starting point for outlining the contributions of different approaches: qualitative approaches reveal what is possible and quantitative approaches demonstrate what is probable. But we can be more specific.

Qualitative and small-scale studies can provide a number of critical ingredients for the research enterprise. Such studies are essential for developing and testing measuring instruments for teaching and learning, including the high-quality measures of student understanding of complex mathematical ideas called for in the National Research Council (2004) report on the effectiveness of the evaluations of mathematics reform curricula. Qualitative and small-scale studies also can support the thorough development of instructional treatments before large-scale studies test their effects with larger samples of students. Additionally, although case studies are not frequently undertaken to test hypotheses, they can, as Yin (2000) argued, rule out important alternate hypotheses if the studies are carefully designed.

Finally, and perhaps most importantly, qualitative smaller scale studies can provide data that deepen our understanding of how systems of teaching work to facilitate learning (Maxwell, 2004). Another brief story from the medical field is instructive. As reported during the December 7, 2004 National Public Radio program "Morning Edition" (Knox, 2004), a very small-scale study had recently tested the effect of a vaccine targeting a pre-leukemia condition called myelodysplasia. The vaccine not only induced immune responses in about 60% of the cases but unexpectedly created remissions in several cancer cases. "Now the challenge," according to Richard Knox, "is to figure out why the vaccine works." Until the researchers understand the reasons for the vaccine's effect, they cannot adapt it to trials under other conditions or on large scales. There are good reasons to apply a similar logic to research on teaching and learning. Qualitative and small-scale studies can detect the existence of teaching effects and then unpack the system of interacting teaching features to explore the reasons for the effects. Understanding why particular features of teaching facilitate particular kinds of learning is essential for creating more refined hypotheses, for adapting the teaching system to other conditions, and for testing its effects on a large scale.

Large-scale quantitative studies play an equally critical role in building a useful knowledge base for teaching. Because the links between teaching and learning are mediated by so many contextual variables, large-scale studies are needed to gauge the size of these effects and estimate the likelihood of finding similar links between teaching and learning across multiple classrooms. Researchers cannot control all of the classroom influences on students' learning and all of the interaction effects among the teaching features

of interest and the systems in which they operate, so replication over large numbers of classrooms and students remains one of the best strategies for detecting links between teaching and learning (Hiebert, Gallimore, & Stigler, 2002). We recommend taking seriously the call for large-scale cause-effect studies that aim to test hypotheses about relationships between teaching and learning. Along with Cohen et al. (2003) and others before them (e.g., Brownell & Moser, 1949), we recommend studying the effects of implementing systems of teaching under different conditions with clear learning goals in mind. To the extent that conditions not of interest can be controlled (e.g., instruction implemented similarly across classrooms, random assignment of students to treatments), the results can be interpreted with clarity and confidence. As we argued earlier, full control is unlikely but need not preclude efforts to design studies with appropriate controls.

Smaller scale qualitative studies and larger scale quantitative studies are essential for building a theoretically driven and practically useful knowledge base for teaching. Neither is sufficient; both are necessary. The confidence with which conclusions can be drawn about the effectiveness of features or systems of teaching increases with the convergence toward similar conclusions from studies using different methods. We agree with the National Research Council (2002), "If a research conjecture or hypothesis can withstand the scrutiny of multiple methods its credibility is enhanced greatly" (p. 64). The issue is not one of choosing between approaches but rather one of using both approaches in a mutually beneficial way.

FINAL THOUGHTS

We began the chapter with the goal of examining directly the issue of teaching effectiveness—why it has been so hard to document, what is known about it, and how the mathematics education community can learn more about the relationships between teaching and learning. We contended that these questions get to the heart of the educational research enterprise because (1) the questions lie at the foundation of education activity and are fundamental research issues and (2) the improvement of classroom practice depends on the answers. This places researchers in a difficult position. The issues they are dealing with are fundamental and complex, but researchers are under pressure from policy and government agencies to produce simple, easy-to-understand evidence-based answers and to produce them quickly. What constitutes an appropriate response to this challenging dilemma?

One response to the challenge, currently popular in the United States, is to address the simple question "What works?" The "What Works Clearinghouse" launched in 2004 (Viadero, 2004) is the latest in a series of efforts by the U.S. government (U.S. Department of Education, 1987), state governments (e.g., Dixon et al., 1998), and professional organizations (e.g., Cawelti, 2004) to provide practitioners with straightforward evidence-based advice for improving classroom practice. The What Works Clearinghouse plans to roll out a series of recommendations on a range of educational issues by conducting reviews of intervention programs, practices, and products. Similar to the review procedure used by Dixon et al. (1998) to identify best practices in mathematics classrooms, the What Works Clearinghouse will screen out most studies for review using a set of selection criteria. In particular, all studies that do not include a cause-effect analysis through the use of comparison groups will be excluded. The studies remaining constitute one segment of the large-scale quantitative studies described earlier. These studies are an important part of the picture. They provide essential information regarding teaching-learning relationships. But, usually, these large-scale comparative studies do not help consumers understand why the documented relationships exist or exactly how particular kinds of teaching facilitated particular kinds of learning. As we have argued, users must understand why particular features of teaching facilitate students' learning in order to deal appropriately with the modifications that inevitably are needed for their own classrooms. Some level of understanding is essential to prevent lethal mutations in teaching (Brown & Campione, 1996).

We applaud the concept of conducting and reviewing research to answer questions of what works. Educators need trustworthy information to inform classroom decisions. Our concerns about the limitations of the process engaged by the What Works Clearinghouse in no way lessen the importance or urgency of the goal toward which the Clearinghouse is working. In fact, we believe this goal must be aggressively pursued by researchers in the mathematics education community. Mathematics education researchers are in the best position to create and sustain a better alternative.

We conclude this chapter by proposing a richer, more useful, but longer term alternative to "what works." We propose the coherent and systematic construction of a knowledge base that documents robust links between teaching and learning *and* provides insights into the mechanisms that are responsible for such links. Researchers will need to participate in efforts to generate knowledge about teaching-learning

relationships that have direct relevance for classroom teachers and to regularly review available evidence to provide recommendations accompanied with estimates of confidence. We expect that these efforts will follow a range of different models for researcher-teacher collaboration (Hiebert et al., 2002; Ruthven, 2002) but will be characterized by the principles outlined in this chapter. Effective teaching will be studied in relation to clearly specified and measured learning outcomes so that conclusions will be phrased, not just as "effective teaching" but as "effective teaching under conditions [X] for learning [Y]." Hypotheses and theories will guide all aspects of the research process so that conjectures can be offered about why and how the targeted teaching features or systems facilitated (or did not facilitate) students' learning. The conditions under which features were effective will be detailed so that users can interpret the conclusions in light of their own local situations. Tests of hypotheses about teaching-learning relationships will be replicated in a variety of contexts to study the influence of contextual factors so that conclusions can be drawn with increasing specificity and confidence. Reviews will look for patterns of consistent findings from studies using different research methods to increase the confidence with which conclusions can be offered.

To build the richer knowledge base we envision, debates about research methods must shift from debates about which methods are appropriate to debates about how to increase the quality of the methods employed. The principle that multiple methods are better than one method must be accepted, and researchers will need to pursue the goal of gathering data with higher levels of trustworthiness, regardless of method. Discussions like those initiated by Hanna (1998), Kilpatrick (1993), Lester and Lambdin (1998), Nissen and Blomhøj (1993), and Simon (2004) on the quality of research in mathematics education should continue and be extended.

A knowledge base of the kind we envision cannot be built without resources. Conducting the large-scale comparisons of teaching that provide critical kinds of information are especially expensive. Assembling networks of researchers who can launch coordinated series of studies and build on one another's work requires long-term investments. No research enterprise that produces important advances is cheaply funded. U.S. mathematics education research funding comes primarily from the National Science Foundation and the U.S. Department of Education, and, as a Rand-conducted study of funding for mathematics education research from 1996 to 2001 pointed out, neither agency has clear strategic long-range planning for mathematics education R&D funding. In fact, funding for research and development in both agencies is a miniscule part of their total education budgets (Lacampagne, Newton, & Markham, 2001). Apparently high-quality evidence-based information on teaching-learning relationships in the mathematics classroom will cost much more than the United States currently is willing to spend.

The knowledge base we envision that links mathematics teaching and learning is beginning to take shape, even with limited resources. But it is difficult to find. It is hidden in the archives of libraries and scattered across time and among countless materials. Teachers (and researchers) must work hard to access it. Recent reviews have succeeded in pulling together some information (English, 2002; Kilpatrick et al., 2003; National Research Council, 2001; this volume), and we hope this chapter extends and contributes to these efforts. But most of this work is post hoc, requiring reviewers to detect patterns across work that was conducted independently and without connections. Much can be gained by researchers launching research agendas that are intentionally and explicitly connected. Through theory development, planned replications, and other research activities that connect research communities, a coherent, rich, accessible, and useful knowledge base can be built. Questions of "What works?" can then be answered with the richness and insight teachers deserve.

REFERENCES

Ausubel, D. P. (1963). *The psychology of meaningful verbal learning.* New York: Grune & Stratton.

Ball, D. L. (1993). Halves, pieces, and twoths: Constructing and using representational contexts in teaching fractions. In T. P. Carpenter, E. Fennema, & T. A. Romberg (Eds.), *Rational numbers: An integration of research* (pp. 157–195). Hillsdale, NJ: Erlbaum.

Ball, D. L., & Rowan, B. (2004). Introduction: Measuring instruction. *Elementary School Journal, 105,* 3–10.

Berliner, D. C. (1976). Impediments to the study of teacher effectiveness. *Journal of Teacher Education, 27,* 5–13.

Bjork, R. A. (1994). Memory and metamemory considerations in the training of human beings. In J. Metcalfe & A. Shimamura (Eds.), *Metacognition: Knowing about knowing* (pp. 185–205). Cambridge, MA: MIT Press.

Boaler, J. (1998). Open and closed mathematics: Student experiences and understandings. *Journal for Research in Mathematics Education, 29,* 41–62.

Borko, H., Stecher, B. M., Alonzo, A., Moncure, S., & McClam, S. (2003). *Artifact packages for measuring instructional practice: A pilot study.* Retrieved June 30, 2004, from http://www.cresst.org/reports/R615.pdf

Brophy, J. E. (1988). Research on teacher effects: Uses and abuses. *Elementary School Journal, 89,* 3–21.

Brophy, J. E. (1997). Effective instruction. In H. J. Walberg & G. D. Haertel (Eds.), *Psychology and educational practice* (pp. 212–232). Berkeley, CA: McCutchan.

Brophy, J. E. (1999). *Teaching* (Education Practices Series No. 1). Geneva, Switzerland: International Bureau of Education. Retrieved June 10, 2002, from http://www.ibe.unesco.org

Brophy, J. E., & Good, T. L. (1986). Teacher behavior and student achievement. In M. C. Wittrock (Ed.), *Handbook of research on teaching* (3rd ed., pp. 328–375). New York: Macmillan.

Brousseau, G. (1997). *Theory of didactical situations in mathematics.* Dordrecht, The Netherlands: Kluwer.

Brown, A. L., & Campione, J. C. (1996). Psychological theory and the design of innovative learning environments: On procedures, principles, and systems. In L. Schauble & R. Glaser (Eds.), *Innovations in learning: New environments for education* (pp. 289–325). Mahwah, NJ: Erlbaum.

Brown, S. I. (1993). Towards a pedagogy of confusion. In A. M. White (Ed.), *Essays in humanistic mathematics* (pp. 107–121). Washington, DC: Mathematical Association of America.

Brownell, W. A. (1935). Psychological considerations in the learning and teaching of arithmetic. In W. D. Reeve (Ed.), *The teaching of arithmetic: Tenth yearbook of the National Council of Teachers of Mathematics* (pp. 1–31). New York: Teachers College, Columbia University.

Brownell, W. A. (1947). The place of meaning in the teaching of arithmetic. *Elementary School Journal, 47,* 256–265.

Brownell, W. A. (1948). Criteria of learning in educational research. *Journal of Educational Psychology, 39,* 170–182.

Brownell, W. A., & Moser, H. E. (1949). Meaningful vs. mechanical learning: A study in grade III subtraction. *Duke University Research Studies in Education* (No. 8). Durham, NC: Duke University Press.

Brownell, W. A., & Sims, V. M. (1946). The nature of understanding. In N. B. Henry (Ed.), *Forty-fifth yearbook of the National Society for the Study of Education: Part I. The measurement of understanding* (pp. 27–43). Chicago: University of Chicago Press.

Campbell, D. T., & Stanley, J. C. (1963). Experimental and quasi-experimental designs for research on teaching. In N. L. Gage (Ed.), *Handbook of research on teaching* (pp. 171–246). Washington, DC: American Educational Research Association.

Capon, N., & Kuhn, D. (2004). What's so good about problem-based learning? *Cognition and Instruction, 22,* 61–79.

Carpenter, T. P., Fennema, E., Peterson, P. L, Chiang, C.-P., & Loef, M. (1989). Using knowledge of children's mathematics thinking in classroom teaching: An experimental study. *American Educational Research Journal, 26,* 499–531.

Carroll, J. (1963). A model for school learning. *Teachers College Record, 64,* 723–733.

Cawelti, G. (Ed.). (2004). *Handbook of research on improving student achievement* (3rd ed.). Arlington, VA: Educational Research Service.

Coalition for Evidence-Based Policy. (2003). *Identifying and implementing educational practices supported by rigorous evidence: A user friendly guide.* U.S. Department of Education, Institute of Education Sciences, National Center for Education Evaluation and Regional Assistance. Retrieved August 5, 2004, from http://www.ed.gov/rschstat/research/pubs/rigorousevid/index.html

Cobb, P. (1994). Where is the mind? Constructivist and sociocultural perspectives on mathematics development. *Educational Researcher, 23*(7), 13–20.

Cobb, P., Wood, T., Yackel, E., Nicholls, J., Wheatley, G., Trigatti, B., et al. (1991). Assessment of a problem-centered second-grade mathematics project. *Journal for Research in Mathematics Education, 22,* 3–29.

Cohen, D. K., McLaughlin, M. W., & Talbert, J. E. (Eds.). (1993). *Teaching for understanding: Challenges for policy and practice.* San Francisco: Jossey-Bass.

Cohen, D. K., Raudenbush, S. W., & Ball, D. L. (2003). Resources, instruction, and research. *Educational Evaluation and Policy Analysis, 25,* 119–142.

Cooney, T. J. (1985). A beginning teacher's view of problem solving. *Journal for Research in Mathematics Education, 16,* 324–336.

Cronbach, L. J. (1986). Social inquiry by and for earthlings. In D. W. Fiske & R. A. Shweder (Eds.), *Metatheory in social science: Pluralisms and subjectivities* (pp. 83–107). Chicago: University of Chicago Press.

Davis, R. B. (1984). *Learning mathematics: The cognitive science approach to mathematics education.* Norwood, NJ: Ablex.

Design-Based Research Collective. (2003). Design-based research: An emerging paradigm for educational inquiry. *Educational Researcher, 32*(1), 5–8.

Dewey, J. (1910). *How we think.* Boston: Heath.

Dewey, J. (1926). *Democracy and education.* New York: Macmillan.

Dewey, J. (1929). *The quest for certainty.* New York: Minton, Balch & Co.

Dixon, R. C., Carnine, D. W., Lee, D.-S., Wallin, J., The National Center to Improve the Tools of Educators, & Chard, D. (1998). *Review of high quality experimental mathematics research: Report to the California State Board of Education.* Eugene, OR: National Center to Improve the Tools of Educators.

Doyle, W. (1978). Paradigms for research on teacher effectiveness. In L. S. Shulman (Ed.), *Review of research in education* (Vol. 5, pp. 163–198). Itasca, IL: Peacock.

Duffy, G. G. (1981). Teacher effectiveness research: Implications for the reading profession. In M. L. Kamil (Ed.), *Directions in reading: Research and instruction. Thirtieth yearbook of the National Reading Conference* (pp. 113–136). Washington, DC: National Reading Conference.

Dunkin, J., & Biddle, B. (1974). *The study of teaching.* New York: Holt, Rinehart, & Winston.

English, L. (Ed.). (2002). *Handbook of international research in mathematics education.* Mahwah, NJ: Erlbaum.

Evertson, C. M., Anderson, C. W., Anderson, L. M., & Brophy, J. E. (1980). Relationships between classroom behaviors and student outcomes in junior high mathematics and English classes. *American Educational Research Journal, 17*(1), 43–60.

Fawcett, H. P. (1938). *The nature of proof: A description and evaluation of certain procedures used in a senior high school to develop an understanding of the nature of proof.* New York: Teachers College, Columbia University.

Fennema, E., & Romberg, T. A. (Eds.). (1999). *Mathematics classrooms that promote understanding.* Mahwah, NJ: Erlbaum.

Festinger, L. (1957). *A theory of cognitive dissonance.* Evanston, IL: Row, Peterson.

Fey, J. (1979). Mathematics teaching today: Perspectives from three national surveys. *Mathematics Teacher, 72,* 490–504.

Fisher, C. W., Berliner, D. C., Filby, N. N., Marliave, R., Cahn, L. S., & Dishaw, M. M. (1980). Teaching behaviors, academic learning time, and student achievement: An overview. In C. Denham & A. Lieberman (Eds.), *Time to learn* (pp. 7–32). Washington, DC: U.S. Department of Health, Education, and Welfare, National Institute of Education.

Fletcher, H. J. (1971). An efficiency reanalysis of the results. *Journal for Research in Mathematics Education, 2,* 143–156.

Floden, R. E. (2001). Research on effects of teaching: A continuing model for research on teaching. In V. Richardson (Ed.), *Handbook of research on teaching* (4th ed., pp. 3–16). New York: Macmillan.

Floden, R. E. (2002). The measurement of opportunity to learn. In A. C. Porter & A. Gamoran (Eds.), *Methodological advances in cross-national surveys of educational achievement* (pp. 231–266). Washington, DC: National Academy Press.

Forman, E. A. (2003). A sociocultural approach to mathematics reform: Speaking, inscribing, and doing mathematics within communities of practice. In J. Kilpatrick, W. G. Martin, & D. Schifter (Eds.), *A research companion to Principles and Standards for School Mathematics* (pp. 333–352). Reston, VA: National Council of Teachers of Mathematics.

Freudenthal, H. (1973). *Mathematics as an educational task.* Dordrecht, The Netherlands: Reidel.

Fuson, K. C., & Briars, D. J. (1990). Using a base-ten blocks learning/teaching approach for first- and second-grade place-value and multidigit addition and subtraction. *Journal for Research in Mathematics Education, 21,* 180–206.

Gage, N. L. (1978). *The scientific basis of the art of teaching.* New York: Teachers College Press.

Gagne, R. M. (1985). *The conditions of learning and theory of instruction* (4th ed.). New York: Holt, Rinehart and Winston.

Gamoran, A. (2001). Beyond curriculum wars: Content and understanding in mathematics. In T. Loveless (Ed.), *The great curriculum debate: How should we teach reading and math?* (pp. 134–162). Washington, DC: Brookings Institution Press.

Gamoran, A., Porter, A. C., Smithson, J., & White, P. A. (1997). Upgrading high school mathematics instruction: Improving learning opportunities for low-achieving, low income youth. *Educational Evaluation and Policy Analysis, 19,* 325–328.

Good, T. L., & Grouws, D. A. (1977). A process-product study in fourth-grade mathematics classrooms. *Journal of Teacher Education, 28*(3), 49–54.

Good, T. L., & Grouws, D. A. (1979). The Missouri mathematics effectiveness project: An experimental study in fourth-grade classrooms. *Journal of Educational Psychology, 71* (3), 355–362.

Good, T. L., Grouws, D. A., & Ebmeier, H. (1983). *Active mathematics teaching.* New York: Longman.

Good, T. L., Mulryan, C., & McCaslin, M. (1992). Grouping for instruction in mathematics: A call for programmatic research on small-group processes. In D. A. Grouws (Ed.), *Handbook of research on mathematics teaching and learning* (pp. 165–196). New York: Macmillan.

Gravemeijer, K. (1994). *Developing realistic mathematics education.* Culemborg, The Netherlands: Technipress.

Greenwald, A., Pratkanis, A., Lieppe, M., & Baumgardner, M. (1986). Under what conditions does theory obstruct research progress? *Psychological Review, 93,* 216–229.

Grouws, D. A. (2004). Mathematics. In G. Cawelti (Ed.), *Handbook of research on improving student achievement* (3rd ed., pp. 160–178). Arlington, VA: Educational Research Service.

Hamilton, L. S., McCaffrey, D. F., Stecher, B. M., Klein, S. P., Robyn, A., & Bugliari, D. (2003). Studying large-scale reforms of instructional practice: An example from mathematics and science. *Educational Evaluation and Policy Analysis, 25,* 1–29.

Handa, Y. (2003). A phenomenological exploration of mathematical engagement: Approaching an old metaphor anew. *For the Learning of Mathematics, 23,* 22–28.

Hanna, G. (1998). Evaluating research papers in mathematics education. In A. Sierpinska & J. Kilpatrick (Eds.), *Mathematics as a research domain: A search for identity* (ICMI Studies Series, Vol. 4, pp. 399–407). Dordrecht, The Netherlands: Kluwer.

Hatano, G. (1988). Social and motivational bases for mathematical understanding. In G. B. Saxe & M. Gearhart (Eds.), *Children's mathematics* (pp. 55–70). San Francisco: Jossey-Bass.

Heath, R. W., & Nielson, M. A. (1974). The research basis for performance-based teacher education. *Review of Educational Research, 44,* 463–484.

Heaton, R. M. (2000). *Teaching mathematics to the new standards: Relearning the dance.* New York: Teachers College Press.

Heid, M. K. (1988). Resequencing skills and concepts in applied calculus using the computer as a tool. *Journal for Research in Mathematics Education, 19,* 3–25.

Hiebert, J. (2003). What research says about the NCTM Standards. In J. Kilpatrick, W. G. Martin, & D. Schifter (Eds.), *A research companion to Principles and Standards for School Mathematics* (pp. 5–23). Reston, VA: National Council of Teachers of Mathematics.

Hiebert, J., & Carpenter, T. P. (1992). Learning and teaching with understanding. In D. A. Grouws (Ed.), *Handbook of research on mathematics teaching and learning* (pp. 65–97). New York: Macmillan.

Hiebert, J., Carpenter, T. P., Fennema, E., Fuson, K., Human, P., Murray, H., et al. (1996). Problem solving as a basis for reform in curriculum and instruction: The case of mathematics. *Educational Researcher, 25*(4), 12–21.

Hiebert, J., Gallimore, R., Garnier, H., Givvin, K. B., Hollingsworth, H., Jacobs, J., et al. (2003). *Teaching mathematics in seven countries: Results from the TIMSS 1999 Video Study* (NCES 2003-013). Washington, DC: U.S. Department of Education, National Center for Education Statistics.

Hiebert, J., Gallimore, R., & Stigler, J. W. (2002). A knowledge base for the teaching profession: What would it look like and how can we get one? *Educational Researcher, 31*(5), 3–15.

Hiebert, J., & Stigler, J. W. (2000). A proposal for improving classroom teaching: Lessons from the TIMSS video study. *Elementary School Journal, 101,* 3–20.

Hiebert, J., & Wearne, D. (1993). Instructional tasks, classroom discourse, and students' learning in second-grade arithmetic. *American Educational Research Journal, 30,* 393–425.

Hoetker, J., & Ahlbrand, W. P., Jr. (1969). The persistence of the recitation. *American Educational Research Journal, 6,* 145–167.

Husen, T. (1967). *International study of achievement in mathematics: A comparison of twelve countries (Vols. 1–2).* New York: John Wiley.

Inagaki, K., Hatano, G., & Morita, E. (1998). Construction of mathematical knowledge through whole-class discussion. *Learning and Instruction, 8,* 503–526.

Karmiloff-Smith, A., & Inhelder, B. (1974). If you want to get ahead, get a theory. *Cognition, 3,* 192–212.

Kawanaka, T., & Stigler, J. W. (1999). Teachers' use of questions in eighth-grade mathematics classrooms in Germany, Japan, and the United States. *Mathematical Thinking and Learning, 1*(4), 255–278.

Kilpatrick, J. (1993). Beyond face value: Assessing research in mathematics education. In G. Nissen & M. Blomhøj (Eds.), *Criteria for scientific quality and relevance in the didactics of mathematics* (pp. 15–34). Roskilde, Denmark: Danish Research Council for the Humanities.

Kilpatrick, J., Martin, W. G., & Schifter, D. (Eds.). (2003). *A research companion to Principles and Standards for School Mathematics.* Reston, VA: National Council of Teachers of Mathematics.

Knox, R. (Reporter). (2004, December 7). *New report suggests that researchers have had some success in making a cancer vaccine* [Radio broadcast]. Washington, DC: National Public Radio.

Lacampagne, C.B., Newton, E., & Markham, K. (2001). Federal funding of mathematics education research. Project Memorandum (PM-1238) Prepared for Office of Educational Research and Improvement (OERI). Arlington, VA: RAND.

Lampert, M. (2001). *Teaching problems and the problems of teaching.* New Haven, CT: Yale University Press.

Lampert, M., & Cobb, P. (2003). Communication and language. In J. Kilpatrick, W. G. Martin, & D. Schifter (Eds.), *A research companion to Principles and Standards for School Mathematics* (pp. 237–249). Reston, VA: National Council of Teachers of Mathematics.

Lave, J. (1988). *Cognition in practice.* Cambridge, U.K.: Cambridge University Press.

Leinhardt, G. (1986). Expertise in math teaching. *Educational Leadership, 43(7),* 28–33.

Leinhardt, G., & Greeno, J. G. (1986). The cognitive skill of teaching. *Journal of Educational Psychology, 78*(2), 75–95.

Lester, F. K., Jr., & Lambdin, D. (1998). The ship of Theseus and other metaphors for thinking about what we value in mathematics education research. In A. Sierpinska & J. Kilpatrick (Eds.), *Mathematics as a research domain: A search for identity* (ICMI Studies Series, Vol. 4, pp. 415–425). Dordrecht, The Netherlands: Kluwer.

Malara, N. A., & Zan, R. (2002). The problematic relationship between theory and practice. In L. English (Ed.), *Handbook of international research in mathematics education* (pp. 553–580). Mahwah, NJ: Erlbaum.

Maxwell, J. A. (2004). Causal explanation, qualitative research, and scientific inquiry in education. *Educational Researcher, 33*(2), 3–11.

Mayer, D. P. (1999). Measuring instructional practice: Can policymakers trust survey data? *Educational Evaluation and Policy Analysis, 21,* 29–45.

McDonald, F. (1976). Report on phase II of the beginning teacher evaluation study. *Journal of Teacher Education, 27,* 39–42.

McDonald, F., & Elias, P. (1976). The effects of teaching performance on pupil learning. Beginning Teacher Evaluation Study, Phase II, 1974–1976 (Final report, Vols. 1–5). Princeton, NJ: Educational Testing Service.

National Advisory Committee on Mathematics Education. (1975). *Overview and analysis of school mathematics, grades K–12.* Washington, DC: Conference Board of the Mathematical Sciences.

National Commission on Mathematics and Science Teaching for the 21st Century. (2000). *Before it's too late: A report to the nation from the National Commission on Mathematics and Science Teaching for the 21st Century.* Washington, DC: Department of Education.

National Research Council. (1999). *How people learn: Brain, mind, experience, and school.* J. D. Bransford, A. L. Brown, & R. R. Cocking (Eds.). Committee on Developments in the Science of Learning, Commission on Behavioral and Social Sciences and Education. Washington, DC: National Academy Press.

National Research Council. (2001). *Adding it up: Helping children learn mathematics.* J. Kilpatrick, J. Swafford, & B. Findell (Eds.). Mathematics Learning Study Committee, Center for Education, Division of Behavioral and Social Sciences and Education. Washington, DC: National Academy Press.

National Research Council. (2002). *Scientific research in education.* R. J. Shavelson & L. Towne (Eds.). Committee on Scientific Principles for Educational Research, Center for Education, Division of Behavioral and Social Sciences and Education. Washington, DC: National Academy Press.

National Research Council. (2004). *On evaluating curricular effects: Judging the quality of K–12 mathematics evaluations.* J. Confrey & V. Stohl (Eds.). Committee for a Review of the Evaluation Data on the Effectiveness of NSF-Supported and Commercially Generated Mathematics Curricular Materials. Mathematical Sciences Education Board, Center for Education, Division of Behavioral and Social Sciences and Education. Washington, DC: National Academy Press.

Nissen, G., & Blomhøj, M. (Eds.). (1993). *Criteria for scientific quality and relevance in the didactics of mathematics.* Roskilde, Denmark: Danish Research Council for the Humanities.

Nye, B., Konstantopoulos, S., & Hedges, L. V. (2004). How large are teacher effects. *Educational Evaluation and Policy Analysis, 26,* 237-257.

Olson, L. (1997, January 22). Keeping tabs on quality. *Education Week Supplement,* pp. 7–11, 14–17.

Oser, F. K., & Baeriswyl, F. J. (2001). Choreographies of teaching: Bridging instruction to learning. In V. Richardson (Ed.), *Handbook of research on teaching* (4th ed., pp. 1031–1065). New York: Macmillan.

Paige, R. (2004). *Meeting the highly qualified teachers challenge: The Secretary's annual report on teacher quality, 2004.* Retrieved October 5, 2004, from http://www.title2.org/secReport04.htm

Piaget, J. (1960). *The psychology of intelligence.* Totowa, NJ: Littlefield, Adams.

Polya, G. (1957). *How to solve it* (2nd ed.). Garden City, NY: Doubleday Anchor Books.

Porter, A., Floden, R., Freeman, D., Schmidt, W., & Schwille, J. (1988). Content determinants in elementary school mathematics. In D. A. Grouws, T. J. Cooney, & D. Jones (Eds.), *Effective mathematics teaching* (pp. 96–113). Reston, VA: National Council of Teachers of Mathematics.

Powell, M. (1980). The beginning teacher evaluation study: A brief history of a major research project. In C. Denham & A. Lieberman (Eds.), *Time to learn* (pp. 1–5). Washington, DC: National Institute of Education.

Resnick, L. B., & Ford, W. W. (1981). *The psychology of mathematics for instruction.* Hillsdale, NJ: Erlbaum.

Reusser, K. (2001, September). *Bridging instruction to learning: Where we come from and where we need to go.* Paper presented at the Ninth Conference of the European Association for Research on Learning and Instruction, Fribourg, Switzerland.

Romberg, T. A., & Carpenter, T. P. (1986). Research on teaching and learning mathematics: Two disciplines of scientific inquiry. In M. C. Wittrock (Ed.), *Handbook of research on teaching* (3rd ed., pp. 850–873). New York: Macmillan.

Rosenshine, B., & Stevens, R. (1986). Teaching functions. In M. C. Wittrock (Ed.), *Handbook of research on teaching* (3rd ed., pp. 376–391). New York: Macmillan.

Ross, J. A., McDougall, D., Hogaboam-Gray, A., & LeSage, A. (2003). A survey measuring elementary teachers' implementation of standards-based mathematics teaching. *Journal for Research in Mathematics Education, 34,* 344–363.

Rowan, B., Correnti, R., & Miller, R. J. (2002). *What large-scale, survey research tells us about teacher effects on student achievement: Insights from the Prospects study of elementary schools* (CPRE Research Report Series PR-051). Philadelphia: University of Pennsylvania, Consortium for Policy Research in Education.

Rowan, B., Harrison, D. & Hayes, A. (2004). Using instructional logs to study mathematics curriculum and teaching in the early grades. *Elementary School Journal, 105,* 103–127.

Ruthven, K. (2002). Linking researching with teaching: Towards a synergy of scholarly and craft knowledge. In L. English (Ed.), *Handbook of international research in mathematics education* (pp. 581–598). Mahwah, NJ: Erlbaum.

Sanders, W. L., & Rivers, J. C. (1996). *Cumulative and residual effects of teachers on future student academic achievement.* Knoxville: The University of Tennessee Value-Added Research and Assessment Center.

Sanders, W. L., Saxton, A. M., & Horn, S. P. (1997). The Tennessee value-added assessment system: A quantitative outcomes-based approach to educational assessment. In J. Millman (Ed.), *Grading teachers, grading schools: Is student achievement a valid evaluation measure?* Thousands Oaks, CA: Corwin Press.

Schoen, H. L., Cebulla, K. J., Finn, K. F., & Fi, C. (2003). Teacher variables that relate to student achievement when using a standards-based curriculum. *Journal for Research in Mathematics Education, 34,* 228-259.

Schoenfeld, A. (1985). *Mathematical problem solving.* Orlando, FL: Academic Press.

Schoenfeld, A. (1988). When good teaching leads to bad results: The disasters of "well taught" mathematics courses. *Educational Psychologist, 23,* 145–166.

Schoenfeld, A. (1998). Toward a theory of teaching-in-context. *Issues in Education: Contributions from Educational Psychology, 4,* 1–94.

Sfard, A. (1998). On two metaphors for learning and the dangers of choosing just one. *Educational Researcher, 27*(2), 4–13.

Shuell, T. J. (1996). Teaching and learning in a classroom context. In D. C. Berliner & R. C. Calfee (Eds.), *Handbook of educational psychology* (pp. 726–764). New York: Macmillan.

Silver, E. A., & Stein, M. K. (1996). The QUASAR Project: The "revolution of the possible" in mathematics instructional reform in urban middle schools. *Urban Education, 30,* 476–522.

Simon, M. A. (2004). Raising issues of quality in mathematics education research. *Journal for Research in Mathematics Education, 35,* 157–163.

Simon, M. A., Tzur, R., Heinz, K., & Kinzel, M. (2004). Explicating a mechanism for conceptual learning: Elaborating the construct of reflective abstraction. *Journal for Research in Mathematics Education, 35,* 305–329.

Skemp, R. R. (1971). *The psychology of learning mathematics.* Middlesex, England: Penguin.

Skemp, R. R. (1976). Relational understanding and instrumental understanding. *Arithmetic Teacher, 26*(3), 9–15.

Smith, M. (2000). *A comparison of the types of mathematics tasks and how they were completed during eighth-grade mathematics instruction in Germany, Japan, and the United States.* Unpublished doctoral dissertation, University of Delaware, Newark.

Stake, R., & Easley, J. (Eds.). (1978). *Case studies in science education.* Urbana: University of Illinois.

Stecher, B., & Borko, H. (2002). Integrating findings from surveys and case studies: Examples from a study of standards-based educational reform. *Journal of Education Policy, 17,* 547–569.

Stein, M. K., Grover, B. W., & Henningsen, M. (1996). Building student capacity for mathematical thinking and reasoning: An analysis of mathematical tasks used in reform classrooms. *American Educational Research Journal, 33,* 455–488.

Stein, M. K., & Lane, S. (1996). Instructional tasks and the development of student capacity to think and reason: An analysis of the relationship between teaching and learning in a reform mathematics project. *Educational Research and Evaluation, 2*(1), 50–80.

Stigler, J. W., Gallimore, R., & Hiebert, J. (2000). Using video surveys to compare classrooms and teaching across cultures: Examples and lessons from the TIMSS video studies. *Educational Psychologist, 35,* 87–100.

Stigler, J. W., Gonzales, P., Kawanaka, T., Knoll, S., & Serrano, A. (1999). *The TIMSS videotape classroom study: Methods and findings from an exploratory research project on eighth-grade mathematics instruction in Germany, Japan, and the United States.* (NCES 1999-074). Washington, DC: U.S. Department of Education, National Center for Education Statistics.

Stigler, J. W., & Hiebert, J. (1999). *The teaching gap: Best ideas from the world's teachers for improving education in the classroom.* New York: Free Press.

Thorndike, E. L. (1906). *The principles of teaching based on psychology.* New York: A. G. Seiler.

Thorndike, E. L. (1912). *Education: A first book.* New York: Macmillan.

Traub, J. (2002, November 10). Does it work? *New York Times.* Retrieved November 18, 2002, from http://www.nytimes.com/2002/11/10/edlife/10CHILD.html

U.S. Department of Education. (1987). *What works: Research about teaching and learning* (2nd ed.). Washington, DC: U.S. Government Printing Office.

Viadero, D. (2004, July 14). "What Works" research site unveiled. *Education Week, 23*(42), 1, 33.

Vygotsky, L. S. (1978). *Mind in society: The development of higher psychological processes.* M. Cole, V. John-Steiner, S. Scribner, & E. Souberman (Eds.). Cambridge, MA: Harvard University Press.

Webb, N. M. (1991). Task-related verbal interaction and mathematics learning in small groups. *Journal for Research in Mathematics Education, 22,* 366–389.

Webb, N. M., Troper, J. D., & Fall, R. (1995). Constructive activity and learning in collaborative small groups. *Journal of Educational Psychology, 87,* 406–423.

Weinert, F. E., Schrader, F.-W., & Helmke, A. (1989). Quality of instruction and achievement outcomes. *International Journal of Educational Research, 13,* 895–914.

Wenger, E. (1998). *Communities of practice: Learning, meaning, and identity.* Cambridge, U.K.: Cambridge University Press.

Weiss, I. R., Pasley, J. D., Smith, P. S., Banilower, E. R., & Heck, D. J. (2003). *Looking inside the classroom: A study of K–12 mathematics and science education in the United States.* Retrieved July 19, 2004, from http://horizon-research.com

Welch, W. (1978). Science education in Urbanville: A case study. In R. Stake & J. Easley (Eds.), *Case studies in science education* (pp. 5-1–5-33). Urbana: University of Illinois.

Winne, P. H., & Marx, R. W. (1980). Matching students' cognitive responses to teaching skills. *Journal of Educational Psychology, 72,* 257–264.

Wittrock, M. C. (1986). Students' thought processes. In M. C. Wittrock (Ed.), *Handbook of research on teaching* (3rd ed., pp. 297–314). New York: Macmillan.

Wubbels, T., Brekelmans, M., & Hooymayers, H. (1992). Do teacher ideals distort the self-reports of their instructional behavior? *Teaching and Teacher Education, 8,* 47–58.

Yin, R. K. (2000). Rival explanations as an alternative to reforms as "experiments." In L. Bickman (Ed.), *Validity & social experimentation: Donald Campbell's legacy* (pp. 239–266). Thousand Oaks, CA: Sage.

AUTHOR NOTE

We thank a number of individuals for their helpful comments on earlier drafts of the chapter. Hilda Borko and Thomas Cooney provided insightful and constructive comments at two critical points during the development of the chapter. Faculty and doctoral students in mathematics education at the University of Delaware offered two rounds of comments during lively seminar discussions. Colleagues at Michigan State University shared generously of their time and advice during a colloquium presentation of these ideas. Finally, we thank Frank Lester, Jr., for providing helpful feedback and coordinating the review of the chapter.

10

CULTURE, RACE, POWER, AND MATHEMATICS EDUCATION

Diversity in Mathematics Education Center for Learning and Teaching

UNIVERSITY OF CALIFORNIA, LOS ANGELES
UNIVERSITY OF WISCONSIN, MADISON
UNIVERSITY OF CALIFORNIA, BERKELEY

For their tireless efforts toward equity and justice, we dedicate this chapter to
Rosa Parks
&
Coretta Scott King

INTRODUCTION

Considerable attention in mathematics education research has been paid to understanding and confronting differential mathematics achievement. Such research is complex, as it is situated within and framed by broader educational, social, and political contexts. Embedded in these contexts are issues of race, class, gender, language, culture, and power. Mathematics education researchers not only have to grapple with issues of mathematics teaching and learning, but also come to understand the ways in which these broader educational and cultural contexts shape opportunities for all students to learn mathematics (Cobb & Nasir, 2002; Martin, 2000; Moschkovich, 2002a). Moreover, such research brings into question the role of mathematics education by situating it within these broader contexts (Apple, 1992; Martin, 2003).

Race is one of the most, if not the most, salient framing characteristics for differential achievement in the discourse that surrounds the "achievement gap" (Schoenfeld, 2002; Tate, 1997). Additionally we, the authors of this chapter, were aware of the significance of race in our everyday work with teachers and students and noted that its impact in mathematics education went beyond issues of differential achievement. Thus, in this chapter we specifically examine research aimed at understanding the various intersections between issues of race and mathematics teaching and learning. We follow the field in rejecting perspectives of race as biologically determined and adopted a perspective of race as constructed, contested, and reified through social activity. Additionally, our synthesis of the research around race and mathematics education provides a context, as Perry (Perry, Steele, & Hilliard, 2003) suggested, to "grapple with the notion that not all racial minorities occupy the same political position in this society" (p. 9). This acknowledgment brings relations of power into our analysis. More specifically, as discussed later in this chapter, the racialization of mathematics education is intertwined with issues of power and authority, making it difficult, if not impossible, to examine issues of race without also considering the impact of power dynamics on mathematics teaching and learning.

We also note that the field often treats culture as race and vice versa (see special issue of *Educational Researcher*, 2004). Some work treats cultural characteristics of people from a racial group as homogenous and immutable; in this model, a static view of race determines culture. Other work opts to deal with local cultural practices as a way of avoiding seemingly static racial categories, often missing broader patterns of behavior that socially construct racial experience. Both of these trends speak to the difficult intersection between race and culture. Thus, we expanded our work to include research examining issues of culture and mathematics teaching and learning. We focus on research in mathematics education that has conceptualized culture not as a static individual characteristic, but as constantly negotiated by individuals in their everyday activities. Conceptualizing culture in this way means that cultural knowledge is an integrated system of learned patterns of behavior, ideas, and products with a focus on people's everyday lived practices (K. Gutiérrez, 2002; Moll & Gonzalez, 2004; Rogoff, 2003). This chapter attempts to highlight and suggest areas for future research that explores the complexity in students' cultural activity, racial experience, and mathematics education.

Individual Mathematical Performance

Before turning to a discussion of future research directions, we begin with a brief examination of the concept of individual mathematical performance as it relates to differential achievement and solutions to inequitable educational situations. Researchers and the media often report achievement gaps in mathematics across broad groups of students. Explanations for these gaps in mathematics achievement both structure larger agendas for addressing equity in schools and affect individual teacher-student interactions. Many of the explanations, however, both in the media and in research, explicitly and implicitly frame the problem as one of individual student learning, asserting that some students "just don't try hard enough" or "fail because they are disadvantaged." Although these explanations often stem from research that examines the success or failure of individual students, they are also mistakenly generalized to the differential patterns of achievement for *groups* of students.

Research that focuses on the mathematical characteristics and performance of individual students frames equity as an issue of equal access to high-quality mathematics education that allows *all* students to

succeed. This notion of equity as equality suggests that equity can be realized when students have equal access to resources, high-quality teachers, and appropriate instructional support regardless of race, class, gender, and so on. However, researchers concerned with issues of social justice and longstanding social inequities, or what Ladson-Billings (2006) referred to in her American Educational Research Association Presidential Address as the "Education Debt," argue that "equality" in educational opportunities does not necessarily address the histories of groups and how those histories have shaped the social structures and beliefs at the base of opportunities in and outside of school.

Rochelle Gutierrez (2002) described the goal of equity in mathematics education as "being unable to predict student patterns (e.g., achievement, participation, and the ability to critically analyze data or society) based solely on characteristics such as race, class, ethnicity, sex, beliefs and creeds, and proficiency in the dominant language" (p. 153). She proposed a twofold approach to achieving this condition: first, providing students with high-quality mathematics education, or what she called *dominant* mathematics, and second, supporting students in using mathematics to perceive and confront inequitable situations in their lives, or *critical* mathematics. In this chapter, we take up Gutierrez's call to the field, to explore the processes involved in differential mathematics achievement and to consider mathematics education as a means of and for a socially just society.

In support of this task, this chapter reviews research within and beyond the field of mathematics education that conjoins accounts of individual-level processes with critical social and cultural processes involved in learning and teaching (Allexsaht-Snider & Hart, 2001; Boaler, 2002; Greeno, 2004; Martin, 2000; Moschkovich, 2002a). Further, instead of making comparisons of individuals on the basis of what has been considered "normative" or "adaptive" (Baratz & Baratz, 1970; Leacock, 1971; Nieto, 2004; Secada, 1992), the research reviewed is concerned with elaborating and deepening conceptions of learning to account for differences and regularities both within and across social and racial groups as well as social and cultural contexts. As Secada (1992) and others noted, over the last 20 years, group comparisons have typically been developed within research with a majority of White children, primarily by White researchers, whose perspectives were shaped by their sociocultural experiences in the dominant culture. A charge to the field of mathematics education, then, is to conduct research within non-dominant populations of students who experience marginalization, and to attend to the positioning of these groups vis-à-vis their White counterparts and to broader sociopolitical structures and forces. *White* as a racial category also needs explication as it entails certain privileges and status within current educational systems (Lee, 2004). For these reasons, we focus this chapter on research that conceptualizes larger constructs such as culture, power, and race as ways to understand issues in mathematics education.

Organization of the Chapter

To help make explicit current understandings, research, and theory around culture, race, power, and mathematics education, we first review the literature that uses culture as a way of conceptualizing achievement differences within and between groups. This review focuses on literature that portrays culture as fluid, socially negotiated, and "lived in the everyday." This review attempts to understand the influence of what we summarize as *theories of cultural activity* on perspectives of learning, development, and culture and the relations between them with respect to differences in opportunities to learn mathematics for non-dominant students. Then we turn to mathematics education research that suggests it is not enough to examine culture without attending to broader issues of power and race. This section reviews research that takes into consideration the constructs of power and race at the classroom level and in larger contexts. This literature stresses the need to consider these constructs in order to fully understand and reconceptualize differential achievement in mathematics education. Finally, we discuss issues around and implications of researching culture, race, and power in mathematics education.

CULTURAL ACCOUNTS OF MATHEMATICS TEACHING AND LEARNING

Culture has long been a focus of research on individual differences in cognition, learning, and development (Lave, Murtaugh, & de la Rocha, 1984; Nunes, Schliemann, & Carraher, 1993; Rogoff, 2003; Saxe, 1988a, 1988b). However, over the past 10 years studies in this area have gained prominence and offered alternative perspectives on the role of culture in learning, in terms of both what counts as learning and who has access to it. One reason for this heightened interest is that research drawn from various theories of cultural activity has afforded an understanding of knowing and learning as a function of what an individual accomplishes over time and across the various communities and practices in which he or she participates.[1] These theories encompass such theoretical approaches as situated cognition, activity theory, cultural historical activity theory, and sociocultural theory. Findings across these lines of research point to the fact that mathematics classrooms are necessarily cultural and social spaces that can perpetuate social inequities by privileging certain forms of discourse and ways of reasoning or reorganize them by positioning multiple forms of learning and knowing as "having clout" (Cobb & Hodge, 2002; Gutierrez, Baquedano-Lopez, & Tajeda, 1999; Gutstein, in press-a).

Understanding the cultural entailments of mathematics learning requires complicated analyses of how people live and learn culturally both within and outside of the mathematics classroom. This chapter reports on (1) the links that researchers have uncovered between the diverse ways that students from different backgrounds negotiate learning within and outside of mathematics classrooms, (2) the structures for mathematics learning within mathematics classrooms, and (3) the relations students develop with the domain of mathematics. All told, this work suggests that differences in mathematical achievement among groups do not rest solely upon students' cultural/mathematical backgrounds, but also in the sociopolitical organization of mathematics classrooms (and mathematics education in general). Moreover, although culture is commonly characterized as synonymous with group membership, this perspective is broadened by an emphasis in this work on how individuals create, contest, and reconfigure roles and relationships within the communities in which they participate.

Research that takes a cultural activity perspective aimed wholly or in part at persistent inequities in mathematics education has found traction in focusing on the mathematics learning of nondominant students both inside and outside of the mathematics classroom. We examine contributions of this research to the field of mathematics education to improving understanding of the processes through which students negotiate mathematics learning and doing across a variety of contexts. One major contribution of this research is the extensive accounts of the participation of students and teachers in various *communities* of mathematics classrooms, the curricular and participation structures within these communities that guide the class's joint work together, and the relation of these structures to the identities students develop as

---

[1] *Participation* in this case refers broadly to the way that individuals interact within a community of practice—whether they play a central role in perpetuating what goes on inside of it or actively contest or challenge from the margins.

learners and doers of mathematics (see, e.g., Cobb & Hodge, 2002; Martin, 2000; Nasir, 2002).

The recognition of the mathematics classroom as a functioning community where teacher and student activity in it is shaped by (and shapes) a set of norms and practices for learning mathematics highlights the importance of issues such as competence, ownership, and alignment in engaging in this community. In particular, alignment between the practices and identities of home and school has implications for whether students negotiate ways of participating that serve their individual goals, as well as the goals of the classroom community (Cobb & Hodge, 2002; Hand, 2003).

A second and growing contribution of this research is documentation of the wide variety of mathematical practices and identities that students bring to the classroom from their home and local communities, which has expanded conceptions of competent classroom participation (see, e.g., Carraher, Carraher, & Schliemann, 1985; Lipka, 1994; Moll, Amanti, Neff, & Gonzalez, 1992; Nasir, 2002; Taylor, 2004). Students' practices and ways of reasoning are often marginalized in mathematics classrooms where teachers rely on traditional scripts of, and formats for, classroom instruction. Broadening mathematical activity to recognize and value the multiple ways that students participate in mathematics can draw in students who may normally be sidelined. Finally, this research offers promising models of classroom learning environments that begin to address issues of race and power in the mathematics classroom by focusing squarely on issues of cultural relevancy and social justice (see, e.g., Gutstein, 1997, 2003, in press-b; Ladson-Billings, 1994; Moses & Cobb, 2001a).

We begin the next section by explicating the key constructs of participation and identity that undergird the research from a cultural activity perspective. Next, we examine how these constructs have been employed to recenter analyses of student achievement to consider differences in students' *opportunities to learn* mathematics both in classrooms and more broadly. Examining opportunities to learn in mathematics classrooms for diverse groups of students has prompted new ways to think about differential access to engagement in mathematics in relation to the social and cultural resources that students bring to the learning environment. As a framework, opportunity to learn orients the review of the literature by highlighting particular themes that run across research with a perspective of culture as activity. These themes include (a) creating equitable opportunities to learn in the classroom, (b) broadening classroom discourse practices, (c) expanding conceptions of mathematical competence, (d) bridging in-school and out-of-school

mathematics knowledge, and (e) reforming the culture of mathematics.

Participation

Participation has become an important construct in situated and sociocultural analyses of students' school and classroom experiences (Boaler, 1999; Boaler & Greeno, 2000; Cobb & Hodge, 2002; Cobb, Stephan, McClain, & Gravemeijer, 2001; Engle & Conant, 2002; Gonzalez, Andrade, Civil, & Moll, 2001; Lave & Wenger, 1991; Martin, 2000, in press-b; Moschkovich, 2002a). As Sfard (1998) has observed, analytical frameworks that embrace the *participation metaphor* regard learning as a constant process of doing and becoming within a context populated by practices and activities in which individuals engage. An analytic focus on participation in studies of learning can highlight in what practices and activities students engage and how, as well as recenter analyses of student achievement and performance on the actual practices and activities in which students, teachers, and mathematics interact. Several researchers concerned with issues of equity in education broadly have achieved this recentering in different ways.

Cobb and Hodge (2002) focused on practice and participation in their analytical perspective on diversity and equity in mathematics education. They proposed a *relational perspective* that highlights "the relations between the specifically mathematical practices in which students participate in the classroom and the practices of the out-of-school communities of which students are members" (p. 251). They emphasized the continuities and discontinuities between students' ways of reasoning, talking, and interacting rooted in their out-of-school communities and those of their mathematics classroom. They argued that these relations, as they play out in classroom interactions, are the locus of the successes and inequities that arise in mathematics classrooms. In this formulation, what students are motivated to do in mathematics, how they engage in doing it, and, hence, what they are learning are understood through investigating patterns of students' participation in and across different contexts, in relation to the normative practices of the mathematics classroom. This analytical frame locates student achievement in the relationships between students' ways of participating in mathematics—which may be shaped by students' histories of engagement in their out-of-school communities—and the norms and valued practices of their mathematics classrooms.

Gutierrez and Rogoff (2003) used practice and participation to understand culturally related approaches to learning. They argued that although research on cultural learning styles was an important move away from deficit-model explanations, it led to an "overly static and categorical" approach to the relation between culture and individual participation and engagement in school-related practices (p. 19). They proposed that instead of attributing cultural differences to individual or group traits, researchers concerned with understanding how students' culture and communities influence their school achievement should also focus on "variations in individuals' and groups' histories of engagement in cultural practices" (p. 19). By distinguishing between membership in a particular group and participation in practices of cultural communities, researchers can attend both to regularities in the practices and organization of cultural communities as well as to how students differently engage in and make meaning of those practices.

A number of researchers have explored the implications of situations in which students' everyday practices are misaligned with classroom practices for students' social and academic positioning in the classroom (Diamondstone, 2002; Hand, 2003; Nasir, 2004). For example, Hand (2003) found that the *a priori* distinctions between social and intellectual activity that often get organized in mathematics classrooms can provoke tension between the ways of being, talking, and reasoning that students from diverse backgrounds bring to the mathematics classroom and the normative aspects of the mathematics classroom. She argued that "open" participation structures, or those that afford negotiation around the framing and positioning of participation, are more likely to encourage broad-based participation among a range of students, versus "closed" or rigid participation structures, which tend to foster student resistance. This perspective attends not only to the relation of students' participation practices in and out of the classroom, but also to the relation of this alignment to students' deep engagement in or opposition to classroom mathematical activity. Both Cobb and Hodge (2002) and Hand (2003) pointed to the relation of students' participation and engagement in school and mathematics to ongoing negotiation and development of who they are and who they want to become—in other words, students' developing and constantly negotiated identities.

Identity

The notion of identity is of central concern in studies of participation as it is intimately linked to social practice and is concerned with both what is made available to individuals in the various social and cultural communities they inhabit and how they enact their participation across them. Although the field of

mathematics education (and education research more broadly) has yet to agree upon a working definition of *identity* (Sfard & Prusak, 2005), the theories of cultural activity that frame this review generally follow Wenger (1998) in proposing that, "identity serves as the pivot between the social and the individual" (p. 145). The working definition that the field eventually arrives at will need to attend to multiple aspects of identity formation and negotiation. For example, the concept of identity must account for the perceptions that individuals hold about themselves and for those held by others about them, and the relation of these multiple perceptions to an individual's social positioning in interaction. This definition will also need to acknowledge multiple identities (or dimensions of identity), such as being an African American and being a mathematics learner (Martin, in press-b), that individuals manage within and across contexts. What a perspective of identity of this kind theorizes is an *identity in practice* that is constantly being formed and reformed in dialectic between social structures and individual lived experiences. In this way, identity captures both the histories of engagement of various sociocultural communities and institutions and the practices they develop over time, as well as the trajectories of individuals as they negotiate and adapt these practices in and across local communities.

A number of researchers have examined identity negotiation and development within and beyond the mathematics classroom, particularly for nondominant students (Boaler & Greeno, 2000; Cobb & Hodge, 2002; Gresalfi, 2004; Martin, 2000, 2006, in press-a, in press-b; Nasir, 2002; Nasir & Hand, 2006; Sfard & Prusak, 2005). Martin's (in press-b) research, for example, focuses on the challenge many African Americans face in negotiating positive identities as mathematics learners both inside and outside of school. He argues that "any analysis of identity construction and students' becoming doers of mathematics must simultaneously consider African American identities as well" (p. 6). This is because mathematics education is shaped by a master narrative within society that positions African Americans as less capable in mathematics than their White peers. In response to and in contestation of the master narrative, Martin focuses on the success of some African American mathematics learners in overcoming structural and cultural impediments to mathematical achievement in order to forge new identities as Black mathematics learners. Martin's significant contributions to our understanding of racialized mathematical identity are explored further in the section on race and power.

Nasir has also examined the relation between identity and learning among African American youth but has explored this connection across multiple activity settings, including dominoes, basketball (Nasir, 2002), and the mathematics classroom (Nasir & Hand, 2006). Nasir studies the organization of social and intellectual resources for students' identity development and learning in relation to students' goals and strategies for participation in various social practices. Drawing on Wenger's *community of practice* model of learning, she has observed a relationship among students' goals, identities, and learning. Identity formation shapes and is shaped by learning. Similar relationships are found between goals and learning and between goals and identity. For example, in the case of elementary and secondary school students gaining competence in the game of dominoes, shifts in goals and identities accompanied age-related shifts in practice. More specifically, older, more experienced, and skilled players employed more complex strategies and had increasingly complex mathematical goals. Nasir also explored the notion of a *practice-linked identity* to capture the sense of connection that high school basketball players developed to the practice of basketball, in contrast to mathematics. In this study, players incorporated more of themselves into the practice of basketball as they drew heavily on a range of available resources for participation to accomplish their play. Nasir's research in this area has shown how identities involve aspects of both community (e.g., relationships) and learning (e.g., mastery), which both affect and reflect identity.

The literature on participation and identity illustrates the complexity underlying individual activity in any given social context, including a mathematics classroom. It also emphasizes the importance of examining the processes by which individual stakeholders in mathematics education (e.g., students, teachers, parents, policymakers, mathematicians) help to *construct* what "counts" as mathematics learning and what it means to be a mathematics learner, both within the local context of the mathematics classroom and in the broader system of mathematics education.

Assessing Opportunities to Learn

The consideration of participation and identity in research on mathematics learning communities shifts the focus from assessing what an individual student knows to discerning what she or he has an opportunity to learn within classroom mathematical practices (Greeno & Gresalfi, 2006). The construct of *opportunities to learn* (or OTL) initially referred to whether certain content appeared in curriculum materials or was included in the implemented curriculum. Recently the construct has been revived by a number of researchers to challenge a common assumption that learning is dictated by the curriculum, instead of being constructed within a classroom system of teachers, students, mathematical tasks, and material and ideological resources (Greeno, 2003; Lampert, 2001). This perspective highlights the enactment of mathematical content within a classroom culture, which affords particular opportunities for engaging with mathematical practices and ideas.

For example, it has been argued that mathematics instruction that prioritizes finding solutions or performing rapid computation over making sense of mathematical ideas and the connections between them can serve to constrain students' opportunities to engage deeply in mathematics (Boaler, 1998; Bransford, Brown, & Cocking, 1999; Carpenter, Franke, & Levi, 2003; Hiebert et al., 1997; Jacobs, 2002; Kilpatrick, Swafford, & Findell, 2001). Although this finding does not necessarily speak directly to the gap in achievement scores, researchers like Boaler (1997, in press) have linked meaningful participation in inquiry-based mathematics classrooms to greater affiliation with, persistence, and achievement in mathematics for women and students of color. There is general consensus in the field of mathematics education that mathematical practices that are guided by principles of mathematics reform, if implemented with fidelity to the curricular goals and intentions, can support powerful engagement in rich mathematics for *all* students (Kilpatrick et al., 2001; National Council of Teachers of Mathematics, 2000; Schoenfeld, 2002).

Opportunities to learn are related to identity in recognizing the diverse ways that opportunities can be *taken up* by students as a function of their cultural and mathematical histories. Student participation in the mathematics classroom is related to affiliation with and membership in local and broader communities (Nasir, 2002) as well as negotiation in moment-to-moment interaction about positioning oneself and being positioned by others (Wortham, 2006) with respect to these communities and the local classroom practices. Thus, differences in students' perceptions of and participation practices with opportunities to learn ultimately shape the nature of their mathematical experiences (Gresalfi, 2004).

To date, the research on opportunities to learn mathematics has focused primarily on comparing the experiences of dominant and nondominant students in classrooms, with less attention to the broader social and political structures shaping these learning environments. Issues of equity, however, run across multiple levels and social contexts and thus suggest that analyses of opportunities to learn be expanded beyond current research efforts.

Creating Equitable Opportunities to Learn

Researchers in mathematics education have investigated the relations between opportunities to learn mathematics for dominant (i.e., Whites and Asian Americans) versus nondominant students (i.e., African Americans, Latino/as, and women) in a variety of ways. The work of Jo Boaler and her colleagues, in particular, has figured prominently in this effort by identifying curricular strategies and classroom mathematical practices that serve to level the playing field in classrooms of students from diverse backgrounds.

Boaler's (1997, 1998, 2002, 2003, 2006) longitudinal studies of traditional versus reform mathematics curricula in the United Kingdom and the United States have focused on the nature of classroom mathematical practices under these different approaches, focusing specifically on what these curricula afford for the mathematical understandings that students develop and the views students construct of mathematics and themselves as mathematics learners. In her early work, Boaler (1997) found that reform mathematics classrooms were more likely to engage students in practices that focused on mathematical problem solving, developing a variety of solution paths, and providing justification for their mathematics work, and that those practices inherently offered opportunities for more students to participate in them. The students in the reform classrooms developed a conceptual understanding rather than just procedural knowledge, enabling them to apply this knowledge to a variety of tasks. Boaler (2002) has taken these findings further, demonstrating how reform mathematics can promote equity. Boaler and Greeno (2000) found that students taking Calculus developed significantly different perspectives on the role of learning mathematics depending on the nature of the classroom's mathematical activity (e.g., discussion-based format versus lecture-based format). Although these findings do not necessarily address issues of equity directly, Boaler and Greeno contended that students who developed a more "connected" relation to the learning of mathematics—meaning that they view mathematics as a meaningful aspect of their lives and important to who they see themselves becoming—were more apt to pursue mathematics over the long run. Thus, the nature of a classroom's mathematical activity could provide opportunities for more students to see themselves as mathematics learners.

Boaler (2003) has continued to pursue these issues in her work and has recently reported a variety of findings in which issues of equity are fore grounded. She highlights the work of the Railside mathematics department, which was highly successful at producing strong mathematics learners in a diverse urban high school.[2] Boaler argued that the success of this department was largely dependent on the ability of the teachers to foster what she calls *relational equity* among all students in their classrooms. She described relational equity as focusing on how "students learn to treat each other and the respect they learn for people from different circumstances to their own" (Boaler, in press, p. 5). This perspective of equity links students' mathematical engagement directly with the deep sense of commitment and respect students develop for one another within the classroom mathematics community. To promote relational equity among students, the teachers at this school drew on strategies of complex instruction (Cohen & Lotan, 1997), which minimizes status differences among students and promotes group accountability (Boaler & Staples, in press). They also utilized an internally developed reform mathematics curriculum, which offered a rich set of group-worthy mathematics tasks for students to grapple with together. Boaler's work over the years has helped paint a nuanced picture of mathematics classrooms that focuses on meaningful and respectful mathematical engagement, where all students' mathematical ideas are respected and their mathematical identities are cultivated.

In-depth case studies of mathematics learning within classrooms and across schools, like the ones presented by Jo Boaler and her colleagues, offer critical perspectives on how teachers, departments, and school administrations create conditions for relational equity among their students. Rich and comprehensive analysis of the practices of teachers and departments that play a role in fostering these relations can be described as design principles for learning environments. Although these principles can help guide the development of better opportunities for students in mathematics within their local schools, they do not necessarily address the systematic and structural aspects of inequity in mathematics education. Additional research needs to be conducted on the application of these principles across a range of schools and classrooms, and the broader social and structural barriers that shape, support, and constrain their proper implementation.

---

[2] At the time of Boaler's (2002) study, Railside student demographics were approximately 38% Latino/a, 23% African American, 20% White, 16% Asian or Pacific Islanders, and 3% other groups.

Broadening Classroom Discourse Practices

One aspect of classrooms that support relational equity is that they utilize a discussion-based format, where students work in groups and as a whole class to engage deeply with mathematical concepts and procedures. Some researchers concerned with English language learners (ELL) and students from nondominant backgrounds suggest that organizing mathematics classrooms around inquiry-based discussion may sometimes serve to perpetuate inequities among students if language and status differences are not taken into account. For example, in her work as a teacher-researcher, Lubienski (2000, 2002) found that her high and low socioeconomic-status (SES) students experienced mathematical discussions differently. While the higher SES students claimed to feel confident in contributing to and sorting out the ideas presented, the lower SES students found discussions frustrating and wanted the teacher to show them how to do problems. Using examples from her classroom, Lubienski (2002) argued that one cannot assume that particular practices, such as open-ended discussions, will necessarily work for all students. Although she does not suggest an alternative practice, a possible interpretation of her finding is that lower SES students will be more successful if taught by traditional methods. This conclusion is inconsistent with Boaler's findings. A more nuanced interpretation of Lubienski's findings is that the practices she studied did not take into account the everyday discursive practices or ways of reasoning of the low SES students. For example, others have found that when students' diverse ways of communicating are taken up as part of classroom practices, all students involved benefit (Brice-Heath, 1982; Staples & Hand, 2004; Warren, Rosebery, & Conant, 1994).

Research that closely examines the role of language in mathematics classrooms has yielded significant insights into the linguistic resources that are often missed in classrooms with students from diverse cultural backgrounds. Moschkovich (1999, 2002a), in studying the participation of ELL students, has examined how classroom practices create opportunities for students to demonstrate their competence in mathematics. She has argued that teachers and researchers should not focus on what students lack (in this case mathematics vocabulary) but instead focus on how students participate in everyday and mathematical discourses as well as on how they draw on multiple resources to communicate mathematically. Specifically, by analyzing students' participation, Moschkovich (1999) found that students brought different ways of talking about mathematical objects (e.g., narrative, predictive, and argumentative) and points of view of

mathematics situations to the discussions (e.g., standard definition of parallelogram versus dynamic view of trapezoid as half of a parallelogram). None of these ways of talking is privileged over the other; instead, each way of talking "can contribute in its own way to the mathematical discussion and bring resources to the conversation" (p. 12). Also, by examining the point of view that the students brought to the problem, Moschkovich illustrated how instead of concluding that the students were wrong or lacked vocabulary, they were simply bringing a different point of view than the teacher expected. Moschkovich concluded that taking a discourse approach to mathematical learning means considering the different ways of talking and different points of view students bring to discussions. This kind of approach shifts the focus of math instruction for ELL students away from vocabulary development toward mathematical content. In concluding her analysis of middle school math discussions, Moschkovich (2002a) provided further implications for mathematics instruction for bilingual and ELL students:

- Instruction should support engagement in conversations that go beyond vocabulary translation, and involve students in communicating about concepts.
- Teachers should support ALL students in participating in discussions. They can move toward this goal by providing opportunities for bilingual students to participate in discussions and by learning to recognize the resources that these students use.
- Instruction should also support students' use of resources from the everyday register, as well as resources such as gestures, objects, and students' first languages.
- Assessment must consider more than vocabulary, expanding to include how students use the situation, the everyday register, and their first languages as resources.
- Determining if a student's error is a conceptual misunderstanding or a language problem is not as important as listening to students and uncovering their competence, which requires a complex perspective. (Moschkovich, 2002a, p. 207)

Issues of language and discursive practices have implications for a large group of students. Mathematical and scientific language, with its particular precision, must be learned by all students. This is often accomplished as students are apprenticed into the community of mathematics and science during their elementary and secondary schooling. The discourse of the mathematical community seems to align with the discourse patterns of the dominant society (R. Gutiérrez, 2002), privileging students from that socioeconomic and language group. In advocating for a mathematics community that is more inclusive, attention may need to be focused in two directions: (a) to apprenticing students from nondominant groups into the mathematics community so that they develop the necessary discourse proficiency at the same time that the resources they bring to the classroom to explore concepts are validated and valued; and (b) to opening the domains of science and mathematics to reflect more of the diversity of the United States and the world. This may be a task that requires more attention from teachers who come from discourse communities that are different from those of the students. The work of Moschkovich (2002a), Lubienski (2002), and others has begun to define the issues and suggests questions that need to be addressed.

Expanding Conceptions of Mathematical Competence

Examining opportunities to learn in mathematics classrooms for diverse groups of students also prompts new ways to think about student achievement in mathematics. The recentering of analyses of student learning on student experience and practice allows for the identification of various student competencies not easily captured by traditional forms of assessment or, for that matter, those that were not previously valued as mathematical competence (Cohen & Lotan, 1997). Departing from traditional notions of *competence*, competence from a cultural perspective is recognized as being coconstructed by teachers and students in relation to classroom opportunities to learn and to what students are held accountable. This analytic shift from *achievement* to *competence* distinguishes a culturally biased deficit approach to assessment from a more culturally inclusive notion of mathematical understanding. Focusing on what students *are* doing opens conceptions of mathematical competence to the possibility of accommodating the diversity in students' ways of knowing and paths to learning.

One approach to expanding what it means to be mathematically competent is to investigate students' mathematical activities outside of school, challenging existing beliefs that pathologize or construct as deficient the cultural practices of nondominant populations. Since Gay and Cole's (1967) landmark ethnography on mathematics learning in a Liberian Kpelle society that focused on community-based out-of-school activities, sociocultural studies of mathematics learning have proliferated (see, e.g., Lave, 1988; Saxe, 1991; Scribner, 1984). Important conclusions of researchers comparing the everyday use of mathematics (e.g., selling candy or fruit, purchasing items at a local store) with performance in school mathematics contexts not only found students to be competent in mathematics in out-of-school contexts, but also suggest the importance of analyzing and drawing on children's informal strategies into the community of practice of the mathematics classroom (Carraher et al.,1985; Lave et al., 1984; Saxe, 1988a, 1988b; Taylor, 2004).

In the first of many studies of young Brazilian street vendors, Carraher et al. (1985) analyzed the everyday use of mathematics by children selling fruit on the streets of Recife, Brazil. The results of the study indicated that students competent in out-of-school settings could not reach equally accurate solutions to the same problem when posed as an in-school task, suggesting that the context in which problem solving is happening cannot be separated from the problem and the act of problem solving itself. Similarly, Lave et al. (1984), in a study comparing the problem solving done by grocery shoppers at the supermarket to their performance on paper-and-pencil tests, concluded that the supports present in the supermarkets could not be divorced from the problem solving process of making purchases. In other words, the environment of the supermarket provided supports for calculations that did not exist when the task was one given in a school-like format. The importance of the setting to the problem-solving process is indicated by this study as "people and settings together create problems and solution shapes, and moreover, they do so simultaneously" (p. 94). In both of these studies not only were people found to be competent in mathematics in out-of-school contexts, but this research also suggests that the environment in which problem solving takes place contributes to mathematical accuracy.

In studying how school mathematics learning is related to out-of-school mathematical practices, Saxe (1988a) found no statistically significant difference between schooled and nonschooled children for problems of currency arithmetic and ratio comparison in out-of-school contexts. On the other hand, in looking at differences in performance on school arithmetic problems, he observed that the children who were sellers did solve more problems drawing on the informal strategies developed in the marketplace than did the nonsellers. Saxe argued for drawing on informal strategies such as these in classroom practice.

In a study of how school mathematics learning might be supported by out-of-school mathematical practices, Taylor (2004) explored the purchasing done by children before and after school at local convenience stores in a northern California city. He documented conventions, artifacts, and other supports that assisted children in making purchases in these real-life situations. Taylor found that accessing the mathematics necessary to complete purchases required children to draw on knowledgeable others, such as older children and store employees. Taylor also found currency to be a powerful context to draw on in school. In working with young children on noncontextualized addition and modeling problems in a school context, Taylor has shown that the use of currency as a base-ten model provided more support in arriving at correct solutions to problems than did base-ten blocks. This study illustrated how students' everyday practices may support the learning of fundamental mathematics concepts in mathematics classrooms. It is important to point out that although these studies illustrate the mathematical fluency with which children engage everyday tasks, they do not suggest that these youth are incapable of success in school mathematics. Rather, studies of students' out-of-school mathematical activities expose a misalignment between the opportunities to learn in school mathematics and students' mathematical practices developed outside of the classroom.

The work of both Saxe (1988a) and Taylor (2004) suggests that expanding notions of mathematical competence to include the valuation of students' informal strategies developed in out-of-school contexts can support students' mathematical knowledge development within school contexts. Uncovering students' mathematical competency both in and out of school broadens the field's understanding of student mathematics achievement and begins to change conceptions of mathematical competence. However, broadening conceptions of mathematical competence by looking outside of the classroom provokes the question of how these insights can be applied generatively to classroom mathematics curriculum and instruction. Research examining the features of out-of-school mathematics activities in comparison to school practices may help bridge students' participation in mathematical activity, and mathematical activity that supports this engagement.

Bridging In-School and Out-of-School Mathematics Knowledge

One project that attempts to forge links between in-school and out-of-school knowledge is the Yup'ik project. The Yup'iks are a group of indigenous people of Alaska whose language and culture has been marginalized within the educational system by its emphasis on English-only language and middle-class, White American cultural practices (Lipka, 1994, 2002; Lipka, Wildfeuer, Wahlberg, George, & Ezran, 2001). The Yup'ik Project is a collaboration among university researchers, teachers (both indigenous and White) and Yup'ik elders to explore and describe the traditional number system used by the Yup'ik as well as the mathematics inherent in Yup'ik cultural practices. The project has begun to publish a culturally based curriculum (*Math in a Cultural Context,* Detselig Enterprises, Calgary, Canada) and has documented that the achievement of students whose teachers use the curriculum after receiving professional development surpasses that of comparable groups of students (Civil, 2001; Lipka, 1998, 2005; Webster, Wiles, Civil, & Clark, 2005).

Moll and his colleagues have conducted seminal work on bridging the in-school and out-of-school knowledge of students from households in working-class Mexican communities (Gonzalez, Moll, & Amanti, 2005; Moll et al. 1992; Moll & Greenberg, 1990). These researchers developed a conceptual framework known as *funds-of-knowledge* to begin to account for the nature and structure of knowledge and skills organized in and across these households. Funds-of-knowledge are "historically accumulated and culturally developed bodies of knowledge and skills essential for household or individual functioning and well-being" (Moll et al., 1992, p. 133). The funds-of-knowledge work explicitly rejects the notion that the problem of underachievement either is located within the students or is attributable to their cultures and communities. Instead the locus of responsibility for underachievement shifts to the school and acknowledges the complexities of students' lives in and out of school, seeking to support the students' scholastic achievement by drawing on the wealth of their home and community experiences. Moll (1992) argued "that these families and their funds-of-knowledge represent a *potential* major social and intellectual resource for the schools" (p. 22).

Whereas the initial funds-of-knowledge work of Moll and his colleagues focused on language and literacy, the BRIDGE project extends the work to mathematics (Civil, 1995a, 1995b). An explicit goal of the project was "the development of mathematics teaching that builds on these students' backgrounds and experiences" (Civil, 1995a, p. 2), and it incorporated the same three basic components of the original project: (a) ethnographic household visits, (b) teacher-researcher study groups, and (c) classroom implementation and curriculum development (Civil, 1994). Views of the families as "somehow disorganized socially and deficient intellectually" were exploded during the data gathering and study-group analysis in the BRIDGE project (Moll et al., 1992, p. 134). When faced with caring and interested parents on the other side of the home interview table, teachers had to confront their notions of the homes these children came from as dysfunctional and begin to examine the potential that existed there (Moll et al., 1992). Civil (1998) suggested that this type of direct connection made between parent and teacher could have an impact on the teacher's thinking and could ultimately influence his or her teaching. The incorporation of mathematical funds-of-knowledge into school mathematics presented particular challenges, however, within this project. A question that evolved from the study of household funds-of-knowledge was the extent to which it was possible for schools to build on everyday mathematical situations (Civil, 1995b). Researchers found, for example, that frequently the mathematics in everyday activities was hidden, such that people engaging in the activity did not acknowledge what they were doing as mathematics and at times even directly rejected it being characterized as math (Gonzalez et al., 2001).

Although the BRIDGE project does seem to have positively affected teacher and student attitudes about school, and although learning clearly was happening in the BRIDGE classrooms, it is not as clear what mathematics the children were learning. As early as 1995, Civil was raising issues regarding the depth of mathematics the children were learning, stating, "unless we have a clearer picture of the mathematical opportunities in a module and on how to push for these, the modules risk to present only surface applications of mathematics, often not challenging enough for the students" (1995a, p. 15). Civil's (2001) recent work documents the explicit connections that students made between a gardening project that she designed with an intermediate grades mathematics teacher and area measurement. The activity was designed to foreground the mathematics by "looking for opportunities in which mathematics will occur naturally" (p. 404).

Moschkovich (2002b) summarized much of the work done in the area of integrating everyday mathematical practices into school mathematics in the interest of making "the mathematical practices of different groups accessible to more students" (p. 8). Her own recommendations for implementing such practices, however, carry a caveat about the overuse of everyday mathematics at the expense of academic mathematics. She is particularly concerned about the emphasis on practical mathematics being implemented or interpreted as part of a tracking policy or as a disguise for vocational education. "It is not merely using everyday mathematics that is important but making connections between the familiar practices of everyday activities and academic mathematical practices" (Moschkovich, 2002b, p. 9). In the end, Moschkovich is asking for a balance between everyday and academ-ic mathematics that will not only motivate students to engage in the study of mathematics but will also provide them with the discursive practices they will need to pursue more advanced studies such as documenting and constructing narratives about their solution processes and reflecting on the "efficiency or generality of different approaches to a problem" (Moschkovich, 2002b, p. 9).

Reforming the "Culture of Mathematics"

Balancing the need for individuals to draw on practices, identities, and realities of their everyday, lived experience with the goal to master formal, domain-related practices and identities within the context of schooling has been a reoccurring theme in the field of education (Dewey, 1902, 1938). R. Gutiérrez (2002) framed this longstanding tension in a particular way with respect to equity in mathematics. Progress on equity cannot be made, she argued, as long as deficit perspectives continue to permeate research and practice, and the field of mathematics perpetuates a culture that excludes individuals with different perspectives (such as minorities and women). To create conditions for equitable mathematics education, she proposed that the field coordinate two different approaches to mathematics instruction. One is mathematics that "reflects the status quo in society, that gets valued in high-stakes and credentialing, that privileges a static formalism in mathematics, and is involved in making sense of a world that favors views and perspectives of a relatively elite group" (p. 151), or what Gutierrez called *dominant mathematics* (in which she included "reform" mathematics). She contrasted this with "mathematics that squarely acknowledges students are members of a society rife with issues of power and domination" (p. 151), or *critical mathematics,* that empowers students to challenge the structures, perspectives, and processes through which they are marginalized. The combination of *dominant* and *critical* mathematics, she argued, "serve as an entrance for students to critically analyze the world with mathematics, and being able to critically analyze the world with mathematics may be an entrance to engage dominant mathematics" (p. 152).

The contention that mathematics education is embedded in a broader "culture of mathematics" that is privileged and privileges a certain few is echoed by researchers who take a critical stance on mathematics education (Delpit, 1988; Ladson-Billings, 1997; Tate, 1994). This critical perspective is serving to push mathematics education research to take seriously issues of power, race/racism, and "White privilege" in mathematics education. In the section that follows, we highlight some of the contributions that the theories of cultural activity have made to the conceptualization and consideration of differences in opportunities to learn between dominant and nondominant students in mathematics education.

Summary

Researchers who study mathematics education with a perspective of learning as cultural activity have made significant strides in disentangling the issues involved in differential achievement—reconceptualized as differences in opportunities to learn. Due to the space constraints of this chapter, we have reviewed the research of only a select group of researchers who are recognized within the field as pioneering this work. This review is by no means exhaustive, and the research has suffered in general from complicated methodological and political challenges. However, the research reviewed here offers important insights into the processes by which a *participation gap* (Hand, 2003) between students from diverse social, cultural, and racial backgrounds can arise within mathematics classrooms and how classrooms can be structured to better afford opportunities to participate in mathematics by a wider range of students.

From a theoretical perspective, research based on a perspective of culture as constituted by and through activity provides new constructs for examining the interplay of culture and the individual. By focusing on participation and identity as key aspects of individual and joint social practice, this perspective acknowledges the role of both individual agency (as individuals choose, adapt, and reject practices) and sociocultural processes (as broader social, political, and racial practices and identities are remade in local classrooms) in shaping the learning experience. Thus, while acknowledging that students bring perspectives, values, and routines to the mathematics classroom from their home and local communities, this perspective emphasizes the processes of negotiation, reconciliation, and rejection that students manage with respect to their activities across multiple communities.

The complex nature of examining these processes within and beyond the mathematics classroom has been aided by the development of complementary methodological tools and techniques. Tools such as video interaction analysis and qualitative data analysis software and techniques such as documenting trajectories of participation and repertoires of practice help researchers capture, analyze, and identify features of learning communities and various aspects of participation within them. These features and aspects can then be linked back to broader cultural practices, discourses, tools and artifacts, and identities, which implicate issues of power and access in the classroom. Locating practices, scripts, and norms found in the local context of the mathematics classrooms within global social and cultural hierarchies can reveal the danger of "neutral" mathematics instruction in perpetuating inequitable processes that marginalize nondominant students.

Attending to the variety of ways individuals engage in mathematical tasks and other activities in different contexts, this perspective has also expanded the notion of what it means to do mathematics and to be a learner and doer of mathematics. On the one hand, definitions of mathematical thinking and reasoning have been widened to include the processes by which people solve mathematical problems informally with the material and social resources around them. Thus, part of calculating the statistics of a pro basketball player involves being part of a community of players and spectators who find this activity meaningful. On the other hand, learning mathematics in school and identifying oneself as a mathematics learner is bound up in social and cultural discourses that position particular mathematical practices as being "competent" or "rigorous" and certain individuals as being "smart" and "capable." Thus, what this research has revealed is how mathematics learning itself is organized within a cultural practice that can serve to either enfranchise or marginalize different groups of individuals.

Finally, one common thread across the research presented in these sections is the rejection of deficit and cultural-deficit thinking. Culture is acknowledged as a critical element in trying to understand the academic disparities between economically advantaged populations and impoverished ones. But instead of viewing the cultures of these families and communities as pathological and the source of academic school failure, culture is viewed as an area of students' lives that can contribute to academic success if appropriately understood. At the same time, however, this research has tended to shy away from making explicit links between opportunities to learn found in mathematics classrooms, cultural practices both in and out of the mathematics classroom, and persistent racial inequities (Cobb & Nasir, 2002; Martin, in press-b).

A potential tension in research aimed at identifying and valuing the cultural practices and identities of nondominant groups of students is the possibility of reifying what are basically essentialist accounts of racial and ethnic communities. This way of thinking, or essentializing of traits of groups of students, has potentially negative consequences for students subject to its application.

The obvious problem with essentializing is that, by treating cultural behaviors, values, and practices as fixed and immutable, it slips perilously close to committing at least two types of error that were observed with cultural deficit thinking. First, in associating and universalizing cultural practices and patterns to all members of a group, the space for the member to be treated as an individual shrinks. Essentializing contributes to a form of *cognitive reductionism* where cognitive complexity is simplified and diminished (Gutierrez & Rogoff, 2003). A person, as an individual, is rendered secondary in the process of teaching and learning. Second, as in the case of cultural deficit thinking, with essentializing there is a tendency to reify culture, that is, to see it as isolated, immune from impact of other cultures and the dominant social and economic forces. If the role of culture is going to be given a central place in educational research and practice, researchers must learn how to think and talk about cultural patterns and practices and student learning without slipping into essentializing and cultural-deficit thinking (Gutierrez & Rogoff, 2003). In order for culture to remain a viable explanatory construct for understanding and improving the persistent underachievement of certain student populations, much care and attention must be exercised in developing theories and pedagogies that do not essentialize cultural traits.

Much of the research reviewed in this section has focused on the conditions of individual classrooms or schools, out-of-school practices, or communities, with only limited attention to how historically based processes and structures that involve race and power have shaped these local contexts and practices over time. Thus, perhaps one of the most important challenges to arise out of research based on a cultural activity perspective is the need to better understand the relation between race and culture. On the one hand, research that isolates the individual from the cultural processes in which she develops has merged the two constructs, treating race and culture as if they were one and the same. In this way, the characteristics of all people from a racial group are treated as being homogenous. In this model race determines culture. On the other hand, the research just reviewed has tended to completely divorce the two and focus only on culture in its most apolitical sense. In this model, culture is all that matters. However, there are a growing number of researchers who have critiqued a focus on culture that excludes consideration of race and power.

CONSIDERATIONS OF RACE AND POWER AT THE CLASSROOM LEVEL

Thus, we turn to mathematics education research that has considered explicitly issues of culture *and* issues of race and power as they intersect with mathematics teaching and learning. Specifically, this section highlights research that provides models of classroom learning environments that focus squarely on issues of cultural relevancy while also taking into consideration issues of race and power (see, e.g., Frankenstein, 1990, 1995, 1997; Gutstein, 2003, 2006; Ladson-Billings, 1994, 1995, 1997; Moses & Cobb, 2001a, 2001b), thus making race and power central constructs in confronting differential achievement in mathematics education.

The Algebra Project

The Algebra Project, for example, is a mathematics literacy effort with a grassroots implementation process (Moses & Cobb, 2001a, 2001b; Moses, Kamii, Swap, & Howard, 1989; Silva, Moses, Rivers, & Johnson, 1990). The program is designed to make Algebra available to all seventh- and eighth-grade students despite their previous levels of academic achievement. The curriculum is created to develop algebraic thinking using projects that engage students with concrete experiences supported by a culture of mutual inquiry. These experiences are drawn from or build on practices of the communities in which students live. Addition and subtraction of negative numbers, for example, is taught within the context of trips. But the subway context that is used for inner-city kids in Boston is exchanged for a bus trip for students living in rural areas of Mississippi. The curriculum explicitly acknowledges that the mathematics must connect with the lived experiences of students.

The work of Moses and colleagues also makes central issues of race and power, equating the need for mathematical literacy in today's society with the need for Black registered voters in Mississippi in 1961. Moses and Cobb (2001a) have argued that differential access to algebra, which disproportionately excludes African Americans, Latinos, and poor White students from college preparatory mathematics classes, is serving as a form of structural discrimination resembling the use of literacy tests in the '60s. Moreover, algebra is the forum where students learn the symbolism necessary for developing technological knowledge demanded in today's high-tech job market. Thus, mathematical literacy is not just needed for access to college preparatory mathematics classes but is also necessary to meaningfully participate, with economic viability, in today's society. Mathematical literacy, then, is the key to citizenship; it becomes a civil-rights issue, and a necessary component in promoting economic and civic equality.

More recent work of the Algebra Project, the Young People's Project, encourages students to take up this right (Kirkland, 2002). Their work organizes young people to work to change their education and the way they relate to it, that is, encouraging students to "demand to understand" in an effort to challenge and transform their marginalization. Specifically, the Young People's Project is dedicated to the creation of mathematically literate communities through the recruitment and training of core high school and college Math Literacy Workers. These Math Literacy Workers come from the communities the Algebra Project seeks to serve, and they focus on developing their knowledge capacity so that they are able to organize and manage math literacy work independently. This work might consist of mentoring middle and elementary school students, providing ongoing after-school workshops for younger students, providing community events for families and community members, and facilitating team organizing for mathematics competitions. In the process, the Math Literacy Workers collectively contribute to the development of ever-expanding networks of mathematically literate young people. For this to occur, though, young people themselves must demand the right to receive a quality education, and the Young People's Project is working to create that demand.

The work of the Algebra Project, and the extended work of the Young People's Project, focuses squarely on issues of cultural relevancy by building a curriculum that explicitly connects with students' lived experiences. Additionally, this work considers the effects of power structures on mathematics teaching and learning by acknowledging the gate-keeping role of mathematics. Their work explicitly confronts this gate-keeping role of mathematics education by helping *all* students, particularly those traditionally denied opportunities, pass through the gates. It has the potential to confront differential mathematics achievement and provide quality mathematics instruction to all students. More research is needed that looks at the effects of the Algebra Project and the Young People's Project on student achievement in mathematics.

Culturally Relevant Pedagogy

Other compelling models of classroom learning environments consider both the multiple constructs of culture, race, and power, and how students can use their knowledge (of mathematics and other subjects) to challenge current oppressive and inequitable structures (see, e.g., Frankenstein, 1995, 1997; Gutstein, 2003, 2006; Ladson-Billings, 1994, 1995, 2001; Tate, 1994, 1995). Ladson-Billing's (1994, 1995, 1997,

2001) work, although not specific to mathematics education, provides a theory of culturally relevant pedagogy that speaks to issues in mathematics education, and as such is important to consider here. From her study of eight successful teachers of African American students, Ladson-Billings developed a theory of culturally relevant pedagogy. Specifically, she argued for the importance of a three-pronged approach to culturally relevant teaching that proposes to (a) produce students who can achieve academically where achievement is not limited to standardized assessment, (b) produce students who demonstrate cultural competence, and (c) develop students who can both understand and critique the existing social order. Two of these three components will be expanded upon here: cultural competence and understanding and critiquing the existing social order (or developing a sociopolitical consciousness).

As a pioneer of culturally relevant teaching, Ladson-Billings argued that teachers should develop in students a cultural competence. That is, students should be provided with "a way to maintain their cultural integrity while succeeding academically" (Ladson-Billings, 1995, p. 476). In classrooms that promotes cultural competence, teachers must first be aware of their own culture and its role in their lives (Ladson-Billings, 2002). They can then work to effectively respond to students from different cultures and classes while valuing and preserving the dignity of cultural differences and similarities between individuals, families, and communities. A classroom that promotes cultural competence acknowledges that everyone has a cultural history that shapes their identity. Moreover, in such a classroom, the academic and cultural assets students bring to the classroom are seen as enriching the community, and the students and teacher together continuously strive to learn about one another and the assets each person brings. Additionally, Ladson-Billings (2001) argued that in classrooms that promote cultural competence in students, the teacher "uses culture as a basis for learning" (p. 98). That is, teachers aim to capitalize on students' prior knowledge, and they view students' culture as a means through which they can acquire new knowledge.

In addition to promoting students' cultural competence, teachers enacting a culturally relevant pedagogy provide students with the tools they need to understand the social structures around them, see how those social structures (such as institutional racism) may affect their lives, and teach students how to challenge those structures. In classrooms where teachers employ culturally relevant pedagogy, teachers and students together create knowledge "in conjunction with the ability (and the need) to be critical of content" (Ladson-Billings, 1994, p. 93). To be critical of content means culturally relevant teachers attempt to make knowledge problematic, challenging students to view knowledge as a means for transforming the world in which they live. For example, in a more contextualized examination of two classroom teachers in her study, Ladson-Billings (1994) noted that both teachers held exceptionally high expectations (demonstrated through both words and actions) of their students while simultaneously helping students understand that societal expectations for them are generally low. In this way, students recognized the teacher's act of holding high expectations and their own efforts to meet those expectations as acts to challenge and defy prevailing societal beliefs.

Although Ladson-Billings is not referring specifically to the teaching of mathematics, mathematics education researchers argue similarly, noting that mathematics teaching entails a shift from thinking of preparing students to live within the world, as it currently exists, to thinking about how to prepare students to restructure "those social systems . . . in order to remove barriers that women, minorities, and others experience" (Secada, 1989, p. 47). Mathematics educators need to work toward using mathematical knowledge to empower students to work for social justice and to confront issues of unequal power relationships that exist in the world in which we live (Martin, 2003). Thus, mathematics education faces a twofold imperative: It needs to provide students with mathematics instruction that includes the mathematics deemed necessary for success in the current system (a similar component to Ladson-Billings *academic achievement*) while simultaneously providing students an opportunity to use mathematics to expose and confront obstacles to their success (Apple, 1992; R. Gutierrez, 2002; Gutstein, 2003, 2006; Martin, 2003; Secada, 1989; Tate, 1994, 1995).

Teaching Mathematics for Social Justice

Teaching mathematics for social justice addresses both of these components. Math teachers employing social-justice pedagogies address the first imperative by recognizing the necessity of mathematical knowledge and including mathematics-specific goals for their students (see, e.g., Frankenstein, 1990, 1997; Gutstein, 2003, in press-a). They address the second imperative by engaging students in using mathematics to analyze their world critically in an effort to ultimately promote a democratic society in which all have an opportunity to participate fully (see, e.g., Frankenstein, 1995; Gutstein, 2003; Skovsmose, 1994).

Frankenstein (1990, 1995, 1997), for example, described her attempts to teach mathematics for social justice with working class adults in basic college mathematics courses. In these courses, Frankenstein employed a critical mathematics pedagogy focused explicitly on using statistical tools to analyze social issues critically (e.g., income data, wealth distribution, home mortgage distribution, the tax system). Her critical mathematics pedagogy involved promoting "the ability to ask basic statistical questions in order to deepen one's appreciation of particular issues [and] it also involved the ability to present data to change people's perceptions of those issues" (Frankenstein, 1990, p. 336). The goal of this critical analysis was to prompt students to question their assumptions about how society is structured and to enable them to act from a more informed position on social structures and processes.

Gutstein (2003, 2006, in press-a, in press-b) also described his enactment of a social- justice pedagogy in a Chicago public middle school with a predominately low-SES Mexican and Mexican American population. His work further demonstrates how such a pedagogy addresses issues of culture, race, and power and works to promote students' use of mathematics to transform oppressive structures. Gutstein (in press-b) conceptualized the pedagogy of teaching mathematics for social justice as developing four main components: "a) academic 'success' (i.e., both mathematical power and what is needed to pass gate keeping tests); b) sociopolitical consciousness; c) a sense of social agency; and d) positive social and cultural identities" (p. 8).

To achieve his goals of teaching for social justice, Gutstein created 17 real-world mathematics projects that connected to students' lives (e.g., examining wealth distribution; analyzing SAT and ACT exam scores by race, class, and gender; questioning whether racism is a factor in mortgage loan opportunities). To achieve the mathematics-related objectives in the classroom, Gutstein used *Mathematics in Context*, a curriculum developed by the National Center for Research in Mathematical Sciences Education and the Freudenthal Institute (1997-1998). More specifically, Gutstein employed a *pedagogy of questioning* in his classroom. He created a classroom environment where students posed their own meaningful questions, engaged in understanding their own realities in sociopolitical context, discussed interrelationships and complexities among questions, engaged and analyzed multiple perspectives, and interacted with questions that connect to actions and social movements. Mathematics played a central role in this pedagogy of questioning. Students used mathematics to develop sociopolitical understanding of their life conditions and broader society. For example, students used mathematics to uncover what the money for one B-2 bomber would mean in terms of college education for thousands of Latino/Latinas. In this project, students

> used mathematics to investigate and calculate the various costs for different ways to use public money; they considered ramifications, shared their views with others, and wrote about them; put their mathematical analyses into sociopolitical, historical context; and built community, a shared sense of purpose, and a dramatically different orientation toward mathematics and its use in understanding reality. (Gutstein, in press-b, p. 24)

The results of Frankenstein's and Gutstein's work are promising. Frankenstein (1997) found that the use of critical mathematics pedagogy changed her students' perceptions of mathematics and their ability to understand mathematics, both important factors in helping students reach the mathematics-specific goals of the course. At the same time, she engaged her students in using mathematics to analyze their world. For Gutstein's students, many developed mathematical power, and the cocreation of a classroom environment where students discussed significant issues of justice and equity, and where students actively and consistently raised their own questions, seemed to foster agency among students.

Although teaching mathematics for social justice holds potential for addressing differential achievement in mathematics because of its simultaneous consideration of issues of culture, race, and power and their intersection with mathematics teaching and learning, more research is needed that examines the effects of the implementation of such pedagogy on students' mathematical learning. It is important, however, to note that when determining whether this pedagogy is effective in terms of students' learning of mathematics, researchers must also consider the purpose of mathematics education. In the cases described here, the purpose of mathematics education is not functional literacy, or "at best, to generate a few more individual successes" (Gutstein, 2006, p. 211); rather, the goal is to conceive of mathematics knowledge as the ability to use mathematics to critique and transform oppressive structures—math literacy is "knowledge for liberation from oppression" (Gutstein, 2006, p. 211).

Frakenstein's and Gutstein's work describes the practices of teacher researchers who are committed to teaching for social justice. The next question is how to build on that work to study what it means for the "average" teacher to learn to teach mathematics for social justice. Gau (2005), in a study of eight secondary mathematics teachers engaged in learning to teach for social justice through the creation, implementation, observation, and revision of a mathematics lesson that incorporated social-justice goals, found that teachers struggled to find a balance between the mathematics and the social justice. For one group, this meant that the mathematics took priority over the social justice, and in four separate lesson implementations they never addressed the social justice goals of their lesson. For the other group of teachers, this meant that mathematics need not be tied in all the time. Instead, they focused on developing the social-justice goal of their lesson, to the detriment of the mathematics. In fact, in this latter case, teachers seemed to be so focused on having students come to a particular conclusion, that they inappropriately interpreted data, suggesting to students that the data supported conclusions that were, in fact, not founded by the data.

Additionally, although teachers in both groups conceptualized teaching mathematics for social justice similarly to that expressed in the literature, they did not ever express that it include the goal of students' learning mathematics. Rather, they conceptualized it as students' using known mathematics to analyze and confront inequities, suggesting that the social-justice component is always an "add on" to the curriculum. More work is needed in this area to see what teachers struggle with as they learn to teach mathematics for social justice, and the implications this has for the potential of such a pedagogy to affect both students' learning of mathematics and students' ability to use mathematics to critique and transform oppressive structures.

Researchers who work in the area of mathematics teaching for social justice (see, e.g., Esmonde, 2006; Gau, 2005; Gutstein, 2003, 2006) also note that such teaching is not just a collection of supplemental projects that could be "dropped into" any context. Rather, this pedagogy permeates all aspects of the classroom. It is a pedagogy that is "forged *with*, not *for*" students and is continually negotiated as students' understandings of their own sociopolitical contexts grow and as the questions they wish to explore evolve (Freire, 1970/1993, p. 30). Additional research on the implications of teaching mathematics for social justice on achievement is still needed, but this work, along with the work of Moses and colleagues in the Algebra Project and Ladson-Billing's theory of culturally relevant pedagogy serves to highlight work that places culture, race, and power at its center.

RACE AND POWER

In the previous section, we examined research that places culture, race,[3] and power[4] at the center in understanding issues of mathematics teaching and learning. Here that examination continues. However, whereas the previous section examined these constructs within the context of the classroom, this section will be situated within a broader, structural context. More closely, this section highlights larger policies and mandates that shape and structure mathematics education. The focus is on two questions. First, how do issues of race and power manifest themselves within the structures that provide mathematics education to students? Second, how might an analysis of race and power within this broader context help in understanding differential achievement in mathematics?

It is important to mention that two issues focus the discussion in this section. First, presently little research centers itself within mathematics education and provides a structural analysis of race and power. Where this literature does exist, we have sought to cite it and its contribution to the field. However, because of the limited amount of such research, in some instances we cite researchers outside of mathematics education that examine race and power. A second reason flows directly from the first. Because there is limited research within the field, this section raises issues and questions with respect to race and power in relation to mathematics education. Many of the issues raised here draw on the experiences and work of the authors in schools and districts that serve poor and minoritized communities. Such communities provide a rich context within which to explore the impact of racialized experiences and power dynamics on differential achievement in mathematics.

Structuring Opportunities to Learn Mathematics

Orfield, Frankenberg, and Lee (2003) stated that the level of segregation of schools is worse now than in 1968. Students of color and Whites are increasingly not in the same schools. Moreover, only 15% of the intensely segregated White schools have populations in which more than half are poor enough to receive free and reduced lunches. For Black and Latino students the percentage is 86%. Schools in communities predominantly consisting of Blacks and Latinos

are poorer, and they generally have fewer AP courses, fewer credentialed teachers (Darling-Hammond & Sykes, 2003), more out-of-field teachers (Ingersoll, 1999; Rogers, Jellison-Holmes, & Silver, 2005), and buildings in worse conditions (Oakes & Saunders, 2002). Kozol (2005) and others (Frankenberg, Lee & Orfield, 2003; Hunter & Donahoo, 2003) have ascribed this situation to a new form of apartheid in the U.S. school system, where low-income public schools have become hypersegregated with populations of up to 99% students of color.

Along with the material conditions in "apartheid" schools, Kozol (2005) notes that in urban schools there are another set of conditions around how we talk about students and the ways they are expected to participate. Kozol points out that he has heard hypnotic slogans like "I'm smart! I know that I'm smart," repeated everyday, "but rarely in suburban schools where potential is assumed" (p. 36). These non-material conditions shape the opportunities of students of color—often blaming them for their own failure. At the same time that these students are blamed for their failure, the system of mathematics education continually fails them. As such, even if the material conditions were equitable, the non-material normative aspects of schools would still construct failure for students of color (Oakes & Lipton, 1999).

The literature on access and opportunity to learn mathematics documents how experiences differ along racial lines. Overall, segregated minority schools offer less access to upper-level math and science courses, many not offering courses beyond Algebra II. Oakes, Muir, and Joseph (2000) wrote that,

> A student can only take a high level class in science and mathematics if his or her school offers such classes or if his or her school opens up access to these courses to all students. In other words, how far a student can go down either the mathematics or science pipeline depends on his or her access to particular courses. (p. 12)

On the basis of a student's race, he or she can expect to experience mathematics education differently (Hunter & Donahoo, 2003). Students of color often experience a lesser form of education, in mathematics and otherwise. In contrast, adequate mathematics course offerings (Lee, Burkham, Chow-Hoy, Smerdon, & Geverdt, 1998), qualified mathematics instructors (Rogers et al., 2005), quality mathematics curricula, and mathematics teachers who respect their culture

---

[3] By *race*, we mean the very real ways in which a student's skin color (and the social significance that society attaches to that skin color) frames that student's educational opportunities and experiences.

[4] In terms of *power*, we refer to the set of relationships and hierarchies that frame interactions.

and hold high expectations of them is the minimum expectation for White children. Whereas a Latina student can expect to attend a school with a large number of courses below Algebra, a White student can expect, as a matter of course, that the classes necessary for her to prepare herself for college will be present at her school. In this sense, statistically speaking, whiteness has a higher property or currency value. With whiteness comes advantage, more valued cultural practices as well as property, educational buildings in better neighborhoods that draw higher taxes, and therefore more funding.

As described in earlier sections, White culture often determines what is "normal" and also constructs the dialogue or ideology for understanding the "other." This dialogue is constructed and reinforced in mathematics education, for example, when achievement scores are reported in terms of race, and lower test scores are ascribed to race (ignoring the fact that "White" is also a racial category). Educators fail to ask how the racial and cultural entailments of whiteness provide opportunities for large groups of White students to be consistently ahead of their Black and Latino/a counterparts. Instead, the success or failure of a White student often gets framed as an individual act, acclaiming or pathologizing the individual rather than the race.

Ideologies are embedded within language and ways of talking that perpetuate stereotypes of the "other." These broad Discourses, as Gee (1990) and colleagues call them, structure the ways of talking about children of color, communities of color, and structure our individual actions (Gee, 1990; Gee, Hull, & Lankshear, 1996). Gee (1990) calls this dialogue Discourse with a big *D* because it contains ideologies, beliefs, practices, and ways of being that further the power of the dominant culture. There is more going on in individual success or failure, or individual interaction, than what is actually seen in front of us. Individual interaction sits inside of a historical reality; it sits within history, within a context, and within a relation of power. The stories embedded in these Discourses limit the ways of talking and thinking about people of color and can limit how one thinks about their intelligence and abilities, quality of family life, and cultural resources (Kana"iaupuni, 2005; Ryan, 1971; Warren, 2005).

Counterstorytelling is a method that "aims to cast doubt on the validity of accepted premises or myths, especially ones held by the majority" (Delgado & Stefanic, 2001, p. 144). Counterstories or counternarratives challenge the privileged discourse, or way of talking and relating, giving voice to people of color and critiquing racialized stereotypes. These counternarratives construct alternative Discourses to the mainstream, counteracting the essentializing that exists within the dominant narrative. The stories that researchers like Kozol (2005) and Martin (2003, 2006, in press-a, in press-b) bring to the education community serve as counternarratives to the myth that schooling and mathematics education are neutral and color blind. In the following sections, race and power are central in creating counternarratives about the neutrality and color blindness of policies both inside and outside mathematics education.

The increasing segregation, decreasing access, and pervading Discourses place race and educational structures as central in educational opportunity. We see these areas as important for the field to explore in relation to mathematics education in particular. Though much of this work has been done outside mathematics education, we think the work on counternarratives provides an area for future research to challenge prevalent Discourses and to open access to those not in power.

Mathematics for All

In the 1980s, several national reports were released that called attention to serious problems in mathematics and science education; one of them was *A Nation at Risk* (National Commission on Excellence in Education, 1983). Among the findings reported in *A Nation at Risk* were that from 1963 to 1983 average mathematics SAT scores dropped nearly 40 points, only one third of 17 year-olds could solve a mathematics problem requiring several steps, and between 1975 and 1980 remedial mathematics courses in public 4-year colleges increased by 72%. The business and industry sector also provided impetus for improving mathematics education by demanding an improvement in workplace proficiency. Criticism had been directed at public education because employees failed to demonstrate, beyond the use of computational algorithms, proficiency levels in reasoning and problem solving (Vandegrift & Dickey, 1993).

In response, national organizations produced documents that advocated changes in mathematics curricula. The National Council of Teachers of Mathematics (NCTM) produced the *Curriculum and Evaluation Standards* (NCTM, 1989), and the Mathematics Sciences Education Board (MSEB) published *Everybody Counts* (MSEB, 1989) and *Reshaping School Mathematics* (MSEB, 1990). These documents described the goals for mathematics education as problem solving, mathematical power, access to technology, and constructivist learning (Huetinck, Munshin, & Murray-Ward, 1995). These documents were part of a national movement reconstructing how educators think about the teaching and learning of mathematics.

The NCTM *Standards* (1989) took a particular stand on equity as well:

- In developing the standards, we considered the content appropriate for *all* students.
- The mathematical content outlined in the standards is what we believe *all* students will need if they are to be productive citizens of the twenty-first century. (italics added)
- We believe that *all* students should have an opportunity to learn the important ideas of mathematics expressed in these standards. (italics added) (p. 9)

The statement made by these excerpts is that all students should have the opportunity to learn high-level mathematics and all students need to learn mathematics. There is an implicit counter-Discourse within this movement; namely that African-American, Latino, and poor children can and should learn mathematics. This national movement intended to bring about greater levels of mathematics achievement for all students. It was born out of a desire to increase the level of mathematics literacy of Americans and to help prepare more students for mathematics-dependent fields such as engineering and computer technology.

The 2000 NCTM *Principles and Standards* pushed notions of equity in mathematics education further by having one of six principles focus specifically on equity. The equity principle states,

> *All* students, regardless of their personal characteristics, backgrounds, or physical challenges, must have opportunities to study—and support to learn—mathematics.... Equity does not mean that every student should receive identical instruction.... All students need access each year to a coherent, challenging mathematics curriculum taught by competent and well-supported mathematics teachers.... Well-documented examples demonstrate that all children, including those who have been traditionally underserved, can learn mathematics when they have access to high-quality instructional programs that support their learning. (italics added, pp. 11–13)

The 2000 *Principles and Standards* show a marked increase in attention to equity in comparison to a decade earlier. Again, the idea embedded in the *Standards* is that everyone can learn mathematics, though this is spelled out in more specificity in the 2000 *Standards*. Both sets of standards, though there is a shift in attention, frame the goal of mathematics education as a problem "for all."

The mathematics education community has taken several steps in moving schools towards the goal of greater equity in the mathematics education of students. Such efforts include comprehensive reviews of the mathematics achievement progress of traditionally underserved students such as English language learners, girls, and African Americans (Lee, 2002; Reyes & Stanic, 1988; Schoenfeld, 2002; Secada, 1992; Tate, 1997) and research studies that have revealed gross inequities in the course-availability, course-placement, and learning opportunities of these underserved students (Oakes, 1990; Oakes, Quartz, Ryan, & Lipton, 2000; Paul, 2003). The field of mathematics education is deeply indebted to the work of these scholars-many of whom are outside of the field of mathematics education research. However, we want to argue for increased efforts—within the mathematics education community—that attend to issues around the structures of schools in general and to issues of race, racism and the racialized experiences of students of color that are enforced through these structures. We argue that the lack of attention to these issues may account for the lack of progress towards equity. We posit that one cannot overcome the inequity experienced by students of color and in schools that serve these students without addressing how and why those inequities came to be and are held in place. Race and the Discourses around race continue to bring inequities into being and hold them into place. Furthermore, power—the set of relationships and hierarchies that frame interactions—is often ignored in the efforts at reform in mathematics education. As with race, power both brings into being and holds into place the inequities that we presently see in schools. By focusing our attention on equity, we deal only with effects while ignoring the causes of the inequity that we see. The issue is not only that underserved children have access to far fewer rigorous mathematics courses, but that they attend schools where such disparities are not questioned or critiqued. The realities of race (the real ways in which students' skin color and the social significance assigned to that skin color) and power (implicit and explicit ways that larger structures, institutions, and normative ways of talking and thinking shape the access, opportunity, and experiences of individuals and groups) hold such inequities in place—making them accepted, acceptable, and normal.

Dismantling the inequities that these scholars bring to the fore such as underfunding certain schools (Kozol, 2005; Oakes, Rogers, Silver, Horng, & Goode, 2004; Rothstein, 2000), a shortage of certified mathematics teachers (Darling-Hammond & Sykes, 2003; Ingersoll, 1999), tracking (Oakes, 1985), cultural conflicts, and standardization (Rogers et al., 2005), we believe, is a matter of wrestling with and confronting Discourses, practices and the ideologies (such as racism) which hold these structures in place. We use this next section to raise issues and questions as to how the mathematics education community deals with equity, race, and systematic power relationships.

Moving from Rhetoric to Reality

A number of researchers have offered critiques of the NCTM standards and the "Mathematics for All" movement, asking whether the call to provide quality mathematics instruction *for all* is simply rhetoric (Apple, 1992/1999; Martin, 2003). Martin (2003) cited several events that have shaped and tested the mathematics education community's commitment to equity including specific calls for greater equity (Reyes & Stanic, 1988) and seminal documents such as the NCTM *Standards*. After nearly 2 decades of equity-minded reform, he argued there have been very few appreciable outcomes for African American, Latino, and Native American students (see, e.g., Lee, 2002; Schoenfeld, 2002; Tate, 1997). The color-blind discourse prevalent in NCTM's seminal documents (1989, 2000) may be a contributing factor in the mathematics education community's inability to make more headway towards equity.

The 1989 standards use equality rather than equity as a frame. Such a framing does not recognize that students have different needs and that the same instruction will not necessarily produce equitable results. The 2000 standards show a shift from equality to equity. The document notes specifically that there is no one-size-fits-all program for students. ("Equity does not mean that every student should receive identical instruction. . . . All students need access each year to a coherent, challenging mathematics curriculum taught by competent and well-supported mathematics teachers," p. 11.) However, this document still makes no mention of race or power and the ways in which these factors make access to mathematics education inequitable (Apple, 1992/1999). Instead this latter document frames inequity as an issue of personal characteristics, background, and instruction only ("All students, regardless of their personal characteristics, backgrounds, or physical challenges, must have opportunities to study—and support to learn—mathematics. . . . all children, including those who have been traditionally underserved, can learn mathematics when they have access to high-quality instructional programs that support their learning," pp. 11–13). Such a framing makes this document, and its consumers, vulnerable to ideologies of defi-

ciency (Cuban & Tyack, 1988; Hull, Rose, Fraser, & Castellano, 1991). Furthermore, this lack of attention to race, gender, and SES allows historical, social, and economic reasons for underachievement to be cast as individual deficiencies (Cuban & Tyack, 1988; Hull et al., 1991).

Martin (2003) has argued that the color-blind discourse (i.e., "for all") found in these policies and documents glosses over the complexities of race and power so present in schools and school systems. For instance, Martin asked who are the students currently not receiving quality mathematics instruction (i.e., who are the *all* spoken of in these documents)? Why are *all* of these students currently not receiving quality mathematics instruction? For whom do we traditionally not consider this content appropriate? What backgrounds keep students from access to high-quality mathematics? These questions bring texture and complexity to the current efforts at equity and force educators to confront the great effort that will be required to achieve it.

Grappling with the realities posed in the previous questions focuses attention on the racialization of mathematics education. More closely, these questions bring light to how the opportunities of students within mathematics are distributed differentially based upon their race (and the social significance assigned to their race). These opportunities are not only technical (e.g., course offerings), but normative such as the conceptualizations of, the talk about, and beliefs around non-White students, their families, and their communities. It is often these norms (that embed themselves in schooling structures, opportunities, and differential resources) that make the attainment of equity so problematic. Although the call for "Mathematics for All" is noble, if it does not address these normative ways of talking and thinking, it will not change the realities of education for students of color.

Next, we present two examples from our work in schools to raise structural and policy issues for mathematics educators to address and document in future research. We situate both cases within the contexts that they arise and use them to encourage different kinds of questions for mathematics educators to research.

Algebra for All[5]

In California, the call for "Mathematics for All" has played out specifically in policies that require all of its eighth-grade students to take a 1st-year Algebra course. The California mandate, "Algebra for All,"

---

[5] Details about this study can be found in Spencer (2006).

serves as an instantiation of the larger "Mathematics for All" movement, and we use this mandate to draw out questions that the mathematics education community needs to address in more depth to better understand issues of equity, race, and power in schools. The structures of schools are reified.

In recent years, Algebra has been declared a necessity for all students (Moses & Cobb, 2001a; Paul, 2003). Part of the outcry for greater access to Algebra is based upon the gate-keeping power of this course to college preparatory mathematics (Moses & Cobb, 2001a). Students who finish Algebra in middle school are positioned to take mathematics courses in high school that are necessary for 4-year college and university admissions. Research revealed that a large number of African American, Latino, and Native American students were not engaging in this course in middle school (Oakes, Joseph, & Muir, 2003). As a result, many in the mathematics education community supported and endorsed a middle-school requirement of Algebra. California took up this decree, requiring its districts to enroll all of its eighth graders in Algebra. On the surface, this decree appears quite egalitarian. Yet, the implementation of such a decree has had significant problems.

The same beliefs, conceptions, labels, and Discourses about urban students that kept them enrolled in non-Algebra courses before the mandate have kept them enrolled in inferior "Algebra" courses after the mandate. Mandates and decrees constantly get interpreted through existing lenses and Discourses within schooling practice. The more the field can understand the lenses, Discourses, and beliefs embedded in the structure of schooling, the more researchers can document and develop successful future reforms to address inequities. These Discourses, in the passing on of ideologies and beliefs, are a form of power acting on the educational opportunity of students of color. In the following section, we pose how using the lens of Discourse can help us research race and power in relation to the "Algebra for All" mandate.

Labels: Recurring Ideologies

Oakes et al. (2003) report on the implementation of the "Algebra for All" mandate in urban districts in California. The mandate, adopted in 2000, has led to all eighth-grade students being enrolled in "Algebra." Subsequent to the mandate, Paul (2003) documented the proliferation of Algebra and pseudo-Algebra courses. Courses were titled with names such as pre-Algebra, 2-year Algebra, 1-year double-dose Algebra, Math Essentials, and Honors Algebra. Typically, only one of these courses, for example Honors Algebra,

actually fulfilled the Algebra requirement that makes students eligible for the college-preparatory mathematics track in high school. Because only a small number of students are allowed to enroll in Honors Algebra, these iterations of Algebra courses work to recreate the inequity that they profess to correct.

This re-naming of courses serves as one example of how a potential change in structure gets reified because the people and their ideologies around learning mathematics have not changed. Despite the changes in course titles, the dispositions towards students changed little (Oakes et al., 2003). Prior to this decree, teachers' conceptions of "remedial" students focused their instruction on math facts because of perceived mathematical ability associated with lower track classes (Raudenbush, Rowan, & Cheong, 1993). After the decree, teachers conceptions remained the same, however, this time, they were referring to their "Algebra Essentials" students (Spencer, 2006). This Discourse about "remedial" students keeps teachers from engaging students in complex or nonprocedural work, and maintains talk about homework not being turned in and a general lack of parental involvement. Algebra students were given a host of intellectually stimulating, complex, and real-world mathematics problems to engage with (Raudenbush, Rowan, & Cheong, 1993). In opposition to the remedial students, the Discourse in reference to Algebra students constructs them as capable, hardworking, well-raised, and advanced. In addition to reinforcing ideological norms, labels reinforce structural inequities. The new titles further disenfranchised students- giving them the illusion that they were actually engaging in a college-preparatory mathematics course- without changing the actual content and goal of their courses (Oakes et al., 2003; Spencer, 2006). The labels have power in that they route students into or away from college-bound trajectories. Furthermore, these routes shape students' relationships to mathematics. Put another way, how students see themselves as thinkers and learners of mathematics.

Historically, the same rationales given for why particular students could not think, learn, or achieve 50 years ago are the same rationales presently given for why they cannot think, learn, or achieve (Cuban & Tyack, 1988; Hull et al., 1991; Spencer, 2006). Labels such as Algebra Essentials, lazy, low skilled and remedial serve as a proxy for African Americans and Latinos without dealing with racial implications. The labels signify underlying ideologies about students of color (Cuban & Tyack, 1988; Hull et al., 1991; Spencer, 2006). Not explicitly attending to race allows these ideologies to remain with superficial changes, no matter how well thought-out the policy.

This perpetuation of inequity and deficit views suggests that the work of mathematics education is to find ways to maintain concern with mathematics content, teaching, and learning, while attending to race and power to understand how the arrangement of mathematical opportunities inside and outside of school interact and contribute to current and historical inequities (Atweh, Forgasz, & Nebres, 2001; Gutstein, 2003; Martin, 2000; Oakes, 1990). In addition, mathematics educators need to develop meaningful interventions, inside and outside of school, to empower marginalized students with mathematics so that they can change the conditions that contribute to inequities (Abraham & Bibby, 1988; Anderson, 1990; Apple, 1992/1999; D'Ambrosio, 1990; de Arbreu, 1995; Frankenstein, 1990, 1994; Gutstein, 2002, 2003; Martin, 2000; Moses & Cobb, 2001a).

For example, instead of the state imposing a blanket eighth-grade Algebra mandate, what if the large urban district or a researcher posed the question: Why are more African American and Latino/a students not enrolled in college-preparatory mathematics courses? Educators could start to answer this question by interviewing African American and Latino/a students, asking them about their experiences in mathematics, the reasons for their course choices, and their future goals. Next, researchers may interview African American and Latino/a parents and ask about the goals that they have for their children. Finally, observing mathematics classrooms and talking with teachers about why they do not recommend more of their Black and Latino/a students for college-preparatory mathematics courses could inform the field about how to design interventions to change classroom practices to open opportunities for students of color. After this work, which is only the beginning, mathematics educators would have more knowledge to inform curricular, administrative, and policy decisions as to how mathematics instruction can better serve African American and Latino/a students and teachers of these students.

Such an approach demystifies the *all* and places a face on those students whom the district is currently not serving. It also gives credence to the lives and experiences of minoritized children and communities, and it forces schools and teachers to confront dynamics of power and race that exist within their school. This approach shares power with students, parents, and communities rather than subjecting them to the mandates of politicians and administrators. Instead, the current "Algebra for All" mandate reifies the stratified system of education existent in these schools. It misrepresents the reality to students and community members who believe that their children are engaged in high-quality Algebra coursework.

Standardized Labels: The Continued Process of Racialization

Many of the same issues previously discussed have manifested themselves in national and local movements towards greater standardization. Policies that establish accountability systems, high stakes standardized tests, and mandated curricula also work to reify labels and more importantly ideologies about children and communities of color.

The culture of standardization places students in groups based on judgments about their academic abilities. These judgments often fall into deficit modes of thinking and are solidified in the schooling institution through state-, district-, and school-sanctioned labels. In California, these labels evolved into terms like *far below basic, below basic, strategic,* or *intensive.* Grouping students by test performance sanctions these labels, making them more acceptable, and creates an institutional Discourse of deficit thinking tied to linking ability with individual standardized test scores (Perry et al., 2003; Rose, 1988). This deficit Discourse in schools, as before, is more often associated with students of color.

Labeling is one way of grouping by race while never making explicit that these labels construct how students of color are seen. Put another way, these labels become proxies for speaking about students of color without referring to race (Perry et al., 2003; Pollock, 2004; Rose, 1988). For example, whereas a teacher or administrator may be reticent to say that their African American boys have trouble sitting still in class, they might state that their "kinesthetic" students have trouble keeping still in class. Taken a step further, they might add that their kinesthetic students need "hands on" lessons to learn. *Kinesthetic,* unlike *African American,* is an ahistorical and seemingly neutral term. It does not carry with it centuries of oppression and denied rights. It therefore fits with a color-blind ideology (DeCuir & Dixson, 2004) and is an acceptable way of talking about students of color. Similar terms such as *Attention Deficit Disorder (ADD), hyperactive, remedial,* and more recently *far below basic* and *math essentials* operate in the same manner. Such labels avoid and cover up the complex and real issues of racialized experience that we as educators need to begin to grapple with. By not documenting, understanding, or deconstructing these Discourses we do not produce work that adequately counters long standing racial and power differentials and the ways that they shape the mathematical experiences of minoritized students.

Summary

Our aim in this section on race and power was to illustrate how not using power and race as central analytical tools in the work of mathematics education reproduces current inequitable practices. Although these cases are preliminary and the descriptions of them largely based on the experiences of the authors involved in this work, they help to illustrate how Discourses and schooling practices are taken for granted, normal, and neutral because they are part of the schooling institution. Better understanding and unpacking the social systems, policies, and narratives that structure classroom learning can help mathematics educators implement new reforms, navigate the political system, and develop policies that work for, rather than against, African Americans, Latinos, and the poor. This would require us to develop new understandings about how power employed through policy, states, districts, Discourses and Whiteness influences the learning that goes on in classrooms. It also means listening and taking up the concerns of those not empowered by the current system in researching mathematics education.

CONCLUSION

Over the last 10 years, there has been a growing concern within education and mathematics education that we need to examine what is happening for groups of students. Although there is recognition that teaching matters, research suggests that as we consider teaching, we cannot ignore cultural histories or essentialize them, and we cannot ignore that learning occurs within schools and communities that are shaped by cultural histories. Research is accumulating that demonstrates that the structures of schooling and society serve to both support and limit student opportunities. However, as we found throughout this chapter, we still know little about the details around how cultural histories and social structures shape and are shaped by communities and their histories. In mathematics education research, we know even less about these details.

The lack of available literature in mathematics education suggests that race, racism, and power remain undertheorized in the field. In other words, the literature does not sufficiently address how race interacts with the experiences of students of color or White students in mathematics education. The high value placed on mathematics education, the career opportunities open to those with strong mathematics backgrounds, and the qualifications for entrance into elite universities situate mathematics in a different way than many other fields. Therefore, it is possible that how racism and the normative ways of talking about students of color play out is different with re-

spect to mathematics from other disciplines. The ways that these normative Discourses that build themselves into the institution of mathematics education through course names such as Math Essentials and labels like *far below basic*, are one example of enacting power on students in systematic ways. The literature does little to make explicit how the schooling institution enacts power, in fairly implicit ways, on students of color. Hopefully, this chapter raises issues for mathematics education researchers to further explore how issues of race, racism, and power structure opportunities and experiences for all students so the field can find ways to counteract inequities.

Standardized tests serve as an example of how social structures and cultural histories shape access, opportunities, and experiences of groups of students. This review suggests that these tests often hold cultural, racial, and social-class biases, making them better assessors of students' cultural practices or their SES, revealing little about students' mathematical understanding. Moreover, standardized-test results have deep consequences for students of color, as the results are often used to inappropriately sort students. As long as differential achievement is associated predominantly with standardized-test scores, students of color will continually be harmed by an inequitable distribution of individual sanctions due to a limited understanding of achievement and neglect of the history of underserving them. Additionally, such sanctions place responsibility on the students for overcoming disparities in achievement, funding, and opportunity to learn that society-at-large and the institution of schools create and perpetuate.

The characterization of inequities in mathematics education synthesized throughout this chapter underscores the benefits of and necessity for (a) researching the concerns of those who have been disenfranchised, (b) questioning the privilege of the powerful, (c) recognizing classrooms as racialized spaces, even all-White classrooms, and (d) looking outside the field of mathematics education for theoretical perspectives and methodologies that can contribute to our understanding of culture, race, power, and mathematics teaching and learning.

Researching Culture, Race, and Power in Mathematics Education

The examination of race, culture, and power with respect to student achievement and learning in mathematics raises different questions for our current system of mathematics educators. Research questions are needed that can help guide studies of mathematics education in both untangling and challenging processes that perpetuate current inequity and injustice in mathematics education and in society writ large. A few researchers have already proposed research questions that attempt to address issues of culture, race, and power in mathematics education (see, e.g., R. Gutierrez, 2002; Gutstein, 2006; Martin, in press). These questions stem from the perspectives that these researchers hold about the relations of race, culture, and power in mathematics, the nature of mathematics teaching and learning, as well as the role of mathematics education in society. These researchers share a concern with analyzing what counts as mathematics learning, in whose eyes, and how these culturally bound distinctions afford and constrain opportunities for students of color to have access to mathematical trajectories in school and beyond. The questions offered by Martin, Gutstein, and Gutierrez share a common theme in observing that mathematics education, and assumptions within it about mathematics teaching and learning, have historical, cultural, and political underpinnings, that have privileged White, middle- and upper-class students over students of color.

We have also embedded questions within the chapter in an effort to reveal subtle power relations, racialized experiences, and implicit ideologies and practices that constrain the development of a more equitable system of mathematics education. This is also an attempt to challenge the field to move beyond studies of mathematics teaching and learning that reduce culture to race, race to non-White students, and that inadvertently strengthen narratives and Discourses that marginalize and oppress. In proposing new areas of research, we do not seek to minimize the work that has been done to date to bring about greater equity in mathematics education. We also recognize the difficulty of doing such research in a contentious social and political climate, and simultaneously with the development of new methodological tools and frameworks.

However, different kinds of questions often require different methodologies. Asking questions about systematic inequities leads to methodologies that allow the researcher to look at multiple levels simultaneously. This means that mathematics education research should take a multifaceted approach, aimed at multiple levels from the classroom to broader social structures, within a variety of contexts both in and out of school, and at a broad span of relationships including researcher to study participants, teachers to schools, schools to districts, and districts to national policy. It is important, then, that researchers understand that policies *do* play out as well as *the ways in which* they play out at the classroom, school, district, and state levels.

Methodologies that capture the relationships between individuals, groups, classrooms, schools, communities, and social structures are needed. In other words, research that avoids looking at these elements in isolation has much more potential for informing the field and policy decisions about culture, race, and power in relation to teaching and learning mathematics. For instance, Spencer (2006) provides a multi-level framework for studying the mathematics achievement of African American students. In this framework, Black student participation and identity, the mathematics content present in classrooms, teacher's dispositions towards and explanations about the participation of their African American students, schools' responses to the needs of African American students, and the historical context within which the school is seated each take a central role. The potential of such a framework is in its ability to address the complex and relational nature of a phenomenon such as "student achievement." For example, such an analysis could help answer how the historical significance assigned to a particular group of students impacts the relation of a school to that group as well as the relation of that group to the school. More central to our work, such an analysis may help us understand how a teacher's perceptions of students as doers of mathematics impacts how that group views themselves in the enterprise of learning mathematics.

Both qualitative and quantitative research have strengths and limitations. Work using narratives, ethnographies, and historical analyses allow research to speak to multiple levels of practice in order to see nuanced details. Although these methods are sometimes discredited, not counted as research, or not given the same respect as other forms, their multilevel nature situates them as particularly powerful in understanding the details of relationships discussed in this chapter. Standards of quality and thoroughness can be achieved just as in quantitative analyses. Similarly, quantitative methods such as multilevel modeling can uncover systematic issues of inequity. Although multilevel modeling such as Hierarchical Linear Modeling (HLM) is already respected as a form of research, when using such techniques we must be just as thoughtful and careful that we are actually measuring what we intend. The measures must be sensitive enough to allow for the subtle ways that culture, race, and power can influence teaching and learning in mathematics classrooms. When this is the case, multilevel modeling allows researchers to understand complex causal relationships that can uncover power dynamics within social structures that shape the experiences of groups. This quantitative work allows researchers to understand complex relationships on much broader scales than by using qualitative methods. We recognize that quality research of this sort is difficult and challenging to do. However, if we are to take equity seriously, we cannot avoid engaging with these issues in new ways.

In addition to different kinds of questions and methodologies, this work will push the field to develop new ways of understanding results. We attempt to provide some different frameworks for understanding the relationships between culture, race, and learning in this chapter. This is not an argument for a particular framework; rather, multiple lenses and theoretical perspectives will be needed to understand our work in relation to social structures and cultural and racial histories. If mathematics education researchers were all to challenge our assumptions about why we have the results we do, it might open up new insights into the complex relationships discussed in this chapter. New frameworks for understanding the interactions between culture, race, and power would shape how we discuss and understand the work of mathematics education.

The seemingly color-blind, neutral polices in mathematics education and in education in general, and discourses about students and student achievement, serve to privilege some students and cultural practices over others. Our discussions in this chapter reveal that not paying explicit attention to the ways race and power manifest themselves in such policies can lead to a masking of the racialization of mathematics education and of the damaging effects of recurring ideologies for students of color. These challenges considered, if the nation is to take seriously current inequities in math education, then it is unacceptable for the field of mathematics education to ignore something that so obviously defines the relationships and realities of many students in the United States.

REFERENCES

Abraham, J., & Bibby, N. (1988). Mathematics and society. *For the Learning of Mathematics, 8*(2), 2–11.

Allexsaht-Snider, M. & Hart, L.E. (2001). "Mathematics for All": How do we get there? *Theory into Practice, 40*(2), 93-101.

Anderson, S. (1990). Worldmath curriculum: Fighting Eurocentrism in mathematics. *Journal of Negro Education, 59*(3), 348–359.

Apple, M. W. (1992). Do the standards go far enough? Power, policy, and practice in mathematics education. *Journal for Research in Mathematics Education, 23*(5), 412–431.

Apple, M. (1999). Do the Standards go far enough? Power, policy, and practice in mathematics education. Reproduced in *Power, meaning, and identity: Essays in critical educational studies.* New York: Peter Lang (Original work published 1992).

Atweh, B., Forgasz, H., & Nebres, B. (2001). *Sociocultural research on mathematics education: An international perspective.* Mahwah, NJ: Erlbaum.

Baratz, S. S., & Baratz, J. C. (1970). Early childhood intervention: The social science base of institutional racism. *Harvard Education Review, 40*(1), 29–50.

Boaler, J. (1997). *Experiencing school mathematics: Teaching styles, sex, and setting.* Philadelphia: Open University Press.

Boaler, J. (1998). Open and closed mathematics: Student experiences and understandings. *Journal for Research in Mathematics Education, 29*(1), 41–62.

Boaler, J. (1999). Participation, knowledge and beliefs: A community perspective on mathematics learning. *Educational Studies in Mathematics, 40*(3), 259–281.

Boaler, J. (2002). Learning from teaching: Exploring the relationship between reform curriculum and equity. *Journal for Research in Mathematics Education, 33*(4), 239–258.

Boaler, J. (2003). When learning no longer matters: Standardized testing and the creation of inequality. *Phi Delta Kappan, 84*(7), 502–507.

Boaler, J. (2006). Urban success: A multidimensional mathematics approach with equitable outcomes. *Phi Delta Kappan, 87*(5), 364–369.

Boaler, J. (in press). *Promoting relational equity: The mixed ability mathematics approach that taught students high levels of responsibility, respect, and thought.* Unpublished manuscript, Stanford University, CA.

Boaler, J., & Greeno, J. G. (2000). Identity, agency, and knowing in mathematics worlds. In J. Boaler (Ed.), *Multiple perspectives on mathematics teaching and learning.* Westport, CT: Ablex Publishers.

Boaler, J., & Staples, M. (in press). Creating mathematical futures through an equitable teaching approach: The case of Railside school. *Teachers College Record.*

Bransford, J. D. E., Brown, A. L. E., & Cocking, R. R. E. (1999). *How people learn: Brain, mind, experience, and school.* Washington, DC: National Academy of Sciences. National Research Council.

Brice-Heath, S. (1982). Questioning at home and at school: A comparative study. In G. Spindler (Ed.), *Doing the ethnography of schooling: Educational anthropology in action* (pp. 102–131). Prospect Heights, IL: Waveland Press.

Carpenter, T. P., Franke, M. L., & Levi, L. (2003). *Thinking mathematically.* Portsmouth, NH: Heinemann.

Carraher, T. N., Carraher, D. W., & Schliemann, A. D. (1985). Mathematics in the streets and in schools. *British Journal of Developmental Psychology, 3,* 21–29.

Civil, M. (1994, April). *Connecting the home and the school: Funds of knowledge for mathematics teaching and learning.* Paper presented at the annual meeting of the American Educational Research Association, New Orleans, LA.

Civil, M. (1995a, April). *Bringing the mathematics to the foreground.* Paper presented at the annual meeting of the American Educational Research Association, San Francisco, CA.

Civil, M. (1995b, April). *Everyday mathematics, "mathematicians' mathematics," and school mathematics: Can we (and should we) bring these three cultures together?* Paper presented at the Annual Meeting of the American Educational Research Association, San Francisco, CA.

Civil, M. (1998, April). *Bridging in-school mathematics and out-of-school mathematics.* Paper presented at the American Educational Research Association Annual Meeting, San Diego, CA.

Civil, M. (2001). Mathematics instruction developed from a garden theme. *Teaching Children Mathematics, 7*(7), 400–405.

Cobb, P., & Hodge, L. L. (2002). A relational perspective on issues of cultural diversity and equity as they play out in the mathematics classroom. *Mathematical Thinking and Learning, 4*(2/3), 249–284.

Cobb, P., & Nasir, N. (2002). Diversity, equity, and mathematical learning. *Mathematical Thinking and Learning, 4*(2/3), 91–102.

Cobb, P., Stephan, M., McClain, K., & Gravemeijer, K. (2001). Participating in classroom mathematical practices. *The Journal of the Learning Sciences, 10*(1/2), 113–163.

Cohen, E. G., & Lotan, R. A. (Eds.). (1997). *Working for equity in heterogeneous classrooms: Sociological theory in practice.* New York: Teachers College Press.

Cuban, L., & Tyack D. (1988). "Dunces," "shirkers," and "forgotten children": Historical descriptions and cures for low achievers. Paper presented at the Conference for Accelerating the Learning of At-Risk Students. Stanford, CA.

D'Ambrosio, U. (1990). The role of mathematics in building a democratic and just society. *For the Learning of Mathematics, 10*(3), 20–23.

Darling-Hammond, L., & Sykes, G. (2003, September 17). Wanted: A national teacher supply policy for education: The right way to meet the "Highly Qualified Teacher" challenge. *Education Policy Analysis Archives, 11*(33). Retrieved April 10, 2006 from http://epaa.asu.edu/epaa/v11n33/.

de Abreu, B. (1995). Understanding how children experience the relationship between home and school mathematics. *Mind, Culture, and Activity, 2,* 119–142.

DeCuir, J. T., & Dixson, A. D. (2004). "So when it comes out, they aren't that surprised that it is there": Using critical race theory as a tool of analysis of race and racism in education. *Educational Researcher, 33*(5), 26–31.

Delgado, R., & Stefanic, J. (2001). *Critical race theory.* New York: New York University Press.

Delpit, L. D. (1988). The silenced dialogue: Power and pedagogy in educating other people's children. *Harvard Educational Review, 58*(3), 280–298.

Dewey, J. (1902). *The school and society: The child and the curriculum.* Chicago: The University of Chicago Press.

Dewey, J. (1938). *Experience and education.* New York: Simon & Schuster.

Diamondstone, J. (2002). Keeping resistance in view in an activity theory analysis. *Mind, Culture, and Activity, 9*(1), 2–21.

Engle, R. A., & Conant, F. R. (2002). Guiding principles for fostering productive disciplinary engagement: Explaining an emergent argument in a community of learners' classroom. *Cognition and Instruction, 20*(4), 399-483.

Esmonde, I. (2006). *"How are we supposed to, like, learn it, if none of us know?": Opportunities to learn and equity in mathematics cooperative learning structures.* Unpublished doctoral dissertation, University of California, Berkeley.

Frankenberg, E., Lee, C., & Orfield, G. (2003). *A multiracial society with segregated schools: Are we losing the dream?* Cambridge, MA: The Civil Rights Project, Harvard University.

Frankenstein, M. (1990). Incorporating race, gender, and class issues into a critical mathematical literacy curriculum. *Journal of Negro Education, 59*(3), 336–347.

Frankenstein, M. (1994). Understanding the politics of mathematical knowledge as an integral part of becoming critically numerate. *Radical Statistics, 56,* 22–40.

Frankenstein, M. (1995). Equity in mathematics education: Class in a world outside of class. In W. G. Secada, E.
Fennema, & L. B. Adajian (Eds.), *New directions for equity in mathematics education* (pp. 165–190). New York: Cambridge University Press.

Frankenstein, M. (1997). In addition to the mathematics: Including equity issues in the curriculum. In J. Trentacosta & M. Kenny (Eds.), *Multicultural and gender equity in the mathematics classroom* (pp. 10–22). Reston, VA: National Council of Teachers of Mathematics.

Freire, P. (1970/1993). *Pedagogy of the oppressed.* New York: The Continuum Publishing Company.

Gau, T. (2005). *Learning to teach mathematics for social justice.* Unpublished doctoral dissertation, University of Wisconsin, Madison.

Gay, J., & Cole, M. (1967). *The new mathematics and an old culture: A study of learning among the Kpelle of Liberia.* New York: Holt, Rinehart, and Winston.

Gee, J. (1990). *Social linguistics and literacies: Ideologies in Discourses.* New York: Falmer Press.

Gee, J., Hull, G., & Lankshear, C. (1996). *The new work order: Behind the language of the new capitalism.* Boulder, CO: Westview Press.

Gonzalez, N., Andrade, R., Civil, M., & Moll, L. (2001). Bridging funds of distributed knowledge: Creating zones of practices in mathematics. *Journal of Education for Students Placed at Risk (JESPAR), 6*(1–2), 115–132.

Gonzalez, N., Moll, L., & Amanti, C. (2005). *Funds of knowledge: Theorizing practices in households and classrooms.* Mahwah, NJ: Erlbaum.

Greeno, J. G. (2003). Situative research relevant to standards for school mathematics. In J. Kilpatrick, G. Martin, & D. Schifter (Eds.), *A research companion to principles and standards for school mathematics* (pp. 304–332). Reston, VA: National Council of Teachers of Mathematics.

Greeno, J. G. (2004). Theoretical and practical advances through research on learning. In J. L. Green, G. Camilli & P. B. Elmore (Eds.), *Handbook of complementary methods in education research* (3rd ed.). Washington, DC & Mahwah, NJ: American Educational Research Association and Erlbaum.

Greeno, J. G., & Gresalfi, M. (2006). Opportunities to learn in practice and identity. Unpublished manuscript.

Gresalfi, M. (2004). *Taking up opportunities to learn: Examining the construction of participatory mathematical identities in middle school students.* Unpublished doctoral dissertation, Stanford University, CA.

Gutiérrez, K. (2002). Studying cultural practices in urban learning communities. *Human Development, 45,* 312–321.

Gutierrez, K.D., Baquedano-Lopez, P., & Tajeda, C. (1999). Rethinking diversity: Hybridity and hybrid language practices in the third space. *Mind, Culture, and Activity, 6*(4), 286–303.

Gutierrez, K.D. & Rogoff, B. (2003). Cultural ways of learning: Individual traits or repertoires of practice. *Educational Researcher, 32*(5), 19–25.

Gutiérrez, R. (2002). Enabling the practice of mathematics teachers in context: Toward a new equity research agenda. *Mathematical Thinking and Learning, 4*(2/3), 145–187.

Gutstein, E. (1997). Culturally relevant mathematics teaching in a Mexican American context. *Journal for Research in Mathematics Education, 28*(6), 709–738.

Gutstein, E. (2002, April). *Roads to equity in mathematics education.* Paper presented at the annual meeting of the American Educational Research Association, New Orleans, LA.

Gutstein, E. (2003). Teaching and learning mathematics for social justice in an urban, Latino school. *Journal for Research in Mathematics Education, 34*(1), 37–73.

Gutstein, E. (2006). *Reading and writing the world with mathematics: Toward a pedagogy for social justice.* New York: Routledge.

Gutstein, R. (in press-a). "And that's just how it starts": Teaching mathematics and developing student agency.

Gutstein, E. (in press-b). "So one question leads to another": Using mathematics to develop a pedagogy of questioning. In N. S. Nasir & P. Cobb (Eds.), *Improving access to mathematics: Diversity and equity in the classroom.* New York: Teachers College Press.

Hand, V. (2003). *Reframing participation: Meaningful mathematical activity in diverse classrooms.* Unpublished doctoral dissertation, Stanford University, CA.

Hiebert, J., Carpenter, T. P., Fennema, B., Fuson, K. C., Wearne, D., & Murray, H. (1997). *Making sense: Teaching and learning mathematics with understanding.* Portmouth, NH: Heinemann.

Huetinck, L., Munshin, S., & Murray-Ward, M. (1995). Eight methods to evaluate support reform in the secondary-level mathematics classroom. *Evaluation Review, 19*(6), 646–662.

Hull, G., Rose, M., Fraser, K. L., & Castellano, M. (1991). Remediation as social construct: Perspectives from an analysis of classroom discourse. *College Composition and Communication, 42*(3), 299–329.

Hunter, R. C., & Donahoo, S. (2003). The nature of urban school politics after Brown: The need for new political knowledge, leadership, and organizational skills. *Education and Urban Society, 36*(1), 3–15.

Ingersoll, R.M. (1999). The problem of underqualified teachers in American secondary schools. *Educational Researcher, 28*(2), 26–37.

Jacobs, V. R. (2002). A longitudinal study of invention and understanding: Children's multidigit addition and subtraction. In J. Sowder & B. Schappelle (Eds.), *Lessons learned from research* (pp. 93–100). Reston, VA: National Council of Teachers of Mathematics.

Kana'iaupuni, S. M. (2005). Ka'aka–lai Ku–Kanaka: A call for strengths-based approaches from a native Hawaiian perspective. *Educational Researcher, 34*(5), 32–38.

Kilpatrick, J., Swafford, J., & Findell, B. (Eds.). (2001). *Adding it up: Helping children learn mathematics.* Washington, DC: National Academy Press.

Kirkland, E. (2002). Do the math. *Teacher Magazine, 13*(6), 14–17.

Kozol, J. (2005). *The shame of the nation: The restoration of apartheid schooling in America.* New York: Crown.

Ladson-Billings, G. (1994). *The dreamkeepers: Successful teachers of African American children.* San Francisco: Jossey-Bass.

Ladson-Billings, G. (1995). Toward a theory of culturally relevant pedagogy. *American Educational Research Journal, 32*(3), 465–491.

Ladson-Billings, G. (1997). It doesn't add up: African American students' mathematics achievement. *Journal for Research in Mathematics Education, 28*(6), 697–708.

Ladson-Billings, G. (2001). *Crossing over to Canaan: The journey of new teachers in diverse classrooms.* San Francisco: Jossey Bass.

Ladson-Billings, G. (2002, Summer). Teaching and cultural competence—What does it take to be a successful teacher in a diverse classroom? *Rethinking Schools Online.*

Ladson-Billings, G. (2006, April). *From the achievement gap to the education debt: Understanding achievement in US schools.*

Paper presented at the annual meeting of the American Educational Research Association, San Francisco. Retrieved June 16, 2006, from http://www.cmcgc.com/media/WMP/260407/49_010_files/Default.htm.

Lampert, M. (2001). *Teaching problems and the problems of teaching.* New Haven: Yale University Press.

Lave, J. (1988). *Cognition in practice: Mind, mathematics, and culture in everyday life.* New York: Cambridge University Press.

Lave, J., Murtaugh, M., & de la Rocha, O. (1984). The dialectic of arithmetic in grocery shopping. In B. Rogoff & J. Lave (Eds.), *Everyday cognition* (pp. 67–94). Cambridge, MA: Harvard University Press.

Lave, J., & Wenger, E. (1991). *Situated learning and legitimate peripheral participation.* Cambridge, MA: Cambridge University Press.

Leacock, E.B. (1971). Introduction. In E.B. Leacock (Ed.), *The culture of poverty: A critique* (pp. 9–37). New York: Simon & Schuster.

Lee, C. (2004). Why we need to re-think race and ethnicity in educational research. *Educational Researcher, 32*(5), 3–5.

Lee, J. (2002). Racial and ethnic achievement gap trends: Reversing the progress toward equity? *Educational Researcher, 31*(1), 3–12.

Lee, V. E., Burkham, D. T., Chow-Hoy, T., Smerdon, B. A., & Geverdt, D. (1998). High school curriculum structure: Effects on coursetaking and achievement in mathematics for high school graduates—An examination of data from the National Education Longitudinal Study of 1988 (Working Paper No. 98). Washington, DC: U.S. Department of Education National Center for Education Statistics.

Lipka, J. (1994). Culturally negotiated schooling: Toward a Yup'ik mathematics. *Journal of American Indian Studies, 33*(3), 1–12.

Lipka, J. (1998). *Transforming the culture of schools: Yup'ik Eskimo examples.* Mahwah, NJ: Erlbaum.

Lipka, J. (2002). *Schooling for self-determination: Research on the effects of including native language and culture in the schools.* ERIC Digest.

Lipka, J. (2005). Math in a cultural context: Two case studies of a successful culturally based math project. *Anthropology & Education Quarterly, 36*(4), 367–385.

Lipka, J., Wildfeuer, S., Wahlberg, N., George, M., & Ezran, D. R. (2001). Elastic geometry and storyknifing: A Yup'ik Eskimo example. *Teaching Children Mathematics, 7*(6), 337–343.

Lubienski, S. T. (2000). Problem solving as a means toward mathematics for all: An exploratory look through a class lens. *Journal for Research in Mathematics Education, 31*(4), 454–482.

Lubienski, S. T. (2002). Research, reform, and equity in U.S. mathematics education. *Mathematical Thinking and Learning, 4*(2/3), 103–125.

Martin, D. B. (2000). *Mathematics success and failure among African-American youth.* Mahwah, NJ: Erlbaum.

Martin, D. B. (2003). Hidden assumptions and unaddressed questions in Mathematics for All rhetoric. *The Mathematics Educator, 13*(2), 7–21.

Martin, D. B. (in press-a). Mathematics learning and participation as racialized forms of experience: African American parents speak on the struggle for mathematics literacy. *Mathematical Thinking and Learning.*

Martin, D. (in press-b). Mathematics learning and participation in African American context: The co-construction of identity in two intersecting realms of experience. In N. S.

Nasir & P. Cobb (Eds.), *Improving access to mathematics: Diversity and equity in the classroom*. New York: Teachers College Press.

Martin, D. (2006). *Researching race in mathematics education*. Manuscript submitted for publication.

Mathematical Sciences Education Board (MSEB) (1989). *Everybody Counts: A report to the nation on the future of mathematics education*. Washington, DC: National Academy Press.

Mathematics Sciences Education Board (MSEB) (1990). *Reshaping school mathematics: A philosophy and framework of curriculum*. Washington DC: National Academy Press.

Moll, L. (1992). Bilingual classroom studies and community analysis: Some recent trends. *Educational Researcher, 21*(2), 20–24.

Moll, L. C., Amanti, C., Neff, D., & Gonzalez, N. (1992). Funds of knowledge for teaching: A qualitative approach to connect homes and classrooms. *Theory into Practice, 31*(1), 132–141.

Moll, L., & Gonzalez, N. (2004). Engaging life: A funds-of-knowledge approach to multicultural education. In J. A. Banks & C. A. M. Banks (Eds.), *Handbook of research on multicultural education* (2nd ed., pp. 699–715). San Francisco: Jossey-Bass.

Moll, L., & Greenberg, J. (1990). Creating zones of possibilities: Combining social contexts for instruction. In L. C. Moll (Ed.), *Vygotsky and Education* (pp. 319–348). Cambridge, England: Cambridge University Press.

Moschkovich, J. (1999). Supporting the participation of English language learners in mathematical discussions. *For the Learning of Mathematics, 19*(1), 11–19.

Moschkovich, J. N. (2002a). A situated and sociocultural perspective on bilingual mathematics learners. *Mathematical Thinking and Learning, 4*(2/3), 189–212.

Moschkovich, J. N. (2002b). An introduction to examining everyday and academic mathematical practices. In M. E. Brenner & J. N. Moschkovich (Eds.), *Everyday and academic mathematics in the classroom* (pp. 1–11). Reston, VA: National Council of Teachers of Mathematics.

Moses, R. P., & Cobb, C.E. (2001a). *Radical equations: Civil rights from Mississippi to the Algebra Project*. Boston: Beacon Press.

Moses, R. & Cobb, C. (2001b). Organizing algebra: The need to voice a demand. *Social Policy,* Summer, 4–12.

Moses, R., Kamii, M., Swap, S. M., & Howard, J. (1989). The Algebra Project: Organizing in the spirit of Ella. *Harvard Educational Review, 59*(4), 423–443.

Nasir, N. S. (2002). Identity, goals, and learning: Mathematics in cultural practice. *Mathematical Thinking and Learning, 4*(2/3), 213–248.

Nasir, N. S. (2004). Halal-ing the child: Deconstructing identities of resistance in an urban Muslim school. *Harvard Educational Review,* summer, 153–174.

Nasir, N. S., & Hand, V. (2006). *From the court to the classroom: Managing identities as learners in basketball and classroom mathematics*. Manuscript submitted for publication.

National Commission on Excellence in Education (1983). *A nation at risk: The imperative for educational reform*. Washington, DC: United States Department of Education.

National Council of Teachers of Mathematics. (1989). *Curriculum and evaluation standards for school mathematics*. Reston, VA: Author.

National Council of Teachers of Mathematics. (2000). *Principles and standards for school mathematics*. Reston, VA: Author.

Nieto, S. (2004). *Affirming diversity: The sociopolitical context of multicultural education*. Boston: Pearson.

Nunes, T., Schliemann, A. D., & Carraher, D. W. (1993). *Street mathematics and school mathematics*. Cambridge, MA: Cambridge University Press.

Oakes, J. (1985). *Keeping track: How schools structure inequality*. New Haven, CT. Yale Press.

Oakes, J. (1990). *Multiplying inequalities: The effects of race, social class, and tracking on opportunities to learn mathematics and science*. Santa Monica, CA: RAND.

Oakes, J., Muir, K., & Joseph, R. (2000, July). *Coursetaking and achievement in mathematics and science: Inequalities that endure and change*. [Paper prepared for the National Institute of Science Education].

Oakes, J., Joseph, R. & Muir, K. (2003). Access and achievement in mathematics and science: Inequalities that endure and change. In J. A. Banks & C. A. Banks, (Eds.), *Handbook of research on multicultural education* (2nd ed., pp. 69–90). San Francisco: Jossey-Bass.

Oakes, J. & Lipton, M. (1999). *Teaching to change the world*. Boston, MA: McGraw-Hill.

Oakes, J., Quartz, K., Ryan, S., & Lipton, M. (2000). *Becoming good American schools: The struggle for civic virtue in education reform*. San Francisco: Jossey-Bass.

Oakes, J., Rogers, J., Silver, D., Horng, E., & Goode, J. (2004). *Separate and unequal 50 years after* Brown: *California's racial "opportunity gap."* UCLA/IDEA publication series. Retrieved May 6, 2004, from http://www.idea.gseis.ucla.edu/publications/index.html.

Oakes, J. & Saunders, M. (2002). *Access to textbooks, instructional materials, equipment, and technology: Inadequacy and inequality in California's public schools*. [Expert Report for *Williams v. State of California*]. Retrieved May 6, 2004, from http://www.decentschools.org/experts

Orfield, G., Frankenberg, E. D., & Lee, C. (2003). The resurgence of school segregation. *Educational Leadership, 60*(4), 16–20.

Paul, F. G. (2003, October). *Re-tracking within algebra one: A structural sieve with powerful effects for low-income, minority, and immigrant students*. Paper presented at the Harvard University of California Policy Conference. Sacramento, CA.

Perry, T., Steele, C., & Hilliard, A., (2003). *Young, gifted, and Black: Promoting high achievement among African American students*. Boston: Beacon Press.

Pollock, M. (2004). *Colormute: Race talk dilemmas in an American school*. Princeton, NJ: Princeton University Press.

Reyes, L. H., & Stanic, G. (1988). Race, sex, socioeconomic status, and mathematics. *Journal for Research in Mathematics Education, 19*(1), 26–43.

Rogers, J., Jellison-Holmes, J., & Silver, D. (2005). *More questions than answers: CAHSEE results, opportunities to learn, and the class of 2006*. UCLA/ IDEA publication series. Retrieved August 22, 2005, from http://www.idea.gseis.ucla.edu/publications/index.html.

Rogoff, B. (2003). *The cultural nature of human development*. Oxford, UK: Oxford University Press.

Rose, M. (1988). Narrowing the mind and page: Remedial writers and cognitive reductionism. *College Composition and Communication, 39*(3), 267–302.

Rothstein, R. (2000, January 5). Closing the gap in state school spending. *The New York Times*.

Ryan, W. (1971). *Blaming the victim*. New York: Vintage Books.

Saxe, G. B. (1988a). Candy selling and math learning. *Educational Researcher, 17*(6), 14–21.

Saxe, G. B. (1988b). The mathematics of child street vendors. *Child Development, 59*(5), 1415–1425.

Saxe, G. B. (1991). *Culture and cognitive development: Studies in mathematical understanding.* Hillsdale, NJ: Erlbaum

Schoenfeld, A. H. (2002). Making mathematics work for all children: Issues of standards, testing, and equity. *Educational Researcher, 31*(1), 13–25.

Scribner, S. (1984). Studying working intelligence. In B. Rogoff & J. Lave (Eds.), *Everyday cognition: Its development in social context* (pp. 9–40). Cambridge, MA: Harvard University Press.

Secada, W. G. (1989). Agenda setting, enlightened self-interest, and equity in mathematics education. *Peabody Journal of Education, 66,* 22–56.

Secada, W. G. (1992). Race, ethnicity, social class, language, and achievement in mathematics. In D. Grouws (Ed.), *Handbook of research on mathematics teaching and learning* (pp. 623–660). Reston, VA: National Council of Teachers of Mathematics.

Sfard, A. (1998). On two metaphors for learning and the dangers of choosing just one. *Educational Researcher, 27*(2), 4–13.

Sfard, A., & Prusak, A. (2005). Telling identities: In search of an analytic tool for investigating learning as a culturally shaped activity. *Educational Researcher, 34*(4), 14–22.

Silva, C. M., Moses, R. P., Rivers, J., & Johnson, P. (1990). The Algebra Project: Making middle school mathematics count. *Journal of Negro Education, 59*(3), 375–391.

Skovsmose, O. (1994). *Towards a philosophy of critical mathematics education.* Dordrecht, The Netherlands: Kluwer.

Spencer, J. (2006). *Balancing the equations: African American students' opportunity to learn mathematics with understanding in two central city schools.* Unpublished doctoral dissertation, University of California, Los Angeles.

Staples, M., & Hand, V. (2004, April). *Co-constructing contributions: Effectively managing the social and intellectual aspects of secondary mathematics classrooms.* Paper presented at the American Educational Research Association, Montreal, Canada.

Tate, W. F. (1994). Race, retrenchment, and the reform of school mathematics. *Phi Delta Kappan, 75,* 447–485.

Tate, W. F. (1995). Returning to the root: A culturally relevant approach to mathematics pedagogy. *Theory into Practice, 34,* 166–173.

Tate, W. F. (1997). Race-ethnicity, SES, gender, and language proficiency trends in mathematics achievement: An update. *Journal for Research in Mathematics Education, 28*(6), 652–679.

Taylor, E. V. (2004, April). *Engagement in currency exchange as support for multi-unit understanding in African-American children.* Paper presented at the annula meeting of the American Educational Research Association, San Diego, CA.

Vandegrift, J. A. & Dickey, L. (1993). *Improving mathematics and science education in Arizona: Recommendations for the Eisenhower Higher Education Program.* Morrison Institute for Public Policy, School of Public Affairs, Arizona State University, Tempe, AZ.

Warren, M. (2005). Communities and schools: A new view of urban education reform, *Harvard Educational Review, 75* (2), 133–139.

Warren, B., Rosebery, A., & Conant, F. (1994). Discourse and social practice: Learning science in language minority classrooms. In D. Spener (Ed.), *Adult biliteracy in the United States* (pp. 191–210). McHenry, IL and Washington, DC: Delta Systems and Center for Applied Linguistics.

Webster, J. P., Wiles, P., Civil, M., & Clark, S. (2005). Finding a good fit: Using MCC in a "Third Space." *Journal of American Indian Education, 44*(3), 9–30.

Wenger, E. (1998). *Communities of practice: Leaning, meaning, and identity.* Cambridge, MA: Cambridge University Press.

Wortham, S. (2006). *Learning identity: The joint emergence of social identification and academic learning.* New York: Cambridge University Press.

AUTHOR NOTE

The material in this chapter is based in part on work supported by the National Science Foundation under Grant No. ESI-0119732 to the Diversity in Mathematics Education Center for Learning and Teaching (DiME). Any opinions, findings, and conclusions or recommendations expressed in this material are those of the author(s) and do not necessarily reflect the position, policy, or endorsement of the National Science Foundation.

We would like to thank the following individuals for their thoughtful comments and suggestions on earlier drafts of this chapter: Patricia Campbell, Eric Gutstein, Gloria Ladson-Billings, Danny Martin, Na'ilah Nasir, and William Tate.

DiME is based at University of Wisconsin-Madison, University of California-Los Angeles, and University of California-Berkeley and is devoted to furthering the study of diversity and equity in mathematics education. This chapter was written by a team consisting of Vanessa Pitts Bannister, Tonya Gau Bartell, Dan Battey, Victoria M. Hand, and Joi Spencer with contributions from Filiberto Barajas, Rozy Brar, Kyndall Brown, Indigo Esmonde, Mary Q. Foote, Charles Hammond, Carolee Koehn, Mara G. Landers, Mariana Levin, Shi-uli Mukhopadhyay, Ann Ryu, Marian Slaughter, and Anita A. Wager. DiME directors Thomas Carpenter, Megan Franke, and Alan Schoenfeld provided oversight and guidance for the writing of the chapter.

11

THE ROLE OF CULTURE IN TEACHING AND LEARNING MATHEMATICS

Norma Presmeg

ILLINOIS STATE UNIVERSITY

Mathematics education has experienced a major revolution in perceptions (cf. Kuhn, 1970) comparable to the Copernican revolution that no longer placed the earth at the center of the universe. This change has implicated beliefs about *the role of culture* in the historical development of mathematics (Eves, 1990), in the practices of mathematicians (Civil, 2002; Sfard, 1997), in its political aspects (Powell & Frankenstein, 1997), and hence necessarily in its teaching and learning (Bishop, 1988a; Bishop & Abreu, 1991; Bishop & Pompeu, 1991; Nickson, 1992). The change has also influenced methodologies that are used in mathematics education research (Pinxten, 1994). Researchers now increasingly concede that mathematics, long considered value- and culture-free, is indeed a cultural product, and hence that the role of culture—with all its complexities and contestations—is an important aspect of mathematics education.

Diverse aspects are implicated in a cultural formulation of mathematics teaching and learning. Other chapters in this volume focus more centrally on some of these aspects, for instance the following:

- the role of language and communication in mathematics education;
- equity, diversity, and learning in multicultural mathematics classrooms.

These topics will thus not be a focus in this chapter, although some issues from these fields will enter spontaneously. Topics that are central in this chapter are those arising from and extending the notions of ethnomathematics and everyday cognition (Nunes, 1992, 1993). Various broad theoretical fields are relevant in addressing these topics. Some of the theoretical notions that are apposite are rooted in—but not confined to—situated cognition (Kirshner & Whitson, 1997; Lave & Wenger, 1991; Watson, 1998), cultural models (Holland & Quinn, 1987), notions of cultural capital (Bourdieu, 1995), didactical phenomenology (Freudenthal, 1973, 1983), and Peircean semiotics (Peirce, 1992, 1998). From this partial list, the breadth of this developing field may be recognized. This chapter does not attempt to treat the general theories in detail: The interested reader is referred to the original authors. Instead, from these fields this chapter highlights some key notions that have explanatory power or usefulness in the central focus, which is *the role of culture in learning and teaching mathematics.* The seminal work of Ascher (1991, 2002), Bishop (1988a, 1988b), D'Ambrosio (1985, 1990) and Gerdes (1986, 1988a, 1988b, 1998) on ethnomathematics is still centrally relevant and thus is treated in some detail in later sections.

In the last decade, the field of research into the role of culture in mathematics education has evolved from "ethnomathematics and everyday cognition" (Nunes, 1992), although both ethnomathematics and everyday cognition are still important topics of investigation. The developments have rather consisted in a broadening of the field, clarification and evolution of definitions, recognition of the complexity of the constructs and issues, and inclusion of social, critical, and political dimensions as well as those from cultural psychology involving valorization, identity, and agency (Abreu, Bishop, & Presmeg, 2002). This chapter in its scope cannot do full justice to political and critical views of mathematics education (see Mellin-Olsen, 1987; Skovsmose, 1994; Vithal, 2003, for a full treatment), but some of the *landscapes* from these fields—as Vithal calls them—are used in this chapter to deepen and problematize aspects of the treatment of culture in mathematics education.

This chapter has four sections. The first addresses an introduction to issues and definitions of key notions involving culture in mathematics education. The organization of the second and third sections uses the research framework of Brown and Dowling (1998). In this framework, resonating theoretical and empirical fields surround and enclose the central research topic, and the description involves layers of increasing specificity as it zooms in on details of the problematic and problems of the research issues, the empirical settings, and the results of the studies, only to zoom out again at the end in order to survey the issues in a broader field, informed by the results of the research studies examined, in order to see where further research on culture might be headed. Thus section 2 addresses theoretical fields that incorporate culture specifically in mathematics education; section 3 addresses salient empirical fields, settings, and some details of results of research on culture in mathematics education, and their implications for the teaching and learning of mathematics. Issues relating to technology and the culture of mathematics and its teaching and learning are included in this section. Using a broader perspective, the fourth (final) section collects and elaborates suggestions for possible directions for future research on culture in mathematics education.

DEFINITIONS AND SIGNIFICANCE OF CULTURE IN MATHEMATICS EDUCATION

Why is it important to address definitions carefully in considering the role of culture in teaching and learning mathematics? As in other areas of reported research in mathematics education, different authors use terminology in different ways. Particularly notions such as *culture, ethnomathematics,* and *everyday mathematics* have been controversial; they have been contested and given varied and sometimes competing interpretations in the literature. Thus these constructs must be problematized, not only for the sake of clarification, but more importantly for the roles they have played, and for their further potential to be major

focal points in mathematics education research and practices. Further, especially in attempts to bridge the gap between the formal mathematics taught in classrooms and that used out-of-school in various cultural practices, *what counts as mathematics* assumes central importance (Civil, 1995, 2002; Presmeg, 1998a). Thus ontological aspects of the nature of mathematics itself must also be addressed.

Culture

In 1988, when Bishop published his book, *Mathematical Enculturation: A Cultural Perspective on Mathematics Education* (1988a) and an article that summarized these ideas in *Educational Studies in Mathematics* (1988b, now reprinted as a classic in Carpenter, Dossey, & Koehler, 2004), the prevailing view of mathematics was that it was the one subject in the school curriculum that was value- and culture-free, notwithstanding a few research studies that suggested the contrary (e.g., Gay & Cole, 1967; Zaslavsky, 1973). Along with those of a few other authors (notably, D'Ambrosio), Bishop's ideas have been seminal in the recognition that culture plays a pivotal role in the teaching and learning of mathematics, and his insights are introduced repeatedly in this chapter.

Whole books have been written about definitions of *culture* (Kroeber & Kluckhohn, 1952). Grappling with the ubiquity yet elusiveness of culture, Lerman (1994) confronted the need for a definition but could not find one that was entirely satisfactory. Yet, as he pointed out, culture is "ordinary. It is something that we all possess and that possesses us" (p. 1). Bishop favored the following definition of culture in analyzing its role in mathematics education:

> Culture consists of a complex of shared understandings which serves as a medium through which individual human minds interact in communication with one another. (1988a, p. 5, citing Stenhouse, 1967, p. 16)

This definition highlights the communicative function of culture that is particularly relevant in teaching and learning. However, it does not focus on the continuous renewal of culture, the dynamic aspect that results in cultural change over time. This dynamic aspect caused Taylor (1996) to choose the "potentially transformative view" (p. 151) of cultural anthropologist Clifford Geertz (1973), for whom culture consists of "webs of significance" (p. 5) that we ourselves have spun.

This potentially transformative view assumes particular importance in the light of the necessity for negotiating social norms and sociomathematical norms in mathematics classrooms (Cobb & Yackel, 1995).

The culture of the mathematics classroom, which was brought to our attention as being significant (Nickson, 1992), is not monolithic or static but continuously evolving, and different in different classrooms as these norms become negotiated. The mathematics classroom itself is one arena in which culture is contested, negotiated, and manifested (Vithal, 2003), but there are various levels of scale. Bishop's (1988a) view of mathematics education as a social process resonates with a transformative, dynamic notion of culture. He suggested that five significant levels of scale are involved in the social aspects of mathematics education. These are the cultural, the societal, the institutional, the pedagogical, and the individual aspects (1988a, p. 14). Culture here is viewed as an all-encompassing umbrella construct that enters into all the activities of humans in their communicative and social enterprises. In addition to a view of culture in this macroscopic aspect, as in these levels of scale, culture as webs of significance may be central also in the societal, institutional, and pedagogical aspects of mathematics education considered as a social process. Thus researchers may speak of the culture of a society, of a school, or of an actual classroom. Culture in all of these levels of scale impinges on the mathematical learning of individual students.

Notwithstanding these general definitions of culture, the word with its various characterizations does not have meaning in itself. Vithal and Skovsmose (1997) illustrated this point starkly by pointing out that interpretations of *culture* (and by implication also anthropology) were used in South Africa to justify the practices of apartheid. They extended the negative connotations still attached to this word in that context to suggest that ethnomathematics is also suspect (as suggested in the title of their article: "The End of Innocence: A Critique of 'Ethnomathematics'"). Some aspects of their critique are taken up in later sections.

How are *culture* and *society* interrelated? The words are not interchangeable (Lerman, 1994), although connections exist between these constructs:

> One would perhaps think of gender stereotypes as cultural, but of 'gender' as socially constructed. One would talk of the culture of the community of mathematicians, treating it as monolithic for a moment, but one would also talk, for example, of the social outcomes of being a member of that group. (p. 2)

A. J. Bishop stated succinctly, on several occasions (personal communication, e.g., July, 1985), that *society involves various groups of people, and culture is the glue that binds them together.* This informal characterization resonates with Stenhouse's definition of *culture* as a complex of shared understandings, and also with that of Geertz, as webs of significance that we ourselves have spun. Considering the cultures of mathematics classrooms, Nickson (1994) wrote of the "invisible and apparently shared meanings that teachers and pupils bring to the mathematics classroom and that govern their interaction in it" (p. 8). Values, beliefs, and meanings are implicated in these "shared invisibles" (p. 18) in the classroom. Nickson saw socialization as a universal process, and culture as the *content* of the socialization process, which differs from one society to another, and indeed, from one classroom to another.

Part of culture as webs of significance, taken at various levels, are the prevailing notions of what counts as mathematics.

Mathematics

As Nickson (1994) pointed out, "one of the major shifts in thinking in relation to the teaching and learning of mathematics in recent years has been with respect to the adoption of differing views of the nature of mathematics as a discipline" (p. 10). Nickson characterized this cultural shift as moving from a *formalist* tradition in which mathematics is absolute—consisting of "immutable truths and unquestionable certainty" (p. 11) without a human face, to one of *growth and change*, under persuasive influences such as Lakatos's (1976) argument that "objective knowledge" is subject to proofs and refutations and thus that mathematical knowledge has a strong social component. That this shift is complex and that both views of mathematics are held simultaneously by many mathematicians was argued by Davis and Hersh (1981). The formalist and socially mediated views of mathematics resonate with the two categories of absolutist and fallibilist conceptions discussed by Ernest (1991). Contributing the notion of mathematics as problem solving, the Platonist, the problem-solving, and the fallibilist conceptions are categories reminiscent of the teachers' conceptions of mathematics that Alba Thompson, already in 1984, gave evidence were related to instructional practices in mathematics classrooms.

Both Civil (1990, 1995, 2002; Civil & Andrade, 2002) in her "Funds of Knowledge" project and in her later research with colleagues into ways of linking home and school mathematical practices, and Presmeg (1998a, 1998b, 2002b) in her use of ethnomathematics in teacher education and research into semiotic chaining as a means of building bridges between cultural practices and the teaching and learning of mathematics in school, described the necessity of broadening conceptions of the nature of mathematics in these endeavors. Without such broadened definitions, high-school and university students alike are naturally inclined to characterize mathematics according to what they have experienced in learning *institutional* mathematics—more often than not as "a bunch of numbers" (Presmeg, 2002b). On the one hand, such limited views of the nature of mathematics inhibit the recognition of mathematical ideas in out-of-school practices. On the other hand, if definitions of what counts as mathematics are too broad, then the "everything is mathematics" notion may trivialize mathematics itself, rendering the definition useless. In examining the mathematical practices of a group of carpenters in Cape Town, South Africa, Millroy (1992) expressed this tension well, as follows:

> [It] became clear to me that in order to proceed with the exploration of the mathematics of an unfamiliar culture, I would have to navigate a passage between two dangerous areas. The foundering point on the left represents the overwhelming notion that 'everything is mathematics' (like being swept away by a tidal wave!) while the foundering point on the right represents the constricting notion that 'formal academic mathematics is the only valid representation of people's mathematical ideas' (like being stranded on a desert island!). Part of the way in which to ensure a safe passage seemed to be to openly acknowledge that when I examined the mathematizing engaged in by the carpenters there would be examples of mathematical ideas and practices that I would recognize and that I would be able to describe in terms of the vocabulary of conventional Western mathematics. However, it was likely that there would also be mathematics that I could not recognize and for which I would have no familiar descriptive words. (pp. 11–13)

Some definitions of mathematics that achieve a balance between these two extremes are as follows. *Mathematics* is "the language and science of patterns" (Steen, 1990, p. iii). Steen's definition has been taken up widely in reform literature in the USA (National Council of Teachers of Mathematics [NCTM], 1989, 2000). Opening the gate to recognizing a human origin of mathematics, Saunders MacLane called mathematics "the study of formal abstract structures arising from human experience" (as cited in Lakoff, 1987, p. 361). According to Ada Lovelace, mathematics is the systematization of relationships (as described by Noss, 1997). All of these definitions strike some sort of balance between the human face of mathematics and its formal aspects. Going beyond Steen's well-known pattern definition in the direction of stressing abstraction, in a critique of ethnomathematics, Thomas (1996) defined mathematics as "the science of detachable relational insights" (p. 17). He suggested a useful distinction between *real mathematics* (as

characterized in his definition), and *proto-mathematics* (the category in which he placed ethnomathematics). In the next section, Barton's (1996) characterization of ethnomathematics, which resolves many of these issues and clarifies this dualism, is presented along with some evolving definitions of ethnomathematics. (For details, the reader should consult Barton's original article.)

Ethnomathematics

What is *ethnomathematics*? In his illuminating article, Barton (1996) wrote as follows:

> In the last decade, there has been a growing literature dealing with the relationship between culture and mathematics, and describing examples of mathematics in cultural contexts. What is not so well-recognised is the level at which contradictions exist within this literature: contradictions about the meaning of the term 'ethnomathematics' in particular, and also about its relationship to mathematics as an international discipline. (p. 201)

Barton pointed out that difficulties in defining ethnomathematics relate to three categories: epistemological confusion, "problems with the meanings of words used to explain ideas about culture and mathematics" (p. 201); philosophical confusion, the extent to which mathematics is regarded as universal; and confusion about the nature of mathematics. The nature of mathematics is part of its ontology, and because both ontology and epistemology are branches of philosophy, all of these categories may be regarded as philosophical difficulties. The strength of Barton's resolution of the difficulties lies in his creation of a preliminary framework (he admitted that it might need revision) whereby the differing views can be seen in relation to each other. His triadic framework is an "Intentional Map" (p. 204) with the three broad headings of *mathematics*, *mathematics education*, and *society* (cf. the whole day of sessions dedicated to these broad areas at the 6th International Congress on Mathematical Education held in Budapest, Hungary, in 1988). The seminal writers whose definitions of ethnomathematics he considered in detail and placed in relation to this framework were Ubiratan D'Ambrosio in Brazil, Paulus Gerdes in Mozambique, and Marcia Ascher in the USA.

As Barton (1996) pointed out, D'Ambrosio's prolific writings on the subject of ethnomathematics have influenced the majority of writers in this area. Thus on the Intentional Map, although D'Ambrosio's work (starting with his 1984 publication) falls predominantly in the socio-anthropological dimension between *society* and *mathematics*, some aspects of his concerns can be found in all of the dimensions. In his later work, he increasingly used his model to analyze "the way in which mathematical knowledge is colonized and how it rationalizes social divisions within societies and between societies" (Barton, 1996, p. 205). In his early writing, D'Ambrosio (1984) defined ethnomathematics as the way different cultural groups mathematize—count, measure, relate, classify, and infer. His definition evolved over the years, to include a changing form of knowledge manifest in practices that change over time. In 1985, he defined ethnomathematics as "the mathematics which is practiced among identifiable cultural groups" (p. 18). Later, in 1987, his definition of ethnomathematics was "the codification which allows a cultural group to describe, manage, and understand reality" (Barton, 1996, p. 207).

D'Ambrosio's (1991) well-known etymological definition of ethnomathematics is given in full in the following passage.

> The main ideas focus on the concept of *ethnomathematics* in the sense that follows. Let me clarify at the beginning that this term comes from an etymological abuse. I use *mathema(ta)* as the action of explaining and understanding in order to transcend and of managing and coping with reality in order to survive. Man has developed throughout each one's own life history and throughout the history of mankind *technê's* (or *tics*) of *mathema* in very different and diversified cultural environments, i.e., in the diverse *ethno's*. So, in order to satisfy the drive towards survival and transcendence in diverse cultural environments, man has developed and continuously develops, in every new experience, **ethno-mathema-tics.** These are *communicated* vertically and horizontally in time, respectively throughout history and through conviviality and education, relying on memory and on sharing experiences and knowledges. For the reasons of being more or less effective, more or less powerful and sometimes even for political reasons, some of these different *tics* have lasted and spread (ex.: counting, measuring) while others have disappeared or been confined to restricted groups. This synthesizes my approach to the history of ideas. (p. 3)

As in some of his other writings (1985, 1987, 1990), D'Ambrosio is in this definition characterizing ethnomathematics as a dynamic, evolving system of knowledge—the "process of knowledge-making" (Barton, 1996, p. 208), as well as a research program that encompasses the history of mathematics.

Returning to Barton's Intentional Map, the work of Paulus Gerdes is "practical, and politically explicit," concentrated in the *mathematics education* area of the Map (Barton, 1996, p. 205). Gerdes's definition of ethnomathematics evolved from the mathematics implic-

it or "frozen" in the cultural practices of Southern Africa (1986), to that of a mathematical movement that involves research and anthropological reconstruction (1994). The work of mathematician Marcia Ascher (1991, 1995, 2002), while overlapping with that of Gerdes to some extent, falls closer to the *mathematics* area on the Map, concerned as it is with cultural mathematics. Her definition is that ethnomathematics is "the study and presentation of the mathematical ideas of traditional peoples" (1991, p. 188). When Ascher (1991) worked out the kinship relations of the Warlpiri, say, in mathematical terms, she acknowledged that she was using her familiar "Western" mathematics. In that sense her ethnomathematics is subjective: The Warlpiri would be unlikely to view their kinship system through her lenses. Referring to mathematics and ethnomathematics, she stated, "They are both important, but they are different. And they are linked" (Ascher & D'Ambrosio, 1994, p. 38). In this view, there is no need to view ethnomathematics as "proto-mathematics" (Thomas, 1996), because it exists in its own right.

Finally, Barton (1996) found a useful metaphor to sum up the similarities and differences between the views of ethnomathematics held by these three proponents: "For D'Ambrosio it is a window on knowledge itself; for Gerdes it is a cultural window on mathematics; and for Ascher it is the mathematical window on other cultures" (p. 213). These three windows are distinguished by the standpoint of the viewer, and by what is being viewed, in each case. Although not eliminating the duality of ethnomathematics as opposed to mathematics (of mathematicians), these three distinct windows represent approaches each of which has something to offer. Taken together, they contribute a broadened lens on the role of culture in teaching and learning mathematics.

Several other writers in the field of ethnomathematics have acknowledged the need and attempted to define ethnomathematics. Scott (1985) regarded ethnomathematics as lying at the confluence of mathematics and cultural anthropology, "mathematics in the environment or community," or "the way that specific cultural groups go about the tasks of classifying, ordering, counting, and measuring" (p. 2). Several definitions of ethnomathematics highlight some of the "environmental activities" that Bishop (1988a) viewed as universal, and also "necessary and sufficient for the development of mathematical knowledge" (p. 182), namely counting, locating, measuring, designing, playing, and explaining. One further definition brings back the problem, hinted at in the foregoing account, of *ownership* of ethnomathematics. Whose mathematics is it?

Ethnomathematics refers to any form of cultural knowledge or social activity characteristic of a social and/or cultural group, that can be recognized by other groups such as 'Western' anthropologists, but not necessarily by the group of origin, as mathematical knowledge or mathematical activity. (Pompeu, 1994, p. 3)

This definition resonates with Ascher's, without fully solving the problem of ownership. The same problem appears in definitions of everyday mathematics, considered next.

Everyday Mathematics

Following on from the description of *everyday cognition* (Nunes, 1992, 1993) and important early studies that examined the use of mathematics in various practices, such as mathematical cognition of candy sellers in Brazil (Carraher, Carraher, & Schliemann, 1985; Saxe, 1991), constructs and issues are still being questioned. In this area, too, clarification of definitions is being sought, along with deeper consideration of the scope of the issues and their potential and significance for the classroom learning of mathematics.

Brenner and Moschkovich (2002) raised the following questions.

What do we mean by *everyday mathematics*? How is everyday mathematics related to *academic mathematics*? What particular everyday practices are being brought into mathematics classrooms? What impact do different everyday practices actually have in classroom practices?" (p. v)

In a similar vein, and with the benefit of two decades of research experience in this area, Carraher and Schliemann (2002) examined how their perceptions had evolved, as they explored the topic of their chapter, "Is Everyday Mathematics Truly Relevant to Mathematics Education?"

All the authors of chapters in the monograph edited by Brenner and Moschkovich (2002) in one way or another set out to explore these and related questions. Several of these authors pointed out that it is problematic to oppose everyday and academic mathematics, for several reasons. For one thing, for mathematicians academic mathematics *is* an everyday practice (Civil, 2002; Moschkovich, 2002a). For another, studies of everyday mathematical practices in workplaces reveal a complex interplay with sociocultural and technological issues (FitzSimons, 2002). In the automobile production industry, variations in the mathematical cognition required of workers have less to do with the job itself than with the decisions of management concerning production procedures and organization of the workplace. Highly skilled machinists display spatial and geometric knowledge that goes beyond what is commonly taught in school: In contrast, assembly-line workers and some machine operators find few if any mathematical demands in their work, which is deliberately stripped of the need for decisions involving knowledge of mathematics beyond elementary counting (Smith, 2002). The complex relationship between use of technology and the demand for mathematical thinking in the workplace is a theme that is explored in a later section of this chapter.

Another aspect that is again apparent in all the chapters of Brenner and Moschkovich's monograph is the importance of perceptions and beliefs about the nature of mathematics, both in the microculture of classroom practices (Brenner, 2002; Masingila, 2002) and in the broader endeavor to bridge the gap between mathematical thinking in and out of school (Arcavi, 2002; Civil, 2002; Moschkovich, 2002a). An essential element in all of these studies is the concern to connect knowledge of mathematics in and out of school. (This issue is revisited later in this chapter.) Because of the difficulties surrounding the construct *everyday mathematics* the terminology that will be adopted in this chapter follows Masingila (2002), who referred to in-school and out-of-school mathematics practices (p. 38).

The developments described in this section parallel the genesis of the movement away from purely psychological cognitive and behavioral frameworks for research in mathematics education, towards cultural frameworks that embrace sociology, anthropology, and related fields, including political and critical perspectives. The following section introduces some relevant theoretical issues and lenses that have been used to examine some of these developments.

THEORETICAL FIELDS THAT INCORPORATE CULTURE IN MATHEMATICS EDUCATION

The notion of theoretical and empirical fields is drawn from Brown and Dowling (1998) and provides a useful framework for characterizing components of research. This section addresses some theoretical fields pertinent to culture in the teaching and learning of mathematics. Their instantiation in empirical studies is described in the next section. The reader is reminded again to consult the original authors for a full treatment of theoretical fields that are introduced in this section, which has as its purpose a wide but by no means exhaustive view of the scope of theories that are available for work in this area.

As suggested in Barton's (1996) sense-making article introduced in the previous section, in the last 2 decades there has already been considerable movement in theoretical fields regarding the interplay of culture and mathematics. One such movement is discernible in the definitions of ethnomathematics given by D'Ambrosio, Gerdes, and Ascher, as their theoretical formulations moved from more static definitions of ethnomathematics as the mathematics of different cultural groups, to characterizations of this field as an anthropological research program that embraces not only the history of mathematical ideas of marginalized populations, but the history of mathematical knowledge itself (see previous section). D'Ambrosio (2000, p. 83) called this enterprise *historiography*.

Historiography

Moving beyond earlier theoretical formulations of ethnomathematics, its importance as a catalyst for further theoretical developments has been noted (Barton, 1996). D'Ambrosio played a large role and served as advisory editor in the enterprise that resulted in Helaine Selin's (2000) edited book, *Mathematics Across Cultures: The History of Non-Western Mathematics*. The chapters in this book are global in scope and record the mathematical thinking of cultures ranging from those of Iraq, Egypt, and other predominantly Islamic countries; through the Hebrew mathematical tradition; to that of the Incas, the Sioux of North America, Pacific cultures, Australian Aborigines, mathematical traditions of Central and Southern Africa; and those of Asia as represented by India, China, Japan, and Korea. As can be gleaned from the scope of this work, D'Ambrosio's original concern to valorize the mathematics of colonized and marginalized people (cf. Paolo Freire's *Pedagogy of the Oppressed* in 1970/1997) has broadened to encompass a movement that is both archeological and historical in nature, based on the theoretical field of "historiography" and visions of world knowledge through the "sociology of mathematics" (D'Ambrosio, 2000, pp. 85–87).

Although these antedated D'Ambrosio's program, earlier studies such as Claudia Zaslavsky's (1973) report on the counting systems of Africa and Glendon Lean's (1986) categorization of those of Papua New Guinea (see also Lancy, 1983), could also be thought of as historiography, as could anthropological research such as that of Pinxten, van Dooren, and Harvey (1983) who documented Navajo conceptions of space. Also in the cultural anthropology tradition, Crump's (1994) research on the anthropology of numbers is another fascinating example of historiography. More recent studies such as some of those collected as *Ethnomathematics* in the book by Powell and Frankenstein (1997) and the work of Marcia Ascher (1991, 1995, 2002) also fall into this category. Many of the cultural anthropological studies of various mathematical aspects of African practices, such as work on African fractals (Eglash, 1999); *lusona* of Africa (Gerdes, 1997); and women, art, and geometry of Africa (Gerdes, 1998), may also be regarded as historiography. As part of ethnomathematics conceived as a research program, this ambitious undertaking of historiography is designed to address lacunae in the literature on the history of mathematical thought through the ages. Much of the work of members of the International Study Group on Ethnomathematics (founded in 1985), and of the North American Study Group on Ethnomathematics (founded in 2003)—including research by Lawrence Shirley, Daniel Orey, and many others—could be placed in the category of historiography (see the list of some of the available web sites following the references).

Because historiography addresses some of the world's mathematical systems that have been ignored or undervalued in mathematics classrooms, it reminds us that there is *cultural capital* involved in the power relations associated with access in school mathematics. Some relevant theories are outlined in the following paragraphs.

Cultural Capital and Habitus

The usefulness of the theoretical field outlined by Bourdieu (1995) for issues of culture in the learning of mathematics has been indicated in research on learner's transitions between mathematical contexts (Presmeg, 2002a). Resonating with D'Ambrosio's original issues of concern (although D'Ambrosio did not use this framework), Bourdieu's work belongs in the field of sociology. The relevance of this work consists in "the innumerable and subtle strategies by which words can be used as instruments of coercion and constraint, as tools of intimidation and abuse, as signs of politeness, condescension, and contempt" (Bourdieu, 1995, p. 1, editor's introduction). This theoretical field serves as a useful lens in examining empirical issues related to the social inequalities and dilemmas faced by mathematics learners as they move between different cultural contexts, for example in the transitions experienced by immigrant children learning mathematics in new cultural settings (Presmeg, 2002a). This field embraces Bourdieu's notions of *cultural capital*, *linguistic capital*, and *habitus*. Bourdieu (1995) used the ancient Aristotelian term *habitus* to refer to "a set of dispositions which incline agents to act and react in certain ways" (p.12). Such dispositions are part of culture viewed as a set of shared understandings. Various forms of capital are *economic capital* (material wealth), *cultural capital* (knowledge, skills, and other cultural acquisitions, as exemplified by educational or technical qualifications), and *symbolic capital* (accumulated prestige or honor) (p. 14). *Linguistic capital* is not only the capacity to produce expressions that are appropriate in a certain social context, but it is also the expression of the "correct" accent, grammar, and vocabulary. The *symbolic power* associated with possession of cultural, symbolic, and linguistic capital has a counterpart in the *symbolic violence* experienced by individuals whose cultural capital is devalued. Symbolic violence is a sociological construct. In that capacity it is a powerful lens with which to examine actions of a group and ways in which certain types of knowledge are included or excluded in what the group counts as knowledge (for examples embracing the learning of mathematics, see the chapters in Abreu, Bishop, & Presmeg, 2002).

Borderland Discourses

The notion of symbolic violence leaves a possible theoretical gap relating to the ways in which individuals choose to construct, or choose *not* to construct, particular knowledge—mathematical or otherwise. Bishop (2002a) gave examples both of immigrant learners of mathematics in Australia who chose not to accept the view of themselves as constructed by their peers or their teacher—and of others who chose to accept these constructions. One student "shouted back" when peers in the mathematics class shouted derogatory names.

Discourse is a construct that is wider than mere use of language in conversation (Philips, 1993; Wood, 1998): It embraces all the aspects of social interaction that come into play when human beings interact with one another (Dörfler, 2000; Sfard, 2000). The notion of *Discourses* formulated by Gee (1992, his capitalization), and in particular his extension of the construct to *borderland Discourses*, those "community-based secondary Discourses" situated in the "borderland" between home and school knowledge (p. 146), are in line with Bourdieu's ideas of habitus and symbolic violence. Borderland Discourses take place in the borderlands between primary (e.g., home) and secondary (e.g., school) cultures of the diverse participants in social interactions. In situations where the secondary culture (e.g., that of the school) is conceived as threatening because of the possibility of symbolic violence there, the borderland may be a place of solidarity with others who may share a certain habitus. These ideas go some way towards closing the theoretical gap

in Bourdieu's characterization of symbolic violence by raising some issues of individuals' choices, because individuals choose the extent to which they will participate in various forms of these Discourses (see also Bishop's, 2002b, use of Gee's constructs).

Gee's work was in the context of second language learning but is also useful in the analysis of meanings given to various experiences by mathematics learners in cultural transition situations. Bishop (2002a) used this theoretical field in moving from the notion of cultural conflict to that of cultural mediation, in analyzing these experiences of learners of mathematics. If one considers the primary Discourse of school mathematics learners to be the home-based practices and conversations that contributed to their socialization and enculturation (forming their habitus) in their early years, and continuing to a greater or lesser extent in their present home experiences, then the secondary Discourse, for the purpose of learning mathematics, could be designated as the formal mathematical Discourse of the established discipline of mathematics. The teacher is more familiar with this Discourse than are the students and thus has the responsibility of introducing students to this secondary Discourse. As students become familiar with this field, their language and practices may approximate more closely those of the teacher. But in this transition the borderland Discourse of interactive classroom practices provides an important mediating space.

The enculturating role of the mathematics teacher was suggested in the foregoing account. However, as Bishop (2002b) pointed out, the learning of school mathematics is frequently more of an acculturation experience than an enculturation. The difference between these anthropological terms is as follows. *Enculturation* is the induction, by the cultural group, of young people into their own culture. In contrast, *acculturation* is "the modification of one's culture through continuous contact with another" (Wolcott, 1974, p. 136, as quoted in Bishop, 2002b, p. 193). The degree to which the culture of mathematics, as portrayed in the mathematics classroom, is viewed as their own or as a foreign culture by learners would determine whether their experiences there would be of enculturation or acculturation.

Cultural Models

Allied with Gee's (1992) notion of different Discourses is his construct of *cultural models*. This construct, defined as "'first thoughts' or taken for granted assumptions about what is 'typical' or 'normal'" (Gee, 1999, p. 60, quoted in Setati, 2003a, p. 153), was used by Setati (2003a) as an illuminating theoretical lens in her research on language use in South African multilingual mathematics classrooms. Already in 1987, D'Andrade defined a cultural model—which he also called a *folk model*—as "a cognitive schema that is intersubjectively shared by a social group," and he elaborated, "One result of intersubjective sharing is that interpretations made about the world on the basis of the folk model are treated as if they were obvious facts about the world" (p. 112). The transparency of cultural models may help to explain why mathematics was for so long considered to be value- and culture-free. The well-known creativity principle of making the familiar strange and the strange familiar (e.g., De Bono, 1975) is necessary for participants to become aware of their implicit cultural beliefs and values, which is why the anthropologist is in a position to identify the beliefs that are invisible to many who are within the culture.

In the context of mathematics learning in multilingual classrooms, Adler (1998, 2001) pointed out aspects of the use of language as a cultural resource that relate to the transparency of cultural models. Particularly in classrooms where the language of instruction is an *additional language*—not their first language—for many of the learners (Adler, 2001), teachers must at times focus on the language itself, in which case the artifact of language no longer serves as a "window" of transparent glass through which to view the mathematical ideas (Lave & Wenger, 1991). In this case the language used is no longer *invisible*, and the focus on the language itself may detract from the conceptual learning of the mathematical content (Adler, 2001).

Valorization in Mathematical Practices

If transparency of culture necessitates making the familiar strange before those sharing that culture become aware of the lenses through which they are viewing their world, then this principle points to a reason for the neglect of issues of valorization in the mathematics education research community until Abreu's (1993, 1995) research brought the topic to the fore. The *value* of formal mathematics as an academic subject was for so long taken for granted that it became a *given* notion that was not culturally questioned. Especially in its role as a gatekeeper to higher education, this status in education is likely to continue. But if ethnomathematics as a research program is to have a legitimate place in broadening notions both of what counts as mathematics and of which people have originated these forms of knowledge, then issues of valorization assume paramount importance.

Working from the theoretical fields of cultural psychology and sociocultural theory, Abreu and colleagues investigated the effects of valorization of various mathematical practices on Portuguese children (Abreu, Bishop, & Pompeu, 1997), Brazilian children (Abreu, 1993, 1995), and British children from Anglo and Asian backgrounds (Abreu, Cline, & Shamsi, 2002). As confirmed also in the research of Gorgorio, Planas, and Vilella (2002), many of these children denied the existence of, or devalued, mathematics as used in practices that they associated with their home- or out-of-school settings.

Valorization, the social process of assigning more value to certain practices than to others, is closely allied to Wertsch's (1998) notion of *privileging*, defined as "the fact that one mediational means, such as social language, is viewed as being more appropriate or efficacious than another in a particular social setting" (Wertsch, 1998, p. 64, in Abreu, 2002, p. 183). Abreu (2002) elaborated as follows.

> From this perspective, cultural practices become associated with particular social groups, which occupy certain positions in the structure of society. Groups can be seen as mainstream or as marginalised. In a similar vein individuals who participate in the practices will be given, or come to construct, identities associated with certain positions in these groups. The social representation enables the individual and social group to have access to a 'social code' that establishes relations between practices and social identities. (p. 184)

Thus Abreu argued strongly that in the learning of mathematics, valorization operates not only on the societal plane but also on the personal plane, because it impacts the construction of social identities. At this psychological level, the construction of mathematical knowledge may be subordinated to the construction of social identity by the individual learner in cases of cultural conflict, as suggested by Presmeg (2002a).

Abreu's ideas are embedded in the field of cultural psychology. Also taking the individual and society into account, another field that has grown in influence in mathematics education research in recent decades is that of situated cognition.

Situated Cognition

The theoretical field of situated cognition explores related aspects of the interplay of knowledge on the societal and psychological planes. Hence it has been a useful lens in research that takes culture into account in mathematical thinking and learning (Watson, 1998). As Ubiratan D'Ambrosio by his writings founded and influenced the field of ethnomathematics, so Jean Lave in analyzing and reporting her anthropological research has influenced the theoretical field of situated cognition. From her early theorizing following ethnography with tailor's apprentices in Liberia (Lave, 1988) to her more recent writings, following research with grocery shoppers and weight-watchers (Lave, 1997), the notion of transfer of knowledge through abstraction in one context, and subsequent use in a new context, was questioned and problematized. In collaboration with Etienne Wenger, her theorizing led to the notion of cognitive apprenticeship and *legitimate peripheral participation* (Lave & Wenger, 1991; Wenger, 1998). In this view, learning consists in a centripetal movement of the apprentice from the periphery to the center of a practice, under the guidance of those who are already masters of the practice. This theory was not originally developed in the context of or for the purpose of informing mathematics education. However, in its challenge to the cognitive position that abstract learning of mathematics facilitates transfer and that this knowledge may be readily applied in other situations than the one in which it was learned (not born out by empirical research), the theory has been powerful and influential. The research studies inspired by Lave and reported by Watson (1998) bear witness to this strong influence.

In her later writings, Lave attempted to bring the theory of socially situated knowledge to bear on the classroom teaching and learning of mathematics. But issues of intentionality and recontextualization separate apprenticeship and classroom situations, although enough commonality exists in the two situations for both legitimately to be called *practices* (Lerman, 1998). Mathematics learners in school are not necessarily aiming to become either mathematicians or mathematics teachers. As Lerman pointed out in connection with the issue of voluntary and nonvoluntary participation, students' presence in the classroom may be nonvoluntary, creating a very different situation from that of apprenticeship learning, and calling into question the assumption of a goal of movement from the periphery to the center. At the same time, the teacher has the intention to teach her students mathematics, notwithstanding Lave's claim that teaching is not a precondition of learning. The learning of mathematics is the goal of the enterprise, and teaching is the teacher's job. In contrast, in the apprenticeship situation the learning is not a goal in itself, for example, in the case of the tailors the goal is to make garments efficiently. Thus, as Adler (1998) suggests, "It is in the understanding of the aims of school education that Lave and Wenger's seamless web of practices entailed in moving from peripheral to full participation in a community of practice is problematic" (p. 174).

Although the theory of legitimate peripheral participation may not translate easily into the classroom teaching and learning of mathematics, the view of learning as a social practice has powerful implications for this learning and has been an influence in changes that have taken place in the practice of teaching mathematics, such as an increased emphasis on communication (NCTM, 1989, 2000) in the mathematics classroom. Even more, situated cognition as a theory has given a warrant to attempts to bridge the gap between in- and out-of-school mathematical practices.

Use of Semiotics in Linking Out-of-School and In-School Mathematics

In the USA, the *Principles and Standards for School Mathematics* (NCTM, 2000) continued the earlier call (NCTM, 1989) for teachers to make connections, in particular between the everyday practices of their students and the mathematical concepts that are taught in the classroom. But various theoretical lenses of situated cognition (Kirshner & Whitson, 1997; Lave & Wenger, 1991) remind us that these connections can be problematic. There are at least three ways in which the activities of out-of-school practices differ from the mathematical activities of school classrooms (Walkerdine, 1988), as follows.

- The goals of activities in the two settings differ radically.
- Discourse patterns of the classroom do not mirror those of everyday practices.
- Mathematical terminology and symbolism have a specificity that differs markedly from the useful qualities of ambiguity and indexicality (interpretation according to context) of terms in everyday conversation.

A semiotic framework that uses chains of signification (Kirshner & Whitson, 1997) has the potential to bridge this apparent gap through a process of chaining of signifiers in which each sign "slides under" the subsequent signifier. In this process, goals, discourse patterns, and use of terms and symbols all move towards that of classroom mathematical practices in a way that has the potential to preserve essential structure and some of the meanings of the original activity.

This theoretical framework resonates with that of *Realistic Mathematics Education* (RME) developed by Freudenthal, Streefland, and colleagues at the Freudenthal Institute (Treffers, 1993). *Realistic* in this sense does not necessarily mean out-of-school in the real world: The term refers to problem situations that learners can imagine (van den Heuvel-

Panhuizen, 2003). A theory of semiotic chaining in mathematics education resonates with RME in that the starting points for the learning are realistic in this sense. But more specifically the chaining model is a useful tool for linking out-of-school mathematical practices with the formal mathematics of school classrooms (an example of such use is presented in the next paragraph).

In brief, the theory of semiotic chaining used in mathematics education research, as it was initially presented by Whitson (1994, 1997), Cobb, Gravemeijer, Yackel, McClain, & Whitenack (1997), and Presmeg (1997, 1998b), followed the usage of Walkerdine (1988). Although she was working in a poststructural paradigm, Walkerdine found the Swiss structural approach of Saussure, as modified by Lacan, useful in building chains of signifiers to link a home practice, such as that of a daughter and her mother pouring drinks for guests, with more formal learning of mathematics, in this case the system of whole numbers. Walkerdine's chain had the following structure.

Signified 1: the actual people coming to visit;

Signifier 1 (*standing for* signified 1): the names of the people.

Signifier 1 and signified 1 form **sign 1.**

The daughter is asked to raise a finger for each name.

Signified 2: the names of the guests;

Signifier 2: the raised fingers.

Signifier 2 and signified 2 form **sign 2,** the second link.

The daughter is asked to count her raised fingers.

Signified 3: the raised fingers;

Signifier 3: the numerals, one two, three, four, five.

Signifier 3 and signified 3 form **sign 3,** the third link.

In this process, signifier 3 actually comes to stand for all of the preceding links in the chain: Five guests are coming to tea. Note that each signifier in turn becomes the next signified, and that at any time any of the links in the chain may be revisited conceptually. As mother and daughter move through the links of the chain, the discourse shifts successively from actual people to their names, to the fingers of one hand, and finally to the more abstract discourse of the numerals of the whole number counting system.

In her distinctive style, Adler (1998) characterized the associated recontextualizing process as that of crossing a discursive bridge:

That there is a bridge to cross between everyday and educated discourses is at the heart of Walkerdine's (1988) argument for "good teaching" entailing chains of signification in the classroom where everyday notions have to be prised out of their discursive practice and situated in a new and different discursive practice. (p. 174)

The theory may be summarized as follows.

In his *semiology* Saussure defined the sign as a combination of a "signified" together with its "signifier" (Saussure, 1959; Whitson, 1994, 1997). Lacan inverted Saussure's model, which gave priority to the *signified* over the signifier, to stress the *signifier* over the signified, and thus to recognize "far ranging autonomy for a dynamic and continuously productive play of signifiers that was not so easily recognized when it was assumed tacitly that a signifier was somehow constrained under domination by the signified" (Whitson, 1994, p. 40). This formulation allows for the chaining process in which a signifier in a previous sign-combination becomes the signified in a new sign-combination, and so on. The chain has at its final link some mathematical concept that may be taught in school.

Using this process, a teacher can use the chain as an instructional model that develops a mathematical concept starting with an out-of-school situation and linking it in a number of steps with formal school mathematics (Hall, 2000; Hall & Presmeg, 2000). Building on the work of Presmeg (1997, 1998b) in his dissertation research, Hall taught three 4th-grade teachers to build semiotic chains appropriate to the practices of their students and the instructional needs of their individual classrooms. Using a similar semiotic chaining model, Cobb et al. (1997) reported on the emergence (using their *emergent* theoretical perspective) of chains of signification in one 1st-grade classroom. In addition to these sources, examples of mathematical chaining at elementary school, high school, and college levels may be found in Presmeg's (1997, 1998b, 2006) writings. In later research on building bridges between out-of-school and in-school mathematics, Presmeg (2006) preferred to use a nested triadic model based on the writings of Charles Sanders Peirce (1992, 1998). In addition to an *object* (which could be an abstract concept) and a *representamen* that stands for the object in some way, each nested sign has a third component that Peirce called the *interpretant*. This triadic model explicitly allows for learners' individual construction of meaning (through the interpretant) in a way that linear chains of signification can do only implicitly.

In closing this section, three more of Peirce's constructs that are relevant to the role of culture and historiography in mathematics education are introduced, for the purpose of showing the potential of these theoretical notions to provide a foundation for ethnomathematics characterized as a research program.

PEIRCE'S NOTIONS OF SYNECHISM, COMMUNITY, AND COMMENS IN RECOGNIZING THE ROLE OF CULTURE IN MATHEMATICS EDUCATION

In this final part of the section, I want to return to ideas in the beginning of this section by introducing some Peircean constructs that have relevance in the *historiography* of D'Ambrosio (2000) with which this section started. Peirce's (1992, 1998) philosophical writings are useful in attending to the historical development of mathematical thought and thus are relevant to the role of culture in mathematics education (Presmeg, 2003). In particular, the construct of *community*—which he used without definition—and his definitions of *commens* and *synechism*, including his "law of mind," are apposite to explorations into the place of culture in mathematical historiography. In the light of these constructs (discussed in the following paragraphs), the role of community in the public institution of mathematics education is an issue of fundamental practical importance. The significance of the community of thinkers in the evolution of mathematical knowledge is indicated in Peirce's (1992) somewhat negative designation of the individual—uninformed by the sociocultural milieu—as ignorant and in error: "The individual man, since his separate existence is manifested only by ignorance and error, so far as he is anything apart from his fellows, and from what he and they are to be, is only a negation" (p. 55).

With regard to the genesis and evolution of mathematics, a point that has relevance in Peirce's epistemology is the continuity of past, present, and future. Continuity is central in Peirce's definition of *synechism* as "the tendency to regard continuity . . . as an idea of prime importance in philosophy" (Peirce, 1992, p. 313). Synechism involves the startling notion that knowledge in its real essence depends on *future* thought and how it will evolve in the community of thinkers:

> Finally, as what anything really is, is what it may finally be come to be known to be in the ideal state of complete information, so that reality depends on the ultimate decision of the community; so thought is what it is, only by virtue of its addressing a future thought which is in its value as thought identical with it, though more developed. In this way, the existence of thought now, depends on what is to be hereafter;

so that it has only a potential existence, dependent on the future thought of the community. (Peirce, 1992, pp. 54–55)

Whether "the ideal state of complete information" is ever an attainable goal is a matter of doubt, but the relevance of synechism for the history of mathematics lies in the role attributed to future generations of thinkers in assessing the achievements of the past and present. The notion of synechism is further explicated in connecting individual and community ideation through the role of convention in semiosis (activity with signs). Because the semiosis of the individual is mediated by the community through the adoption of certain ways of thinking and representing ideas as conventional, the growth of (mathematical) knowledge manifests continuity. Peirce (1992) cast further light on what he meant by continuity in his *law of mind*:

> Logical analysis applied to mental phenomena shows that there is but one law of mind, namely, that ideas tend to spread continuously and to affect certain others which stand to them in a peculiar relation of affectability. In this spreading they lose intensity, and especially the power of affecting others, but gain generality and become welded with other ideas. (p. 313)

Because of the importance of personal interpretations in forging a community of thinkers with its conventions, and thus in the continuity of ideas, Peirce formulated three kinds of interpretant in his semiotic model. Accordingly, he used triads not only in his semiotic model including object, representamen (sometimes called the sign), and interpretant, but also in the types of each of these components. This model includes the need for expression or communication: "Expression is a kind of representation or signification. A sign is a third mediating between the mind addressed and the object represented" (Peirce, 1992, p. 281). In an act of communication, then, there are three kinds of interpretant, as follows:

- the "*Intentional* Interpretant, which is a determination of the mind of the utterer";
- the "*Effectual* Interpretant, which is a determination of the mind of the interpreter"; and
- the "*Communicational* Interpretant, or say the *Cominterpretant*, which is a determination of that mind into which the minds of utterer and interpreter have to be fused in order that any communication should take place." (Peirce, 1998, p. 478)

It is the latter fused mind that Peirce designated the *commens*. There is a clear resonance of the commens with culture defined as a set of shared understandings (in the previous section).

For the continuity of mathematical ideas and their evolution in the history of mathematics, a central requirement is a community of thinkers who share a "fused mind" sufficiently to communicate effectively with one another—and with posterity through their artifacts—through this commens. Both the intentional and the effectual interpretants are important for communication (and have the implicit potential for miscommunication). But the third member of this triad, the interpretant generated by the commens, leads to the adoption of conventional signs by an intellectual community.

D'Ambrosio and others following the tradition of historiography that he has inaugurated (Selin, 2000) have introduced a different commens. Future generations, through synechism, will judge the impact of this historiography on the ontology and epistemology of mathematics in general, and of mathematics education in particular.

EMPIRICAL FIELDS THAT INCORPORATE CULTURE IN MATHEMATICS EDUCATION

The last section dealt with some theoretical fields that have been useful, or have the potential to be useful, in research on the role of culture in mathematics education. In the current section, some associated empirical fields are discussed. Empirical fields entail broad methodologies of empirical research, but in addition to these, in this section specific methods of research and also the participants, data, and results of selected studies are introduced. These details of participants, time, and place in the research are termed the *empirical settings* (Brown & Dowling, 1998). Thus appropriate empirical fields and settings are the focus of this section, along with some results of research on the role of culture in mathematics education.

Because ethnography is the special province of holistic cultural anthropological research (Eisenhart, 1988), which has as a broad goal the understanding of various cultures, it is natural to expect that ethnography (sometimes in modified form) would be used by researchers who are interested in the role of culture in teaching and learning mathematics. This is indeed the case. However, researchers in mathematics education in general sometimes call their studies ethnography when they have not spent long enough in the field to warrant that term for their research (Eisen-

hart, 1988). For instance, valuable as studies such as that of Millroy (1992) are for mathematics educators to learn more about the use of mathematical ideas in out-of-school settings such as that of Millroy's carpenters in Cape Town, South Africa, the $5\frac{1}{2}$ month duration of her fieldwork might be considered brief for an ethnography, and most ethnographic studies in mathematics education have a shorter duration than Millroy's. For this reason, the field of such research is often more appropriately called *case studies* (Merriam, 1998), because they satisfy the criterion of being bounded in particular ways. If such studies are investigating a particular topic, such as ways of bringing out-of-school mathematics into the classroom, and using several cases to do so, they could be called *instrumental case studies* (Stake, 2000). Where the individual cases themselves are the focus, case studies are *intrinsic* (ibid.).

Most of the studies mentioned in this section may be regarded as instrumental case studies in the sense explained in the foregoing. Research that concerns in-school and out-of-school mathematical practices falls into two main categories. In the first category are studies that attempt to build bridges between informal or nonformal mathematical practices (Bishop, Mellin-Olsen, & van Dormolen, 1991) and those of formal school mathematics. In the second category are studies that link with formal mathematics education less directly but with the potential to deepen understanding of this education, by exploring the culture of mathematics in various workplaces. The studies in both of these categories are too numerous for a summary chapter such as this one to do full justice to the aims, research questions, methodologies and methods for data collection, and the subtleties unearthed in the results of such research. (Again, the reader is reminded to read the original literature for depth of coverage.) What follows is an introduction to the range of research in each of these two categories, with discussion of some of the empirical settings, a sampling of issues addressed, and some of the results.

Linking Mathematics Learning Out-of-School and In-School

In the first category, studies that specifically investigated linking out-of-school and in-school mathematics are of several types. Some researchers interviewed parents of minority learners and their teachers to investigate relationships between home and school learning of mathematics. In England, Abreu and colleagues conducted such interviews for the purpose of investigating how the parents of immigrant learners are able to support their children in their transitions to learning mathematics in a new school culture (Abreu, Cline & Shamsi, 2002). Civil and colleagues in the USA interviewed such parents for the purpose of finding out which home practices of immigrant families might be suitable for pedagogical purposes in mathematics classrooms (Civil, 1990, 1995, 2002; Civil & Andrade, 2002). Related empirical studies in which researchers investigated the transitions experienced by immigrant children in their learning of mathematics have been reported by Bishop (2002a) in Australia and by Gorgorio, Planas, and Vilella (2002) in Catalonia, Spain.

In the next type of study in this category, research was conducted in mathematics classrooms to investigate the effects of bringing out-of-school practices into that arena. Using video recording as a data collection method, Brenner (2002) investigated how four teachers and their junior high school students went about an activity in which the students were required to decide cooperatively which of two fictitious pizza companies should be given the contract to supply pizza to the school cafeteria. In a different classroom video study, Moschkovich's (2002b) research addressed the mathematical activities of seventh-grade students during an architectural design project: Students working collaboratively tried to design working and living quarters for a team of scientists in Antarctica in such a way that space would be maximized while still taking into account the heating costs of various designs. These studies emphasized the point that through their pedagogy and instructional decisions in the classroom, teachers are an important component of the success (or lack of success) of attempts to use out-of-school practices effectively in mathematics classrooms. Beliefs of teachers about the nature of mathematics, what culture is, and its role in the classroom learning of mathematics are strong factors in teachers' decisions in this regard (Civil, 2002; Presmeg, 2002b).

Like the researchers in the preceding paragraphs, Presmeg (1997, 1998b, 2002b, 2006), Hall (2000), and Adeyemi (2004) recognized the lack of congruence of out-of-school cultural practices and those "same" practices when brought into the school mathematics classroom (Walkerdine, 1988, 1990, 1997). Their research used semiotic chaining as a theoretical tool (see previous section) in an attempt to bridge the gaps, not only between these practices in- and out-of-school, but also between mathematical ideas implicit in these activities and the formal mathematics of the syllabi that teachers are expected to use in their practices. The fourth-grade teachers with whom Hall worked built chains of signifiers that they implemented with their classes, starting with a cultural practice of at least some of their students and aimed at linking mathematical ideas in this practice in a number of steps with a formal school mathematics topic. Chains constructed by the teachers were designated as either intercultural—bridging two or more cultures, or intracultural—having a chain that remained within a single culture. Examples of the first type involved number of children in students' families, pizzas, coins, and measurement of students' hands, linking in a series of steps with classroom mathematical concepts. These are intercultural because the cultures and discourse of students' homes or activities are linked with the different discourse and culture of classroom mathematics, for instance, the making of bar graphs. Manipulatives were frequently used as intermediate links in these chains. The intracultural type, involving chains that were developed within the culture of a single activity, was evidenced in a chain involving baseball team statistics. The movement along the chain could be summarized as follows:

Baseball Game → Hits vs. At Bats → Success Fraction → Batting Average

It was not the activity that was preserved throughout the chain, but merely the culture of baseball (at least as it was imported into the classroom practice) within which the chain was developed. The need for interpretation at each link in the chain led to further theoretical developments using a triadic nested model that was a better lens for interpreting the results (Presmeg, 2006).

One study that is ethnographic in its scope and methodology has less explicit claims to link out-of-school and in-school mathematics, although implicit ties between the two contexts are present. This study is a thorough investigation of mathematical elements in learning the practice of selling newspapers in the streets by young boys (called in Portuguese *ardinas*) on the island of Cabo Verde in the Atlantic ocean (Santos & Matos, 2002). Using Lave and Wenger's (1991) theoretical framework of legitimate peripheral participation, Santos and Matos described the goal of their research as follows.

> Our goal was to look into the ways (mathematics) learning relates to forms of participation in social practice in an environment where mathematics is present but that escapes the characteristics of the school environment. Because we believe that culture is an unavoidable fact that shapes our way of seeing and analyzing things, we decided to look at a culturally distinct practice and that constituted a really strange domain for us: the practice of the *ardinas* at Cabo Verde islands in Africa. (p. 81)

In addition to interesting explicit and implicit mathematical aspects of the changing practice of the *ardinas* as they moved from being newcomers to full participants, methodological difficulties in this kind of research were foregrounded by Santos and Matos (2002):

> The fact that the research is studying a phenomena [*sic*] which was almost totally strange to us in most of its aspects, led us to realize that we had to go through a process which should involve, to a certain extent, our participation in the (*ardinas'*) practice with the explicit (for us and for them) goal of learning it but not in order to be a full member of that community of practice. This starting point (more in terms of knowing that there are more things that we don't know than that we know) opened that community of practice to us but also gave us consciousness that methodological issues were central in this research. (p. 120)

Many of the difficulties that Santos and Matos described with sensitivity and vividness are common to most anthropological research and thus also impact investigations of cultural practices in the field of mathematics education. They were required to become part of the culture sufficiently to be able to interpret it to others who are not participants, but at the same time not to become so immersed in the culture that it would be transparent—a lens that is not the focus of attention because one looks *through* it. They faced the difficulty that entering the group as an outsider might in fact change the culture of that group to a greater or lesser extent. On several occasions there was evidence that the mathematical reporting by key informants among the *ardinas* was influenced by the fact that the fieldworker was a Portuguese-speaking woman, whom these boys might have associated with their schoolteachers. Finally, the researchers recognized the not inconsiderable difficulties associated with practical matters concerned with collecting data in the street.

The investigation of the *ardinas'* mathematical thinking could be classified as an ethnographic workplace study. Thus this short account of this research provides a transition to the second aspect of this empirical section of the chapter, mathematics in the workplace.

Culture of Mathematics in the Workplace

In the second category of studies in this section, the culture of mathematics in various workplaces was investigated intensively by FitzSimons (2002), by Noss and colleagues (Noss & Hoyles, 1996a; Noss, Hoyles & Pozzi, 2000; Noss, Pozzi & Hoyles, 1999; Pozzi, Noss, and Hoyles, 1998), and by several other researchers (see later). FitzSimons investigated the culture and epistemology of mathematics in Technical and Further Education in Australia, and how this education related or failed to relate to the cultures of mathematics as used in workplaces. Noss, Hoyles, and Pozzi examined workplace mathematics in more specific detail. *Inter alia* their studies addressed mathematical aspects of banking and nursing practices. A common thread in these studies is the *demathematization* of the workplace: "As the seminal work of researchers Richard Noss and Celia Hoyles among others indicates, mathematics actually used in the workplace is contingent and rarely utilizes 'school mathematics' algorithms in their entirety, if at all, or necessarily correctly" (FitzSimons, 2002, p. 147).

The same point was brought out strongly, but contingently, in an investigation of mathematical activity in automobile production work in the USA (Smith, 2002). One result of Smith's study was that "the organizational structure and management of automobile production workplaces directly influenced the level of mathematics expected of production workers" (p. 112). This level varied from a minimal expectation on assembly lines—designed to be "worker-proof"—to "a surprisingly high level of spatial and geometric competence, which outstripped the preparation that most K–12 curricula provide" (p. 112), e.g., as manifested by skilled machinists who translated between two- and three-dimensional space with sophistication and accuracy in creating products that were sometimes one-of-a-kind. Organization that used "lean manufacturing principles" (p. 124) to move away from assembly lines and give more autonomy to workers was more likely to enhance and require these highly developed forms of mathematical thinking. Workers are not usually allowed to decide what level of mathematics will be required in their work:

> The fact that the organization of production systems and work practices mediates and in some cases limits workers' mathematical expectations and their access to workplace mathematics is a reminder that the everyday mathematics of work is inseparable from issues of power and authority. In large measure, someone other than the production workers themselves decides when, how often, and how deeply they will be called on to think mathematically. (Smith, 2002, p. 130)

Other sources of insights into the mathematical ideas involved in practices of various workplaces are detailed in FitzSimons's (2002) book. These include the following:

> operators in the light metals industry (Buckingham, 1997); front-desk motel and airline staff (Kanes, 1997a,

1997b); landless peasants in Brazil... (Knijnik, 1996, 1997, 1998); carpet layers (Masingila, 1993); commercial pilots (Noss, Hoyles, & Pozzi, 2000);... draughtspersons (Strässer, 1998); semi-skilled operators (Wedege, 1998b, 2000a, 2000b) and swimming pool construction workers (Zevenbergen, 1996). Collectively, these reports highlight not only the breadth and depth of mathematical concepts encountered in the workplace, but underline the complex levels of interactions in the broad range of professional competencies as outlined above, where mathematical knowledge can come into play—when permitted by (or in spite of) management. (p. 72)

In addition to this broad range of reported research, in an early study Mary Harris (1987) investigated "women's work," challenging her readers to "derive a general expression for [knitting] the heel of a sock" (p. 28). The more recent mathematics education conferences have included presentations on the topic of the mathematics of the workplace (e.g., Strässer, 1998). Strässer and Williams (2001) organized a Discussion group on "Work-Related Mathematics Education" at the 25th conference of the International Group for the Psychology of Mathematics Education. They presented the aims of this Discussion Group in the following terms.

Discussions in the group should aim at better understanding the contradictory trends of hiding or revealing workplace maths by means of artefacts and discourses, and identify problems and potentials for teaching and learning work-related mathematics. Current research into the use of mathematics at work (with the spectrum from traditional statistics to ethnomethodology) seems to favour case studies in a participatory style, while different ways of "stimulated recall" are also in use. (p. 265)

From the foregoing, the breadth of this field, the variety of practices that have been investigated from a mathematical point of view, and the scope of methodologies employed are apparent. Many of these studies may be characterized as investigations in ethnomathematics, for example, those of Masingila (1993, 1994) with the art of carpet laying, and Harris (1987). This overlap in classification also applies to research on mathematics in the world of work that is concerned with its artifacts and tools, and with its technology. The confluence of technology and culture in mathematics education is the topic of the last part of this section.

Influences of Technology

A study that stands at the intersection of workplace mathematics studies and those that are concerned with the role and influence of technology in mathematics education is that by Magajna and Monaghan (2003). In their research they investigated both the mathematical elements and the use of computer aided design (CAD) and manufacture (CAM) in the work of six skilled technicians who designed and produced moulds for glass factories. The technicians specialized in "moulds for containers of intricate shapes" (p. 102), such as bottles in the form of twisted pyramids, stars, or guitars. The study is interesting because the mathematics used by the CAD/CAM technicians in various stages of their work (for instance, calculating the interior volume of a mould) was not elementary, and in theory school-learned mathematics could have been used. However, the researchers reported as follows.

Although the technicians did not consider their activity was related to school mathematics there is evidence that in making sense of their practice they resorted to (a form of) school mathematics. The role of technology in technicians' mathematical activity was crucial: not only were the mathematical procedures they used shaped by the technology they used but the mathematics was a means to achieve technological results. Further to this the mathematics employed by the technicians must be interpreted within the goal-oriented behaviour of workers who 'live' the imperatives and constraints of the factory's production cycle. (p. 101)

An implication of results such as these, resonating with the conclusions of other workplace studies, is that there is no direct path from school mathematics to the mathematics used, sometimes indirectly, in various occupations, where the specific practices are learned on the job. It may not be possible then to gear a mathematics curriculum, even in vocational education (FitzSimons, 2002), to the specific requirements of a number of vocations simultaneously. However, as technology has developed in the last 6 decades, its influence has been felt in mathematics curricula of various time periods, as illustrated by Kelly (2003) and described next.

Each of the mathematics curriculum movements of the last century felt the impact of the state of the current technology of that period. Building on the assertion that "the tool defines the skill," Kelly (2003, p. 1041) described the influence of technological developments on mathematics curricula in four different time periods, overlapping with the following crucial years:

- 1942—the mainframe computer;
- 1967—the first four-function calculator;
- 1978—the personal computer (microcomputer);
- 1985—the graphing calculator.

In a broad sense, then, technology of various types has influenced considerably the culture of mathematics education taken as a complex of shared understandings (Stenhouse, 1967) about the nature of mathematics and its pedagogy. However, this influence does not imply unanimity about *what* technology should be brought into classrooms, or *how* it should be used. Contestation and conflict accompanied the growing use of computers and calculators of increasing sophistication in mathematics education, as Kelly's account showed vividly. He pointed out that the influence has not only permeated the culture of mathematics education through various curricula, but that as the technology changed and became more interactive through graphical user interfaces, the potential for a different kind of learning of mathematics became possible, and hence the need arose for a different kind of instruction. However, two impediments to the incorporation of new technology into mathematics classrooms have been the inaccessibility of the technology—exacerbated particularly in developing countries (Setati, 2003b)—and a lack of teacher readiness (Kelly, 2003).

Kelly's (2003) portrayal also suggests the importance of the avenues opened up by the graphing calculator, especially in—but not limited to—secondary and post-secondary schooling. Use of graphing calculators and computers impacts the nature of mathematics learned at all levels. The research in this growing field is beyond the scope of this chapter, but two examples highlight the potential of graphing calculators and computers to change the culture of learning mathematics. First, the research of Ricardo Nemirovsky (2002) and his colleagues (Nemirovsky, Tierney, & Wright, 1998) demonstrates how the use of motion detectors and associated technology changes the culture of learning from one of fostering *formal* generalizations in mathematics (e.g., "all *x* are *y*"), to one of constructing *situated* generalizations. The latter kind of generalization is "embedded in how people relate to and participate in tasks, events, and conversations" (Nemirovsky, 2002, p. 250)—and it is "loaded" with the values of "grasping the circumstances and transforming aspects that appear to be just ordinary or incidental into objects of reflection and significance" (p. 251). Nemirovsky illustrated this kind of generalization in the informal mathematical constructions of Clio, an 11-year-old girl working with the problem of trying to predict, based on the graph created on a computer screen by a motion detector, where a toy train is situated in a tunnel.

A second example is from the doctoral research of Paul Yu (2004), who investigated middle grades students' learning of geometry via the interactive geometric program called Shape Makers (Battista, 1998). Yu's research demonstrated how use of an interactive computer program such as this one has the potential to change the order of acquisition of the van Hiele (1986) levels of learning geometry. For instance, for these learners, a trapezoid was *what the trapezoid maker makes* in comparison with other shapes, and only later would they focus on individual appearance and properties of particular shapes.

The influence of dynamic geometry systems has been considerable, and research on the impact of such interactive software on the culture of school learning of geometry is important and ongoing. As Kelly (2003) pointed out, not only does the tool define the skill, but the development of new tools changes the culture and practices of mathematics itself, for instance in the impact of computer algebra systems on mathematics, or in changes made possible by the ease with which the computer can perform large numbers of routines that are beyond human capabilities.

In summary, then, technology, by entering all avenues of life, influences not only the mathematics of the curriculum and the ontology of mathematics itself, but also the culture of the future, through its children.

This section has examined in broad detail some of the empirical research relating to the role of culture in teaching and learning mathematics. Aspects that were considered included the linking of out-of-school and in-school mathematics learning, both from the point of view of comparing these different cultural practices and from the point of view of building bridges between cultural practices and the classroom learning of mathematics—of "bringing in the world" (Zaslavsky, 1996). A related strand that was outlined was the culture of mathematics in the workplace, and finally, some aspects of the cultural influence of technology on the learning of mathematics were suggested. Clearly in all of these strands ongoing research is needed and in progress. However, perhaps just as significantly, research is needed that will increase understanding of and highlight how these strands are related amongst themselves. For instance, as the use of technology both in the workplace and in the mathematics classroom changes the culture of learning mathematics in these broad arenas, hints of avenues to be explored in future work appear, for example in whether and how technology has the potential to bridge or decrease the gap between formal and informal mathematical knowledge. The theme of technology and culture in future research is elaborated, along with other significant areas, in the final section.

FUTURE DIRECTIONS FOR RESEARCH ON CULTURE IN MATHEMATICS EDUCATION

The final section uses a wider lens to zoom out and consider areas of research on culture in mathematics education that require development or are likely to assume increasing importance in the years ahead. By its nature, this section is speculative, but trends are already apparent that are likely to continue. The influence of technology is one such trend (Morgan, 1994; Noss, 1994).

Technology

Already in 1988, Noss pointed out the cultural entailments of the computer in mathematics education, as follows: "Making sense of the advent of the computer into the mathematics classroom entails a cultural perspective, not least because of the ways in which children are developing the computer culture by appropriating the technology for their own ends" (p. 251). He elaborated, "The key point is that children see computer screens as 'theirs,' as a part of a predominantly adult culture which they can appropriate and use for their own ends" (p. 257). As these children become adults with a facility with technology beyond that of their parents (Margaret Mead's *prefigurative enculturation* comes into play as children *teach* their parents), the culture of mathematics education in all its aspects is likely to change in fundamental ways. Noss (1988) and Noss and Hoyles (1996b) put forward a vision for the role that computer microworlds have played and might play in the future of mathematics education. With accelerating changes in platforms, designs, and software, research must keep pace and inform resulting changes in the cultures of teaching and learning of mathematics in schools. An aspect of potential use for the computer stems from its ability to mimic reality in a world of virtual reality. Noting the complex relationship between formal and informal mathematics (ethnomathematics), Noss (1988) suggested, "I propose that the technology itself—specifically the computer—can be the instrument for bridging the gap between the two" (p. 252). This area remains one in which research is needed, both in its own right and in ways that the change of context of bringing the virtual reality of computer images into mathematics classrooms changes the culture of teaching and learning in those classrooms.

Bridging Mathematics In-School and Out-of-School

Resonating with the Realistic Mathematics Education viewpoint that *real* contexts are not confined to those concrete situations with which learners are familiar (van den Heuvel-Panhuizen, 2003), Carraher and Schliemann (2002) made a useful distinction between *realism* and *meaningfulness*:

> What makes everyday mathematics powerful is not the concreteness of the objects or the everyday realism of the situations, but the meaning attached to the problems under consideration (Schliemann, 1995). In addition meaningfulness must be distinguished from realism (D. W. Carraher & Schliemann, 1991). It is true that engaging in everyday activities such as buying and selling, sharing, or betting may help students establish links between their experience and intuitions already acquired and topics to be learned in school. However, we believe it would be a fundamental mistake that schools attempt to emulate out-of-school institutions. After all, the goals and purposes of schools are not the same as those of other institutions. (p. 137)

The dilemma, then, is *how* to incorporate out-of-school practices in school mathematics classrooms in ways that are meaningful to students and that do not trivialize the mathematical ideas inherent in those practices. This issue remains a significant one for mathematics education research.

Reported research (Civil, 2002; Masingila, 2002; Presmeg, 2002b) has shown that learners' beliefs about the nature of mathematics affect what mathematics they identify *as* mathematics in out-of-school settings. Students' and teachers' conceptions of mathematics as decontextualized and abstract may limit what can be accomplished in terms of meaningfulness derived from out-of-school practices (Presmeg, 2002b). And yet, abstract thinking is not necessarily antagonistic to the idea of reasoning in particular contexts, as Carraher and Schliemann (2002) pointed out, which led them to propose the construct of *situated generalization*. They summarized these issues as follows.

> Research sorely needs to find theoretical room for contexts that are not reducible to physical settings or social structures to which the student is passively subjected. Contexts can be imagined, alluded to, insinuated, explicitly created on the fly, or carefully constructed over long periods of time by teachers and students. Much of the work in developing flexible mathematical knowledge depends on our ability to recontextualize problems—to see them from diverse and fresh points of view and to draw upon our former experience, including formal mathematical learning. Mathematization is not to be opposed to contextualization, since it always involves thinking in contexts. Even the apparently context-free activity of applying syntax transformation rules to algebraic expressions can involve meaningful contexts, particularly for experienced mathematicians. (p. 147)

They mentioned the irony that the mechanical following of algorithms characterizes the approaches of both highly unsuccessful and highly successful mathematical thinkers.

Taking into account both the need for meaningful contexts in the learning of mathematics and the necessity of developing mathematical ideas in the direction of abstraction and generalization (in the flexible sense, not to be confused with decontextualization), at least two extant fields of research have the potential to address these issues in significant ways. The notions of *horizontal* and *vertical* mathematizing that have informed Realistic Mathematics Education research for several decades (Treffers, 1993) could resolve a seeming conflict between abstraction and context. In harmony with these ideas, recent attempts to use semiotic theories in linking out-of-school and in-school mathematics also have the potential for further development (Hall & Presmeg, 2000; Presmeg, 2006; Yackel, Stephan, Rasmussen, & Underwood, 2003).

As Moschkovich (1995) pointed out, a tension exists between educators' attempts to engage learners in "real world" mathematics in classrooms and movements to make mathematics classrooms reflect the practices of mathematicians. Multicultural mathematics materials for use in classrooms have been available for some time (Krause, 1983; Zaslavsky, 1991, 1996). However, the tensions, the contradictions, and the complexity of trying to incorporate practices for which "making change" serves as a metonymy at the same time that students are "making mathematics" (Moschkovich, 1995) will engage researchers in mathematics education for some time to come. Bibliographies such as that compiled by Wilson and Mosquera (1991) will continue to be necessary, to inform both researchers and practitioners what has already been accomplished as the field of culture in mathematics education continues to grow.

Teacher Education

In all of the foregoing areas of potential cultural research in mathematics education, the role of the teacher remains important. Noting that concrete and abstract domains in mathematical thinking are not necessarily disparate, Noss (1988) suggested,

The key idea is that of focusing attention on the important relationships involved, a role in which—as Weir (1987) points out—the computer is rather well cast; but not without the conscious intervention of educators, and the careful development of an ambient learning culture. (p. 263)

Bishop (1988a, 1988b) was also intensely aware of the role of the *mathematical enculturators* in personifying the values that are inherent in the teaching and learning of mathematics. In the final chapter of his seminal book (1988a) he suggested requirements for the education—rather than the more restricted notion of "training"—of those who will be mathematics teachers, at both the elementary and secondary levels. He did not distinguish between these levels, for teachers at both elementary school and secondary school have important mathematical enculturation roles. Bishop (1988a) summarized the necessary criteria as follows.

I propose, then, these four criteria for the selection of suitable Mathematical enculturators:

- ability to "personify" the mathematical culture
- commitment to the Mathematics enculturation process
- ability to communicate Mathematical ideas and values
- acceptance of accountability to the Mathematical cultural group. (p. 168)

These ideas still seem timely; in fact the literature on discourse and communication has broadened in the decades since these words were written, to suggest that communication amongst all involved in the negotiation of the cultures of mathematics classrooms (teacher and learners) plays a significant role in the learning of mathematics in those arenas (Cobb et al., 1997; Dörfler, 2000; Sfard, 2000). *Language and Communication in Mathematics Education* was the title of a Topic Study Group at the 10th International Congress on Mathematical Education (Copenhagen, 2004), and the literature in this field is already extensive. But *how* to educate future teachers of mathematics to satisfy Bishop's four criteria is still an open field of research.

As Arcavi (2002) acknowledged, much has already been accomplished in curriculum development, research on teachers' beliefs and practices, and "the development of a classroom culture that functions in ways inspired by everyday practices of academic mathematics" (p. 27). However, open questions still exist concerning ways of using the recognition that the transition from out-of-school mathematical practices to those within school is sometimes not straightforward, in order to inform the practices of mathematics teaching. Arcavi (2002) gave examples of such questions.

However, much remains to be researched. For example, is it always possible to smooth the transition between familiar and everyday contexts, in which students use ad hoc strategies to solve problems, and academic contexts in which more general, formal, and decontextualized mathematics is to be learned?

Are there breaking points? If so, what is their nature? Studies in everyday mathematics and in ethnomathematics are very important contributions, not only because of their inherent value but also because of the reflection they provoke in the mathematics education community at large. There is much to be gained from those contributions. (p. 28)

Clearly ethnomathematics conceived as a research program (D'Ambrosio, 2000) has already permeated the cultures of mathematics education research and practice in various ways. The influence, and the need for research that addresses the complexities of the issues involved, are ongoing.

REFERENCES

Abreu, G. de (1993). *The relationship between home and school mathematics in a farming community in rural Brazil.* Unpublished doctoral dissertation, University of Cambridge, England.

Abreu, G. de (1995). Understanding how children experience the relationship between home and school mathematics. *Mind, Culture and Activity, 2*(2), 119–142.

Abreu, G. de (2002). Towards a cultural psychology perspective on transitions between contexts of mathematical practices. In G. de Abreu, A. J. Bishop, & N. C. Presmeg (Eds.), *Transitions between contexts of mathematical practices* (pp. 173–192). Dordrecht, The Netherlands: Kluwer.

Abreu, G. de, Bishop, A. J., & Pompeu, G. (1997). What children and teachers count as mathematics. In T. Nunes & P. Bryant (Eds.), *Learning and teaching mathematics: An international perspective* (pp. 233–264). Hove, East Sussex, England: Psychology Press.

Abreu, G. de, Bishop, A. J., & Presmeg, N. C. (Eds.). (2002). *Transitions between contexts of mathematical practices.* Dordrecht, The Netherlands: Kluwer.

Abreu, G. de, Cline, T., & Shamsi, T. (2002). Exploring ways parents participate in their children's school mathematical learning: Case studies in multiethnic primary schools. In G. de Abreu, A. J. Bishop, & N. C. Presmeg (Eds.), *Transitions between contexts of mathematical practices* (pp. 123–147). Dordrecht, The Netherlands: Kluwer.

Adeyemi, C. (2004). *The effect of knowledge of semiotic chaining on the beliefs, mathematical content knowledge, and practices of pre-service teachers.* Unpublished doctoral dissertation, Illinois State University.

Adler, J. (1998). Lights and limits: Recontextualising Lave and Wenger to theorise knowledge of teaching and of learning school mathematics. In A. Watson (Ed.), *Situated cognition and the learning of mathematics* (pp. 162–177). Oxford, England: University of Oxford Department of Educational Studies.

Adler, J. (2001). *Teaching mathematics in multilingual classrooms.* Dordrecht, The Netherlands: Kluwer.

Arcavi, A. (2002). The everyday and the academic in mathematics. In M. E. Brenner & J. N. Moschkovich (Eds.), *Everyday and academic mathematics in the classroom* (pp. 12–29). *Journal for Research in Mathematics Education* Monograph No. 11. Reston, VA: National Council of Teachers of Mathematics.

Ascher, M. (1991). *Ethnomathematics: A multicultural view of mathematical ideas.* New York: Chapman and Hall.

Ascher, M. (1995). Models and maps from the Marshall Islands: A case in ethnomathematics. *Historia Mathematica, 22,* 347–370.

Ascher, M. (2002). *Mathematics elsewhere: An exploration of ideas across cultures.* Princeton, NJ: Princeton University Press.

Ascher, M., & D'Ambrosio, U. (1994). Ethnomathematics: A dialogue. *For the Learning of Mathematics, 14*(2), 36–43.

Barton, B. (1996). Making sense of ethnomathematics: Ethnomathematics is making sense. *Educational Studies in Mathematics, 31,* 201–233.

Battista, M. (1998). *Shape Makers: Developing Geometric Reasoning with The Geometer's Sketchpad.* Emeryville, CA: Key Curriculum Press.

Bishop, A. J. (1988a). *Mathematical enculturation: A cultural perspective on mathematics education.* Dordrecht, The Netherlands: Kluwer.

Bishop, A. J. (1988b). Mathematics education in its cultural context, *Educational Studies in Mathematics, 19*(2), 179–191.

Bishop, A. J. (2002a). The transition experience of immigrant secondary school students: dilemmas and decisions. In G. de Abreu, A. J. Bishop, & N. C. Presmeg (Eds.), *Transitions between contexts of mathematical practices* (pp. 53–79). Dordrecht, The Netherlands: Kluwer.

Bishop, A. J. (2002b). Mathematical acculturation, cultural conflicts, and transition. In G. de Abreu, A. J. Bishop, & N. C. Presmeg (Eds.), *Transitions between contexts of mathematical practices* (pp. 193–212). Dordrecht, The Netherlands: Kluwer.

Bishop, A. J. & Abreu, G. de (1991). Children's use of outside school knowledge to solve mathematical problems at school. In F. Furinghetti (Ed.), *Proceedings of the 15th International Conference for the Psychology of Mathematics Education,* Assisi, Italy, *1,* 128–135.

Bishop, A. J., Mellin–Olsen, S., & van Dormolen, J. (Eds.). (1991). *Mathematical knowledge: Its growth through teaching.* Dordrecht, The Netherlands: Kluwer.

Bishop, A. J. & Pompeu, G. (1991). Influences of an ethnomathematical approach on teacher attitudes to mathematics education. In F. Furinghetti (Ed.) *Proceedings of the 15th International Conference for the Psychology of Mathematics Education,* Assisi, Italy, *1,* 136–143.

Bourdieu, P. (1995). *Language and symbolic power.* Cambridge, MA: Harvard University Press.

Brenner, M. E. (2002). Everyday problem solving and curriculum implementation: An invitation to try pizza. In M. E. Brenner & J. N. Moschkovich (Eds.), *Everyday and academic mathematics in the classroom* (pp. 63–92). *Journal for Research in Mathematics Education Monograph No. 11.* Reston, VA: National Council of Teachers of Mathematics.

Brenner, M. E. & Moschkovich, J. N. (Eds.) (2002). *Everyday and academic mathematics in the classroom. Journal for Research in Mathematics Education Monograph No. 11.* Reston, VA: National Council of Teachers of Mathematics.

Brown, A. & Dowling, P. (1998). *Doing research/reading research: A mode of interrogation for education.* New York: Falmer Press.

Carpenter, T. P., Dossey, J. A., & Koehler, J. L. (Eds.). (2004). *Classics in mathematics education research.* Reston, VA: National Council of Teachers of Mathematics.

Carraher, T. N., Carraher, D. W., & Schliemann, A. D. (1985). Mathematics in the streets and in schools. *British Journal of Developmental Psychology, 3,* 21–29.

Carraher, D. W., & Schliemann, A. D. (2002). Is everyday mathematics truly relevant to mathematics education? In

M. E. Brenner & J. N. Moschkovich (Eds.), *Everyday and academic mathematics in the classroom* (pp. 131–153). *Journal for Research in Mathematics Education Monograph No. 11.* Reston, VA: National Council of Teachers of Mathematics.

Civil, M. (1990). "You only do math in math": A look at four prospective teachers' views about mathematics. *For the Learning of Mathematics, 10*(1), 7–9.

Civil, M. (1995). Connecting home and school: Funds of knowledge for mathematics teaching. In B. Denys & P. Laridon (Eds.), *Working group on cultural aspects in the learning of mathematics: Some current developments* (pp. 18–25). Monograph following the 19th Annual Meeting of the International Group for the Psychology of Mathematics Education, Recife, Brazil.

Civil, M. (2002). Everyday mathematics, mathematicians' mathematics, and school mathematics: Can we bring them together? In M. E. Brenner & J. N. Moschkovich (Eds.), *Everyday and academic mathematics in the classroom* (pp. 40–62). *Journal for Research in Mathematics Education Monograph No. 11.* Reston, VA: National Council of Teachers of Mathematics.

Civil, M. & Andrade, R. (2002). Transitions between home and school mathematics: Rays of hope amidst the passing clouds. In G. de Abreu, A. J. Bishop, & N. C. Presmeg, (Eds.), *Transitions between contexts of mathematical practices* (pp. 149–169). Dordrecht, The Netherlands: Kluwer.

Cobb, P., Gravemeijer, K., Yackel, E., McClain, K., & Whitenack, J. (1997). Mathematizing and symbolizing: The emergence of chains of signification in one first grade classroom. In. D. Kirshner & J.A. Whitson (Eds.), *Situated Cognition: Social Semiotic, and Psychological Perspectives* (pp. 151–233). Hillsdale, NJ: Erlbaum.

Cobb, P., & Yackel, E. (1995). Constructivist, emergent, and sociocultural perspectives in the context of developmental research. In D. T. Owens, M. K. Reed, & G. M. Millsaps (Eds.), *Proceedings of the 17th Annual Meeting of the North American Chapter of the International Group for the Psychology of Mathematics Education* (pp. 3–29). Columbus, OH: ERIC Clearinghouse for Science, Mathematics, and Environmental Education.

Crump, T. (1994). *The anthropology of numbers.* Cambridge, England: Cambridge University Press.

D'Ambrosio, U. (1984). Socio–cultural bases for mathematics education. Plenary presentation, *Proceedings of the 5th International Congress on Mathematical Education,* Adelaide, Australia.

D'Ambrosio, U. (1985). Ethnomathematics and its place in the history and pedagogy of mathematics. *For the Learning of Mathematics, 5*(1), 44–48.

D'Ambrosio, U. (1987). Ethnomathematics, what it might be. *International Study Group on Ethnomathematics Newsletter, 3*(1).

D'Ambrosio, U. (1990). The role of mathematics education in building a democratic and just society. *For the Learning of Mathematics, 10*(3), 20–23.

D'Ambrosio, U. (1991). *On ethnoscience.* Campinas, Brazil: Interdisciplinary Center for the Improvement of Science Education.

D'Ambrosio, U. (2000). A historiographical proposal for non-Western mathematics. In H. Selin (Ed.), *Mathematics across cultures: The history of non-Western mathematics* (pp. 79–92). Dordrecht, The Netherlands: Kluwer.

D'Andrade, R. (1987). A folk model of the mind. In D. Holland & N. Quinn (Eds.), *Cultural models in language and thought.* Cambridge, England: Cambridge University Press.

Davis, P. J., & Hersh, R. (1981). *The mathematical experience.* Boston: Houghton Mifflin.

De Bono, E. (1975). *Mechanisms of mind.* Harmondsworth, England: Penguin.

Dörfler, W. (2000). Means for meaning. In P. Cobb, E. Yackel, & K. McClain (Eds.), *Symbolizing and communicating in mathematics classrooms: Perspectives on discourse, tools, and instructional design* (pp. 99–131). Mahwah, NJ: Erlbaum.

Eglash, R. (1999). *African fractals: Modern computing and indigenous design.* New Brunswick, NJ: Rutgers University Press.

Eisenhart, M. A. (1988). The ethnographic research tradition and mathematics education research. *Journal for Research in Mathematics Education, 19*(2), 99–114.

Ernest, P. (1991). *The philosophy of mathematics education.* London: Falmer Press.

Eves, H., (1990). *An introduction to the history of mathematics* (6th ed.). Philadelphia: Saunders College Publishing.

FitzSimons, G. E. (2002). *What counts as mathematics? Technologies of power in adult and vocational education.* Dordrecht, The Netherlands: Kluwer.

Freire, P. (1970/1997). *Pedagogy of the oppressed.* New York: Continuum.

Freudenthal, H. (1973). *Mathematics as an educational task.* Dordrecht, The Netherlands: Kluwer.

Freudenthal, H. (1983). *Didactical phenomenology of mathematical structures.* Dordrecht, The Netherlands: Kluwer.

Gay, J., & Cole, M. (1967). *The new mathematics in an old culture.* New York: Holt, Rinehart and Winston.

Gee, J. P. (1992). *The social mind: Language, ideology and social practice.* New York: Bergin & Garvey.

Geertz, C. (1973). *The interpretation of culture.* New York: Basic Books.

Gerdes, P. (1986). How to recognise hidden geometrical thinking: A contribution to the development of anthropological mathematics. *For the Learning of Mathematics, 6*(2), 10–17.

Gerdes, P. (1988a). On possible uses of traditional Angolan sand drawings in the mathematics classroom. *Educational Studies in Mathematics, 19*(1), 3–22.

Gerdes, P. (1988b). On culture, geometrical thinking and mathematics education. *Educational Studies in Mathematics, 19*(2), 137–162.

Gerdes, P. (1994). Reflections on ethnomathematics. *For the Learning of Mathematics, 14*(2), 19–22.

Gerdes, P. (1997). *Lusona—Geometrical recreations of Africa.* Montréal, Canada: L'Harmattan.

Gerdes, P. (1998). *Women, art, and geometry in Southern Africa.* Trenton, NJ: Africa World Press.

Gorgorio, N., Planas, N., & Vilella, X. (2002). Immigrant children learning mathematics in mainstream schools. In G. de Abreu, J. Bishop, & N. Presmeg (Eds.), *Transitions between contexts of mathematical practices* (pp. 23–52). Dordrecht, The Netherlands: Kluwer.

Hall, M. (2000). *Bridging the gap between everyday and classroom mathematics: An investigation of two teachers' intentional use of semiotic chains.* Unpublished doctoral dissertation, Florida State University.

Hall, M. & Presmeg, N. C. (2000). Teachers' uses of semiotic chaining to link home and middle school mathematics in classrooms. Research Report, in M. L. Fernández

(Ed.), *Proceedings of the 22nd Annual Meeting of the North American Chapter of the International Group for the Psychology of Mathematics Education* (Vol. 2, pp. 427–432). Columbus, OH: ERIC Clearinghouse for Science, Mathematics, and Environmental Education.

Harris, M. (1987). An example of traditional women's work as a mathematics resource. *For the Learning of Mathematics, 7*(3), 26–28.

Holland, D. & Quinn, N. (Eds.). (1987). *Cultural models in language and thought.* Cambridge, England: Cambridge University Press.

Kelly, B. (2003). The emergence of technology in mathematics education. In G. M. A. Stanic & J. Kilpatrick (Eds.), *A history of school mathematics.* Reston, VA: National Council of Teachers of Mathematics.

Kirshner, D. & Whitson, J. A. (Eds.). (1997). *Situated cognition: Social, semiotic, and psychological perspectives.* Mahwah, NJ: Erlbaum.

Krause, M. (1983). *Multicultural mathematics materials.* Reston, VA: National Council of Teachers of Mathematics.

Kroeber, A. L., & Kluckhohn, C. (1952). *Culture: A critical review of concepts and definitions.* New York: Vintage Books.

Kuhn, T. S. (1970). *The structure of scientific revolutions.* Chicago: University of Chicago Press.

Lakatos, I. (1976). *Proofs and refutations.* Cambridge, England: Cambridge University Press.

Lakoff, G. (1987). W*omen, fire and dangerous things: What categories reveal about the mind.* Chicago: University of Chicago Press.

Lancy, D. (1983). *Cross cultural studies in cognition and mathematics.* New York: Academic Press.

Lave, J. (1988). *Cognition in practice: Mind, mathematics and culture in everyday life.* Cambridge, England: Cambridge University Press.

Lave, J. (1997). The culture of acquisition and the practice of understanding. In D. Kirshner & J. A. Whitson (Eds.), *Situated cognition: Social, semiotic, and psychological perspectives* (pp. 17–35). Mahwah, NJ: Erlbaum.

Lave, J. & Wenger, E. (1991). *Situated learning: Legitimate peripheral participation.* Cambridge, England: Cambridge University Press.

Lean, G. A. (1986). *Counting systems of Papua New Guinea.* Lae, Papua New Guinea: Research Bibliography, Department of Mathematics, Papua New Guinea University of Technology.

Lerman, S. (Ed.) (1994). *Cultural perspectives on the mathematics classroom.* Dordrecht, The Netherlands: Kluwer.

Lerman, S. (1998). Learning as social practice: An appreciative critique. In A. Watson (Ed.), *Situated cognition and the learning of mathematics* (pp. 34–42). Oxford, England: University of Oxford Department of Educational Studies.

Magajna, Z., & Monaghan, J. (2003). Advanced mathematical thinking in a technological workplace. *Educational Studies in Mathematics, 52*(2), 101–122.

Masingila, J. O. (1993). Learning from mathematics practice in out-of-school situations. *For the Learning of Mathematics, 13*(2), 18–22.

Masingila, J. O. (1994). Mathematics practice in carpet laying. *Anthropology and Education Quarterly, 25*(4), 430–462.

Masingila, J. O. (2002). Examining students' perceptions of their everyday mathematical practice. In M. E. Brenner & J. N. Moschkovich (Eds.), *Everyday and academic mathematics in the classroom* (pp. 30–39). *Journal for Research in Mathematics*

Education Monograph No. 11. Reston, VA: National Council of Teachers of Mathematics.

Mellin-Olsen, S. (1987). *The politics of mathematics education.* Dordrecht, The Netherlands: D. Reidel.

Merriam, S. B. (1998). *Qualitative research and case study applications in education.* San Francisco: Jossey-Bass.

Millroy, W. L. (1992). *An ethnographic study of the mathematics of a group of carpenters. Journal for Research in Mathematics Education Monograph No. 5.* Reston, VA: National Council of Teachers of Mathematics.

Morgan, C. (1994). The computer as a catalyst in the mathematics classroom? In S. Lerman (Ed.), *Cultural perspectives on the mathematics classroom* (pp. 115–131). Dordrecht, The Netherlands: Kluwer.

Moschkovich, J. (1995, April). *"Making change" and "making mathematics" as proposals for mathematics classroom practices.* Paper presented at the Annual Meeting of the American Educational Research Association, San Francisco.

Moschkovich, J. N. (2002a). An introduction to examining everyday and academic mathematical practices. In M. E. Brenner & J. N. Moschkovich (Eds.), *Everyday and academic mathematics in the classroom* (pp. 1–11). *Journal for Research in Mathematics Education Monograph No. 11.* Reston, VA: National Council of Teachers of Mathematics.

Moschkovich, J. N. (2002b). Bringing together workplace and academic mathematical practices during classroom assessments. In M. E. Brenner & J. N. Moschkovich (Eds.), *Everyday and academic mathematics in the classroom* (pp. 93–110). *Journal for Research in Mathematics Education Monograph No. 11.* Reston, VA: National Council of Teachers of Mathematics.

National Council of Teachers of Mathematics (1989). *Curriculum and evaluation standards for school mathematics.* Reston, VA: Author.

National Council of Teachers of Mathematics (2000). *Principles and standards for school mathematics.* Reston, VA: Author.

Nemirovsky, R. (2002). On guessing the essential thing. In K. Gravemeijer, R. Lehrer, B. van Oers, & L. Verschaffel (Eds.), *Symbolizing, modeling and tool use in mathematics education* (pp. 233–256). Dordrecht, The Netherlands: Kluwer.

Nemirovsky, R., Tierney, C., & Wright, T. (1998). Body motion and graphing. *Cognition and Instruction, 16*(2), 119–172.

Nickson, M. (1992). The culture of the mathematics classroom: An unknown quantity? In D. A. Grouws (Ed.) *Handbook of research on mathematics teaching and learning* (pp. 101–114). New York: Macmillan.

Nickson, M. (1994). The culture of the mathematics classroom: An unknown quantity? [Modified reprint of the 1992 paper], In S. Lerman (Ed.), *Cultural perspectives on the mathematics classroom* (pp. 7–35). Dordrecht, The Netherlands: Kluwer.

Noss, R. (1988). The computer as a cultural influence in mathematics learning. *Educational Studies in Mathematics, 19*(2), 251–268.

Noss, R. (1994). Sets, lies and stereotypes. In S. Lerman (Ed.), *Cultural perspectives on the mathematics classroom* (pp. 37–49). Dordrecht, The Netherlands: Kluwer.

Noss, R. (1997). Meaning mathematically with computers. In T. Nunes & P. Bryant (Eds.), *Learning and teaching mathematics: An international perspective* (pp. 289–314). Hove, East Sussex, England: Psychology Press.

Noss, R., & Hoyles, C. (1996a). The visibility of meanings: Modelling the mathematics of banking. *International Journal of Computers for Mathematical Learning, 1*(1), 3–31.

Noss R., & Hoyles, C. (1996b). *Windows on mathematical meanings: Learning cultures and computers.* Dordrecht, The Netherlands: Kluwer.

Noss, R., Hoyles, C,. & Pozzi, S. (2000). Working knowledge: Mathematics in use. In A. Bessot & J. Ridgway (Eds.), *Education for mathematics in the workplace* (pp. 17–35). Dordrecht, The Netherlands: Kluwer.

Noss, R., Pozzi, S., & Hoyles, C. (1999). Teaching epistemologies: Meanings of average and variation in nursing practice. *Educational Studies in Mathematics, 40*(1), 25–51.

Nunes, T. (1992). Ethnomathematics and everyday cognition. In D. A. Grouws (Ed.), *Handbook of research on mathematics teaching and learning* (pp. 557–574). New York: Macmillan.

Nunes, T. (1993). The socio-cultural context of mathematical thinking: research findings and educational implications. In A. J. Bishop, K. Hart, S. Lerman & T. Nunes, *Significant influences on children's learning of mathematics.* (UNESCO Publcation No. 44, pp. 27–42). Paris.

Peirce, C. S. (1992). *The essential Peirce: Selected philosophical writings.* Volume 1 (1967–1893). (Edited by N. Houser & C. Kloesel.) Bloomington: Indiana University Press.

Peirce, C. S. (1998). *The essential Peirce: Selected philosophical writings.* Volume 2 (1893–1913). (Edited by the Peirce Edition Project.) Bloomington: Indiana University Press.

Philips, S. U. (1993). *The invisible culture: Communication in classroom and community on the Warm Springs Indian Reservation.* Prospect Heights, IL: Waveland Press.

Pinxten, R. (1994). Anthropology in the mathematics classroom? In S. Lerman (Ed.), *Cultural perspectives on the mathematics classroom* (pp. 85–97). Dordrecht, The Netherlands: Kluwer.

Pinxten, R., van Dooren, I., & Harvey, F. (1983). *Anthropology of space: Explorations into the natural philosophy and semantics of Navajo Indians.* Philadelphia: University of Pennsylvania Press.

Pompeu, G. Jr. (1994). Another definition of ethnomathematics? *Newsletter of the International Study Group on Ethnomathematics, 9*(2), 3.

Powell, A. B. & Frankenstein, M. (Eds.) (1997). *Ethnomathematics: Challenging Eurocentrism in mathematics education.* Albany, NY: SUNY Press.

Pozzi, S., Noss, R., & Hoyles, C. (1998). Tools in practice, mathematics in use. *Educational Studies in Mathematics, 36*(2), 105–122.

Presmeg, N. C. (1997). A semiotic framework for linking cultural practice and classroom mathematics. In J. Dossey, J. Swafford, M. Parmantie, & A. Dossey (Eds.), *Proceedings of the 19th Annual Meeting of the North American Chapter of the International Group for the Psychology of Mathematics Education,* (Vol. 1, pp. 151–156). Columbus, OH: ERIC Clearinghouse for Science, Mathematics, and Environmental Education.

Presmeg, N. C. (1998a). Ethnomathematics in teacher education. *Journal of Mathematics Teacher Education, 1*(3), 317–339.

Presmeg, N. C. (1998b). A semiotic analysis of students' own cultural mathematics. Research Forum Report. In A. Olivier & K. Newstead (Eds.), *Proceedings of the 22nd Conference of the International Group for the Psychology of Mathematics Education,* (Vol. 1, pp. 136–151). Stellenbosch, South Africa: University of Stellenbosch.

Presmeg, N. C. (2002a). Shifts in meaning during transitions. In G. de Abreu, A. J. Bishop, & N. C. Presmeg (Eds.), *Transitions between contexts of mathematical practices* (pp. 213–228). Dordrecht, The Netherlands: Kluwer.

Presmeg, N. C. (2002b). Beliefs about the nature of mathematics in the bridging of everyday and school mathematical practices. In G. Leder, E. Pehkonen, & G. Törner (Eds.), *Beliefs: A hidden variable in mathematics education?* (pp. 293–312). Dordrecht, The Netherlands: Kluwer.

Presmeg, N. C. (2003, July). *Ancient areas: A retrospective analysis of early history of geometry in light of Peirce's "commens."* Paper presented in the Discussion Group on Semiotics in Mathematics Education, 27th Annual Conference of the International Group for the Psychology of Mathematics Education, Honolulu, HI. (Available at http://www.math.uncc.edu/~sae/.)

Presmeg, N. C. (2006). Semiotics and the "connections" standard: Significance of semiotics for teachers of mathematics. *Educational Studies in Mathematics, 61,* 163–182.

Santos, M. & Matos, J. F. (2002). Thinking about mathematical learning with Cabo Verde *ardinas*. In G. de Abreu, A. J. Bishop, & N. C. Presmeg (Eds.), *Transitions between contexts of mathematical practices* (pp. 81–122). Dordrecht, The Netherlands: Kluwer.

Saussure, F. de (1959). *Course in general linguistics.* New York: McGraw-Hill.

Saxe, G. (1991). *Culture and cognitive development: Studies in mathematical understanding.* Hillsdale, NJ: Erlbaum.

Scott, P. R. (1985). Ethnomathematics: What might it be? *Newsletter of the International Study Group on Ethnomathematics, 1*(1), 2.

Selin H. (Ed.). (2000). *Mathematics across cultures: The history of non-Western mathematics.* Dordrecht, The Netherlands: Kluwer.

Setati, M. (2003a). Language use in a multilingual mathematics classroom in South Africa: A different perspective. In N. A. Pateman, B. J. Dougherty, & J. T. Zilliox (Eds.), *Proceedings of the 27th Conference of the International Group for the Psychology of Mathematics Education,* (Vol. 4, pp. 151–158). Honolulu: University of Hawai'i Press.

Setati, M. (2003b). Availability and (non-) use of technology in and for mathematics education in poor schools in South Africa. In N. A. Pateman, B. J. Dougherty, & J. T. Zilliox (Eds.), *Proceedings of the 27th Conference of the International Group for the Psychology of Mathematics Education,* (Vol. 1, pp. 149–152). Honolulu: University of Hawai'i Press.

Sfard, A. (1997). The many faces of mathematics: Do mathematicians and researchers in mathematics education speak about the same thing? In A. Sierpinska & J. Kilpatrick (Eds.), *Mathematics education as a research domain: A search for identity* (pp. 491–511). Dordrecht, The Netherlands: Kluwer.

Sfard, A. (2000). Symbolizing mathematical reality into being—or how mathematical discourse and mathematical objects create each other. In P. Cobb, E. Yackel, & K. McClain (Eds.), *Symbolizing and communicating in mathematics classrooms: Perspectives on discourse, tools, and instructional design* (pp. 37–98). Mahwah, NJ: Erlbaum.

Skovsmose, O. (1994). *Towards a critical philosophy of mathematics education.* Dordrecht, The Netherlands: Kluwer.

Smith, J. P. III (2002). Everyday mathematical activity in automobile production work. In M. E. Brenner & J. N. Moschkovich (Eds.), *Everyday and academic mathematics in the classroom* (pp. 111–130). *Journal for Research in Mathematics Education Monograph No. 11.* Reston, VA: National Council of Teachers of Mathematics.

Stake, R. E. (2000). Case studies. In N. K. Denzin & Y. S. Lincoln (Eds.), *Handbook of qualitative research* (pp. 435–454). Thousand Oaks, CA: Sage.

Steen, L. A. (1990). *On the shoulders of giants: New approaches to numeracy.* Washington, DC: National Academy Press

Stenhouse, L. (1967). *Culture and education.* New York: Weybright & Talley.

Strässer, R. (1998). Mathematics for work: A didactical perspective. In C. Alsina, J. M. Alvarez, B. Hodgsen, C. Laborde, & A. Pérez (Eds.), *8th International Congress on Mathematical Education: Selected lectures* (pp. 427–441). Seville, Spain: S. A. E. M 'THALES'.

Strässer R. & Williams, J. (2001). Discussion group on work-related mathematics education. In M. van den Heuvel-Panhuizen (Ed.), *Proceedings of the 25th Conference of the International Group for the Psychology of Mathematics Education,* (Vol. 1, p. 265). Utrecht, The Netherlands: Freudenthal Institute.

Taylor, P. C. (1996). Mythmaking and mythbreaking in the mathematics classroom. *Educational Studies in Mathematics, 31*(1–2), 151–173.

Thomas, R. S. D. (1996). Proto-mathematics and/or real mathematics. *For the Learning of Mathematics, 16*(2), 11–18.

Thompson, A. G. (1984). The relationship of teachers' conceptions of mathematics and mathematics teaching to instructional practice. *Educational Studies in Mathematics, 15,* 105–127.

Treffers, A. (1993). Wiskobas and Freudenthal realistic mathematics education. *Educational Studies in Mathematics, 25,* 89–108.

Van den Heuvel-Panhuizen, M. (2003). The didactical use of models in Realistic Mathematics Education: An example from a longitudinal trajectory on percentage. *Educational Studies in Mathematics, 54*(1), 9–35.

Van Hiele, P. M. (1986). *Structure and insight: A theory of mathematics education.* London: Academic Press.

Vithal, R. (2003). *Towards a pedagogy of conflict and dialogue for mathematics education.* Dordrecht, The Netherlands: Kluwer.

Vithal, R. & Skovsmose, O. (1997). The end of innocence: A critique of 'ethnomathematics.' *Educational Studies in Mathematics, 34*(2), 131–157.

Walkerdine, V. (1988). *The mastery of reason: Cognitive developments and the production of rationality.* New York: Routledge.

Walkerdine, V. (1990). Difference, cognition, and mathematics education. *For the Learning of Mathematics, 10*(3), 51–56.

Walkerdine, V. (1997). Redefining the subject in situated cognition theory. In D. Kirshner & J. A. Whitson (Eds.), *Situated cognition: Social, semiotic, and psychological perspectives* (pp. 57–70). Mahwah, NJ: Erlbaum.

Watson, A. (Ed.) (1998). *Situated cognition and the learning of mathematics.* Oxford, England: University of Oxford Department of Educational Studies.

Wenger, E. (1998). *Communities of practice: Learning, meaning, and identity.* Cambridge, England: Cambridge University Press.

Wertsch, J. V. (1998). *Mind as action.* Oxford, England: Oxford University Press.

Whitson, J. A. (1994, November). Elements of a semiotic framework for understanding situated and conceptual learning. In D. Kirshner (Ed.), *Proceedings of the 16th Annual Meeting of the North American Chapter of the International Group for the Psychology of Mathematics Education,* (Vol. 1, pp. 35–50). Baton Rouge: Louisiana State University.

Whitson, J. A. (1997). Cognition as a semiosic process: From situated mediation to critical reflective transcendance. In D. Kirshner & J. A. Whitson (Eds.), *Situated cognition: Social, semiotic, and psychological perspectives* (pp. 97–149). Mahwah, NJ: Lawrence Erlbaum Associates.

Wilson, P. S. & Mosquera, P. (1991, October). A challenge: Culture inclusive research. *Proceedings of the 13th Annual Meeting of the North American Chapter of the International Group for the Psychology of Mathematics Education,* (Vol. 2, pp. 22–28). Blacksburg, VA.

Wood, L. N. (1998). Communicating mathematics across culture and time. In H. Selin (Ed.), *Mathematics across cultures: The history of non-Western mathematics* (pp. 1–12). Dordrecht, The Netherlands: Kluwer.

Yackel, E., Stephan, M., Rasmussen, C., & Underwood, D. (2003). Didactising: Continuing the work of Leen Streefland. *Educational Studies in Mathematics, 54*(1), 101–126.

Yu, P. W. (2004). *Prototype development and discourse among middle school students in a dynamic geometric environment.* Unpublished doctoral dissertation, Illinois State University.

Zaslavsky, C. (1973). *Africa counts: Number and pattern in African cultures.* Chicago: Lawrence Hill Books.

Zaslavsky, C. (1991). World cultures in the mathematics class. *For the Learning of Mathematics, 11*(2), 32–36.

Zaslavsky, C. (1996). *The multicultural mathematics classroom: Bringing in the world.* Portsmouth, NH: Heinemann.

Useful Web Sites on Culture and the Learning and Teaching of Mathematics:

http://www.csus.edu/indiv/o/oreyd/once/once.htm

http://www.geometry.net/pure_and_applied_math/ethnomathematics.html

http://www.dm.unipi.it/~jama/ethno/

http://phoenix.sce.fct.unl.pt/GEPEm/

http://www.fe.unb.br/etnomatematica/

http://www2.fe.usp.br/~etnomat

http://web.nmsu.edu/~pscott/spanish.htm

http://etnomatematica.univalle.edu.co

http://www.rpi.edu/~eglash/isgem.htm

http://chronicle.com/colloquy/2000/ethnomath/ethnomath.htm

http://chronicle.com/free/v47/i06/06a01601.htm

http://chronicle.com/colloquy/2000/ethnomath/re.htm

http://www.ecsu.ctstateu.edu/depts/edu/projects/ethnomath.html

AUTHOR NOTE

I wish to thank Ubiratan D'Ambrosio and Joanna Masingila for helpful comments on an earlier draft of this chapter, and Ubiratan D'Ambrosio for supplying the list of useful web sites, including those of relevant dissertations.

PART IV

Students and Learning

12

EARLY CHILDHOOD MATHEMATICS LEARNING

Douglas H. Clements and Julie Sarama

UNIVERSITY AT BUFFALO, STATE UNIVERSITY OF NEW YORK

It seems probable that little is gained by using any of the child's time for arithmetic before grade 2, though there are many arithmetic facts that he [sic] can learn in grade 1.

—Thorndike, 1922, p. 198

Children have their own preschool arithmetic, which only myopic psychologists could ignore.

—Vygotsky, 1978/1935, p. 84

For over a century, views of young children's mathematics have differed widely. The recent turn of the century has seen a dramatic increase in attention to the mathematics education of young children. Although mathematics researchers have studied young children's learning for some time, much of the research corpus originated in fields such as developmental psychology. Our goal is to synthesize relevant research on the learning of mathematics before first grade from multiple fields. The synthesis reveals a nascent field that nonetheless is a potential harbinger of theoretical advancement. For example, genetic epistemology, as the study of the origins of knowledge itself, has never been more deeply developed or empirically tested than in recent research on early mathematics learning. Thus, this body of research illuminates foundational issues on the learning of mathematics with theoretical implications for all ages.

We begin with a brief overview of mathematics in early childhood and young children's learning of mathematics. In the bulk of the review, we address children's learning of ideas within the five topical domains determined to be significant for young children's learning (Clements & Conference Working Group, 2004; NCTM, 2000), because it is most fruitful to focus on the big ideas of mathematics (Bowman, Donovan, & Burns, 2001; Clements, 2004; Fuson, 2004; Griffin, Case, & Capodilupo, 1995; Tibbals, 2000; Weiss, 2002). We then briefly discuss issues of mathematical processes and learning and teaching contexts, including early childhood school settings and equity issues. Space constraints limit discussion of educational implications to those with direct empirical evidence (but see other sources, Clements, Sarama, & DiBiase, 2004).

MATHEMATICS IN EARLY CHILDHOOD

There are at least seven reasons for the recent surge of attention to mathematics in early childhood. First, increasing numbers of children attend early care and education programs. In 1999, 70% of 4-year-olds and 93% of 5-year-olds were enrolled in preprimary education, up from 62 and 90%, respectively, in 1991 (U.S. Department of Education, 2000, p. 7). Several states are instituting universal pre-K,[1] with about 1 million children enrolled in 1999, and that number is increasing (Hinkle, 2000). In 2001, about two thirds of all 4-year-olds were enrolled, with that ratio increasing (Loeb, Bridges, Bassok, Fuller, & Rumberger, in press; Magnuson, Meyers, Rathbun, & West, 2004). Various government agencies, federal and state, provide financial support for pre-K programs designed to facilitate academic achievement, particularly in low-income children.

Second, there is an increased recognition of the importance of mathematics (Kilpatrick, Swafford, & Findell, 2001). In a global economy with the vast majority of jobs requiring more sophisticated skills than in the past, American educators and business leaders have expressed strong concern about students' mathematics achievement. These concerns are echoed in international comparisons of mathematics achievement (Mullis et al., 1997).

Third, the mathematics achievement of American students compares unfavorably with the achievement of students from several other nations, even as early as kindergarten. Some cross-national differences in informal mathematics knowledge appear as early at 4 to 5 years of age (Starkey et al., 1999; Yuzawa, Bart, Kinne, Sukemune, & Kataoka, 1999).

Fourth, the knowledge gap is most pronounced in the performance of U.S. children living in economically deprived urban communities (Griffin, Case, & Siegler, 1994; Saxe, Guberman, & Gearhart, 1987; Siegler, 1993). That is, differences are not just between nations, but also between socioeconomic groups within countries.

Fifth, researchers have changed from a position that young children have little or no knowledge of or capacity to learn mathematics (e.g., Piaget, Inhelder, & Szeminska, 1960; Piaget & Szeminska, 1952; Thorndike, 1922) to theories that posit competencies that are either innate or develop in the first years of life (Baroody, Lai, & Mix, 2006; Clements, Sarama, & DiBiase, 2004; R. Gelman & Gallistel, 1978; Perry & Dockett, 2002). We will discuss these issues in more depth; here, it suffices to say that young children clearly can engage with substantive mathematical ideas.

Sixth, specific quantitative and numerical knowledge in the years before first grade has been found to

be a stronger predictor of later mathematics achievement than tests of intelligence or memory abilities (Krajewski, 2005).

Seventh, research indicates that knowledge gaps appear in large part due to the lack of connection between children's informal and intuitive knowledge (Ginsburg & Russell, 1981; Hiebert, 1986) and school mathematics. This is especially detrimental when this informal knowledge is poorly developed (Baroody, 1987; Griffin, Case, & Siegler, 1994).

For these reasons, there has been much recent interest in, and attention to, the learning and teaching of mathematics before first grade, the focus of this chapter.

YOUNG CHILDREN AND MATHEMATICS LEARNING

Young children possess an informal knowledge of mathematics that is surprisingly broad, complex, and sophisticated (Baroody, 2004; Clements, Swaminathan, Hannibal, & Sarama, 1999; Fuson, 2004; Geary, 1994; Ginsburg, 1977; Kilpatrick, Swafford, & Findell, 2001; Piaget & Inhelder, 1967; Piaget, Inhelder, & Szeminska, 1960; Steffe, 2004). For example, preschoolers engage in substantial amounts of foundational free play.[2] They explore patterns, shapes, and spatial relations; compare magnitudes; and count objects. Importantly, this is true for children regardless of income level and gender (Seo & Ginsburg, 2004). They engage in mathematical thinking and reasoning in many contexts, especially if they have sufficient knowledge about the materials they are using (e.g., toys), if the task is understandable and motivating, and if the context is familiar and comfortable (Alexander, White, & Daugherty, 1997).

In a similar vein, most entering kindergartners, and even entering preschoolers, show a surprisingly high level of mathematical skills, with much individual variation. For example, in 1999, 94% of entering U.S. kindergartners could count to 10 and recognize numerals and shapes, and 58% could also read numerals, count beyond 10, make patterns, and use non-

---

[1] Terms for the preprimary years are not used consistently. We use *pre-K* for prekindergarten, the year before kindergarten entrance (in the United States pre-K children are usually 4 years of age); *pre-K children* and *preschoolers* are often used interchangeably, with plural phrases such as *the preschool years* explicitly meaning the pre-K year and the year before this year (3-year-olds); *toddlers* refers to 1- to 2-year-olds and infants to children below 1 year of age.

[2] Such everyday foundational experiences form the intuitive, implicit conceptual foundation for later mathematics. Later, children represent and elaborate these ideas—creating models of an everyday activity with mathematical objects, such as numbers and shapes; mathematical actions, such as counting or transforming shapes; and their structural relations. We call this process mathematization. A distinction between foundational and mathematized experiences is necessary to avoid confusion about the type of activity in which children are engaged (Kronholz, 2000).

standard units of length to compare objects (National Center for Education Statistics [NCES], 2000). Similarly, about 68% of a sample of low-income children entering preschool in two states in the United States could count verbally to 5, 44% to 10 (Clements, Sarama, & Gerber, 2005). From 37 to 45% could count small groups of objects (2 to 7). However, only 2 to 16% of the children could perform various nonverbal addition and subtraction problems and 2 to 12% could solve verbal problems with sums of 5 or less. Between 75 and 86% of the children could identify prototypical examples of squares, triangles, and rectangles, and 39 to 74% could identify palpable nonexamples (palpable distractors) or categorize nonprototypical members and nonpalpable distractors (26–59%).

Such findings indicate that foundational mathematical knowledge begins during infancy and undergoes extensive development over the first 5 years of life. It is just as natural for young children to think premathematically and then mathematically as it is for them to use language, because "humans are born with a fundamental sense of quantity" (Geary, 1994, p. 1), as well as spatial sense, and a propensity to search for patterns.

In summary, young children have the interests and ability to engage in significant mathematical thinking and learning. Their abilities extend beyond what is introduced in most programs (Aubrey, 1997; Clements, 1984c; Geary, 1994; Griffin & Case, 1997; A. Klein & Starkey, 2004).

Most of this review is organized around the *big ideas of mathematics* (Bowman, Donovan, & Burns, 2001; Clements, 2004; Fuson, 2004; Griffin, Case, & Capodilupo, 1995; Tibbals, 2000; Weiss, 2002), the overarching concepts that are mathematically central and coherent, consistent with children's thinking, and generative of future learning (Clements & Conference Working Group, 2004). This organization reflects the idea that children's early competencies are organized around several large conceptual areas.

In addition, in developing each of these big ideas, children often pass through a sequence of levels of thinking. These developmental progressions can underlie hypothetical learning trajectories. Learning trajectories have three parts: a goal, a learning path or trajectory through which children move through levels of thinking, and instruction that helps them move along that path. Formally, learning trajectories are descriptions of children's thinking as they learn to achieve specific goals in a mathematical domain and a related, conjectured route through a set of instructional tasks designed to engender those mental processes or actions hypothesized to move children through a developmental progression of levels of thinking (Clements & Sarama, 2004c).

Learning trajectories are useful pedagogical as well as theoretical constructs (Bredekamp, 2004; Clements & Sarama, 2004b; M. A. Simon, 1995). In one study, the few teachers who actually led in-depth discussions in reform mathematics classrooms saw themselves not as moving through a curriculum, but as helping students move through levels of understanding (Fuson, Carroll, & Drueck, 2000). Further, researchers suggest that professional development focused on developmental progressions increases not only teachers' professional knowledge but also their students' motivation and achievement (D. M. Clarke et al., 2001; D. M. Clarke et al., 2002; Fennema et al., 1996; G. Thomas & Ward, 2001; R. J. Wright, Martland, Stafford, & Stanger, 2002). A focus on both big ideas and the "conceptual storylines" of curricula in the form of hypothetical learning trajectories is supported by research on systemic reform initiatives (Heck, Weiss, Boyd, & Howard, 2002). For these reasons, this review attempts to explicate the developmental progressions of the big ideas.

This emphasis on content is not meant to deny the importance of processes, which we discuss throughout and focus on in the second last section.

> As important as mathematical content are general mathematical processes such as problem solving, reasoning and proof, communication, connections, and representation; specific mathematical processes such as organizing information, patterning, and composing, and habits of mind such as curiosity, imagination, inventiveness, persistence, willingness to experiment, and sensitivity to patterns. All should be involved in a high-quality early childhood mathematics program. (Clements & Conference Working Group, 2004, p. 57)

This structure of goals is consistent both with recommendations of the National Council of Teachers of Mathematics (NCTM, 2000) and with research on young children's development of a network of logical and mathematical relations (Kamii, Miyakawa, & Kato, 2004). Finally, it is consistent with the conclusions of another review that "the overriding premise of our work is that throughout the grades from pre-K through 8 all students can and should be mathematically proficient" (Kilpatrick, Swafford, & Findell, 2001, p. 10), including conceptual understanding, procedural fluency, strategic competence, adaptive reasoning (capacity for logical thought, reflection, explanation, and justification), and a productive disposition.

Theoretical Frameworks

Three main theoretical frameworks for understanding young children's mathematical thinking are empiricism, (neo)nativism, and interactionalism. The introductory quote by Thorndike illustrates an empiricist framework. In traditional empiricism, the child is seen as a "blank slate," truth lies in correspondences between children's knowledge and reality, and knowledge is received by the learner via social transmission or abstracted from repeated experience with a separate ontological reality. An extension, traditional information processing theory, uses the computer as a metaphor for the mind and moves slightly toward an interactionalist perspective.

In contrast, nativist theories, in the traditional of philosophical rationalism (e.g., Plato and Kant), emphasize the inborn or early developing capabilities of the child. For example, quantitative or spatial cognitive structures present in infancy support the development of later mathematics, and thus innate structures are fundamental to mathematical development. Many theorists build on Gelman and Gallistel's *privileged domains hypothesis* (Rittle-Johnson & Siegler, 1998). In this view, a small number of innate or early-developing mathematical competencies are privileged and easy to learn. These are hypothesized to have evolutionary significance and be acquired or displayed by children in diverse cultures at approximately the same age. This is in contrast to other perspectives that would explain relative competence as resulting from frequency of experience. As we shall show in more detail, neither the empiricist nor nativist position fully explains children's learning and development. An intermediate position seems warranted, such as interactionalist theories that recognize the interacting roles of nature and nurture (Newcombe, 2002).

In interactionalist, constructivist theories, children actively and recursively create knowledge. Structure and content of this knowledge are intertwined and each structure constitutes the organization and components from which the child builds the next, more sophisticated structure. In comparison to nativism's initial representational cognition, children's early structures are prerepresentational.

Constructivist theories take several forms. In *trivial constructivism*, children are not passive receivers of knowledge, but knowledge is seen as learning about an objective reality. In contrast, in *radical constructivism* cognition is adaptive and serves to help children organize their experiential world, rather than uncover or discover a separate, objective ontology. Radical constructivism often uses an evolutionary view of cognitive development, in which children generate cognitive schemes to solve perceived problems and test them to see how well they fit the experiential world. *Social constructivism*, building on Vygotsky's seminal work, holds that individuals and the society are inter-connected in fundamental ways; there is no isolated cognition (Ernest, 1995).

We believe that the research reviewed here supports a synthesis of aspects of previous theoretical frameworks that we call *hierarchic interactionalism*. The term indicates the influence and interaction of global and local (domain specific) cognitive levels and the interactions of innate competencies, internal resources, and experience (e.g., cultural tools and teaching). Mathematical ideas are represented intuitively, then with language, then metacognitively, with the last indicating that the child possesses an understanding of the topic and can access and operate on those understandings. The tenets of hierarchic interactionalism follow; research supporting these tenets will be developed throughout the chapter.

1. *Developmental progression.* Most content knowledge is acquired along developmental progressions of levels of thinking. These progressions play a special role in children's cognition and learning because they are particularly consistent with children's intuitive knowledge and patterns of thinking and learning at various levels of development (at least in a particular culture, but guided in all cultures by "initial bootstraps"—see below), with each level characterized by specific mental objects (e.g., concepts) and actions (processes), (e.g., Clements, Wilson, & Sarama, 2004; Steffe & Cobb, 1988).

2. *Domain specific progression.* These developmental progressions often are most propitiously characterized within a specific mathematical domain or topic (see also Karmiloff-Smith, 1992; cf. Resnick's "conceptual rationalism," 1994). Children's kowledge (i.e., the objects and actions they have developed with that domain) are the *main determinant of the thinking within each progression, although hierarchic interactions occur at multiple levels* within and between topics, as well as with general cognitive processes (e.g., metacognitive processes, potentialities for general reasoning and learning-to-learn skills).

3. *Hierarchic development.* This key tenet contains two ideas. First, development is less about the emergence of entirely new processes and products and more an interactive interplay among specific components of knowledge and processes (both general and specific; Minsky, 1986). Second, each level builds hierarchically on the concepts and processes of the previous levels. Levels are coherent and

often characterized by increased sophistication, complexity, abstraction, power, and generality. However, the learning process is more often incremental and gradually integrative than intermittent and tumultuous (i.e., occurring mainly between stable levels). Various models and types of thinking grow in tandem to a degree, but a critical mass of ideas from each level must be constructed before thinking characteristic of the subsequent level becomes ascendant in the child's thinking and behavior (Clements, Battista, & Sarama, 2001). Successful application leads to the increasing use of a particular level. However, under conditions of increased task complexity, stress, or failure this probability level decreases, and an earlier level serves as a fallback position (Hershkowitz & Dreyfus, 1991; Siegler, 1986). No level of thinking is deleted from memory (Davis, 1984). That is, rerecording a mental representation at a more explicit level does not erase the earlier representation. (Indeed, replacing or deleting any successful representations would eliminate "fall-back" strategies that arguably are essential when new, untested knowledge is being formed, cf. Minsky, 1986; Vurpillot, 1976). The schemes that constitute the representation are instantiated dynamically (i.e., their form is interactively shaped by the situational context, including the task demands), and they can be modified at this time (or new schemes can be formed), although schemes that are basic subschemes of many other schemes become increasingly resistant to alteration because of the disruptive effect alteration would have on cognitive functioning. The continued existence of earlier levels and the role of intentionality and social influences in their instantiation explain why in certain situations even adults fall back to earlier levels, for example, failing to conserve in some contexts.

4. *Cyclic concretization.* Development progressions often proceed from sensory-concrete levels at which perceptual concrete supports are necessary and reasoning is restricted to limited cases (such as small numbers) to verbally based (or enhanced) generalizations and abstractions that are tenuous to integrated-concrete understandings relying on internalized mental representations that serve as mental models for operations and abstractions that are increasingly sophisticated and powerful.

5. *Co-mutual development of concepts and skills.* Concepts constrain procedures, and concepts and skills develop in constant interaction (Baroody, Lai, & Mix, 2005; Greeno, Riley, & Gelman, 1984). (In imbalanced cultural or educational environments, one may take precedence, to an extent that can be harmful.) Effective instruction often places initial priority on conceptual understanding, including children's creations of solution procedures (e.g., Carpenter, Franke, Jacobs, Fennema, & Empson, 1998), the importance of autonomy (Kamii & Housman, 1999), and mathematical dispositions and beliefs ("Once students have learned procedures without understanding, it can be difficult to get them to engage in activities to help them understand the reasons underlying the procedure," Kilpatrick, Swafford, & Findell, 2001, p. 122). The constructs of concepts and skills include symbolic representations, utilization competence (Greeno, Riley, & Gelman, 1984), and general cognitive skills.

6. *Initial bootstraps.* Children have important but often inchoate premathematical and general cognitive competencies and predispositions at birth or soon thereafter that support and constrain but do not absolutely direct subsequent development of mathematics knowledge. These have been called *experience-expectant processes* (Greenough, Black, & Wallace, 1987), in which universal experiences lead to an interaction of inborn capabilities and environmental inputs that guide development in similar ways across cultures and individuals (cf. Karmiloff-Smith, 1992).

7. *Different developmental courses.* Different developmental courses are possible within those constraints, depending on individual and social confluences (Clements, Battista, & Sarama, 2001). Within any developmental course, at each level of development, children have a variety of cognitive tools—concepts, strategies, and skills—that coexist.

8. *Progressive hierarchization.* Within and across developmental progressions, children gradually make connections between various mathematically relevant concepts and procedures, weaving ever more robust understandings that are hierarchical (i.e., they employ generalizations while maintaining differentiations). These generalizations and metacognitive abilities eventually connect

to form logical-mathematical structures that virtually compel children toward decisions in certain domains, such as those on traditional Piagetian conservation of number tasks, that are resistant to confounding via misleading perceptual cues. Children provided with high-quality educational experiences build similar structures across a wide variety of mathematical domains (again, in contrast to instrumental knowledge).

9. *Environment and culture.* Environment and culture affect the pace and direction of the developmental courses. For example, the degree of experience children have to observe and use number and other mathematical notions and to compare these uses will affect the rate and depth of their learning along the developmental progressions. The degree to which children learn mathematical words, exposure to which varies greatly across cultural groups (Hart & Risley, 1995), affects developmental courses. Words alert children to the class of related words to be learned and to specific mathematical properties, laying the foundation for learning mathematical concepts and language (cf. Sandhofer & Smith, 1999) by providing a nexus on which to build their nascent constructs (Vygotsky, 1934/1986).

10. *Consistency of developmental progressions and instruction.* Instruction based on learning consistent with natural developmental progressions is more effective, efficient, and generative for the child than learning that does not follow these paths.

11. *Learning trajectories.* An implication of the tenets to this point is that a particularly fruitful instructional approach is based on hypothetical learning trajectories (Clements & Sarama, 2004c). On the basis of the hypothesized specific mental constructions (mental actions-on-objects) and patterns of thinking that constitute children's thinking, curriculum developers design instructional tasks that include external objects and actions that mirror the hypothesized mathematical activity of children as closely as possible. These tasks are sequenced, with each corresponding to a level of the developmental progressions, to complete the hypothesized learning trajectory. Such tasks will theoretically constitute a particularly efficacious educational program; however, there is no implication that the task sequence is the only

path for learning and teaching, only that it is hypothesized to be one fecund route.

12. *Instantiation of hypothetical learning trajectories.* Hypothetical learning trajectories must be interpreted by teachers and are only realized through the social interaction of teachers and children around instructional tasks. Societally determined values and goals are substantive components of any curriculum (Confrey, 1996; Hiebert, 1999; National Research Council [NRC], 2002; R. W. Tyler, 1949); research cannot ignore or determine these components (cf. Lester & Wiliam, 2002). There is no understanding without reflection, and reflection is an activity students have to carry out themselves. No one else can do it for them. Yet a teacher who has some inkling as to where a particular student is in his or her conceptual development has a better chance of fostering a further reflective abstraction than one who merely follows the sequence of a preestablished curriculum. (von Glasersfeld, 1995)

NUMBER AND QUANTITATIVE THINKING

For early childhood, number and operations is arguably the most important area of mathematics learning. In addition, learning of this area may be one of the best developed domains in mathematics research, especially in the early primary grades (Baroody, 2004; Fuson, 2004; Kilpatrick, Swafford, & Findell, 2001; Steffe, 2004). Preschool mathematics education, although an emergent area, benefits from a wealth of psychological and early childhood studies.

We discuss children's numerical concepts and operations separately, although they and their components are highly interrelated. We do not limit "operations" to standard arithmetic operations of adding, subtracting, multiplying, and dividing but include counting, comparing, unitizing, grouping, partitioning, and composing. The importance of these ideas is highlighted by their close correspondence to the components of number sense, including composing and decomposing numbers, recognizing the relative magnitude of numbers, dealing with the absolute magnitude of numbers, using benchmarks, linking representations, understanding the effects of arithmetic operations, inventing strategies, estimating, and possessing a disposition toward making sense of numbers (Sowder, 1992). The competencies rest on early quantitative reasoning that begins to develop as early as the first 2 years of life. Preceding cognitive psychology research concerning early numbers in infants by half a century, Vygotsky said

> The first stage is formed by the natural arithmetical endowment of the child, i.e., his operation of quantities before he knows how to count. We include here the immediate conception of quantity, the comparison of greater and smaller groups, the recognition of some quantitative group, the distribution into single objects where it is necessary to divide, etc.(Vygotsky, 1929/1994, p. 67)

Introduction to the Early Development of Quantity

In 1917, Warren McCulloch's academic advisor asked him to describe his research interests. He replied, "What is number, that a man [*sic*] may know it, and a man that he may know number?" His advisor responded, "Friend, thee will be busy as long as thee lives" (McCulloch, 1963, p. 1).

Over more than 100 years, research on early number knowledge has passed through four broad phases (cf. Clements, 1984b). After a prescient analysis of early number by Dewey (1898), researchers studied subitizing (the direct and rapid perceptual apprehension of the numerosity of a group, from the Latin "to arrive suddenly," named by Kaufman, Lord, Reese, & Volkmann, 1949, who referred to verbal naming of the numerosity), counting, and the relationship between subitizing and counting (e.g., Douglass, 1925; Freeman, 1912).

In the second phase, the wide influence of Piagetian theory redirected theoretical and empirical study, as it explained the development of number concepts based on underlying logical operations.

> Our hypothesis is that the construction of number goes hand-in-hand with the development of logic, and that a pre-numerical period corresponds with the pre-logical level. Our results do, in fact, show that number is organized, stage after stage, in close connection with the gradual elaboration of systems of inclusions (hierarchy of logical classes) and systems of asymmetrical relations (qualitative seriations), the sequence of numbers thus resulting from an operational synthesis of classification and seriation. In our view, logical and arithmetical operations therefore constitute a single system that is psychologically natural, the second resulting from generalization and fusion of the first, under the two complementary headings of inclusion of classes and seriation of relations, quality being disregarded (Piaget & Szeminska, 1952, p. viii).

Counting was viewed as ineffectual, with "no connection between the acquired ability to count and the actual operations of which the child is capable" (Piaget & Szeminska, 1952, p. 61), with the child *capable* of meaningful counting only upon reaching the level of reversible operations (p. 184). Studies in this tradition tended to disregard subitizing "the intuitive numbers" as well as counting, which had been the major focus of earlier studies of number concepts. Instead, Piagetians focused on number conservation, which they considered the sine qua non of all rational (i.e., based on reason) number activity. Number conservation is the ability to state quickly and assuredly that the numerosity of a collection above subitizing range (e.g., more than 7) had not changed after a change in the relationship among its elements, such as increasing the distance between them. In the Piagetian view, children do not acquire a notion of quantity and then conserve it; they discover true quantification only when they become capable of conservation. This develops in three stages: Stage 1, gross quantity, in which children make global perceptual judgments without one-to-one correspondence; Stage 2, intensive quantity, in which they can make one-to-one correspondences perceptually; and Stage 3, in which they construct the notion of unit and numerical correspondence and they understand that inverse changes in length and density can compensate each other and thus that changes are reversible. Spatial qualities no longer determine number. The elements become equivalent and interchangeable units (i.e., equivalent members of a class), differing only by their relative order (i.e., seriation). Each successive element creates a category containing all previous classes (i.e., hierarchical inclusion).

The result was a broad belief, in research and educational practice arenas, that children could not reason logically about quantities until the early elementary school years. This view seemed to be supported by hundreds of investigations of the validity of the Piagetian stages. Studies generally supported the existence of the stages (Clements, 1984a). For example, children scoring lower on tests of conservation scored significantly lower on tests of problem solving (Harper & Steffe, 1968; Steffe, 1966).

The third phase began when new perspectives and research methodologies challenged critical aspects of the Piagetian view. Researchers questioned whether the Piagetian tasks measured number knowledge or other competencies, such as attending to relevant attributes (R. Gelman, 1969). Others doubted the Piaget position that without logical operations there is no possibility of meaningful quantitative reasoning and therefore attempting to develop children's number abilities, such as counting, was indefensible.

Supporting the new perspective, seminal research showed numerical and arithmetic competencies in young children who were not operational in Piagetian assessments (R. Gelman, 1969; Harper & Steffe, 1968). For example, in the "magic" experiments, two plates where shown, one with, for example, 2 mice, designated "the loser," and the other with three mice, designated "the winner." After a series of identification tasks, the experimenter surreptitiously altered the winner, changing the spatial arrangement in some experiments, altering the identity of the items or size of the collection in others. The result we emphasize here is that children as young as 3 and sometimes 2.5 years seem to know that transformations involving displacements do not change the numerical value of a display—an early form of conservation of number (R. Gelman & Gallistel, 1978). Children preferred to make such decisions based on equivalence or nonequivalence of verbal numbers, rather than on the 1-to-1 object correspondences emphasized in Piagetian studies. Such studies formed the basis for a nativist view of early number development.

Other researchers built number skills-integration models. Taking an information-processing approach, these empiricist-oriented models used task and scalogram analyses to describe sequences of learning numerals, counting and numeration of small before larger collections, and independence of counting and one-to-one correspondence (M. Wang, Resnick, & Boozer, 1971). Other models postulated hierarchic skills-integration sequences (Klahr & Wallace, 1976; Schaeffer, Eggleston, & Scott, 1974). Numerous studies supported the contention of a multidimensional concept of number in young children (Clements, 1984a).

The fourth phase is an extension of the third phrase. Debates continue about the meaning of early (and lack of) competence (Baroody, Lai, & Mix, 2006; Cordes & Gelman, 2005; Mix, Huttenlocher, & Levine, 2002; T. J. Simon, 1997; Spelke, 2003; Uller, Carey, Huntley-Fenner, & Klatt, 1999). Some researchers have attempted to synthesize components of the various theoretical perspectives or create intermediate positions. Nativist, empiricist, and interactionalist accounts have been evaluated both philosophically and empirically. Number knowledge similarly has been subjected to analysis, resulting in more sophisticated conceptualizations. For example, number and quantitative knowledge apparently develops substantially earlier than Piagetian logical operations but nonetheless appears to be built upon inchoate foundations, such as unitizing or creating an image of a collection of objects (classification) and tracking and ordering (seriation).

Issues concerning nativist, empiricist, and interactionalist accounts arise in each of the following main sections. In the remainder of this section, we illustrate these issues by focusing on children's initial competence with number. For example, using new techniques, some researchers claimed that infants possessed understanding of cardinality. Many of these researchers built on Gelman and Gallistel's nativist position, including the privileged domains of whole-number counting and simple arithmetic, which necessitated an underlying foundation of both cardinality and ordinality. The hypothesis predicts that conceptual understanding precedes skilled execution of the relevant procedures in these domains (R. Gelman & Gallistel, 1978). In contrast, the empiricist skills-integration approach would predict that exposure frequency determines development, and thus procedural skill would emerge before conceptual development.

Providing evidence for their position, nativists reported that babies in the first 6 months of life can discriminate one object from two, and two objects from three (Antell & Keating, 1983; Starkey, Spelke, & Gelman, 1990). This was determined via a habituation paradigm in which infants "lose interest" in a series of displays that differ in size, density, color, and so forth, but have the same number of objects. Renewed interest when shown a display with a different number of objects provides evidence that they are sensitive to number (Wynn, Bloom, & Chiang, 2002). Thus, infants can discriminate among and match small configurations (1–3) of objects. The experiments also indicate that children's discrimination is limited to such small numbers. Children do not discriminate four objects from five or six until the age of 3 or 4 years (Starkey & Cooper, 1980). Some researchers have therefore suggested that infants use subitizing, defined as an automatic perceptual process that all people, including adults, can apply only to small collections up to around four objects (Chi & Klahr, 1975). Children can also construct quantitative equivalence relations by 6 to 8 months of age through exchange operations (substituting or commuting objects; Langer, Rivera, Schlesinger, & Wakeley, 2003).

Some initial competence is generally accepted. What this competence *means*, however, remains in dispute. Recent findings suggest that infants in "number" experiments may be responding to overall contour length, area, mass, or density rather than discrete number (Feigenson, Carey, & Spelke, 2002; Tan & Bryant, 2000). For example, infants dishabituated to changes in contour length when the number of objects was held constant, but they did not dishabituate to changes in number when contour length was held constant (Clearfield & Mix, 1999). One model proposes that infants encode object files that store data on each object's properties, so that situations can be processed in terms of individuation or analog properties of these objects such as contour, depending on the situation (Feigenson, Carey, & Spelke, 2002). For example, parallel-processed individuation is used for very small collections, with continuous extent used when storage for individuation is exceeded.

The strongest claim that infants are processing number, rather than perceptual information, is based on two types of evidence. First is the claim that infants discriminate not only collections of objects, but also temporal sequences such as sounds or events like puppet jumps as early as 6 months of age (Wynn, 1998). Second is the claim that infants show cross-modal number abilities; that is, they can match visual representations of certain numbers and auditory sequences consisting of the same number of sounds (Starkey, Spelke, & Gelman, 1990). Such evidence has led to strong nativist claims. More recent evidence, however, has called these results and claims into question. Even if children do sometimes make these connections, general, nonnumerical processes of comparing inputs from different modalities (e.g., from sounds and their locations) may be used (T. J. Simon, 1997). More important, one team found opposing results—either no effect of auditory information and merely a tendency to prefer the larger contour (consistent with several previous studies), or longer looking at *non*corresponding displays (Moore, Benenson, Reznick, Peterson, & Kagan, 1987). In another study, children as old as 3 years performed at chance on an auditory-visual equivalence test (Mix, Huttenlocher, & Levine, 1996). There is no theoretical reason to justify why children more than 3 years old would have difficulty with tasks on which neonates ostensibly show competence. Consistent with these findings, more recent data found no evidence that infants notice quantitative equivalence between auditory sequences and visual displays (Mix, Levine, & Huttenlocher, 1997). Thus, there is little reason to believe that infants have the ability to deal with cross-modal quantitative correspondence or even discrete quantity (i.e., number).

This is not to say that empiricist positions are supported. Early quantitative competencies have been demonstrated. However, many, if not all, of the earliest competencies may be explicable through other frameworks. Further, how infants may use a variety of processes and how they may be weighted or combined, as well as how they may be evoked differently in various situations, have not been thoroughly studied (Baroody, Lai, & Mix, 2005). Researchers also need to learn more about what role such processes, even if they are numerical, play in children's early cognition

or development of later numerical activity (Nunes & Bryant, 1996).

Part of the problem is, of course, that infant studies are particularly difficult to perform and interpret. For example, the frequently employed habituation methodology, although useful, was developed to study perceptual, not conceptual issues, leaving open the question of what results mean (Haith & Benson, 1998). Nevertheless, research suggests that children discriminate collections on some quantitative bases from birth. Furthermore, most accounts suggest that these limited capabilities, with as yet undetermined contributions of maturation and experience, form a foundation for later connection to culturally based cognitive tools such as number words and the number word sequence, to develop exact and extended concepts and skills in number. We elaborate these issues in each of several areas of numerical and quantitative competence.

Subitizing

Early Work on Subitizing

In the first half of the 20th century, researchers believed that counting did not imply a true understanding of number but subitizing did (e.g., Douglass, 1925). Many saw the role of subitizing as a developmental prerequisite to counting. Freeman (1912) suggested that whereas measurement focused on the whole and counting focused on the unit item, only subitizing focused on both the whole and the unit—thus, subitizing underlies number ideas. Carper (1942) agreed subitizing was more accurate than counting and more effective in abstract situations. (Recall that the term "subitizing" was not created until later, Kaufman, Lord, Reese, & Volkmann, 1949.)

In the second half of the century, educators developed several models of subitizing and counting. Some models were based on the notion that subitizing was a more "basic" skill than counting (Klahr & Wallace, 1976; Schaeffer, Eggleston, & Scott, 1974). One reason was that children can subitize directly through interactions with the environment, without social interactions. Supporting this position, Fitzhugh (1978) found that some children could subitize sets of one or two but were unable to count them. None of these very young children, however, were able to count any sets that they could not subitize. She concluded that subitizing is a necessary precursor to counting. Research with infants similarly suggested that young children possess and spontaneously use subitizing to represent the number contained in small sets and that subitizing emerges before counting (A. Klein & Starkey, 1988).

As logical as this position seems, counterarguments exist. In 1924, Beckmann found that young children used counting rather than subitizing (cited in Solter, 1976). Others agreed that children develop subitizing later, as a shortcut to counting (Beckwith & Restle, 1966; Brownell, 1928; Silverman & Rose, 1980). In this view, subitizing is a form of rapid counting (R. Gelman & Gallistel, 1978).

Evidence for subitizing as a separate process from "fast counting" includes reaction time data showing that recognition of 1–3 objects is faster and more accurate than larger numerosities. Reaction times increase with the number of items in a set to be quantified; it takes adults about 250 to 350 ms longer to quantify a set of 7 than 6, or 6 than 5. In contrast, the increase is slight when numbers are small, about 40 to 100 ms longer to respond to 3 than 2, or 2 than 1, and this may be response choice time rather than time to process the stimuli. At least some data indicate that adults can process 3 within the same exposure time as 2 or 1 (Trick & Pylyshyn, 1994).

Nature of the Subitizing Process and the Early Development of Quantity

Researchers still debate the basis for subitizing ability. Some claim that there are two distinct types of enumeration: numerical subitizing and nonnumerical (or figural) subitizing. In addition, some models assume that subitizing is possible only for sets that are simultaneously displayed, whereas others allow for sequential enumeration (Canfield & Smith, 1996).

Recognition of spatial patterns, such as triangular arrays, and attentional mechanisms are the main explanations for those who posit an underlying nonnumerical process (Chi & Klahr, 1975; Klahr & Wallace, 1976; Mandler & Shebo, 1982; von Glasersfeld, 1982). For Mandler and Shebo, these abstract geometric patterns are mapped to specific numerosities using processes like those for recognizing colors. For von Glasersfeld, the patterns are constituted by motion, either physical or attentional, forming scanpaths. They become numerical after reflective abstraction that focuses attention on their iterative structure raises them to "pure" abstraction. In another model, objects are first individuated by a limited but parallel processing preattentive mechanism associated with object tracking, and then markers are enumerated in a serial process (Trick & Pylyshyn, 1994). There is a neural mechanism that distinguishes single, dual, or triple incidences of a given evident or object, based on inherent temporal parameters.

Other models consider subitizing to be a numerical process. In the Meck and Church model (1983) subitizing is a numerical process enabled by the availability of the functional equivalent of a number line in the brain that operates on both simultaneous and sequential items. A pacemaker emits equivalent pulses at a constant rate. When a unitized item is encountered, a pulse is allowed to pass through a gate, entering an accumulator. The gradations on the accumulator estimate the number in the collection of units, similar to height indicating numerosity in a stack of blocks. An alternative model postulates an evolutionarily based, abstract mental component dedicated to number perception; that is, the child perceives numbers directly (Dehaene, 1997). This counting-like process guides development of whole number counting, hypothesized to be a privileged domain. Findings from both humans and nonhuman animals support this position (Gallistel & Gelman, 2005).

Research using different methodologies confirms that infants rapidly recognize and use the number of sequentially presented pictures to predict the location of the next picture in the sequence (Canfield & Smith, 1996). There was no evidence they had to reaccumulate items by scanning working memory, supporting subitizing rather than fast-counting models. (G. A. Miller, 1956, agreed that his famous working-memory limit was not related to human subitizing limits.) Thus, the researchers (Canfield & Smith, 1996) judged the Meck and Church model (1983) most consistent with their results. Because infants never saw more than a single stimulus at any moment during the session and the items were presented in a manner that was neither rhythmic nor temporally predictable, the process of combining sequential events was viewed as numerical.

Recent reviews (Baroody, Lai, & Mix, 2005; Mix, Huttenlocher, & Levine, 2002) have concluded that children younger than 3 years tend not to represent any numbers except 1 and 2 precisely. For example, 3-day-old to 5-month-old infants could discriminate between collections of one and two, two and three but not between collections of three and four (Antell & Keating, 1983), and 10- to 16-month-olds successfully discriminated between 2 and 3 but not between 3 and 4 or even 2 and 4 or 3 and 6 (Feigenson, Carey, & Hauser, 2002). Even in studies that did show success, results are equivocal; for example, 10–12-month-olds could discriminate between 2 and 3 items, but not 4 and 5, with mixed results for 3 vs. 4 (Strauss & Curtis, 1984); performance of even 2-year-olds drops off on tasks involving 3 items, with above-chance levels not necessarily indicating exact representations (Starkey & Cooper, 1995).

Developmental differences may exist in the processes that underlie subitizing (cf. Resnick & Singer, 1993). The mental models view (Huttenlocher, Jordan, & Levine, 1994; Mix, Huttenlocher, & Levine, 2002) postulates that children represent numbers nonverbally and approximately, then nonverbally but exactly, and eventually via verbal, counting-based processes. (This in contrast to the accumulator model, which is seen as unable to account for the greater difficulty of sequential presentations, use of overall amount in early quantification, and representation of cardinality, Mix, Huttenlocher, & Levine, 2002.) In the mental models view, children cannot initially differentiate between discrete and continuous quantities but represent both approximately using one or more perceptual cues such as contour length (Mix, Huttenlocher, & Levine, 2002). Children develop the ability to individuate objects, providing the ability to build notions of discrete number. About the age of 2 years, they develop representational, or symbolic, competence (such as shown in symbolic play), allowing them to create mental models of collections, which they can retain, manipulate (move), add to or subtract from, and so forth (although the model does not adequately describe how cardinality is ultimately cognized and how comparisons are made). Early nonverbal capabilities then provide a basis for the development of verbally based numerical and arithmetic knowledge (young children are more successful on nonverbal than verbal versions of number and arithmetic tasks, Huttenlocher, Jordan, & Levine, 1994; Jordan, Hanich, & Uberti, 2003; Jordan, Huttenlocher, & Levine, 1992; Jordan, Huttenlocher, & Levine, 1994; Levine, Jordan, & Huttenlocher, 1992). (One difficulty with this argument is that calling both "arithmetic tasks" may unintentionally hide potential differences in children's schemes.) In this view, there is no reason to consider early quantitative development solely a number competence, much less assume that number is a privileged domain (Mix, Huttenlocher, & Levine, 2002).

A recent version of this theory (Baroody, Lai, & Mix, 2005) suggests that during the time when children are gaining representational precision, they are also moving toward more generalized but deeper (more interconnected) concepts. Approximate mental models serve as a transition between number based on perceptual cues and one based on an exact, abstract, mental model. Implicit distinctions between discrete and continuous quantities lie on a continuum from clearly distinct for collections of one to indistinct in the case of "many"—more than three or four. Meaningful learning of number words (in contrast to symbolic ability) causes the transition to exact numerical representations. Such development may apply to certain numbers at different times, even among very small numbers. The child can then begin to verbally represent numbers (beyond one and two, which may

be represented nonverbally and exactly and verbally at about the same time). This may provide the basis for understanding cardinality and other counting principles, as well as arithmetic ideas (Baroody, Lai, & Mix, 2006). Anywhere from 3 to 6 years, depending on the home and preschool environments, children also make the transition to using written representations, which help them develop numerical or abstract reasoning.

Research reviewed here is generally consistent with this position, with only a few modifications and additions. What infants quantify are collections of rigid objects; that is, sequences of sounds and events, or materials that are nonrigid and noncohesive (e.g., water), are not quantified (Huntley-Fenner, Carey, & Solimando, 2002). Such quantifications, including number, begin as an undifferentiated, innate notion of amount of objects. Object individuation, which occurs early in preattentive processing (and is a general, not numerical, process), helps lay the groundwork for differentiating discrete from continuous quantity. Multiple systems are employed, including an object-file system that stores information about the objects, some or all of which is used depending on the situation, and an estimator mechanism that stores analogue quantitative information only (Feigenson, Carey, & Spelke, 2002; Gordon, 2004; Johnson-Pynn, Ready, & Beran, 2005). This estimator may be a set of filters (a cognitive scheme detects the numerosity of a group), each tuned to an approximate number of objects (e.g., 3) but overlapping (Nieder, Freedman, & Miller, 2002). The child encountering small sets opens object-files for each in parallel. If the situation elicits quantitative comparisons, continuous extent is retrieved and used except in extreme circumstances. For example, by about 6 months of age, infants may represent very small numbers (1 or 2) as individuated objects. Conversely, large numbers in which continuous extent varies or is otherwise unreliable (McCrink & Wynn, 2004) may be processed by the analog estimator as a collection of binary impulses (as are event sequences later in development), but not by exact enumeration (Shuman & Spelke, 2005).

To compare quantities, correspondences are processed. Initially, these are inexact estimates comparing the results of two estimators, depending on the ratio between the sets (Johnson-Pynn, Ready, & Beran, 2005). Once the child can represent objects mentally, that child can also make exact correspondences between these nonverbal representations and eventually develop a quantitative notion of that comparison (Baroody, Lai, & Mix, 2005).

Even these correspondences, however, do not imply a cardinal representation of the collection (a representation of the collection qua a *numerosity* of a *group* of items). That is, one still must distinguish between noncardinal representations of a collection and an explicit cardinal representation. Only for the latter does the individual apply an integration operation to create a composite with some numerical index. Some claim that the accumulator yields a cardinal output; however, it may be quantitative, constituted as an indexing of a collection using an abstract, cross-modality system for numerical magnitude (Shuman & Spelke, 2005) without an explicit cardinality. Comparisons, such as correspondence mapping, might still be performed, but only at an implicit level (cf. Sandhofer & Smith, 1999). In this view, only with experience representing and naming collections is an explicit cardinal representation created. This is a prolonged process. Children may initially make word-word mappings between requests for counting or numbers (e.g., "how many?") to number words until they have learned several (Sandhofer & Smith, 1999). Then they label some (small number) cardinal situations with the corresponding number word, that is, map the number word to the numerosity property of the collection. They begin this phase even before 2 years of age, but for some time this applies mainly to the word *two*, a bit less to *one*, and with considerable less frequency, *three* and *four* (Fuson, 1992). Only after many such experiences do children abstract the numerosities from the specific situations and begin to understand that the situations named by *three* correspond; that is, they begin to establish what adults would term a numerical equivalence class. Counting-based verbal systems are then more heavily used and integrated, as described in the following section, eventually leading to explicit, verbal, mathematical abstractions.

In summary, early quantitative abilities exist, but they may not initially constitute systems that can be said to have an explicit number concept. Instead, they may be prematerial, foundational abilities (cf. Clements, Sarama, & DiBiase, 2004) that develop and integrate slowly, in a piecemeal fashion (Baroody, Benson, & Lai, 2003). For example, object individuation must be stripped of perceptual characteristics and understood as a perceptual unit item through abstracting and unitizing to be mathematical (Steffe & Cobb, 1988), and these items must be considered simultaneously as individual units and members of a collection whose numerosity has a cardinal representation to be numerical, even at the lowest levels. The explicit, cultural, numeral-based sense of number develops in interaction with, but does not replace (indeed, may always be based on, Gallistel & Gelman, 2005), the analog sense of number. Regardless of its origins in continuous or discontinuous processing, number rather than amount of substances does achieve a core status in quantitative reasoning for preschoolers to adults (K. F. Miller, 1995). The human facility with language probably plays a central role in linking relations between different representations and thus making early premathematical cognition numeric (Wiese, 2003).

Types of Subitizing

Regardless of the precise mental processes, subitizing appears to be phenomenologically distinct from counting and other means of quantification and deserves differentiated educational consideration. *Perceptual subitizing* (Clements, 1999b) is closest to the original definition of subitizing: recognizing a number without consciously using other mental or mathematical processes and then naming it. Perceptual subitizing employs a preattentional quantitative process but adds an intentional numerical process; that is, infant sensitivity to number is not (yet) perceptual subitizing. The term *perceptual* applies only to the quantification mechanism as phenomenologically experienced by the person; the intentional numerical labeling, of course, makes the complete cognitive act conceptual.

Perceptual subitizing also plays the primitive role of *unitizing*, or making single "things" to count from the stream of perceptual sensations (von Glasersfeld, 1995). "Cutting out" pieces of experience, keeping them separate, and eventually coordinating them with number words are not trivial tasks for young children. For example, a toddler, to recognize the existence of a plurality, must focus on the items such as apples and repeatedly apply a template for an apple *and* attend to the *repetition* of the template application.

How is it that people see an 8-dot domino and "just know" the total number, when evidence indicates that this lies above the limits of perceptual subitizing? They are using the second type of subitizing (a distinction for which there is empirical evidence, Trick & Pylyshyn, 1994). *Conceptual subitizing* (Clements, 1999b) plays an advanced organizing role. People who "just know" the domino's number recognize the number pattern as a composite of parts and as a whole. They see each side of the domino as composed of four individual dots and as "one four." They see the domino as composed of two groups of four and as "one eight." These people are capable of viewing number and number patterns as units of units (Steffe & Cobb, 1988). (Some research suggests that only the smallest numbers, perhaps up to 3, are actually perceptually recognized; thus, sets of 1 to 3 may be perceptually recognized, sets of 3 to about 6 may be decomposed and recomposed without the person being aware of the process. Conceptual subitizing as we use the term here refers to recognition in which the person consciously uses such partitioning strategies.)

Spatial patterns such as those on dominoes are just one kind. Other patterns are temporal and kinesthetic, including finger patterns, rhythmic patterns, and spatial-auditory patterns. Creating and using these patterns through conceptual subitizing helps children develop abstract number and arithmetic strategies (Steffe & Cobb, 1988). For example, children use temporal patterns when counting on. "I knew there were 3 more so I just said, nine . . . *ten, eleven, twelve*" (rhythmically gesturing three times, one "beat" with each count). They use finger patterns to figure out addition problems. Children who cannot subitize conceptually are handicapped in learning such arithmetic processes. Children who can, may only do so with small numbers at first. Such actions, however, can be "stepping stones" to the construction of more sophisticated procedures with larger numbers.

Factors Affecting Difficulty of Subitizing Tasks

In contrast to what might be expected from a view of innate subitizing ability, subitizing ability *develops*, in a stepwise fashion. That is, in laboratory settings, children can initially differentiate "one" from "more than one" at about 33 months of age (Wynn, 1992b). Between 35 and 37 months, they differentiate between one and two, but not larger numbers. A few months later, at 38 to 40 months, they identify three as well. After about 42 months, they can identify all numbers that they can count, four and higher, at about the same time. However, research in natural, child-initiated settings shows that the development of these abilities occurs much earlier, with children working on one and two around their second birthdays or earlier (Mix, Sandhofer, & Baroody, 2005). Further, some children many begin with "two" rather than "one." These studies intimate that language and social interactions interact with internal factors in development, as well as showing that number knowledge develops in levels, over time. Most studies suggest that children begin recognizing "one," then "one" and "two," then "three" and then "four," whereupon they learn to count and know other numbers (see R. Gelman & Butterworth, 2005, for an opposing view concerning the role of language; Le Corre, Van de Walle, Brannon, & Carey, 2005).

The spatial arrangement of sets also influences how difficult they are to subitize. Children usually find rectangular arrangements easiest, followed by linear, circular, and scrambled arrangements (Beckwith & Restle, 1966; M. Wang, Resnick, & Boozer, 1971). This is true for students from kindergarten to college.

Certain arrangements are easier for specific numbers. Arrangements yielding a better "fit" for a given number are easier (Brownell, 1928). Children make fewer errors for 10 dots than for 8 with the "domino five" arrangement, but fewer errors for 8 dots for the "domino four" arrangement.

For young children, however, neither of these arrangements is easier for any number of dots. Indeed, children 2- to 4-years-old show no differences between any arrangements of four or fewer items (Potter & Levy, 1968). For larger numbers, the linear arrangements are easier than rectangular arrangements. It may be that many preschool children do not use decomposing (conceptual subitizing). They can learn to conceptually subitize, though older research indicated that first graders' limit for subitizing scrambled arrangements is about four or five (Dawson, 1953).

Textbooks often present sets that discourage subitizing. Their pictures combine many inhibiting factors, including complex embedding, different units with poor form (e.g., birds that are not compact as opposed to squares), lack of symmetry, and irregular arrangements (Carper, 1942; Dawson, 1953). Such complexity hinders conceptual subitizing, increases errors, and encourages simple one-by-one counting.

Role of Spontaneous Subitizing in Early Mathematics Development

A series of five studies indicated that the child's tendency to spontaneously focus on numerosity is a distinct, mathematically significant process (Hannula, 2005). Such tendency at 3 years predicted development of cardinality knowledge a year later. Focusing on numerosity was also related to counting and arithmetic skills, even when nonverbal IQ and verbal comprehension were controlled. Results suggested that spontaneous focus builds subitizing ability, which in turn supported the development of counting and arithmetic skills. Thus, children low in the tendency to spontaneously focus on numerosity in the early years are at risk for later mathematical failure (Hannula, 2005).

A quasi-experimental study (Hannula, 2005) showed that it is possible to enhance 3-year-old children's spontaneous focusing on numerosity and thus catalyze children's deliberate practice in numerical skills (cf. Ericsson, Krampe, & Tesch-Römer, 1993). Children in the experimental group showed increased tendency to focus on numerosity and develop cardinality competencies on a delayed posttest compared to children in the control group.

The most effective technique may be to begin naming very small collections with numbers after children have established names and categories for some physical properties such as shape and color (Sandhofer & Smith, 1999). Numerous experiences naming such collections help children build connections between quantity terms (number, how many) and number words, then build word-cardinality connections (•• is "two") and finally build connections among the representations of a given number.

In summary, subitizing small numbers appears to proceed and support the development of counting ability (Le Corre, Van de Walle, Brannon, & Carey, 2005). Thus, it forms a foundation for all learning of number. Across development, numerical knowledge initially develops qualitatively and becomes increasingly mathematical. In subitizing, children's ability to "see small collections" grows from preattentive but quantitative, to attentive perceptual subitizing, to imagery-based subitizing, to conceptual subitizing (Clements, 1999b; Steffe, 1992). Perceptual patterns are those the child can, and must, immediately see or hear, such as domino patterns, finger patterns, or auditory patterns (e.g., three beats). A significant advance is a child's focusing on the exact number in these patterns, attending to the cardinality. Finally, children develop conceptual patterns, which they can operate on, as when they can mentally decompose a five pattern into two and three and then put them back together to make five again. These types of patterns may "look the same" on the surface but are qualitatively different. All can support mathematical growth and thinking, but conceptual patterns are the most powerful.

Verbal and Object Counting

As previously described, historically some (Dewey, 1898; Thorndike, 1922) have argued that initial mathematics education should emphasize counting, whereas others (e.g., Piaget, Inhelder, & Szeminska, 1960; Piaget & Szeminska, 1952; B. Russell, 1919) contended that counting was a rote skill until logical foundations were acquired. Even recent accounts treated counting, at least verbal counting, as a rote skill (Fuson, 1992; Ginsburg, 1977). However, number words can be meaningful in some contexts and can orient children to numerical meanings (Mix, Huttenlocher, & Levine, 2002). Further, without language, development of number appears to be severely restricted. For example, members of a tribe with only a "one-two-many" system of counting had remarkably poor performance on number tasks above three (Gordon, 2004). Even with language, without a *verbal counting system,* exact naming of and operations on number do not appear (Pica, Lemer, Izard, & Dehaene, 2004).

Verbal Counting

By 24 months of age, many toddlers have learned their first number word (typically "two"). Words for larger collections usually appear after children use verbal counting. Depending on their early environment, children begin to try to count using verbal number names at age 2 or 3 years. Important developments in counting continue during the preschool years. Children from ages 2 to 5 years learn more of the system of number words ("one, two, three, ...") due to a desire to count larger collections and a curiosity about the number word system itself (Baroody, 2004; Fuson, 1988, 2004; Griffin, 2004; Steffe, 2004). In this developmental progression, children learn verbal counting, reproducing the forward number word sequence.

We do not use the common phrase "rote counting" for verbal counting, because children have to learn the principles and patterns in the number system as coded in their natural language, at least for numbers words above 20 (Baroody, 1987; Fuson, 1992). Supporting evidence for this position is that children who could count when given a new point from which to start performed better on all numerical tasks, suggesting that fluent verbal counting does not depend primarily on rote factors, but rather on the recognition that the system is rule-governed (Pollio & Whitacre, 1970). That is, children learn that numbers derive order and meaning from their embeddedness in a system, and they learn a set of rules that allows the generation, not recall, of the appropriate sequence.

Further, cross-cultural studies indicate that learning of counting varies with the language in which the number system is learned (Ginsburg, Choi, Lopez, Netley, & Chi, 1997; Han & Ginsburg, 2001; K. F. Miller, Major, Shu, & Zhang, 2000; K. F. Miller, Smith, Zhu, & Zhang, 1995). For example, Chinese, like many East Asian languages, has a more regular sequence of number words than does English. In both English and Chinese, the numbers 1 through 10 are arbitrary and the numbers after 20 follow a regular pattern of naming the decade name and then the digit name (e.g., "twenty-one"). However, in Chinese (and in many Asian languages rooted in ancient Chinese), there are two important differences. The tens numbers directly mirror the single-digit number names ("two-tens" rather than "twenty; "three-tens" rather than "thirty"), and the numbers from 11 to 20 also follow a regular pattern (comparable to "ten-one," "ten-two," etc.) instead of the obscure "eleven, twelve. . . ." Through 3 years of age, children in the various cultures learn 1 through 10 similarly; however, those learning English learn the "teens" more slowly and with more errors. Further, Asian number words can be pronounced more quickly, providing another significant cognitive advantage (Geary & Liu, 1996).

Overall, children learning Asian languages show a substantial advantage in learning verbal counting.

Support for the notion that this advantage is a direct consequence of language is provided by parallel findings that Asian children do *not* differ in other aspects of number knowledge, including counting small sets and solving simple numerical problems (Fuson, 1992; K. F. Miller, Major, Shu, & Zhang, 2000). One group of researchers abandoned studying U.S. kindergartners because they could not count consistently and accurately from 1 to 50, but Korean kindergartners had no such problems (Miura & Okamoto, 2003). Only U.S. children made errors such as "twenty-nine, twenty-ten, twenty-eleven…"; Chinese children do not make that kind of mistake (K. F. Miller, Major, Shu, & Zhang, 2000). Although all children tend to make more mistakes at decade boundaries, for Chinese-speaking children, these are largely limited to infrequent mistakes with numbers above 60. Conclusions must be made with caution, however, as other cultural factors, such as Asian parents' emphasize on counting in the early years, may affect these results (Towse & Saxton, 1998; J. Wang & Lin, 2005).

The deleterious effects of less mathematically coherent languages are not limited to counting larger numbers but hamper children's development of place value, multidigit arithmetic, and other concepts (Fuson, 1992). The effects are detrimental in surprising ways: Even the counting words from 1 to 9 are learned better by Chinese children because they practice these exact numbers when learning to count from 11 to 19 (Miura & Okamoto, 2003).

Fuson and colleagues traced the development of the number word sequence from its beginnings at age 2 to its general extension to the notion of base-ten systems at about the age of 8 (Fuson & Hall, 1982; Fuson, Richards, & Briars, 1982). Most middle-class children less than 3.5 years of age are working on learning the sequence to 10 (Fuson, 1992; Saxe, Guberman, & Gearhart, 1987). For the next year, they develop the sequence from 10 to 20. From 4.5 to 6 years, they still make errors in the teens, but most also develop the decades to 70, although a substantial number count to 100 and higher (Bell & Bell, 1988; Fuson, Richards, & Briars, 1982).

Fuson and colleagues also reported two distinct, overlapping phases: an acquisition phase of learning the conventional sequence of the number words, and an elaboration phase, during which this sequence is decomposed into separate words and relations upon these words are established. During the first phase, the sequence is a single, connected serial whole from which interior words cannot be produced singly. The most common overall structure of the sequences is (a) stable conventional, an initial group that is the beginning of the conventional sequence; (b) stable noncon-ventional, a group that deviates from convention but is produced with some consistency, and (c) nonstable, a final group with little consistency. The acquisition of longer sequences consists of the extension of the sequence and the consolidation of the extension so that it is produced reliably.

At first, children can only produce the sequence by starting at the beginning. In the elaboration phase, there is a gradual process of constructing relations among these words. Five levels were differentiated and empirically supported (Fuson & Hall, 1982; Fuson, Richards, & Briars, 1982): (a) string level—words are not objects of thought nor "heard" as separate words, (b) unbreakable list level—words are heard and become objects of thought, (c) breakable chain level—parts of the chain can be produced starting from an arbitrary number (enabling certain counting on strategies), (d) numerical chain level—words become units that themselves can be counted, and (e) bidirectional chain level—words can be produced in either direction and the unitized seriated embedded numerical sequence allows part-whole relations and a variety of flexible strategies to be employed. These developments precede and enable changes in addition and subtraction solution strategies (Fuson, 1992).

Fuson and colleagues (1982) also claimed that backward verbal counting is learned in a slow and laborious manner, based on an already-mastered forward sequence. However, specific instruction may enable children to develop backward counting to almost the same extent as forward counting (B. Wright, 1991).

Object Counting

To count objects, children learn to coordinate the production of counting words with indicator actions, such as pointing to or moving objects (Dewey, 1898; Fuson, 1988; Ginsburg, 1977; Judd, 1927). More elaborately, a fully developed object counting scheme has four components. First, there is a situation that is recognized, say a collection of countable items. Second is a goal, to find out how many. Third is an activity, counting, and fourth, the result, a unitary whole of counted items (Steffe & Cobb, 1988).

The easiest type of collection for 3-year-olds to count has only a few objects arranged in a straight line and can be touched as children proceed with their counting. Between 3 and 5 years of age, children acquire more skill as they practice counting, and most become able to cope with numerically larger collections in different arrangements, without needing to touch or move objects during the act of counting.

As discussed previously, experiences in which the configurations are labeled with a number word by adults or older children ("Here are two blocks.") enable children to build meaning for number words as telling how many items in a configuration or collection of items. The capstone of early numerical knowledge, and the necessary building block for all further work with number and operations, is connecting the counting of objects in a collection to the number of objects in that collection. Initially, children may not know how many objects there are in a collection after counting them. If asked how many are there, they typically count again, as if the "how many?" question is a directive to count rather than a request for how many items are in the collection. Children must learn that the last number word they say when counting refers to how many items have been counted. Many 3-year-olds do learn this result of counting: "One, two, three, four. There are four olives."

How do such competencies develop? Gelman and Gallistel (1978) hypothesized that the ease and rapidity that even very young children display in learning to count indicated that such development was guided by knowledge of counting principles. In particular, they hypothesized that young children know five principles. The three how-to-count principles included the stable order principle (always assign the numbers in the same order), one-one principle (assign one and only one number word to each object), and cardinal principle (the last count indicates the number of objects in the set). The two what-to-count principles were the order irrelevance principle (the order in which objects are counted is irrelevant) and the abstraction principle (all the other principles can apply to any collection of objects). The researchers presented evidence indicating that children understand—explicitly or implicitly—all these principles by age 5, and many by age 3 years. For example, in the "magic" experiments described previously, all three how-to-count principles were followed (not perfectly, as the question was competence, not performance) by children as young as 2.5 years, at least with homogeneous, linearly arranged collections of no more than five items (R. Gelman & Gallistel, 1978). The one-one errors tended to be over- or undercounting on all set sizes, or skipping or double counting on the larger of the set sizing. Almost no children even used the same number word more than once.

As another example, even when children make mistakes, they show knowledge of the principle. For example, if they do not know the standard order of number words, the idiosyncratic order they *do* use is usually stable. Or, they may make mistakes in executing the one-one rule (Fuson, 1988) but assign exactly one number word to most of the objects (about 75% of the objects even for 2- and 3-year-olds). Thus, these might reflect performance errors, not conceptual limitations. Preschoolers give evidence of using the cardinal principle by repeating or emphasizing the last number. They count events, collections that include different categories of objects, and even "missing" objects (e.g., eggs not in a carton). The order irrelevance principle is not demonstrated as widely, but most 5-year-olds will start counting in the middle of a row of objects and count each object.

The principles were expanded and implemented within a computational model in which it was assumed that implicit understanding of principles was characterized as conceptual competence, hypothesized to be a set of action schemata corresponding to those principles (Greeno, Riley, & Gelman, 1984). Theoretically, a main thrust of the nativist argument is that such principles guide children's acquisition of counting skills and understandings. This was in contrast to Piagetians (e.g., Elkind, 1964; Kamii & DeVries, 1993), who maintained that counting errors occurred because the child lacked fundamental understandings, for example, recognizing the logical need to order objects.

Other researchers gave more importance to repeated, often massive, experience and demonstrations, modeling, or scaffolding from adults in learning counting competencies (e.g., Fuson, 1988). Supporting this view are findings that suggest children count skillfully before they understand the principles that underlie counting (Bermejo, 1996; Briars & Siegler, 1984; Frye, Braisby, Lowe, Maroudas, & Nicholls, 1989; Wynn, 1992b). From this perspective, children may abstract features that are common to all acts of counting (e.g., counting each object exactly once) and distinguish them from those that are incidental (e.g., starting at the end of a row). Another possibility is that children build on their verbal number recognition (subitizing) skill to abstract the counting principles from their own counting experience (Le Corre, Van de Walle, Brannon, & Carey, 2005) or to make sense of adult efforts to model counting procedures (Baroody, Lai, & Mix, 2006).

Additional research has clarified these issues, especially the main principles. Children improve on counting in a "stable order" but do not show signs of understanding the principle (Fuson & Hall, 1982; Fuson, Richards, & Briars, 1982). Researchers agree that children as young as 3 years usually display one-to-one correspondences in counting (Fuson, 1988). However, they do not do so in all situations. There are actually several related but distinct competencies, including coordinating number words with objects and keeping track of counted objects (Alibali & DiRusso, 1999). Even 5-year-olds show difficulty sequencing large disorganized collections. Further, coordination of words with objects actually involves a chain of two types of correspondence, that between the production of a number word and an indicating act such as pointing, and another between that act and each item to be counted. At age 3 or 4, children violate the first type of correspondence in two main ways, by pointing to an object without saying a word and by pointing to an object and saying more than one number word. They violate the second by skipping over objects or by double-counting (see Fuson, 1988, for an elaboration of many less frequent errors). Gesturing helps children maintain both types of correspondences, keeping track and coordinating number words with objects (Alibali & DiRusso, 1999). The rhythmic, physical motions may focus children's attention on the individual items, aiding segmentation. Further, the use of touching may lessen the working memory demands of counting by externally representing some of the contents of working memory, such as marking the child's place in the set of objects. In summary, for these two principles, evidence does not provide strong support for the nativist view but does not disprove it; children could understand the principle and have difficulty carrying out the procedures.

Evidence regarding the cardinality principle is perhaps most inconsistent with a strong nativist view. Cardinality does not appear to be a component of many children's early counting in most situations (Fuson, 1988; Ginsburg & Russell, 1981; Linnell & Fluck, 2001; Schaeffer, Eggleston, & Scott, 1974). As previously stated, to these children, the purpose of counting is the action of enumeration (Fluck, 1995; Fluck & Henderson, 1996). Showing an even more limited understanding, some children, asked why they count, seem perplexed by the notion of counting as a purposeful activity (Munn, 1997). Fuson and colleagues posited that repeating the last number of a counting act is initially merely the last step in a chain of responses that children learn to meet adult requests (Fuson & Hall, 1982; Fuson, Richards, & Briars, 1982). Even when children respond correctly to the cardinality question, they may not fully understand it (Sophian, 1988). For example, asked to indicate which object(s) to which their last-number-word response referred, children often indicate only the last object. As another example, children give the last-number-word response even when it is incorrect (e.g., when they were asked to begin counting with a number other than one, Bermejo, 1996). Children may have to represent, or "redescribe," the final number word at an explicit level before it is available to serve the purpose of establishing a relationship between the counting act and cardinality (Karmiloff-Smith, 1992). This theory emphasizes internal processes, including visuo-spatial abilities, which play a greater role than language ability

in the development of cardinality in normally developing young children. However, social interaction probably aids this development (Linnell & Fluck, 2001).

This does not mean that cardinality is not understood, only that it is not an innate notion in counting. Instead, young children's earliest meaning for number words may be rooted in recognition or subitizing (Le Corre, Van de Walle, Brannon, & Carey, 2005). This may be generalized to counting, so that by 3.5 years, many children begin to consciously use and understand the cardinal notion in counting (Wynn, 1990, 1995), at least for small numbers, and most 5-year-olds do so consistently in simple counting situations. However, the idea still undergoes development. For example, younger children may not maintain cardinal and ordinal meanings in Piagetian conservation of number tasks in which the child builds a row of objects equivalent to a given row of objects. However, different tasks reveal greater competence. Many preschoolers, told the cardinal value of each of two sets, can determine whether the items would be put in one-to-one correspondence (and the inverse), providing evidence that they understand the relation between the number words and quantity and can use number words to reason numerically about one-to-one correspondences that are not perceptually available (Becker, 1989; Sophian, 1988). This provides evidence that preschoolers have at least initial integration of the cardinality of the collection as a whole and the individual items in the collection. If asked to count, many 4- and 5-year-olds count and use the information to correctly judge equivalence (Fuson, 1988; Fuson, Secada, & Hall, 1983; Michie, 1984a). Thus, counting can be a meaningful quantifier for children.

The abstraction principle addresses what units children can count. Research indicates that a significant development is to understand fully the unit that one is counting (K. F. Miller, 1989). For example, asked to count whole forks when shown some intact forks and some broken in half, children 4 to 5 years of age were reluctant to count two halves as a single item (Shipley & Shepperson, 1990). Although performance improves on such tasks from 4 to 7 years, a substantial portion of each age had such difficulties (Sophian & Kailihiwa, 1998). Thus, the abstraction principle appears only partially valid.

The nativist position is therefore neither definitively supported nor disproved. Addressing similar issues, a review of research generally supported the skills-before-concepts position, although there was a lack of studies with a strong design (Rittle-Johnson & Siegler, 1998). The argument and evidence against a strictly nativist (privileged domains) position is more compelling than evidence in support of the skills-before-concepts position. This is not to say the concepts-before-skills position is supported, but rather that the evidence reviewed may provide the most support for an interactionalist position, in which growth in each of concepts and skills supports growth in the other (Baroody, 1992; Fuson, 1988). For example, a skills-first perspective may suggest that very early use of number or counting is mathematically meaningless (Ginsburg, 1977; von Glasersfeld, 1982); however, early developing quantitative and numerical concepts may imbue these words and procedures with meaning (Baroody, Lai, & Mix, 2005; Baroody & Tiilikainen, 2003), and the words may help explicate and differentiate the initially syncretic quantities (Clements, Battista, & Sarama, 2001). Further, humans are sufficiently flexible to permit different paths to learning (e.g., mostly skills or concepts first), even if they are differentially generative of future learning (as hierarchical interactionalism posits, see also Clements, Battista, & Sarama, 2001). Finally, the extant literature offers many cases in which conceptual knowledge leads to construction of procedures (i.e., concepts result in skills). Moreover, that literature often does not examine the types of experiences that have been offered to children; we contend that especially in school, but also at home, simplistic and reductionistic views of mathematics and mathematics learning have biased research examining relationships between concepts and skills. Researchers need better evidence, including solid causal evidence (Rittle-Johnson & Siegler, 1998), on these issues.

Zero and Infinity

Situations similar to those that engender counting may also lead to children's initial encounters with zero and infinity. Because touching objects is important to young children's initial counting, zero is often not encountered in the act of counting. (Resistance to the ideas is not merely a developmental limitation, of course. Historically, societies have taken considerable time to confront, represent, and incorporate ideas of zero and infinity.) Kindergartners solve problems with zero well, although there is a separation between their knowledge of zero and their knowledge of the counting numbers. In one study, for example, they showed no evidence of applying counting strategies to zero problems and performed better on zero problems than on problems involving one or two, suggesting that such problems were solved differently (Evans, 1983). Preschoolers often have a more limited understanding of zero and infinity (Evans, 1983; R. Gelman & Meck, 1983). However, even 3-year-olds represent zero as the absences of objects more than half the time, and by 4 years, children did as well with zero as

with small whole numbers (Bialystok & Codd, 2000; similar results are reported by Hughes, 1986).

Counting played a larger role in children's early ideas related to infinity (Evans, 1983). Some children showed little or no knowledge of numbers greater than one hundred or the structure of counting. Other children at level showed knowledge of this structure, including generating new, larger numbers. The most advanced children understood that there was no greatest number because it was always possible to create a larger number.

Summary

Accurate, effortless, meaningful, and strategic counting is an essential early numerical competence. Initial level of achievement and subsequent growth in number and arithmetic from preschool into the elementary years are both predicted better by early counting ability than by other abilities, such as visual attention, metacognitive knowledge, and listening comprehension (Aunola, Leskinen, Lerkkanen, & Nurmi, 2004).

Synthesizing across this and the previous sections, we summarize that early numerical knowledge includes four *interrelated* aspects (as well as others): recognizing and naming how many items are in a small configuration (small number recognition and, when done quickly, subitizing), learning the names and eventually the ordered list of number words to ten and beyond, enumerating objects (i.e., saying number words in correspondence with objects), and understanding that the last number word said when counting refers to how many items have been counted. Children learn these aspects, often separately by different kinds of experiences, but gradually connect them during the preschool years (cf. Linnell & Fluck, 2001). Each of the aspects begins with the smallest numbers and gradually includes larger numbers. In addition, each includes significant developmental levels. For example, small number recognition moves from nonverbal recognition of one or two objects, to quick recognition and discrimination of one to four objects, to conceptual subitizing of larger (composed) groups. Development shares similarity across these aspects.

This describes the progression from sensory-concrete to integrated-concrete knowledge. Sensory-concrete refers to knowledge that demands the support of concrete objects and children's knowledge of manipulating these objects (Clements, 1999a). *Integrated-concrete* refers to concepts that are "concrete" at a higher level because they are connected to other knowledge, both physical knowledge that has been abstracted and thus distanced from concrete objects and abstract knowledge of a variety of types (Clements, 1999a). Ultimately, these are descriptions of changes in the configuration of knowledge as children develop. Consistent with other theoreticians (Anderson, 1993), this distinction assumes that there are not fundamentally different, incommensurable types of knowledge, such as "concrete" versus "abstract."

Comparing and Ordering

Comparing and Equivalence

Infants begin to construct equivalence relations between sets by establishing correspondences (spatial, temporal, or numerical) as early as 10 months and at most by 24 months of age (Langer, Rivera, Schlesinger, & Wakeley, 2003). Studies of early discrimination and conservation of number, previously discussed, also address children's initial abilities to compare quantities. To briefly highlight relevant findings, children appear to base comparisons on number and may use nonverbal representations of cardinality (Mix, Huttenlocher, & Levine, 2002). For example, they may share with monkeys a set of mental filters, each tuned to an approximate number of objects (e.g., 3), but overlapping with adjacent filters (e.g., that detect 2 or 4), which explains why discrimination improves with greater numerical distance (Nieder, Freedman, & Miller, 2002). At 3 years of age, children can identify as equivalent or nonequivalent static (simultaneously presented) collections consisting of a few (1 to about 4) highly similar items (e.g., Huttenlocher, Jordan, & Levine, 1994; Mix, 1999). For instance, they can identify ▥ ▥ ▥ and ▥ ▥ ▥ as equal and different from ▥ ▥ or ▥ ▥ (the researchers describe this as a nonverbal competence, but children may have subitized the arrays). At 3.5 years, they can match different homogeneous visual sets and match sequential and static sets that contain highly similar items. At 4.5, they can nonverbally match equivalent collections of random objects and dots—a heterogeneous collection and a collection of dissimilar items.

Children's success with sequentially presented objects emerges later than with simultaneously presented objects, and sequential events are the most difficult (Mix, 1999). These abilities seem to depend on counting abilities, with sequential events requiring a mastery of counting.

Both infants (R. G. Cooper, Jr., 1984) and preschoolers (Sophian, 1988) do better comparing sets of equal, rather than unequal, number, presumably because there are many ways for collections to be unequal. In brief, children show competence in comparing simultaneously present, equivalent collections as early as 2 or 3 years of age in everyday, spontaneous situations but show only the beginnings of such com-

petence on clinical tasks at 2.5 to 3.5 years of age, with success across a wide range of tasks only appearing at ages suggested by Piaget (Baroody, Lai, & Mix, 2005; Mix, Huttenlocher, & Levine, 2002).

Clinical studies of equivalence also highlight specific difficulties in number comparison. For example, some children will count the sets, recognize the number is the same, but still maintain that one set has more (Piaget & Szeminska, 1952). Children from 4 to 5 years of age do compare set sizes on the basis of misleading length cues, even when the situation is set up to encourage counting (Fuson, 1988).

Conversely, children cannot always infer number from that of an equivalent set. In one study, fair sharing was established by dealing out items to two puppets, and the experimenter counted one set aloud and asked children how many the other puppet had. No 4-year-old children made the correct inference that the number was the same; instead, they tried to count the second set. The experimenter stopped them and asked again, but less than half made the correct inference (Frydman & Bryant, 1988).

Researchers have posited several explanations for preschoolers' reluctance or inability to use counting (Curtis, Okamoto, & Weckbacher, 2005). Some prefer an explanation that relies on available memory resources (Case & Okamoto, 1996; Pascual-Leone, 1976). However, others state that children believe that counting-based strategies are too difficult or unreliable (Cowan, 1987; Michie, 1984b; Siegler, 1995; Sophian, 1988). Finally, others say that counting is an activity embedded in situated tasks, and that use of it in other tasks develops only over time (Nunes & Bryant, 1996; Steffe, Cobb, von Glasersfeld, & Richards, 1983), as children come to trust the results of counting (e.g., relative to conflicting perceptions of length, Cowan & Daniel, 1989). Most agree that preschoolers do not often use counting to compare numerosities of collections.

However, presented different tasks, preschoolers can perform such inferences, such as from matching-based to counting-based equivalence. Recall studies showed that, told the cardinal value of each of two sets, preschoolers determined whether the items would be put in one-to-one correspondence and the inverse (Becker, 1989; Sophian, 1988). Thus, success is dependent on the task, with success appearing first when one-to-one correspondence (rather than sharing) is used to establish equivalence. Between the ages of 3 and 5, children develop from counting only single collections to being able to use counting to compare the results of counting two collections (Saxe, Guberman, & Gearhart, 1987) and reason about these comparisons across different situations.

Further, counting is not without positive influence. Even when children use misleading length cues, from 70 to 80% turned to counting when asked to "do something" to justify their response (Fuson, 1988). Further, although encouraging 4.5- to 5-year-old children to match and count increases their correct judgments of equivalence, children only spontaneously adopt counting for subsequent problems of the same type (Fuson, Secada, & Hall, 1983). When the counting is performed for them, preschoolers are more likely to base their comparison decisions on the results of counting (Curtis, Okamoto, & Weckbacher, 2005), which may imply that counting imposes a large processing load on children of this age, or that they do not appreciate the relevance of the counting strategy.

Apparently children have to learn that the same number implies the same numerosity and that different numbers imply (and necessitate) different numerosities (Cowan, 1987). That is, many children up to first grade need to learn about the significance of the results of counting in different situations, such as comparing sets or producing equivalent sets (Nunes & Bryant, 1996). Further, they may not have a concept of space that is sufficiently articulated and unitized to reconcile the perceptual evidence and numerical interpretation of the situation (Becker, 1989).

Thus, as we previously concluded, counting can be a meaningful quantifier for children before they reach the Piagetian levels of operational thought about number conservation. However, children's ability to solve problems using counting develops slowly, as they learn counting skills and about the application of counting to various tasks.

Ordering and Ordinal Numbers

When the topic of *ordinality* is discussed, many authors and some researchers, often assume that all ordinal notions must involve the terms *first, second...*, and so forth. This is a limited view. As was previously discussed, numbers have an ordinal property in that they are sequenced. A number may be first or second in an order without considering cardinality, or it may be considered to have a greater or lesser magnitude than another number. We first address young children's ability to order collections by number, then their ability to deal explicitly with traditional ordinal numbers.

In most studies of conservation discussed previously, children needed only to decide whether collections were equivalent or not, but not to explicitly order them (even determining which contained more). Evidence indicates that human beings can make perceptual judgments of relative quantities early, but somewhat later than they evince other quantitative abilities. As an example, 16- to 18-month-olds were reinforced for selecting a square containing two dots but not another square containing one dot, regardless of which side it was on and regardless of how big or bright the dots were (Strauss & Curtis, 1984). Then, presented two new squares with three and four dots, respectively, the infants more often chose the square with more dots, indicating sensitivity to the ordinal concept of "more numerous." Other studies confirm that children differentiate between "less than" and "greater than" relations by 14 months of age (R. G. Cooper, Jr., 1984; Feigenson, 1999; Haith & Benson, 1998), and possible earlier, at least with collections no larger than two (Sophian & Adams, 1987). Even 11-, but not 9-month-olds, can distinguish between ascending and descending sequences of quantities (Brannon, 2002). That is, following habituation to three-item sequences of decreasing numerical displays, in which area was not confounded with number, only 11-month-olds could then distinguish between new displays of decreasing vs. increasing sequences.

Seminal research on equivalence, Gelman's "magic experiments," showed that children explicitly recognized which of two sets was the "winner" even if the length, density, or even properties (color) of one set were altered. However, when the change involved an addition or subtraction that created equivalent sets, children correctly responded that there was no winner (R. Gelman & Gallistel, 1978). These experiments also revealed that children about 3 years of age are sensitive to ordinal relations. Shown that 2 was the winner compared to 1, children selected 4 as the winner compared to 3. In both experiments, this competence was just emerging at 2 to 3 years of age. By 3 years of age, children show knowledge of order in comparing collections, separate from their counting skills (Huntley-Fenner & Cannon, 2000; Mix, Huttenlocher, & Levine, 2002).

As with cardinal number, children develop the ability to order numbers over several years by learning the cultural tools of subitizing (Baroody, Lai, & Mix, 2006), matching, and counting (Fuson, Secada, & Hall, 1983). For example, children can answer questions such as "Which is more, 6 or 4?" only by age 4 or 5 years (Siegler & Robinson, 1982). Tasks involving smaller numbers and numbers farther apart are easier (Cowan & Daniel, 1989), suggesting that counting skills are relevant with numbers larger than 4. Unlike middle-income children, low-income 5- and 6-year-olds were unable to tell which of two numbers, such as 6 or 8, is bigger, or which number, 6 or 2, is closer to 5 (Griffin, Case, & Siegler, 1994). They may not have developed the "mental number line" representation of numbers as well as their more advantaged peers. These representations have several aspects, such as two additional principles of counting. The "plus and minus one" principle involves understanding the generative pattern that relates adjacent numerical values (e.g., a collection of 4 is a collection of 3 with one added). The related comparison principle involves understanding the consequences of each successive number representing a collection that contains more objects (Griffin, Case, & Siegler, 1994). Children who use and understand these principles can reason that if the counts of two collections are 9 and 7, the collection with 9 has more because 9 comes later in the counting sequence than 7.

Finding out *how many more* (or fewer) there are in one collection than another is more demanding than simply comparing two collections to find which has more. Children have to understand that the number of elements in the collection with fewer items is contained in the number of items in the collection with more items. That is, they have to mentally construct a "part" of the larger collection (equivalent to the smaller collection) that is not visually present. They then have to determine the "other part" or the larger collection and find out how many elements are in this "left-over amount."

Ordinal numbers, usually (but not necessarily) involving the words *first, second* . . . , indicate position in a series or ordering. As such, they have different features (e.g., their meaning is connected to the series they describe). Of children entering their first year of school in Australia, 29 and 20% could identify the third and fifth items in a series (B. A. Clarke & Clarke, 2004). In the U.S., performance for first was 60 and 72%, and for third was 9 and 13% for low- and middle-income groups, respectively (Clements, Sarama, & Gerber, 2005).

As in other areas, different languages differentially affect children's acquisition and use of ordinal words. Differences in ordinal words are even more distinct across languages. Beginning right with *first* and *second* compared to *one* and *two*, English ordinal and counting words often have little or no relation. In comparison, Chinese speakers form ordinal words simply by adding a predicate (*di*) to the cardinal number name. English-, compared to Chinese-, speaking children show dramatically lower performance on naming ordinal words from the first words and from the earliest ages, persisting into elementary school (K. F. Miller, Major, Shu, & Zhang, 2000). A precipitous decline is between 19th and 21st, with only 30% of the English-speaking sample counting to 21st, versus nearly perfect performance in Chinese-speaking children. However, this research also reveals complexities in the relationships between concepts and language. Although few children understood the unique features of ordinal numbers, English speakers seem to understand the distinctions between ordinal and cardinal number concepts sooner, perhaps because struggles with the different names engendered reflection on these differences (K. F. Miller, Major, Shu, & Zhang, 2000).

Beginning Addition and Subtraction, and Arithmetic Counting Strategies

For most people in the United States, arithmetic is arguably the most salient topic in elementary mathematics education. However, most people would also argue that this was an inappropriate topic in preprimary education. We begin our discussion considering research on the age at which knowledge of arithmetic seems to emerge.

The Earliest Arithmetic

Educational practice, past and present, often takes a skills hierarchy view that arithmetic follows counting and other simple work with number and therefore is beyond children's grasp until about first grade. However, decades ago, researchers illustrated the possibility of arithmetic competence in children before kindergarten (R. Gelman & Gallistel, 1978; Groen & Resnick, 1977; Hughes, 1981).

More recently, nativist researchers claimed that even infants have knowledge of simple arithmetic. For example, there is evidence that infants and toddlers notice the effects of increasing or decreasing small collections by one item. For example, after seeing a screen hide 1 or 2 dolls, then a hand place another doll behind the screen, then the screen retracted, 5-month-olds look longer when an incorrect result is revealed (a violation-of-expectations procedure, Wynn, 1992a).

In another study, children placed balls in a box placed high enough that they could not see into the box. They then watched as balls were added or removed (Starkey, 1992). Finally, they were asked to remove all the balls. The question was whether the number of times children reached was the same as the number of balls they placed. To prevent children from feeling whether or not any remained in the box, the balls were secretly removed and replaced one-by-one before each reach. Children did reach the same number of times above chance. However, a flaw in the design was that these removals might have inadvertently kept children from reaching in again when they would have (mistakenly) done so (Mix, Huttenlocher, & Levine, 2002). This would favor tasks with an answer of one. Note that children were evaluated as correct on 64% of the 4 − 3 trials but only 14% on the 4 − 1 trials, even though the latter would be considered the easier of the two tasks.

Despite such problems, authors of the original studies have made strong nativist claims, such as "infants can discriminate between different small numbers . . . determine numerical equivalence across perceptual modalities," and "calculate the results of simple arithmetical operations on small numbers," all of which indicates that they "possess true numerical concepts, and suggests that humans are innately endowed with arithmetic abilities" (Wynn, 1992a, p. 749). They have "access to the ordering of and numerical relationships between small numbers, and can manipulate these concepts in numerically meaningful ways" (p. 750). A recent study grounded in that perspective suggested that very young children, viewing dots representing 5 + 5 and 10 − 5 (controlled for area and contour length), could discriminate between outcomes of 5 and 10 (McCrink & Wynn, 2004).

Possible alternative explanations have been addressed. For example, it is possible that infants merely track locations, but evidence indicates they can track objects or number of objects (Koechlin, Dehaene, & Mehler, 1997). A different possible confound is that every outcome in Wynn's work that was arithmetically impossible was also physically impossible; however, evidence indicated that 3- to 5-month-old babies were more upset if an arithmetically impossible result occurred than if one doll changed to another (T. J. Simon, Hespos, & Rochat, 1995).

Still, there are concerns about the nativist claims. In some studies, using standardized procedures, infants were not found to discriminate between correct and incorrect results of addition and subtraction (Wakeley, Rivera, & Langer, 2000). In addition, multiple studies have found that 3- but not 2-year-olds are successful on nonverbal addition and subtraction tasks (Houdé, 1997; Huttenlocher, Jordan, & Levine, 1994). If toddlers do not have the ability, it may be unreasonable to suggest that infants do (see Huttenlocher, Jordan, & Levine, 1994, for a discussion of these and other problems with research claiming that infants perform exact arithmetic).

Further, early behaviors often are explicable through other frameworks (Haith & Benson, 1998). For example, the children might be processing continuous amounts. Using Wynn's tasks but different size puppets, infants were more likely to attend to the display with the unexpected change in amount and the expected number than the opposite (Feigenson, Carey, & Spelke, 2002; Mix, Huttenlocher, & Levine, 2002). Children's success with larger number arithmetic (e.g., 10 − 5, McCrink & Wynn, 2004) may involve their use of an estimator such as in the "large number exception" situation discussed in the subitizing section. This is supported by cultures that have number words only up to five but can compare and add large *approximate* numbers beyond that range (Pica, Lemer, Izard, & Dehaene, 2004). It is also consistent with the view that infants represent small collections as individual objects but not as groups and large numbers as groups but not as individual objects (Spelke, 2003). Thus, analogue estimators may be innate or early developing and may mediate (Gallistel & Gelman, 2005), but not directly determine, later developing explicit, accurate arithmetic.

Another possible framework for smaller number problems postulates that infants are tracking objects via "mental tokens" using general processes of simple categorization (an early "object concept") and comparison (Koechlin, Dehaene, & Mehler, 1997; T. J. Simon, 1997; T. J. Simon, Hespos, & Rochat, 1995). Children may possess neural traces of visual information that are not consistent with the final display, and thus their "representations" could be literal and knowledge-based. That is, these babies may *not* be *quantifying* a *collection* of objects but instead individuating and tracking separate objects, reacting if one disappears or appears mysteriously (Koechlin, Dehaene, & Mehler, 1997; Sophian, 1998). For example, infants' surprising greater sensitivity to subtraction than addition across multiple studies (Koechlin, Dehaene, & Mehler, 1997; Wynn, 1992a, 1995) may indicate that the smaller the result, the more likely the child will track accurately. Further, experiments directly comparing two models, numeric/symbolic and object file, were consistent with the object file model's predictions (Uller, Carey, Huntley-Fenner, & Klatt, 1999). For example, infants are much more likely to succeed if the first object in a 1 + 1 task is seen on the floor before the screen is introduced, compared to tasks in which the screen is introduced first and thus the objects must be imagined there (Uller, Carey, Huntley-Fenner, & Klatt, 1999). Consistent results from other experiments (Uller, Carey, Huntley-Fenner, & Klatt, 1999) and the mixed results of various researchers imply that adults should remain aware that we often attribute numerosity and arithmetic operations to reactions to changes in number, but that may be our inability to decenter and understand that this is our projection of number onto the situation, deeply embedded in our adult cognition (cf. von Glasersfeld, 1982). This does not mean competencies do not exist, or that they are irrelevant to the foundations of numbers development; in contrast, relevant competencies are supported in most studies. However, even comparing via an early version of one-to-one correspondence may be at the service of general goals and not be numerical *for the infant in the sense that it would be for adults.* Probably the most we can confidently say is that infants react to situations that older children and adults would experience as arithmetical.

Semantic analysis is consistent with this latter interpretation (von Glasersfeld, 1982). For example, children may discover, perhaps by putting up two fingers on each hand, that they can make a configuration of four fingers, but perhaps only at the sensory-motor level. Through application of subitizing or counting and reflective abstraction on the attentional patterns whose unconscious "automatic" application to sensory material made the figurative compositions possible, children eventually rebuild these experiences as arithmetical relationships.

Supporting this constructivist approach is evidence that young children's active sensorimotor production may develop before their reactive perceptual discrimination of adding and subtracting. For example, when exposed to a standardized violation-of-expectations procedure (Wynn's were not standardized), 11-, 16-, and 21-month-olds did not look longer at results for 1 + 1 or 2 − 1 that were incorrect than those that were correct (Langer, Gillette, & Arriaga, 2003). However, 21-month-olds did reach into a box correctly in about three fourths of the trials. Thus, pragmatic knowledge of what "works" with objects may precede "understanding how" or "what."

When do children explicitly understand the order relations in arithmetic? An early study showed that children from 3 to 6 years show sensitivity to the effects of adding or subtracting one or two marbles to one or more containers, initially established to hold an equal (but uncounted) number of marbles, when exact computations were not required (Brush, 1978). Children 14 to 28 months showed similar sensitivity to insertions and deletions from hidden sets of numerosity no more than 2, with a caveat that results were affected by children's bias toward choosing the transformed set (Sophian & Adams, 1987). Thus, as early as 14 to 24 months of age, children appear to understand some sense of addition increasing, and subtraction decreasing, the size of collections, before their counting skills are well established (Bisanz, Sherman, Rasmussen, & Ho, 2005; Cowan, 2003; Mix et al., 2002).

Preschoolers develop in reasoning about the effects of increasing or decreasing the items of two collections of objects. When two collections are created simultaneously by placing items one-for-one in separate locations, many preschoolers correctly judge that the items in the two collections are equal even though they do not know exactly how many objects are in the collections. If items of one collection are then increased or decreased, children as young as age 3 correctly judge that the collection added to contains more or the collection subtracted from contains fewer than the collection that was not changed. However, problems involving two collections that are initially unequal present a difficulty for 3-year-olds. For example, if two collections initially differ by two objects, and one object is added to the smaller collection, many 3-year-olds will incorrectly say that the collection to which the object was added has more. In contrast, some 5-year-olds know that this collection still has fewer. They know that both the addition and the initial inequality must be taken into account in reasoning about the effect of the addition on the collection (R. G. Cooper, Jr., 1984).

Children's accuracy is limited until about 3 years of age. Children are initially more accurate with small numbers (operands from 1 to 3) and nonverbal problems (Hughes, 1981; see also Huttenlocher, Jordan, & Levine, 1994). For example, children solve nonverbal arithmetic problems at about 3 years of age. In such tasks, the experimenter might show a number of disks, move them behind a screen, add more disks to the now-hidden collection, and ask the child to make a collection showing the number hidden (Huttenlocher, Jordan, & Levine, 1994). Even when preschoolers develop this ability, their competencies are approximate and are strongly related to their general intellectual competence. This is inconsistent with a perspective that such arithmetical competence is innate. Addition is easier than subtraction for larger problems (operands 5 to 8; (Hughes, 1981). Verbal-based addition develops later, as counting becomes the key basis for understanding verbal computation above small numbers (Cowan, 2003).

Thus, most agree that children have built an initial, explicit understanding of addition and subtraction situations by about 3 years of age. However, it is not until 4 years of age that children can solve addition problems involving even slightly larger numbers with accuracy (Huttenlocher, Jordan, & Levine, 1994).

Most children do not solve larger number problems without the support of concrete objects until 5.5 years of age (Levine, Jordan, & Huttenlocher, 1992). They have, apparently, not only learned the counting sequence and the cardinal principle (usually about 3.5 years) but also have developed the ability to convert verbal number words to quantitative meaning (cf. the ordinal-to-cardinal shift in Fuson, 1992).

However, this limitation is not so much developmental as experiential. With experience, preschoolers and kindergartners can learn "count all" and even "count on" strategies (Clements & Sarama, in press; Groen & Resnick, 1977; Hughes, 1986), a point to which we will return.

Arithmetic Strategies

Once children show dependable signs of understanding addition and subtraction tasks, they also show a range of strategies for solving them. Most people of any age can show diversity in solution strategies, as was recognized for young children more than 50 years ago (Ilg & Ames, 1951). Solutions of children as young as preschool (perhaps because school has not yet negatively affected them, cf. Kamii & Lewis, 1993) are notably creative and diverse (Bisanz et al., 2005; Geary, Bow-Thomas, Fan, & Siegler, 1993). For example, preschool to first-grade children use a variety of covert and overt strategies, including counting fingers, finger patterns, verbal counting, retrieval, derived facts, and covert strategies, some slower and some faster (Siegler & Jenkins, 1989; Siegler & Shrager, 1984; Steffe, 1983). Children are flexible strategists, using different strategies on problems they perceive to be easier or harder. Preschoolers can invent surprisingly sophisticated arithmetic strategies without instruction (Groen & Resnick, 1977).

Strategies may emerge from children's modeling the problem situation. Kindergartners can solve a wide range of addition and subtraction problem types when they represent the objects, actions, and relationships in the situations (Carpenter, Ansell, Franke, Fennema, & Weisbeck, 1993). About 90% used a valid strategy even for the basic subtraction and multiplication problems, with half the children using a valid strategy on every problem. About 62% solved more than three fourths of the problems accurately.

Although researchers cannot always ascertain their strategies, children from all income levels are first able to respond to the simplest tasks, nonverbal addition and subtraction problems with small numbers (no number greater than 4; Huttenlocher, Jordan, & Levine, 1994; Jordan, Huttenlocher, & Levine, 1994). Middle-income children can also respond verbally to these problems; low-income children find this, and conventional skills such as counting, more difficult. Thus, early solutions may use manipulation of nonverbal images when possible, and conventional skills may be requisite for wider types of response types and solution procedures, especially for problems that involve larger numbers. Automatic (fast and accurate) counting skills, such as naming the counting number immediately after a given number, predicts accuracy of arithmetic performance, which predicts overall mathematics achievement for kindergartners (Penner-Wilger et al., 2005). An open question is whether the distinction between nonverbal and verbal responses is mainly a difference in language competencies or also includes the children's possession of a figural collection (which may not have a cardinal representation) versus a numerosity or protonumerosity (Steffe & Cobb, 1988).

Confirming cultural or experiential influences, Chinese kindergartners answer three times as many addition problems as U.S. children (Geary, Bow-Thomas, Fan, & Siegler, 1993). Language, again, appears to be a substantial factor. Quicker pronunciation of Chinese number words allows a greater memory span for numbers that in turn supports more sophisticated verbal strategies (compared to use of fingers) that in turn support learning number combinations (because addends and sum are simultaneously in working memory).

The structure and abstruseness of English (compared to Asian) counting words slows early learning of counting, as shown earlier. Given the importance of counting in learning subsequent skills such as addition and subtraction, the complexity of the number system may slow learning in those areas as well (Fuson, 1992; Miller, Smith, Zhu, & Zhang, 1995). In addition, languages such as Japanese add coherence to problem statements by integrating items into a set using numeral classifiers (Miura & Okamoto, 2003). A word problem such as "Joe has 6 marbles, 2 more than Tom. How many does Tom have?" in Japanese would be "Joe has 6-round-small-things, 2-round-small-things more than Tom. How many round-small-things does Tom have?" This may help children build images and keep the notion of the sets of objects explicit.

Average 4-year-olds in the United States integrate counting with their arithmetic knowledge and, in doing so, invent counting strategies to compute the effects of addition and subtraction operations on sets of objects. They often initially use a "count all" procedure. For example, given a situation of 5 + 2, they count objects for the initial collection of 5 items, count 2 more items, and then count the items of the two collections together and report "7." These children naturally use such counting methods to solve story situations as long as they understand the language in the story. After children develop such methods, they eventually curtail them (Fuson, 1992). In an early study (Groen & Resnick, 1977), researchers were surprised that, given only practice on addition problems and demonstrations of the count-all strategy, about half of the 5-year-olds moved from a counting-all strategy to a counting-on-from-larger strategy by themselves. That is, given 2 + 4, they might start with 4 and count 2 more. This illustrates that young children can invent efficient strategies without direct instruction, or even despite being taught alternative methods (see also Siegler & Jenkins, 1989). They follow different developmental paths, some illustrating awareness of the new strategies, others not, and most using a variety of more and less advanced strategies at any time. This research also illustrates that failing to solve tasks is not requisite to inventing new strategies; instead, success is often the catalyst (DeLoache, Miller, & Pierrout-sakos, 1998; Karmiloff-Smith, 1984, 1986), especially if one is not changing a conception but only changing to a more efficient or elegant approach to producing the same answer. After concentrated experience, 4- to 5-year-olds showed a variety of strategies, often the counting-on-from-larger (also called "min") strategy, but their successes, rather than failures, appeared to catalyze the invention of new strategies. Transitional strategies were important, such as the "shortcut-sum" strategy, which appears similar to a count-all strategy but involves only one count; for example, to solve 4 + 3, 1, 2, 3, 4, 5, 6, 7 and answer 7. This often preceded the min strategy, which was preferred by most children once they invented it, especially on problems such as 2 + 23 where it saved the most work (Siegler & Jenkins, 1989).

Children continue developing and curtailing their counting methods. For example, given a task in which two collections are hidden from view but are said to contain 6 and 2 items, respectively, children may put up fingers sequentially while saying, "1, 2, 3, 4, 5, 6" and then continue on, putting up two more fingers, "7, 8. Eight!" Such counting on is a landmark in children's numerical development. It is not a rote step, as is true of all such curtailments of processes. It requires conceptually embedding the 6 inside the total (Steffe & Cobb, 1988; Steffe, Cobb, von Glasersfeld, & Richards, 1983).

Counting on when increasing collections and the corresponding counting-back-from when decreasing collections are powerful numerical strategies for children. However, they are only beginning strategies. In the case where the amount of increase is unknown, children count-up-to to find the unknown amount. If six items are increased so that there are now nine items, children may find the amount of increase by counting, "Six; 7, 8, 9. Three." And if nine items are decreased so that six remain, children may count from nine down to six to find the unknown decrease, as follows: "Nine; 8, 7, 6. Three." However, counting backwards, especially more than three counts, is difficult for most children (Fuson, Smith, & Lo Cicero, 1997).

When children fully realize that they can find the amount of decrease by putting the items back with the six and counting from six up to nine, they establish that subtraction is the inversion of addition and can use addition instead of subtraction. This understanding develops over several years. There is little sign that children as young as 2 years use or understand the principle (Vilette, 2002), but 3- and 4-year-old children use procedures consistent with the inversion

principle (J. S. Klein & Bisanz, 2000; Rasmussen, Ho, & Bisanz, 2003), showing use of arithmetical (quantitative, not just qualitative) principles before formal schooling. This use is inconsistent, however, perhaps reflecting children's creative tendency to try out a variety of procedures (Shrager & Siegler, 1998), limits on their information-processing capacities (J. S. Klein & Bisanz, 2000), their nascent understanding of the concept (Vilette, 2002), or some combination. Given a problem with larger numbers, no 4-, some 5-, and most 6-year-olds used the addition-subtraction inverse principle (Baroody & Lai, 2003). In summary, only in the primary grades do many children understand and use the inversion principle in a conscious and interrelated fashion (P. Bryant, Christie, & Rendu, 1999). At that time, they often use addition to solve certain subtraction problems. However, many kindergartners do use this mathematically significant principle consistently in a quantitative manner in some situations.

Summary

Early in life, babies are sensitive to some situations that older people view as arithmetic. They may be using an innate subitizing ability that is limited to very small numbers, such as 2 + 1. Alternatively, they may be individuating and tracking individual objects. In any case, they possess the beginnings of invariance and transformations and thus have far richer foundations for arithmetic than traditional Piagetian accounts suggested.

The contents of minds can be categorized as data, information, knowledge, understanding, and wisdom (Ackoff, 1989). Infants appear to have data, and perhaps information, but knowledge and understanding develop over considerable time. As for the other areas involving number, calculation first emerges with very small numbers, then extends to larger numbers. Exact calculation is preceded by a period of approximations that are more accurate than random guessing. Only years later do they extend their abilities to larger collections that, although still small (e.g., 3 + 2), are amenable to solution with other methods. These other methods, usually using concrete objects and based on subitizing and counting, play a critical developmental role, as the sophisticated counting and composition strategies that develop later are all curtailments of these early solution strategies (Carpenter & Moser, 1984; Fuson, 1992). That is, children learn to count from a given number (rather than starting only from one), generate the number before or after another number, and eventually embed one number sequence inside another. They think about the number sequence, rather than just saying it (Fuson, 1992). Such reflection empowers counting to be an effective and efficient representational tool for problem solving. Thus, educators must study the processes children use as well as the problems they can solve to understand both their strengths and limitations at various ages.

Education

Children can learn arithmetic from 3 years of age and, in some situations, even earlier. Yet most pre-K teachers do not believe arithmetic is appropriate (Sarama, 2002; Sarama & DiBiase, 2004). In multiple countries, professionals in multiple educational roles vastly underestimate beginning students' abilities (Heuvel-Panhuizen, 1990).

Given the abilities of even preschool children to directly model different types of problems using concrete objects, fingers, and other strategies (Carpenter, Ansell, Franke, Fennema, & Weisbeck, 1993), these restricted educational beliefs and strategies are not only unnecessary but also deleterious. That is, children unfortunately constrain the symbols + and − to these limited interpretations and lose the flexibility of their intuition-based solution strategies (Carpenter, 1985), and their confidence in themselves as problem solvers decreases. Textbooks also present symbols and number sentences too soon and in the wrong ways (i.e., before meaningful situations); do little with subitizing or counting, automatization of which aids arithmetical reasoning; and de-emphasize counting strategies (Fuson, 1992). The younger the children, the more problematic these approaches become. No wonder that American schooling has a positive effect on children's accuracy on arithmetic but an inconsistent effect on their use of strategies (Bisanz et al., 2005; Naito & Miura, 2001).

In addition, textbooks offer an inadequate presentation of problems with anything but small numbers. In one kindergarten text, only 17 of the 100 addition combinations were presented, and each of these only a small numbers of times (Hamann & Ashcraft, 1986). Primary students encountered combinations with larger numbers, but less frequently than small-number combinations. Correlational evidence supported that this accounts for difficulty students of all ages have with addition with larger sums (Hamann & Ashcraft, 1986).

Several benefits stem from children's invention and discussion of diverse strategies for more demanding arithmetic problems. The number of different strategies children show predicts their later learning (Siegler, 1995). For example, when learning to solve problems involving mathematical equivalence, children were most successful when they had passed through a stage of considering multiple solution strategies (Alibali & Goldin-Meadow, 1993; Siegler, 1995). Also, it is too seldom recognized that these processes are themselves worthwhile mathematical goals (Steffe & Cobb, 1988). Teachers might encourage children to advance in their sophistication, but effective advances usually do not involve replacing initial strategies with school-based algorithms but instead helping children curtail and adapt their early creations. When asked to explain a concept they are acquiring, children often convey one procedure in speech and a different procedure in gesture. Such "discordant" children are in a transitional state and are particularly receptive to instruction (Alibali & Goldin-Meadow, 1993; Goldin-Meadow, Alibali, & Church, 1993).

For what period are manipulatives necessary? For children at any age, they are necessary at certain stages of development, such as counters of perceptual unit items (Steffe & Cobb, 1988). Preschoolers can learn nonverbal and counting strategies for addition and subtraction (Ashcraft, 1982; Clements & Sarama, in press; Groen & Resnick, 1977), but they need concrete objects to give meaning to the task, the count words, and the ordinal meanings embedded in the situations.

However, once children have established that level of thinking, they can often solve simple arithmetic tasks without explicitly provided manipulatives. In one study of kindergartners, there were no significant differences between those given and not given manipulatives in accuracy or in the discovery of arithmetic strategies (Grupe & Bray, 1999). The similarities go on: Children without manipulatives used their fingers on 30% of all trials, whereas children with manipulatives used the bears on 9% of the trials but used their fingers on 19% of trials for a combined total of 28%. Finally, children stopped using external aids approximately halfway through the 12-week study.

Composing and Decomposing

Composing and decomposing numbers is another approach to addition and subtraction, one that often overlaps with counting strategies. Phenomenologically it can be experienced similarly to subitizing; indeed, conceptual subitizing, previously discussed, is a special case of composition of number. Composing and decomposing numbers also contributes to developing part-whole relations, one of the most important accomplishments in arithmetic (Kilpatrick, Swafford, & Findell, 2001).

Composing and decomposing are combining and separating operations that allow children to build concepts of parts and wholes. Toddlers may first learn that wholes consist of parts, then that a whole is more than its parts and that parts make a larger whole. Next, they may learn that nonverbally represented parts make a specific whole (e.g., •• and •• make ••••). Later, they may learn that "two" and "two" make "four" (Baroody, Lai, & Mix, 2005). At that point, children can develop the ability to recognize that the numbers 2 and 3 are "hiding inside" 5, as are the numbers 4 and 1. Such thinking continues to develop. As previously shown, most preschoolers can "see" that 2 items and 1 item make 3 items; even 3-year-olds can solve problems such as one and one more, at least nonverbally. However, 4-year-olds do not appreciate the part-whole structure of initial-unknown change problems (Sophian & McCorgray, 1994). They would choose a small number as the answer for both addition and subtraction problems. In contrast, children a year or two older responded with a number that was larger than that given for the final set more often on addition than on subtraction problems, and they responded with a smaller number than that given for the final set more often on subtraction than on addition problems. This is not to say that children always apply such understandings in finding precise answers to arithmetic tasks. The findings are less pessimistic than the view that this level of part-whole thinking is not accessible to children until the primary grades. For example, Piagetian theory might be interpreted as excluding missing-addend tasks until after first or even second grade (Kamii, 1985). However, children appear to understand the part-whole relationships of tasks by kindergarten, although they may not know how or think to apply it to all arithmetic tasks (Sophian & McCorgray, 1994).

At least by 5 years of age, children are ready to engage in tasks that put a substantive demand on their explicit understanding of part–whole relationships, such as initial-unknown change problems. However, they may not apply these understandings in all relevant arithmetic tasks. One study suggests that teachers may need to help children see the relevance in and apply their understandings of part-whole relationships to these types of problems (Sophian & McCorgray, 1994).

Language, Numerals, and Other Symbols

As the brief history of research on early number knowledge indicated, the hypothesized role of language has changed through the four historical phases described previously. In the first phase, counting and number words played a definite, if underspecified, role. The second, Piagetian, phase severely limited the role of language. In the theories of the nativists and skills-integration theorists of the third phase, language was important in building more sophisticated number concepts but was not central to development per se, especially in its earliest foundations. In the fourth phase, several theorists proposed that language plays a critical role in developing the concept of cardinal numbers. For example, they claimed that language sensitizes children to cardinality and helps them move from approximate to exact representations of number (Spelke, 2002; Van de Walle, Carey, & Prevor, 2000). That is, the word *three* may enable children to create an explicitly cardinal representation that applies to various triads of objects (Baroody, Lai, & Mix, 2006).

Empirical evidence indicates that basic verbal number competencies precede some nonverbal tasks, contrary to the mental-models view (Baroody, Benson, & Lai, 2003). This supports the view that learning number words facilitates the development of exact representations. Thus, as stated previously, human language plays an important role in connecting different representations, making early premathematical cognition numeric (Wiese, 2003). Much of the research discussed in this section supports this view, as does research on learning other concepts, such as color words (Sandhofer & Smith, 1999). Not all language involves the term for the concept. Often, introduction of the term is better delayed until concepts are developed with informal language and experiences (Lansdell, 1999). Further, there is still an open debate about the exact role of language, with some arguing that it is critical for the development of number concepts above four, and others citing evidence of competence in cultures with limited counting words and other evidence that language is not the cause of the development of these concepts (R. Gelman & Butterworth, 2005). For example, there is evidence that when verbal counting is too slow to satisfy time constraints, nonverbal mental magnitudes mediate the production of a number word that approximates that numerosity of a set. It also mediates the ordering of symbolic numbers (Gallistel & Gelman, 2005). In general, research indicates that the interactions between language on the one hand, and number concepts and skills on the other, are more bidirectional, fluid, and interactive than previous accounts allowed (Mix, Sandhofer, & Baroody, 2005).

Children come to elementary school, and often pre-K, with initial knowledge of reading and even writing numerals (Bell & Bell, 1988; Bialystok & Codd, 2000; Fuson, 1992). For example, kindergartners from a range of populations were generally competent with the numerals from 1 to 20, and many from 1 to 100. The early autumn percentages for reading and writing were for 3, reading, 95%, writing, 80%; for 9, 74% and 53%. By the end of the year, the percentages for 3 and 9 for both reading and writing were above 93%. Reversing the digits for numbers in the teens was a frequent error, due to the reversal in English pronunciation. (Reversals in writing digits, e.g., writing 3 backwards, were ignored in scoring as common errors that children outgrow.)

Children's ability to read or produce numerals or other representations does not imply a commensurate understanding of what they represent (Bialystok & Codd, 2000). In one study, about a fourth of the children entering school were unaware of the function of numerals (Munn, 1998). When asked to represent how many objects there were, children may produce idiosyncratic, pictographic (representing the appearance of objects as well as their numerosity), iconic (one-to-one correspondence between objects and marks), or symbolic (numerals) responses (Ewers-Rogers & Cowan, 1996; Hughes, 1986; Munn, 1998). Similarly, in other studies preschoolers showed an understanding that written marks on paper can preserve and communicate information about quantity (Ewers-Rogers & Cowan). For example, 3- and 4-year-olds invented informal marks on paper, such as tally marks and diagrams, to show how many objects are in a set (Allardice, White, & Daugherty, 1977). However, children this age did not notice when numerals were missing from pictures of such objects and rarely said number words in giving fast-food orders; thus, understanding preceded use of numerals (Ewers-Rogers & Cowan, 1996).

Just as important as learning to read and write representations of number is learning to use such representations functionally. Except for the idiosyncratic representations, children use their symbols to remember the amount in collections, one level of functional use. Children continue to develop competence in functional use of numerals. For example, children were shown several collections labeled with symbols. They were told a teddy bear added one object to only one collection and asked to figure out which one had the extra object. Only some young children could use the symbol to complete the task (Munn, 1998). Functional use is more likely when children use numerals rather than pictographic or iconic representations, and nonconventional representations may not be important precursors to conventional symbol use.

Final Words

Research on number and quanty supports the tenets of the *hierarchic interactionalism* framework. For example, cognitive progressions through levels in the specific domains of number, such as counting, comparing, and arithmetic, are evident, building on initial bootstraps, yet undergoing significant development through the early years. There are different courses

for such development. Two examples, emphasizing experiences in the social environment, briefly illustrate these possibilities. First, children from low-income homes may engage in informal premathematical activity and possible premathematical knowledge (Ginsburg, Inoue, & Seo, 1999) but, due to different experiences, may still lack components of the conceptual structures possessed by children from more affluent homes (Griffin, Case, & Siegler, 1994). Second, educational emphases on skills, for example, can lead to instrumental knowledge of, and beliefs about, mathematics (Kamii & Housman, 1999; Skemp, 1976). More specifically, experiences with sophisticated counting strategies, composition/decomposition, and other imagistic approaches lead children to develop different strategies and understandings of number than, for example, experiences limited to simple counting strategies and memorization. Relevant here is the notion, perhaps expressed best by Wittgenstein (1953/1967), but expressed previously by several others (e.g., Douglass, 1925; von Glasersfeld, 1982), of weaving together various threads to form number concepts, in which various different weavings create different number concepts of various strength and applicability. The tenet of progressive hierarchization is illustrated by the connections children come to make between various mathematically relevant concepts and procedures (e.g., connecting experiences with nonverbal subitizing, hearing number words applied to particular situations, and counting to learn cardinality, Klahr & Wallace, 1976; Sophian & Adams, 1987), creating more robust understandings that are hierarchical in that they employ generalizations (e.g., of additive composition) while maintaining differentiations (e.g., discrete vs. continuous compositions).

The domains of number and geometry share commonalities that emerge from the loosely differentiated competencies of the young child. For example, Mix, Huttenlocher, and Levine suggested that "the quantification of infants and young children could be accurately termed 'spatial quantification'" (Mix, Huttenlocher, & Levine, 2002, p. 139). In this view, spatial thinking is essential in its role in the development of number and quantification. It is important for many other reasons.

GEOMETRY AND SPATIAL THINKING

Geometry and spatial thinking are important because they involve "grasping that space in which the child lives, breathes and moves . . . that space that the child must learn to know, explore, conquer, in order to live, breathe and move better in it" (Freudenthal in NCTM, 1989, p. 48). In this section, we begin with just that space in which the child lives, then examine how the child learns to know it better, and then turn to issues of smaller scale geometric shape, including composition and transformation of shapes. The separation of these two basic geometric domains is based on distinct systems in the primate brain for object perception, or recognizing *what* an object is (inferior temporal cortex, ventral pathway), and for spatial perception, or *where* an object is (posterior parietal cortex, ventral stream; Stiles, 2001; Ungerleider & Mishkin, 1982).

Spatial Thinking

Spatial thinking is important because it is an essential human ability that contributes to mathematical ability. Spatial reasoning is also a process that is distinct from verbal reasoning (Shepard & Cooper, 1982) and functions in distinct areas of the brain (Newcombe & Huttenlocher, 2000). Further, mathematics achievement is related to spatial abilities (Ansari et al., 2003; Fennema & Sherman, 1977; Fennema & Sherman, 1978; Guay & McDaniel, 1977; Lean & Clements, 1981; Stewart, Leeson, & Wright, 1997; Wheatley, 1990). Although researchers do not fully understand why and how, children who have strong spatial sense do better at mathematics. This relationship, however, is not straightforward. Sometimes, "visual thinking" is "good," but sometimes it is not. For example, many studies have shown that children with specific spatial abilities are more mathematically competent. However, other research indicates that students who process mathematical information by verbal-logical means outperform students who process information visually (for a review, see Clements & Battista, 1992).

Similarly, limited imagery in mathematical thinking can cause difficulties. An idea can be *too* closely tied to a single image. For example, connecting the idea of *triangles* to a single image such as an equilateral triangle with a horizontal base restricts young children's thinking. Therefore, spatial ability is important in learning many topics of mathematics. The role it plays, however, is elusive and, even in geometry, complex. Two major abilities are spatial orientation and spatial visualization (Bishop, 1980; Harris, 1981; McGee, 1979). We first discuss spatial orientation, which involves an extensive body of research, then spatial visualization and imagery.

Spatial Orientation

Spatial orientation is knowing where you are and how to get around in the world, that is, understanding and operating on relationships between different positions in space, at first with respect to your own position and your movement through it, and eventually from a more abstract perspective that includes maps and coordinates at various scales. Like number, spatial orientation has been postulated as a core domain, for which competencies, including the ability to actively and selectively seek out pertinent information and certain interpretations of ambiguous information, are present from birth (R. Gelman & Williams, 1997). Infants focus their eyes on objects and then begin to follow moving objects (Leushina, 1974/1991). Toddlers ignore other cues and instead use geometric information about the overall shape of their environment to solve location tasks (Hermer & Spelke, 1996). Again, however, evidence supports the interaction of inborn endowments, (possibly) maturation, experience, and sociocultural influences. What, then, can young children understand and represent about spatial relationships and navigation? When can they represent and ultimately mathematize this knowledge?

Piaget maintained that children are born without knowledge of space, or even permanent objects (Piaget & Inhelder, 1967). His topological primacy thesis posited that they move through stages of egocentric spatial constructions (e.g., objects within reach, or those in front of the child, including topological relations of connectedness, enclosure, and continuity) to allocentric constructions (e.g., objects farther away, including having relationships to one another). Studies can be interpreted in that light. For example, if 6- to 11-month-old children are placed in a maze and repeatedly find a toy by crawling straight and then turning left at an intersection, then are moved to another end, most will incorrectly turn in the same direction that originally led to the toy (Acredolo, 1978). At 16 months of age, children correctly compensate for the change in their position. Piaget claimed that children's first notions are of topological space (e.g., understanding closure and connectedness) and later projective (relations between the child and objects, establishing a "point of view") and Euclidean, or coordinate space.

Research supports Piaget's prediction about development of near space before far space (see Haith & Benson, 1998, for a review). However, Piaget's topological primacy thesis seems of limited usefulness. In addition, research suggests that young children are more, and adults less, competent than the Piagetian position indicated, although substantial development does occur. Young children can reason about spatial perspectives and spatial distances, although their abilities develop considerably throughout the school years. In the first year of life, infants can perceive the shape and size of objects and can represent the location of objects in a three-dimensional space (Haith & Benson, 1998; Kellman & Banks, 1998). As another example, "egocentrism" is not displayed if landmarks provide cues (Rieser, 1979). Between 5 and 9 months, they develop a geometric mechanism that allows them to identify an object to which another person is pointing (Butterworth, 1991). Older students and adults still display biases and errors in spatial reasoning (Fischbein, 1987; Uttal & Wellman, 1989) and do not always perform successfully on tasks designed by Piaget to assess an underlying Euclidean conceptual system (Liben, 1978; Mackay, Brazendale, & Wilson, 1972; H. Thomas & Jamison, 1975). Thus, both the topological primacy thesis and the traditional egocentric-to-allocentric theory should be replaced. However, Piaget's constructivist and interactionist positions remain viable, especially that the representation of space is not a perceptual "reading off" of the spatial environment but is built up from active manipulation of that environment (Piaget & Inhelder, 1967). We review research on spatial orientation in four categories: spatial location and intuitive navigation, spatial thought, models and maps, and coordinates and spatial structuring.

Spatial Location and Intuitive Navigation

What kind of "mental maps" do young children possess? Neither children nor adults actually have "maps in their heads"—that is, their mental maps are not like a mental picture of a paper map. Instead, they consist of private knowledge and idiosyncrasies. Children learn about space by developing two types of self-based reference systems and two types of external-based reference systems (Newcombe & Huttenlocher, 2000). The younger the child, the more loosely linked these systems are.

Self- and external-based systems. Self-based spatial systems are related to the Piagetian construct of egocentric space and encode the position of the moving self. The most primitive—both in its early emergence and its limited power—is response learning, or sensorimotor coding. The child records a location or route to a location by a pattern of movements that have been associated with a goal (e.g., looking to the left from a high chair to see a parent cooking). The second is path integration, in which locations are coded based on the distance and direction of one's own movement. One's location is continually updated based on input regarding movement (as well as from landmarks encountered—integrated with external-based systems). Such automatic processes could serve as the foundation for explicit mathematics (Newcombe & Huttenlocher, 2000).[3]

External-based reference systems are based on environmental structures and landmarks. The landmarks are usually salient, familiar, or important objects. Cue learning associates an object with a coincident landmark, such as a toy on a couch. As with response learning, cue learning is the more limited and less powerful of two systems. The more powerful, place learning, comes closest to people's intuition of mental maps (albeit, as we shall show, this is a limited metaphor), as it builds locations from distances and directions among environmental landmarks. One example of place learning, taking the edges of a region or walls of a room as a frame of reference, illustrates a possible early, implicit foundation for later learning of coordinate systems.

Finally, when information from these four systems is combined, it is combined hierarchically (e.g., chair in a room, school, city . . .), with different precisions at each level. This provides a best estimate of locations but can also introduce systematic biases (Newcombe & Huttenlocher, 2000). For example, knowing an object was dropped and lost in the left half of the back yard might be combined with a memory of a specific location via path integration, but it may bias the estimate toward the center of the region.

When do these systems emerge? Once they do, do they develop? Development consists of two interrelated aspects. First, children learn through experience which coding systems are more effective and accurate, and in which situations. (Following Newcombe & Huttenlocher, 2000, we use the term coding to mean a memory trace of some type in the information-processing system that supports action in the environment without any commitment to a cognitively accessible "representation" that will, eventually, support thinking about space.) Secondly, each of these coding systems becomes more effective, although the simple systems of response and cue learning only extend the situations to which they apply. We begin by describing the latter development.

---

[3] We use R. F. Wang and Spelke's (2002) term rather than Newcombe and Huttenlocher's (2000) term *dead reckoning* because the root of the latter is "deductive reasoning." With magnetic compasses, sailors could take compass bearings, or headings. Bearings on two landmarks were used to construct two intersecting lines, which determined a location. This starting point was then iteratively updated based on movement—direction and distance (speed times time). In both its implicit beginnings and navigational application, the difficulty is "drift"—small errors lead to increasingly inaccurate calculations. The common written abbreviation of deductive reckoning, "ded. reckoning," was misread by an early mariner and the name stuck as *dead reckoning*. As a term, *path integration* has no implication of conscious deduction.

Response learning and cue learning. In the first year of life, stationary infants rely on response learning to locate objects in their environment. An example is children who incorrectly turn in the same direction that originally led to a toy even after being physically reoriented (Acredolo, 1978; see also Acredolo & Evans, 1980).

In addition to response learning, two other systems, cue learning and path integration, emerge by at least 6 months of age (Newcombe & Huttenlocher, 2000). As an example of cue learning, 7-month-olds can remember which of two containers holds an object, even after a minute's delay filled with distractions (McDonough, 1999). Landmarks can help children depress an incorrect response after they are rotated, but 6-month-olds were uncertain of the correct choice; 9- and 11-month-olds used the cues successfully (Acredolo & Evans, 1980). As another example, infants associate objects as being adjacent to a parent (Presson, 1982; Presson & Somerville, 1985) but cannot associate objects to distant landmarks. By the age of 1 year, they can use a different colored cushion among an array of cushions to locate a toy (Bushnell, McKenzie, Lawrence, & Com, 1995). Toddlers and 3-year-olds can place objects in prespecified locations near distant landmarks but "lose" locations that are not specified ahead of time once they move. R. F. Wang and Spelke (2002) described a *view-dependent place recognition* system that operates similarly in animals such as ants and bees and argued that in humans as well, determining location and navigating are based on view-specific representations.

Path integration ("dead reckoning"). Infants use response and cue learning when they are stationary. They ignore movement when they are carried (Acredolo, Adams, & Goodwyn, 1984) but use path integration as early as 6 months of age when they have actively moved themselves (Newcombe, Huttenlocher, Drummey, & Wiley, 1998). By 1 year of age, they can encode both distance and direction with some degree of accuracy during self-movement (Bushnell, McKenzie, Lawrence, & Com, 1995). Thus, self-movement appears important, although infants as young as 6 but not 4 months of age can demonstrate above-chance performance after training (to localize a fixed location from two orientations; see also Haith & Benson, 1998, for a review and statement of methodological challenges; D. Tyler & McKenzie, 1990). They can internally represent the amount of rotational movement about their own axis by at least 8 months and code amount of movement along a straight line by at least 9 months. By 16 months, children are likely to perform path integration following movements involving both translation and rotation. They sometimes use this system, even more than other systems for certain tasks (Bremner, Knowles, & Andreasen, 1994). A general point is that spatial systems that produce motor activity and spatial representations are intimately connected (Rieser, Garing, & Young, 1994). Path integration has been identified in insects, birds, and mammals, and findings support that aspects of it are fundamental inborn endowments in humans as well (R. F. Wang & Spelke, 2002). The fine calibration of the system, and the ability to ignore distracting visual information, improves from 4 years of age to adulthood, probably on the basis of fine-tuning from the senses, including proprioception and kinesthesia (Newcombe & Huttenlocher, 2000).

Place learning. During their second year, children develop the ability to code locations using objects in their external environment. They also become capable of spatial reasoning, in that they can solve problems with that information. They continue to grow in their abilities in spatial coding, reasoning, and symbolizing through their elementary school years, as they develop spatial visualization abilities such as maintaining and operating on mental images (e.g., mental rotation) and learn to use such tools as language and maps.

The first of these abilities is place learning, which involves creating a frame of reference by coding objects' positions with respect to perceptually available landmarks, using information of their distances and directions. In contrast to Piaget's topological primacy thesis, in which children build a "Euclidean system" only by age 9 or 10, research indicates that toddlers are able to code distance information and use that to locate objects. For example, infants as young as 5 months use spatiotemporal information to track and even individuate objects and from 12 months of age can code distance to support the search for hidden objects (Bushnell, McKenzie, Lawrence, & Com, 1995; Newcombe & Huttenlocher, 2000); however, in one of these studies they could not use separate indirect landmarks at 1 year of age (Bushnell, McKenzie, Lawrence, & Com, 1995). They can also use simple geometric properties of a room to guide a search; for example, looking more often in two corners of a rectangular room with the long wall on the right and the short wall on the left (Hermer & Spelke, 1996).

Place learning using multiple objects may appear after 21 months of age (Newcombe, Huttenlocher, Drummey, & Wiley, 1998). For example, after shown an object hidden in a sandbox, children were taken around the other side of it. Children 21 months or younger were not helped by the availability of landmarks around the sandbox (that is, they did no worse when these landmarks were hidden by a curtain), but older children were significantly more accurate with the visible landmarks. By 5 years of age, children can represent an object's position relative to multiple landmarks, such as an object midway between two other objects (Newcombe, 1989). From 5 to 7 years, children increase their ability to keep track of their locations in mazes or open areas.

Functional use of such spatial knowledge for searching, which requires coding spatial information and forming and utilizing spatial relationships, develops over the toddler and preschool years (Newcombe & Sluzenski, 2004). Children show significant growth in the ability to search for multiple objects between 18 and 24 months of life and an increase in the ability to use relations among objects between 24 and 42 months (Newcombe & Sluzenski, 2004). In addition, toddlers can systematically check an array by exhaustively searching within groupings at three sites. Preschoolers can plan comprehensive searches in a small area, including memory of sites that have been checked, often using a circular search path (Cornell, Heth, Broda, & Butterfield, 1987).

Selection of systems. As mentioned, the second aspect of development involves learning to select the coding systems that are more effective and accurate in different situations. Evidence of early appearance of all systems and gradual growth favor the view that children change the relative importance they assign to different types of spatial information when these types provide conflicting information (Newcombe & Huttenlocher, 2000), possibly through the formation of metacognitive strategies (Minsky, 1986). Infants' choices among conflicting spatial systems depend on a combination of cue salience, complexity of movement, whether response learning has been recently reinforced, and whether the infant is emotionally secure or under stress (Newcombe & Huttenlocher, 2000). Further development depends on experience. For example, when children attain self-mobility by crawling between 6 and 9 months of age, they can experience failures in response learning. Infants who crawl, and even those with extensive experience with a walker, succeed more often in locating objects' spatial positions (Bertenthal, Campos, & Kermoian, 1994). The longer they have been moving themselves, the greater the advantage, probably because they learn to attend to relevant environmental information and update their spatial codings as they move. This is supported by research showing that when 12-month-old children walk to the other side of a layout and have the opportunity to look at all times, they both look more than children who are guided and also subsequently do better in turning toward the object from the new position (Acredolo, Adams, & Goodwyn, 1984). There is great variability on the onset of crawling (Bertenthal,

Campos, & Kermoian, 1994), which, along with failures in the codes from response learning, probably leads to developmental advances via the variation-and-selection process.

This perspective does not deny the possibility of an apparently qualitative, or general, shift in default propensities (Newcombe & Huttenlocher, 2000), which may be the result of newly available experiences (e.g., crawling) or of the gradual growth in the effectiveness of a new system that has achieved ascendance in the overlapping stages of development theory (Minsky, 1986, see also a later but more developed version in Siegler, 1996). Such shifts may also result from biological maturation, although some argue that experience may be the cause of the development of the nervous system (Thelen & Smith, 1994).

Hierarchical combination of systems. Before their second birthday, at about 16 months, children show the beginnings of a hierarchical combination of spatial reference frames, including categorical coding of regions and fine-grained information, although the categories and combinations of categories develop considerably after that age (Newcombe & Huttenlocher, 2000). For example, children form categories that are not physically demarcated (e.g., the left half of a field), beginning in a limited way at 4–6 years of age (in small rectangles) and expanding to larger regions at about 10 years. Second, they develop the ability to code hierarchically along two dimensions simultaneously. For example, to recall the location of a point in a circular region, people 10 years of age and older code locations for both the distance from the center of a circle and angle, as well as categorical information about the quadrant. Children younger than 7 years do not code the categorical information about quadrants in such tasks, whereas 7-year-olds do if angle information is requested. In general, as they age, children divide regions into smaller and more abstract categorically coded regions. Such changes, rather than a qualitative shift to Euclidean space, appear to explain changes in performance. In summary, hierarchical coding begins at 16 months, but not until 10 years of age do children reach sophisticated levels (Newcombe & Huttenlocher, 2000).

Summary. Children are born with potential abilities in response learning, cue learning, path integration, and place learning. During the 1st year of life, these systems are further integrated, based on feedback from experiences with their physical environment. During the 2nd year, starting at about 21 months, children develop place learning proper; that is, they begin to use distance information from multiple environmental landmarks to define location. Development of symbolic skills may play an important role. From that point forward, the path integration and place learning systems develop in effectiveness and accuracy. Path integration improves from 4 years on. By 5 years of age, children can represent a location in terms of multiple landmarks, and from 5 to 7 years, develop in their ability to maintain locations in challenging circumstances such as open areas. All four systems are further reconfigured in their application and integration. Finally, there are changes in the size and nature of categories used in hierarchical spatial coding, but these remain limited in the preschool years.

Further, the systems are used with increasing efficiency. At 6 months of age, children are more likely to rely on response learning, but older children, with more visual and especially self-produced movement experiences, rely more on cue learning and path integration (Newcombe & Huttenlocher, 2000). This variation-and-selection approach is similar to Siegler's (1996) approach, as previously discussed regarding the learning of number. Again, the phenomenon of increased variability leads eventually to greater stability and generalizability.

We now turn to what Piaget and Inhelder's (1967) research focused upon, representation that would support reflection. They separated representation that would support action in the world (what we called *spatial coding*) and that would support reflection about the world (*spatial thought*).

Spatial Thought

As children gain the capacity for symbolic thought in their 2nd year, they begin to gain access to their spatial knowledge and thus build upon spatial codings to create accessible representations, supporting the emergence of spatial thought. We first address the question of whether spatial reasoning is an innate or developed ability and then consider spatial perspective taking, navigation through large-scale environments, and the language of space.

Development of spatial thought. Some have argued that abilities such as spatial inference are innate, on the basis of, for example, the finding that children who are blind can infer paths that they have not been taught (Landau, Gleitman, & Spelke, 1981). However, blind children performed less accurately in the key aspect of the task (accuracy at final position) than age-matched sighted but blindfolded children performed in similar tasks. This is noteworthy given that the blindfolding created an artificial task for the latter children (Morrongiello, Timney, Humphrey, Anderson, & Skory, 1995). Thus, at least some visual experience appears important for full development of spatial knowledge (Arditi, Holtzman, & Kosslyn, 1988; Morrongiello, Timney, Humphrey, Anderson,

& Skory, 1995; Newcombe & Huttenlocher, 2000). Nevertheless, spatial representations are spatial, not strictly visual. Visual input is important for full development.

Spatial perspective taking. A component of Piaget's topological primacy thesis was that "projective relations" did not develop until elementary school age. When they developed, children considered figures and locations in terms of a "point of view." As an example, the concept of the straight line results from the child's act of "taking aim" or "sighting." Children perceive a straight line from their earliest years, but they initially cannot independently place objects along a straight path. They realize, based on perception, that the line is not straight but cannot construct an adequate conceptual representation to make it so. At about 7 years of age, children spontaneously "aim" or sight along a trajectory to construct straight paths.

The Piagetians confirmed these theoretical claims with other experiments, such as the "three mountains" task in which children had to construct a scene from the perspective of a doll. For each new position of the doll, young children methodically went about their task of re-creating the appropriate viewpoint, but it always turned out to be from the same perspective—their own! Thus, Piaget and Inhelder inferred that children must construct systems of reference, not from familiarity born of experience, but from operational linking and coordination of all possible viewpoints, each of which they are conscious. They concluded that such global coordination of viewpoints is the basic prerequisite in constructing simple projective relations. For although such relations are dependent upon a given viewpoint, nevertheless a single point of view cannot exist in an isolated fashion but necessarily entails the construction of a complete system linking together all points of view.

However, other studies have shown conflicting results. For example, young children do recognize that other observers see something different and develop in their ability to construct those viewpoints (Pillow & Flavell, 1986). They may use a "line-of-sight" idea, in which they reason that people can see any object for which one can imagine an unobstructed line between their eyes and the object. Some ability to coordinate perspectives mentally is present in children as young as 18 months. For example, children performed better than chance in retrieving a reward after it was hidden randomly in one of two identical left-right locations on a turntable, then its location reversed via either a 180-degree rotation of the turntable by the experimenter or a move by the child to the opposite side of the table (Benson & Bogartz, 1990).

In summary, the development of projective space may involve not just the coordination of viewpoints but also the establishment of an external framework. That is, a key to solving the three mountains task may be, again, the conflict between frames of reference, experienced and imagined. Full competence in dealing with such conflicts is achieved at approximately the ages Piaget originally claimed. Simultaneously, children are developing additional knowledge of projective relations. At 4 years of age, but not 3, children begin to understand that moving objects nearer or farther increases and decreases its apparent size; they can also indicate how a circular object should be rotated to make it appear circular or elliptical (Pillow & Flavell, 1986).

Navigation through large-scale environments. Navigation in large environments requires integrated representations, because one can see only some landmarks at any given point. Some researchers believe that people learn to navigate using landmarks; then routes, or connected series of landmarks; then scaled routes; and finally survey knowledge, a kind of mental map that combines many routes and locations (Siegel & White, 1975). Only older preschoolers learn scaled routes for familiar paths; that is, they know about the relative distances between landmarks (Anooshian, Pascal, & McCreath, 1984). Even young children, however, can put different locations along a route into some relationship, at least in certain situations. For example, they can point to one location from another even though they never walked a path that connected the two (Uttal & Wellman, 1989).

Children as young as 3.5 years were able, like adults, to accurately walk along a path that replicated the route between their seat and the teacher's desk in their classroom (Rieser, Garing, & Young, 1994). Self-produced movement is again important. Kindergarten could neither imagine similar movements nor point to various objects accurately, but they could recreate the route and point accurately when they actually walked and turned. Thus, children can build imagery of locations and use it, but they must physically move to show their competence.

Comparing routes, as in finding the shortest of several routes, is a difficult task. Children as young as 1 or 2 years can plan shortest routes only in simple situations. For example, 1.5-year-olds will choose the shortest route around a wall separating them from their mothers (Lockman & Pick, 1984). Such early planning may depend on direct sighting of the goal and on an obvious choice of a shorter or more direct route to the goal (Wellman, Fabricius, & Sophian, 1985). Children 3.5 to 4.5 years of age show a mixture of sighting and planning, that is, considering extended courses of action taking into account the overall distances of competing routes, probably on the basis of qualitative dis-

tance-relevant aspects of routes, such as the necessity of backtracking (Wellman, Fabricius, & Sophian, 1985). There is some evidence that young children can compute shortest routes even when the goal is not visible (Newcombe & Huttenlocher, 2000). A significant proportion (40%) of 4-year-olds can not only identify that a direct and indirect route to a given location are not the same distance but can explain why the direct route was shorter (Fabricius & Wellman, 1993). However, ability to plan routes in situations in which the optimal route involves locations that are not close or in sight (when others are) appears at 5–6 years of age.

Considering these results in the context of the broad research on spatial orientation, it is unsurprising that children grow over a number of years in their ability to form integrated spatial representations. However, evidence does not support a simple, qualitative shift from landmark to route to integrated, 2-dimensional representations (Newcombe & Huttenlocher, 2000). Instead, development results from substantive refinements in the effectiveness and connectedness of already-existing representational systems, including hierarchical categorization (e.g., children create more precise embedded categories as they age). Young children undoubtedly rely on cues (landmarks) and partially connected landmark-and-route representations when they have limited experience in an environment and when information-processing demands, such as those on memory; use of strategies (e.g., turning around to visualize how a route will look on the return trip); and inferences required to combine spatial knowledge from different systems overwhelm their cognitive abilities. They develop abilities such as recognition-in-context memory that facilitate acquisition of landmark knowledge (Allen & Ondracek, 1995).

The language of space. The development of both geometric domains, the "what" and "where" systems (Ungerleider & Mishkin, 1982), begins early in life as children represent objects at a detailed level of shape, and simply as a set of axes (a representation of an object allowing a region in front and a region in back; Landau, 1996). However, spatial relations are not perceived "automatically"; they require attention (Logan, 1994; Regier & Carlson, 2002). Children learning English show strong biases to ignore fine-grained shape when learning novel spatial terms such as "on" or "in front of" or when interpreting known spatial terms, and equally strong biases to attend to fine-grained shape when learning novel object names. For example, 3-year-olds will ignore shape and generalize primarily on the basis of an object's demonstrated location when shown an unusual object and told, "This is acorp my box" [sic] (Landau, 1996).

Children represent objects in terms of axes by the second year of life; contrary to Piagetian theory, the development of regions appears to begin with the axis and broaden afterward. However, spatial terms are acquired in a consistent order, even across different languages (Bowerman, 1996), one of the few sequences that is consistent with the Piagetian topological primacy thesis. The first terms acquired are *in, on,* and *under,* along with such vertical directionality terms as *up* and *down.* These initially refer to transformations of one spatial relationship into other ("on" not as a smaller object on top of another, but only as making an object become physically attached to another, Gopnik & Meltzoff, 1986). Second, children learn words of proximity, such as *beside* and *between.* Third, children learn words referring to frames of reference such as *in front of, behind.* The words *left* and *right* are learned much later and are the source of confusion for many years.

The consistency of acquisition order, along with early learning of space and the reliance on spatial understanding in learning new words (e.g., predicting what new words mean and extending their use to new situations) argue that spatial language builds upon already-constructed spatial concepts (Bowerman, 1996; Regier & Carlson, 2002; Spelke, 2002). However, that does not mean that children merely map spatial words directly onto extant spatial concepts. The concepts may be just forming, and develop from attempts to change spatial relationships of objects (*down* as *get me down*) to references to all such changes, and finally to static spatial relationships. The words children use encode concepts that are problematic for them—that they are developing—and are used as a cognitive tool to support that development. In addition, adult language helps children consolidate their emerging understandings (Gopnik & Meltzoff, 1986). Further, semantic organization of language appears to influence children's development of spatial concepts (Bowerman, 1996). For example, while English uses *on* for contact with and support by a surface (*on* a table and handle *on* a door) and *in* for containment, Finnish categorizes the handle on a door and apple in a bowl together; the horizontal support of "on a table" requires a different construction. In Dutch, the hanging attachment of "handle on a door" is a separate construct (Bowerman, 1996). Consistent with the notion that these differences affect children's special learning, cross-cultural studies show that, for example, children learning English acquire concepts referring to verticality faster than children learning a language such as Korean for which those terms are less central, whereas Korean children learn different meanings for terms that mean *in* as in *fit tightly* and as in *inside a larger container* whereas English-speaking children are slower to learn that differentiation, with these differences occurring both in production (Choi & Bowerman, 1991) and in comprehension (Choi, Mc-Donough, Bowerman, & Mandler, 1999) as early as 18 to 23 months. Thus, language appears to affect conceptual growth by affecting what kind of spatial relationships and categories children attend to and build.

By 2 years of age, children have considerable spatial competence on which language might be based. Moreover, the use of even a single-word utterance by a 19-month-old, such as "in" may reflect more spatial competence than it first appears when the contexts differ widely, such as saying "in" when about to climb into the child seat of a shopping cart and saying "in" when looking under couch cushions for coins she just put in the crack between cushions (Bowerman, 1996). However, notions such as *left* and *right,* whose relative understanding may require mental rotation (Roberts & Aman, 1993), may not be fully understood until about 6 to 8 years of age. Between these ages, children also learn to analyze what others need to hear to follow a route through a space. To a large degree, however, development past pre-K depends on sociocultural influences on children's understanding of conventions, such as negotiating which frame of reference is used (Newcombe & Huttenlocher, 2000). Such influences have strong effects. For example, achieving flexible spatial performance is correlated and may be caused by acquisition-appropriate spatial vocabulary (R. F. Wang & Spelke, 2002). Indeed, it may be the unique way that people combine concepts from different inborn spatial systems into mature and flexible spatial understandings. Finally, although language supports simple representations such as cue learning, place learning is difficult to capture verbally because of the multiple simultaneous relationships. This leads us to more apropos external representations, such as models and maps.

Models and Maps

Young children can represent and, to an extent, mathematize their experiences with navigation. They begin to build mental representations of their spatial environments and can use and create simple maps. Children as young as 2 years of age can connect oblique and eye-level views of the same space, finding their mother behind a barrier after observing the situation from above (Rieser, Doxsey, McCarrell, & Brooks, 1982). In another study, 2.5- but not 2-year-olds could locate a toy shown in a picture of the space, even when 2-year-olds were given help (DeLoache, 1987; DeLoache & Burns, 1994).

To make sense of maps, children have to create relational, geometric correspondences between elements, as these vary in scale and perspective (Newcombe & Huttenlocher, 2000). By 3 years, children may be able to build simple but meaningful models with landscape toys such as houses, cars, and trees (Blaut & Stea, 1974), although this ability is limited through the age of 6 years (Blades, Spencer, Plester, & Desmond, 2004). However, researchers know less about what specific abilities and strategies they use to do so. For example, kindergarten children making models of their classroom cluster furniture correctly (e.g., they put the furniture for a dramatic play center together) but may not relate the clusters to each other (Siegel & Schadler, 1977).

Thus, preschoolers have some impressive initial competencies, but these are just starting to develop. We begin a more in-depth look at this development by considering children's use, rather than production, of models and maps. One study confirms that children can use both models and maps by 2.5 to 3 years of age, but with a twist. Children were shown a location on a scale model of a room then asked to find same object in the actual room (DeLoache, 1987). For example, a miniature dog was hidden behind a small couch in the model, and the child was asked to find a larger stuffed dog hidden behind a full-sized couch. Interestingly, raising a point to which we shall return, 3- but not 2.5-year-olds could find the corresponding object (the authors do not state, but we assume, that the two were in alignment). However, both ages were successful with line drawings or photographs of the room. It may be that the younger children saw the model as an interesting object, but not as a symbol for another space, leading to the counterintuitive result that the more "concrete" model was less useful to them. In support of this notion, having these children play with the model decreased their success using it as a symbol in the search task, and eliminating any interaction increased their success. The 3s were successful with either, revealing cognitive flexibility in their use of the model, as an object per se and a symbol for another space.

In a similar vein, beginning about 3 and more so at 4 years of age, children can interpret arbitrary symbols on maps, such as a blue rectangle standing for blue couch, or "x marks the spot" (Dalke, 1998). Their abilities lack sophistication; for example, preschoolers recognized roads on a map but suggested that the tennis courts were doors (Liben & Downs, 1989). In another study, 4- and 5-year-olds criticized symbols that lacked features (e.g., tables without legs) but could recognize a plane view of their classroom, so findings such as these may be the result of children merely voicing preferences. All could distinguish between representational (in room) and nonrepresentational (outside of room) paper space (Liben & Yekel, 1996). In any case, by age 5 or 6, children can consistently interpret the arbitrary symbolic relationships used in maps (Newcombe & Huttenlocher, 2000). Yet, children may understand that symbols on maps represent objects but have limited understanding of the geometric correspondence between maps and the referent space. As we shall show, both understandings are developing but have far to go by the end of the preschool years (Liben & Yekel, 1996).

Shortly after 3 years of age, children are able to scale distance across simple spatial representations, a fundamental competence (Huttenlocher, Newcombe, & Vasilyeva, 1999). However, they perform better with symmetric than asymmetric configurations (Uttal, 1996). Children also understand that a map represents space (Liben & Yekel, 1996). By 4 years of age, they can build upon these abilities and begin interpreting maps, planning navigation, reasoning about maps, and learning from maps, at least in simple situations (Newcombe & Huttenlocher, 2000). For example, 4-year-olds benefit from maps and can use them to guide navigation (i.e., follow a route) in simple situations (Scholnick, Fein, & Campbell, 1990). In a similar study, 4- to 7-year-olds had to learn a route through a six-room playhouse with a clear starting point. Children who examined a map beforehand learned a route more quickly than those who did not (Uttal & Wellman, 1989). Children younger than 6, however, have trouble knowing where they are in the space; therefore, they have difficulty using information available from a map relevant to their own present position (Uttal & Wellman, 1989). Preschoolers also have difficulty aligning maps to the referent space, a skill that improves by age 5 (Liben & Yekel, 1996). Competencies in geometric distances and scaling are underway by age 6 or 7, and primary grade children can recognize features on aerial photographs and large-scale plans of the same area (Blades, Spencer, Plester, & Desmond, 2004; Boardman, 1990), but these abilities continue to improve into adulthood. However, even adults do not attain perfect competence.

The ability to use a map to plan routes is more challenging than following specified routes. This ability is forming at about age 5, although the spaces researchers use are usually simple and rectangular. By 6 years, children can plan routes in more complex environments with multiple alternatives, using distance information. For example, by 5 to 6 years of age, children can use maps to navigate their way around a school but are less successful navigating complex streets or a cave (Jovignot, 1995). More research is needed on naturalistic spaces, as well as on children's ability to plan efficient routes (Newcombe & Huttenlocher, 2000).

Young preschoolers show some ability to create models of spaces such as their classrooms (Blaut & Stea, 1974; Siegel & Schadler, 1977). Preschoolers, like older people, could preserve the configuration of objects when reconstructing a room depicted on a map. However, preschoolers placed objects far from correct locations and performed worse with asymmetric than symmetric configurations (Uttal, 1996). Most 4-year-olds can locate clusters of model furniture items in a scale model of their classroom but get confused when they must position the items, getting only about half the items correct (Golbeck, Rand, & Soundy, 1986; Liben, Moore, & Golbeck, 1982). Much of the difficulty may be not in coding and producing locations, but rather in scaling distances, especially as that difficulty is compounded with multiple elements (Newcombe & Huttenlocher, 2000).

There are individual differences in such abilities. In one study, most preschoolers rebuilt a room better using real furniture than toy models. For some children, however, the difference was slight. Others placed real furniture correctly but grouped the toy models only around the perimeter. Some children placed the models and real furniture randomly, showing few capabilities (Liben, 1988). Even children with similar mental representations may produce quite different maps due to differences in drawing and map-building skills (Uttal & Wellman, 1989). Nevertheless, by the primary grades, most children are able to draw simple sketch-maps of the area around their home from memory (Boardman, 1990).

What accounts for differences and age-related changes? Maturation and development are significant. Children need mental processing capacity to update directions and location. The older they get, the more spatial memories they can store and transformations they can perform. Such increase in processing capacity, along with general experience, determines how a space is represented more than the amount of experience with the particular space (Anooshian, Pascal, & McCreath, 1984). Learning is also important, as will be discussed in a following section.

Fundamental is the connection of primary to secondary uses of maps (Presson, 1987). Even young children can form primary relations to spaces on maps, once they see them as representing a space at about 2.5 to 3 years of age (DeLoache, 1987). They must learn to treat the spatial relations as separate from their immediate environment. These secondary meanings require people to take the perspective of an abstract frame of reference ("as if you were there") that conflicts with the primary meaning. You no longer imagine yourself "inside" but rather must see yourself at a distance, or "outside," the information. Show-ing children several models and explicitly comparing them using language, and possibly visual highlights, can help (Loewenstein & Gentner, 2001), probably because it helps children notice common relationships on subsequent tasks. Such meanings of maps challenge people from preschool into adulthood, especially when the map is not aligned with the part of the world it represents (Uttal & Wellman, 1989). For example, successful map use and mental rotation abilities are correlated in 4- to 6-year-olds (Scholnick, Fein, & Campbell, 1990). Ability to use misaligned maps, especially those 180° misaligned, shows considerable improvement up to about 8 years of age. In addition, children may learn other strategies to deal with the basic problem of conflicting frames of reference (Newcombe & Huttenlocher, 2000). These findings re-emphasize that researchers must be careful how they interpret the phrase *mental (or cognitive) map*. Spatial information may be different when it is garnered from primary and secondary sources such as maps.

Coordinates and Spatial Structuring

As stated, young children can learn to relate various reference frames, and they appear to use, implicitly, two coordinates in remembering direction, either polar or Cartesian. This seems inconsistent with the Piagetian account of the development of two-dimensional space, in which only in later years do children come to "see" objects as located in a two-dimensional frame of reference. That is, Piaget and Inhelder (1967) challenged the claim that there is an innate tendency or ability to organize objects in a two- or three-dimensional reference frame. Spatial awareness does not begin with such an organization; rather, the frame itself is a culminating point of the development of Euclidean space.

To test their theory in the case of horizontality, Piaget and Inhelder showed children jars half-filled with colored water and asked them to predict the spatial orientation of the water level when the jar was tilted. For verticality, a plumb line was suspended inside an empty jar, which was similarly tilted, or children were asked to draw trees on a hillside. Children initially were incapable of representing planes; a scribble, for example, represented water in a tilted jar. At the next state, the level of the water was always drawn perpendicular to the sides of the jar, regardless of tilt. Satisfaction with such drawings was in no way undermined even when an actual water-filled tilted jar was placed next to the drawing. Striking is "how poorly commonly perceived events are recorded in the absence of a schema within which they may be organized" (1967, p. 388). Sometimes, sensing that the water moves towards the mouth of the jar, children raised the level of the water, still keeping the surface perpendicular to the sides. Only at the final stage—at about 9 years of age—did children ostensibly draw upon the larger spatial frame of reference (e.g., tabletop) for ascertaining the horizontal.

Ultimately, the frame of reference constituting Euclidean space is analogical to a container, made up of a network of sites or positions. Objects within this container may be mobile, but the positions are stationary. From the simultaneous organization of all possible positions in three dimensions emerges the Euclidean coordinate system. This organization is rooted in the preceding construction of the concept of straight line (as the maintenance of a constant direction of travel), parallels, and angles, followed by the coordination of their orientations and inclinations. This leads to a gradual replacement of relations of order and distance between objects by similar relations between the positions themselves. It is as if a space were emptied of objects to organize the space itself. Thus, intuition of space is not a "reading" or innate apprehension of the properties of objects, but a system of relationships borne in actions performed on these objects.

Young children's grasp of Euclidean spatial relationships is apparently more adequate than the theory posits. Very young children can orient a horizontal or vertical line in space (Rosser, Horan, Mattson, & Mazzeo, 1984). Similarly, in very simple situations, 4- to 6-year-old children (a) can extrapolate lines from positions on both axes and determine where they intersect, (b) are equally successful going from point to coordinate as going from coordinate to point, and (c) extrapolate as well with or without grid lines (Somerville & Bryant, 1985). Piagetian theory seems correct in postulating that the coordination of relations develops after such early abilities. Young children fail on double-axis orientation tasks even when misleading perceptual cues are eliminated (Rosser, Horan, Mattson, & Mazzeo, 1984). Similarly, the greatest difficulty is in coordinating two extrapolations, which has its developmental origins at the 3- to 4-year-old level, with the ability to extrapolate those lines developing as much as a year earlier (Somerville, Bryant, Mazzocco, & Johnson, 1987).

These results suggest an initial inability to utilize a conceptual coordinate system as an organizing spatial framework (Liben & Yekel, 1996). Some 4-year-olds can use a coordinate reference system, whereas most 6-year-olds can (Blades & Spencer, 1989), at least in scaffolded situations. However, most 4-year-olds can coordinate dimensions if the task is set in a meaningful, guided context in which the orthogonal dimensions are cued by the lines of sight of imaginary people (Bremner, Andreasen, Kendall, & Adams, 1993).

Conceptual integration of coordinates is not limited to two orthogonal dimensions. Children as young as 5 years can metrically represent spatial information in a polar coordinate task, using the same two dimensions as adults, radius and angle, although children do not use categorizations of those dimensions until age 9 (Sandberg & Huttenlocher, 1996).

However, performance on coordinate tasks is influenced by a variety of factors at all ages. Performance on horizontality and verticality tasks may reflect bias toward the perpendicular in copying angles, possibly because this reference is learned early (Ibbotson & Bryant, 1976). Representations of figures also are distorted either locally by angle bisection, or by increasing symmetry of the figure as a whole (Bremner & Taylor, 1982). Finally, performance on these Piagetian spatial tasks correlates with disembedding as well as with general spatial abilities (Liben, 1978). Such results indicate a general tendency to produce symmetry or simplicity in constructions that confound the traditional Piagetian interpretation (Bremner & Taylor, 1982; Mackay, Brazendale, & Wilson, 1972).

Thus, young children have nascent abilities to structure two-dimensional space (which they often have to be prompted to use), but older students often have not developed firm conceptual grounding in grid and coordinate reference systems. To consciously structure space with such systems requires considerable conceptual work. Spatial structuring is the mental operation of constructing an organization or form for an object or set of objects in space (Battista, Clements, Arnoff, Battista, & Borrow, 1998). Structuring is a form of abstraction, the process of selecting, coordinating, unifying, and registering in memory a set of mental objects and actions. Structuring takes previously abstracted items as content and integrates them to form new structures. Spatial structuring precedes meaningful use of grids and coordinate systems. On the one hand, grids that are provided to children may aid their structuring of space; however, children face additional hurdles in understanding grid and coordinate systems. The grid itself may be viewed as a collection of square regions, rather than as sets of perpendicular lines. In addition, order and distance relationships within the grid must be constructed and coordinated across the two dimensions. Labels must be related to grid lines and, in the form of ordered pairs of coordinates, to points on the grid, and eventually integrated with the grid's order and distance relationships so that they constitute mathematical objects and ultimately can be operated upon.

In summary, even young children can use coordinates that adults provide for them. However, when facing traditional tasks, they and their older peers may not yet be able or predisposed to spontaneously make and use coordinates for themselves.

As a final note, we argue that the term *spatial structuring* be reserved for this specific construct of organizing such two- or three-dimensional concepts, the context in which the term was created (Battista & Clements, 1996; Battista, Clements, Arnoff, Battista, & Borrow, 1998; Sarama, Clements, Swaminathan, McMillen, & González Gómez, 2003). Although one could think of all geometric and spatial activity discussed in this section as "structuring space," this would enervate the construct. Furthermore, better, specific local theories exist for other areas of geometry and spatial thinking. Building stronger and more detailed local theories is a superior approach for psychology and education (Newcombe & Huttenlocher, 2000).

Spatial Visualization and Imagery

Spatial visualization is the ability to generate and manipulate images, which are internally experienced, holistic representations of objects that are to a degree isomorphic to their referents. Kosslyn (1983) defined four classes of image processes: generating an image, inspecting an image to answer questions about it, maintaining an image in the service of some other mental operation, and transforming and operating on an image. Thus, spatial visualization involves understanding and performing imagined movements of two- and three-dimensional objects. To do this, one must be able to create a mental image and manipulate it. An image is not a "picture in the head," although the mental processes are similar to those that underlie the perception of objects (Shepard, 1978). For example, they are integrated and can be scanned or rotated as one would do to perceptually available objects, with transitional images and times in proportion to those of perceptual activity (Eliot, 1987; Kosslyn, Reiser, & Ball, 1978). Images are more abstract, more malleable, and less crisp than pictures. They are often segmented into parts and represent relationships among those parts (Shepard, 1978). Some images can cause difficulties, especially if they are too inflexible, vague, or filled with irrelevant details.

People's first images are static, not dynamic. They can be mentally re-created and even examined but not transformed. Having a dynamic image allows one to, for example, mentally "move" the image of one shape (such as a book) to another place (such as a bookcase, to see if it will fit) or mentally move (slide) and turn an image of one shape to compare that shape to another one.

In Piagetian theory (Piaget & Inhelder, 1967, 1971), children up to 4 years of age cannot construct an entire image of a two-dimensional shape after only tactile-kinesthetic experience (visual experiences were thought to rely overly on perceptual thinking) because preschoolers are too passive, touching one part of a shape only. Children aged 4 to 7 would touch another part and regulate their actions by establishing relations among them, building a more accurate representation of the shape. Such processes allow them to accurately distinguish between rectilinear (e.g., a triangle) and curvilinear (e.g., circle) shapes and build images of simple shapes.

Although Piaget argued that most children cannot perform full dynamic motions of images until the primary grades (Piaget & Inhelder, 1967, 1971), pre-K children show initial transformational abilities. Some researchers have reported that second graders learned only manual procedures for producing transformation images but could not mentally perform such transformations (Williford, 1972). In contrast, other studies indicate that even young children can learn something about these motions and appear to internalize them, as indicated by increases on spatial-ability tests (Clements, Battista, Sarama, & Swaminathan, 1997; Del Grande, 1986). Slides appear to be the easiest motions for children, then flips and turns (Perham, 1978); however, the direction of transformation may affect the relative difficulty of turn and flip (Schultz & Austin, 1983). Results depend on specific tasks, of course; even 4- to 5-year-olds can do turns if they have simple tasks and orientation cues (Rosser, Ensing, Glider, & Lane, 1984). For children of ages 4 to 8, there were no significant effects for slides and flips (with the trend being a negative effect for younger children) but a dramatic beneficial effect for turns. A slide task was at least as easy as a flip, and turns were most difficult (Moyer, 1978).

Transformations and perspective taking appear to follow the development of perceptual and imagistic reproduction. There is a hierarchical developmental sequence of reproduction of geometric figures requiring only encoding (i.e., building a matching configuration of shapes, with the original constantly in sight), reproduction requiring memory (building a matching configuration from recall), and transformation involving rotation and visual perspective-taking (building a matching configuration either from recall after a rotation or from another's perspective), with pre-K children able to perform at only the first two levels (Rosser, Lane, & Mazzeo, 1988). In a similar vein, a framework of imagery for early spatial mathematics learning that is generally consistent with the research reviewed here has been proposed by Owens (1999). In each of three categories, orientation and motion, part-whole relationships, and classification and language, children develop strategies in five sub-categories: emergent (begin to attend, manipulate, and explore), perceptual (attend to features and make comparisons, rely on what they can see or do), pictorial (use mental images and standard language), pattern and dynamic imagery (understand conceptual relationships), and efficient strategies. Preliminary evidence suggests the validity and usefulness of the framework, for researchers and teachers. Using a similar framework, researchers found that 11% of Australian kindergartners were unable to visualize simple shapes at the beginning of the year (2% at the end of the year in the experimental group); 70% were at the level of forming static, pictorial images in conjunction with models or manipulatives (37%), with only 19% at the level of visualizing effects of motions (52%) and 1% using dynamic imagery (10%; D. M. Clarke et al., 2002). Thus, there is much room for growth in the earliest year, but helping teachers understand developmental progressions and learning trajectories can promote that growth.

Education

Little is known, in the spatial sphere, about specific cultural and educational experiences and their impact on these capabilities. Experience-expectant processes (Greenough, Black, & Wallace, 1987) seem to account for much of children's development. Universal experiences, such as the physical world provides, lead to an interaction of inborn capabilities and environmental inputs that guide development in similar ways across cultures and individuals. However, other competencies, such as spatial reasoning and the use and creation of external spatial representations like language, models, and maps, probably develop via experience-dependent processes (Greenough, Black, & Wallace, 1987), and thus capability differs across cultures and individuals. For example, preschool teachers spend more time with boys than girls and usually interact with boys in the block, construction, sand play, and climbing areas and with girls in the dramatic play area (Ebbeck, 1984). Boys engage in spatial activities more than girls at home do, both alone and with caretakers (N. Newcombe & Sanderson, 1993). Such differences may interact with biology to account for early spatial skill advantages for boys (note that some studies find no gender differences, e.g., Brosnan, 1998; Ehrlich, Levine, & Goldin-Meadow, 2005; Jordan, Kaplan, Oláh, & Locuniak, 2005; Levine, Huttenlocher, Taylor, & Langrock, 1999; Rosser, Horan, Mattson, & Mazzeo, 1984). Thus, there are important questions not only for cognitive psychology, but also for mathematics education research.

The psychological research once again indicates that active experiences, here emphasizing both physical and mental activity, are appropriate and, in some cases, critical for children, especially as they are just developing a skill. For example, 1-year-olds who walk themselves around a display are more active observers and better locators than those who are carried (Acredolo, Adams, & Goodwyn, 1984). Similarly, self-directed movement at 5 years of age led to superior recall of distances in a spatial layout of a room (Poag, Cohen, & Weatherford, 1983). As children develop, they become able to perform well under a greater variety of conditions (e.g., outcomes did not differ among self- or other-directed movement and viewing conditions for 7-year-olds). These and other studies (Benson & Bogartz, 1990; Newcombe & Huttenlocher, 2000; Rieser, Garing, & Young, 1994) emphasize the importance of self-produced movement for success in spatial tasks and suggest the benefit of maximizing such experience for all young children.

Given the early competence in foundational spatial representational systems, there is every reason to assume that rich environments will contribute to spatial competence at the intuitive and explicit levels. For example, adults responsible for young children know that they learn practical navigation early. Channeling that experience is valuable. For example, when nursery-school children tutor others in guided environments, they build geometrical concepts (Filippaki & Papamichael, 1997). Such environments might include interesting layouts inside and outside classrooms, incidental and planned experiences with landmarks and routes, and frequent discussion about spatial relations on all scales, including distinguishing parts of their bodies (Leushina, 1974/1991) and spatial movements (forward, back), finding a missing object ("under the table that's next to the door"), putting objects away, and finding the way back home from an excursion. Verbal interaction is important. For example, parental scaffolding of spatial communication helped both 3- and 4-year-olds perform direction-giving tasks in which they had to disambiguate by using a second landmark ("it's in the bag *on the table*"), which children are more likely to do the older they are. Both age groups benefited from directive prompts, but 4-year-olds benefited more quickly than younger children did from nondirective prompts (Plumert & Nichols-Whitehead, 1996). Control children never disambiguated, showing that interaction and feedback from others is critical to certain spatial communication tasks.

Representing spatial environments with models and maps, as well as implicitly or verbally, is less likely to be spontaneously developed by children. Specific teaching strategies may therefore yield as yet largely unexplored benefits. Models and maps are sociocultural tools (Gauvain, 1991), but ones whose development is not well supported in the U.S. culture. For example, school experiences are limited and fail to connect map skills with other curriculum areas, including mathematics (Muir & Cheek, 1986). Most students do not become competent users of maps even beyond their early childhood years.

Such deficits may have negative ramifications past map use. Using and thinking about maps may contribute to spatial development by helping children acquire abstract concepts of space, the ability to think systematically about spatial relationships that they have not experienced directly, and the ability to consider multiple spatial relations among multiple locations (Uttal, 2000). As we have documented, children benefit from exposure to maps (Uttal & Wellman, 1989) and to overhead views (Rieser, Doxsey, McCarrell, & Brooks, 1982), at least when children *use* the information (cf. Liben & Yekel, 1996). Indeed, using maps to teach children locations before they entered the space helped children to identify rooms out of the learned sequence, thus improving their mental representation of that space (Uttal & Wellman, 1989).

Instruction on spatial ability, symbolization, and metacognitive skills (consciously self-regulated map-reading behavior through strategic map referral) can increase 4- to 6-year-olds' competence with reading route maps, although it does not overcome age-related differences (Frank, 1987). Using oblique maps (e.g., tables are shown with legs) aids preschoolers' subsequent performance on plan ("bird's-eye view") maps, perhaps because symbolic understanding was developed (Liben & Yekel, 1996). Telling very young children that a model was the result of putting a room in a "shrinking machine" helped the children see the model as a symbolic representation of that space (DeLoache, Miller, Rosengren, & Bryant, 1997). Using structured maps that help preschoolers match the elements on the map with the corresponding elements in the space facilitated their use of maps (DeLoache, Miller, & Pierroutsakos, 1998). These may be good beginnings, but models and maps should eventually move beyond overly simple iconic picture maps and challenge children to use geometric correspondences. Teachers need to help children connect the abstract and concrete meanings of map symbols.

According to some research, navigation activities that combine physical movement, paper-and-pencil, and computer work can facilitate learning of mathematics and map skills. Such spatial learning can be particularly meaningful because it can be consistent with young children's ways of moving their bodies (Papert, 1980). For example, young children can abstract and generalize directions and other map concepts by working with the Logo turtle (Borer, 1993; Clements,

Battista, Sarama, Swaminathan, & McMillen, 1997; Clements & Meredith, 1994; Goodrow, Clements, Battista, Sarama, & Akers, 1997; Kull, 1986; Try, 1989; Watson, Lange, & Brinkley, 1992; Weaver, 1991), although results are not guaranteed (Howell, Scott, & Diamond, 1987). The interface must be appropriate, and activities must be well planned (Watson & Brinkley, 1990/91). By giving the turtle directions such as forward 10 steps, right turn, forward 5 steps, they learn orientation, direction, and perspective concepts, among others. Walking paths and then recreating those paths on the computer help them abstract, generalize, and symbolize their experiences navigating. For example, one kindergartner abstracted the geometric notion of *path* saying, "A path is like the trail a bug leaves after it walks through purple paint" (Clements, Battista, & Sarama, 2001). Logo can also control a floor turtle robot, which may have special benefits for certain populations. For example, blind and partially sighted children using a computer-guided floor turtle developed spatial concepts such as right and left and accurate facing movements (Gay, 1989). Other simple (non-Logo) navigational programs may have similar benefits. For example, using such software (with on-screen navigation) has shown to increase kindergartners' understanding of the concepts of left and right (Carlson & White, 1998).

As early as the preschool years, through to first grade, U.S. children perform lower than children in countries such as Japan and China on spatial visualization and imagery tasks (Starkey et al., 1999; Stigler, Lee, & Stevenson, 1990). There is more cultural support for activities in these other countries (e.g., using visual representations, expecting competency in drawing), and, similarly, within cultural groups in the United States, and within higher SES families in all countries (Starkey et al., 1999; Thirumurthy, 2003). Research shows that even pre-K and kindergarten children show initial transformational abilities in certain settings (Clements, Battista, Sarama, & Swaminathan, 1997; Del Grande, 1986). While incomplete, research suggests that all children benefit from developing their ability to create, maintain, and represent mental images of mathematical objects.

Shape

Shape is a fundamental construct in cognitive development in and beyond geometry. For example, young children form artifact categories characterized by similarity among instances in shape (Jones & Smith, 2002). Even very young children show strong biases to attend to fine-grained shape when learning novel object names, at least when directed to a rigid object, contrary to the topological primacy thesis of Piagetian theory (Landau, 1996). How do children come to understand shape? We first introduce several theoretical perspectives, then present the hierarchic interactionist position that is consistent with extant evidence.

Theories of Young Children's Perception and Knowledge of Shape

Piaget and Inhelder (1967) claimed that young children initially discriminate objects on the basis of "topological" features, such as being closed or otherwise topologically equivalent, especially when given only tactile, rather than visual, perceptions of the shapes. Only older children could discriminate rectilinear from curvilinear forms and, finally, among rectilinear closed shapes, such as squares and diamonds, via systematic and coordinated explorations. However, as previously discussed with regard to spatial thinking, support for this view has decreased. Decades ago, researchers reported that even at the earliest ages (2–3 years), children can distinguish between curvilinear and rectilinear shapes, contrary to the theory (Lovell, 1959; Page, 1959).

A difficulty in designing such experiments is in quantifying the degree of equivalence of the shapes. Geeslin and Shar (1979) modeled figures via a finite set of points on a grid. Degree of distortion was defined as the sum of displacements of these points. The authors postulated that children compare two figures in terms of the amount of distortion necessary to transform one figure into another, after an attempt at superimposition using rigid motions and dilations. This model received strong support; however it is more predictive than explanatory.

An overarching criticism has been that Piaget and Inhelder's use of the terms such as *topological, separation, proximity*, and *Euclidean*, as well as the application of these and related concepts to the design of their studies, are not mathematically accurate (Darke, 1982; Kapadia, 1974; Martin, 1976). For example, proximity, in Piaget and Inhelder's interpretation of "nearbyness," is not topological, for it involves distance. Piaget and Inhelder also claimed that only by the stage of formal operations (11 or 12 years) do children synthesize notions of proximity, separation, order, and enclosure to form the notion of continuity. However, continuity is not the synthesis of these concepts; it is a defining concept of topology. If it does not develop until a late stage, the argument for the primacy of topological concepts is weakened (Darke, 1982; Kapadia, 1974). In a similar vein, classifying figures as topological or Euclidean is problematic, as all figures possess both these characteristics.

In sum, results of many of the Piagetian studies may be an artifact of the particular shapes chosen and the abilities of young children to identify and name these shapes (G. H. Fisher, 1965). This does not support a strong version of the topological primacy thesis. The class of topological properties may not enable young children to identify certain shapes. Visually salient properties such as holes, curves, and corners; simplicity; and familiarity—rather than topological versus Euclidean properties—may underlie children's discrimination.

Although the topological primacy thesis is not supported, Piaget and Inhelder's theory included a second theme. They claimed that children's representation of space is not a perceptual "reading off" of their spatial environment but is constructed from prior active manipulation of that environment. From this perspective, abstraction of shape is the result of a coordination of children's actions. Children

> can only "abstract" the idea of such a relation as equality on the basis of an action of equalization, the idea of a straight line from the action of following by hand or eye without changing direction, and the idea of an angle from two intersecting movements. (p. 43)

The Piagetian's tactile-kinesthetic experiments appeared to support this view.

In contrast, Gibson (1969) stated that motor activity plays at most an indirect role in perception. She also disagreed with the empiricists who believed that perception was a matter of association. Instead, she claimed that perception involves both learning and development, increasing the person's ability to (selectively) extract information from the stimulation in the environment. Her theory is similar to Gestalt theory, which emphasizes (a) wholes, irreducible to parts, that drive toward "best structures" and (b) developmental processes including articulation and differentiation, rather than accretion by association. However, Gibson emphasized learning in sensory reorganization through processes of filtering and abstraction. Perception begins as only crudely differentiated and grossly selected. With growth and exposure, perception becomes better differentiated and more precise, as the person learns detection of properties, patterns, and distinctive features. Gibson suggested that learning is facilitated by instructors' emphasizing distinctive features and by beginning with broad differences in those features and moving toward finer distinctions.

From this perspective, perception is active, adaptive, internally directed, and self-regulated. In contrast to a radical constructivist orientation, Gibson assumed an external reality, including structure and information already existing in the stimulus, without the need for cognitive mediation. Consistent with the constructivist interpretation, other research indicates that, although components such as edge-lines may already be discrete, and basic perceptual competencies such as size constancy (correcting for variations in distance) inborn (Granrud, 1987; Slater, Mattock, & Brown, 1990), one cannot claim that ontological reality is divided into invariants (E. Wright, 1983). There is no "information" in the environment separate from the individual.

Also in the constructivist tradition, the theory of Pierre and Dina van Hiele posits that students progress through qualitatively distinct levels of thought in geometry (van Hiele, 1986; van Hiele-Geldof, 1984). At Level 1, the visual level, students can only recognize shapes as wholes and cannot form mental images of them. A given figure is a rectangle, for example, because "it looks like a door." They do not think about the defining attributes, or properties,[4] of shapes. At Level 2, descriptive/analytic, students recognize and characterize shapes by their properties. For instance, a child might think of a square as a figure that has four equal sides and four right angles. Properties are established experimentally by observing, measuring, drawing, and model making.

The hierarchic interactionalism perspective builds most directly on elaborations and revisions of the most educationally relevant theory, that of the van Hieles, introducing other theoretical and empirical contributions. To begin, although both theories posit domain-specific developmental progressions, hierarchic interactionalism hypothesizes several revisions of the van Hiele levels (Clements & Battista, 1992). These revisions have been subsequently supported (Clements, Battista, & Sarama, 2001; Clements, Swaminathan, Hannibal, & Sarama, 1999; Yin, 2003). A level antecedent to the visual level is required to describe the empirical corpus. At Level 0, pre-recognition, children cannot reliably distinguish circles, triangles, and squares from nonexamples of those classes. Children at this level are just starting to form visual schemes for the shapes. These early, unconscious schemes are formed through several initial bootstrap competencies. For example, pattern matching through some type of feature analysis (Anderson, 2000; Gibson, Gibson, Pick, & Osser, 1962) is conducted after the visual image of the shape is transformed by heuristics built into the visual system that imposes an intrinsic frame of reference on the shape, possibly using symmetry (Palmer, 1989).

Research indicates that even infants can perceive wholes as well as parts of geometric patterns (Bornstein & Krinsky, 1985) and that children as young as 3 years engage in active spatial analysis in both construction (Tada & Stiles, 1996) and perception (Feeney & Stiles, 1996), a process that changes with development. That is, children will draw a + using 4 separate line segments, treating intersections as junctions of separate parts and will even choose 4 small segments as being most like the goal object. Older children produce one long vertical and two short horizontal segments, and adults two long segments. As another example, when copying a square divided into fourths (all lines parallel or perpendicular), some young children use all short segments, and others use multiple closed forms (e.g., four squares drawn inside the larger square). As early as 4.5 years, most children's constructions are similar to those of adults. Thus, the youngest children parse out simple, well-formed, spatially independent parts and use simple combination rules such as seriation and adjacency, then organization around a central point, to connect these parts. Older children relate parts across intersections and parse out continuous unsegmented parts, coordinating relations across boundaries (Tada & Stiles, 1996). These first units may be inborn, well-defined, rigid, unarticulated primary structures that may not follow Gestalt principles. Indeed, Gestalt processes themselves, and what adults perceive, may be a result of a developmental process. Development includes both components of *spatial analysis*—identification of the parts of a geometric form and integration of those parts into a coherent whole. Whereas 4-year-olds use fragmented strategies, such as drawing segmented forms radiating around a central point, children as young as 6 years of age possess multiple spatial-analytic strategies and can use strategies like those of adults, at least in simple situations. The strategy they use is a function of both their capabilities and the complexity of the pattern they are copying (Akshoomoff & Stiles, 1995). Throughout development, children process more parts and more difficult sets of relations, such as intersections and oblique segments, in increasingly higher order hierarchical units (Akshoomoff & Stiles, 1995).

Simple, closed shapes may initially tend to form undifferentiated, cohesive units in children's phenomenological perceptual experiences (c.f. Smith, 1989). (Research indicates that shapes are perceived holistically with a separate subsystem; Ganel & Goodale, 2003.) For example, contour is a salient characteristic for young children (e.g., Tada & Stiles, 1996)

---

[4] We reserve the term *property* for those attributes that indicate a relationship between parts, or components, of shapes. Thus, parallel sides, or equal sides, are properties. We use *attributes* and *features* interchangeably to indicate any characteristic of a shape, including properties, other defining characteristics (e.g., straight sides) and nondefining characteristics (e.g., "right-side up").

that, when suggestive of a spatially continuous object, may direct the perception of the shape as a whole unit. Nascent schemes may ascertain the presence of the features of closed and "rounded" to match circles, "pointy" features along a figural scan-path to match triangles (without necessarily attaching numerical significance to this path, cf. Piaget & Szeminska, 1952; von Glasersfeld, 1982), four near-equal sides with approximately right angles to match squares, and approximate parallelism of opposite "long" sides to match rectangles.

As they develop triangle schemes (cognitive networks of relationships connecting geometric concepts and processes in specific patterns), children develop, tacitly at first, the ability to "see" both the parts *and* the whole so that they do not focus so much on the angles ("pointy parts") that they consider an angle (e.g., upside-down *V*) to be a triangle. Similarly, they learn not to focus only on the enclosing, simple contour of the shape that initially led them to consider shapes such as chevrons to be triangles (Owens, 1999). Typical "geometry deprived" (Fuys, Geddes, & Tischler, 1988) young children have a limited number of imagistic schemes for triangle, for example, an equilateral triangle and a right triangle, both with a horizontal base (Hershkowitz et al., 1990; Vinner & Hershkowitz, 1980). These prototype schemes are not rigid, but they have constraints. For example, the more the lengths of the legs of a right triangle differ in length, the less likely it will be assimilated to that prototype. Such prototypes can be thought of as having multivariate distributions of possible values (e.g., for the relationships between the side lengths and for degree of the base's rotation from the horizontal) in which the nearer to the mode of the distribution the perceived figure is (equal for the side lengths and 0° rotation of the base for this example), the more likely it will be assimilated to that prototype (parallel distributed processing, or PDP, networks and other structural theories model this type of system, see Clements & Battista, 1992; McClelland, Rumelhart, & the PDP Research Group, 1986).

What is the role of language in this process? As the initial schemes are developing, so do mappings similar to those of number, color, and other properties (Sandhofer & Smith, 1999). That is, children first learn that the question "what shape?" is answered by words such as *circle* and *square*. They then map these shape words to a few concrete examples. Next, they combine these abilities to produce correct shape names for prototypical examples of common shapes. Only after these experiences do they build shape categories (probably because the PDP-type networks of each of these examples contains much more information than geometric shape and thus property-to-prop-

erty matching and shape category creation is a slower developmental process, cf. Sandhofer & Smith, 1999). Thus, shape words and names help organize and direct attention to the relevant features of objects.

This analysis reveals that the nature of the van Hiele levels also requires clarification. The "visual" level includes visual/imagistic and verbal declarative knowledge ("knowing what") about shapes. That is, it is not viable to conceptualize a purely visual level (1), followed and replaced by a basically verbal descriptive level (2) of geometric thinking—a common interpretation. Instead, different types of reasoning—those characterizing different levels—coexist in an individual and can be developing simultaneously but at different rates. Consistent with the tenet of *progressive hierarchization*, levels do not consist of unadulterated knowledge of only one type. This view is consistent with literature from Piagetian and cognitive traditions (e.g., Minsky, 1986; Siegler, 1996; Snyder & Feldman, 1984), as well as reinterpretations of van Hielie theory (Clements & Battista, 1992; Gutiérrez, Jaime, & Fortuny, 1991; Lehrer, Jenkins, & Osana, 1998; Pegg & Davey, 1998).

Certain conditions, such as a request to explain one's decision, may prompt children, even at early levels, to abstract, attend to, and describe certain features ("this is pointy"). They might also consciously attend to a subset of the shape's visual characteristics and use such a subset to identify geometric shapes. Thus, their descriptions of shapes may include a variety of terms and attributes (Clements, Battista, & Sarama, 2001; Lehrer, Jenkins, & Osana, 1998). However, these are "centrations" in the Piagetian sense, not integrated with other components of the shape. This supports the recognition of both comparison-to-prototype and attention-to-attributes in young children's geometric categorization (Clements, Swaminathan, Hannibal, & Sarama, 1999; see also Lehrer, Jenkins, & Osana, 1998). Conditions also affect how such features and prototype matching are processed. For example, young children's overall acceptance rate of both examples and nonexamples increases with the inclusion of palpable distractors (Hannibal & Clements, 2000).

Thus, even at a very early age children attend to some attribute of their imagistic scheme, even if it is only its "prickliness" for triangles (c.f. Lehrer, Jenkins, & Osana, 1998). Then, as the scheme becomes better formed (and as the child gains the ability and predisposition to give increased selective attention to single dimensions when comparing objects, Smith, 1989), the child is able to discern more attributes (both defining and nondefining) that s/he uses to construct her or his definition of a triangle. Children can match and identify many shapes (most reliably, prototypical forms determined by both biology and culture). However, they often attend only to a proper subset of a shape's visual characteristics and are unable to identify many common shapes. In tactile contexts, they can distinguish between figures that are curvilinear and those that are rectilinear but not among figures within those classes. Even in visual contexts, they may not be able to construct an image of shapes, or a re-presentation of the image. They are unable to rotate shapes and place them into part-whole relationships (Wheatley & Cobb, 1990). Thus, before Level 1, children lack the ability to construct and manipulate visual images of geometric figures. This example of *cyclic concretization* is consistent with the Piagetian tradition of the construction of geometric objects on the "perceptual plane" before a reconstruction on the "representational" or "imaginal plane" (Piaget & Inhelder, 1967).

Later in development, additional visual-spatial elements, such as the right angles of squares, are incorporated into these schemes, and thus traditional prototypes may be produced. Further, older children can attend to these features separately, whereas younger children are not able or predisposed to focus on single features analytically (Smith, 1989; Vurpillot, 1976). Therefore, younger children can produce a prototype in identifying rectangles without necessarily attending to the components or specific features that constitute these prototypes. For children of all ages, the prototypes may be overgeneralized or undergeneralized compared to mathematical categorization, of course, depending on the examples and nonexamples and teaching acts children experience. Also, progress to strong Level 1 and eventually to Level 2 understanding is protracted. For example, primary grades continue to apply many different types of cognitive actions to shapes, from detection of features like fat or thin, to comparison to prototypical forms, to the action-based embodiment of pushing or pulling on one form to transform it into another (Clements, Battista, & Sarama, 2001; Lehrer, Jenkins, & Osana, 1998).

Thus, young children operating at Levels 0 and 1 show evidence of recognition of components and attributes of shapes, although these features may not be clearly defined (e.g., sides and corners). Some children operating at Level 1 appear to use both matching to a visual prototype (via feature analysis) and reasoning about components and properties to solve these selection tasks. Thus, Level 1 geometric thinking as proposed by the van Hieles is more syncretic (a fusion of differing systems) than visual, as Clements (1992) suggested. That is, this level is a synthesis of verbal declarative and imagistic knowledge, each interacting with and enhancing the other. We therefore suggest

the term *syncretic level (indicating a fusion of different perspectives)*, rather than *visual level*, because we wish to signify a global combination of verbal and imagistic knowledge without an analysis of the specific components and properties of figures. At the syncretic level, children more easily use declarative knowledge to explain why a particular figure is not a member of a class, because the contrast between the figure and the visual prototype provokes descriptions of differences (Gibson, 1985).

Children making the transition to the next level sometimes experience conflict between the two parts of the combination (prototype matching vs. component and property analysis), leading to mathematically incorrect and inconsistent task performance. For example, one girl started pre-K with a stable concept, a scheme of the triangle as a visual whole. However, when introduced instructionally to the attributes of triangles, she formed a separate scheme for a "three-sided shape." For some time thereafter, she held complex and unstable ideas about triangles, especially when the two schemes conflicted (Spitler, Sarama, & Clements, 2003). As another example, many young children call a figure a square because it "just looked like one," a typical holistic, visual response. However, some attend to relevant attributes; for them, a square has "four sides the same and four points." Because they have not yet abstracted perpendicularity as another relevant and critical attribute, some accept certain rhombi as squares (Clements, Battista, & Sarama, 2001). That is, even if their prototype has features of perpendicularity (or aspect ratio near 1), young children base judgments on similarity (in this case, near perpendicularity) rather than on identity (perpendicularity), and therefore they accept shapes that are "close enough" (Smith, 1989). The young child's neglect of such relevant (identity) attributes or reliance on irrelevant attributes leads to categorization errors. This is consistent with early findings that preschoolers show a slow development of skills, sudden insight, and regression (Fuson & Murray, 1978).

Mervis and Rosch (1981) theorized that generalizations based on similarity to highly representative exemplars will be the most accurate. This theory would account for the higher number of correct categorizations by those children who appeared to be making categorization decisions based on comparison to a visual prototype without attention to irrelevant attributes. Finally, strong feature-based schemes and integrated declarative knowledge, along with other visual skills, may be necessary for high performance, especially in complex, embedded configurations. To form useful declarative knowledge, especially robust knowledge supporting transition to Level 2, children must construct and consciously attend to the components and properties of geometric shapes as cognitive objects (a learning process that requires mediation and is probably aided by physical construction tasks using manipulatives as well as reflection often prompted by discussion, points to which we shall return).

Thus, the *hierarchic interactionalism* theory predicts that children are developing stronger imagistic prototypes and gradually gaining verbal declarative knowledge. Those figures that are more symmetric and have fewer possible imagistic prototypes (circles and squares) are more amenable to the development of imagistic prototypes and thus show a straightforward improvement of identification accuracy. Rectangles and triangles have more prototypes that are possible. Rectangle identification may improve only over substantial periods. Similarly, shapes such as triangles, the least definable by imagistic prototypes discussed here, may show complex patterns of development as the scheme widens to accept more forms, overwidens, and then must be further constrained.

Children's variegated responses (some visual, some verbal declarative) may be another manifestation of this syncretic level. Further, they substantiate Clements's (1992) claim that geometric levels of thinking coexist, as previously discussed. Progress through such levels is determined more by social influences, specifically instruction, than by age-linked development. Although each higher level builds on the knowledge that constitutes lower levels, the nature of the levels does not preclude the instantiation and application of earlier levels in certain contexts (not necessarily limited to especially demanding or stressful contexts). For each level, there exists a probability for evocation for each of numerous different sets of circumstances, but this process is codetermined by conscious metacognitive control, control that increases as one moves up through the levels, so people have increasing choice to override the default probabilities (*progressive hierarchization*). The use of different levels is environmentally adaptive; thus, the djective *higher* indicates a higher level of abstraction and generality, without implying either inherent superiority or the abandonment of lower levels as a consequence of the development of higher levels of thinking. Nevertheless, the levels would represent veridical qualitative changes in behavior, especially the construction of mathematical representations (i.e., construction of geometric objects) from action.

In summary, geometric knowledge at every level maintains nonverbal, imagistic components; that is, every mental geometric object includes one or more image schemes—recurrent, dynamic patterns of kinesthetic and visual actions (Johnson, 1987). Imagistic knowledge is not left untransformed and merely "pushed into the background" by higher levels of thinking. Imagery has a number of psychological layers, from more primitive to more sophisticated (each connected to a different level of geometric thinking), which play different (but always critical) roles in thinking depending on which layer is activated. Even at the highest levels, geometric relationships are intertwined with images, though these may be abstract images. Thus, images change over development. The essence of Level 2 thinking, for example, lies in the integration and synthesis of properties of shapes, not merely in their recognition. Through the process of *progressive hierarchization*, children at this level have transcended the perceptual and have constructed the properties as singular mental geometric objects that can be acted upon, not merely as descriptions of visual perceptions or images (cf. Steffe & Cobb, 1988). Ideally, however, these objects are "neither words nor pictures" (Davis, 1984), but a synthesis of verbal declarative and rich imagistic knowledge, each interacting with and supporting the other. The question, therefore, should not be whether geometric thinking is visual or not visual but rather whether imagery is limited to unanalyzed, global visual patterns or includes flexible, dynamic, abstract, manipulable imagistic knowledge (Clements, Swaminathan, Hannibal, & Sarama, 1999).

Thinking and Learning About Specific Shapes

With these processes in mind, it is useful to consider children's learning about specific shapes. Infants may be born with a tendency to form certain mental prototypes. When people in a Stone Age culture with no explicit geometric concepts were asked to choose a "best example" of a group of shapes, such as a group of quadrilaterals and near-quadrilaterals, they chose a square and circle more often, even when close variants were in the group (Rosch, 1975). For example, the group with squares included square-like shapes that were not closed, had curved sides, and had nonright angles. So, people might have innate preferences for closed, symmetric shapes (c.f. Bornstein, Ferdinandsen, & Gross, 1981). Further, symmetry affects shape perception in two ways, on the global level (e.g., preference for symmetry about the vertical, and to lesser extent, horizontal axis) and local level (symmetries within a shape even if the axes are not vertical or horizontal). For example, global symmetries support the recognition of a prototypical "diamond" (with the long diagonal vertical), but the local symmetry still affects the recognition of any rhombus, even if oriented with a horizontal side (Palmer, 1985).

Culture influences these preferences. We conducted an extensive examination of materials that teach children about shapes from books, toy stores, teacher supply stores, and catalogs. With few exceptions (and with signs that this is changing in recent years), these materials introduce children to triangles, rectangles, and squares in rigid ways. Triangles are usually equilateral or isosceles and have horizontal bases. Most rectangles are horizontal, elongated shapes about twice as long as they are wide. No wonder so many children, even throughout elementary school, say that a square turned is "not a square anymore, it's a diamond" (Clements, Swaminathan, Hannibal, & Sarama, 1999; Lehrer, Jenkins, & Osana, 1998). In one study, 4- to 5-year-olds considered rotated squares no longer the same shape or even size, 6- to 7-year-olds retained its characteristics but lost its category and name—it was no longer a square, and 8- to 9-year-olds achieved invariance (Vurpillot, 1976). This may reflect systematic bias for horizontal and vertical lines and a need for perceptual learning and flexibility, but restricted experiences exacerbate such limitations. Research indicates that such rigid visual prototypes can rule children's thinking throughout their lives (Burger & Shaughnessy, 1986; N. D. Fisher, 1978; Fuys, Geddes, & Tischler, 1988; Kabanova-Meller, 1970; Vinner & Hershkowitz, 1980; Zykova, 1969).

Specifically, what visual prototypes and ideas do young children form about common shapes? Decades ago, Fuson and Murray (1978) reported that by 3 years of age over 60% of children could name a circle, square, and triangle. More recently, A. Klein, Starkey, and Wakeley (1999) reported shape-naming accuracy of middle-income 5-year-olds: circle, 85%; triangle, 80% (note all were close to isosceles); square, 78%; rectangle, 44% (note squares were not scored as correct choices).

A study of young children used the same line drawings previously used with elementary students for comparison purposes (Clements, Swaminathan, Hannibal, & Sarama, 1999); replication studies have been conducted in Singapore (Yin, 2003) and Turkey (Aslan, 2004). Children identified circles quite accurately: 92%, 96%, and 99% for 4-, 5-, and 6-year-olds, respectively, in the United States. Only a few of the youngest children chose the ellipse and curved shape. Most children described circles as "round," if they described them at all. Thus, the circle was easily recognized but relatively difficult to describe for these children. Evidence suggests that they matched the shapes to a visual prototype. Turkish children showed the same pattern (Aslan, 2004).

Children also identified squares fairly well: 82%, 86%, and 91% for 4-, 5-, and 6-year-olds, respectively. Younger children tended to mistakenly choose nonsquare rhombi; 25% of 6-year-olds and 5% of 7-year-

olds did so in Singapore. However, they were no less accurate in classifying squares without horizontal sides. In Singapore, 7-year-olds were less likely to correctly identify these as squares than were 6-year-olds (Yin, 2003). Children in all three countries were more likely to be accurate in their square identification when their justifications for selection were based on the shape's attributes (e.g., number and length of sides).

Children were less accurate at recognizing triangles and rectangles (except in Turkey, where children identified these shapes only slightly less accurately, 68% and 71%, than squares, 73%, Aslan, 2004). However, their scores were not low; about 60% correct for triangles. Although ages of children in the studies differed, both the U.S. and Singapore data revealed a phase in which children chose more triangle examples and distractors, then "tightened" their criteria to omit some distractors but also some examples. The children's visual prototype seems to be of an isosceles triangle.

Young children tended to accept "long" parallelograms or right trapezoids as rectangles. Thus, children's visual prototype of a rectangle seems to be a four-sided figure with two long parallel sides and "close to" square corners.

Additionally interesting are cross-cultural and longitudinal comparisons. Similarly aged children in the United States performed slightly better than Singapore children on circles and squares, but the reverse was true for triangles, rectangles, and embedded figures (Yin, 2003)—the more complex tasks. However, these differences lacked statistical significance.

In the follow-up study, children ages 3 to 6 were asked to sort a variety of manipulative forms (Hannibal & Clements, 2000). Certain mathematically irrelevant characteristics affected children's categorizations: skewness, aspect ratio, and, for certain situations, orientation. With these manipulatives, orientation had the least effect. Most children accepted triangles even if their base was not horizontal, although a few protested. Skewness, or lack of symmetry, was more important. Many rejected triangles because "the point on top is not in the middle." For rectangles, on the other hand, many children accepted nonright parallelograms and right trapezoids. Also important was aspect ratio, the ratio of height to base. Children preferred an aspect ratio near one for triangles, that is, about the same height as width. Other forms were "too pointy" or "too flat." Children rejected both triangles and rectangles that were "too skinny" or "not wide enough." These same factors, with one additional factor of size, similarly affected children's judgments in Turkey (Aslan, 2004).

Another study addressed recognition of squares at various orientations. The percentage of primary children recognizing squares was 96.1 for squares with a horizontal base, and 93.2, 84.5, and 73.8 for rotated squares, the last 45° from the horizontal (Kerslake, 1979). Not until the age of 8 did children begin to generalize the concept of square. Again, the explanation for such phenomena may be that shapes are perceived relative to a reference-frame structure in which the orientation of the axes is taken as the descriptive standard. The visual system has heuristics for assigning the frame, which usually, but not always, allows the detection of shape equivalence. However, the reference orientation results from an interaction of intrinsic structure (e.g., symmetry) and orientation relative to the environment (verticality or gravity) and the observer (Palmer, 1989).

A study not designed to test the van Hiele theory also provides evidence on classification schemes. Children of 3, 4, 5, 7, and 9 years of age, and adults, were asked to sort shapes including exemplars, variants, and distractors (Satlow & Newcombe, 1998). A substantial change occurred between 4 and especially 5 years of age to 7 years, consistent with Keil's description of characteristic-to-defining shifts, in which older children relied more on rule-based definitions and less on perceptual similarity than younger children. Younger children were more likely to accept distractors with characteristic features and reject variants. Development regarding recognition of variants was incremental, but identification of distractors showed sudden improvement. Consistent with research discussed, shapes with multiple variants, such as triangles, were more difficult. The authors stated that this evidence disconfirms theories of general development, including Piaget's cognitive and van Hiele's mathematical ability. However, the shift itself is consistent with our *hierarchic interactionalism* reinterpretations of van Hielie theory (Clements, Battista, & Sarama, 2001).

In their play, children's behaviors were coded as "pattern and shape" more frequently than six other categories (Seo & Ginsburg, 2004). About 47% of these behaviors involved recognizing, sorting, or naming shapes. However, children's capabilities, in play and other situations, exceed naming and sorting shapes. This continues into the primary grades, as we have shown (Lehrer, Jenkins, & Osana, 1998). As a final example from the U.K., 7-year-olds were given a set of shapes in random order and asked to match and name each. More than 90% easily matched the shapes, but the percentage naming each was circle, 97.4; square, 96.4; equilateral triangle, 92.8; rectangle, 78.1; regular hexagon, 55.3; and regular pentagon, 31.1. They were also less confident about drawing shapes they had more difficulty naming. The percentage able to name properties of the shapes varied

from 92 for the square to 80 for the pentagon. These data substantiate the conclusion, consistent across the studies reviewed here, that children can easily distinguish the shapes but are exposed to a limited number of shape names. Although their performance is lower for unfamiliar shapes, these children can still name properties, further evidence that they are in "geometrically deprived" environments. We now turn to additional aspects of children's knowledge of shape and spatial structure.

3-D Figures

There is a paucity of research on very young children's knowledge of 3-D figures. Babies only 1 or 2 days old can maintain object size despite changes in distance (and thus change in size of the retinal image, Slater, Mattock, & Brown, 1990). That is, they habituate in looking at a sphere of constant size that changes in distance from the newborn, but not when both this distance and the size of the sphere is changed so that the sphere subtends the same angle on the retina (Granrud, 1987). In addition, infants can perceive 3-D shapes; this is limited, however, to continuously moving objects, rather than single or even multiple static views of the same object (Humphrey & Humphrey, 1995).

Two related studies required children to match solids with their nets. Kindergartners had reasonable success when the solids and nets both were made from the same interlocking materials (Leeson, 1995). An advanced kindergartner had more difficulty with drawings of the nets (Leeson, Stewart, & Wright, 1997), possibly because he was unable to visualize the relationship with the more abstract materials.

Congruence, Symmetry, and Transformations

Young children develop beginning ideas not just about shapes, but also about symmetry, congruence, and transformations. The ability to detect symmetry develops early (Vurpillot, 1976). Infants as young as 4 months dishabituate more quickly to symmetric figures than asymmetric figures, at least for vertical symmetry (Bornstein, Ferdinandsen, & Gross, 1981; Bornstein & Krinsky, 1985; Ferguson, Aminoff, & Gentner, 1998; C. B. Fisher, Ferdinandsen, & Bornstein, 1981; Humphrey & Humphrey, 1995). A preference for vertical symmetry seems to develop between 4 and 12 months of age, and vertical bilateral symmetry remains easier for children than horizontal symmetry, which in turn is easier than diagonal symmetries (Genkins, 1975; Palmer, 1985; Palmer & Hemenway, 1978). Extreme spatial separation of components of the pattern caused infants to lose the advantage for vertical symmetry (Bornstein & Krinsky, 1985). Further, they do not dishabituate to horizontal symmetry or different vertical patterns (C. B. Fisher, Ferdinandsen, & Bornstein, 1981), so initial competence is limited to "goodness of organization."

There appear to be two phases in recognition of symmetry (Palmer & Hemenway, 1978). The system first selects potential axes of symmetry defined by symmetric components via a crude but rapid analysis of symmetry in all orientations simultaneously. It maps alignment relations, using detection of closed loops, (non)perpendicularity, intersections, and protrusions, to produce structured representations (i.e., structurally consistent matches between identical attributes and relations of the objects in each). Vertical and, to a lesser extent, horizontal axes are preferred, because orientation relationships such as *above* and *beside* develop early; vertical is preferred because *beside* is bidirectional (Ferguson, Aminoff, & Gentner, 1998). The system then evaluates specific axes sequentially in a detailed comparison (Palmer & Hemenway, 1978), which has a significant influence on shape recognition (Palmer, 1989).

Children often use and refer to rotational symmetry as much as they do line symmetry in working with pattern blocks, such as remarking that an equilateral triangle was "special, because when you turn it a little it fits back on itself" (Sarama, Clements, & Vukelic, 1996). They also produce symmetry in their play (Seo & Ginsburg, 2004). For example, preschooler Jose puts a double unit block on the rug, two unit blocks on the double unit block, and a triangle unit on the middle, building a symmetrical structure. Play in this category, "pattern and shape," was the most frequent of the six categories coded. For people of all ages, symmetric shapes are detected faster, discriminated more accurately, and often remembered better than asymmetrical ones (Bornstein, Ferdinandsen, & Gross, 1981). However, many explicit concepts of symmetry are not firmly established before 12 years of age (Genkins, 1975).

Many young children judge congruence (Are these two shapes "the same"?) based on whether they are, overall, more similar than different (Vurpillot, 1976). As with symmetry, the comparison of two figures may evoke a pair of structured representations, and the comparison is represented as a mapping between sets of relations between components of the representations (Ferguson, Aminoff, & Gentner, 1998). Children younger than 5.5 years may not do an exhaustive comparison and until about 7 years of age may not attend to the spatial relationships of all the parts of complex figures (Vurpillot, 1976). Shown pairs of figures, some congruent, but all rotated, kindergartners tend to judge all pairs as "different," considering orientation a significant feature (Rosser, 1994).

With guidance, however, even 4-year-olds and some younger children can generate strategies for verifying congruence for some tasks. Preschoolers often try to judge congruence using an edge-matching strategy, although only about 50% can do it successfully (Beilin, 1984; Beilin, Klein, & Whitehurst, 1982). They gradually develop a greater awareness of the types of differences between figures that are considered relevant and move from considering various parts of shapes to considering the spatial relationships of these parts (Vurpillot, 1976). Thus, strategies supersede one another in development (e.g., motion-based superposition), becoming more powerful, sophisticated, geometrical, and accurate (recall that rigid motions were discussed in the section on spatial thinking).

The origins of the symmetry and equivalence concepts, however, may lie in early actions. Evidence indicates that children as young as 18 months will pick up two blocks and bang them together and are more likely to choose equivalent blocks than other available blocks. Children may progress through levels, from experiencing similar input between two hands (symmetry in action), to bilateral banging, to relating the block forms (at about 2 years of age), stacking equivalent blocks (an early explicit attempt to create a static expression of equivalence), placing blocks side by side, and eventually spacing equivalent blocks (Forman, 1982). Thus, human sensitivity to symmetry and equivalences may be prefigured in the bilateral symmetry of our anatomy, prefigured in the sense that particular types of object manipulatives are more likely to occur as a result of our having hands bilaterally opposed, and the feedback that is thus pleasing, as having identical objects in two hands (Forman, 1982), perhaps reflecting perception of the equivalence of our own bodies and even neurological symmetries.

Under the right conditions, children of all ages can apply similarity transformations to shapes. Even 4- and 5-year-olds can identify similar shapes in some circumstances (Sophian & Crosby, 1998). The coordination of height and width information to perceive the proportional shape of a rectangle (fat vs. skinny, wide or tall) might be a basic way of accessing proportionality information. This may serve as a foundation for other types of proportionality, especially fractions.

Education

Research indicates the importance of educational experiences. If the examples and nonexamples children experience are rigid, not representing the range of the shape category, so will be their concepts. For example, many children learn to accept only isosceles triangles, often only with a horizontal base, as "triangles." Others learn richer concepts, even at a young age; for example, one of the youngest 3-year-olds scored higher than every 6-year-old on shape-recognition tasks (Clements, Swaminathan, Hannibal, & Sarama, 1999). Concepts of two-dimensional shapes begin forming in the pre-K years and stabilize as early as age 6 (Gagatsis & Patronis, 1990; Hannibal & Clements, 2000). It is therefore critical that children be provided better opportunities to learn about geometric figures between 3 and 6 years of age. This learning will be more effective if it includes a full range of examples and distractors to build valid and strong concept images, including dynamic and flexible imagery (Owens, 1999).

Unfortunately, U.S. educational practice does not reflect these recommendations. Children often know as much about shapes entering school as their geometry curriculum "teaches" them in the early grades. This is due to teachers' and curriculum writers' assumptions that children in early childhood classrooms have little or no knowledge of geometric figures. Further, teachers have had few experiences with geometry in their own education or in their professional development. Thus, it is unsurprising that most classrooms exhibit limited geometry instruction. One early study found that kindergarten children had a great deal of knowledge about shapes and matching shapes before instruction began. Their teacher tended to elicit and verify this prior knowledge but did not add content or develop new knowledge. That is, about two thirds of the interactions had children repeat what they already knew in a repetitious format as in the following exchange: Teacher: "Could you tell us what type of shape that is?" Children: "A square." Teacher: "Okay. It's a square" (B. Thomas, 1982).

A more recent study confirmed that current practices in the primary grades also promote little conceptual change: First-grade students in one study were more likely than older children to differentiate one polygon from another by counting sides or vertices (Lehrer, Jenkins, & Osana, 1998). Over time, children were less likely to notice these attributes, given conventional instruction of geometry.

Such comparisons may be present even among preschoolers in various countries (Starkey et al., 1999). On a geometry assessment, 4-year-olds from America scored 55% compared to those from China at 84%. Thus, cultural supports are lacking from the earliest years in the United States. Computer environments can catalyze children's thinking about squares and rectangles for young children. In one large study (Clements, Battista, & Sarama, 2001), some kindergartners formed their own concept (e.g., "it's a square rectangle") in response to their work with Logo mi-

croworlds. This concept was applied only in certain situations: Squares were still squares, and rectangles were still rectangles, unless one formed a square while working with procedures—on the computer or in drawing—that were designed to produce rectangles. The concept was strongly visual in nature, and no logical classification per se, such as class inclusion processes, could be inferred. The creation, application, and discussion of the concept, however, were arguably a valuable intellectual exercise.

Less is known about teaching concepts of 3-D shapes. Certainly, consistent experience with building materials such as building blocks seems warranted (Leeb-Lundberg, 1996), especially as children engage in considerable premathematical play with these materials (Seo & Ginsburg, 2004) and can build general reasoning skills in this geometric context (Kamii, Miyakawa, & Kato, 2004). However, it is critical that teachers mathematize such activity. They can engage in fruitful discussions of blocks and other solids, and using specific terminology for solids, faces, and edges makes such discussion particularly beneficial (Nieuwoudt & van Niekerk, 1997).

Children can develop theories-in-action involving congruence as toddlers, stacking congruent blocks to make towers (Kamii, Miyakawa, & Kato, 2004). Block building can be used to develop such early notions throughout the preschool years. Beginning as early as 4 years of age, children can create and use strategies for judging whether two figures are "the same shape," and over the next couple of years, they can develop sophisticated and accurate mathematical procedures for determining congruence. Using puzzles and matching games, with task selection and scaffolding adjusted to children's strategies, caretakers can effectively move children through this learning trajectory.

There is mixed evidence regarding young children's ability with geometric motions. Pre-K to kindergarten children may be limited in their ability to mentally transform shapes, although there is evidence that even these sophisticated processes are achievable (Ehrlich, Levine, & Goldin-Meadow, 2005; Levine, Huttenlocher, Taylor, & Langrock, 1999). Further, these children can learn to perform rotations on objects (physical or virtual), and a rich curriculum, enhanced by manipulatives and computer tools, may reveal that knowledge and mental processes are valid educational goals for most young children. For example, interventions improve the spatial skills of low-income kindergartners, especially when embedded in a story context (Casey, 2005). Computers are especially helpful, as the screen tools make motions more accessible to reflection, and thus bring them to an explicit level of awareness for children (Clements & Sarama, 2003; Sarama, Clements, & Vukelic, 1996).

Computer environments also can be helpful in learning congruence and symmetry (Clements, Battista, & Sarama, 2001). Indeed, the effects of Logo microworlds on symmetry were particularly strong for kindergarten children. Writing Logo commands for the creation of symmetric figures, testing symmetry by flipping figures via commands, and discussing these actions apparently encouraged children to build richer and more general images of symmetric relations (with possibly some overgeneralization). Children had to abstract and externally represent their actions in a more explicit and precise fashion in Logo activities than, say, in freehand drawing of symmetric figures. There is undeveloped potential in generating curricula that seriously consider children's intuitions, preference, and interest in symmetry.

Research supports the use of manipulatives in developing geometric and spatial thinking in young children (Clements & McMillen, 1996). Using a greater variety of manipulatives is beneficial (Greabell, 1978). Such tactile-kinesthetic experiences as body movement and manipulating geometric solids help young children learn geometric concepts (Gerhardt, 1973; Prigge, 1978). Children also fare better with solid cutouts than printed forms, the former encouraging the use of more senses (Stevenson & McBee, 1958). Pictures also can be important; even children as young as 5 or 6 years (but not younger) can use information in pictures to build a pyramid, for example (Murphy & Wood, 1981). Thus, pictures can give children an immediate, intuitive grasp of certain geometric ideas. However, pictures need to be sufficiently varied so that children do not form limited ideas. Further, research indicates that pictures are rarely superior to manipulatives. In fact, in some cases, pictures may not differ in effectiveness from instruction with symbols (Sowell, 1989). But the reason may not lie in the "nonconcrete" nature of the pictures as much as it lies in their "nonmanipulability"—that is, that children cannot act on them as flexibly and extensively. Research shows that manipulatives on computers can have real benefit.

Instructional aids help because they are manipulable and meaningful. In providing these features, computers can provide representations that are just as real and helpful to young children as physical manipulatives. In fact, they may have specific advantages (Clements & McMillen, 1996). For example, children and teachers can save and later retrieve any arrangement of computer manipulatives. Similarly, computers allow one to store more than static configurations. They can record and replay *sequences* of actions on ma-

nipulatives. This helps young children form dynamic images. Computers can help children become aware of and mathematize their actions. For example, very young children can move puzzle pieces into place, but they do not think about their actions. Using the computer, however, helps children become aware of and describe these motions (Clements & Battista, 1991; Johnson-Gentile, Clements, & Battista, 1994).

Composition of Shapes

Disembedding Shapes

Children develop over years in learning how to separate structures within embedded figures. Visual discrimination, including figure-ground discrimination, seems to be innate, and visual stimuli are perceptually organized in the 1st year of life (Vurpillot, 1976). The primary perceptual structures operate as rigid, indivisible, unanalyzable, and unarticulated up to about 4 years of age, but between 6 and 8 years they become flexible, decomposable, and composable. Eventually these concepts and operations operate hierarchically (i.e., units can be combined, each one serving as a whole to the smaller units of which it is composed and, at the same time, as a component part of the comprehensive structure). Thus, in this age range, children develop the ability to break down line figures and reassemble them in new forms (as in embedded-figures problems), link up isolated perceptual units by means of imaginary lines to identify the more complex structures of incomplete figures, and pass from one structure to another when these are reversible figures. Gestalt theory leads to the following principles.

1. All the lines of a figure are involved in construction of the primary contour structures (PCS).

2. No line or part can belong to more than one primary structure.

3. A line belongs in its entirety to a single primary contour structure.

4. The PCS are preferably symmetrical or at least as regular as possible.

5. The number of PCS must be the fewest possible.

6. Each area entirely surrounded by the lines of a figure and not crossed by another line constitutes a primary area structure (PAS).

Solving embedded-figures problems requires going beyond the primary structure to create secondary structures, borrowed from one or more PCS.

The preschooler can thus find a figure identical to a given PCS but in other cases will incorrectly designate a PCS as the solution, as empirical findings suggest. The following problems present increasing degrees of difficulty, the first three primary, the second three secondary (in which parts can belong simultaneously to several structures, allowing the construction of new units) levels of organization.

1. The PCS that is the frame, or external contour (e.g., largest square in the 3 by 3 arrangement of squares).
2. A PCS of a complex figure (triangle of Star of David).
3. A PCS other than the frame (e.g., the tic-tac-toe of the arrangement of squares).
4. A figure that is not a PCS but encloses a PAS (e.g., smallest square of the arrangement of squares).
5. Similar, but a sum of several PAS.
6. A nonclosed figure that is not a PCS.

In finding a secondary structure, children less than 6 years of age often identify a PCS instead. In sum, before 6 years of age, what children perceive is organized in a rigid manner into structures whose form is determined by Gestalt principles. Children grow in the flexibility of the perceptual organizations they can create. They eventually integrate parts and use imaginary components, with anticipation. Beyond a certain level of complexity, people of any age cannot perceive secondary structures but must construct them piece-by-piece.

Composition and Decomposition

The ability to describe, use, and visualize the effects of composing and decomposing geometric regions is significant in that the concepts and actions of creating and then iterating units and higher order units in the context of constructing patterns, measuring, and computing are established bases for mathematical understanding and analysis (Clements, Battista, Sarama, & Swaminathan, 1997; Reynolds & Wheatley, 1996; Steffe & Cobb, 1988). Additionally, there is empirical support that this type of composition corresponds with and supports children's ability to compose and decompose numbers (Clements, Sarama, Battista, & Swaminathan, 1996).

Block building provides a view of children's initial abilities to compose 3-D objects (as well as their formation of a systemic of logic, cf. Forman, 1982). As they do with 2-D shapes, children initially build structures from simple components and later explicitly synthesize 3-D shapes into higher order 3-D shapes. Children either engage in little systematic organization of objects or show little interest in stacking in their 1st year (Forman, 1982; Kamii, Miyakawa, & Kato, 2004; Stiles & Stern, 2001). Stacking begins at 1 year, thus showing use of the spatial relationship of *on* (Kamii, Miyakawa, & Kato, 2004). Occasionally, children balance blocks intuitively, but often they place blocks off center or on an edge of a triangular prism. In the latter case, they could recognize lack of success when the block fell but made no attempt to understand it. The *next-to* relation develops at about 1.5 years (Stiles-Davis, 1988). At 2 years, children place each successive block congruently on or next to the one previously placed (Stiles-Davis, 1988). They appear to recognize that blocks do not fall when so placed. Children begin to reflect (think back) and anticipate (Kamii, Miyakawa, & Kato, 2004). At 3 to 4 years of age, children regularly build vertical and horizontal components within a building (Stiles & Stern, 2001). They can use the relations only in sequence at age 3, creating multicomponent structures, but within a limited range. When asked to build "a tall tower," they use long blocks vertically, because they have added to the goal of making a stable tower, making a stable tall tower, first using only one block in this fashion, then several (Kamii, Miyakawa, & Kato, 2004). At 4 years, they can use multiple spatial relations, extending in multiple directions and with multiple points of contact among components, showing flexibility in how they generate and integrate parts of the structure. A small number of children will build a tower with all blocks; for example, by composing the triangular blocks, making subparts to coordinate with the whole (Kamii, Miyakawa, & Kato, 2004).

This developmental progression was used to study children's processes in relation to task complexity (Stiles & Stern, 2001). Three-year-olds could not produce the more complex constructions, including an enclosure, +, and horizontal corner. Their strategy was simple and unsystematic accretion of parts. Children of 4 or 5 years could use a more sophisticated strategy organized around main construction components, but they regressed to simpler strategies for the most difficult tasks. Children of 6 years did not differ from adults on these tasks. Thus, spatial processing in young children is not qualitatively different from that of older children or adults. However, with age, children produce progressively more elaborate constructions. The way they analyze spatial arrays also changes; they segment out different elements and relational structures. For example, to make a +, adults often complete one long segment and add two short segments, but children build four components around a point. Finally, as they develop, children choose more advanced strategies, but this is influenced by the task.

In this and other studies, Kamii argued for focusing on children's general reasoning with a Piagetian perspective. From a different perspective, children's performance on a block design task was consistent with Vygotskian theory, predicted by both preschoolers' initial abilities and a dynamic assessment of their learning potential (Day, Engelhardt, Maxwell, & Bolig, 1997). Both these areas were domain-specific, supporting the position that cognition is multidimensional.

A related study required children to make cube buildings from 2-D representations or determine the number of cubes in a pictured cube building. A case study indicated than an advanced kindergartner could do the former, but not the latter. Perhaps he was unable to visualize the shapes and needed manipulatives (Leeson, Stewart, & Wright, 1997). Thus, less advanced children in pre-K and kindergarten may require hands-on materials to help them interpret 2-D representations of 3-D shapes.

These 3-D studies only touched on children's abilities to actually compose shapes. Research on 2-D shape composition provides a more detailed picture. The following is a research-based developmental progression, which approximately spans ages 4 to 8 years (Clements, Sarama, & Wilson, 2001; Sarama, Clements, & Vukelic, 1996). From lack of competence in composing geometric shapes, children gain abilities to combine shapes—initially through trial and error and gradually by attributes—into pictures and finally synthesize combinations of shapes into new shapes (composite shapes). Three sources of evidence support the validity of this theory (Clements, Sarama, & Wilson, 2001; Clements, Wilson, & Sarama, 2004). First, the original hypothetical learning trajectory and the developmental progression underlying it emerged from naturalistic observations of young children composing shapes (Clements, Battista, Sarama, & Swaminathan, 1997; Mansfield & Scott, 1990; Sales, 1994; Sarama, Clements, & Vukelic, 1996, Sarama & Clements, in preparation). Second, the levels of the developmental progression were tested iteratively in formative research that involved researchers and teachers. Their case studies indicated that about four fifths of the children studied evinced behaviors consistent with the developmental progression (using an early version of the instrument). By the end of this phase, all participants believed that the developmental progression and the items retained to measure levels in the progression were valid and that they could reliably classify children as exhibiting thinking on the progression. Third, a summative study employed the final instrument with 72 randomly selected children from pre-K to grade 2. Analyses revealed that the level scores fit the hypothesized structure in which scores from one level would be more highly correlated with scores immediately adjacent to that level than to scores on levels nonadjacent to the given level. Further, the developmental progression showed development across ages, with children at each grade scoring significantly higher than those at the preceding grade did. In addition, support for this theory lies in its consistency with previously discussed research on children's perceptions of shape. In both, parts are related to wholes, with each part initially playing a single functional role in the pattern structure (cf. Tada & Stiles, 1996). Indeed, children spatially isolate parts at first, then arrange them contiguously, and later combine them in an integrative manner, eventually creating more complex units within different structural layers. In both cases, mature cognition is a result of a developmental process in which parts and wholes are interrelated across hierarchical levels.

Summary and Issues

Theoretical Framework

Our theoretical framework, hierarchic interactionalism, is supported by the research on geometric and spatial thinking. Foundational geometric competencies are inborn or developed early, become represented with language, and eventually are reflected upon metacognitively (probably following phases of representational redescription, cf. Clements, Battista, & Sarama, 2001; Karmiloff-Smith, 1984, 1986, 1990). Developmental progressions, along with specific mental objects and actions defining each level, have been identified and, especially in the domain of knowledge of traditional topics such as geometric shapes, have been generally confirmed through a variety of research methods, which describe the *hierarchic development* of those topics separately, as well as their progressive hierarchization (Clements, Battista, & Sarama, 2001; Lewellen, 1992). Further, these progressions are topic specific, as the theory indicates, rather than general (van Hiele, 1986). Environment and culture affect spatial reasoning, although many are experience-expectant processes, but they are particularly important in the learning of geometry, with both children and adults showing low levels of geometric thinking if they are deprived of high-quality education. Learning trajectories appear to be an effective basis for curriculum and teaching (Clements, 2003; Clements, Wilson, & Sarama, 2004).

Although the section "Theories of Young Children's Perception and Knowledge of Shape" elaborated on the hierarchic interactionalism framework, a few additional points can be made. There is particular support for hierarchic development and the existence of different developmental courses in geometry (Clements, Battista, & Sarama, 2001). That is, levels of thinking about geometric shapes, for example, do not consist of unadulterated knowledge of one type only. Levels of geometric knowledge are not strictly visual, followed by descriptive/analytic (as described in van Hiele, 1986). Instead, children at both levels possess shape schemes that include visual/imagistic and nonvisual or verbal declarative knowledge ("knowing what") about shapes.

Research has substantiated that children possess several different prototypes for figures (e.g., a vertically and a horizontally oriented rectangle) without accepting the "middle" case (e.g., an obliquely oriented rectangle). In one study, participants studied a preponderance of rectangles with extreme values and few intermediate values of variables such as size (Neumann, 1977). Participants were presented with test items and asked to rate their confidence that they had already studied that particular item. Interestingly, they rated the extremes (e.g., large or small rectangles) much higher than items created by using the mean of these values (e.g., middle-sized rectangles), showing that they extracted multiple foci of centrality, thereby creating several visual prototypes. Thus, they did not cognitively "average" what they had studied. This finding is consistent with studies on the van Hiele theory (Burger & Shaughnessy, 1986; Clements & Battista, 1992; Clements, Battista, & Sarama, 2001; Fuys, Geddes, & Tischler, 1988).

Initially, children have constrained imagistic prototypes for triangles but also possess verbal declarative knowledge. With repeated exposure to exemplars in the culture, these prototypes grow stronger; to the extent these exemplars are limited (e.g., mostly equilateral for many shape categories), acceptance decreases radically, leading to the child's rejection of both distractors and variants that do not closely match the visual prototype, which may have rigid constraints regarding aspect ratio, skewness, and orientation. This rejection may be particularly noticeable in situations in which shapes are drawn (or otherwise cannot be manipulated) and are presented in a canonical rectangular frame. Nascent declarative knowledge, while developing, does not gain transcendence in the scheme (consistent with the *hierarchic development* tenet of the theory). (Note this provides an elaboration and mechanism for the construct of concept images—a combination of all the mental pictures and properties that have been associated with the concept, Vinner & Hershkowitz, 1980.) Even people who know a correct verbal description of a concept and possess a specific visual image (or concept image) associated strongly with the concept may have difficulty applying the verbal description correctly. Eventually, exposure to a wider variety of examples and a strengthening of declarative knowledge ("3 straight sides") leads to a wider acceptance of varieties of geometric figures while rejecting nonmembers of the class. Each type of knowledge increasingly constrains the other (*co-mutual development of concepts and skills*).

Consistent with the tenet of *different developmental courses* and the evidence that children possess multiples types of geometric knowledge, we suggest that children's knowledge of geometry might be enhanced in different ways. First, their imagistic prototypes might be vastly elaborated by the presentation of a variety of variants, via the systematic variation of irrelevant and relevant attributes (e.g., through dynamic media such as the computer). Children must actively attend to the examples and connect them as through verbal labeling. Such enhancement is accomplished mainly through a usually unconscious visual-induction process. Second, through the presentation of particular tasks and engagement in dialogue about them, children's verbal declarative knowledge might be refined to extend, elaborate, and constrain their visual knowledge. Preschoolers exposed to such discussions consciously used and were excited about their ability to tell if any figure was a triangle just by counting to see if it had three straight sides (Clements & Sarama, in press).

We hypothesize that each of these two paths toward enhancing knowledge of geometry can be followed separately or together. If separately, knowledge of one type can "substitute" for knowledge of another type on certain tasks, within certain limits (including a performance decrement in accuracy or speed of execution). For example, a rich and varied exposure to various examples, could allow near-perfect performance on triangle discrimination tasks such as those discussed previously. Performance would suffer only if a figure were presented that fell outside the range of any of the multiple imagistic prototypes developed by the children, or if the task demanded reasoning based on properties not supported by such imagistic-oriented schemes (e.g., calculations regarding angle relationships), in which case the schemes would be inadequate to the task.

Similarly, well-developed verbal declarative knowledge, even in the absence of exposure to various examples, could allow perfect performance on such simple discrimination tasks. However, performance would suffer (at least in speed of execution) because a chain of reasoning would be required for every figure that could not be immediately assimilated into the (hypothesized-limited) range of imagistic prototypes. In each of these cases, the characteristics of the figures would have to be "read off" the figure and compared to those that are held in verbal declarative knowledge as characteristics of the class. This type of verbal knowledge would support more sophisticated analysis of geometric figures. Without the imagistic knowledge, however, the range and flexibility of the application of this knowledge would be limited. Research on the learning of geometry shows that students limit their conceptualizations to studied exemplars and often consider common but nondefining features (e.g., an altitude of a triangle is always located inside the triangle) as essential features of the concept (Burger & Shaughnessy, 1986; N. D. Fisher, 1978; Fuys, Geddes, & Tischler, 1988; Kabanova-Meller, 1970; Zykova, 1969). Further, students who know a correct verbal definition of a concept but also have a limited visual prototype associated with it may have difficulty applying the verbal description correctly (Clements & Battista, 1989; Hershkowitz et al., 1990; Vinner & Hershkowitz, 1980). This is consistent with the theory of simultaneous and sequential processing. In this view, pictorially presented materials may be more likely to evoke the visual/holistic/simultaneous processing, whereas aurally presented materials would be more likely to evoke verbal/sequential/linear processing. The type of presentation may interact with individual differences in abilities in these two domains.

This view is consistent with the work of Karmiloff-Smith (1984; , 1986; , 1990), who postulated a repeating 3-phase cycle of representational redescription. Phases, in contrast to structurally constituted stages or levels, are recurrent, general (across-domain) processes that people work through during development or learning. At Phase 1, the building of mental representations is predominantly driven by interactions of the children's goal-directed schemes with the environment. The children's goal is behavioral success, or the reaching of that goal, which sometimes is evaluated by consistency with adult responses and feedback.[5] Such success leads to the recording of isolated correspondences between environmental situations that do or do not allow or aid the attainment of the goal in a form inaccessible to the system (i.e., input-output correspondences along with their contexts in compiled form). In this form, any relationship between bits of knowledge is, at best, implicit.

When behavioral success is achieved, a Phase 2 metaprocess (a procedure that operates on internal knowledge structures) evaluates the knowledge base. Now, the goal is not to behave successfully, but to gain control over the representational forms (Vygotsky, 1934/1986, similarly stated that the development of thought cannot be derived from the failure of thought and postulated a genetic predisposition to gain control of mental representations). The first operation of Phase 2 is to rerecord Phase 1 representations in a form that can be accessed, though not yet consciously. The implicit representations are analyzed into semanticized components, linking them into a simplified but growing network structure that is predicated on the usefulness of the initial correspondences to goal attainment. The second operation is to form relationships between bits of knowledge. These two new operations place demands on cognitive processing, which, together with the (over)simplified structure and the need to test the mental representation in new situations, often leads to new "errors"—ostensibly a step backward to Phase 1 from an adult's perspective—that mask the progress in explicating representations of the domain. Finally, these two operations constitute an internalization of relationships and processes that were previously only implicit (cf. Steffe & Cobb, 1988).

Once successful rerepresentation is achieved, children develop Phase 3 control mechanisms that integrate and balance consideration of the external environment and the new internal representational connections forged in Phase 2. At the end of Phase 3, these connections are rerecorded again in abstract symbolic form, the first form accessible to conscious thought. Now, performance improves beyond that which was achieved at either of the two earlier phases.

At the syncretic level of geometric thinking, children are implicitly recognizing the properties of shapes. For example, their schemes for squares and for rectangles both contain patterns for right angles. However, these are patterns in spatial subsystems that emerge when instantiated; they are not conceptual objects (mental entities that can be manipulated or scrutinized, Davis, 1984). Right angles are not represented explicitly, and therefore no relationship is formed between them. In general, this type of representation explains how operation at one level can presuppose knowledge from the succeeding level, without allowing access to knowledge at the higher level because

---

[5] This sentence could be misconstrued to mean that students are controlled by the external environment. We assume that students are always sense-making beings; however, during this phase in building a representation, they are actively making sense of their social and physical environments, rather than their representations, of which they are not yet conscious. This active sense-making is critical, in that it allows the students to differentiate between environments that do and do not assist goal attainment.

the form of such knowledge is proceduralized or schematized behaviorally and is, therefore, inaccessible to the rest of the cognitive system.

When shapes are dealt with successfully on the level of behavior, metaprocesses rerecord the mental representations, creating a mental geometric object for the visual image of the right angles and a link between these mental objects for different cases of right angles, including links between those in rectangles and those in squares. Because children in Phase 2 are seeking control over their representations of these geometric forms, some increases in "errors" (from an adult's point of view) may occur (e.g., maintaining the notion that a nonrectangular parallelogram is a rectangle "because you're only looking at it from the side").

Eventually, the properties become conceptual objects that are available as data to conscious processes. That is, visual features become salient in isolation and are linked to a verbal label, and the child becomes capable of reflecting on the visual features and thus can explicitly recognize the shape's properties. At first, however, the flexibility of application is limited. Indeed, children can be expected not only to think at different levels for different topics (Clements & Battista, 1992) but also to think at different phases for different topics. Problem solving and discussion involving the geometric objects help build connections between the constructed knowledge (e.g., of right angles) and other similarly accessible knowledge (e.g., of parallelism and side lengths of rectangles and squares). Eventually, connections are built between properties of figures such as rectangles and squares and of properties of other geometric objects. In this manner, children's geometric knowledge can become increasingly abstract, coherent, and integrated, because it is freed from the constraints of compiled, and thus inflexible, mental representations. At each level, the degree of integration increases as the connections span greater numbers of geometric and eventually nongeometric topics (cf. Gutiérrez & Jaime, 1988). Here again, instruction has a strong influence. Ideally, it encourages unification; however, the instruction of isolated bits of knowledge at low levels retards such development. Unfortunately, the latter is pervasive in both curriculum materials (Fuys, Geddes, & Tischler, 1988) and teaching (Clements & Battista, 1992; Porter, 1989; B. Thomas, 1982).

To place instructional implications in a different light, only after the third phase do children become explicitly aware of their geometric conceptualiza-

tions; therefore, it is after Phase 3 that the last three instructional steps in the van Hiele model (explicitation, free orientation, and integration) can begin.[6] An implication is that short-circuiting this developmental sequence (e.g., by beginning with explicitation) is a serious pedagogical mistake. Deprived of the initial construction of their own mental geometric objects and relationships (images), children construct Phase 1 (behaviorally "correct") verbal responses based on "rules without reason" (Skemp, 1976). A more viable goal is the construction of mathematical meaning from actions on geometric objects and subsequent reflections on those actions.

In summary, development of geometric properties as conceptual objects leads to pervasive Level 2 thinking. This development represents a reconstruction on the abstract/conscious/verbal plane of those geometric conceptualizations that Piaget and Inhelder (1967) hypothesized were first constructed on the perceptual plane and then reconstructed on the representational/imaginal plane. Thus, Level 2 thinking requires what Piaget called the construction of articulated mental imagery, which develops most fully via the combined enhancement of both imagistic and verbal declarative knowledge, as previously discussed. We posit that the same recurrent phases explain the reconstruction on each new plane (level), but that the role of social interaction and instruction increases in importance with higher levels.

The theory of reiterated phases of rerepresentation applies equally to children and adults, regardless of their overall stage of cognitive development (Karmiloff-Smith, 1990). This is consonant with reports of low van Hiele levels among high school students and preservice teachers (Burger & Shaughnessy, 1986; Denis, 1987; Gutiérrez & Jaime, 1988; Mayberry, 1983; Senk, 1989).

Final Words

As stated in the introduction to this section and supported by the research, spatial and geometric thinking are essential human abilities that contribute to mathematical ability. Their importance is highlighted by findings that infants assign greater weight to spatiotemporal information than color or form in their definition of what an object is (Newcombe & Huttenlocher, 2000). Thereafter, toddlers use the shape of objects as the essential cue in learning the identity and names of objects. For example, training 17-month-olds to attend to shape led them to general-

---

[6] The first two steps, Information and guided orientation, of the van Hiele model's five-step instructional sequence would be, with some modification, consonant with the three phases of development described here. This topic is related to, but different from, the topics addressed here and will not be examined.

ize that objects with similar shapes have the same name and engendered a dramatic increase of 350% in learning new words outside of the laboratory (Smith, Jones, Landau, Gershkoff-Stowe, & Samuelson, 2002). As important is spatial thinking, which supports geometry and creative thought in all mathematics. Given their importance, it is essential that geometry and spatial sense receive greater attention in instruction and in research. Unfortunately, U.S. children's performance in geometry and spatial reasoning is woefully lacking. Educators need to understand more about learning and teaching geometry and spatial sense.

Although far less developed than our knowledge of number, research provides guidelines for developing young children's learning of geometric and spatial abilities. However, researchers do not know the potential of children's learning if a conscientious, sequenced development of spatial thinking and geometry were provided throughout their earliest years. This can also be seen as a caveat. Research on the learning of shapes and certain aspects of visual literacy suggest the inclusion of these topics in the early years. Insufficient evidence exists on the effects (efficacy and efficiency) of including topics such as congruence, similarity, transformations, and angles in curricula and teaching at specific age levels. Such research, and longitudinal research on many such topics, is needed.

Finally, competencies in these first two major realms, spatial/geometric and quantitative/number, are connected, probably at deep levels, throughout development. The earliest competencies may share common perceptual and representational origins (Mix, Huttenlocher, & Levine, 2002). Infants are sensitive to both amount of liquid in a container (Gao, Levine, & Huttenlocher, 2000) and distance a toy is hidden in a long sandbox (Newcombe, Huttenlocher, & Learmonth, 1999). Visual-spatial deficits in early childhood are detrimental to children's development of numerical competencies (Semrud-Clikeman & Hynd, 1990; Spiers, 1987). Other evidence shows specific spatially related learning disabilities in arithmetic, possibly more so for boys than girls (Share, Moffitt, & Silva, 1988). Children with Williams Syndrome, who show impairment on visuo spatial construction tasks, can learn reading and spelling but have difficulty learning mathematics (Bellugi, Lichtenberger, Jones, Lai, & St. George, 2000; Howlin, Davies, & Udwin, 1998). Primary school children's thinking about "units" and "units of units" was found to be consistent in both spatial and numerical problems (Clements, Battista, Sarama, & Swaminathan, 1997). In this and other ways, specific spatial abilities appear to be related to mathematical competencies (D. L. Brown & Wheatley, 1989; Clements & Battista, 1992; Fennema

& Carpenter, 1981; G. H. Wheatley, Brown, & Solano, 1994). Geometric measurement connects the spatial and numeric realms explicitly and is the topic of the following section.

GEOMETRIC MEASUREMENT

Geometric measurement can serve as a bridge between the two critical domains of geometry and number, with each providing conceptual support to the other. Indications are, however, that this potential is usually not realized, as measurement is not taught well. Many children use measurement instruments or count units in a rote fashion and apply formulas to attain answers without meaning (Clements & Battista, 1992). In international comparisons, U.S. students' performance in measurement was low (NCES, 1996).

Children's understanding of measurement has its roots in infancy and the preschool years but grows over many years, as the work of Piaget and his collaborators has shown (Piaget & Inhelder, 1967; Piaget, Inhelder, & Szeminska, 1960). As with number, however, Piagetians underestimated the abilities of the youngest children. For example, shown an object that was then occluded by a drawbridge, infants looked longer when the drawbridge rotated past the point where the object should have stopped it (Baillargeon, 1991). Thus, even infants are sensitive to continuous quantity (Gao, Levine, & Huttenlocher, 2000), and even comparisons (Spelke, 2002) and accumulations (Mix, Huttenlocher, & Levine, 2002) of continuous quantity, at least in some conditions (cf. Huntley-Fenner, Carey, & Solimando, 2002). These studies show that early cognitive foundations of mathematics are not limited or unique to number. As with number, however, these abilities have limits. Infants and toddlers can discriminate between lengths of dowels, but only when a salient standard (a same-length dowel or container) was present; 4-year-olds could discriminate with or without such as standard (Huttenlocher, Duffy, & Levine, 2002). Infants and toddlers may lack the ability to create and maintain a mental image of a length.

As we discussed, some believe that humans may be sensitive to amount of continuous quantity and not discrete number (Clearfield & Mix, 1999). However, they do make distinctions. For example, when a small number of objects or portions of a substance is hidden and then revealed, 12- to 13-month-old infants expect that the former cannot be combined to make a larger object but that the latter can be coalesced into larger portions (Huntley-Fenner, 1999a). Recall that 11- but not 9-month-old infants could discriminate sequences of number (Brannon, 2002). Noteworthy here is that the 9-month-olds could discriminate sequences of size; therefore, nonnumerical ordinal judgments may develop before the capacity for numerical ordinal judgments.

Preschool children know that continuous attributes such as mass, length, and weight exist, although they cannot quantify or measure them accurately. Children 2 to 4 years of age can use the same three types of standards when judging "big" and "little" that adults use: perceptual (object is compared to another physically present object), normative (object is compared to a class standard stored in memory, such as a chihuahua is small for a dog), and functional (Is this hat the right size for this doll?; Ebeling & Gelman, 1988). Children also can coordinate these, preferring the perceptual to the normative if there is a conflict. For example, when asked if an object is larger than an egg, but then given an even larger egg, children will compare the object to the larger egg. They prefer either the perceptual or normative standards to the functional (which must be brought to the attention of 3 year olds, but not 4 year olds) (S. A. Gelman & Ebeling, 1989). Finally, children can switch within contexts and from a normative context to the others but have difficulty switching to a normative context. The normative context may differ from the perceptual and functional contexts in that stimuli are not physically present and thus may be accessible only when no other context has been recently experienced (Ebeling & Gelman, 1994).

As young as 3 years of age, children know explicitly that if they have some clay and then are given more clay, they have more than they did before. However, preschoolers cannot reliably make judgments about which of two amounts of clay is more; they use perceptual cues such as which is longer. Children do not reliably differentiate between continuous and discrete quantity, for example, basing equal sharing on the number of cookie pieces rather than the amount of substance (K. F. Miller, 1984; Piaget, Inhelder, & Szeminska, 1960).

Further, they have not yet integrated their counting (e.g., of discrete entities) with measurement (counting units of continuous quantity). For example, 4- and 5-year-olds were more likely to count when asked which group had "more glasses" than if asked which had "more sand," even though in each comparison the same sand was in the same glasses (Huntley-Fenner, 1999b). Younger children responded inaccurately when asked to compare sand that was not in glasses. When children observed cupfuls of sand poured into boxes they again did not use counting and used the rate, not duration, of pouring as the basis of judgment (if researchers hid the pouring act, younger children performed the worst). Thus, children must learn to ap-

ply their counting skills to unitized measures of continuous quantities. They appear to do so first for small numerosities. For example, 6-year-olds were more accurate in addition tasks when red blocks were placed in cylinders instead of liquid, but only for small numerosities (5 or less). To measure, children have to overcome a natural inclination to quantify continuous substances with mental processes that are analogously continuous (i.e., that do *not* involve discrete units and counting). In a striking example, 3- to 5-year-old children were no less successful comparing amounts of sand in piles than when the same amount was shown in 3 vs. 2 discrete glasses (Huntley-Fenner, 2001).

Despite such challenges, young children can be guided to have appropriate measurement experiences. They naturally encounter and discuss quantities (Seo & Ginsburg, 2004). They initially learn to use words that represent quantity or magnitude of a certain attribute. Then they compare two objects directly and recognize equality or inequality (Boulton-Lewis, Wilss, & Mutch, 1996). At age 4–5 years, most children can learn to overcome perceptual cues and make progress in reasoning about and measuring quantities. They are ready to learn to measure, connecting number to the quantity (even though the average U.S. child, with limited measurement experience, exhibits limited understanding of measurement until the end of the primary grades). We next examine this development in more detail for the case of length.

Length Measurement

Length is a characteristic of an object found by quantifying how far it is between the endpoints of the object. *Distance* is often used similarly to quantify how far it is between any two points in space. Measuring length or distance consists of two aspects, identifying a unit of measure and *subdividing* (mentally and physically) the object by that unit and placing that unit end to end (*iterating*) alongside the object. Subdividing and unit iteration are complex mental accomplishments that are too often ignored in traditional measurement curriculum materials and instruction. Therefore, many researchers go beyond the physical act of measuring to investigate children's understandings of measuring as covering space and quantifying that covering.

We discuss length in the following three sections. First, we identify several key concepts that underlie measuring (Clements & Stephan, 2004; Stephan & Clements, 2003). Second, we discuss early development of some of these concepts. Third, we describe research-based instructional approaches that were designed to help children develop concepts and skills of length measurement.

Concepts in Linear Measurement

At least eight concepts form the foundation of children's understanding of length measurement. These concepts include understanding of the attribute, conservation, transitivity, equal partitioning, iteration of a standard unit, accumulation of distance, origin, and relation to number.

Understanding of the attribute of length includes understanding that lengths span fixed distances ("Euclidean" rather than "topological" conceptions in the Piagetian formulation).

Conservation of length includes understanding that lengths span fixed distances (Euclidean rather than topological conceptions in the Piagetian formulation) and the understanding that as an object is moved, its length does not change. For example, if children are shown two equal-length rods aligned, they usually agree that they are the same length. If one is moved to project beyond the other, children 4.5 to 6 years often state that the projecting rod is longer. At 5 to 7 years, many children hesitate or vacillate; beyond that, they quickly answer correctly. Conservation of length develops as the child learns to measure (Inhelder, Sinclair, & Bovet, 1974).

Transitivity is the understanding that if the length of object X is equal to (or greater/less than) the length of object Y and object Y is the same length as (or greater/less than) object Z, then object X is the same length as (or greater/less than) object Z. A child with this understanding can use an object as a referent by which to compare the heights or lengths of other objects.

Equal partitioning is the mental activity of slicing up an object into the same-sized units. This idea is not obvious to children. It involves mentally seeing the object as something that can be partitioned (or "cut up") before even physically measuring. Asking children what the hash marks on a ruler mean can reveal how they understand partitioning of length (Clements & Barrett, 1996; Lehrer, 2003). Some children, for instance, may understand "five" as a hash mark, not as a space that is cut into five equal-sized units. As children come to understand that units can also be partitioned, they come to grips with the idea that length is continuous (e.g., any unit can itself be further partitioned).

Iteration of a unit is the ability to think of the length of a small unit such as a block as part of the length of the object being measured and to place the smaller block repeatedly along the length of the larger object (Kamii & Clark, 1997; Steffe, 1991), counting these iterations.

Accumulation of distance is the understanding that as you iterate a unit along the length of an object and count the iteration, the number words signify the space covered by all units counted up to that point (Petitto, 1990). Piaget et al. (1960) characterized children's measuring activity as an accumulation of distance when the result of iterating forms nesting relationships to each other. That is, the space covered by three units is nested in or contained in the space covered by four units.

Origin is the notion that any point on a ratio scale can be used as the origin. Young children often begin a measurement with "1" instead of zero.

Relation between number and measurement requires children to reorganize their understanding of the items they are counting to measure continuous units. They make measurement judgments based upon counting ideas, often based on experiences counting discrete objects. For example, Inhelder and others (1974) showed children two rows of matches that were the same length, but each row comprised a different number of matches. Although, from the adult perspective, the lengths of the rows were the same, many children argued that the row with more matches was longer because it had more matches. Note that this context differs from the cardinal situations in that the order-irrelevance principle does not apply and every element (e.g., units on a ruler) should not necessarily be counted (Fuson & Hall, 1982).

Researchers debate the order of the development of these concepts and the ages at which they are developed; perhaps education and experience have a large effect on both. Researchers generally agree that these ideas form the foundation for various aspects of measurement. Traditional measurement instruction is insufficient for helping children build these conceptions.

Early Development of Length Measurement Concepts

The same landmarks that aid children in cue or place learning also can affect their representations of the distances separating objects. Piaget, Inhelder, and Szeminska (1960) reported that after placing a third object between two objects, young children claim that the distance is smaller or larger than before. In another study, children judged that two routes, one direct and one indirect, cover the same distance (i.e., straight path is the shortest distance between two points). Subsequent studies have confirmed that most 4-year-olds and about half of 5- and 6-year-olds show such patterns (Fabricius & Wellman, 1993; K. F. Miller & Baillargeon, 1990). Thus, the Piagetian position is that young children do not possess understanding of distance and length.

Mistakes on tasks may not be due to the misconceptions of space that Piagetian theory assumed. First, children can encode and apply distance information. For example, preschoolers do well at simple distance judgment tasks, with this competence appearing as early as 12 to 16 months of age (Huttenlocher, Newcombe, & Vasilyeva, 1999).

Second, there are inconsistencies in the literature on Piagetian tasks. About 40% of 4-year-olds could avoid errors on direct and indirect routes, including giving correct explanations (Fabricius & Wellman, 1993). Further, children 3.5 to 5 years of age appear to understand both the direct-indirect principle and the same-plus principle (if two routes are the same up to a point, but only one continues, it is longer) in a task modification in which the items were screened, so responses would not be the result of perceptual scanning (Bartsch & Wellman, 1988).

In a variation of the conservation-of-length task, children were asked which of five boxes a stick would fit into (Schiff, 1983). Children's judgments remained consistent after sliding. Thus, they correctly judged that a stick would go into the same box after it was moved (e.g., apparently believing that sliding the stick across the table did not change the physical dimensions of the stick). In another study, children were asked to choose a stick to bridge a gap. They appeared to understand that occlusion of the stick did not affect the length (K. F. Miller & Baillargeon, 1990). Children first understand affordances, such as "will this stick fit here," and later integrate knowledge between length and distance (K. F. Miller, 1984).

Across several experiments, then, there is little empirical support for the notion, such as in the Piagetian topological primacy thesis, that conceptualizations underlying children's reasoning about distance and length differ from those of adults. Preschoolers understand that lengths span fixed distances. Still, some researchers hold that complete conservation is essential for, but not equivalent to, a full conception of measurement. Piaget, Inhelder, and Szeminska (1960) argued that transitivity is impossible for children who do not conserve lengths because once they move a unit, it is possible, in the child's view, for the length of the unit to change. Most researchers agree that children develop the notion of conservation before transitivity (Boulton-Lewis, 1987). Although researchers agree that conservation is essential for a complete understanding of measurement, children do not necessarily need to develop conservation before they can learn some measurement ideas (Boulton-Lewis, 1987; Clements, 1999c; Hiebert, 1981; Petitto, 1990). Two measurement ideas that seem to depend on conservation and transitivity are the inverse relation between the size of the unit and the number of those units and the need to use equal length units when measuring. However, in several anecdotal reports preschoolers understood the inverse relation in reform curricula contexts, and one study (Sophian, 2002) showed an increase in the understanding of effect of object size on measurement of volume in 3- and 4-year-olds, when children are given the opportunity to compare the result of measurements made with different units.

Most researchers argue that children must reason transitively before they can understand measurement adequately (Boulton-Lewis, 1987). Some researchers conclude that the ruler is useless as a measuring tool if a child cannot yet reason transitively (Kamii & Clark, 1997). As with conservation, this may only be true for some tasks. Further, as we have shown before, understanding is not a dichotomous phenomenon. Children as young as pre-K and kindergarten age use transitivity in simple measurement tasks. For example, given two holes and a marked stick, they can compare the depth of the holes by inserting the stick and comparing the marks (Nunes & Bryant, 1996). Such abilities appear in even younger children on some tasks (K. F. Miller, 1989).

On many tasks that *appear* to require general logical reasoning, children find their own strategy to measure, and they do so correctly. These solution strategies do not necessarily match the structural logic of the task. For example, children use intermediate measurements to compare two lengths without explicitly asking the transitivity question. They move a unit to measure the length of an object and do not worry about whether the length is being conserved. Finally, children of all developmental levels solve simple measurement tasks that do not appear to rely heavily on general reasoning.

In summary, children have an intuitive understanding of length on which to base reasoning about distance and length, but that reasoning develops considerably. They may have difficulty mapping words such as *long* onto the adult concept, instead assuming it means endpoint comparison (Schiff, 1983). They need to learn to coordinate and resolve perceptual and conceptual information when it conflicts. Finally, they need to learn to use measurement, understanding that units of lengths can be iterated along successive distances and these iterations counted to determine length. Thus, young children know early that properties such as length (as well as area, volume, mass, and weight) exist, but they do not initially know how to reason about these attributes or to measure them accurately. Nevertheless, when there are no perceptually conflicting cues, preschoolers can accurately compare objects directly.

Before kindergarten, many children lack measurement rules such as lining up an end when comparing the lengths of two objects (Piaget & Inhelder, 1967; Piaget, Inhelder, & Szeminska, 1960), although they can learn about such ideas. Even 5- to 6-year-olds, given a demarcated ruler, wrote in numerals haphazardly with little regard to the size of the spaces. Few used zero as a starting point, showing a lack of understanding of the origin concept. At age 4–5 years, however, many children can, with opportunities to learn, become less dependent on perceptual cues and thus make progress in reasoning about or measuring quantities. From kindergarten to Grade 2, children significantly improve in measurement knowledge (Ellis, Siegler, & Van Voorhis, 2000). They learn to represent length with a third object, using transitivity to compare the length of two objects that are not compared directly in a wider variety of situations (Hiebert, 1981). They can also use given units to find the length of objects and associate higher counts with longer objects (Hiebert, 1981, 1984). Some 5-year-olds, but most 7-year-olds, can use the concept of unit to make inferences about the relative size of objects; for example, if the numbers of units are the same, but the units are different, the total size is different (Nunes & Bryant, 1996). However, even 7-year-olds found tasks demanding conversion of units challenging.

Even kindergartners are fairly proficient with a conventional ruler and understand quantification in measurement contexts, but their skill decreases when features of the ruler deviate from the convention. Thus, measurement is supported by characteristics of measurement tools, but children still need to develop understanding of key measurement concepts. In one study, all K–2 children understood several measurement concepts. However, there were significant age differences on understanding concepts such as iterating a standard unit and the cardinality principle (Ellis, Siegler, & Van Voorhis, 2000). Children initially may iterate a unit leaving gaps between subsequent units or overlapping adjacent units (Horvath & Lehrer, 2000; Lehrer, 2003); therefore, it is a physical activity of placing units along a path in some manner, not an activity of covering the space/length of the object with no gaps. Furthermore, students often begin counting at the numeral "1" on a ruler (Lehrer, 2003) or, when counting paces heel-to-toe, start their count with the movement of the first foot, missing the first foot and counting the second foot as "one" (Lehrer, 2003; Stephan, Bowers, Cobb, & Gravemeijer, 2003). Students probably are not thinking about measuring as covering space. Rather, the numerals on a ruler (or the placement of a foot) signify when to start counting, not an amount of space that has already been cov-

ered (i.e., "one" is the space from the beginning of the ruler to the hash mark, not the hash mark itself). Many children initially find it necessary to iterate the unit until it "fills up" the length of the object and will not extend the unit past the endpoint of the object they are measuring (Stephan, Bowers, Cobb, & Gravemeijer, 2003). Finally, many children do not understand that units must be of equal size. They will even measure with tools subdivided into different size units and conclude that quantities with more units are larger (Ellis, Siegler, & Van Voorhis, 2000). This may be a deleterious side effect of counting, in which children learn that the size of objects does not affect the result of counting (Mix et al., 2002, although we disagree with the authors' claim that units are always "given" in counting situations—along with most teachers, Mix et al. do not consider counting situations, such as counting whole toy people constructed in two parts, top and bottom, when some are fastened and some are separated, cf. Sophian & Kailihiwa, 1998).

Children are also learning accumulation of distance. Some, for example, measured the lengths of objects by pacing heel to toe and counting their steps (Stephan, Bowers, Cobb, & Gravemeijer, 2003). As one child paced the length of a rug, the teacher stopped the child midmeasure and asked her what she meant by "8." Some children claimed that "8" signified the space covered by the eighth foot whereas others argued that it was the space covered from the beginning of the first foot to the end of the eighth. These latter children were measuring by accumulating distances. This type of interpretation may not appear until students are 9 or 10 years old (Clements, 1999c; Kamii & Clark, 1997). However, with meaningful instruction, children as young as 6 years old can learn to measure by accumulating distance (Stephan, Bowers, Cobb, & Gravemeijer, 2003).

Finally, young children are developing the foundational ideas of origin and relation between number and measurement. As Piagetian research indicated, they draw on their counting experiences to interpret their measuring activity, to which the "starting at 1" error may be related. If children understand measuring only as "reading the ruler," they may not understand this idea (Lehrer, 2003). Children also have to understand and apply counting concepts, including one-to-one correspondence and the cardinality principle.

Thus, significant development occurs in the early childhood years. However, the foundational length ideas are usually not integrated. For example, children may still not understand the importance of, or be able to create, equal size units, even into the primary grades (Clements, Battista, Sarama, Swaminathan, & McMillen, 1997; Lehrer, Jenkins, & Osana, 1998;

K.F. Miller, 1984). This indicates that children have not necessarily differentiated fully between counting discrete objects and measuring.

Education

Young children naturally encounter and discuss quantities in their play (Ginsburg, Inoue, & Seo, 1999). They first learn to use words that represent quantity or magnitude of a certain attribute. Facilitating this language is important not only to develop communication abilities, but for the development of mathematical concepts. Simply using labels such as "Daddy/Mommy/ Baby" and "big/little/tiny" helped children as young as 3 years to represent and apply higher order seriation abilities, even in the face of distracting visual factors, an improvement equivalent to a 2-year gain. Language provides an invitation to form comparisons and a method to remember the newly represented relational structure (Rattermann & Gentner, 1998). Thus, language can modify thought (cf. Vygotsky, 1934/1986). Along with progressive alignment, in which children are presented with easy literal similarity matches before difficult matches, language provides powerful scaffolding potential (Kotovsky & Gentner, 1996).

Next, children compare two objects directly and recognize equality or inequality, for example, of the length of two objects (Boulton-Lewis, Wilss, & Mutch, 1996). Following this, children can learn to measure, connecting number to length. Again, language, such as the differences between counting-based terms (e.g., "a toy" or "two trucks") and mass terms (e.g., "some sand"), can help children form relationships between counting and continuous measurement (Huntley-Fenner, 2001).

Traditionally, the goal of measurement instruction has been to help children learn the skills necessary to use a conventional ruler. In contrast, research and recent curriculum projects suggest that developing the conceptual foundation for such skills is critical to develop understanding and procedures. Kamii and Clark (1997) argued that comparing lengths is at the heart of developing the notions of conservation, transitivity, and unit iteration, but most textbooks do not have these types of tasks. Textbooks tend to ask questions such as "How many paper clips does the pencil measure?" rather than "How much longer is the blue pencil than the red pencil?" Although Kamii and Clark advocate beginning instruction by comparing lengths with nonstandard or standard units (not a ruler), they caution that such an activity is often done by rote.

Many recent curricula promote a sequence of instruction in which children compare lengths, measure with nonstandard units, incorporate the use of manipulative standard units, and measure with a ruler (Clements, 1999c; Kamii & Clark, 1997). The basis for this sequence is, explicitly or implicitly, Piaget et al.'s (1960) theory of measurement. The argument is that this approach motivates children to see the need for a standard measuring unit. Challenges to this sequence have arisen from studies suggesting that standard units and even rulers might lay a better foundation for initial understanding of measurement, with nonstandard and different standards units introduced later (Clements, 1999c).

Whatever the specific instructional approach taken, four implications can be abstracted from the research. First, measurement is not a simple skill, but rather a complex combination of concepts and skills that develops over years (Clements, Battista, Sarama, Swaminathan, & McMillen, 1997; Sophian, 2002; Steffe, 1991; Stephan, Bowers, Cobb, & Gravemeijer, 2003). Second, initial informal activities can establish the attribute of length and develop concepts such as *longer*, *shorter*, and *equal in length* and strategies such as direct comparison. Third, emphasis on solving real measurement problems and, in so doing, building and iterating units, as well as units of units, helps children develop strong concepts and skills. Fourth, children need to closely connect the use of manipulative units and rulers. When conducted in this way, measurement tools and procedures become tools for mathematics and tools for thinking about mathematics (Clements, 1999c; K. F. Miller, 1984, 1989). Well before first grade, children have begun the journey toward that end.

Area Measurement

Understanding of area measurement also involves learning and coordinating many ideas (Clements & Stephan, 2004). Eventually, children must understand that decomposing and rearranging shapes does not affect their area.

Some researchers report that preschoolers use only one dimension or one salient aspect of the stimulus (Bausano & Jeffrey, 1975; Maratsos, 1973; Mullet & Paques, 1991; Piaget, Inhelder, & Szeminska, 1960; Raven & Gelman, 1984; J. Russell, 1975; Sena & Smith, 1990). For example, 4- and 5-year-olds may match only one side of figures when attempting to compare their areas. Others claim that children can integrate more than one feature of a region but judge areas with additive combination, for example, making area judgments based on the longest single dimension (Mullet & Paques, 1991) or height + width rules (Cuneo, 1980; Rulence-Paques & Mullet, 1998). Children from 6 to 8 years use a linear extent rule, such as the diagonal of a rectangle. Only after this age do most children move to explicit use of multiplicative rules. This leads to our next concept.

In most of these studies, children did not interact with the materials. Doing so often changes their strategies. Children as young as 3 years are more likely to make estimates consistent with multiplicative rules when in a problem-solving setting (count out the right number of square tiles to cover a floor and put them in a cup) than when just asked to make a perceptual estimation (K. F. Miller, 1984).

Children of 5 to 6 years of age were more likely to use strategies consistent with multiplicative rules after playing with the stimulus materials (Wolf, 1995). Wolf argued that more complex rules are often used when people are more familiar with the materials involved in a task. This may be so; however, children did better when their manipulation followed their estimating sizes. Thus, small numbers of objects and familiarity with materials may be beneficial, but familiarity with the task (the conceptual goal) and the physical and cognitive actions applied to the materials may also encourage more accurate strategies such as scanning one length through another.

Although some researchers imply multiplicative thinking on the part of the child, we take the conservative position that there is little evidence of true 2-dimensional spatial structuring in these studies, although children may be using linear or additive strategies that are more consistent with the result of accurate multiplicative rules.

Another more accurate strategy is superimposition. Children as young as 3 years have a rudimentary concept of area based on placing regions on top of one another, but it is not until 5 or 6 years that their strategy is accurate and efficient. Asked to manipulate regions, preschoolers used superimposition instead of the less precise strategies of laying objects side-by-side or comparing single sides, both of which use one dimension at best in estimating the area (Yuzawa, Bart, & Yuzawa, 2000). Multiplicative thinking may be a result of internalization of such procedures as placing figures on one another, which may be aided by cultural tools (manipulatives) or scaffolding by adults (cf. Vygotsky, 1934/1986). For example, kindergartners who were given origami practice increased the spontaneous use of the procedure of placing one figure on another for comparing sizes (Yuzawa, Bart, Kinne, Sukemune, & Kataoka, 1999). Because origami practice includes the repeated procedure of folding one sheet into two halves, origami practice might facilitate the development of an area concept, which is related with the spontaneous use of the procedure.

Equal partitioning is the mental act of cutting two-dimensional space with a two-dimensional unit. Teachers often assume that the product of two lengths structures a region into an area of two-dimensional units for older students. However, the construction of a two-dimensional array from linear units is nontrivial. Young children often cannot partition and conserve area and instead use counting as a basis for comparing. For example, when it was determined that one share of pieces of a paper cookie was too little, 6 preschoolers cut one of that share's pieces into two and handed them both back (K. F. Miller, 1984).

Conclusion

Measurement is one of the principal real-world applications of mathematics. It bridges two critical realms of mathematics, geometry and real numbers. Number and operations are essential elements of measurement. The measurement process subdivides continuous quantities such as length to make them countable. Measurement provides a model and an application for both number and arithmetic operations. In this way, measurement helps connect the two realms of number and geometry, each providing conceptual support to the other.

Research indicates that measuring in general is more complex than learning the skills or procedures for determining a measure. Learning specific measurement concepts and skills is intrinsically important and also helps children differentiate between two basic types of quantity, discrete and continuous, which are often confused by young children (K. F. Miller, 1984; Piaget, Inhelder, & Szeminska, 1960). That is, representing exact discrete quantities by counting and more precise continuous quantities by measuring may be instrumental in children's development of the distinction between and attributes of these two basic categories (Mix, Huttenlocher, & Levine, 2002). This gradual integration and differentiation across domains supports the previously described principle of progressive hierarchization, which appears consistently in studies of measurement. For example, recall that children who observed agreement between measurement and direct comparison were more likely to use measurement later than were children who observed both procedures but did not have the opportunity to compare them (P. E. Bryant, 1982). Finally, we repeat the caveat that although young children can develop early ideas in measurement of various attributes, such as area and angle, there is little research on how valuable it would be to invest instructional time in these areas rather than others.

PATTERNS AND ALGEBRAIC THINKING

One key foundational ability for all topics in mathematics is the algebraic insight—the conscious understanding that one thing can represent another. Children develop this key idea at about 3 years of age in the simplest form; for example, as described earlier, understanding that a map or picture can represent another space. This ability develops considerably over the next several years, as children interact with symbols and with other people using these symbols in a variety of situations, from maps to spatial patterns to numerical patterns.

From this perspective, algebraic thinking can permeate much of the instruction of these main areas (and the smaller amount of research done on such thinking should not diminish its importance to the curriculum). One common, albeit often underappreciated, route to algebra begins with a search for patterns. Identifying patterns helps bring order, cohesion, and predictability to seemingly unorganized situations and allows one to generalize beyond the information directly available. Recognition and analysis of patterns are important components of the young child's intellectual development because they provide a foundation for the development of algebraic thinking.

Although pre-K children engage in pattern-related activities and recognize patterns in their everyday environment, research has revealed that an abstract understanding of patterns develops gradually during the early childhood years (B. A. Clarke & Clarke, 2004; A. Klein & Starkey, 2004). Of children entering their first year of school in Australia, 76% could copy a repeating color pattern, but only 31% could extend or explain it (B. A. Clarke & Clarke, 2004). In the pre-K years, children can learn to copy simple patterns. In kindergarten, they can learn to extend and create patterns. Further, children learn to recognize the relationship between patterns with nonidentical objects or between different representations of the same pattern (e.g., between visual and motoric, or movement, patterns). This is a crucial step in using patterns to generalize and to reveal common underlying structures. Through kindergarten and the primary grades, children must learn to identify the core unit (e.g., AB) that either repeats (ABABAB) or "grows" (ABAABAAAB) and then use it to generate both these types of patterns. Note that by the pre-K or kindergarten year, many children can name patterns with conventions such as "ABAB." This is another step to algebra, as it involves using variable names (letters) to label or identify patterns that involve different physical embodiments. Such naming helps children recognize

that mathematics focuses on underlying structure, not physical appearances. Little else is known, except that pattern identification is one of many elements of teaching visual literacy with positive long-term impact in the Agam program (Razel & Eylon, 1990).

In the later months of kindergarten, it is possible that with guidance children can begin analyzing numerical patterns, just as primary grade students can do. These students learn to find and extend numerical patterns— extending their knowledge of patterns to thinking algebraically about arithmetic (Baroody, 1993). Two central themes are generalizing and using symbols to represent mathematical ideas and to represent and solve problems (Carpenter, Fennema, Franke, Levi, & Empson, 1999). For example, children might generalize that when you add zero to a number the sum is always that number or that when you add three numbers it does not matter which two you add first (Carpenter, Fennema, Franke, Levi, & Empson, 1999). Thus, students in the primary grades can learn to formulate, represent, and reason about generalizations and conjectures, although their justifications do not always adequately validate the conjectures they create.

This body of research on young children's understanding of patterns may be used to establish developmentally appropriate learning trajectories for pattern instruction in early mathematics education. However, the main conclusion of this section is that far too little research has been conducted as a basis for curriculum development in the domain of patterns and algebraic thinking. This is equally true of the final content area, to which we now turn.

DATA ANALYSIS

There is a dearth of research on children's learning of data analysis in the preprimary years. From a small amount of research conducted with older students, we can describe in broad strokes that the developmental continuum for data analysis includes growth in classifying and counting and in data representations. (The development of classification per se, a major consideration in the domain, is described in the following section.) For example, children initially learn to sort objects and quantify their groups. They might sort a collection of buttons into those with one to four holes and count to find out how many they have in each of the four groups. To do this, they focus on and describe the attributes of objects, classifying according to those attributes, and quantify the resulting categories. Children eventually became capable of simultaneously classifying and counting, for example, counting the number of colors in a group of objects, as described previously.

After gathering data to answer questions, children's initial representations often do not use categories. Their interest in data is on the particulars (S. J. Russell, 1991). For example, they might simply list each child in their class and each child's response to a question. They then learn to classify these responses and represent data according to category. Finally, young children can use physical objects to make graphs (objects such as shoes or sneakers, then manipulatives such as connecting cubes), then picture graphs, then line plots, and finally bar graphs that include grid lines to facilitate reading frequencies (Friel, Curcio, & Bright, 2001).

As stated, however, research on data analysis before the primary years is sparse. Most consist of anecdotal reports. They are promising, but limited. There is some evidence that preschoolers can understand discrete graphs as representations of numerosity based on one-to-one correspondence (Solomon, 2003). Similarly, although 5- to 6-year-olds were more accurate than 4- to 5-year-olds on most, but not all graphing tasks, all children could interpret the graphs and use them to solve mathematical problems (J. Cooper, Brenneman, & Gelman, 2005). Children performed better than in previous research, possibly because they were provided examples and given feedback and because they were motivated by the task, graphing their progress toward gathering items for a scavenger hunt. Children performed better with discrete than continuous formats.

Across all the available literature, the consensus of a group attempting to create research-based standards for young children concluded that the main role for data analysis would lie in supporting the areas of number and spatial sense (Clements & Conference Working Group, 2004). For example, gathering data to answer a question or make a decision is potentially an effective means to develop applied problem solving and number sense. To do so, curricula and teachers might focus on one big idea: Classifying, organizing, representing, and using information to ask and answer questions.

MATHEMATICAL PROCESSES

We have woven discussion of mathematical processes throughout this chapter but make a few additional points focused on specific processes in this section.

Reasoning and Problem Solving

Reasoning is not a radically different sort of force operating against habit but the organization and coop-

eration of many habits, thinking facts together. Reasoning is not the negation of ordinary bonds, but the action of many of them, especially of bonds with subtle elements of the situation. . . . Almost everything in arithmetic should be taught as a habit that has connections with habits already acquired and will work in an organization with other habits to come. The use of this organized hierarchy of habits to solve novel problems is reasoning. (Thorndike, 1922, p. 193–194)

Thorndike accurately reflects turn-of-the-century perspectives that learning mathematics is like mental exercising, building the mind like a muscle. This perspective was perhaps most problematic in generalizing that reasoning could be explained as bonds and habits.

In contrast, others have thought that reasoning is merely logic, with uneducated people less knowledgeable of that domain. George Boole considered his research on pure mathematics and symbolic logic a central contribution to psychology, entitling his 1854 book, *An Investigation of the Laws of Thought*. The Piagetians agreed. "In short, reasoning is nothing more than the propositional calculus itself" (Inhelder & Piaget, 1958, p. 305). Instead, as has already been illustrated in previous topically oriented sections, young children's reasoning is a complex cognitive process that depends on local and global knowledge and procedures. Space constraints prevent our doing more than the briefest of overviews of young children's mathematical processes, including reasoning, problem solving, and communication, with the goal of illustrating their complexity, impressive extent, and importance.

As some vignettes throughout this chapter have illustrated, even very young children can reason and solve problems if they have a sufficient knowledge base, if the task is understandable and motivating, and if the context is familiar (Alexander, White, & Daugherty, 1997; DeLoache, Miller, & Pierroutsakos, 1998). Consistent with hierarchic interactionalism, there is no cogent evidence that these processes undergo substantive qualitative shifts. Instead, development occurs in interactive interplay among specific components of both general and specific knowledge and processes, together with an increasing effectiveness of these components.

Multiple types of reasoning processes can be identified in young children, although the various types overlap. Perceptual reasoning begins early, as similarity between objects is the initial relation from which children draw inferences, but it remains an important relation throughout development (DeLoache, Miller, & Pierroutsakos, 1998, on which this summary is largely based). Once concepts are formed, children as young as toddlers build structural-similarity relations, even between perceptually dissimilar objects, giving rise to analogies (Alexander, White, & Daugherty, 1997). Analogies in turn increase knowledge acquisition, which increases information-processing capacity through unitizing and symbolizing. Myriad represented relations give birth to reasoning with rules and symbolic relations, freeing reasoning from concrete experience and encouraging the appropriation of cultural tools that vastly expand the possible realms of reasoning. Again, then, we see the mutual development of concepts and skills, with knowledge catalyzing new processes, which in turn facilitate knowledge growth.

In solving problems, children may often use most of these types of reasoning, highlighting the overlap and even lack of real distinction (except for adult purposes of classification or research) among them. One illustration researchers use is the task, described in the "Models and Maps" section, in which children were shown a location of a toy on a scale model of a room and then asked to find the toy in the actual room (DeLoache, 1987). Perceptual similarity and concepts supported children's relating the model and room. Symbols were important—indeed, the picture was more useful to the youngest children because it served as a symbol. Children's inferring the toy's location was analogical, mapping a relationship in one space onto a similar relationship in the other.

Reasoning is, of course, used in solving problems. There are also additional strategies possessed by young children. Luke, 3 years old, watched his father unsuccessfully looking under the van for a washer that had fallen and suggested, "Why don't you just roll the car back, so you can find it?" Luke employed means-end analysis better than his father did. This strategy involves determining the difference between the current state and the goal and then taking action that reduces the difference between them, reasoning backwards from the goal to set subgoals. Means-end problem solving may emerge between 6 and 9 months when, for example, children learn to pull on a blanket to bring a toy into their reach.

We discussed in previous sections that even young children have multiple problem-solving strategies at their disposal and the ability to choose among them. Means-ends analysis is a general strategy, as are several others. Children know and prefer cognitively easier strategies. For example, in hill climbing, you reason forward from the current state in the direction of the desired goal (DeLoache, Miller, & Pierroutsakos, 1998). Trial and error, with light cognitive requirements, begins early, with Piagetian circular reactions.

These strategies develop in generality and flexibility during the toddler and preschool years, enabling children to address problems of increasing complex-

ity. For example, recall that kindergartners can solve a wide range of addition, subtraction, multiplication, and division problems when they are encouraged to represent the objects, actions, and relationships in those situations (Carpenter, Ansell, Franke, Fennema, & Weisbeck, 1993). The researchers argued that modeling is a parsimonious and coherent way of thinking about children's mathematical problem solving that is accessible to teachers and children. An Australian study supported the notion that kindergartners could learn to solve a variety of quite difficult word problems (Outhred & Sardelich, 1998). Children used concrete objects but were also required to draw and explain their representations, including the relationships among the elements. By the end of the year, they could pose and solve complex problems using a variety of strategies to represent aspects of the problem, including showing combining and partitioning groups and using letters and words to label elements of sets.

The ability to choose among alternative problem-solving strategies also emerges early in life; for example, in early spatial planning (Wellman, Fabricius, & Sophian, 1985). At 18 months or earlier, for example, children can use multiple strategies to pull a toy into reach with a third object (DeLoache, Miller, & Pierroutsakos, 1998). As an example in a different domain, recall that young children can make adaptive choices among arithmetic strategies, especially if the situations and strategies are meaningful for them (Siegler, 1993). In another study (Outhred & Sardelich, 1997), by the end of their kindergarten year, all children modeled arithmetic problems using concrete materials and accurately solved them. Their drawings showed a variety of strategies for representing the situations, including displaying properties (e.g., size), separating groups or crossing out items, partitioning sets, drawing lines to indicate sharing relationships, drawing array structures to show equal groups, and using letters and words to label items in collections. They could write their own problems and represent problem situations symbolically (they only struggled to accurately represent multistep problems with symbols). All this was accomplished despite the small amount of time the teacher engaged the children in problem solving.

Problem posing appears to be an effective way for children to express their creativity and integrate their learning (S. I. Brown & Walter, 1990; Kilpatrick, 1987; van Oers, 1994). Few empirical studies have been conducted that verify effects of problem posing, however, and none involved young children.

In summary, considering their minimal experience young children are impressive problem solvers. They are learning to learn and learning the rules of the "reasoning game." Research on problem solving and reasoning again reveals that children are more skilled, and adults less skilled, than conventionally thought. Finally, although domain-specific knowledge is essential, researchers should not fail to recognize that reasoning from domain-specific knowledge builds upon the basis of mindful general problem-solving and reasoning abilities that are evident from the earliest years.

Classification and Seriation

The complex relationship between seriation and classification and the development of number concepts (Piaget & Szeminska, 1952) has already been discussed. Piagetian theory held that these operations also underlie logic and reasoning (Piaget, 1964). For example, the claim that the abilities to seriate, construct a correspondence between two series, and insert an element into a series developed synchronically was an argument for the Piagetian operational theory of intelligence. The research literature on these constructs is vast (Clements, 1984a); we will only highlight some relevant findings here.

Although research militates against any simple, direct causal relationship, there is also evidence that these processes do play important roles in the development of mathematical reasoning and learning (cf. Kamii, Rummelsburg, & Kari, 2005; Piaget, 1971/1974). For example, children who do not acquire basic competencies in classification (*oddity*—which one is not like the others), seriation, and conservation by kindergarten do not perform as well in mathematics in later schooling (Ciancio, Rojas, McMahon, & Pasnak, 2001; Lebron-Rodriguez & Pasnak, 1977; Pasnak et al., 1987). Similarly, assessments of Piagetian reasoning tasks in kindergarten were related to children's mathematics concepts years later (Silliphant, 1983).

At all ages, children classify informally as they intuitively recognize objects or situations as similar in some way (e.g., differentiating between objects they suck and those they do not at 2 weeks of age) and eventually label what adults conceive of as classes. Often, functional relationships (the cup goes with a saucer) are the bases for sorting (Piaget, 1964; Vygotsky, 1934/1986). In addition, even infants place objects that are different (6 months), then alike (12 months), on some attribute together (Langer, Rivera, Schlesinger, & Wakeley, 2003). By 18 months, they form sets in which objects in each set are identical and objects in other set are different and, by 2 years, form sets with objects that are similar on some properties but not necessarily identical. Some 2-year-olds and all

3-year-olds will substitute elements to reconstruct misclassified sets (Langer, Rivera, Schlesinger, & Wakeley, 2003). These are often partial arrangements with fluid criteria; nevertheless, they play an essential role in number, through the unitizing process.

Not until age 3 can most children follow verbal rules for sorting. For example, told two simple rules for sorting pictures, children aged 36 months could sort regardless of the type of category, but children even a few months younger could not (Zelazo & Reznick, 1991). Having relevant knowledge and having fewer memory demands were not significant in this case.

In the preschool ages, many children learn to sort objects according to a given attribute, forming categories, although they may switch attributes during the sorting (Kofsky, 1966; Vygotsky, 1934/1986). The result may appear to reflect adult categorizations but often has a different conceptual basis, such as general resemblance (Vygotsky, 1934/1986, calls these "pseudoconcepts"). Preschoolers appear to encode examples holistically, distributing their attention nonselectively across many stimulus features, and then generalize to new stimuli on the basis of their overall similarity to the stored examples (Krascum & Andrews, 1993).

Not until age 5 or 6 years do children usually sort consistently by a single attribute and reclassify by different attributes. At this point, they can sort consistently and exhaustively by an attribute, given or created, and use the terms *some* and *all* in that context (Kofsky, 1966). Young children also learn seriation from early in life. Preverbal infants are able to make perceptual size comparisons and, from the age of 18 months or so, are able to respond to and use terms such as *big*, *small*, and *more* in ways that show they appreciate quantity differences (Resnick & Singer, 1993). By 2 or 3 years of age, children can compare numbers and number pairs on the basis of a common ordering relation (Bullock & Gelman, 1977). At 3 years, children can make paired comparisons, and 4-year-olds can make small series but do not seriate all objects (Clements, 1984a, Piaget & Inhelder reported that 55% make no attempt at seriation and 47% build partial uncoordinated series, Inhelder & Piaget, 1958; 1967). At age 5, 18% make no seriation, 61% build partial series; and 12% solve the problem, but only by trial and error. In a recent study, 43% of 5-year-olds put six lengths in order by length (A. Klein, Starkey, & Wakeley, 1999, but note that Piaget's lengths were more difficult to distinguish perceptually). Most 5-year-olds can insert elements into a series.

Children exhibit many strategies in seriation. Some choose the smallest (or largest) object to begin, then continue to select the next smallest (largest). Some place randomly selected objects in place. Others begin with the largest block, then select a proximate block, accepting it only if order is preserved, switching if it is not. Some analyses have claimed that those with less sophisticated, or even incorrect, strategies are not qualitatively different from seriators; they are just missing one or more rules (such as accepting any monotonic increase, rather than only unit differences).

There is also evidence that children can make transitive, deductive inferences, often considered the most difficult of the seriation tasks (if A is longer than B and B is longer than C, then A is longer than C). For example, 4-year-olds do so if they are trained to code and retrieve the relevant information (Trabasso, 1975). The authors claimed, as we did with reasoning and problem solving previously, that the cognitive processes of even young children are not that much different from those of adults, at least in the nature of the strategies used.

As mentioned, Piaget claimed that seriation and serial correspondence (between two series) and reconstruction of ordinal correspondences (a similarity that has become numerical) are solved at approximately the same time (Piaget & Szeminska, 1952). Young children make global comparisons with seriation or correspondences. They then develop the ability to construct series and correspondences intuitively but often fall back on perceptual correspondences. Only at the stage of concrete operations do they solve all the problems. They understand correspondences numerically; that is, the element n represents both the position n and the cardinal value n. Seriation and multiple seriation follow the development of classification and multiple classification, respectively. Generally, the developmental progressions within a domain (such as seriation) are supported, but few developmental concurrences across domains have been found (Clements, 1984a).

Studies involving both classification and seriation suggest that young children are more likely to abstract a property represented through literal comparisons. For example, both 4- and 6-year-olds can match three circles increasing in size with three squares that similarly increased in size, but only 6-year-olds could match the circles to three circles that increased in color saturation, that is, match seriations across dimensions (Kotovsky & Gentner, 1996). If the 4-year-olds are trained to mastery on the former, they can match across dimensions. Thus, as with counting and comparing number, we see a relational shift, in which children's early reliance on physical similarity develops into the ability to perceive purely relational commonalities (Kotovsky & Gentner, 1996).

Education

A strong argument can be made that developing reasoning and problem-solving abilities be the main focus in any attempt to help babies and toddlers develop in mathematics (see, e.g., Kamii, Miyakawa, & Kato, 2004). Encouraging language can support the growth of reasoning abilities (Rattermann & Gentner, 1998). Presenting children with easy literal similarity matches also helps them solve difficult analogical matches. Similarly, helping preschoolers learn the properties and relationships in a mathematical situation allows them to solve analogies of various types, such as geometric analogy items (Alexander, White, & Daugherty, 1997).

Problem solving can also be facilitated. Even for the very young, there are substantial benefits of varied situations, encouragement of diverse strategies, discussions, simple justifications, and prompts and hints as needed (DeLoache, Miller, & Pierroutsakos, 1998). Children make progress when they solve many problems over the course of years. Children as young as kindergartners, and perhaps younger, benefit from planned instruction, but not prescribed strategies, from a teacher who believes problem solving is important. They benefit from modeling a wide variety of situations (problem types, including addition, subtraction, and, at least from kindergarten on, multiplication and division) with concrete objects, from drawing a representation to show their thinking, from explaining and discussing their solutions, and from connecting representations (Carpenter, Ansell, Franke, Fennema, & Weisbeck, 1993; Outhred & Sardelich, 1997; van Oers, 1994). A final point is that concrete objects often make an important contribution but are not guaranteed to help (Baroody, 1989; Clements, 1999a). Children must see the structural similarities between any representation and the problem situation to use objects as tools. (These issues are discussed in Clements, 1999a.)

All young children need opportunities to achieve at least a minimal level of competence with classification (e.g., oddity) and seriation before they reach the primary grades. Simple teaching strategies can have a significant effect, especially for children with special needs. For example, Pasnak and colleagues (Ciancio, Rojas, McMahon, & Pasnak, 2001; Kidd & Pasnak, 2005; Lebron-Rodriguez & Pasnak, 1977) have taught seriation and classification via a simple learning set procedure, including demonstration, practice, and feedback with many varied concrete examples to a variety of children. These children, including blind and sighted, at-risk pre-K children and at-risk kindergarten children, and older blind children, increased their scores on instruments measuring intelligence quotient and mathematics achievement. (Other instructional approaches may also have these and other benefits as we shall see, Clements, 1984c; Kamii, Rummelsburg, & Kari, 2005.) Similarly, although solving oddity problems is difficult for children below the age of 6, telling the children the oddity rule quickly leads to success (DeLoache, Miller, & Pierroutsakos, 1998). Thus, children do not have difficulty following the rule, but inducing it. Game-like instruction may help children learn to induce simple rules. Again, evidence indicates that giving children clues about a rule or discussing it with them enables them to represent and follow it.

When should such instruction begin? Before 3 years of age, informal, child-centered experiences are indicated. Many 2.5-year-old children know a rule and have relevant conceptual knowledge but fail to use it to regulate their behavior. Seemingly impervious to efforts to improve their rule use, 32-month-olds could not label pictures in terms of appropriate categories, even with varieties of extra help in sorting, including feedback and reinforcement (Zelazo, Reznick, & Piñon, 1995). Improvements in sorting by rules may require emerging control over actions.

Materials to think with, to sort, or to order are important. Again, the meaningfulness of the representations and tasks is more important than the form of the materials; thus, for example, computer materials may be as or more useful than physical materials (Clements, 1999a). In one study, children working with computer manipulatives learned classification and other topics as well as children learning from physical materials, but only the computer group gained significantly on seriation (Kim, 1994). Further, the computer manipulatives provided children with a more interesting learning environment that generated more time on task.

Beyond the youngest ages, many educational approaches have been tried, including direct verbal instruction, contingent feedback, and modeling by peers just one level above the target child's assessed level. Researchers disagreed as to the theoretical position their results supported, but all methods were found to be effective in at least some situations (Clements, 1984a). For example, between 28% and 66% of kindergarten children in a transitional level learned all classification including class inclusion following six training sessions on various classification competencies (P. Miller, 1967). The researcher claimed that the training might have catalyzed cognitive reorganization.

Several educational approaches have also been found effective in teaching seriation, which is often easier to teach than classification or number concepts such as conservation (Clements, 1984a). These approaches include televised modeling; developmentally sequenced lessons; use of Montessori materials; a combination of corrective feedback, attention to task

stimuli, and cuing and cue fading; and discussion of children's own seriation strategies. Some studies reported that only children in Piagetian transition stages make substantial gains.

Focused, teacher-centered activities to develop these processes are thus effective, but there are reasons to believe they should be supplemented with occasions in which the processes are used to solve meaningful everyday problems for the child. As Piaget (1971/1974) stated,

> The child may on occasion be interested in seriating for the sake of seriating, in classifying for the sake of classifying, etc., but, in general, it is when events or phenomena must be explained and goals attained through an organization of causes that operations [logico-mathematical knowledge] will be used [and developed] most. (p. 17)

For example, although many types of activities may support the learning of classification, a pedagogical guideline of "classify with good causation" (Forman & Hill, 1984) indicates that children will benefit less from sorting shapes according to a teachers' directions and more from sorting three-dimensional objects to distinguish which will and will not roll down a ramp.

A final issue is whether training in Piagetian foundational areas such as classification and seriation aids the development of number. Evidence is mixed. Several studies show positive effects. For example, teaching classification and seriation to kindergarten children improved their performance on tests of number concepts (Lesh, 1972). Similarly, kindergartners taught classification, seriation, and conservation with simple teaching procedures (Pasnak, 1987) made twice the gains of children receiving "traditional mathematics instruction" on measures of general learning and reasoning ability and matched their gains on reading and mathematics achievement. These gains have held longitudinally (Pasnak, Madden, Malabonga, & Holt, 1996).

Findings similar to those of Lesh and Pasnak were reported for pre-K children. Four-year-old children were randomly assigned to one of three educational conditions for 8 weeks: logical foundations (classification and seriation), number (counting), and control (Clements, 1984c). After engaging in activities teaching classification, multiple classification, and seriation operations, the logical foundations group significantly outperformed the control group both on measures of conservation and on number concepts and skills. However, inconsistent with Piagetian theory, the number group also performed significantly better than the control group on classification, multiple classification, and seriation tasks as well as on a wide variety of number tasks. Further, there was no significant difference between the experimental groups on the logical operations test, and the number group significantly outperformed the logical foundations group on the number test. Thus, the transfer effect from number to classification and seriation was stronger than the reverse. The domains of classes, series, and number seem to be interdependent, but experiences in number have priority (Clements, 1984c). Note that some number and logical foundation activities were structurally isomorphic, so children received implicit experience with classification and seriation in number activities. For example, in one activity, children counted the blue cars, the red cars, and all the cars. Thus, in meaningful counting situations, the child may have a familiar cognitive tool with which to construct the logical structures required.

LEARNING AND TEACHING CONTEXTS

Early Childhood Mathematics Education

How much mathematics is done in early childhood settings? Observations of the full day of 3-year-olds' lives, across all settings, revealed remarkably few mathematics activities, lessons, or episodes of play with mathematical objects, with 60% of the children having no experience across 180 observations (Tudge & Doucet, 2004). Factors such as race-ethnicity, SES, and setting (home or child care) did not significantly affect this low frequency. Teacher reports from the large Early Childhood Longitudinal Study (ECLS, Hausken & Rathbun, 2004) indicate that kindergarten teachers spend an average of 39 minutes each session, 4.7 days per week, for a total of 3.1 hours each week of mathematics instruction. This is about half of what they spend on reading. Direct instruction was more commonly used with girls and higher SES, and "constructivist" approaches used more often with children of low ability. Each week teachers engage children in manipulative activities 1–2 times and problem-solving and practice (worksheet) activities 2–3 times. Individual skills were not related to the frequency of exposure to different types of mathematics problems. The authors warned that uniformity of activities across the classrooms may account for these findings. Especially given contrasting data from smaller but more focused studies described later, additional caveats concerning the limitation of post hoc teacher self-report data and the lack of specific information on curricula are warranted.

Research is beginning to describe corresponding preschool practices. The large National Center for Early Development and Learning (NCEDL) studies report that children are not engaged in learning or constructive activities during a large proportion of the pre-K day (Early et al., 2005). About 8% of the day on the average involves mathematics activities. A survey (Sarama, 2002; Sarama & DiBiase, 2004) asked teachers from a range of preschool settings at what age children should start large-group mathematics instruction. Family and group care providers chose ages 2 or 3 most often, whereas the other group felt large-group instruction should not start until age 4. Most teachers professed to use manipulatives (95%), number songs (84%), basic counting (74%), and games (71%); few used software (33%) or workbooks (16%). They preferred children to "explore math activities" and engage in "open-ended free play" rather than participate in "large group lessons" or be "doing math worksheets." When asked about mathematics topics, 67% taught counting, 60% sorting, 51% numeral recognition, 46% patterning, 34% number concepts, 32% spatial relations, 16% making shapes, and 14% measuring. Geometry and measurement concepts were the least popular.

A small observational study of four pre-K teachers from two settings revealed that little mathematics was presented in any of the classrooms, either directly or indirectly (Graham, Nash, & Paul, 1997). Researchers observed only one instance of informal mathematical activity with concrete materials and few instances of informal or formal mathematics teaching. Teachers stated that they believed that mathematics was important and that they engaged in mathematical discussions. Apparently the selection of materials and activities such as puzzles, blocks, games, songs, and finger plays constituted mathematics for these teachers.

Several findings support the traditional emphasis on play and child-centered experiences. In one study, children made more progress overall and specifically on mathematics when they attended child-initiated, compared to strictly academically oriented, programs (Marcon, 1992). There was some evidence that these children's grades were higher at the end of elementary school (6th, but not 5th, grade; Marcon, 2002). This may be consistent with practices in some Asian countries. For example, Japanese pre-K and kindergarten education places emphasis on social-emotional rather than academic goals. Preschoolers engage in free play most of the day. Parents interact with their children in mathematics, usually in real-life, such as counting down elevator numbers. Few mention workbooks (Murata, 2004). However, Macron's studies have been criticized on methodological grounds (Lonigan, 2003) and are correlational. Further, exposure to mathematics instruction explained a substantive portion of the greater gains of young Chinese compared to U.S. children (Geary & Liu, 1996).

Although fewer studies have been conducted specifically of small- or whole-group pedagogical strategies, evidence suggests these approaches too can be effective. Small-group work can significantly increase children's scores on tests aligned with that work (A. Klein & Starkey, 2004; A. Klein, Starkey, & Wakeley, 1999). Small-group activities can also transfer to knowledge and abilities that have not been taught (Clements, 1984c). Combinations of whole-group, small-group, everyday, and computer activities were employed in the *Building Blocks* project (Clements & Sarama, 2004a, in press). Large effect sizes supported the strategy of designing a curriculum built on comprehensive research-based principles, including an emphasis on hypothesized learning trajectories (Clements & Sarama, in press). Finally, observations in countries that use far more whole-group instruction with young children suggest its advantages may be overlooked in the United States. For example, the teacher-directed, whole-class Korean approach provides a positive, nurturing environment that offers children the opportunity to develop essential preacademic skills (French & Song, 1998).

In summary, in this new educational arena, several approaches, if performed in high-quality settings, can be effective. Most successful pedagogical strategies, even those with focused goals, include play or play-like activities. All approaches have a shared core of concern for children's interest and engagement as well as content matched to children's cognitive level. Although some studies support general, play-oriented approaches, learning mathematics seems to be a distinct process, even in preschool (Day, Engelhardt, Maxwell, & Bolig, 1997), and approaches focused on mathematics have been successful. We return to this issue in the following section.

Issues of Equity and Individual Differences

Children who live in poverty and who are members of linguistic and ethnic minority groups demonstrate significantly lower levels of achievement (Bowman, Donovan, & Burns, 2001; Brooks-Gunn, Duncan, & Britto, 1999; P. F. Campbell & Silver, 1999; Denton & West, 2002; Entwisle & Alexander, 1990; Halle, Kurtz-Costes, & Mahoney, 1997; Mullis et al., 2000; Natriello, McDill, & Pallas, 1990; Rouse, Brooks-Gunn, & McLanahan, 2005; Secada, 1992; Sylva, Melhuish, Sammons, Siraj-Blatchford, & Taggart, 2005). Ethnic gaps widened in the 1990s (Lee, 2002). The achievement

gaps have origins in the earliest years, with low-income children possessing less extensive mathematics knowledge than middle-income children of pre-K and kindergarten age (Arnold & Doctoroff, 2003; Denton & West, 2002; Ginsburg & Russell, 1981; Griffin, Case, & Capodilupo, 1995; Jordan, Huttenlocher, & Levine, 1992; Saxe, Guberman, & Gearhart, 1987; Sowder, 1992). The SES gap is broad and encompasses several aspects of mathematical knowledge: numerical, arithmetic, spatial/geometric, patterning, and measurement knowledge (Clements, Sarama, & Gerber, 2005; A. Klein & Starkey, 2004). The reason for this gap seems to be that children from low-income families receive less support for mathematics development in their home and school environments (Blevins-Knabe & Musun-Miller, 1996; Holloway, Rambaud, Fuller, & Eggers-Pierola, 1995; Saxe, Guberman, & Gearhart, 1987; Starkey et al., 1999). Also, public pre-K programs serving low-income, compared to those serving higher income, families provide fewer learning opportunities and supports for mathematical development, including a narrower range of mathematical concepts (D. M. Bryant, Burchinal, Lau, & Sparling, 1994; Farran, Silveri, & Culp, 1991).

Differences in specific aspects of young children's mathematical knowledge have been reported in two types of comparisons. First, there are cross-national differences. Some mathematical knowledge is more developed in East Asian children than in American children (Geary, Bow-Thomas, Fan, & Siegler, 1993; Ginsburg, Choi, Lopez, Netley, & Chi, 1997; K. F. Miller, Smith, Zhu, & Zhang, 1995; Starkey et al., 1999). Second, there are differences related to socioeconomic status. Some mathematical knowledge is more developed in children from middle-income, compared to lower-income, families (Clements, Sarama, & Gerber, 2005; Griffin & Case, 1997; Jordan, Huttenlocher, & Levine, 1992; Kilpatrick, Swafford, & Findell, 2001; Saxe, Guberman, & Gearhart, 1987; Starkey & Klein, 1992). The key factors in one study were the educational level attained by the child's mother and the level of poverty in the child's neighborhood (Lara-Cinisomo, Pebley, Vaiana, & Maggio, 2004). These are distinct factors, with income having a direct effect on the child and an effect mediated by the parents' interaction with children (e.g., higher, compared to lower, income parents providing more support for problem solving, Brooks-Gunn, Duncan, & Britto, 1999; Duncan, Brooks-Gunn, & Klebanov, 1994). Low-income, compared to middle-income parents, believe that mathematics education is the responsibility of the preschool and that children cannot learn aspects of mathematics that research indicates they can learn (Starkey et al., 1999). In addition, low-income families more strongly endorsed a skills perspective than middle-income families, and the skills and entertainment perspectives were not predictive of later school achievement, as was the "math in daily living" perspective adopted by more middle-income parents (Sonnenschein, Baker, Moyer, & LeFevre, 2005). These deleterious effects are more prevalent and stronger in the United States than other countries and stronger in early childhood than for other age ranges.

As an example of this gap, one large survey of U.S. kindergartners found that 94% of first-time kindergartners passed their Level 1 test (counting to 10 and recognizing numerals and shapes) and 58% passed their Level 2 test (reading numerals, counting beyond 10, sequencing patterns, and using nonstandard units of length to compare objects). However, 79% of children whose mothers had a bachelor's degree passed the Level 2 test, compared to 32% of those whose mothers had less than a high school degree. Large differences were also found between ethnic groups on the more difficult Level 2 test (NCES, 2000). Differences appear even in the preschool years. Another study showed that mathematical knowledge is greater in 3- and 4-year-old Chinese children than in American middle-class children and greater in American middle-class children than in 3- and 4-year-olds from low-income families (Starkey et al., 1999). As a caveat, what mechanisms account for all these cross-national differences is unknown. Some factors seem to be nationally situated, such as teacher' knowledge, formal teaching practice, and curriculum standards. Others are transnational, such as language differences, and yet others may be cultural without reflecting national boundaries, such as family values (J. Wang & Lin, 2005). Japanese kindergartners perform better in mathematics than those from the United States, but neither families nor schools in Japan emphasize academics for this age group (Bacon & Ichikawa, 1988). Their lower but perhaps more realistic expectations and reliance on informal instruction at the child's level, including eliciting interest and providing examples rather than direct teaching of procedures, may account for their success.

Children from low-income families show specific difficulties. They do not understand the relative magnitudes of numbers and how they relate to the counting sequence (Griffin, Case, & Siegler, 1994). They have more difficulty solving addition and subtraction problems. Working-class children in the U.K. are a year behind in simple addition and subtraction as early as 3 years of age (Hughes, 1981). Similarly, U.S. low-income children begin kindergarten behind middle-income children, and, although they progressed at the same rate on most tasks, they ended behind and made no progress in some tasks. For example, although they performed adequately on nonverbal arithmetic tasks, they made no progress over the entire kindergarten year on arithmetic story problems (Jordan, Kaplan, Oláh, & Locuniak, 2005). Further, lower class children were more likely to show a "flat" growth curve for the year.

Similarly, lower income preschoolers lag behind higher income peers in addition, comparing, and subitizing (Clements, Sarama, & Gerber, 2005), as well as on the earliest form of subitizing, spontaneous recognition of numerosity (Hannula, 2005). Further, the subtests on which the gaps were significant were those that assess more sophisticated mathematical concepts and skills, such as composing numbers or shapes (Clements, Sarama, & Gerber, 2005). They often lack foundational abilities to classify and seriate (Pasnak, 1987). Older children entering first grade showed a smaller effect of familial factors on computation than on mathematics concepts and reasoning. Majority-minority contrasts were small, but parents' economic and psychological resources were strong influences (Entwisle & Alexander, 1990).

Into kindergarten and the primary grades, lower income children more than middle class children, use less adaptive and even maladaptive strategies, probably revealing a deficit in intuitive knowledge of numbers and different strategies (Griffin, Case, & Siegler, 1994; Siegler, 1993). Most 5- and 6-year-old low-income children are unable to answer the simplest arithmetic problems, where most middle-income kindergartners could do so (Griffin, Case, & Siegler, 1994). In one study, 75% of children in an upper-middle-class kindergarten were capable of judging the relative magnitude of two different numbers and performing simple mental additions, compared to only 7% of lower income children from the same community (Case, Griffin, & Kelly, 1999; Griffin, Case, & Siegler, 1994). As another example, about 72% of high-, 69% of middle- and 14% of low-SES groups can answer the orally presented problem, "If you had 4 chocolate candies and someone gave you 3 more, how many chocolates would you have altogether?" Low-income children often guess or use other maladaptive strategies such as simple counting (e.g., 3 + 4 = 5). They often do this because they lack knowledge of strategies and understandings of why they work and what goal they achieve (Siegler, 1993) However, given more experience, lower income children use multiple strategies with the same accuracy, speed, and adaptive reasoning as middle-income children.

Thus, there is an early developmental basis for later achievement differences in mathematics: Children from different sociocultural backgrounds are provided different foundational experiences (Starkey et al., 1999). This is important, as knowledge of mathematics in preschool predicts later school success (Jimerson, Egeland, & Teo, 1999; H. W. Stevenson & Newman, 1986; Young-Loveridge, 1989). Specific quantitative and numerical knowledge is more predictive of later achievement than are tests of intelligence or memory abilities (Krajewski, 2005). Those with low mathematics knowledge in the earliest years fall farther behind each year (Arnold & Doctoroff, 2003; Aunola, Leskinen, Lerkkanen, & Nurmi, 2004; R. J. Wright, Stanger, Cowper, & Dyson, 1994).

Research shows that providing educational support to children at risk results in greater school readiness upon entry into kindergarten (Bowman, Donovan, & Burns, 2001; Magnuson & Waldfogel, 2005; Shonkoff & Phillips, 2000), because such support helps young children develop a foundation of informal mathematics knowledge (Clements, 1984c). Early knowledge has been shown to support later school mathematics achievement, and lack of it places minorities on a path away from engagement in mathematics and science (F. A. Campbell, Pungello, Miller-Johnson, Burchinal, & Ramey, 2001; Oakes, 1990). Longitudinal research indicates that attendance in center-based (but not other types of) care in the pre-K year is associated with higher mathematics scores in kindergarten and (to a lesser extent) in first grade (Turner & Ritter, 2004), and that achievement in preschool is related to differences in elementary school achievement for Hispanic children (Shaw, Nelsen, & Shen, 2001). In another study, African American, Hispanic, and female children who attended an intervention preschool program had a significantly greater probability of achieving high scores in fourth grade than their peers who did not attend school (Roth, Carter, Ariet, Resnick, & Crans, 2000). Childcare in general can help, with greater number of hours in childcare correlating with greater quantitative skills in low-income children (Votruba-Drzal & Chase, 2004). High-quality preschool experience is predictive of later school success in mathematics (Broberg, Wessels, Lamb, & Hwang, 1997; F. A. Campbell, Pungello, Miller-Johnson, Burchinal, & Ramey, 2001; Peisner-Feinberg et al., 2001). However, results are sometimes moderate, and there is not always a significant effect for quality.

In comparison, children living in poverty and those with special needs increase in mathematics achievement after high-quality interventions *focused on mathematics* (P. F. Campbell & Silver, 1999; Fuson, Smith, & Lo Cicero, 1997; Griffin, 2004; Griffin, Case, & Capodilupo, 1995; Ramey & Ramey, 1998), which can be sustained into first (Magnuson, Meyers, Rathbun, & West, 2004) *to* third grade (Gamel-McCormick & Amsden, 2002). For example, the *Rightstart* (now *Number Worlds*) program (Griffin, Case, & Siegler,

1994), featuring games and active experiences with different models of number, led to substantial improvement in children's knowledge of number. Across five studies, almost all children failed the number pretest, and the majority in the comparison groups failed the posttest as well, whereas the vast majority of those in the program passed the posttest. Children in the program were better able to employ reasonable strategies and were able to solve arithmetic problems even more difficult than those included in the curriculum. Program children also passed five far-transfer tests that were hypothesized to depend on similar cognitive structures (e.g., balance beam, time, money). The foundation these children received supported their learning of new, more complex mathematics through first grade. In a 3-year longitudinal study in which children received consistent experiences through the grades from kindergarten through primary, children gained and surpassed both a second low-SES group and a mixed-SES group who showed a higher initial level of performance and attended a magnet school with an enriched mathematics curriculum. The children also compared favorably with high-SES groups from China and Japan (Case, Griffin, & Kelly, 1999). (A caution is that other research indicates that certain components of the curriculum are difficult to implement, Gersten et al., 2006.)

A series of studies have similarly indicated that the *Building Blocks* curriculum (Clements & Sarama, 2007) significantly and substantially increases the mathematics knowledge of low-SES preschool children. Formative, qualitative research indicated that the curriculum raised achievement in a variety of mathematical topics (Clements & Sarama, 2004a; Sarama & Clements, 2002). Summative, quantitative research confirmed these findings, with effect sizes ranging from .85 (Cohen's *d*) for number to 1.47 for geometry in a small-scale study (Clements & Sarama, in press). In a larger study involving random assignment of 36 classrooms, the *Building Blocks* curriculum increased the quantity and quality of the mathematics environment and teaching and substantially increased scores on a mathematics achievement test. The effect size compared to the control group score was very large ($d = 2.8$), and the effect size compared to a group receiving a different, extensive mathematics curriculum was large ($d = 1.3$). There was no significant interaction by program type (Head Start vs. public preschool).

The *Number Worlds* and *Building Blocks* programs share several characteristics. Both use research to include a comprehensive set of cognitive concepts and processes (albeit *Number Worlds* only does so in the domain of numbers). Both curricula are based on developmentally sequenced activities and help teachers become aware of, assess, and remediate based on those sequences (projects around the world that use research-based developmental progressions help raise achievement of all children, e.g., G. Thomas & Ward, 2001; R. J. Wright, Martland, Stafford, & Stanger, 2002). Both use a mix of instructional methods.

Higher quality programs result in learning benefits upon entering elementary school, including in mathematics (Fuson, 2004; Griffin, 2004; Karoly et al., 1998). Unfortunately, most American children are not in high-quality programs, much less in programs that use research-based mathematics curricula (Hinkle, 2000). Further, children whose mothers had college degrees were nearly twice as likely to be in higher quality, center-based care as those whose mothers had not completed high school (Magnuson, Meyers, Rathbun, & West, 2004). As another example, Hispanic children are less likely to be enrolled in preschool, even though the benefit they receive from attending is double that for non-Hispanic White children in mathematics and prereading (Loeb, Bridges, Bassok, Fuller, & Rumberger, in press). Similarly, children from extremely poor families show a .22 *SD* (compared to .10 *SD* average for all children) advantage in mathematics concepts compared to peers who remain at home. Further, these average gains would be predictably higher if focused mathematics programs were in place. Thus, children from low-income and minority communities are provided fewer educational opportunities. Implications include providing high-quality programs for all children and their families. Even controlling for parents' occupations and education, family practices such as playing with numbers at home have significant impact on children's mathematical development (Sylva, Melhuish, Sammons, Siraj-Blatchford, & Taggart, 2005).

There seems to be a contradiction between two pictures of the mathematical knowledge and competencies of children of different SES groups. On the one hand, the evidence suggests a substantive and widening gap. On the other hand, there are few or no differences between low- and middle-income children in the amount of mathematics they exhibit in their free play (Ginsburg, Ness, & Seo, 2003; Seo & Ginsburg, 2004). The authors often conclude that low-income children are more mathematically competent than expected. This contradiction may have several explanations. Low-income children may not have the same kind of informal opportunities at home (although there is only weak support for this hypothesis, Tudge & Doucet, 2004). Researchers often observe them in a school setting, and there is evidence that low-income families provide less support for mathematical thinking (Thirumurthy, 2003). Thus, it may be that they

evince mathematics in their play in school but are still engaged in such play far less than higher income children. Another explanation is that children have not been provided with the opportunities to reflect on and discuss their premathematical activity. There are huge, meaningful differences in the amount of language in which children from different income levels engage (Hart & Risley, 1995, 1999). Low-income children may engage in premathematical play but be unable to connect this activity to school mathematics because to do so requires the children to bring the ideas to an explicit level of awareness. This is supported by the finding that a main difference between children is not their ability to perform with physical objects, but to solve problems verbally (Jordan, Huttenlocher, & Levine, 1992) or explain their thinking (Sophian, 2002). Consider a child who turned 4 years of age. When asked to solve "how much is ten and one more," she used physical blocks, added 1 to 10, and answered, "eleven." Five minutes later, asked several times using the same wording, "How many is two and one more," the child did not respond and, asked again, said, "fifteen"in a couldn't-care-less tone of voice (Hughes, 1981, pp. 216–217).

In summary, although there is little direct evidence on this, we believe the pattern of results suggest that, although low-income children have premathematical knowledge, they do lack important components of mathematical knowledge. They lack the ability—because they have been provided less support to learn—to connect their informal premathematical knowledge to school mathematics. Children must learn to mathematize their informal experiences, abstracting, representing, and elaborating them mathematically, and using mathematical ideas and symbols to create models of their everyday activities. This includes the ability to generalize, connecting mathematical ideas to different situations and using the ideas and processes adaptively. In all its multifaceted forms, they lack the language of mathematics.

We believe the significance of this conclusion needs to be highlighted. Some authors find that low-income children perform similarly to middle-income children on nonverbal calculation tasks, but significantly worse on verbal calculation tasks (Jordan, Hanich, & Uberti, 2003), consistent with our review. These authors also state their agreement with others (Ginsburg & Russell, 1981) that the differences are associated with language and approaches to problem solving rather than "basic mathematical abilities" (p. 366). We prefer to call these *not* "basic mathematical," but rather foundational abilities. We emphasize that mathematization—including redescribing, reorganizing, abstracting, generalizing, reflecting upon, and giving language to that which is first understood on an intuitive and informal level—*is requisite to basic mathematical ability*. This distinction goes beyond semantics to involve a definition of the construct of mathematics and then to critical ramifications for such practical decisions as allocation of resources on the basis of equity concerns.

Gender equity also remains a concern. In some cultures, females are socialized to view mathematics as a male domain and themselves as having less ability. Teachers show more concern when boys, rather than girls, struggle. They call on and talk to boys more than girls. Finally, they believe success in mathematics is due to high ability more frequently for boys than girls and view boys as the most successful students in their class. All these may unintentionally undermine girls' achievement motivation (Middleton & Spanias, 1999). In more than one study, boys were more likely than girls to score in the lowest and highest ranges in mathematics (Callahan & Clements, 1984; Rathbun & West, 2004). In addition, there was evidence of a faster growth rate for high-achieving boys (Aunola, Leskinen, Lerkkanen, & Nurmi, 2004). Reasons for this are still unclear, but there are practical ramifications. There are also some indications that boys outperform girls as early as kindergarten on some tasks, such as number sense, estimation, and nonverbal estimation, all of which may have a spatial component (Jordan, Kaplan, Oláh, & Locuniak, 2005). However, in the U.K., preschool girls scored higher than boys (Sylva, Melhuish, Sammons, Siraj-Blatchford, & Taggart, 2005). Thus, the problems are complex, and there are distinct concerns about boys and girls.

Some children show specific learning mathematical difficulties (MD) at young ages. Unfortunately, they are often not identified or categorized broadly with other children as "developmentally delayed." This is especially unfortunate because focused mathematical interventions at early ages are effective (Dowker, 2004; Lerner, 1997).

Foundational abilities in subitizing, counting and counting strategies, simple arithmetic, and magnitude comparison are critical for young children with MD (Aunola, Leskinen, Lerkkanen, & Nurmi, 2004; Geary, Hoard, & Hamson, 1999; Gersten, Jordan, & Flojo, 2005, note that these studies often ignore mathematics topics other than number). Some children with MD may have difficulty keeping one-to-one correspondence when counting or matching. They may need to physically grasp and move objects, as grasping is an earlier skill than pointing in development (Lerner, 1997). They often understand counting as a rigid, mechanical activity (Geary, Hamson, & Hoard, 2000). These children also may count objects in small sets one-by-one for long after their peers are strategically subitizing these

amounts. Emphasizing their ability to learn to subitize the smallest number, perhaps representing them on their fingers, may be helpful. (Children who have continued difficulty perceiving and distinguishing even small numbers are at risk for severe general mathematical difficulties, Dowker, 2004).

Some children may have difficulty with magnitude comparisons (e.g., knowing which of two digits is larger) and in learning and using more sophisticated counting and arithmetic strategies (Gersten, Jordan, & Flojo, 2005). Their lack of progress in arithmetic, especially in mastering arithmetic combinations, causes consistent problems; thus, early and intensive intervention is indicated.

In another domain, some children have difficulty with spatial organization across a wide range of tasks. Children with certain mathematics learning difficulties may struggle with spatial relationships, visual-motor and visual-perception, and sense of direction (Lerner, 1997). They may not perceive a shape as a complete and integrated entity as children without learning disabilities do. For example, a triangle may appear to them as 3 separate lines, as a rhombus, or even as an undifferentiated closed shape (Lerner, 1997). Children with different brain injuries show different patterns of competence. Those with right hemispheric injuries have difficulty organizing objects into coherent spatial groupings, whereas those with left hemispheric injuries have difficulty with local relations within spatial arrays (Stiles & Nass, 1991). Teaching with learning trajectories based on the developmental sequences described here is even more important for children with learning disabilities, as well as children with other special needs.

Spatial weakness may underlie children's difficulties with numerical magnitudes (e.g., knowing that 5 is a greater than 4, but only by a little, whereas 12 is a lot greater than 4) and rapid retrieval of numeral names and arithmetic combinations (Jordan, Hanich, & Kaplan, 2003). These children may not be able to manipulate visual representations of a number line.

Children diagnosed as autistic need structured interventions from the earliest years. They must be kept engaged with their world, including mathematics. Many children with autism are visually oriented. Manipulatives and pictures can aid children's learning of most topics, in geometry, number, and other areas. Children benefit from illustrating even verbs with dramatizations. About a tenth of children with autism exhibit savant (exceptional) abilities, often spatial in nature, such as art, geometry, or a particular area of arithmetic. These abilities are probably due not to a mysterious talent, but from massive practice, the reason and motivation for which remains unknown (Ericsson, Krampe, & Tesch-Römer, 1993).

In conclusion, there are substantial inequities in mathematics experiences in the early years. Some children not only start behind but also begin a negative and immutable trajectory in mathematics (Case, Griffin, & Kelly, 1999). Low mathematical skills in the earliest years are associated with a *slower growth rate*— children without adequate experiences in mathematics start behind and lose ground every year thereafter (Aunola, Leskinen, Lerkkanen, & Nurmi, 2004). Interventions should start in pre-K and kindergarten (Gersten, Jordan, & Flojo, 2005). There is substantial evidence that this lag can be avoided or ameliorated, but also evidence that our society has not taken the necessary steps to do either. Without such interventions, children in special need are often relegated to a path of failure (Baroody, 1999; Clements & Conference Working Group, 2004; Jordan, Hanich, & Uberti, 2003; B. Wright, 1991; R. J. Wright, Stanger, Cowper, & Dyson, 1996).

FINAL WORDS

Research in early mathematics has and continues to lead the way in investigating fundamental issues in epistemology, psychology, and education. Number and space are common topics of investigations by psychologists engaged in a debate among empiricist, nativist, and interactionalist positions (Haith & Benson, 1998; Spelke, 2000). Further, researchers in mathematics education have emphasized the need to specify children's abilities to learn, and learn to learn, as well as the ecological influences on such learning, from sociocultural background to school learning experiences. There is as yet no consensus about exactly when knowledge begins, what it consists of, how it manifests itself, what causes it to emerge, or how it changes with growth and experience in the earliest years of life. Furthermore, reminiscent of introspective psychology of a century ago (Anderson, 2000), researchers' empirical evidence is consistent with their own theoretical orientation with uncomfortable frequency.[7] This is another reason why mathematics education research that emphasizes curriculum research will be so valuable (Clements, in press). Because it is result-centered, rather than theory-centered, curriculum research minimizes seductive theory-confirming strategies that tend to insidiously replace the intended theory-testing strategies and maximizes strategies that attempt to produce specified patterns of data and thus mitigate confirmation bias, stimulating creative development of theory (Greenwald, Pratkanis, Leippe, & Baumgardner, 1986). Nevertheless, we already have a growing body of knowledge that is at the least suggestive about early competencies on which to build mathematics learning.

Through more than a century, research has moved from a cautious assessment of the number competencies of children entering school, to a Piagetian position that children were not capable of true numeric thinking, to the discovery of infant sensitivity to mathematical phenomena, to the present debate about the meaning of these contradictions and an attempt to synthesize ostensibly opposing positions. Frequent in the last two of these phases is the paradox of contradictions to Piagetian findings and confirmation of the basic constructivist Piagetian framework, the influence of which has been so fundamental that even substantive new theories were born in reactions to the monumental Piagetian corpus.

Often due to their reactions against certain Piagetian functions and the nature of their methods, researchers tended to create increasingly specialized and local theories. Recently, similarities among those theories have laid the groundwork for new hybrid theories that provide general frameworks but are replete with local detail, including specific innate predispositions as well as competencies, conceptualizations, and strategies along developmental progressions. These progressions are at various levels of detail, both topical (e.g., "number" vs. "counting" vs. "specific counting competencies and errors at the *nth* level of development") and social-psychological (e.g., a broad level for goals, deeper for teachers, deeper for curriculum developers, deeper for researchers).

To develop such theories, researchers need to synthesize psychological and clinical approaches with others, such as those used in mathematics education research. The theoretical framework of *hierarchic interactionalism* that we proposed here is one such attempt. This framework connects *initial bootstraps* with a gradual development of conscious knowledge of systems of mathematical knowledge, depending on educational experiences. Thus, intellectual development results from an interplay between internal and external fac-

---

[7] A similar, contributing problem is the theoretical and empirical insularity of the various research communities. Mathematics education researchers are aware of some, but not all, of research from other fields, but those in various branches of psychology are not aware of relevant work in mathematics education research, even making discoveries or inventing "new" research methods that have a long history in mathematics education.

tors, including innate competencies and dispositions, maturation, experience with the physical environment, sociocultural experiences (as opposed to only "social transmission"), and self-regulatory processes (reflective abstraction). To be useful educationally, the roles and interactions of each of these factors must be described in detail within specific domains. The field of early mathematics is fortunate to include several research programs in which the development of such descriptions is underway. Similar proposals and increased clinical, longitudinal, and educational studies of the usefulness of the theories are needed. Studies and specific theories must avoid eclecticism (Newcombe & Huttenlocher, 2000), in which "everything matters," and garner empirical evidence that details development and the factors that influence it. Evaluating the theories' usefulness requires well-designed studies that connect specific pedagogical processes and contexts to outcomes to identify moderating and mediating variables and to compare the immediate and long-term outcomes of different approaches. We caution researchers interested in studying young children's mathematical thinking against restricting their attention to numbers and arithmetic. The mathematical world of young children is much richer.

REFERENCES

Ackoff, R. L. (1989). From data to wisdom. *Journal of Applied Systems Analysis, 16*, 3–9.

Acredolo, L. P. (1978). The development changes of spatial orientation in infancy. *Developmental Psychology, 14*, 224–234.

Acredolo, L. P., Adams, A., & Goodwyn, S. W. (1984). The role of self-produced movement and visual tracking in infant spatial orientation. *Journal of Experimental Child Psychology, 38*, 312–327.

Acredolo, L. P., & Evans, D. (1980). Developmental changes in the effects of landmarks on infant spatial behavior. *Developmental Psychology, 16*, 312–318.

Akshoomoff, N. A., & Stiles, J. (1995). Developmental trends in visuospatial analysis and planning: I. Copying a complex figure. *Neuropsychology, 9*, 364–377.

Alexander, P. A., White, C. S., & Daugherty, M. (1997). Analogical reasoning and early mathematics learning. In L. D. English (Ed.), *Mathematical reasoning: Analogies, metaphors, and images* (pp. 117–147). Mahwah, NJ: Erlbaum.

Alibali, M. W., & DiRusso, A. A. (1999). The function of gesture in learning to count: More than keeping track. *Cognitive Development, 14*, 37–56.

Alibali, M. W., & Goldin-Meadow, S. (1993). Gesture-speech mismatch and mechanisms of learning: What the hands reveal about a child's state of mind. *Cognitive Psychology, 25*, 468–573.

Allardice, P. A., White, C. S., & Daugherty, M. (1977). The development of written representations for some mathematical concepts. *Journal of Children's Mathematical Behavior, 1*(4), 135–148.

Allen, G. L., & Ondracek, P. J. (1995). Age-sensitive cognitive abilities related to children's acquisition of spatial knowledge. *Developmental Psychology, 31*, 934–945.

Anderson, J. R. (2000). *Cognitive psychology and its implications* (5th ed.). New York: W. H. Freeman.

Anderson, J. R. (Ed.). (1993). *Rules of the mind.* Hillsdale, NJ: Erlbaum.

Anooshian, L. J., Pascal, V. U., & McCreath, H. (1984). Problem mapping before problem solving: Young children's cognitive maps and search strategies in large-scale environments. *Child Development, 55*, 1820–1834.

Ansari, D., Donlan, C., Thomas, M. S. C., Ewing, S. A., Peen, T., & Karmiloff-Smith, A. (2003). What makes counting count? Verbal and visuo-spatial contributions to typical and atypical number development. *Journal of Experimental Child Psychology, 85*, 50–62.

Antell, S. E., & Keating, D. P. (1983). Perception of numerical invariance in neonates. *Child Development, 54*, 695–701.

Arditi, A., Holtzman, J. D., & Kosslyn, S. M. (1988). Mental imagery and sensory experience in congenital blindness. *Neuropsychologia, 26*, 1–12.

Arnold, D. H., & Doctoroff, G. L. (2003). Early education of socioeconomically disadvantaged children. *Annual Review of Psychology, 54*, 517–545.

Ashcraft, M. H. (1982). The development of mental arithmetic: A chronometric approach. *Developmental Review, 2*, 213–236.

Aslan, D. (2004). *Anaokuluna devam eden 3–6 yas grubu çocuklarina temel geometrik sekilleri tanimalari ve sekilleri ayirtetmede kullandiklari kriterlerin incelenmesi* [The investigation of 3 to 6 year-olds preschool children's recognition of basic geometric shapes and the criteria they employ in distinguishing one shape group from the other]. Unpublished Masters, Cukurova University, Adana, Turkey.

Aubrey, C. (1997). Children's early learning of number in school and out. In I. Thompson (Ed.), *Teaching and learning early number* (pp. 20–29). Philadelphia, PA: Open University Press.

Aunola, K., Leskinen, E., Lerkkanen, M.-K., & Nurmi, J.-E. (2004). Developmental dynamics of math performance from pre-school to grade 2. *Journal of Educational Psychology, 96*, 699–713.

Bacon, W. F., & Ichikawa, V. (1988). Maternal expectations, classroom experiences, and achievement among kindergartners in the United States and Japan. *Human Development, 31*, 378–383.

Baillargeon, R. (1991). Reasoning about the height and location of a hidden object in 4.5 and 6.5 month-old children. *Cognition, 38*, 13–42.

Baroody, A. J. (1987). *Children's mathematical thinking.* New York: Teachers College.

Baroody, A. J. (1989). Manipulatives don't come with guarantees. *Arithmetic Teacher, 37*(2), 4–5.

Baroody, A. J. (1992). The development of preschoolers' counting skills and principles. In J. Bideaud, C. Meljac, & J.-P. Fischer (Eds.), *Pathways to number: Developing numerical abilities* (pp. 99–126). Mahwah, NJ: Erlbaum.

Baroody, A. J. (1993). *Problem solving, reasoning, and communicating (K–8): Helping children think mathematically.* New York: Merrill/Macmillan.

Baroody, A. J. (1999). The development of basic counting, number, and arithmetic knowledge among children classified as mentally handicapped. In L. M. Glidden (Ed.), *International review of research in mental retardation* (Vol. 22, pp. 51–103). New York: Academic Press.

Baroody, A. J. (2004). The developmental bases for early childhood number and operations standards. In D. H. Clements, J. Sarama, & A.-M. DiBiase (Eds.), *Engaging young children in mathematics: Standards for early childhood mathematics education* (pp. 173–219). Mahwah, NJ: Erlbaum.

Baroody, A. J., Benson, A. P., & Lai, M.-l. (2003). Early number and arithmetic sense: A summary of three studies.

Baroody, A. J., & Lai, M.-l. (2003, April). *Preschoolers' understanding of the addition-subtraction inverse principle.* Paper presented at the Society for Research in Child Development Tampa, FL.

Baroody, A. J., Lai, M.-L., & Mix, K. S. (2005, December). *Changing views of young children's numerical and arithmetic competencies.* Paper presented at the National Association for the Education of Young Children, Washington, DC.

Baroody, A. J., Lai, M.-l., & Mix, K. S. (2006). The development of young children's number and operation sense and its implications for early childhood education. In B. Spodek, & O. N. Saracho (Eds.), *Handbook of research on the education of young children* (pp. 187–221). Mahwah, NJ: Erlbaum.

Baroody, A. J., & Tiilikainen, S. H. (2003). Two perspectives on addition development. In A. J. Baroody, & A. Dowker (Eds.), *The development of arithmetic concepts and skills: Constructing adaptive expertise* (pp. 75–125). Mahwah, NJ: Erlbaum.

Bartsch, K., & Wellman, H. M. (1988). Young children's conception of distance. *Developmental Psychology, 24*(4), 532–541.

Battista, M. T., & Clements, D. H. (1996). Students' understanding of three-dimensional rectangular arrays of cubes. *Journal for Research in Mathematics Education, 27*, 258–292.

Battista, M. T., Clements, D. H., Arnoff, J., Battista, K., & Borrow, C. V. A. (1998). Students' spatial structuring of 2D arrays of squares. *Journal for Research in Mathematics Education, 29*, 503–532.

Bausano, M. K., & Jeffrey, W. E. (1975). Dimensional salience and judgments of bigness by three-year-old children. *Child Development, 46*, 988–991.

Becker, J. (1989). Preschoolers' use of number words to denote one-to-one correspondence. *Child Development, 60*, 1147–1157.

Beckwith, M., & Restle, F. (1966). Process of enumeration. *Journal of Educational Research, 73*, 437–443.

Beilin, H. (1984). Cognitive theory and mathematical cognition: Geometry and space. In B. Gholson, & T. L. Rosenthanl (Eds.), *Applications of cognitive-developmental theory* (pp. 49–93). New York: Academic Press.

Beilin, H., Klein, A., & Whitehurst, B. (1982). *Strategies and structures in understanding geometry.* New York: City University of New York.

Bell, M. S., & Bell, J. B. (1988). *Assessing and enhancing the counting and numeration capabilities and basic operation concepts of primary school children.* Unpublished manuscript, University of Chicago, Chicago, IL.

Bellugi, U., Lichtenberger, L., Jones, W., Lai, Z., & St. George, M. (2000). The neurocognitive profile and Williams Syndrome: A complex pattern of strengths and weaknesses. *Journal of Cognitive Neuroscience, 12: Supplement,* 7–29.

Benson, K. A., & Bogartz, R. S. (1990). Coordination of perspective change in preschoolers. New Orleans, LA: Society for Research in Child Development.

Bermejo, V. (1996). Cardinality development and counting. *Developmental Psychology, 32*, 263–268.

Bertenthal, B. I., Campos, J. J., & Kermoian, R. (1994). An epigenetic perspective on the development of self-produced locomotion and its consequences. *Current Directions in Psychological Sciences, 5*, 140–145.

Bialystok, E., & Codd, J. (2000). Representing quantity beyond whole numbers: Some, none, and part. *Canadian Journal of Experimental Psychology, 54*, 117–128.

Bisanz, J., Sherman, J., Rasmussen, C., & Ho, E. (2005). Development of arithmetic skills and knowledge in preschool children. In J. I. D. Campbell (Ed.), *Handbook of mathematical cognition* (pp. 143–162). New York: Psychology Press.

Bishop, A. J. (1980). Spatial abilities and mathematics achievement—A review. *Educational Studies in Mathematics, 11*, 257–269.

Blades, M., & Spencer, C. (1989). Young children's ability to use coordinate references. *The Journal of Genetic Psychology, 150*, 5–18.

Blades, M., Spencer, C., Plester, B., & Desmond, K. (2004). Young children's recognition and representation of urban landscapes: From aerial photographs and in toy play. In G. L. Allen (Ed.), *Human spatial memory: Remembering where* (pp. 287–308). Mahwah, NJ: Erlbaum.

Blaut, J. M., & Stea, D. (1974). Mapping at the age of three. *Journal of Geography, 73*(7), 5–9.

Blevins-Knabe, B., & Musun-Miller, L. (1996). Number use at home by children and their parents and its relationship to early mathematical performance. *Early Development and Parenting, 5*, 35–45.

Boardman, D. (1990). Graphicacy revisited: Mapping abilities and gender differences. *Educational Review, 42*, 57–64.

Borer, M. (1993). *Integrating mandated Logo computer instruction into the second grade curriculum.* Unpublished Practicum Report, Nova University.

Bornstein, M. H., Ferdinandsen, K., & Gross, C. G. (1981). Perception of symmetry in infancy. *Developmental Psychology, 17*, 82–86.

Bornstein, M. H., & Krinsky, S. J. (1985). Perception of symmetry in infancy: The salience of vertical symmetry and the perception of pattern wholes. *Journal of Experimental Child Psychology, 39*, 1–19.

Boulton-Lewis, G. M. (1987). Recent cognitive theories applied to sequential length measuring knowledge in young children. *British Journal of Educational Psychology, 57*, 330–342.

Boulton-Lewis, G. M., Wilss, L. A., & Mutch, S. L. (1996). An analysis of young children's strategies and use of devices of length measurement. *Journal of Mathematical Behavior, 15*, 329–347.

Bowerman, M. (1996). Learning how to structure space for language: A crosslinguistic perspective. In P. Bloom, M. A. Peterson, L. Nadel, & M. F. Garrett (Eds.), *Language and space* (pp. 385–436). Cambridge, MA: MIT Press.

Bowman, B. T., Donovan, M. S., & Burns, M. S. (Eds.). (2001). *Eager to learn: Educating our preschoolers.* Washington, DC: National Academy Press.

Brannon, E. M. (2002). The development of ordinal numerical knowledge in infancy. *Cognition, 83*, 223–240.

Bredekamp, S. (2004). Standards for preschool and kindergarten mathematics education. In D. H. Clements, J. Sarama, & A.-M. DiBiase (Eds.), *Engaging young children in mathematics: Standards for early childhood mathematics education* (pp. 77–82). Mahwah, NJ: Erlbaum.

Bremner, J. G., Andreasen, G., Kendall, G., & Adams, L. (1993). Conditions for successful performance by 4-year-olds in a dimensional coordination task. *Journal of Experimental Child Psychology, 56*(2), 149–172.

Bremner, J. G., Knowles, L., & Andreasen, G. (1994). Processes underlying young children's spatial orientation during movement. *Journal of Experimental Child Psychology, 57,* 355–376.

Bremner, J. G., & Taylor, A. J. (1982). Children's errors in copying angles: Perpendicular error or bisection error? *Perception, 11,* 163–171.

Briars, D., & Siegler, R. S. (1984). A featural analysis of preschoolers' counting knowledge. *Developmental Psychology, 20,* 607–618.

Broberg, A. G., Wessels, H., Lamb, M. E., & Hwang, C. P. (1997). Effects of day care on the development of cognitive abilities in 8-year-olds: A longitudinal study. *Developmental Psychology, 33,* 62–69.

Brooks-Gunn, J., Duncan, G. J., & Britto, P. R. (1999). Are socieconomic gradients for children similar to those for adults? In D. P. Keating, & C. Hertzman (Eds.), *Developmental health and the wealth of nations* (pp. 94–124). New York: Guilford.

Brosnan, M. J. (1998). Spatial ability in children's play with Lego blocks. *Perceptual and Motor Skills, 87,* 19–28.

Brown, D. L., & Wheatley, G. H. (1989). Relationship between spatial knowledge and mathematics knowledge. In C. A. Maher, G. A. Goldin, & R. B. Davis (Eds.), *Proceedings of the eleventh annual meeting, North American Chapter of the International Group for the Psychology of Mathematics Education* (pp. 143–148). New Brunswick, NJ: Rutgers University.

Brown, S. I., & Walter, M. I. (1990). *Art of problem posing.* Mahwah, NJ: Erlbaum.

Brownell, W. A. (1928). *The development of children's number ideas in the primary grades.* Chicago: Department of Education, University of Chicago.

Brush, L. R. (1978). Preschool children's knowledge of addition and subtraction. *Journal for Research in Mathematics Education, 9,* 44–54.

Bryant, D. M., Burchinal, M. R., Lau, L. B., & Sparling, J. J. (1994). Family and classroom correlates of Head Start children's developmental outcomes. *Early Childhood Research Quarterly, 9,* 289–309.

Bryant, P., Christie, C., & Rendu, A. (1999). Children's understanding of the relation between addition and subtraction: Inversion, identify, and decomposition. *Journal of Experimental Child Psychology, 74,* 194–212.

Bryant, P. E. (1982). The role of conflict and of agreement between intellectual strategies in children's ideas about measurement. *British Journal of Psychology, 73,* 242–251.

Bullock, M., & Gelman, R. (1977). Numerical reasoning in young children: The ordering principle. *Child Development, 48,* 427–434.

Burger, W. F., & Shaughnessy, J. M. (1986). Characterizing the van Hiele levels of development in geometry. *Journal for Research in Mathematics Education, 17,* 31–48.

Bushnell, E. W., McKenzie, B. E., Lawrence, D. A., & Com, S. (1995). The spatial coding strategies of 1-year-old infants in a locomotor search task. *Child Development, 66,* 937–958.

Butterworth, G. (1991). Evidence for the "geometric" comprehension of manual pointing. Seattle, WA: Society for Research in Child Development.

Callahan, L. G., & Clements, D. H. (1984). Sex differences in rote counting ability on entry to first grade: Some observations. *Journal for Research in Mathematics Education, 15,* 378–382.

Campbell, F. A., Pungello, E. P., Miller-Johnson, S., Burchinal, M., & Ramey, C. T. (2001). The development of cognitive and academic abilities: Growth curves from an early childhood educational experiment. *Developmental Psychology, 37,* 231–242.

Campbell, P. F., & Silver, E. A. (1999). *Teaching and learning mathematics in poor communities.* Reston, VA: National Council of Teachers of Mathematics.

Canfield, R. L., & Smith, E. G. (1996). Number-based expectations and sequential enumeration by 5-month-old infants. *Developmental Psychology, 32,* 269–279.

Carlson, S. L., & White, S. H. (1998). The effectiveness of a computer program in helping kindergarten students learn the concepts of left and right. *Journal of Computing in Childhood Education, 9*(2), 133–147.

Carpenter, T. P. (1985). Learning to add and subtract: An exercise in problem solving. In E. A. Silver (Ed.), *Teaching and learning mathematical problem solving* (pp. 17–40). Hillsdale, NJ: Erlbaum.

Carpenter, T. P., Ansell, E., Franke, M. L., Fennema, E. H., & Weisbeck, L. (1993). Models of problem solving: A study of kindergarten children's problem-solving processes. *Journal for Research in Mathematics Education, 24,* 428–441.

Carpenter, T. P., Fennema, E. H., Franke, M. L., Levi, L., & Empson, S. B. (1999). *Children's mathematics: Cognitively guided instruction.* Portsmouth, NH: Heinemann.

Carpenter, T. P., Franke, M. L., Jacobs, V. R., Fennema, E. H., & Empson, S. B. (1998). A longitudinal study of invention and understanding in children's multidigit addition and subtraction. *Journal for Research in Mathematics Education, 29,* 3–20.

Carpenter, T. P., & Moser, J. M. (1984). The acquisition of addition and subtraction concepts in grades one through three. *Journal for Research in Mathematics Education, 15,* 179–202.

Carper, D. V. (1942). Seeing numbers as groups in primary-grade arithmetic. *The Elementary School Journal, 43,* 166–170.

Case, R., Griffin, S., & Kelly, W. M. (1999). Socieconomic gradients in mathematical ability and their responsiveness to intervention during early childhood. In D. P. Keating, & C. Hertzman (Eds.), *Developmental health and the wealth of nations* (pp. 125–149). New York: Guilford.

Case, R., & Okamoto, Y. (1996). The role of central conceptual structures in the development of children's thought. *Monographs of the Society for Research in Child Development, 61*(1–2, Serial No. 246).

Casey, M. B. (2005, April). *Evaluation of NSF-funded mathematics materials: use of storytelling contexts to improve kindergartners' geometry and block-building skills.* Paper presented at the National Council of Supervisors of Mathematics, Anaheim, CA.

Chi, M. T. H., & Klahr, D. (1975). Span and rate of apprehension in children and adults. *Journal of Experimental Child Psychology, 19,* 434–439.

Choi, S., & Bowerman, M. (1991). Learning to express motion events in English and Korean: The influence of language-specific lexication patterns. *Cognition, 41,* 83–121.

Choi, S., McDonough, L., Bowerman, M., & Mandler, J. (1999). Comprehension of spatial terms in English and Korean. *Cognitive Development, 14,* 241–268.

Ciancio, D. S., Rojas, A. C., McMahon, K., & Pasnak, R. (2001). Teaching oddity and insertion to Head Start children; an economical cognitive intervention. *Journal of Applied Developmental Psychology, 22,* 603 - 621.

Clarke, B. A., & Clarke, D. M. (2004). *What mathematical knowledge and understanding do young children bring to school?* Paper presented at the Australian Research in Early Childhood Education, Melbourne, Australia.

Clarke, D. M., Cheeseman, J., Clarke, B., Gervasoni, A., Gronn, D., Horne, M., et al. (2001). Understanding, assessing and developing young children's mathematical thinking: Research as a powerful tool for professional growth. In J. Bobis, B. Perry, & M. Mitchelmore (Eds.), *Numeracy and beyond (Proceedings of the 24th Annual Conference of the Mathematics Education Research Group of Australasia, Vol. 1)* (pp. 9–26). Reston, Australia: MERGA.

Clarke, D. M., Cheeseman, J., Gervasoni, A., Gronn, D., Horne, M., McDonough, A., et al. (2002). *Early Numeracy Research Project final report*: Department of Education, Employment and Training, the Catholic Education Office (Melbourne), and the Association of Independent Schools Victoria.

Clearfield, M. W., & Mix, K. S. (1999). Infants use contour length—not number—to discriminate small visual sets. Albuquerque, NM: Society for Research in Child Development.

Clements, D. H. (1984a). Foundations of number and logic: Seriation, classification, and number conservation from a Piagetian perspective. *Psychological Documents (Ms. No. 2607), 14*(4).

Clements, D. H. (1984b). The development of counting and other early number knowledge: A review of research and psychological models. *Psychological Documents (Ms. No. 2644), 14*(2).

Clements, D. H. (1984c). Training effects on the development and generalization of Piagetian logical operations and knowledge of number. *Journal of Educational Psychology, 76,* 766–776.

Clements, D. H. (1992). Elaboraciones sobre los niveles de pensamiento geometrico [Elaborations on the levels of geometric thinking]. In A. Gutiérrez (Ed.), *Memorias del Tercer Simposio Internacional Sobre Investigatcion en Educacion Matematica* (pp. 16–43). València, Spain: Universitat De València.

Clements, D. H. (1999a). 'Concrete' manipulatives, concrete ideas. *Contemporary Issues in Early Childhood, 1*(1), 45–60.

Clements, D. H. (1999b). Subitizing: What is it? Why teach it? *Teaching Children Mathematics, 5,* 400–405.

Clements, D. H. (1999c). Teaching length measurement: Research challenges. *School Science and Mathematics, 99*(1), 5–11.

Clements, D. H. (2003). Teaching and learning geometry. In J. Kilpatrick, W. G. Martin, & D. Schifter (Eds.), *A research companion to Principles and Standards for School Mathematics* (pp. 151–178). Reston, VA: National Council of Teachers of Mathematics.

Clements, D. H. (2004). Geometric and spatial thinking in early childhood education. In D. H. Clements, J. Sarama, & A.-M. DiBiase (Eds.), *Engaging young children in mathematics: Standards for early childhood mathematics education* (pp. 267–297). Mahwah, NJ: Erlbaum.

Clements, D. H. (in press). Curriculum research: Toward a framework for "research-based curricula." *Journal for Research in Mathematics Education.*

Clements, D. H., & Barrett, J. (1996). Representing, connecting and restructuring knowledge: A micro-genetic analysis of a child's learning in an open-ended task involving perimeter, paths and polygons. In E. Jakubowski, D. Watkins, &

H. Biske (Eds.), *Proceedings of the 18th annual meeting of the North America Chapter of the International Group for the Psychology of Mathematics Education* (Vol. 1, pp. 211–216). Columbus, OH: ERIC Clearinghouse for Science, Mathematics, and Environmental Education.

Clements, D. H., & Battista, M. T. (1989). Learning of geometric concepts in a Logo environment. *Journal for Research in Mathematics Education, 20,* 450–467.

Clements, D. H., & Battista, M. T. (Artist). (1991). *Logo geometry*

Clements, D. H., & Battista, M. T. (1992). Geometry and spatial reasoning. In D. A. Grouws (Ed.), *Handbook of research on mathematics teaching and learning* (pp. 420–464). New York: Macmillan.

Clements, D. H., Battista, M. T., & Sarama, J. (2001). Logo and geometry. *Journal for Research in Mathematics Education Monograph Series, 10.*

Clements, D. H., Battista, M. T., Sarama, J., & Swaminathan, S. (1997). Development of students' spatial thinking in a unit on geometric motions and area. *The Elementary School Journal, 98,* 171–186.

Clements, D. H., Battista, M. T., Sarama, J., Swaminathan, S., & McMillen, S. (1997). Students' development of length measurement concepts in a Logo-based unit on geometric paths. *Journal for Research in Mathematics Education, 28*(1), 70–95.

Clements, D. H., & Conference Working Group. (2004). Part one: Major themes and recommendations. In D. H. Clements, J. Sarama, & A.-M. DiBiase (Eds.), *Engaging young children in mathematics: Standards for early childhood mathematics education* (pp. 1–72). Mahwah, NJ: Erlbaum.

Clements, D. H., & McMillen, S. (1996). Rethinking "concrete" manipulatives. *Teaching Children Mathematics, 2*(5), 270–279.

Clements, D. H., & Meredith, J. S. (1994). Turtle math [Computer software]. Montreal, Quebec: Logo Computer Systems, Inc. (LCSI).

Clements, D. H., & Sarama, J. (2003). Young children and technology: What does the research say? *Young Children, 58*(6), 34–40.

Clements, D. H., & Sarama, J. (2004a). *Building Blocks* for early childhood mathematics. *Early Childhood Research Quarterly, 19,* 181–189.

Clements, D. H., & Sarama, J. (2004b). Hypothetical learning trajectories. *Mathematical Thinking and Learning, 6*(2).

Clements, D. H., & Sarama, J. (2004c). Learning trajectories in mathematics education. *Mathematical Thinking and Learning, 6,* 81–89.

Clements, D. H., & Sarama, J. (2007). *Building Blocks—SRA Real Math Teacher's Edition, Grade PreK.* Columbus, OH: SRA/McGraw-Hill.

Clements, D. H., & Sarama, J. (in press). Effects of a preschool mathematics curriculum: Summary research on the *Building Blocks* project. *Journal for Research in Mathematics Education.*

Clements, D. H., Sarama, J., Battista, M. T., & Swaminathan, S. (1996). Development of students' spatial thinking in a curriculum unit on geometric motions and area. In E. Jakubowski, D. Watkins, & H. Biske (Eds.), *Proceedings of the 18th annual meeting of the North America Chapter of the International Group for the Psychology of Mathematics Education* (Vol. 1, pp. 217–222). Columbus, OH: ERIC Clearinghouse for Science, Mathematics, and Environmental Education.

Clements, D. H., Sarama, J., & DiBiase, A.-M. (2004). *Engaging young children in mathematics: Standards for early childhood mathematics education*. Mahwah, NJ: Erlbaum.

Clements, D. H., Sarama, J., & Gerber, S. (2005). Mathematics knowledge of low-income entering preschoolers. *Manuscript submitted for publication.*

Clements, D. H., Sarama, J., & Wilson, D. C. (2001). Composition of geometric figures. In M. v. d. Heuvel-Panhuizen (Ed.), *Proceedings of the 21st Conference of the International Group for the Psychology of Mathematics Education* (Vol. 2, pp. 273–280). Utrecht, The Netherlands: Freudenthal Institute.

Clements, D. H., & Stephan, M. (2004). Measurement in pre-K–2 mathematics. In D. H. Clements, J. Sarama, & A.-M. DiBiase (Eds.), *Engaging young children in mathematics: Standards for early childhood mathematics education* (pp. 299–317). Mahwah, NJ: Erlbaum.

Clements, D. H., Swaminathan, S., Hannibal, M. A. Z., & Sarama, J. (1999). Young children's concepts of shape. *Journal for Research in Mathematics Education, 30,* 192–212.

Clements, D. H., Wilson, D. C., & Sarama, J. (2004). Young children's composition of geometric figures: A learning trajectory. *Mathematical Thinking and Learning, 6,* 163–184.

Confrey, J. (1996). The role of new technologies in designing mathematics education. In C. Fisher, D. C. Dwyer, & K. Yocam (Eds.), *Education and technology, reflections on computing in the classroom* (pp. 129–149). San Francisco: Apple Press.

Cooper, J., Brenneman, K., & Gelman, R. (2005, April). *Young children's use of graphs for arithmetic.* Paper presented at the Biennial Meeting of the Society for Research in Child Development, Atlanta, GA.

Cooper, R. G., Jr. (1984). Early number development: Discovering number space with addition and subtraction. In C. Sophian (Ed.), *Origins of cognitive skills* (pp. 157–192). Mahwah, NJ: Erlbaum.

Cordes, S., & Gelman, R. (2005). The young numerical mind: When does it count? In J. Campbell (Ed.), *Handbook of mathematical cognition* (pp. 127–142). New York: Psychology Press.

Cornell, E. H., Heth, C. D., Broda, L. S., & Butterfield, V. (1987). Spatial matching in 1 1/2- to 4 1/2-year-old children. *Developmental Psychology, 23,* 499–508.

Cowan, R. (1987). When do children trust counting as a basis for relative number judgments? *Journal of Experimental Child Psychology, 43,* 328–345.

Cowan, R. (2003). Does it all add up? Changes in children's knowledge of addition combinations, strategies, and principles. In A. J. Baroody, & A. Dowker (Eds.), *The development of arithmetic concepts and skills: Constructing adaptive expertise* (pp. 35–74). Mahwah, NJ: Erlbaum.

Cowan, R., & Daniel, H. (1989). Children's use of counting and guidelines in judging relative number. *British Journal of Educational Psychology, 59,* 200–210.

Cuneo, D. (1980). A general strategy for quantity judgments: The height + width rule. *Child Development, 51,* 299–301.

Curtis, R. P., Okamoto, Y., & Weckbacher, L. M. (2005). Preschoolers' use of counting information to judge relative quantity, *American Educational Research Association.* Montreal, CA.

Dalke, D. E. (1998). Charting the development of representational skills: When do children know that maps can lead and mislead? *Cognitive Development, 70,* 53–72.

Darke, I. (1982). A review of research related to the topological primacy thesis. *Educational Studies in Mathematics, 13,* 119–142.

Davis, R. B. (1984). *Learning mathematics: The cognitive science approach to mathematics education.* Norwood, NJ: Ablex.

Dawson, D. T. (1953). Number grouping as a function of complexity. *The Elementary School Journal, 54,* 35–42.

Day, J. D., Engelhardt, J. L., Maxwell, S. E., & Bolig, E. E. (1997). Comparison of static and dynamic assessment procedures and their relation to independent performance. *Journal of Educational Psychology, 89*(2), 358–368.

Dehaene, S. (1997). *The number sense: How the mind creates mathematics.* New York: Oxford University Press.

Del Grande, J. J. (1986). Can grade two children's spatial perception be improved by inserting a transformation geometry component into their mathematics program? *Dissertation Abstracts International, 47,* 3689A.

DeLoache, J. S. (1987). Rapid change in the symbolic functioning of young children. *Science, 238,* 1556–1557.

DeLoache, J. S., & Burns, N. M. (1994). Early understanding of the representational function of pictures. *Cognition, 52,* 83–110.

DeLoache, J. S., Miller, K. F., & Pierroutsakos, S. L. (1998). Reasoning and problem solving. In D. Kuhn, & R. S. Siegler (Eds.), *Handbook of Child Psychology (5th Ed.): Vol. 2. Cognition, Perception, & Language* (pp. 801–850). New York: Wiley.

DeLoache, J. S., Miller, K. F., Rosengren, K., & Bryant, N. (1997). The credible shrinking room: Very young children's performance with symbolic and nonsymbolic relations. *Psychological Science, 8,* 308–313.

Denis, L. P. (1987). Relationships between stage of cognitive development and van Hiele level of geometric thought among Puerto Rican adolescents. *Dissertation Abstracts International, 48,* 859A.

Denton, K., & West, J. (2002). Children's reading and mathematics achievement in kindergarten and first grade. 2002, from http://nces.ed.gov/pubsearch/pubsinfo.asp? pubid=2002125

Dewey, J. (1898). Some remarks on the psychology of number. *Pedagogical Seminary, 5,* 426–434.

Douglass, H. R. (1925). The development of number concept in children of preschool and kindergarten ages. *Journal of Experimental Psychology, 8,* 443–470.

Dowker, A. (2004). *What works for children with mathematical difficulties?* Nottingham, UK: University of Oxford/DfES Publications.

Duncan, G. J., Brooks-Gunn, J., & Klebanov, P. K. (1994). Economic deprivation and early childhood development. *Child Development, 65,* 296–318.

Early, D., Barbarin, O., Burchinal, M. R., Chang, F., Clifford, R., Crawford, G., et al. (2005). *Pre-Kindergarten in Eleven States: NCEDL's Multi-State Study of Pre-Kindergarten & Study of State-Wide Early Education Programs (SWEEP)* Chapel Hill, NC: University of North Carolina.

Ebbeck, M. (1984). Equity for boys and girls: Some important issues. *Early Child Development and Care, 18,* 119–131.

Ebeling, K. S., & Gelman, S. A. (1988). Coordination of size standards by young children. *Child Development, 59,* 888–896.

Ebeling, K. S., & Gelman, S. A. (1994). Children's use of context in interpreting "big" and "little." *Child Development, 65,* 1178–1192.

Ehrlich, S. B., Levine, S. C., & Goldin-Meadow, S. (2005, April). *Early sex differences in spatial skill: The implications of spoken and gestured strategies.* Paper presented at the Biennial Meeting of the Society for Research in Child Development, Atlanta, GA.

Eliot, J. (1987). *Models of psychological space.* New York: Springer-Verlag.

Elkind, D. (1964). Discrimination, seriation, and numeration of size and dimension of differences in young children: Piaget replication study VI. *Journal of Genetic Psychology, 104,* 275–296.

Ellis, S., Siegler, R. S., & Van Voorhis, F. E. (2000). Developmental changes in children's understanding of measurement procedures and principles.

Entwisle, D. R., & Alexander, K. L. (1990). Beginning school math competence: Minority and majority comparisons. *Child Development, 61,* 454–471.

Ericsson, K. A., Krampe, R. T., & Tesch-Römer, C. (1993). The role of deliberate practice in the acquisition of expert performance. *Psychological Review, 100,* 363–406.

Ernest, P. (1995). The one and the many. In L. P. Steffe, & J. Gale (Eds.), *Constructivism in education* (pp. 459–486). Mahwah, NJ: Erlbaum.

Evans, D. W. (1983). *Understanding infinity and zero in the early school years.* Unpublished doctoral dissertation, University of Pennsylvania.

Ewers-Rogers, J., & Cowan, R. (1996). Children as apprentices to number. *Early Child Development and Care, 125,* 15–25.

Fabricius, W. V., & Wellman, H. M. (1993). Two roads diverged: Young children's ability to judge distances. *Child Development, 64,* 399–414.

Farran, D. C., Silveri, B., & Culp, A. (1991). Public preschools and the disadvantaged. In L. Rescorla, M. C. Hyson, & K. Hirsh-Pase (Eds.), *Academic instruction in early childhood: Challenge or pressure? New directions for child development* (pp. 65–73). San. Francisco: Jossey-Bass.

Feeney, S. M., & Stiles, J. (1996). Spatial analysis: An examination of preschoolers' perception and construction of geometric patterns. *Developmental Psychology, 32,* 933–941.

Feigenson, L. (1999). An anticipatory-looking paradigm for examining infants' ordinal knowledge. Albuquerque, NM: Society for Research in Child Development.

Feigenson, L., Carey, S., & Hauser, M. (2002). The representations underlying infants' choice of more: Object files versus analog magnitudes. *Psychological Science, 13,* 150–156.

Feigenson, L., Carey, S., & Spelke, E. S. (2002). Infants' discrimination of number vs. continuous extent. *Cognitive Psychology, 44,* 33–66.

Fennema, E. H., & Carpenter, T. P. (1981). Sex-related differences in mathematics: Results from National Assessment. *Mathematics Teacher, 74,* 554–559.

Fennema, E. H., Carpenter, T. P., Frank, M. L., Levi, L., Jacobs, V. R., & Empson, S. B. (1996). A longitudinal study of learning to use children's thinking in mathematics instruction. *Journal for Research in Mathematics Education, 27,* 403–434.

Fennema, E. H., & Sherman, J. (1977). Sex-related differences in mathematics achievement, spatial visualization and affective factors. *American Educational Research Journal, 14,* 51–71.

Fennema, E. H., & Sherman, J. A. (1978). Sex-related differences in mathematics achievement and related factors. *Journal for Research in Mathematics Education, 9,* 189–203.

Ferguson, R. W., Aminoff, A., & Gentner, D. (1998). Early detection of qualitative symmetry. Submitted for publication.

Filippaki, N., & Papamichael, Y. (1997). Tutoring conjunctions and construction of geometry concepts in the early childhood education: The case of the angle. *European Journal of Psychology of Education, 12*(3), 235–247.

Fischbein, E. (1987). *Intuition in science and mathematics.* Dordrecht, Holland: D. Reidel.

Fisher, C. B., Ferdinandsen, K., & Bornstein, M. H. (1981). The role of symmetry in infant form discrimination. *Child Development, 52,* 457–462.

Fisher, G. H. (1965). Developmental features of behaviour and perception. *British Journal of Educational Psychology, 35,* 69–78.

Fisher, N. D. (1978). Visual influences of figure orientation on concept formation in geometry. *Dissertation Abstracts International, 38,* 4639A.

Fitzhugh, J. I. (1978). The role of subitizing and counting in the development of the young children's conception of small numbers. *Dissertation Abstracts International, 40,* 4521B–4522B.

Fluck, M. (1995). Counting on the right number: Maternal support for the development of cardinality. *Irish Journal of Psychology, 16,* 133–149.

Fluck, M., & Henderson, L. (1996). Counting and cardinality in English nursery pupils. *British Journal of Educational Psychology, 66,* 501–517.

Forman, G. E. (1982). A search for the origins of equivalence concepts through a microanalysis of block play. In G. E. Forman (Ed.), *Action and thought* (pp. 97–135). New York: Academic Press.

Forman, G. E., & Hill, F. (1984). *Constructive play: Applying Piaget in the preschool (revised edition).* Menlo Park, CA: Addison Wesley.

Frank, R. E. (Cartographer). (1987). The emergence of route map reading skills in young children. Lida Lee Tall Learning Resources Center Research in Brief: A Newsletter for Early Childhood and Elementary Educators, 3–4. Towson, MD: Towson State University.

Freeman, F. N. (1912). Grouped objects as a concrete basis for the number idea. *Elementary School Teacher, 8,* 306–314.

French, L., & Song, M.-J. (1998). Developmentally appropriate teacher-directed approaches: Images from Korean kindergartens. *Journal of Curriculum Studies, 30,* 409–430.

Friel, S. N., Curcio, F. R., & Bright, G. W. (2001). Making sense of graphs: Critical factors influencing comprehension and instructional implications. *Journal for Research in Mathematics Education, 32,* 124–158.

Frydman, l., & Bryant, P. (1988). Sharing and the understanding of number equivalence by young children. *Cognitive Development, 3,* 323–339.

Frye, D., Braisby, N., Lowe, J., Maroudas, C., & Nicholls, J. (1989). Young children's understanding of counting and cardinality. *Child Development, 60,* 1158–1171.

Fuson, K. C. (1988). *Children's counting and concepts of number.* New York: Springer-Verlag.

Fuson, K. C. (1992). Research on learning and teaching addition and subtraction of whole numbers. In G. Leinhardt, R. Putman, & R. A. Hattrup (Eds.), *Analysis of arithmetic for mathematics teaching* (pp. 53–187). Hillsdale, NJ: Erlbaum.

Fuson, K. C. (2004). Pre-K to grade 2 goals and standards: Achieving 21st century mastery for all. In D. H. Clements, J. Sarama, & A.-M. DiBiase (Eds.), *Engaging young children in mathematics: Standards for early childhood mathematics education* (pp. 105–148). Mahwah, NJ: Erlbaum.

Fuson, K. C., Carroll, W. M., & Drueck, J. V. (2000). Achievement results for second and third graders using the *Standards*-based curriculum *Everyday Mathematics*. *Journal for Research in Mathematics Education, 31,* 277–295.

Fuson, K. C., & Hall, J. W. (1982). The acquisition of early number word meanings: A conceptual analysis and review. In H. P. Ginsburg (Ed.), *Children's mathematical thinking* (pp. 49–107). New York: Academic Press.

Fuson, K. C., & Murray, C. (1978). The haptic-visual perception, construction, and drawing of geometric shapes by children aged two to five: A Piagetian extension. In R. Lesh, & D. Mierkiewicz (Eds.), *Concerning the development of spatial and geometric concepts* (pp. 49–83). Columbus, OH: ERIC Clearinghouse for Science, Mathematics, and Environmental Education.

Fuson, K. C., Richards, J., & Briars, D. (1982). The acquisition and elaboration of the number word sequence. In C. J. Brainerd (Ed.), *Children's logical and mathematical cognition: Progress in cognitive development research* (pp. 33–92). New York: Springer-Verlag.

Fuson, K. C., Secada, W. G., & Hall, J. W. (1983). Matching, counting, and conservation of number equivalence. *Child Development, 54,* 91–97.

Fuson, K. C., Smith, S. T., & Lo Cicero, A. (1997). Supporting Latino first graders' ten-structured thinking in urban classrooms. *Journal for Research in Mathematics Education, 28,* 738–760.

Fuys, D., Geddes, D., & Tischler, R. (1988). The van Hiele model of thinking in geometry among adolescents. *Journal for Research in Mathematics Education Monograph Series, 3.*

Gagatsis, A., & Patronis, T. (1990). Using geometrical models in a process of reflective thinking in learning and teaching mathematics. *Educational Studies in Mathematics, 21,* 29–54.

Gallistel, C. R., & Gelman, R. (2005). Mathematical cognition. In K. Holyoak, & R. Morrison (Eds.), *Cambridge handbook of thinking and reasoning* (pp. 559–588). Cambridge, U.K.: Cambridge University Press.

Gamel-McCormick, M., & Amsden, D. (2002). *Investing in better outcomes: The Delaware early childhood longitudinal study.* Delaware Interagency Resource Management Committee and the Department of Education.

Ganel, T., & Goodale, M. A. (2003). Visual control of action but not perception requires analytical processing of object shape. *Nature, 426,* 664–667.

Gao, F., Levine, S. C., & Huttenlocher, J. (2000). What do infants know about continuous quantity? *Journal of Experimental Child Psychology, 77,* 20–29.

Gauvain, M. (1991). The development of spatial thinking in everyday activity. Seattle, WA: Society for Research in Child Development.

Gay, P. (1989). Tactile turtle: Explorations in space with visually impaired children and a floor turtle. *British Journal of Visual Impairment, 7*(1), 23–25.

Geary, D. C. (1994). *Children's mathematical development: Research and practical applications.* Washington, DC: American Psychological Association.

Geary, D. C., Bow-Thomas, C. C., Fan, L., & Siegler, R. S. (1993). Even before formal instruction, Chinese children outperform American children in mental addition. *Cognitive Development, 8,* 517–529.

Geary, D. C., Hamson, C. O., & Hoard, M. K. (2000). Numerical and arithmetical cognition: A longitudinal study of process and concept deficits in children with learning disability. *Journal of Experimental Child Psychology, 77,* 236–263.

Geary, D. C., Hoard, M. K., & Hamson, C. O. (1999). Numerical and arithmetical cognition: Patterns of functions and deficits in children at risk for a mathematical disability. *Journal of Experimental Child Psychology, 74,* 213–239.

Geary, D. C., & Liu, F. (1996). Development of arithmetical competence in Chinese and American children: Influence of age, language, and schooling. *Child Development, 67*(5), 2022–2044.

Geeslin, W. E., & Shar, A. O. (1979). An alternative model describing children's spatial preferences. *Journal for Research in Mathematics Education, 10,* 57–68.

Gelman, R. (1969). Conservation acquisition: A problem of learning to attend to relevant attributes. *Journal of Experimental Child Psychology, 7,* 167–187.

Gelman, R., & Butterworth, B. (2005). Number and language: How are they related? *Trends in Cognitive Sciences, 9*(1), 6–10.

Gelman, R., & Gallistel, C. R. (1978). *The child's understanding of number.* Cambridge, MA: Harvard University Press.

Gelman, R., & Meck, E. (1983). Preschoolers' counting: Principles before skill. *Cognition, 13,* 343–359.

Gelman, R., & Williams, E. M. (1997). Enabling constraints for cognitive development and learning: Domain specificity and epigenesis. In D. Kuhn, & R. Siegler (Eds.), *Cognition, perception, and language. Volume 2: Handbook of Child Psychology* (5th ed., pp. 575–630). New York: John Wiley & Sons.

Gelman, S. A., & Ebeling, K. S. (1989). Children's use of nonegocentric standards in judgments of functional size. *Child Development, 60,* 920–932.

Genkins, E. F. (1975). The concept of bilateral symmetry in young children. In M. F. Rosskopf (Ed.), *Children's mathematical concepts: Six Piagetian studies in mathematics education* (pp. 5–43). New York: Teaching College Press.

Gerhardt, L. A. (1973). *Moving and knowing: The young child orients himself in space.* Englewood Cliffs, NJ: Prentice-Hall.

Gersten, R., Chard, D., Baker, S., Jayanthi, M., Flojo, J. R., & Lee, D. (2006). *Experimental and quasi-experimental research on instructional approaches for teaching mathematics to students with learning disabilities: A research synthesis.* Center on Instruction: Technical Report 2006-2, Signal Hill, CA: RG Research Group.

Gersten, R., Jordan, N. C., & Flojo, J. R. (2005). Early identification and interventions for students with mathematical difficulties. *Journal of Learning Disabilities, 38,* 293–304.

Gibson, E. J. (1969). *Principles of perceptual learning and development.* New York: Appleton-Century-Crofts, Meredith Corporation.

Gibson, E. J., Gibson, J. J., Pick, A. D., & Osser, H. (1962). A developmental study of the discrimination of letter-like forms. *Journal of Comparative and Physiological Psychology, 55,* 897–906.

Gibson, S. (1985). The effects of position of counterexamples on the learning of algebraic and geometric conjunctive concepts. *Dissertation Abstracts International, 46,* 378A.

Ginsburg, H. P. (1977). Children's arithmetic. In. Austin, TX: Pro-ed.

Ginsburg, H. P., Choi, Y. E., Lopez, L. S., Netley, R., & Chi, C.-Y. (1997). Happy birthday to you: The early mathematical thinking of Asian, South American, and U.S. children. In T. Nunes, & P. Bryant (Eds.), *Learning and teaching mathematics: An international perspective* (pp. 163–207). East Sussex, England: Psychology Press.

Ginsburg, H. P., Inoue, N., & Seo, K.-H. (1999). Young children doing mathematics: Observations of everyday activities. In J. V. Copley (Ed.), *Mathematics in the early years* (pp. 88–99). Reston, VA: National Council of Teachers of Mathematics.

Ginsburg, H. P., Ness, D., & Seo, K.-H. (2003). Young American and Chinese children's everyday mathematical activity. *Mathematical Thinking and Learning, 5,* 235–258.

Ginsburg, H. P., & Russell, R. L. (1981). Social class and racial influences on early mathematical thinking. *Monographs of the Society for Research in Child Development, 46*(6, Serial No. 193).

Golbeck, S. L., Rand, M., & Soundy, C. (1986). Constructing a model of a large scale space with the space in view: Effects of guidance and cognitive restructuring. *Merrill-Palmer Quarterly, 32,* 187–203.

Goldin-Meadow, S., Alibali, M. W., & Church, R. B. (1993). Transitions in concept acquisition: Using the hand to read the mind. *Psychological Review, 100,* 279–297.

Goodrow, A., Clements, D. H., Battista, M. T., Sarama, J., & Akers, J. (1997). *How long? How far? Measurement.* Palo Alto, CA: Dale Seymour Publications.

Gopnik, A., & Meltzoff, A. N. (1986). Words, plans, things, and locations: Interactions between semantic and cognitive development in the one-word stage. In S. A. Kuczaj II, & M. D. Barrett (Eds.), *The development of word meaning* (pp. 199–223). Berlin, Germany: Springer-Verlag.

Gordon, P. (2004). Numerical cognition without words: Evidence from Amazonia. *Science, 306,* 496–499.

Graham, T. A., Nash, C., & Paul, K. (1997). Young children's exposure to mathematics: The child care context. *Early Childhood Education Journal, 25,* 31–38.

Granrud, C. E. (1987). Visual size constancy in newborn infants. *Investigative Ophthalmology & Visual Science, 28*((Suppl. 5)).

Greabell, L. C. (1978). The effect of stimuli input on the acquisition of introductory geometric concepts by elementary school children. *School Science and Mathematics, 78*(4), 320–326.

Greeno, J. G., Riley, M. S., & Gelman, R. (1984). Conceptual competence and children's counting. *Cognitive Psychology, 16,* 94–143.

Greenough, W. T., Black, J. E., & Wallace, C. S. (1987). Experience and brain development. *Children Development, 58,* 539–559.

Greenwald, A. G., Pratkanis, A. R., Leippe, M. R., & Baumgardner, M. H. (1986). Under what conditions does theory obstruct research progress? *Psychological Review, 93,* 216–229.

Griffin, S. (2004). Number Worlds: A research-based mathematics program for young children. In D. H. Clements, J. Sarama, & A.-M. DiBiase (Eds.), *Engaging young children in mathematics: Standards for early childhood mathematics education* (pp. 325–342). Mahwah, NJ: Erlbaum.

Griffin, S., & Case, R. (1997). Re-thinking the primary school math curriculum: An approach based on cognitive science. *Issues in Education, 3*(1), 1–49.

Griffin, S., Case, R., & Capodilupo, A. (1995). Teaching for understanding: The importance of the Central Conceptual Structures in the elementary mathematics curriculum. In A. McKeough, J. Lupart, & A. Marini (Eds.), *Teaching for transfer: Fostering generalization in learning* (pp. 121–151). Mahwah, NJ: Erlbaum.

Griffin, S., Case, R., & Siegler, R. S. (1994). Rightstart: Providing the central conceptual prerequisites for first formal learning of arithmetic to students at risk for school failure.

In K. McGilly (Ed.), *Classroom lessons: Integrating cognitive theory and classroom practice* (pp. 25–49). Cambridge, MA: MIT Press.

Groen, G., & Resnick, L. B. (1977). Can preschool children invent addition algorithms? *Journal of Educational Psychology, 69,* 645–652.

Grupe, L. A., & Bray, N. W. (1999). What role do manipulatives play in kindergartners' accuracy and strategy use when solving simple addition problems? Albuquerque, NM: Society for Research in Child Development.

Guay, R. B., & McDaniel, E. (1977). The relationship between mathematics achievement and spatial abilities among elementary school children. *Journal for Research in Mathematics Education, 8,* 211–215.

Gutiérrez, A., & Jaime, A. (1988). *Globality versus locality of the van Hiele levels of geometric reasoning.* Unpublished manuscript, Universitat De València, Valencia, Spain.

Gutiérrez, A., Jaime, A., & Fortuny, J. M. (1991). An alternative paradigm to evaluate the acquisition of the van Hiele levels. *Journal for Research in Mathematics Education, 22,* 237–251.

Haith, M. M., & Benson, J. B. (1998). Infant cognition. In W. Damon, D. Kuhn, & R. S. Siegler (Eds.), *Handbook of child psychology (5th edition), Cognition, perception, and language* (Vol. 2, pp. 199–254). New York: Wiley.

Halle, T. G., Kurtz-Costes, B., & Mahoney, J. L. (1997). Family influences on school achievement in low-income, African American children. *Journal of Educational Psychology, 89,* 527–537.

Hamann, M. S., & Ashcraft, M. H. (1986). Textbook presentations of the basic addition facts. *Cognition and Instruction, 3,* 173–192.

Han, Y., & Ginsburg, H. P. (2001). Chinese and English mathematics language: The relation between linguistic clarity and mathematics performance. *Mathematical Thinking and Learning, 3,* 201–220.

Hannibal, M. A. Z., & Clements, D. H. (2000). Young children's understanding of basic geometric shapes. *Manuscript submitted for publication.*

Hannula, M. M. (2005). *Spontaneous focusing on numerosity in the development of early mathematical skills.* Turku, Finland: University of Turku.

Harper, E. H., & Steffe, L. P. (1968). *The effects of selected experiences on the ability of kindergarten and first-grade children to conserve numerousness. Technical Report No. 38, Research and Development Center for Cognitive Learning, C-03, OE 5-10-154:* Madison, WI: University of Wisconsin.

Harris, L. J. (1981). Sex-related variations in spatial skill. In L. S. Liben, A. H. Patterson, & N. Newcombe (Eds.), *Spatial representation and behavior across the life span* (pp. 83–125). New York: Academic Press.

Hart, B., & Risley, T. R. (1995). *Meaningful differences in the everyday experience of young American children.* Baltimore, MD: Paul H. Brookes.

Hart, B., & Risley, T. R. (1999). *The social world of children: Learning to talk.* Baltimore, MD: Paul H. Brookes.

Hausken, E. G., & Rathbun, A. (2004). Mathematics instruction in kindergarten: Classroom practices and outcomes. San Diego, CA: American Educational Research Association.

Heck, D. J., Weiss, I. R., Boyd, S., & Howard, M. (2002). *Lessons learned about planning and implementing statewide systemic initiatives in mathematics and science education.* New Orleans, LA: American Educational Research Association.

Hermer, L., & Spelke, E. (1996). Modularity and development: The case of spatial reorientation. *Cognition, 61,* 195–232.

Hershkowitz, R., Ben-Chaim, D., Hoyles, C., Lappan, G., Mitchelmore, M., & Vinner, S. (1990). Psychological aspects of learning geometry. In P. Nesher, & J. Kilpatrick (Eds.), *Mathematics and cognition: A research synthesis by the International Group for the Psychology of Mathematics Education* (pp. 70–95). Cambridge, U.K.: Cambridge University Press.

Hershkowitz, R., & Dreyfus, T. (1991). Loci and visual thinking. In F. Furinghetti (Ed.), *Proceedings of the fifteenth annual meeting International Group for the Psychology of Mathematics Education* (Vol. II, pp. 181–188). Genova, Italy: Program Committee.

Heuvel-Panhuizen, M. v. d. (1990). Realistic arithmetic/mathematics instruction and tests. In K. P. E. Gravemeijer, M. van den Heuvel, & L. Streefland (Eds.), *Contexts free productions tests and geometry in realistic mathematics education* (pp. 53–78). Utrecht, The Netherlands: OW&OC.

Hiebert, J. C. (1981). Cognitive development and learning linear measurement. *Journal for Research in Mathematics Education, 12,* 197–211.

Hiebert, J. C. (1984). Why do some children have trouble learning measurement concepts? *Arithmetic Teacher, 31*(7), 19–24.

Hiebert, J. C. (1986). *Conceptual and procedural knowledge: The case of mathematics.* Hillsdale, NJ: Lawrence Erlbaum.

Hiebert, J. C. (1999). Relationships between research and the NCTM Standards. *Journal for Research in Mathematics Education, 30,* 3–19.

Hinkle, D. (2000). *School involvement in early childhood:* National Institute on Early Childhood Development and Education, U.S. Department of Education Office of Educational Research and Improvement.

Holloway, S. D., Rambaud, M. F., Fuller, B., & Eggers-Pierola, C. (1995). What is "appropriate practice" at home and in child care?: Low-income mothers' views on preparing their children for school. *Early Childhood Research Quarterly, 10,* 451–473.

Horvath, J., & Lehrer, R. (2000). The design of a case-based hypermedia teaching tool. *International Journal of Computers for Mathematical Learning, 5,* 115–141.

Houdé, O. (1997). Numerical development: From infant to the child. Wynn's (1992) paradigm in 2- and 3-year-olds. *Cognitive Development, 12,* 373–391.

Howell, R. D., Scott, P. B., & Diamond, J. (1987). The effects of "instant" Logo computing language on the cognitive development of very young children. *Journal of Educational Computing Research, 3*(2), 249–260.

Howlin, P., Davies, M., & Udwin, U. (1998). Syndrome specific characteristics in Williams syndrome: To what extent do early behavioral patterns persist into adult life? *Journal of Applied Research in Intellectual Disabilities, 11*(207–226).

Hughes, M. (1981). Can preschool children add and subtract? *Educational Psychology, 1,* 207–219.

Hughes, M. (1986). *Children and number: Difficulties in learning mathematics.* Oxford, U.K.: Basil Blackwell.

Humphrey, G. K., & Humphrey, G. K. (1995). The role of structure in infant visual pattern perception. *Canadian Journal of Psychology, 43*(2), 165–182.

Huntley-Fenner, G. (1999a). Infants' expectation of quantity varies with material kind. Albuquerque, NM: Society for Research in Child Development.

Huntley-Fenner, G. (1999b). The effect of material kind on preschoolers' judgments of quantity. Albuquerque, NM: Society for Research in Child Development.

Huntley-Fenner, G. (2001). Why count stuff?: Young preschoolers do not use number for measurement in continuous dimensions. *Developmental Science, 4,* 456–462.

Huntley-Fenner, G., & Cannon, E. (2000). Preschoolers' magnitude comparisons are mediated by a preverbal analog mechanism. *Psychological Science, 11,* 147–152.

Huntley-Fenner, G., Carey, S., & Solimando, A. (2002). Objects are individuals but stuff doesn't count: Perceived rigidity and cohesiveness influence infants' representations of small groups of discrete entities. *Cognition, 85,* 203–221.

Huttenlocher, J., Duffy, S., & Levine, S. C. (2002). Infants and toddlers discriminate amount: Are they measuring? *Psychological Science, 13,* 244–249.

Huttenlocher, J., Jordan, N. C., & Levine, S. C. (1994). A mental model for early arithmetic. *Journal of Experimental Psychology: General, 123,* 284–296.

Huttenlocher, J., Newcombe, N. S., & Vasilyeva, M. (1999). Spatial scaling in young children. *Psychological Science, 10,* 393–398.

Ibbotson, A., & Bryant, P. E. (1976). The perpendicular error and the vertical effect in children's drawing. *Perception, 5,* 319–326.

Ilg, F., & Ames, L. B. (1951). Developmental trends in arithmetic. *Journal of General Psychology, 79,* 3–28.

Inhelder, B., & Piaget, J. (1958). *The growth of logical thinking from childhood to adolescence.* New York: Basic Books.

Inhelder, B., Sinclair, H., & Bovet, M. (1974). *Learning and the development of cognition.* Cambridge, MA: Harvard University Press.

Jimerson, S., Egeland, B., & Teo, A. (1999). A longitudinal study of achievement trajectories: Factors associated with change. *Journal of Educational Psychology, 91,* 116–126.

Johnson, M. (1987). *The body in the mind.* Chicago: The University of Chicago Press.

Johnson-Gentile, K., Clements, D. H., & Battista, M. T. (1994). The effects of computer and noncomputer environments on students' conceptualizations of geometric motions. *Journal of Educational Computing Research, 11*(2), 121–140.

Johnson-Pynn, J. S., Ready, C., & Beran, M. (2005, April). *Estimation mediates preschoolers: numerical reasoning: evidence against precise calculation abilities.* Paper presented at the Biennial Meeting of the Society for Research in Child Development, Atlanta, GA.

Jones, S. S., & Smith, L. B. (2002). How children know the relevant properties for generalizing object names. *Developmental Science, 2,* 219–232.

Jordan, N. C., Hanich, L. B., & Kaplan, D. (2003). A longitudinal study of mathematical competencies in children with specific mathematics difficulties versus children with comorbid mathematics and reading difficulties. *Child Development, 74,* 834–850.

Jordan, N. C., Hanich, L. B., & Uberti, H. Z. (2003). Mathematical thinking and learning difficulties. In A. J. Baroody, & A. Dowker (Eds.), *The development of arithmetic concepts and skills: Constructing adaptive expertise* (pp. 359–383). Mahwah, NJ: Erlbaum.

Jordan, N. C., Huttenlocher, J., & Levine, S. C. (1992). Differential calculation abilities in young children from middle- and low-income families. *Developmental Psychology, 28,* 644–653.

Jordan, N. C., Huttenlocher, J., & Levine, S. C. (1994). Assessing early arithmetic abilities: Effects of verbal and nonverbal response types on the calculation performance of middle- and low-income children. *Learning and Individual Differences, 6,* 413–432.

Jordan, N. C., Kaplan, D., Oláh, L. N., & Locuniak, M. N. (2005). Number sense growth in kindergarten: A longitudinal investigation of children at risk for mathematics difficulties.

Jovignot, F. (1995). Can 5–6 year old children orientate themselves in a cave? *Scientific Journal of Orienteering, 11*(2), 64–75.

Judd, C. H. (1927). *Psychological analysis of the fundamentals of arithmetic. Supplementary Educational Monographs, No. 32.* Chicago: Department of Education, University of Chicago.

Kabanova-Meller, E. N. (1970). The role of the diagram in the application of geometric theorems. In J. Kilpatrick, & I. Wirszup (Eds.), *Soviet studies in the psychology of learning and teaching mathematics (Vols. 4)* (pp. 7–49m). Chicago: University of Chicago Press.

Kamii, C., Miyakawa, Y., & Kato, Y. (2004). The development of logico-mathematical knowledge in a block-building activity at ages 1–4. *Journal of Research in Childhood Education, 19,* 13–26.

Kamii, C. (1985). *Young children reinvent arithmetic: Implications of Piaget's theory.* New York: Teaching College Press.

Kamii, C., & Clark, F. B. (1997). Measurement of length: The need for a better approach to teaching. *School Science and Mathematics, 97,* 116–121.

Kamii, C., & DeVries, R. (1993). Piaget for early education. In M. C. Day, & R. Parker (Eds.), *The preschool in action* (2nd ed.). Boston: Allyn and Bacon.

Kamii, C., & Housman, L. B. (1999). *Young children reinvent arithmetic: Implications of Piaget's theory* (2nd ed.). New York: Teaching College Press.

Kamii, C., & Lewis, B. (1993). The harmful effects of algorithms . . . in primary arithmetic. *Teaching K–8, 23*(4), 36–38.

Kamii, C., Rummelsburg, J., & Kari, A. R. (2005). Teaching arithmetic to low-performing, low-SES first graders. *Journal of Mathematical Behavior, 24,* 39–50.

Kapadia, R. (1974). A critical examination of Piaget-Inhelder's view on topology. *Educational Studies in Mathematics, 5,* 419–424.

Karmiloff-Smith, A. (1984). Children's problem solving. In M. E. Lamb, A. L. Brown, & B. Rogoff (Eds.), *Advances in developmental psychology (Vol. 3)* (pp. 39–90). Mahwah, NJ: Erlbaum.

Karmiloff-Smith, A. (1986). From meta-processes to conscious access: Evidence from children's metalinguistic and repair data. *Cognition, 23,* 95–147.

Karmiloff-Smith, A. (1990). Constraints on representational change: Evidence from children's drawing. *Cognition, 34,* 57–83.

Karmiloff-Smith, A. (1992). *Beyond modularity: A developmental perspective on cognitive science.* Cambridge, MA: MIT Press.

Karoly, L. A., Greenwood, P. W., Everingham, S. S., Houbé, J., Kilburn, M. R., Rydell, C. P., et al. (1998). *Investing in our children: What we know and don't know about the costs and benefits of early childhood interventions.* Santa Monica, CA: Rand Education.

Kaufman, E. L., Lord, M. W., Reese, T. W., & Volkmann, J. (1949). The discrimination of visual number. *American Journal of Psychology, 62,* 498–525.

Kellman, P. J., & Banks, M. S. (1998). Infant visual perception. In W. Damon, D. Kuhn, & R. S. Siegler (Eds.), *Handbook of child psychology (5th ed.), Cognition, perception, and language* (Vol. 2, pp. 103–146). New York: Wiley.

Kerslake, D. (1979). Visual mathematics. *Mathematics in the School, 8,* 34–35.

Kidd, J. K., & Pasnak, R. (2005, April). *Increasing learning by promoting early abstract thought.* Paper presented at the American Educational Research Association, Montréal, Québec.

Kilpatrick, J. (1987). Problem formulating: Where do good problems come from? In A. H. Schoenfeld (Ed.), *Cognitive science and mathematics education* (pp. 123–147). Hillsdale, NJ: Erlbaum.

Kilpatrick, J., Swafford, J., & Findell, B. (2001). *Adding it up: Helping children learn mathematics.* Washington, DC: National Academy Press.

Kim, S.-Y. (1994). The relative effectiveness of hands-on and computer-simulated manipulatives in teaching seriation, classification, geometric, and arithmetic concepts to kindergarten children. *Dissertation Abstracts International, 54/09,* 3319.

Klahr, D., & Wallace, J. G. (1976). *Cognitive development: An information-processing view.* Mahwah, NJ: Erlbaum.

Klein, A., & Starkey, P. (1988). Universals in the development of early arithmetic cognition. In G. B. Saxe, & M. Gearhart (Eds.), *Children's mathematics* (pp. 27–54). San Francisco: Jossey-Bass.

Klein, A., & Starkey, P. (2004). Fostering preschool children's mathematical development: Findings from the Berkeley Math Readiness Project. In D. H. Clements, J. Sarama, & A.-M. DiBiase (Eds.), *Engaging young children in mathematics: Standards for early childhood mathematics education* (pp. 343–360). Mahwah, NJ: Erlbaum.

Klein, A., Starkey, P., & Wakeley, A. (1999, April). *Enhancing prekindergarten children's readiness for school mathematics.* Paper presented at the American Educational Research Association. Montreal, Canada.

Klein, J. S., & Bisanz, J. (2000). Preschoolers doing arithmetic: The concepts are willing but the working memory is weak. *Canadian Journal of Experimental Psychology, 54*(2), 105–115.

Koechlin, E., Dehaene, S., & Mehler, J. (1997). Numerical transformation in five-month-old human infants. *Mathematical Cognition, 3,* 89–104.

Kofsky, E. (1966). A scalogram study of classificatory development. *Child Development, 37,* 191–204.

Kosslyn, S. M. (1983). *Ghosts in the mind's machine.* New York: W. W. Norton.

Kosslyn, S. M., Reiser, B. J., & Ball, T. M. (1978). Visual images preserve metric spatial information: Evidence from studies of image scanning. *Journal of Experimental Psychology: Human Perception and Performance, 4*(1), 47–60.

Kotovsky, L., & Gentner, D. (1996). Comparison and categorization in the development of relational similarity. *Child Development, 67,* 2797–2822.

Krajewski, K. (2005, April). *Prediction of mathematical (dis-)abilities in primary school: A 4-year German longitudinal study from Kindergarten to grade 4* Paper presented at the Biennial Meeting of the Society for Research in Child Development, Atlanta, GA.

Krascum, R., & Andrews, S. (1993). Feature-based versus exemplar-based strategies in preschoolers' category learning. *Journal of Experimental Child Psychology, 56,* 1–48.

Kronholz, J. (2000, May 16). See Johnny jump! Hey, isn't it math he's really doing? *The Wall Street Journal*, p. A1; A12.

Kull, J. A. (1986). Learning and Logo. In P. F. Campbell, & G. G. Fein (Eds.), *Young children and microcomputers* (pp. 103–130). Englewood Cliffs, NJ: Prentice-Hall.

Landau, B. (1996). Multiple geometric representations of objects in languages and language learners. In P. Bloom, M. A. Peterson, L. Nadel, & M. F. Garrett (Eds.), *Language and space* (pp. 317–363). Cambridge, MA: MIT Press.

Landau, B., Gleitman, H., & Spelke, E. (1981). Spatial knowledge and geometric representation in a child blind from birth. *Science, 213*, 1275–1277.

Langer, J., Gillette, P., & Arriaga, R. I. (2003). Toddlers' cognition of adding and subtracting objects in action and in perception. *Cognitive Development, 18*, 233–246.

Langer, J., Rivera, S. M., Schlesinger, M., & Wakeley, A. (2003). Early cognitive development: Ontogeny and phylogeny. In J. Valsiner, & K. J. Connolly (Eds.), *Handbook of developmental psychology* (pp. 141–171). London: Sage.

Lansdell, J. M. (1999). Introducing young children to mathematical concepts: Problems with 'new' terminology. *Educational Studies, 25*, 327–333.

Lara-Cinisomo, S., Pebley, A. R., Vaiana, M. E., & Maggio, E. (2004). *Are L.A.'s children ready for school?* Santa Monica, CA: RAND Corporation.

Le Corre, M., Van de Walle, G. A., Brannon, E. M., & Carey, S. (2005). Re-visiting the competence/performance debate in the acquisition of counting as a representation of the positive integer.

Lean, G., & Clements, M. A. (1981). Spatial ability, visual imagery, and mathematical performance. *Educational Studies in Mathematics, 12*, 267–299.

Lebron-Rodriguez, D. E., & Pasnak, R. (1977). Induction of intellectual gains in blind children. *Journal of Experimental Child Psychology, 24*, 505–515.

Lee, J. (2002). Racial and ethnic achievement gap trends: Reversing the progress toward equity? *Educational Researcher, 31*, 3–12.

Leeb-Lundberg, K. (1996). The block builder mathematician. In E. S. Hirsh (Ed.), *The block book* (3rd ed., pp. 35–60). Washington, DC: National Association for the Education of Young Children.

Leeson, N. (1995). Investigations of kindergarten students' spatial constructions. In B. Atweh, & S. Flavel (Eds.), *Proceedings of 18th Annual Conference of Mathematics Education Research Group of Australasia* (pp. 384–389). Darwin, AU: Mathematics Education Research Group of Australasia.

Leeson, N., Stewart, R., & Wright, R. J. (1997). Young children's knowledge of three-dimensional shapes: Four case studies. In F. Biddulph, & K. Carr (Eds.), *Proceedings of the 20th Annual Conference of the Mathematics Education Research Group of Australasia* (Vol. 1, pp. 310–317). Hamilton, New Zealand: MERGA.

Lehrer, R. (2003). Developing understanding of measurement. In J. Kilpatrick, W. G. Martin, & D. Schifter (Eds.), *A research companion to Principles and Standards for School Mathematics* (pp. 179–192). Reston, VA: National Council of Teachers of Mathematics.

Lehrer, R., Jenkins, M., & Osana, H. (1998). Longitudinal study of children's reasoning about space and geometry. In R. Lehrer, & D. Chazan (Eds.), *Designing learning environments for developing understanding of geometry and space* (pp. 137–167). Mahwah, NJ: Erlbaum.

Lerner, J. (1997). *Learning disabilities*. Boston: Houghton Mifflin Company.

Lesh, R. A. (1972). The generalization of Piagetian operations as it relates to the hypothesized functional interdependence between class, series, and number concepts. *Dissertation Abstracts International, 32*, 4731B.

Lester, F. K., Jr., & Wiliam, D. (2002). On the purpose of mathematics education research: Making productive contributions to policy and practice. In L. D. English (Ed.), *Handbook of International Research in Mathematics Education* (pp. 489–506). Mahwah, NJ: Erlbaum.

Leushina, A. M. (1974/1991). *The development of elementary mathematical concepts in preschool children* (Vol. 4). Reston, VA: National Council of Teachers of Mathematics.

Levine, S. C., Huttenlocher, J., Taylor, A., & Langrock, A. (1999). Early sex differences in spatial skill. *Developmental Psychology, 35*(4), 940–949.

Levine, S. C., Jordan, N. C., & Huttenlocher, J. (1992). Development of calculation abilities in young children. *Journal of Experimental Child Psychology, 53*, 72–103.

Lewellen, H. (1992). *Description of van Hiele levels of geometric development with respect to geometric motions*. Unpublished doctoral dissertation, Kent State University.

Liben, L. S. (1978). Performance on Piagetian spatial tasks as a function of sex, field dependence, and training. *Merrill-Palmer Quarterly, 24*, 97–110.

Liben, L. S. (1988). Conceptual issues in the development of spatial cognition. In J. Stiles-Davis, M. Kritchevsky, & U. Bellugi (Eds.), *Spatial cognition: Brain bases and development* (pp. 145–201). Mahwah, NJ: Erlbaum.

Liben, L. S., & Downs, R. M. (1989). Understanding maps as symbols: The development of map concepts in children. In H. W. Reese (Ed.), *Advances in child development and behavior: Vol. 22* (pp. 145–201). San Diego: Academic Press.

Liben, L. S., Moore, M. L., & Golbeck, S. L. (1982). Preschooler's knowledge of their classroom environment. Evidence from small-scale and life-size spatial tasks. *Child Development, 53*, 1275–1284.

Liben, L. S., & Yekel, C. A. (1996). Preschoolers' understanding of plan and oblique maps: The role of geometric and representational correspondence. *Child Development, 67*(6), 2780–2796.

Linnell, M., & Fluck, M. (2001). The effect of maternal support for counting and cardinal understanding in pre-school children. *Social Development, 10*, 202–220.

Lockman, J. J., & Pick, H. L., Jr. (1984). Problems of scale in spatial development. In C. Sophian (Ed.), *Origins of cognitive skills* (pp. 3–26). Mahwah, NJ: Erlbaum.

Loeb, S., Bridges, M., Bassok, D., Fuller, B., & Rumberger, R. (in press). How much is too much? The influence of preschool centers on children's development nationwide. *Economics of Education Review*.

Loewenstein, J., & Gentner, D. (2001). Spatial mapping in preschoolers: Close comparisons facilitate far mappings. *Journal of Cognition and Development, 2*, 189–219.

Logan, G. (1994). Spatial attention and the apprehension of spatial relations. *Journal of Experimental Psychology: Human Perception and Performance, 20*, 1015–1036.

Lonigan, C. J. (2003). Comment on Marcon (ECRP, Vol. 4, No. 1, Spring 2002): "Moving up the grades: Relationship between preschool model and later school success". *Early Childhood Research & Practice, 5*(1).

Lovell, K. (1959). A follow-up study of some aspects of the work of Piaget and Inhelder on the child's conception of space. *British Journal of Educational Psychology, 29,* 104–117.

Mackay, C. K., Brazendale, A. H., & Wilson, L. F. (1972). Concepts of horizontal and vertical: A methodological note. *Developmental Psychology, 7,* 232–237.

Magnuson, K. A., Meyers, M. K., Rathbun, A., & West, J. (2004). Inequality in preschool education and school readiness. *American Educational Research Journal, 41,* 115–157.

Magnuson, K. A., & Waldfogel, J. (2005). Early childhood care and education: Effects on ethnic and racial gaps in school readiness. *The Future of Children, 15,* 169–196.

Mandler, G., & Shebo, B. J. (1982). Subitizing: An analysis of its component processes. *Journal of Experimental Psychology: General, 111,* 1–22.

Mansfield, H. M., & Scott, J. (1990). Young children solving spatial problems. In G. Booker, P. Cobb, & T. N. deMendicuti (Eds.), *Proceedings of the 14th annual conference of the Internation Group for the Psychology of Mathematics Education* (Vol. 2, pp. 275–282). Oaxlepec, Mexico: Internation Group for the Psychology of Mathematics Education.

Maratsos, M. P. (1973). Decrease in the understanding of the word "big" in preschool children. *Child Development, 44,* 747–752.

Marcon, R. A. (1992). Differential effects of three preschool models on inner-city 4-year-olds. *Early Childhood Research Quarterly, 7,* 517–530.

Marcon, R. A. (2002). Moving up the grades: Relationship between preschool model and later school success. *Early Childhood Research & Practice.* Retrieved 1, 4, from http://ecrp.uiuc.edu/v4n1/marcon.html

Martin, J. L. (1976). An analysis of some of Piaget's topological tasks from a mathematical point of view. *Journal for Research in Mathematics Education, 7,* 8–24.

Mayberry, J. (1983). The van Hiele levels of geometric thought in undergraduate preservice teaching. *Journal for Research in Mathematics Education, 14,* 58–69.

McClelland, J. L., Rumelhart, D. E., & the PDP Research Group. (1986). *Parallel distributed processing: Explorations in the microstructure of cognition. Volume 2: Psychological and biological models.* Cambridge, MA: MIT Press.

McCrink, K., & Wynn, K. (2004). Large number addition and subtraction by 9-month-old infants. *Psychological Science, 15,* 776–781.

McCulloch, W. (1963). *Embodiments of mind.* Cambridge, MA: MIT Press.

McDonough, L. (1999). Early declarative memory for location. *British Journal of Developmental Psychology, 17,* 381–402.

McGee, M. G. (1979). Human spatial abilities: Psychometric studies and environmental, genetic, hormonal, and neurological influences. *Psychological Bulletin, 86,* 889–918.

Meck, W. H., & Church, R. M. (1983). A mode control model of counting and timing processes. *Journal of Experimental Psychology: Animal Behavior Processes, 9,* 320–334.

Mervis, C. B., & Rosch, E. (1981). Categorization of natural objects. *Annual Review of Psychology, 32,* 89–115.

Michie, S. (1984a). Number understanding in preschool children. *British Journal of Educational Psychology, 54,* 245–253.

Michie, S. (1984b). Why preschoolers are reluctant to count spontaneously. *British Journal of Developmental Psychology, 2,* 347–358.

Middleton, J. A., & Spanias, P. (1999). Motivation for achievement in mathematics: Findings, generalizations, and criticisms of the research. *Journal for Research in Mathematics Education, 30,* 65–88.

Miller, G. A. (1956). The magical number seven, plus or minus two: Some limits on our capacity for processing information. *The Psychological Review, 63,* 81–97.

Miller, K. F. (1984). Child as the measurer of all things: Measurement procedures and the development of quantitative concepts. In C. Sophian (Ed.), *Origins of cognitive skills: The eighteenth annual Carnegie symposium on cognition* (pp. 193–228). Hillsdale, NJ: Erlbaum.

Miller, K. F. (1989). Measurement as a tool of thought: The role of measuring procedures in children's understanding of quantitative invariance. *Developmental Psychology, 25,* 589–600.

Miller, K. F. (1995). Origins of quantitative competence. In R. Gelman, & T. Au (Eds.), *Handbook of perception and cognition (Vol. 13: Perceptual and cognitive development)* (Vol. 13, pp. 213–241). Orlando, FL: Academic Press.

Miller, K. F., & Baillargeon, R. (1990). Length and distance: Do preschoolers think that occlusion brings things together? *Developmental Psychology, 26,* 103–114.

Miller, K. F., Major, S. M., Shu, H., & Zhang, H. (2000). Ordinal knowledge: Number names and number concepts in Chinese and English. *Canadian Journal of Experimental Psychology, 54,* 139–149.

Miller, K. F., Smith, C. M., Zhu, J., & Zhang, H. (1995). Preschool origins of cross-national differences in mathematical competence: The role of number-naming systems. *Psychological Science, 6,* 56–60.

Miller, P. (1967). The effects of age and training on children's ability to understand certain basic concepts. *Dissertation Abstracts International, 28,* 2161B–2162B.

Minsky, M. (1986). *The society of mind.* New York: Simon and Schuster.

Miura, I. T., & Okamoto, Y. (2003). Language supports for mathematics understanding and performance. In A. J. Baroody, & A. Dowker (Eds.), *The development of arithmetic concepts and skills: Constructing adaptive expertise* (pp. 229–242). Mahwah, NJ: Erlbaum.

Mix, K. S. (1999). Preschoolers' recognition of numerical equivalence: Sequential sets. *Journal of Experimental Child Psychology, 74,* 309–332.

Mix, K. S., Huttenlocher, J., & Levine, S. C. (1996). Do preschool children recognize auditory-visual numerical correspondences? *Child Development, 67,* 1592–1608.

Mix, K. S., Huttenlocher, J., & Levine, S. C. (2002). *Quantitative development in infancy and early childhood.* New York: Oxford University Press.

Mix, K. S., Levine, S. C., & Huttenlocher, J. (1997). Numerical abstraction in infants: Another look. *Developmental Psychology, 33,* 423–428.

Mix, K. S., Sandhofer, C. M., & Baroody, A. J. (2005). Number words and number concepts: The interplay of verbal and nonverbal processes in early quantitative development. In R. Kail (Ed.), *Advances in Child Development and Behavior* (Vol. 33, pp. 305–345). New York: Academic Press.

Moore, D., Benenson, J., Reznick, J. S., Peterson, M., & Kagan, J. (1987). Effects of auditory numerical information on infants' looking behavior: Contradictory evidence. *Developmental Psychology, 23,* 665–670.

Morrongiello, B. A., Timney, B., Humphrey, G. K., Anderson, S., & Skory, C. (1995). Spatial knowledge in blind and sighted children. *Journal of Experimental Child Psychology, 59,* 211–233.

Moyer, J. C. (1978). The relationship between the mathematical structure of Euclidean transformations and the spontaneously developed cognitive structures of young children. *Journal for Research in Mathematics Education, 9,* 83–92.

Muir, S. P., & Cheek, H. N. (1986). Mathematics and the map skill curriculum. *School Science and Mathematics, 86,* 284–291.

Mullet, E., & Paques, P. (1991). The height + width = area of a rectangle rule in five-year-olds: effects of stimulus distribution and graduation of the response scale. *Journal of Experimental Child Psychology, 52*(3), 336–343.

Mullis, I. V. S., Martin, M. O., Beaton, A. E., Gonzalez, E. J., Kelly, D. L., & Smith, T. A. (1997). *Mathematics achievement in the primary school years: IEA's third international mathematics and science study (TIMSS).* Chestnut Hill, MA: Center for the Study of Testing, Evaluation, and Educational Policy, Boston College.

Mullis, I. V. S., Martin, M. O., Gonzalez, E. J., Gregory, K. D., Garden, R. A., O'Connor, K. M., et al. (2000). *TIMSS 1999 international mathematics report.* Boston: The International Study Center, Boston College, Lynch School of Education.

Munn, P. (1997). Children's beliefs about counting. In I. Thompson (Ed.), *Teaching and learning early number* (pp. 9–19). Philadelphia, PA: Open University Press.

Munn, P. (1998). Symbolic function in pre-schoolers. In C. Donlan (Ed.), *The development of mathematical skills* (pp. 47–71). East Sussex, UK: Psychology Press.

Murata, A. (2004). Paths to learning ten-structured understanding of teen sums: Addition solution methods of Japanese Grade 1 students. *Cognition and Instruction, 22,* 185–218.

Murphy, C. M., & Wood, D. J. (1981). Learning from pictures: The use of pictorial information by young children. *Journal of Experimental Child Psychology, 32,* 279–297.

Naito, M., & Miura, I. T. (2001). Japanese children's numerical competencies: Age- and schooling-related influences on the development of number concepts and addition skills. *Developmental Psychology, 37,* 217–230.

Natriello, G., McDill, E. L., & Pallas, A. M. (1990). *Schooling disadvantaged children: Racing against catastrophe.* New York: Teachers College Press.

National Center for Education Statistics. (1996). Pursuing excellence, NCES 97–198 (initial findings from the Third International Mathematics and Science Study). Washington DC: National Center for Education Statistics, U.S. Government Printing Office.

National Center for Education Statistics. (2000). *America's kindergartners (NCES 2000070).* Washington, DC: National Center for Education Statistics, U.S. Government Printing Office.

National Council of Teachers of Mathematics. (1989). *Curriculum and evaluation standards for school mathematics.* Reston, VA: Author.

National Council of Teachers of Mathematics. (2000). *Principles and standards for school mathematics.* Reston, VA: Author.

Neumann, P. G. (1977). Visual prototype formation with discontinuous representation of dimensions of variability. *Memory and Cognition, 5,* 187–197.

Newcombe, N. (1989). The development of spatial perspective taking. In H. W. Reese (Ed.), *Advances in Child Development and Behavior* (Vol. 22, pp. 203–247). New York: Academic Press.

Newcombe, N., & Sanderson, H. L. (1993). The relation between preschoolers' everyday activities and spatial ability. New Orleans, LA: Society for Research in Child Development.

Newcombe, N. S. (2002). The nativist-empiricist controversy in the context of recent research on spatial and quantitative development. *Psychological Science, 13*(5), 395–401.

Newcombe, N. S., & Huttenlocher, J. (2000). *Making space: The development of spatial representation and reasoning.* Cambridge, MA: MIT Press.

Newcombe, N. S., Huttenlocher, J., Drummey, A. B., & Wiley, J. G. (1998). The development of spatial location coding: Place learning and dead reckoning in the second and third years. *Cognitive Development, 13,* 185–200.

Newcombe, N. S., Huttenlocher, J., & Learmonth, A. (1999). Infants' coding of location in continuous space. *Infant Behavior and Development, 22,* 483–510.

Newcombe, N. S., & Sluzenski, J. (2004). Starting points and change in early spatial development. In G. L. Allen (Ed.), *Human spatial memory: Remembering where* (pp. 25–40). Mahwah, NJ: Erlbaum.

Nieder, A., Freedman, D. J., & Miller, E. K. (2002). Representation of the quantity of visual items in the primate prefrontal cortex. *Science, 297,* 1708–1711.

Nieuwoudt, H. D., & van Niekerk, R. (1997). The spatial competence of young children through the development of solids. Chicago: American Educational Research Association.

National Research Council. (2002). *Scientific research in education.* Washington, DC: National Research Council, National Academy Press.

Nunes, T., & Bryant, P. (1996). *Children doing mathematics.* Cambridge, MA: Balckwell.

Oakes, J. (1990). Opportunities, achievement, and choice: Women and minority students in science and mathematics. In C. B. Cazden (Ed.), *Review of research in education* (Vol. 16, pp. 153–222). Washington, DC: American Educational Research Association.

Outhred, L. N., & Sardelich, S. (1997). Problem solving in kindergarten: The development of representations. In F. Biddulph, & K. Carr (Eds.), *People in Mathematics Education. Proceedings of the 20th Annual Conference of the Mathematics Education Research Group of Australasia* (Vol. 2, pp. 376–383). Rotorua, New Zealand: Mathematics Education Research Group of Australasia.

Outhred, L. N., & Sardelich, S. (1998). Representing problems in the first year of school. In A. Olivier, & K. Newstead (Eds.), *Proceedings of the 22nd conference for the International Group for the Psychology of Mathematics Education* (Vol. 3, pp. 319–327). Stellenbosch, South Africa: University of Stellenbosch.

Owens, K. (1999). The role of visualization in young students' learning. In O. Zaslavsky (Ed.), *Proceedings of the 23rd Conference of the International Group for the Psychology of Mathematics Education* (Vol. 1, pp. 220–234). Haifa, Isreal: Technion.

Page, E. I. (1959). Haptic perception: A consideration of one of the investigations of Piaget and Inhelder. *Educational Review, 11,* 115–124.

Palmer, S. E. (1985). The role of symmetry in shape perception. *Acta Psychologica, 59*(1), 67–90.

Palmer, S. E. (1989). Reference frames in the perception of shape and orientation. In B. E. Shepp, & S. Ballesteros (Eds.), *Object perception: Structure and process* (pp. 121–163). Hillsdale, NJ: Erlbaum.

Palmer, S. E., & Hemenway, K. (1978). Orientation and symmetry: Effects of multiple, rotational, and near symmetries. *Journal of Experimental Psychology, 4*, 691–702.

Papert, S. (1980). *Mindstorms: Children, computers, and powerful ideas.* New York: Basic Books.

Pascual-Leone, J. (1976). A view of cognition from a formalist's perspective. In K. F. R. J. A. Meacham (Ed.), *The developing individual in a changing world: Vol. I Historical and cultural issues.* The Hague: The Netherlands: Mouton.

Pasnak, R. (1987). Accelerated cognitive development of kindergartners. *Psychology in the Schools, 28*, 358–363.

Pasnak, R., Brown, K., Kurkjian, M., Mattran, K., Triana, E., & Yamamoto, N. (1987). Cognitive gains through training on classification, seriation, conservation. *Genetic, Social, and General Psychology Monographs, 113*, 295–321.

Pasnak, R., Madden, S. E., Malabonga, V. A., & Holt, R. (1996). Persistence of gains from instruction in classification, seriation, and conservation. *Journal of Educational Research, 89*, 1–6.

Pegg, J., & Davey, G. (1998). Interpreting student understanding in geometry: A synthesis of two models. In R. Lehrer, & D. Chazan (Eds.), *Designing learning environments for developing understanding of geometry and space* (pp. 109–135). Mahwah, NJ: Erlbaum.

Peisner-Feinberg, E. S., Burchinal, M. R., Clifford, R. M., Culkins, M. L., Howes, C., Kagan, S. L., et al. (2001). The relation of preschool child-care quality to children's cognitive and social developmental trajectories through second grade. *Child Development, 72*, 1534–1553.

Penner-Wilger, M., Fast, L., Smith-Chant, B. L., Skwarchuck, S.-L., LeFevre, J.-A., Arnup, J. S., et al. (2005, April). *What's next? Performance on the next number task as a predictor of primary children's math achievement.* Paper presented at the Biennial Meeting of the Society for Research in Child Development, Atlanta, GA.

Perham, F. (1978). An investigation into the effect of instruction on the acquisition of transformation geometry concepts in first grade children and subsequent transfer to general spatial ability. In R. Lesh, & D. Mierkiewicz (Eds.), *Concerning the development of spatial and geometric concepts* (pp. 229–241). Columbus, OH: ERIC Clearinghouse for Science, Mathematics, and Environmental Education.

Perry, B., & Dockett, S. (2002). Young children's access to powerful mathematical ideas. In L. D. English (Ed.), *Handbook of International Research in Mathematics Education* (pp. 81–111). Mahwah, NJ: Erlbaum.

Petitto, A. L. (1990). Development of numberline and measurement concepts. *Cognition and Instruction, 7*, 55–78.

Piaget, J. (1964). *The early growth of logic in the child.* New York: W.W. Norton & Company, Inc.

Piaget, J. (1971/1974). *Understanding causality.* New York: Norton.

Piaget, J., & Inhelder, B. (1967). *The child's conception of space* (F. J. Langdon, & J. L. Lunzer, Trans.). New York: W. W. Norton.

Piaget, J., & Inhelder, B. (1971). *Mental imagery in the child.* London, U.K.: Routledge and Kegan Paul.

Piaget, J., Inhelder, B., & Szeminska, A. (1960). *The child's conception of geometry.* London: Routledge and Kegan Paul.

Piaget, J., & Szeminska, A. (1952). *The child's conception of number.* London: Routledge and Kegan Paul.

Pica, P., Lemer, C., Izard, V., & Dehaene, S. (2004). Exact and approximate arithmetic in an Amazonian indigene group. *Science, 306*, 499–503.

Pillow, B. H., & Flavell, J. H. (1986). Young children's knowledge about visual perception: Projective size and shape. *Child Development, 57*, 125–135.

Plumert, J. M., & Nichols-Whitehead, P. (1996). Parental scaffolding of young children's spatial communication. *Developmental Psychology, 32*(3), 523–532.

Poag, C. K., Cohen, R., & Weatherford, D. L. (1983). Spatial representations of young children: The role of self- versus adult-directed movement and viewing. *Journal of Experimental Child Psychology, 35*, 172–179.

Pollio, H. R., & Whitacre, J. D. (1970). Some observations on the use of natural numbers by preschool children. *Perceptual and Motor Skills, 30*, 167–174.

Porter, A. C. (1989). A curriculum out of balance: The case of elementary school mathematics. *Educational Researcher, 18*, 9–15.

Potter, M., & Levy, E. (1968). Spatial enumeration without counting. *Child Development, 39*, 265–272.

Presson, C. C. (1982). Using matter as a spatial landmark: Evidence against egocentric coding in infancy. *Developmental Psychology, 18*, 699–703.

Presson, C. C. (1987). The development of spatial cognition: Secondary uses of spatial information. In N. Eisenberg (Ed.), *Contemporary topics in developmental psychology* (pp. 77–112). New York: John Wiley & Sons.

Presson, C. C., & Somerville, S. C. (1985). Beyond egocentrism: A new look at the beginnings of spatial representation. In H. M. Wellman (Ed.), *Children's searching: The development of search skill and spatial representation* (pp. 1–26). Mahwah, NJ: Erlbaum.

Prigge, G. R. (1978). The differential effects of the use of manipulative aids on the learning of geometric concepts by elementary school children. *Journal for Research in Mathematics Education, 9*, 361–367.

Ramey, C. T., & Ramey, S. L. (1998). Early intervention and early experience. *American Psychologist, 53*, 109–120.

Rasmussen, C., Ho, E., & Bisanz, J. (2003). Use of the mathematical principle of inversion in young children. *Journal of Experimental Child Psychology, 85*, 89–102.

Rathbun, A., & West, J. (2004). *From kindergarten through third grade: Children's beginning school experiences.* Washington, DC: U.S. Department of Education, National Center for Education Statistics.

Rattermann, M. J., & Gentner, D. (1998). The effect of language on similarity: The use of relational labels improves young children's performance in a mapping task. In K. Holyoak, D. Gentner, & B. Kokinov (Eds.), *Advances in analogy research: Integration of theory & data from the cognitive, computational, and neural sciences* (pp. 274–282). Sophia, Bulgaria: New Bulgarian University.

Raven, K. E., & Gelman, S. A. (1984). Rule usage in children's understanding of "big" and "little". *Child Development, 55*, 2141–2150.

Razel, M., & Eylon, B.-S. (1990). Development of visual cognition: Transfer effects of the Agam program. *Journal of Applied Developmental Psychology, 11*, 459–485.

Regier, T., & Carlson, L. (2002). Spatial language: Perceptual constraints and linguistic variation. In N. L. Stein, P. J. Bauer, & M. Rabinowitz (Eds.), *Representation, memory, and development. Essays in honor of Jean Mandler* (pp. 199–221). Mahwah, NJ: Erlbaum.

Resnick, L. B. (1994). Situated rationalism: Biological and social preparation for learning. In H. L. A., & S. A. Gelman

(Eds.), *Mapping the mind. Domain-specificity in cognition and culture* (pp. 474–493). Cambridge, MA: Cambridge University Press.

Resnick, L. B., & Singer, J. A. (1993). Protoquantitative origins of ratio reasoning. In T. P. Carpenter, E. H. Fennema, & T. A. Romberg (Eds.), *Rational numbers: An integration of research* (pp. 107–130). Hillsdale, NJ: Erlbaum.

Reynolds, A., & Wheatley, G. H. (1996). Elementary students' construction and coordination of units in an area setting. *Journal for Research in Mathematics Education, 27*(5), 564–581.

Rieser, J. J. (1979). Spatial orientation of six-month-old infants. *Child Development, 50,* 1078–1087.

Rieser, J. J., Doxsey, P. A., McCarrell, N. S., & Brooks, P. H. (1982). Wayfinding and toddlers' use of information from an aerial view. *Developmental Psychology, 18,* 714–720.

Rieser, J. J., Garing, A. E., & Young, M. F. (1994). Imagery, action, and young children's spatial orientation: It's not being there that counts, it's what one has in mind. *Child Development, 65,* 1262–1278.

Rittle-Johnson, B., & Siegler, R. S. (1998). The relation between conceptual and procedural knowledge in learning mathematics: A review. In C. Donlan (Ed.), *The development of mathematical skills* (pp. 75–110). East Sussex, UK: Psychology Press.

Roberts, R. J., Jr., & Aman, C. J. (1993). Developmental differences in giving directions: Spatial frames of reference and mental rotation. *Child Development, 64,* 1258–1270.

Rosch, E. (1975). Cognitive representations of semantic categories. *Journal of Experimental Psychology: General, 104,* 192–233.

Rosser, R. A. (1994). Children's solution strategies and mental rotation problems: The differential salience of stimulus components. *Child Study Journal, 24,* 153–168.

Rosser, R. A., Ensing, S. S., Glider, P. J., & Lane, S. (1984). An information-processing analysis of children's accuracy in predicting the appearance of rotated stimuli. *Child Development, 55,* 2204–2211.

Rosser, R. A., Horan, P. F., Mattson, S. L., & Mazzeo, J. (1984). Comprehension of Euclidean space in young children: The early emergence of understanding and its limits. *Genetic Psychology Monographs, 110,* 21–41.

Rosser, R. A., Lane, S., & Mazzeo, J. (1988). Order of acquisition of related geometric competencies in young children. *Child Study Journal, 18,* 75–90.

Roth, J., Carter, R., Ariet, M., Resnick, M. B., & Crans, G. (2000, April). *Comparing fourth-grade math and reading achievement of children who did and did not participate in Florida's statewide Prekindergarten Early Intervention Program.* Paper presented at the American Educational Research Association, New Orleans, LA.

Rouse, C., Brooks-Gunn, J., & McLanahan, S. (2005). Introducing the issue. *The Future of Children, 15,* 5–14.

Rulence-Paques, P., & Mullet, E. (1998). Area judgment from width and height information: The case of the rectangle. *Journal of Experimental Child Psychology, 69*(1), 22–48.

Russell, J. (1975). The interpretation of conservation instructions by five-year-old children. *Journal of Child Psychology and Psychiatry, 16,* 233–244.

Russell, S. J. (1991). Counting noses and scary things: Children construct their ideas about data. In D. Vere-Jones (Ed.), *Proceedings of the Third International Conference on Teaching Statistics.* Voorburg, The Netherlands: International Statistical Institute.

Sales, C. (1994). *A constructivist instructional project on developing geometric problem solving abilities using pattern blocks and tangrams with young children.* Unpublished Masters, University of Northern Iowa, Cedar Falls, Iowa.

Sandberg, E. H., & Huttenlocher, J. (1996). The development of hierarchical representation of two-dimensional space. *Child Development, 67*(3), 721–739.

Sandhofer, C. M., & Smith, L. B. (1999). Learning color words involves learning a system of mappings. *Developmental Psychology, 35,* 668–679.

Sarama, J. (2002). Listening to teachers: Planning for professional development. *Teaching Children Mathematics, 9,* 36–39.

Sarama, J., & Clements, D. H. (2002). *Building Blocks* for young children's mathematical development. *Journal of Educational Computing Research, 27*(1&2), 93–110.

Sarama, J., Clements, D. H., Swaminathan, S., McMillen, S., & González Gómez, R. M. (2003). Development of mathematical concepts of two-dimensional space in grid environments: An exploratory study. *Cognition and Instruction, 21,* 285–324.

Sarama, J., Clements, D. H., & Vukelic, E. B. (1996). The role of a computer manipulative in fostering specific psychological/mathematical processes. In E. Jakubowski, D. Watkins, & H. Biske (Eds.), *Proceedings of the 18th annual meeting of the North America Chapter of the International Group for the Psychology of Mathematics Education* (Vol. 2, pp. 567–572). Columbus, OH: ERIC Clearinghouse for Science, Mathematics, and Environmental Education.

Sarama, J., & DiBiase, A.-M. (2004). The professional development challenge in preschool mathematics. In D. H. Clements, J. Sarama, & A.-M. DiBiase (Eds.), *Engaging young children in mathematics: Standards for early childhood mathematics education* (pp. 415–446). Mahwah, NJ: Erlbaum.

Satlow, E., & Newcombe, N. (1998). When is a triangle not a triangle? Young children's developing concepts of geometric shape. *Cognitive Development, 13,* 547–559.

Saxe, G. B., Guberman, S. R., & Gearhart, M. (1987). Social processes in early number development. *Monographs of the Society for Research in Child Development, 52*(2, Serial #216).

Schaeffer, B., Eggleston, V. H., & Scott, J. L. (1974). Number development in young children. *Cognitive Psychology, 6,* 357–379.

Schiff, W. (1983). Conservation of length redux: A perceptual-linguistic phenomenon. *Child Development, 54,* 1497–1506.

Scholnick, E. K., Fein, G. G., & Campbell, P. F. (1990). Changing predictors of map use in wayfinding. *Developmental Psychology, 26,* 188–193.

Schultz, K. A., & Austin, J. D. (1983). Directional effects in transformational tasks. *Journal for Research in Mathematics Education, 14,* 95–101.

Secada, W. G. (1992). Race, ethnicity, social class, language, and achievement in mathematics. In D. A. Grouws (Ed.), *Handbook of research on mathematics teaching and learning* (pp. 623–660). New York: Macmillan.

Semrud-Clikeman, M., & Hynd, G. W. (1990). Right hemispheric dysfunction in nonverbal learning disabilities: Social, academic, and adaptive functioning in adults and children. *Psychological Bulletin, 107*(196–209).

Sena, R., & Smith, L. B. (1990). New evidence on the development of the word big. *Child Development, 61,* 1034–1052.

Senk, S. L. (1989). Van Hiele levels and achievement in writing geometry proofs. *Journal for Research in Mathematics Education, 20,* 309–321.

Seo, K.-H., & Ginsburg, H. P. (2004). What is developmentally appropriate in early childhood mathematics education? In D. H. Clements, J. Sarama, & A.-M. DiBiase (Eds.), *Engaging young children in mathematics: Standards for early childhood mathematics education* (pp. 91–104). Mahwah, NJ: Erlbaum.

Share, D. L., Moffitt, T. E., & Silva, P. A. (1988). Factors associated with arithmetic and reading disabilities and specific arithmetic disability. *Journal of Learning Disabilities, 21*(313–320).

Shaw, K., Nelsen, E., & Shen, Y.-L. (2001, April). *Preschool development and subsequent school achievement among Spanish-speaking children from low-income families.* Paper presented at the American Educational Research Association, Seattle, WA.

Shepard, R. N. (1978). The mental image. *American Psychologist, 33*, 125–137.

Shepard, R. N., & Cooper, L. A. (1982). *Mental images and their transformations.* Cambridge, MA: The MIT Press.

Shipley, E. F., & Shepperson, B. (1990). Countable entities: Developmental changes. *Cognition, 34*, 109–136.

Shonkoff, J. P., & Phillips, D. A. (Eds.). (2000). *From neurons to neighborhoods: The science of early childhood development.* Washington, DC: National Academy Press.

Shrager, J., & Siegler, R. S. (1998). SCADS: A model of children's strategy choice and strategy discoveries. *Psychological Science, 9*, 405–410.

Shuman, M., & Spelke, E. S. (2005, April). *The development of numerical magnitude representation.* Paper presented at the Biennial Meeting of the Society for Research in Child Development, Atlanta, GA.

Siegel, A. W., & Schadler, M. (1977). The development of young children's spatial representations of their classrooms. *Child Development, 48*, 388–394.

Siegel, A. W., & White, S. H. (1975). The development of spatial representations of large-scale environments. In H. W. Resse (Ed.), *Advances in Child Development and Behavior* (Vol. 10, pp. 9–55). New York: Academic Press.

Siegler, R. S. (1986). *Children's thinking.* Englewood Cliffs, NJ: Prentice-Hall.

Siegler, R. S. (1993). Adaptive and non-adaptive characteristics of low income children's strategy use. In L. A. Penner, G. M. Batsche, H. M. Knoff, & D. L. Nelson (Eds.), *Contributions of psychology to science and mathematics education* (pp. 341–366). Washington, DC: American Psychological Association.

Siegler, R. S. (1995). How does change occur: A microgenetic study of number conservation. *Cognitive Psychology, 28*, 255–273.

Siegler, R. S. (1996). *Emerging minds: The process of change in children's thinking.* New York: Oxford University Press.

Siegler, R. S., & Jenkins, E. (1989). *How children discover new strategies.* Mahwah, NJ: Erlbaum.

Siegler, R. S., & Robinson, M. (1982). The development of numerical understandings. In H. W. Reese, & L. P. Lipsitt (Eds.), *Advances in child development and behavior* (Vol. 16, pp. 241–312). New York: Academic Press.

Siegler, R. S., & Shrager, J. (1984). Strategy choices in addition and subtraction: How do children know what to do? In C. Sophian (Ed.), *The origins of cognitive skills* (pp. 229–294). Hillsdale, NJ: Erlbaum.

Silliphant, V. (1983). Kindergarten reasoning and achievement in grades K–3. *Psychology in the Schools, 20*, 289–294.

Silverman, I. W., & Rose, A. P. (1980). Subitizing and counting skills in 3-year-olds. *Developmental Psychology, 16*, 539–540.

Simon, M. A. (1995). Reconstructing mathematics pedagogy from a constructivist perspective. *Journal for Research in Mathematics Education, 26*(2), 114–145.

Simon, T. J. (1997). Reconceptualizing the origins of numerical knowledge: A "nonnumerical" account. *Cognitive Development, 12*, 349–372.

Simon, T. J., Hespos, S. J., & Rochat, P. (1995). Do infants understand simple arithmetic? A replication of Wynn (1992). *Cognitive Development, 10*, 253–269.

Skemp, R. (1976, December). Relational understanding and instrumental understanding. *Mathematics Teaching, 77*, 20–26.

Slater, A., Mattock, A., & Brown, E. (1990). Size constancy at birth: Newborn infants' responses to retinal and real size. *Journal of Experimental Child Psychology, 49*, 314–322.

Smith, L. B. (1989). A model of perceptual classification in children and adults. *Psychological Review, 96*(1), 125–144.

Smith, L. B., Jones, S. S., Landau, B., Gershkoff-Stowe, L., & Samuelson, L. (2002). Object name learning provides on-the-job training for attention. *Psychological Science, 13*, 13–19.

Snyder, S. S., & Feldman, D. H. (1984). Phases of transition in cognitive development: Evidence from the domain of spatial representation. *Child Development, 55*, 981–989.

Solomon, T. (2003, April). *Early development of the ability to interpret graphs graphs.* Paper presented at the Biennial Meeting of the Society for Research in Child Development, Tampa, FL.

Solter, A. L. J. (1976). Teaching counting to nursery school children. *Dissertation Abstracts International, 36*(8-A), 5844B.

Somerville, S. C., & Bryant, P. E. (1985). Young children's use of spatial coordinates. *Child Development, 56*, 604–613.

Somerville, S. C., Bryant, P. E., Mazzocco, M. M. M., & Johnson, S. P. (1987). *The early development of children's use of spatial coordinates.* Baltimore, MD: Society for Research in Child Development.

Sonnenschein, S., Baker, L., Moyer, A., & LeFevre, S. (2005, April). *Parental beliefs about children's reading and math development and relations with subsequent achievement.* Paper presented at the Biennial Meeting of the Society for Research in Child Development, Atlanta, GA.

Sophian, C. (1988). Early developments in children's understanding of number: Inferences about numerosity and one-to-one correspondence. *Child Development, 59*, 1397–1414.

Sophian, C. (1998). A developmental perspective on children's counting. In C. Donlan (Ed.), *The development of mathematical skills* (pp. 27–46). East Sussex, UK: Psychology Press.

Sophian, C. (2002). Learning about what fits: Preschool children's reasoning about effects of object size. *Journal for Research in Mathematics Education, 33*, 290–302.

Sophian, C., & Adams, N. (1987). Infants' understanding of numerical transformations. *British Journal of Educational Psychology, 5*, 257–264.

Sophian, C., & Crosby, M. E. (1998). Ratios that even young children understand: The case of spatial proportions. Ireland: Cognitive Science Society of Ireland.

Sophian, C., & Kailihiwa, C. (1998). Units of counting: Developmental changes. *Cognitive Development, 13*, 561–585.

Sophian, C., & McCorgray, P. (1994). Part-whole knowledge and early arithmetic problem solving. *12*, 3–33.

Sowder, J. T. (1992). Making sense of numbers in school mathematics. In G. Leinhardt, R. Putman, & R. A. Hattrup

(Eds.), *Analysis of arithmetic for mathematics teaching.* Mahwah, NJ: Erlbaum.

Sowell, E. J. (1989). Effects of manipulative materials in mathematics instruction. *Journal for Research in Mathematics Education, 20,* 498–505.

Spelke, E. S. (2000). Nativism, empiricism, and the origns of knowledge. In D. Muir, & A. Slater (Eds.), *Infant development: The essential readings* (pp. 36–51). Malden, MA: Blackwell Publishers.

Spelke, E. S. (2002). Conceptual development in infancy: The case of containment. In N. L. Stein, P. J. Bauer, & M. Rabinowitz (Eds.), *Representation, memory, and development. Essays in honor of Jean Mandler* (pp. 223–246). Mahwah, NJ: Erlbaum.

Spelke, E. S. (2003). What makes us smart? Core knowledge and natural language. In D. Genter, & S. Goldin-Meadow (Eds.), *Language in mind* (pp. 277–311). Cambridge, MA: MIT Press.

Spiers, P. A. (1987). Alcalculia revisited: Current issues. In G. Deloche, & X. Seron (Eds.), *Mathematical disabilities: A cognitive neuropyschological perspective.* Hillsdale, NJ: Erlbaum.

Spitler, M. E., Sarama, J., & Clements, D. H. (2003). *A Preschooler's Understanding of "Triangle:" A Case Study.* Paper presented at the 81th Annual Meeting of the National Council of Teachers of Mathematics. Retrieved.

Starkey, P. (1992). The early development of numerical reasoning. *Cognition, 43,* 93–126.

Starkey, P., & Cooper, R. G., Jr. (1980). Perception of numbers by human infants. *Science, 210,* 1033–1035.

Starkey, P., & Cooper, R. G., Jr. (1995). The development of subitizing in young children. *British Journal of Developmental Psychology, 13,* 399–420.

Starkey, P., & Klein, A. (1992). Economic and cultural influence on early mathematical development. In F. L. Parker, R. Robinson, S. Sombrano, C. Piotrowski, J. Hagen, S. Randoph, & A. Baker (Eds.), *New directions in child and family research: Shaping Head Start in the 90s* (pp. 440). New York: National Clouncil of Jewish Women.

Starkey, P., Klein, A., Chang, I., Qi, D., Lijuan, P., & Yang, Z. (1999). Environmental supports for young children's mathematical development in China and the United States. Albuquerque, NM: Society for Research in Child Development.

Starkey, P., Spelke, E. S., & Gelman, R. (1990). Numerical abstraction by human infants. *Cognition, 36,* 97–128.

Steffe, L. P. (1966). *The performance of first grade children in four levels of conservation of numerousness and three I.Q. groups when solving arithmetic addition problems (Technical Report No. 14), Research and Development Center for Cognitive Learning, C-03, OE 5-10-154:* Madison, WI: University of Wisconsin.

Steffe, L. P. (1983). Children's algorithms as schemes. *Educational Studies in Mathematics, 14,* 109–125.

Steffe, L. P. (1991). Operations that generate quantity. *Learning and Individual Differences, 3,* 61–82.

Steffe, L. P. (1992). Children's construction of meaning for arithmetical words: A curriculum problem. In D. Tirosh (Ed.), *Implicit and explicit knowledge: An educational approach* (pp. 131–168). Norwood, NJ: Ablex Publishing Corporation.

Steffe, L. P. (2004). *PSSM* From a constructivist perspective. In D. H. Clements, J. Sarama, & A.-M. DiBiase (Eds.), *Engaging young children in mathematics: Standards for early childhood mathematics education* (pp. 221–251). Mahwah, NJ: Erlbaum.

Steffe, L. P., & Cobb, P. (1988). *Construction of arithmetical meanings and strategies.* New York: Springer-Verlag.

Steffe, L. P., Cobb, P., von Glasersfeld, E., & Richards, J. (1983). *Children's counting types: Philosophy, theory, and application.* New York: Praeger Scientific.

Stephan, M., Bowers, J., Cobb, P., & Gravemeijer, K. P. E. (2003). Supporting students' development of measuring conceptions: Analyzing students' learning in social context. *Journal for Research in Mathematics Education Monograph Series, 12.*

Stephan, M., & Clements, D. H. (2003). Linear, area, and time measurement in prekindergarten to grade 2. In D. H. Clements (Ed.), *Learning and teaching measurement: 65th Yearbook* (pp. 3–16). Reston, VA: National Council of Teachers of Mathematics.

Stevenson, H. W., & McBee, G. (1958). The learning of object and pattern discrimination by children. *Journal of Comparative and Psychological Psychology, 51,* 752–754.

Stevenson, H. W., & Newman, R. S. (1986). Long-term prediction of achievement and attitudes in mathematics and reading. *Child Development, 57,* 646–659.

Stewart, R., Leeson, N., & Wright, R. J. (1997). Links between early arithmetical knowledge and early space and measurement knowledge: An exploratory study. In F. Biddulph, & K. Carr (Eds.), *Proceedings of the Twentieth Annual Conference of the Mathematics Education Research Group of Australasia* (Vol. 2, pp. 477–484). Hamilton, New Zealand: MERGA.

Stigler, J. W., Lee, S.-Y., & Stevenson, H. W. (1990). *Mathematical knowledge of Japanese, Chinese, and American elementary school children.* Reston, VA: National Council of Teaching of Mathematics.

Stiles, J. (2001). Spatial cognitive development. In C. A. Nelson, & M. Luciana (Eds.), *Handbook of developmental cognitive neuroscience* (pp. 399–414). Cambridge, MA: Bradford.

Stiles, J., & Nass, R. (1991). Spatial grouping activity in young children with congenital right or left hemisphere brain injury. *Brain and Cognition, 15,* 201–222.

Stiles, J., & Stern, C. (2001). Developmental change in spatial cognitive processing: Complexity effects and block construction performance in preschool children. *Journal of Cognition and Development, 2,* 157–187.

Stiles-Davis, J. (1988). Developmental change in young children's spatial grouping ability. *Developmental Psychology, 24,* 522–531.

Strauss, M. S., & Curtis, L. E. (1984). Development of numerical concepts in infancy. In C. Sophian (Ed.), *The origins of cognitive skills* (pp. 131–155). Hillsdale, NJ: Erlbaum.

Sylva, K., Melhuish, E., Sammons, P., Siraj-Blatchford, I., & Taggart, B. (2005). *The effective provision of pre-school education [EPPE] project: A longitudinal study funded by the DfEE (1997–2003).* London: EPPE Project, Institute of Education, University of London.

Tada, W. L., & Stiles, J. (1996). Developmental change in children's analysis of spatial patterns. *Developmental Psychology, 32,* 951–970.

Tan, L. S. C., & Bryant, P. (2000). The cues that infants use to distinguish discontinuous quantities: Evidence using a shift-rate recovery paradigm. *Child Development, 71,* 1162–1178.

Thelen, E., & Smith, L. B. (1994). *A dynamic systems approach to the development of cognition and action.* Cambridge, MA: MIT Press.

Thirumurthy, V. (2003). *Children's cognition of geometry and spatial thinking—A cultural process.* Unpublished doctoral dissertation, University of Buffalo, State University of New York.

Thomas, B. (1982). *An abstract of kindergarten teachers' elicitation and utilization of children's prior knowledge in the teaching of shape concepts.* Unpublished manuscript, School of Education, Health, Nursing, and Arts Professions, New York University.

Thomas, G., & Ward, J. (2001). *An evaluation of the Count Me In Too pilot project.* Wellington, New Zealand: Ministry of Education.

Thomas, H., & Jamison, W. (1975). On the acquisition of understanding that still water is horizontal. *Merrill-Palmer Quarterly, 21,* 31–44.

Thorndike, E. L. (1922). *The psychology of arithmetic.* New York: Macmillan.

Tibbals, C. (2000). Standards for preschool and kindergarten mathematics education. Arlington, VA: Conference on Standards for Preschool and Kindergarten Mathematics Education.

Towse, J., & Saxton, M. (1998). Mathematics across national boundaries: Cultural and linguistic perspectives on numerical competence. In C. Donlan (Ed.), *The development of mathematical skills* (pp. 129–150). East Sussex, UK: Psychology Press.

Trabasso, T. (1975). Representation, memory and reasoning: How do we make transitive inferences. *Minnesota Symposium on Child Psychology, 9,* 135–172.

Trick, L. M., & Pylyshyn, Z. W. (1994). Why are small and large numbers enumerated differently? A limited-capacity preattentive stage in vision. *Psychological Review, 101,* 80–102.

Try, K. M. (1989). *Cognitive and social change in young children during Logo activities: A study of individual differences.* Unpublished Doctorial dissertation, The University of New England, Armidale, New South Wales, Australia.

Tudge, J. R. H., & Doucet, F. (2004). Early mathematical experiences: Observing young Black and White children's everyday activities. *Early Childhood Research Quarterly, 19,* 21–39.

Turner, R. C., & Ritter, G. W. (2004, April). *Does the impact of preschool childcare on cognition and behavior persist throughout the elementary years?* Paper presented at the American Educational Research Association, San Diego, CA.

Tyler, D., & McKenzie, B. E. (1990). Spatial updating and training effects in the first year of human infancy. *Journal of Experimental Child Psychology, 50,* 445–461.

Tyler, R. W. (1949). *Basic principles of curriculum and instruction.* Chicago: University of Chicago Press.

U.S. Department of Education, N. C. E. S. (2000). *The condition of education 2000.* Washington, DC: U.S. Government Printing Office.

Uller, C., Carey, S., Huntley-Fenner, G., & Klatt, L. (1999). What representations might underlie infant numerical knowledge? *Cognitive Development, 14,* 1–36.

Ungerleider, L. G., & Mishkin, M. (1982). Two cortical visual systems. In D. J. Ingle, M. A. Goodale, & R. J. W. Mansfield (Eds.), *Analysis of visual behavior* (pp. 549–586). Cambridge, MA: MIT Press.

Uttal, D. H. (1996). Angles and distances: Children's and adults' reconstruction and scaling of spatial configurations. *Child Development, 67*(6), 2763–2779.

Uttal, D. H. (2000). Seeing the big picture: Map use and the development of spatial cognition. *Developmental Science, 3*(3), 247–286.

Uttal, D. H., & Wellman, H. M. (1989). Young children's representation of spatial information acquired from maps. *Developmental Psychology, 25,* 128–138.

Van de Walle, G. A., Carey, S., & Prevor, M. (2000). Bases for object individuation in infancy: Evidence from manual search. *Journal of Cognition and Development, 1,* 249–280.

van Hiele, P. M. (1986). *Structure and insight: A theory of mathematics education.* Orlando, FL: Academic Press.

van Hiele-Geldof, D. (1984). The didactics of geometry in the lowest class of secondary school (M. Verdonck, Trans.). In D. Fuys, D. Geddes, & R. Tischler (Eds.), *English translation of selected writings of Dina van Hiele-Geldof and Pierre M. van Hiele* (pp. 1–214). Brooklyn, NY: Brooklyn College, School of Education. (ERIC Document Reproduction Service No. 289 697).

van Oers, B. (1994). Semiotic activity of young children in play: The construction and use of schematic representations. *European Early Childhood Education Research Journal, 2,* 19–33.

Vilette, B. (2002). Do young children grasp the inverse relationship between addition and subtraction? *Cognitive Development, 17,* 1365–1383.

Vinner, S., & Hershkowitz, R. (1980). Concept images and common cognitive paths in the development of some simple geometrical concepts. In R. Karplus (Ed.), *Proceedings of the Fourth International Conference for the Psychology of Mathematics Education* (pp. 177–184). Berkeley, CA: Lawrence Hall of Science, University of California.

von Glasersfeld, E. (1982). Subitizing: The role of figural patterns in the development of numerical concepts. *Archives de Psychologie, 50,* 191–218.

von Glasersfeld, E. (1995). Sensory experience, abstraction, and teaching. In L. P. Steffe, & J. Gale (Eds.), *Constructivism in education* (pp. 369–383). Mahwah, NJ: Erlbaum.

Votruba-Drzal, E., & Chase, L. (2004). Child care and low-income children's development: Direct and moderated effects. *Child Development, 75,* 296–312.

Vurpillot, E. (1976). *The visual world of the child.* New York: International Universities Press.

Vygotsky, L. S. (1929/1994). The problem of the cultural development of the child. In R. van der Veer, & J. Valsiner (Eds.), *The Vygotsky reader* (pp. 57–72). Oxford, U.K.: Blackwell.

Vygotsky, L. S. (1934/1986). *Thought and language.* Cambridge, MA: MIT Press.

Vygotsky, L. S. (1978). *Mind in society: The development of higher psychological processes.* Cambridge, MA: Harvard University Press.

Wakeley, A., Rivera, S. M., & Langer, J. (2000). Can young infants add and subtract? *Child Development, 71,* 1525–1534.

Wang, J., & Lin, E. (2005). Comparative studies on U.S. and Chinese mathematics learning and the implications for standards-based mathematics teaching reform. *Educational Researcher, 34*(5), 3–13.

Wang, M., Resnick, L., & Boozer, R. F. (1971). The sequence of development of some early mathematics behaviors. *Child Development, 42,* 1767–1778.

Wang, R. F., & Spelke, E. S. (2002). Human spatial representation: Insights from animals. *Trends in Cognitive Sciences, 6,* 376–382.

Watson, J. A., & Brinkley, V. M. (1990/91). Space and premathematic strategies young children adopt in initial Logo problem solving. *Journal of Computing in Childhood Education, 2,* 17–29.

Watson, J. A., Lange, G., & Brinkley, V. M. (1992). Logo mastery and spatial problem-solving by young children: Effects of Logo language training, route-strategy training, and learning styles on immediate learning and transfer. *Journal of Educational Computing Research, 8,* 521–540.

Weaver, C. L. (1991). *Young children learn geometric and spatial concepts using Logo with a screen turtle and a floor turtle.* Unpublished doctoral dissertation, State University of New York at Buffalo.

Weiss, I. R. (2002). Systemic reform in mathematics education: What have we learned? Las Vegas, NV: Research presession of the 80th annual meeting of the National Council of Teachers of Mathematics.

Wellman, H. M., Fabricius, W. V., & Sophian, C. (1985). The early development of planning. In H. M. Wellman (Ed.), *Chiildren's searching: The development of search skill and spatial representation* (pp. 123–149). Hillsdale, NJ: Erlbaum.

Wheatley, G., & Cobb, P. (1990). Analysis of young children's spatial constructions. In L. P. Steffe, & T. Wood (Eds.), *Transforming early childhood mathematics education: International perspectives* (pp. 161–173). Mahwah, NJ: Erlbaum.

Wheatley, G. H. (1990). Spatial sense and mathematics learning. *Arithmetic Teacher, 37*(6), 10–11.

Wheatley, G. H., Brown, D. L., & Solano, A. (1994). Long term relationship between spatial ability and mathematical knowledge. In D. Kirshner (Ed.), *Proceedings of the sixteenth annual meeting North American Chapter of the International Group for the Psychology of Mathematics Education* (Vol. 1, pp. 225–231). Baton Rouge, LA: Louisiana State University.

Wiese, H. (2003). Iconic and non-iconic stages in number development: The role of language. *Trends in Cognitive Sciences, 7,* 385–390.

Williford, H. J. (1972). A study of transformational geometry instruction in the primary grades. *Journal for Research in Mathematics Education, 3,* 260–271.

Wittgenstein, L. (1953/1967). *Philosophical investigations* (G. E. M. Anscombe, Trans. Vol. 3rd). New York: Macmillan.

Wolf, Y. (1995). Estimation of Euclidian quantity by 5- and 6-year-old children: Facilitating a multiplication rule. *Journal of Experimental Child Psychology, 59,* 49–75.

Wright, B. (1991). What number knowledge is possessed by children beginning the kindergarten year of school? *Mathematics Education Research Journal, 3*(1), 1–16.

Wright, E. (1983). Pre-phenomenal adjustments and Sanford's Illusion Objection against sense-data. *Pacific Philosophical Quarterly, 64,* 266–272.

Wright, R. J., Martland, J., Stafford, A. K., & Stanger, G. (2002). *Teaching number: Advancing children's skills and strategies.* London: Paul Chapman Publications/Sage.

Wright, R. J., Stanger, G., Cowper, M., & Dyson, R. (1994). A study of the numerical development of 5-year-olds and 6-year-olds. *Educational Studies in Mathematics, 26,* 25–44.

Wright, R. J., Stanger, G., Cowper, M., & Dyson, R. (1996). First-graders' progress in an experimental mathematics recovery program. In J. Mulligan, & M. Mitchelmore (Eds.), *Research in early number learning* (pp. 55–72). Adelaide, Australia: AAMT.

Wynn, K. (1990). Children's understanding of counting. *Cognition, 36,* 155–193.

Wynn, K. (1992a). Addition and subtraction by human infants. *Nature, 358,* 749–750.

Wynn, K. (1992b). Children's acquisition of the number words and the counting system. *Cognitive Psychology, 24,* 220–251.

Wynn, K. (1995). Origins of numerical knowledge. *Mathematical Cognition, 1,* 35–60.

Wynn, K. (1998). Numerical competence in infants. In C. Donlan (Ed.), *The development of mathematical skills* (pp. 3–25). East Sussex, UK: Psychology Press.

Wynn, K., Bloom, P., & Chiang, W.-C. (2002). Enumeration of collective entities by 5-month-old infants. *Cognition, 83,* B55–B62.

Yin, H. S. (2003). Young children's concept of shape: van Hiele visualization level of geometric thinking. *The Mathematics Educator, 7*(2), 71–85.

Young-Loveridge, J. M. (1989). The relationship between children's home experiences and their mathematical skills on entry to school. *Early Child Development and Care, 43,* 43–59.

Yuzawa, M., Bart, W. M., Kinne, L. J., Sukemune, S., & Kataoka, M. (1999). The effects of "origami" practice on size comparison strategy among young Japanese and American children. *Journal of Research in Childhood Education, 13*(2), 133–143.

Yuzawa, M., Bart, W. M., & Yuzawa, M. (2000). Development of the ability to judge relative areas: Role of the procedure of placing one object on another. *Cognitive Development, 15,* 135–152.

Zelazo, P. D., & Reznick, J. S. (1991). Age-related asynchrony of knowledge and action. *Child Development, 62,* 719–735.

Zelazo, P. D., Reznick, J. S., & Piñon, D. E. (1995). Response control and the execution of verbal rules. *Developmental Psychology, 31,* 508–517.

Zykova, V. I. (1969). Operating with concepts when solving geometry problems. In J. Kilpatrick, & I. Wirszup (Eds.), *Soviet studies in the psychology of learning and teaching mathematics (Vols. 1)* (pp. 93–148). Chicago: University of Chicago.

AUTHOR NOTE

This chapter is based upon work supported in part by the National Science Foundation under Grant ESI-9730804 to D. H. Clements and J. Sarama, "Building Blocks—Foundations for Mathematical Thinking, Pre-Kindergarten to Grade 2: Research-Based Materials Development," and Grant ESI-9817540 to D. H. Clements, "Conference on Standards for Preschool and Kindergarten Mathematics Education." Work on the research was also supported in part by the Interagency Educational Research Initiative (NSF, DOE, and NICHHD) Grant REC-0228440, to D. H. Clements, J. Sarama, and J. Lee "Scaling Up the Implementation of a Pre-Kindergarten Mathematics Curricula: Teaching for Understanding With Trajectories and Technologies." Any opinions, findings, and conclusions or recommendations expressed in this material are those of the authors and do not necessarily reflect the views of the funding agencies. We thank Art Baroody, Frank Lester, and Erna Yackel for their comments on an early draft.

13

WHOLE NUMBER CONCEPTS AND OPERATIONS

Lieven Verschaffel

UNIVERSITY OF LEUVEN, BELGIUM

Brian Greer

PORTLAND STATE UNIVERSITY

Erik De Corte

UNIVERSITY OF LEUVEN, BELGIUM

Whole number concepts and operations constitute, universally, a dominant part of elementary mathematics education. The volume of research devoted to them within mathematics education and other fields such as instructional psychology, experimental psychology, developmental psychology, and (increasingly) neurosciences, is correspondingly vast; hence we have been necessarily selective in constructing this chapter. We intend it to be a systematic and balanced overview, but in the process of selecting and thematically arranging the material, it is impossible not to be influenced by our value systems about mathematics education (research). For example, we consider it relevant to situate the research and theory that we have surveyed within a broader cultural context.

Of course, this chapter overlaps to some extent other chapters, especially that on early childhood mathematics (by Clements and Sarama). To minimize overlap with that chapter, we have narrowed ours to the elementary level (i.e., from about 6 years old), whereas the Clements and Sarama chapter addresses the same topics (as well as others) at the preschool to kindergarten level (i.e., from birth to age 6).

By reference to earlier reviews, we can briefly summarize material that is fully dealt with elsewhere. Thus, we refer the reader to relevant chapters of the previous NCTM *Handbook of Research on Mathematics Teaching and Learning* (Grouws, 1992), in particular those by Fuson, Greer, and Sowder, which we explicitly take as our point of departure (see also De Corte, Greer, & Verschaffel, 1996).

Our review surveys studies and analyses inspired by various theories about the nature of arithmetic cognition and how it develops in students' learning activities. In our characterizations of these underlying theories, we largely follow the terminology used by Greeno, Collins, and Resnick (1996). These authors distinguished three general perspectives that frame the nature of knowing and learning in contrasting and complementary ways, which they labeled behaviorist/empiricist, cognitive/rationalist, and situative/pragmatist-sociohistoric. Representative traditions for the first view, according to which knowing is an organized accumulation of associations and skills, are associationism, behaviorism, and connectionism. Gestalt psychology, symbolic information processing, and constructivism are traditions of research they considered to be illustrative of the second perspective, because of their emphasis on the organization of information in cognitive structures and procedures. Although symbolic information processing and constructivism share the above-mentioned common feature, there are significant differences (of emphasis) between these research traditions. Among the traditions that have contributed to the situative/pragmatist-sociohistoric perspective are ethnography, ecological psychology, and situation theory.

In terms of the trichotomy of Greeno et al. (1996), the first view is scarcely represented here, as it plays little part in contemporary work in the field. However, as pointed out by De Corte et al.

> Conceptions of learning as incremental (with errors to be avoided or immediately stamped out), of assessment as appropriately implemented by reference to atomistic behavioral objectives, of teaching as the reinforcement of behavior, of motivation as directly mediated by rewards and punishments, and of mathematics as precise, unambiguous, and yielding uniquely correct answers through the application of specific procedures remain prevalent in folk psychology and, as such, represent the legacy of behaviorism. (1996, p. 492)

Sfard (1998, 2003) suggested a broadly similar contrast between the second and third perspectives as "acquisitionist" versus "participationist." According to Sfard, the "acquisitionist" grouping consists of those (traditional) cognitivist approaches that explain learning and knowledge in terms of mental entities such as cognitive schemes, tacit models, concept images, and misconceptions, whereas the participationist framework embraces all those relatively new theories that prefer to view learning as a reorganization of activity accompanying the integration of an individual learner within a community of practice. So, the latter framework focuses on the process through which the learner becomes a skillful participant in a given (mathematical) discourse.

For a related, but slightly different, classification of the psychological roots of various theories of (mathematical) cognition and learning, including the important differences between information processing theory and constructivism, see Baroody (2003).

One historical characterization of the contrast between the cognitive/rationalist and situative/pragmatist-sociohistoric views is that they constitute two "waves" of the "Cognitive Revolution" (De Corte et al., 1996), the second wave being a reaction to the perceived limitations of the first (Gardner, 1985; Greer & Verschaffel, 1990). However, as argued by Sfard (1998), the two perspectives cannot be completely separated, and both are present in almost every current theory (although one perspective is usually much more prominent than the other).

OUTLINE OF THE REST OF THE CHAPTER

The body of the chapter consists of four sections.

In the *first section*, we update ascertaining studies reported in earlier reviews on aspects of individual children's performance on, and understanding of, a range of key topics, namely:

1. Single-digit computation
2. Multidigit computation
3. Estimation and number sense
4. Word problems
5. The structure of the whole-number system

An overarching theme of earlier research and theoretical analysis, which receives a lot of attention in this first section, was the appropriate balance between procedural competence and conceptual understanding. Particularly since the seminal book edited by Hiebert (1986), the prevailing view has been that procedural and conceptual aspects are not oppositional, but rather complementary and interacting (see also Baroody & Dowker, 2003; Hiebert et al., 1997; Rittle-Johnson & Siegler, 1998; Romberg, Carpenter, & Kwako, 2005; Siegler, 2003). However, as argued by Baroody (2003), some influential thinkers (e.g., Brownell, 1945; Resnick & Ford, 1981; Thorndike, 1922) came to this conclusion earlier.

Section 2 deals with the design, implementation, and evaluation of (experimental) environments for learning and teaching whole number arithmetic. Looking towards the future at the end of her chapter for the Grouws handbook, Fuson (1992, p. 268) described a vision of future mathematics classrooms "as places where children construct meanings for mathematical concepts, words, and written marks and carry out, discuss, and justify solution procedures for mathematical situations." Further (p. 269), they are places where (a) children engage in mathematical situations that are meaningful and interesting to them; (b) the emphasis is on sustained engagement in mathematical situations, not on rapidly obtaining answers; (c) alternative solution procedures are accepted, discussed, and justified; and (d) errors are just expected way stations on the road to solutions and should be analyzed in order to increase everyone's understanding.

Exemplary studies reported in the second section reflect that, in the intervening years, substantial strides have been made towards actualizing this vision through a major development of that period, namely a shift from mainly ascertaining studies of individual children to extensive classroom-based research. The major approaches/topics exemplified in these classroom-based studies are

1. Learning arithmetic as problem solving
2. Teaching arithmetic as inherently algebraic
3. Creating a social-constructivist classroom culture
4. Using emergent models as a design heuristic
5. Teaching word problems from a modeling perspective
6. Changing teachers' knowledge and beliefs

Collectively, these studies represent a richer and more complex view of whole number arithmetic in relation to mathematics as a whole, and of the socially mediated processes within the culture of the elementary mathematics classroom.

In the *third section*, we reflect on the wider contexts of the research—historical, cultural, social, and political. Whole numbers and operations on them are pan-cultural concepts and activities (Bishop, 1988). Within this universality, cultural differences impact teaching and learning, in terms of cognitive tools (most importantly, language) and societal attitudes towards, and beliefs about, both arithmetic and learning/teaching arithmetic. Arithmetic in school constitutes a particular form of activity, a *situated practice* in Lave's (1988) term. In this section, we consider two theoretical perspectives that link with recent important trends that characterize cognition as both situated and distributed. The first is the relatively new field of Ethnomathematics, defined by D'Ambrosio (1985, p. 45) as "the mathematics which is practised among identifiable cultural groups, such as national-tribal societies, labor groups, children of a certain age bracket, professional classes, and so on." The second is Activity Theory, a development of the Russian school of Vygotsky and others. In particular, we consider the characterization by Engestrom (1987, p. 103) of the activity of school-going. These perspectives also focus attention on the lack of connections between arithmetic in school, outside of school, and prior to formal education. Furthermore, it is becoming increasingly obvious—particularly, but by no means exclusively in the United States—that mathematics education is politically situated. Associated fundamental philosophical differences are particularly clearly focused in early arithmetic.

In the concluding *fourth section*, drawing on the research reviewed, we endorse a balanced view of proficiency as implying intertwined strands of conceptual understanding, strategic competence, adaptive reasoning, productive disposition, and procedural fluency (Kilpatrick, Swafford, & Findell, 2001). This conception clearly implies that students should develop *adaptive expertise*, defined by Hatano (2003, p. xi) as "the ability to apply meaningfully learned procedures flexibly and creatively" rather than merely *routine expertise*, which he defined as "being able to complete school exercises quickly and accurately without understanding." We further propose a vision of early learning of arithmetic as an experience that is foundational in establishing a positive disposition towards mathematics and introduces the young child to defining characteristics of mathematical activity: sense-making, problem-solving, modeling of aspects of the real world, structural analysis, constructive use of symbolic and physical tools, and logical argumentation.

ASCERTAINING STUDIES OF INDIVIDUAL CHILDREN LEARNING AND DOING WHOLE-NUMBER MATHEMATICS

Here we review and illustrate the last decade's developmental studies, expert-novice studies, studies about children with learning problems, neuropsychological research, and so on, on five key topics related to number and arithmetic, always on a necessarily highly selective basis and with reference to existing surveys. Most of the highlighted studies in this section are exemplary for the continuation of the cognitive/rationalist (predominantly information-processing) approach based on experiments or interviews with individual children, with little or no attention to the broader cultural or instructional contexts in which these studies take place. However, at some places, results from classroom-based research and (inter)cultural studies will be discussed insofar as their results are directly relevant.

Single-Digit Computation

Children's Single-Digit Conceptual Structures

The conceptual field of whole numbers is central in elementary school mathematics and foundational for an individual's mathematical education. (We deliberately use Vergnaud's, 1996, term *conceptual field* to draw attention to the complexity of the constellation of ideas relating to whole numbers, rather than the frequently—and, in our view, misleadingly—used term *concept*.) The development of children's understanding and use of natural numbers has attracted an enormous (arguably disproportionate) amount of literature. For a long time, the view and practice relating to the development of understanding of the natural numbers were heavily influenced by Piaget's "logical operations" framework, including classification, seriation, and conservation as the foundations of the natural number understanding. Accordingly, many educators, aware of Piaget's findings and claims, believed that it was impossible for children to learn about natural numbers in a meaningful way until they had reached the concrete-operational stage, roughly at the age of 6–8, wherein the above-mentioned logical operations become part of their cognitive structures in an integrated way (Kilpatrick et al., 2001; Verschaffel & De Corte, 1996). Research over the last 25 years, however, has questioned the pivotal role of these logical operations in the development of natural number understanding and has shown at the same time the importance of children's declarative and procedural knowledge of counting for this development. In general, this research has convincingly documented that the development of counting ability is intertwined with the development of the understanding of counting principles. Moreover, it has been shown that by the time they enter elementary school most children understand the rules that underlie counting, can perform conventional counting correctly with sets of objects even greater than 10, can use counting to solve some simple mathematical problems, and even know the Arabic symbols for some, if not for most, of the numbers up to 10 (Dehaene, 1993; Kilpatrick et al., 2001; see chapter by Clements and Sarama, this volume).

Besides developmental studies relating to basic concepts of natural number, there are also many analyses of the conceptual structures for addition and subtraction. In an attempt to integrate the work of Steffe and his colleagues on counting and cardinal conceptual units (Steffe & Cobb, 1988), as well as her own work in this area, Fuson (1992) proposed a model of the conceptual structures for addition and subtraction consisting of three levels. At Level 1, children are limited to seeing sums as an aggregate of single-unit items and to seeing only a single presentation of a quantity at a time, with the consequence that they can apply only the most primitive procedures to solve addition and subtraction problems (see below). At Level 2, all three quantities can be simultaneously represented by embedding entities for the addends within the sum, allowing the application of more efficient and more abbreviated counting methods. At Level 3 children have created "ideal chunk items," enabling them to construct simultaneous embedded mental presentations of numbers within a sum. This ability permits recomposition of the addends so that a

problem can be transformed into an easier sum of different addends or be solved using other derived-fact strategies. For an extensive and systematic review of the research about the early conceptual development of numerosity and of the concepts relating to natural numbers, see the chapter by Clements and Sarama in this handbook.

Besides these developmental analyses, a very important area of cognitive-psychological and neuroscientific research describes how people internally represent and process numbers. Among the most influential models for number processing are those of McCloskey, Caramazza, and Basili (1985) and Dehaene and Cohen (1995). The former model assumes that all numerical inputs are initially translated, via notation-specific comprehension modules, into an amodal abstract representation of numbers. The model of Dehaene abandons the idea of a unique number concept and clusters numerical abilities in different groups according to the format in which numbers are manipulated (verbal, Arabic symbols, or quantity). This model posits an analog magnitude code (a kind of mental number line) as the main, and perhaps the only, semantic representation of a number. Dehaene and Cohen also speculated about the anatomical substrates of the different processes involved in their model (see also Delazer, 2003).

Children's Strategies for Single-Digit Computation

As children begin to learn mathematics in elementary school, much of their number activity is designed to help them become proficient with single-digit arithmetic, namely mastery of the sums and products of single-digit numbers and their companion differences and quotients, the necessity of which is not disputed in the current reform documents (Kilpatrick et al., 2001; Thompson, 1999a; Treffers, 2001a, 2001b). However, the view on what it means to have knowledge of these facts and, even more importantly, how these number facts should be learnt and taught has drastically changed in the past decades. Whereas learning single-digit arithmetic was for a long time based mainly on memorizing those facts through drill-and-practice to the point of automatization (with early exceptions such as, notably, Brownell, 1945), current instructional approaches put great emphasis on the gradual development of these number facts from children's invented and informal strategies (Kilpatrick et al., 2001; Verschaffel & De Corte, 1996).

The domain of single-digit addition and subtraction is undoubtedly one of the most frequently investigated areas of numerical cognition and school mathematics. Much work in the domain has been done from a cognitive/rationalist—and, especially, information-processing—perspective. Numerous older and more recent studies provide detailed descriptions of the progression in children of mastery of orally stated single-digit additions, such as 3 + 4:

- from the earliest concrete counting-all-with-materials strategy, for example, finding 3 + 4 by making a set of 1, 2, 3 blocks, then adding 1, 2, 3, 4 blocks to this set of 3 blocks, and then counting the total,

- over several types of progressively more streamlined counting strategies, such as counting-all-without-materials (1, 2, 3, . . . 4, 5, 6, 7), counting-on-from-first (3, . . . 4, 5, 6, 7), and counting-on-from-larger (4, . . . 5, 6, 7), and derived-fact strategies (3 + 4 = [3 + 3] + 1 = 6 + 1 = 7), which take advantage of certain arithmetic principles to shorten and simplify the computation,

- to the final state of fact mastery (for extensive reviews of this research, see Baroody & Tiilikainen, 2003; Fuson, 1992; Kilpatrick et al., 2001; Thompson, 1999a).

Similar levels for subtraction have been described, although this developmental sequence is somewhat less clearly defined (Thompson, 1999a).

These and other studies document at the same time how, at any given time during this development, an individual child uses a variety of addition strategies, even within the same session and even for the same item (for an overview see Siegler, 1998). Even older pupils and adults do not always perform at the highest developmental level of fact mastery but still demonstrate a range of different procedures even for simple addition problems (Siegler, 1998).

Proposing and testing models framed in terms of storage in the brain of arithmetic facts in learners having reached the final stage of the above developmental process is a special area of research in numerical cognition (for overviews see Ashcraft, 1995; Dehaene, 1993; Verguts, 2003). Most of this research has been done with adults. Most such models share the notion that in the "expert fact retriever" arithmetic facts such as 2 + 3 = 5 are memorized in, and (semi) automatically retrieved from, a stored associative network or lexicon (Ashcraft, 1995; Verguts, 2003). Well-known "problem size effects" (i.e., the fact that the time needed to answer single-digit addition problems increases slightly with the size of the operands) and "tie effects" (i.e., the fact that response time for ties such as 2 + 2 remain constant or increases very little with operand size) are considered in this common view as reflecting the duration and the difficulty of memory

retrieval. According to Ashcraft (1995; Ashcraft & Christy, 1995), both effects faithfully reflect the frequency with which arithmetic facts are acquired and practiced by individuals (but see Baroody, 1994, for a thorough critique of such evidence). However, it is quite generally accepted that not all experts' knowledge of single-digit arithmetic is mentally represented in separate and independent units. Part of their knowledge about simple addition seems to be stored as rules (e.g., $N + 0 = N$) rather than as isolated facts. A related assumption is that because of relational knowledge not all problems need be represented. For instance, for each commutative pair (e.g., 3 + 5 and 5 + 3) there may be only one representational unit in the network.

It is well known that the above-mentioned developmental process from counting to fact retrieval does not proceed smoothly for all children. Single-digit arithmetic among children with mathematical difficulties or learning problems has also attracted a lot of research. Generally speaking, this research shows that children with learning disabilities, and others having difficulty with mathematics, do not use procedures that differ from the progression described above, but they are slower in moving through it (Kilpatrick, Martin, & Schifter, 2003; Torbeyns, Verschaffel, & Ghesquière, 2004). Central but still unresolved research questions are whether these difficulties are associated with general weaknesses in linguistic processes related to representing and retrieving semantic information from long-term memory (Geary, 2003; Jordan, Hanich, & Kaplan, 2003) and how these persistent problems with fact retrieval are related to developmental delays or more fundamental deficits in particular parts of the brain (Geary, 2003). Although according to some authors these persistent retrieval difficulties appear to reflect a highly persistent, perhaps lifelong, deficit (Geary, 2003; Jordan, Hanich, & Kaplan, 2003; Ostad, 1997), others (see, e.g., Baroody, in press) argue that poor instruction may be the basis of most of these problems.

Another very important line of research addresses the relationship between the development of these strategies or procedures for doing simple addition and subtraction, and children's conceptual development. In an attempt to account for this relationship, Fuson (1992) proposed a model of the conceptual structures for addition consisting of three developmental levels, across which both the conceptual structures for quantities change (declarative aspect) and the addition and subtraction strategies become increasingly abbreviated, abstract, internalized (procedural aspect). This model was briefly described in the previous section.

Several other studies have addressed the question about the relationship between declarative and procedural knowledge by investigating whether one kind of knowledge develops before the other. As far as single-digit addition is concerned, this question has focused on the relationship between children's understanding of the commutativity of addition principle and their progression towards more efficient counting strategies (like the counting-on-from-larger strategy) based on this mathematical principle (for an extensive review of this literature see Baroody, Wilkins, & Tiilikainen, 2003; Rittle-Johnson & Siegler, 1998). Rittle-Johnson and Siegler claimed that their findings showed that most children understand the commutativity concept before they generate the procedures based on it, but this claim has been disputed by Baroody (2003). In any case, whereas in previous decades the debate about the relation between conceptual and procedural knowledge was dominated by proponents of two camps (i.e., the "skills-first" versus the "concepts first" view), most researchers now adopt a more nuanced perspective wherein the relationship between procedural and conceptual knowledge is assumed to develop more concurrently and iteratively than suggested in either of the older views, and wherein it is accepted that the nature of this relationship may differ among different mathematical (sub)domains (Baroody, 2003; Baroody & Ginsburg, 1986; Rittle-Johnson & Siegler, 1998).

While many cognitive psychologists continue to investigate the growth of children's procedures for single-digit arithmetic with little or no concern for the impact of the instructional and cultural environments in which this development is situated, a growing number of studies belonging to the situative/pragmatist-sociohistoric perspective make this instructional and broader cultural impact the focus of their work. For instance, several researchers who examined the developmental paths of addition solution methods used by Japanese children have reported that they typically move more quickly from counting-all methods to derived-fact and known-fact methods—without passing through a clearly identifiable stage of more efficient counting strategies—than United States children do. Interestingly, many Japanese children use the number 5 as an intermediate anchor number to think about numbers and to do additions and subtractions before starting to do sums by means of retrieval or using 10 as an anchor in their derived-fact strategies. Arguably, these developmental characteristics of Japanese children are closely related to a number of cultural and instructional supports and practices, such as the emphasis that is put on using groups of 5 in the early arithmetic instruction in general and in abacus instruc-

tion in particular (Hatano, 1988; Kuriyama & Yoshida, 1995; Murata, 2004). Similarly, Torbeyns, Verschaffel, and Ghesquière (2005) found a remarkably quick and good mastery of ties with totals above 10 (such as $8 + 8$) and, correspondingly, an unusually frequent, efficient, and adaptive use of a so-called tie strategy on sums above 10 (i.e., solving almost-tie sums such as $7 + 8$ by means of $[7 + 7] + 1$, rather than by the "decomposition-to-10" strategy $[7 + 3] + 5$) among classes of Flemish children who used a (new) textbook series that put great emphasis on the deliberate and flexible use of multiple solution strategies (rather than on the mastery of the decomposition-to-10 strategy as the only acceptable approach to sums above 10).

Although the research concerning the development of children's strategies for multiplying and dividing single-digit numbers is less extensive than for single-digit addition and subtraction, there is a growing body of studies in this domain (see, e.g., Anghileri, 1989, 1999; Baroody, 1999; Butterworth, Marscchesini, & Girelli, 2003; Cooney, Swanson, & Ladd, 1988; Kouba, 1989; LeFevre, Smith-Chant, Hiscock, Daley, & Morris, 2003; Lemaire & Siegler, 1995; Mulligan & Mitchelmore, 1997; Rickard & Bourne, 1996; Steffe & Cobb, 1998). Similarly to the progression for single-digit addition and subtraction, this research documents how, generally speaking, children progress from (material-, fingers-, or paper-based) concrete counting-all strategies, through additive-related calculations (repeated adding and additive doubling), pattern-based (e.g., multiplying $\times$ 9 as by $10 - 1$), and derived-fact strategies (e.g., deriving 7×8 from $7 \times 7 = 49$) to a final mastery of learned multiplication products. However, consistency between the names and the characterizations of the different categories is less than for addition and subtraction. As for addition and subtraction, research on multiplication and division has documented how multiplicity and flexibility of strategy use are basic features of people doing simple number combinations, even for older children and for adults (LeFevre et al., 2003). Here, too, the exact features of the organization and the functioning of the postulated multiplication facts store and, more particularly, to what extent (part of) experts' knowledge about the multiplication table is stored as rules ($0 \times N = 0$, $1 \times N = N$, $10 \times N = N0$, etc.) rather than as strengthened associative links between particular mathematical expressions and their correct answers has been, and still is, an important research issue (for a review see Verguts, 2003). On the basis of a recent study with third graders and fifth graders solving multiplication items with the larger operand placed either first (7×3) or second (3×7), Butterworth et al. (2003, p. 201) concluded with respect to that issue that

> the child learning multiplication facts may not be passive, simply building associative connections between an expression and its answer as a result of practice. Rather, the combinations held in memory may be reorganized in a principled way that takes into account a growing understanding of the operation, including the commutativity principle, and, perhaps, other properties of multiplication.

(See also Baroody, 1999, for a similar conclusion based on a study wherein he found transfer of mastery to unpracticed but commuted versions of practiced multiplication combinations).

Probably the most ambitious and most influential attempt to model this development and this variety of strategy use in single-digit arithmetic from an information-processing perspective has been the model of strategy choice and strategy change in the domain of simple addition developed by Siegler and associates (Siegler & Jenkins, 1989; Siegler & Shipley, 1995; Siegler & Shrager, 1984), which is described and critiqued in the next section.

Strategy Choice and Discovery Simulation (SCADS)

SCADS is a computer model, the latest in a series of models constructed by Siegler and his colleagues (e.g., Shrager & Siegler, 1998). Central to SCADS is a database, with information about problems (in the form of associations between individual problems and potential answers to these problems that differ in strength and the strength of which changes continuously as a result of reinforced practice) and about strategies (including global, featural, problem-specific, and novelty data about each strategy available in the database). Whenever SCADS is presented with a simple addition problem, such as 3 + 6, it activates the global, featural, and problem-specific data about the speed and accuracy of each of the available strategies, weights these data as a function of the amount of information they reflect and how recently they were generated, and chooses the strategy with the highest outcome generated by this complex weighting process. If a retrieval strategy is chosen, the model specifies the minimum strength of the problem-answer association that must be reached before it will be retrieved as well as the maximum number of attempts that the model will make to retrieve the answer. If this retrieval strategy does not work, a so-called back-up strategy is chosen. The execution of a back-up strategy is represented in the model as a modular sequence of operators. The first time SCADS executes a particular back-up strategy (e.g., counting on from first), the model devotes its attentional resources to ensuring the correct execution of the strategy, but the more frequently the model executes a strategy, the less attentional resources SCADS needs to monitor this strategy's execution. SCADS allocates these freed attentional resources to the built-in discovery heuristics, which sometimes propose new strategies, some valid, others flawed. SCADS evaluates these proposed new strategies (e.g., counting on from larger) for consistency with conceptual constraints. If the proposed strategy violates these constraints, it is abandoned. If not, SCADS adds it to its strategy repertoire.

According to the developers of SCADS, its performance on single-digit additions and on additions with one addend above 20 is highly consistent with the strategy choice and discovery phenomena that can be observed in 4- and 5-year-old children (as in the microgenetic study by Shrager & Siegler, 1998; see also Siegler, 2001).

Siegler's strategy-choice model has been tested not only for simple addition, but also, although to a much less fine-grained extent, for multiplication. Siegler and Lemaire (1997) reported a longitudinal investigation of French second graders' acquisition of single-digit multiplication skill. Speed, accuracy, and strategy data were assessed three times within the year when children learned multiplication. The data showed improvements in speed and accuracy that reflected four different aspects of strategic changes (namely origin of new strategies, more frequent use of more efficient strategies, more efficient execution of each strategy, and more adaptive choices among available strategies) that generally accompanied learning. According to the authors, these findings supported a number of predictions of the SCADS model.

Although the model of Siegler and his colleagues is considered by many as providing exemplary support for the success of the information-processing paradigm, and although it has influenced and still influences a lot of research in the domain of single-digit arithmetic, the model also has its critics. Starting from a constructivist and social-learning theoretical framework and from a broader data set, Baroody and Tiilikainen (2003) performed a very critical analysis of Siegler's model of early addition performance and its underlying assumptions. Their major criticisms are that the theoretical advance that has resulted from this computer simulation is very limited (because most of what the model does was known and understood already) and that the model in many respects does not behave and learn as real children do. According to these authors, the computer model looks impressively sophisticated, comprehensive, and detailed, but close inspection indicates that it does not provide an adequate account of, and even is at odds with, how children's "expanding web of conceptual and relational knowledge" (Baroody & Tiilikainen, 2003, p. 82) promotes the development of increasingly advanced arithmetical procedures. For instance, SCADS does not model key constructs such as assimilation, integration, and developmental readiness, or key phenomena such as why order of addends initially constrains children's strategy choice, the evolution of children's estimation ability, or their invention of thinking strategies to reason out unknown number combinations (A.J. Baroody, personal communication, May 24, 2004). Moreover, in the model, little or no attention is given to the social, and in particular instructional, context in which the development of arithmetic skills takes place. Indeed, it seems incontrovertible to assume that the occurrence and the frequency, efficiency, and adaptivity with which certain strategies are used by children will depend heavily on the nature of instruction. By instruction, we mean more than the pattern of frequencies of arithmetic facts in textbooks (Ashcraft & Christy, 1995), the number of times a particular item has been shown, or the number of times a child has received positive or negative feedback to a particular item.

Commenting on Baroody and Tiilikainen's (2003) very critical analysis of SCADS and their own "schema-based view" (Baroody & Ginsburg, 1986) that they presented as a more valuable alternative, Bisanz (2003) remarked that, although it is quite clear how SCADS works, this schema-based view, which speaks of "a web of conceptual, procedural, and factual knowledge," is not described in equal detail.

While SCADS exemplifies one of the major virtues claimed for information-processing theories, namely that they encourage precise theorizing and provide a clear basis for evaluating alternative accounts (Cowan, 2003), we certainly do not go as far as many adherents of this approach, such as Klahr and MacWhinney (1998, p. 670), saying that "computational modeling appears to be a necessary condition for advancing our understanding of cognitive development." As we (De Corte & Verschaffel, 1988; Greer & Verschaffel, 1990) and many others (see, e.g., Baroody and Tiilikainen, 2003) have argued, for a variety of reasons, using computer simulation does not *guarantee* clarity or rigor over theories expressed (only) in nonformal language. Even if the model of Baroody and Tiilikainen (2003) lacks the specificity of SCADS, it certainly has pointed to the complex mutual relationship between different kinds of knowledge (conceptual and procedural) in the development of single-digit arithmetic as well as to the crucial role of the broader sociocultural and instructional contexts within which this development occurs.

Conclusion and Discussion

Although psychologists' and mathematical educators' views of arithmetical expertise and of the method of achieving mastery have changed, the importance of mastery of number combinations has certainly remained highly prominent in elementary school mathematics. Clearly, a child who cannot efficiently produce results for the basic combinations is at a great disadvantage in multidigit written and oral arithmetic (Cowan, 2003; Kilpatrick et al., 2001; Thompson, 1999a).

The available research over the past 10 years has provided further documentation that acquiring proficiency with single-digit computations involves much more than rote memorization. This domain of whole number arithmetic demonstrates (a) how the different components of arithmetic skill (strategies, principles, and number facts) contribute to each other, (b) how children begin with understanding of the meaning of operations and how they gradually develop more efficient methods, and (c) how they choose quite adaptively among different strategies depending on the numbers involved (Kilpatrick et al., 2001). Although researchers have made considerable progress in describing these phenomena and although there are now sophisticated (computer) models that fit (to some extent) with these data, we are still some way from a full understanding of the development of expertise in this subdomain (Cowan, 2003).

A first important focus for future research efforts is the need for increased methodological rigor. For instance, a very quick response or the absence of any conscious awareness of the answering process is no guarantee that an answer is generated by retrieval (Baroody, 1993, 2003; Cowan, 2003; LeFevre et al., 2003). Highly practiced routines can become (quasi-)automatized to a point that they can be carried out very quickly and without any conscious awareness. So, neither self-report nor speed will distinguish automatic retrieval from rapid nonconscious computation or application of a rule. With the development of models of number processing that specify both functional components and neuoranatomical circuits (Dehaene & Cohen, 1995), it might become possible to discriminate retrieval of memorized facts from unconscious computation using event-related potentials or functional magnetic resonance imaging techniques (Baroody, 2003; Cowan, 2003). Evidently, this neuroscientific work may be helpful to further our theorizing. Another important methodological problem is that some of the obtained findings may partly be artifacts of the methodologies used, which have different strengths but also potential weaknesses. For example, data about the speed and the accuracy of strategies obtained in many studies wherein participants are simply asked to solve the problems (by means of whatever strategy they want) can be seriously biased by selection effects. To control for these effects, researchers have developed the "choice/no-choice method", wherein strategy data collected in a "(free) choice condition" are compared with those gathered in one or more "no-choice conditions," wherein participants are (experimentally) forced to solve all items by means of a particular strategy (Siegler & Lemaire, 1997; Torbeyns, Arnaud, Lemaire, & Verschaffel, 2004). However, as argued by Torbeyns, Arnaud et al. (2004), this methodology, like any technique, is not without problems.

Second, although researchers now generally accept that the concepts and skills involved in single-digit arithmetic develop iteratively or simultaneously, rather than in a fixed order (either "concepts-first" or "skills-first," Baroody, 2003, pp. 10–11), how these two components interact and, more precisely, exactly when and how the development of a particular (sub)skill promotes the discovery of a particular principle (or vice versa) is still a matter of serious debate. As argued convincingly by Siegler and others, further research on this issue requires the application of so-called microgenetic methods, which involve the repeated examination of children's conceptual and procedural knowledge during the whole learning process (Siegler, 1998; Torbeyns, Arnaud, et al., 2004).

Third, while a lot of research has been done on addition and multiplication, relatively little attention has been given to a *comparative* analysis of the two operations. This raises the question to what extent the role of the different components (namely: strategies, principles, and number facts) in the development of arithmetic skill are the same for single-digit addition and single-digit multiplication. Moreover, compared with the operations of addition and multiplication, subtraction and division have attracted much less research, although clinical and neuroscientific evidence suggests that single-digit arithmetic around these latter operations proceeds quite differently than for their inverse counterparts, especially with respect to performance at the highest level of development (namely the level of fact retrieval).

Finally, although this line of research has unraveled the role of self-invention or self-discovery in the development of addition skill, the impact of cultural and instructional factors is beyond the scope of most of the (cognitive/rationalist) psychological work being discussed in this section. Many of these (computer) models seem to assume that there is a kind of universal taxonomy or developmental sequence of computational strategies that is fundamentally independent of the nature of instruction or the broader cultural environment. Although it seems plausible that some elements of this development are strongly constrained by general factors other than the instructional and cultural context wherein this development occurs (such as the inherent structure of mathematics, some characteristics of young children's everyday experiences, the unfolding of certain cognitive capacities in early childhood), others are less constrained and much more dependent on children's experiences with early mathematics at home and at school (such as the provision of cultural supports and practices as sources to move quickly beyond counting-based methods or the involvement in a [classroom] climate that nurtures and values flexibility).

Multidigit Arithmetic

After they have been introduced to single-digit arithmetic in the first year(s) of elementary school, children spend several years learning multidigit arithmetic. Children in the United States, but also in many other countries, receiving regular classroom instruction have been shown to experience considerable difficulty constructing appropriate concepts of multidigit numeration and appropriate procedures for multidigit arithmetic (Brown, 1999; Fuson, 1992; Rittle-Johnson & Siegler, 1998; Thompson, 1999c).

Illustrations of lack of conceptual understanding are elementary school children's inability to identify the tens, hundreds, etc. digit in a multidigit number, to relate ungrouped and grouped objects to the different digits in a multidigit number, and to demonstrate or explain 10-for-1 trading with concrete representations (Fuson, 1992; Fuson, Smith, & Lo Cicero, 1997; Fuson, Wearne, et al., 1997; Thompson, 1999c). Lack of procedural knowledge has been evidenced in findings that children of these ages frequently err while solving multidigit arithmetic problems with paper and pencil (see also Fuson, 1992). As a typical example, we refer to the numerous documentations of children applying incorrect variations of the correct procedures involved in the execution of the standard algorithms ("buggy procedures"; see the section *Mental Calculation Strategies Versus Written Algorithms*), because they misremembered one step or reversed two steps of the standard algorithm (Thompson, 1999c).

According to Fuson (1992, p. 263), many errors children in the United States make on numeration indicate that they interpret and treat multidigit numbers "as single-digit numbers placed adjacent to each other, rather than using multidigit meanings for the digits in different positions." Thus, they seem to be using a concatenated single-digit conceptual structure for multidigit numbers. This pattern of behavior may stem from classroom experiences

> that do not sufficiently support children's construction of multiunit meanings, do require children to add and subtract multidigit numbers in a procedural, rule-directed fashion, and do set expectations that school mathematics activities do not require to think or to access meanings. (Fuson, 1992, p. 263)

A similar analysis of children in the U.K. was provided by Anghileri (1999), who cited the anecdotal example of a child who, after having struggled with the (long) division problem $432 \div 15$ put up his hand and asked: "Could I add the 2 and the 43 because I know that 15 goes into 45 three times?" So, an important focus of the current worldwide reform movement is to help children develop good understanding of the decimal number system as a basis for developing procedural fluency in multidigit arithmetic (Brown, 1999; Kilpatrick et al., 2001).

In this section, we review recent literature on the development of children's conceptual structures and procedures for multidigit arithmetic. First, the research on children's conceptual structures for multidigit numbers will be reviewed. Next, we discuss recent work on the development of children's (invented, informal) strategies for, respectively, multidigit addition and subtraction, and multiplication and division. We categorize these strategies, which typically show up in reform-based classroom environments that support children's invented strategies (discussed later in this chapter), as *mental arithmetic* and contrast them with the standard algorithms for (written) computation. When we use the term "mental arithmetic," we do not use it in the traditional sense of "calculating an exact arithmetic operation without the aid of an external calculation or recording device," but rather in the sense of inventing and applying handy and flexible means of calculation, based on one's understanding of the basic features of the number system and of the arithmetic operations, as well as on a well-developed feeling for numbers and a sound knowledge of the elementary number facts (see Anghileri, 1999; Reys, Reys, Nohda, & Emori, 1995; Thompson, 1999a). A typical characteristic of mental arithmetic thus conceived is that it involves the selection and adaptive use of strategies, possibly idiosyncratic but appropriate for a particular problem and always linked to the conceptual underpinnings. In this view, mental arithmetic does not exclude the use of intermediate notes or "jottings" (as an *aide-mémoire*) if the situation or the problem allows or necessitates it (Anghileri, 1999;

Buys, 2001; Thompson, 1999a). In short, according to this second view, mental arithmetic is "thinking *with* your head" rather than as "thinking *in* your head" (Buys, 2001, p. 122). Thus, it is not the presence or absence of paper and pencil, but rather the nature of the mathematical entities and actions that is crucial in our differentiation between mental arithmetic and (written) algorithms.

Children's Conceptual Structures for Multidigit Numbers

Although the conceptual structures needed to do single-digit arithmetic can, in principle, also be used to solve multidigit problems, these procedures quickly become inefficient when the numbers get larger. Consequently, children must develop conceptual structures that reflect the basic principles underlying our decimal number system as well as the mutual relationships between the different ways in which the numbers of that system can be represented (i.e., the number words, the written marks, and the quantities they represent). So, the fundamental building blocks for children's conceptual structures for multidigit numbers are the single-digit combinations and rules for combining them within the decimal system. Some scholars, like Wu (1999), suggest that the most appropriate way to think about multidigit arithmetic is to regard it as nothing more than a sequence of single-digit arithmetic operations, which is odd, because one needs to understand how to orchestrate the sequence. Wu's view, which fits with an information-processing perspective, may be adequate for routine expertise, but not adaptive expertise as defined by Hatano (2003, and see introduction).

Hereafter we summarize a framework of conceptual structures developed by Fuson, Smith, et al. (1997) and Fuson, Wearne, et al. (1997) for understanding (English) number words for multidigit numbers and multidigit written marks. This framework extends and clarifies the theoretical analysis presented in Fuson (1992) and integrates the theoretical perspectives of several recent projects designed to help children learn number concepts and operations with understanding, which will be reviewed in the second part of this chapter (Carpenter, Franke, Jacobs, Fennema, & Empson, 1997; Fuson & Kwon, 1992; Hiebert & Wearne, 1996; Murray & Olivier, 1989). Fuson, Smith, et al. (1997) and Fuson, Wearne, et al. (1997) called it the UDSSI triad model after the names of the five conceptual structures (unitary, decade, sequence, separate, integrated). According to the authors, several different conceptions may be available to a given child and be used in different situations. Thus, new conceptions are added to, rather than replacing, older conceptions. A sixth conception, the incorrect concatenated single-digit conception (discussed above) is also included in their framework. Each conception involves a triad of two-way relationships among number words, written number marks, and quantities.

According to the model, the development begins with an established triad structure for single-digit numbers. Multidigit numbers build on and use the unitary single-digit triads of knowledge. Thus, before children can learn about two-digit numbers, they must have learned for 1 to 9 how to read and say number words corresponding to each number mark, write the numeral corresponding to each number word, and count or count out quantities for each mark and number word up to 9. Because the number words for single-digit numbers in English, as in most languages, and the corresponding written marks are arbitrary, most children learn most of the unitary single-digit triads as rote associations.

Unitary multidigit conception. Children begin with a unitary multidigit conception, in which quantities are not differentiated into groupings, and the number word and number marks are not differentiated into parts. So, for 15 doughnuts, for example, the 1 is not related to "teen" in "fifteen," and the quantities are not meaningfully separable into 10 doughnuts and 5 doughnuts.

Decade-and-ones conception. In the second phase children begin to separate the decade and the ones parts of a number word and start to relate each part separately to the quantity to which it refers, e.g., "fifty" to 50 objects and "three" to 3 objects. They may make the same separation and try to link the decade and the ones parts of a number word to written marks. A typical error made by children in this stage of conceptual development is to write 503 as they hear "fifty-three."

Sequence-tens-and-ones conception. In the next phase children construct a sequence-tens-and-ones conception, in which each decade is structured into groups of ten. This conception requires the skill of being able to count by tens, but also of "seeing" the groups of ten within a quantity and choosing to count these by tens.

Separate-tens-and-ones conception. In the next stage the child begins to think of a two-digit quantity as comprising two separate kinds of units—units of ten and units of one. When asked to make a two-digit quantity, the child counts the groups of ten using single-digit numbers (e.g., "one ten, two tens, three tens, . . .) or sometimes even omits the word *tens* while counting, leaving it as understood ("one, two, three, four, five tens"). While using this terminology, the child understands that each ten is composed of 10 ones and can switch to thinking of 10 ones if that approach becomes useful.

Integrated sequence-separate tens conceptions. In the last stage, children may eventually relate the sequence-tens and separate-tens conceptions of two-digit numbers to each other so that they are able to switch between them rapidly. In this integrated sequence-separate tens conception, bidirectional relationships are established between the tens and the ones component of each of the three parts (number words, marks, quantities) of the sequence-tens and the separate-tens conceptions. This integrated conception allows children considerable flexibility in approaching and solving problems using two-digit numbers.

Concatenated single-digit conception. Even when children have one of the adequate multidigit conceptions and use this conception to add or subtract numbers meaningfully and accurately when these are presented in a word problem or horizontally, they may use a concatenated single-digit conception for the same computation presented vertically and make an error. The vertical presentation elicits an orientation of vertical slots on the multidigit numbers that partitions these numbers into single digits. (The physical appearance of the written multidigit marks as single digits and the nonintuitive use of the left-right position as a signifier may combine to seduce children to use a concatenated single-digit conceptual structure even if they have a more meaningful conception available.)

So far, we have only discussed two-digit numbers, but according to Fuson, Smith, et al. (1997) and Fuson, Wearne, et al. (1997) the conceptual structures for numbers with three or more digits are extensions of these conceptual structures, except that the hundreds and thousands (in English) have more "separate" than "sequence" characteristics because of their more regular named structure (e.g., *three hundred* and *four thousand* by comparison with *thirty* and *forty*).

Fuson, Smith, et al. (1997) acknowledged that the developmental model presented above is deceptively neat in several respects. First, there are qualitative and quantitative differences depending on the language used. Second, children learn the six relationships for a given number (or set of numbers) at different times and may not construct the last triad relationship for all numbers up to 99 for one kind of conception before the first triad relationships for another conception are construed. Third, not all children construct all conceptions; these constructions depend on the conceptual supports experienced by individual children in their classroom and outside of school. Finally, children who have more than one multidigit conception may use different conceptions in different situations or combine parts of different triads in a single situation. So, children's multiunit conceptions definitively do not conform to a uniform and stage-like model (Fuson, Smith, et al., 1997).

We now turn to some comments on the above framework. First, the empirical basis of the latest version of the model, as presented above, seems somewhat unclear. Clearly, it is partly based on the ascertaining studies that formed the basis of the earlier version of this developmental model described in Fuson (1992). But, for the latest version, the authors state that it showed up "throughout their work with children in the various design studies around more meaningful, constructive, problem-based learning environments aimed at a better proficiency with (operations with) multidigit numbers" (Fuson, Wearne, et al., 1997, p. 138), without giving much further explication of the precise nature of the empirical underpinnings of the theoretical framework and without much clarification about its generalizibility. In other words, it remains unclear what aspects of this development are shaped by specific characteristics of the (innovative) learning environments in which it was observed and what aspects of it are shaped by more general factors that are largely independent of instruction. In an attempt to submit Fuson, Smith, et al.'s (1997) and Fuson, Wearne, et al.'s (1997) UDSSI model to a rigorous empirical test, Collet (2004) has designed a series of tests allowing her to assess every aspect of the triad relationships at each developmental level of the model. This test battery was administered individually to 120 first- and second-grade children from different but traditional learning environments. A provisional analysis of the results suggests that, whatever the instruction received, most (but not all) children construct the unitary conception first. However, once the unitary conception is in construction, the individual pathways seem to become more divergent. Some children construct the decades-and-ones conception before being able to use the sequence-tens-and-ones conceptions (as postulated in the UDSSI model), but some other children can use these latter two conceptions (and are able to count by tens and the tens by ones) without being able to use the decades-and-ones conception. Collet's (2004) results suggest identification of the aspects of the development of the base-ten system that are independent of instruction (i.e., the unitary conception constructs first) and clarify the developmental aspects that vary from one child to another. Further research is needed to clarify the exact contribution of the learning environments in the diversity of the developmental pathways observed.

Second, Fuson, Wearne, et al.'s (1997) model focuses on one aspect of children's growing understanding of numbers and number relationships when they start exploring and operating on multidigit numbers (Fuson, 1992; Jones, Thornton, & Putt, 1994; Treffers, 2001b), namely their base-ten structure. Fuson (1992, p. 265) herself pointed to the fact that besides this "collection-based" interpretation of numbers, there is also a "sequence-based" interpretation wherein the focus is on the number's position within the number-word sequence. Treffers (2001b) referred to both conceptions as, respectively, the "structuring" and "positioning" representation of numbers. He (Treffers, 2001b, p. 109) defined *structuring* as "seeing numbers as a junction in a network of number relations," as in decimal splitting (breaking down a number into its compound positional values) or in factorization (relating numbers to each other by multiplication and division), whereas *positioning* is defined as "being able to place whole numbers on an empty number line with a fixed start and end point. . . . Positioning enables students to gain a general idea of the sizes of numbers to be placed" (p. 104). As such, Treffers's (2001b) positioning interpretation shows some alignment with Dehaene's theory about how numbers are internally represented in the human mind (and brain), wherein an analog magnitude code (a kind of mental number line) is assumed as the main, if not only, semantic representation of a number (see earlier section on single-digit conceptual structures). Although several mathematics educators working in the domain of multidigit arithmetic give this positioning interpretation an increasingly prominent place in their (experimental) curricula, textbooks, and instructional materials (see, e.g., Beishuizen, 1999; Gravemeijer & Stephan, 2002; Selter, 1998; Treffers, 2001b), we are not aware of any (ascertaining) study documenting the development of this latter aspect of children's growing conceptual knowledge of numbers (and its relation to the other aspects of multidigit number development) in a broad and systematic way.

Third, the conceptual development described in this section builds on a development that started much earlier, namely at the time of the very first signs of perception of numerosity and understanding of and operating on natural numbers in the early years of childhood (see the chapter on early childhood mathematical thinking), and continues when children start to learn about other numbers that are included in the upper grades of school mathematics (rational numbers, integers, etc.) (e.g., Greer, 2004; Moss & Case, 1999; Vosniadou & Verschaffel, 2004). With respect to this continuation, there is a great deal of agreement that the extension of number concepts beyond the domain of natural numbers is a serious conceptual obstacle requiring drastic conceptual change (Greer, 2004; Vosniadou & Verschaffel, 2004) in the mathematical development of children. Making use of their knowledge of natural numbers when interpreting new information about rational numbers gives rise to numerous well-documented misconceptions such as the misbeliefs that "the more digits a number has, the bigger it is" (leading to the misbelief that, for instance, 1.3 is smaller than 1.25) or that "multiplication always makes bigger and division always makes smaller" (leading to the expectation that, for instance, 45 divided by 0.9 will result in a quotient smaller than 45). These misconceptions typically result from the unwarranted overgeneralizations of rules that are (only) valid in the domain of natural numbers (Greer, 1992, 2004; Verschaffel & De Corte, 1996; Vosniadou & Verschaffel, 2004; see also the chapter on rational numbers and proportional reasoning in this volume).

Finally, although we have separated this section from the next ones, it is, of course, difficult to separate the development of base-ten number concepts from learning the procedures for doing multidigit arithmetic. In their review of the relation between conceptual and procedural knowledge of multidigit arithmetic, Rittle-Johnson and Siegler (1998) reported several types of evidence for this relationship. A first type comes from evidence from several large-scale studies that individual children's success on measures of conceptual knowledge and procedural knowledge are positively correlated. For instance, Hiebert and Wearne (1996) followed children from first grade to fourth grade. They assessed conceptual understanding by asking children to identify the number of tens in a number, to represent the value of each digit in a number with concrete materials, and to make different concrete representations of multidigit numbers. Procedural knowledge was assessed through performance on two-digit addition and subtraction (story) problems, which could be solved either by the standard algorithm or by an invented procedure. The size of the numbers used in the tests differed as the children grew older. Across assessment periods, children who demonstrated higher levels of conceptual understanding also obtained higher scores on the procedural measures. As a second kind of support, Rittle-Johnson and Siegler (1998) referred to another finding from Hiebert and Wearne's (1996) study, namely that early conceptual understanding predicted not only concurrent, but also future procedural skill. Thirdly, as will be reviewed in later, recent design experiments aimed at improving teaching of multidigit arithmetic by emphasizing concepts of place value and how they relate to steps in a procedure usually led to successful increases in both conceptual and procedural knowledge compared to traditional forms of instruction that are characterized by a gap between the two kinds

of knowledge (Hiebert & Wearne, 1996). At the same time, Rittle-Johnson and Siegler (1998) referred to some research evidence (Resnick & Omanson, 1987) that in conventional instruction that emphasized practicing procedures without linking this practice to conceptual understanding, the links between conceptual and procedural development are much weaker.

Strategies for Multidigit Addition and Subtraction

Whereas existing theory and research offer a rather comprehensive picture of how children learn to add and subtract with small numbers (discussed earlier), the literature about what strategies should be distinguished, and how they develop over time, is much more limited in the number domain beyond 20. At the end of her overview of the research on whole number addition and subtraction, Fuson (1992, p. 269) concluded that, except for research on buggy algorithms, our knowledge of the development of conceptual structures and solution procedures for multidigit addition and subtraction "lags far behind" that for single-digit concepts and skills.

However, since Fuson's (1992) review, many studies exploring this new domain have been conducted in the United States (Jones et al., 1994), the U.K. (Thompson, 1994, 1999a), Japan (Reys et al., 1995), Australia (Cooper, Heirdsfield, & Irons, 1998), the Netherlands (Beishuizen, 1993, 1999; Beishuizen, Van Putten, & Van Mulken, 1993), Germany (Selter, 1993), Belgium (Torbeyns, Verschaffel, & Ghesquière, 2006; Verschaffel, 1997), and many other countries. These studies have documented the frequent and varied nature of children's and adults' use of (informal) strategies for mental computation that depart from the formal written algorithms taught in school.

Fuson, Wearne, et al. (1997) provided a comprehensive analysis of the mental calculation strategies children use to solve multidigit addition and subtraction problems, involving two primary classes of strategies, namely the *decompose-tens-and-ones* method and the *begin-with-one-number* method, as well as a third category of mixed strategies. In the first type of strategy, namely decompose-tens-and-ones, the tens and units of both numbers are split off and handled separately (e.g., 46 + 47 is determined by taking 40 + 40 = 80 and 6 + 7 = 13, answer 80 + 13 = 93). In the second type of strategy, the begin-with-one-number method, the different values of the second number are counted up or down from the first unsplit number (e.g., 46 + 47 is determined by taking 46 + 40 = 86, 86 + 7 = 93). In the literature, these two methods have a variety of names such as *combining units separately, decomposition, partitioning* or *the split method* (for the decompose-tens-and-ones method), versus *sequential strategy, cumulative method* or *jump method* (for the begin-with-one-number strategy). Besides these two primary types of strategies (and their mixed versions), some researchers (e.g., Buys, 2001; Treffers, 2001b) have identified a third type, called *compensating* or *varying* strategies, wherein the given numbers are adjusted in a clever and flexible way to simplify the calculation (e.g., 46 + 47 = [45 + 45] + 1 + 2 = 93 or 46 + 47 = 46 + 50 – 3 = 96 – 3 = 93), whereas other researchers (such as Fuson, Wearne, et al., 1997) consider such clever strategies as a subcategory of one of their two major categories of solution methods. Evidently, these categorizations apply also for addition and subtraction with multidigit numbers larger than two.

The first two types of strategies (for which we will from now on use the short but telling terms *split* and *jump*) in fact appear to be consistent with the two fundamentally different ways of interpreting two-digit numerals and number words, which Fuson (1992, p. 265) called "collection-based" and "sequence-based" and Treffers (2001b) called "structuring" and "positional" interpretations of numbers (discussed earlier; see also Buys, 2001; Cobb, Gravemeijer, Yackel, McClain, & Whitenack, 1997; Treffers, 2001b). Whereas the idea implicit in the split strategy is that of numbers as collections of items that can be partitioned and recomposed, the view implicit in the jump strategy is that of a counting activity that can be curtailed or chunked. Accordingly, whereas base-ten blocks (or corresponding square/bar/point drawings) have been used typically as concrete referents to teach and to assess the collection-based interpretation and the corresponding split strategy, educators and researchers focusing on the sequence-based interpretation and the jump strategy will often use other representational tools, such as the hundred square or the hundred row and, more recently, the (empty) number line (Beishuizen, 1999; Treffers, 2001b; Wittmann, 2001).

Although these two strategies have a widespread use worldwide, there seem to be great international variations as consequence of differences in the (school) mathematics culture and practice. For instance, in most studies with second- to fourth-grade children in the United States and the U.K. the split method is the dominant one, probably because of an educational tradition that has emphasized written "column arithmetic" and the use of base-ten blocks (sharing with the split method the same conceptual number structure). In most studies involving continental European children, by contrast, the jump method seems more dominant among children who have received instruction in addition and subtraction in the number domain up to 100, probably because of the greater historical attention to mental calcula-tion in general and the greater instructional focus on the sequential method in textbooks and on materials (such as the hundred board, the number line…) supporting this method (Beishuizen et al., 1993).

Beishuizen's research program aimed at the characteristics of the solution and learning process of both strategies in the domain 20–100, basically from a viewpoint of cognitive (developmental) psychology (Beishuizen, 1993, 1999; Beishuizen et al., 1993). Like other Dutch researchers, he found that, although Dutch mathematics education in Grade 2 focuses on the jump method, the strategies of Dutch pupils showed a mixed picture. In general, only about half of the third graders systematically used the (explicitly taught) jump method, whereas the other half preferred the split method (as a nontaught alternative). Only a small minority used both strategies in a flexible or adaptive way (e.g., the split strategy for addition and the jump strategy for subtraction problems). However, contrary to what occurs in this early stage, in the long run the split method turns out to be the more complicated and error-prone computation method, whereas the jump method becomes the more efficient and effective one. Especially in subtraction-with-carry problems, pupils using the split method show many errors, such as the well-known "smaller-from-larger bug" (e.g., 63 – 38 solved as 60 – 30 = 30 and 8 – 5 = 3, answer 33). In line with this argument, Beishuizen (1993) observed more errors for the split as compared to the jump strategy when competent and consistent users of either method were compared on mental computation tasks.

Following these exploratory studies on the characteristics of the jump and the split strategies, Beishuizen et al. also conducted a series of intervention studies (Blöte, Van der Burg, & Klein, 2001; Klein, Beishuizen, & Treffers, 1998) wherein they compared the impact of reform-based instruction, focusing on the flexible application of different strategies from the start of the learning process onwards, with traditional instruction, focusing on the mastery of one single strategy, namely the jump strategy. The results of these intervention studies showed that, after instruction, both groups of children applied the jump strategy quite frequently, especially on subtractions (Blöte et al., 2001; Klein et al., 1998). But the children who used the reform-based instruction fitted their strategy choices more often to the number characteristics of the items than the traditionally taught children and relied more frequently on so-called varying strategies. For instance, they solved items with an integer ending with the number 9 (like 35 + 49) more frequently with a compensation or varying strategy (e.g., 35 + 50 – 1) than children who had received traditionally-oriented instruction (Blöte et al., 2001; Klein et al., 1998).

Although the research programs of Beishuizen and his colleagues deepened our understanding of the developmental changes in children's strategy competencies in the number domain up to 100, their results are weakened because of some methodological issues. For example, their operationalization of adaptivity only takes into account item characteristics. As discussed by several authors (Payne, Bettman, & Johnson, 1993; Siegler, 1998), an adaptive strategy choice process takes into account not only item characteristics, but also characteristics of the strategies (e.g., accuracy, speed), the subject (e.g., ability level), and the context (e.g., instruction). In an attempt to overcome these shortcomings, Torbeyns et al. (2006) analyzed the development of children's strategy competencies for doing additions and subtractions up to 100 in terms of the model of strategy change (Lemaire & Siegler, 1995), using the choice/no-choice method (Siegler & Lemaire, 1997). Second graders of high, average, or low mathematical achievement level solved four types of additions and subtractions before and after (traditional) instruction in the jump method. At both measurement times, children answered the items in one choice condition in which they could choose between the jump and the split strategy on each item, and in two no-choice conditions in which they had to solve all items with, respectively, the jump and the split strategy. As in Beishuizen et al.'s studies, at the first time of measurement, most children solved all problems by one strategy, mostly splitting, in the choice condition. Second, although the no-choice data showed that most children became quite accurate in applying the taught jump method, many of them (especially the weaker ones) continued to opt for the split strategy (especially on the addition problems) in the choice condition. Finally, although the children in this study, as in the studies of Beishuizen et al., seemed to adapt their strategy choices to item characteristics (e.g., by using the split strategy much less frequently on subtraction than on addition items), a more comprehensive (but at the same time more demanding) measure of adaptivity, involving also strategy and subject characteristics, revealed that, both before and after instruction, children were remarkably nonadaptive in their strategy choices. Although the reasons for this nonadaptivity remain obscure, it seems evident that this lack of strategy adaptivity is at least partly due to an orientation towards routine rather than adaptive expertise in the instructional environment.

As part of a broader ascertaining study, Reys et al. (1995) assessed mental computation performance, as well as attitude and computational preferences, of large groups of Japanese students in Grades 2, 4, 6, and 8 on problems involving whole-number additions and subtractions ranging from single-digit to three-digit problems such as $182 + 97$ and $264 - 99$. Problems were presented visually or orally, and students could not make use of paper and pencil. A sample of students in Grades 4 and 8 scoring in the upper and middle quintiles on the mental computation test was interviewed to identify strategies used to mentally compute. A wide range of performance on mental computation was found with respect to the different number types and operations involved at every grade level. The mode of presentation affected performance level, with visual items generally producing higher performance. The most important finding was that the range of strategies used for mental computation was quite narrow, with the most popular approach reflecting a "mental version" of the learned right to left paper-and-pencil algorithm. This tendency to work algorithmically, even in the absense of paper and pencil, was more pronounced for the middle-quintile students than for the high-quintile students.

In the United States, Carpenter et al. (1997) did a longitudinal study investigating the development of children's multidigit addition and subtraction in relation to their understanding of multidigit concepts in Grade 1–3. Students were individually interviewed five times on a variety of tasks involving straightforward, result-unknown addition and subtraction word problems with two-digit numbers for the first three interviews and three-digit numbers in the last two interviews. During the same interviews, children were individually administered five tasks measuring their knowledge of base-ten number concepts, together with a task wherein they had to apply a specific invented strategy to solve another problem and two unfamiliar (missing addend) problems that required some flexibility in calculation. Contrary to the above-mentioned study with Japanese pupils, all students were in classes of teachers who were participating in a 3-year intervention study designed to help teachers understand and build on children's mathematical thinking. The emphasis of this program was on how children's intuitive mathematical ideas emerge to form the basis for the development of more formal concepts and procedures. Teachers learned about how children solve problems using base-ten materials and the various invented strategies children often construct. The researchers identified the following categories of strategies:

- modeling or counting by ones,
- modeling with tens materials,
- sequential strategies (jump strategies),

- combining-units strategies (decomposition or split strategies),
- compensating strategies,
- other invented strategies,
- algorithms (correct as well as buggy ones).

First, the study showed that, under favorable circumstances, children can invent mental calculation strategies for addition and subtraction problems. Second, buggy algorithms occurred more frequently among children who started out working algorithmically early than among children who used invented mental strategies before or at the same time they used standard algorithms. Third, students who used mental calculation strategies before they started using standard algorithms also demonstrated better knowledge of base-ten number concepts and were more successful in extending their knowledge to new situations. Finally, the data suggest that there is no explicit sequence in which the three basic categories of mental calculation strategies (sequential/jump, combining units/split, and compensating/varying) develop for addition. The majority of students applied all three, and the order in which they occurred was mixed. For subtraction, the sequential method was mostly used, but for this kind of problem some compensation strategies were also observed. For other studies reporting similar findings about the development of mental calculation strategies for multidigit addition and subtraction in nonconventional classrooms in close relation to the development of the concepts that support them, we refer to Fuson, Wearne, et al. (1997) and to Hiebert and Wearne (1996).

Strategies for Multidigit Multiplication and Division

Although several researchers have documented that children can also invent strategies for multiplying and dividing multidigit numbers and have described some strategies they use, less progress has been made in characterizing such inventions than for the domain of multidigit addition and subtraction. Here, we summarize the main findings from a recent analysis by Ambrose, Baek, and Carpenter (2003; see also Baek, 1998) of children's invented multidigit multiplication and division procedures (and the concepts and skills on which they depend). We stress again that these constructions did not take place in a vacuum, but in an instructional environment that stimulated children to construct, elaborate, and refine their own mental strategies rather than forcing them to follow a uniform, standardized trajectory for mental or written arithmetic (Ambrose et al., 2003, p. 309). As such, the constructions described by these authors show great similarities with those in Treffers's (1987, Chapter VI)

analysis of how elementary school children gradually build up the algorithms for column multiplication and division starting from their invented informal strategies in an instructional environment based on the principles of realistic mathematics education (RME), and to similar analyses by Anghileri (1999) and Thompson (1999c). For a description of these environments, we refer to the Classroom-Based Research section.

Children's mental calculation strategies for multiplication problems were classified into four categories: direct modeling, complete number strategies, partitioning number strategies, and compensating strategies (Ambrose et al., 2003; Baek, 1998). A child using a *direct modeling strategy* models each of the groups using concrete manipulatives or drawings. A second category, *complete number strategies*, are those based on progressively more efficient techniques for adding and doubling, the most basic one being simply repeated addition. A child using the *partitioning number strategy* will split the multiplicand or multiplier into two or more numbers and create multiple subproblems that are easier to deal with. Finally, a child using a *compensating strategy* will adjust one or both of the multiplicand and multiplier on the basis of special characteristics of the number combination to make the calculation easier. Each of these mental calculation strategies is illustrated in Figure 13.1 (Baek, 1998).

Many children in the study developed their mental calculation strategies for multidigit numbers in a sequence from direct modeling to the complete number strategy to the partitioning numbers strategy (Ambrose et al., 2003; Baek, 1998). Children's strategies for solving multidigit multiplication problems varied with their conceptual knowledge of addition, units, grouping by ten, place value, and properties of the four basic operations.

Ambrose et al. (2003) presented a similar taxonomy for division, distinguishing between the following strategies. First, *Working with one group at a time*, whereby the child repeatedly subtracts the smaller given number from the larger (solving "180 divided into groups of 6" by doing $180 - 6 = 174$, $174 - 6 = 168$, etc.), or repeatedly adds up the smaller given number until the larger one is reached ($0 + 6 = 6$, $6 + 6 = 12$, etc.), and then counts how many subtractions or additions were made. Repeated subtraction or addition occurred regularly for quotitive-division problems, but very rarely for partitive-division problems, because in the latter case children do not know what number to subtract or to add repeatedly. However, these problems can, and are, solved at a comparable basic level by effectively distributing the larger given number in small parts over the different groups, usually with help of

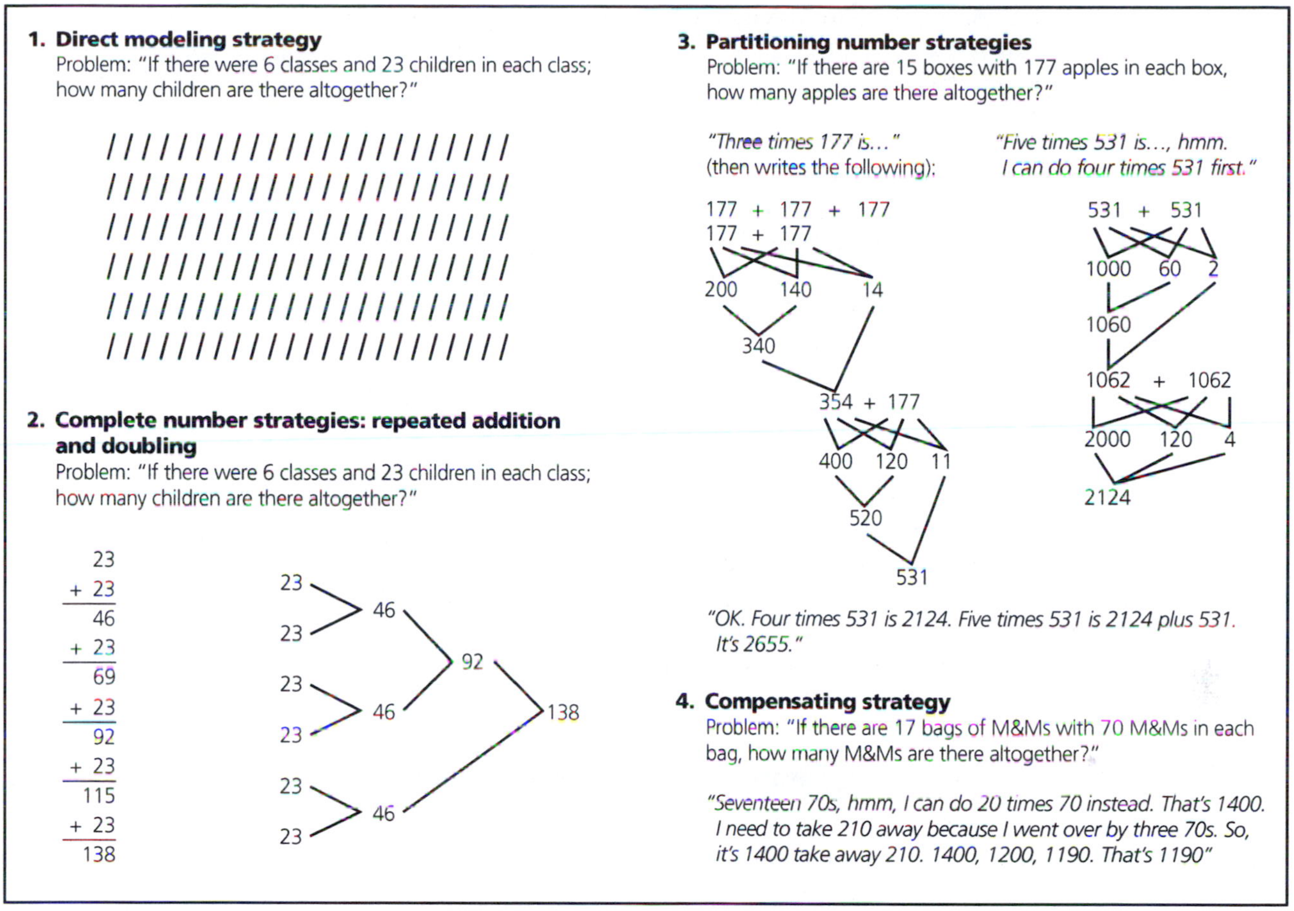
1. Direct modeling strategy
Problem: "If there were 6 classes and 23 children in each class; how many children are there altogether?"
2. Complete number strategies: repeated addition and doubling
Problem: "If there were 6 classes and 23 children in each class; how many children are there altogether?"
23
+ 23
46
+ 23
69
+ 23
92
+ 23
115
+ 23
138
23
23
46
23
23
46
23
23
46
92
138
3. Partitioning number strategies
Problem: "If there are 15 boxes with 177 apples in each box, how many apples are there altogether?"
"Three times 177 is…"
(then writes the following):
177 + 177 + 177
177 + 177
200
140
14
340
354 + 177
400
120
11
520
531
"Five times 531 is…, hmm. I can do four times 531 first."
531 + 531
1000
60
2
1060
1062 + 1062
2000
120
4
2124
"OK. Four times 531 is 2124. Five times 531 is 2124 plus 531. It's 2655."
4. Compensating strategy
Problem: "If there are 17 bags of M&Ms with 70 M&Ms in each bag, how many M&Ms are there altogether?"
"Seventeen 70s, hmm, I can do 20 times 70 instead. That's 1400. I need to take 210 away because I went over by three 70s. So, it's 1400 take away 210. 1400, 1200, 1190. That's 1190"

Figure 13.1. Examples of (invented) strategies for multidigit multiplication problems (Baek, 1998, pp. 153–158).

some kind of visual representations (e.g., solving "228 divided into 12 groups" by drawing 12 circles—one for each group—and then distributing the 228 units over the 12 groups by first putting a 10 in each circle, then a 2, then a 5, and finally another 2, while always keeping track how much remains after each distribution round). A second category involves *more abstract strategies that do not involve decomposition of the dividend,* wherein the child subtracts (or divides) in a more efficient and systematic way by relying explicitly on the ten-structure (e.g., solving "544 divided into 17 groups" by subtracting 544 three times by 10 × 17 = 170, and, having arrived at 34, by 2 × 17 = 34, leading to the solution of 30 + 2 = 32). A third category consists of *strategies wherein the dividend is split into parts* (e.g., dividing 896 by 35 by dividing first the units [6], then the tens [90] and finally the hundreds from the dividend [800] by 35 while keeping track of the number of times 35 goes into each of these numbers as well as of the remainder of each of these three partial divisions). A final category involves strategies wherein both partitive and quotitive division problems are solved by means of *building up strategies* (e.g., solving 544 divided into 17 groups by adding up 170 + 170 = 340, 340 + 170 = 510, 510 + 34 = 544, so there are 32 candies/friend). For similar inventories of informal strategies for multidigit division, see, for example, Anghileri (1999, 2001), Treffers (1987), and, most recently, Van Putten, van den Brom-Snijders, and Beishuizen (2005). Using a 10-category system of strategies and a sophisticated multivariate approach, the latter authors were able to describe in a very detailed and systematic way the evolution of Grade 4 children's strategies for solving multidigit division problems under a realistic mathematics approach in Dutch primary schools. Just after the introduction of division problems with larger numbers, the strategies varied from no-chunking to high-level chunking, but after 5 months, the diversity was reduced, not only for the high-ability pupils but also for pupils with a rather weak mathematics level, to mainly high-level chunking.

Underlying these multidigit mental calculation strategies for multiplication and division are the commutative, associative, and distributive laws, although they may not be known to pupils by name and may not be applied deliberately (Anghileri, 1999). As such, they constitute what Vergnaud (1996) termed *theorems-in-action*. A possible question for future research is whether making the laws (more) explicit would enhance the adaptivity based on their implicit application.

Mental Calculation Strategies versus Written Algorithms

Besides the kind of (invented) mental calculation strategies described in the two previous sections, there is, of course, another well-known type of procedure for doing multidigit arithmetic, namely the standard algorithm. Algorithms are finite, well-defined step-by-step procedures for accomplishing (familiar) tasks (Plunkett, 1979; Usiskin, 1998). More specifically, in this context, they are recipes for performing calculations with the digits of the numbers in the problem to find the result of an addition, subtraction, multiplication, or division (Thompson, 1999c; Treffers, Nooteboom, & De Goeij, 2001). The purpose of any arithmetical algorithm is to reduce a complex calculation to a series of more elementary calculations that can be performed using well-established processes, together with the rules for co-ordinating these smaller calculations. As explained earlier, these standard algorithms are typically done by means of paper and pencil, but they can be, and sometimes are, done mentally.

Standard algorithms have evolved over centuries for efficient, accurate calculation and for the most part are far removed from their conceptual underpinnings. Their compact notation tends to mask the underlying principles that make them work. For instance, the principled basis of the standard algorithm for multiplication and especially for division in many countries is very difficult to unpack. Actually, standard practice in the use of algorithms even demands that one does not think about what the digits represent if one does not want to become confused (Thompson, 1999c). According to the same author, algorithms encourage "cognitive passivity" on behalf of the person using them: "The decision as to how to set out the calculation, where to start, what value to assign to the digits, etc. are all taken out of the individual's hands" (Thompson, 1999c, p. 173). Taught or invented mental calculation strategies like those discussed in the two previous sections, by contrast, are derived directly and perspicuously from the underlying concepts of base-ten numbers, the fundamental principles of the operations, and the relations among them (Ambrose et al., 2003; Anghileri, 1999; Thompson, 1999a; Treffers, 1987).

Evidently, there are different standard algorithms available for whole-number operations. For instance, Figure 13.2 shows three main written methods for subtraction that have been and are still being taught in

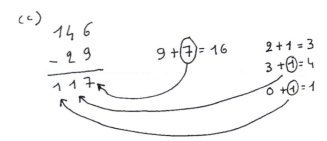

Figure 13.2 Alternative algorithms for whole-number subtraction.

different Western countries. For an overview of different algorithms for arithmetic, we refer to Morrow and Kenney (1988), Thompson (1999c), and the Web site of Orey (2005).

Ascertaining research on algorithms is rather scarce, except for the extensive (older) work on buggy algorithms by Brown, Burton, and Van Lehn (Brown & Van Lehn, 1982; Burton, 1982) in the early '80s. Drawing on computer-based metaphors, they developed a theory that characterizes observed patterns of errors as "bugs" within correct procedures for computations. A finite collection of such bugs, often operating in combinations, have been shown to account for many of the errors made by pupils. This work has often been held up as an example of the contribution that the information-processing approach can make to mathematics education, being cited by Boden (1988, p. 263) as one of the few examples of a domain modeled in enough detail to be useful to pupils or teachers, by Rissland (1985) as a good example of "principled" modeling of human mathematical behavior, and by Newell (1990, p. 368) as a rare case in which cognitive theory can predict long stretches of behavior. Mathematics educators, by contrast, have been much less enthusiastic (e.g., Cobb, 1990; Fischbein, 1990), a fundamental basis of their criticism being that skill in computation involves much more than procedural competence. On the one hand, local repair of bugs in children's procedures amounts to treating symptoms rather than the underlying causes of imperfect understanding. On the other hand, it is appropriate to exploit systematicity in errors to diagnose conceptual weaknesses (e.g., Resnick et al., 1989).

Other researchers, particularly mathematics educators, have also studied algorithms. Results show that after having completed time-consuming courses in written algorithms, many students still have not mastered them. For instance, Treffers (1987) reported that after between 50 and 100 lessons, only 2 out of 3 pupils in the Netherlands had mastered the long division algorithm. Moreover, traditionally schooled pupils find it very difficult to relate the algorithm to situations in which it is nevertheless applicable, and they also perform very weakly on test items asking for insight into the algorithm rather than for procedural mastery (Treffers, 1987; see also Verschaffel & De Corte, 1996). Some mathematics educators went further in the '90s and documented the harmful effects of algorithms in elementary school children (Kamii & Dominick, 1998). These manifestations may be interpreted as evidence that teaching with the narrow aim of establishing routine expertise may not be the best way to achieve even that limited goal.

As explained before, algorithmic calculation is given a less prominent place in current educational reform—or rather, to be more precise, its prominence is placed against the background of emphasizing other aspects of proficiency. Although the importance of mental arithmetic and estimation is stressed in all current reform documents, the view is still generally accepted that children should learn in elementary school at least one standard written method of calculation for each of the four arithmetic operations. However, it is also agreed, first, that premature teaching of standard written algorithms should be avoided and that an introduction into written algorithms can only be given after prolonged work with concrete and mental arithmetic. Second, with respect to the way of teaching, there is wide agreement that children should be actively involved in devising the algorithms for written computation, starting from their knowledge about number and single-digit and multidigit operations. Third, more and more people in the mathematics education community accept that there is no absolute need that all pupils reach the highest level of all formal written standard algorithms (Thompson, 1999a, 1999c; Treffers et al., 2001).

Illustrative, and supportive, for this latter reform-based approach are the design experiments realized by the Freudenthal Institute for the development of multiplication and long division, wherein the hypo-thetical learning trajectories for the four algorithms involve a gradual progression from the kind of informal and invented mental calculation strategies described above through a stage of "column arithmetic" strategies to the culturally-agreed algorithmic calculations (Treffers, 1987; Treffers et al., 2001; see also De Corte et al., 1996; Van Putten et al., 2005; Verschaffel & De Corte, 1996). For similar design experiments done recently in the United States and the U.K., see the Classroom-Based Research section.

Conclusion and Discussion

Whereas in the '70s and '80s research focused on children's solutions of arithmetic problems involving relatively small whole numbers, researchers more recently have paid more attention to problems that involve multidigit calculations and have made significant progress in identifying and characterizing the different conceptual elements and strategies that children construct to calculate with multidigit numbers besides the regularly taught standard algorithms for written computation.

This research has helped us to illustrate another meaning of "mental arithmetic" besides the traditional one wherein mental arithmetic is seen (merely) as doing sums in one's head instead of with paper and pencil. Most classifications of children's procedures for operating on multidigit numbers distinguish among three basic categories of strategies of mental arithmetic (which seem to be closely linked to different conceptions of numbers):

- Jump strategies in which the numbers are seen primarily as objects in the counting row and for which the operations are movements along the counting row—further (+) or back (−) or repeatedly further (×) or repeatedly back (÷).
- Split strategies in which the numbers are seen primarily as objects with a decimal structure and in which operations are performed by splitting and processing the numbers on the basis of this structure.
- Varying strategies based on arithmetic properties in which the numbers are seen as objects that can be structured in all sorts of ways and in which operations take place by exploiting a suitable structure and using the appropriate arithmetic properties (Buys, 2001).

For each of the basic arithmetic operations, each of these three basic forms can be performed at different levels of internalization, abbreviation, abstraction, and formalization (e.g., at a lower level by using models such as the base-ten blocks or the "empty number line," at a higher level by noting intermediate steps in formal-arithmetic language, or purely internally).

As in the *Single-Digit Computation* section, the above review of the last decade's research on multidigit mental arithmetic has pointed to (a) the invented nature of several of these procedures of mental arithmetic; (b) the flexible or adaptive use of different strategies as a basic characteristic of expertise in mental arithmetic (Hatano, 2003); (c) the close relationship between procedural and conceptual knowledge in the domain of multidigit mental arithmetic (Rittle-Johnson & Siegler, 1998); (d) the "dispositional nature" of mental arithmetic, as is convincingly documented, albeit in a negative way, by many traditionally taught children's inclination to apply their standard algorithms in a stereotyped, inflexible way, even in cases in which adaptive mental arithmetic seems much more appropriate, such as problems like $24,000 \div 6,000$ and $4,002 - 3,998$ (Buys, 2001; Treffers, 1987, 2001b), by their lack of conceptually based "critics" to detect errors in computation, and by their lack of self-confidence to "have a go" and take risks when leaving the safe path of standard algorithms (Thompson, 1999a).

Although we have focused mainly on findings about the development of individual children's concepts and strategies in multidigit arithmetic and postponed the role of instructional and cultural issues to the later section on classroom-based research, the research reported so far leads already to some important instructional implications. On the one hand, there is a well-documented research tradition on the many kinds of errors ("bugs") that students make when the algorithmic methods for multidigit addition, subtraction, multiplication, and division are not connected to place value meanings (Kilpatrick et al., 2001), while at the same time there is increasing evidence of the variety of strategies and the gradual progression in these strategies when children are allowed and even stimulated to invent and use their own methods. Typically, these invented strategies can be ordered on a developmental line beginning with concrete methods that model the problem rather directly, towards more efficient methods, which are less transparent and more problem independent (Kilpatrick et al., 2001). So, these informal mental calculation strategies seem to be a perfect stepping-stone to "informal column arithmetic" and, afterwards, to the standard formal algorithms for written computation (Treffers et al., 2001).

Whether these algorithmic methods still need to be mastered perfectly by all children, and (thus) deserve so much instructional time, is a question that elicits hot debates. The idea that drilling students for long periods on solving problems involving large numbers by means of these written algorithms "seems to be more appropriate to the twentieth century than the twenty-first" (Kilpatrick et al., 2001, p. 214) is gaining ground. Nevertheless, internationally, most elementary school mathematics curricula continue to pay, for mixed reasons, considerable attention to the teaching and learning of these algorithms.

Finally, an important concern, with serious theoretical, methodological, and instructional implications—not only for the research on multidigit arithmetic but also for the previous and the next sections—is what is really meant by "a student is using strategy x" or "strategy x is being taught." According to authors such as Gravemeijer (2004), it is inappropriate to think of strategies as ready-made techniques that are available in the repertoire of the children and are consciously or deliberately chosen and applied in a particular situation. Especially with young children, a more plausible way of thinking is that they (initially) base their computations on their familiarity with certain number relations. Take, for instance, a second grader who, when solving the problem $29 + 5$, activates a variety of number relations involving 29 and 5 to solve the problem at hand. One option would be to combine $29 + 1 = 30$ and $4 + 1 = 5$ to conclude that you can take 1 from the 5, add that 1 to 29 to get 30, and add the remaining 4 to the 30 to get 34. As Gravemeijer commented (2004, p. 114),

> Then, from the observer's point of view, it could look like the student is using a building-up-to-10s strategy [or, in our terminology, a jump strategy]. For the student, however, this strategy may not be at the horizon yet. Only after reflecting on substantial experience with similar problems, the student could start to notice a pattern and construe the building-up-to-10s strategy. Even then, it could take a while before the student starts to use this strategy as an a priori guidance for choosing a solution procedure.

This distinction recalls Vergnaud's (1996) term *theorem-in-action*. If this view on (the development of) strategies is true, the instructional goal, then, becomes, according to Gravemeijer (2004, p. 115), "not to teach the students a set of strategies. Instead, our primary goal is for the students to develop a framework of number relations that offers the building blocks for flexible mental computation."

This idea fits perfectly with the "schema-based view" of arithmetic learning proposed by Baroody and Ginsburg (1986, see also Baroody, Cibulskis, Lai, & Li, 2004).

Estimation and Number Sense

In the lives of both children and adults, many types of questions and situations call for estimation, yet, until recently, curriculum developers have given it short shrift. Before 1980, only isolated voices spoke in support of its role in the mathematics curriculum. Current reform documents, however, call for incorporating estimation activities "into all areas of the program on a regular and sustaining basis" (Sowder, 1992, p. 373; see also Siegler & Booth, 2005). According to these documents, ability to estimate is not only extremely important in many kinds of employment and in the ordinary activities of life, but estimation experience is also claimed to have beneficial effects on many other mathematical abilities such as mental computation, spatial visualization, and measurement (Brade, 2003; Dowker, 2003; Van den Heuvel-Panhuizen, 2001). A third reason why estimation is important is that many types of estimation require going beyond routine application of procedures and applying mathematical knowledge and skills in a flexible way. So it is a privileged vehicle for developing adaptive expertise in mathematics (Siegler & Booth, 2005).

Before these pleas in favor of estimation in curricular documents and recommendations, it had prompted very little research. The scarce research that had been done offered ample evidence that most school-age children are remarkably bad at it and even many adults lacked basic estimation skills (Sowder, 1992), a finding echoed in Siegler and Booth's (2005) more recent review. In her review of the literature, Sowder (1992, p. 371) referred to three types of estimation: computational, numerosity, and measurement estimation, "each requiring different kinds of understanding and different kinds of skills." In this section we focus on computational and numerosity estimation. Because of its close relation to estimation, we have decided to review the research on number sense in this section too. As in the other parts concerning ascertaining studies, we pay more attention to the psychological (and psychometric, and neuroscientific) research on estimation and number sense and their development than to studies dealing with the question of how the skill should be addressed in school mathematics. Measurement estimation (and "sense for measures"), the third component of estimation distinguished by Sowder (1992), will not be discussed, as it lies beyond the scope of this chapter. Although numerosity estimation is clearly the first to evolve, we prefer to start our review with computational estimation, as this has, by far, elicited most of the research on estimation.

Computational Estimation

Computational estimation can be defined as "finding an approximate answer to arithmetic problems without actually (or before) computing the exact answer" (Lemaire & Lecacheur, 2002, p. 282). Within computational estimation, a distinction can be drawn between (a) calculating with rounded-off numbers with the intention of finding a global answer to an arithmetic problem given in exact terms (instead of, before, or after making the exact computation), and (b) calculating with estimated values because (some of) the exact data are unavailable (Van den Heuvel-Panhuizen, 2001).

Most of the research on estimation has dealt with children over 10 years old or with adults. This means that estimation has usually been studied after it is at least partially established, rather than charting its emergence (Dowker, 2003). However, some researchers have investigated how estimation abilities develop (in the absence of instruction) before the age of 10 (see Sowder, 1992, for an overview of the studies up to 1992, and Brade, 2003; Dowker, 2003; and Siegler & Booth, 2005, for reviews of the later research).

A comprehensive study on computational estimation done by Reys, Rybolt, Bestgen, and Wyatt (1982) on a large group of pupils from Grades 7 to 12 and out-of-school adults led to the identification among good estimators of three key processes or procedures that are still encountered in recent work, namely reformulation, translation, and compensation.

- Reformulation can be described as the process of altering numerical data to produce a more mentally manageable form while leaving the structure of the problem intact (e.g., changing $[6 \times 347] \div 42$ into $[6 \times 350] \div 42$).
- Translation involves changing the mathematical structure of the problem into a more mentally manageable form (e.g., changing $[347 \times 6] \div 43$ into $347 \times [6 \div 43]$ and then into $350 \div 7$).
- Compensation involves making adjustments to reflect changes that come about as a result of transformation or reformulation of the problem (e.g., $\$21,319,908 \div 26$ is estimated by dividing $26,000,000$ by 26, yielding $1,000,000$, then compensating downward to $\$850,000$).

The investigators noted that good estimators used a variety of strategies within these three key processes.

Many investigators have characterized good estimators as individuals who, besides having the ability to use these three procedures, "have a good grasp of basic facts, place value, and arithmetic properties, are skilled at mental computation, are self-confident and tolerant of error, and can use a variety of strategies and switch easily between strategies" (Sowder, 1992, p. 374). Poor estimators, by contrast, seem to be bound, with only slight variations, to one strategy—that of applying algorithms more suitable for finding the exact answer. They have only a vague notion of the nature and purpose of estimation, believe it to be inferior to exact calculation, and equate it with guessing. According to Sowder (1992), Hatano's distinction between adaptive and routine expertise seems particularly relevant when thinking about how to characterize good and weak estimators and about how to plan instruction on estimation. Indeed, many estimation problems are novel in nature, making it difficult, if not impossible, to delineate a list of knowledge components and skills that, if learned, would constitute a comprehensive knowledge base for computational estimation (Sowder, 1992).

Since Sowder's (1992) landmark review, interesting work in this domain has been done by LeFevre, Greenham, and Waheed (1993); Dowker (1997, 2003); and Lemaire and Lecacheur (2002). In their cross-sectional study with Canadian participants, ranging from elementary school children to adults, LeFevre et al. (1993) found that until Grade 4 multidigit computational estimation develops very slowly, followed by a leap in estimation ability between Grade 4 and Grade 6, and further improvements beyond Grade 6. Second, using observation of ongoing behavior and immediate retrospective self-reports as their data-gathering techniques, these authors observed a wide strategic variability among children as well as adults, with rounding being the most common approach. Third, both children and adults seemed to adapt their strategy choices to problem characteristics. Finally, the range, the sophistication, and the adaptiveness of the strategy choices increased with age. For instance, a sophisticated strategy like post-compensation (whereby the person corrects for distortion after a computation has been done on rounded or truncated numbers) was used (efficiently) more often by adults than by children. To account for the observed (development in) variety, sophistication, and adaptivity in people's strategy choices, these authors proposed a process model of computational estimation, consisting of retrieval, calculation, and compensation as the basic strategic components, heavily inspired by Siegler and Shrager's (1984) model of strategy choices in one-step arithmetic (see *SCADS* section).

Dowker (1997, 2003) investigated the development of arithmetic estimation with special attention to the question how estimation changes in relation to the difficulty level of the items and development of ar-

ithmetical skill. A large group of children between 4–5 and 9–10 years old was divided into five levels according to their performance on a set of mental addition tasks. For instance, a typical Level 2 child could solve $5 + 3$ but not $8 + 6$, whereas a typical Level 5 child could solve $52 + 39$ but not $523 + 168$. Depending on the level at which they were placed, the children then were given one or more sets of estimation items that went beyond the capacities of the child's arithmetic level. For instance, children categorized in Level 2 received estimation items beyond the capabilities of the addition-to-10 level (e.g., $8 + 6$). The child's task was not to work out the answers, but to make good guesses about them. Dowker found that estimation performance improved with the calculation performance level of the children and that, for each performance level, estimation accuracy deteriorated as the children moved further and further from the number domain corresponding to their level of arithmetical skill (Dowker, 2003).

In an attempt to overcome some methodological problems of many of the previous studies, Lemaire and Lecacheur (2002) recently investigated people's strategies in computational estimation of three-by-three-digit addition problems with small units ($263 + 471$) or with large units (e.g., $249 + 743$) using the "choice/no choice" method. Grade 4 and Grade 6 students, and adults, were asked to provide estimates for such calculations either in a choice condition, wherein they could choose between a "round-down" (rounding both operands to the closest smaller decades) and a "round-up" strategy (rounding both operands to the closest larger decades) or in one of two no-choice conditions (either always rounding-down or rounding-up). Results showed that (a) participants of all age groups used the round-down strategy more often and executed it more quickly than the round-up strategy; (b) strategy use and execution were influenced by participants' age, problem features, and relative strategy performance; and (c) age-related changes in computational estimation included also changes in the relative influences of problem and strategy characteristics on strategy choices.

In their search for factors that might underlie the above-mentioned developments in estimation performance, strategy use, and strategy choice, Siegler and Booth (2005) disinguished three factors: (a) improved conceptual understanding of estimation and its underlying principles, (b) working memory growth, and (c) improved computational skills.

Numerosity Estimation

Of the three types of estimation distinguished by Sowder (1992), numerosity estimation has received the least research attention. But since the early 1990s, numerosity estimation has received growing attention in developmental and cognitive psychological research. Numerosity estimation is typically described as gauging the cardinal number of a discrete quantity (Brade, 2003), such as beans on a string, people in a football stadion, or candies in a jar.

Subitizing, a process whereby very small numerosities are identified instantaneously, had attracted the attention of experimental psychologists already a long time age and also leads to the determination of the cardinality of a discrete quantity. However, most of the research on subitizing has been done with children of a very young age, so we will not review this (for an extensive review, see Whalen, Gallistel, & Gelman, 1999).

As for the research on computational estimation, research on numerosity estimation has shown that (a) accuracy of numerosity estimation increases over a prolonged age range, (b) there is a variety in strategies, (c) strategy choices in numerosity estimation are sensitive to problem characteristics, and (d) the age-related differences in accuracy of numerosity estimation stem from the adoption of new, superior strategies, improvement in strategy execution with age, and improvement in the adaptivity of the choices among the existing strategies. Hereafter, we briefly review a few representative studies illustrating the above general findings.

Crites (1992) analyzed the strategies used by third, fifth, and seventh graders who were judged skilled and less skilled in making numerosity estimates on the basis of their results on a paper-and-pencil estimation test. Students' strategies were identified by means of a structured interview. Three strategies accounted for two thirds of the strategies used by the students of all grade levels: (a) benchmark comparison (estimating the numerosity of a new set by application of a known standard), (b) decomposition/recomposition (decomposing the to-be-estimated quantity into small samples and recomposing these small samples to arrive at an estimate), and (c) eyeball (a perceptually based strategy accompanied with an explanation like "looks like there are that many"). Further, the first two kinds of strategies were associated with higher performance scores and the third kind of strategy with lower performance scores. Finally, skilled estimators outperformed less skilled estimators, especially on questions that involved large quantities (greater than 1000).

Brade (2003) analyzed the numerosity estimation strategies and skills in a class of Kindergarten, Grade 1, and Grade 2 children who were involved in an intervention program (consisting mainly of a computer program involving estimation of numerosity tasks arranged in different levels according to the size of the to-be-estimated quantity) and comparable control classes. All children from each class were pretested and posttested using a free-response task requiring children to estimate the number of objects in drawings containing up to about 80 identical and randomly arranged objects. Using a classification scheme consisting of eight (developmental) levels, each child was scored into one of these levels based on his or her performance on each test. Contrary to his expectations, the author did not find a significant effect of the computer-based intervention. Moreover, a significant relationship between children's ages and their estimation strategies was found, showing, for instance, that most kindergartners did not go beyond a level where they must count at least a portion of the to-be-estimated numerosity before an estimate can be made, whereas Grade 1 and 2 children already started to respond (appropriately) by means of typical small, medium, and large numbers (without ostensive use of counting by ones). However, more sophisticated strategies, like using the above-mentioned benchmark and decomposition/recomposition methods, were almost completely absent. Finally, fairly strong correlations were found between strategy and accuracy data, although some strikingly discrepant cases were observed too.

Camos (2003) examined the development of the strategies people use to determine the numerosity of large sets, from childhood to adulthood. Using Lemaire and Siegler's (1995) framework for describing strategic change, Camos analyzed her data in terms of the repertoire, the distribution, and the efficiency of execution of numerosity judgment strategies. Four main strategies were identified in this range: counting by ones ("1, 2, 3, 4, . . ."), by ns (e.g., "5, 10, 15, 20 . . ."), with additions (e.g., "one set of 4, plus a set of 6 and another one of 5 is 15"), and with multiplications (e.g., "6 sets of 5 is 30"). Although counting by ones continued to be present at all ages, counting by ns and with additions appeared at 7 years, and counting with multiplications only at 9. The frequency of use of the three new strategies increased with age, and counting by ns became the dominant strategy (only) after 11 years. The evaluation of the strategies (in terms of speed, accuracy, and use of manual pointing behavior) showed that all of these strategies became more efficient with age, except for the counting with multiplications (which was the most efficient as of Age 9, but also the least used). Finally, the frequency of use of strategies was examined in relation to the characteristics of the arrays to be counted. Although the manipulated characteristics rendered the arrays more or less difficult to count (by changing size, arrangement, density of subgroups, and so on), none of them affected the use of strategies until 11 years, suggesting little signs of adaptivity before that age.

In the studies reviewed so far, the objects were arranged randomly. By contrast, in a series of recent studies Verschaffel, De Corte, Lamote, and Dhert (1998), and Luwel, Verschaffel, Onghena, and De Corte (2000, 2003a, 2003b) used tasks wherein the objects were presented within a regularly spaced grid form. Participants (ranging from second graders to university students) were asked to judge numerosities of colored blocks that were randomly arranged within rectangular grid structures (e.g., a grid of 7×7, 6×9, or 10×10). Siegler's model of strategic change was used to predict and analyze differences in strategic competence among different age groups and between different tasks. A scrutinized analysis of individuals' response-time and error-rate data, combined with individual interview data, revealed that besides the above-mentioned addition strategies from Camos's classification, many participants applied another strategy, namely a subtraction strategy, in which the number of empty squares was subtracted from the total number of squares in the grid. Second, even from the early years of elementary school on, many participants switched adaptively between these two kinds of strategies. Thus, for the items with few blocks and many empty squares (like a numerosity of 10 blocks in a 10×10 grid) they applied (different versions of) the addition strategy, whereas for items with many blocks and few empty squares (like a numerosity of 90 blocks in a 10×10 grid), most participants applied a subtraction strategy. Third, as expected, the occurrence, frequency, efficiency, and adaptivity with which this subtraction strategy was used increased with age.

All above-mentioned studies investigated people's numerosity estimation strategies in rather artificial contexts. Sowder (1992) referred to a couple of studies dealing with estimation by elementary school children of (large) numerosities embedded in a meaningful context, such as estimating the number of names on a page of a telephone book or how many sugar cubes would fill a matchbox, revealing that the above-mentioned benchmark and decomposition/recomposition methods reported by Camos (2003) were also observed in these realistic contexts. A recent study by Verschaffel, Luwel, and Torbeyns (2006) revealed that the same holds for the so-called subtraction strategy. Groups of second, fourth, and sixth graders were individually confronted with numerosities smaller and larger than 100 embedded in one-, two-, or three-dimensional realistic contexts. Whereas one third of these contexts were unstructured (e.g., a representation of an irregular piece of land with 72 cars unsystematically arranged on it), another third had a clear structure (e.g., a 16×4 rect-

angular parking lot completely filled with 64 cars), and the final third had a semistructure that made the use of the subtraction strategy highly efficient (e.g., a 13×7 rectangular parking lot containing 83 cars). Overall, almost 20% of all identifiable strategies for the semistructured estimation tasks could be classified as subtraction strategies. As expected, this percentage increased significantly with age.

In their review of numerical estimation, Siegler and Booth (2005) also reviewed another line of research, which they called *number line estimation*, which has become the focus of these reviewers' own research attention. In a typical number-line estimation experiment, such as the study by Siegler and Opfer (2003), a person is asked to estimate the position of a set of numbers on a particular scale (e.g., a 0–20, a 0–100, or a 0–1000 scale). Generally speaking, the results of these studies (in terms of the development of estimation performance as well as of strategy variety, efficiency, and adaptiveness) are the same as for the previously reviewed domains of numerical estimation. However, a very interesting specific aspect of this latter study is that it also revealed how people internally represent numerical magnitude. For instance, by means of fine-grained analyses of the patterns of percent absolute error for varied groups of numbers within a particular scale, Siegler and Opfer (2003) found distortions in the numbers used to label collections of objects, due to nonlinear mappings between numbers and their magnitudes.

Number Sense

In many analyses, estimation is connected to number sense (or to numeracy). For instance, Van den Heuvel-Panhuizen (2001, p. 173) started her didactical treatise of estimation as follows: "Estimation is one of the fundamental aspects of numeracy. It is the pre-eminent calculation form in which numeracy manifests itself most explicitly." McIntosh, Reys, and Reys (1992, p. 3) described number sense as follows:

> Number sense refers to a person's general understanding of number and operations along with the ability and inclination to use this understanding in flexible ways to make mathematical judgements and to develop useful strategies for handling numbers and operations. It reflects an inclination and an ability to use numbers and quantitative methods as a means of communicating, processing and interpreting information. It results in, and reciprocally derives from, an expectation that numbers are useful and that mathematics has a certain regularity.

Number sense is highlighted in current mathematics education reform documents as it typifies the theme of learning mathematics as a sense-making activity. Discussions around this "flag term" have resulted in listings of the essential components of number sense (McIntosh et al., 1992; Sowder, 1992) as well as descriptions of students displaying (lack of) number sense (Reys & Yang, 1998).

Probably the most comprehensive and most influential attempt to articulate a structure that clarifies, organizes, and interrelates some of the generally agreed components of basic number sense has been provided by McIntosh et al. (1992). In their model, distinction is made among three areas where number sense plays a key role, namely number concepts, operations with number, and applications of number and operation.

- The first component, knowledge of and facility with numbers, involves subskills such as sense of orderliness of number (e.g., indicate a number on an empty number line, given some benchmarks), multiple representations for numbers (e.g., 3/4 = 0.75), sense of relative and absolute magnitude of numbers (e.g., have you lived more or less than 1000 days?), and system of benchmarks (e.g., recognizing that the sum of two 2-digit numbers is less than 200).

- The second component, called knowledge of, and facility with, operations, involves understanding the effect of operations (e.g., knowing that multiplication does *not* always make bigger), understanding mathematical properties (such as commutativity, associativity, and distributivity and intuitively applying these properties in inventing procedures for [mental] computation), and understanding the relationships between operations (e.g., the inverse relationship between addition and subtraction, and multiplication and division).

- The third component, applying knowledge and facility with numbers and operations to computational settings, involves subskills such as understanding the relationship between problem contexts and the necessary computation (e.g., understanding that a problem like "Skip spent $2.88 for apples, $2.38 for bananas, and $3.76 for oranges; could Skip pay for this fruit with $10?" can be solved quickly and confidently by adding the three estimated quantities rather than by exact calculation), awareness that multiple strategies exist for a given problem, and inclination to utilize an efficient representation or method (e.g., not solving

[375 + 375 + 375 + 375 + 375]/ 5 by first adding the five numbers and then dividing the result by 5), and inclination to review the data and result (having a natural tendency to examine one's answer in the light of the original problem).

Several researchers have documented the problems children experience with the different aspects of number sense mentioned before. For instance, Reys and Yang (1998) investigated the relationship between computational performance and number sense among sixth-grade and eight-grade students in Taiwan. Students were interviewed about their knowledge of the different aspects of number sense from the theoretical framework mentioned above. Their performance on number sense was lower than that on similar questions requiring written computation. There was little evidence that identifiable components of number sense, such as use of benchmarks, were naturally used by Taiwanese students in their decision making. This is not surprising because, at least until recently, Taiwanese schools were typically very traditional and computationally driven in their instructional approach (Reys & Yang, 1998).

Given the above conceptualization of number sense, it is evident that, according to many authors, the development of number sense "is not a finite entity that a student has or does not have but rather a process that develops and matures with experience and knowledge" (Reys & Yang, 1998, p. 227) and that this development results from a whole range of activities of mathematics education on a day-by-day basis within each mathematics lesson, rather than a designated subset of specially designed activities (Reys & Yang, 1998).

Conclusion and Discussion

Our review of studies on estimation has revealed how difficult it is for students who receive traditional instruction, with its overemphasis on routine paper-and-pencil calculation, to understand that besides counting precisely and calculating exact answers, there is also something like estimating and developing appropriate procedures and strategies for making appropriate estimations (Kilpatrick et al., 2001; Siegler & Booth, 2005; Van den Heuvel-Panhuizen, 2001). Besides showing that the development of estimation begins rather late and proceeds rather slowly, the available research has documented, as in most other domains, (a) that estimating is a complex activity wherein all strands of mathematical proficiency are involved; (b) that children use several (invented) strategies to find estimates of arithmetic problems, of numerosities or of positions on a number line; (c) that these estimation strategies vary in frequency of use and efficacy; (d) that estimation strategy choices are influenced by problem characteristics; and (e) that estimation strategy use, execution, and adaptivity improve with age.

At a more general level, the research reviewed in this section has shown that the domain of estimation is, *par excellence*, an area where the recurrent basic features of arithmetical thinking and learning are manifest, notably (a) the interrelationship between declarative and procedural knowledge, (b) flexibility rather than routine skill as the quintessence of expertise in the domain, and (c) the dispositional nature of number sense, involving not only aspects of capacity but also aspects of inclination and sensitivity (see also Dowker, 2003).

The above review also prompts some questions, such as why estimation skill develops so slowly. Besides a number of well-known sources of difficulty, which have been raised in relation to limitations of conceptual understanding, component skills, and working memory, the recent work on number-line estimation suggests an additional source of the slow and incomplete development of estimation skills, namely the nonlinear representations of numerical magnitudes: "If representations of the magnitudes of different numbers are largely indistinguishable, children are unlikely to gain much of a sense of the types of numbers that are plausible answers to numerical problems or that typically accompany real world values" (Siegler & Booth, 2005, p. 211). This idea not only suggests further ascertaining studies on the relationships between number line, computational, and numerosity estimation, but also design experiments aimed at the improvement of estimation accuracy over a wide range of numerical estimation by helping (young) children develop linear representations of a wide range of numerical magnitudes.

We end this section with a conceptual comment on the notion of *number sense*. As suggested already, there is a serious problem with it and, more specifically, with its relation to other aspects of arithmetic competence described in this and other chapters. Number sense seems to be somewhat like *intuition* in the sense that the mathematics education community commonly agrees on its importance, but it is hard to define and even harder to operationalize in order to do research on it. Also, recently there has been a rash of—more or less—related terms such as *numeracy* and *mathematical literacy*, with little definitional precision. Indeed, number sense is, in most cases where we have encountered it, defined so broadly that it includes not only problem solving, but also most, if not all, other

skills and dispositions that have been discussed in the other sections. Although we acknowledge its power in curricular reforms (Thompson, 1999b), we have some doubts about its usefulness for scientific research, unless and until its specific meaning has been defined in a clear and consistent way.

Understanding and Solving Arithmetic Word Problems

Arithmetic word problems constitute an important part of the mathematics program at the elementary school. Historically, they had mainly an application function, that is, they were used to train children to apply the formal-mathematical knowledge and skills (previously) learned at school in real-world situations. Later, word problems were given other functions, in particular as a vehicle for developing students' general problem-solving skills. According to the current reform documents, word problems should also be mobilized in the earliest stages of the teaching and learning of whole-number calculation to promote a thorough and broad understanding of, and proficiency with, whole-number arithmetic (Kilpatrick et al., 2001; Verschaffel & De Corte, 1997). Indeed, in solving word problems young children get ample opportunities to build a rich and broad understanding of the four basic operations, to display their most advanced levels of counting performance, and to build a gradually more efficient and more abstract repertoire of procedures for whole-number computation.

Word problems have already for a very long time attracted the attention of researchers, both cognitive psychologists and mathematics educators. Here we will review only the work on word problems involving one or more basic arithmetic operations on whole numbers. Moreover, the focus is on studies that have been carried out since the reviews by Fuson (1992) and Greer (1992). Although some issues of instructional and cultural factors will be discussed already here, these issues are mainly dealt with later in this chapter.

Understanding and Solving One-Step Addition and Subtraction Word Problems

Throughout the 1980s and 1990s, concentrated research was carried out on how children learn to do one-step arithmetic word problems. Undoubtedly, the most popular subdomain has been the solution of one-step addition and subtraction problems involving small whole numbers or collections of discrete objects. In the early 1980s a basic distinction emerged that guided much of the subsequent research, distinguishing three classes of problem situations modeled by addition and subtraction. These are situations involving a change from an initial state to a final state through the application of a transformation (change problems), situations involving the combination of two discrete sets or splitting of one set into two discrete sets (combine problems), and situations involving the quantified comparison of two discrete sets of objects (compare problems). Within each of these three major semantic categories, further distinctions were made resulting in 14 different types of one-step addition and subtraction problems (Riley, Greeno, & Heller, 1983; see also Fuson, 1992; Reed, 1999; Verschaffel & De Corte, 1993, 1997).

Numerous empirical studies carried out during this period with children between 5 and 8 demonstrated the psychological significance of this classification scheme, specifically that word problems that can be solved by the same arithmetic operation (i.e., a direct addition or a subtraction with the two given numbers in the problem), but that belong to different semantic problem types, yield different degrees of difficulty, different ways of representing and solving these problems, and different error categories (for reviews of this research, see Fuson, 1992; Reed, 1999; Verschaffel & De Corte, 1997).

On the basis of the results of this research program, researchers have also developed experimental programs for teaching elementary arithmetic word problems, wherein children were taught to materialize or schematize the semantic structure of the addition and subtraction word problems before actually solving them. Some of these experimental programs focused on learning to (re-)represent all 14 semantic problem types in terms of the part-whole structure (Rathmell, 1986; Tamburino, 1982; Wolters, 1983), whereas others are characterized by the use of different material or schematic representations for the distinct basic categories of addition and subtraction word problems (see, e.g., De Corte & Verschaffel, 1985a; Fuson & Willis, 1989; Jaspers & Van Lieshout, 1994). A somewhat different kind of instructional application is the Cognitively Guided Instruction approach, wherein researchers tried to improve in-service teachers' instructional practice and their students' learning outcomes by confronting these teachers with the above-mentioned research-based knowledge of the different types of word problems and of (the development of) children's skills and processes for solving them (Carpenter & Fennema, 1992; see also the section on *Teaching Word Problems*).

Computer simulations were developed concurrently with the empirical research (for a more recent example of the genre, see Leblanc & Weber-Russell, 1996, and for more recent reviews of these computer models, see Fuson, 1992; Reed, 1999; Verschaffel & De Corte, 1997). Many studies compared the performance, strategy, and error data of students with the behavior of these computer models (for reviews of these empirical studies see Fuson, 1992; Reed, 1999; Verschaffel & De Corte, 1997). For example, empirical support for the assumed central role of part-whole knowledge in children's addition and subtraction word-problem solving comes from the work of Sophian (Sophian & McCorgray, 1994; Sophian & Vong, 1995).

Until the early 1990s, the computer models and empirical findings were much more elaborated for change and combine problems than for problems about comparison. Therefore, it is not surprising that in the last decade that last problem category seems to have especially attracted the continued attention of researchers. This research has focused on the most difficult compare problem types, namely compare problems with an unknown reference set, such as "Pete has 9 apples. He has 3 apples more than Connie. How many apples does Connie have?" and "John has 5 marbles. He has 3 fewer marbles than Ann. How many marbles does Ann have?" In an attempt to explain students' difficulties with these types of compare problems, Lewis and Mayer (1987) presented a model that simulates the comprehension processes when solving these problems in quite a different way than according to the hypothetical processes implemented in the above-mentioned computer models. Their model has been tested in several empirical studies with a variety of data-gathering techniques, including the registration of eye movements while reading and solving the problems (Hegarty, Mayer, & Monk, 1995; Verschaffel, De Corte, & Pauwels, 1992), asking participants to retell the problems before answering them (Verschaffel, 1994), asking them to formulate a hypothesis about the required mathematical operation while reading every problem sentence (Kintsch & Lewis, 1993), and doing individual interviews wherein subjects are confronted with a variety of carefully chosen diagnostic tasks (Stern, 1993). Whereas some of the findings support the hypotheses underlying the Lewis and Mayer (1987) model, there are also others for which this model does not provide a proper explanation and for which alternative explanatory frameworks (e.g., Kintsch & Lewis, 1993; Okamoto, 1996) seem more appropriate. In any case, which (combination) of the different available theoretical explanations provides the best explanation for the remarkable difficulty of compare problems with an unknown reference set and for the extra complicating effect of the presence of the "marked" relational term is still a matter of hot debate among (cognitive) researchers.

As with the above-mentioned research programs of Siegler and associates (Shrager & Siegler, 1998;

Siegler, 1998; Siegler & Jenkins, 1989; see *SCADS* section) and Brown, Burton, and Van Lehn (Brown & Van Lehn, 1982; Burton, 1982; see section *Mental Calculation Strategies* section), the cross-fertilization between these empirical studies and computer simulation models is considered, in some circles, as one of the most impressive demonstrations of the cognitive-rationalist, and especially the information-processing, paradigm (De Corte et al., 1996; Greer & Verschaffel, 1990). As a result of the extended joint effort of computer model builders and empirical researchers, Fuson (1992, p. 269) stated that "there is a fairly clear consensus about the development of the skill in solving one-step addition and subtraction word problems over time, both in terms of the underlying conceptual structures and the kind of solution strategies."

However, it was clear from an early stage that there were still several problematic issues and remaining questions. By no means are all empirical findings consistent with the (computer) models (for systematic comparisons of these empirical data with the predictions of the computer models, see Carpenter & Moser, 1984; De Corte & Verschaffel, 1988; Fuson, 1992). Many of these inconsistencies have to do with the fact that the models do not address important aspects, such as (a) the influence of textual variables on the construction of children's problem representations and solutions, and (b) the fact that the solution of a word problem does not take place in a sociocultural vacuum but within a particular sociocultural setting, namely a mathematics classroom. To illustrate the first complicating element, we refer to a great number of studies that have convincingly shown how small changes in the wording of the problems may have a dramatic impact on children's solution processes and skills (Coquin-Viennot & Moreau, 2003; Cummins, Kintsch, Reusser, & Weimer, 1988; De Corte, Verschaffel, & De Win, 1985). Taking into account such findings, some researchers have started to question the basic assumption underlying the computer models that children's competence to solve word problems depends basically on the development of logico-mathematical knowledge structures, including the general part-whole conceptual schema (= logico-mathematical models), and proposed instead "text-comprehension models" that pay much more attention to the importance of children's understanding of the statements referring to the situation described in the problem (Cummins et al., 1988). Illustrative for the second complicating factor, namely the impact of the classroom culture on children's problem-solving processes and skills, are the well-documented examples of children's errors on arithmetic word problems due to the use of superficial coping strategies they have developed within the particular culture of school

mathematics, rather than to their failure to represent or solve the problem because of the lack of the necessary logico-mathematical or linguistic knowledge (see the next *Conclusion and Discussion* section for a further elaboration of this theme).

A general limitation of the theoretical models that have been developed through a combination of empirical studies and computer simulations is that this work has been almost entirely confined to word problems that can be "unproblematically" modeled by addition or subtraction of small positive whole numbers. In other words, rather than covering the whole range of addition and subtraction situations that (can) occur in the real world, most of this research has instead used a very restricted set of scholastic word problem versions of the real world, without much reflection on the complexities from a sociocultural perspective.

Understanding and Solving Word Problems About Multiplication and Division

Partly because of the complexity of the domain, partly because of the lack of computer models that have tried to provide an integrative picture of the whole representation and solution process, the field of multiplication and division word problems has developed somewhat differently, and with less integration, than that of addition and subtraction word problems. We consider here three prominent lines of research.

First, research on semantic types is concerned with categorizing the situations described in word problems according to how they are or can be schematized (prior to solution). Having reviewed the work done throughout the 1980s and early 1990s in this first line of research, Greer (1992) proposed a synthesis of semantic types—which he termed *models of situations*—for multiplication and division. An important difference from the dominant categorizations for addition and subtraction is that this taxonomy also pays attention to the kind of numbers involved (whole numbers or positive rationals). A distinction is made between situations that are "psychologically commutative" (or "symmetric") and those that are "psychologically non-communicative" (or "asymmetric"). In the asymmetric classes the multiplier and multiplicand can be distinguished—for example, in the class labeled "equal groups" the number of objects in each group is the multiplicand and the number of groups is the multiplier. By contrast, for Cartesian product (e.g., "How many different combinations can be assembled if you have 3 shirts and 4 trousers?") the numbers multiplied have symmetric roles. With respect to division for asymmetric cases, two types of division can be distinguished, namely division by the multiplier (partitive division) and division by the multiplicand (quotitive division). By contrast, for a symmetric class of situations, there is no clear distinction between multiplier and multiplicand, and, consequently, only one type of division. This type of structural analysis provides a valuable framework for research, but it is not very clear to what extent this is useful in considering children's thinking or as a guide to pedagogical strategies (Greer, 1992). Contrary to research on addition and subtraction, in which the categories of problem types also represent hypothetical cognitive structures that are assumed to drive pupils' understanding and solving of particular problems, this does not seem to be so clearly the case for the different classes in the above scheme (Anghileri, 2001; Greer, 1992; Verschaffel & De Corte, 1997). Although this line of research is basically conceptual/epistemological in nature, researchers also point to empirical research (e.g., about levels of difficulty or about typical strategies or errors) to support the validity of certain distinctions or characterizations in their models.

A second line of inquiry that has attracted a lot of research, and that is also extensively reviewed by Greer (1992), concerns the role played by so-called "intuitive models" in the solution of word problems. According to Fischbein, Deri, Nello, and Marino's (1985) theory of primitive models, each arithmetic operation is linked to a primitive, intuitive model, even long after that operation has acquired a formal status. Identification of the operation needed to solve a word problem is mediated by these intuitive models. The model affecting the meaning and use of multiplication is repeated addition, in which a number of collections of the same size are put together. One intuitive model for division is equal sharing of a group of objects (partition), and a second less dominant interpretation relates division to finding the number of equal groups in a given total (the quotitive model). Many research findings, such as the amount and the kind of errors made when confronting students with multiplicative problems involving different kinds of decimals in the role of multiplier or multiplicand (for an extensive overview see Greer, 1992), children's problem solving and reasoning processes in individual interviews around such problems (see also Greer, 1992), and, more recently, the kind of correct as well as incorrect word problems that pupils generate when being asked to construct problems corresponding to particular multiplications and divisions (e.g., 4×1.5, 0.75×8, $4 \div 0.5$; De Corte & Verschaffel, 1996) have been successfully interpreted within this theory. However, several questions still exist about how to account for the origin and the further development of these intuitive models (e.g., whether the primitive models reflect natural features of human thinking or rather

how the relevant concepts were taught in school, or both), as well as their precise role in the solution of different types of word problems involving multiplication and division (Greer, 1992; Verschaffel & De Corte, 1997).

The third line of research, not systematically reviewed by Greer (1992), pertains to analyses of computational strategies, that is, the kinds of strategies that children apply to arrive at an answer of a multiplication or division word problem. Generally speaking, this research yielded similar outcomes as for addition and subtraction (see also Verschaffel & De Corte, 1997). First, the results indicate that before pupils have been instructed in multiplication or division, many of them are able to solve problems involving these operations by means of a widespread variety of informal strategies (Kouba, 1989; Levain, 1992; Mulligan & Mitchelmore, 1997; Nunes & Bryant, 1995; Treffers, 1987). An overview of these strategies, ordered in terms of increasing levels of abbreviation, internalization, and formalization, has already been given in the sections on *Single-Digit Computation* and on *Multidigit Arithmetic*. This development towards higher levels of abbreviation, internalization, and formalization proceeds differently for the different categories of problems. For instance, use of derived-fact or known-fact strategies seems to occur later for problems about Cartesian product than for equal group problems. Second, in the early stages of development, children's strategies tend to reflect the action or relationship described in the problem. For instance, they use different material strategies for solving partitive and quotitive division problems that fit very well with the meaning of the problem. Afterwards, children start to apply strategies that are more efficient from a computational point of view but that no longer necessarily match this semantic problem structure. Third, compared to addition and subtraction, the research clearly shows that many children do not progress to secure multiplicative thinking in the primary school years (Anghileri, 2001; Greer, 1992). Or, as Clark and Kamii (1996, p. 48) stated it, "multiplicative thinking appears early but develops slowly." Indeed, in their study wherein a very large group of children in Grades 1–5 were interviewed using a Piagetian task to study the development of their multiplicative thinking, 45% of the second graders demonstrated already evidence of multiplicative thinking, but only 49% of the fifth graders could be classified as solid multiplicative thinkers.

Strategic Aspects of Word Problem Solving

Most of the word problem solving research described above starts from the assumption that a competent problem solver tries to represent the problem situation and to base the choice of an appropriate solution (whether it be a material or verbal counting action, or a formal-arithmetical operation) on that understanding. However, research has convincingly shown that the process of solving word problems for many pupils is along the lines of another model, whereby the intermediate representation is bypassed and the problem text immediately guides the choice of an operation—a choice that may be based on superficial features such as the presence of certain key words in the text (for instance, the word *less* in the problem text automatically results in a decision to perform a subtraction). The directly evoked operation is then applied to the numbers given in the problem text, and the result of the calculation is found and given as the answer. For a review of these coping strategies we refer to Verschaffel and De Corte (1997) and Verschaffel, Greer, and De Corte (2000).

Moreover, researchers have documented that when confronted with a word problem for which they have no ready-made solution, many pupils do not spontaneously apply valuable heuristic strategies, such as making a sketch or a drawing of the problem, or decomposing the problem into parts, or guessing-and-checking (De Bock, Verschaffel, & Janssens, 1998; De Corte & Somers, 1982; Garofalo & Lester, 1985; Van Essen, 1991). Although many researchers have reported differences between pupils who are good and weak in word problem solving with respect to the use of heuristics, the outcomes of teaching experiments focusing on the development of these heuristic skills have generally been rather disappointing (Schoenfeld, 1992).

These failures of efforts to improve students' problem solving by teaching them heuristics led researchers to study word problem solving from the perspective of metacognition (Garofalo & Lester, 1985; Mevarech, 1999; Schoenfeld, 1992). Several studies have revealed that in many pupils' solution attempts of word problems, moments of metacognitive awareness or self-regulatory activities such as analyzing the problem, monitoring the solution process, and evaluating the outcome are completely absent. A typical approach used by many pupils can be summarized as follows: The student glimpses the problem, quickly deciding what calculations to perform with the numbers given in the problem statement and then proceeds with these calculations without considering any alternatives even if no progress is made at all (Schoenfeld, 1992; Verschaffel & De Corte, 1997). Several researchers reported differences between children who are more and less able in solving arithmetic word problems with respect to self-monitoring and self-control (see, e.g., Garofalo & Lester, 1985;

Nelissen, 1987). Starting from these findings from ascertaining studies, several scholars have started to design and evaluate instructional environments aimed at the development of metacognitive skills for solving word problems (sometimes in combination with heuristic training). In these studies a detailed cognitive model of the heuristic and metacognitive abilities underlying competent problem solving in mathematics is specified and used to guide the development of teacher and student activities that will lead to the acquisition and use of these (meta)cognitive structures. In some studies this training was done on an individual (computer-aided) basis (Teong, 2000), whereas in others the metacognitive instruction was employed in cooperative settings where small groups of 4–6 students studied together. Typical examples of such intervention studies are Lester, Garofalo, and Kroll (1989), Cardelle-Elawar (1995), and Mevarech (1999). Most of the above studies suggest that providing metacognitive training has a significant impact on students' mathematical performance. As will be explained in the section *Emergent Models in Design Heuristics*, in more recent design experiments, the cognitive-rationalist perspective behind these intervention studies has been replaced, or at least enriched, by a more situative orientation, a socioconstructivist orientation, or both, as evidenced by a more critical attitude towards traditional word problems, a greater awareness of the complex relationship between mathematics and reality, and a greater emphasis on authentic mathematical tasks than in the studies belonging to the cognitive-rationalist tradition. The rationale for this shift of attention is given in the next *Conclusion and Discussion* section.

Pupils' Lack of Sense-Making When Doing Word Problems

As already announced before, in recent years the characteristics, use, and rationale of word problems have been critically analyzed from multiple perspectives, including linguistic, cultural, and sociological. In particular, many mathematics educators have argued that the stereotyped and artificial nature of word problems typically represented in mathematics textbooks and the discourse and activity around these problems in traditional mathematics lessons have detrimental effects on children's disposition towards realistic mathematical modeling and sense-making. Many observations have led to the conclusion that pupils answer word problems without taking into account realistic considerations about the situations described in the text, or even whether the question and the answer make sense.

The most dramatic example comes from French researchers who posed (lower) elementary school children nonsensical questions such as: "There are 26 sheep and 10 goats on a ship. How old is the captain?" It was found that the majority of pupils were prepared to offer an answer to such questions (Baruk, 1985).

Inspired by this and several other striking examples of suspension of sense-making in school mathematics, Greer (1993) in Northern Ireland and Verschaffel, De Corte, and Lasure (1994) in Flanders carried out in parallel a pencil-and-paper study with 11–13-year-old students, using a set of problems including the following:

- A man wants to have a rope long enough to stretch between two poles 12 meters apart, but he only has pieces of rope 1.5 meters long. How many of these would he need to tie together to stretch between the poles?

- Bruce and Alice go to the same school. Bruce lives at a distance of 17 km from the school and Alice at 8 km. How far do Bruce and Alice live from each other ?

- John's best time to run 100 meters is 17 seconds. How long will it take him to run 1 kilometer?

They termed these items *problematic* (P-items, for short) in the sense that a proper answer requires (at least from the researchers' point of view) the application of judgment based on real-world knowledge and assumptions rather than the routine application of arithmetical operations on the given numbers. A response was classified as a "nonrealistic reaction" if a pupil's response sheet contained no apparent trace of a judgment based on real-world knowledge. A reaction was scored as a "realistic reaction" if either the answer given revealed that realistic considerations had been taken into account or a nonrealistic answer was accompanied by a comment indicating that the pupil was aware that the problem was not straightforward. Both in Greer's (1993) and in Verschaffel et al.'s (1994) study, pupils demonstrated a very strong overall tendency to exclude real-world knowledge and realistic considerations when confronted with these P-items. For instance, in Verschaffel et al.'s study, only 17% of all reactions to the P-items were realistic. (The percentages for the three examples of P-items given above were between 0.5% and 3%!)

These studies were replicated in many other countries using the same P-items, the same testing conditions, and the same scoring criteria as in the original studies. Findings have been strikingly consistent with the first results, sometimes to the great surprise and disappointment of these other researchers, who had anticipated that the "disastrous" picture of the Irish and Flemish pupils would not apply to their students (see Verschaffel et al., 2000, for a review of the earlier replications).

Several follow-up studies have tested the effectiveness of variations in the experimental setting on children's tendency to solve problems in a realistic way. A first group of studies assessed the effectiveness of making students more alert to consider aspects of reality and to legitimize alternative forms of answering. For instance, Yoshida, Verschaffel, and De Corte (1997) gave half of the pupils an explicit warning at the top of their test sheet (a translation of Verschaffel et al.'s, 1994, test) that some of the problems in the test were problematic and invited them explicitly to write down and explain any lack of clarity or complexities. However, this manipulation did not result in a significant increase in the number of realistic reactions to the P-items in the test. In a different set of studies, one or more categories of P-items were presented in more authentic, performance-based settings. Contrary to the first kind of experimental manipulation, increasing the authenticity of the experimental setting (see, e.g., De Franco & Curcio, 1997) yielded great improvements in students' inclination to include the real-world knowledge in their problem-solving endeavors (see Palm, 2002, and Verschaffel et al., 2000, for a full account of these studies). However, Van Dooren, De Bock, Janssens, and Verschaffel (in press) showed that the facilitating authenticity effect had only marginal impact on students' future problem-solving behavior with respect to traditional word problems.

Besides inferring how students think about word problems from their responses to word problems in a paper-and-pencil test, it is, of course, possible to ask them directly. Interviews carried out by Caldwell (1995) and Hidalgo (1997) and Palm (2002) suggest that, although unfamiliarity with the contexts involved in the problems and lack of appropriate heuristic and metacognitive skills may provide contributory explanations, (mis)beliefs about school arithmetic word problems constitute the major reason why so many students solve the P-items in a nonrealistic way. For instance, a 10-year-old interviewed by Caldwell (1995, p. 39) commented as follows in response to the interviewer's question as to why she had answered a P-item in a nonrealistic way: "I know all these things, but I would never think to include them in a maths problem. Maths isn't about things like that. It's about getting sums right and you don't need to know outside things to get sums right."

How can the results of some years of mathematics education be the willingness of children to collude in negating their knowledge of reality? This apparent "suspension of sense-making" can be construed as sense-making of a different sort, namely a strategic decision to play the "word problem game." As expressed by Schoenfeld (1991, p. 340)

> such behavior is sense-making of the deepest kind. In the context of schooling, such behavior represents the construction of a set of behaviors that results in praise for good performance, minimal conflict, fitting in socially etc. What could be more sensible than that?

Students' strategies and beliefs develop from their perceptions and interpretations of the didactical contract (Brousseau, 1997) or sociomathematical norms (Yackel & Cobb, 1996) that determine—mainly implicitly—how to behave in a mathematics class, how to think, how to communicate with the teacher, and so on. More specifically, this enculturation seems to be strongly influenced by two aspects of current instructional practice, namely the nature of the problems given and the ways in which these problems are conceived and treated by teachers (Verschaffel et al., 2000). Support for the second factor comes from a study by Verschaffel, De Corte, and Borghart (1997), wherein preservice elementary school teachers were asked, first, to solve a set of problems themselves and, second, to evaluate alternative answers from (imaginary) pupils to the same set of problems. The results indicated that these future teachers shared, though in a less extreme form, students' tendency to suspend sense-making. As a result of these factors, students learn to play the Word Problem Game (De Corte & Verschaffel, 1985b), the rules of which include

- Any problem presented by the teacher or in a textbook is solvable and makes sense.
- There is a single, correct, and precise numerical answer that must be obtained by performing one or more arithmetical operations with numbers given in the text.
- Violations of knowledge about the everyday world may be ignored.

Conclusion and Discussion

In this section we have reviewed research on the development of children's arithmetic word problem solving processes and skills, focusing on that done since the thorough and extensive reviews by Fuson (1992) and Greer (1992). More precisely, we have described what we see as the continuation of the cognitive-rationalist line of research, although we have, especially in the last section, already touched on a considerable part of the last decade's research on word problems wherein this theoretical perspective has been enriched by others, in particular representatives of the situative/pragmatist-sociohistoric perspective.

This research has shown that most children entering school can solve some word problems, not only with an additive but also with a multiplicative structure, by means of solutions based on modeling the actions and relations described in them. Moreover, researchers have documented how children's proficiency gradually develops in two significant directions. One direction is from informal, external, laborious strategies towards more efficient (i.e., more abbreviated, internalized, formalized) calculation strategies. The other direction is from having a different solution method for each type of problem that is a direct reflection of the problem situation, to a single general method that can be used for classes of problems with a similar underlying mathematical structure (see also Kilpatrick et al., 2001).

Notwithstanding all these accomplishments, there is no shortage of issues and tasks for further research. Besides the specific questions for future research that we raised at the end of each part of this section, there are some broader issues. A first general limitation of past research is that it has largely focused on the initial and middle stages of the development of the concepts and skills needed for solving arithmetic word problems. Therefore, a general research priority should be to investigate as well the upper stages of the development of these concepts and skills, in which the additive and multiplicative concepts extend beyond the domain of positive integers (Greer, 1992, 1994; Vergnaud, 1988). Second, although separate analyses of the conceptual fields of additive and multiplicative structures will doubtless continue, there is a strong need for a comparative analysis between or a synthesis of these two hitherto rather separate bodies of research (as was already pleaded for by Greer, 1992). Such a synthesis should involve both theoretical analyses of the similarities, dissimilarities, and relationships between the conceptual fields of additive and multiplicative structures and empirical studies in which data are simultaneously collected for all four arithmetic operations. Some general comments about differences between addition/subtraction and multiplication/division are appropriate. For the first pair of operations, extension to non-whole-numbers seems to be assumed nonproblematic (there appears to be essentially no research on it)—instead the main problem is extension to negative numbers. By contrast, for the second pair, the extension to positive rational numbers provides major difficulties (e.g., Greer, 1994). Secondly, it may be that by the time students get to multiplication and division, they have already developed, through their work with addition and subtraction, some dissociation between calculations and situations, that is to say they have learned that, for the purposes of finding the answer, the calculation, once identified, can be done "off line."

Finally, there is a strongly growing awareness among researchers that the level of our understanding of how children learn to solve arithmetic application problems drops enormously if we allow the domain to expand beyond the restricted range of traditional school word problems that can be "unproblematically" mapped onto an arithmetic operation with the given numbers and if we are prepared to take seriously into account the complexities of the instructional and cultural context wherein this learning process takes place.

The Structure of the Whole-Number System

Formal Structure of the Natural Numbers and Associated Arithmetic

At the time of "New Math" it was joked that children knew that 6×7 and 7×6 are equal but did not know what they are equal *to*. More seriously, there was a belief, stemming largely from mathematicians, that the structural edifice of mathematics as seen from an advanced standpoint could be communicated directly to children (Kilpatrick, 2003, p. 474). In retrospect, the effort was founded on the error of "replacing the learner's insight with the adult mathematician's" (Freudenthal, 1991, p. 112). Perhaps the time is right to reevaluate the role of structural understanding *from the perspective of the learner.*

The name *natural numbers* for the set of positive integers 1, 2, 3, . . . reflects their original conceptualization as Platonic objects and their ubiquitous usefulness for modeling aspects of the real world that afford counting. The four basic operations of arithmetic may be regarded as naturally emergent from reflection on counting acts and manipulations of sets of objects, physical or mental. As a structure, the set of whole numbers, which appends zero to the natural numbers, has been analyzed in great depth and formalized in several variants. Here we deal with this formalization briefly and informally—in standard mathematical texts, the reader can find as much detail and rigor as required.

The ordinal aspects of natural numbers, intuitively grounded in acts of counting, are formalized in the Peano axioms (anticipated by Dedekind) from which the arithmetic operations can be defined, and their properties derived. The cardinal aspects have been formalized in set-theoretic terms, notably by Cantor, Frege, Russell, and Whitehead (Freudenthal, 1983, p. 76). One formulation is to define a whole number, n, as the class of all sets with cardinal n. This formula-

tion corresponds to the idea that sets of 3 sheep, 3 colors, 3 rectangles, and so on. all have "threeness" in common. Alternatively, the whole number, n, can be defined as the set of all the whole numbers (including 0) less than n. As Freudenthal wryly commented (1983, p. 80), "with each profundity we get further away from the phenomenology of number as it is naively expressed." Nor has history come to an end in this respect, as new formulations continue to appear (e.g., Conway & Guy, 1996).

The set of whole numbers, with the operations of addition and multiplication, forms a structure with properties such as commutativity and associativity of addition and multiplication and distributivity of multiplication over addition (Kilpatrick et al., 2001, Chapter 3). In addition, there are numerous properties relating to order.

It is characteristic of the notion of *number* that it has historically and logically been extended. In particular, the set of whole numbers is extended to the set of integers (positive and negative) and to the set of rationals (all those numbers of the form a/b where $b \neq 0$). An important mathematical motivation for these extensions is the urge for closure. The whole numbers are not closed under either subtraction or division (i.e., it is not always possible to subtract one whole number from another, or divide one whole number by another, and get a whole number as the result), leaving gaps that are closed by the extensions to the integers and to rational numbers, respectively. Although the focus of this chapter is on whole-number arithmetic, we draw attention to the need to anticipate the difficulties students commonly experience in reconceptualizing arithmetic within larger and larger domains as the concept of *number* itself is extended (Lakoff & Nunez, 2000).

Straddling the development of natural numbers as, on the one hand, a practical tool for dealing with calculations on sometimes very large numbers and, on the other hand, the infinite set of natural numbers as a mathematical object is the invention of finite (hence, necessarily generative, like language) representations. A reasonable case can be argued that the decimal system that has emerged as the evolutionarily fittest is optimal insofar as it concatenates polynomial expressions with a constant base; however, the choice of 10 as that base (rather than, say, 8 or 12) appears to be a biological accident, and the specific choice of place-value representation (with the key step that zero could be seen as a number in its own right) is a cultural decision. A great deal of what we call early arithmetic is more precisely described as arithmetic using decimal representation, as one solution to the generative requirement of a finite representation.

Research on Understanding of Structural Properties

There has been considerable research on children's awareness of the structural properties of whole numbers and operations upon them. In accordance with our principled division between this section and the next on *Classroom-Based Research*, here we concentrate on ascertaining studies, which are generally associated with the first level, whereby awareness is inferred from behavior that, in some sense, indicates implicit understanding of the principle—what Vergnaud (1996) termed a *theorem-in-action*. It is unusual for children taught traditionally to spontaneously formulate explicit generalizations and provide general arguments in their support; hence these manifestations are rare in ascertaining studies. Work relating to these higher levels of understanding and the explicitation of structural properties, which is mainly situated within teaching experiments, is dealt with in the second part of the *Classroom-Based Research* section.

At the simplest level, exploitation of properties may be applied in the mastery of the basic number facts. For example, the task of learning or constructing the 100 multiplication facts may be drastically simplified by flexible and strategic use of a wide variety of properties and regularities (Kilpatrick et al., 2001; see earlier section on *Single-Digit Computation*). In the case of multidigit computations, algorithms mostly consist of systematic methods to simplify the computation to a coordination of basic number facts, and these methods depend on several properties, notably the associative and distributive laws in the case of multiplication. As pointed out earlier, the compact notation of standard algorithms tends to mask the underlying principles that make them work (although these principles may be unmasked, as shown by Wu, 1999). On the other hand, other algorithms and taught or invented mental strategies are derived directly from the underlying properties of decimally represented whole numbers and the relations among them (Resnick, 1992).

So-called "smart arithmetic" tasks may be used to probe students' awareness of such networks of relationships. For example, starting from the given equation

$$86 + 57 = 143$$

students may be asked to use it in finding: $86 + 56$, $57 + 86$, $860 + 570$, . . . and so on (Selter, 1996; Van den Heuvel-Panhuizen, 1996). Children's "flexible arithmetic" (Gravemeijer, 1994, p. 37) is based on (often implicit) use of the formal properties of arithmetic, plus others derived therefrom, such as

$$a + (a + 1) = (a + a) + 1$$
$$(a + p) - (b + p) = a - b$$

Many studies have been carried out that have probed students' understanding of specific properties. As an example of research in this spirit, we consider one of the most basic structural principles, the commutativity of addition and multiplication.

Within a more general theory of the emergence of arithmetical cognition from its rudimentary origins, Resnick (1992) presented a theory of how children's understanding of commutativity of addition develops (see Table 13.1).

Baroody et al. (2003, p. 130) quoted a useful clarification of Resnick's position as follows to the effect that

> additive commutativity is reconstructed at each successive level, using both what the child knows from the previous level and new knowledge (e.g., the experience of counting and computing). . . . For a child to understand [for example] that 3 apples + 6 apples = 6 apples + 3 apples [requires] more than just applying an existing [part-whole] schema. . . . It would involve also noticing . . . that in both cases the sum is 9 apples.

They presented evidence that understanding of commutativity is correlated with computational experience and expertise. However, there is a general point that needs to be raised, namely what evidence might

Table 13.1 Developmental Levels of Commutativity (Resnick, 1992)

Level of additive commutativity	Example
Protoquantitative (prenumerical)	apples + oranges = oranges + apples (Part1 + Part2 = Part2 + Part1)
Quantitative (specific numbers in context)	3 apples + 5 apples = 5 apples + 3 apples
Numerical (specific numbers in the abstract)	3 + 5 = 5 + 3
Operator (general arithmetic principle)	$a + b = b + a$

we take to show that a child understands commutativity of addition, and in what sense?

- If the child computes the answer to 3 + 5 by saying "five . . . six, seven, eight" (perhaps accompanied by finger actions), or if the child rapidly and confidently answers the question "do the additions 6 + 4 and 4 + 6 have the same answer?", we infer that the child, in some sense, understands commutativity of addition.

- If the child gives an explicit statement implying that commutativity holds for any two whole numbers, then we infer that the property is understood at some level of generality, at least as the result of empirical observation ("every time I've calculated an addition both ways, I've got the same result").

- At the highest level an explicit statement is accompanied by a (more or less) convincing argument, often based on a physical or visual representation (Kilpatrick et al., 2001, p. 77). A key distinction here is that made by Piaget (e.g., Piaget, 2001) between an empirical abstraction, whereby a student notices and generalizes a pattern, and a sense of the logical necessity underlying that pattern, which is the product of reflective abstraction (cf. the comment by Sawyer, 1955, that a mathematician always wants to know the reason for a pattern).

Thus, as with mathematical concepts in general, we suggest it is problematic to speak of understanding of commutativity as something a student either does or does not have.

A key idea Baroody et al. (2003) emphasized is that addition may be treated as a unary or binary operation. Within, say, the whole numbers, addition as a binary operation is a rule that assigns to any two whole numbers another whole number. On the other hand 5 + 3 can be interpreted as the application of the unary operation + 3 to 5. As discussed below (see previous section on arithmetic word problems), this distinction is reflected in the different classes of situations modeled by addition and subtraction. Moreover, although the binary conception is more conducive to an emergent grasp of commutativity, the unary conception facilitates development of understanding of subtraction as the inverse of addition (Siegler, 2003; though the latter may also be facilitated by experience of working with a part-part-whole binary schema, see also Bryant, Christie, & Rendu, 1999).

Synthesizing a wide range of experimental evidence and theoretical analysis, Baroody et al. (2003,

p. 155) proposed a revised model for development of additive commutativity (Table 13.2).

Table 13.2 Development of Additive Commutativity Understanding (Baroody et al., 2003)

Level 0:	Unary conception + noncommutativity
Level 1:	Unary conception + noncommutativity and binary conception + protocommutativity
Level 2:	Unary conception + protocommutativity and binary conception + part-whole ("true") commutativity
Level 3:	Unary conception + change-add-to commutativity ("pseudocommutativity") and binary conception + part-whole ("true") commutativity

Despite the amount of work surveyed, Baroody et al. (2003) finished by acknowledging that more research is needed on this topic. The fact that the interpretation remains controversial and unclear, despite a considerable effort in experimentation, in itself suggests that there is an inherent messiness arising from individual differences, teaching, differential experience, artefacts of the experimental contract . . . and so on. In a later section the discussion will be resumed in the context of classrooms in which children are encouraged, through strategically chosen tasks and discussions, to enunciate and even explain commutativity as a general principle.

A highly important point, with general relevance, is the suggestion by Baroody et al. (2003) that children's early understanding of arithmetic should be viewed "in terms of multiple, loosely connected weak schemes" (p. 156). As such, it seems inevitable that a neat linear description of developmental stages will always be complicated by the exigencies of individual differences in cognition and in experience. This reality underlies one major advantage of teaching experiments, in that they afford access to at least partial knowledge of the nature of instruction the child has experienced and how that experience has affected the child's learning. The model presented in Table 13.2 may therefore be more usefully seen as a framework for interpreting children's performance.

Understanding of multiplication from a purely computational point of view has also been extensively studied from the aspect of looking for evidence of use of structure, particularly commutativity. For example Baroody and Tiilikainen (2003, p. 105) proposed a possible schema underlying mental multiplication that invokes several rules, such as if 1 and a number are multiplied, the answer is that number, that if 2 and a number are multiplied, the answer is that number added to itself (see also the earlier section *Children's Strategies for Single-Digit Computation*). An overriding principle is that the order of the numbers multiplied is irrelevant and accordingly may be switched strategically to facilitate the calculation. Interestingly, as Baroody and Tiilikainen pointed out, the research of Butterworth et al. (2003) suggests that Italian children convert small-factor-first combinations to large-factor-first, whereas that of LeFevre et al. (2003) suggests the converse for Chinese experts.

Modeling Different Classes of Situations

The research exemplified thus far is based on counting acts, operations on numbers represented by sets of taken-as-countable objects, or purely symbolic computational tasks. Further complications arise if the task involves the use of arithmetic to model a situation.

For example, consider the awareness of subtraction as the inverse of addition for each of the three main semantic structures: combine, change, and compare. Splitting a superset into subsets as the inverse of combining subsets to form a superset is grounded in actions, as Piaget stressed, and related to his concept of reversibility. So is undoing a change—but in a different way. Conceptualizing "Peter is 4 inches taller than Anne" and "Anne is 4 inches shorter than Peter" is different again, and particularly difficult for young children (Stern, 1993). Inverseness also plays a role in cases in which the operation that yields the answer is incompatible with the "story-line," as in a change-unknown problem such as "John had 5 apples and found some more. Now he has 9 apples. How many apples did he find?" The computation 9 − 5 represents an inversion, in a sense, of the direct modeling of the story as 5 + ? = 9 (note the algebraic nature of this equation).

Likewise, for a simple additive word problem, the arithmetic structure of the operation involved may depend on which class of problems—combine, change, or compare—is involved (De Corte & Verschaffel, 1987). Thus, as mentioned already in the SCADS section, the computationally more efficient counting-on-from-larger strategy, which for a purely numerical task exploits the commutativity of addition, is much more prevalent for a combine problem than for a change problem. The explanation is that the addends in the former case play "symmetrical" functions, whereas in the latter case, the functions are asymmetrical and imply order (a distinction closely related to Weaver's, 1982, distinction between unary and binary conceptions of addition).

The concept of symmetry also applies in the case of situations modeled by multiplication and division (Greer, 1992). In many classes, it is possible to distinguish a multiplier that acts upon the multiplicand, implying asymmetric functions, as in the simple case of equal groups, where the multiplicand is the number in each group and the multiplier is the number of groups. As a consequence, there are two types of division, namely partition (division by the multiplier) and quotition (division by the multiplicand) for such situations (Greer, 1992, p. 286; see also elsewhere in this chapter). For this reason, there are interactions between the semantics of modeled situations and computational solutions. For example, Ambrose et al. (2003, pp. 318–319) reported that for an equal groups problem involving 153 bags with 37 balloons in each "[the students] were not inclined to use the commutative property to switch the numbers around, even though it might have made calculation easier" because their thinking was linked to the context. Many observations confirm that, in this sense, an equal-groups-of meaning of multiplication is psychologically non-commutative.

Number Theory

Number theory, in general, is the study of properties of number systems. Originally, its scope was patterns and relations among the natural numbers, to which we restrict ourselves here. Typical topics within this domain are "figurate numbers, whole number patterns and sequences, multiples, factors, divisors, primes, composites, prime decomposition, relatively prime numbers, divisibility, and divisibility rules" (Campbell & Zazkis, 2002b, p. 3). Campbell and Zazkis (2002a) argued for a greater recognition of number theory within the curriculum at all levels. Justifications for this recommendation include the following points:

- The topic is of historical interest.
- It offers a context for fostering structural awareness and algebraic thinking.
- It affords mathematical sense-making, reasoning, argumentation, and proof about familiar mathematical objects.
- It reinforces the idea that mathematics is, at heart, concerned with patterns and relations.

Campbell and Zazkis included examples of studies with preservice teachers, but none with school students. We suggest number theory as a possible area for future research with children, for the reasons cited above. Zazkis (2001) made an interesting suggestion that using problems with specific large numbers could force generalizable thinking that could serve as a bridge to algebra.

CLASSROOM-BASED RESEARCH

In this section, we reflect a major development of the research since the early 1990s, namely a shift from mainly ascertaining studies of individual children to extensive classroom-based intervention research. This methodological shift is, first of all, related to the theoretical shift from the (strictly) cognitive-rationalist analysis of individual children's mathematical thinking and learning, and the task and subject variables influencing these thinking and learning processes, to other (socioconstructivist and sociocultural) frameworks wherein the complex social and cultural context is considered as truly constitutive for what and how children think and learn mathematically. Second, this methodological shift towards classroom-based intervention research is also a consequence of researchers' growing awareness that a general priority for them should be to complement ascertaining studies, which only document and diagnose existing ills, with research efforts aimed towards designing, implementing, and assessing innovative mathematics curricula, programs, teaching/learning activities, and so on, as recommended both by Fuson (1992, p. 269) and by Greer (1992, p. 293). Or, as Selter (2002, p. 221) put it: "mathematics education has a fundamental and clear-cut responsibility to develop practical suggestions for teaching within the capacity of 'average teachers' in a systematic process involving researchers, teachers, designers, textbook authors and children."

This plea for such "design experiments" can be found both in the research field of instructional psychology and in that of mathematics education. As far as the former is concerned, we refer to Greeno et al. (1996) and Schauble and Glaser (1996), who describe design experiments as a new kind of interdisciplinary teamwork that is evolving among practitioners, researchers, teacher educators, and community partners around the design, implementation, and analysis of changes in practice. Results provide case studies that can serve as instructive models about conditions that need to be satisfied for reforms of the same kind to be successful, and about conditions that impede success. Researchers believe that the design of such learning environments and the development of theory must proceed in a mutually supportive fashion. Among the mathematics educators who have pleaded strongly for design experiments as the main type of research in the field are Freudenthal (1991), Rouche (1992), and Wittmann (1995). In his seminal article entitled *Mathematics Education as a Design Science*, Wittmann (1995, p. 356) described the main task of researchers in mathematics education as follows:

Scientific knowledge about the teaching of mathematics, however, cannot be gained by simply combining results from other fields; rather it presupposes a specific didactic approach that integrates different aspects into a coherent and comprehensive picture of mathematics teaching and learning and then transposing it into practical use in a constructive way. . . . In my view, the specific tasks of mathematics education can only be actualized if research and development have specific linkages with practice at their core and if the improvement of practice is merged with the progress of the field as a whole.

According to Wittmann (1995, p. 363), the core of research in mathematics education is "the construction of 'artificial objects,' namely teaching units, sets of coherent teaching units and curricula as well as the investigation of their possible effects in different educational 'ecologies.'" And he continued: "Indeed, the quality of these constructions depends on the theory-based constructive fantasy, the 'ingenium', of the designers, and on systematic evaluation, both typical activities for design sciences." According to Wittmann (1995) both the so-called "developmental research" projects of the Dutch Freudenthal Institute (see, e.g., Gravemeijer, 1994; Treffers, 1987; see also De Corte et al., 1996) and his own team's work on the MATHE 2000 project (see, e.g., Becker & Selter, 1996) are typical examples of this approach.

Design research basically encompasses three phases: developing a preliminary design, conducting a teaching experiment, and carrying out a retrospective analysis. The first phase starts with the formulation of a "conjectured local instruction theory" (Gravemeijer, 2004, p. 109) or a "hypothetical learning trajectory" (Simon, 1995, p. 133) that involves conjectures about (a) the learning goals, (b) the instructional tasks and activities and possible tools that will be used, and (c) the thinking and learning processes in which the students might engage in this instructional environment. Most typically, these conjectures are based on the historical development of mathematics or knowledge about children's informal strategies. These conjectured goals, tasks and activities, and envisaged thinking and learning processes are often not worked out in detail during this first phase. In the second phase, the teaching experiment, those conjectures are put to the test. In the course of the teaching experiment, the hypothetical teaching/learning trajectory is gradually and cyclicly adapted, corrected, refined, and so on on the basis of the input of the students and assessments of their actual understandings. However, we should acknowledge that the nascency and complex nature of design experiments has led to a variety of interpretations and applications, both about the methodological requirements of this approach to research and about the constituents of the learning trajectories (for a thorough discussion, see Baroody et al., 2004; Clements & Sarama, 2004b).

Rather than structuring our review of this classroom-based research around the different curricular subdomains of elementary school arithmetic distinguished in the previous section, we judge it more productive to highlight and illustrate some typical features for the design studies realized in these different curricular subdomains during the last decade. The selection of examples represents both a variety of subject-matter content and age levels as well as a diversity in theoretical perspective and background. However, these examples also share some "family resemblances": (a) the active and constructive view on mathematics learning; (b) the recognition of the crucial role of students' prior knowledge for future learning; (c) the orientation towards conceptual understanding and problem solving; (d) a reluctance to "impose" a single representational model or solution strategy on all learners and the pursuit of adaptive expertise (Baroody et al., 2004); (e) the importance of social interaction and collaboration in doing and learning mathematics, and, more generally, of conceiving learning mathematics as a community activity (Romberg et al., 2005); (f) the efforts to embed mathematics learning in authentic and meaningful contexts; (g) the seeking of an appropriate balance between "horizontal and vertical mathematisation" (Treffers, 1987); (h) the focus on basic principles (or, as Baroody et al., 2004, or Romberg et al., 2005, would say: on "big ideas" that underlie numerous concepts and procedures across topics) while leaving most of the details to students (Paulos, 1991); (i) the belief that the above instructional design principles do not hold only for teaching mathematics to mathematically able children, but also to low achievers; and (j) the attention to the teacher's role in the learning process. This implies that all these examples represent quite a radical departure from a "traditional skills approach" to mathematics instruction based on the view that mathematics learning is a highly individual activity consisting mainly in absorbing and memorizing a fixed body of decontextualized and fragmented knowledge and procedural skills transmitted by the teacher (Baroody, 2003). But these examples can also be differentiated from a so-called "laissez-faire problem-solving approach," which was especially influential in the initial period of the current reform movement in countries like the United States (Baroody, 2003; Romberg et al., 2005) and the Netherlands (Van den Heuvel-Panhuizen, 2001). They generally fit with what Baroody (2003) called the investigative approach, that is, a blend of a conceptual and a problem-solving approach, aiming at the mastery of basic skills, conceptual learning, and mathematical thinking and characterized by both meaningful and inquiry-based instruction and by purposeful learning and practice. From a methodological perspective, the selected examples can be considered as prototypes of the new generation of ecologically valid design experiments (as defined above) rather than the traditional kind of teaching experiments, which focused more on internal rather than external or ecological validity.

Problem-Solving as a Basis for Learning Multidigit Calculation

According to many researchers and policy makers, students can learn about number through problem solving—including both standard aspects of computation, as well as more puzzle-like problems that promote understanding of the structure of the whole-number system, elicit algebraic modes of reasoning prior to the use of formal algebraic notation, and so on. Taking the description of problem solving in NCTM's *Principles and Standards for School Mathematics* (2000, p. 116) as "a major means for developing mathematical knowledge" as a starting point, a number of related research projects in the United States and South Africa have addressed children's well-known deficiencies with multidigit numbers and arithmetic by trying new approaches that are hypothesized to support children's construction of accurate and robust conceptual structures for multidigit numbers and facilitate the use of these conceptual structures in multidigit calculation. All these programs take a so-called problem-solving approach to teaching multidigit number concepts and operations (see Carpenter et al., 1997; Fuson, Wearne, et al., 1997; Hiebert et al., 1996). In a joint manifest in the *Educational Researcher*, Hiebert et al. (1996, p. 12) explained this problem-solving principle as follows:

> The principle is this: students should be allowed to make the subject *problematic*. We argue that this single principle captures what is essential for instructional practice. . . . Allowing the subject to be problematic means allowing students to wonder why things are, to inquire, to search for solutions, and to resolve incongruities. It means that both curriculum and instruction should begin with problems, dilemmas, and questions for students. . . . We use "problematic" in the sense that students should be allowed and encouraged to problematize what they study, to define problems that elicit curiosities and sense-making skills.

These authors justified the practice of problematizing the subject by claiming "that it is this activity that most likely leads to the construction of understanding" (p. 15). Examples of design experiments wherein this basic principle is applied are contained in a large number of case-study descriptions of classrooms by these researchers in which the open and constructive examination of methods of inquiry and instruction is emphasized. For instance, Fuson, Smith, et al. (1997) carried out a year-long classroom teaching experiment in two predominantly Latino low socioeconomic status (SES) urban classrooms (one English-speaking and one Spanish-speaking) that sought to support first graders' thinking of two-digit quantities as tens and ones, starting from the developmental sequence of conceptual structures for two-digit numbers (the so-called UDSSI triad model) described earlier (Fuson, Smith, et al., 1997). By the end of the year, most children could accurately add and subtract two-digit numbers that require trading (regrouping) by using drawings or Dienes blocks and gave answers by using tens and ones on various tasks. Their performance was substantially above that reported in other studies for U.S. first graders of higher SES and for older U.S. children. According to the authors, children's responses looked more like those of East Asian children than of U.S. children in other (ascertaining) studies.

According to Hiebert et al. (1996), the available evidence from design experiments such as the one by Fuson, Smith, et al. (1997) shows that students who are presented with these kinds of tasks and engage in the kinds of discussions described and illustrated above do develop, first of all, deeper structural understandings of the number system than their peers who move through a more traditional skills-based curriculum. Second, two kinds of strategies are produced by working through problematic situations: (a) the particular arithmetic strategies that can be used for solving particular problems and (b) the general metastrategic approaches or ways of thought that are needed to construct these procedures. As far as the specific strategic aspect is concerned, the research evidence from these authors' studies suggests that students who are allowed to problematize arithmetic procedures perform as well on routine tasks as their more traditionally taught peers. With respect to the metastrategic aspect, there is evidence that pupils who have been encouraged to treat situations as problematic and develop their own strategies can adapt them later, or invent new ones, to solve new problems. A third important observation is that when students experience curricula that treat mathematics as problematic, such as those described above, the traditional separation between conceptual knowledge and procedural skill is weakened. Finally, the authors stress that what students take away from these problem-based environments cannot be

adequately described in (meta)cognitive terms only. There is evidence, according to Hiebert et al. (1996), that students who are allowed and stimulated to treat mathematics as problematic develop also appropriate beliefs and positive attitudes towards mathematics that, in turn, influence their orientation toward future activities.

Whereas the experimental program of Fuson, Smith, et al. (1997) aimed at the development of the integration of declarative and procedural knowledge for doing multidigit addition and subtraction from a basically ten-base and decomposition approach, several mathematics educators in the Netherlands and Germany have recently designed and evaluated experimental training programs for mental addition and subtraction up to 100 or up to 1000 wherein the empty number line, and, consequently, the sequential concept and the jump method, play a crucial role (Beishuizen, 1999; Gravemeijer, 2004; Menne, 2001; Selter, 1998). By way of example, we review here Selter's (1998) design experiment.

As in the above-mentioned design experiments from the United States, Selter (1998) started from the idea that powerful instruction should involve a gradual progression from highly contextualized, externalized, less elegant, and inefficient methods toward more general, internalized, more elegant, and efficient methods (vertical mathematization in Treffers's, 1987, terminology), typically (but not necessarily!) starting from the exploration of realistic problems. Pupils' different solution strategies are discussed, explained, and elaborated via pupil-pupil and pupil-teacher interactions. The teacher's role is to encourage the pupils to develop their own ways of working and to reflect on them, much in the same way as in the design experiments mentioned above. During these moments of reflection, the role of the teacher is to act as a mediator between the "inventions" of the pupils and the "conventions" of mathematics (Lampert, 1990).

Selter's (1998) teaching experiment was conducted with 27 German 8- and 9-year-old third graders. The course consisted of 15 lessons on mental addition and subtraction (as defined in the earlier section on arithmetic word problems) in the domain between 100 and 1000. The aim of the study was not to prove the superiority of this instructional approach to other, more traditional ways of teaching mental arithmetic. Rather, it was more modest, namely to describe in a qualitative way how he tried to realize in one class the idea of vertical mathematization for this part of the curriculum and to document the kind of student learning resulting from it.

In the first stage, the children solved word problems (mostly around the context of a big new cinema center recently built in their city) with their own informal methods. Especially during this first stage, children were stimulated to use the empty number line as an informal and approximate way to support both their thinking and their communication. As il-lustrated in Figure 13.3, Selter observed a wide variety of solutions to problems such as "In cinema 5, 216 persons can be seated. 148 persons are already sitting there. How many empty seats are there?" From the first stage on, all children had a diary wherein they

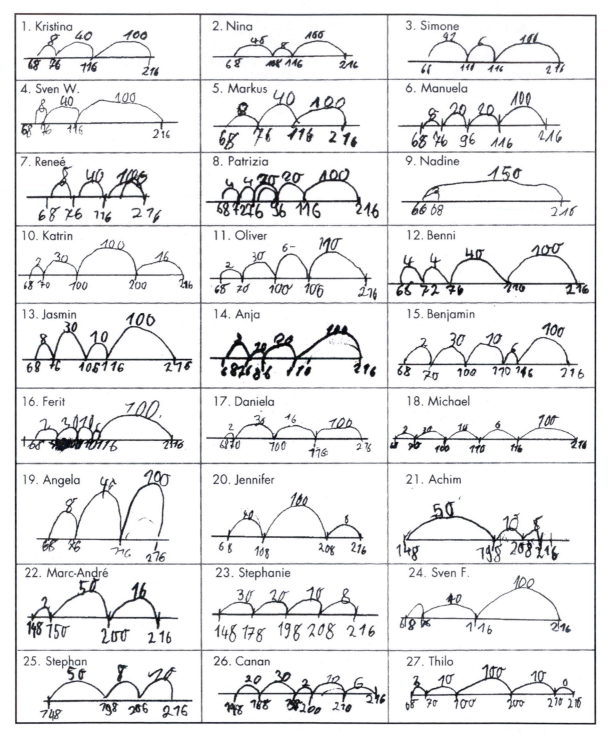

Figure 13.3. All solutions to a subtraction problem ("In cinema 5, 216 persons can be seated. 148 persons are already sitting there. How many empty seats are there?") (Selter, 1998, p. 16).

had not only to document their strategy by means of the empty number line (on the right half of the page), but also to write a brief verbal description of their solution and, if possible, to put down their method in symbolic expression (on the left half of the page). Many children's descriptions point to a high degree of articulation and reflection.

In the second stage, vertical mathematization was achieved by means of "math conferences" among small groups of pupils as well as whole-class discussions, wherein they explicitly discussed the different methods in terms of elegance and efficiency. An interesting instructional technique that was used in that stage, called "Do it like . . . ," involved the explicit instruction to try to solve a given addition or subtraction problem according to a particular strategy that had appeared to be effective during previous lessons. For instance, in a worksheet called "Doing it like Ferit," pupils were asked to apply a compensation strategy (see *Strategies for Multidigit Addition and Subtraction* section) to solve the problem 128 + 96. Although more elegant and more efficient methods were favored through these activities, children were never forced to use the most efficient methods in their own work (not even at the end of the course). In other words, strategies were never imposed upon children.

During the third part of the course, children focused on the advantages and disadvantages of the different strategies and were stimulated to develop new ones. Selter (1998) documented the generation of several new strategies in this stage, such as the use of a sort of "adding down to" strategy to solve subtraction of close numbers like 312 − 278.

Finally, children were asked to reflect on the whole course by explaining to other people (i.e., their parents and children from another class) what they had done and learnt in the course.

Looking back, Selter (1998) pointed to a number of important didactical problems that were revealed through this design experiment, such as the handling of the decomposition or split strategies, which are difficult, if not impossible, to visualize by means of the empty number line. Moreover, as the study was basically qualitative and exploratory in nature, he acknowledged the necessity of further, more systematic evaluation.

Laying Foundations for Algebra in Early Arithmetic

One important aspect of algebra, though by no means the only one (Kaput, 1999), is the generalization of arithmetic, and researchers commonly refer to problems student encounter during the transition from arithmetic to algebra. More recently, however, the emphasis has shifted to recognizing the aspects of algebraic thinking that are inherent within arithmetic and to teaching arithmetic accordingly. Note that by *algebraic thinking* is meant such basic processes as the recognition and exploitation of structure and, above all, generalization. Such processes are not dependent on algebra as the intensive study of the last three letters of the alphabet or, as expressed by Engestrom (1987), making what should be a tool into the object of study itself. Pre-algebra can also entail a focus on analyzing process instead of generating answers.

The refocus on algebra as latent within arithmetic affords complementary advantages, in that it lays foundations for symbolic algebra and at the same time consolidates arithmetical knowledge by exploiting its structural coherence (refer back to the *Structure of the Whole-Number System* section). Furthermore, it exposes children at an early age to one important view of mathematics as "the classification and study of all possible patterns" (Sawyer, 1955, p. 12) and thus contributes to the foundations of a positive mathematical disposition.

Thinking Ahead, Laying Groundwork

By strategic choices in the types of arithmetic activities, the foundations for later symbolic algebra may be established early. Adumbrating the use of letters to denote specific unknowns is simple and familiar—for example, through a question such as " times what equals 20?" The more advanced use of a letter to denote a variable can also be foreshadowed by posing a problem such as "describe how the store owner would figure the amount of refund for any number of returned cans," given that the refund is $5 per can (Swafford & Langrall, 2000, p. 93). Thus, paralleling the historical development from rhetoric to syncopated algebra, the groundwork is done for the major conceptual advance whereby generalized quantitative relationships are symbolically expressed—which really begins to pay off when those expressions are conceived of as mathematical entities that can be operated upon. Careful choice of tasks that strain the limits of arithmetical treatment and demonstrate the power and elegance of algebra can provide motivation and rationale for this conceptual advance.

A specific difference between algebra and arithmetic is in the use of the equals sign. Taught without regard to their future, many children come to have a very limited notion of what this symbol means in arithmetic, something like "makes"—which indeed is its major role in arithmetic. This limitation is a natural result of their experience being only of seeing it come between two arithmetic expressions, with that on the right being a single number. As a result, they have no basis for dealing with arithmetic equations in noncanonical form, with or without unknowns, such as

$$8 + 4 = \_ + 5$$

(Carpenter, Franke, & Levi, 2003; Falkner, Levi, & Carpenter, 1999; Seo & Ginsburg, 2003; Van Amerom, 2002). To interpret such equations requires a relational view of the equals sign, namely that the expressions on either side are equal.

Seo and Ginsburg (2003) carried out a study (using several research methods including analysis of materials, participant observations, and clinical interviews) of a second-grade classroom with 29 students in which the teacher was attempting to foster such a relational view. In their study, they specifically addressed the following questions:

Issue 1: How do textbooks present the equals sign?

Issue 2: How did the teacher present the equals sign?

Issue 3: How did context affect the children's interpretation of the equals sign?

In the textbooks used, they found little effort to promote the relational view. The teacher did include a variety of activities promoting it, such as linking the equals sign with the signs for inequality, writing a variety of expressions with the same numerical value, using Cuisenaire rods, currency equivalence (e.g., 1 nickel = 5¢, 1 dime = 10¢), and everyday examples linking "equals" with "same number." Children's interpretation varied widely depending on the context. One student carefully explained that the meaning of the equals sign depended on the mathematical sentence, and most of the others agreed. Also the children made distinctions of meaning that mathematicians would not! They considered that the meaning is different for an addition and a subtraction, also for, for example, $2 + 3 = 5$ (means "answer") and $3 = 3$ (means "the same").

In passing, we note that a lot of the recent work on the equals sign implies that *the* mathematical meaning of the equals sign is equivalence, which is wrong. There are many other meanings encountered later, such as expressing an identity ($a + b = b + a$) or a definition ($f(x) = 3x - 4$) and subtleties such as $17 \div 3 = 5$ remainder 2 and $22 \div 4 = 5$ remainder 2, but $17 \div 3 \neq 22 \div 4$.

Falkner et al. (1999) reported that of 145 sixth-grade students posed the problem

$$8 + 4 = \_ + 5$$

every single student thought that either 12 or 17 should go in the box. Such findings serve particularly well to illustrate the dangers of short-termism. If the aim is to get the children to a high level of proficiency in arithmetic for the next test, then the "makes" meaning will serve its purpose. Unfortunately, such a temporary gain is more than offset by repercussions in the future when students need to work with algebraic uses of the equals sign, notably in equations.

In any case, the relational meaning of the equals sign is needed to express equivalence in the course of exploring the structure of numbers or doing mental arithmetic (in the sense described earlier in the arithmetic word problems section). Thus, according to Falkner et al. (1999, p. 233)

> A child who has had many opportunities to express and reflect on such number sentences as $17 - 9 = 17 - 10 + 1$ might be able to use the same mathematical principle to solve more difficult problems, such as $45 - 18$, by expressing $45 - 18 = 45 - 20 + 2$. This example shows the advantages of integrating the teaching of arithmetic with the teaching of algebra. By doing so, teachers can help children increase their understanding of arithmetic at the same time that they learn algebraic concepts.

Structural Insight, Problem-Solving, Argumentation, and Proof

In the *Structure of the Whole-Number System* section, we reviewed evidence that children spontaneously—and mostly implicitly—exploit the structure of arithmetic, in the process demonstrating considerable adaptive expertise. One focus of recent studies is how enquiry-based classrooms can support making the structural properties explicit, through processes of conjecture and refutation, argumentation and (informal) proof. For example, Carpenter, Franke, and Levi (2003, chapter 4) gave several examples of students constructing and testing generalizations about calculations involving 0 and 1, about inverse operations, and about computations with even and odd numbers. There are several accounts of emergent convincing arguments about the commutativity of addition and multiplication. Carpenter and Franke (2001) reported a case study of a sixth-grade class that illustrates the progression of forms of argument used by students to explain commutativity of multiplication, culminating in general argument based on rotating a rectangular array of dots. A similar explanation from a third-grade student was provided by Schifter and Fosnot (2003).

Another set of studies addressed how children tackle puzzle-like situations that depend on insight into structural relationships. An example of an extremely rich problem is "arithmagons" (or "arithmogons"; Becker, Sawada, & Shimizu, 1999; Becker & Selter, 1996; Driscoll, 1999; Mason, 1996; Mason,

Burton, & Stacey, 1982; Wittmann, 2001). The core problem may be stated generally as follows: "A secret number is assigned to each vertex of a triangle. On each side of the triangle is written the sum of the secret numbers at its ends. Given the values on the three sides, find a simple rule for revealing the secret numbers." For example, the numbers 11, 18, 27 are produced by the secret numbers 1, 10, 17 (see Figure 13.4). A specific problem such as this may be solved by many methods—from undirected trial and error, to systematic trial and improvement, to arguments such as one that begins "the two bottom secret numbers must differ by 7, since added to the top secret number they produce 11 and 18, respectively," to formal algebra, assigning a, b, c as the secret numbers to produce a system of three equations.

$$a + b = 11 \qquad b + c = 27 \qquad c + a = 18$$

An insightful step (based on the symmetry of the system of equations) is to add the three equations to find that $(a + b) + (b + c) + (c + a) = 2(a + b + c) = 11 + 27 + 18 = 56$, hence $a + b + c = 28$. Now c can be found by subtracting 11 ($= a + b$) from 28 ($= a + b + c$), and so on.

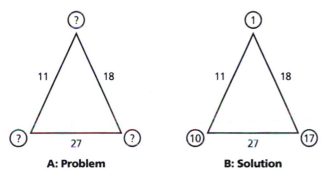

A: Problem   **B: Solution**

Figure 13.4   Arithmogons (Becker et al., 1999).

Various modifications of the task making it suitable for first-grade students have been used in the MATHE 2000 project (Becker & Selter, 1996, pp. 538–540; Wittmann, 2001).

The problem can also be generalized to other polygons (Mason et al., 1982, p. 160) with interesting results. For any even number of secret numbers, it turns out that the solution is indeterminate, whereas for any odd number of secret numbers, the solution that begins by finding the sum of all the secret numbers by taking half of the sum of all their pairwise sums generalizes. The comparison of odd and even cases leads naturally to a discussion of dependent versus independent systems of equations.

Another rich problem is that of finding which whole numbers can be expressed as the sum of two or more consecutive whole numbers in at least one way

(e.g., $15 = 7 + 8 = 4 + 5 + 6 = 1 + 2 + 3 + 4 + 5$; Driscoll, 1999; Mason et al., 1982). Exploration very naturally leads to the conjecture that all numbers have this property except for powers of 2 (including $2^0 = 1$). As Sawyer (1955, p. 23) commented, "for every pattern that appears, a mathematician feels he [sic] ought to know why it appears." The search for a reason in this case is likely to throw up many conjectures and generalizations, such as that 3 times any number can be expressed as the sum of that number plus the preceding and succeeding numbers, that prime numbers (apart from 2) can be expressed as the sum of two consecutive numbers but not otherwise, and so on. It is perhaps worth pointing out that a lot of practice in computation is a by-product of such investigations.

Blanton and Kaput (2000) described one strand of a year-long teaching experiment with a socioeconomically and ethnically diverse third-grade class in an urban school, part of a district-based project. The focus of the description was generalizing and formalizing the students' thinking about odd and even numbers, with the class as the grain-size for analysis. The researchers documented progression in complexity of generalizations and supporting justifications about even and odd numbers over the course of the year. Specifically, they found evidence of multiple forms of reasoning (including representational, numerical, and pattern-based), algebraic use of the terms *even* and *odd* as placeholders, use of a generalized referent (e.g., *odd*) to produce a generalization, extension of generalizations across mathematical operations, validation of generalizations through peer argumentation, and construction of sophisticated justifications. For large-scale implementation of an "algebrafication" strategy in early grades, Kaput and Blanton (2001, pp. 344–345) called for classroom-grounded teacher development in three dimensions:

1. Building algebraic reasoning opportunities, especially generalization and progressive formalization opportunities, from available instructional materials, including the algebrafication of existing arithmetic problems by transforming them from one-answer arithmetic problems to opportunities for pattern-building, conjecturing, generalizing, and justifying mathematical facts and relationships.

2. The building of teachers' "algebra eyes and ears" so that they can spot opportunities for generalization and systematic expression of that generality . . . and then exploit these opportunities as they occur across mathematical topics.

3. The process of creating classroom practice and culture to encourage and support active student generalization and formalization within the context of purposeful conjecture and argument, so that algebraic reasoning opportunities occur frequently and are viable when they do occur.

Creating a Social-Constructivist Classroom Culture

Whereas many researchers focus their analysis of classroom environments on the development of arithmetical thinking and how it can be improved through innovative didactical materials, tools, techniques, and so on, in another influential set of classroom-based studies of the past decade, the focus is on creating, from a very young age, an appropriate classroom culture for arithmetic lessons along lines suggested by a social-constructivist perspective. In reflecting upon their longstanding developmental research project on mathematical learning in the early grades of elementary school, Yackel and Cobb (1996, p. 459) wrote,

> we began the project intending to focus on learning primarily from a cognitive perspective, with constructivism as our guiding framework. However, as we attempted to make sense of our experiences in the classroom, it was apparent that we needed to broaden our interpretative stance by developing a socio-cultural perspective on mathematical activity.

According to this sociocultural perspective, the development of individuals' arithmetic concepts and skills cannot be separated from their participation in the interactive constitution of "taken-as-shared" mathematical meanings. An extremely important proactive role in the development of such a classroom microculture is played by the teacher. According to Yackel and Cobb (1996), one of the teacher's major tasks, which has been neglected for too long, is to establish sociomathematical norms. These norms, which are specifically mathematical (and are therefore distinguished from the social norms that relate to classroom actions and interactions in general), regulate classroom discourse and influence the learning opportunities that arise for both the students and the teacher. Typical examples of sociomathematical norms include what counts as a different mathematical solution, a sophisticated mathematical solution, and an efficient mathematical solution. Others relate to what counts as a good mathematical problem, a proper explanation, or an acceptable justification. The importance attributed to sociomathematical norms stems from the contention that students reorganize their specifi-

cally mathematical beliefs and values as they participate in and contribute to the establishment of these norms. This claim implies that teachers can support students' development of appropriate dispositions toward mathematics by guiding the development of sociomathematical norms (Yackel & Cobb, 1996).

Starting from Yackel and Cobb's (1996) analysis, McClain and Cobb (2001) collaborated intensively with one teacher to consciously attempt to guide the development of sociomathematical norms and thus influence her students' beliefs about what it means to know and do mathematics, in the context of a teaching experiment. Together with the teacher, they traced the emergence of sociomathematical norms in one classroom over a 4-month period while she was teaching mental computation and estimation with numbers up to 100. The data consist of videotaped recordings from two cameras of about 100 lessons, supplemented with copies of students' written work, field notes, several videotaped clinical interviews with each student and with the teacher. In their article, they documented in great detail the development of sociomathematical norms in this classroom over a 12-week period, wherein the focus was on becoming familiarized with the quantitative environment structured by relationships between numbers up to 20. Ample use was made of the "arithmetic rack," a device with two rows of five white and five colored beads each, which enhances, more than others, multiple approaches to single-digit arithmetic (see Figure 13.5, which illustrates how, with the help of the arithmetic rack, a sum like 6 + 7 can be solved easily either as 6 + 4 + 3 or as 6 + 6 + 1; Treffers, 1991, 2001a). This rich and detailed analysis led to several interesting conclusions and issues for further research.

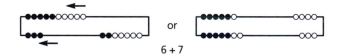

Figure 13.5 Solving a sum like 6 + 7 in two different ways using the arithmetic rack (Treffers, 2001a, p. 50).

The first issue concerns the complexity and challenges that arise for teachers as they attempt to reorganize their instructional practices along the lines of the curriculum reforms. In the beginning of the experiment, the teacher accepted all students' contributions equally and rejected the view that she should act in any way as a mathematical authority in the class. According to the researchers, she had developed this form of practice as a reaction to traditional instructional practices. In other words, she had developed an instructional approach that could be defined negatively in terms of what she did not do compared with traditional teaching. When she started to guide the development of the norm of mathematical difference at the researchers' urging, she did it in a rather directive and nonargued way. The teacher's learning during the rest of the teaching experiment consisted of improved attempts to cope with the tension between direction and structure, on the one hand, and openness to students' interpretations and interests, on the other.

A second issue concerns the way in which the emergence of the sociomathematical norm of what counted as a different solution formed the foundation for the emergence of the norms of what counted as a sophisticated and as an efficient solution, and the role that notating and symbolizing plays in supporting the emergence of these norms.

The third issue concerns the manner in which the establishment of these norms could directly influence the mathematical agenda. For instance, McClain and Cobb (2001) observed instances in which the students initiated reflective shifts in classroom discussions by conjecturing about the number of different ways in which a task could be solved by grouping, "which appear to involve an initial step in the transition from arithmetic to algebraic reasoning" (p. 264).

One major goal of this section was to demonstrate exemplarily how the teaching and learning of number and arithmetic can, and should, be interwoven with establishing appropriate social and sociomathematical norms from a young age. A second was to show how careful attention to these (subtle) sociocultural aspects of school mathematics can lead to significant learning outcomes related to children's mathematical reasoning and communication skills, and intellectual autonomy, without losing significant learning effects for number and arithmetic "strictu sensu"(see also Gravemeijer, 2004).

Emergent Models as a Design Heuristic

In many approaches to elementary mathematics learning, both older and recent, tactile and visual models play a very important role. However, there are different views about the use of these models. In a traditional skills approach (Baroody, 2003), they are not used because meaningful instruction is not an important goal. In a conceptual approach, the goal of meaningful learning of concepts and skills is typically reached by the imposition of a "ready-made" model (e.g., MAB material, Cuisenaire blocks) by the teacher that concretizes or embodies a given concept or technique. Recent theoretical developments, especially the growing appreciation of social constructivism and sociocultural theories of learning and instruction and of investigative approaches

to mathematics education, have led to strong calls to find an alternative for the teacher-imposed use of tactile and visual models as (unproblematical) "embodiments" of mathematical concepts, as in the above-mentioned conceptual approach.

In a number of (theoretically closely related) recent design experiments dealing with the teaching of certain arithmetic concepts or skills the option of building on symbolizing, modeling, and tool use as personally meaningful activities of students, in the context of social practices, is explored. The traditional metaphor of "transmission of knowledge" with the help of symbols that function as "carriers of meaning" is replaced by the image of students "constructing their own ways of symbolizing as part of their mathematical activity" (Gravemeijer, Lehrer, Van Oers, & Verschaffel, 2002, p. 1; see also Gravemeijer, 2002, 2004). What is called symbolizing and modeling in these studies differs significantly from the use of manipulative materials and visual models that has been common practice for a long time in mathematics education (research). In contrast, within the conception that is paramount in these design experiments, modeling is seen primarily as a form of organizing, in the course of which both the symbolic means and the model itself emerge (Gravemeijer, 2002, 2004; Gravemeijer & Stephan, 2002). This "emergent modeling as an instructional design" approach to the teaching and learning of arithmetic can be illustrated best by referring to several developmental research studies realized by the Freudenthal Institute in the Netherlands (Gravemeijer, 1994; Treffers, 1987), and to other studies conceptualized and realized along similar lines (Cobb et al., 1997; Izsak, 2004). These studies, which deal with the use of models *of* certain meaningful situations transitioning into models *for* arithmetic operations, show how these models emerge as context-specific models that refer to concrete and paradigmatic situations that are "experientially real" for the students. Gradually, the contextual meaning may shift to the background and the visual expression may change, but the idea is that the roots of the model are preserved, and that the students can always fall back on this original meaning (Gravemeijer & Stephan, 2002). Although the proponents of the emergent modeling approach claim that, ideally, the pupils should invent the necessary modeling tools for themselves, they acknowledge (partly based on experiences wherein they have tried to implement this design principle quite radically) that this ideal is mostly not really feasible and that as a reasonable compromise the pupils should at least "experience an involvement in the invention process, even though they do not invent the tools for themselves" (Gravemeijer, 2004, p. 122).

Rather than summarizing Gravemeijer's description of design studies based on the empty number line (Gravemeijer, 2002, 2004; see also Gravemeijer & Stephan, 2002) as a prototypical example of the emergent modeling approach, we refer to a recent study by Iszak (2004), dealing with a different curricular topic. Iszak's (2004) study combined an in-depth analysis of two-digit multiplication instruction in one U.S. fourth-grade classroom with an analysis of learning accomplished by a cross-section of students from the same classroom. More specifically, the study compares how taken-as-shared classroom mathematical practices and individual students used features of rectangular area representations for accomposing conceptual, procedural, and problem-solving goals. Except for a chapter in Treffers's (1987) book, describing an experimental program for multidigit multiplication based on the principle of "progressive schematization" starting from area representations (but without providing any empirical analysis of classroom interactions with these experimental materials), Iszak (2004) did not find any previous research on the use of rectangular areas to develop students' understanding of multidigit multiplication. Starting from these ideas of Treffers (1987) and the results of ascertaining studies that fourth-grade students have good (intuitive) understanding of arrays, Iszak (2004) designed a learning environment wherein students' understanding of arrays was used as the starting point to develop understandings of multidigit multiplication, as a kind of alternative to the more traditional instructional programs wherein multiplication is introduced as repeated addition (as discussed earlier).

The study was conducted in the context of the Children's Math Worlds (CMW) project, the main objective of which is to make the goals of the U.S. Standards accessible to urban and suburban students and teachers. The CMW two-digit multiplication block builds on another CMW unit aimed at developing connections among single-digit multiplication, equal groups, and areas of rectangles understood as arrays of unit squares. The two-digit multiplication unit scaffolds the development of more efficient strategies with three area representations and a numeric method to break apart factors and products into smaller, easier-to-handle pieces (Figure 13.6): (a) a unit squares representation, (b) a 100s/10s/1s representation, (c) a quadrant representation, and (d) the expanded algorithm representation.

The experimental course consisted of 13 two-digit lessons, spread over a period of 8 weeks, wherein (groups of) pupils' representations and solutions to a small number of two-digit multiplication problems were intensively discussed with the whole class and

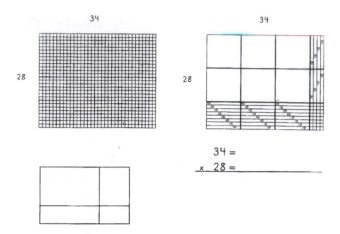

Figure 13.6 The unit squares, 100s/10s/1s, quadrants, and expanded algorithm representations for 28 × 34 (Iszak, 2004, p. 50).

connected to each other as well as to representations and solutions proposed by the teacher.

A scrutinized analysis of the videotapes of these lessons revealed, first, how taken-as-shared class strategies for finding groups and determining areas in unit squares and 100s/10s/1s representations emerged as the teacher and her students discussed alternatives for problems such as 14 × 13 and how, afterwards, increasingly more complex problems and accompanying new instructions led to the gradual development of more efficient strategies of the types shown in Figure 13.6. Ultimately, class discussion converged on essentially one multiplication strategy, the expanded algorithm representation shown in Figure 13.6, that coordinated magnitudes of partial products, expanded forms for factors, and the distributive property. As a consequence of this convergence, alternative choices for equal-groups strategies gradually faded from class discussions. The analysis of the whole-class solutions also revealed the challenging decisions that the teacher had to make about the range of strategies to pursue with the students, especially with respect to finding a good balance between providing too little guidance (in which case students might not develop understandings of efficient and general multiplication methods) and too much guidance (leading to too quick removal of alternative methods from the class discussions).

The results on a paper-and-pencil posttest that asked pupils to solve problems ranging between 17 × 12 and 92 × 78 and also "to draw any pictures that they need" demonstrated that by the end of the lesson unit a large majority of students could compute correctly with the expanded algorithm, and many could use alternative methods as well.

To gain deeper insight into pupils' strategies, the researcher also conducted videotaped semi-structured interviews with two pairs of low-, middle- and high-achieving pupils at the end of the unit, wherein they were administered two nonroutine tasks (e.g., a visually presented problem about the size of a 6×17 rectangular pattern and a word problem wherein children had to compute the total number of stores owned by a chain that had 26 stores in each of 38 states). These qualitative data made it clear that the pupils understood class strategies to very different degrees, that some of them did not readily make the required connections, and that more time and support might be necessary if all students are to find an approach to two-digit multiplication based on areas of rectangles. According to the author's analysis, this moderate overall outcome is at least partly due to the sociomathematical norms for justifying solutions applied by the teacher:

> Had norms been established in the teacher's classroom that afforded opportunities for students to explain strategies in full, and that required students to take responsibility for asking questions when they did not understand, then the standard for taken-as-shared might have been higher and taken more time to achieve (Iszak, 2004, p. 78).

In a position paper about "developmental research" (unfortunately not available in English), Treffers (1993) extensively discussed the advantages and disadvantages of an instructional approach for multidigit multiplication based on the rectangular area model (as in some older developmental studies from the Freudenthal Institute and as in Iszak's study), and an alternative teaching/learning trajectory based on the repeated addition model (as in the study of Carpenter et al.,1997, described earlier and in other design experiments realized at the Freudenthal Institute; Dekker, Ter Heege, & Treffers, 1982; Treffers, 1987). Although he acknowledged that both approaches have their advantages and disadvantages, he preferred the latter approach based on the repeated addition model, mostly because it fits better with children's intuitive understanding of, and informal strategies for, multiplication. An empirical comparison of the strengths and weaknesses of both instructional approaches to multidigit multiplication seems necessary.

Teaching Word Problems From a Modeling Perspective

As discussed earlier, in recent years, the characteristics, use, and rationale of word problems have been critically analyzed from multiple perspectives, including linguistic, cultural, and sociological (Verschaffel et al., 2000). In particular, many mathematics educators have argued that the stereotyped and artificial nature of word problems typically represented in mathematics textbooks, and the discourse and activity around these problems in traditional mathematics lessons, have detrimental effects on students' disposition towards mindful and realistic mathematical modeling and problem solving.

One reaction to criticisms of traditional practice surrounding word problems in schools is to undermine the approach that allows students to succeed using superficial strategies based on the "rules of the game." This is done by breaking up the stereotypical nature of the problems posed. For example, by including problems that do not make sense or contain superfluous or insufficient data, students can be guided to interpret word problems critically. A more radical suggestion is to treat word problems as genuine exercises in mathematical modeling.

Without any attempt to be exhaustive, we list a few recent design studies that have been set up recently according to this modeling perspective:

- the Jasper studies of the Cognition and Technology Group at Vanderbilt (1997),
- the numerous design experiments with model-eliciting activities summarized in the recent book by Lesh and Doerr (2003),
- the modeling curriculum for mathematics and science by Lehrer and Schauble (2005),
- English's (1998) design studies on mathematical problem posing,
- the learning environment for mathematical modeling and problem solving in upper elementary school children that Verschaffel et al. (1999) developed and tested a few years ago.

Hereafter we exemplarily describe the last study from that list. Verschaffel et al. (1999) carried out a design experiment aiming at the design and evaluation of a powerful learning environment that elicits in upper primary school children the appropriate learning processes for acquiring the intended competence in mathematical modeling and applied problem solving as well as positive mathematics-related beliefs. The learning environment in the classroom was fundamentally changed, and its design, implementation, and evaluation were done in close cooperation with the teachers of the four participating experimental classrooms and their principals. The intervention consisted of a series of 20 lessons that were taught by the regular classroom teachers (for a more detailed report, see Verschaffel et al., 1999).

The learning environment focused on the acquisition by the pupils of an overall cognitive self-regulation strategy for solving mathematical problems consisting of five stages (representation, planning, execution, interpretation and evaluation, and communicating the answer), and embedding a set of eight heuristic strategies that are especially functional in the first two stages of that strategy (e.g., make a drawing, make a table, use your real-world knowledge, simplify the numbers, make a guess). An additional aim was to foster positive mathematics-related beliefs.

To elicit and support the intended constructive learning processes in *all* pupils, an intervention characterized by the following three basic features was developed. First, a varied set of carefully designed meaningful, complex, and open problems was used differing substantially from traditional textbook tasks. Moreover, these problems were presented in different formats—a text, a newspaper article, a brochure, a comic strip, a table, or a combination of several of these formats. An example is described below.

School Trip Problem

The teacher told the children about a plan for a school trip to visit the Efteling, a well-known amusement park in the Netherlands. If that turned out to be too expensive, one of the other amusement parks might be an alternative.

Each group of four pupils received copies of folders with entrance prices for the different parks. The lists mentioned various prices depending on the period of the year, the age of the visitors, and the kind of party (individuals, families, groups).

In addition, each group received a copy of a fax from a local bus company addressed to the principal of the school. The fax gave information about the prices for buses of different sizes (with a driver) for a 1-day trip.

The first task of the groups was to check whether it was possible to make the school trip to the Efteling given that the maximum price per child was limited to 12.50 euro. After finding out that this was not possible, the groups received a second task: They had to find out which of the other parks could be visited for the maximum amount of 12.50 euro per child.

Second, a learning community was created through the application of a varied set of stimulating and interactive instructional techniques. The basic instructional model for each lesson period consisted of the following sequence of classroom activities: (a) a short whole-class introduction; (b) two group assignments solved in fixed heterogeneous groups of three to four pupils, each of which was followed by a whole-class discussion; (c) an individual task also with a subsequent whole-class discussion. Throughout the whole lesson the teacher encouraged and scaffolded pupils to engage in, and to

reflect upon, the kinds of cognitive and metacognitive activities involved in the model of skilled problem solving. These instructional supports were gradually faded out as pupils became more competent in and aware of their problem-solving activity and, thus, took more responsibility for their own learning and problem-solving processes.

Third, an innovative classroom culture was created through the establishment of new social and sociomathematical norms about learning and teaching problem solving, aiming at fostering positive mathematics-related attitudes and beliefs in children. Typical aspects of this classroom culture were (a) stimulating pupils to articulate and reflect upon their solution strategies, (mis-)conceptions, beliefs, and feelings relating to mathematical problem solving; (b) discussing about what counts as a good problem, a good response, and a good solution procedure (e.g., "there are often different ways to solve a problem"; "for some problems a rough estimate is a better answer than an exact number"); and (c) reconsidering the role of the teacher and the pupils in the mathematics classroom (e.g., "the class as a whole, under the guidance of the teacher, will decide which of the generated solutions is the optimal one after an evaluation of the pros and cons of the different alternatives").

As mentioned above, the learning program consisted of a series of 20 lessons. One introductory lesson was followed by 15 lessons that systematically addressed each step and each heuristic of the five-step problem-solving model. At the end of each of these 15 lessons pupils completed a sheet for their personal notebook summarizing the essentials about the what, the how, and the why of each step of the model, or of each of the embedded heuristics. At the end of the series four project lessons built around more complex application problems were organized wherein the pupils learnt to use all the elements of the problem-solving model in an integrated and flexible way. The School Trip Problem presented above is an example of such a problem.

To enhance a reliable and powerful implementation of the learning environment, the teachers of the four experimental classes received a great variety of support materials. It is important to mention here that this lesson series as well as the above-mentioned elements of teacher support were elaborated in partnership with the teachers of the participating experimental classes and their principals.

During the period of about 4 months during which the intervention was implemented in the four experimental classes, pupils from seven comparable control classes followed the regular maths curriculum.

Students' progress toward the major goals of the learning environment was assessed using a variety of instruments. Three parallel versions of a written test consisting of 10 difficult nonroutine application problems were used as pretest, posttest and retention test, respectively. A questionnaire on pupils' beliefs and attitudes relating to the teaching and learning of mathematical word-problem solving and a standardized mathematics achievement test that covers the entire mathematics curriculum were both applied as pretest and posttest. In addition, in each of the four experimental classes the solution processes of three pairs of children for five nonroutine problems were videotaped and analyzed before and after the intervention. Finally, in order to assess the implementation of the learning environment by the teachers, a sample of representative lessons was videotaped in each experimental class, and these lessons were analyzed by means of an observation scale. Moreover, the four experimental teachers were interviewed at the end of the whole lesson series about (their personal appreciation of) all aspects of the learning environment.

The results can be summarized as follows. First, whereas no significant difference was found between the experimental and control groups on the word-problem test during the pretest, the former group significantly outperformed the latter during the posttest, and this difference in favor of the experimental group continued to exist on the retention test. Second, the learning environment had also a significant, albeit small, positive impact on children's pleasure and persistence in solving mathematics problems, and on their mathematics-related beliefs and attitudes. Third, the results on a standardized achievement test showed that the experimental classes also performed significantly better than the control classes on the standard achievement test. Fourth, the analysis of pupils' written notes on their response sheets of the word problem test showed that the better results of the experimental group were paralleled by a very substantial increase in the spontaneous use of the heuristic strategies taught in the learning environment. This finding was confirmed by a qualitative analysis of the videotapes of the problem-solving processes of three groups of two children from each experimental class before and after the intervention. Indeed, this revealed that on the posttest these dyads made nearly twice as much spontaneous use of the eight heuristics as during the pretest, and that they also clearly progressed in terms of the spontaneous use of four metacognitive skills (i.e., orientation, planning, monitoring, and evaluation). Fifth, not only the high- and the medium-ability pupils, but also those of low ability benefited significantly—albeit to a lesser degree—from the intervention in all aspects just mentioned. Indeed, there was no significant interaction between the variables experimental group, time of test, and ability level. Finally, the analysis of the videotapes of the lessons of the four teachers indicated that all four experimental teachers implemented the learning environment in (at least) a satisfactory way.

The preceding results show that an intervention, combining a set of carefully designed word problems with highly interactive teaching methods and the introduction of new sociomathematical classroom norms, can lead to the creation of a powerful learning environment that significantly boosts pupils' cognitive and metacognitive competency in solving mathematical word problems. In most of the design experiments in the domain of mathematical modeling, promisingly positive outcomes have been obtained in terms of performance, underlying concept formation, heuristic, metacognitive and communicative skills, and motivational and affective aspects of learning, although these outcomes are sometimes less impressive or convincing than one would have hoped, both in terms of the size of the obtained effects and their impact on all groups of pupils including low-ability children and in terms of the methodological rigor with which the reported effects have been obtained (see also Baroody et al., 2004). However, taken as a whole, the available research evidence shows that, to quote Niss (2001, p. 80), "application and modeling capability can be learnt, and according to the above-mentioned findings has to be learnt, but at a cost, in terms of effort, complexity of task, time consumption, and reduction of syllabus in the traditional sense."

Changing Teachers' Knowledge and Beliefs

Another typical characteristic of the last decade's research in the domain of number and arithmetic is the central place accorded to the teacher's knowledge and beliefs and the general acknowledgment of the importance of the teacher, not only in children's learning but also, for example, in curriculum reform and in research (Sfard, 2005). Researchers generally agree that improvement of mathematics education starts with improvement of the mathematical knowledge and the pedagogical content knowledge of teachers (Ma, 1999).

On the other hand, several studies have documented the weak subject-matter and pedagogical content knowledge of teachers. Subject-matter knowledge includes mastery of the key facts, concepts, principles and explanatory frameworks, procedures, and problem-solving techniques and strategies within the given domain of instruction. Crucial in this respect is also the level of teachers' understanding of the domain. There is empirical evidence to support the alarming statement about (future) teachers' insufficient mathematical competence. For instance, in the domain of multiplicative structures, Graeber and Tirosh (1988) studied preservice elementary school teachers' knowledge and skills with respect to multiplication and division involving decimals larger and smaller than 1. They found that a considerable number of preservice teachers made the same errors and shared the same misconceptions as observed in 10- to 12-year-olds. In another study that focused on the connectedness rather than the correctness of prospective teachers' knowledge of division, Simon (1993) found that their knowledge base was weak with respect to several types of connections, such as the conceptual underpinnings of the familiar algorithm of division, the relationship between partitive and quotitive division and between division and subtraction, and the connection between symbolic division and real-world situations to which it is applicable. Verschaffel, Janssens, and Janssen (2005) conducted a large-scale longitudinal study in which the elementary school mathematical knowledge and skills of a large group of Flemish preservice elementary school teachers from 15 different institutes was assessed by means of a paper-and-pencil test that was administered both at the beginning and at the end of their 3-year training. The 30-items test covered the new standards for mathematics in the elementary school curriculum in Flanders. The test was divided into six subsets differing in terms of the curricular subdomain and of the cognitive operations being addressed by the item. The results revealed that at the beginning of their training preservice elementary school teachers have rather weak mathematical competencies. At the end of their 3-year training, the overall test performance had become substantially better, although there were still reasons to be seriously concerned about the readiness of some student teachers to teach mathematics to elementary school children, especially with respect to their modeling and applied problem-solving skills. Finally, Ma (1999) found that the majority of the U.S. teachers (versus only a minority of the Chinese teachers) involved in her study displayed only procedural knowledge of topics of whole number and arithmetic such as subtraction with regrouping and multidigit multiplication. In fact, the knowledge gap between the U.S. and Chinese teachers paralleled the learning gap between U.S. and Chinese students revealed by scholars like Stevenson and Stigler (1992). According to Ma (1999), this limitation in their knowledge confined their expectations of student learning as well as their capacity to promote conceptual learning in the classroom. Given that the parallel of the two gaps is, according to Ma (1999), no coincidence, it follows that "while we want to work on improving students' mathematics education, we also need to improve their teachers' knowledge of school mathematics" (p. 144).

The second category of teachers' knowledge—teachers' pedagogical content knowledge—includes several subsystems such as knowledge of mathematics lesson scripts and mathematics teaching routines, knowledge about the kinds of problem types, graphical representations, and so on. that are best suited to introduce particular mathematical notions and skills to pupils, and knowledge of instructional materials available for teaching various mathematical topics. Several studies have documented particular weaknesses in (future) teachers' pedagogical content knowledge, mostly in close relation to weaknesses in their mathematical knowledge base (Ma, 1999; Van Dooren, Verschaffel, & Onghena, 2002; Verschaffel et al., 1997).

Other studies have focused on the impact of teachers' mathematical knowledge on students' learning processes and outcomes. Taken as a whole, the majority of the results are in favor of the old adage, which is also included in the recent publication *Adding It Up* (Kilpatrick et al., 2001) and is echoed in Ma's (1999) book, namely that you cannot properly teach what you do not know yourself.

Given the well-documented relationship between these aspects of teachers' professional knowledge, on the one hand, and students' learning processes and outcomes, on the other, clearly without an improvement in the state of preparedness of the teaching body, calls for teaching with, or for, understanding such as those contained in the U.S. *Standards* (NCTM, 2000) and in similar reform documents are simply doomed. Therefore, researchers have begun to address the question of whether and how (preservice) teachers' cognitions and conceptions can be influenced. These intervention studies differ widely in

- the aspects of (preservice) teachers' cognition addressed (e.g., mathematical content knowledge, pedagogical content knowledge, beliefs about mathematics and mathematics teaching)
- the amount and kind of training provided (e.g., short interventions, interactive video, intensive workshops, workshops followed by classroom visits, and afterschool working sessions)
- the number of preservice teachers involved
- the way training effects were determined (e.g., teacher cognitions, teaching behavior, student outcomes).

In an unusually systematic and sustained intervention study of this kind, Carpenter, Fennema, Peterson, Chiang, and Loef (1989; see also Carpenter & Fennema, 1992) investigated whether it is possible to improve teachers' instructional practice and their students' learning outcomes by confronting them with the available research-based knowledge of the development of children's skills and processes for solving addition and subtraction (word) problems described in ascertaining studies. They called their approach *cognitively guided instruction* (CGI). Basic ideas underlying the CGI approach are that the teaching/learning process is too complex to specify in advance, and that teaching as problem solving is mediated by teachers' thinking and decision making. Consequently, significant changes in educational practice can best be pursued "by helping teachers to make informed decisions rather than by attempting to train them to perform in a specified way" (Carpenter & Fennema, 1992, p. 460). Twenty first-grade teachers participated in a 1-month long workshop that attempted to familiarize them with the available research findings on learning and the development of addition and subtraction concepts and skills. Teachers learned to classify problems, to distinguish levels of mastery, to identify strategies that young children use to solve different problem types, and to relate those strategies to the mastery levels and problem types in which they are commonly used by children. Afterwards teachers discussed principles of instruction derived from research and designed their own programs of instruction based on those principles. Based on a large set of data-gathering techniques, including classroom observations, measurement of teachers' cognitions and beliefs, and tests of student learning outcomes, the following findings were obtained:

- CGI teachers taught problem solving significantly more and number facts significantly less than control teachers
- CGI teachers encouraged pupils to use a variety of strategies, and they listened to the processes their students used significantly more than did control teachers
- CGI teachers believed that mathematics instruction should build on students' existing knowledge more than did control teachers, and they knew more about individual students' problem-solving processes
- Pupils of CGI teachers outperformed control pupils in number-fact knowledge, problem solving, reported understanding, and reported confidence in problem solving.

Notwithstanding that several unanswered questions remain, these studies demonstrate that sharing with teachers research-based knowledge about students' thinking and problem solving can profoundly affect teachers' cognitions and beliefs about arithmetic classroom learning and instruction, their classroom practices, and, most important, their students' learning outcomes and beliefs. As such, these positive findings provide evidence for the relevance of the constructive conception and the dispositional view of mathematics learning and teaching, which is complementary to one already presented earlier. At the same time, these studies demonstrate the feasibility of this new conception in the setting of a typical elementary mathematics classroom. In light of the extremely time-consuming nature of the knowledge-sharing approach followed in CGI and other similar projects, the question remains whether such an approach can be applied on a large scale. New information technology, such as simulation programs and interactive video, can probably play an important role in helping inservice and preservice teachers interact with CGI or related ideas.

A quite different approach to professional development is followed in the Lesson Study project, the core form of professional development for Japanese teachers (Lewis, 2002). In the Lesson Study cycle, teachers work together to

- formulate goals for student learning and long-term development;
- collaboratively plan a "research lesson" designed to bring to life these goals;
- conduct the lesson, with one team member teaching and others gathering evidence on students' learning and development;
- discuss the evidence gathered during the lesson, using it to improve the lesson, the unit, and instruction more generally;
- teach the revised lesson in another classroom, if desired, and study and improve it again.

Research lessons occur in many settings in Japan. Lesson Study is credited for the shift from "teaching as telling" to "teaching for understanding" in Japanese mathematics education and is highly valued by both Japanese teachers and administrators. However, except for Yoshida's (1999) case study of the impact of the Lesson Study approach on teacher development and improved mathematics instruction in one Japanese school, we know of no research-based evidence (of the type described for CGI) for the impact of this approach to improving instruction through teacher development.

CULTURE, SOCIETY, SCHOOLING, AND ARITHMETIC

Introduction

The purpose of this section is to situate the research summarized in the first two sections in wider human contexts—historical, social, cultural, and political. Although these perspectives are, of course, applicable to mathematics education as a whole, their application to early arithmetic is of particular interest and importance for many reasons, including the following:

- the human (near-)universality of natural numbers, operations on them, and applications
- the importance of early arithmetic in a child's (mathematical) education—not just in the sense that it forms the basis for later mathematics, but also that the child's experience in the early years of formal schooling is foundational for the development of her or his conception of, and disposition towards, mathematics
- the popular perception of arithmetic as the central aspect of mathematics
- the philosophical and ideological debates that are often fought out over this terrain and that influence both mathematics education policy and research on mathematics education.

To begin with, it is appropriate to make some comments from this perspective on the studies summarized in the first two sections of this chapter. In respect of the types of experimental investigations described as ascertaining studies, and as stressed several times during that section, we should avoid any suggestion of a universal developmental trajectory for the development of numerical and arithmetic cognition. As with Piagetian theory, this implies careful acknowledgment that any results are relative to a specific cultural environment, particularly, but not limited to, the nature of instruction in arithmetic and, more generally, schooling and to the cognitive tools (including language) available. This point may seem obvious if, for example, one does a thought experiment on computation with Roman numerals, yet arguably its implications are frequently not given enough weight.

These remarks are clearly applicable to cognitive models such as SCADS, discussed earlier. As commented there, a major shortcoming of this and related models is that "little or no attention is given to the social, especially instructional context in which the development of arithmetic skills takes place." More-over, as stated on p. 565, "many of these (computer) models seem to assume that there is a kind of universal taxonomy and/or developmental sequence of computational strategies which is fundamentally independent of the nature of instruction or the broader cultural environment." In this respect, the model of Baroody and Tiilikainen (2003), although lacking the precision highly valued by many in cognitive science, does acknowledge "the crucial role of the broader socio-cultural and instructional contexts." As one very straightforward example among several mentioned by Fuson (2003), the emphasis on commutativity in Chinese teaching of multiplicative single-digit combinations is reflected in reaction-time patterns of Chinese adults (LeFevre & Liu, 1997).

Although work within neurosciences is clearly a growing area for the scientific study of mathematical cognition (see Campbell, 2004, for a recent overview of this research), it is limited in several respects. Though there are aspects of the physical environment common to all humans, such as affordances for counting, and although there are common evolutionarily produced phenomena in the development of the brain, the aspects of mathematical cognition that are the products of biological evolution are minimal relative to those that are the products of cultural evolution. The conclusion by Delazer (2003, p. 404) that "recent neuropsychological studies provided clear evidence that different types of knowledge are involved in arithmetical processing and that they are functionally independent and separately implemented in the human brain" may be interesting, but its implications and utility for teaching arithmetic have as yet not been demonstrated. Moreover, the neuropsychological work is largely limited to context-free arithmetic executed with routine expertise, whereas the research on even the addition and subtraction of small whole numbers in context shows the semantic and cognitive complexity of such operations. As Lakoff and Nunez (2000, p. 26) put it, we know something about what parts of the brain are active when we use different arithmetical capacities, and we have some idea of which of these capacities are innate (Butterworth, 1999), but "compared to the huge edifice of mathematics, [that] is almost nothing."

We may suggest, therefore, that the explanatory power and educational utility of such cognitive and neuroscientific studies would be very significantly increased if they took into account the specifics of instructional environments and aimed to show how those differentially affect mathematical thinking and performance and how they align (or not) with hard-wired cognition. For example, is it possible to relate instruction based on representations of the number line—and the historical record of such representations and their use—to a theory that number is innately so represented in the brain?

Another inherent limitation of the work from cognitive science is that it is typically limited to the simplest arithmetic, while often making claims that it is—in principle—extendable. Yet it seems obvious that the more complex the arithmetic the more divergence may be expected due to cultural/instructional diversity. By way of example—and still without penetrating at all far into the edifice of mathematics—we may cite the profound effects of linguistic regularity/irregularity for number names from 11–100 documented by Fuson (2003).

A common theme of recent research (as in many of the ascertaining studies reviewed in the first section, such as those of Carpenter et al., 1997) is the invention by children of strategies that they are not directly taught. This phenomenon may be taken as a strong indication of the pervasive influence of cultural context beyond direct instruction, such as the representation of arithmetic operations and applications in children's social worlds outside of school. In the *Classroom-Based Research* section, several examples show how the accumulation of these discoveries has led to increasing consideration of how instruction should take account of, encourage, and be aligned with these inventions.

Generally, in the second section, examples of specific approaches made clear that the nature of the instructional environment is constitutive of mathematical cognition and that what happens in the mathematics classroom is about many more and much more profound things than knowledge of number facts and computational competency, including laying foundations for the student's view of mathematics, as conveyed through sociomathematical norms, in particular (in the third part of that section).

Beyond the physical and cultural environments, it is necessary to consider *designed* environments. The shift from ascertaining to intervention studies provides case studies of precisely how the culture of the mathematics classroom (Seeger, Voigt, & Waschescio, 1998) influences both what mathematics is learned by students and what they learn *about mathematics*. Many of the characteristics common to the representative design experiments discussed in the *Classroom-Based Research* section reflect a cultural perspective, particularly the fostering of mathematical discourse and sociomathematical norms as forms of enculturation and attention to sense-making connected to students' lived experiences outside of school.

Cultural Perspectives on Arithmetic

Sociohistorical Aspects of Number and Arithmetic

Crump (1990, p. 146) summarized the overview of his book *The Anthropology of Numbers* by concluding that "the series of natural numbers, together with the basic arithmetical operations of addition and subtraction, multiplication and division, are a resource open to use in almost any culture." However, this comment should be clarified by the recognition that the application of the "resource" is highly variable across cultures and throughout history. Many of the uses of arithmetic that are taken for granted in U.S. and European societies are actually reflections of a particular contemporary worldview. In 1985, the Brazilian scholar Ubiratan D'Ambrosio introduced the term and concept of *Ethnomathematics*, a field that he defined as lying on the borderline between the history of mathematics and cultural anthropology. Subsequent work on ethnomathematics has had a variety of foci, notably the demonstration of the pan-cultural nature of mathematics (Ascher, 2002; Bishop, 1988; Selin, 2000), in particular to counterbalance the prevailing Eurocentric view of the development of mathematics (Joseph, 1992; Powell & Frankenstein, 1997).

Although much of ethnomathematical work has been concerned with the mathematics of non-Western peoples, the breadth of the conception of ethnomathematics is clear from the definition given by D'Ambrosio (1985, p. 45): "the mathematics which is practised among identifiable cultural groups, such as national-tribal societies, labor groups, children of a certain age bracket, professional classes, and so on." Gerdes (1994, p. 20) offered an equally broad characterization as "the field of research that tries to study mathematics (or mathematical ideas) in its (their) relationship to the whole of cultural and social life." These broad definitions clearly suggest strong theoretical links between ethnomathematics and varieties of sociocultural theory, notably situated cognition and activity theory.

The combination of historical and pan-cultural perspectives offered by ethnomathematics provides a useful corrective to tendencies to assume stability and universality of mathematics discernible in Western mathematics education; however, these issues lie beyond the scope of this chapter. Staying more focused, we find several reasons why the issues raised here are relevant to the teaching of whole-number arithmetic:

First, mathematics is a human activity. The study of arithmetic across time and cultures (Swetz, 1994) and stories about insights into the structure of natural numbers by gifted individuals such as Gauss and Ramanujan (Wells, 1997, p. 44 and p. 77, respectively) afford an early introduction for children to an understanding of mathematical creativity and mathematics as part of culture.

Second, the historical and pan-cultural lore of number and arithmetic, including representations, algorithms, pattern analysis, games, and puzzles is a rich source for interesting mathematics (e.g., Bazin, Tamez, & The Exploratorium Teacher Institute, 2002; Conway & Guy, 1996; Selter, 2001; Wells, 1997).

Third, an appreciation of the contributions of all cultures to arithmetic (more generally, mathematics) is part of culturally appropriate instruction in countries around the world (including those in a post-colonial phase) that acknowledges the contributions of all peoples (e.g., Stillman & Balatti, 2001; Zaslavsky, 1973).

Fourth, similarly, an appreciation of the contributions of other cultures and cultural variation in doing arithmetic is part of appropriate instruction for the increasingly multicultural classes of the United States, many European countries, and elsewhere (e.g., Orey, 2005; Zaslavsky, 1991).

Arithmetic in School and Out

Arithmetic is not something children do only in school. Before they go to school, children are engaged in many activities with number and arithmetic, which they learn in ways that are in certain respects similar to how they learn their native tongue, namely in context and, very often, with an older person to provide instant feedback. So, by the time a child goes to school, he or she already has a lot of foundational knowledge in place (Clements & Sarama, this volume; Kilpatrick et al., 2001, chapter 5).

Encounters with number and arithmetic in children's lives outside of school are semantically and contextually much richer and varied than the narrow focus of formal education in arithmetic. Consider what a child entering school (in the United States, for the purposes of our example) typically knows about the number 5—what the numeral looks like, how to say the word for 5 in their language, five fingers on each hand, 5 objects, 5 people, 5 inches, what it means to be 5 years old (which is very complex inasmuch as there is "I am five today," perhaps represented by 5 candles on a cake, as opposed to "I am five," which probably means between 5 and 5:6, because a child of 5:6 is likely to say "I am five and a half"), five cents/dollars as opposed to a nickel/five dollar bill and their equivalence for some purposes, 5 as part of a date or a time (think about the complexity of interpretations of the number 5 on an analog clock), 5 as the number of a house or a bus, fifth in a series, . . . and on and on. In short, 5 provides a model for a very wide variety of situations familiar to a child.

As formal education proceeds, the connections between school mathematics and children's lived experiences and sense-making become weaker. A dramatic example was discussed in the *Word Problems* section. Children in France were prepared to give answers to questions such as "There are 26 sheep and 10 goats on a ship. How old is the captain?" (Baruk, 1985). Many other examples, in many countries (Verschaffel et al., 2000), point to an apparent effect whereby many children, after some years of formal education in mathematics, are inhibited from saying "that does not make sense" and learn to play the word problem game. Many mathematics educators are making attempts to tackle the perceived lack of connection between math in and out of school; several examples were discussed earlier. Increasing emphasis on data handling, and on modeling and applications, in elementary school is another development contributing to this movement.

Theoretical frameworks have also been brought to bear. Lave (1988, p. 74) used word problems as "a microcosm of theories of learning" in order to suggest "a theory of situated learning as a possible next step in understanding the learning and teaching of mathematics," She argued (p. 81) that

> math in school is situated practice: school is the site of children's everyday activity. If school activities differ from the activities children and adults engage in elsewhere, the view of schooling must be revised accordingly; it is a site of specialized everyday activity—not a privileged site where universal knowledge is transmitted.

A related theoretical framework leads us to consider school mathematics as an activity system (Engestrom, 1987; Engestrom & Miettinen, 1999). Starting from Vygotsky's theory of mediated acts, Engestrom elaborated a triangular structure that he applied to the analysis of many and varied activity systems, in particular that of school-going (Figure 13.7, Engestrom, 1987, p. 103). Although this figure presents an analysis of school-going in general, it applies very appositely to mathematics education in particular. It offers a powerful theoretical lens for analyzing the teaching/learning of elementary arithmetic, including the use of physical and symbolic tools (Cobb, Yackel, & McClain, 2000; Gravemeijer, Lehrer, Van Oers, & Verschaffel, 2002). Moreover, much of the work described in the *Classroom-Based Research* section exactly fits the suggestion by Engestrom and Miettinen (1999, p. 3) that

> the type of methodology I have in mind requires that general ideas of activity theory be put to the acid test of practical validity and relevance in interventions

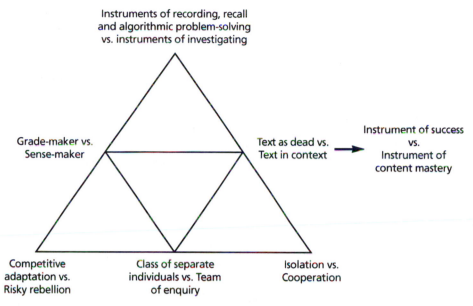

Figure 13.7 The activity of school going (Engestrom, 1987, p. 103).

that aim at the construction of new models of activity jointly with the local participants.

Although as yet only limited research directly uses this theory, it may be suggested as a powerful resource for future work, as a perspective that, for example, "overcomes and transcends the dualism between individual subject and objective societal circumstances" (Engestrom & Miettinen, 1999, p. 3). Moreover, it is in alignment with recent interest in contextual and cultural theories in cognitive science, as shown by the concepts of situated cognition and distributed cognition (Salomon, 1993), and in education in the idea of situated learning in communities of practice (again, as reflected in many of the case studies described in the second section).

Cultural and Educational System Differences

As exemplified at various points in the first two sections of this chapter, many research findings deal with differences in arithmetic teaching and learning that can be attributed to specific or general cultural differences. Pervasive aspects of the culture milieu, such as attitude to learning, apply more broadly to all education, and we will not attempt to cover them here. On the other hand, it is appropriate to deal with language, insofar as there are clear and specific ways in which its influence on early arithmetic teaching and learning can be traced. Then we will turn to aspects of different educational systems, without attempting the complex task of analyzing in depth the varying extents to which these differences are reflections of cultures *per se.*

Effects of Linguistic Differences on Early Arithmetic

A strong and specific effect on learning and teaching whole-number arithmetic is the observation that languages differ in the extent to which the structure of the whole numbers is mirrored in the sytem of words used to name them. In particular, Chinese, Japanese, and Korean (all derived from ancient Chinese) systematically name the whole numbers from 11 to 100, whereas European languages do not (Fuson & Kwon, 1992; Miura & Okamoto, 2003). Consequent differences have been noted, as early as kindergarten, in development of verbal counting (Miura & Okamoto, 2003, p. 233). In another comparison, early first-graders (who had no prior experience with base-ten blocks or instruction in using them) were asked to represent 42. The children from the United States, Sweden, and France were most likely to use 42 unit blocks, whereas those from China, Japan, and Korea generally used 4 ten-blocks and 2 units (Miura & Okamoto, 2003, p. 231). It is at least highly plausible that the linguistic differences, at least partially, underlie better understanding of place value by Chinese, Japanese, and Korean first-graders as opposed to those from the United States, Sweden, and France (Miura, Okamoto, Kim, Steere, & Fayol, 1993).

The systematicity of the naming system may be one reason for, and certainly facilitates the use of, early calculating methods using the "make a ten" method. Combined with making the connections beween 6, 7,

8, 9 and 5 + 1, 5 + 2, etc., this method is taught in first grade in China, Japan, Korea, and Taiwan (Fuson & Kwon, 1992). As a result, most children in these countries, by the end of first grade, can rapidly carry out all single-digit additions. The method is also taught in some European countries, and Fuson strongly advocated using it in first grade in U.S. schools, instead of restricting almost all of first grade to the learning of addition and subtraction below 10. Further strong motivation for adopting this approach is that it lays a strong conceptual foundation for two-digit addition and subtraction in general, and for the understanding of place value in relation to multidigit addition and subtraction. In Liping Ma's (1999) study of Chinese and American elementary teachers, she reported that most of the Chinese teachers who were asked about two-digit subtraction sums referred to the principle of "decomposing a unit of higher value" which is related to abacus use (Ma, 1999, p. 7) and which formed a coherent "knowledge package" (p. 19) relevant in many arithmetical procedures. By contrast, the majority of American teachers showed only procedural knowledge of subtraction with regrouping.

Differences in Educational Systems

As commented by Towse and Saxton (1998), a clear pattern of findings seems to show that children from Asian countries such as Japan, Korea, China, and Singapore outperform those in the United States and Europe (even though the picture is complicated by the methodological quagmire of cross-cultural comparisons and often contradictory results). A considerable amount of research has been undertaken to try to find reasons for these differences and thereby generate ways to improve mathematics education. In addition to specific linguistic factors, possible explanations considered by Towse and Saxton include the influence of preschool education, quality and quantity of teaching, differences in problem-solving strategies, and cultural attitudes to learning mathematics.

However, as remarked by Kilpatrick et al. (2001, p. 31), "In every country, the complex system of school mathematics is situated in a cultural matrix." Thus, as stated by Stigler and Hiebert (1999, p. 103), "teaching is a complex cultural activity that is highly determined by beliefs and habits that work partly outside the realm of consciousness." Awareness of this cultural embedding is needed, accordingly, to temper the temptation to uncritically import aspects of other systems that appear to contribute to success. For example, describing the outcomes of the Third International Mathematics and Science Study (TIMSS), Kilpatrick et al. (2001, p. 32) warned that comparisons of practices, programs, and policies in the United States with those of high-achieving countries "at best can only be suggestive of the sources of achievement differences."

On the other hand, the example of linguistic factors discussed above, with ramifications spreading upwards to the learning of multidigit computation, suggests that a careful analysis may lead to beneficial results. Another line of research that may well have had tangible benefit is the comparative studies of textbook treatments of whole-number arithmetic, of which an early and influential example was Stigler, Fuson, Ham, and Kim (1986).

An interesting study of the effects of different teaching approaches was reported by Anghileri, Beishuizen, and Van Putten (2002). They explored the written calculation methods for division used by pupils in England and the Netherlands at two points in the same school year. The test involved a series of five calculations and five word problems involving both single-digit and two-digit dividers. Informal strategies were analyzed, and progression towards more structured procedures were identified. Whereas the English approach was characterized by a discontinuity between the formal computational procedure being currently taught and informal solution methods used earlier, the Dutch approach was based on careful progression from informal strategies towards more formal and efficient procedures. Comparison of the methods used by Year 5 pupils in the two countries showed greater success in the Dutch approach on the tests. With respect to strategies, the traditional algorithm was widely used during both tests by English children (with rather low level of success), whereas Dutch children most frequently applied high-level chunking strategies (with rather high levels of success). Although some methodological problems warrant caution, this comparative study can be considered as evidence that learning is more effective when written methods build upon pupils' intuitive understanding and informal strategies in a progressive way rather than by abruptly replacing them with standard procedures.

Political/Ideological Aspects

Towards Evidence-Based Reform: Standards-Based Elementary Curricula

Research on mathematics education, to an unprecedented extent, has become a political issue especially, but by no means exclusively, in the United States (see, e.g., Ginsburg, Klein, & Starkey, 1998). Thus, Kilpatrick (2003, p. 471) pointed out that "research on teaching and learning has become a weapon in the so-called math wars over *Standards*-based school mathematics curricula." As mentioned in the introduction to this section, elementary arithmetic, for several reasons, is one of the major battlegrounds in that war.

One response to the vociferous criticism in some quarters following publication of the *Curriculum and Evaluation Standards for School Mathematics* (National Council of Teachers of Mathematics, 1989) was to encourage the evaluation of a number of elementary, middle, and high school programs based on the *Standards* (Senk & Thompson, 2003) in the spirit of "evidence-based policy making." Given the importance of textbooks (Kilpatrick et al., 2001, pp. 36–37), the relative paucity of evaluative research may be considered surprising.

Part II of *Standards-Based School Mathematics Curricula* (Senk & Thompson, 2003) includes evaluations of four elementary grades curriculum projects: *Math Trailblazers*, *Everyday Mathematics*, *Investigations*, and *Number Power*. These evaluations are of textbooks in widespread use, in contrast with most of the examples in the second section of this chapter (with the exception of the work stemming from the Freudenthal Institute). There are inevitably problems in conducting tightly controlled experimentation within the constraints of real schools (Hiebert, 1999). Nevertheless, in an overview of the studies Putnam (2003) concluded that they provided convincing evidence of beneficial effects in that the students in the experimental programs performed at least as well on conventional standardized tests, and better on custom-made tests assessing various other aspects targeted by the projects, such as mental arithmetic and number sense. Moreover, collectively these curricula cover a wider range of topics than has historically been the case, in particular data analysis, modeling, and applications, including in science.

Looking to the future, Kilpatrick (2003, p. 485) identified as major concerns for further development of such evaluative research "a deeper examination of how teachers are using the curriculum, the inclusion of comparisons across new curricula, and the development of more instruments to measure ambitious learning goals." The difficulties of evaluating curricula have been starkly exposed in a report prepared for the National Research Council (Confrey & Stohl, 2004), which concluded (p. 189) that

> The corpus of evaluation studies across the 19 programs studied does not permit one to determine the effectiveness of individual programs with high degree of certainty, due to the restricted number of studies for any particular curriculum, limitations in the array of methods used, and the uneven quality of the studies.

The report makes recommendations that prompted one commentator (Steen, 2004) to remark that "the standards that the committee set for future evaluations to meet the test of 'scientific validity' are generally sound, although taken together they pose hurdles that are both administratively and financially impossible to achieve."

Philosophical Basics: What Does 2 + 2 = 4 Mean?

$$2 + 2 = 4$$

"There is a mathematically correct answer"

(Masthead for website of Mathematically Correct)

However, it is not so simple. Hersh (1997, p. 16) offered a clarification that we find helpful:

> So "two" and "four" have double meanings: as Counting Numbers or as pure numbers. The formula:
>
> $$2 + 2 = 4$$
>
> has a double meaning. It's about counting—about how discrete, reasonably permanent, noninteracting objects behave. And it's a theorem in pure arithmetic.

The quotation from "Mathematically Correct" implies that only the second meaning suggested by Hersh counts as part of mathematics. However, even theorems in pure arithmetic are not such a simple matter. Again, we can turn to Hersh (1997, p. 16) for enlightenment:

> *Fact 1:* Mathematical objects are created by humans. Not arbitrarily, but from activity with existing mathematical objects, and from the needs of science and daily life.
>
> *Fact 2:* Once created, mathematical objects can have properties that are difficult for us to discover.

Thus $2 + 2 = 4$ may be thought of as a property of the whole numbers that is easy to discover, in contrast to, say, Fermat's last theorem. The set of properties discovered is, of course, time-dependent as are interpretations of those properties (straying outside the positive whole numbers momentarily, the history of interpretations of the equation $-1 \times -1 = 1$ is particularly interesting). Also varying across time are the many different formal interpretations of the system of whole numbers (discussed earlier).

Our preferred description of the first meaning of $2 + 2 = 4$ suggested by Hersh is that it is a mathematical statement that may or may not provide an adequate model for some aspect of reality. In relation to statements about natural numbers used to model aspects of the real world, clearly 2 + 2 = 4 is *not* an appropriate model for many situations that can be differentiated from situations for which it *is* an appropriate model only by extramathematical interpretations of the real world (Usiskin, 2004, and see Verschaffel et al., 2000, chapter 7, for more examples), and such interpretations are certainly culturally dependent, as the ethnomathematical literature in particular shows (e.g., Ferreira, 1997). For example, Ascher and Ascher (1997, p. 29) related an oft-told anecdote about an African sheepherder and "someone else variously described as an explorer, trader, scientist, anthropometrist, or ethnologist." The herder agreed to trade two sticks of tobacco for one sheep but became upset when offered four sticks of tobacco for two sheep. As pointed out by Ascher and Ascher, from one perspective the herder does not understand that 2 + 2 = 4, whereas from another, the scientist/trader does not understand sheep, which are not standardized units.

The above is not merely an abstract discussion about philosophy; it has implications for the realpolitik of teaching arithmetic, because the beliefs of those who exercise power in educational policy-making and influence public opinion impact what happens in schools. By way of illustration, at the end of his book *Entering the Child's Mind*, Ginsburg (1997, p. 228) quoted an article from the *San Francisco Chronicle* (Saunders, 1994):

> If the proposed framework is approved, California's third graders could be taught that even math is relative. [The program] would teach children that there are different "strategies" for solving simple equations. Instead of memorizing 5 + 4 = 9, the program would tell children to find creative ways to find the answer, such as "I think of 5 + 5, because I know that 5 + 5 = 10. And 5 + 4 is 1 less, so it's 9." . . .
>
> Maureen DiMarco, Governor Wilson's point person on education and candidate for state schools superintendent, dismisses new-new math as "fuzzy crap."

The point of view represented here fits clearly the philosophical stance labeled "dualism" by Baroody et al. (2004, p. 228). It is beyond the scope of this chapter to unpack the many implications of the rhetoric (note the title of the article) and the underlying ideologies (of both the reporter and the candidate for state schools superintendent). The reader will see that the dismissal of encouraging *thinking* is incompatible with the views expressed in this chapter, yet the expression and influence of such views are part of the political reality of mathematics education today, and not just in the United States. In his comments, Ginsburg (1997,

p. 228) starkly characterized the choice as "between fundamentalism and modernity."

SALIENT THEMES AND CHALLENGES

In this final section, we highlight overarching themes of our review, followed by what we see as major challenges facing the field. From many points of view— what kind of research is needed, what it means to have a profound understanding of whole-number arithmetic, how arithmetic fits within the edifice of mathematics, the need to improve understanding of the issues among the general public and those wielding power, and the appropriate preparation and continuing professional development of elementary teachers—the central messages are that the learning and teaching of whole-number arithmetic are extremely complex, culturally embedded, and foundational for every person's mathematical education.

Major Themes of the Research

Complementary Approaches

The contrast between the first two sections of our review of recent research reflects a major development to complement studies predominantly carried out with individual children, largely within the broad family of cognitive perspectives and methodologies, with extended classroom-based action research and design studies within the family of sociocultural theoretical frameworks. Note that the use of the word *complement* is intended to signal our belief in the continuing valuable contribution of cognitive analyses (cf. Greer, 1996).

This contrast corresponds to that between the two broad theoretical orientations, with associated research methodologies, identified by Sfard (2003, p. 355), namely the aquisitionist and the participationist group, that were characterized as follows in the introductory section of this chapter (see p. 558).

> The "acquisitionist" group consists of those traditional cognitivist approaches that explain learning and knowledge in terms of such entities as cognitive schemes, tacit models, concept images, or misconceptions. The "participationist" framework embraces all of those relatively new theories that view learning as a reorganization of activity accompanying the integration of an individual learner with a community of practice.

The idea of enculturation into "*the* mathematical community" had a vogue, but, if interpreted too narrowly, is problematic because what children do in school is very different from what practicing and research mathematicians do. However, "integration with a community of practice" is a more general conception, consistent with D'Ambrosio's (1985) definition of ethnomathematics as the mathematics of cultural groups, and Lave's (1992) characterization of school mathematics as a form of situated practice.

Research within the first framework follows the traditional approach that, through systematic enquiry, yields evidence for theoretical descriptions of the development of cognitive processes underlying the identified elements of competence in the domain. The aim and expectation is that this theoretical knowledge can subsequently be applied by teachers, curriculum developers, policymakers and, indeed, those engaged in classroom-based research. Thus it follows what Greeno (2003, p. 312) called the RDDE (research, development, dissemination, evaluation) model in which the components follow a linear temporal trajectory.

By contrast, much of the work surveyed in the second section follows a cyclical or intermeshed pattern of research and development that has been called "developmental research" (Gravemeijer, 1994, chapter 4; Gravemeijer, 2004). Related terms include *design experiments* (Greeno et al., 1996), *interactive research and design* (Greeno & the Middle-School Mathematics Through Applications Project Group, 1998), and *hypothetical learning trajectories* (Clements & Sarama, 2004a). A major advantage of such work is that it greatly reduces the gap between research and practice. However, in most cases the huge task remains of scaling up—from existence proofs with exceptional teachers given strong support—to the generality of teachers, as Selter (2002) commented (see Burkhardt & Schoenfeld, 2003). The work of the Freudenthal Institute stands out in this regard as perhaps the best example of a long-term and large-scale implementation that has substantially impacted mathematics education at a national level.

The contrast between the first and second sections is correlated with divergent aims between psychologists and mathematics educators (De Corte et al., 1996, p. 492). In particular, psychologists who take mathematical cognition as a domain for study tend to accept mathematics education as it is, whereas mathematics educators are likely to want to change it.

The Multifaceted Nature of Proficiency

Perhaps the most salient observation about research on learning/teaching arithmetic in recent years is that it reflects a much more complex and sophisticated view of the domain in many respects. For example, Kilpatrick et al. (2001) characterized mathematical proficiency, with particular emphasis on whole-number arithmetic, using the metaphor of intertwined strands representing different facets of a complex whole: conceptual understanding, procedural fluency, strategic competence, adaptive reasoning, and productive disposition. They stressed that (p. 116): "the five strands are interwoven and interdependent in the development of proficiency in mathematics." This conception represents an enrichment of the earlier recognition that conceptual understanding and procedural competence interact in complex ways (Hiebert, 1986)—including the case when understanding is retrospective and facilitated by previous routine use of procedures (Sfard, 2003).

Much of the enrichment, relative to educational policies and goals predominantly restricted to the second strand of procedural fluency, is captured by Hatano's distinction between routine and adaptive expertise (see earlier in this chapter). In the first section, we reported on numerous observations of children's spontaneous adaptive expertise—successively streamlined methods for calculations, increasing sophistication of strategy choice, flexible mental arithmetic and number sense, contextual interpretation of outcomes of computational work, awareness and exploitation of structure. . . . Correspondingly, many of the examples in the second section amount to the construction and evaluation of designed instructional environments that foster adaptive expertise, in particular by harnessing the creativity implicit in Hatano's characterization.

Long-Term Developmental Perspective

Long-term developmental has two meanings with respect to the cultural/historical development (phylogeny) of whole-number arithmetic and with respect to the development for an individual child in an instructional environment and a social environment outside of school (ontogeny).

Using natural numbers and operations upon them is a human activity that is pan-cultural and reaches several millennia back into the past. Although it may be psychologically difficult to avoid a Platonic view of the natural numbers (the name itself is indicative and the mathematician Kronecker famously said that the natural numbers are the work of God, whereas all else is the work of man), the creation of symbol systems to record counted numbers efficiently, and the invention of physical devices and algorithms for calculating with them, represent a cultural activity of constructing and using tools. This cultural-historical perspective is a necessary backdrop when considering development of understanding of arithmetic in today's student, as Freudenthal, in particular, has argued (1991, p. 47): "Traditionally, mathematics is taught as a ready made subject. Students are given definitions, rules and algorithms, according to which they are expected to proceed. Only a small minority learn mathematics this way."

As we have described, the long arc of children's experience with whole-number concepts begins before school and continues in their social lives outside of school, in parallel with formal instruction. Even if we restrict attention to whole-number arithmetic, the complexity of this development demands a coherent long-term theoretical framework. Further, when we consider this period in a wider context, a comprehensive conception is needed, such as Vergnaud's (1996) characterization of the additive and multiplicative conceptual fields, and their synthesis, with a view to the further development of the concept of number. Further elements include the conceptual and not merely procedural extension of the operations to domains beyond the natural numbers (Greer, 1994; Verschaffel & Vosniadou, 2004), acts of generalization that foreshadow formal algebra (Kaput, 1998), and understanding of the mathematical structures in which numbers and operations are embedded.

In terms of the instruction that such a theoretical framework should inform, we have seen, through several examples, that aligning instruction with the child's spontaneous creations, problem-solving, and sense-making—under guidance, rather than presenting it as "ready-made"—has beneficial results. An example already mentioned is the study by Anghileri et al. (2002) in which Dutch students who were taught in accordance with the principle of progressive schematization outperformed English students who were directly taught standard algorithms.

A common feature of such approaches is that they do not proceed by systematic incremental changes in superficial features such as the size of the numbers. Rather, they operate at the level of larger ideas, in the spirit of the alternative to traditional scope and sequence termed a *distributed curriculum* by Resnick (1992, p. 423), which calls for and fosters well-connected knowledge and adaptive expertise (cf. Ambrose et al., 2003, p. 330, and see the notion of "big ideas" in Baroody et al., 2004).

Sense-Making and Modeling

One kind of sense-making in this context is when doing disembedded calculations where the sense to be made is of the structures that relate numbers and operations upon them, without explicitly modeling aspects of the real world (Hiebert et al.,1997, predominantly addresses this form of sense-making). However, modeling with a different meaning (Greer, 1994, p. 76; Greer, 2004) is often involved in that those structural relationships may be represented using physical (including symbolic) embodiments, such as blocks (unitary or structured), fingers, number line, abacus (see *Multidigit Arithmetic* section). These embodiments fall into two broad classes, discrete and continuous. The use of collections of uniform, small, hard, permanent objects has the advantage of being immediately accessible to counting expertise, and such a collection becomes internalized as a generalized image for natural numbers and operations on them. Further, the objects can be structured in order to accommodate the decimal system for the whole numbers. On the other hand, the number line has the great advantage that it extends naturally to rational numbers, directed numbers, and real numbers (and eventually, with an orthogonal number line, to both complex numbers and Cartesian graphs). The decimal system has the same design strength in that the extension of computations to rational numbers, decimally represented, is relatively straightforward (but the conceptual restructuring is certainly not; Greer, 1994), although the procedural as well as conceptual problems of extension to rationals represented as fractions are well documented.

As reviewed in the *Single-Digit Computation* section, children first learning to compute, as an extension of counting, are heavily dependent on external devices including their fingers. We have argued and presented evidence for the assertion that the trajectory from that point to the point at which they can do multidigit computations should not require the abandonment of sense-making.

What we might call the mathematician's approach is exemplified by Wu (1999) who characterized conceptual understanding of multidigit computations in terms of realizing how they are based on laws of arithmetic such as associativity and commutativity. He also showed how those laws themselves can be grasped as intuitively accessible generalizations through the use of arrays of dots. From this perspective, the title of his paper (rightly, in our opinion) labels "basic skills versus conceptual understanding" as "a bogus dichotomy in mathematics education" (Wu, 1999, p. 1).

However, as Wu (1999, p. 5) pointed out, the standard algorithms represent "efficient compression of a valuable piece of *mathematical reasoning* into a compact shorthand" and many mathematics educators believe that a more promising route—with much the same destination in view—is to reach this level of compression more gradually and more in concert with children's intuitive sense-making. Thus, according to Treffers (1987, p. 267) "The fact that the standard algorithm is taken as the final product of a process of algorithmisation rather than as starting point is, in a way, paralleled by the historical development . . . [through] processes of schematizing and shortening." As Freudenthal (1991, p. 61) pointed out, the process of "progressive schematization" may be highly facilitated by presenting problems as word problems. The work of Ambrose et al. (2003), summarized earlier follows the same principle (and see also Hiebert et al., 1997). For example, a partitive division problem is "Hattie had 544 candies to share with her friends. She gave each of her 17 friends the same amount of candy. How many candies did each friend get?" Here the story provides a simple semantic scaffolding for the logic of the calculation, a kind of "mental manipulative" (Toom, 1999); in particular, this particular example appeals to children's strong social understanding of fair sharing. At the same time, for the students being studied, this was a genuine problem in that they did not have a routine solution to hand, certainly not the standard algorithm.

The same distinction between sense-making in terms of the structural properties of the system of whole numbers and sense-making in relation to a described real-world or imaginable situation applies also to estimation and number sense.

In what has been described above, word problems are assumed to map unproblematically onto arithmetic operations, reflecting their main purpose, namely to support computation with understanding. We have argued (Verschaffel et al., 2000) that sense-making involves a second aspect whereby the relationship between a description of a situation and the mathematics appropriate for modeling that situation needs to be considered more critically and not always accepted as unproblematic (thereby calling for adaptive expertise—see Van Dooren, Verschaffel, Greer, & De Bock, 2006). Further, as argued by Selter (2002, p. 224), beyond the discussion of a range of particular models is the higher level activity of understanding the nature of modeling activity (Greer & Verschaffel, in press).

A related issue is the common concern about the lack of connection between school mathematics and out-of-school life. The long-term perspective referred to above also extends backwards in time, with a heightened attention to the experience children have prior to formal education, from the counting they have done, from hearing number words and seeing number symbols used in everyday life, and from various experiences in judging, comparing quantities, sharing collections, and so on (see chapter 5 of Kilpatrick et al., 2001). More generally, according to Becker and Selter (1996, p. 516),

> The informal knowledge that children possess is valuable and should be used as the basis for mathematics teaching and learning. . . . Connecting or relating formal mathematics to these informal approaches is

not an easy task . . . however, it has potential for making school mathematics more meaningful and less threatening and overwhelming to children.

Challenges Ahead

Conducting Research in a Political Context

In the third section of this chapter, we commented briefly on the ideological and political forces that are, to a considerable extent, shaping mathematics education—most extremely, perhaps, but certainly not only, in the United States (e.g. Dowling & Noss, 1990; Ellerton & Clements, 1994; Thomas, 2001). The same forces are acting on research in mathematics education. The *No Child Left Behind Act* of 2001 calls for the use of "scientifically based research" as the foundation for many education programs and for classroom instruction. On February 6, 2002, a seminar was held at the U.S. Department of Education where "leading experts in the fields of education and science discussed the meaning of scientifically based research and its status across various disciplines" (U.S. Department of Education, 2002). In this seminar, a talk entitled *Math Education and Achievement: Scientifically Based Research* by Russell Gersten, a specialist on reading comprehension and special education, began thus: "This is actually an easy topic to be brief on because there isn't a lot of scientific research in math [education]."

The validity of research, including a great deal of that reported in this chapter, and the credentials of those in the discipline of mathematics education are, to an unprecedented extent, under fire. Although these comments apply to this volume as a whole, for reasons set out earlier elementary arithmetic is at the epicenter. Yet, educational research is "the hardest science of all" (Berliner, 2002), and research in mathematics education is characterized by "reasonable ineffectiveness" (Kilpatrick, 1981). Recent shifts within the field have led to a proliferation of theories and methodologies that may be regarded as liberating or anarchic, depending on one's perspective (De Corte et al., 1996, p. 499). (For the anarchic interpretation, see Steen's, 1999, review of the ICMI Study *Mathematics Education as a Research Domain: A Search for Identity*, Sierpinska & Kilpatrick, 1998).

Consider the following key issue. We have suggested that the teaching of whole-number concepts and operations could be viewed as an opportunity to lay the foundations for many aspects of higher order mathematical thinking—problem-solving, algebraic forms of reasoning, argumentation, symbolizing and representational acts, sense-making, modeling. Examples cited constitute existence proofs demonstrating the feasibility of classroom teaching that addresses these higher level, long-term aims and the positive results that can be obtained. It hardly seems possible, however, to envisage research that could definitively validate or invalidate the belief of proponents of these approaches that the laying of such early foundations pays off in the long term in older students who are better at problem solving, understanding proof, mathematizing both vertically and horizontally, and so on. The promotion of such approaches is a matter of beliefs and values—just as is the statement in the Mathematics Framework for California Public Schools (California Department of Education, 2000, p. 18 and oft-repeated) that "all students are capable of learning rigorous mathematics and learning it well." (This example was chosen because the creators of that framework are firmly in the camp of those who call for decision-making based on rigorous scientific evidence.)

Another belief shared by most, if not all, of the mathematics educators cited in this chapter is that, by contrast with traditional teaching that has less ambitious aims of computational fluency, conceptual understanding, and perhaps problem solving (often narrowly conceived), classrooms organized according to the progressive principles exemplified in the *Classroom-Based Research* section will be more intellectually enjoyable and motivating and lay the foundations of positive mathematical dispositions. Again, taking into account the time frame, is this a researchable hypothesis?

In response to the critique of the NCTM-led reform, there have been notable efforts to bring interdisciplinary groups together to try to "balance the unbalanceable" (Sfard, 2003) and to produce consensual interpretations of research. Examples include the report *Adding It up* (Kilpatrick et al., 2001) that concentrates on elementary arithmetic and has been cited at numerous points in this chapter, and *A Research Companion to Principles and Standards for School Mathematics* (Kilpatrick et al., 2003), several relevant chapters of which we have drawn on. Further, in response to criticism of reform-based curricula, an effort has been made to demonstrate empirically their efficacy (Senk & Thompson, 2003) as reviewed in relation to the elementary curricula described earlier (but see the conclusions reached by Confrey & Stohl, 2004). Insofar as these represent a determined effect to improve the interface between research and practice, such reactions may be considered a positive by-product of the ideological conflict.

The Central Importance of Teachers and Instructional Environments

As Selter (2002, p. 222) stated (in commenting on three specific pieces of research, but clearly with more general relevance), "It is a challenging but necessary task to transform the research and the theoretical background . . . in such a way that it can reach and have an impact on a considerable number of teachers." As in mathematics education as a whole, there is a wide gap between what can be demonstrated by researchers who have had the benefit of many years of reflection, working in atypically supportive circumstances, and the propagation of the same effective approaches through the ranks of day-by-day teachers. As Carroll and Isaacs (2003, p. 81) put it, "reforms must take account of the working lives of teachers."

A coherent attempt to improve the teaching of whole-number concepts and operations along the lines suggested above is possible only with a radical change in the preparation and professional development of elementary teachers (Kilpatrick et al., 2001; and see the last part of section 2), in concert with corresponding changes in the nature of curricular materials and assessment. A major obstacle to such reform is the common perception that elementary mathematics and the teaching of elementary mathematics are simple. For this reason and others, the political will to devote the necessary resources is lacking. Thus, although we may indulge here in laying out a utopian solution, it might be more effective to consider what could be achieved in reality.

Building Strong Foundations for Mathematics

If the last 100 years of school mathematics is viewed as a long-running experiment, how much more evidence do we need that radical change is required? The learning of whole-number arithmetic is foundational for a person's mathematical education. Throughout this chapter, we have sought to emphasize ways in which this is so, including the long-term perspective that takes into account the extensions of the number concept. Further, arithmetic cannot be considered in isolation. We argued, in particular, that reconceptualizing arithmetic as inherently algebraic would serve the complementary aims of laying the groundwork for formal algebra and strengthening the understanding of arithmetic as structured. Arithmetic is foundational for other parts of mathematics—most clearly data handling.

Even more fundamentally, the early years of formal education in mathematics provide the opportunity to establish the forms of discourse, reasoning, and creativity that are definitive of mathematics—appreciation of pattern, problem solving, informal proof, use of symbolic and physical tools, modeling, communication. From this perspective, the importance of procedural competence is in no way diminished but is envisaged as a natural component of a much richer early mathematical education. In the second section of the chapter, examples have been described that represent promising steps in actualizing such a program. (It should be acknowledged that although our emphasis has been on recent work, there are many noteworthy precursors going back several decades.)

The ideal conception of early learning of arithmetic is as an intellectually and culturally rich experience that connects with children's personal and social lives and lays a firm foundation for mathematical proficiency as defined by Kilpatrick et al. (2001), including a positive and productive disposition towards mathematics as a human activity.

REFERENCES

Ambrose, R., Baek, J.-M., & Carpenter, T. P. (2003). Children's invention of multidigit multiplication and division algorithms. In A. J. Baroody & A. Dowker (Eds.), *The development of arithmetic concepts and skills* (pp. 305–336). Mahwah, NJ: Erlbaum.

Anghileri, J. (1989). An investigation of young children's understanding of multiplication. *Educational Studies in Mathematics, 20,* 367–385.

Anghileri, J. (1999). Issues in teaching multiplication and division. In I. Thompson (Ed.), *Issues in teaching numeracy in primary schools* (pp. 184–194). Buckingham, U.K.: Open University Press.

Anghileri, J. (2001). Development of division strategies for year 5 pupils in ten English schools. *British Educational Research Journal, 27*(1), 85–103.

Anghileri, J., Beishuizen, M., & Van Putten, K. (2002). From informal strategies to structured procedures: Mind the gap. *Educational Studies in Mathematics, 49,* 149–170.

Ascher, M. (2002). *Mathematics elsewhere: An exploration of ideas across cultures.* Princeton, NJ: Princeton University Press.

Ascher, M., & Ascher, R. (1997). Ethnomathematics. In A. B. Powell & M. Frankenstein (Eds.), *Ethnomathematics: Challenging Eurocentrism in mathematics education* (pp. 25–50). Albany, NY: SUNY Press.

Ashcraft, M. H. (1995). Cognitive psychology and simple arithmetic: A review and summary of new directions. *Mathematical Cognition, 1*(1), 3–34.

Ashcraft, M. H., & Christy, K. S. (1995). The frequency of arithmetic facts in elementary texts. Addition and multiplication in grades 1–6. *Journal for Research in Mathematics Education, 26,* 396–421.

Baek, J.-M. (1998). Children's invented algorithms for multidigit multiplication problems. In L. J. Morrow & M. J. Kenney (Eds.), *The teaching and learning of algorithms in school mathematics* (pp. 151–160). Reston, VA: National Council of Teachers of Mathematics.

Baroody, A. J. (1993). Early multiplication performance and the role of relational knowledge in mastering combinations involving "two." *Learning and Instruction, 3,* 93–112.

Baroody, A. J. (1994). An evaluation of evidence supporting fact-retrieval models. *Learning and Individual Differences, 6,* 1–36.

Baroody, A. J. (1999). The roles of estimation and the commutativity principle in the development of third-graders' mental multiplication. *Journal of Experimental Psychology, 74,* 153–193.

Baroody, A. J. (2003). The development of adaptive expertise and flexibility: The integration of conceptual and procedural knowledge. In A. J. Baroody & A. Dowker (Eds.), *The development of arithmetic concepts and skills* (pp. 1–34). Mahwah, NJ : Erlbaum.

Baroody, A. J. (in press). Chapter 2: Framework. In F. Fennell (Ed.), *Special education and mathematics: Helping children with learning difficulties achieve mathematical proficiency.* Reston, VA: National Council of Teachers of Mathematics.

Baroody, A. J., Cibulskis, M., Lai, M.-L., & Li, X. (2004). Comments on the use of learning trajectories in curriculum development and research. *Mathematical Thinking and Learning, 6,* 227–260.

Baroody, A. J., & Dowker, A. (Eds.). (2003). *The development of arithmetic concepts and skills.* Mahwah, NJ: Erlbaum.

Baroody, A. J., & Ginsburg, H. P. (1986). The relationship between initial meaningful learning and mechanical knowledge of arithmetic. In J. Hiebert (Ed.), *Conceptual and procedural knowledge: The case of mathematics* (pp. 75–112). Hillsdale, NJ : Erlbaum.

Baroody, A. J., & Tiilikainen, S. H. (2003). Two perspectives on addition development. In A. J. Baroody & A. Dowker (Eds.), *The development of arithmetic concepts and skills* (pp. 75–126). Mahwah, NJ : Erlbaum.

Baroody, A. J., Wilkins, J. L. M., & Tiilikainen, S. H. (2003). The development of children's understanding of additive commutativity: From protoquantitive concept to general concept? In A. J. Baroody & A. Dowker (Eds.), *The development of arithmetic concepts and skills* (pp. 127–160). Mahwah, NJ: Erlbaum.

Baruk, S. (1985). *L'âge du capitaine. De l'erreur en mathématiques.* [The captain's age. About errors in mathematics]. Paris: Seuil.

Bazin, M., Tamez, M., & The Exploratorium Teacher Institute (2002). *Math and science across cultures: Activities and investigations from the exploratorium.* New York: New Press.

Becker, J. P., Sawada, T., & Shimizu, Y. (1999). Some findings of the U.S.-Japan cross-cultural research on students' problem-solving behaviors. In G. Kaiser, E. Luna, & I. Huntley (Eds.), *International comparisons in mathematics education* (pp. 121–150). London: Falmer.

Becker, J. P., & Selter, C. (1996). Elementary school practices. In A. J. Bishop, K. Clements, C. Keitel, J. Kilpatrick, & C. Laborde (Eds.), *International handbook of mathematics education* (pp. 511–564). Dordrecht, The Netherlands: Kluwer.

Beishuizen, M. (1993). Mental strategies and materials or models for addition and subtraction up to 100 in Dutch second grades. *Journal for Research in Mathematics Education, 24,* 294–323.

Beishuizen, M. (1999). The empty number line as a new model. In I. Thompson (Ed.), *Issues in teaching numeracy in primary schools* (pp. 157–168). Buckingham, U.K.: Open University Press.

Beishuizen, M., Van Putten, K., & Van Mulken, F. (1993). Mental arithmetic and strategy use with indirect number problems up to one hundred. *Learning and Instruction, 7,* 87–106.

Berliner, D. C. (2002). Educational science: The hardest science of all. *Educational Researcher, 31*(8), 18–20.

Bisanz, J. (2003). Arithmetic development. Commentary on chapters 1 through 8 and reflections on directions. In A. J. Baroody & A. Dowker (Eds.), *The development of arithmetic concepts and skills* (pp. 435–452). Mahwah, NJ: Erlbaum.

Bishop, A. J. (1988). Mathematics education in its cultural context. *Educational Studies in Mathematics, 19*, 179–191.

Blanton, M., & Kaput, J. (2000). Generalizing and progressively formalizing in a third grade mathematics classroom: Conversations about even and odd numbers. In M. Fernández (Ed.), *Proceedings of the 22nd Annual Meeting of the North American Chapter of the International Group for the Psychology of Mathematics Education* (pp. 115–119). Columbus, OH: ERIC Clearinghouse.

Blöte, A. W., Van der Burg, E., & Klein, A. S. (2001). Students' flexibility in solving two-digit addition and subtraction problems: Instruction effects. *Journal of Educational Psychology, 93*, 627–638.

Boden, M. A. (1988). *Computer models of mind*. New York: Cambridge University Press.

Brade, G. A. (2003). *The effect of a computer activity on young children's development of numerosity estimation skills*. Unpublished doctoral dissertation, University of New York at Buffalo.

Brousseau, G. (1997). *Theory of didactical situations in mathematics*. (N. Balacheff, M. Cooper, R. Sutherland, & V. Warfield, Eds. & Trans.). Dordrecht, The Netherlands: Kluwer.

Brown, J. S., & Van Lehn, K. (1982). Toward a generative theory of bugs. In T. P. Carpenter, J. M. Moser, & T. Romberg (Eds.), *Addition and subtraction. A cognitive perspective* (pp. 117–135). Hillsdale, NJ: Erlbaum.

Brown, M. (1999). Swings of the pendulum. In I. Thompson (Ed.), *Issues in teaching numeracy in primary schools* (pp. 3–16). Buckingham, U.K.: Open University Press.

Brownell, W. A. (1945). When is arithmetic meaningful? *Journal of Educational Research, 38*, 481–498.

Bryant, P., Christie, C., & Rendu, A. (1999). Children's understanding of the relation between addition and subtraction: Inversion, identity, and decomposition. *Journal of Experimental Child Psychology, 74*, 194–212.

Burkhardt, H., & Schoenfeld, A. H. (2003) Improving educational research: Toward a more useful, more influential, and better-funded enterprise. *Educational Researcher, 32*(9), 3–14.

Burton, R. B. (1982). Diagnosing bugs in a simple procedural skill. In D. H. Sleeman & J. S. Brown (Eds.), *Intelligent tutoring systems* (pp. 157–183). New York: Academic Press.

Butterworth, B. (1999). *What counts: How every brain is hardwired for math*. New York: Free Press.

Butterworth, B., Marschesini, N., & Girelli, L. (2003). Young adults' strategic choices in simple arithmetic: Implications for the development of mathematical representations. In A. J. Baroody & A. Dowker (Eds.), *The development of arithmetic concepts and skills* (pp. 189–202). Mahwah, NJ: Erlbaum.

Buys, K. (2001). Mental arithmetic. In M. Van den heuvel (Ed.), *Children learn mathematics* (pp. 121–146). Utrecht, The Netherlands: Freudenthal Institute, University of Utrecht.

Caldwell, L. (1995). *Contextual considerations in the solution of children's multiplication and division word problems*. Unpublished master's thesis, Queen's University, Belfast, Northern Ireland.

California Department of Education. (2000). *Mathematics framework for California public schools*. Sacramento, CA: Author.

Camos, V. (2003). Counting strategies from 5 years to adulthood: Adaptation to structural features. *European Journal of Psychology of Education, 18*, 251–265.

Campbell, J. I. D. (2004). *Handbook of mathematical cognition*. New York: Psychology Press.

Campbell, S. R., & Zazkis, R. (Eds.). (2002a). *Learning and teaching number theory*. Westport, CT: Ablex.

Campbell, S. R., & Zazkis, R. (2002b). Toward number theory as a conceptual field. In S. R. Campbell & R. Zazkis (Eds.), *Learning and teaching number theory* (pp. 1–14). Westport, CT: Ablex.

Cardelle-Elawar, M. (1995). Effects of metacognitive instruction on low achievers in mathematics problems. *Teaching and Teacher Education, 11*(1), 81–95.

Carpenter, T. P., & Fennema, E. (1992). Cognitively guided instruction: Building on the knowledge of students and teachers. *International Journal of Educational Research, 17*, 457–470.

Carpenter, T. P., Fennema, E., Peterson, P. L., Chiang, C. P., & Loef, M. (1989). Using knowledge of children's mathematical thinking in classroom teaching: An experimental program. *American Educational Research Journal, 26*, 499–532.

Carpenter, T. P., & Franke, M. (2001). Developing algebraic reasoning in the elementary grades: Generalization and proof. In H. Chick, K. Stacey, J. Vincent, & J. Vincent (Eds.), *Proceedings of the 12th ICMI Study Conference* (Vol. 1, pp. 155–162). Melbourne, Australia: University of Melbourne.

Carpenter, T. P., Franke, M. L., Jacobs, V., Fennema, E., & Empson, S. B. (1997). A longitudinal study of intervention and understanding in children's multidigit addition and subtraction. *Journal for Research in Mathematics Education, 29*, 3–30.

Carpenter, T. P., Franke, M., & Levi, L. (2003). *Thinking mathematically: Integrating arithmetic and algebra in elementary school*. Portsmouth, NH: Heinemann.

Carpenter, T. P., & Moser, J. M. (1984). The acquisition of addition and subtraction concepts in grades one through three. *Journal for Research in Mathematics Education, 15*, 179–202.

Carroll, W. M., & Isaacs, A. (2003). Achievement of students using the University of Chicago School Mathematics Project's Everyday Mathematics. In S. L. Senk & D. R. Thompson (Eds.), *Standards-based school mathematics curricula* (pp. 79–108). Mahwah, NJ: Erlbaum.

Clark, F. B., & Kamii, C. (1996). Identification of multiplicative thinking in children in grades 1–5. *Journal for Research in Mathematics Education, 27*, 41–51.

Clements, D. H., & Sarama, J. (Eds.). (2004a). Hypothetical learning trajectories [Special issue]. *Mathematical Thinking and Learning, 6*(2).

Clements, D. H., & Sarama, J. (2004b). Learning trajectories and mathematics education. *Mathematical Thinking and Learning, 6*, 81–90.

Cobb, P. (1990). A constructivist perspective on information-processing theories of mathematical activity. *International Journal of Educational Research, 14*, 67–92.

Cobb, P., Gravemeijer, K., Yackel, E., McClain, K., & Whitenack, J. (1997). Mathematizing and symbolizing: The emergence of chains of signification in one first-grade classroom. In D. Kirschner & J. A. Whitson (Eds.), *Situated cognition theory: Social, semiotic and neurological perspectives* (pp. 151–233). Mahwah, NJ: Erlbaum.

Cobb, P., Yackel, E., & McClain, K. (Eds.). (2000). *Communicating and symbolizing in mathematics education: Perspectives on discourse, tools, and instructional design*. Mahwah, NJ: Erlbaum.

Coquin-Viennot, D., & Moreau, S. (2003). Highlighting the role of the episodic situation model in solving of arithmetical problems. *European Journal of Psychology of Education, 18,* 267–279.

Cognition and Technology Group at Vanderbilt. (1997). *The Jasper project: Lessons in curriculum, instruction, assessment, and professional development.* Mahwah, NJ: Erlbaum.

Collet, M. (2004). *Le développement du système en base-dix chez les enfants de 1ière et 2ème primaire.* [The development of the ten base system in first and second graders.] Unpublished doctoral dissertation, Université Catholique de Louvain, Faculté de Psychologie et des Sciences de l'Éducation, Louvain-la-Neuve, Belgium.

Confrey, J., & Stohl, V. (Eds.). (2004). *On evaluating curricular effectiveness: Judging the quality of K–12 mathematics evaluations.* Washington, DC: National Academies Press.

Conway, J. H., & Guy, R. K. (1996). *The book of numbers.* New York: Springer-Verlag.

Cooney, J. B., Swanson, H. L., & Ladd, S. F. (1988). Acquisition of mental multiplication skill: Evidence for the transition between counting and retrieval strategies. *Cognition and Instruction, 5,* 323–345.

Cooper, T. J., Heirdsfield, A. M., & Irons, C. J. (1998, July). *Years 2 and 3 children's mental addition and subtraction strategies for 2- and 3-digit word problems and algorithmic exercices.* Paper presented at Topic Group 1 at the 8th International Congress on Mathematics Education, Sevilla, Spain.

Cowan, R. (2003). Does it all add up? Changes in children's knowledge of addition combinations, strategies and principles. In A. J. Baroody & A. Dowker (Eds.), *The development of arithmetic concepts and skills* (pp. 35–74). Mahwah, NJ: Erlbaum.

Crites, T. (1992). Skilled and less skilled estimators' strategies for estimating discrete quantities. *The Elementary School Journal, 92,* 601–619.

Crump, T. (1990). *The anthropology of numbers.* Cambridge, England: Cambridge University Press.

Cummins, D., Kintsch, W., Reusser, K., & Weimer, R. (1988). The role of understanding in solving word problems. *Cognitive Psychology, 20,* 405–438.

D'Ambrosio, U. (1985). Ethnomathematics and its place in the history and pedagogy of mathematics. *For the Learning of Mathematics, 5*(1), 44–48.

De Bock, D., Verschaffel, L., & Janssens, D. (1998). Solving problems involving length and area of similar plane figures and the illusion of linearity: An inquiry of the difficulties of secondary school students. *Educational Studies in Mathematics, 35,* 65–83.

De Corte, E., Greer, B., & Verschaffel, L. (1996). Learning and teaching mathematics. In D. Berliner & R. Calfee (Eds.), *Handbook of educational psychology* (pp. 491–549). New York: Macmillan.

De Corte, E., & Somers, R. (1982). Estimating the outcome of a task as a heuristic strategy in arithmetic problem solving. A teaching experiment with sixth-graders. *Human Learning, 1,* 105–121.

De Corte, E., & Verschaffel, L. (1985a). Working with simple word problems in early mathematics instruction. In L. Streefland (Ed.), *Proceedings of the Ninth International Conference for the Psychology of Mathematics Education: Vol. 1. Individual contributions* (pp. 304–309). Utrecht, The Netherlands: Research Group on Mathematics Education

and Educational Computer Center, Subfaculty of Mathematics, University of Utrecht.

De Corte, E., & Verschaffel, L. (1985b). Beginning first graders' initial representation of arithmetic word problems. *Journal of Mathematical Behavior, 4,* 3–21.

De Corte, E., & Verschaffel, L. (1987). The effect of semantic structure on first graders' solution strategies of elementary addition and subtraction word problems. *Journal for Research in Mathematics Education, 18,* 363–381.

De Corte, E., & Verschaffel, L. (1988). Computer simulation as a tool in research on problem solving in subject-matter domains. *The International Journal of Educational Research, 12,* 49–69.

De Corte, E., & Verschaffel, L. (1996). An empirical test of the impact of primitive intuitive models of operations on solving word problems with a multiplicative structure. *Learning and Instruction, 6,* 219–243.

De Corte, E., Verschaffel, L., & De Win, L. (1985). The influence of rewording verbal problems on children's problem representations and solutions. *Journal of Educational Psychology, 77,* 460–470.

De Franco, T. C., & Curcio, F. R. (1997). A division problem with a remainder embedded across two contexts: Children's solutions in restrictive versus real-world settings. *Focus on Learning Problems in Mathematics, 19*(2), 58–72.

Dehaene, S. (1993). Varieties of numerical abilities. In S. Dehaene (Ed.), *Numerical cognition* (pp. 1–42). Cambridge, MA: Blackwell.

Dehaene, S., & Cohen, L. (1995). Towards an anatomical and functional model of number processing. *Mathematical Cognition, 1,* 83–120.

Dekker, A., Ter Heege, H., & Treffers, A. (1982). *Cijferend vermenigvuldigen en delen volgens Wiskobas.* [Teaching written multiplication and division according to Wiskobas]. Utrecht, The Netherlands: Ontwikkeling van de Wiskunde & Onderwijs Computercentrum, University of Utrecht.

Delazer, M. (2003). Neuropsychological findings on conceptual knowledge of arithmetic. In A. J. Baroody & A. Dowker (Eds.), *The development of arithmetic concepts and skills* (pp. 385–408). Mahwah, NJ: Erlbaum.

Dowker, A. (1997). Young children's addition estimates. *Mathematical Cognition, 3,* 141–154.

Dowker, A. (2003). Young children's estimates for addition: the zone of partial knowledge and understanding. In A. J. Baroody & A. Dowker (Eds.), *The development of arithmetic concepts and skills* (pp. 243–266). Mahwah, NJ: Erlbaum.

Dowling, P., & Noss, R. (Eds.). (1990). *Mathematics versus the National Curriculum.* London: Falmer.

Driscoll, M. (1999). *Fostering algebraic thinking.* Portsmouth, NH: Heinemann.

Ellerton, N. F., & Clements, M. A. (1994). *The National Curriculum debacle.* West Perth, Australia: Meridian Press.

Engestrom, Y. (1987). *Learning by expanding: An activity-theoretical approach to developmental research.* Helsinki, Finland: Orienta-Konsultit Oy.

Engestrom, Y., & Miettinen, R. (1999). Introduction. In Y. Engestrom, R. Miettinen, & R.-J. Punamaki (Eds.), *Perspectives on activity theory* (pp. 1–18). Cambridge, England: Cambridge University Press.

English, L. (1998). Children's problem posing within formal and informal contexts. *Journal for Research in Mathematics Education, 29,* 83–106.

Falkner, K. P., Levi, L., & Carpenter, T. P. (1999). Children's understanding of equality: A foundation for algebra. *Teaching Children Mathematics, 6,* 232–236.

Ferreira, M. (1997). When $1 + 1 \neq 2$: Making mathematics in central Brazil. *American Ethnologist, 24,* 132–147.

Fischbein, E. (1990). Introduction. In P. Nesher & J. Kilpatrick (Eds.), *Mathematics and cognition. A research synthesis by the International Group for the Psychology of Mathematics Education: ICME Study Series* (pp. 1–13). Cambridge, England: Cambridge University Press.

Fischbein, E., Deri, M., Nello, M. S., & Marino, M. S. (1985). The role of implicit models in solving verbal problems in multiplication and division. *Journal for Research in Mathematics Education, 16,* 3–17.

Freudenthal, H. (1983). *Didactical phenomenology of mathematical structures.* Dordrecht, The Netherlands: Reidel.

Freudenthal, H. (1991). *Revisiting mathematics education.* Dordrecht, The Netherlands: Kluwer.

Fuson, K. C. (1992). Research on whole number addition and subtraction. In D. A. Grouws (Ed.), *Handbook of research on mathematics teaching and learning* (pp. 243–275). New York: MacMillan.

Fuson, K. C. (2003). Developing mathematical power in whole number operations. In J. Kilpatrick, W. G. Martin, & D. Schifter (Eds.), *A research companion to Principles and Standards for School Mathematics* (pp. 68–94). Reston, VA: National Council of Teachers of Mathematics.

Fuson, K. C., & Kwon, Y. (1992). Korean children's understanding of multidigit addition and subtraction. *Child Development, 63,* 491–506.

Fuson, K. C., Smith, S., & Lo Cicero, A. (1997). Supporting Latino first graders' ten-structured thinking in urban classrooms. *Journal for Research in Mathematics Education, 28,* 738–760.

Fuson, K. C., Wearne, D., Hiebert, J. C., Murray, H. G., Human, P. G., Olivier, A. I., et al. (1997). Children's conceptual structures for multidigit numbers and methods of multidigit addition and subtraction. *Journal for Research in Mathematics Education, 28,* 130–162.

Fuson, K. C., & Willis, G. B. (1989). Second graders' use of schematic drawings in solving addition and subtraction word problems. *Journal of Educational Psychology, 81,* 514–520.

Gardner, H. (1985). *The mind's new science.* New York: Basic Books.

Garofalo, J., & Lester, F. K. (1985). Metacognition, cognitive monitoring and mathematical persformance. *Journal for Research in Mathematics Education, 16,* 163–176.

Geary, D. C. (2003). Arithmetic development. Commentary on chapters 9 through 15 and future directions. In A. J. Baroody & A. Dowker (Eds.), *The development of arithmetic concepts and skills* (pp. 453–464). Mahwah, NJ : Erlbaum.

Gerdes, P. (1994). Reflections on ethnomathematics. *For the Learning of Mathematics, 14*(2), 19–22.

Ginsburg, H. (1997). *Entering the child's mind: The clinical interview in psychological research and practice.* Cambridge, England: Cambridge University Press.

Ginsburg, H. P., Klein, A., & Starkey, P. (1998). The development of children's mathematical knowledge: Connecting research with practice. In I. E. Sigel & K. A. Renninger (Eds.), *Handbook of child psychology: Vol. 4. Child psychology in practice* (5th ed., pp. 401–476). New York: Wiley.

Graeber, A., & Tirosh, D. (1988). Multiplication and division involving decimals: Preservice elementary teachers'

performance and beliefs. *Journal of Mathematical Behavior, 7,* 263–280.

Gravemeijer, K. (1994). *Developing realistic mathematics education.* Utrecht, The Netherlands: Freudenthal Institute, University of Utrecht.

Gravemeijer, K. (2002). Preamble: From models to modeling. In K. Gravemeijer, R. Lehrer, B. Van Oers, & L. Verschaffel (Eds.), *Symbolizing, modeling, and tool use in mathematics education* (pp. 7–24). Dordrecht, The Netherlands: Kluwer.

Gravemeijer, K. (2004). Local instruction theories as means of support for teachers in reform mathematics education. *Mathematical Thinking and Learning, 6,* 105–128.

Gravemeijer, K., Lehrer, R., Van Oers, B., & Verschaffel, L. (2002). Introduction and overview. In K. Gravemeijer, R. Lehrer, B. Van Oers, & L. Verschaffel (Eds.), *Symbolizing, modeling and tool use in mathematics education* (pp. 1–5). Dordrecht, The Netherlands: Kluwer.

Gravemeijer, K., & Stephan, M. (2002). Emergent models as an instructional design heuristic. In K. Gravemeijer, R. Lehrer, B. Van Oers, & L. Verschaffel (Eds.), *Symbolizing, modeling, and tool use in mathematics education* (pp. 145–170). Dordrecht, The Netherlands: Kluwer.

Greeno, J. G. (2003). Situative research relevant to Standards for School Mathematics. In J. Kilpatrick, W. G. Martin, & D. Schifter (Eds.), *A research companion to Principles and Standards for School Mathematics* (pp. 304–332). Reston, VA: National Council of Teachers of Mathematics.

Greeno, J. G., Collins, A. M., & Resnick, L. B. (1996). Cognition and learning. In D. C. Berliner & R. C. Calfee (Eds.), *Handbook of educational psychology* (pp. 15–46). New York: Macmillan.

Greeno, J. G., & The Middle-School Mathematics Through Applications Project Group. (1998). The situativity of cognition, learning, and research. *American Psychologist, 53,* 5–26.

Greer, B. (1992). Multiplication and division as models of situations. In D. A. Grouws (Ed.), *Handbook of research on mathematics teaching and learning* (pp. 276–295). New York: Macmillan.

Greer, B. (1993). The modeling perspective on wor(l)d problems. *Journal of Mathematical Behavior, 12,* 239–250.

Greer, B. (1994). Extending the meaning of multiplication and division. In G. Harel & J. Confrey (Eds.), *The development of multiplicative reasoning in the learning of mathematics* (pp. 61–85). Albany, NY: SUNY Press.

Greer, B. (1996). Theories of mathematics education: The role of cognitive analyses. In L. Steffe, P. Nesher, P. Cobb, G. Goldin, & B. Greer (Eds.), *Theories of mathematics education* (pp. 179–196). Hillsdale, NJ: Erlbaum.

Greer, B. (2004). The growth of mathematics through conceptual restructuring. *Learning and Instruction, 14,* 541–548

Greer, B., & Verschaffel, L. (1990). Introduction to the special issue on mathematics as a proving ground for information-processing theories. *International Journal for Educational Research, 14,* 3–12.

Greer, B., & Verschaffel, L. (in press). Characterizing modeling competencies. In W. Blum, W. Henne, & M. Niss (Eds.), *Applications and modeling in mathematics education (ICMI Study 14).* Dordrecht, The Netherlands: Kluwer.

Grouws, D. A. (Ed.). (1992). *Handbook of research on mathematics teaching and learning.* New York: Macmillan.

Hatano, G. (1988). Social and motivational bases for mathematical understanding. In G.B. Saxe & M. Gearhart

(Eds.), *Children's mathematics* (pp. 55–70). San Fransisco: Jossey-Bass.

Hatano, G. (2003). Foreword. In A. J. Baroody & A. Dowker (Eds.), *The development of arithmetic concepts and skills* (pp. xi–xiv). Mahwah, NJ : Erlbaum.

Hegarty, M., Mayer, R. E., & Monk, C. A. (1995). Comprehension of arithmetic word problems: A comparison of successful and unsuccessful problem solvers. *Journal of Educational Psychology, 87*, 18–32.

Hersh, R. (1997). *What is mathematics, really?* New York: Oxford University Press.

Hidalgo, M. C. (1997). *L'activation des connaissances à propos du monde réel dans la résolution de problèmes verbaux en arithmétique.* [The activation of real world knowledge in the solution of arithmetic word problems.] Unpublished doctoral dissertation, Université Laval, Québec, Canada.

Hiebert, J. (Ed.). (1986). *Conceptual and procedural knowledge: The case of mathematics.* Hillsdale, NJ: Erlbaum.

Hiebert, J. (1999). Relationships between research and the NCTM Standards. *Journal for Research in Mathematics Education, 30*, 3–19.

Hiebert, J., Carpenter, T. P., Fennema, E., Fuson, K., Human, P., Murray, H., et al. (1996). Problem solving as a basis for reform in curriculum and instruction. The case of mathematics. *Educational Researcher, 25*, 12–21.

Hiebert, J., Carpenter, T. P., Fennema, E., Fuson, K. C., Wearne, D., Murray, H., et al. (1997). *Making sense: Teaching and learning mathematics with understanding.* Portsmouth, NH: Heinemann.

Hiebert, J., & Wearne, D. (1996). Instruction, understanding, and skill in multidigit addition and subtraction. *Cognition and Instruction, 14*, 251–284.

Iszak, A. (2004). Teaching and learning two-digit multiplication: Coordinating analyses of classroom practices and individual student learning. *Mathematical Thinking and Learning, 6*, 37–79.

Jaspers, M. W. M., & Van Lieshout, E. C. D. M. (1994). The evaluation of two computerised instruction programs for arithmetic word problem solving by educable mentally retarded children. *Learning and Instruction, 4*, 193–215.

Jones, G., Thornton, C. A., & Putt, I. J. (1994). A model for nurturing and assessing multidigit number sense among first grade children. *Educational Studies in Mathematics, 27*, 117–143.

Jordan, N. C., Hanich, L. B., & Kaplan, D. (2003). Arithmetic fact mastery in young children. A longitudinal investigation. *Journal of Experimental Child Psychology, 85*, 103–119.

Joseph, G. G. (1992). *The crest of the peacock: Non-European roots of mathematics.* London: Penguin.

Kamii, C., & Dominick, A. (1998). The harmful effects of algorithms in Grades 1–4. In L. J. Morrow & M. J. Kenney (Eds.), *The teaching and learning of algorithms in school mathematics* (pp. 130–140). Reston, VA: National Council of Teachers of Mathematics.

Kaput, J. (1998). Transforming algebra from an engine of inequity to an engine of mathematical power by "algebrafying" the K–12 curriculum. In S. Fennel (Ed.), *The nature and role of algebra in the K–14 curriculum: Proceedings of a national symposium* (pp. 25–36). Washington, DC: National Research Council, National Academy Press.

Kaput, J. (1999). Teaching and learning a new algebra. In E. Fennema & T. Romberg (Eds.), *Mathematics classrooms that promote understanding* (pp. 133–155). Mahwah, NJ: Erlbaum.

Kaput, J., & Blanton, M. (2001). Algebrafying the elementary experience: Part I. Transforming task structures. In H. Chick, K. Stacey, J. Vincent, & J. Vincent (Eds.), *Proceedings of the 12th ICMI Study Conference* (Vol. 1, pp. 344–351). Melbourne, Australia: University of Melbourne.

Kilpatrick, J. (1981). The reasonable ineffectiveness of research in mathematics education. *For the Learning of Mathematics, 2*(2), 22–29.

Kilpatrick, J. (2003). What works? In S. L. Senk & D. R. Thompson (Eds.), *Standards-based school mathematics curricula* (pp. 471–488). Mahwah, NJ: Erlbaum.

Kilpatrick, J., Martin, W. G., & Schifter, D. (Eds.). (2003). *A research companion to Principles and Standards for School Mathematics.* Reston, VA: National Council of Teachers of Mathematics.

Kilpatrick, J., Swafford, J., & Findell, B. (2001). *Adding it up. Helping children learn mathematics.* Washington, DC: National Academy Press.

Kintsch, W., & Lewis, A. B. (1993). The time course of hypotheses formulation in solving arithmetic word problems. In M. Denis & G. Sabah (Eds.), *Modèles et concepts pour la science cognitive: Hommage à Jean-Francois Le Ny* [Models and concepts for the cognitive sciences. A tribute to Jean-Francois Le Ny.] (pp. 11–23). Grenoble, France: Presses Universitaires de Grenoble.

Klahr, D., & MacWhinney, B. (1998). Information processing. In W. Damon (Ed.), *Handbook of Child Psychology, Volume 2: Cognition, perception, and language* (pp. 631–678). New York: Wiley.

Klein, A. S., Beishuizen, M., & Treffers, A. (1998). The empty number line in Dutch second grades: Realistic versus gradual program design. *Journal for Research in Mathematics Education, 29*, 443–464.

Kouba, V. (1989). Children's solution strategies for equivalent set multiplication and division word problems. *Journal for Research in Mathematics Education, 20*, 147–158.

Kuriyama, K., & Yoshida, H. (1995). Representational structure of numbers in mental addition. *Japanese Journal of Educational Psychology, 43*, 402–410.

Lakoff, G., & Nunez, R. E. (2000). *Where mathematics comes from: How the embodied mind brings mathematics into being.* New York: Basic Books.

Lampert, M. (1990). Connecting inventions with conventions. In L. P. Steffe, & T. Wood (Eds.), *Transforming children's mathematics education* (pp. 253–265). Mahwah, NJ: Erlbaum.

Lave, J. (1988). *Cognition in practice: Mind, mathematics, and culture in everyday life.* Cambridge, England: Cambridge University Press.

Lave, J. (1992). Word problems: A microcosm of theories of learning. In P. Light & G. Butterworth (Eds.), *Context and cognition: Ways of learning and knowing* (pp. 74–92). New York: Harvester Wheatsheaf.

Leblanc, M., & Weber-Russell, S. (1996). Text integration and mathematical connections: A computer model of arithmetic word problem solving. *Cognitive Science, 20*, 357–407.

LeFevre, J., Greenham, S.L., & Waheed, N. (1993). The development of procedural and conceptual knowledge in computational estimation. *Cognition and Instruction, 11*, 95–148.

LeFevre, J., & Liu, J. (1997). The role of experience in numerical skill: Multiplication performance in adults from Canada and China. *Mathematical Cognition, 3,* 31–62.

LeFevre, J.-A., Smith-Chant, B. L., Hiscock, K., Daley, K. E., & Morris, J. (2003). Young adults' strategic choices in simple arithmetic: Implications for the development of mathematical representations. In A. J. Baroody & A. Dowker (Eds.), *The development of arithmetic concepts and skills* (pp. 203–228). Mahwah, NJ : Erlbaum.

Lehrer, R., & Schauble, L. (2005). Modeling and argument in the elementary grades. In T. A. Romberg, T. P. Carpenter, & F. Dremock (Eds.), *Understanding mathematics and science matters* (pp. 29–54). Mahwah, NJ: Erlbaum.

Lemaire, P., & Lecacheur, M. (2002). Children's strategies in computational estimation. *Journal of Experimental Child Psychology, 82,* 281–304.

Lemaire, P., & Siegler, R. S. (1995). Four aspects of strategic change: Contributions to children's learning of multiplication. *Journal for Experimental Psychology: General, 124,* 83–97.

Lesh, R., & Doerr, H. M. (Eds.). (2003). *Beyond constructivism. Models and modeling perspectives on mathematical problem solving, learning and teaching.* Mahwah, NJ: Erlbaum.

Lester, F., Garofalo, J., & Kroll, D. (1989). *The role of metacognition in mathematical problem solving: A case study of two grade seven classes.* Final report to the National Science Foundation of NSF project MDR 85–50346.

Levain, J. P. (1992). La résolution des problèmes multiplicatifs à la fin du cycle primaire. [Solutions to multiplication problems at the end of the primary school]. *Educational Studies in Mathematics, 23,* 139–161.

Lewis, A., & Mayer, R. (1987). Students' misconceptions of relational statements in arithmetic word problems. *Journal of Educational Psychology, 74,* 199–216.

Lewis, C. C. (2002). *Lesson study. A handbook of teacher-led instructional change.* Philadelphia: Research for Better Schools.

Luwel, K., Verschaffel, L., Onghena, P., & De Corte, E. (2000). Children's strategies for numerosity judgement in square grids of different sizes. *Psychologica Belgica, 40,* 183–209.

Luwel, K., Verschaffel, L., Onghena, P., & De Corte, E. (2003a). Analyzing the adaptiveness of strategy choices using the choice/no-choice method. The case of numerosity judgement. *European Journal of Cognitive Psychology, 15,* 511–537.

Luwel, K., Verschaffel, L., Onghena, P., & De Corte, E. (2003b). Flexibility in strategy use: Adaptation of numerosity judgement strategies to task characteristics. *European Journal of Cognitive Psychology, 15,* 247–266.

Ma, L. (1999). *Knowing and teaching elementary mathematics. Teachers' understanding of fundamental mathematics in China and the United States.* Mahwah, NJ: Erlbaum.

Mason, J. (1996). Expressing generality and roots of algebra. In N. Bednarz, C. Kieran, & L. Lee (Eds.), *Approaches to algebra: Perspectives for teaching and research* (pp. 65–86). Dordrecht, The Netherlands: Kluwer.

Mason, J., Burton, L, & Stacey, K. (1982). *Thinking mathematically.* New York: Addison-Wesley.

McClain, K., & Cobb, P. (2001). An analysis of the development of sociomathematical norms in one first-grade classroom. *Journal for Research in Mathematics Education, 32,* 236–266.

McCloskey, M., Caramazza, A., & Basili, A. G. (1985). Cognitive mechanisms in number processing and calculation: Evidence from dyscalculia. *Brain and Cognition, 4,* 171–196.

McIntosh, A., Reys, B. J., & Reys, R. E. (1992). A proposed framework for examining basic number sense. *For the Learning of Mathematics, 12*(3), 2–8.

Menne, J. J. M. (2001). *Met sprongen vooruit. Een productief oefenprogramma voor zwakke rekenaars in het getallengebied tot 100—een onderwijsexperiment.* [A productive training program for mathematically weak children in the number domain up to 100—a design study]. Utrecht, The Netherlands: CD-beta Press.

Mevarech, Z. (1999). Effects of metacognitive training embedded in cooperative setting on mathematical problem solving. *Journal of Educational Research, 92,* 195–205.

Miura, I. T., & Okamoto, Y. (2003). Language supports for mathematics understanding and performance. In A. J. Baroody & A. Dowker (Eds.), *The development of arithmetic concepts and skills* (pp. 229–242). Mahwah, NJ: Erlbaum.

Miura, I. T., Okamoto, Y., Kim, C. C., Steere, M., & Fayol, M. (1993). First graders' cognitive representation of number and understanding of place value: Cross-national comparisons – France, Japan, Korea, Sweden, and the United States. *Journal of Educational Psychology, 85,* 24–30.

Morrow, L. J., & Kenney, M. J. (1988). *The teaching and learning of algorithms in school mathematics. (NCTM Yearbook 1988).* Reston, VA: National Council of Teachers of Mathematics.

Moss, J., & Case, R. (1999). Developing children's understanding of the rational numbers. A new model and an experimental curriculum. *Journal for Research in Mathematics Education, 30,* 122–147.

Mulligan, J., & Mitchelmore, M. (1997). Young children's intuitive models of multiplication and division. *Journal for Research in Mathematics Education, 28,* 309–330.

Murata, A. (2005). Paths to learning 10-structured understandings of tens sums: Addition solution methods of Japanese Grade 1 students. *Cognition and Instruction, 22,* 185–218.

Murray, H., & Olivier, A. (1989). A model of understanding of two-digit numeration and computation. In G. Vergnaud, J. Rogalski, & M. Artigue (Eds.), *Proceedings of the 13th International Conference for the Psychology of Mathematics Education* (Vol. 3, pp. 3–10). Paris: Laboratoire de Psychologie du Developpement.

National Council of Teachers of Mathematics. (2000). *Principles and standards for school mathematics.* Reston, VA: Author.

Nelissen, J. M. C. (1987). *Kinderen leren wiskunde. Een studie over constructie en reflectie in het basisonderwijs.* [Children learning mathematics. A study on construction and reflection in elementary school children]. Gorinchem, The Netherlands: Uitgeverij De Ruiter.

Newell, A. (1990). *Unified theories of cognition.* Cambridge, MA: Harvard University Press.

Niss, M. (2001). Issues and problems of research on the teaching and learning of applications and modelling. In J. F. Matos, W. Blum, S. K. Houston, & S. P. Carreira (Eds.), *Modelling and mathematics education. ICTMA 9: Applications in science and technology* (pp. 72–89). Chichester, U.K.: Horwood.

Nunes, T., & Bryant, P. (1995). Do problem situations influence children's understanding of the commutativity of multiplication? *Mathematical Cognition, 1,* 245–260.

Okamoto, Y. (1996). Modelling children's understanding of quantitative relations in texts: A developmental perspective. *Cognition and Instruction, 14,* 409–440.

Orey, D. (2005). *The algorithm collection project.* Retrieved, June 26 2006, from: http://www.csus.edu/indiv/o/oreyd/ACP.htm_files/Alg.html

Ostad, S. A. (1997). Developmental differences in addition strategies. A comparison of mathematically disabled and mathematically normal children. *British Journal of Educational Psychology, 67,* 345–357.

Palm, T. (2002). *The realism of mathematical school tasks.* Unpublished doctoral thesis, Department of Mathematics, Umea University, Sweden.

Paulos, J.A. (1991). *Beyond numeracy: Ruminations of a numbers man.* New York: Knopf.

Payne, J.W., Bettman, J.R., & Johnson, E.J. (1993). *The adaptive decision maker.* Cambridge, England: Cambridge University Press.

Piaget, J. (2001). *Studies in reflecting abstraction.* Hove, UK: Psychology Press.

Plunkett, S. (1979). Decomposition and all that rot. *Mathematics in School, 8*(3), 2–5.

Powell, A. B., & Frankenstein, M. (Eds.). (1997). *Ethnomathematics: Challenging Eurocentrism in mathematics education.* Albany, NY: SUNY Press.

Putnam, R. T. (2003). Commentary on four elementary mathematics curricula. In S. L. Senk & D. R. Thompson (Eds.), *Standards-based school mathematics curricula* (pp. 161–180). Mahwah, NJ: Erlbaum.

Rathmell, E. C. (1986). Helping children learn to solve story problems. In A. Zollman, W. Speer, & J. Meyer (Eds.), *The Fifth Mathematics Methods Conference papers.* Bowling Green, OH: Bowling Green State University.

Reed, S. K. (1999). *Word problems. Research and curriculum reform.* Mahwah, NJ: Erlbaum.

Resnick, L. B. (1992). From protoquantities to operators: Building mathematical competence on a foundation of everyday knowledge. In G. Leinhardt, R. Putnam, & R. A. Hattrup (Eds.), *Analysis of arithmetic for mathematics teaching* (pp. 373–425). Hillsdale, NJ: Erlbaum.

Resnick, L.B., & Ford, W. W. (Eds.). (1981). *The psychology of mathematics for instruction.* Hillsdale, NJ: Erlbaum.

Resnick, L. B., Nesher, P., Leonard, F., Magone, M., Omanson, S., & Peled, I. (1989). Conceptual bases of arithmetic errors: The case of decimal fractions. *Journal for Research in Mathematics Education, 20,* 8–27.

Resnick, L. B., & Omanson, S. F. (1987). Learning to understand arithmetic. In R. Glaser (Ed.), *Advances in instructional psychology.* (Vol. 3, pp. 41–95). Hillsdale, NJ: Erlbaum.

Reys, R. E., Reys, B. J., Nohda, N., & Emori, H. (1995). Mental computation performance and strategy use of Japanese students in grades 2, 4, 6, and 8. *Journal for Research in Mathematics Education, 26,* 304–326.

Reys, R. E., Rybolt, J. F., Bestgen, B. J., & Wyatt, J. W. (1982). Processes used by good computational estimators. *Journal for Research in Mathematics Education, 13,* 183–201.

Reys, R. E., & Yang, D.-C. (1998). Relationship between computational performance and number sense among sixth- and eighth-grade students in Taiwan. *Journal for Research in Mathematics Education, 29,* 225–237.

Rickard, T. C., & Bourne, L. E., Jr. (1996). Some tests of an identical elements model of basic arithmetic skills. *Journal of Experimental Psychology: Learning, Memory, and Cognition, 22,* 1281–1295.

Riley, M. S., Greeno, J. G., & Heller, J. I. (1983). Development of children's problem-solving ability in arithmetic. In H. P. Ginsburg (Ed.), *The development of mathematical thinking* (pp. 153–196). New York: Academic Press.

Rissland, E. L. (1985). Artificial intelligence and the learning of mathematics. A tutoring sampling. In E. A. Silver (Ed.), *Teaching and learning mathematical problem solving: Multiple research perspectives* (pp. 147–176). Hillsdale, NJ: Erlbaum.

Rittle-Johnson, B., & Siegler, R. S. (1998). The relation between conceptual and procedural knowledge in learning mathematics. A review. In C. Donlan (Ed.), *The development of mathematical skills* (pp. 75–110). Hove, U.K.: Psychology Press.

Romberg, T. A., Carpenter, T. P., & Kwako, J. (2005). Standards-based reform and teaching for understanding. In T. A. Romberg, T. P. Carpenter, & F. Dremock (Eds.), *Understanding mathematics and science matters* (pp. 3–28). Mahwah, NJ: Erlbaum.

Rouche, N. (1992). *Le sense de la mesure.* [The sense of measurement.] Bruxelles: Didier Hatier.

Salomon, G. (Ed.). (1993). *Distributed cognitions: Psychological and educational considerations.* Cambridge, U.K.: Cambridge University Press.

Saunders, D. (1994, September 26). Duck, it's the New-New Math. *San Francisco Chronicle,* p. A20.

Sawyer, W. W. (1955). *Prelude to mathematics.* Harmondsworth, England: Penguin.

Schauble, L., & Glaser, R. (Eds.). (1996). *Innovations in learning: New environments for education.* Mahwah, NJ: Erlbaum.

Schifter, D., & Fosnot, C. T. (2003). *Reconstructing mathematics education: Stories of teachers meeting the challenge of reform.* New York: Teachers College Press.

Schoenfeld, A. H. (1991). On mathematics as sense-making: An informal attack on the unfortunate divorce of formal and informal mathematics. In J. F. Voss, D. N. Perkins, & J. W. Segal (Eds.), *Informal reasoning and education* (pp. 311–343). Hillsdale, NJ: Erlbaum.

Schoenfeld, A. H. (1992). Learning to think mathematically. Problem solving, metacognition and sense-making in mathematics. In D. A. Grouws (Ed.), *Handbook of research on mathematics teaching and learning: A project of the National Council of Teachers of Mathematics* (pp. 334–370). New York: MacMillan.

Seeger, F., Voigt, J., & Waschescio, U. (Eds.). (1998). *The culture of the mathematics classroom.* Cambridge, England: Cambridge University Press.

Selin, H. (Ed.). (2000). *Mathematics across cultures: The history of non-Western mathematics.* Dordrecht, The Netherlands: Kluwer.

Selter, C. (1993). *Eigenproduktionen im Arithmetik der Primarstufe.* [Children's own productions in primary arithmetic teaching]. Wiesbaden, Germany: Deutscher Universitätsverlag.

Selter, C. (1996). Doing mathematics while practicing skills. In C. Van den Boer & M. Dolk (Eds.), *Modellen, meten en meetkunde. Paradigma's van adaptief onderwijs* [Models, mensuration, and geometry. Paradigms for adaptive education] (pp. 31–43). Utrecht, The Netherlands: Panama/HvU & Freudenthal Institute.

Selter, C. (1998). Building on children's mathematics—A teaching experiment in grade 3. *Educational Studies in Mathematics, 36,* 1–27.

Selter, C. (2001). Chapter 2: Understanding – the underlying goal of teacher education. In M. Van den Heuvel-Panhuizen (Ed.), *Proceedings of the 25th International Conference of the International Group for the Psychology of Mathematics Education,*

1, 198–202. Utrecht, The Netherlands: Freudenthal Institute.

Selter, C. (2002). Taking into account different views. Three brief comments on papers by Gravemeijer & Stephan, Cobb, and Thompson. In K. Gravemeijer, R. Lehrer, B. Van Oers, & L. Verschaffel (Eds.), *Symbolizing, modeling, and tool use in mathematics education* (pp. 221–230). Dordrecht, The Netherlands: Kluwer.

Senk, S. L., & Thompson, D. R. (2003). *Standards-based school mathematics curricula*. Mahwah, NJ: Erlbaum.

Seo, K-H., & Ginsburg, H. P. (2003). "You've got to carefully read the math sentence . . .": Classroom context and children's interpretations of the equals sign. In A. J. Baroody & A. Dowker (Eds.), *The development of arithmetic concepts and skills* (pp. 161–188). Mahwah, NJ: Erlbaum.

Sfard, A. (1998). Two metaphors for learning mathematics: Acquisition metaphor and participation metaphor. *Educational Researcher, 27*(2), 4–13.

Sfard, A. (2003). Balancing the unbalanceable: The NCTM Standards in light of theories of mathematics. In J. Kilpatrick, W. G. Martin, & D. Schifter (Eds.), *A research companion to Principles and Standards for School Mathematics* (pp. 353–392). Reston, VA: National Council of Teachers of Mathematics.

Sfard, A. (2005). What could be more practical than good research? *Educational Studies in Mathematics, 58*, 393–413.

Shrager, J., & Siegler, R. S. (1998). SCADS: A model of children's strategy choices and strategy discoveries. *Psychological Sciences, 9*, 405–410.

Siegler, R. S. (1998). *Children's thinking*. Saddle River, NJ: Prentice Hall.

Siegler, R. S. (2001). Children's discoveries and brain-damaged patients' rediscoveries. In J. L. McClelland & R. S. Siegler (Eds.), *Mechanisms of cognitive development. Behavioral and neural perspectives* (pp. 33–63). Mahwah, NJ: Erlbaum.

Siegler, R. S. (2003). Implications of cognitive science research for mathematics education. In J. Kilpatrick, W. G. Martin, & D. Schifter (Eds.), *A research companion to Principles and Standards for School Mathematics* (pp. 289–303). Reston, VA: National Council of Teachers of Mathematics.

Siegler, R. S., & Booth, J. L. (2005). Development of numerical estimation: A review. In J. I. D. Campbell (Ed.), *Handbook of mathematical cognition* (pp. 197–212). New York: Psychology Press.

Siegler, R. S., & Jenkins, E. A. (1989). *How children discover new strategies*. Hillsdale, NJ: Erlbaum.

Siegler, R. S., & Lemaire, P. (1997). Older and younger adults' strategy choices in multiplication: Testing predictions of ASCM using the choice/no-choice method. *Journal of Experimental Psychology: General, 126*, 71–92.

Siegler, R. S., & Opfer, J. (2003). The development of numerical estimation. Evidence for multiple representations of numerical quantity. *Psychological Science, 14*, 237–243.

Siegler, R. S., & Shipley, C. (1995). Variation, selection and cognitive change. In T. Simon & G. Halford (Eds.), *Developing cognitive competence: New approaches to process modelling* (pp. 31–76). Hillsdale, NJ: Erlbaum.

Siegler, R. S., & Shrager, J. (1984). Strategy choices in addition and subtraction: How do children know what to do? In C. Sophian (Ed.), *Origins of cognitive skills*. Hillsdale, NJ: Erlbaum.

Sierpinska, A., & Kilpatrick, J. (Eds.). (1998). *Mathematics education as a research domain: A search for identity*. Dordrecht, The Netherlands: Kluwer.

Simon, M. (1993). Prospective elementary teachers' knowledge of division. *Journal for Research in Mathematics Education, 24*, 233–254.

Simon, M. A. (1995). Reconstructing mathematical pedagogy from a constructivist perspective. *Journal for Research in Mathematics Education, 26*, 114–145.

Sophian, C., & McCorgray, P. (1994). Part-whole knowledge and early arithmetic problem solving. *Cognition and Instruction, 12*, 3–33.

Sophian, C., & Vong, K. I. (1995). The parts and wholes of arithmetic story problems: Developing knowledge in the preschool years. *Cognition and Instruction, 13*, 469–477.

Sowder, J. (1992). Estimation and number sense. In D. A. Grouws (Ed.), *Handbook of research on mathematics teaching and learning* (pp. 371–389). New York: MacMillan.

Steen, L. (1999). Theories that gyre and gimble in the wabe. *Journal for Research in Mathematics Education, 30*, 235–241.

Steen, L. (2004). Cited in *Focus* [Newsletter of the Mathematical Association of America], *24*(6), 29.

Steffe, L. P., & Cobb, P. (1988). *Construction of mathematical meanings and strategies*. New York: Springer.

Steffe, L. P., & Cobb, P. (1998). Multiplicative and division schemes. *Focus on Learning Problems in Mathematics, 20*(1), 45–61.

Stern, E. (1993). What makes certain arithmetic word problems involving the comparison of sets so difficult for children? *Journal of Educational Psychology, 85*, 7–23.

Stevenson, H. W., & Stigler, J. W. (1992). *The learning gap. Why our schools are failing and what can we learn from Japanese and Chinese education*. New York: Summit Books.

Stigler, J. W., Fuson, K. C., Ham, M., & Kim, M. S. (1986). An analysis of addition and subtraction word problems in American and Soviet elementary mathematics textbooks. *Cognition and Instruction, 3*, 153–171.

Stigler, J. W., & Hiebert, J. (1999). *The teaching gap: Best ideas from the world's teachers for improving education in the classroom*. New York: Free Press.

Stillman, G., & Balatti, J. (2001). Contribution of ethnomathematics to mainstream mathematics classroom practice. In B. Atweh, H. Forgasz, & B. Nebres (Eds.), *Sociocultural research on mathematics education: An international perspective*. Mahwah, NJ: Erlbaum.

Swafford, J. O., & Langrall, C. W. (2000). Grade 6 students' preinstructional use of equations to describe and represent problem situations. *Journal for Research in Mathematics Education, 31*, 89–112.

Swetz, F. J. (Ed.). (1994). *From five fingers to infinity*. Chicago: Open Court.

Tamburino, J. L. (1982). *The effects of knowledge-based instruction on the abilities of primary grade children in arithmetic word problem solving*. Unpublished doctoral dissertation, University of Pittsburgh, PA.

Teong, S. K. (2000). *The effect of metacognitive training on the mathematical word problem solving of Singapore 11–12 years old in a computer environment*. Unpublished doctoral thesis, Centre for Studies in Science and Mathematics Education, University of Leeds, United Kingdom.

Thomas, J. (2001). Globalization and politics of mathematics education. In B. Atweh, H. Forgasz, & B. Nebres (Eds.),

Sociocultural research on mathematics education (pp. 95–112). Mahwah, NJ: Erlbaum.

Thompson, I. (1994). Young children's idiosyncratic written algorithms for addition. *Educational Studies in Mathematics, 26*, 323–345.

Thompson, I. (1999a). Getting your head around mental calculation. In I. Thompson (Ed.), *Issues in teaching numeracy in primary schools* (pp. 145–156). Buckingham, U.K.: Open University Press.

Thompson, I. (Ed.). (1999b). *Issues in teaching numeracy in primary schools.* Buckingham, U.K.: Open University Press.

Thompson, I. (1999c). Written methods of calculation. In I. Thompson (Ed.), *Issues in teaching numeracy in primary schools* (pp. 169–183). Buckingham, U.K.: Open University Press.

Thorndike, E. L. (1922). *The psychology of arithmetic.* New York: Macmillan.

Toom, A. (1999). Word problems: Applications or mental manipulatives. *For the Learning of Mathematics, 19*(1), 36–38.

Torbeyns, J., Arnaud, L., Lemaire, P., & Verschaffel, L. (2004). Cognitive change as strategy change. In A. Demetriou & A. Raftopoulos (Eds.), *Cognitive developmental change: Theories, models and measurement* (pp. 186–216). Cambridge, U.K.: Cambridge University Press.

Torbeyns, J., Verschaffel, L., & Ghesquière, P. (2004). Strategy development in children with mathematical disabilities: Insights from the choice/no-choice method and the chronological-age/ability-level-match design. *Journal of Learning Disabilities, 37*, 119–131.

Torbeyns, J., Verschaffel, L., & Ghesquière, P. (2005). Simple addition strategies in a first-grade class with multiple strategy instruction. *Cognition and Instruction, 23*, 1–21.

Torbeyns, J., Verschaffel, L., & Ghesquière, P. (2006). Development of children's strategy competencies in the number domain from 20 up to 100. *Cognition and Instruction, 24*, 439–465.

Towse, J., & Saxton, M. (1998). Mathematics across national boundaries: Cultural and linguistic perspectives on numerical competence. In C. Donlan (Ed.), *The development of mathematical skills* (pp. 129–150). Hove, England: Psychology Press.

Treffers, A. (1987). *Three dimensions. A model of goal and theory description in mathematics education. The Wiskobas project.* Dordrecht, The Netherlands: Reidel.

Treffers, A. (1991). Didactical background of a mathematics program for primary education. In L. Streefland (Ed.), *Realistic mathematics education in primary school* (pp. 21–57). Utrecht, The Netherlands: CD Press.

Treffers, A. (1993). Ontwikkelingsonderzoek in eerste aanzet. [Developmental research: A first attempt.] In R. De Jong & M. Wijers (Eds.), *Ontwikkelingsonderzoek: Theorie en praktijk* [Developmental research: Theory and practice] (pp. 17–34). Utrecht, The Netherlands: Freudenthal Institute.

Treffers, A. (2001a). Grade 1 (and 2): Calculation up to twenty. In M. Van den heuvel (Ed.), *Children learn mathematics* (pp. 43–60). Utrecht, The Netherlands: Freudenthal Institute, University of Utrecht.

Treffers, A. (2001b). Numbers and number relationships. In M. Van den heuvel (Ed.), *Children learn mathematics* (pp. 101–120). Utrecht, The Netherlands: Freudenthal Institute, University of Utrecht.

Treffers, A., Nooteboom, A., & De Goeij, E. (2001). Column arithmetic and algorithms. In M. Van den heuvel (Ed.),

Children learn mathematics (pp. 147–172). Utrecht, The Netherlands: Freudenthal Institute, University of Utrecht.

United States Department of Education (2002). *Proven methods: Scientifically based research.* Retrieved, July 7 2006, from: http://www.ed.gov/nclb/methods/whatworks/research/index.html

Usiskin, Z. (1998). Paper-and-pencil algorithms in a calculator-and-computer age. In L. J. Morrow & M. J. Kenney (Eds.), *The teaching and learning of algorithms in school mathematics* (pp. 7–20). Reston, VA: National Council of Teachers of Mathematics.

Usiskin, Z. (2004). The arithmetical operations as mathematical models. In H.-W. Henn & W. Blum (Eds.), *Pre-conference volume, ICMI Study 14: Applications and modelling in mathematics education* (pp. 279–284). Dortmund, Germany: University of Dortmund.

Van Amerom, B. A. (2002). *Reinvention of early algebra.* Utrecht, The Netherlands: Utrecht University.

Van den Heuvel-Panhuizen, M. (1996). *Assessment and realistic mathematics education.* Utrecht, The Netherlands: Utrecht University.

Van den Heuvel-Panhuizen, M. (Ed.). (2001). *Children learn mathematics.* Utrecht, The Netherlands: Freudenthal Institute, University of Utrecht.

Van Dooren, W., De Bock, D., Janssens, D., & Verschaffel, L. (in press). Pupils' overreliance on linearity: A scholastic effect? *British Journal of Educational Psychology.*

Van Dooren, W., De Bock, D., Hessels, A., Janssens, D., & Verschaffel, L. (2005). Not everything is proportional: Effects of age and problem type on propensities for overgeneralization. *Cognition and Instruction, 23*, 57–86.

Van Dooren, W., Verschaffel, L., Greer, B., & De Bock, D. (2006). Modelling for life: Developing adaptive expertise in mathematical modelling from an early age. In L. Verschaffel, F. Dochy, M. Boekaerts, & S. Vosniadou (Eds), *Instructional psychology: Past, present and future trends* (pp. 91–112). Oxford: Elsevier.

Van Dooren, W., Verschaffel, L., & Onghena, P. (2002). The impact of preservice teachers' content knowledge on their evaluation of students' strategies for solving arithmetic and algebra. *Journal for Research in Mathematics Education, 33*, 319–351.

Van Essen, G. (1991). *Heuristics and arithmetic word problems.* Unpublished doctoral dissertation, State University, Amsterdam.

Van Putten, C. M., van den Brom-Snijders, P., & Beishuizen, M. (2005). Progressive mathematization of long division in Dutch primary schools. *Journal for Research in Mathematics Education, 36*, 44–73.

Vergnaud, G. (1988). Multiplicative structures. In J. Hiebert & M. Behr (Eds.), *Research agenda in mathematics education: Number concepts and operations in the middle grades* (pp. 141–161). Hillsdale, NJ: Erlbaum.

Vergnaud, G. (1996). The theory of conceptual fields. In L. P. Steffe, P. Nesher, P. Cobb, G. A. Goldin, & B. Greer (Eds.), *Theories of mathematical learning* (pp. 219–239). Mahwah, NJ: Erlbaum.

Verguts, T. (2003). *Retrieval in single-digit multiplication arithmetic.* [Internal report]. Ghent, Belgium: Laboratory for Experimental Psychology, University of Ghent.

Verschaffel, L. (1994). Using retelling data to study elementary school children's representations and solutions of compare problems. *Journal for Research in Mathematics Education, 25,* 141–165.

Verschaffel, L. (1997). Young children's strategy choices for solving elementary arithmetic wor(l)d problems: The role of task and context variables. In M. Beishuizen, K. Gravemeijer, & E. Van Lieshout (Eds.), *The role of contexts and models in the development of mathematical strategies and procedures* (pp. 113–126). Utrecht, The Netherlands: Center for Science and Mathematics Education, Freudenthal Institute, University of Utrecht.

Verschaffel, L., & De Corte, E. (1993). A decade of research on word-problem solving in Leuven: Theoretical, methodological and practical outcomes. *Educational Psychology Review, 5,* 239–256.

Verschaffel, L., & De Corte, E. (1996). Number and arithmetic. In A. Bishop, K. Clements, C. Keitel, & C. Laborde (Eds.), *International handbook of mathematics education.* (Vol. 1, pp. 99–138). Dordrecht, The Netherlands: Kluwer.

Verschaffel, L., & De Corte, E. (1997). Word problems. A vehicle for promoting authentic mathematical understanding and problem solving in the primary school. In T. Nunes & P. Bryant (Eds.), *Learning and teaching mathematics: An international persepctive* (pp. 69–97). Hove, U.K.: Psychology Press.

Verschaffel, L., De Corte, E., & Borghart, I. (1997). Pre-service teachers' conceptions and beliefs about the role of real-world knowledge in mathematical modelling of school word problems. *Learning and Instruction, 4,* 339–359.

Verschaffel, L., De Corte, E., Lamote, C., & Dhert, N. (1998). Development of a flexible strategy for the estimation of numerosity. *European Journal for Psychology of Education, 13,* 347–270.

Verschaffel, L., De Corte, E., & Lasure, S. (1994). Realistic considerations in mathematical modelling of school arithmetic word problems. *Learning and Instruction, 4,* 273–294.

Verschaffel, L., De Corte, E., Lasure, S., Van Vaerenbergh, G., Bogaerts, H., & Ratinckx, E. (1999). Design and evaluation of a learning environment for mathematical modeling and problem solving in upper elementary school children. *Mathematical Thinking and Learning, 1,* 195–229.

Verschaffel, L., De Corte, E., & Pauwels, A. (1992). Solving compare problems: An eye-movement test of Lewis and Mayer's consistency hypothesis. *Journal of Educational Psychology, 84,* 85–94.

Verschaffel, L., Greer, B., De Corte, E. (2000). *Making sense of word problems.* Lisse, The Netherlands: Swets & Zeitlinger.

Verschaffel, L., Janssens, S., & Janssen, R. (2005). Development of mathematical competence in pre-service elementary school teachers in Flanders. *Teaching and Teacher Education, 22,* 49–63.

Verschaffel, L., Luwel, K., & Torbeyns, J. (2006). *Development of estimation of 'real' numerosities in elementary school children.* (Manuscript submitted for publication.)

Verschaffel, L., & Vosniadou, S. (Eds.). (2004). The conceptual change approach to mathematics learning and teaching. *Learning and Instruction, 14,* 445–548.

Vosniadou, S., & Verschaffel, L. (2004). Extending the conceptual change approach to mathematics learning and teaching. *Learning and Instruction, 14,* 445–451.

Weaver, J. F. (1982). Interpretations of number operations and symbolic representations of addition and subtraction. In T. P. Carpenter, J. M. Moser, & T. A. Romberg (Eds.), *Addition and subtraction: A cognitive perspective* (pp. 60–66). Hillsdale, NJ: Erlbaum.

Wells, D. (1997). *The Penguin book of curious and interesting mathematics.* London: Penguin.

Whalen, J., Gallistel, C.R., & Gelman, R. (1999). Nonverbal counting in humans: The psychophysics of number representation. *Psychological Science, 10,* 130–137.

Wittmann, E. Ch. (1995). Mathematics education as a design science. *Educational Studies in Mathematics, 29,* 355–374.

Wittmann, E. Ch. (2001). Drawing on the richness of elementary mathematics in designing substantial learning environments. In M. Van den Heuvel-Panhuizen (Ed.), *Proceedings of the 25th International Conference of the International Group for the Psychology of Mathematics Education, 1,* 198–202. Utrecht, The Netherlands: Freudenthal Institute.

Wolters, M. A. D. (1983). The part-whole schema and arithmetical problems. *Educational Studies in Mathematics, 14,* 127–138.

Wu, H. (1999, Fall). Basic skills versus conceptual understanding: A bogus dichotomy in mathematics education. *American Educator/American Federation of Teachers,* 1–7.

Yackel, E., & Cobb, P. (1996). Sociomathematical norms, argumentations, and autonomy in mathematics. *Journal for Research in Mathematics Education, 27,* 458–477.

Yoshida, H., Verschaffel, L., & De Corte, E. (1997). Realistic considerations in solving problematic word problems: Do Japanese and European children have the same difficulties? *Learning and Instruction, 7,* 329–338.

Yoshida, M. (1999). *Lesson study: A case study of a Japanese approach to improving instruction through school-based teacher development.* Unpublished doctoral dissertation, University of Chicago.

Zaslavsky, C. (1973). *Africa counts: Number and pattern in African culture.* Brooklyn, NY: Lawrence Hill Books.

Zaslavsky, C. (1991). Multicultural mathematics education in the middle grades. *Arithmetic Teacher,* 8–13.

Zazkis, R. (2001). From arithmetic to algebra via big numbers. In H. Chick, K. Stacey, J. Vincent, & J. Vincent (Eds.), *Proceedings of the 12th ICMI Study Conference* (Vol. 1, pp. 676–681). Melbourne, Australia: University of Melbourne.

AUTHOR NOTE

We wish to thank a number of individuals for their helpful comments on earlier drafts of the chapter. Art Baroody and Karen Fuson provided insightful and constructive comments at critical points during the development of the chapter. The work herein was partly supported from grant GOA/2006/01 of the Research Fund of the K.U.Leuven, Belgium.

14

RATIONAL NUMBERS AND PROPORTIONAL REASONING

Toward a Theoretical Framework for Research

Susan J. Lamon

MARQUETTE UNIVERSITY

Traditionally, ratio and proportion have been considered middle school topics, corresponding to Piaget's theory that children achieve understanding of these topics in adolescence, but when viewed in a larger context, their tight connections with fractions and other multiplicative concepts suggest a body of knowledge that spans the entire curriculum—elementary school through university-level mathematics. Of all the topics in the school curriculum, fractions, ratios, and proportions arguably hold the distinction of being the most protracted in terms of development, the most difficult to teach, the most mathematically complex, the most cognitively challenging, the most essential to success in higher mathematics and science, and one of the most compelling research sites. In the last decade or more, researchers have made little progress in unraveling the complexities of teaching and learning these topics. Worse yet, the number of scholars pursuing long-term research agendas in the field of rational numbers is disproportionate to the mathematical richness of the domain and to the amount of research that needs to be accomplished.

Currently, research in the domain is at a turning point, a propitious time to take stock of where research in this domain has been and where it is going. With the hindsight afforded by the passage of approximately 15 years since the last handbook chapter was begun, I will look back at the major themes and messages addressed in that chapter, and, taking into account some changes that have occurred since then, reevaluate the research area, identify persistent unresolved issues, and add to the research agenda the new questions that have arisen. This chapter will attempt to explicate the mathematical complexities of the domain so that it is more inviting to researchers who may be interested in studying related topics. I begin by revisiting the three major areas addressed by Behr, Harel, Post, and Lesh (1992): their semantic/mathematical analysis of rational numbers, their principles for qualitative reasoning, and the teaching and learning of rational number concepts.

LOOKING BACK

Behr, Harel, Post, and Lesh wrote the chapter on rational number, ratio, and proportion in the 1992 *Handbook of Research on Mathematics Teaching and Learning* shortly after the Research Agenda Project Conference, at which the leading scholars on rational number topics evaluated the state of the art and set an agenda for future research in middle school mathematics. The consensus of the research community was that "the differences between early and later number concepts need to be recognized and more adequately understood before instructional programs can be developed to enhance number learning in the middle grades" (Sowder, 1988, p. ix). Rational number construct theory was paramount; multiple attempts were made to capture the essence of rational numbers for pedagogical purposes. Our tradition of teaching for computational ability in the industrial era had left us pedagogically bankrupt for an age that values connections and meaning, and basic questions needed exploration. What does it mean to understand the rational numbers? What are the conceptual hurdles for children as they make the transition from whole number topics to rational number topics? Thus, Behr et al. began with a discussion of the most important work of that era: Behr, Lesh, Post, and Silver (1983), Kieren (1976), and Ohlsson (1987).

These independent analyses represented different perspectives and resulted in different conclusions about how many subconstructs or personalities of the rational numbers show up at the phenomenological level (Freudenthal, 1983), that is, in real situations for which rational numbers are useful. When these analyses were reconciled, *measure, quotient, ratio,* and *operator* were recognized as distinct subconstructs (Kieren, 1988), and there was some disagreement about whether *part-whole* was distinct from the *measure* subconstruct. Whatever the number of subconstructs, the overriding issue was that current instruction, which addresses only one of them, is inadequate.

Behr et al. (1992) also presented a semantic/ mathematical analysis of rational numbers and an accompanying mathematical notation system based on units and the ways in which they may be regrouped to represent the subconstructs of the rational numbers. Their formal microanalysis of the possible unit conversions did not take into account whether these operations have any basis in real problems, whether they represent mental actions that children actually perform, or whether teachers should be advised to teach every possible (reasonable) conversion. Nonetheless, this work served the authors' stated purpose of emphasizing the critical deficiencies in the mathematics curriculum in terms of the kinds of opportunities it affords students to develop flexibility in unitizing and reunitizing quantities. (*Unitizing* refers to the process of mentally chunking or restructuring a given quantity into familiar or manageable or conveniently sized pieces in order to operate with that quantity. For example, one might think of a case of 24 colas as 24 cans, 2 (12-packs), or 4 (6-packs).)

For several reasons, the Behr et al. (1992) semantic analysis and notational system has not proved useful in ensuing research. The authors stated that "it remains for curriculum development and research to create problem situations whose solutions would involve these manipulations and to determine whether such solutions can be created by children" (p. 305). The notion of beginning with the mathematics and inventing problems to create a purpose for the mathematics is antithetical to current conceptions of doing mathematics, especially from a mathematical modeling perspective, in which one begins with a real or realistic situation and formulates a mathematical description for the purpose of explaining, predicting, drawing conclusions, or answering questions (e.g., Lamon, Parker, & Houston, 2003).

A second objection relates to the authors' emphasis on manipulating physical representations. The authors claimed that the manipulation of objects should facilitate the construction of the units given in their analysis (p. 305) and that "a goal of our analysis is to provide a theoretical context that will guide research into the cognition underlying students' manipulations in transforming physical representations" (p. 315). Building upon Dienes's (1960) principles of perceptual variability, they advocated the teaching of fractions using a model that entails multiple translations between real-world situations and written symbols, pictures, spoken symbols, and manipulative aids. In fact, Cramer (2003) and Cramer, Post, and delMas (2002) have had success with this translation model in helping students to develop understanding of initial fraction concepts. Nevertheless, Behr et al.'s (1992) formal quantity analysis of rational number constructs deals only with *discrete* representations of rational number and does not address the more important instructional question of how to facilitate children's conceptual leap to quantities that defy physical representation, those that result from reflective abstraction, and those that do not result from counting and correspondence mechanisms, or direct measurement—the intensive or "per" quantities (Schwartz, 1988). These include such quantities as speed (miles per hour), price (cents per gram), density (grams per cubic centimeter) and tire pressure (pounds per square inch).

The second section of the Behr et al. chapter addressed the intuitive qualitative reasoning that might be applied to rational number and proportional situations, specifically, in determining fraction order or equivalence. Essentially, the concepts of equivalence and ordering of rational numbers by size rely on determining whether a multiplicative relationship varies or remains invariant. In some cases, but not all, qualitative reasoning is sufficient to determine order relationships.

A *quantity* is a measurable quality of an object— whether that quality is actually quantified or not. For example, you can compare the heights of two people in your family without measuring them. When one is standing beside or near the other, you can tell which is taller. If you are in New York, you can safely say that Philadelphia is closer than Los Angeles without checking a mileage table to get the distance between the cities. Relating quantities that are not quantified is an important kind of reasoning that is easily accessible to children when posed in contexts that they understand (Lamon, 2006). For example:

> Yesterday you shared some cookies with some friends. Today, you share fewer cookies with more friends. Will everyone get more, less, or the same amount as they received yesterday?

Of course, you know that everyone will receive less today. Both the number of people and the number of cookies might be quantified by counting; however, even without quantifying either quantity, the question is easily answered. In fact, the question invites you to determine the direction of change between yesterday's ratio of the number of cookies to the number of people and today's ratio.

Suppose that today fewer people share more cookies. Again, you can tell that everyone would receive more than they did yesterday. Suppose that today more people shared more cookies. In this case, it could be more or less, depending on the numbers chosen.

This type of reasoning is easy for children because it builds on their prior knowledge and experience with sharing. It is, in fact, intuitive. There are nine different situations, only two of which are indeterminate, in which students can think about the change in the size of the ratio of cookies to children without quantifying either quantity and simply by knowing the direction of change in each quantity. Letting + represent an increase, −, a decrease, and ?, an indeterminate change, the nine cases are summarized in the Table 14.1.

Table 14.1 Direction of Change in a Ratio

Change in quantity _cookies per person_

Change in number of cookies	Change in number of people		
	+	−	0
+	?	+	+
−	−	?	−
0	−	+	0

This thinking is useful for some comparisons, but not sufficient for all comparisons. For example, children can rely on their intuitive knowledge to compare

$\frac{4}{7}$ and $\frac{5}{6}$ (today fewer people share more cookies), but other methods are needed to compare, say, $\frac{4}{7}$ and $\frac{5}{8}$.

Behr et al. analyzed a second type of multiplicative task, one that entails the invariance of a product. The example they used was that of the balance-beam task. However, this type of task is more difficult than a simple direct proportion (Vergnaud, 1988). The product of measures is a structure that consists of a Cartesian composition of two measure spaces into a third measure space and is useful in problems involving area, volume, Cartesian products, work, and many other physical concepts. For example, when you multiply (number of men) × (number of hours they work), you get # of man-hours; when you multiply area of the base of a right prism by height, you get neither area nor height, but rather a new measure, volume. Thus, on several accounts, this type of task is more cognitively demanding: a) It requires the coordination at least three quantities simultaneously; b) knowledge of physical principles is involved; and c) as research has consistently demonstrated (Case, 1992; Inhelder & Piaget, 1958; Marini, 1992; Siegler, 1976), children's understanding of this task and others like it (e.g., projection-of-shadow problems) shows clear developmental trends, both in terms of the M factor (Pascual-Leone, 1969), which limits the amount of information to which a child can attend, and in terms of the level of understanding of the situation.

As is common in neo-Piagetian theories of conceptual development (Case, 1992), it is important to distinguish the types of competence a task requires and the epistemic level of that competence. Any competence can be demonstrated to be present in some form at almost any age (Bruner, 1960). However, the level of functioning with that concept changes noticeably with development; that is, children in certain age groups reach a common developmental ceiling across a variety of tasks (Marini, 1992). Cross-task asynchrony occurs, however, as children solve each task independently, using whatever task-specific background they might have and heavily relying on the contextual factors present in the task.

Although people with a mature understanding of physical situations such as the balance-beam task can determine when the order relation between products is determinate or indeterminate for each of the nine change conditions given in the table above, stating these relations as principles or rules for which initial instruction should help children "construct principles to apply qualitatively to questions of order" (Behr et al., 1992, p. 322) goes well beyond what is typically classified as intuitive knowledge or "knowledge that is self-evident and obvious to the person who has it" (Resnick, 1986, p. 188). Thus, qualitative reasoning is not necessarily intuitive.

In the first handbook chapter, the authors named the clarification of concepts and standardization of vocabulary as prerequisites to progress in research on rational numbers. Their observations remain as compelling today as they were then. Specifically, we still need to "agree on the concepts of fraction and rational number" (Behr et al., 1992, p. 296). Other distinctions such as the difference between ratios and rates, and the difference between reasoning proportionally and understanding proportionality still need clarification. Another persistent issue is the confusion in naming student strategies for determining the equivalence of ratios—the *within* strategy or the *between* strategy—that resulted from the different uses of the term in early research, depending on whether the researcher came from a mathematics background or a science background. Several issues were not addressed in that chapter. What counts as proportional reasoning? What is the connection between proportional reasoning and fractions? What is the connection of proportional reasoning to the proportionality schema (Inhelder & Piaget, 1958)? As work in the field progresses, these questions continue to present themselves for clarification and will be addressed in this chapter.

In the final section of the chapter, Behr et al. discussed their original charge to address research on teaching rational numbers. They noted that, unfortunately, they were "unable to find much research that specifically targeted teaching rational number concepts" (1992, p. 300). The full significance of that statement was that the research domain including fractions, rational numbers, and ratios and proportions had not yet reached a level of maturity from which it could offer empirical propositions for teaching, that is, generalizations that derive directly from empirical findings. Nevertheless, they reported on a different kind of knowledge that informs teaching, a conceptual invention. Conceptual inventions (Shulman, 1986a) do not derive directly from findings in a straightforward way. They are acts of scholarly invention based on theoretical understanding, practical wisdom, and generalizations from empirical work. They explained Lesh's model for translations between modes of representation for teaching initial fraction concepts and the manner in which it exploits perceptual variability to aid mathematical abstraction. In this chapter, I will report on research that has finally taken rational number concepts into the classroom.

SCOPE AND ORGANIZATION OF THIS CHAPTER

The nature of this chapter is influenced by three important developments. Chief among them is the fact that six major contributors to rational number research have died since the last handbook chapter (including Merlyn Behr), and several other contributing scholars have retired. These were great thinkers to whom all future researchers owe a debt of gratitude. Unfortunately, new researchers, committed to long-term agendas in this research area, are not stepping up to replace those pioneers. Opportunistic treatment of rational numbers and multiplicative ideas in the course of pursuing other research topics is clearly not sufficient to make progress in this domain. The work that needs to be done is abundant enough to provide career-long research agendas for many scholars.

As the reader will note, the number of references in this chapter predating 1992 is far greater than the number appearing since the last handbook. This crisis—I do not exaggerate to call it that—stems from the complex nature of the mathematics as well as from the difficulty of conducting research because a) teachers are not prepared to teach content other than part-whole fractions; b) long-term commitment is needed because rational number topics are learned over many years; c) the nonlinear development of content does not mesh well with scope and sequence charts currently prescribing mathematics instruction in schools; and d) in comparison to a domain such as early addition and subtraction, little research progress is evident. Indeed, the rational number domain, however critical, has been perceived as a difficult research site. One goal of this chapter is to share recent exciting developments in this research area, to summarize the history and work in this field in understandable terms and to inspire new researchers to adopt personal research agendas based on multiplicative topics.

Second, the diversification of research approaches can drastically change the status of a research domain by making feasible and legitimate work that was previously impossible. In recent years, the design experiment (Brown, 1992, also deceased since the last handbook; Kelly & Lesh, 2000) has come into vogue and has opened the door to new possibilities for research in complex domains such as rational numbers. Brown originally envisioned a dynamic relationship between classroom-based and laboratory-based research, a dialectic in which substantive action aims to make changes in classroom practice but at the same time tries to discover the means to enact those changes most effectively. Just as researchers several decades ago were able to begin research in children's thinking in addition and subtraction because of the cognitive revolution and the acceptance of Piaget's clinical interview methods, so has the design experiment provided the same impetus to research in the rational number domain. Short-term and cross-sectional studies have not been particularly useful in studying a conceptual domain that is subject to developmental processes. Although Piagetian interviews reveal something about the changes underlying children's thought processes, they fail to capture the dynamic processes underlying such change, including some of the fleeting transitional stages, the full range of contexts and decisions that might affect those changes, and the full range of reactions of teachers and students in real classrooms, rather than under highly controlled research conditions. The design experiment allows the researcher to influence, to study, and to adjust the course of the longitudinal study based on the unstaged and serendipitous dynamics of a real classroom.

The third significant development, related to the last, is that research in this field has met a milestone since the last handbook chapter in terms of beginning the longitudinal research in real classrooms (Lamon, in progress). A nonclassical framework for this research was based on top-down analyses of content, bottom-up investigations of student thinking, and neo-Piagetian cognitive psychology, and was implemented in two longitudinal studies using a design-experiment research paradigm. Two 4-year studies with the same children tracked their development from the start of fraction instruction in the third grade until the end of sixth grade, yielding new insights into the teaching and learning of rational numbers. Because the work in this field is reaching a new level of maturity, this is a propitious time to look back and to analyze the research field in terms of the broad stages through which it has progressed. Hopefully taking stock of where we as researchers have been from a vantage point that also allows us to see new directions for future work will give new impetus to research in this domain.

This chapter is divided into three sections, the first of which addresses some persistent language and definition issues, with the goal of promoting better communication. The second section will review the evolution of research in this field, looking at the development of rational number research in the broad stages through which work has progressed. The third section will present a theoretical framework that provided the basis for several longitudinal studies, as well as a summary of the major results of those studies. Research generally raises more questions than it answers. Therefore, the chapter will conclude with research questions that may provide inspiration for those who wish to take up the challenge of research in this field.

To keep this chapter tractable, I will restrict discussion to the content and issues most appropriately addressed in the school years 3–8, spanning the time that children begin fraction instruction until they finish middle school. Understanding fractions, ratios, and proportions is but a small piece of the huge system of interrelated multiplicative concepts, operations, and associated contexts whose construction extends from the elementary school years through undergraduate and graduate school. However, with an eye toward later usefulness and connections, this chapter will focus on the early years because research into the learning of advanced multiplicative concepts is a moot issue when we educators are failing miserably at teaching the most elementary multiplicative concepts and operations.

Neither will this chapter address fractions and proportional reasoning within the context of inservice or preservice teacher education. Many adults, including middle school teachers (for example, Harel, Behr, Post, & Lesh, 1994; Post, Harel, Behr, & Lesh, 1988) and preservice teachers (e.g., Graeber, Tirosh, & Glover, 1989; Harel et al., 1994; Simon & Blume, 1994), struggle with the same concepts and hold the same primitive ideas and misconceptions as students do. This may be attributed to the fact that the mathematics curriculum has never appropriately addressed the multiplicative conceptual field, and teachers themselves have had the same school experiences as current students receive.

DEFINING TERMS AND CLARIFYING CONSTRUCTS

Several mathematical constructs are ill-defined, and their relationships to one another have become obfuscated as terminology has been used indiscriminately. In other cases, concepts and language need to be periodically updated because meanings have evolved as part of the natural development of a research area. In this section, I will discuss *within* and *between* strategies, the difference between ratios and rates, the difference between fractions and rational numbers, what it means to understand fractions as numbers, the difference between proportional reasoning and proportionality, and the connection between the rational numbers and proportional reasoning.

Within (Internal) and Between (External) Ratios

Freudenthal (1973, 1978) classified a ratio as an internal or within ratio if its constituent magnitudes shared the same measure space, or in Freudenthal's terminology, came from the same "system." Similarly, an external or between ratio is composed of magnitudes from different measure spaces.

The two ratios in the following proportion involve an internal comparison by Freudenthal's definition:

$$4 \text{ people} : 20 \text{ people} = 1 \text{ car} : 5 \text{ cars}$$

whereas this proportion uses external or between ratios:

$$4 \text{ people} : 1 \text{ car} = 20 \text{ people} : 5 \text{ cars}.$$

Confusion arose, however, because research originating in a science tradition (Karplus, Pulos, & Stage, 1983a, 1983b; Noelting 1980a, 1980b) used an alternate definition of a system: a set of interacting elements. According to this alternate conception, the example above shows two systems with people and cars interacting within each of them. The scientist's notion of internal ratio (corresponding to Freudenthal's external ratio) is a comparison of elements within one scientific state or system. An external ratio is one involving elements from two different systems. Thus, the following proportions involve internal and external ratios, respectively, according to the scientist's definition:

$$1 \text{ mile} : 4 \text{ hours} = 3 \text{ miles} : 12 \text{ hours}$$
$$1 \text{ mile} : 3 \text{ miles} = 4 \text{ hours} : 12 \text{ hours}.$$

The confusion is easily eliminated by using the terminology "within or between *systems*" or "within or between *measure spaces*." For example, consider the problem:

Richard buys 4 cakes for $8.95 each.
How much does he have to pay?

You could analyze the problem using, say, Vergnaud's (1988) terminology, *isomorphism of measures* where Measure 1 (M1) refers to numbers of cakes and Measure 2 (M2) refers to costs. Measure spaces usually refer to different sets of objects, different types of quantities, or different units of measure.

M_1	M_2
1	$8.95
4	$ x

In this case, $1 : 4 = \$ 8.95 : \$ x$ is a proportion whose ratios originate *within measure spaces*.

If you compare the "oranginess" of the liquid in pitchers containing cans of orange concentrate and cans of water, such as the following:

Pitcher 1	Pitcher 2
3 o.j.	5 o.j.
5 H_2O	11 H_2O

then ratios comparing parts orange concentrate to parts water, such as 3 o.j. : 5 H_2O and 5 o.j. : 11 H_2O are *within-system ratios*.

Ratios and Rates

In the last decade, progress has been made in more clearly specifying the essential differences between ratios and rates. Early definitions of ratio and rate were linked to comparisons *within* and *between* measure spaces. A ratio was considered a comparison between like quantities (e.g., pounds : pounds) and a rate a comparison of unlike quantities (e.g. distance : time), although, as noted by Lesh, Post, and Behr (1988), there was "disagreement about the essential characteristics" that distinguish the two. Ohlsson (1988), for example, defined a ratio as a numerical expression of how much there is of one quantity in relation to another quantity, and a rate as a ratio between a quantity and a period of time.

Schwartz's (1988) distinction between ratio and rate was based upon the nature of the entity whose attribute is described by a multiplicative comparison. If the material is homogeneous (such as a mixture) or a speed is constant, then the quantity (a rate) characterizes the material or the trip as a whole, but if the referent is not homogeneous, the quantity is a more local property of the referent entity and is a ratio. Building on this distinction, Kaput and West (1994) associated ratios with the accumulation of attributes and rates with homogeneity of attributes. A ratio may be iterated in a series of static and independent instances, describing a local property of the referent, so that amounts of each quantity composing it accumulate. An example might be the *building up* strategy often observed in children's work: 2 pencils for $.10, 4 pencils for $.20, 6 pencils for $.30, and so on. In a rate, however, constituent elements always exist in a constant ratio to one another, regardless of how much of the original entity is sampled. For example, if you take samples of a mixture such as paint or orange juice, they are always composed of constituent elements in the same ratio to one another, regardless of the amount of the mixture sampled.

After working extensively with teachers and students on the concept of speed (Thompson & Thompson, 1994, 1996), P. Thompson (1994) suggested that the inherent difference lies not so much in the situations and contexts per se, as in the mental operations by which people apprehend rate and ratio situations. He defined a rate as a "reflectively abstracted constant ratio" (p. 192). This means that one conceives a ratio when he/she conceives two quantities in a multiplicative comparison. However, a ratio *becomes* a rate when that person is able to conceive that the ratio applies beyond the particular situation in which it was conceived and can think about it as a characteristic of a whole class of covarying quantities. Thompson's distinction between ratio and rate is attractive for those who study children's thinking because, in a neo-Piagetian sense, it applies the words to levels of epistemic competence.

Fractions and Rational Numbers

Often the word *fraction* is used when rational number is intended, and vice versa. The word *fraction* is used in a variety of ways inside the classroom as well as outside the classroom. The many uses of the word are bound to cause confusion. In particular, the word *fraction* has several meanings, not all of which are mathematical. For example, a fraction might be a piece of undeveloped land, whereas in church, it would refer to the breaking of the Eucharistic bread. In the statement "All but a fraction of the townspeople voted in the presidential election," the word *fraction* means *a small part*. When you hear that "the stock rose fractionally," it means *less than one dollar*. In math class, students are disconcerted by the need to learn the technical definition of a part-whole fraction when they already know from their everyday experience of the word *fraction* that it means *any little bit*. When a fraction such as $\frac{4}{3}$ refers to more than one whole unit, the interpretation *a little bit* does not apply very well.

The colloquial usage of the word is complicated enough, but multiple uses of the word in mathematics education are also problematic. Some people use the word *fraction* to refer specifically to a part-whole comparison. It may also refer to any number written in the symbolic form $\frac{a}{b}$. Often the words *fraction* and *rational number* are used interchangeably.

What are fractions? First, *fractions* are bipartite symbols, a certain form for writing numbers: $\frac{a}{b}$. This sense of the word fraction refers to a notational system, a symbol, two integers written with a bar between them. Second, *fractions* are non-negative rational numbers. That is, in the set of rational numbers, $\frac{a}{b}$, *a* and *b* may be positive or negative integers, with the restriction that $b \neq 0$. Traditionally, because students begin to study fractions long before they are introduced to the integers, *a* and *b* are restricted to the set of whole numbers, so that fractions are only a subset of the rational numbers.

I reject the use of the word *fraction* as referring exclusively to one of the interpretations of the rational numbers, namely, part-whole comparisons. Because the part-whole comparison was the only meaning ever used in instruction, it is understandable that *fraction* and *part-whole fraction* became synonymous. However, educators now realize that restricting instruction to the part-whole interpretation has left students with an impoverished notion of the rational numbers, and increasingly, teachers are becoming aware of the alternate interpretations or subconstructs of the rational numbers and referring to them as *operator, measure, ratio,* and *quotient*. Part-whole comparisons are on equal ground with the other interpretations and no longer merit the distinction of being synonymous with *fractions*.

Therefore, the terms *fractions* and *rational numbers* are not synonymous and it is more accurate to think of fractions as a subset of the rational numbers. Eventually, it is important for students to understand the full set of rational numbers as well as to grasp the following distinctions:

- All rational numbers may be written in fraction form.
  $\frac{3}{4}$, $\frac{\sqrt{4}}{3}$ (usually written as $\frac{2}{3}$),
  $\frac{2.1}{4.1}$ (usually written as $\frac{21}{41}$) and
  $\frac{\frac{1}{2}}{\frac{1}{1}}$ (usually written as $\frac{2}{1}$)
  are all fractions and rational numbers.
- All numbers written in fraction form are not rational.
  $\frac{\pi}{2}$ is not a rational number although it is written in fraction form.
- Each fraction does not correspond to a different rational number.

A different rational number does not exist for each of the three fractions $\frac{2}{3}$, $\frac{6}{9}$, and $\frac{10}{15}$. Just as one and the same woman might be addressed as Mrs. Jones, Mom, Mother, Maggie, Dear, Aunty Meg, and Margaret, these fractions are different numerals designating the same rational number. A single rational number underlies all of the equivalent forms of a fraction.

- Rational numbers may be written in fraction form, but they may be written in other forms as well.
- Terminating decimals are rational numbers. Nonterminating, repeating decimals are rational numbers. Percents are rational numbers. Nonterminating, nonrepeating decimals are not rational numbers.

Fractions as Numbers

When I speak of a fraction as a number, I am really referring to the underlying rational number. Understanding a fraction as a number entails realizing, for example, that $\frac{1}{4}$ refers to the same relative amount in each of the following pictures. There is but one rational number underlying all of these relative amounts.

Whether I call it $\frac{1}{4}$, $\frac{4}{16}$, $\frac{3}{12}$, or $\frac{2}{8}$ is not as important as the fact that a single relationship is conveyed. When considering fractions as single numbers, rather than focusing on the two parts used to write the fraction symbol, the focus is on the relative amount conveyed by those symbols (Figure 14.1).

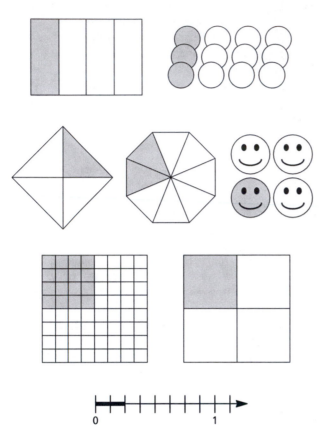

Figure 14.1 Illustrations for a single rational number.

Regardless of the size of the pieces, their color, their shape, their arrangement, or any other physical characteristic, the same relative amount and the same rational number is indicated in each picture. Nevertheless, psychologically, for the purpose of instruction, when fractions are connected with pictorial representations, which fraction name you connect with which picture *is* an important issue. For example, you would not call the first picture $\frac{2}{8}$ or $\frac{4}{16}$.

Understanding Rational Numbers

The rational numbers and proportionality are logical mathematical constructs. These constructs are not terribly useful for instructional and learning purposes. Knowing rational numbers is certainly more than reciting the axioms of an ordered quotient field. Similarly, the definition of *proportionality* as a condition in which two quantities or variables are related in a linear manner offers little information as to what

it might mean to understand proportionality. These two constructs are not disjoint. The rational numbers have a proportional nature; that is, within each equivalence class, each element is a constant multiple of every other. Nevertheless, as expected, understanding the rational numbers and understanding proportionality are not equivalent.

The work of Kieren (1976, 1980, 1983, 1993) and Behr et al. (1983) has highlighted five subconstructs of the rational numbers: part-whole, measure, quotient, operator, and ratio. These subconstructs suggest learning sites from which children might gain insights into the essential nature of rational numbers. Freudenthal (1983) called fractions the "phenomenological source of the rational number" (p. 134). The didactical phenomenology are a set of situations in which important mathematical ideas are grounded in real phenomena; these sites provide the context in which the learning of the mathematics takes place so that both the meaningfulness and the utility of the mathematical ideas are ensured. They are points at which a student may enter into the learning process and reconstruct important ideas for himself because the real phenomena he is exploring beg to be organized by the mathematics of interest. Although much needs to be done in this direction, in recent years progress has been made in building didactical phenomenology (Lamon, 2006).

When educators say that the goal of fraction instruction is to *understand the rational numbers*, we mean that by providing students with experiences with nonnegative rational numbers (fractions), we want them to recognize nuances in meaning; to associate each meaning with appropriate situations and operations; and, in general, to develop insight, comfort, and flexibility in dealing with the rational numbers. The goals of fraction instruction are best summarized in the language of number sense (Greeno, 1991; Sowder, 1989). Students who have developed *rational number sense* have an intuitive feel for the relative sizes of rational numbers and the ability to estimate, to think qualitatively and multiplicatively, to solve proportions and to solve problems, to move flexibly between interpretations and representations, to make sense, and to make sound decisions and reasonable judgments. In general, they should be comfortable in reasoning, computing, and problem solving in the domain of rational numbers.

What Is Proportional Reasoning?

Educational researchers have always perceived that something called *proportional reasoning* is a consequence of understanding the nature of rational numbers. Unsurprisingly, for the purposes of textbook instruction, large-scale assessment (Lamon, 1988), and for research (Tourniaire & Pulos, 1985) the domain of ratio and proportion has been defined traditionally in terms of two problem types similar to fraction comparison and equivalence tasks: comparison problems and missing-value problems. In a comparison problem, four values are given (a, b, c, and d) and the goal is to determine the order relation between the ratios $\frac{a}{b}$ and $\frac{c}{d}: \frac{a}{b} \{<>=\} \frac{c}{d}$. The following example is a typical ratio comparison problem:

> John makes lemonade concentrate by using 3 spoonfuls of sugar and 12 spoonfuls of lemon juice. Mary makes concentrate by using 5 spoonfuls of sugar and 20 spoonfuls of lemon juice. Whose lemonade concentrate is sweeter, John's or Mary's, or do they taste the same? (Karplus et al., 1983b, p. 53)

A missing-value problem provides three of the four values in the proportion $\frac{a}{b} = \frac{c}{d}$ and the goal is to find the missing value. This is a typical missing-value problem:

> John makes lemonade concentrate by using 3 spoonfuls of sugar and 12 spoonfuls of lemon juice. How much lemon juice would Mary need with 5 spoonfuls of sugar to make her concentrate taste just like John's? (Karplus et al., 1983b, pp. 53–54)

In spite of the implicit definition of the domain of ratio and proportion in terms of these two problem types, the terms *proportionality* and *proportional reasoning* are often used interchangeably (e.g., Karplus et al., 1983b). As a result, both have become ill-defined umbrella terms, referring to anything and everything related to ratio and proportion. In the 1989 Standards, the National Council of Teachers of Mathematics claimed that *reasoning proportionally* "is of such great importance that it merits whatever time and effort must be expended to assure its careful development" (p. 82), and in 2000, the Council claimed that *proportionality* "is an important integrative thread that connects many of the mathematics topics studied in grades 6–8" (p. 217). Although no suggestions were offered for teaching either proportional reasoning or proportionality, these strong statements express a pervasive sense that this reasoning ability is critical to mathematical and scientific thinking, and a sense of urgency about the consistent failure of students and adults to reason proportionally. Even using the basic definition of understanding the structural relationships in comparison and missing-value problems, my own estimate is that more than 90% of adults do not reason proportionally—compelling evidence that this reasoning process entails more than

developmental processes and that instruction must play an active role in its emergence. Yet, without appropriate instructional goals, purposeful teaching for well-structured knowledge is impossible. Clearly, our failure to define terms is a deterrent to progress in this research domain.

Many scholars share the view that proportional reasoning is a long-term developmental process in which the understanding at one level forms a foundation for higher levels of understanding (Case, 1985, 1992; Hart, 1981, 1984; Inhelder & Piaget, 1958; Kieren, 1988, 1993; Lesh et al., 1988; Noelting, 1980a, 1980b). "The domain represents a critical juncture at which many types of mathematical knowledge are called into play and a point beyond which a student's understanding in the mathematical sciences will be greatly hampered if the conceptual coordination of all the contributing domains is not attained" (Lamon, 1994, p. 90). The ability to solve comparison and missing-value problems is assumed knowledge in the secondary school curriculum (Post, 1986) and as the knowledge of multiplicative topics deepens through the study of algebra, geometry, trigonometry, statistics, and sciences in the secondary curriculum, it becomes the basis for understanding fields of application such as inflation, force, density, the physics of lenses, and the physics of sound (Gray, 1979). This view is consistent with other philosophical perspectives (Bohm & Peat, 1987) and mechanisms described in computer science and psycholinguistics (Pinker, 1997) that view the mind as a recursive processor, constantly embedding knowledge in progressively more complex thoughts.

This perspective suggests that proportional reasoning, defined by one's ability to understand the structural relationships in comparison and missing-value problems, might be conceived as a necessary but not sufficient prerequisite to understanding proportionality. The issue becomes one of where to draw lines (benchmarks) in a long-term process that is not easily segmented; for example, others have referred to some stages of the reasoning process as preproportional reasoning (Inhelder & Piaget, 1958), protoratio reasoning (Resnick & Singer, 1993), ethnomathematical knowledge (Kieren, 1988, 1993), and theorems-in-action (Vergnaud, 1983, 1988). One approach is to define proportional reasoning in terms that present reasonable expectations for students by the end of middle school, keeping in mind that the study of variables, functions, linear equations, vectors, and other topics studied in high school mathematics will continue to broaden and deepen their perspectives on multiplicative relationships. Therefore, I propose that *proportional reasoning* means supplying reasons in support of claims made about the structural relationships among four quantities, (say *a, b, c, d*) in a context simultaneously involving covariance of quantities and invariance of ratios or products; this would consist of the ability to discern a multiplicative relationship between two quantities as well as the ability to extend the same relationship to other pairs of quantities.

The critical structural relationships, both scalar and functional, are best illustrated using Vergnaud's (1983) framework for analyzing multiplicative structures. The following example shows a simple direct proportion between two measure spaces. A linear functional relationship exists between corresponding elements of the measure spaces, and a scalar operator transforms quantities of the same type.

If Maria can sew 5 team shirts with $7\frac{1}{2}$ yards of material, how many yards will she need to make a T-shirt for each of the 15 boys on the soccer team?

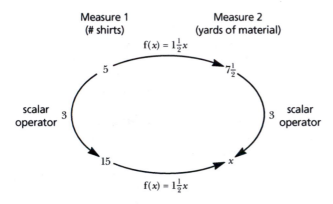

The ability to give correct answers is no guarantee that proportional reasoning is taking place. Often, proportions may be solved using mechanical knowledge about equivalent fractions or about numerical relationships, or by applying algorithmic procedures (for example, the cross-multiply rule) that circumvent the use of the constant of proportionality.

> Which is sweeter, a mixture of 2 teaspoons of sugar with 6 teaspoons of lemon juice, or a mixture of 8 teaspoons of sugar with 24 teaspoons of lemon juice?
>
> Student: They are the same because 8 goes into 24 three times and 2 goes into 6 three times. (Karplus et al., 1983b, p. 55).

In such contexts, multiplicative explanations have "counted" as proportional reasoning; however an observer cannot say much about these students' understanding of proportionality. Indeed, most adults do not associate comparison problems with trying to find out which situation has the greater constant of proportionality. As researchers, we must conclude that

proportionality is a much larger construct than proportional reasoning.

The *reasoning* aspect of proportional reasoning entails recognition of the constant ratio between elements of the same measure space and recognition of the functional relationship between measure spaces (recognition of the constant ratio $M_2 : M_1 = 3 : 2$). In the case of a simple inversely proportional relationship, recognizing the structural aspects of the situation entails understanding that there are two scalar operators, one of which is the multiplicative inverse of the other, and that product of corresponding measures is constant.

If 3 people can mow and trim a lawn in 2 hours, how long will it take 2 people to do the same work?

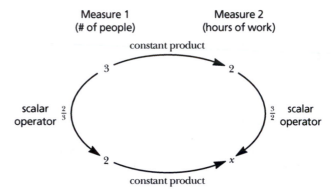

Understanding Proportionality

Proportionality is a mathematical construct referring to the condition or the underlying structure of a situation in which a special invariant (constant) relationship exists between two covarying quantities (quantities that are linked and changing together). In this section, I explore some of the critical aspects of understanding proportionality, both mathematical and psychological, arguing that it is a larger construct than that of proportional reasoning.

The constant of proportionality. The mathematical model for proportional relationships is a linear function of the form $y = kx$, where k is called the constant of proportionality. Thus, *y* is a constant multiple of *x*. Equivalently, two quantities are proportional when they vary in such a way that they maintain a constant ratio: $\frac{y}{x} = k$. The constant k plays an essential role in understanding proportionality.

Pedagogically speaking, k is a slippery character, because its guise changes in each particular context and representation involving proportional relationships. It frequently does not appear explicitly in the context in which it plays a role, but rather is a structural element waiting to be discovered beneath the obvious details. In symbols, it is a constant. In a graph, it is the slope. In a tabular representation, it may be the difference between any entry and the one before it,

# of stacked wooden cubes	1	2	3	4	5	6
height of the stack in inches	3	6	9	12	15	18

or, equivalently, it may be the rate at which one quantity changes with respect to the other expressed as a unit rate.

# of stacked wooden cubes	2	5	9	12	15
height of the stack in inches	6	15	27	36	45

In general, in rate situations, k is the constant rate. In reading maps, it is the scale. In shrinking or enlarging contexts, or in similar figures, it is the scale factor. It may be a percentage if you are discussing sales tax, or a theoretical probability if you are rolling dice. These examples suggest that the interpretation of the constant of proportionality is situation-specific.

Even when a student reasons proportionally, the task of interpreting the constant of proportionality in relatively simple contexts can be challenging. For example, consider this recent discussion that I had with a calculus student (S) as he struggled with many issues related to number versus quantity, working and interpreting within a context, and understanding the constant of proportionality. He was considering the following problem:

> Suppose that you are going to take a trip by car on interstate highways. Estimate the amount of gas you will need.

S: I know that the amount of gas I will need (let's say g) depends on the length of the trip I am taking (let's call it d for distance). In fact, if I drive 1000 miles, I will use about twice as much gas as if I drive 500 miles and about three times as much if I drive 1500 miles. So I know that the amount of gas I use (g) is proportional to my distance traveled:

$$g \propto d.$$

This relationship means that g is a constant multiple of d, so I can write

$$g = kd.$$

What is that constant k? I suspect that it has something to do with my car's fuel economy. I know that on the highway, my car's fuel economy rating is 30 mpg, meaning that it can go about 30 miles per gallon of gas. So I said my equation

would be $g = 30d$. But here is where I get stuck. That equation doesn't make sense because if I plug in 1 mile for my distance, it says I need 30 gallons.

I: Tell me the way you *think* it works.

S: The amount of gas it takes is your miles divided by 30, which is $\frac{1}{30}$ times the number of miles. So I would say $g = \frac{1}{30}d$. So it looks like k = $\frac{1}{30}$. But how was I supposed to know that?

I: Go back to the equation $g = kd$. Solve for k.

S: k = $\frac{g}{d}$

I: What label is on that quantity?

S: Gas over distance—that's gallons over miles. OK! Let me back up. I can write 30 mpg as $\frac{30 \text{ miles}}{1 \text{ gallon}}$. But that's the same as $\frac{1 \text{ gallon}}{30 \text{ miles}}$. So my equation has to be $g = \frac{1}{30}d$.

I: So what does k = 30 mpg mean?

S: To drive a distance of 1 mile, I need about $\frac{1}{30}$ gallon of gas.

This scenario suggests that students should be able to

- express the meanings of quantities and variables and the constant of proportionality in the contexts in which they are used.

Additionally, it seems reasonable to expect that an understanding of proportionality entails

- the ability to use proportionality as a mathematical model to organize in appropriate real-world contexts;
- the ability to distinguish situations in which proportionality is not an appropriate mathematical model from situations in which it is useful;
- development and use of the language of proportionality;
- use of functions to express the covariation of 2 quantities;
- the ability to explain the difference between functions of the form $y = mx$ and functions of the form $y = mx + b$. In the latter function, y is not proportional to x, but rather, Δy is proportional to Δx.
- knowing that the graph of a direct proportional situation is a straight line through the origin;
- knowing that the graph of $y = mx + b$ is a straight line intersecting the y axis b units above the origin;
- the ability to distinguish different types of proportionality and to associate each of them with appropriate real-world situations in which they are applicable: direct proportions,

inverse proportions, square proportionalities ($y = kx^2$), and cubic proportionalities ($y = kx^3$);

- knowing that k is the constant ratio between two quantities in a direct proportional situation;
- knowing that k, the constant in an inversely proportional situation, is simply the product of any pair of values of the two quantities;
- knowing that the graph of an inversely proportional situation is a hyperbola.

The Proportionality and Proportional Reasoning Hypothesis

Educators are accustomed to mentioning *rational numbers, ratios*, and *proportions* together in the same phrase simply because the terms are so close mathematically and psychologically. Proportional reasoning develops (or should develop) as one studies fractions, a subset of the rational numbers. Thus, proportional reasoning is, to some degree, an indicator of one's rational number sense. However, understanding proportionality is a far more demanding goal than either rational number sense or proportional reasoning. In particular, the ability to use proportionality as a mathematical model of real world situations develops only as one studies higher mathematics and science.

Few would disagree that proportions have always played a key role in quantitative relations in science. Furthermore, recall that Piaget and his colleagues (Inhelder & Piaget, 1958) referred to the "proportionality schema" (p. 173). In general, a schema has always been taken to mean something more complex and encompassing than the phenomenon defined above as *proportional reasoning*. Using clinical interview methods, the Genevans investigated children's development of the proportionality schema in complex situations (e.g., within physical systems such as the balance beam or with an oscillating pendulum). Later, when researchers wished to work in classrooms and test larger numbers of students, they developed tasks amenable to paper-and-pencil large-scale testing (see Karplus, Karplus, Formisano, & Paulsen, 1979; Karplus et al., l983b). Piaget's tasks were criticized because they required some knowledge of physical principles in addition to proportional reasoning, and tasks such as orange juice and lemonade mixture tasks and Mr. Tall and Mr. Short emerged. Notwithstanding the confusing use of the terms *proportionality* and *proportional reasoning*, implicit in the kinds of tasks that have been devised to study proportional reasoning is an effort to draw the line long before what researchers might call *application problems*.

My guess is that Piaget and his colleagues were correct and that proportionality is a much larger concept than proportional reasoning as we have come to know it. Due to the context-bound interpretations illustrated above, understanding proportionality is tied to investigating and discerning principles and regularities in problem situations before using proportions to express those relationships symbolically, and then interpreting the symbols and what one has learned in terms of the specific context to which they apply (as in the gas consumption problem). Mathematicians recognize this process as mathematical modeling.

My hypothesis is that proportions arise in the study of rational numbers as a natural expression of their equivalence and that as one develops rational number sense through various experiences with many personalities of the rational numbers, one learns to reason proportionally. However, understanding the larger concept of proportionality comes about later, through interaction with mathematical and scientific systems that involve the invariance of a ratio or of a product.

EVOLUTION OF THE FIELD

In any research domain, it is useful to step back periodically and consider the state of the field. In this section, I analyze the evolution of research in the domain of rational numbers and ratio and proportion. In every field of endeavor, research tends to proceed through regular stages, each of which produces distinctive types of research, fueled by beliefs, values, questions, and approaches shared by a scholarly community. Looking at the field in developmental stages provides a useful organization for those who may be considering a research agenda in the field, and at the same time, it foretells "next steps." The following broad stages, named for their characteristic types of research, do not necessarily proceed sequentially; sometimes parallel development occurs: empirical analysis, rational task analysis, clinical interviews, interventions, and longitudinal studies. To borrow a metaphor, this section will not be concerned with rebuilding brick by brick, as much as it will expose the current edifice (Begle & Gibb, 1980). Therefore, rather than summarizing extant literature, only the contributions that have most directly influenced current work will be reviewed.

Empirical Task Analysis: Defining the Problem

Standardized tests, especially those that compare student performance to national and international audiences, are often a stimulus for research and change. The tests, not withstanding criticism of their validity and reliability, identify areas in which student performance is weak. Ensuing research examines student performance to better understand the nature and extent of children's preconceptions and misconceptions, often verifying the results in a broad range of contexts with a variety of age groups. In general, the initial focus is on children's incompetencies, on what they *cannot* do. This research tends not to offer explanations, but rather results in conjectures about the variables that affect problem difficulty.

Almost 50 years ago, Inhelder and Piaget (1958) drew attention to proportional reasoning when they designated it the hallmark of their formal operations stage of development. For a long time, research was piecemeal; it was conducted in several different disciplines, including science education, psychology, and mathematics education, and it lacked an underlying framework. Years after Piaget and his associates posited the centrality of proportional reasoning in their developmental theory, a large-scale study of science education in 19 countries (Comber & Keeves, 1973) showed a low success rate on questions involving proportional reasoning, and many researchers began to study the phenomenon (e.g., Karplus et al., 1979; Karplus, Karplus, & Wollman, 1974; Karplus et al., 1983a, 1983b; Noelting, 1980a, 1980b; Wollman, 1974). In 1979, the Rational Number Project (Behr, Lesh, & Post, 1984; Behr, Lesh, Post, & Silver, 1979) began to investigate fractions, ratios, and proportions from a mathematics education perspective.

A large body of research, roughly from 1970 to 1985, produced an extensive inventory of factors that influence the difficulty of proportion problems. Among these are context (Tourniaire, 1983, 1986); how familiar students are with using proportions in a given context (Tourniaire, 1983); the location of the missing element in a proportion in relation to the other three numbers (Bezuk, 1986; Rupley, 1981); whether a problem concerns discrete or continuous quantities (Behr et al., 1983; Pulos, Karplus, & Stage, 1981); whether the number to be found is the largest of the four terms (Abramowitz, 1974; Rupley, 1981); the presence of integral ratios as opposed to nonintegral ratios (Bezuk, 1986; Hart, 1984, 1988; Karplus et al., 1983b; Noelting, 1980a, 1980b; Rupley, 1981); the presence of unit ratios, especially 1:2 (Hart, 1981, 1988; Noelting, 1980a, 1980b); and whether perceptual cues are consistent or inconsistent (Behr et al., 1983; Cramer, Post, & Behr, 1989; Lesh, Landau, & Hamilton, 1983; Novillis, 1976). In addition, researchers noted that children frequently use additive strategies where multiplicative comparisons are required (Hart, 1981, 1988; Karplus et al., 1983b). Many children lack understanding of the conventional notation (e.g., 12:1) for expressing a ratio (Silver, 1981), and, in general, they show an overdependence on standard textbook representations when trying to model a given situation (Lesh et al., 1983; Silver, 1981). Schwartz (1987) found that multiplication and division problems involving intensive quantities (for example, "per" quantities) are difficult, even for above-average 12th-grade students. The mode of representation used by students is distinctly problem-specific and in some rational number tasks, when students were encouraged to use concrete aids alone, performance declined (Lesh et al., 1983).

Children experience difficulties linking underlying problem structure and operation when presented with problems in multiplication and division. Faced with problems involving the same content, they change their minds about the operation needed to solve the problem, depending on the numerical data used. Greer (1987) called this phenomenon *nonconservation of operations* (p. 37). A line of research in multiplication and division, the collective work of Bell, Swan, and Taylor (1981), Fischbein, Deri, Nello, and Marino (1985); Bell, Fischbein, and Greer (1984); Luke (1988); and Graeber et al. (1989), hypothesized that the fundamental operations of arithmetic are linked to implicit, unconscious, and primitive intuitive models that constrain students' ability to predict appropriate operations and that these models lead to the assumption that multiplication makes bigger and division makes smaller.

Rational Task Analysis: Understanding the Content

Researchers use empirical analysis to study children's mathematics, that is, children's actual performance on tasks, but they use rational task analysis to examine the content from a mature mathematical perspective, making assumptions about the ways of thinking that are necessary to solve problems. On a theoretical level, rational task analysis identifies the essential mathematical components of the domain, the cognitive objectives of instruction, both conceptual and procedural, with an eye toward the future usefulness of those ideas in higher mathematics. Often, in the process of carefully analyzing the mathematics, implicit processes become explicit for the first time, and a vocabulary develops by which researchers can discuss the various phenomena they are studying. Especially in a domain where the mathematics is so complex this stage is critical for devising a framework by which the research can proceed. That framework provides a rationale that helps the researcher to ask appropriate questions, to devise activities and problems, and to interpret children's thinking.

By the late 1970s, researchers were convinced of the complexity of the connections among rational numbers, proportional reasoning, and many other mathematical concepts of a multiplicative nature. A number of task analyses examined the huge domain from many different angles and enhanced our understanding of the mathematics and its cognitive demands.

Kieren (1976, 1980) proposed that to understand rational numbers, one must have adequate experience with multiple interpretations, not merely with fractions as objects of computation. For five major interpretations or personalities of rational numbers—part-whole fractions, ratios, operators, quotients, and measures—he analyzed the kinds of mathematical understanding unique to each, the mechanisms such as equivalence and partitioning that help to build competence in each interpretation, and some classroom activities appropriate for building knowledge of each interpretation. Other analyses (e.g., Behr et al, 1983; Ohlsson, 1987, 1988) resulted in different numbers of subconstructs, but, whether one admits five or six interpretations or more, the compelling issue was that the mathematics curriculum has not been providing an adequate foundation for understanding rational numbers.

Freudenthal's (1983) didactical phenomenology helped researchers to relate very complex mathematical ideas (among these, the personalities of rational numbers) to the mental objects, human activities, and real phenomena that are appropriately organized by those mathematical ideas. He suggested many everyday examples accessible to children through which instruction could provide an inroad to the world of rational numbers, from what he described as a "wealth," a veritable "ocean" of examples, "a source that never dries up" (p. 134).

Another contribution was the introductory chapter to a monograph summarizing the presentations at the National Council of Teachers of Mathematics Research Agenda Project Conference on concepts and operations in the middle grades. Hiebert and Behr's (1988) synthesis of the themes of the conference and the status of research on mathematical learning in the middle grades was, in its own right, an expert treatise on the mathematical and psychological dimensions of change in moving from whole numbers to rational numbers.

Schwartz (1976, 1987, 1988) and Kaput (1985) emphasized the role of the mathematics of quantity in multiplicative operations and enhanced researchers' understanding of the cognitive complexity of those operations. Their semantic analysis, which distinguished extensive and intensive quantities, emphasized the need to always link numbers with their referents and to carefully consider the *relation* between the referents in a problem. An intensive quantity is an abstraction from the comparison of two other quantities. It is not easily attained because its referent is different from that of either entity that composes it, it has no adequate representation, and operations with such quantities do not conform well to dominant cognitive models of multiplication as repeated addition or of (partitive) division as equal sharing.

Vergnaud (1983, 1988) introduced the construct of a *conceptual field*. The multiplicative conceptual field refers to a complex system of interrelated concepts, students' ideas (both competencies and misconceptions), procedures, problems, representations, objects, properties, and relationships that cannot be studied in isolation. The content of the multiplicative conceptual field is extensive, including such topics as multiplication, division, fractions, ratios, simple and multiple proportions, rational numbers, dimensional analysis, and vector spaces, and understanding of this field develops over many years. This construct suggests that learning is not linear, that it is content-based, and that a piece-by-piece conception of research will prove to be inadequate because in a web-like model of the structure of knowledge, contact with one strand will have ramifications across the entire structure. These views are consistent with the conclusions of others (e.g, Bereiter, 1990; Lesh & Lamon, 1992) who have argued that research capable of informing the teaching and learning enterprise in schools (complex systems) must take into account the full complexity of the content, and the classroom, and the children's worlds of experience.

Clinical Interviews: Understanding Students' Thinking

Whereas analyses of rational number knowledge, such as those just discussed, begin with the mathematics and view the content as a prelude to more advanced mathematical concepts, a bottom-up approach to research in the domain begins with children and the knowledge they bring to the task of learning rational numbers. Clinical interviews deeply probe not only children's current understanding of a topic, but how the researcher might affect change in that understanding. Beginning with questions based upon a conceptual analysis, the researcher opportunistically revises or devises questions as the child's primitive conceptions or misconceptions surface and thereby can study the student's constructions in the face of these improvisations. Assuming that the small number of children who are interviewed demonstrate thinking that is characteristic of the larger population, clinical interviews fuel theories about how certain ideas develop and about how material might be psychologized for instruction. Two kinds of research using clinical interviews have been influential in rational number research: those that study children's construction of knowledge and studies that seek to identify children's competencies, useful intuitive knowledge, and informal strategies.

Steffe and his colleagues study the functioning schemes (repeatable, generalizable actions) that children bring to a specific task and how the children adapt their scheme in response to the specific task at hand. On the basis of a careful study of counting and number sequences, Steffe has articulated a theory concerning the way in which children's formation and use of units is progressively elaborated from early counting through multiplication (Steffe, 1988, 1992; Steffe & Cobb, 1988; Steffe, von Glasersfeld, Richards, & Cobb, 1983). The centrality of the unit in fraction instruction, especially the role of composite units (units of numerosity greater than one) and the fact that ratios and rates may be viewed as complex types of units (Lamon, 1993a, 1994) suggest that unit building may be an important mechanism in accounting for the development of increasingly sophisticated mathematical ideas. Research in the natural development of language hierarchies (Callanan & Markman, 1982; Markman, 1979) substantiates this perspective, suggesting that more sophisticated thinking results when one reframes a situation in terms of a more collective unit. When that happens, a part-whole schema comes in to play and the individual is able to think about both the aggregate and the individual parts that compose it. In turn, the ability to compose and to decompose a unit into its constituent parts adds the flexibility to one's reasoning that is needed in the field of rational numbers.

Olive & Steffe (1994) have embodied these mechanisms by which children build more complex units in a microworld that allows children to experiment with operations on discrete and continuous objects that can help them to understand rational numbers: composing and decomposing, iterating, partitioning, measuring. It further serves as a dynamic medium by which teachers and researchers can investigate the itinerary children take as they construct meaning for rational numbers.

In spite of the fact that many whole number strategies and ways of thinking are not useful for dealing with fractions, research in the last decade has proved that there is not an unbridgeable divide between the two number domains. In particular, Mack (1990, 1995)

demonstrated in her teaching experiments that students' informal and whole-number knowledge can be extended and connected to rational number symbols and procedures. Others have found competencies, even in very young students, that could be exploited in instruction. Student representations of rational numbers and of proportions occur on an informal, qualitative basis long before these students are capable of treating these topics quantitatively (Tourniaire, 1986; Treffers & Goffree, 1985). Children naturally employ a form of mathematical intuition, an informal knowledge system (Kieren, 1983; Streefland, 1984, 1985; Treffers & Goffree, 1985). This informal knowledge includes a visual understanding of ratio and proportion (Steffe & Parr, 1968), especially of congruence and similarity (van den Brink & Streefland, 1979). Some strategies are more primitive and depend on counting, adding, and halving (Hart, 1984, 1988). However, many fairly sophisticated strategies and processes develop independent of instruction (Lo & Watanabe, 1997; Post, Behr, Lesh, & Wachsmuth, 1985; Rupley, 1981; Saenz-Ludlow, 1994).

Partitioning or fair sharing has long been recognized both as an important mechanism for building rational number understanding (Kieren, 1976, 1980; Piaget, Inhelder, & Szeminska, 1960; Pothier & Sawada, 1983; Streefland, 1991) and as a part of students' informal knowledge base beginning in their preschool days. Because the rational numbers are a quotient field, partitioning is an operation that plays an important role in generating the rational number subconstructs (Lamon, 1996) as well as in constructing meaning for operations with the rational numbers (Lamon, 2006; Mack, 2001). Partitioning activities merit a place of greater prominence than they currently receive in the elementary curriculum because children's partitioning strategies develop in efficiency over a long period of time and share important connections with equivalence and the formation of complex unit structures (Lamon, 1996).

Among the early, correct strategies children use in situations involving proportions is the *building-up* strategy as documented by Piaget, Grize, Szeminska, and Bang (1968), Karplus and Peterson (1970), and Hart (1984). In this strategy, a student establishes a ratio and extends it to a second ratio by addition:

> 2 pencils cost $.65
> $.65 for 2 more makes $1.30
> $.65 for 2 more makes $1.95

Although it is a primitive strategy by which children reason up to some desired quantity by using pattern recognition and replication, it is useful in solving some proportion problems. Nevertheless, without additional information, it would not be considered proportional reasoning because it does not take into consideration the constant ratio between the two measure spaces.

Children also demonstrate interesting and productive strategies in ratio comparison problems. Lamon (1989, 1993a, 1993b) found that students are able to use a double counting strategy to successfully solve some problems. For example, consider this comparison task:

Who gets more pizza, a girl or a boy?

Spontaneously, before instruction in rational numbers, children construct complex units when solving problems such as this. They choose one of the ratios (in this case, 1 pizza: 3 boys) and use it to reinterpret the other ratio (3:7). This process is called *norming*, or reinterpreting a situation in terms of some chosen unit. Essentially, it is a double counting process: Measure the ratio 1:3 out of the ratio 3:7 by counting both parts of the pizza-to-boy ratio out of corresponding parts of the pizza-to-girl ratio. It is a fairly easy task for students who use this strategy to represent it in symbols. The ratio arithmetic is

$$(3:7) - 3(1:3) = (3:7) - (3:9) = (0:-2).$$

Thus, the student explains, if the pizza were served so that there was always 1 pizza for 3 people, the first group would actually get more because they could have fed 2 more people. Like the building-up strategy, norming is not a good indicator of proportional reasoning because students who use it often fail to recognize all of the structural relationships in a proportion. Although students who use the strategy do not yet conceptualize a ratio as a single number (a divided ratio or an index), the norming strategy appears to be a natural step in the process of conceptualizing a ratio as a unit of units of units (Lamon, 1993a, 1993b, 1994). The child coordinates a) number of pizzas, b) number of people, and further recognizes the new intensive quantity (pizzas per person) composed of a and b. This strategy and the next build upon the unit construction processes identified by Les Steffe and his colleagues (e.g., Steffe, 1988, 1992; Steffe et al., 1983), who documented the role of counting in children's construction of number sequences, and the

mechanisms by which children build up increasingly complex unit structures.

Lo and Watanabe (1997) have documented other variations of a double matching strategy. They asked a fourth-grade student to solve this proportion:

Yesterday I bought 28 candies with 12 quarters. Today, if I go to the same store with 15 quarters, how many candies can I buy?

The child made 7 groups of 4 candies and put quarters in groups of 2:

This did not prove satisfactory because there were 6 groups of quarters, not 7. After experimenting with several regroupings of candies and quarters, he arrived at this matching:

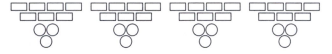

Then he was able to verbalize the relationship 3 quarters for 7 candies and imagine one more such matching to arrive at the answer 15 quarters for 35 candies.

Another important intuitive strategy is that of *unitizing*. Recall that unitizing is the cognitive chunking or regrouping of a given quantity into manageable or conveniently sized parcels (Lamon, 1996). One of the most salient differences between those who think proportionally and those who do not is that the proportional reasoners are adept at building and using composite units when the context suggests that it is more efficient to use composite units than to use singleton units (Lamon, 1993a, 1993b). For example, decide which of these cereals is the best buy:

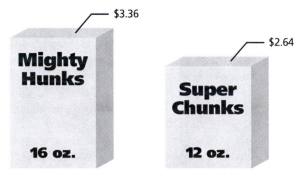

In this case, without pencil or calculator, comparing the prices of 4 ounces of each cereal is more convenient than mentally using a two-digit divisor to compute the cost of 1 ounce of each cereal. Although third-grade students rarely construct such composite units, fourth and fifth graders begin to develop thinking strategies based on unitizing (Lamon, 1996).

As a whole, the studies in which young children were interviewed before they had had substantial fraction instruction show that powerful, intuitive strategies are available to children at an early age. Even Inhelder and Piaget (1958), speaking of proportional reasoning, observed, "We have seen subjects construct the notion in the experimental situation before they have learned it in school" (p. 314). On the other hand, research with older children who have had 5 or more years of traditional mathematics instruction (Karplus et al., 1983b) suggests that rules and algorithms replace reasoning strategies, and that these often fail the student when integral ratios do not exist. In the Karplus et al. research, a proportion such as one solved by a much younger child in the Lo and Watanabe (1997) study, $\frac{28}{12} = \frac{x}{15}$, would have been one of the most difficult for eighth graders. Researchers must begin to suspect that instruction played a role in the poor performance of the older students. One conjecture is that informal strategies were replaced with rules for creating equivalent fractions (multiply numerator and denominator by the same number) that failed to serve the students well in the absence of integral ratios. Thus, instruction that builds upon children's preinstructional strengths should be a high priority.

Interventions: Correcting Problems

If one suspects that understanding in a particular field has to do both with development and instruction, the question naturally arises as to whether learning can occur more rapidly in older children who, for whatever reason, have misconceptions or deficits in their knowledge. Intervention studies are limited-term teaching experiments that target specific problems with the goal of producing more acceptable outcomes. Unfortunately, in the domain of fractions, ratios, and proportions, these studies have been largely ineffective.

Greer (1987) devised a teaching experiment for children whose problem-solving strategies varied on two juxtaposed problems that differed only in the numbers involved. His goal was to see if they could overcome their nonconservation of operation and recognize the invariance of operation. He concluded that "this conservation can be as difficult to induce as the standard Piagetian conservations" (p. 44).

Hart (1984) described teaching experiments designed to help children overcome the incorrect use of addition in multiplicative situations, only to conclude that "children have misconceptions of a more funda-

mental nature than can be readily cured by a brief series of lessons" (p. 81).

Similar disappointing results came from a number of teaching experiments conducted by the Rational Number Project. In a study of children's understanding of fraction size, Behr, Wachsmuth, and Post (1985) reported that "given the amount of instructional time, the degree of special attention, the extensive use of manipulative aids, and the amount of time devoted to developing rational number concepts," they were surprised that 49% of the students in the study showed "little or no comprehension of fraction addition or rational number size" (p. 128). In a study of children's understanding of order and equivalence, Post, Wachsmuth, Lesh, and Behr (1985) reported, "Even under the conditions of instruction in the teaching experiment—which employed a rich diversity of manipulative aids, placed a heavy emphasis on concept development, and took more time than traditional instruction—many rational number concepts remained exceedingly difficult for these children" (p. 23). Similarly, in an 18-week study in which instruction focused on multiple concrete teaching materials, Behr, Wachsmuth, Post, and Lesh (1984) reported, "Even after extensive instruction, the performance of a significant number of fourth graders demonstrates a substantial lack of understanding" (p. 337). Collectively, these intervention studies indicate that building fraction, ratio, and proportion knowledge will involve a long-term learning process. There will be no quick fixes.

Experimental Materials

In spite of the lack of clear direction from research, experimental curriculum materials have been developed to help students build understanding of rational numbers and other multiplicative content. These materials vary in scope from full middle school curricula to units of study to less comprehensive materials. All have been developed from slightly different perspectives. In the United States, the so-called reform curricula, including Connected Mathematics (Lappan, Fey, Fitzgerald, Friel, & Phillips, 1998) and Mathematics in Context (Romberg & de Lange, 1998), and in South Africa, materials produced by the MALATI Fraction Group at Stellenbosch (Murray, Olivier, & Human, 1996) have incorporated multiple rational number subconstructs (ratio, quotient, measure, operator, and part-whole comparison). In the Netherlands, an extensive study (Streefland, 1991) used materials that developed fractions and ratios simultaneously through realistic problem-centered approaches. Other interesting materials were developed for a study at the Institute of Child Study at the University of Toronto (Moss & Case, 1999) and in work at Cornell University (Confrey, 1994; Lachance & Confrey, 1995; Scarano, 1996; Smith & Confrey, 1994). The Canadian study, originating in a psychological tradition, began with percents and decimals and used the proportional nature of the rational numbers to help children build number sense. Confrey and her colleagues emphasized a measurement interpretation, using continuous quantities and splitting. All of these materials and approaches were significant in that they took full advantage of children's informal knowledge and encouraged intuitive, sense-making strategies from the onset of instruction, while going well beyond the traditional content of fraction instruction. Unfortunately, the work begun in The Netherlands and in Canada was cut short by the untimely deaths of their chief investigators, Leen Streefland and Robbie Case, respectively.

Questions arise as to whether the new curricula that have incorporated various rational number subconstructs will be effective in increasing students' understanding of the rational numbers and whether students' proportional reasoning ability will be enhanced. The answers to these questions may require long-term follow-up of students who have used the curricula as they study mathematics in high school and college. Questions concerning the effectiveness of the curricula present challenges to the research community to more explicitly define what it means to understand rational numbers and to develop intrinsic measures of understanding. We habitually look to standardized examinations to measure success in spite of the fact that they reflect a different set of values, include more procedural than conceptual content, are often developed for different purposes, and are problematic in that the structure and arrangement of content they assess is inconsistent with the nature and growth of knowledge in the multiplicative conceptual field.

A Temporary Standstill

Research can sometimes seem to come to a halt when, after years of studying a complex domain, scholars have uncovered a complex web of questions and misconceptions and compelling needs, only to realize that they do not have the tools and methods and classrooms needed to answer the challenge. Certainly this was true at the time Behr et al. wrote the first handbook chapter. It was difficult to see where the field was going. Even as late as 1992, Davis, Hunting, and Pearne characterized the field as one that is not producing any definite conclusions, one in which "no real progress is being made" (p. 2). Traditional experimental methods and interventions can fail to

be productive, especially when dealing with complex content in the multifaceted world of the classroom. A research community may need to wait for the evolution and diversification of research paradigms; sometimes they must diverge from classical frameworks and incorporate new perspectives; sometimes, both. In the next section, I discuss a new research framework that incorporates nontraditional perspectives and its use in the new design experiment research paradigm for conducting longitudinal classroom research.

A FRAMEWORK FOR RESEARCH IN THE MULTIPLICATIVE CONCEPTUAL FIELD

In this section, I discuss a framework that is currently being used in longitudinal studies to facilitate rational number understanding and proportional reasoning. This framework rests upon the assumption that pedagogical content knowledge in a complex domain such as the rational numbers does not concern target concepts and operations in the domain, as much as it does core or central cognitive structures, composed of concepts, ways of thinking, and mechanisms for growth that are foundational to a broad range of multiplicative topics. Later in this section the conclusions of that research will be presented, as will questions that might launch additional longitudinal studies using the framework.

Reconsidering Pedagogical Content Knowledge

Pedagogical content knowledge (Shulman, 1986b, 1987) is the kind of knowledge a teacher needs to be able to teach a particular strand of content. One needs to know the students and their current capabilities and, beyond a personal knowledge of mathematics, the teacher needs to know the particular ingredients of understanding that can facilitate children's construction of knowledge about that content. The pertinent issue is *what* content should be psychologized for instruction. Do we in education teach to some goal state, or do we seek some foundational kind of knowledge that can grow to support larger, more complex content?

Historically, rational number education has taken adult knowledge of mathematics and simplified it so that it is palatable for children. As confirmed by a survey reported by Hart (1984), teachers equate understanding of ratio and proportion with the ability to solve problems involving scale drawing, similarity, percentages, trigonometry, probability, enlargement, and so on, or to find equivalent fractions or ratios, or to solve for missing numbers in a proportion. Nevertheless, teaching definitions, algorithms, and applications of rational number knowledge has not facilitated the development of rational number sense and the ability to reason proportionally.

Rather than teaching to goal states, seeking initial, front-end ideas and processes may be more important. When considered from the perspective of emergent concepts and processes, pedagogical content knowledge in the field of rational numbers and ratios and proportions may not appear to be related to textbook definitions and the problems one uses may not look like the problems or the applications for which proportional reasoning is useful. Therefore, before asking what kinds of problems we educators want students to be able to solve, we might begin by looking for fundamental perspectives on the field, including the cognitive processes, focal points, and big ideas that form the basis of thought and knowledge in rational numbers and related multiplicative topics. This is generally referred to as *conceptual analysis* (Lesh, 1985). Conceptual analyses may be performed at a variety of levels. The previous section in which proportional reasoning and proportionality were discussed is a kind of conceptual analysis. For another example, see Thompson and Saldanha (2003). Identifying student ideas (useful ways of thinking or mechanisms that contribute to the development of more powerful processes) is another form of conceptual analysis (Lesh, 1985). For example, unitizing, norming and partitioning (fair sharing) are primitive student capabilities that underlie rational number ideas. The following analysis, for the purpose of psychologizing critical content in the multiplicative field for instruction, examines additional basic concepts and processes and includes consideration of student thinking about them.

Continuing the Conceptual Analysis

Reasoning: Beyond Mechanization

Proportional *reasoning* refers to detecting, expressing, analyzing, explaining, and providing evidence in support of assertions about proportional relationships. The word *reasoning* further suggests that one uses common sense, good judgment, and a thoughtful approach to problem solving, rather than plucking numbers from word problems and blindly applying rules and operations. Researchers typically do not associate reasoning with rule-driven or mechanized procedures, but rather with mental, free-flowing processes that require conscious analysis of the relationships among quantities. Consider these problems and an eighth grader's approach to them:

a. If a bag of topsoil weighs 40 pounds, how much will 3 identical bags weigh?

b. If a football player weighs 225 pounds, how much will 3 players weigh?

c. If Ed can paint the bedroom by himself in 3 hours, and his friend, Jake, works at the same pace as Ed does, how long will it take to paint the room if the boys work together?

d. I was charged $1.30 for sales tax when I spent $20. How much sales tax would I pay on a purchase of $50?

e. Bob and Marty like to run laps together because they run at the same pace. Today, Marty started running before Bob came out of the locker room. Marty had run 7 laps by the time that Bob ran 3. How many laps had Marty run by the time that Bob had run 12?

Mason

a. $\dfrac{1\ bag}{40\ p.} = \dfrac{3\ b}{?\ p.}$ $3 \times 40 = 120\ p.$

b. $\dfrac{1\ f}{225\ p.} = \dfrac{3f}{?\ p.}$ $3 \times 225 = 675\ p.$

c. $\dfrac{1\ man}{3\ hrs} = \dfrac{2\ men}{?\ hrs}$ $2 \times 3 = 6\ hrs$

d. $\dfrac{\$1.20\ tax}{\$20} = \dfrac{?}{\$30}$ $\$1.20 \times 50 = 60$
   $60 \div 20 = \$3$

e. $\dfrac{M}{7\ laps}{?\ laps} = \dfrac{B}{3\ laps}{12\ laps}$ $7 \times 12 = 84$
   $84 \div 3 = 28$

Problems a and d can be correctly solved with the student's algorithmic approach. The student's blind application of proportions is a reminder that part of understanding a concept is knowing what it is *not* and when it does *not* apply. Two quantities may be unrelated (as in problem b); one quantity may be related in an inversely proportional way to another quantity, as in problem c (increasing the number of people decreases the time needed to do a job); or the *change* in one quantity might be proportional to the *change* in the other quantity, as in problem e. Many other important relationships do not entail proportional relationships at all, and students must learn to recognize the difference.

Another nuance of the word *reasoning* is that it relies on headwork and logical reasoning patterns to draw conclusions or inferences, rather than written computations.

Colloquially, proportional reasoning is "reasoning up and down." For example, the following problem can be solved by reasoning up, without using the form $\frac{a}{b} = \frac{c}{d}$ or doing any writing at all.

> If a box of detergent contains 80 cups of powder and your washing machine recommends $1\frac{1}{4}$ cups per load, how many loads can you do with one box?

Think: $1\frac{1}{4}$ cups do 1 load

5 cups do 4 loads

40 cups do 32 loads

80 cups do 64 loads

> It takes 6 men 4 days to complete a job. How long will it take 8 men to do the same job?

Think: 6 men take 4 days

1 man doing the work of 6 men all by himself takes 24 days

8 men, dividing up the work that 1 man did, take 3 days

Reasoning up and down can replace pencil work from the initial stages of fraction instruction. For example, consider the following fraction question, in which the unit is implicit.

> These are $\frac{2}{3}$ of Joan's pennies. How many pennies are $\frac{1}{2}$ of all she has?

The problem gives information about $\frac{2}{3}$ of the unit; from there, one can reason down to $\frac{1}{3}$, then to 1 (because three $\frac{1}{3}$s make the whole set of pennies), then to $\frac{1}{2}$.

Think: 8 pennies are $\frac{2}{3}$

4 pennies are $\frac{1}{3}$

12 pennies are 1 whole set

6 pennies are $\frac{1}{2}$ of Joan's pennies

Unfortunately, students rarely reason without a pencil and paper. Yet, from my experience with children, I consider reasoning aloud the most profound and compelling constructivist activity. Not only does the reasoning aloud enable students in the class to engage in interactive, critical communication in which they influence each other's interpretations and actions (Cobb, Yackel, &Wood, 1992), but in addition, the student who is reasoning aloud becomes his own best critic. Hearing his own message constitutes instantaneous reflective abstraction and serves to develop his metacognitive skills.

Covariance and Invariance

One of the most ubiquitous ways of thinking and operating in mathematics entails the transformation of quantities or equations in such a way that the mathematical structure is invariant (unchanged). As I have already mentioned, proportional relationships involve one of the simplest forms of covariation. Two quantities are linked to each other in such a way that when one changes, the other one also changes in a precise way with the first quantity. How does one know precisely how the changes in the quantities should be linked? How would one recognize an error in reasoning in either of the problems just solved (laundry detergent and pennies)?

Part of what it means to understand the proportional nature of the rational numbers is to recognize valid and invalid transformations. Two major types of invariance play a role in proportional reasoning. One is invariance of the ratio of two quantities. The second is invariance of the product of two quantities.

In a direct proportion, the direction of change in the related quantities is the same. One says that "y is directly proportional to x" or that "y varies as x." For example,

> For every 2 cups of water in a punch recipe, you add $\frac{1}{2}$ cup of sugar. If you quadruple the recipe so that you are using 8 cups of water, you should use 2 cups of sugar.

The amount of water as compared to the amount of sugar is always the same, no matter how large a batch you make. In any size batch that is made with this recipe, there is always 4 times as much water as there is sugar. The two quantities may be increased or decreased as long as the relationship between the quantities is preserved, that is, as long as their original ratio is maintained. Symbolically,

$$\frac{y}{x} = k, \frac{n_1 y}{n_1 x} = k, \frac{n_2 y}{n_2 x} = k, \frac{n_3 y}{n_3 x} = k,$$

where n_i is a natural number.

For example, in the laundry detergent problem, the ratio of cups to loads was $1\frac{1}{4}$:1. The constant of proportionality, k, is the divided ratio: $1\frac{1}{4}:1=1.25$. That ratio was maintained throughout the reasoning process: $5:4 = 1.25$, $40:32 = 1.25$, and $80:64 = 1.25$. Similarly, in the pennies problem, the ratio (number of pennies):(fractional part of the set) was preserved:

$$8:\tfrac{2}{3}=12:1$$

In a second type of invariance two quantities change together, but the direction of change is not the same for both. In the work problem used above, increasing the number of people working on the job should mean that the job can be completed in less time. Conversely, decreasing the number of people working on the job should increase the number of days needed to complete the job. In this situation, "*y* is inversely proportional to *x*" or "*y* varies inversely as *x*." The invariance occurs in the product of the two quantities, number of men (*m*) and number of days (*d*). Symbolically, $m \cdot d = k$.

# men	# days	# man-days
6	4	24
8	3	24
1	24	24
4	6	24
2	12	24

Multiple layers of development precede the ability to detect the invariant quantity when two quantities are changing together. First of all, students do not always focus on the quantifiable characteristics of a situation. For example, middle school students were asked to think about this situation:

When you get on your bicycle in the front of your house and you ride your bicycle down the street, what changes?

Their answers included the following observations:

- The pedals go up and down
- The trees go by
- I move away from where I started
- My wheels go around
- I pass my friend's house
- I get tired

Students must be encouraged to direct their thinking beyond obvious, surface-level observations toward more significant, quantifiable characteristics such as distance traveled, speed of the bike, height of one's

foot above the ground, and so on. If children are not thinking about quantities, it is pointless to talk to them about the ways that quantities change together.

Very young students have a tendency to assume that related quantities change together: If one quantity increases, so does the other to which it is related. For example, in a study of ratio sense, well over 50% of third graders said that if an orchestra can play a composition in 15 minutes, then two orchestras could play it in 30 minutes. Some fourth and most fifth graders observed that 2 orchestras playing would not take longer; it would just be louder. Students often need reminders to think about common sense statements that can help them to think about the way that quantities work together. For example, "If fewer people are working on a task, it should take longer to complete it" or, "If I travel at the same speed but I take a longer bike ride than I did yesterday, then it should take me longer today."

However, using common sense and recognizing the explicit quantities in a problem are not enough. One must look beyond the given quantities to construct a new quantity. A third (implicit) quantity derives from the relationship of the two changing and connected quantities; this new quantity remains constant. Conceiving of a relationship between two quantities (a ratio) as a quantity is difficult, but an intensive quantity is precisely such an abstraction. It is not explicit in the situation described; it has a different referent than either of the quantities that are present; and it is difficult to represent (Kaput, Luke, Poholsky, & Sayer, 1986).

Multiplicative Reasoning

In the previous problems about the laundry detergent and men working, why did the transformations consist of *multiplying* (or dividing) both quantities by the same whole number? Some degree of mathematical maturity is required to understand the difference between adding and multiplying and contexts in which each operation is appropriate. One of the most difficult tasks for children is to understand the multiplicative nature of the rational numbers. Children who reason additively indiscriminately employ additive transformations, but whether additive reasoning is an invariant stage in the development of proportional reasoning, as Inhelder and Piaget (1958) suspected, is unclear. Other problems, including imprecise colloquial language, often interfere with children's apprehension of mathematical ideas. For example, when asked to compare the sizes of the cats in the following picture, young children will say that the cat on the left "shrunk into" the cat on the right, or that the cat on the right was enlarged. These children simply do not know that shrinking or enlarging entails more than just adding to or subtracting from the height of the cats.

Proportional reasoners are able to differentiate between additive and multiplicative situations and to apply whichever transformation is appropriate. The process of addition is associated with situations that entail adding, joining, subtracting, separating, and removing—actions with which children are familiar because of their experiences with counting and whole number operations. The process of multiplication is associated with situations that involve such processes as shrinking, enlarging, scaling, duplicating, exponentiating, and fair sharing. As students interact with multiplicative situations, analyzing the quantitative relationships, they eventually understand why additive transformations do not work. However, this takes time and experience, and it cannot happen until the student can detect intensive quantities. For example, consider this problem:

For your party, you had planned to purchase 2 pounds of mixed nuts for 8 people, but now 10 people are coming. How many pounds should you purchase?

# People	Pounds of nuts
8	2
10	?

Having 10 people is 2 more than 8 people, but adding 2 pounds of nuts for 2 more people suggests that a whole pound is needed for each person, when originally you were not figuring on a pound per person. Hence, an additive transformation does not preserve the ratio (pounds of nuts : number of people). As this example illustrates, a student is not going to be able to analyze this situation until he "sees" in it the implicit third quantity "pounds per person."

Abstraction: Beyond Observation and Direct Measurement

Multiplicative reasoning moves beyond the realm of concrete operations and into the stage of formal reasoning. It deals with abstractions, and Piaget considered it the hallmark of formal reasoning. In this computer age, students are accustomed to a barrage

of sense data; understanding comes from perceptually based data, but in logico-mathematical knowledge, understanding consists in grasping abstractions imposed upon sense data. Abstraction entails imposing a relational structure, a classification scheme apart from the object, event, or image in which it is exemplified. It is not a *per*ception as much as it is a *con*ception. The construction and use of ratios as measures is one example.

In the previous example, the quantities 8 people and 2 pounds of nuts relied on counting and weighing, two activities with which most people are comfortable. However, the quantity *pounds per person* is both a relationship between the other two quantities, people and pounds, and a new measure derived from the other two measures by considering their ratio. The underlying concept of ratio (a relationship between a pair of quantities) is fundamental to fractions (in arithmetic), to rates (in physics), to similarity transformations (in geometry), and to probabilities (in statistics), just to name a few mathematical applications. Density, slope, and speed, for example, are all called intensive quantities and they differ from extensive quantities (those that tell numerosity, length, height, time, etc.). They are, in fact, not directly observable and not directly measurable. Intensive quantities are relationships (comparisons or ratios) between two other quantities. In short, relative or comparative thinking, as opposed to absolute thinking, is needed to move beyond the sense data.

Relative thinking activities provide the opportunity for students to further expand the range of applicability of certain words that were formerly only associated with additive concepts (Pimm, 1987). For example, the word *more*. Although most children will be familiar with the word as a signal word for addition/subtraction, the word can have a proportional or relational meaning:

Which family has more girls?

The Jones Family

The King Family

Being able to see that 2 girls comprise a larger proportion of the King family ($\frac{1}{2}$ or 50%) than they do of the Jones family ($\frac{2}{5}$ or 40%) is related to understanding percent increase/decrease problems.

Understanding intensive quantities sometimes requires more than intuition and experience; content knowledge sometimes plays a role. In turn, without knowledge of intensive quantities, one cannot reason proportionally. In some cases, mathematical knowledge is needed to quantify some of these characteristics. For example, the property *squareness* relies on a mathematical definition of a square as a quadrilateral with congruent sides and angles.

Which of these rectangles (not drawn to scale) is most square?

Early elementary instruction includes such topics as counting, telling time, making change, measuring length with a ruler, and, perhaps, massing objects. Later topics include perimeter, area, and volume. On more than one account, researchers might argue that this is an impoverished treatment of measurement. Measurement lies at the very heart of human activity; humans have always been preoccupied with measuring their universe, and the units and methods of measurement—which have a long history of development (Klein, 1974)—are essentials of science. Yet, the elementary mathematics curriculum does not address the fact that some quantities cannot be measured directly. Students need time to analyze such characteristics as color intensity, sourness, roundness, the oranginess of a drink, and the crowdedness of an elevator, for example, and to engage in argumentation and justification about how to measure these characteristics and distinguish them from quantities with extensive measures.

In addition, the treatment of measurement in the elementary curriculum does not allow time for students to develop *concepts* of measurement, as opposed to the acts of measuring. For example, by the start of third grade, very few students understand the inverse relationship between the size of a unit of measure and the number of times you can measure it out of some fixed quantity. Very few understand that no matter what the unit of measure, you can break it into smaller and smaller subunits to make your measurement as precise as you need. Unit conversions, in which the same quantity is expressed in terms of different-sized chunks, is related to unitizing. Until measurement

becomes a more reflective process, and one that is connected to other aspects of mathematical thinking, children are inadequately prepared for rational number understanding.

Need for Longitudinal Research

By the time students begin to study fractions, they have had 3 or more years of schooling in addition to 7 or more years of diverse, individual experiences that contribute to their intuitive knowledge. Knowing the path that any particular child has taken in coming to grasp a particular strand of knowledge is impossible because, in fact, many different routes can lead to the same useful understanding. A conceptual field, much like the theorized structure of the human mind (Pinker, 1997), is not linear. Thus, a single best way to reach a goal or to prescribe instruction to a desired end (à la Thorndike) is unlikely to exist. Nor could students possibly be taught each and every context and application that is useful. The process of achieving a flexible number sense in the rational number realm is partly developmental, so that instruction alone is not the full solution. Multiplicative ideas, in particular, fractions, ratios, and proportions, are difficult and develop over a long period of time; brief teaching experiments have had disappointing results. There seems to be no substitute for longitudinal research that takes into account the full complexity of the classroom, the full complexity of students' experiential backgrounds and intuitive knowledge, and the full complexity of the mathematics.

Nevertheless, conducting research has been difficult because the web-like organization of this knowledge defies the linear organization of scope and sequence charts, and the organization of schools into grade levels and terms of fixed length are not conducive to the kind of research that is needed. Time constraints, unstable communities, and other realities of public schools have made longitudinal research very difficult.

Over time, the nature of schools is changing and obstacles to doing long-term research in classrooms are diminishing. Many charter schools, for example, are not bound by traditional rules and organization, and opportunities are increasing for researchers to work intensively over 4 or 5 years with charter schools that are seeking excellence in mathematics. Teacher and student turnover in these schools is not the problem that it is in traditional, urban public schools, and some use mathematics specialists.

A Longitudinal Study

The preceding rational analysis of proportional reasoning, many studies of children's thinking in the field of rational numbers, and a rare bit of serendipity occasioned a longitudinal study of 6 classes of children as they developed rational number meanings and operations from the start of Grade 3 until they finished Grade 6. A glimpse into the frameworks, curriculum, instruments, and results of that study (Lamon, in progress) are offered here with the hope that they can stimulate research in the area in many new directions.

Central Multiplicative Structures

In my longitudinal study, I used a 7-point model for building mathematics instruction that grounded children in one of the subconstructs of rational numbers as well as six other ideas considered to be central to the development of multiplicative thinking. Following Case (1992), I refer to the six topics as *central multiplicative structures.* Case formulated the construct of a *central conceptual structure* in response to the debate about whether a general intelligence results from a developmental process, as proposed by Piaget (e.g., Flavell, 1963), or whether intelligence is modular or domain-specific as proposed by Gardner (1983). Case and his colleagues presented evidence (Case, 1992) that the real case is somewhere in between these two positions. Central conceptual structures are critical systems—built-in and subject to developmental ceilings, and partly nurtured through instruction—including content, action, associations with appropriate contexts, representations, and a web of conceptual relations both within and between them. They are built up by a complex interaction of knowledge and experience over a long period of time and are called *central* because they constitute part of the very backbone of mathematics, ideas and processes and representations that are recurrent, recursive, and of increasing complexity across mathematical and scientific domains. When these structures are sufficiently mature and connected, their owner develops control structures so that they are under her command (reasoning is conscious and deliberate), they enable higher order thinking that was previously impossible, and they affect knowledge and performance across a variety of domains.

Paradigm and Methodology

On the basis of the preceding rational task and conceptual analyses, the following 7 central structures were hypothesized as critical components of instruction if children are to eventually develop a rich understanding of multiplicative concepts.

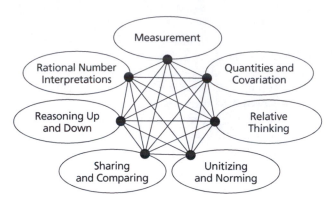

For 4 years, in Grades 3 through 6, five cohorts of students in two urban schools in different parts of the country each built their understanding of rational numbers on a different initial interpretation of the symbol $a\text{–}b$: as a part/whole comparison with unitizing, as a measure, as a quotient, as an operator, and as a ratio. A sixth class that received traditional fraction instruction served as a control group. Instruction in all of the classes except the control group entailed elements from the other 6 hypothesized central multiplicative structures. The rationale was this: If these are, indeed, central conceptual structures, then in time (and expectations were that this was going to take a *long* time), children's knowledge built in each of the "nodes" should grow together. As a result, students should demonstrate a usable knowledge of multiple rational number interpretations, not only the one in which their instruction was grounded. Their thinking should be flexible enough to support proportional reasoning in problem-solving situations. This study provided a unique opportunity to compare unconventional approaches with traditional instruction and to document the sequencing and growth of ideas and the breadth and depth of the understanding that developed.

This research was conducted in the paradigm that has become known as *design experiment* research. The chief characteristic of design research (Brown, 1992) is that the product is being designed in a series of systematic trials and adaptations. The data collection, analysis, and refinement of the product being developed (in this case, effective instruction in rational numbers and proportional reasoning) go through an extensive test-and-revise cycle. One's well-researched framework is the starting point, but daily activities are designed and changed in response to classroom observations to better meet the needs of the students that they are purportedly serving. That is, based on data (e.g., observations, students' written work, classroom discussions), the next day's lessons reflected what the teachers and I learned from that data. Rather than empirically proving that some theory "works," a design experiment actually studies the synergy, the ecology of a fully functioning classroom in all of its complexity. Rather than "controlling" for variables, it allows the researcher and teacher to regulate what they do in response to the diversity they encounter. Simply put, the teachers and I meddled in the research. We brought about innovative forms of learning so that we could study them and tweak them and make them as effective as possible for as many students as possible.

This design research was based upon the following questions:

1. What understanding might develop if fraction instruction were based on a subconstruct other than part-whole comparisons?

2. What sorts of instructional activities support understanding of each of the nontraditional rational number constructs?

3. How long do students take to develop a useful understanding of any single subconstruct of rational number?

4. Because understanding is not an all-or-none affair, what are some of the benchmarks along the path to understanding?

5. Are the subconstructs equally good alternatives for learning concepts and computation?

6. Which subconstructs are powerful enough to connect to other interpretations without direct instruction?

7. Are any developmental processes or mechanisms operating that connect whole number knowledge and rational number knowledge?

8. Will the set of hypothesized central conceptual structures be necessary and sufficient to facilitate proportional reasoning?

Because I was concerned that my results generalize to other public school classrooms, I was careful not to idealize the study in any regard. Results include all students, including learning disabled students, students with behavioral disabilities, one student with a truancy problem, and students for whom English is a second language—all the realities normally found in a mainstream public school classroom. The reality principle applied equally to the teachers. Teachers studied to learn the material and the teaching methods immediately before applying the content and methods in their own classrooms. Although they were experienced teachers, they were not mathematics specialists.

Children were given the freedom and the encouragement to express their thinking in whatever manner they wished. The two teachers who taught the five classes facilitated learning through the kinds of activities they encouraged and through the problems they posed. None of the groups were taught any rules or operations. Teachers sometimes used whole-class instruction, and occasionally, whole-group discussions of the ideas arose from the activities. However, on most days, the mathematical activity consisted of group problem solving, reporting, and then individually writing and revising solutions for homework.

Three measures were used to compare the five classes with the control group: number of rational number subconstructs the student was able to use by the end of Grade 6, proportional reasoning, and fraction computation ability.

Problem types and children's strategies. A summary of the subconstructs of rational numbers and some of the kinds of classroom activities that supported each of them is given in Table 14.2, using the fraction $\frac{3}{4}$ as an example.

Children have a tremendous capacity to create ingenious solutions when they are sufficiently challenged and when they do not feel expected to follow rules. To give the reader both a sense of the kinds of nontraditional questions explored during instruction and a sense of the kind of student thinking those questions elicited, several examples are given here. More detail about instructional activities can be found in Lamon (in progress).

- Here, a student uses "reasoning up and down" to determine the fractional part of a set when the unit is given implicitly: If 6 books are $\frac{2}{3}$ of all the books on Robert's shelf, figure out how many books are $\frac{5}{9}$ of the books on his shelf.

Table 14.2 Alternatives to Part–Whole Fraction Instruction

Interpretations of 3/4	Meaning	Selected Classroom Activities
Part–Whole Comparisons with Unitizing "3 parts out of 4 equal parts"	$\frac{3}{4}$ means three parts out of four equal parts of the unit, with equivalent fractions found by thinking of the parts in terms of larger or smaller chunks. $$\frac{3 \text{ (whole pies)}}{4 \text{ (whole pies)}} = \frac{12 \text{ (quarter pies)}}{16 \text{ (quarter pies)}} = \frac{1\frac{1}{2} \text{ (pair of pies)}}{2 \text{ (pair of pies)}}$$	Unitizing to produce equivalent fractions and to compare fractions
Measure "$3(\frac{1}{4}\text{-units})$"	$\frac{3}{4}$ means a distance of $3(\frac{1}{4}\text{-units})$ from 0 on the number line or $3(\frac{1}{4}\text{-units})$ of a given area.	Successive partitioning, reading meters and gauges
Operator "$\frac{3}{4}$ of something"	$\frac{3}{4}$ is a rule that tells how to operate on a unit (or on the result of a previous operation): multiply by 3 and divide the result by 4 or divide by 4 and multiply the result by 3. This results in multiple meanings for $\frac{3}{4}$: $3(\frac{1}{4}\text{-units})$, $1(\frac{3}{4}\text{-unit})$, and $\frac{1}{4}$ (3-unit).	Machines, paperfolding, xeroxing, discounting, area models for multiplication and division
Quotient "3 divided by 4"	$\frac{3}{4}$ is the amount each person receives when 4 people share a 3-unit of something.	Partitioning
Ratios "3 of A are compared to 4 of B"	3:4 is a relationship in which 3 A's are compared, in a multiplicative rather than an additive sense, to 4 B's.	Bi-color chip activities

- This problem entails identification of the unit and the notion of operator: Here is one day's supply of juice. You have taken only one drink. How much of your day's supply do you still have left?

Terry

I measure it out and get ¹/₅ 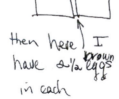 but it is ¹/₅ of ¹/₂ of all my juice that I drank.

¹/₅ of ¹/₂ = ¹/₁₀ drank

⁹/₁₀ left

- This is a relative thinking question that entails consideration of the how one might measure a quality such as "brownness" of a carton of eggs: Here are two egg cartons, each containing some brown and some white eggs. Which container is more brown?

Bert

I put them into 6 packs like this

then here I have 2¹/₂ brown eggs in each

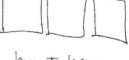

here I have 2¹/₃ brown eggs in each

¹/₂ is more, so the 12 pack is browner.

- How would you compare rates of growth? The first picture shows Jeb and Sarah Smart when they were younger. The second shows them as they look now. Who grew faster between the first and second pictures?

Grade 5 Student:

Well they both grew taller, but Jeb grew *taller*. Get it? He was shorter then he grew taller. Well, actually, Sarah was shorter then she grew taller, too. But he grew TALLER (loud and with emphasis).

(The student realized that his words were not conveying what he meant and showed some frustration in not being able to express himself. There was a long pause as he thought.)

Look. I can show you. (Taking a ruler, he aligned it with the top of Sally's head then and the top of her head now. Then he did the same for Jeb.)

See. The ruler is more slanty for Jeb. That proves he grew faster.

- Rational numbers as measures were introduced using successive partitioning of a number line.
  Locate $\frac{17}{24}$ on this number line:

Students were taught to use arrow notation to keep track of the size of the subdivisions after each time they partitioned.

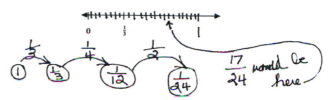

This was a particularly powerful interpretation of rational numbers. After a time, students stopped partitioning the number lines and were able to visualize and reason about the fractions without them.

- Find two fractions between $\frac{1}{8}$ and $\frac{1}{9}$.
  Martin, end of Grade 5

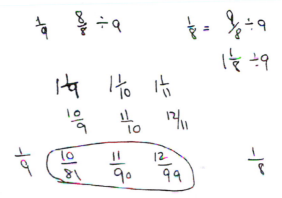

- Find 3 fractions between $\frac{7}{12}$ and $\frac{7}{11}$.
  Jonathan, end of Grade 5

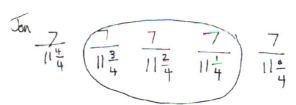

Annie, Grade 6, midyear

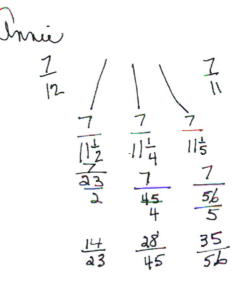

- Next, two students use unitizing to answer the question, "Which is more pizza, $\frac{1}{2}$ pizza or $\frac{3}{5}$ pizza?

Tommy R.

$$\frac{1 \text{ piz}}{2 \text{ piz}} = \frac{2 \left(\frac{1}{2} \text{ piz}\right)}{4 \left(1\frac{1}{2} \text{ piz}\right)} = \frac{3 \left(\frac{1}{3} \text{ piz}\right)}{6 \left(\frac{1}{3} \text{ piz}\right)} =$$

$$\frac{4 \left(\frac{1}{4} \text{ piz}\right)}{8 \left(\frac{1}{4} \text{ piz}\right)} = \boxed{\frac{5 \left(\frac{1}{5} \text{ piz}\right)}{10 \left(\frac{1}{5} \text{ piz}\right)}}$$

$$\frac{3 \text{ piz}}{5 \text{ piz}} = \boxed{\frac{6 \left(\frac{1}{2} \text{ piz}\right)}{10 \left(\frac{1}{2} \text{ piz}\right)}} = \frac{9 \left(\frac{1}{3} \text{ piz}\right)}{15 \left(\frac{1}{3} \text{ piz}\right)}$$

$\frac{3}{5}$ of a pizza is more.

Angela

$$\frac{1 \text{ pizza}}{2 \text{ pizzas}} = \frac{5 \left(\frac{1}{5}\text{-pizzas}\right)}{10 \left(\frac{1}{5}\text{-pizzas}\right)}$$

$$\boxed{\frac{3 \text{ pizzas}}{5 \text{ pizzas}} = \frac{6 \left(\frac{1}{2}\text{-pizzas}\right)}{10 \left(\frac{1}{2}\text{-pizzas}\right)}}$$

- How would you divide 1 by $\frac{2}{3}$?

Grace

$\frac{2}{3}$ is twice as big as $\frac{1}{3}$ and I know $\frac{1}{3}$ goes into 1 three time. So $\frac{2}{3}$ can go in only half as many times.

Troy

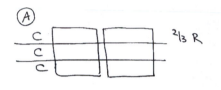

$1 = 6 \left(\frac{1}{6} \text{ pieces}\right)$

$\frac{2}{3} = 4 \left(\frac{1}{6} \text{ pieces}\right)$

After I take out 4 from the 6, I have $2 \left(\frac{1}{6} \text{ pieces}\right)$ left. This is $\dfrac{2 \left(\frac{1}{6} \text{ pieces}\right)}{4 \left(\frac{1}{6} \text{ pieces}\right)}$

- Here is a question that entails relative thinking and ratio comparison: The 5 people at Table A drank 2 root beers and 3 colas. The 7 people at Table B drank 3 root beers and 4 colas. Which table could be called the cola drinkers?

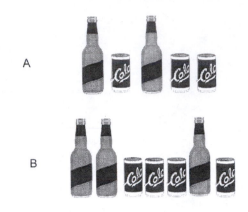

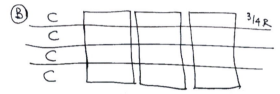

A, a cola is matched with $\frac{2}{3}$ of a root beer.

B, a cola is matched with $\frac{3}{4}$ of a root beer.

Cola is stronger at table A because it is matched with less root beer.

- Another ratio comparison question: Which group, the girls or the boys, got the better deal? 2 girls got into the theatre on State Street for $3. 5 boys got into the theatre on Main Street for $6.

I did $(5:6) - 2(2:3) = (1:0)$

$(5:6) - (4:6) = (1:0)$

The boys price is better because for the same amount of money they can get 1 more person in.

Selected Results

After 4 years, all five groups of students had developed deeper understanding of rational numbers as measured by the number of subconstructs they were using than did the students in the control group. The numbers of proportional reasoners far exceeded the number in the control group, and even in computation, achievement was greater in all five groups.

Under different rational number interpretations, children acquired meanings and processes in different sequences, to different depths of understanding, and at different rates. Even children who were taught the same initial interpretation of fractions showed different learning profiles. The time-honored learning principle of transferability was robust. Not only did children transfer their knowledge to unfamiliar circumstances, but to other interpretations that they were not directly taught. By the end of sixth grade, more than 50% of the children had demonstrated the ability to apply their knowledge to at least two of the rational number subconstructs. Table 14.3 shows the numbers of students from each class who were able to reason proportionally at the end of 4 years, the number from each class who could pass a computation test at the 80% level, and the number from each class who had demonstrated the ability to apply their knowledge to 1, 2, 3, or 4 of the rational number subconstructs.

Student reasoning and problem solving in the five special classes were extraordinary. Their abilities brought into question the conclusion from former (now classic) research studies that number relationships affect the solution of fraction or ratio comparison tasks. For example, consider this comparison task:

Are these fractions equivalent? Show how you figure it out.

$$\frac{3}{5} \quad \frac{7}{11}$$

If you are like most adults, you either used a common denominator or checked the cross products. When elementary school students in the beginning stages of fraction instruction are asked this question, most conclude that the fractions are not equivalent.

Grade 5 student: They are not equivalent because there is no number I can multiply the 3 and the 5 by to get 7 and 11.

Table 14.3 Achievement of the Four-Year Participants from Each Class on Three Criteria

Class	Proportional Reasoning	Computation[a]	Number of Rational Number Subconstructs				
			1	2	3	4	5
Unitizing[b]	8	13	19	8	7	1	0
Measures[c]	8	12	17	11	9	3	0
Operators[d]	3	9	15	9	0	0	0
Quotients[e]	2	9	16	7	0	0	0
Ratio/Rate[f]	12	11	18	10	6	0	0
Control[g]	1	6	11	2	0	0	0

(Header spanning: "Numbers of Students Achieving Each Criterion")

[a] 80% accuracy or better; [b] $n = 19$; [c] $n = 17$; [d] $n = 19$; [e] $n = 18$; [f] $n = 18$; [g] $n = 20$

Students will, in fact, arrive at the same conclusion for the same reason when asked to compare $\frac{4}{6}$ and $\frac{6}{9}$. They have no notion of equivalence, but they remember their teachers saying that "to get equivalent fractions, you multiply the numerator and the denominator by 1 or 2 or 3, and so on." Older children who have learned how to reduce or to find common denominators do a little better. However, middle school students generally perform very poorly on this type of comparison task (for example, see Karplus et al., 1983b). Either children do not know how to proceed, or they resort to incorrect additive strategies. Researchers have concluded that the lack of integral multiples between the given ratios and within the given ratios greatly contributes to their difficulty. Students can more easily compare fractions or ratios such as $\frac{3}{12}$ and $\frac{5}{24}$, $\frac{6}{8}$ and $\frac{12}{15}$, or $\frac{3}{9}$ and $\frac{6}{15}$.

Children, who were never given rules, did not perform as the *integral multiples* hypothesis would predict. In fact, when they were given all of the tasks used in the Karplus et al. study, the students made no more errors on comparison tasks without integral multiples than on tasks involving integral relationships. This suggests that the nature of the instruction children receive may have more to do with their ability to compare these fractions than with the kind of numbers involved. The children in the Karplus study were in sixth and eighth grades and presumably had rule-based instruction. When rules and procedures are not learned with meaning, students forget them or do not always realize when to use them. When children are accustomed to thinking and reasoning without rules, which numbers they are given makes little difference.

The question naturally arises as to the kind of learning experiences that will result in a well-rounded understanding of the rational numbers. One way that question is asked is: Should children be taught one interpretation in depth or should instruction include all of the interpretations? This seems to be the wrong question. All of the interpretations do not provide equal access to a deep rational number understanding, and no single interpretation is a panacea. However, interpretations are tightly intertwined, and with the proper attention to the central multiplicative structures, students will naturally develop an understanding of multiple interpretations. Results are briefly summarized by class.

Part-Whole Comparisons with Unitizing

Unitizing added a dynamic aspect to the traditional part-whole interpretation so that children developed a very strong notion of the unit, and of equivalent fractions, which definitely facilitated the operations of addition and subtraction. This interpretation has strong, natural connections to the measure interpretation, to ratios, and to operators. By the end of sixth grade, some students had a working knowledge of four interpretations.

The part-whole interpretation of rational numbers does not provide a particularly intuitive or insightful path to fraction multiplication. For example, using part-whole language or thinking to multiply $1\frac{1}{3} \cdot 2\frac{1}{2}$ is difficult. Nevertheless, for division, it led to the common denominator algorithm.

$$1\frac{1}{4} \div \frac{2}{3}$$

$$1\frac{1}{4} = \frac{5}{4} = \frac{15}{12}\left(\frac{1}{3}\text{-parts}\right) \text{ and } \frac{2}{3} = \frac{8}{12}\left(\frac{1}{4}\text{-parts}\right)$$

How many times can you measure 8 parts out of 15 parts? $\frac{15}{8}$

$$\text{This tells us that } \frac{5}{4} \div \frac{2}{3} = \frac{15}{12} \div \frac{8}{12} = \frac{15}{8}$$

In this case, both the divisor and the dividend were reinterpreted in terms of equal-sized pieces of the unit so that the division was essentially a whole number division.

Quotients

Quotients shared natural connections with ratios and rates, and there was a great deal of power in being able to interchange quotients and ratios in sharing and comparing contexts. Students could easily choose and use equivalent fractions, compare fractions, and add and subtract fractions. However, the thinking used for addition and subtraction did not prove as easy to use for multiplication and division. The model broke down and became too complicated for multiplication and division, and area models were used to provide a meaningful introduction to multiplication and division. The children were already using the language and ideas of operators, which had evolved during their partitioning activities. Thus, introducing the area models to the children was quite natural and easy.

Operators

Although the operator interpretation provided a useful context for multiplication and division, scaling, and general fraction sense, it did not lead naturally to fraction addition and subtraction. This is because rational numbers have a dual nature. In mathematical terms, they are a field; they obey two sets of group axioms, those for addition and those for multiplication. These operations are defined independently from one another, and in the operator construct, perhaps moreso than in the other interpretations, the separateness of those operations was evident.

However, the representations used for multiplication and division, area models, could be extended quite easily to compare fractions and to build meaning for symbolic addition and subtraction.

$$\text{Add } \frac{2}{3} + \frac{3}{4}$$

Look at $\frac{2}{3}$ of a unit and $\frac{3}{4}$ of the same unit by partitioning a unit area into thirds in one direction and into fourths in the other direction.

It can then be seen that $\frac{2}{3}$ (2 rows) is $\frac{8}{12}$ of that area and that $\frac{3}{4}$ (3 columns) is $\frac{9}{12}$ of that area.

$$\text{So } \frac{3}{4} > \frac{2}{3} \text{ and } \frac{2}{3} + \frac{3}{4} = \frac{17}{12}.$$

Measures

Students developed strong notions of the unit and subintervals, equivalence, the order and density of the rational numbers, and the operations of addition and subtraction. Most children naturally extended their knowledge to the operator interpretation. For example, from their operations on number lines and the supporting use of arrow notation, they already knew that $\frac{1}{2}$ of $\frac{1}{6}$ is $\frac{1}{12}$ and $\frac{1}{3}$ of $\frac{1}{7}$ is $\frac{1}{21}$. They were well prepared for the language of operators and had no difficulty connecting to that division model. Some of these children demonstrated a knowledge of four interpretations by the end of sixth grade.

Ratios

The greatest difference between ratios and the other interpretations of rational numbers is in the way they combine through arithmetic operations. The other interpretations of rational numbers are all different conceptually, but they are indistinguishable once they are written symbolically. They add, subtract, multiply, and divide according to the same rules. However, one does not operate on ratios in the same way as on fractions.

Students who studied ratios and rates as their primary interpretation of rational numbers developed a very strong notion of equivalence classes and of proportional reasoning in general. They easily switched between ratio and part-whole comparisons and had no trouble with fraction addition and subtraction. Most of them developed their own ways of reasoning about

multiplication and division. For example, to multiply $\frac{3}{8} \cdot \frac{2}{3}$, they used proportional reasoning:

$$\frac{1}{3} \text{ of } \frac{3}{8} = \frac{1}{8}, \text{ so } \frac{2}{3} \text{ of } \frac{3}{8} = \frac{2}{8} \text{ or } \frac{1}{4}.$$

Most of the ratio class developed a good working knowledge of ratio, part-whole, and operator.

Conclusions and Implications

The deep connectivity of the rational number subconstructs made teaching only one interpretation impossible. The rational number personalities that I have discussed highlight different and essential characteristics of the rational numbers, but they are inextricably connected. In the end, the measure subconstruct seemed to be the strongest because it connected most naturally with the other subconstructs; in fact, I agree with Kieren that part-whole is not a separate construct, but really a case of the measure subconstruct (Kieren, 1988). Not all of the subconstructs were equally good starting points; operators and quotients were less powerful than the measures, ratios, and part-whole subconstructs.

Children need a starting place in one of the rational number interpretations for their fraction instruction. In this space, they need to develop the ideas of unit and equivalence of fractions. They need to develop techniques for comparison so that they can judge the relative size of fractional numbers. They need sufficient time to work in that interpretation without being given rules so that they develop fraction sense, a comfort and flexibility in fraction thinking. Although the measure subconstruct and the ratio subconstruct were especially powerful, pretesting suggested that informal experiences and intuitive knowledge better prepared some children for being sharers or comparers and others for being concatenators or composers when performing multiple operations. Thus, children were placed into their groups to best take advantage of their preinstructional strengths, and clearly not all children were equally prepared to deal with the measure and ratio interpretations. A more significant factor in overall success, no matter where a child was placed, was the development of the central multiplicative structures.

Will Incorporating More Rational Number Subconstructs Be Effective?

My research suggests that incorporating rational number interpretations other than the traditional part-whole meaning, will, in itself, be insufficient to facilitate meaningful learning. Children need in-depth grounding in a specific rational number interpretation (part-whole comparisons, ratios, operators, measures, or quotients) to which they can return to center their thinking when they face a new problem. In addition, I question whether studying all of the interpretations will allow students the time to get to know any of the subconstructs indepth.

This observation is supported by evidence, even when working in-depth with a single interpretation, of a long-term learning process. As an example, in another study (Lamon, 1996), many students whose instruction began with partitioning (the quotient subconstruct) took up to 2 years before they could answer the following questions:

If the girls each get an equal share of the pizza, how much pizza will each girl receive?

How much of the total pizza will each girl receive?

Because of all the unnecessary marking and cutting they did, students were unable to reassemble the many pieces that comprised one person's share to name the amount of pizza in one share (3/7 of a pizza) until they had some notion of equivalence and could distribute larger chunks to each person. Understanding the task also demands that one know that when 7 people are sharing something, each share will be 1/7 of the unit, regardless of the unit, whether or not it is composite, and what size the components are. Many adults were instructed long ago that the fraction symbol 3/7 means "3 divided by 7;" however, they fail to realize that that fact has anything to do with the sharing situation presented here, or that 1/7 of the total pizza is 3/7 of a pizza. Thus, partitioning and the quotient interpretation of a rational number are not easily grasped, and a superficial treatment of the process and concepts does not result in understanding. This perspective is further substantiated by the small numbers of children who understand part-whole comparisons in spite of their historical emphasis in instruction (Lesh et al., 1983).

Current part–whole fraction instruction likely could be improved merely by allowing children the time and opportunity to build understanding of the central multiplicative structures and not directly presenting rules and algorithms. In fact, in a 4-year follow-up study in which instruction in the part-whole subconstruct was supplemented with materials emphasizing the central multiplicative structures, results were similar to those in the first study. Details concerning the pre- and post-tests and the nature of instruction under each subconstruct are given in Lamon (in progress).

Building meaning took a long time, and during the first 2 years, students in the rational number study were no competition for other students who were already performing fraction computations. However, in the long run, all five classes of students surpassed the rote learners, and their representations showed that they were performing meaningful operations. Of course, this delayed gratification has many implications for teacher accountability, state and national assessment, and many other aspects of schools and schooling.

Developing Research Agendas

Shulman (1987) suggested that not all research sites are of equal importance. Some areas represent more critical sites because they can result in dramatic reconceptions with far-reaching results. As he pointed out, all too often, researchers choose their research sites because of their amenability to available concepts and research methodologies. However, the complexity of the multiplicative conceptual field has been impenetrable with traditional perspectives and research methodologies. Researchers in this area have had to find new ways to think about the content, new vocabulary by which to discuss it, and new methodologies by which to study student thinking (Harel & Confrey, 1994).

Mathematics education has a compelling need for longitudinal studies with the same children in the domain of rational numbers and proportional reasoning. Many of the generalizations about stages and order of learning in mathematics education have been made by taking snapshots of children's thinking. A persistent problem in snapshot research is that a child's actions at any particular time really depend on many causes—constraints and affordances—in his immediate environment. What the child might have done under different conditions, over longer periods of time, goes undiscovered. Particularly in biological research, models suggest that two of the prerequisites for natural selection are enough variation and enough time (Lamon, 2003; Pinker, 1997). This was true for the children in the longitudinal study. For the first couple of years, in comparison to the children who received traditional part-whole fraction instruction, the special classes appeared to make little progress. However, once children found processes that worked and these were internalized, they had more powerful ways of thinking than the children who had been taught to execute standard algorithms. They had been given the time to develop more diverse perspectives that, in the end, provided them with great flexibility.

Another important consideration is that, given the difficulty and the complexity of the multiplicative conceptual field, progress is unlikely to be made if researchers do not adopt personal research agendas, life-long areas of interest that keep them focused on a significant issue. Researchers are encouraged to carve a niche in which they will focus in-depth on a particular piece of the mathematics and on children's thinking about that mathematics. Research that moves from topic to topic in "opportunity" studies is not as powerful in advancing knowledge in the field.

In the interest of encouraging new researchers to devote their time and talent to this engaging domain, I will conclude with some ideas for building research agendas. Usually studies answer some questions and unearth many others, and this was certainly the case in the longitudinal research reported in this chapter. Some general directions as well as a sampling of more pointed questions are presented.

Adapting the Framework

The framework used for the study reported in this chapter provides a useful starting point and can be used as a springboard for additional studies. It could be used in a number of ways, in whole or in part, for future research.

Research in Microworlds

The work of Les Steffe and his colleagues is an example of conceptual microanalysis. The work of studying in fine detail how children construct units and operate with them is critical in psychologizing mathematical content. Much more work of this nature is needed. In particular, given the importance of measurement ideas to understanding not only rational numbers, but all of mathematics, a microanalysis of the development of measurement could provide another long-term research agenda with broad impact.

The Reversibility of Processes

The concept of multiplicative inverses becomes important when a child enters the world of rational numbers. Children do not always construct operations and inverse operations at the same time. Similarly, Lamon (1994) found that when children first constructed complex units, they were not immediately able to decompose those units into their constituent parts. Researchers know very little about reversibility or about multiplicative operations and inverses, and

these could be subjects for a valuable microanalysis research agenda.

Focus on a Single Central Multiplicative Structure

Another way to conceive of doing research in the domain is to form a research agenda based on one of the central multiplicative structures that seeks answers to the many questions surrounding that single structure. For example, how do children use unitizing in coming to understand each of the rational number subconstructs? Alternatively, building an agenda around *quantities and covariation,* one might study the process by which children recognize the requirements of a ratio-preserving mapping and later, construct such mappings. Very little is also known about children's thinking in inversely proportional contexts.

Focus on Children's Thinking at the Beginning of Third Grade

One might investigate the possibility that different subconstructs might provide better inroads to the rational numbers for different students. Some evidence from pretests given to students in the longitudinal study reported in this chapter, indicated that children who are oldest in their families are more inclined to be sharers (partitioners) and children with older siblings, comparers (ratio thinkers). My hypothesis is that the need for younger siblings to hold their own within their respective families might make them better comparers. Mothers frequently give snacks to the older child with directions to share them with younger brothers or sisters. After being shortchanged a few times, the younger child begins to monitor the division of the goods. In the same vein, some third-grade children were better at concatenating than composing and vice versa. Apparently by third grade, or the beginning of fraction instruction, children may have already developed a propensity for ways of thinking that might make one interpretation of rational numbers a more logical starting point for them in order to build upon their informal knowledge.

Generative Ordering

Kieren's work (e.g., Kieren, 1993) is an excellent example of research that considers the generative nature of knowledge. Intelligence bootstraps itself both by enlarging its domain of effective operation but also by building its own depth or *interiority* by embedding previous thinking and elaborating it into more sophisticated forms. Kieren's work used the recursive nature of knowledge building to try to understand how one can build rational number knowledge from basic concepts and the actions they support. The goal of the framework presented in this chapter was to capture the germs of the vast system of relationships that interact and eventually build upon one another to facilitate proportional reasoning, but much more generative ordering research (but on a smaller scale than Kieren uses) is needed to more fully understand the power of various elementary structures and mechanisms. For example, beginning with a single idea or action—say, unitizing and norming, or partitioning—use a deep mathematical analysis to determine how far the idea or the process can be pushed to build increasingly sophisticated forms of rational number knowledge. Then, test that theoretical analysis in empirical research to see if or how it corresponds to what children do or can do.

Other Outstanding Questions

Many other researchable questions lie in new areas, or in areas where much fertile ground still exists:

- What are the links between additive and multiplicative structures?
- What mechanisms drive the change from additive to multiplicative reasoning?
- What intuitive (informal) strategies are available to children before fraction instruction?
- What are the limits of a child's intuitive knowledge in building fraction concepts and operations?
- What knowledge beyond fractions is needed to understand rational numbers?
- How does one measure rational number sense?
- What are the developmental ceilings for understanding multiplicative concepts at various ages?
- Which aspects of whole number knowledge are useful in building fraction knowledge?
- How does a child come to understand a rational number as a single quantity as opposed to regarding it as a pair of numbers?
- How is the quantification of motion best taught?
- What do children of various ages understand about distance-rate-time relationships?
- How do children come to understand this triad?
- What is the relationship between the development of increasingly complex unit structures and unit splitting?
- What is the connection of this splitting that leads to exponentiation and the successive partitioning that was used to study the measurement interpretation of rational numbers?

- What are the connections between rational number learning and algebra?
- What are the connections between rational number learning and calculus?
- As children study each of the rational number interpretations, what are the benchmarks by which to judge that their knowledge is moving in a desirable direction?
- What constitutes understanding of each of the rational number subconstructs? Specifically, how is understanding operationalized? (What are the observable effects of understanding a particular rational number subconstruct?)
- How can we assess depth of understanding of the rational numbers?
- How many subconstructs should a student understand before high school algebra? Before calculus?
- Instruction must build upon children's intuitive knowledge but push them beyond it as well. What are the limits of children's intuitive knowledge and qualitative reasoning? How far can that intuitive knowledge be pushed, and what critical mathematical content and ways of thinking will not occur without instruction?
- How can unit analysis help to connect fraction learning with other conceptual areas?
- What are the connections between proportional reasoning (as defined in this chapter) and proportionality?
- What constitutes understanding of proportionality?
- What are some of the benchmarks of understanding between proportional reasoning and proportionality?
- Is it better to teach one rational number subconstruct or all five?
- If one, which should it be?
- What are the conceptual difficulties (learning problems) associated with each rational number subconstruct?
- Do algorithms preclude reasoning, or can older students develop useful knowledge about the central multiplicative structures?

REFERENCES

Abramowitz, S. (1974). *Investigation of adolescent understanding of proportionality.* Unpublished doctoral dissertation, Stanford University.

Begle, E. G., & Gibb, E. G. (1980). Why do research? In R. J. Shumway (Ed.), *Research in mathematics education* (pp. 3–19). Reston, VA: National Council of Teachers of Mathematics.

Behr, M., Harel, G., Post, T., & Lesh, R. (1992). Rational number, ratio, and proportion. In D. A. Grouws (Ed.), *Handbook of research on mathematics teaching and learning* (pp. 296–333). New York: Macmillan.

Behr, M. J., Lesh, R., Post, T. R., & Silver, E. A. (1979). *The role of manipulative materials in the learning of rational number concepts: The rational number project.* (NSF SED 79-20591). Washington, DC: National Science Foundation.

Behr, M. J., Lesh, R., Post, T. R., & Silver, E. A. (1983). Rational-number concepts. In R. Lash & M. Landau (Eds.), *Acquisition of mathematics concepts and processes* (pp. 91–126). Orlando, FL: Academic Press.

Behr, M. J., Lesh, R., & Post, T. R. (1984). *The role of rational number concepts in the development of proportional reasoning skills* (NSF DPE-8470077). Washington, DC: National Science Foundation.

Behr, M. J., Wachsmuth, I., & Post, T. R. (1985). Construct a sum: A measure of children's understanding of fraction size. *Journal for Research in Mathematics Education, 16*(2), 120–131.

Behr, M. J., Wachsmuth, I., Post, T. R., & Lesh, R. (1984). Order and equivalence of rational numbers: A clinical teaching experiment. *Journal for Research in Mathematics Education, 15*(5), 323–341.

Bell, A., Fischbein, E., & Greer, B. (1984). Choice of operation in verbal arithmetic problems: The effects of number size, problem structure, and context. *Educational Studies in Mathematics, 15,* 129–147.

Bell, A., Swan, M., & Taylor, G. (1981). Choice of operations in verbal problems with decimal numbers. *Educational Studies in Mathematics, 12,* 399–420.

Bereiter, C. (1990). Aspects of an educational learning theory. *Review of Educational Research, 60*(4), 603–624.

Bezuk, N. (1986). *Variables affecting seventh grade students' performance and solution strategies on proportional reasoning word problems.* Unpublished doctoral dissertation, University of Minnesota.

Bohm, D., & Peat, F. D. (1987). *Science, order, and creativity.* New York: Bantam Books.

Brown, A. L. (1992). Design experiments: Theoretical and methodological challenges in creating complex interventions. *Journal of the Learning Sciences, 2*(3), 141–178.

Bruner, J. (1960). *The process of education.* New York: Vintage Books.

Callanan, M. A., & Markman, E. M. (1982). Principles of organization in young children's natural language hierarchies. *Child Development, 53,* 1093–1101.

Case, R. (1985). *Intellectual development: Birth to adulthood.* New York: Academic Press.

Case, R. (Ed.) (1992). *The mind's staircase: Exploring the conceptual underpinnings of children's thought and knowledge.* Hillsdale, NJ: Erlbaum.

Cobb, P., Yackel. E., & Wood, T. (1992). A constructivist alternative to the representational view of mind in mathematics education. *Journal for Research in Mathematics Education, 23*(1), 2–33.

Comber, L. C., & Keeves, J. P. (1973). *Science education in 19 countries.* Stockholm: Almquist & Wiksell.

Confrey, J. (1994). Splitting, similarity, and the rate of change: A new approach to multiplication and exponential functions. In G. Harel & J. Confrey (Eds.), *The development of multiplicative reasoning in the learning of mathematics* (pp. 291–330). Albany: State University of New York Press.

Cramer, K. (2003). Using a translation model for curriculum development and classroom instruction. In R. Lesh & H. Doerr (Eds.), *Beyond constructivism: Models and modeling perspectives on mathematics problem solving, learning, and teaching* (pp. 449–463). Mahwah, NJ: Erlbaum.

Cramer, K., Post, T, & Behr, M. (1989). Cognitive restructuring ability, teacher guidance, and perceptual distractor tasks: An aptitude treatment interaction study. *Journal for Research In Mathematics Education, 20,* 103–110.

Cramer, K. A., Post, T. R., & delMas, R. C. (2002). Initial fraction learning by fourth- and fifth-grade students: A comparison of the effects of using commercial curricula with the effects of using the rational number project curriculum. *Journal for Research in Mathematics Education, 33*(2), 111–144.

Davis, G., Hunting, R., & Pearne, C. (1992, November). *What might a fraction mean to a child and how would a teacher know?* Paper presented at the joint meeting of the Australian and New Zealand Associations for Research in Education, Deakin University, Geelong, Victoria, Australia.

Dienes, Z. (1960). *Building up mathematics.* London: Hutchinson Educational Ltd.

Fischbein, E., Deri, M., Nello, M. S., & Marino, M. S. (1985). The role of implicit models in solving verbal problems in multiplication and division. *Journal for Research in Mathematics Education, 16*(1), 3–17.

Flavell, J. H. (1963). *The developmental psychology of Jean Piaget.* New York: Van Nostrand.

Freudenthal, H. (1973). *Mathematics as an educational task.* Dordrecht, The Netherlands: D. Reidel.

Freudenthal, H. (1978). *Weeding and sowing: Preface to a science of mathematical education.* Dordrecht, The Netherlands: D. Reidel.

Freudenthal, H. (1983). *Didactical phenomenology of mathematical structures.* Dordrecht, The Netherlands: D. Reidel.

Gardner, H. (1983). *Frames of mind: The theory of multiple intelligences.* New York: Basic Books.

Graeber, A., Tirosh, D., & Glover, R. (1989). Preservice teachers' misconceptions in solving verbal problems in multiplication and division. *Journal for Research in Mathematics Education, 20*(1), 95–102.

Gray, R. L. (1979). Toward observing that which is not directly observable. In J. Lochhead & J. Clement (Eds.), *Cognitive process instruction: Research on teaching thinking skills* (pp. 217–227). Philadelphia: Franklin Institute Press.

Greeno, J. G. (1991). Number sense as situated knowing in a conceptual domain. *Journal for Research in Mathematics Education, 22*(13), 170–218.

Greer, B. (1987). Nonconservation of multiplication and division involving decimals. *Journal for Research in Mathematics Education, 18*(1), 37–45.

Harel, G., Behr, M, Post, T., & Lesh, R. (1994). The impact of the number type on the solution of multiplication and division problems. In G. Harel & J. Confrey (Eds.), *The development of multiplicative reasoning in the learning of mathematics* (pp. 363–384). New York: State University of New York Press.

Harel, G., & Confrey, J. (Eds.). (1994). *The development of multiplicative reasoning in the learning of mathematics.* New York: State University of New York Press.

Hart, K. (1981). Strategies and errors in secondary mathematics: The addition strategy in ratio. *Proceedings of the fifth conference of the International Group for the Psychology of Mathematics Education* (pp. 199–202). Columbus, OH:

ERIC Clearinghouse for Science, Mathematics, and Environmental Education.

Hart, K. M. (1984). *Ratio: Children's strategies and errors. A report of the strategies and errors in secondary mathematics project.* London: NFER-Nelson.

Hart, K. (1988). Ratio and proportion. In J. Hiebert & M. Behr (Eds.), *Number concepts and operations in the middle grades* (pp. 198–219). Reston, VA: National Council of Teachers of Mathematics and Erlbaum.

Hiebert, J., & Behr, M. (Eds.). (1988). Introduction: Capturing the major themes. In J. Hiebert & M. Behr (Eds.), *Number concepts and operations in the middle grades* (pp. 1–18). Reston, VA: National Council of Teachers of Mathematics and Erlbaum.

Inhelder, B, & Piaget, J. (1958). *The growth of logical thinking from childhood to adolescence.* New York: Basic Books.

Kaput, J. J. (1985). *Multiplicative word problems and intensive quantities: An integrated software response.* (Tech. Rep.). Cambridge, MA: Harvard Graduate School of Education, Educational Technology Center.

Kaput, J. J., Luke, C., Poholsky, J., & Sayer, A. (1986). *The role of representations in reasoning with intensive quantities: Preliminary analysis* (Tech. Rep. No. 86-9). Cambridge,MA: Harvard Graduate School of Education, Educational Technology Center.

Kaput, J., & West, M. M. (1994). Missing value proportional reasoning problems: Factors affecting informal reasoning patterns. In G. Harel & J. Confrey (Eds.), *The development of multiplicative reasoning in the learning of mathematics* (pp. 235–287). New York: State University of New York Press.

Karplus, K., Karplus, E., Formisano, M., & Paulsen, A. C. (1979). Proportional reasoning and control of variables in seven countries. In J. Lochhead & J. Clement (Eds.), *Cognitive process instruction: Research on teaching thinking skills* (pp. 47–103). Philadelphia: Franklin Institute Press.

Karplus, K., Karplus, E., & Wollman, W. (1974). Intellectual development beyond elementary school IV: Ratio, the influence of cognitive style. *School Science and Mathematics, 74*, 476–482.

Karplus, K., & Peterson, R. (1970). Intellectual development beyond elementary school II: Ratio, a survey. *School Science and Mathematics, 70*, 735–742.

Karplus, R., Pulos, S., & Stage, E. K. (l983a). Early adolescents' proportional reasoning on 'rate' problems. *Educational Studies in Mathematics, 14*, 219–233.

Karplus, R., Pulos, S., & Stage, E. K. (l983b). Proportional reasoning of early adolescents. In R. Lesh & M. Landau (Eds.), *Acquisition of mathematics concepts and processes* (pp. 45–90). Orlando, FL: Academic Press.

Kelly, A. E., & Lesh, R. (2000). *Handbook of research design in mathematics and science education.* Mahwah, NJ: Erlbaum.

Kieren, T. (1976). On the mathematical, cognitive, and instructional foundations of rational numbers. In R. A. Lesh (Ed.), Number and measurement (pp. 101–144). Columbus: OH: ERIC/SMEAC.

Kieren, T. (1980). The rational number construct–Its elements and mechanisms. In T. Kieren (Ed.), *Recent research on number learning* (pp. 125–149). Columbus: OH: ERIC/SMEAC.

Kieren, T. (1983). Axioms and intuition in mathematical knowledge building. In J. C. Bergeron & N. Herscovics (Eds.), *Proceedings of the fifth annual meeting of the North American chapter of the International Group for the Psychology of Mathematics Education* (pp. 67–73). Columbus, OH:

ERIC Clearinghouse for Science, Mathematics, and Environmental Education.

Kieren, T. (1988). Personal knowledge of rational numbers: Its intuitive and formal development. In J. Hiebert & M. Behr (Eds.), *Number concepts and operations in the middle grades* (pp. 162–181). Reston, VA: National Council of Teachers of Mathematics and Erlbaum.

Kieren, T. (1993). Rational and fractional numbers: From quotient fields to recursive understanding. In T. P. Carpenter, E. Fennema, & T. A. Romberg (Eds.), *Rational numbers: An integration of research* (pp. 49–84). Hillsdale, NJ: Erlbaum.

Klein, H. A. (1974). *The world of measurements.* New York: Simon and Schuster.

Lachance, A., & Confrey, J. (1995). Introducing fifth graders to decimal notation through ratio and proportion. In D. T. Owens, M. K. Reed, & G. M. Milsaps (Eds.), *Proceedings of the seventeenth annual meeting of the North American chapter of the International Group for the Psychology of Mathematics Education* (Vol. 1, pp. 395–400). Columbus, OH: ERIC Clearinghouse for Science, Mathematics, and Environmental Education.

Lamon, S. J. (1988). *Ratio and proportion: Preinstructional cognitions.* Unpublished research proposal, University of Wisconsin, Madison.

Lamon, S. J. (1989). *Ratio and proportion: Preinstructional cognitions.* Unpublished doctoral dissertation, University of Wisconsin, Madison.

Lamon, S. J. (1993a). Ratio and proportion: Children's cognitive and metacognitive processes. In T. P. Carpenter, E. Fennema, & T. A. Romberg (Eds.), *Rational numbers: An integration of research* (pp. 131–156). Hillsdale, NJ: Erlbaum.

Lamon, S.J. (1993b). Ratio and proportion: Connecting content and children's thinking. *Journal for Research in Mathematics Education, 24*(1), 41–61.

Lamon, S. J. (1994). Ratio and Proportion: Cognitive Foundations in Unitizing and Norming. In G. Harel & J. Confrey (Eds.), *The development of multiplicative reasoning in the learning of mathematics* (pp. 89–121). New York: State University of New York Press.

Lamon, S. J. (1996). The development of unitizing: Its role in children's partitioning strategies. *Journal for Research in Mathematics Education, 27*(2), 170–193.

Lamon, S.J. (2003). Beyond constructivism: An improved fitness metaphor for the acquisition of mathematical knowledge. In R. Lesh & H. M. Doerr (Eds.), *Beyond constructivism: Models and modeling perspectives on mathematics problem solving, learning, and teaching* (pp. 435–447). Mahwah, NJ: Erlbaum.

Lamon, S.J. (2006). *Teaching fractions and ratios for understanding; Essential content knowledge and instructional strategies for teachers.* Mahwah, NJ: Erlbaum.

Lamon, S. J. (in progress). *Understanding rational numbers.* Mahwah, NJ: Erlbaum.

Lamon, S. J., Parker, W. A., & Houston, S. K. (Eds.) (2003). *Mathematical modelling: A way of life.* West Sussex, UK: Ellis Horwood.

Lappan, G., Fey, J. T., Fitzgerald, W. M., Friel, S. N., & Phillips, E. D. (1998). *Connected mathematics.* Palo Alto, CA: Dale Seymour.

Lesh, R. (1985). Conceptual analyses of problem-solving performance. In E. A. Silver (Ed.), *Teaching and learning mathematical problem solving: Multiple research perspectives* (pp. 309–329). Hillsdale, NJ: Erlbaum.

Lesh, R., & Lamon, S. J. (1992). Assessing authentic mathematical performance. In R. Lesh & S. J. Lamon (Eds.), *Assessment of authentic performance in school mathematics* (pp. 17–62). Washington, DC: American Association for the Advancement of Science.

Lesh, R., Landau, M., & Hamilton, E. (1983). Conceptual models and applied mathematical problem-solving research. In R. Lesh & M. Landau (Eds.), *Acquisition of mathematics concepts and processes* (pp. 263–343). Orlando, FL: Academic Press.

Lesh, R., Post, T. R., & Behr, M. (1988). Proportional Reasoning. In J. Hiebert & M. Behr (Eds.), *Number concepts and operations in the middle grades*, (pp. 93–118). Reston, VA: National Council of Teachers of Mathematics.

Lo, J. J., & Watanabe, T. (1997). Developing ratio and proportion schemes: A story of a fifth grader. *Journal for Research in Mathematics Education, 28*(2), 216–236.

Luke, C. (1988). The repeated addition model of multiplication and children's performance on mathematical word problems. *Journal of Mathematical Behavior, 7*(3), 217–226.

Mack, N. (1990). Learning fractions with understanding: Building on informal knowledge. *Journal for Research in Mathematics Education, 21*(1), 16–32.

Mack, N. (1995). Confounding whole-number and fraction concepts when building on informal knowledge. *Journal for Research in Mathematics Education, 26*(5), 422–441.

Mack, N. (2001). Building on informal knowledge through instruction in a complex content domain: Partitioning, units, and understanding multiplication of fractions. *Journal for Research in Mathematics Education, 32*(3), 267–295.

Marini, Z. (1992). Synchrony and asynchrony in the development of children's scientific reasoning. In R. Case (Ed.), *The mind's staircase: Exploring the conceptual underpinnings of children's thought and knowledge* (pp. 55–73). Hillsdale, NJ: Erlbaum.

Markman, E. M. (1979). Classes and collections: Conceptual organization and numerical abilities. *Cognitive Psychology, 11*, 395–411.

Moss, J., & Case, R. (1999). Developing children's understanding of rational numbers: A new model and an experimental curriculum. *Journal for Research in Mathematics Education, 30*(2), 122–147.

Murray, H., Olivier, A., & Human, P. (1996). Young students' informal knowledge of fractions. In L. Puig & A. Gutiérrez (Eds.), *Proceedings of the 20th conference of the International Group for the Psychology of Mathematics Education* (Vol. 4, pp. 43–50).

National Council of Teachers of Mathematics. (1989). *Curriculum and evaluation standards for school mathematics.* Reston, VA: Author.

National Council of Teachers of Mathematics. (2000). *Principles and standards for school mathematics.* Reston, VA: Author.

Noelting, G. (1980a). The development of proportional reasoning and the ratio concept. Part 1–Differentiation of stages. *Educational Studies in Mathematics, 11*, 217–253.

Noelting, C. (1980b). The development of proportional reasoning and the ratio concept. Part 2–Problem-structure at successive stages; Problem-solving strategies and the mechanism of adaptive restructuring. *Educational Studies in Mathematics, 11*, 331–363.

Novillis, C. (1976). An analysis of the fraction concept into a hierarchy of selected subconcepts and the testing of the hierarchical dependencies. *Journal for Research in Mathematics Education, 7*(3), 131–144.

Ohlsson, S. (1987). Sense and reference in the design of iterative illustrations for rational numbers. In R. W. Lawler & M. Yazdani (Eds.), *Artificial intelligence and education*, (pp. 307–344). Norwood, NJ: Ablex.

Ohlsson, S. (1988). Mathematical meaning and applicational meaning in the semantics of fractions and related concepts. In J. Hiebert & M. Behr (Eds.), *Number concepts and operations in the middle grades*, (pp. 53–92). Reston, VA: National Council of Teachers of Mathematics.

Olive, J., & Steffe, L. P. (1994). *TIMA Bars* [Computer software]. Acton, MA: William K. Bradford.

Pascual-Leone, J. (1969). *Cognitive development and cognitive style.* Unpublished doctoral dissertation, University of Geneva, Geneva, Switzerland.

Piaget, J., Grize, J. B., Szeminska, A., & Bang, V. (1968). *Epistemologie etpsychologie de la fonction* [Epistemology and psychology of the function]. Paris: Presses Universitaires de France.

Piaget, J., Inhelder, B., & Szeminska, A. (1960). *The child's conception of geometry.* London: Routledge & Kegan Paul.

Pimm, D. (1987). *Speaking mathematically.* London: Routledge & Kegan Paul.

Pinker, S. (1997). *How the mind works.* New York: W. W. Norton & Company.

Post, T. R. (1986). The learning and assessment of proportional reasoning. In G. Lappan. & R. Even (Eds.), *Proceedings of the eighth annual meeting of the North American Chapter of the International Group for the Psychology of Mathematics Education. Vol. 1: Individual contributions* (pp. 349–353). Columbus, OH: ERIC Clearinghouse for Science, Mathematics, and Environmental Education.

Post, T. R., Behr, M., Lesh, R., & Wachsmuth, I. (1985). Selected results from the Rational Number Project. In L. Streefland (Ed.), *Proceedings of the ninth International Conference for the Psychology of Mathematics Education: Vol I. Individual contributions* (pp. 342–351). Columbus, OH: ERIC Clearinghouse for Science, Mathematics, and Environmental Education.

Post, T. R., Harel, C., Behr, M., & Lesh, R. (1988). Intermediate teachers' knowledge of rational number concepts. In E. Fennema, T. P. Carpenter, & S. J. Lamon (Eds.), *Integrating Research on Teaching and Learning Mathematics* (pp. 194–217). Madison: Wisconsin Center for Education Research.

Post, T. R., Wachsmuth, I., Lesh, R., & Behr, M. (1985). Order and equivalence of rational numbers: A cognitive analysis. *Journal for Research in Mathematics Education, 16*(1), 18–36.

Pothier, Y., & Sawada, D. (1983). Partitioning: The emergence of rational number ideas in young children. *Journal for Research in Mathematics Education, 14*(4), 307–317.

Pulos, S., Karplus, R., & Stage, E. K. (1981). Generality of proportional reasoning in early adolescence: Content effects and individual differences. *Journal of Early Adolescence, 1*, 257–264.

Resnick, L. B. (1986). The development of mathematical intuition. In M. Perlmutter (Ed.), *Perspectives on intellectual development: The Minnesota Symposia on Child Psychology.* (Vol. 19, pp. 159–194). Hillsdale, NJ: Erlbaum.

Resnick, L. B., & Singer, J. A. (1993). Protoquantitative origins of ratio reasoning. In T. P. Carpenter, E. Fennema, & T. A. Romberg (Eds.), *Rational numbers: An integration of research.* Hillsdale, NJ: Erlbaum.

Romberg, T. A., & de Lange, J. (1998). *Mathematics in context.* Chicago: Encyclopaedia Brittanica.

Rupley, W. H. (1981). *The effects of numerical characteristics on the difficulty of proportional problems.* Unpublished doctoral dissertation, University of California, Berkeley.

Saenz-Ludlow, A. (1994). Michael's fraction schemes. *Journal for Research in Mathematics Education, 25*(1), 50–85.

Scarano, G. H. (1996, April). *Results from a 3-year longitudinal teaching experiment designed to investigate splitting, ratio and proportion.* Paper presented at the annual meeting of the American Educational Research Association, New York.

Schwartz, J. (1976). *Semantic aspects of quantity.* Unpublished manuscript, MIT, Cambridge.

Schwartz, J. (1987). *Intensive quantity and referent transforming arithmetic operations.* Unpublished manuscript, MIT and Harvard Graduate School of Education, Cambridge.

Schwartz, J. (1988). Intensive quantity and referent transforming arithmetic operations. In J. Hiebert & M. Behr (Eds.), *Number concepts and operations in the middle grades,* (pp. 41–52). Reston, VA: National Council of Teachers of Mathematics.

Shulman, L. S. (1986a). Paradigms and research programs in the study of teaching: A contemporary perspective. In M.C. Wittrock (Ed.), *Handbook of research on teaching* (3rd Ed.) (pp. 3–36). New York: Macmillan.

Shulman, L. S. (1986b). Those who understand: Knowledge growth in teaching. *Educational Researcher, 15*(1) 4–14.

Shulman, L. S. (1987). Knowledge and teaching: Foundations of the new reform. *Harvard Educational Review, 57*(1), 1–22.

Siegler, R. S. (1976). Three aspects of cognitive development. *Cognitive Psychology, 8,* 481–520.

Silver, E. A. (1981). Young adults' thinking about rational numbers. In T. R. Post & M. P. Roberts (Eds.), *Proceedings of the third annual meeting of the North American Chapter of the International Group for the Psychology of Mathematics Education* (pp. 149–159). Columbus, OH: ERIC Clearinghouse for Science, Mathematics, and Environmental Education.

Simon, M. A., & Blume, G. W. (1994). Building and understanding multiplicative relationships: A study of prospective elementary teachers. *Journal for Research in Mathematics Education, 25*(5), 472–494.

Smith, E., & Confrey, J. (1994). Multiplicative structures and the development of logarithms: What was lost by the invention of function? In G. Harel & J. Confrey (Eds.), *The development of multiplicative reasoning in the learning of mathematics* (pp. 331–360). Albany: State University of New York Press.

Sowder, J. (1988). Series forward. In J. Hiebert & M. Behr (Eds.), *Number concepts and operations in the middle grades* (pp. vii–x). Reston, VA: National Council of Teachers of Mathematics and Erlbaum.

Sowder, J. (1989). *Setting a research agenda.* Reston, VA: National Council of Teachers of Mathematics and Erlbaum.

Steffe, L. P. (1988). Children's construction of number sequences and multiplying schemes. In J. Hiebert & M. Behr (Eds.), *Number concepts and operations in the middle grades* (pp. 41–52). Reston, VA: National Council of Teachers of Mathematics.

Steffe, L. P. (1992). Schemes of action and operation involving composite units. *Learning and Individual Differences, 4*(3), 259–309.

Steffe, L. P. & Cobb, P. (1988). *Construction of arithmetical meanings and strategies.* New York: Springer-Verlag.

Steffe, L. P., & Parr, R. B. (1968). *The development of the concepts of ratio and fraction in the fourth, fifth, and sixth years of elementary school* (Tech. Rep. No. 49). Madison, WI: Wisconsin Research and Development Center for Cognitive Learning.

Steffe, L. P., von Glasersfeld, E., Richards, E., & Cobb, P. (1983). *Children's counting types: Philosophy, theory, and application.* New York: Praeger.

Streefland, L. (1984). Search for the roots of ratio: Some thoughts on the long term learning process (Towards . . . A Theory). Part I: Reflections on a teaching experiment. *Educational Studies in Mathematics, 15,* 327–348.

Streefland, L. (1985). Search for the roots of ratio: Some thoughts on the long term learning process (Towards . . . A Theory). Part II: The outline of the long term learning process. *Educational Studies in Mathematics, 16,* 75–94.

Streefland, L. (1991). *Fractions in realistic mathematics education: A paradigm of developmental research.* Dordrecht, The Netherlands: D. Reidel.

Thompson, A. G., & Thompson, P. W. (1994). Talking about rates conceptually, Part I: A teacher's struggle. *Journal for Research in Mathematics Education, 25*(3), 279–303.

Thompson, A. G., & Thompson, P. W. (1996). Talking about rates conceptually, Part II: Mathematical knowledge for teaching. *Journal for Research in Mathematics Education, 27*(1), 2–24.

Thompson, P. W. (1994). The development of the concept of speed and its relationship to concepts of rate. In G. Harel & J. Confrey (Eds.), *The development of multiplicative reasoning in the learning of mathematics* (pp. 179–234). New York: State University of New York Press.

Thompson, P. W., & Saldanha, L. (2003). Fractions and multiplicative reasoning. In J. Kilpatrick, G. Martin, & D. Schifter (Eds.), *Research companion to the Principles and Standards for School Mathematics* (pp. 95–114). Reston, VA: National Council of Teachers of Mathematics.

Tourniaire, F. (1983). Some aspects of proportional reasoning in young children. In J. C. Bergeron & N. Herscovics (Eds.), *Proceedings of the fifth annual meeting of the North American chapter of the International Group for the Psychology of Mathematics Education* (pp. 319–324). Columbus, OH: ERIC Clearinghouse for Science, Mathematics, and Environmental Education.

Tourniaire, F. (1986). Proportions in elementary school. *Educational Studies in Mathematics, 17,* 410–412.

Tourniaire, F., & Pulos, S. (1985). Proportional reasoning: A review of the literature. *Educational Studies in Mathematics, 16,* 181–204.

Treffers, A., & Goffree, F. (1985). Rational analysis of realistic mathematics education—The Wiskobas Program. In L. Streefland (Ed.), *Proceedings of the ninth International Conference on the Psychology of Mathematics Education* (pp. 97–121). Columbus, OH: ERIC Clearinghouse for Science, Mathematics, and Environmental Education.

van den Brink, J., & Streefland, L. (1979). Young children (6–8)—Ratio and proportion. *Educational Studies in Mathematics, 10,* 403–420.

Vergnaud, G. (1983). Multiplicative structures. In R. Lesh & M. Landau (Eds.), *Acquisition of mathematics concepts and processes* (pp. 127–174). Orlando, FL: Academic Press.

Vergnaud, G. (1988). Multiplicative structures. In J. Hiebert & M. Behr (Eds.), *Number concepts and operations in the middle grades* (pp. 141–161). Reston, VA: National Council of Teachers of Mathematics and Erlbaum.

Wollman, W. (1974). Intellectual development beyond elementary school V: Using ratio in differing tasks. *School Science and Mathematics, 74,* 593–611.

AUTHOR NOTE

I gratefully acknowledge the contributions of Bill Parker (Marquette University), Nancy K. Mack (Grand Valley State University) and Kathleen Cramer (University of Minnesota), who commented on earlier drafts of this chapter.

G

M

D

CPSIA information can be obtained at www.ICGtesting.com
Printed in the USA
BVOW040536140312

284591BV00007B/1/P